CHARACTERIZATION TECHNIQUES AND TABULATIONS FOR ORGANIC NONLINEAR OPTICAL MATERIALS

OPTICAL ENGINEERING

Series Editor

Brian J. Thompson

Distinguished University Professor
Professor of Optics
Provost Emeritus
University of Rochester
Rochester, New York

1. Electron and Ion Microscopy and Microanalysis: Principles and Applications, *Lawrence E. Murr*
2. Acousto-Optic Signal Processing: Theory and Implementation, *edited by Norman J. Berg and John N. Lee*
3. Electro-Optic and Acousto-Optic Scanning and Deflection, *Milton Gottlieb, Clive L. M. Ireland, and John Martin Ley*
4. Single-Mode Fiber Optics: Principles and Applications, *Luc B. Jeunhomme*
5. Pulse Code Formats for Fiber Optical Data Communication: Basic Principles and Applications, *David J. Morris*
6. Optical Materials: An Introduction to Selection and Application, *Solomon Musikant*
7. Infrared Methods for Gaseous Measurements: Theory and Practice, *edited by Joda Wormhoudt*
8. Laser Beam Scanning: Opto-Mechanical Devices, Systems, and Data Storage Optics, *edited by Gerald F. Marshall*
9. Opto-Mechanical Systems Design, *Paul R. Yoder, Jr.*
10. Optical Fiber Splices and Connectors: Theory and Methods, *Calvin M. Miller with Stephen C. Mettler and Ian A. White*
11. Laser Spectroscopy and Its Applications, *edited by Leon J. Radziemski, Richard W. Solarz, and Jeffrey A. Paisner*
12. Infrared Optoelectronics: Devices and Applications, *William Nunley and J. Scott Bechtel*
13. Integrated Optical Circuits and Components: Design and Applications, *edited by Lynn D. Hutcheson*
14. Handbook of Molecular Lasers, *edited by Peter K. Cheo*
15. Handbook of Optical Fibers and Cables, *Hiroshi Murata*
16. Acousto-Optics, *Adrian Korpel*
17. Procedures in Applied Optics, *John Strong*
18. Handbook of Solid-State Lasers, *edited by Peter K. Cheo*
19. Optical Computing: Digital and Symbolic, *edited by Raymond Arrathoon*
20. Laser Applications in Physical Chemistry, *edited by D. K. Evans*
21. Laser-Induced Plasmas and Applications, *edited by Leon J. Radziemski and David A. Cremers*
22. Infrared Technology Fundamentals, *Irving J. Spiro and Monroe Schlessinger*
23. Single-Mode Fiber Optics: Principles and Applications, Second Edition, Revised and Expanded, *Luc B. Jeunhomme*
24. Image Analysis Applications, *edited by Rangachar Kasturi and Mohan M. Trivedi*
25. Photoconductivity: Art, Science, and Technology, *N. V. Joshi*
26. Principles of Optical Circuit Engineering, *Mark A. Mentzer*

27. Lens Design, *Milton Laikin*
28. Optical Components, Systems, and Measurement Techniques, *Rajpal S. Sirohi and M. P. Kothiyal*
29. Electron and Ion Microscopy and Microanalysis: Principles and Applications, Second Edition, Revised and Expanded, *Lawrence E. Murr*
30. Handbook of Infrared Optical Materials, *edited by Paul Klocek*
31. Optical Scanning, *edited by Gerald F. Marshall*
32. Polymers for Lightwave and Integrated Optics: Technology and Applications, *edited by Lawrence A. Hornak*
33. Electro-Optical Displays, *edited by Mohammad A. Karim*
34. Mathematical Morphology in Image Processing, *edited by Edward R. Dougherty*
35. Opto-Mechanical Systems Design: Second Edition, Revised and Expanded, *Paul R. Yoder, Jr.*
36. Polarized Light: Fundamentals and Applications, *Edward Collett*
37. Rare Earth Doped Fiber Lasers and Amplifiers, *edited by Michel J. F. Digonnet*
38. Speckle Metrology, *edited by Rajpal S. Sirohi*
39. Organic Photoreceptors for Imaging Systems, *Paul M. Borsenberger and David S. Weiss*
40. Photonic Switching and Interconnects, *edited by Abdellatif Marrakchi*
41. Design and Fabrication of Acousto-Optic Devices, *edited by Akis P. Goutzoulis and Dennis R. Pape*
42. Digital Image Processing Methods, *edited by Edward R. Dougherty*
43. Visual Science and Engineering: Models and Applications, *edited by D. H. Kelly*
44. Handbook of Lens Design, *Daniel Malacara and Zacarias Malacara*
45. Photonic Devices and Systems, *edited by Robert G. Hunsberger*
46. Infrared Technology Fundamentals: Second Edition, Revised and Expanded, *edited by Monroe Schlessinger*
47. Spatial Light Modulator Technology: Materials, Devices, and Applications, *edited by Uzi Efron*
48. Lens Design: Second Edition, Revised and Expanded, *Milton Laikin*
49. Thin Films for Optical Systems, *edited by Francoise R. Flory*
50. Tunable Laser Applications, *edited by F. J. Duarte*
51. Acousto-Optic Signal Processing: Theory and Implementation, Second Edition, *edited by Norman J. Berg and John M. Pellegrino*
52. Handbook of Nonlinear Optics, *Richard L. Sutherland*
53. Handbook of Optical Fibers and Cables: Second Edition, *Hiroshi Murata*
54. Optical Storage and Retrieval: Memory, Neural Networks, and Fractals, *edited by Francis T. S. Yu and Suganda Jutamulia*
55. Devices for Optoelectronics, *Wallace B. Leigh*
56. Practical Design and Production of Optical Thin Films, *Ronald R. Willey*
57. Acousto-Optics: Second Edition, *Adrian Korpel*
58. Diffraction Gratings and Applications, *Erwin G. Loewen and Evgeny Popov*
59. Organic Photoreceptors for Xerography, *Paul M. Borsenberger and David S. Weiss*
60. Characterization Techniques and Tabulations for Organic Nonlinear Optical Materials, *edited by Mark G. Kuzyk and Carl W. Dirk*

Additional Volumes in Preparation

CHARACTERIZATION TECHNIQUES AND TABULATIONS FOR ORGANIC NONLINEAR OPTICAL MATERIALS

EDITED BY

MARK G. KUZYK
Washington State University
Pullman, Washington

CARL W. DIRK
University of Texas at El Paso
El Paso, Texas

CRC Press
Taylor & Francis Group
Boca Raton London New York

CRC Press is an imprint of the
Taylor & Francis Group, an **informa** business

First published 1998 by Marcel Dekker, Inc.

Published 2022 by CRC Press
Taylor & Francis Group
6000 Broken Sound Parkway NW, Suite 300
Boca Raton, FL 33487-2742

ISBN 13: 978-0-8247-9968-7 (hbk)

Visit the Taylor & Francis Web site at
http://www.taylorandfrancis.com

and the CRC Press Web site at
http://www.crcpress.com

Library of Congress Cataloging-in-Publication Data

Characterization techniques and tabulations for organic nonlinear optical materials / edited by Mark G. Kuzyk, Carl W. Dirk.
p. cm.— (Optical engineering; v. 60)
Includes bibliographical references and index.
ISBN 0-8247-9968-2
1. Organic compounds—Optical properties. 2. Optical materials. 3. Nonlinear optics—Materials. I. Kuzyk, Mark G. II. Dirk, Carl William. III. Series: Optical engineering (Marcel Dekker, Inc.); v. 60.
QD473.C47 1998
620.1'1795—dc21 98-16706
CIP

From the Series Editor

As I was sitting down to write this, I happened to be reading the current issue of *Physics World* [1] and was struck by two pertinent and related items. The first was a book review entitled "Do You Know Your Stuff?" The "stuff" came from the title of the book being reviewed, which was *Stuff: Materials the World Is Made of* by Ivan Amato. Optical "stuff," i.e., optical materials, is clearly something we do need to know about.

Only a few decades ago optical materials meant the wide variety of glasses available for the fabrication of optical elements such as lenses and prisms. The use of glass in optical components and systems was supplanted by some natural materials including quartz, calcite, and mica that could be used for a variety of special purposes. In summary, these materials were used to produce controlled refraction and polarization.

Times have changed as today we deal with materials for acousto-optics, electro-optics, magneto-optics, and quantum optics. My second observation in *Physics World* was a short description of the use of a (man-made) crystal of beta barium borate with its nonlinear optical properties interacting with an ultraviolet beam of light from a laser. A few decades ago, most optical materials were used for their linear properties. Linear materials are clearly of critical importance today, but nonlinear materials have opened up many new and exciting areas of optics.

The current volume is devoted to discussion of both linear and nonlinear optical organic materials with a focus on those materials pertinent to photonic devices—how the materials are characterized and cataloged, and used.

Brian J. Thompson

REFERENCE

1. Colliex, Christian. "Do you know your stuff?" *Physics World, 11*(3), The Institute of Physics, London, (March 1998).

Preface

With the recent explosion in the growth of applications that incorporate photonic devices, an insatiable thirst has developed for materials that are better able to perform these functions. The last two decades of synthetic work has resulted in a broad range of new materials whose optical and nonlinear-optical properties have been characterized. Many of these new materials are organic and include crystals, polymers, composites, liquids, and liquid crystals. This book therefore primarily focuses on optical properties of organic materials germane to photonic devices.

Current research in optical properties of organics can be viewed as three highly interconnected activities: synthesis, characterization, and theoretical modeling. Most research in this field is necessarily highly interdisciplinary and includes materials synthesis, optical characterization, a comparison of experimental results to theory, refinement of theory, and synthesis of new materials based on the better models. This iterative process has been applied through several generations, resulting in improvements of materials, better theoretical understanding, and new characterization techniques. The diversity of roles of contributors to this field means that information is often compartmentalized in widely disparate sources, and it is the role of books like this to centralize the latest work of the field.

The aim of this book is to provide tabulations of measured optical properties of organic materials and characterization techniques used in measuring these properties. As preliminaries, Chapter 1 introduces conventions, standards, and notation. There are many different systems of units and conventions available, and several different ones are commonly used by researchers. As such, values reported in the literature may appear to vary considerably when a consistent set of conventions is not used. This problem is often compounded by the common practice of theorists using a different convention than is used in most experiments. In addition, experimentalists often use known materials as standards that are not themselves always accurately measured. These three issues are carefully examined in Chapter 1.

Because materials are made of atoms and molecules, an understanding of the molecular optical properties can often be used to comprehend the optics of a bulk material. Chapter 2 provides a description of various computational techniques that give results that are consistent with measurements. The applicability of these techniques to various molecules with regard to size and computational efficiency is also discussed.

Chapter 3 gives a formal derivation of how the molecular optical properties of a material are related to the bulk response. Models for molecular crystals, liquid crystals, and polymers are presented. Local field models are discussed and limits of applicability defined. Most importantly, these models show how various mechanisms of nonlinear-optical response arise from the arrangement of molecules in a material.

Chapters 1 to 3 serve as a bridge to the rest of the book, which gives tabulations of optical properties of organic materials. Chapters 4 to 10 are each divided into two main categories: characterization techniques and tabulations. The characterization section concentrates on providing a tutorial on how to measure a particular property of a material along with possible pitfalls—at a level of detail not often found in the research literature. This is followed by extensive tabulations of materials properties.

Chapter 4 discusses linear optical properties of organics and concentrates on measurement techniques. In particular, the question of how a limited number of measurements can be used to approximate the linear properties of a material at other wavelengths is treated. Examples comparing measurement and theory are provided to show the accuracy of these techniques. Because the amount of measurements of linear properties is too expansive to fit even in several volumes, tabulations are limited to example cases.

Chapters 5 and 6 focus on what can be considered mostly second-order nonlinear effects. Chapter 5 surveys several experimental methods used to measure electro-optic effects in thin films, waveguides, and crystals. Chapter 6 discusses second harmonic generation in thin films and waveguides along with the infinite dilution technique used to obtain the molecular response. Current topics such as hyper-Raleigh scattering and poled polymers are treated. Most importantly, the relationship between the electro-optic effect and second harmonic generation is described.

Chapters 7–10 primarily deal with third-order nonlinear-optical processes. Chapter 7 discusses many experimental techniques that can be used to measure two photon absorption processes. Nuances of the effects of excited state properties and two photon processes are presented. Chapter 8 describes the z-scan technique, which has become popular because of its ease of use. Of importance to the experimentalist is an analysis that relates z-scan data to the third-order susceptibility. In particular, a simple approximation to the more complex solutions of the wave equation is shown to give accurate results. Chapter 9 treats the general case of four-wave mixing, which include self-focusing, pump-probe techniques, and phase-conjugation. Chapter 10 provides a rigorous analysis of third harmonic and shows how subtle air and substrate effects can invalidate a measurement.

This book should serve interdisciplinary scientists who make compounds, characterize materials, do calculations, and design devices using nonlinear optical materials. The contents of this book, along with the references, should serve to bring outsiders up to speed in the field of organic nonlinear optics, as well as to help researchers keep abreast of measurements and techniques.

We are indebted to many funding agencies who made the writing of this book possible. MGK would like to thank the Air Force Office of Scientific Research (AFOSR), the Army Research Office, the National Science Foundation, Washington State University, and the state of Washington. CWD would like to thank AFOSR, BMDO, NSF, and the state of Texas for kind support over the course of developing this book. We thank Lori Bruce for her dedication in pulling together the index. We are indebted to our families (Lexa, Mark, and Pat Kuzyk, and Katie, Kimi, Kristin, and Kathy Dirk) for being supportive and putting up with our absence while we worked on this project. Finally CWD would like to thank MGK (and MGK to CWD) for the opportunity to work with him on this project and as a valued colleague over the past 13 years.

Mark G. Kuzyk
Carl W. Dirk

Contents

Contributors

Israel D. L. Albert Northwestern University, Evanston, Illinois

Carl W. Dirk University of Texas at El Paso, El Paso, Texas

L. Michael Hayden University of Maryland Baltimore County, Baltimore, Maryland

Steven F. Hubbard Case Western Reserve University, Cleveland, Ohio

A. F. Garito University of Pennsylvania, Philadelphia, Pennsylvania

Kennita Johnson University of Maryland Baltimore County, Baltimore, Maryland

F. Kajzar LETI, CEA, Technologies Advancées, DEIN/SPE, Saclay, France

Stephen Kershaw BT Laboratories, Martlesham Heath, Suffolk, England

Mark G. Kuzyk Washington State University, Pullman, Washington

George F. Lipscomb Akzo Nobel Electronic Products Inc., Sunnyvale, California

Rick Lytel* Akzo Nobel Electronic Products, Inc., Sunnyvale, California

Tobin J. Marks Northwestern University, Evanston, Illinois

C. R. Moylan Lightwave Microsystems Corporation, Santa Clara, California

R. A. Norwood AlliedSignal, Morristown, New Jersey

Mark A. Ratner Northwestern University, Evanston, Illinois

Andrew Schober Case Western Reserve University, Cleveland, Ohio

Mansoor Sheik-Bahae University of New Mexico, Albuquerque, New Mexico

R. F. Shi University of Pennsylvania, Philadelphia, Pennsylvania

Kenneth D. Singer Case Western Reserve University, Cleveland, Ohio

J. D. Swalen University of California at Santa Cruz, Santa Cruz, California

John I. Thackara Akzo Nobel Electronic Products, Inc., Sunnyvale, California

Eric W. Van Stryland CREOL (Center for Research and Education in Optics and Lasers), University of Central Florida, Orlando, Florida

---

**Current affiliation*: Amp Inc., Palo Alto, California

1

Introduction: Conventions and Standards for Nonlinear Optical Processes

R. F. Shi and A. F. Garito
University of Pennsylvania, Philadelphia, Pennsylvania

I. INTRODUCTION

Nonlinear optics is the basis of all the fledgling photonics technologies,[1,2] where light works with, or even replaces, electrons in applications traditionally carried out by microelectronics. Over the past decade, the study of nonlinear optical properties in organic and polymer systems has enjoyed rapid and sustained growth.[3–7] Because of their large optical nonlinearities and mechanical, chemical, thermal and photo stabilities, organic nonlinear optical materials are the leading practical materials for fabricating optoelectronic devices. They also have proven to be excellent subjects in which to study the physics of many-body electron correlation effects, excitation dynamics and materials relaxations, and exotic electronic excitations such as exiton strings.[8]

In the search for understanding physical mechanisms that determine the nonlinear optical responses in organic structures and for new materials that may qualify for device applications, we are constantly required to know accurately the values of microscopic and macroscopic nonlinear optical properties of the system under consideration. As a result of an extraordinary expansion in the study of nonlinear optical responses of organic systems, different conventions, reference standards, and units have been used, making it difficult to compare theoretical results with experimental values, or to evaluate nonlinear optical properties of different systems or sometimes even to understand the results on the same system from different research groups. The best example is perhaps the value of the molecular second order susceptibility $\beta_x(-2\omega; \omega, \omega)$ for the heavily studied system, paranitroaniline (PNA).[39,47,48] A careful survey shows that at the same wavelength, different research groups using different conventions obtained approximately the same value, while in fact they should differ by a factor of $3/2$. This case is one of several major findings that will be made more clear later in this Chapter.

One possible explanation for the above difference is the inconsistency in the use of glass $\chi^{(3)}$ values that are referenced in the data analysis. The analysis by Teng and Garito[39] adopted a reference value for BK-7 glass which is only half of the value recommended by Buchalter and Meredith.[55] The analysis by Burland and coworkers,[47,48] however, is based on the reference $\chi^{(3)}$ value for fused silica directly from Buchalter and Meredith. It was pointed out recently that this $\chi^{(3)}$ value used by Buchalter and Meredith is too large by a factor of two.[40] Adding to these questions is the recent finding that the d_{11} value for quartz is smaller than what is commonly used.[79,83] It was realized that d_{11} values measured by parametric fluorescence technique were consistently higher than those from direct second harmonic generation measurements. Furthermore, Mito *et al.*[79] measured the $\chi^{(3)}$ value for fused silica, with the results being more than a factor of two smaller than those obtained by Buchalter and Meredith, as recommended by Heflin *et al.*[40]

This Chapter tries to clarify issues and some confusions surrounding various conventions that are in common use. In section II, we briefly review the physical origin of the nonlinear optical processes in organic and polymeric systems, focusing on second and third order optical processes. In section III, we list and compare several different conventions that are being used, especially in the second and third harmonic generation measurements where possible confusions are present. The discussions surrounding the d_{11} values for quartz and, separately but related, the $\chi^{(3)}$ values for BK-7 glass and fused silica will be laid out in section IV. The commonly used units, along with their conversion relations, will also be presented in section IV.

II. OVERVIEW OF NONLINEAR OPTICAL PROCESSES FOR ORGANIC SYSTEMS

Nonlinear optical properties in organic and polymer systems are under intensive study because, unlike inorganic systems, where nonlinear optical phenomena arise from band structure effects, in organic and polymer systems, nonlinear optical effects originate in the virtual electron excitations occurring on the individual molecular, or polymer chain, units. This allows researchers to employ ultrastructure synthesis, a new level of molecular design in which molecular assembly is controlled down to the placement of individual substituent groups, to optimize the nonlinear optical properties of organic systems while preserving the thermal, mechanical and chemical properties that have made organic materials so useful in industry.

The origin of the outstanding nonlinear optical properties of organic systems is the relative ease with which optical fields affect the motions of the electrons that take part in the delocalized, multicenter bonds characteristic of unsaturated organic compounds: the π electrons. Because π electrons are not tightly bound to the individual positive nuclear sites, their paths, or orbitals, extend over long distances, spanning an entire molecule or even a macroscopic solid.

Macroscopically, the nth order optical susceptibility tensor $\chi^{(n)}_{ijk...}(-\omega; \omega_1, \omega_2, \ldots, \omega_n)$ (of rank $n + 1$) of a medium relates the applied electric field to the polarization. Assuming there is no spatial dispersion and only electric dipole approximation is considered,[9] the ω Fourier component of the polarization is

$$P_i^\omega = P_i + \chi_{ij}^{(1)}(-\omega;\ \omega)\ E_j^\omega + \chi_{ijk}^{(2)}(-\omega;\ \omega_1,\ \omega_2)\ E_j^{\omega_1}\ E_k^{\omega_2}$$
$$+\ \chi_{ijkl}^{(3)}(-\omega;\ \omega_1,\ \omega_2,\ \omega_3)\ E_j^{\omega_1} E_k^{\omega_2} E_l^{\omega_3} + \ldots, \tag{1}$$

where summation over repeated indices is assumed, **P** and **E** are the polarization and the electric field in the medium, respectively, and all the susceptibilities are defined in the frequency domain. We emphasize that the above equation is only valid in certain conventions, as we will see in the next section. The first term on the right side of the equation is the permanent spontaneous polarization which is the dipole moment per unit volume. $\chi_{ij}^{(1)}(-\omega;\ \omega)$, the first-order susceptibility, is responsible for phenomena such as refraction and absorption. $\chi_{ijk}^{(2)}(-\omega;\ \omega_1,\ \omega_2)$, the second order optical susceptibility, includes second harmonic generation ($2\omega = \omega + \omega$) and linear electro-optic effect ($\omega = \omega + 0$). Examples of the third order optical susceptibility $\chi_{ijkl}^{(3)}(-\omega;\ \omega_1,\ \omega_2,\ \omega_3)$ are third harmonic generation ($3\omega = \omega + \omega + \omega$), electric field induced second harmonic generation ($2\omega = \omega + \omega + 0$), degenerate four-wave mixing and Kerr effect ($\omega = \omega + \omega - \omega$), and D. C. Kerr effect ($\omega = \omega + 0 + 0$). Higher order terms have also been discussed recently. For example, one phase-conjugation experiment[62] by nondegenerate six-wave mixing, actually involves $\chi^{(5)}(-2\omega;\ \omega,\ \omega,\ \omega,\ \omega,\ -2\omega)$, a fifth order optical process.

It is important to keep in mind that, because of the symmetry requirement, all even order susceptibilities vanish when the material is centrosymmetric; therefore, for example, second harmonic generation in a centrosymmetric solution is not possible. Molecules in a material must therefore be polar-aligned to achieve macroscopic noncentrosymmetry. In comparison, odd order optical susceptibilities are possible in all systems. An interesting anomaly is a solution of randomly oriented chiral molecules which is noncentrosymmetric with symmetry D_∞. Although second harmonic generation and linear electro-optic effects are not present, sum-frequency generation is possible.[63,64]

Similarly, the microscopic susceptibilities relate the local electric field to the induced dipole moment of the molecule:

$$p_i^\omega = \mu_i + \alpha_{ij}(-\omega;\ \omega)\ f_j^\omega + \beta_{ijk}(-\omega;\ \omega_1,\ \omega_2) f_j^{\omega_1} f_k^{\omega_2}$$
$$+\ \gamma_{ijkl}(-\omega;\ \omega_1,\ \omega_2,\ \omega_3) f_j^{\omega_1} f_k^{\omega_2} f_l^{\omega_3} + \ldots, \tag{2}$$

where summation over repeated indices is assumed, **p** and **f** are the induced dipole moment of the molecule and local electric field, respectively, $\boldsymbol{\mu}$ is the molecular dipole moment, α_{ij} is the linear polarizability, $\beta_{ijk}(-\omega;\ \omega_1,\ \omega_2)$ is the microscopic second order susceptibility (or hyperpolarizability), and $\gamma_{ijkl}(-\omega;\ \omega_1,\ \omega_2,\ \omega_3)$ is the microscopic third order susceptibility. Once again we caution the reader that Eq. (2) is only valid in certain conventions.

Under nonresonant conditions, the frequencies of the input and output optical fields ($\approx 10^{14}$ Hz) are far away from material resonances. Typically, the frequencies are below π-electron resonances but well above vibrational and rotational modes. In this regime, the nonlinear optical response is due predominantly to the electronic response. Because the nonlinear optical responses in organic materials originate primarily from π-electron excitations on individual molecular or polymer chain units, each unit can be viewed essentially as an independent source of nonlinear optical response. One can calculate the optical response of a macroscopic assembly like a solid by simply summing these individual responses, averaged over their orientations. specifically, we have the following equations relating macroscopic susceptibilities to microscopic susceptibilities:

$$\chi_{ij}^{(1)} = N \langle a_{ii'}^{s} a_{jl'}^{s} \alpha_{j'k'} (L_s^{\omega})_{i'j'} (L_s^{\omega_1})_{k'l'} \rangle, \tag{3}$$

$$\chi_{ijk}^{(2)} = N \langle a_{ii'}^{s} a_{jm'}^{s} a_{kn'}^{s} \beta_{j'k'l'} (L_s^{\omega})_{i'j'} (L_s^{\omega_1})_{k'm'} (L_s^{\omega_2})_{l'n'} \rangle, \tag{4}$$

and

$$\chi_{ijkl}^{(3)} = N \langle a_{ii'}^{s} a_{jn'}^{s} a_{ko'}^{s} a_{lp'}^{s} \gamma_{j'k'l'm'} (L_s^{\omega})_{i'j'} (L_s^{\omega^1})_{R'n'} (L_s^{\omega_2})_{l'o'} (L_s^{\omega_3})_{m'p'} \rangle, \tag{5}$$

where N is the number density, $\langle . . . \rangle$ refers to an angular integration over the orientational distribution, $a_{ii'}^{s}$ is the orthogonal transformation matrix that transforms the molecule-fixed frame (primed frame) to the laboratory frame (unprimed frame), and L_s^{ω} is the local field factor at frequency ω. Chapter 3 treats local field models and microscopic origins of nonlinearity in more detail.

Over the last decade, it has become apparent that second order nonlinear optical materials have nonlinearities that are large enough for practical applications. Ongoing research in third order nonlinear optical materials will answer the question whether organics are appropriate for all-optical switching devices.

A. Overview of Second Order Nonlinear Optical Processes

The hyperpolarizability $\beta_{ijk}(-\omega_3; \omega_1, \omega_2)$ is accurately described by quantum many-electron theory that explicitly accounts for electron correlation effects due to electron-electron repulsions. The motion of any one π-electron depends on the motion of all the other π-electrons. For many structures, theoretical calculations reveal that the large values of $\beta_{ijk}(-\omega_3; \omega_1, \omega_2)$ result from highly asymmetric, charge correlated excited states. Once one understands the underlying physical mechanisms of second order processes, we can use this understanding to optimize the desired effect. Generally, β depends on transition energies (difference between state energies) of all states, dipole moments, and transition dipole moments. As described later in this Chapter, β can be enhanced by designing molecules with low transition energy (to the dominant state), and large dipole moment differences and large transition moments. Typical ways to do this are to increase the strength of the electron donor and acceptor substituents and to increase the conjugation length of the π bridge that spans from donor to acceptor.

Bulk materials having large macroscopic second order nonlinear optical susceptibilities $\chi_{ijk}^{(2)}$ require not only molecular constituents with large microscopic second order susceptibilities β_{ijk}, but also macroscopic noncentrosymmetric assemblies aligned so that the individual tensor components of β_{ijk} add constructively. Other than standard crystal growth, the most common approach for attaining such structures is electric-field-induced alignment of glassy amorphous polymer films. Such materials are technologically relevant because of the ease of film processing, the ability to form multilayers and the compatibility with standard semiconductor device manufacturing steps. In this approach,[14,15] one applies a dc electric field across a polymer film sample that is heated above its glass transition temperature (the temperature above which a material tends to become rubber-like) to allow the dipolar chromophores to align with the field. One then lowers the temperature below the glass transition temperature to room temperature to lock in the noncentrosymmetric orientational order. The thermodynamic model of poling is described in detail in Chapter 3.

There exist four major methods by which chromophores are incorporated into polymers.[16] In the main-chain approach, chromophores are chemically attached to the polymer backbone itself. In the side-chain approach, chromophores are chemically attached as part of the side chain to the polymer backbone. These systems have the advantage that a high concentration for the chromophores can be obtained without crystallization, phase separation, or the formation of concentration gradients. In the cross-linking approach, chromophores act as cross-linking bridges between two polymer chains. This method effectively destroys the mobilization of segments of the polymer chains. And since polar or octopolar alignment of the dipole moments is required, most of the cross-linking must occur during or after the poling process is completed. In the guest-host system approach, chromophores form a solid solution in a polymer host. This approach has the simplicity that the chromophores do not need to be chemically bound to the polymer.

For each of the four approaches mentioned, a poled polymer system in the absence of an applied static electric field is not in thermodynamic equilibrium. The order parameter $\langle\cos(\theta)\rangle$, where θ is the angle between the dipole moment of a molecule and the poling field, therefore decays over time, resulting in the decay of the macroscopic second order susceptibility. One can greatly reduce this intrinsic thermal relaxation by using polymer systems with intrinsically high glass transition temperatures. One elegant example currently under development[17,19] is the class of optical polyimides with glass transition temperatures greater than 300–400°C.

The first applications of π-conjugated organic materials will likely be based on the linear electro-optic (EO) effect. The EO effect is a second order nonlinear optical process in which an externally applied DC or low AC frequency electric field couples to the optical field to change the refractive index of the polymer material. The resulting phase shift is

$$\Delta\phi = \frac{n^3 rEL}{2}, \tag{6}$$

where r is the EO coefficient of the material, n is the refractive index with no field applied, E is the applied electric field, and L is the sample length where the electric field is applied. A phase change of the input light can be converted to an intensity change in the output light using interference effects. The net effect is that light is modulated by the applied electric field. Since the driving power required to modulate light depends on the dielectric constant ϵ, an EO figure of merit (FOM) can be defined by

$$FOM \equiv \frac{n^3 r}{\epsilon}. \tag{7}$$

In addition, the bandwidth of an electro optic device is limited by the velocity mismatch between the optical wave (whose velocity is proportional to $1/n(\omega)$) and the driving wave (usually microwave, whose velocity is proportional to $1/\sqrt{\epsilon_{\text{eff}}}$, where ϵ_{eff} is the microwave dielectric constant). Because $\sqrt{\epsilon_{\text{eff}}} \sim n(\omega)$ in organic materials, they are excellent material candidates for high speed waveguide electro optic device applications.

An attractive technological opportunity presented by high performance EO polymers is the ability to design and build novel optoelectronic integrated devices and circuits using existing microelectronic IC manufacturing lines. Various EO modulation and waveguide coupling mechanisms are being developed and tested in EO polymer modulators.[18,19] Recently, researchers using novel EO polymer materials have demonstrated impressive high speed EO modulation at 40 GHz driven by relatively low IC level voltages.

Second harmonic generation (SHG), or frequency doubling, is another important second order nonlinearity that is the subject of intense research and development. Efficient SHG must meet not only the same material requirements as in the case of EO modulators, but also additional ones associated with the coherent nature of the SHG process. For maximum SHG, the material should possess propagation directions where the material birefringence cancels the natural optical dispersion, leading to a condition of equal refractive indices at the fundamental and second harmonic frequencies. This condition, known as phase matching, results in a high percentage of conversion of the input light to the output SHG. The second harmonic output power $P^{2\omega}$ is related to the fundamental input power P^{ω} by[20]

$$P^{2\omega} \propto \frac{d_{\mathrm{eff}}^2}{n^3} \frac{\mathrm{Sin}^2 \dfrac{\Delta kL}{2}}{\left(\dfrac{\Delta kL}{2}\right)^2} (P^{\omega})^2, \tag{8}$$

where d_{eff} is the effective second-harmonic coefficient, n is the refractive index, $\Delta k = 2k_{\omega} - k_{2\omega}$ where k_{ω} and $k_{2\omega}$ are the wavevectors at the fundamental and second harmonic frequencies, respectively, and L is the interaction length. The second harmonic figure of merit (FOM) can then be defined as

$$FOM \equiv \frac{d_{\mathrm{eff}}^2}{n^3}. \tag{9}$$

The fact that low index values (1.3–1.8) characterize organic materials is a distinct advantage over the relatively large values of inorganic materials.

Various approaches are being pursued for achieving efficient SHG. They include filling hollow-core optical fibers with crystalline organic material, or forming spatially periodic polymer films in which layers are alternately poled in opposite directions to achieve quasi phase matching.[21] In approaches based on Langmuir-Blodgett films, waveguides consisting of alternating layers of opposite $\chi^{(2)}$ across the film are used to simultaneously achieve phase matching and improve the optical field overlap between the propagating waveguide modes.[22] In each of these approaches, the second harmonic efficiency is greatly enhanced.

Currently, applications of SHG are centered on frequency doublers for laser diode sources operating in the 800 nm range to the blue spectral region near 400 nm. In optical data storage applications, the shorter wavelengths enable higher packing densities of optical data bits. The main issue associated with blue or violet light generation is optical transparency at both the fundamental and harmonic wavelengths. Most materials with high SHG efficiency show significant absorption in the blue or violet spectral region. Researchers have recently demonstrated efficient blue light generation below 400 nm using highly efficient organic single crystals.[23]

The purpose is to demonstrate the feasibility of using organic materials in terabit optical data storage applications.

B. Third Order Nonlinear Optical Processes in Organic Structures

Third order nonlinear optical effects occur naturally in all optical media since there are no symmetry restrictions as in the case of the second order susceptibility. The large, ultrafast, nonresonant third order nonlinear optical susceptibility $\chi^{(3)}$ of π-conjugated organic and polymeric systems are determined by electron correlation effects among the π-electrons,[24] a finding that has been confirmed in a number of third order nonlinear optical studies.[25] In linear chains, explicit consideration of the coulomb repulsion between electrons leads to results for the magnitude, sign, dispersion, and chain length for the nonresonant response $\chi^{(3)}$ in agreement with experiment. Most importantly, virtual π-electron transitions that involve previously unknown, strongly correlated, two-photon states energetically high lying are essential to the determination of $\gamma_{ijkl}(-\omega; \omega_1, \omega_2, \omega_3)$.[24–28] Descriptions based on independent particle models, which entirely neglect electron-electron repulsions and correlation, miss these states and thus, the underlying mechanism for $\gamma_{ijkl}(-\omega; \omega_1, \omega_2, \omega_3)$. In contrast, the proper account of electron correlation effects correctly predict the $\gamma_{ijkl}(-\omega; \omega_1, \omega_2, \omega_3)$ responses in terms of the sign, magnitude, and dispersion.

For short linear chains such as the six-site all-*trans*-hexatriene (HT)[24] there are two types of competing third order virtual excitation processes that almost exclusively determine $\gamma_{ijkl}(-\omega; \omega_1, \omega_2, \omega_3)$. In each transition, the parity alternates as required by selection rules.[32] In the first type (I) of the form $S_0 \rightarrow S_1 \rightarrow S_0 \rightarrow S_1 \rightarrow S_0$, the intermediate state is S_0 itself, and it makes a negative contribution to γ. In the second type (II) of $S_0 \rightarrow S_1 \rightarrow S_2 \rightarrow S_1 \rightarrow S_0$, the intermediate state S_2 is an energetically high-lying π-electron state that is a strongly electron correlated, two-photon state. The type II process makes a larger, positive contribution to γ and therefore determines the overall sign of γ to be positive, a result which is confirmed by experiment.

For all chain lengths, the electron correlation mechanism holds for γ which, in the finite chain limit, increases supralinearly with increased chain length L.[34] The good fit for polyene chains with different conformations and corresponding ring structures indicates that $\gamma_g(-3\omega; \omega, \omega, \omega)$ possesses a power law dependence on L with an exponent of 3.5. γ is much more sensitive to the physical length of the chain than to the conformation. The rapid increase with L originates in the increased transition moments between the principal virtual states, the decreased excitation energies of those states, and an increased number of significant virtual excitation processes with increased chain lengths. The power law dependence can be generalized to two- and three-dimensional structures as well. The strong dependence of γ on L is the primary reason for the intense interest in π-conjugated linear chains as third order optical materials. The supralinear behavior of γ with chain length, however, continues only to lengths of approximately 50–100 Å, or equivalently 50–100 carbon atom sites, and then saturates in magnitude. Thus, in general, the large nonresonant values of γ, and correspondingly $\chi^{(3)}$, require centrosymmetric chains of only intermediate length of order 50–100 Å with little to be gained beyond this limit. In addition, even assuming little or no optical loss, current available third order susceptibilities are too small to make 1 cm-long devices. The largest value of γ, observed in poly-diacetylenes[20] was observed over a decade ago. Recent searches for materials with much larger molecular third order susceptibilities have proved futile.

A number of approaches in fundamentally new directions have been tried to enhance γ. Among the mechanisms being studied are populated excited states,[29–31] new collective excitations,[8] and bond-length alternation.[33]

Compared with the ground state S_0, the nonresonant $\chi^{(3)}$ of a π-conjugated structure can be enhanced by orders of magnitude or even change sign when a metastable first or second electronic excited state (S_1 or S_2) is optically pumped and then populated for times long enough to permit nonresonant measurements of $\chi^{(3)}$. Such enhancements have been observed for nonresonant third harmonic generation and degenerate four-wave mixing in phthalocyanines[31] and polyene chains.[30]

Our understanding of nonlinear optical processes in organic systems was advanced by the recent discovery of entirely new elementary excitations in an organic crystal in which electron charge transfer occurs in one dimension along the molecular stacking axis.[8] These experiments have shown the existence of coherently propagating strings of excitons, which are the counterparts of excited states in individual molecules and polymers. As the exciton number density increases, strings of two, three, and even more excitons form and propagate coherently through the material. The nonlinear optical responses associated with the new exciton strings, as well as those associated with more familiar collective excitations such as polarons and bipolarons, are being explored as possible sources for increasing γ of linear chains.

Researchers have also recently proposed that the nonresonant γ can be further increased by controlling bond-length alternation, the difference in length between carbon-carbon single and double bonds in a π-conjugated molecular structure. Changing the alternation may markedly alter the electronic charge density distribution of the molecular ground state. For optimized alternation, enhancements of γ by up to a factor of five are possible.

The past 15 years have witnessed an extraordinary expansion of the nonlinear optical studies in organic systems. One result has been that to describe these properties, sometimes different conventions have been adopted, leading to confusing statements in the literature. The next section discusses various conventions that are frequently used and their relations, with a focus on certain important aspects which are often neglected.

III. CONVENTIONS

A. Definition of Nonlinear Optical Susceptibilities

In general, the nonlinear optical susceptibilities are nonlocal in both time and spatial domains. Throughout this Chapter, we will neglect the spatial dispersion, thus adopting the so called electric-dipole approximation. With this approximation, the nonlinear optical susceptibilities can be defined in either the time or frequency domain. In the time domain, the polarization **P** can be written as[10]

$$\mathbf{P}(t) = \mathbf{P}^{(0)}(t) + \mathbf{P}^{(1)}(t) + \mathbf{P}^{(2)}(t) + \cdots + \mathbf{P}^{(n)}(t) + \cdots, \tag{10}$$

where $\mathbf{P}^{(0)}(t)$ is the static polarization independent of the external electric field $\mathbf{E}(t)$, $\mathbf{P}^{(1)}(t)$ is linear in $\mathbf{E}(t)$, $\mathbf{P}^{(2)}(t)$ is quadratic in $\mathbf{E}(t)$, and so on. $\mathbf{P}^{(n)}(t)$, which is proportional to the nth power of $\mathbf{E}(t)$, can be expressed as

$$P_i^{(n)}(t) = \sum_{jk\ldots m} \int_{-\infty}^{+\infty} dt_1 dt_2 \cdots dt_n \chi^{(n)}_{ijk\ldots m}(t - t_1, t - t_2, \cdots, t - t_n) \times E_j(t_1)E_k(t_2) \cdots E_m(t_m), \tag{11}$$

where the time-invariance principle has been applied.[10] The above expression can be rewritten as

$$P_i^{(n)}(t) = \sum_{jk\ldots m} \int_{-\infty}^{+\infty} dt_1 dt_2 \ldots dt_n \chi^{(n)}_{ijk\ldots m}(t_1, t_2, \ldots, t_n) \times E_j(t - t_1)E_k(t - t_2) \ldots E_m(t - t_n). \tag{12}$$

When the dispersion of $\chi^{(n)}$ can be neglected, $P^{(n)}(t)$ can simply be written as

$$P^{(n)}(t) = \chi^{(n)}(E(t))^n. \tag{13}$$

This expression is easy to understand physically and is useful for order of magnitude estimates of $\chi^{(n)}$.

To define the susceptibility tensor in the frequency domain, recall that

$$\mathbf{E}(t) = \int_{-\infty}^{+\infty} d\omega\, \mathbf{E}(\omega) e^{-i\omega t}, \tag{14}$$

and

$$\mathbf{E}(\omega) = \frac{1}{2\pi} \int_{-\infty}^{+\infty} dt \mathbf{E}(t) e^{i\omega t}, \tag{15}$$

where $\mathbf{E}(\omega)$ is the Fourier transform of $\mathbf{E}(t)$. Substituting Eq. (14) into Eq. (12), we have

$$P_i^{(n)}(t) = \sum_{jk\ldots m} \int_{-\infty}^{+\infty} d\omega_1 d\omega_2 \ldots d\omega_n \chi^{(n)}_{ijk\ldots m}(-\omega_{n+1}; \omega_1, \omega_2, \ldots, \omega_n) \times E_j(\omega_1)E_k(\omega_2) \ldots E_m(\omega_n) e^{-i\omega_{n+1}t}, \tag{16}$$

where $\omega_{n+1} = \omega_1 + \omega_2 + \cdots + \omega_n$ and

$$\chi^{(n)}_{ijk\ldots m}(-\omega_{n+1}; \omega_1, \omega_2, \ldots, \omega_n) = \int_{-\infty}^{+\infty} dt_1 dt_2 \ldots dt_n \chi^{(n)}_{ijk\ldots m}(t_1, t_2, \ldots, t_n) \exp\left[i \sum_{\alpha=1}^{n} \omega_\alpha t_\alpha \right] \tag{17}$$

We now have defined the $\chi^{(n)}$ tensor in the frequency domain. From the above equation, we easily get

$$\chi^{(n)}_{ijk\ldots m}(t_1, t_2, \ldots, t_n). = \left(\frac{1}{2\pi}\right)^n \int_{-\infty}^{+\infty} d\omega_1 d\omega_2 \ldots d\omega_n \chi^{(n)}_{ijk\ldots m}(-\omega_{n+1}; \omega_1, \omega_2, \ldots, \omega_n) \exp\left[-i \sum_{\alpha=1}^{n} \omega_\alpha t_\alpha \right]. \tag{18}$$

When the dispersion is neglected,

$$\chi^{(n)}_{ijk\ldots m}(t_1, t_2, \ldots, t_n) = \chi^{(n)}_{ijk\ldots m}\delta(t_1)\delta(t_2)\ldots\delta(t_n). \tag{19}$$

The Fourier transform of $\mathbf{P}^{(n)}(t)$ is

$$\mathbf{P}^{(n)}(\omega) = \frac{1}{2\pi}\int_{-\infty}^{+\infty} dt\mathbf{P}^{(n)}(t)e^{i\omega t}. \tag{20}$$

Since

$$\delta(x - y) = \frac{1}{2\pi}\int_{-\infty}^{+\infty} dt\, e^{it(x-y)}, \tag{21}$$

we can obtain

$$P_i^{(n)}(\omega) = \sum_{jk\ldots m}\int_{-\infty}^{+\infty} d\omega_1 d\omega_2 \ldots d\omega_n \chi^{(n)}_{ijk\ldots m}(-\omega_{n+1}; \omega_1, \omega_2, \ldots, \omega_n) \times E_j(\omega_1)E_k(\omega_2)\ldots E_m(\omega_n)\delta(\omega - \omega_{n+1}). \tag{22}$$

The above expression is in a somewhat unfamiliar form. When the electric field oscillates at several discrete frequencies ω', the field can be expressed by

$$\mathbf{E}(t) = \frac{1}{2}\sum_{\omega'\geq 0}(\mathbf{E}^{\omega'}e^{-i\omega' t} + \mathbf{E}^{-\omega'}e^{i\omega' t}), \tag{23}$$

where $\mathbf{E}^{\omega'}$ is the amplitude of the electric field oscillating at frequency ω' and the summation is only over the non-negative frequency components. In addition, $\mathbf{E}^{-\omega'} = (\mathbf{E}^{\omega'})^*$. Often $\mathbf{E}^{\omega'}$ is real so that $\mathbf{E}^{-\omega'} = \mathbf{E}^{\omega'}$, therefore Eq. (23) is equivalent to

$$\mathbf{E}(t) = \sum_{\omega'\geq 0}\mathbf{E}^{\omega'}\cos(\omega' t). \tag{24}$$

The factor of 1/2 in front of Eq. (23) is our preferred convention for defining $\mathbf{E}^{\omega'}$, but is not included in some of the other conventions.

The Fourier transform of $\mathbf{E}(t)$ is then

$$\mathbf{E}(\omega) = \frac{1}{2}\sum_{\omega'\geq 0}[\mathbf{E}^{\omega'}\delta(\omega - \omega') + \mathbf{E}^{-\omega'}\delta(\omega + \omega')]. \tag{25}$$

Notice that $\mathbf{E}(\omega)$ is different from $\mathbf{E}^{\omega'}$. When the electric field contains discrete frequencies, so should the induced polarization $\mathbf{P}^{(n)}(t)$. Therefore,

$$\mathbf{P}^{(n)}(t) = \frac{1}{2}\sum_{\omega'\geq 0}[\mathbf{P}^{(n)\omega'}e^{-i\omega' t} + \mathbf{P}^{(n)-\omega'}e^{i\omega' t}], \tag{26}$$

where $\mathbf{P}^{(n)-\omega'} = (\mathbf{P}^{(n)\omega'})^*$. Note that ω' now spans all possible sum and difference frequencies of the incoming fields. Similar to the Fourier transform of the electric field, the Fourier transform of the polarization is

$$\mathbf{P}(\omega) = \frac{1}{2}\sum_{\omega'\geq 0}[\mathbf{P}^{\omega'}\delta(\omega - \omega') + \mathbf{P}^{-\omega'}\delta(\omega + \omega')]. \tag{27}$$

Substituting Eqs. (25) and (27) into Eq. (22), we can finally get

$$P_i^{(n)\omega_{n+1}} = K(-\omega_{n+1}; \omega_1, \omega_2, \ldots, \omega_n) \sum_{jk\ldots m} \chi^{(n)}_{ijk\ldots m}(-\omega_{n+1}; \omega_1, \omega_2, \ldots, \omega_n) E_j^{\omega_1} E_k^{\omega_2} \ldots E_m^{\omega_n}. \tag{28}$$

Here

$$K(-\omega_{n+1}; \omega_1, \omega_2, \ldots, \omega_n) = \frac{D}{2^{q-p}}, \tag{29}$$

where q is the number of nonzero input frequencies and p is the number of nonzero output frequencies, which is one when ω_σ is nonzero and zero otherwise, and D is the number of distinguishable orderings in the input frequencies. Rewriting $P_i^{(n)\omega_{n+1}}$ as $P_i^{\omega_{n+1}}$ and keeping in mind that K is independent of Cartesian components, we have

$$P_i^{\omega_{n+1}} = K(-\omega_{n+1}; \omega_1, \omega_2, \ldots, \omega_n) \sum_{jk\ldots m} \chi^{(n)}_{ijk\ldots m}(-\omega_{n+1}; \omega_1, \omega_2, \ldots, \omega_n) E_j^{\omega_1} E_k^{\omega_2} \ldots E_m^{\omega_n}. \tag{30}$$

Note that in the above expression, $\omega_1, \omega_2, \ldots, \omega_n$ can be either positive, zero, or negative, in contrast to Eqs. (26) and (27), where only non-negative frequencies are considered.

Another physically much simpler method to calculate K is to ignore the fact that $\chi^{(n)}$ is a tensor and assume the medium to be dispersionless. Under these two approximations, the n-order nonlinear polarization of the medium can be expressed in the time domain as

$$P^{(n)}(t) = \chi^{(n)}[E_\alpha \cos\omega_\alpha t + E_\beta \cos\omega_\beta t + \ldots]^n \tag{31}$$

where the input electric fields are oscillating at frequencies $\omega_\alpha, \omega_\beta, \ldots$. The amplitude of the nonlinear polarization with frequency ω_σ is then

$$P^{(n)\omega_\sigma} = K^{(n)}\chi^{(n)}(E_\alpha)^{n_1}(E_\beta)^{n_2} \ldots, \tag{32}$$

where $\omega_\sigma = n_1\omega_\alpha + n_2\omega_\beta + \ldots$. It can easily be shown that

$$K^{(n)} = \frac{1}{2^{q-p}} \frac{n!}{(n_1)!(n_2)! \ldots}, \tag{33}$$

where q is the number of nonzero input frequencies and p is the number of nonzero output frequencies, which is unity when ω_σ is nonzero and zero otherwise. The first part on the right side of Eq. (33) is due to our definition of the polarization and the electric field. The second part is the number of distinguished orderings in the input frequencies as can be seen when we expand Eq. (31). Therefore, for example, $K(-2\omega; \omega, \omega) = 1/2$ and $K(-2\omega; \omega, \omega, 0) = 3/2$. Table 1 lists values of $K(-\omega_{n+1}; \omega_1, \omega_2, \ldots, \omega_n)$ for various second and third order nonlinear optical processes.

As an example of how to calculate $K(-\omega_4; \omega_1, \omega_2, \omega_3)$ for the cubic susceptibility, consider an optical field at frequency ω with amplitude E^ω and a dc field with amplitude E^0

Table 1 Values of $K(-\omega_{n+1}; \omega_1, \omega_2, \ldots, \omega_n)$ for Various Second and Third Order Optical Processes

NLO process	Frequencies	K
Static	$0; 0, 0$	1
SHG	$-2\omega; \omega, \omega$	1/2
LEO	$-\omega; \omega, 0$	2
SFG	$-(\omega_1 + \omega_2); \omega_1, \omega_2$	1
Static	$0; 0, 0, 0$	1
DCSHG (EFISH)	$-2\omega; \omega, \omega, 0$	3/2
THG	$-3\omega; \omega, \omega, \omega$	1/4
DFWM	$-\omega; \omega, \omega, -\omega$	3/4
Kerr effect	$-\omega; \omega, 0, 0$	3
Rectification	$0; \omega, -\omega, 0$	3/2

$$K(-\omega_{n+1}; \omega_1, \omega_2, \ldots, \omega_n) = \frac{D}{2^{q-p}},$$

where q is the number of nonzero input frequencies and p is the number of nonzero output frequencies, which is one when ω_σ is nonzero and zero otherwise, and D is the number of distinguishable orderings in the input frequencies.
SHG: second harmonic generation
LEO: linear electro-optic effect
SGF: sum frequency generation
DCSHG (EFISH): electric field induced second harmonic generation
THG: third harmonic generation
DFWM: degenerate four wave mixing

$$[E^\omega \cos\omega t + E^0]^3 =$$

$$\frac{1}{4}(E^\omega)^3 \cos 3\omega t + \frac{3}{2}(E^\omega)^2 E^0 \cos 2\omega t + \left[\frac{3}{4}(E^\omega)^3 + 3E^\omega(E^0)^2\right] \cos \omega t$$

$$+ \left[\frac{3}{2}(E^\omega)^2 E^0 + (E^0)^3\right]. \tag{34}$$

The coefficients on the right side describe values of K for various third order optical processes. For example, $K = 1/4$ is for third harmonic generation, $K = 3/2$ for electric-field induced second harmonic generation, and so on.

B. Recommended Conventions

One might anticipate that Eqs. (23), (26), and (30) would naturally define the convention for $\chi^{(n)}$. Many other conventions, however, are possible. In this subsection, we define two conventions that are most frequently used by researchers in the field of organic nonlinear optics, especially those involved with measurements of liquids and solids.

Convention I:

$$P_i^{\omega_{n+1}} = \sum_{jk\ldots m} \chi^{(n)}_{ijk\ldots m}(-\omega_{n+1}; \omega_1, \omega_2, \ldots, \omega_n) E_j^{\omega_1} E_k^{\omega_2} \ldots E_m^{\omega_n}, \tag{35}$$

and

$$\mathbf{A}(t) = \frac{1}{2} \sum_{\omega' \geq 0} [\mathbf{A}^{\omega'} e^{-i\omega' t} + \mathbf{A}^{-\omega'} e^{i\omega' t}], \tag{36}$$

where $\mathbf{A}(t)$ can be either $\mathbf{P}(t)$ or $\mathbf{E}(t)$. In this convention, all the numerical factors are absorbed into $\chi^{(n)}$. This convention has the advantage that in most cases, one does not have to worry about the additional numerical factors K in the equation relating $\mathbf{P}^{\omega}$ to $\mathbf{E}^{\omega}$. The disadvantage is that different susceptibilities of the same order (for example, second harmonic generation $\chi^{(2)}_{ijk}(-2\omega; \omega, \omega)$ and sum frequency generation $\omega^{(2)}_{ijk}(-\omega_3; \omega_1, \omega_3)$) may not have the same value in the zero frequency limit.

Convention II:

$$P_i^{\omega_{n+1}} = \sum_{jk\ldots m} K(-\omega_{n+1}; \omega_1, \omega_2, \ldots, \omega_n) \chi^{(n)}_{ijk\ldots m}(-\omega_{n+1}; \omega_1, \omega_2, \ldots, \omega_n) \times E_j^{\omega_1} E_k^{\omega_2} \ldots E_m^{\omega_n}, \tag{37}$$

and

$$\mathbf{A}(t) = \frac{1}{2} \sum_{\omega' \geq 0} [\mathbf{A}^{\omega'} e^{-i\omega' t} + \mathbf{A}^{-\omega'} e^{i\omega' t}],$$

where again $\mathbf{A}(t)$ represents $\mathbf{P}(t)$ or $\mathbf{E}(t)$. This convention is a natural extension of what we discussed in the previous subsection. It has the advantage that different processes of the same order yield the same value at the zero frequency limit. The disadvantage is that one always has to write down values of K explicitly for a nonlinear optical process.

C. Other Conventions

Convention III:

$$P_i^{\omega_{n+1}} = \sum_{jk\ldots m} \chi^{(n)}_{ijk\ldots m}(-\omega_{n+1}; \omega_1, \omega_2, \ldots, \omega_n) E_j^{\omega_1} E_k^{\omega_2} \ldots E_m^{\omega_n}, \tag{38}$$

$$\mathbf{A}(t) = \sum_{\omega' \geq 0} [\mathbf{A}^{\omega'} e^{-i\omega' t} + \mathbf{A}^{-\omega'} e^{i\omega' t}]. \tag{39}$$

This convention differs from convention I in that the electric field $\mathbf{E}^{\omega}$ and polarization $\mathbf{P}^{\omega}$ are differently defined.

Convention IV:

$$\chi^{(n)}_{ijk\ldots m}(-\omega_{n+1}; \omega_1, \omega_2, \ldots, \omega_n) = \frac{\partial^n P_i^{\omega_{n+1}}}{\partial E_j^{\omega_1} \, \partial E_k^{\omega_2} \ldots \partial E_m^{w_n}}. \tag{40}$$

Convention IV will yield an expression for $\mathbf{P}$ in terms of $\mathbf{E}$ in the form of the so-called Taylor expansion. This convention is heavily used by researchers in finite field calculations and measurements on gas phase atoms and molecules. In fact, it is the convention used in the popular MOPAC semi-empirical calculation package.[71] This definition is most used in calculations of the static limit.

Convention V:

$$P_i^{\omega_{n+1}} = \sum_{jk\ldots m} K'(-\omega_{n+1};\ \omega_1,\ \omega_2,\ \ldots,\ \omega_n)$$

$$\chi^{(n)}_{ijk\ldots m}(-\omega_{n+1};\ \omega_1,\ \omega_2,\ \ldots,\ \omega_n)E_j^{\omega_1}\ E_k^{\omega_2}\ \ldots\ E_m^{\omega_n}, \tag{41}$$

and

$$\mathbf{A}(t) = \sum_{\omega' \geq 0} [\mathbf{A}^{\omega'} e^{-i\omega' t} + \mathbf{A}^{-\omega'}\ e^{iw' t}].$$

The prime on $K'(-\omega_{n+1};\ \omega_1,\ \omega_2,\ \ldots,\ \omega_n)$ indicates that although it is related to $K(-\omega_{n+1};\ \omega_1,\ \omega_2,\ \ldots,\ \omega_n)$ as defined in conventions I and II, they are different in value because of the difference in definitions for the polarizations and electric fields.

Table 2 lists various groups associated with the conventions they adopted in their publications. The assignments of individual works to different conventions have been made solely on their outlined equations. Notice that we placed Kajzar,[56] Hurst,[61] and Neher[58] in both convention III and convention V because the equations outlined in their papers did not uniquely specify the convention they used. In addition, we note that for second harmonic and third harmonic generations, conventions III and V are identical.

The five conventions listed in Table 2 are by no means complete. Conventions other than the five listed have been used, especially in the early days of nonlinear optics research. For example,[66] Maker and Terhune[67] used c_{1111} for third order optical processes, where

$$c_{1111}(-3\omega;\ \omega,\ \omega,\ \omega) = \chi^{(3)I}_{1111}(-3\omega;\ \omega,\ \omega,\ \omega). \tag{42}$$

However, they only included the $1/2^{q-p}$ factor in their expression for c_{1111} and not the other permutation factor. For them,

$$c_{1111}(-\omega;\ \omega,\ \omega,\ -\omega) = c_{1111}(-3\omega;\ \omega,\ \omega,\ \omega), \tag{43}$$

when the dispersion is neglected. Therefore

$$c_{1111}(-\omega;\ \omega,\ \omega,\ -\omega) = \frac{1}{3}\ \chi^{(3)I}_{1111}(-\omega;\ \omega,\ \omega,\ -\omega). \tag{44}$$

In the papers by Hellwarth *et al.*,[68]

Table 2 Various Groups with the Conventions They Adopted

Convention	Group
I	Yariv,[a] Levine,[b] Oudar,[c] Garito,[d] Singer[e]
II	Orr,[f] Ward,[g] Kajzar,[h] Burland,[i] Prasad,[j] Cheng[k]
III	Kajzar,[m] Hermann,[n] Neher[o]
IV	Bishop,[p] Papadopoulos,[q] Hurst,[r] Buckingham[s]
V	Meredith,[l] Boyd,[l] Kajzar,[m] Hermann,[n] Neher[o]

a: Ref.[35]; *b*: Ref.[36]; *c*: Ref.[37]; *d*: Ref.[38–40]; *e*: Ref.[41]; *f*: Ref.[42]; *g*: Ref.[43]; *h*: Ref.[44–46]; *i*: Ref.[47–49]; *j*: Ref.[51]; *k*: Ref.[52]; *l*: Ref.[53,54]; *m*: Ref.[56]; *n*: Ref.[57]; *o*: Ref.[58]; *p*: Ref.[59]; *q*: Ref.[60]; *r*: Ref.[61]; *s*: Ref.[94]; and *t*: Ref.[70].

$$\sigma = 8\chi^{(3)I}_{1111}(-3\omega;\ \omega,\ \omega,\ \omega). \tag{45}$$

In the papers by Soos *et al.*,[69]

$$\gamma^{Soos}_{ijkl}(-3\omega;\ \omega,\ \omega,\ \omega) = \frac{1}{2}\,\gamma^{I}_{ijkl}(-3\omega;\ \omega,\ \omega,\ \omega). \tag{46}$$

D. Relationships between Conventions

We can now turn to the relationships between the various conventions. It can be seen from the definitions that

$$\begin{aligned}
&\chi^{(n)I}_{ij\ldots m}(-\omega_{n+1};\ \omega_1,\ \omega_2,\ \ldots,\ \omega_n)\\
&\quad = K(-\omega_{n+1};\ \omega_1,\ \omega_2,\ \ldots,\ \omega_n)\chi^{(n)II}_{ij\ldots m}(-\omega_{n+1};\ \omega_1,\ \omega_2,\ \ldots,\ \omega_n)\\
&\quad = \frac{1}{2^{q-p}}\,\chi^{(n)III}_{ij\ldots m}(-\omega_{n+1};\ \omega_1,\ \omega_2,\ \ldots,\ \omega_n) = \frac{1}{2^{q-p}}\\
&\qquad K'(-\omega_{n+1};\ \omega_1,\ \omega_2,\ \ldots,\ \omega_n)\chi^{(n)V}_{ij\ldots m}(-\omega_{n+1};\ \omega_1,\ \omega_2,\ \ldots,\ \omega_n),
\end{aligned} \tag{47}$$

where q and p have been defined previously in Eq. (33). Since in convention V $K'(-\omega_{n+1};\ \omega_1,\ \omega_2,\ \ldots,\ \omega_n)$ is equal to the number of distinguished orderings in the input frequencies, we therefore have

$$K(-\omega_{n+1};\ \omega_1,\ \omega_2,\ \ldots,\ \omega_n) = \frac{1}{2^{q-p}}\,K'(-\omega_{n+1};\ \omega_1,\ \omega_2,\ \ldots,\ \omega_n). \tag{48}$$

This indicates that the following relation is true:

$$\chi^{(n)II}_{ij\ldots m}(-\omega_{n+1};\ \omega_1,\ \omega_2,\ \ldots,\ \omega_n) = \chi^{(n)V}_{ij\ldots m}(-\omega_{n+1};\ \omega_1,\ \omega_2,\ \ldots,\ \omega_n), \tag{49}$$

demonstrating that although conventions II and V are different in definition, they generate the same numerical values for all orders of optical susceptibilities. Convention IV is mainly used in the static limit, where it can be easily shown that

$$\chi^{(n)II}_{ij\ldots m}(0;\ 0,\ 0,\ \ldots,\ 0) = \frac{1}{n!}\,\chi^{(n)IV}_{ij\ldots m}(0;\ 0,\ 0,\ \ldots,\ 0). \tag{50}$$

Common examples for macroscopic susceptibilities are the following:

$$\chi^{(2)I}_{ijk}(-2\omega;\ \omega,\ \omega) = \frac{1}{2}\,\chi^{(2)II}_{ijk}(-2\omega;\ \omega,\ \omega) = \frac{1}{2}\,\chi^{(2)III}_{ijk}(-2\omega;\ \omega,\ \omega), \tag{51}$$

$$\chi^{(3)I}_{ijkl}(-2\omega;\ \omega,\ \omega,\ 0) = \frac{3}{2}\,\chi^{(3)II}_{ijkl}(-2\omega;\ \omega,\ \omega,\ 0) = \frac{1}{2}\,\chi^{(3)III}_{ijkl}(-2\omega;\ \omega,\ \omega,\ 0), \tag{52}$$

and

$$\chi^{(3)I}_{ijkl}(-3\omega;\ \omega,\ \omega,\ \omega) = \frac{1}{4}\,\chi^{(3)II}_{ijkl}(-3\omega;\ \omega,\ \omega,\ \omega) = \frac{1}{4}\,\chi^{(3)III}_{ijkl}(-3\omega;\ \omega,\ \omega,\ \omega), \tag{53}$$

as well as

$$\chi_{ijk}^{(2)II}(0; 0, 0) = \frac{1}{4}\chi_{ijk}^{(2)IV}(0; 0, 0), \tag{54}$$

and

$$\chi_{ijkl}^{(3)II}(0; 0, 0, 0) = \frac{1}{6}\chi_{ijkl}^{(3)IV}(0; 0, 0, 0). \tag{55}$$

The corresponding microscopic susceptibilities are

$$\beta_{ijk}^{I}(-2\omega; \omega, \omega) = \frac{1}{2}\beta_{ijk}^{II}(-2\omega; \omega, \omega) = \frac{1}{2}\beta_{ijk}^{III}(-2\omega; \omega, \omega), \tag{56}$$

$$\gamma_{ijkl}^{I}(-2\omega; \omega, \omega, 0) = \frac{3}{2}\gamma_{ijkl}^{II}(-2\omega; \omega, \omega, 0) = \frac{1}{2}\gamma_{ijkl}^{III}(-2\omega; \omega, \omega, 0), \tag{57}$$

and

$$\gamma_{ijkl}^{I}(-3\omega; \omega, \omega, \omega) = \frac{1}{4}\gamma_{ijkl}^{II}(-3\omega; \omega, \omega, \omega) = \frac{1}{4}\gamma_{ijkl}^{III}(-3\omega; \omega, \omega, \omega), \tag{58}$$

as well as

$$\beta_{ijk}^{II}(0; 0, 0) = \frac{1}{4}\beta_{ijk}^{IV}(0; 0, 0), \tag{59}$$

and

$$\gamma_{ijkl}^{II}(0; 0, 0, 0) = \frac{1}{6}\gamma_{ijkl}^{IV}(0; 0, 0, 0). \tag{60}$$

We note that for SHG and THG measurements, conventions II and III yield the same values.

E. Analysis and Conventions

We shall first discuss relations among different processes of the same order when the dispersion can be neglected. This is crucial in convention I where these processes may not have the same value and, as a result, one needs to compare results from different measurements. Typical examples include comparisons between $\chi^{(3)}$ values from electric field induced second harmonic generation and third harmonic generation measurements and between $\chi^{(2)}$ values from linear electro optic and second harmonic generation measurements.

From the definition of convention II, we can easily see that when the dispersion is neglected, different susceptibilities of the same order should have the same value. For example, the susceptibility for sum-frequency generation should have the same value as that for the second harmonic generation. Mathematically,

$$\lim_{\omega_1, \omega_2 \to \omega} \chi_{111}^{(2)II}(-\omega_3; \omega_1, \omega_2) = \chi_{111}^{(2)II}(-2\omega; \omega, \omega). \tag{61}$$

On the other hand, from Eq. (47):

$$\chi^{(2)I}_{111}(-\omega_3;\ \omega_1,\ \omega_2) = \chi^{(2)II}_{111}(-\omega_3;\ \omega_1,\ \omega_2), \tag{62}$$

and Eq. (51), we have

$$\lim_{\omega_1,\omega_2 \to \omega} \chi^{(2)I}_{111}(-\omega_3;\ \omega_1,\ \omega_2) = 2\chi^{(2)I}_{111}(-2\omega;\ \omega,\ \omega). \tag{63}$$

This indicates that when the dispersion is neglected, $\chi^{(2)}_{111}(-\omega_3;\ \omega_1,\ \omega_2)$ is twice as large as $\chi^{(2)}_{111}(-2\omega;\ \omega,\ \omega)$ when they are expressed in convention I. The discontinuity is an intrinsic reflection of convention I. To see physically why this is the case, let us consider sum-frequency generation in a medium with $\chi^{(2)}_{ijk}(-\omega_3;\ \omega_1,\ \omega_2)$ where the optical field is polarized in the x direction with two frequency components, $E^{\omega_1}_1$ and $E^{\omega_2}_2$. In convention I, $\chi^{(2)}_{111}(-\omega_3;\ \omega_1,\ \omega_2)$ is defined by

$$P^{\omega_3}_1 = \chi^{(2)I}_{111}(-\omega_3;\ \omega_1,\ \omega_2)E^{\omega_1}_1\ E^{\omega_2}_1, \tag{64}$$

where I indicates convention I and the polarization $P^{\omega_3}_1$ at the sum-frequency ω_3 is directly related to the measured sum-frequency intensity. This is important in the sense that we have already included the contribution from $\chi^{(2)I}_{111}(-\omega_3;\ \omega_2,\ \omega_1)$, which, because of the intrinsic permutation symmetry, is equal to $\chi^{(2)I}_{111}(-\omega_3;\ \omega_1,\ \omega_2)$. When $\omega_1 = \omega_2 = \omega$, the following relation holds:

$$P^{2\omega}_1 = \chi^{(2)I}_{111}(-2\omega;\ \omega,\ \omega)[E^{\omega_1}_1 + E^{\omega_2}_1]^2. \tag{65}$$

Here we implied that since $E^{\omega_1}_1$ and $E^{\omega_2}_1$ are amplitudes for their respective frequencies, they are frequency independent.

The relation between $\chi^{(2)I}_{111}(-\omega_3;\ \omega_1,\ \omega_2)$ and $\chi^{(2)I}_{111}(-2\omega;\ \omega,\ \omega)$ when $\omega_1 = \omega_2 = \omega$ can be made clear by combining Eq. (64) with the following two equations:

$$P^{2\omega_1}_1 = \chi^{(2)I}_{111}(-2\omega_1,\ \omega_1,\ \omega_1)E^{\omega_1}_1\ E^{\omega_1}_1 \tag{66}$$

$$P^{2\omega_2}_1 = \chi^{(2)I}_{111}(-2\omega_2;\ \omega_2,\ \omega_2)E^{\omega_2}_1\ E^{\omega_2}_1, \tag{67}$$

and then comparing with Eq. (65). One can easily see that Eq. (63) is indeed true.

There are several other worthwhile observations. One has to be careful when making comparisons between theory and experiment, between different theoretical calculations, and between different experimental measurements. Attention must be paid to the conventions being used before drawing comparisons of results. In addition, we also need to be careful of the reliability of susceptibility values of reference materials. Typical examples of reference materials are $\chi^{(2)}$ values for quartz and $\chi^{(3)}$ values for glass. This will be the main subject in the next section where we will discuss standards and references. In the case of third harmonic generation, the concerns surrounding results of different measurements are mainly due to BK-7 glass and fused silica $\chi^{(3)}$ values.

The most common case of inconsistencies in the literature concerns DCSHG measurements, where different $\chi^{(2)}$ values for the same molecule exist not only because of the different reference values used, but also because of the improper use of conventions. The main reason is that in DCSHG measurements of noncentrosymmetric molecules, both second and third order nonlinear optical processes contribute to the measurement. It is often assumed that in DCSHG measurements of noncentrosymmetric structures, $\langle\gamma\rangle + \vec{\mu}.\ \vec{\beta}/(5kT)$ is the expression valid in all conventions, where $\langle\gamma\rangle$ is the scalar part of γ_{ijkl} defined by

$$\langle\gamma(-2\omega;\ \omega,\ \omega,\ 0)\rangle = \frac{1}{5}\left\{\sum_i \gamma_{iiii} + \frac{1}{3}\sum_{i\neq j}(\gamma_{iijj} + \gamma_{ijij} + \gamma_{ijji})\right\}, \tag{68}$$

$\vec{\mu}$ is the electric dipole moment, and $\vec{\beta}$ is the vector part of β_{ijk} defined by the following:

$$\beta_i(-2\omega;\ \omega,\ \omega) = \beta_{iii} + \frac{1}{3}\sum_{j\neq i}[2\beta_{jji} + \beta_{ijj}]. \tag{69}$$

Recently, it was pointed out[49] that in convention II what was actually measured was $\langle\gamma\rangle + \vec{\mu}\cdot\vec{\beta}/(15kT)$. This can be made clear by the following:

The relationship between $\chi^{(2)}_{111}$ and $\vec{\beta}$ is given by

$$\chi^{(2)}_{111}(-2\omega;\ \omega,\ \omega) = NL^{2\omega}(L^{\omega})^2\,\frac{\vec{\mu}\cdot\vec{\beta}(-2\omega;\ \omega,\ \omega)}{5kT}\,L^0E^0_1, \tag{70}$$

while $\chi^{(3)}_{1111}$ is given by

$$\chi^{(3)}_{1111}(-2\omega;\ \omega,\ \omega,\ 0) = NL^{2\omega}(L^{\omega})^2L^0\,\langle\gamma(-2\omega;\ \omega,\ \omega,\ 0)\rangle, \tag{71}$$

where N is the number density, L's are the local field factors, and E^0_1 is the dc field. *The above two equations hold in all conventions.* Therefore, in convention I, the polarization $P^{2\omega}_1$ is

$$\begin{aligned} P^{2\omega}_1 &= \chi^{(2)I}_{111}(-2\omega;\ \omega,\ \omega)(E^{\omega}_1)^2 + \chi^{(3)I}_{1111}(-2\omega;\ \omega,\ \omega,\ 0)(E^{\omega}_1)^2E^0_1 \\ &= NL^{2\omega}(L^{\omega})^2L^0\left[\langle\gamma^I(-2\omega;\ \omega,\ \omega,\ 0)\rangle + \frac{\vec{\mu}\cdot\vec{\beta}^I(-2\omega;\ \omega,\ \omega)}{5kT}\right](E^{\omega}_1)^2E^0_1, \end{aligned} \tag{72}$$

but in convention II, it is given by

$$\begin{aligned} P^{2\omega}_1 &= \frac{1}{2}\chi^{(2)II}_{111}(-2\omega;\ \omega,\ \omega)(E^{\omega}_1)^2 + \frac{3}{2}\chi^{(3)II}_{1111}(-2\omega;\ \omega,\ \omega,\ 0)(E^{\omega}_1)^2E^0_1 \\ &= \frac{3}{2}NL^{2\omega}(L^{\omega})^2L^0\left[\langle\gamma^{II}(-2\omega;\ \omega,\ \omega,\ 0)\rangle + \frac{\vec{\mu}\cdot\vec{\beta}^{II}(-2\omega;\ \omega,\ \omega)}{15kT}\right](E^{\omega}_1)^2E^0_1. \end{aligned} \tag{73}$$

Burland and coworkers[49] came up with a method to recover the factor of 1/5 by changing convention II to convention II*, which is different from convention II only by

$$\beta^{II}_{ijk}(-2\omega;\ \omega,\ \omega) = 3\beta^{II*}_{ijk}(-2\omega;\ \omega,\ \omega). \tag{74}$$

In this manner, the coefficient associated with kT on the right side of Eq. (73) will become 5, which is the same as that associated with kT on the right side of Eq. (72).

We shall now discuss the various DCSHG measurements of p-nitroaniline (PNA) and 2-methyl-4-nitroaniline (MNA)[39,47,48] in 1,4-dioxane. At $\lambda = 1064$ *nm*, the results by Teng and Garito[39] are for PNA, $\beta_x^I(-2\omega; \omega, \omega) = 16.9 \times 10^{-30}$ *esu*, and for MNA, $\beta_x^I(-2\omega; \omega, \omega) = 16.7 \times 20^{-30}$ *esu*. In comparison, the results by Burland *et al.*[47,48] are for PNA, $\beta_x^{II*}(-2\omega; \omega, \omega) = 16.3 \times 10^{-30}$ *esu*, and for MNA, $\beta_x^{II*}(-2\omega; \omega, \omega) = 18.8 \times 10^{-30}$ *esu* at the same wavelength. The two sets of data are not consistent with each other since the measured values are expressed in different conventions, I and II*, where they are related by the following

$$\beta_{ijk}^I(-2\omega; \omega, \omega) = \frac{3}{2}\, \beta_{ijk}^{II*}(-2\omega; \omega, \omega). \tag{75}$$

A careful study reveals that the two sets of experiments adopted the equivalent values of d_{11} for quartz: in the case of Teng and Garito,[39] $d_{11}^I(-2\omega; \omega, \omega) = 1.2 \times 10^{-9}$ *esu*, and in the case of Burland *et al.*,[47,48] $d_{11}^{II}(-2\omega; \omega, \omega) = 2.4 \times 10^{-9}$ *esu*. The problem may lie with the $\chi^{(3)}$ values for glass windows, as we pointed out in the beginning of this Chapter. The confusion surrounding this will be discussed in more detail in the next section.

The second observation relates to the much-cited paper by Cheng *et al.*[52] From the discussion and equations outlined in their paper, it seems that they have used convention II*. But Eq. (7) in the paper is inconsistent with convention II*. The correct form of the equation should be

$$E_b^i = q' \frac{8q\pi\chi_i^{(3)} l_{c_i}}{\lambda(n_i^b + n_i^f)} E_i^{\omega} E_i^{\omega} E_i^{\omega'}, \tag{76}$$

where $q' = 1/4$ for THG and $q' = 3/2$ for DCSHG. The reason is that the above equation comes from the nonlinear polarization which has different coefficients when convention II (II*) is used. It is not clear how the correction will affect their measured values of β and γ for the organic systems they studied.

Table 2 also shows that Kajzar *et al.* adopted two different conventions. Their earlier work[56] was based on convention III (or V), and later they switched to convention II. Fortunately, the change did not cause any trouble since the earlier work[56] was on THG, for which both conventions produce the same value. The main concern is the d_{11} value of quartz at $\lambda = 1064$ *nm*. In another paper by Kajzar *et al.*,[46] they used the value for the second harmonic coefficient of quartz to be 1.2×10^{-9} *esu*, which, in convention II, is clearly too small by a factor of two compared to the value commonly accepted. The inconsistency was corrected in a later paper.[48] But γ values for chloroform,[48] for example, did not change as a result of the correction. So it is unclear how the quartz d_{11} values will affect the nonlinear susceptibilities of chloroform. This is particularly important since the β value of chloroform[46] was found to be -0.49×10^{-49} *esu*, which was later used to calibrate β values of other organic structures by Hyper-Rayleigh Scattering.[50] It is still an open question as to how these values should be rescaled.

A final observation concerns the two definitions of the vector part of β_{ijk}. The first is widely used in DCSHG measurements as defined in Eq. (69). The second differs from the first by a factor of 3/5,

$$\beta_i' = \frac{3}{5} \sum_j \beta_{jji}. \tag{77}$$

β_i' is in fact the true definition of the vector part of β_{ijk} which is often used in many other calculations. For example, β_i' is found in the widely used MOPAC program which does semiemperical calculations of molecular susceptibilities (see Chapter 2 for the concepts behind MOPAC). Also the octopolar part of β_{ijk} is calculated by subtracting β_i' from β_{ijk}.[72]

The microscopic nonlinear optical susceptibilities can be derived from the time-dependent quantum perturbation theory using the Bogoliubov and Mitropolsky method of averages.[65] This method employs the idea of dealing separately with slowly varying and rapidly varying components of the state wave function. The secular and resonant terms are associated with slowly varying amplitudes of state wave function which leads to divergences in the standard perturbation theory. The following expressions can then be obtained[42]:

$$\begin{aligned}
&\beta_{ijk}^{I}(-\omega_3; \omega_1, \omega_2) \\
&= K(-\omega_3; \omega_1, \omega_2) I_{1,2} \left(-\frac{e^3}{\hbar^2}\right) \sum_{m_1, m_2} \\
&\left\{ \frac{r^i_{gm_1} \bar{r}^j_{m_1 m_2} r^k_{m_2 g}}{(\Omega^*_{m_1 g} - \omega_3)(\Omega^*_{m_2 g} - \omega_1)} + \frac{r^i_{gm_1} \bar{r}^k_{m_1 m_2} r^j_{m_2 g}}{(\Omega^*_{m_1 g} + \omega_2)(\Omega_{m_2 g} + \omega_3)} \right. \\
&\left. + \frac{r^k_{gm_1} \bar{r}^i_{m_1 m_2} r^j_{m_2 g}}{(\Omega_{m_1 g} + \omega_2)(\Omega_{m_2 g} - \omega_1)} \right\},
\end{aligned} \tag{78}$$

and

$$\begin{aligned}
&\gamma_{ijkl}^{I}(-\omega_4; \omega_1, \omega_2, \omega_3) \\
&= K(-\omega_4; \omega_1, \omega_2, \omega_3) \left(\frac{e^4}{\hbar^3}\right) I_{1,2,3} \\
&\times \left[\sum_{m_1, m_2, m^3}' \right. \\
&\left\{ \frac{r^i_{gm_3} \bar{r}^l_{m_3 m_2} \bar{r}^k_{m_2 m_1} r^j_{m_1 g}}{(\omega_{m_3 g} - \omega_4)(\omega_{m_2 g} - \omega_1 - \omega_2)(\omega_{m_1 g} - \omega_1)} \right. \\
&+ \frac{r^l_{gm_3} \bar{r}^i_{m_3 m_2} \bar{r}^{k m_2 m_1} r^j_{m_1 g}}{(\omega^*_{m_3 g} + \omega_3)(\omega_{m_2 g} - \omega_1 - \omega_2)(\omega_{m_1 g} - \omega_1)} \\
&+ \frac{r^j_{gm_3} \bar{r}^k_{m_3 m_2} \bar{r}^i_{m_2 m_1} r^l_{m_1 g}}{(\omega^*_{m_3 g} + \omega_1)(\omega^*_{m_2 g} + \omega_1 + \omega_2)(\omega_{m_1 g} - \omega_3)} \\
&\left. + \frac{r^j_{gm_3} \bar{r}^k_{m_3 m_2} \bar{r}^l_{m_2 m_1} r^i_{m_1 g}}{(\omega^*_{m_3 g} + \omega_1)(\omega^*_{m_2 g} + \omega_1 + \omega_2)(\omega^*_{m_1 g} + \omega_4)} \right\} \\
&\sum_{m_1, m_2}' \left\{ \frac{r^i_{gm_2} r^l_{m_2 g} r^k_{gm_1} r^j_{m_1 g}}{(\omega_{m_2 g} - \omega_4)(\omega_{m_2 g} - \omega_3)(\omega_{m_1 g} - \omega_1)} \right.
\end{aligned}$$

$$+ \frac{r^i_{gm_2} r^l_{m_2g} r^k_{gm_1} r^j_{m_1g}}{(\omega_{m_2g} - \omega_3)(\omega^*_{m_1g} + \omega_2)(\omega_{m_1g} - \omega_1)}$$

$$+ \frac{r^l_{gm_2} r^i_{m_2g} r^j_{gm_1} r^k_{m_1g}}{(\omega^*_{m_2g} + \omega_4)(\omega^*_{m_2g} + \omega_3)(\omega^*_{m_1g} + \omega_1)}$$

$$+ \left.\left. \frac{r^l_{gm_2} r^i_{m_2g} r^j_{gm_1} r^k_{m_1g}}{(\omega^*_{m_2g} + \omega_3)(\omega_{m_1g} - \omega_2)(\omega^*_{m_1g} + \omega_1)} \right\} \right], \tag{79}$$

where $r^i_{m_1m_2}$ is the matrix element of the ith-component of the position operator $\langle m_1|r^i|m_2\rangle$ ($\bar{r}^i = r^i - \langle r\rangle_{gg}$), $\hbar\omega_{m_1g}$ is the energy difference between state m_1 and the ground state g. The prime on the summations indicates a sum over all states except the ground state, and $I_{1,2,3}$ denotes an average over all permutations of ω_1, ω_2, and ω_3 with associated indices j, k, and l. The similar definition applies to $I_{1,2}$.

Numerous theoretical and experimental studies have shown that for second order nonlinear optical processes in many organic systems, a two-level model involving the ground state and charge transfer excited state is incomplete and quantitatively incorrect. Nevertheless, the model reveals the physical mechanism for the processes and yields qualitative results. Due to this understanding, the model is still widely used. For a quasi one-dimensional system where charge transfer mainly occurs in the x direction, we have

$$\beta^l_{xxx}(-2\omega;\ \omega,\ \omega) = \frac{3}{2\hbar^2} \frac{(\mu^x_{01})^2(\mu^x_{11} - \mu^x_{00})\omega^2_{10}}{(\omega^2_{10} - \omega^2)(\omega^2_{10} - 4\omega^2)}, \tag{80}$$

$$\beta^l_{xxx}(-\omega;\ \omega,\ 0) = \frac{2}{\hbar^2} \frac{(\mu^x_{01})^2(\mu^x_{11} - \mu^x_{00})(3\omega^2_{10} - \omega^2)}{(\omega^2_0 - \omega^2)^2}, \tag{81}$$

where 0 and 1 indicate the ground and excited states, respectively, μ_{00} and μ_{11} are the electric dipole moments of the ground and excited states, respectively, μ_{01} is the transition moment, and $\hbar\omega_{10}$ is the transition energy between states 1 and 0. The above equations are valid in nonresonant cases. The relation $\lim_{\omega\to 0} \beta^l_{xxx}(-\omega;\ \omega,\ 0) = 4 \lim_{\omega\to 0} \beta^l_{xxx}(-2\omega;\ \omega,\ \omega)$ holds as is expected.

Next we turn to the relationship between $\chi^{(2)}_{ijk}(-\omega;\ \omega,\ 0)$ and the electro optic (EO) coefficient r_{ijk}. In MKS units (see next section), r_{ijk} is defined as[35]

$$\left(\frac{1}{\varepsilon + \delta\varepsilon}\right)_{ij} - \left(\frac{1}{\varepsilon}\right)_{ij} = \sum_k r_{ijk} E^0_k, \tag{82}$$

where ε_{ij} is the dielectric constant tensor defined by $D_i = \varepsilon_0 \sum_j \varepsilon_{ij} E_j$, $(\delta\varepsilon)_{ij} = \sum_k \chi^{(2)l}_{ijk}(-\omega;\ \omega,\ 0)E^0_k$, and $\mathbf{E}$ is the dc electric field. Equation (82) can be rewritten as

$$-(\delta\varepsilon)_{ij} = \sum_{\alpha\beta k} \varepsilon_{i\alpha} r_{\alpha\beta k} E^0_k (\varepsilon + \delta\varepsilon)_{\beta j}. \tag{83}$$

The above equation can be simplified further to

$$\chi^{(2)l}_{ijk}(-\omega;\ \omega,\ 0) = -\sum_{\alpha\beta} \varepsilon_{i\alpha} r_{\alpha\beta k} \varepsilon_{\beta j}. \tag{84}$$

In the principal coordinate system, with the negative sign neglected, the relation becomes

$$\chi^{(2)I}_{333}(-\omega;\ \omega,\ 0) = n^4 r_{33}, \tag{85}$$

therefore

$$r_{33} = \frac{\chi^{(2)I}_{333}(-\omega;\ \omega,\ 0)}{n^4} = \frac{2\chi^{(2)II}_{333}(-\omega;\ \omega,\ 0)}{n^4}, \tag{86}$$

where n is the refractive index of the system determined by $n^2 = \varepsilon$.

Equation (86) is particularly useful in estimating EO r_{33} coefficients in poled material systems where the dc field is along the labelled 3 direction. The equation certainly has to be modified accordingly if a different convention is used. $\chi^{(2)}_{333}(-\omega;\ \omega,\ 0)$ is related to the microscopic $\beta_{ijk}(-\omega;\ \omega,\ 0)$ by the following

$$\chi^{(2)}_{333}(-\omega;\ \omega,\ 0) = N(L^{\omega})^2 L^0\, \frac{\mu_x E_{\text{total}}}{5kT}\, \beta_x(-\omega;\ \omega,\ 0), \tag{87}$$

where N is the number density, L^{ω} and L^0 local field factors, E_{total} the effective poling electric field which is related to the poling field (E_{poling}) by $E_{\text{total}} = L^0 E_{\text{poling}}$, and $\beta_x(-\omega;\ \omega,\ 0)$ is the vector part of $\beta_{ijk}(-\omega;\ \omega,\ 0)$ defined by Eq. (69) along the dipole moment direction. *The above equation is valid in all conventions.* Fortunately, a survey of the literature found wide-spread agreement in the usage of conventions for EO coefficients. Most researchers adopted the definition shown in Eq. (82).

Lastly, we summarize the relationship between $\chi^{(3)}_{1111}(-\omega;\ \omega,\ \omega,\ -\omega)$ and the intensity-dependent refractive index n_2, which is defined as

$$n = n_0 + n_2\,\langle E^2 \rangle, \tag{88}$$

where n is the total refractive index of the material, n_0 is the linear refractive index, and $\langle E^2 \rangle$ is the time average of the squared optical field.[73] It can be easily shown that n_2 is directly related to $\chi^{(3)}_{1111}(-\omega;\ \omega,\ \omega,\ -\omega)$ by

$$n_2 = \frac{4\pi}{n_0}\,\chi^{(3)I}_{1111}(-\omega;\ \omega,\ \omega,\ -\omega) = \frac{3\pi}{n_0}\,\chi^{(3)II}_{1111}(-\omega;\ \omega,\ \omega,\ -\omega). \tag{89}$$

The commonly used relation[73,86] is

$$n_2 = \frac{12\pi}{n_0}\,\chi^{(3)}_{1111}(-\omega;\ \omega,\ \omega,\ -\omega). \tag{90}$$

Equation (90) comes from Eq. (88) and the following equation that defines the nonlinear polarization:

$$P^{\omega}_1 = 3\chi^{(3)}_{1111}(-\omega;\ \omega,\ \omega,\ -\omega) E^{\omega}_1 E^{\omega}_1 E^{-\omega}_1. \tag{91}$$

F. An Overlooked Issue

In the above discussion, both the input and output fields are assumed to be aligned in the same direction. Questions arise as to whether any additional permutation factors should be considered when the input and output fields are not aligned. To illustrate this issue, we turn to the case where the two input fields are not polarized in the same direction; one is in the x direction, the other in the y direction: $\mathbf{E}^{\omega} = E_1^{\omega}\mathbf{e}_1$, and $\mathbf{E}^{\omega+\Delta} = E_2^{\omega+\Delta}\mathbf{e}_2$. In this case, $\chi_{212}^{(2)I}(-2\omega - \Delta; \omega, \omega + \Delta)$ is obtained by measuring the sum-frequency generation from the polarization in the y direction,

$$P_2^{2\omega+\Delta} = \chi_{212}^{(2)I}(-2\omega - \Delta; \omega, \omega + \Delta)E_1^{\omega}E_2^{\omega+\Delta}. \tag{92}$$

The second harmonics at 2ω and $2\omega + 2\Delta$ are generated by

$$P_2^{2\omega} = \chi_{211}^{(2)I}(-2\omega, \omega, \omega)E_1^{\omega}E_1^{\omega} \tag{93}$$

and

$$P_2^{2\omega+2\Delta} = \chi_{222}^{(2)I}(-2\omega + 2\Delta; \omega + \Delta, \omega + \Delta)E_2^{\omega+\Delta}E_2^{\omega+\Delta}. \tag{94}$$

To see how $\chi_{212}^{(2)}(-2\omega; \omega, \omega)$ is defined, let us first consider a case where the dispersion can be neglected. Suppose the total external electric field is $\mathbf{E}^{\omega} = E_1\mathbf{e}_1 + E_2\mathbf{e}_2$, so that the polarization at the second harmonic frequency is

$$P^{2\omega} = \chi^{(2)I}(E^{\omega})^2. \tag{95}$$

This indicates that, for example, for $\chi_{212}^{(2)I}(-2\omega; \omega, \omega)$, there is an additional factor of 2 in the equation relating $P^{2\omega}$ to E^{ω}, in the form of

$$P_2^{2\omega} = 2\chi_{212}^{(2)I}(-2\omega; \omega, \omega)E_1^{\omega}E_2^{\omega}. \tag{96}$$

To show that the above equation is correct, we have to work in a new coordinate system where x' is along the $\mathbf{E} = E_1\mathbf{e}_1 + E_2\mathbf{e}_2$ direction. The new coordinate system is related to the old one by the following:

$$\begin{pmatrix} x' \\ y' \end{pmatrix} = \begin{pmatrix} \cos\alpha & \sin\alpha \\ -\sin\alpha & \cos\alpha \end{pmatrix} \begin{pmatrix} x \\ y \end{pmatrix} \equiv A\begin{pmatrix} x \\ y \end{pmatrix}, \tag{97}$$

where α is the angle between $\mathbf{E}$ and the x axis. The polarizations in the new coordinate system are

$$P_{1'} = \chi_{1'1'1'}^{(2)I}(E_{1'})^2 = \chi_{1'1'1'}^{(2)I}(E_1^2 + E_2^2), \tag{98}$$

$$P_{2'} = \chi_{2'1'1'}^{(2)I}(E_{1'})^2. \tag{99}$$

Here, the frequency arguments are omitted for simplicity. $\chi_{i'j'k'}^{(2)I}$ and $\chi_{ijk}^{(2)I}$ are related by the following:

$$\chi_{i'j'k'}^{(2)I} = \sum_{ijk} A_{i'i}A_{j'j}A_{k'k}\chi_{ijk}^{(2)I}. \tag{100}$$

We recall that $P_2 = \sin\alpha P_{1'} + \cos\alpha P_{2'}$, which, after some algebra, leads to

$$P_2 = [\cos^2\alpha\chi_{211}^{(2)I} + \sin\alpha\cos\alpha\chi_{212}^{(2)I} + \sin\alpha\cos\alpha\chi_{221}^{(2)I} + \sin^2\alpha\chi_{222}^{(2)I}](E_{1'})^2. \tag{101}$$

Therefore, by comparing Eq. (101) with Eqs. (92), (93), and (94), we can see that in the case of SHG, the following is true:

$$P_2^{2\omega} = 2\chi_{212}^{(2)I}(-2\omega; \omega, \omega)E_1^{\omega}E_2^{\omega} = \chi_{212}^{(2)II}(-2\omega; \omega, \omega)E_1^{\omega}E_2^{\omega}. \tag{102}$$

The factor of two in front of $\chi_{212}^{(2)I}$ in Eq. (102) results from the fact that E_1 and E_2 are the coherent components of a single optical field. Great care has to be taken in this case, which applied not only to convention I, but also to convention II and

other conventions. Table 3 lists examples of the correct expressions for $\chi_{ijk}^{(2)}$ in conventions I and II, indicating that one always needs to be careful about the additional degeneracy involved when one input frequency has a component along more than one different direction of the $\chi^{(2)}$ tensor. In such cases, we get the following expressions:
Electric field induced second harmonic generation (DCSHG):

$$P_1^{2\omega} = \chi_{1111}^{(3)I}(-2\omega; \omega, \omega, 0)E_1^{\omega}E_1^{\omega}E_1^{0} = \frac{3}{2}\chi_{1111}^{(3)II}(-2\omega; \omega, \omega, 0)E_1^{\omega}E_1^{\omega}E_1^{0}, \tag{103}$$

$$P_2^{2\omega} = 2\chi_{2121}^{(3)I}(-2\omega; \omega, \omega, 0)E_1^{\omega}E_2^{\omega}E_1^{0} = 3\chi_{2121}^{(3)II}(-2\omega; \omega, \omega, 0)E_1^{\omega}E_2^{\omega}E_1^{0}. \tag{104}$$

Third harmonic generation (THG):

$$P_1^{3\omega} = \chi_{1111}^{(3)I}(-3\omega; \omega, \omega, \omega)E_1^{\omega}E_1^{\omega}E_1^{\omega} = \frac{1}{4}\chi_{1111}^{(3)II}(-3\omega; \omega, \omega, \omega)E_1^{\omega}E_1^{\omega}E_1^{\omega}, \tag{105}$$

$$P_2^{3\omega} = 6\chi_{2123}^{(3)I}(-3\omega; \omega, \omega, \omega)E_1^{\omega}E_2^{\omega}E_3^{\omega} = \frac{3}{2}\chi_{2123}^{(3)II}(-3\omega; \omega, \omega, \omega)E_1^{\omega}E_2^{\omega}E_3^{\omega}. \tag{106}$$

And degenerate four wave mixing (DFWM):

$$P_1^{\omega} = \chi_{1111}^{(3)I}(-\omega; \omega, \omega, -\omega)E_1^{\omega}E_1^{\omega}E_1^{-\omega} = \frac{3}{4}\chi_{1111}^{(3)II}(-\omega; \omega, \omega, -\omega)E_1^{\omega}E_1^{\omega}E_1^{-\omega}, \tag{107}$$

$$P_1^{\omega} = 2\chi_{1122}^{(3)I}(-\omega; \omega, \omega, -\omega)E_1^{\omega}E_2^{\omega}E_2^{-\omega} = \frac{3}{2}\chi_{1122}^{(3)II}(-\omega; \omega, \omega, -\omega)E_1^{\omega}E_2^{\omega}E_2^{-\omega}. \tag{108}$$

The above examples demonstrate that *additional permutation factors must be included when different Cartesian coordinates in a susceptibility tensor are associated with the same input frequency.* This rule applied to all conventions. The reasons is that mathematically, the K factor does not take these into consideration. To the best of our knowledge, few, if any, have expressed Eq. (108) correctly in the literature.

Table 3 Examples of Expressions for $\chi_{ijk}^{(2)}$ in Conventions I and II

Convention	$\chi_{211}^{(2)}$	$\chi_{212}^{(2)}$
I	$P_2^{2\omega+\Delta} = \chi_{211}^{(2)}(-2\omega - \Delta; \omega, \omega + \Delta) \times E_1^{\omega}E_1^{\omega+\Delta}$	$P_2^{2\omega+\Delta} = \chi_{212}^{(2)}(-2\omega - \Delta; \omega, \omega + \Delta) \times E_1^{\omega}E_2^{\omega+\Delta}$
	$P_2^{2\omega} = \chi_{211}^{(2)}(-2\omega; \omega, \omega)\, E_1^{\omega}E_1^{\omega}$	$P_2^{2\omega} = 2\chi_{212}^{(2)}(-2\omega; \omega, \omega)E_1^{\omega}E_2^{\omega}$
II	$P_2^{2\omega+\Delta} = \chi_{211}^{(2)}(-2\omega - \Delta; \omega, \omega + \Delta) \times E_1^{\omega}E_1^{\omega+\Delta}$	$P_2^{2\omega+\Delta} = \chi_{212}^{(2)}(-2\omega - \Delta; \omega, \omega + \Delta) \times E_1^{\omega}E_2^{\omega+\Delta}$
	$P_2^{2\omega} = \frac{1}{2}\chi_{211}^{(2)}(-2\omega; \omega, \omega)E_1^{\omega}E_1^{\omega}$	$P_2^{2\omega} = \chi_{212}^{(2)}(-2\omega; \omega, \omega)E_1^{\omega}E_2^{\omega}$

1. *The Second Harmonic Tensor*

Let us define the commonly used second harmonic d tensors.[35,10] In convention I, $d_{ijk}(-2\omega;\ \omega,\ \omega)$ is defined as

$$P_i^{2\omega} = \sum_{jk} d_{ijk}^{I}(-2\omega;\ \omega,\ \omega)\ E_j^{\omega} E_k^{\omega}. \tag{109}$$

Therefore d_{ijk} is equivalent to χ_{ijk}. Because of the intrinsic permutation symmetry associated with jk, we can replace jk with a single subscript so that

$$\begin{array}{lll} xx = 1, & yy = 2 & zz = 3, \\ yz = zy = 4, & xz = zx = 5, & xy = yx = 6. \end{array}$$

The resultant contracted d_{ij} tensor forms a 3×6 matrix

$$\begin{pmatrix} P_1^{2\omega} \\ P_2^{2\omega} \\ P_3^{2\omega} \end{pmatrix} = \begin{pmatrix} d_{11}^{I} & d_{12}^{I} & d_{13}^{I} & d_{14}^{I} & d_{15}^{I} & d_{16}^{I} \\ d_{21}^{I} & d_{22}^{I} & d_{23}^{I} & d_{24}^{I} & d_{25}^{I} & d_{26}^{I} \\ d_{31}^{I} & D_{32}^{I} & d_{33}^{I} & d_{34}^{I} & d_{35}^{I} & d_{36}^{I} \end{pmatrix} \begin{pmatrix} (E_1^{\omega})^2 \\ (E_2^{\omega})^2 \\ (E_3^{\omega})^2 \\ 2E_2^{\omega}\ E_3^{\omega} \\ 2E_3^{\omega}\ E_1^{\omega} \\ 2E_1^{\omega}\ E_2^{\omega} \end{pmatrix}. \tag{110}$$

In convention II, $d_{ijk}(-2\omega;\ \omega,\ \omega)$ is defined as

$$P_i^{2\omega} = \frac{1}{2} \sum_{jk} d_{ijk}^{II}(-2\omega;\ \omega,\ \omega) E_j^{\omega} E_k^{\omega} \tag{111}$$

so that the contracted d_{ij} yields

$$\begin{pmatrix} P_1^{2\omega} \\ P_2^{2\omega} \\ P_3^{2\omega} \end{pmatrix} = \frac{1}{2} \begin{pmatrix} d_{11}^{II} & d_{12}^{II} & d_{13}^{II} & d_{14}^{II} & d_{15}^{II} & d_{16}^{II} \\ d_{21}^{II} & d_{22}^{II} & d_{23}^{II} & d_{24}^{II} & d_{25}^{II} & d_{26}^{II} \\ d_{31}^{II} & d_{32}^{II} & d_{33}^{II} & d_{34}^{II} & d_{35}^{II} & d_{36}^{II} \end{pmatrix} \begin{pmatrix} (E_1^{\omega})^2 \\ (E_2^{\omega})^2 \\ (E_3^{\omega})^2 \\ 2E_2^{\omega}\ E_3^{\omega} \\ 2E_3^{\omega}\ E_1^{\omega} \\ 2E_1^{\omega}\ E_2^{\omega} \end{pmatrix}. \tag{112}$$

Of course, some researchers use mixed conventions; their $\chi_{ijk}^{(2)}$ is defined in convention II, but d_{ijk} is in convention I (see, for example, the textbook by Butcher and Cotter[10]).

IV. REFERENCES, STANDARDS AND UNITS

While the confusion surrounding various conventions in second and third order nonlinear optical studies as we discussed in the previous section can lead to inconsistencies between researchers, no problems arise as long as the convention used is internally self-consistent between the theoretical and experimental studies. In contrast, uncertainties associated with standard materials used as references are more daunting. For example, the $d_{11}(-2\omega;\ \omega,\ \omega)$ value for crystalline quartz has been accepted to be 1.2×10^{-9} *esu* in convention I over the years. This value is essential to DCSHG and THG measurements in computing absolute $\chi^{(2)}$ and $\chi^{(3)}$ values of the systems under study. Recently, however, several researchers[83,84,79] have concluded from independent meticulous measurements that this commonly accepted value might be too large. Furthermore, there are controversies regarding the $\chi^{(3)}$ values for BK-7 glass and fused silica. Heflin *et al.*[40] pointed out that the accepted $\chi^{(3)}$ values for BK-7 and fused silica, which were derived from previous measurements by Buchalter and Meredith,[55] were too large by a factor of two compared to the values from other measurements. In this paper by Heflin *et al.*,[40] the accuracy of the d_{11} value for quartz was not challenged. If, however, the d_{11} value for quartz is smaller than the currently accepted value, then the $\chi^{(3)}$ value for glass needs to be modified further, since it is derived from the cascading second harmonic process for quartz.[55] This was recently confirmed by Mito *et al.*[79] It is beyond the scope of this Chapter to resolve these issues completely. Therefore, our strategy is to present the evidence and show how absolute measurements may be systematically inaccurate. It is our belief that further measurements are required before any further conclusions can be reached. In comparison, different units used by researchers are fairly transparent, with little associated confusion or controversy.

A. Quartz

Historically, the $d_{11}(-2\omega; \omega, \omega)$ values for quartz were derived by two independent measurements. In the measurement by Jerphagnon[76] using the well-known Maker fringe technique with a Q-switched Nd:YAG laser, it was found that $d_{31}(\text{LiIO}_3) = (15.5 \pm 0.8)d_{11}$ (quartz) at the fundamental wavelength of $\lambda = 1064$ *nm*. The measurements made by Choy and Byer[85] involved the parametric fluorescence of several nonlinear crystals, $LiIO_3$ being among them. It was found that $d_{31}(-2\omega; \omega, \omega) = 7.31 \pm 0.62$ *pm*/*V* for $LiIO_3$. Combining these two measurements, the $d_{11}(-2\omega; \omega, \omega)$ value for quartz was determined to be 0.47 ± 0.04 *pm*/*V*, or $(1.13 \pm 0.11) \times 10^{-9}$ *esu* in Gaussian units. This value has been widely used as a reference standard. Interestingly, it was found that the measured d_{11} values did not agree with those from direct second harmonic generation measurements. For example, it was pointed out that the SHG measurement by Francois[78] would yield a value of 0.41 *pm*/*V* for d_{36} of KDP, which is much smaller than 0.63 *pm*/*V* from the parametric fluorescence measurment. The difference was never reconciled.

Recently, Eckardt *et al.*[83] reexamined this issue by measuring d_{ij} coefficients of several standard crystals using the phase-matched second harmonic generation technique with an injection-seeded, single-mode, Q-switched Nd:YAG laser operating with the spatially filtered output at $\lambda = 1064$ *nm*. Their results showed that, once again, d_{ij} coefficients determined by SHG measurements were consistently smaller than those from parametric fluorescence measurements. For example, the SHG value of d_{31} ($LiIO_3$) is only 58% of the parametric fluorescence value presented by Choy and Byer.[85] According to Eckardt *et al.*, the advantage of parametric fluorescence is that it is only necessary to measure the power ratio of the pump radiation and the generated fluorescence, whereas second harmonic generation requires the measurement of an absolute power and the spatial and temporal distributions of the pump field. While phase-matching is also of critical importance, the second harmonic generation technique has the attraction that it is a direct measurement of the nonlinear coefficient under conditions that are closer to actual applications.

More recently, Mito *et al.*[79] also carried out d coefficient measurements on ADP crystals using SHG technique at the fundamental wavelengths of 633 *nm* and 1064 *nm*. Their main purpose was to determine the $\chi^{(3)}$ values of fused silica. Their measured d_{33} value for the ADP crystal using the absolute SHG phase-matched technique showed that at $\lambda = 633$ *nm*, $d_{33} = 0.57 \pm 0.02$ *pm*/*V*, which is consistent with two previous measurements on ADP crystals, 0.57 ± 0.07 *pm*/*V* by Francois[78] and 0.58 ± 0.09 *pm*/*V* by Bjorkholm *et al.*[77] Table 4 summarizes the nonlinear optical coefficients of crystal ADP from these three independent measurements in both MKS and Gaussian units. The resulting d_{11} values for quartz measured by Mito *et al.*[79] are also shown in the Table. It is clear that they are much smaller than the value obtained from parametric fluorescence measurements,[76,85] also listed in the Table for comparison.

The difference between d values from parametric fluorescence and SHG measurements can be also seen from Table 5, which lists nonlinear coefficients for KDP, $LiIO_3$, and quartz crystals. The parametric fluorescence measurements on KDP made by Campillo and Tang[80] and Choy and Byer[85] showed that the measured d_{36} values are consistently larger than those from second harmonic generation measurements made by Craxton[81] and Eckardt *et al.*[83] d_{31} values for $LiIO_3$ are also much different from these two sets of measurements. And, as a result, d_{11} values for quartz vary widely, depending on the particular measurement. After several decades of nonlinear optics research, many unresolved inconsistencies remain.

A careful analysis of the equations in the second section of the paper by Eckardt *et al.*[83] shows that they adopted convention I. In comparison, the convention adopted by Choy and Byer[85] is not obvious. But the theoretical part of their paper came from the work by Byer and Harris,[89] where their Eq. (3) seems to be valid in convention I. The confusion around the conventions can thus be ruled out in this case. Interestingly, Cheung *et al.*[90] measured the nonlinear optical coefficients of several crystals using the spectral brightness of spontaneous parametric fluorescence. They found that d_{eff} for crystal KTP is 3.9 *pm*/*V* at $\lambda = 1064$ *nm*. This disagrees by a small amount with the value of 3.18 *pm*/*V* obtained by Echardt *et al.*[83] at the same wavelength using second harmonic generation technique.

Table 4 Nonlinear Optical Coefficients of ADP and Quartz Crystals in Convention I

Crystals	Coefficients (*pm*/*V*)	Coefficients (10^{-9} *esu*)	λ (*nm*)	Method	Reference
ADP (d_{33})	0.57 ± 0.07	1.36 ± 0.16	633	SHG (abs)	a
ADP (d_{33})	0.58 ± 0.09	1.38 ± 0.22	633	SHG (abs)	b
ADP (d_{33})	0.57 ± 0.02	1.35 ± 0.05	633	SHG (abs)	c
quartz (d_{11})	0.30 ± 0.01	0.71 ± 0.03	1064	SHG (ADP)	c
quartz (d_{11})	0.32 ± 0.01	0.76 ± 0.03	633	SHG (ADP)	c
quartz (d_{11})	0.47 ± 0.05	1.13 ± 0.11	1064	PF ($LiIO_3$)	d,h

Note: SHG (abs): absolute second harmonic generation measurements; SHG (ADP): relative second harmonic generation measurements of crystal ADP.
a: Ref.[78]; *b*: Ref.[77]; *c*: Ref.[79]; *d*: Ref.[85]; *h*: Ref.[76].

Table 5 Nonlinear Optical Coefficients of KDP, $LiIO_3$, and Quartz Crystals in Convention I

Crystals	Coefficients (*pm/V*)	Coefficients (10^{-9} *esu*)	λ (*nm*)	Method	Reference
KDP (d_{36})	0.63 ± 0.13	1.50 ± 0.31	1064	PF ($LiIO_3$)	d
KDP (d_{36})	0.66 ± 0.10	1.57 ± 0.23	1064	PF ($LiIO_3$)	e
KDP (d_{36})	0.38 ± 0.04	0.91 ± 0.01	1064	SHG (abs)	f
KDP (d_{36})	0.39 ± 0.04	0.93 ± 0.01	1054	SHG (abs)	g
$LiIO_3$ (d_{31})	7.31 ± 0.62	17.4 ± 1.5	1064	PF (abs)	d
$LiIO_3$ (d_{31})	4.1 ± 0.2	9.8 ± 0.5	1064	SHG (abs)	f
quartz (d_{11})	0.47 ± 0.05	1.13 ± 0.11	1064	PF ($LiIO_3$)	d,h
quartz (d_{11})	0.27 ± 0.02	0.64 ± 0.04	1064	SHG ($LiIO_3$)	f,h

Note: PF ($LiIO_3$), relative parametric fluorescence measurements from $LiIO_3$; SHG (abs), absolute second harmonic generation measurements; PF (abs), absolute parametric fluorescence measurements; SHG ($LiIO_3$), relative second harmonic generation measurements from $LiIO_3$; *d*: Ref.[85]; *e*: Ref.[80]; *f*: Ref.[83]; *g*: Ref.[81]; *h*: Ref.[76].

B. BK-7 Glass and Fused Silica

The majority of researchers use either BK-7 glass[36–40,95] or fused silica[45,46,48,52] as sample cell windows for DCSHG and THG measurements of organic liquid and thin film samples. It has widely been accepted that the reference values are $\chi^{(3)}_{1111}(-3\omega;\ \omega,\ \omega,\ \omega) = 4.67 \times 10^{-14}$ *esu* for BK-7 glass and $\chi^{(3)}_{1111}(-3\omega;\ \omega,\ \omega,\ \omega) = 2.80 \times 10^{-14}$ *esu* for fused silica at $\lambda = 1907$ *nm* in convention II.[55] These values were determined by Buchalter and Meredith[55] by comparison with quartz, taken to have $\chi^{(3)II}_{1111}(-3\omega;\ \omega,\ \omega,\ \omega) = (3.81 \pm 0.15) \times 10^{-14}$ *esu* at $\lambda = 1907$ nm, where II indicates that the value was expressed in convention II. This $\chi^{(3)}$ value for quartz was obtained from an analysis of interference fringes between the third harmonic light from a direct third harmonic generation and the third harmonic light from a cascading process that involves two second order optical processes: $\chi^{(2)}(-2\omega;\ \omega,\ \omega)$ and $\chi^{(2)}(-3\omega;\ 2\omega,\ \omega)$.[53] (See Chapter 3 for a more detailed discussion of cascading.) These values are too large by a factor of two, as shown in Tables 6 and 7 listing various measured values of $\chi^{(3)}$ for BK-7 glass[40] and fused silica, respectively, through several nonlinear optical processes: DCSHG,[37,36,39] nondegenerate four-wave mixing,[86] time-resolved interferometry,[87] ellipse rotation,[68,92] and THG.[55]

Careful analysis shows that in Table 6, $\chi^{(3)}$ values for BK-7 glass from DCSHG[37,36,93] and THG[55] were determined based on the assumption that $d^{I}_{11}(-2\omega;\ \omega,\ \omega) = 1.2 \times 10^{-9}$ *esu* for quartz at 1064 *nm*. There are two different kinds of explanations for the disagreement. One explanation is that Levine and Bethea[36] neglected the second harmonic contribution from the surrounding air.[48] This SHG contribution may be significant when the glass slab that forms the DCSHG cell is much smaller than the elctrodes.[48] The other explanation tries to attribute the discrepancy to the incompatibility of the multi-mode laser source in the quartz experiment with the experimental analysis that assumed a single-mode source.[55] The second reason seems to be ruled out by the measurements of Mito *et al.*,[79] who

Table 6 Measured Values of $\chi^{(3)}_{1111}(-\omega_4; \omega_1, \omega_2, \omega_3)$ in Convention II for BK-7 Glass through Several Nonlinear Optical Processes

Process	$\chi^{(3)}_{1111}(-\omega_4; \omega_1, \omega_2, \omega_3)$ (10^{-14} *esu*)	λ (*nm*)	Reference
DCSHG (EFISH)	2.33 ± 0.12	1064	a
DCSHG (EFISH)	2.33 ± 0.36	1064	b
DCSHG (EFISH)	2.12 ± 0.04	1064	c
NDFWM	2.08 ± 0.31	1064, 1071	d
TRI	2.00 ± 0.20	1064	e
Ellipse Rotation	2.20 ± 0.10	694	f
THG	4.67	1907	g

Note: NDFWM: nondegenerate four-wave mixing; TRI: time-resolved interferometry.
a: Ref.[37]; *b*: Ref.[36]; *c*: Ref.[93]; *d*: Ref.[86]; *e*: Ref.[87]; *f*: Ref.[68,92]; *g*: Ref.[55].

performed cascading harmonic measurements on quartz with two lasers, a single-mode cw YAG laser and a multimode Q-switched YAG laser. Taking into account the dispersion of the refractive index of quartz, their results showed that the measured values were in agreement within experimental errors. $\chi^{(3)}$ values obtained from time resolved interferometry and ellipse rotation, however, do not depend on quartz d_{11} values. These values were determined from the change of the refractive index due to $\chi^{(3)}(-\omega; \omega, \omega, -\omega)$.[87,68,92] In comparison, results from nondegenerate four-wave mixing[86] are relative. These form a set of independent evidence in support of results from other absolute measurements.

Excluding possible systematic errors in DCSHG measurements for BK-7 glass, the fact that the results (assuming $d'_{11} = 1.2 \times 10^{-9}$ *esu* for quartz) agree with independent measurement from time resolved interferometry and ellipse rotation seems to indicate that the correct value of d_{11} of quartz was used. But these results are in conflict with the result obtained by Buchalter and Meredith, the validity of whose measurement has been confirmed by Mito *et al.* The source of inconsistency is not clear. Further, Mito *et al* showed that their results were in agreement with the earlier ones by Buchalter and Meredith in terms of the ratio of the third harmonic to the second harmonic coefficients for quartz and in terms of the ratio of

Table 7 Measured Values of $\chi^{(3)}_{1111}(-\omega_4; \omega_1, \omega_2, \omega_3)$ in Convention II for Fused Silica through Several Nonlinear Optical Processes

Process	$\chi^{(3)}_{1111}(-\omega_4; \omega_1, \omega_2, \omega_3)$ (10^{-14} *esu*)	λ (*nm*)	Reference
NDFWM	1.30 ± 0.19	1064, 1071	d
TRI	1.45 ± 0.01	1064	e
Ellipse Rotation	1.45 ± 0.03	694	f
THG	2.80	1907	g
THG	0.95 ± 0.07	1907	h

d: Ref.[86]; *e*: Ref.[87]; *f*: Ref.[68,92]; *g*: Ref.[55]; *h*: Ref.[79].

the third harmonic coefficients between fused silica and quartz. For example, Meredith[53] determined that $\chi^{(3)}(-3\omega; \omega, \omega, \omega)/[\chi^{(2)}(-2\omega; \omega, \omega)\chi^{(2)}(-3\omega; 2\omega, \omega)] = (8.4 \pm 0.03) \times 10^3$ *esu* for quartz at $\lambda = 1907$ *nm*, which is only about 10% higher than the result of $\chi^{(3)}(-3\omega; \omega, \omega, \omega)/[\chi^{(2)}(-2\omega; \omega, \omega)\chi^{(2)}(-3\omega; 2\omega, \omega)] = (7.52 \pm 0.36) \times 10^3$ *esu* at the same wavelength obtained by Mito *et al.* In addition, Buchalter and Meredith[55] showed that $\chi^{(3)}$(fused silica)/$\chi^{(3)}$(quartz) = 0.732, in good agreement with the result of $\chi^{(3)}$(fused silica)/$\chi^{(3)}$(quartz) = (0.723 ± 0.022) by Mito *et al.* The difference in the $\chi^{(3)}$ values of fused silica obtained by Buchalter and Meredith[55] and Mito *et al.*, as shown in Table 7, comes from different d_{11} values for quartz, shown in Table 4.

As a practical matter, although it is important to establish accurate $\chi^{(3)}$ reference values for BK-7 glass and fused silica, the measured solution $\chi^{(3)}(-2\omega; \omega, \omega, 0)$ values from a DCSHG experiment are not sensitive to large uncertainties in the glass $\chi^{(3)}(-2\omega; \omega, \omega, 0)$ if the solution $\chi^{(3)}(-2\omega; \omega, \omega, 0)$ is much larger than $\chi^{(3)}(-2\omega; \omega, \omega, 0)$ for Bk-7 glass or fused-silica. This certainly is true in the case of many π-conjugated second order materials which possess large hyperpolarizability $\beta(-2\omega; \omega, \omega)$ so that the solution containing the chromophore have $\chi^{(3)}(-2\omega; \omega, \omega, 0)$ values on the order of 10×10^{-14} *esu.*

To illustrate the above statement, standard DCSHG measurements were performed on pure nitrobenzene (electronic grade, Eastman Kodak) at $\lambda_\omega = 1064$ *nm* using BK-7 glass windows in the DCSHG cell and referenced to quartz. The fitting formula is from Eq. (14) in the paper by Heflin *et al.*[40] which reads (in convention I):

$$\chi_L^{(3)}(-2\omega; \omega, \omega, 0) = \frac{n_\omega^L + n_{2\omega}^L}{l_c^L T_L [t_\omega^{(1)} t_\omega^{(2)}]^2} \left\{ T_G \frac{l_c^G \chi_G^{(3)}(-2\omega; \omega, \omega, 0)[t_\omega^{(1)}]^2}{n_\omega^G + n_{2\omega}^G} \right.$$
$$\left. \pm \frac{T_Q}{E_0} \left(\frac{1 + n_{2\omega}^G}{2n_{2\omega}^G} \right) \frac{l_c^Q d_{11}^Q}{(n_\omega^Q + n_{2\omega}^Q)} \left(\frac{2}{1 + n_\omega^Q} \right)^2 \left(\frac{\bar{A}_L^{2\omega}}{\bar{A}_Q^{2\omega}} \right)^{1/2} \right\}, \tag{113}$$

where

n_ω^L = 1.52 (refractive index for nitrobenzene at ω),
$n_{2\omega}^L$ = 1.55 (refractive index for nitrobenzene at 2ω),
n_ω^Q = 1.5341 (refractive index for quartz at ω),
$n_{2\omega}^Q$ = 1.5470 (refractive index for quartz at 2ω),
n_ω^G = 1.5067 (refractive index for BK-7 glass at ω),
$n_{2\omega}^G$ = 1.5191 (refractive index for BK-7 glass at 2ω),
l_c^Q = 21.2 μm (quartz coherence length),
l_c^G = 21.37 μm (glass coherence length),
l_c^L = 8.7 μm (nitrobenzene coherence length),
$E_0 = V_0/d$ = 3500 $V/(0.16\ cm)$ = 3500 $Stat.V/(0.16 \times 300\ cm)$ (electric field applied across the sample cell),
$d_{11}^Q = 1.2 \times 10^{-9}$ *esu* (quartz second-order susceptibility),
$\chi_G^{(3)I}(-2\omega; \omega, \omega, 0) = 3.5 \times 10^{-14}$ *esu* (glass DCSHG $\chi^{(3)}$ value),

Fresnel factors are

$$t_\omega^{(1)} = \frac{2}{1 + n_\omega^G}, \qquad t_\omega^{(2)} = \frac{2n_\omega^G}{n_\omega^G + n_\omega^L},$$

$$T_L = \frac{n_\omega^L + n_{2\omega}^L}{n_{2\omega}^G + n_{2\omega}^L}, \qquad T_G = \frac{n_\omega^G + n_{2\omega}^L}{n_{2\omega}^G + n_{2\omega}^L},$$

$$T_Q = \frac{n_\omega^Q + n_{2\omega}^Q}{1 + n_{2\omega}^Q},$$

and $\bar{A}_L^{2\omega}$ and $\bar{A}_Q^{2\omega}$ are the mean Maker fringe[67] heights of nitrobenzene and quartz, respectively. They are obtained from the least squares fit of the Maker fringes to a function in the form of

$$I^{2\omega}(l) = A_1 \sin^2\left(\frac{\pi l}{2A_3} + A_4\right) + A_2, \tag{114}$$

where A_1 is the fringe amplitude, A_2 is the fringe minimum, A_3 is the coherence length, and A_4 is the initial phase. The mean fringe height is defined as $\bar{A}^{2\omega} = A_1/2 + A_2$. In our measurements, they are, for nitrobenzene, $A_1 = 0.164 \times 10^{2.43}$, $A_2 = 0.0007 \times 10^{2.43}$ (in arbitrary units), and for quartz, $A_1 = 0.423 \times 10^{4.71}$, $A_2 = 0.007 \times 10^{4.71}$ (in arbitrary units). With these obtained experimentally, we have the following value for nitrobenzene:

$$\chi^{(3)I}(-2\omega; \omega, \omega, 0) = (185 \pm 18) \times 10^{-14}\ esu.$$

If, however, $\chi_G^{(3)I}(-2\omega; \omega, \omega, 0) = 7.0 \times 10^{-14}$ *esu* for the BK-7 glass windows, is used in Eq. (113), $\chi^{(3)I}(-2\omega; \omega, \omega, 0)$ for nitrobenzene would become $(194 \pm 19) \times 10^{-14}$ *esu*, which is the same as the previous one within the experimental uncertainty.

To illustrate that $\chi^{(3)}$ of glass is important in experimentally determining nonlinear optical susceptibilities, when the liquid cell has a small third order susceptibility, DCSHG measurements were also performed on anhydrous 1,4-dioxane (Aldrich). Parameters which differ from the previous measurement are

$n_\omega^L = 1.41$ (refractive index for dioxane at ω),
$n_{2\omega}^L = 1.42$ (refractive index for dioxane at 2ω),
$l_c^L = 26.7\ \mu m$ (dioxane coherence length).

For dioxane, $A_1 = 0.4816$, $A_2 = 0.020$, and for quartz, $A_1 = 1.484 \times 10^{4.71}$, $A_2 = 0.011 \times 10^{4.71}$ (in arbitrary units). Therefore, we have $\chi_L^{(3)I}(-2\omega; \omega, \omega, 0) = (5.52 \pm 0.55) \times 10^{-14}$ *esu* for 1,4-dioxane if $\chi_G^{(3)I}(-2\omega; \omega, \omega, 0) = 3.5 \times 10^{-14}$ *esu* for BK-7 glass windows. If we double the BK-7 glass value of $\chi^{(3)}$, $\chi_L^{(3)I}(-2\omega; \omega, \omega, 0)$ would become $(8.05 \pm 0.80) \times 10^{-14}$ *esu* for 1,4-dioxane, which is significantly different from the previous result.

C. Units

Different units have been used by researchers in the field of nonlinear optics, as we can see in the discussions presented in the previous sections. For example, the unit associated with the quartz d_{11} value can be either *pm/V* or *esu*. Table 8 lists

Table 8 Unit Conversion from CGS to MKS Units

	CGS (esu)	MKS (I)	MKS (II)
V	1 statvolt	2.9979×10^{2} V	2.9979×10^{2} V
E	1 statvolt cm^{-1}	2.9979×10^{4} Vm^{-1}	2.9979×10^{4} Vm^{-1}
P	1 statvolt cm^{-1}	3.3356×10^{-6} Cm^{-2}	3.3356×10^{-6} Cm^{-2}
$\chi^{(1)}$	1	1.2566×10^{1}	1.1127×10^{-10} $C^2m^{-1}J^{-1}$
$\chi^{(2)}$	1 $statvolt^{-1}$ cm	4.1888×10^{-4} mV^{-1}	3.7088×10^{-15} C^3J^{-2}
$\chi^{(3)}$	1 $statvolt^{-2}$ cm^2	1.3963×10^{-8} m^2V^{-2}	1.2380×10^{-19} C^4mJ^{-3}
μ	1 statvolt cm^2	3.3356×10^{-12} Cm	3.3356×10^{-12} Cm
α	1 cm^3	1.2566×10^{-5} m^3	1.1127×10^{-16} $C^2m^2J^{-1}$
β	1 $statvolt^{-1}$ cm^4	4.1888×10^{-10} m^4V^{-1}	3.7088×10^{-21} $C^3m^3J^{-2}$
λ	1 $statvolt^{-2}$ cm^5	1.3963×10^{-14} m^5V^{-2}	1.2380×10^{-25} $C^4m^4J^{-3}$

three commonly used units and their conversions for several frequently used physical quantities. These units are defined by the following equations:

CGS: $P^{(n)} = \chi^{(n)} = \chi^{(n)}(E)^n$,

MKS(I): $P^{(n)} = \epsilon_0\chi^{(n)}(E)^n$,

and MKS (II): $P^{(n)} = \chi^{(n)}(E)^n$,

where the polarization P and electric field E are in CGS, MKS(I) and MKS(II) units, respectively, $\chi^{(n)}$ refers to the nth order nonlinear susceptibility, and ϵ_0 is the vacuum dielectric constant. In deriving Table 8, the following relations have been used:

$$(10^{-4}c)E_{CGS} = E_{MKS(I)} = E_{MKS(II)}, \tag{115}$$

$$(10^{3}c)P_{CGS} = P_{MKS(I)} = P_{MKS(II)}, \tag{116}$$

$$(4\pi\epsilon_0)(10^{-4}c)^{1-n}\,\chi^{(n)}_{CGS} = \epsilon_0\chi^{(n)}_{MKS(I)} = \chi^{(n)}_{MKS(II)}, \tag{117}$$

where the speed of light $c = 2.9979 \times 10^{8}$ m/s and the vacuum dielectric constant $\epsilon_0 = 8.8542 \times 10^{-12}$ $coul^2/N \cdot m^2$.

V. CONCLUSION

In this Chapter, we have summarized the basic physical properties of nonlinear optical processes in organic and polymer systems and their potential applications. Nonlinear optical phenomena in these systems originate in the virtual excitations occurring on the individual molecular, or polymer chain, units due to the delocalized π-electrons that are easily polarizable. Second order optical properties in π-conjugated systems result from the charge-separated nature of the π-electrons upon virtual excitations. Applications for second order materials will likely be based on the linear EO properties of aligned glassy polymers in integrated optoelectronic circuits using high temperature chromophores and polymers. Electron correlations effects are more pronounced in the third order optical processes of π-conjugated systems. New materials and phenomena are needed to further enhance third order optical properties in these materials for practical applications.

We have also studied in this Chapter several conventions for both macroscopic and microscopic nonlinear optical susceptibilities, with a focus on their definitions and relations among different conventions. We concluded that conventions I and II are recommended because of their simplicity and the fact that they are heavily used in the field of nonlinear optics, especially for those involved in the second and third order nonlinear optical studies of liquid and solid organic materials. We have also identified various researchers with the conventions they adopted. In addition, the nonlinear optical susceptibilities were derived in such a manner that the permutation factor $K(-\omega_{n+1,}, \omega_1, \omega_2, \ldots, \omega_n)$ can easily be defined. Furthermore, one needs to be careful when adopting results from the literature because conventions vary between authors. Also, additional permutation factors have to be included when different Cartesian coordinates in a susceptibility tensor are associated with the same input frequency.

There are still unresolved issues associated with materials that are used as reference standards. For example, the commonly accepted d_{11} value for quartz, which results from parametric fluorescence measurements, was recently found to be larger than those determined by absolute phase-matched second harmonic measurements. Until these issues are completely resolved, we suggest that $d^{I}_{11}(-2\omega; \omega, \omega) = 1.2 \times 10^{-9}$ *esu* and $d^{II}_{11}(-2\omega; \omega, \omega) = 2.4 \times 10^{-9}$ *esu* for quartz at $\lambda\omega = 1064$ *nm* is the best estimate based on literature review. The controversy surrounding $\chi^{(3)}$ values for BK-7 glass and fused silica is also unresolved. For the time being, we recommend that $\chi^{(3)II}_{1111}(-3\omega; \omega, \omega, \omega) = 2.33 \times 10^{-14}$ *esu* for BK-7 glass and $\chi^{(3)II}_{1111}(-3\omega; \omega, \omega, \omega) = 1.40 \times 10^{-14}$ *esu* for fused silica.

ACKNOWLEDGMENTS

This research was generously supported by AFOSR and ARPA (grant F49620-85-C-0105). We gratefully acknowledge the contributions of Drs. Y. M. Cai and J. R. Heflin to the results presented in this Chapter. We also thank Mr. W. D. Chen for his critical reading of the manuscript.

REFERENCES

1. Hornak, L. A., ed., *Polymers for Lightwave and Integrated Optics: Technology and Applications*, Marcel Dekker, Inc., New York, June 1992.
2. Garito, A. F., Shi, R. F., and Wu M. H., *Physics Today*, *47*(5), 51 (1994).
3. Chiang, L. Y., Garito, A. F., and Sandman, D. J. eds., *Electrical, Optical, and Magnetic Properties of Organic Solid State Materials, Mater. Res. Soc. Proc.*, *247* (1992).
4. Messier, J., Kajzar, F., and Prasad, P. N. eds., *Organic Molecules for Nonlinear Optics and Photonics*, NATO ASI Series E, (Kluwer Academic, Boston 1991.
5. Kuzyk, M. G., and Swalen, J. D., eds., *Progress in Nonlinear Optics: Organics and Polymeric Materials*, Nonlinear Opt. *6* (1993).
6. Ashwell, G. J., and Bloor, D. *Organic Materials for Nonlinear Optics* III, *International Symposium on Organic Materials for Nonlinear Optics*, Royal Society of Chemistry, Oxford, England, 1993.

7. Garito, A. F., Jen, A. K. Y., Lee, C. Y. C. and Dalton, L. R., eds., *Mater. Res. Soc. Proc., 328* (1994).
8. Kuwata-Gonokami, M., Peyghambarian, N., Meissner, K., Fluegel, B., Sata, Y., Ema, R., Shimano, R., Mazumdar, S., Guo, F., Tokihiro, T., Ezaki, H., and Hanamura, E., *Nature*, *367*, 47 (1994); Garito and Shi, *Supra*, note 2, *19* (1994).
9. Shen, Y. R., *The Principle of Nonlinear Optics*, Wiley, New York, 1984.
10. Butcher, P. N., and Cotter, D. *The Elements of Nonlinear Optics*, Cambridge University Press, New York, 1990.
11. Charra, F., Kajzar, F., Nunzi, J. M., Raimond, P., and Idiart, E., *Opt. Lett. 18*, 941 (1993).
12. Giordmaine, J. A., *Phys. Rev. 138*, A1600 (1965).
13. Rentzepis, P. M., Giordmaine, J. A., and Wecht, K. W. *Phys. Rev. Lett. 16*, 792 (1966).
14. Havinga, E. E., and van Pelt, P., *Ber Bunsenges Phys. Chem. 83*, 816 (1979).
15. Singer, K. D., Sohn, J. E., and Lalama, S. J. *App. Phys. Lett. 49*, 248 (1986).
16. Burland, D. M., Miller, R. D., and Walsh, C. A., *Chem. Rev. 94*, 31 (1994).
17. Wu, J. W., Binkley, E. S., Kenney, J. T., Lytel, R., and Garito, A. F., *J. Appl. Phys. 69*, 7366 (1991), and refs. therein.
18. Binkley, E. S., and Nara, S., in *Organic Thin Films for Photonic Applications Technical Digest*, vol. 17, (1993) (Optical Society of America, Washington, D.C., 1993) 266.
19. Matsura, T., Ando, S., Matsui, S., Hirata, H., Sasaki, S., and Yamamoto, F. in *Organic Thin Films for Photonic Applications Technical Digest*, vol. 17, (1993) (Optical Society of America, Washington, D.C., 1993) 262.
20. Stegeman, G. I., and Torruellas, W., *Mat. Res. Soc. Sym. Proc. 328*, 397 (1994).
21. Khanarian, G., Norwood, R. A., Haas, D., Feuer, B., and Karim, D., *App. Phys. Lett. 57*, 977 (1990).
22. Penner, T. L., Motschmann, H. R., Armstrong, N. J., Ezenyilimba, M. C., and Williams, D. J. *Nature 367*, 49 (1994).
23. Sagawa, M., Kagawa, H., Kakuta, A., and Kaji, M., *App. Phys. Lett. 63*, 1877 (1993).
24. Heflin, J. R., Wong, K. Y., Zamani Khamiri, O., and Garito, A. F., *Phys. Rev. B38*, 1573 (1988).
25. Bredas, J. L., Adant, C., Tackx, P., Persoons, A., and Pierce, B. M., *Chem. Rev. 94*, 243 (1994), and refs. therein.
26. Zhou, Q. L., Shi, R. F., Zamani Khamiri, O., and Garito, A. F., *Nonlinear Opt. 6*, 145 (1993).
27. Shi, R. F., Zhou, Q. L., and Garito, A. F., *Mat. Res. Soc. Sym. Proc. 328*, 643 (1994).
28. Yu, Y. Z., Shi, R. F., Garito, A. F., and Grossman, C. H. *Opt. Lett. 19*, 786 (1994).
29. Zhou, Q. L., Heflin, J. R., Wong, K. Y., Zamani-Khamiri, O., and Garito, A. F., *Phys. Rev. A43*, 1673 (1991).
30. Rodenberger, D. C., Heflin, J. R., and Garito, A. F., *Nature 359*, 309 (1992).
31. Heflin, J. R., Rodenberger, D. C., Shi, R. F., Wu, M. H., Wang, N. Q., Cai, Y. M., and Garito, A. F., *Phys. Rev. A45*, R4233 (1992).
32. Dirk, C. W., Cheng, L-T., and Kuzyk, M. G., *Int. J. Quant. Chem. 43*, 27 (1992).
33. Marder, S. R., Perry, J. W., Bourhill, G., Gorman, C. B., Tiemann, B. G., and Mansour, K., *Science 261*, 186 (1993).
34. Zhou, Q. L., Ph.D. Dissertation, University of Pennsylvania, Philadelphia, PA, 1993.
35. Yariv, A. M., *Quantum Electronics*, John Wiley & Sons, Inc., New York, 1975.
36. Levine, B. F., and Bethea, C. G., *J. Chem. Phys. 63*, 2666 (1975).
37. Oudar, J. L. J., *Chem. Phys. 67*, 446 (1977).
38. Singer, K. D., and Garito, A. F. *J. Chem. Phys. 75*, 3572 (1981).
39. Teng, C. C., and Garito, A. F., *Phys. Rev. B28*, 6766 (1983).
40. Heflin, J. R., Cai, Y. M., and Garito, A. F., *J. Opt. Soc. Am. B8*, 2132 (1991).
41. Singer, K. D., Kuzyk, M. G., and Sohn, J. E. *J. Opt. Soc. Am. B4*, 968 (1987).

42. Orr, B. J., and Ward, J. F., *Mol. Phys. 20*, 513 (1971).
43. Ward, J. F., and New, G. H. C., *Phys. Rev. 185*, 57 (1969).
44. Kajzar, F., and Messier, J., *Rev. Sci. Instrum. 58*, 2081 (1987).
45. Kajzar, F., and Messier, J., *J. Opt. Soc. Am. B4*, 1040 (1987).
46. Kajzar, F., Ledoux, I., and Zyss, J., *Phys. Rev. A36*, 2210 (1987).
47. M. Stähelin, M., Burland, D., Ebert, M., Miller, R., Smith, B., Twieg, R., Volksen, W., and Walsh, C., *App. Phys. Lett. 61*, 1626 (1991).
48. Burland, D. M., Walsh, C. A., Kajzar, F., and Sentein, C., *J. Opt. Soc. Am. B8*, 2269 (1991).
49. Willetts, A., Rice, J. E., Burland, D. M., and Shelton, D. P., *J. Chem. Phys. 97*, 7590 (1992).
50. Clays, K., and Persoons, A., *Phys. Rev. Lett. 66*, 2980 (1991).
51. Chopra, P., Carlacci, L., King, H. F., and Prasad, P. N., *J. Phys. Chem. 93*, 7120 (1989).
52. Cheng, L. T., Tam, W., Stevenson, S. H., Meredith, G., Rikken, G., and Marder, S. R. *J. Phys. Chem. 95*, 1063 (1991).
53. Meredith, G. R., *Phys. Rev. B24*, 5522 (1981).
54. Meredith, G. R., Buchalter, B., and Hanzlik, C. *J. Chem. Phys. 78*, 1533 (1983).
55. Buchalter, B., and Meredith, J. R., *Appl. Opt. 21*, 3221 (1982).
56. Kajzar, F., and Meissier, J., *Phys. Rev. A32*, 2352 (1985).
57. Hermann, J. R., *Opt. Comm. 9*, 74 (1973).
58. Neher, D., Wolf, A., Bubeck, C., and Wegner, G., *Chem. Phys. Lett. 163*, 116 (1989).
59. Bishop, D. M., *Rev. Mod. Phys. 62*, 343 (1990).
60. Papadopoulos, M. G., Waite, J., and Nicolaides, C. A., *J. Chem. Phys. 77*, 2527 (1982).
61. Hurst, G. J. B., Dupuis, M., and Clementi, E., *J. Chem. Phys. 89*, 385 (1988).
62. Charra, F., Kajzar, F., Nunzi, J. M., Raimond, P., and Idiart, E., *Opt. Lett. 18*, 941 (1993).
63. Giordmaine, J. A., *Phys. Rev. 138*, A1600 (1965).
64. Rentzepis, P. M., Giordmaine, J. A., and Wecht, K. W., *Phys. Rev. Lett. 16*, 792 (1966).
65. Bogoliubov, N. N., and Mitropolsky, Y. A., *Asymptotic Methods in the Theory of Nonlinear Oscillations*, Gorden & Breach, New York, 1961.
66. Heflin, J. R., Ph.D. Dissertation, University of Pennsylvania, Philadelphia, PA, 1990.
67. Maker, P. D., and Terhune, R. W., *Phys. Rev. 137*, A801 (1965).
68. Hellwarth, R. W., Owyoung, A., and George, N., *Phys. Rev. A4*, 2342 (1971).
69. Soos, Z. G., and Ramasesha, S., *Chem. Phys. Lett. 153*, 171 (1988); *J. Chem. Phys. 90*, 1067 (1989).
70. Boyd, R. W., *Nonlinear Optics* (Academic Press, Inc.), Boston, 1992.
71. Kurtz, H. A. Stewart, J. J. P., and Dieter, K. M., *J. Comp. Chem. 11*, 82 (1990).
72. Zyss, J. *Nonlinear Opt. 1*, 3 (1991).
73. Smith, W. L., in *Handbook of Laser Science and Technology*, Vol. III, Optical Materials (M. J. Webber, ed), CRC Press, Inc., 1986.
74. Jerphagnon, J., and Kurtz, S. K., *J. App. Phys. 41*, 1667 (1970).
75. Jerphagnon, J., *App. Phys. Lett. 16*, 298 (1970).
76. Jerphagnon, J., *Phys. Rev. B2*, 1091 (1970).
77. Bjorkholm, J. E., and Siegman, A. E., *Phys. Rev. 154*, 851 (1967).
78. Francois, G. E., *Phys. Rev. 143*, 597 (1966).
79. Mito, A., Hagimoto, K., and Takahashi, C., *International Symposium on Nonlinear Photonics Materials*, Japan, May 1994, 85–99.
80. Campillo, A. J., and Tang, C. L., *App. Phys. Lett. 16*, 242 (1970).
81. Craxton, R. S., *IEEE J. Quant. Electron. QE-17*, 1771 (1981).
82. Boyd, G. D., and Kleinman, D. A., *J. App. Phys. 39*, 3567 (1968).
83. Eckardt, R. C., Masuda, H., Fan, Y. X., and Byer, R. L., *IEEE J. Quant. Electron. QE-26*, 922 (1992).

84. Roberts, D. A., *IEEE J. Quant. Electron. QE-28*, 2057 (1992).
85. Choy, M. M., and Byer, R. L., *Phys. Rev. B14*, 1693 (1976).
86. Adair, R., Chase, L. L., and Payne, S., *J. Opt. Soc. Am. B4*, 875 (1987).
87. Milam, D., and Weber, M. J., *J. App. Phys. 47*, 2497 (1976).
88. Hellwarth, R. W., *Prog. Quantum Electron. 5*, 1 (1977).
89. Byer, R. L., and Harris, S. E., *Phys. Rev. 168*, 1064 (1968).
90. Cheung, E. C., Koch, K., Moore, G. T., and Liu, J. M., *Opt. Lett. 19*, 168 (1994).
91. Skinner, I. M., and Garth, S. J., *Am. J. Phys. 58(2)*, 177 (1990).
92. Hellwarth, R. W., Owyoung, A., and George, N., *Phys. Rev. A4*, 2342 (1971).
93. Teng, C. C., Ph.D. Dissertation, University of Pennsylvania, Philadelphia, PA, 1983.
94. Buckingham, A. D., *Adv. Chem. Phys. 12*, 107 (1967).
95. Andrews, J. H., Khaydarov, J. D. V., and Singer, K. D., *Opt. Lett. 19*, 984 (1994).
96. Stegeman, G. I., Torruellas, W., *Mat. Res.* See Proc. *247*, 397 (1894).

2
Practical Computational Approaches to Molecular Properties

Israel D. L. Albert, Tobin J. Marks, and Mark A. Ratner
Northwestern University, Evanston, Illinois

I. ABSTRACT

Computational schemes for prediction and analysis of the $\chi^{(2)}$ response of molecular chromophore-based systems are overviewed.[a] Emphasis is placed both on normative methods for high accuracy calculations on relatively small molecules, and on more approximate but practical calculations, based on semiempirical electronic structure models, for large chromophores with significant β response.

The discussion is centered on the spectral representation, or sum over states representation, which permits both exact calculation and facile interpretation. The relation of this spectral representation to more efficient methodologies is discussed in the context of ab-initio calculations, whose strengths, development, and shortcomings are presented.

More extensive discussion of semiempirical methods is given, including an overview of some of the problems with these methods, their accuracy and applicability, and some of the important chemical motifs effecting nonlinear response that have been sorted out using the predictive power and wide range of the semiempirical methodologies. Special topics such as geometric distortions, frequency dependencies, environmental and solvent effects are presented.

II. INTRODUCTION

In the design of molecular nonlinear optical materials (for example, molecular crystals or dye-doped polymers), there is a set of challenges that differs substan-

[a]While higher-order susceptibilities are not treated, there are many similarities with calculations of $\chi^{(2)}$. Note, however, that in calculating higher-order susceptibilities, additional subtleties often need to be taken into account.

tially from that associated with solid state nonlinear materials such as lithium niobate. In particular, the fundamental response of molecular materials arises from the molecular subunit, and, if all local field effects are ignored, the overall nonlinear susceptibilities are proportional to number densities of chromophores times the individual chromophore response (see Chapter 3 for details).[1–10] This implies, in turn, that the most important fundamental variables in determining the macroscopic susceptibilities are chromophore number density and the individual chromophore response. Thus, we can understand the nonlinear optical properties of the materials if we know the response of the individual molecules, and can characterize both their number density and the effects of the interaction between them and of the solvent or host environment. Therefore, computation of the bulk nonlinear optical properties of these materials depends critically on the NLO susceptibilities of the individual molecular chromophore.

Accordingly, the majority of effort aimed at understanding a material's nonlinear optical properties (and the dominant emphasis of this Chapter) has been toward molecular calculation and characterization.[6–10] Very extensive reviews of this subject occur in many other places.[6–10] In this chapter, we discuss some of the most popular ways to computationally characterize the molecular nonlinear optical response, and the current state of such calculations.

The aim of this chapter is not *complete* coverage of computational methods or an extensive tabulation of calculations; rather, it is to provide the working scientist with a basic understanding of the role that theory can play in connection with nonlinear response of individual molecules, either alone or in the condensed phase. The reader interested in a more elaborate analysis can consult one of several excellent advanced treatments.[7,8,11–14]

One simple approach to understanding the nonlinear response of molecules is an anharmonic oscillator representation.[11,13,15a] In the simplest such picture, the electronic structure is represented simply as an electron in an anharmonic potential. For small amplitude oscillations about the equilibrium position of the electron in an anharmonic potential, the polarization is a series expansion in the electric field and the nth-order susceptibility is proportional to the coefficient of the nth-power of the electric field. Extensions and modifications to this simple model make it much more useful than it might first appear.[15] Almost all computational work, however, has been based on a very different picture, involving molecular orbital approaches. Here the wave functions of the relevant electrons (valence electrons in semiempirical theories, all electrons in ab-initio calculations) are explicitly determined or estimated, and the nonlinear response is computed from these wave functions using the various perturbation-theoretic formulas describing the nonlinear interaction of a molecule with probe electric fields. In this chapter, we discuss exclusively this molecular-orbital-based analysis.

Although much higher-order susceptibilities have been studied,[16] particularly for atoms, the dominant interest in the chemical and materials communities has been in the second and third order susceptibilities—that is, on the first and second hyperpolarizabilities, usually denoted as $\chi^{(2)}$ and $\chi^{(3)}$ for materials systems or as β and γ for the individual molecular components.[1–15] Both theoretically and experimentally, the first hyperpolarizability β or the second-order susceptibility $\chi^{(2)}$, is better characterized and understood than the second hyperpolarizability, or the third-order susceptibility, $\chi^{(3)}$.[b1–10] Therefore we will concentrate our discussions on the β response.

The remainder of this chapter is organized as follows: Section III discusses the fundamental spectral representation (that is, the perturbation theory result that permits actual calculation of electronic wave functions using molecular orbital methodology). Section IV is devoted to semiempirical electronic structure theory and its application to molecular and nonlinear optical response. Ab-initio-based methods are discussed and illustrated in Section V, while Section VI provides some examples of chemical insight derived from semiempirical calculations. Important special topics, including environmental effects, vibrational features, and particular structural motifs are overviewed in Section VII, and general remarks are made in Section VIII.

Different sets of units are common in discussing all critical properties. For the β response, discussed in detail here, the more common units (for details of unit conversions see Chapter 1 or Ref. 1, page 303), are a.u. and cm^5esu^{-1} (note that in the literature, the designation esu is often used for cm^5/esu, where esu is the unit of charge in gaussian units, the conversion is between these units

$$\beta(cm^5esu^{-1}) = \beta(a.u.) \times 8.6392187 \times 10^{-30} \tag{1a}$$

$$\gamma(cm^7esu^{-2}) = \gamma(a.u) \times 5.03669297 \times 10^{-40}. \tag{1b}$$

III. SPECTRAL REPRESENTATIONS—EXACT SUM OVER STATES FORMULAS

Numerical calculations of polarizabilities and hyperpolarizabilities, at the molecular level are a particularly vexing and taxing aspect of modern electronic structure theory. Because the perturbation operators that are responsible for the interaction with the electromagnetic field can be expressed semiclassically, we can perceive, from simple examination of the Hamiltonian, some of the difficulties entailed in such calculations. The total Hamiltonian,

$$H_{TOT} = H + h^{(1)}, \tag{2a}$$

is the sum of the "dark" molecular Hamiltonian, H, and the interaction Hamiltonian, $h^{(1)}$, which describes the energy of interaction between molecule and light. The classical interaction Hamiltonian, of the form,

$$h^{(1)} = -\sum_j e\vec{E}_j \cdot \vec{r}_j \cos\omega t, \tag{2b}$$

is often used, where e is the electron charge, $\vec{E}_j$ the electric field amplitude at the jth electron, and $\vec{r}_j$ the jth electron's position. The classical interaction Hamiltonian neglects field quantization and the energy associated with the electric field. Note that the interaction potential $h^{(1)}$, which is assumed to be a perturbation, involves a term proportional to the electric displacement $\vec{r}_j$ and leads to a net force that

[b]For a good description of the third order nonlinear optical properties or second hyperpolarizabilities see the recent review by J. L. Bredas et al. in Ref. 7.

moves electrons out of the valence region, toward more diffuse regions of the molecule. While this same interaction Hamiltonian is used in the calculation of optical spectra, the effect of $h^{(1)}$ on the nonlinear optical response involves repeated interactions, so the diffuse regions of electron density may play a more important role in the response. This in turn suggests that the one electron basis set (that is, the set of functions, when added in linear combination describe exactly where in space the electronic density is located) will have to be more extensive for accurate calculations of hyperpolarizabilities (indeed of optical properties in general) than say, computations of such properties as dipole moment, geometry or binding energy.

Given the form of the perturbation in Eq. 2a and 2b, it is a straightforward exercise to show that the polarizability and first hyperpolarizability can be expressed in terms of sums over the excited states of this system:[17–24]

$$\alpha_{ij}(-\omega;\omega) = \sum_{l=0}^{1} \left(\sum_{n} \frac{(r_i)_{gn}\,(r_j)_{ng}}{E_n - E_g + (-1)^l \omega} \right] \tag{4}$$

$$\begin{aligned} &\beta_{ijk}(\omega = \omega_1 + \omega_2) \\ &= -N\frac{e^3}{\hbar^2} \sum_{n,n'}{}' \left[\frac{(r_i)_{gn}(r_j)_{nn'}(r_k)_{n'g}}{(\omega - \omega_{ng})\,(\omega_2 - \omega_{n'g})} + \frac{(r_i)_{gn}(r_k)_{nn'}(r_j)_{n'g}}{(\omega - \omega_{ng})\,(\omega_1 - \omega_{n'g})} \right. \\ &+ \frac{(r_k)_{gn'}(r_j)_{n'n}(r_i)_{ng}}{(\omega + \omega_{ng})\,(\omega_2 + \omega_{n'g})} + \frac{(r_j)_{gn'}(r_k)_{n'n}(r_i)_{ng}}{(\omega + \omega_{ng})\,(\omega_1 + \omega_{n'g})} \\ &- \frac{(r_j)_{ng}(r_i)_{n'n}(r_k)_{gn'}}{(\omega - \omega_{nn'})} \left(\frac{1}{\omega_2 + \omega_{n'g}} + \frac{1}{\omega_1 - \omega_{ng}} \right) \\ &\left. - \frac{(r_k)_{ng}(r_i)_{n'n}(r_j)_{gn}{}'}{(\omega - \omega_{nn'})} \left(\frac{1}{\omega_2 - \omega_{ng}} + \frac{1}{\omega_1 + \omega_{n'g}} \right) \right]. \end{aligned} \tag{5}$$

Equations (4) and (5) are *exact*,[c] ignoring effects of the environment, vibronic states, and excited state dynamics. In these equations, $E_n = \hbar\omega_n$ is the energy of the nth state of the molecule and $\hbar\omega_{ng} = E_n - E_g$, where ψ_g is the groundstate and ψ_n the nth excited state. The r_i and r_j are the coordinate operators for the electrons, ω_1 and ω_2 are the laser frequencies.

When a denominator in the sum vanishes, the polarizability clearly diverges. In the exact spectral representation, one adds a small imaginary part of the energy of each state so that the susceptibility has a pole in the complex energy plane. The real and imaginary parts of the hyperpolarizability can then be shown to be related, which formally arises from demanding causality. On the real energy axis, then, the susceptibility no longer diverges. Physically, the imaginary part of the energy level is the decay width which is the inverse of the lifetime of the state. The net result is a Lorentzian or dispersive frequency dependence of the hyperpolarizability. For electronic structure calculations of the type used in molecular design where only the off-resonant response is of interest, this imaginary component is unnecessary.

[c]Provided that all excited state transition dipole moments, $e(r_i)_{n'n}$ and energy levels, $\omega_n = E_n/\hbar$ are known. Often, the sums are not evaluated over all electronic states.

The forms of Eqs. 4 and 5 are generalized sum over states (SOS) formulas. If the wave functions and energies are computed exactly, and when the energies are made complex with the appropriate line-width, the results are exact provided that the expressions are not evaluated directly on resonance. On the other hand, it is impossible to calculate either wave functions or energies of a many electron system exactly. Additionally, truncation of the states used or neglect of continuum states in the SOS calculation is normally made.[21-27] For those reasons, most calculations give inexact results. Indeed part of the "art" in state-of-the-art computations of electronic structure properties is to overcome this inexactness by improved approximations to the basis set, and the treatment of electron correlation. Note that these spectral representations yield, directly, the frequency dependence of β.

Because electrons have spin ½, they are called fermions (particles of half-integer spin). As such, the total wave function must be antisymmetric upon exchanging a pair of electrons (i.e., the wave function changes sign upon interchange). The wave function of a molecule's electrons can be expressed as a sum (or difference) of a product of single electron wave functions. The antisymmetrization is conveniently expressed as a determinant. For example, the two-particle wave function for a fermion is of the form $\psi = 1/\sqrt{2}\ (\psi_1(r_1)\psi_2(r_2) - \psi_1(r_2)\psi_2(r_1))$. The determinant form is obvious.

If the groundstate is approximated by a single determinant (the Hartree-Fock choice), then, since the operators r_i are of one-electron type, the excited state can only have single excitations from the Hartree-Fock groundstate. This means that they are of single particle/hole type (one electron moves from an occupied to an empty molecular orbital). If all possible monoexcited configurations (named equivalently, mono-excited configuration interaction, (MECI) or singly excited configuration interaction (SCI), or configuration interaction including singles (CIS)) are included, a consistent approximation is well-defined. A very large number of calculations, using both semiempirical and ab-initio model Hamiltonians, have studied the hyperpolarizabilities of molecules using this SOS-MECI approximation.[21-29] In actual practice, it is inefficient to perform explicitly the sum over states of Eq. 5. For most computations at the MECI level, an equivalent computation using linear equation methods or a correction vector approach is used instead.[29-31] Actual SOS calculations are, however, very useful for characterizing β in conceptual, chemical terms (which orbitals dominate in the response).[21-28,32,33] For nomenclature purposes, it is important to remark that these calculations are sometimes referred to as uncoupled Hartree-Fock, Tamm Dancoff (TDA), MECI, SCI, or CIS calculations. Very closely related approximations are called time dependent Hartree-Fock (TDHF)[34-36] or random phase approximation[d] (RPA);[37,38] these include an extra set of terms, not present in MECI and are equivalent to coupled perturbed Hartree-Fock (CPHF) or finite-field methods. The formal advantages of the TDHF over MECI include equivalence of dipole length and dipole velocity calculations for the oscillator strength (f).[e] More important, perhaps, is that these approximations point

[d]The terms TDHF and RPA refer to the approximations used in deriving the wave susceptibility.

[e]The oscillator strength, a dimensionless quantity characterizing the intensity of an absorption peak, is defined by $f_{na} = 2m\omega_{na}|\vec{\mu}_{na}|^2/3\hbar e^2$ where ω_{na} is the frequency difference between states n and a, $\vec{\mu}_{na}$ the transition moment, m the electron mass, and e the electron charge.

in different directions for going beyond the single determinant groundstate: the TDHF approach is used extensively in the calculation of dispersion of correlated wave functions, whereas the monoexcited CI is essentially a convenient stopping point for approximations which, because they neglect electron correlation of the ground state, are clearly never going to be quantitatively accurate.

The spectral representation of Eq. 5 is essentially an exact scheme for calculating the nonlinear response of an isolated molecule. An extremely useful approximation that is often semiquantitatively accurate and nearly always very helpful in understanding molecular behavior can be found by truncating the sum to include only one excited state.[39,40] When this is done, the very simple form of Eq. 6, called the two-level model, results,

$$\beta_{\text{two-state}} = \frac{3e^2}{2\hbar^3} \frac{\omega_{gn} f_{gn} \Delta\mu_{gn}}{[\omega_{gn}^2 - \omega^2][\omega_{gn}^2 - 4\omega^2]}, \tag{6}$$

where g and n represent, respectively, the groundstate and the excited state[f] and f_{gn} and $\Delta\mu_{gn}$ are, respectively, the oscillator strength and the dipole moment difference between state g and n. This formula shows that knowledge of the difference of dipole moments between the ground and excited states, the transition dipole moment between the groundstate and excited state, and the energy difference between the groundstate and excited state, are sufficient to estimate β provided that the state n dominates. Since these properties can be independently measured or calculated on the basis of linear optical behavior, the two-level formula provides a direct link between linear phenomena (such as absorption and solvatochromism) and the first hyperpolarizability.[41–45] Even more important, however, is the utility of the two-level formula in suggesting how to optimize the nonlinear optical response of a particular structure: β is largest for large oscillator strength (i.e., highly absorbing molecule) on resonance (either a one-photon or two-photon resonance) and maximum dipole moment difference between groundstate and excited state. The two-level formula also shows the trade-off between material properties required for making nonlinear optical devices: while coming close to resonance can make the nonlinear response larger, a comparison of the energy denominators of Eqs. 4 and 5 show that linear absorption is also on resonance. This illustrates one of the obvious trade-offs in designing large β chromophores.

From the spectral representations of Eqs. 5 and 6, a numerical calculation of the linear and nonlinear response of a molecule raises the following obvious questions:

1. The one electron basis: In what set of localized electronic basis functions can we expand the wave function describing the electron location?
2. Vibrational and vibronic terms: Are the states in these summation taken as purely electronic, or do we admit vibrational factors, which will change both the matrix elements in the numerator and the energies in the denominator?

---

[f]The usual definition of oscillator strength requires an additional factor of 3/2 on the right of (6); this is usually ignored.

3. Electron correlation effects: if ψ_g is taken not as the Hartree-Fock (uncorrelated) electronic groundstate but rather as a correlated state, which correlation schemes can be used to calculate the groundstate, as well as the excited states, and the associated energies?
4. Dispersion (frequency dependence): Formally, one can calculate α and β for any given set of excitation frequencies using Eqs. 4 and 5, and thereby obtain the frequency dependent response. This becomes substantially less convenient when more sophisticated and advanced methods are used for calculating the wave functions.
5. Truncation: The sums over states in Eqs. 4 and 5 formally include all contributions from all the electronic states of the system: What is the effect of truncating this manifold, and how can the missing states be included, or their effects estimated?

All of these questions are important, and all have been studied in detail in the context of ab-initio electronic structure calculations. Since most studies will use the Hartree-Fock states, and therefore correspond to MECI calculations, we have devoted several sections to this level of theory. A discussion of particular effects calculated at this level of theory, using semiempirical model Hamiltonians, is given in Section VI.

The states indicated in Eqs. 4 and 5 can be calculated from any model Hamiltonian given $h^{(1)}$ in Eq. 1. There has been extensive study using semiempirical electronic structure Hamiltonians, particularly at the MECI level, for the polarizability and the first hyperpolarizability. For the second hyperpolarizability γ, an analogous equation to 3 and 4 can be given; it involves terms of the sort[18,20,46–50]

$$\gamma_{ijkl}(\vec{\omega}) \propto \wp \sum_{nn'n''} (r_i)_{gn} (r_j)_{nn'} (r_k)_{n'n''} (r_1)_{n''g} / D(\vec{\omega}, \omega_{gn}, \omega_{gn'}, \omega_{gn''}), \tag{7}$$

where $\vec{\omega}$, ω_{gs} ($s = n, n', n''$) denote respectively the set of appropriate frequencies and the state energy differences; D is simply a notation for denominator; and $\wp$ is a permutation operator. Notice, since the r's are one electron operators, that if the groundstate ψ_g is taken as Hartree-Fock, then the states ψ_n and $\psi_{n''}$ are of single excitation type, but the state $\psi_{n'}$ can either be a single excitation or double excitation. Therefore it is necessary to include doubly excited states in the summation if one is to be consistent with the choice of the Hartree-Fock groundstate. While the calculation of $\chi^{(3)}$ properties is not really our focus here, at the simplest level one can compute γ in precisely the same way as β, using the so-called doubly excited configuration interaction (DCI). A number of calculations of this kind, particularly using semiempirical models, have appeared in the literature.[46–50] The questions of accuracy in the calculation of frequency-dependent hyperpolarizabilities have been investigated extensively using both ab-initio and semiempirical models. Section IV below describes recent progress in the ab-initio context. While a semiempirical calculation is less precise, it is nevertheless worthwhile to discuss those calculations because they are able to handle larger molecules and because of the insight they provide for chemical interpretation, optimization, screening of candidate chromophores, and for aiding our understanding of the ab-initio results.

IV. SEMIEMPIRICAL CALCULATIONS OF THE β-RESPONSE: SYSTEMATIC ASPECTS

We, and others, have discussed semiempirical models extensively, and much has been written about their application to nonlinear optical properties.[28] In this context, it is important to remember that for any electronic structure calculation of determinantal type (that is, not including such techniques as density functional theory or anharmonic oscillator or Thomas/Fermi models), the Hamiltonian can be written in terms of an orbital basis set.[51,52] It is convenient to do this using a second quantization notation. The functional idea is that a basis set, by virtue of completeness, is used to express the Hamiltonian operator as $H = \Sigma_m \Sigma_n |m\rangle\langle m|H|n\rangle\langle n|$. More explicitly, in second quantized form,

$$H = \sum_{ij} \sum_{\mu} t_{ij}\, a^{\dagger}_{i\mu}\, a_{j\mu} + \frac{1}{2} \sum_{ijkl} \sum_{\mu v} \langle ij|lk\rangle\, a^{\dagger}_{i\mu}\, a^{\dagger}_{jv}\, a_{kv}\, a_{l\mu}, \tag{8}$$

where the operator $a^{\dagger}_{i\mu}$ creates an electron in the ith local basis function with spin μ, and its complex conjugate, $a_{i\mu}$ destroys an electron in that state. The first term in Eq. 8 is the one electron term, arising from nuclear attractions and kinetic energies; the second term is the electronic repulsion term. In the Hückel model,[52–54] the second terms are implicitly included (in a mean field approximation) in the first term, and all explicit inter-electronic interaction is ignored; while some calculations of the nonlinear response of molecules using the Hückel Hamiltonian have appeared,[48,55–58] it is quite clear that electron repulsions are too important to be ignored even for semiquantitative prediction.[48,55]

Elements t_{ij} and $(ij|lk)$ are, respectively, the elements corresponding to the one electron operators and electron repulsion. They have dimensions of energy. If they are calculated properly, using their formal definitions in terms of kinetic energies, nuclear attractions and electron repulsions, one has an ab-initio electronic structure Hamiltonian. If they are partially approximated using some standard numerical data for a particular set of test molecules, one has a semiempirical model Hamiltonian. The sums over *ijkl* are sums over basis functions; in simple semiempirical models[59] such as PPP, INDO or CNDO, these are valence functions of Slater-type, and the basis is minimal (one orbital for hydrogen, four for carbon, nine for iron, and so forth) and only a few dominant integrals are included in the Hamiltonian.

An important observation is that semiempirical and ab-initio theories differ only in the definition of the basis set and the evaluation of the integrals; while there are many advantages of ab-initio theories over semiempirical models, the latter are of substantial use in understanding molecular properties, and have been widely and successfully applied to the problem of nonlinear optical response.[20–32,60–70]

The semiempirical models used in electronic structure theory differ in the choice of valence basis set, and in the value selected for the matrix elements t_{ij} and $(ij|lk)$. The acronyms such as CNDO (complete neglect of differential overlap), INDO (intermediate neglect of differential overlap), MINDO (modified intermediate neglect of differential overlap), and others refer to the definitions involved and which of the two electron repulsion terms, the $(ij|lk)$, are included. In all semiempirical calculations, however, the basis set is small, so that the calculations are fast and interpretation of the calculations in terms of Mulliken population analysis is straightforward.

In applications to compute a nonlinear response, the most commonly used semiempirical methods are the PPP model[23,24,48,60,61] (used for conjugated hydrocarbons and heterocycles), the CNDO/S[21,22,29,30,47,62–64] and INDO/S[25,26,31,35,38a,49,50,65,66] models (the /S refers to screening or spectroscopic, and refers to a choice of values for the semiempirical parameters), and the various NDO models such as MINDO, MNDO, and AM1.[27,67] All of the latter two categories include complete valence sets, rather than just π-orbitals, and therefore can be used for molecules that have atomic sites that are not sp^2-hybridized. In particular, the INDO/S model, parameterized by Zerner[68] to include metals, has been used in extensive calculations of organic and organometallic nonlinear optical response.[26,28,35,38,65,66] Section VI below discusses more extensively actual semiempirical calculations of response at the MECI level. Here we discuss briefly the five issues of improved calculations discussed above, in a semiempirical context.

A. Basis Sets

The issue of basis is not often discussed; semiempirical models are defined in terms of a given (minimal) basis that can generally not be extended. Some work has addressed extended-basis semiempirical models,[69] but for the most part the parameterization involved and the choice of the Hamiltonian are sufficiently complicated that such an approach has not been used in systematic studies.

B. Vibrational Effects

For particular frequency ranges (especially for small molecules) vibrational effects are important. Vibrational motions arise from structural changes and vibronic resonances. (See Section V.) Vibrational effects are not often included within semiempirical models. Soos and collaborators[70] have considered vibrational effects, particularly on $\chi^{(3)}$ for isolated molecules. They find that the effects of vibrations in medium-sized molecules are not important; this might have been anticipated, since the transition matrix elements are generally going to be small for vibrational levels, and therefore contributions to the SOS form of Eq. 5 should not be very large.[g] More extensive studies have been done using ab-initio calculations.

The reason that vibrational effects are generally not included in semiempirical model calculations of nonlinear response is that such studies are often used to understand how variations in molecular geometry and molecular structure effect nonlinear response. The usual experimental value for comparison is the frequency-doubling response in the visible region of the spectrum, as derived from EFISH measurements (described in Chapter 6). Since the aim is not so much quantitative understanding as the understanding of trends, vibrational properties are not relevant, and therefore are not generally studied.

---

[g]Recent work on $\chi^{(3)}$ response by Aramaki et al. (S. Aramaki, W. Torruellas, R. Zanoni, G. I. Stegeman, *Opt. Commun.*, **1991,** *85*, 527) has shown that inclusion of vibrational overlaps can lead to improved quantitative agreement between semiempirical computation and experiment.

C. Correlation

The effects of correlation within semiempirical models can be studied in precisely the same way as within ab-initio models. Indeed, much of the methodology involved in the study of correlation effects in molecules began with studies on semiempirical Hamiltonians.[34] Within semiempirical model calculations of nonlinear optical response, Soos and Ramasesha[48,71] pioneered the use of valence bond representations that give the *exact* states for substitution into Eqs. 4, 5 and 7. Therefore, with a given choice of model Hamiltonian, one can calculate the exact frequency dependent first hyperpolarizability.

Table 1 shows the results for the hyperpolarizability of aniline.[48] Note that inclusion of correlation effects substantially changes the nonlinear response. A much clearer picture that shows the changes in β as a function of increasing order of CI is shown in Table 2.[72] (Here, and elsewhere in this Chapter, we use the term "correlation effects" as is commonly understood in quantum chemistry: The correlation energy is the difference between the exact energy and the energy calculated in the Hartree-Fock approximation with the same basis, and correlation effects are due to electron repulsion which is not included in the averaging that characterizes the Hartree-Fock approximation.[73,74])

Since electron correlation effects can be so large within a given semi-empirical Hamiltonian, one might question the validity of semiempirical model Hamiltonians at the MECI level. Semiempirical calculations are based on the assumption that parameterization of the model Hamiltonian of Eq. 8, and its use within a series of well-defined approximations, can yield an effective model chemistry for that class of molecule and problem. For optical response calculations, the choice of semiempirical model Hamiltonian is generally based on calculations at the MECI level. One therefore can choose, for example, the parameter set for INDO/S or PPP that has been adjusted to fit the optical properties of a set of sample molecules. These parameters differ substantially from those used for calculating groundstate geometries because (as we indicated earlier) the optical properties involve growing dif-

Table 1 Comparison of $\beta_{ijk}(-2\omega;\omega,\omega)$ for Aniline at $\hbar\omega$ = 1.17 eV for Exact PPP, Approximate PPP, and Approximate CNDO Calculations. Adapted from Ref: 48

β_{ijk}	Exact PPP	PPP-MECI[1]	CNDO[2]
xxx	0.468	1.702	1.209
xxy (xyx)	0.227		
yxx	0.234		
xyy	0.037	−0.320	0.250
yxy (yyx)	0.092	−0.268	
β_x^3	0.542	1.417	1.589

[1]From D. Li, M. A. Ratner, and T. J. Marks, *J. Amer. Chem. Soc.*, **110**, 1707 (1988) and references therein.
[2]From S. J., Lalama, A. F. Garito, *Phys. Rev. A.*, **20**, 1179 (1979).
[3]Experimental value at $\hbar\omega$ = 1.17, 0.89 (From B. F. Levine, and C. G. Bethea, *J. Chem. Phys.*, **63**, 2666 (1975).) or 1.09 (From J. Oudar, and D. S. Chemla, *J. Chem. Phys.*, **66**, 2664 (1977).)

Table 2 Linear and Nonlinear Optical Properties of *Para*-Nitroaniline. The NLO Coefficients are Calculated at an Excitation Energy of 0.65 eV. Adapted from Ref: 72

	$\Delta E(^1B_2)$(eV)[1]		$\alpha_{xx}(10^{-23}$ esu)[1]		$\beta_x(10^{-30}$ esu)[1]	
CI order[4]	Γ_C^2	Γ_C^3	Γ_C^2	Γ_C^3	Γ_C^2	Γ_C^3
SCI	3.93		10.42		11.57	
SDCI	4.68	4.59	5.97	6.29	4.58	4.54
SDTCI	4.59	4.52	6.31	6.61	5.54	5.42
QCI	4.59	4.52	6.24	6.55	5.38	5.29
FCI	4.59	4.52	6.25	6.54	5.36	5.26

[1]Experimental values of $\Delta E(^1B_2)$, α_{av} and β_x in acetone are 3.40 eV, 1.70×10^{-23} esu and 9.2×10^{-30} esu (source: Ref: 135a). β_x should be multiplied by 0.58 for comparison (see Ref: 35 for details).
[2]The two-electron one center integral, Γ_C, is taken to be 11.11 eV.
[3]A reduced value of 10.33 eV for the two electron one center integral, Γ_C, is taken. See Ref: 72 for details.
[4]SCI, SDCI, SDTCI, QCI, FCI denote singles, singles and doubles, singles, doubles and triples, quadratic, and full CI.

fuseness in the molecular cloud, upon excitation, corresponding to a screening of the electrons. This reduces the effects of electron repulsion, and therefore the screened parameter set is generally chosen. If, on the other hand, one chooses the groundstate parameterization, one would expect increased errors in the MECI calculation of nonlinear optical response.[72]

It therefore seems inappropriate to use strongly correlated wave functions (in particular, to use exact wave functions) within a semiempirical model Hamiltonian with model parameters based on analysis at the MECI level. If one wishes to do calculations using exact wave functions within the semiempirical context, it would be more appropriate to choose the semiempirical model itself from a set of normative calculations, and comparison with experiment, on sample molecules—and these calculations themselves would have to be at the full CI level. This has not been done.

Calculations such as those of Soos *et al.*[48] are extremely valuable in showing the errors involved in MECI approximations (just as comparable calculations have been extremely valuable in illustrating approaches to the general, difficult problem of electron correlation). Nevertheless, we feel that for consistency one should probably limit the computation of nonlinear optical response using semiempirical methods to the MECI level (for β) or the DCI level (for γ).

D. Dispersion

Semiempirical calculations of dispersion (that is, frequency dependence) use the SOS formula for β: one simply does the calculation at each frequency.[21–31,75] While this is slightly tedious, it also works well. Figure 1 shows a typical frequency-dependent polarizability and first hyperpolarizability response of a generic molecule. Note that the dispersion peak (zero of the denominator in Eq. 4 or 5) of the hyperpolarizability is at twice the frequency of the main peak of the polarizability.

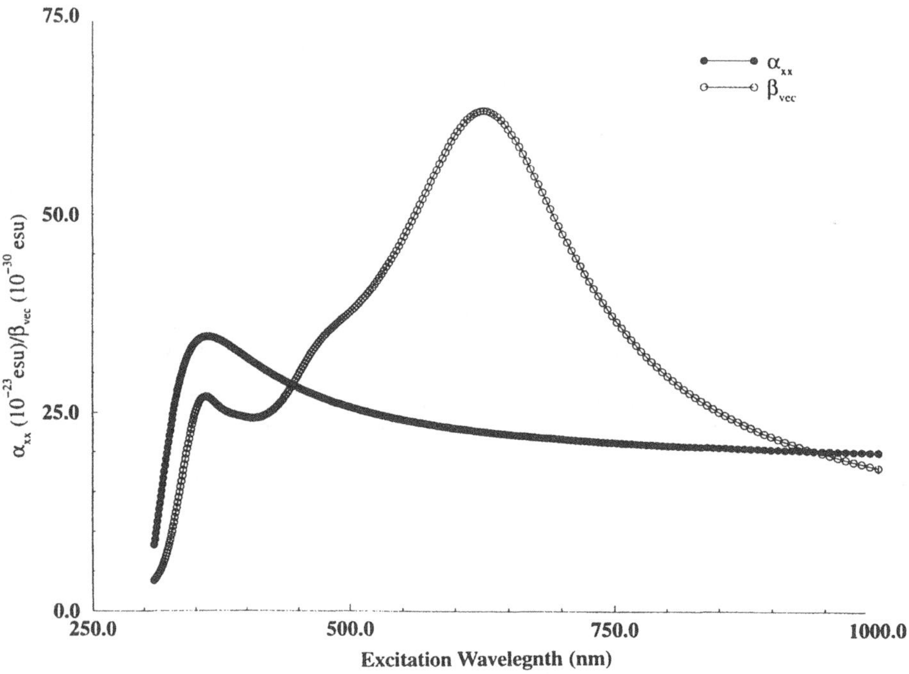

Figure 1 Computed dominant component of the frequency-dependent polarizability and the vector part of the first hyperpolarizability β_{vec}, for para nitroaniline, using the INDO/S model Hamiltonian. Notice that the peak in polarizability occurs at half the wavelength of the hyperpolarizability peak, consistent with the latter being a three-photon process.

Because the hyperpolarizability is a two-photon process, a material resonance is excited by two photons at half the resonance frequency—i.e., frequency doubling. These results also suggest that dispersion would be most pronounced for the highest order hyperpolarizability, because the poles get closer and closer to the frequency origin. This is indeed true. (Damping, lifetime, or vibronic effects that derive from the energy denominators in Eq. 5, will broaden the poles to Lorentzians that are seen in Fig. 1.)

The frequency dependence of the nonlinear response can be measured using frequency tunable sources such as optical parametric amplifiers (OPAs).[76] Such measurements are important for understanding the origin of the hyperpolarizability of a molecule, and as a tool for optimizating molecular design for response at particular frequencies, making materials for communications applications, or fabricating blue light sources. This facile incorporation of dispersion effects is one of the great advantages of the simple SOS formulation, in the MECI approach.

E. Truncation

Finally, the effects of truncation of the sum in Eqs.. 4 and 5 have been extensively studied.[21–30] Figure 2, shows that the largest single contributor is the HOMO/LUMO excitation. This type of behavior is characteristic of many molecules.

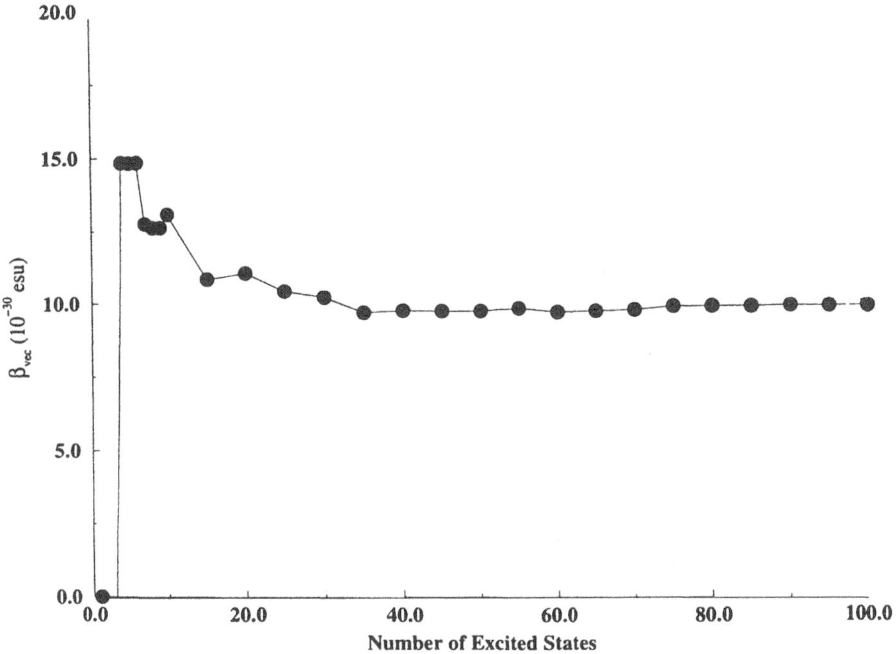

Figure 2 The hyperpolarizability in Eq. 5, for *para*-nitroaniline calculated with MECI, INDO/S as a function of number of states included in the sum. The characteristic behavior (higher excitations reduce the magnitude of the dominant, HOMO → LUMO transition) occurs for most standard chromophores.

Higher excitations (that is, excitations between molecular orbitals whose energies differ by more than the HOMO/LUMO gap) make smaller but nevertheless sizeable contributions. Generally one finds that it is necessary to include in the sum so-called two-level terms and three-level terms, as illustrated in Fig. 3.[28] These differ only in which intermediate states are included; in the two-level terms, only one intermediate state appears. Characteristically, three-level terms contribute an amount that is about half the magnitude of the two-level terms, and, of opposite sign.

For relatively small molecules such as *para*-nitroaniline (*p*NA) or substituted stilbenes, the sum converges after a few dozen excited states are included in the sum.[60–66] One can easily estimate when the sum has converged from a curve similar to the one shown in Fig. 2 (usually ranked or ordered by the energy difference of the molecular orbitals in the sum).

For semiempirical models, then, the effects of the basis set, vibrations, electron correlation, dispersion and truncation on the calculated response are all reasonably well understood; for semiempirical calculations at the MECI level dealing with these effects poses no difficulty. Such calculations will, however, have substantial errors when the minimal basis is used, from the neglect of electron correlation, and from the semiempirical nature of the Hamiltonian. Nevertheless, to the level of approximation, it should be internally consistent, and the deductions based on such

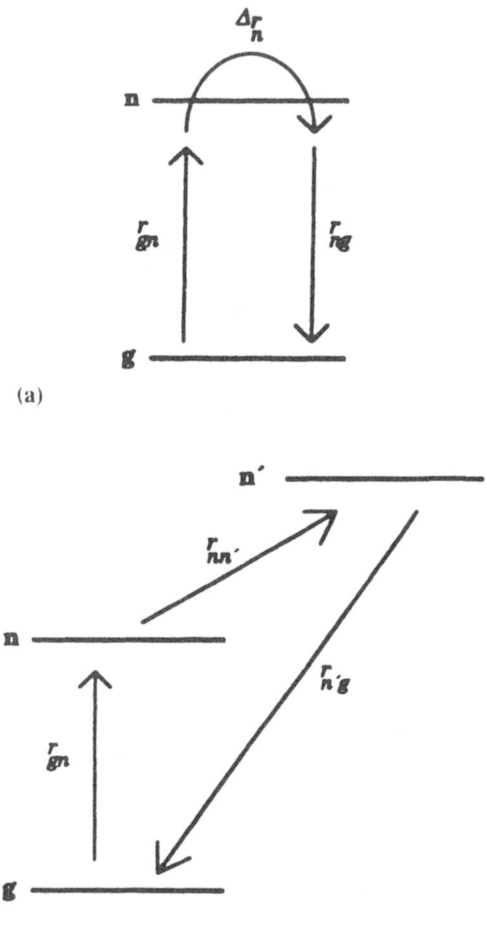

Figure 3 A diagrammatic representation of two dominant terms that enter into the SOS expression of Eq. 5 for the hyperpolarizability. The two-level terms of Fig. 3a correspond to a single excited state acting as the intermediate state, while the three-level terms of Fig. 3b involve two virtual excited states. Very often, the three-level terms differ in sign from the two-level contributions. (From Ref: 28.)

calculations, as extensively discussed both elsewhere[28] and in Section VI below, make such calculations of substantive value in understanding molecular hyperpolarizabilities.

V. AB-INITIO CALCULATIONS: BASIS, CORRELATION AND DISPERSION

The above discussions of semiempirical calculations can also be applied to comparing and improving ab-initio studies. The aim of ab-initio calculations can sometimes be quite different from that of semiempirical models: Often it is the desire to have high accuracy results, both as a test of the computational techniques involved, and to pinpoint such issues as standardization, solvent effects, small changes in geometry of the molecule, or effects of vibrations.

Calculation of the nonlinear response within ab-initio models is one of areas at the forefront of electronic structure studies.[77] Several of the leading groups worldwide have devoted considerable attention to this problem, and a good deal of progress has been made. Nevertheless, the problem is quite a difficult one: (1). The effects of correlation even on the static response (for any many-electron Hamiltonian) are substantial; (2) the extension of the charge cloud arising from photon interaction places severe demands on the one electron basis set; and (3) the inclusion of dispersion requires a complex treatment of the dynamical correlation problem. This area is rapidly evolving, with substantial advances being made quickly in a number of laboratories. Instead of reviewing the field fully (as is done in several other places),[28,77] we will give an overview of the different methods that have been used to attack the problem, and some indication of where the bulk of current research efforts is headed.

A. Basis Sets

This is probably the simplest of complications involved in accurate ab-initio calculations of the hyperpolarizability. As we have already indicated, it is necessary to use one-electron basis functions that describe the diffuse tails of the wave function. One electron basis sets, such as those developed by Sadlej[78] for properties calculations, have been used to give substantially improved results compared to the sort of extended valence plus polarization basis sets generally valid for describing groundstate properties of molecules.[73,74,79] For example, Table 3 shows the results of calculations of the static first hyperpolarizability for a series of heterocyclic species using different basis sets.[79] Clearly, the diffuse functions are important and can affect the calculated results by factors of two or more. Extensive study has

Table 3 Calculated Values of Dipole Moment, μ (Debye), and First Hyperpolarizability, $|\beta|(-2\omega;\omega,\omega)(10^{-30}$ esu), of Heterocyclic Molecules as a Function of Basis Set. Adapted from Ref: 79

		Basis Set				
Molecule		A_1	B^2	C^3	D^4	E^5
Thiophene	μ	0.18	0.99	0.99	1.09	0.91
	β	0.06	0.199	0.111	0.091	0.14
Benzoxazole	μ	1.30	1.52	1.132	1.48	
	β	0.034	0.071	0.073	0.113	
Benzothiazole	μ	1.49	1.62	1.26	1.61	
	β	0.144	1.141	0.095	0.067	

^{1}STO-3G
^{2}4-31G
^{3}B + Semidiffuse, see Ref: 79 for details.
^{4}B + Diffuse, see Ref: 79 for details.
^{5}B + Field-induced polarization functions, see Ref: 79 for details.

shown that an appropriate property basis set, such as that of Sadlej,[78] can be extended by adding more diffuse orbitals (particularly to deal with hydrogenic sites) to yield a satisfactory solution to the basis set problem. Inclusion of these large basis sets, of course, makes the calculation more technically demanding, but an accumulating body of knowledge[74,77] can be applied to limiting appropriately the nature of the basis.

B. Vibrations

For small molecules, vibrational components of the nonlinear response can be important; accurate calculations on small molecules by Bishop and his collaborators have demonstrated this quite clearly.[80,81] A relatively trivial method for calculating the vibrational contribution is to average β over all possible zero point extensions of the molecule. For stable bound molecules, the zero point extensions are relatively small and can be included using harmonic approximations with reasonable levels of accuracy. Vibrational averaging of hyperpolarizabilities of small molecules, particularly diatomics and tetra-atomics, can be (relatively) very large. Indeed, an early calculation on the HF molecule showed that vibrational averaging can change the sign of the first hyperpolarizability, and change the second hyperpolarizability by an order of magnitude.[81f] Part of the reason for this is that these variations are calculated for hyperpolarizabilities that are very small: For example, $\beta(0;0,0)$ of HF is 8.65 atomic units. This corresponds to $\beta(0;0,0) = 0.007 \times 10^{-30}$ esu; only very elaborate first principles calculations can obtain accurate hyperpolarizabilities that are this small. On the other hand, from the point of view of materials properties such hyperpolarizabilities are entirely uninteresting. Small hyperpolarizabilities should therefore not be investigated using such semiempirical models, nor are they of substantive interest in any chemical context.

The more complicated situation involves vibronic intermediate states, previously discussed in connection with semiempirical calculations. It is convenient to examine the sum over states form of Eq. 5. The intermediate states ψ_n are eigenstates of the full electronic and nuclear Schrödinger equation, and therefore should have vibrational as well as electronic components. If the vibrational energies in the denominator are ignored, then the vibrational states do not contribute to the hyperpolarizability (the sum over all the vibrational states is unity). If, however, the energies due to the vibrational states are included, the vibronic states contribute to the response.

One might expect that molecules containing high vibrational frequencies would have the largest vibronic contributions to the hyperpolarizability. The hydrogen molecule has been used as a test case for learning about the importance of the vibronic contribution. Table 4 shows the second hyperpolarizability of the hydrogen molecule[81] as calculated by Bishop and Kirtman in which vibrational effects are demonstrated to contribute strongly to γ. Because the vibrational frequency of the hydrogen molecule is about 4400 cm^{-1}, one cannot ignore the vibrational energies. For organics, vibrational frequencies of skeletal modes, distortions, twists and so forth (those of primary interest) will be of much lower frequency, so that vibrational effects are not expected to be of dominant importance in most cases. Clearly, in any calculation that aims for a quantitative comparison with experiment, vibrational terms must be included.

Table 4 Electronic and Vibrational Hyperpolarizabilities (γ) for H_2 (in arbitrary units). Source: Ref: 80a.

$\lambda/\text{Å}$	$\gamma_{\parallel}^{e1}$	$\gamma_{\parallel}^{v1}$	$\gamma_{\perp}^{e1}$	$\gamma_{\perp}^{v1}$
∞	670.8	183.7	223.6	61.2
4880-Kerr	753.4	63.5	247.8	64.2
4880-ESHG*	969.1	−6.8	337.8	−0.9

¹2nd hyperpolarizability values are in a.u.
*The fundamental wavelength is 4888 nm.

In conjugated systems, a particularly strong vibronic coupling with a collective C-C stretching coordinate effectively modulates the bond-length alternation (see Section VII.B.2). The relative importance of vibrations should depend upon the particular NLO property under study; this is, again, apparent from the denominators of Eq. 5. For the general frequency-mixing response, $\beta(-\omega_1 - \omega_2; \omega_1, \omega_2)$, the relative importance of vibrational contributions is expected to be dominated by near-resonance contributions that appear whenever the real part of a denominator in Eq. 5 vanishes. A small purely imaginary term is added to the energy level to account for the finite lifetime of that state. For the frequency doubling response, $\omega_1 = \omega_2$. In the visible part of the spectrum, the dominant terms would come from electronic excitations, since only these can produce small denominators. For electrooptic effects, however, where $\omega_2 \rightarrow 0$, calculations show that vibrations (especially low-frequency vibrations) are expected to contribute to β.

C. Correlation Effects

The correlation problem is the dominant theme of modern electronic structure theory.[73,74,82–85] Any model Hamiltonian of the type of Eq. 8 is by definition a many-body problem, in which static and dynamic correlations among all the electrons must be considered. In the Hartree-Fock approximation, the "averaged" interelectronic repulsion is included. The effects of correlation (using chemical nomenclature) go beyond Hartree-Fock averaging.

There are many approaches to the correlation problem,[73,74,82–85] and nearly all of those have been used in the context of hyperpolarizability response calculations. Rather than present an extensive discussion and review, we will highlight particularly promising approaches that have been used, and some of the successes that they have achieved.

Correlation effects on optical properties that are clearly closely related to nonlinear optical properties (as can be seen by comparison of Eqs. 4 and 5) are generally studied using one of four approaches.[73,74] The first and most obvious is simply configuration interaction:[73] one constructs the groundstate to include not only the single determinant of Hartree-Fock, but a number of determinants corresponding to excited electronic configurations.[h] The Brillouin theorem[74] guarantees that no singly excited states should mix directly with the groundstate, and therefore such calculations usually begin with a groundstate consisting of the Hartree-Fock state mixed with single and double excitations. Similarly, the excited states are no longer limited to single excitations, but can include singles, doubles, triples, and so forth. This approach is obvious and straightforward, but is not commonly used. One problem is that it is difficult to know the number of multiple levels of excitation that need to be used in the calculation. Another problem is the so-called size consistency problem[73] (a calculation of a response of a pair of *noninteracting* molecules does not give the same results as the calculations on a single molecule). Finally, more efficient and more appropriate methods for dealing with correlation are (deservedly) more popular.

A second approach to the correlation problem in spectroscopy and nonlinear optical response is based on multiconfiguration self consistent field approximations (MCSCF).[86,87] In these one writes the groundstate not as a single determinant, but as a linear combination of single determinants and certain excitations outside of those single determinants. For example, a representation of the sth state of the molecule is given in Eq. 9 in terms of linear combination by the Hartree-Fock state and single excitations (indexed by the $B_{\lambda i}$), double excitations (indexed by $C_{\lambda\kappa ij}$), and higher excitations; here λ and κ label HF unoccupied orbitals, and i, j occupied mo's:

$$|\psi_s> = \{A_o^{(s)} + \sum B_{\lambda i}^{(s)} a_\lambda^+ a_i + \sum_{\lambda\kappa ij} C_{\lambda\kappa ij}^s a_\lambda^+ a_\kappa^+ a_i a_j + \cdots\}|\, \text{HF}>. \tag{9}$$

The orbitals within the Hartree-Fock and the linear expansion coefficients A, B, C are determined simultaneously and self-consistently; this is called a multi-configuration SCF approximation. Often one uses a variant of this in which all possible excitations are allowed, but only within a particular choice of the molecular orbitals. For example, one could allow all excitations between the two highest occupied molecular orbitals and the two lowest empty molecular orbitals. For closed-shell molecules in a singlet state, this would yield 20 excitations (the Hartree-Fock state, four single excitations, ten double excitations, four triple excitations, and one quadruple excitation). The determinantal space that includes these 20 configurations is called the complete active space, and the technique is called the CASSCF approximation.[86–88] The active space is simply the set of molecular orbitals that can be emptied (if occupied) or filled (if empty).

CASSCF calculations are useful for a number of electronic structure properties,[88] and can be used to do optical and nonlinear optical calculations.[88–90] In general, CASSCF itself is not adequate. For accurate representations of optical properties, one generally adds some configuration interaction to the CAS—that is, electrons can be excited by light beyond the levels that the CAS uses. This makes sense in terms of our previous arguments about the nature of the perturbation operator in Eq. 1: This operator is of one electron type, and therefore if a CAS is

[h]For β calculations, finite-field, or coupled-pertubated HF, gives results identical to those from RPA. These are often adequate, though they neglect ground-state correlation. RPA, as stated already, is very similar to MECI. Therefore, neglect of ground-state correlation seems reasonable for relatively accurate β calculations.

used for the groundstate of the molecule, then for the optical response, one would need a manifold that includes (at least) one particle/hole excitation beyond that in the CAS. (This is not true if a response function approach is used for the computation.)[89,90]

Table 5 shows the calculations using CAS for *p*NA.[90] It is quite clear that the CAS calculations are problematic, and not accurate, even for this relatively simple chromophore. Inclusion of configuration interaction beyond the CAS has substantially improved the agreement between theory and experiment of optical properties, and the same would be expected in the calculation of nonlinear optical properties.

A third approach for dealing with correlation effects is the use of perturbation theory.[73,74,79b,91] Essentially, equivalent formulations have been based on the use of diagrammatic expansions (many body perturbation theory or MBPT) or on the use of simple Rayleigh/Schrödinger perturbation theory using as the perturbation the difference between the Hartree-Fock Hamiltonian and the full Hamiltonian:[73,74]

$$V_{MP} = H - H_{HF}. \tag{10}$$

This approach was pioneered by Møller and Plesset, hence the subscript *MP* in Eq. 10. More recently, Pople and his collaborators have applied it extensively to the solution of a large range of electronic structure problems,[92] and it is included in several standard packages for quantum chemical calculations, including GAUSSIAN,[93] GAMESS,[94] and SPARTAN.[95] The MBPT is essentially equivalent to MP, and is found in the ACES II program.[96]

These perturbation theory methods can be brought to arbitrarily high order. Using the Møller/Plesset classification,[74] MP/1 is the Hartree-Fock calculation. MP/2 includes second order perturbation corrections, and similarly MP/3, MP/4 would include higher order corrections.

Use of fairly straightforward perturbation-theoretic corrections results in substantial improvement of the treatment of correlation. For example, Rice and coworkers extensively tested MP/2 calculations for describing the nonlinear response in a number of sample molecules.[97] The results show clearly that MP/2, with a reasonable basis, does very well for predicting nonlinear response. Table 6 shows the results of MP/2 calculations for H_2O.[98] Notice for example, the second harmonic generation result in a larger basis set. The number is −5.4 at the TDHF level, −8.8 at the second order perturbation theory level, −9.1 at the most elaborate computational level, and −9.4 experimentally. Therefore we see from this simple example that the second order perturbation theory indeed eliminates most (in this case 85%) of the correlation error made at the simple Hartree-Fock level.

Based on results of several groups, the MP/2 level is an attractive formalism for dealing with correlation: MP/2 calculations are relatively easy to do, are coded in several accessible packages, and give results that, apparently, capture most of the correlation effects.

A fourth popular approach to dealing with the correlation problem involves the so-called coupled cluster method.[99] In this scheme, the states of the molecule are written as a transformation of the Hartree-Fock state in terms of a coupled cluster operator:

$$|\psi_o^{cc}> = e^T|0> \tag{11}$$

Table 5 The Hyperpolarizability Dispersion of *p*NA (in units of 10^{-30} esu). Adapted from Ref: 89b

Freq.	Level	β_{zzz}	β_{zxx}	β_{zyy}	β_z	Experiment[1]	Experiment[2]
$\omega = 0$							
	RPA	5.10	−0.90	−0.11	4.09		
	MCQR[3]	5.93	0.13	−0.13	5.93		
$\omega = 0.65$ eV							
	RPA	5.58	−0.96	−0.12	4.50		
	MCQR[3]	6.50	0.15	−0.13	6.52	9.6 ± 0.5	28.8 ± 1.5
$\omega = 1.17$ eV							
	RPA	6.94	−1.13	−0.13	5.68		
	MCQR[3]	8.12	0.23	−0.15	8.20	16.9 ± 0.4	50.7 ± 1.2
$\omega = 1.364$ eV							
	RPA	7.88	−1.24	−0.14	6.50		
	MCQR[3]	9.24	0.29	−0.15	9.38	25 ± 1	75 ± 3
$\omega = 1.494$ eV							
	RPA	8.74	−1.35	−0.15	7.24		
	MCQR[3]	10.25	0.34	−0.16	10.43	40 ± 3	120 ± 9

[1]Experimental Values (from Teng, C. C., and Garito, A. F., *Phys. Rev. B.*, *28*, 6766 (1983).
[2]Value in previous column multiplied by a factor of three according to the convention given in Willets, A., Rice, J. E., Burland, D. M., and Shelton, D. P., *J. Chem. Phys.*, *97*, 7590 (1992).
[3]MCQR (Multi-Configurational Quadratic Response) is a CAS-Type Method.

Table 6 Theoretical[1] vs. Experimental Hyperpolarizabilities ($\chi_{\parallel}^{(2)} = \frac{1}{2}\beta_{\parallel}^{SHG}$) of Simple Molecules in Units of esu. Adapted from Ref: 98.

Molecule	TDHF	MBPT(2)	CCSD	CCSD(T)	Experiment[2]
CO	10.5	11.2	11.4	11.7	12.9 ± 1.4
HF	−2.5	−3.3	−3.2	−3.4	−4.7 ± 0.41
HCl	−1.6	−4.0	−3.3	−3.8	−4.22 ± 0.50
H_2O	−5.4	−8.8	−8.2	−9.1	−9.4 ± 0.4
NH_3	−9.3	−20.1	−18.4	−21.2	−20.9 ± 0.5
H_2S	0.6	−5.0	−3.4	−4.5	−4.3 ± 0.9

[1]Value corrected for the dispersion effect at 694.3 nm using Eq. 15. For details on the basis set used in the calculation see Ref: 98.
[2]Value obtained by dc-induced second harmonic generation.

$$T = \sum_{\lambda i} (T_{\lambda i}\, a_{\lambda}^{+}\, a_i + C.\,C.) + \cdots \tag{12}$$

Equations 11 and 12 give the simplest kind of coupled cluster approximation. Here the coupled cluster groundstate is given as a transformation of a determinantal reference configuration, denoted $|0>$ in Eq. 11. The cluster operator T, as indicated in Eq. 12, consists of excitations out of this reference determinant; by increasing the order of excitations, one obtains increasingly sophisticated coupled cluster approximations.

The coupled cluster (*CC*) approximations have some disadvantages. Like the perturbation theory approximations, they do not satisfy the Hellmann/Feynman theorem, and therefore the polarizability tensors must be determined directly from perturbed energies.[100] The *CC* method as ordinarily derived is not variational. Nevertheless, it has substantive advantages in dealing with difficult correlation problems, and a number of techniques based on the coupled cluster approach have been extremely important in analyzing correlation effects in chemistry.[99]

Table 7 shows two more accurate computational methods—the so-called coupled cluster singles and doubles (CCSD) and coupled cluster singles and doubles with partial inclusion of triples (CCSD(T)).[101] These have to do with the order at which the truncation of the cluster operator is made. Generally, the more terms that are included, the more accurate the calculation.

Extensions of the simple coupled cluster scheme have been developed by several research groups. One can, for example, utilize the coupled clusters within an equation of motion technique, and thereby find directly a coupled cluster approximation for the first polarizability, α.[102] By numerical differentiation of α, one can then attain the β and γ responses, but only those for which one frequency is not vanishing such as $\beta(-\omega;\, \omega,\, 0)$.

Several modifications of coupled cluster techniques have been used to analyze correlation effects. For example, Aiga and collaborators[103] have used the Brueckner variant of coupled cluster theory to obtain full-frequency dependent hyperpolarizabilities. In a complex methodological paper, they show relationships between the response formulation of the hyperpolarizabilities and their quasi-energy method. We also note that there remain formal difficulties even with sophisticated coupled

Table 7 Dipole Moment (in units of Debye) and Components of the First Hyperpolarizabilities (in units of esu) of H_2O Molecule and Results from Various Processes. Adapted from Ref: 101.

Property	SCF/TDHF	MBPT(2)	CCSD	CCSD(T)	Experiment
μ_z	0.7804	0.7297	0.7287		
	0.7793	0.7272	0.7273	0.7196	0.721[2]
β_{zzz}	−4.6	−8.7	−7.3		
	−7.1	−13.7	−12.0	−13.7	
β_{zxx}	−0.6	−2.4	−2.0		
	−1.4	−5.9	−5.2	−6.2	
β_{zyy}	−9.7	−10.6	−10.8		
	−9.4	−9.4	−9.9	−10.2	
$\beta_{\parallel}$	−8.3	−13.0	−12.1		
	−10.8	−13.5	−16.2	−18.0	
SHG	−9.860	−15.5	−14.5		
	−12.568	−20.4	−19.0	−21.1	−22 ± 6[3]
OR or	−8.727	−13.8	−12.8		
EOPE	−11.290	−18.4	−17.1	−19.0	−22 ± 6[3]

[1]The z axis is the principal axis of the molecule. For basis set details, see Ref: 101.
[2]From Werner, H.-J., and Meyer, W., *Mol. Phys.*, *31*, 855 (1976).
[3]From Ward, J. F., and Miller, C. K., *Phys. Rev. A.*, *19*, 826 (1979).

cluster approaches. For example, the polarizability response in their coupled cluster technique has the wrong pole structure when compared with the exact results. In an attempt to circumvent this problem, they suggest that one should replace the operator of Eq 12 with a unitary cluster operator. This might be useful, but we are not aware of actual applications of this unitary method to nonlinear optical response.

In favorable cases, with extended basis sets, correlation treatments within the coupled cluster approximation, whether formulated in terms of response properties or equations of motion, can get within roughly 10% of the correct value. Most of these correlation effects can also be captured by simpler, second order perturbation-theory computations. While the results of MECI calculations can be off by factors of two for small molecules, they tend to get better for larger species. The lack of correlation in MECI results is a drawback, and shows up most clearly and importantly in calculations of small components of the hyperpolarizability, or of molecules that have small hyperpolarizabilities.[104] For the most part, however, it appears that for stable small molecules containing first-row elements, correlation treatments at the MP/2 level are probably adequate for semi-quantitative analysis of the hyperpolarizability response. Indeed, because the semiempirical model Hamiltonians are parameterized to fit optical spectra at the TDHF or MECI level, such calculations are accurate for the appropriate molecular response (the dominant β component in a chromophore with large response). For such cases, semiempirical MECI results are usually more accurate than ab-initio treatments, unless the latter includes a large basis set and some appropriate treatment of correlation.

Finally, we should point out that in almost all cases, the effect of correlation is to increase the polarizabilities or hyperpolarizabilities (for example, compare Table 6). This makes sense in terms of the physical nature of the electron correlation: at the Hartree-Fock level, all interelectronic repulsions are replaced by averaged interelectronic repulsion. The effect of specific, dynamic electron correlation is to reduce the average repulsion energies; this in turn makes excitations easier, reducing the energy denominators in the sum over states expression of Eq. 5, and therefore increasing the magnitude of the response.

D. Dispersion

The fundamental expansion in electric field of either the dipole moment or the energy, as detailed elsewhere, shows that one can find the static polarizabilities and hyperpolarizabilities simply by taking derivatives, with respect to the electric field, of the dipole moment or of the energy (if the Hellmann-Feynman theorem holds). Indeed, this is the basis for finite field (CPHF being the simplest) calculations of the hyperpolarizabilities which are appropriate for zero-frequency. With the use of gradient techniques and other analytical methods for finding derivatives directly,[105–109] the finite field is not necessary. Therefore, static (all frequencies vanishing) hyperpolarizabilities can be calculated using various levels of correlation, as discussed in subsection C. Indeed, standard programs directly calculate the static first hyperpolarizability.

When the dispersion or frequency dependence of the hyperpolarizability is required, the situation gets more difficult. The sum over states formula, or spectral representation, shows that in principle one can directly calculate the dispersion simply by performing the calculation as a function of frequency. In almost all cases, however, the sum over states *is not* computed directly. Even in the simplest TDHF or MECI cases, it is generally more efficient to solve a set of linear equations, rather than perform the sum.[35,36,47,48,89,90,97] The results (SOS and linear equations) are equivalent. When correlation effects are included, the situation becomes substantially more complex, because consistency between the levels of treatment of the excited state and the groundstates is required. In the response theory formulation,[87,89–90] one deals directly with the correlation functions or propagators, and explicit excited states do not enter directly.[86–90]

With correlated methods, then, there is currently no straightforward, general, consistent way to include dispersion. It would be preferable to have a set of sophisticated correlated dynamical equations. Often a series of approximate dynamical schemes that do not necessarily treat the full frequency dependence are used under differing levels of approximations to estimate dispersion.

One such dynamical scheme, the multi-configurational linear response technique, has been used by several workers such as the dispersion calculations by Rice and coworkers using MP/2.[36,97] They give reasonable results both for the dispersion and magnitude of β. Quasi-energy methods have been applied by several workers to the calculation of dispersion. In this technique, one calculates the time-dependent quasi-energy, W, as[110]

$$\left(H - \hbar i \frac{\partial}{\partial t}\right) |\phi> = |\phi> W(t). \tag{13}$$

Here, $|\phi>$ is the reference, groundstate wave function from which excitations occur. Once the quasi-energy $W(t)$is calculated, the full-frequency dependent hyperpolarizability can be obtained from the quasi-energy according to:

$$\beta_{ijk}(-\omega_3;\omega_2,\omega_1) = \frac{-\partial^3 W}{\partial E_i(-\omega_3)\partial E_j(\omega_1)\partial E_k(\omega_2)}\Bigg|_o . \tag{14}$$

Note that in this equation no excited states enter directly; knowledge of the quasi-energy itself is adequate to compute the full dynamic response, including dispersion. This technique has been developed by several workers, and some applications are beginning to appear.

Formally, there are some difficulties associated with this method. The function ϕ must be calculated a priori. When ϕ is the Hartree-Fock state, one recovers the TDHF approximation. Alternatively, one can use different coupled cluster approximations for the state ϕ; this yields values for the dipole moment (computed as the first derivative of the quasi-energy) comparable to those obtained using the coupled cluster response formalism. The excitation energies are essentially equivalent to those obtained from the coupled cluster. This quasi-energy formulation is general, but has not been extensively tested in molecular studies.

The frequency-dependent response formalism[89,90] has been used with CAS calculations to deal with correlation, and the method has been applied to a number of organic chromophores. Not surprisingly (given the difficulties associated with using CAS itself for optical properties), results are often unsatisfactory, as seen in Table 5.[89b]

As discussed in subsection C above, it is generally possible to obtain single frequency dependencies, such as $\beta(-\omega;\omega,0)$ or $\gamma(-\omega;\omega\ 0,0)$, by differentiation of the frequency dependent polarizability. This method has been used, with the polarizability calculated from the equations of motion coupled cluster method, by Sekino and Bartlett. In its application to Kerr effect hyperpolarizabilities, they find estimated values for experimentally unknown Kerr response.[111]

Perhaps the most common way to include both correlation and dispersion is to do the zero-frequency calculation using a higher order level of correlation (such as coupled cluster techniques at singles and doubles or singles, doubles and triples; or quadratic CI methods), and then to include the dispersion response calculated at the TDHF level.[112] That is, one assumes

$$\chi_{\text{corr}}(\omega) = \frac{\chi_{\text{TDHF}}(\omega)\cdot\chi_{\text{corr}}(0)}{\chi_{\text{TDHF}}(0)} . \tag{15}$$

Here, χ is any frequency dependent polarizability or hyperpolarizability, where χ_{TDHF} and χ_{corr} are the time dependent Hartree-Fock and correlated, (hyper)polarizabilities respectively. Calculating the frequency dependence at the TDHF level is straightforward (and involves solving linear equations).

The assumptions in Eq. 15 have been tested in a number of cases,[113] and have generally been found to be good. In the particular case of the Kerr effect, Sekino and Bartlett found that the percent dispersion at the frequency $\omega = 0.0656$au,[i] (i.e.,

---

[i] "au" are atomic units with $\hbar = 1$ so that energy and frequency have the same units.

percent difference in β between the stated frequency and zero-frequency) corresponding to the ruby laser, yields an increase of 16.3% in the dominant component β_{xxx} $(-\omega;\ \omega,\ 0)$, at the equation of motion coupled cluster (EOM-CC) level of correlation, and 16.8% using TDHF. In a number of calculations on both butadiene and ammonia, comparable results are found for various components of γ and β. Table 8 shows that for the DC second harmonic generation of ammonia, the effects of correlation and dispersion are to some extent independent (the TDHF hyperpolarizability is about half the CCSD(T) hyperpolarizability but the TDHF dispersion results and coupled cluster results agree within 2%).[114] Furthermore, the final correlated, dynamical result agrees very well with experiment.

A general form for the frequency dependence of the hyperpolarizability at low frequencies was first derived by Bishop:[115]

$$\beta(-\omega_3;\ \omega_1,\ \omega_2) \cong \beta(0;\ 0,\ 0) \cdot (1 + A\omega_L^2), \tag{16}$$

where A is a constant and the characteristic frequency ω_L is

$$\omega_L^2 = \omega_3^2 + \omega_2^2 + \omega_1^2. \tag{17}$$

This is useful for relating different components of the frequency-dependent response to one another, and has been used by several workers to estimate the full frequency dependence.

For systems studied with ab-initio calculations, dispersion seems well approximated by TDHF. This suggests that problems of correlation and dispersion can be treated separately, and that Eq. 15 gives a good estimate of the (hyper)polarizability.

E. Truncation

The SOS formula by direct summation is not commonly used in accurate calculations. This is essentially because the SOS expression is inefficient to implement, and its replacement by techniques such as equations of motion or solution of linear equations substantially speeds up the process, and can therefore produce better results with smaller computational demands. On the other hand, as is discussed

Table 8 DC-SHG (in units of 10^{-30} esu) of NH_3 by Different Dispersion Estimates at an Excitation Energy of 0.0656 au. Adapted from Ref: 111.

Method	β^1	γ^1 (atomic units)
TDHF	−0.190	1.65
CCSD(T) + % TDHF dispersion[2]	−0.424	2.82
CCSD(T) + % EOM-CCSD dispersion[3]	−0.436	2.97
Experiment[4]	−0.422 ± 0.001	3.09 ± 0.055

[1]These TDHF and CCSD(T) results use basis 3 in Ref: 113.
[2]Static CCSD(T) value scaled by % TDHF dispersion.
[3]Value evaluated by EOM-CCSD Kerr effect number assuming the dispersion ratio among the different NLO process mentioned in the text.
[4]From Shelton, D. P., and Rice, J. E., *Chem. Rev.*, *29*, 3 (1993). Ward, J. F., and Miller, C. K., *Phys. Rev. A.*, *19*, 826 (1979).

extensively in Sections VI and VII below, the SOS formulation provides important interpretative advantages that the calculation of simple numerical quantities does not necessarily provide.

We have already discussed the effects of truncation in the MECI calculations using semiempirical models. Comparable calculations, again focusing on the truncation, are reported by Bartlett and his collaborators using ab-initio TDHF schemes, and the results are comparable (Fig. 4).[114]

Table 9 shows the effect of truncation on β of water. Truncation of a limited number of states does not necessarily have a large effect on the TDHF result but drastic truncation (removing more than half the excitations) can result in substantial changes. Further, the response is not monotonic as a function of number of sequential states excluded from the calculation.[116]

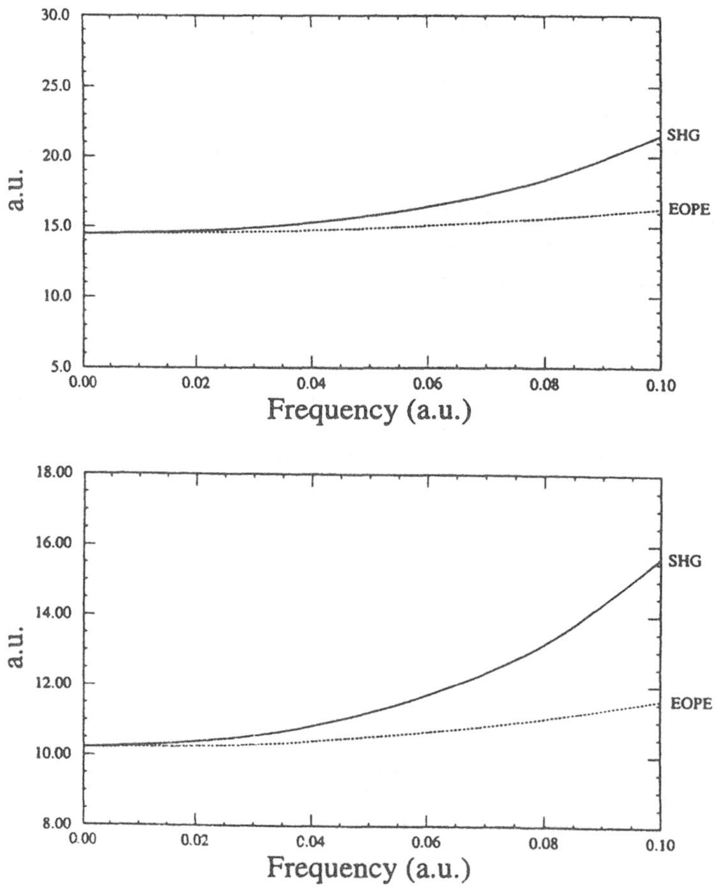

Figure 4 The frequency-dependence of the second harmonic generation coefficient for the water molecule, calculated using an ab-initio Hamiltonian at the MECI level. The curves of Fig. 4a shows a truncated excitation space, while the full excitation space appears in Fig. 4b. As in semi-empirical theories, extension of the excitation space generally lowers the computed response, but often, including only a few dominant transitions gives good quantitative results. (From Ref: 98.)

Table 9 Effect of Truncation on the TDHF-Level Computation of $\beta_{\|}(-2\omega;\omega,\omega)$ for H_2O[1]

# of Excitations	$\beta(\omega = 0)$	$\beta(\omega = 0.06)$	$\beta(\omega = 0.1)$
255	0.088	0.101	0.136
192	0.089	0.102	0.136
129	0.077	0.091	0.124
63	0.172	0.189	0.232
10	0.125	0.142	0.186

[1]From Sekino, H., and Bartlett, R. J., (To be published). The basis is Sadlej's polarization-consistent one, augmented by one diffuse d-type set with a screening constant of 0.1 on each H. Values in (10^{-30} esu). Frequency is in atomic units (27.211 eV).

Three conclusions of interest follow from these calculations. First, the dispersion is reasonably well predicted even when a moderately truncated set of states is used. Second, the largest contributions to the hyperpolarizability (that is, the dominant components) are less sensitive to truncation than the smaller components. Note that with ab-initio, or MECI calculations using semiempirical Hamiltonians, similar conclusions hold: The larger the hyperpolarizability response, the less sensitive it is to approximations in the Hamiltonian, or in the treatment of correlations, or in the treatment of the dispersion process.

The most severe possible degree of truncation is the two-level model of Oudar and Chemla[39,40]: Only one excited state is permitted. While the two-level model is generally not accurate quantitatively, it has been enormously useful in qualitative calculations, in understanding the response of a number of molecular chromophores, and in comparing classes of chromophores with one another. This is another illustration of an instance in which highly accurate calculations are not necessarily required if one is to understand chemical trends, dispersion, or effects of molecular architecture modification on the hyperpolarizability.

VI. SEMIEMPIRICAL CALCULATIONS: SOME RESULTS AND CONCLUSIONS

The primary use of computational studies of chromophore response by chemists, physicists, engineers, and materials scientists interested in nonlinear optical materials is the prediction of the relative susceptibilities of different molecular structures. Semiempirical calculations are ideally suited for this task for several fundamental reasons: They are accurate for medium-size organic and organometallic molecules such as dimethyamino-nitrostilbene (DANS), PNA, disperse orange, or metalloporphyrins.[21–28,35,67,117,118] These types of molecules are not computationally demanding so that large numbers of molecules can be calculated quickly. There already exists a large data base of such calculations, including comparison with appropriate experiments, that demonstrate both the applicability of the models to a wide variety of chromophores and a few exceptions in which these methods fail (Tables 10 and 11). Full frequency dependence is easily obtained. Finally, and

Table 10 $\chi_{\parallel}^{(2)}$ in Units of 10^{-33} esu/molecule. Theoretical Values from Various Molecular Orbital Calculations—Semi-empirical (SE), Uncoupled Hartree-Fock (HF), and Coupled Hartree-Fock (CHF)—and a Single Other Experimental Value are Included for Comparison. The Sign of $\mu\chi_{\parallel}^{(2)}$ is Unambiguously Determined by the Experiment and is Independent of the Sense Chosen for the Molecular z Axis. This Table is Extracted from Ward, J. F., and Miller, C. K. Adapted from Ref: 98.

			Theory			
Molecule	μ^1	$\chi_{\parallel}^{(2)}$ SHG	SE	HF	CHF	Other Exp
C^-O^+	0.112 ± 0.005	129 ± 14	−43.5[2] (+95)[4]	879[3] 420[3] 387[3] −438[5]		
N^-O^+	0.15872 ± 0.00002	147 ± 17				
$H_2^+S^-$	0.974 ± 0.005	−43 ± 9	47.7[2]			
$N^-H_3^+$	1.474 ± 0.009	−209 ± 5	56.4[2]		−44.4[5] −65.1[5] −19.0[5] −15.6[6] −40.8[6]	3500[6]
$H_2^+O^-$	1.86 ± 0.02	−94 ± 4	120[2]	90.6	−52.5[5] −79.2[5] −21.9[5] −51.6[6] −48.0[6]	

[1]Electric dipole moments in Debye units from Landolt-Berstein, Zhahlenwerte und Funktionene, Neue Serie, Vols. II/4 and II/6 (Springer-Verlag, Berlin) and Ref. 9. The sense of the NO moment is suggested by Billingsley, F. P., II, *J. Chem. Phys.*, *63*, 2267 (1975); *62*, 864 (1975).

[2]Hush, N. S., and Williams, M. L., *Theor. Chim. Acta*, *25*, 346 (1972). Signs for NH_3 and H_2O are ambiguous.

[3]O'Hare, J. M., and Hurst, R. P., *J. Chem. Phys.*, *46*, 2356 (1967).

[4]$\chi_{zzz}^{(2)}$ (0;0,0) from McLean, A. D., and Yoshimine, M., *J. Chem. Phys.*, *46*, 3682 (1967).

[5]Liebmann, S. P., and Moskowitz, J. W., *J. Chem. Phys.*, *54*, 3622 (1971).

[6]Lazzeretti, P., and Zanasi, R., *Chem. Phys. Lett.*, *39*, 323 (1976) *Note added in proof.* Recent reconsideration of the sign of these entries yields the negative signs now shown here—Lazzeretti, P., (Private communication). Overall consistency is substantially improved by this change.

perhaps most importantly, the combination of simple minimal basis, valence only semiempirical model Hamiltonians with the sum over states analysis, and the interpretation in terms of contributions to the spectral representation of Eq. 5, or the two-level model of Eq. 6, permits interpretation of why particular molecules exhibit particularly large nonlinear susceptibilities.

Given these strengths, it is not surprising that there is extensive current activity in using semiempirical models at the MECI level to predict and interpret nonlinear optical response. This section reviews what has been learned using this methodology.

Table 11 Dominant Components of the SHG Coefficients (in esu units) of *p*NA Computed Using the Semi-empirical INDO Hamiltonian and From an ab-initio Calculation

Component	INDO/SCI[1]	MP2/HF[2]	MP2/MP2[3]
β_{xxx}	11.37	5.48	8.89
β_{xyy}	−1.50	−0.82	−0.34
β_{xzz}	0.00	−0.21	−0.20
β_{x}	9.84	4.45	8.35

[1]CIS calculated response from AM1 optimized geometry.
[2]NLO properties computed at the HF level using MP2 optimized geometry. Source: Ref: 97.
[3]NLO properties computed at the MP2 level using MP2 optimized geometry. Source: Ref: 97.

A. Limits of Validity

Like all electronic structure computational schemes, the semiempirical model/MECI picture is valid only for certain classes of problems. For example, these methods are expected to fail badly for the response of small molecules such as HF, H_2O, CH_3F or NH_3, for both optical and NLO responses.[98] These responses are dominated by Rydberg-type transitions to diffuse regions of space far away from the valence orbitals, and the response is weak. Table 10 shows an example of the failure of the semiempirical/MECI methodology for such systems.

The approximation is also not expected to hold for the smaller components of the nonlinear response tensor even in molecules for which large components are predicted well.[104] This is due to the semiempirical nature of the model Hamiltonians, that were originally parameterized for the purpose of describing dominant optical transitions. In addition, the smaller components often correspond to repolarizations of the valence density in a region of space for which the effective, single function valence basis set used in semiempirical models is inappropriate. Table 11 shows an example in which, indeed, large components that dominate the response are calculated accurately, but small components are in poor agreement with better calculations.

Additionally, there are other situations in which the semiempirical/MECI models may be questioned. Although computations of open shell groundstates have been given, there is less of a database of measured values to assure the validity of the methodology. Organometallic systems have also been analyzed (compare sub-section F of this Chapter), but analysis of their optical spectra is in general a complicated problem, and the semiempirical MECI theory should be used with caution for such problems (although sub-section F and Table 15 show remarkable agreement with experiment for INDO/S SOS calculations on open-shell organometallics). Finally, if one is interested in very high-frequency response (above the first few resonances), the mono-excited configuration interaction semiempirical models are not expected to be accurate.

While these caveats are important to help understand the accuracy of the calculations, the method is intended to describe nonlinear response of good chromophores, to indicate why they are good chromophores, and to permit comparisons of chromophores with one another in different solvent environments, geometries, and frequency ranges. For these, the evidence suggests that the models are indeed valid and helpful.

B. Dependence on Molecular Dimensions

The two-level model (Eq. 6) has obvious implications for design of large beta chromophores: They should have a high oscillator strength, a relatively low energy resonance, and large dipole moment difference between groundstate and excited state. For this reason, a linear donor/bridge/acceptor geometry, where the bridge is capable of strongly mixing donor and acceptor states, would appear optimal for providing large responses.[64,119–122] Accordingly, this geometry has been the most widely explored for developing chromophores. Computational literature which is aimed at understanding the molecular response is extensive.[21–28,60–67]

As an example, Fig. 5 shows some early results using the PPP model for computing the NLO response of donor/acceptor chromophores as a function of the length of the bridging species.[23] For these characteristic structures, the length dependence in this regime is exponential. Subsequent studies have shown the limitations of this picture that arise both because of geometric modifications (bond length alternation[123–128] and Peierls distortion,[126] twisting around the Z axis causing structure localization,[129] eventual limiting of the spatial extent and energetic modifications of the electronic structure with length) and electron correlation. It has also been shown, as discussed in Section VI.E, that for a given size of the π-electronic system, extended linear geometries exhibit larger responses than rings or polycylic structures; again, this is expected on the basis of the simple two-level model.[64,122]

C. Polarity and Inductive Effects

According to the two-level model (Eq. 6), increasing the polarity difference between donor and acceptor substituents in a simple donor/bridge/acceptor polar structure increases the dipole moment difference, and therefore the size of the β response.[23,130–135] Figure 6 shows results of calculations using the PPP model for a series of conjugated π-systems, and, indeed, the frequency doubling response increases with dipole moment difference. Accordingly, extensive experimentation has shown that very strong electron acceptors, particularly cyano and nitro, produce strong β responses.[135,136] Combining such strong acceptors with strong donors such as dimethylamino and amino has been useful in producing improved chromophores such as 4,4′-DANS. An even stronger response can be found in zwitterionic systems, with donors such as O^- and acceptor substituents such as quaternary ammonium groups.[10,137–139]

Beratan and Marder and their colleagues have stressed that too large a dipole moment difference can eventually result in reduction of the nonlinear response.[57,132] This is, again, reasonably clear from the two-level model: In the limit in which the groundstate wave function is localized entirely on the donor and the excited state entirely on the acceptor, there is no overlap of the wave function and therefore the transition dipole matrix element will vanish. Calculations at the Hückel level,[57] reproduced in Fig. 7, show that, indeed, sharp decreases in the

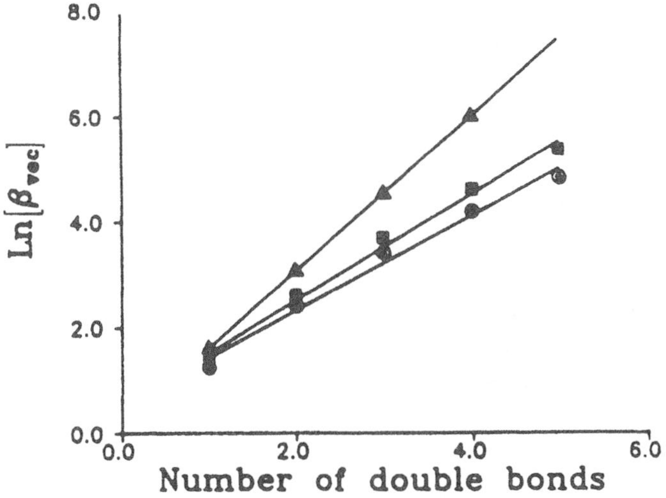

Figure 5 The natural log of the vector part of the first hyperpolarizability as a function of a number of double bonds, calculated at various frequencies for the transoid conjugated polyenes $O_2N(CH{=}CH)NMe_2$. Frequencies are, from steepest to shallowest slopes, 0.0eV, 0.656eV, and 1.17eV. The characteristic growth of the vector response with chain length is quite clear. (From Ref: 23b.)

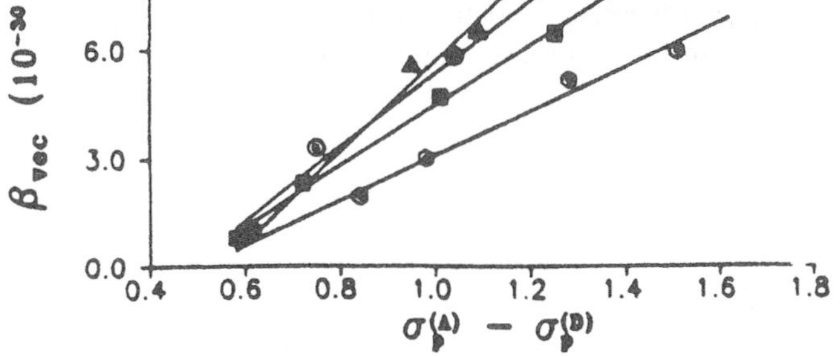

Figure 6 The increase of the vector part of the hyperpolarizability as a function of the difference in Hammett constants, σ, for para-disubstituted benzenes at different frequencies, as the donors change with the given acceptor species. From steepest to shallowest slopes, the acceptors are NO_2, CHO, COOMe and CN, respectively. Calculations are at the PPP-MECI level of theory. (From Ref: 23b.)

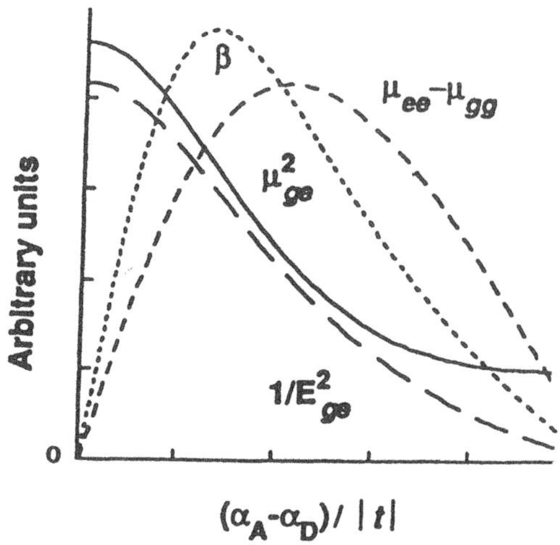

Figure 7 Variation of $\Delta\mu_{ge}$, μ_{ge}^2, $1/E_{ge}^2$ and β as a function of the difference in donor and acceptor strength $(\alpha_A - \alpha_D)/|t|$. (From Refs. 56 and 57.)

transition moment lead to overall reduction in the β response. These results have been more difficult to obtain using Hamiltonians that include electron repulsions; repulsion effects prevent strong charge localization, and therefore even very highly polarized systems do not seem to show the expected strong decrease in transition moment.

An interesting variation on the inductive effect issue involves the actual location of the donor and acceptor sites.[130–140] In the donor/acceptor substituted stilbene structure (Fig. 8) molecular orbital analysis (Fig. 9) shows that the acceptor group is largely the nitro functionality and attached avene ring, whereas the donor functionality includes components from the amino group and the entire carbon π system. Figure 10 shows charge density plots of the HOMO and LUMO for a donor-acceptor stilbene. Again, we see that the charge is transferred largely onto the nitro, but that it comes from the entire molecule (see the HOMO density plot). This suggests that further addition of inductive donor groups (amino) anywhere on the π system should increase nonlinear response, and that substitution by an acceptor (nitro) anywhere in the π system might decrease the response. Figure 11 shows

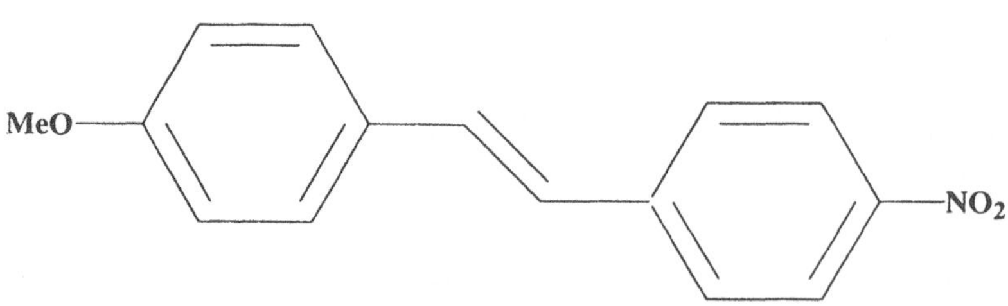

Figure 8 A characteristic donor/acceptor stilbene, whose molecular orbital analysis appears in Fig. 9.

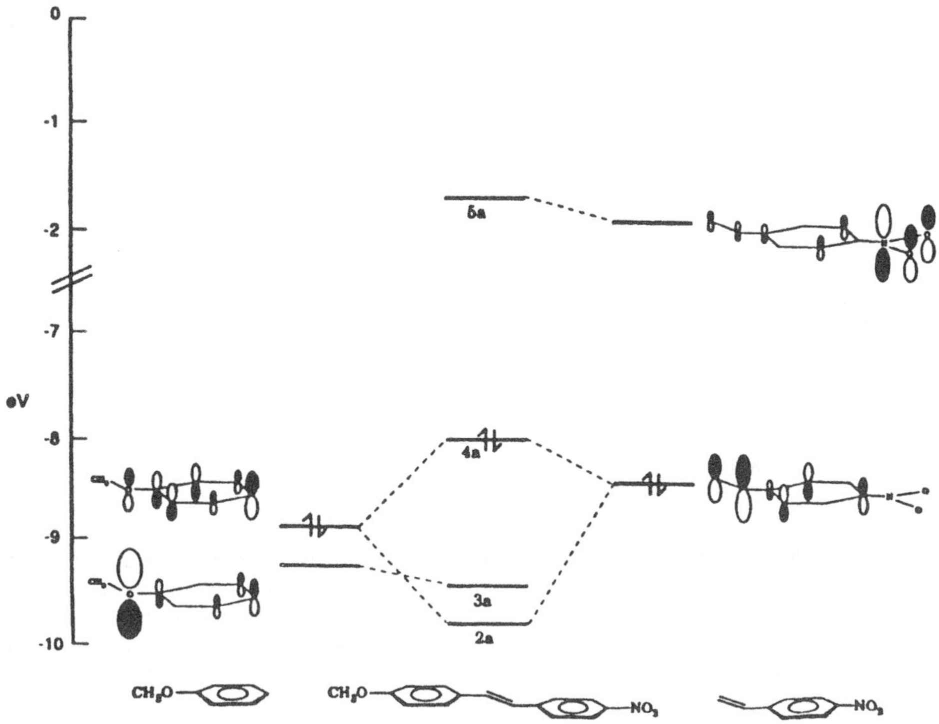

Figure 9 The molecular orbital analysis for the donor/acceptor species of Fig. 8, calculated at the MECI, INDO/S level of theory. Notice that the HOMO is derived from a mixture of methoxy and $p\pi$ systems, while the acceptor is dominantly on the nitro group. Such behavior is characteristic of very strong acceptors with benzenoid frameworks. (From Ref: 28.)

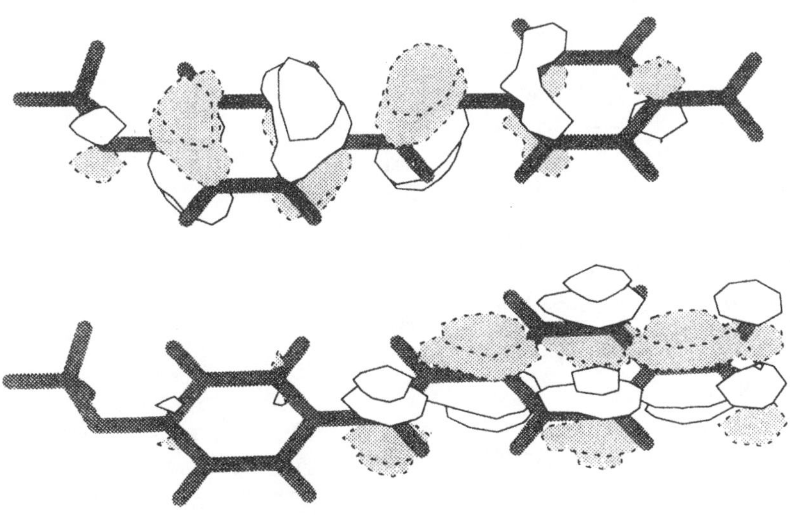

Figure 10 The molecular orbital density plots from an ab-initio calculation using a 3-21G basis set for the donor/acceptor species in Fig. 8.

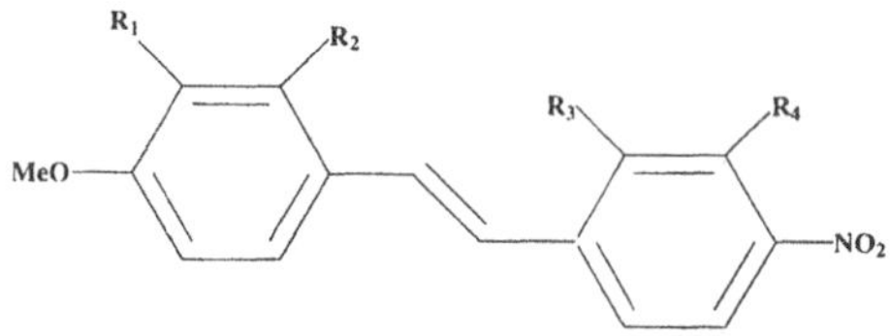

Chromophore	R_1	R_2	R_3	R_4	β_{vec}
a	H	H	H	H	24.4
b	NO_2	H	H	H	16.4
c	H	NO_2	H	H	13.5
d	H	H	H	NO_2	15.6
e	NH_2	H	H	H	27.4
f	H	NH_2	H	H	29.0
g	H	H	H	NH_2	22.4

Figure 11 Inductive affects in the 4,4′-methoxy nitro stilbene structure, calculated at the MECI-INDO/S level of theory. Notice that further substitution with the withdrawing nitro group anywhere on the molecule decreases the β response, whereas the electron releasing amino group generally increases the response. This is consistent with the nitro being the primary acceptor substituent.

Table 12 Experimental $\mu\beta$ Values[1] at a Fundamental Wavelength of 1.907 μm for Benzenoid and Thiophene Compounds with Dialkylamino Donor and Nitro Acceptor Substituents

	Chromophores	β 10^{-30} esu
Ia	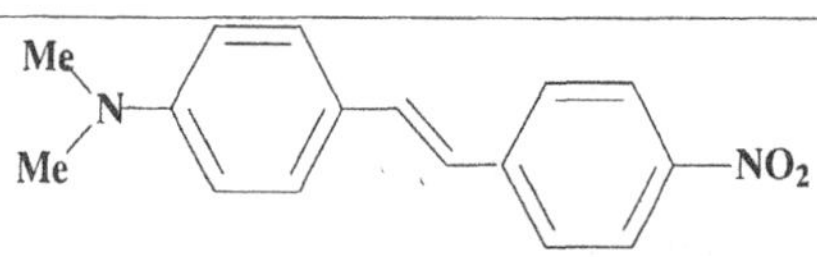	580
Ib	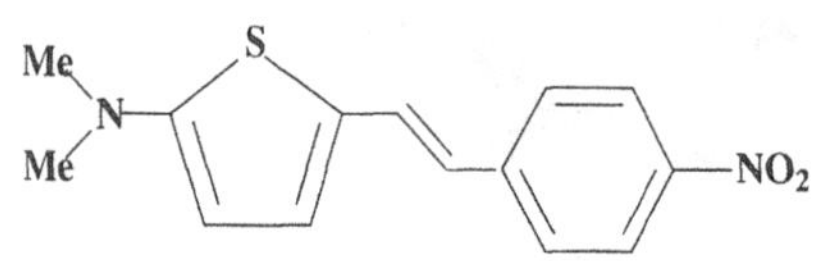	600
Ic	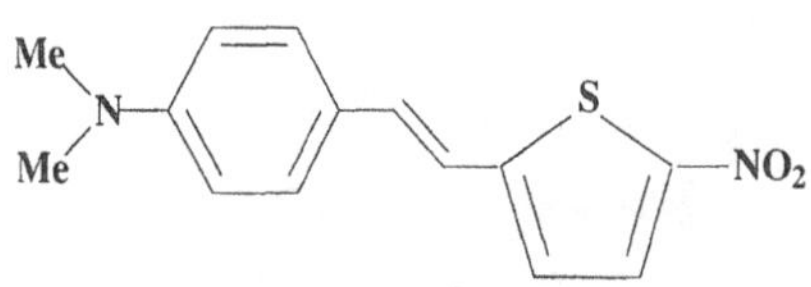	660
Id	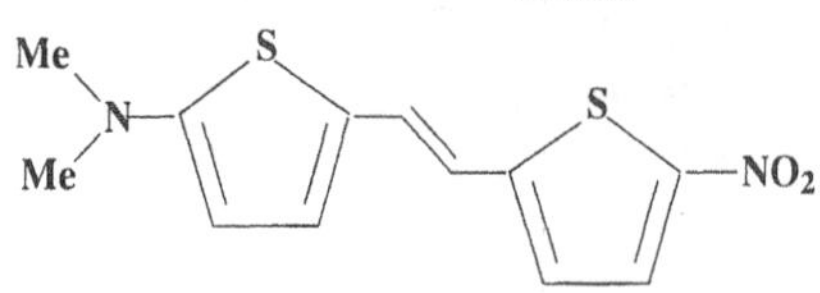	1040

[1]Source: Refs. 142 and 143.

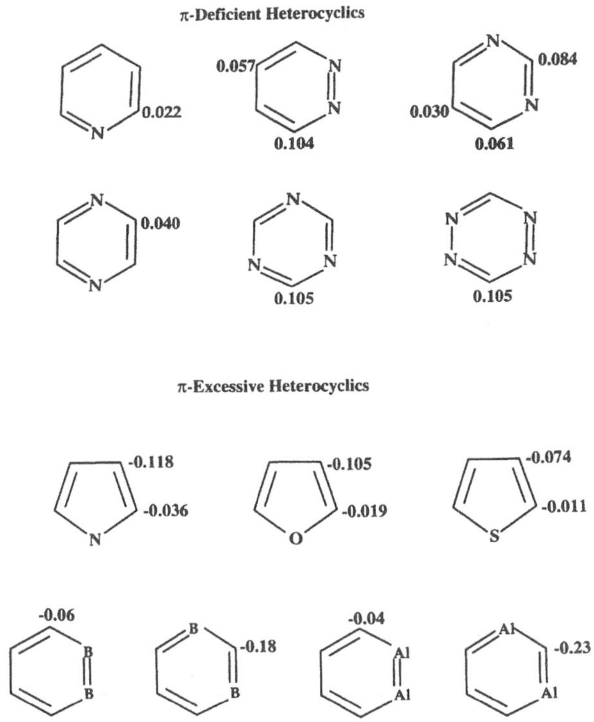

Figure 12 Charge densities on carbon atoms in π deficient and π excessive heterocycles. These results are derived from Mulliken population analysis using simple Hückel-type calculations. (From Ref: 143.)

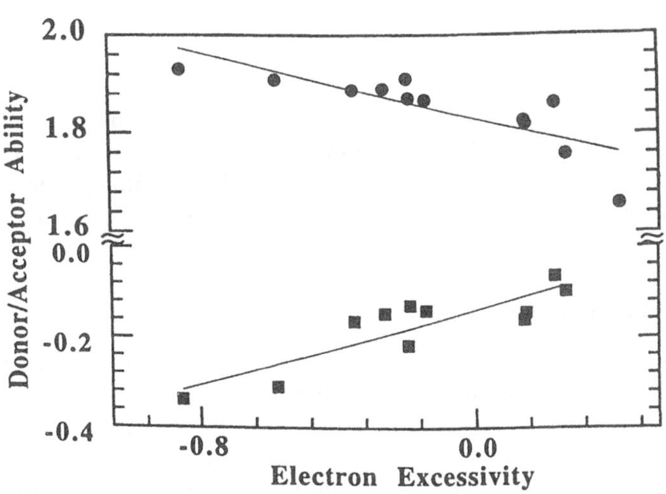

Figure 13 Correlation of electron excessively with donor or acceptor ability of a series of heterocyclic substituents. These results, from MECI-INDO/S calculations, show that the electron excessive/electron deficient heterocycles act, respectively, as auxiliary donor or auxiliary acceptor groups. (From Ref: 143.)

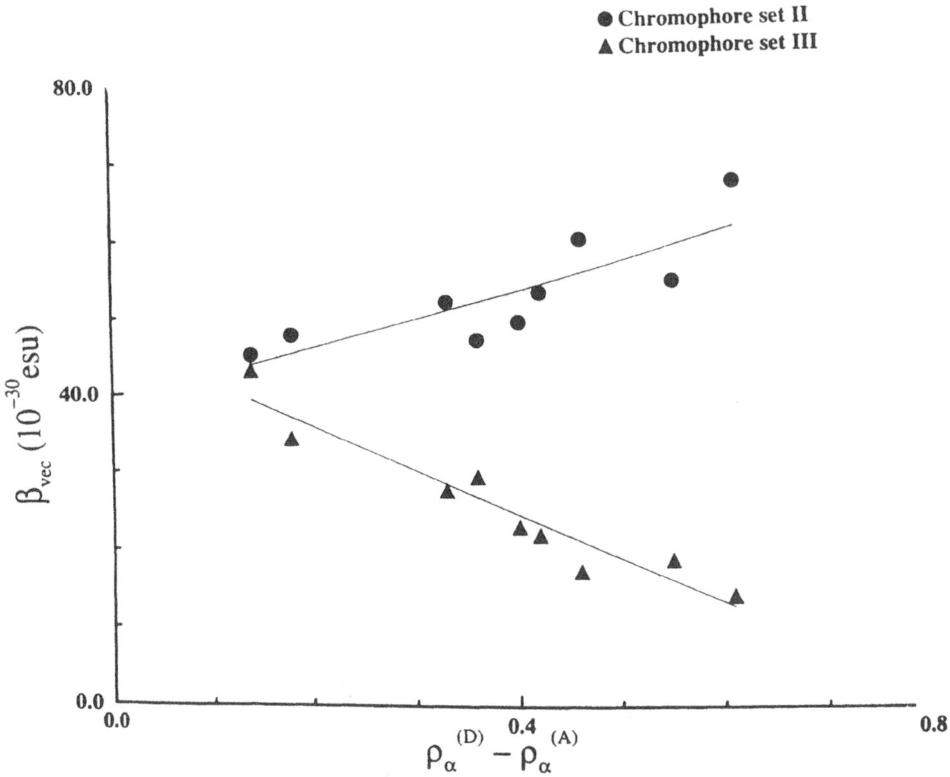

Figure 14 Computed hyperpolarizability, at the MECI, INDO/S level of theory, for a series of chromophores based on the five- and six-membered rings of Fig. 12. The points represent a series of chromophores based on the structures of Fig. 12, plotted against the difference in auxiliary donor and acceptor strengths. (From Ref: 143.)

results of a semiempirical calculation (MECI, ZINDO model Hamiltonian) which illustrates that these substitutions affect β_{vec} as expected. The observation that the acceptor is the nitro group, rather than the π system to which the nitro group is appended, is somewhat surprising, but it suggests new design motifs, as outlined in subsection D below.

D. Auxiliary Donor and Acceptor Effects

As discussed in subsection C, for a typical D/A substituted stilbene, computational study shows that the nitro group acts as an acceptor with the entire remainder of the molecule acting as an extended donor. Adding further donor strength, e.g., by more NH_2 substituents, augments the donor strength.[141] This idea can be extended. Experimental and theoretical studies on donor-acceptor substituted stilbene-like bridges have shown that replacing a homocyclic benzene ring by a heterocyclic ring such as thiophene yields substantial increase in the NLO response (Table 12).[142,145] A plausible explanation of the origin of this enhanced response on the basis of the electron-excessivity and deficiency of heterocyclic rings has been re-

Chromophore Set I

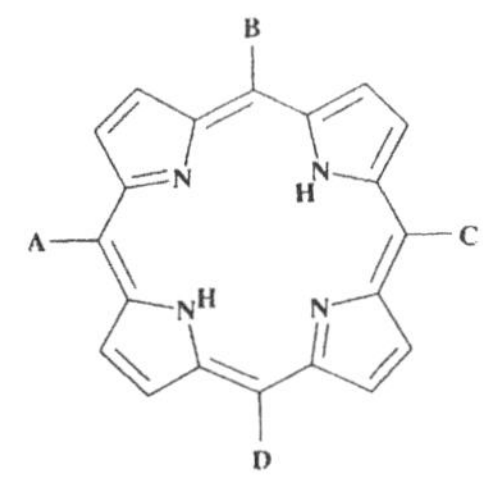

I

Chromophore	A	B	C	D
a	Ph-NH_2	H	Ph-NO_2	H
b	Ph-NH_2	Ph-NH_2	Ph-NH_2	Ph-NO_2
c	Ph-NH_2	Ph-NO_2	Ph-NO_2	Ph-NO_2
d	Ph-NH_2	Ph-NH_2	Ph-NO_2	Ph-NO_2
e	Ph-NH_2	Ph-NH_2	Dz-NO_2	Dz-NO_2
f	Ph-NH_2	Ph-NH_2	Tt-NO_2	Tt-NO_2
g	Fu-NH_2	Fu-NH_2	Ph-NO_2	Ph-NO_2
h	Fu-NH_2	Fu-NH_2	Dz-NO_2	Dz-NO_2
i	Fu-NH_2	Fu-NH_2	Tt-NO_2	Tt-NO_2
j	Th-NH_2	Th-NH_2	Ph-NO_2	Ph-NO_2
k	Th-NH_2	Th-NH_2	Dz-NO_2	Dz-NO_2
l	Th-NH_2	Th-NH_2	Tt-NO_2	Tt-NO_2
m	Py-NH_2	Py-NH_2	Ph-NO_2	Ph-NO_2
n	Py-NH_2	Py-NH_2	Dz-NO_2	Dz-NO_2
o	Py-NH_2	Py-NH_2	Tt-NO_2	Tt-NO_2

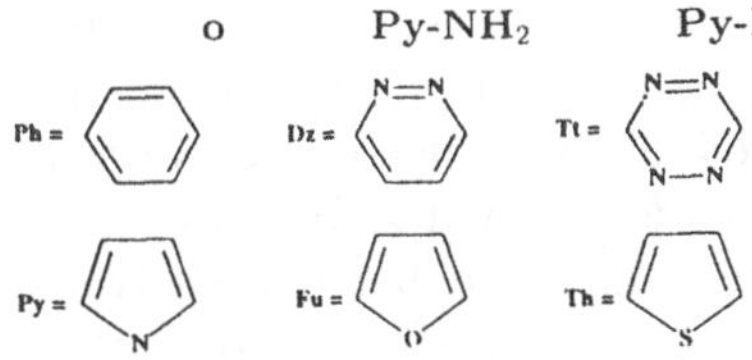

cently provided.[145] It is known that five-membered aromatic heterocyclic rings, depending on electronegativity of the heteroatom with respect to carbon, are electron-excessive and six-membered aromatic rings are electron-deficient.[142–145] The concepts of electron-excessivity and deficiency of heterocyclics arise from a simple Hückel description of the charge densities on the various carbon atoms of the ring (Fig. 12). These concepts have been used by organic chemists to explain the unusual reactivities of aromatic heterocyclics. Recently, these concepts have been used to explain some unusually large NLO responses of donor-acceptor heterocycles.[143,145] These studies show that the electron excessive and deficient

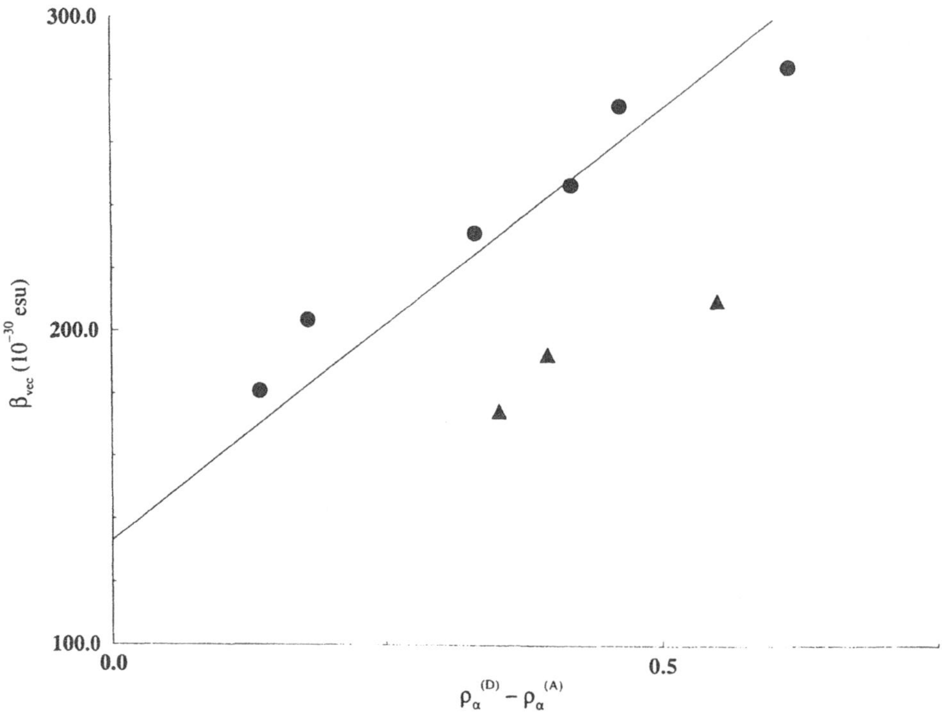

Figure 15 A series of substituted porphyrin species chromophore set I, studied at the MECI-INDO/S level, to analyze the variation of β with different auxiliary donors and acceptor substituents. The three triangles are for pyridazine, that shows a markedly smaller response (see text). (From Ref: 146.)

heterocyclics, through inductive effects, modify the effective strength of donors and acceptors resulting in enhanced donor and acceptor efficiencies (Fig. 13). Thus they are called auxiliary donor and acceptor groups. The enhanced donor and acceptor efficiencies of the groups result in enhanced NLO properties. The effects of electron excessivity and deficiency of five- and six-membered aromatic heterocycles on the first hyperpolarizabilities of a range of chromophores are shown in Fig. 14.

This idea of auxiliary donor and acceptor effects has also been recently extended to design novel porphyrin-based chromophores which afford substantially larger NLO response properties compared to previously examined push-pull porphyrins.[146] In porphyrins, auxiliary donor and acceptor effects can be induced in two ways. First, auxiliary donors and acceptors can be appended to donor or acceptor groups attached to the porphyrin macrocycle. This yields chromophores whose NLO response properties are substantially larger than those previously studied. The variation of β_{vec} for a series of chromophores (chromophore set **I**) with varying auxiliary donor and acceptor strengths is shown in Fig. 15. The linear variation of the SHG coefficient as a function of the auxiliary donor and acceptor strengths is clear. While the variation is uniform in most of the chromophores, chromophores in which the auxiliary acceptor is a pyridazine moiety show a consistently smaller response. Such a behavior has been observed even in simple

DANS-type chromophores. This has been attributed to the possibility of charge transfer reduction by the electronegative nitrogen atoms at the ortho-position.

Second, the electron excessive and deficient sites of the porphyrin macrocycle itself could act as auxiliary donors and acceptors. The INDO/S-computed charge densities of the porphyrin macrocycle show that the β-positions of the porphyrin are electron excessive and the *meso*-positions are electron deficient. Thus, appending the electron-excessive β-sites with donor-groups and electron deficient *meso*-positions with electron acceptors should yield enhanced NLO responses compared to chromophores in which both donor and acceptor groups are attached to the electron excessive *meso*-positions. Computed linear and NLO response properties of a set of chromophores (chromophore set **II**) that support this argument are presented in Table 13.

E. Comparative Optical Nonlinearities in Substituted Polyenes, Polyphenyls, and Polyynes

Most organic π-conjugated chromophores studied to date contain a polyene, a polyphenyl or a polyyne fragment in the bridge. Studies have compared the contrasting NLO response properties of chromophores which contain the above three moieties in varying lengths.[27,64,67,122,140,148] These theoretical and experimental studies have shown that there is a large difference in the NLO response between these bridges and that the NLO responses of chromophores having a polyene bridge are substantially larger than the polyphenyls and the polyynes. The computed λ_{max} and β_{vec} of the three chromophores as a function of length are shown in Fig. 16 and 17.[147] Here, the properties were computed using the CNDO/VSB model Hamiltonian and a CI with all singles and doubles from an active space of 10 orbitals. The dimensionality of the CI space was 325 in all the cases. The substantially larger linear and NLO response of polyenes compared to polyphenyls or polyynes can be seen from the two figures. Such an observation has also been made by the earlier CNDO/VSB-MECI and AM1 calculations.

It is interesting to compare the β_{vec} values of the three series of chromophores as a function of the number of units. In these chromophores, one unit contains four carbon atoms (for polyenes and polyynes) and a phenyl ring (for polyphenyls). While β_{vec} for all three series are comparable at $n = 1$, the β_{vec} value of the polyenes increases much more rapidly with n than the other two series of chromophores. This could be largely attributed to the red shift in λ_{max} values as n increases. The larger excitation energies of the polyphenyls could be attributed to the quinoidal nature of the charge transfer state in polyphenyls: on charge transfer, the geometry of the bridge distorts from a low-energy benzenoidal geometry to a high-energy quinoidal geometry. This increases the excitation energy of the polyphenyls and hence reduces the NLO response. This behavior is not observed in polyene and polyynes. In polyenes, the geometry distorts from a polyenic geometry to a zwitterionic geometry (to be discussed in more detail in Section VII), and one might expect distortion from an acetylenic geometry to a cumulene geometry (this is probably *not* energetically favorable, since the required distortion (bond length difference) is roughtly 0.25Å, compared to 0.12Å in polyenes). It is not clear if the smaller response of polyynes can be attributed to the stabilities of the geometries in the groundstate and in the charge transfer state, as in the case of poly-

Chromophore Set II

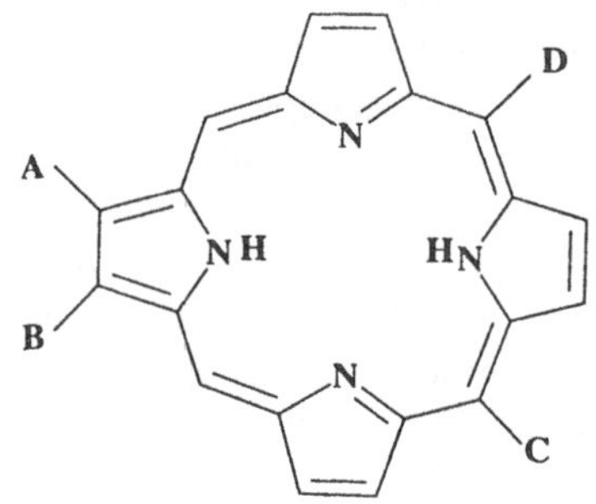

II

Chromophore	A	B	C	D
a	NH_2	NH_2	Ph-NO_2	Ph-NO_2
a	NO_2	NO_2	Ph-NH_2	Ph-NH_2
c	C=C-Ph-NH_2	C=C-Ph-NH_2	C=C-Ph-NO_2	C=C-Ph-NO_2
d	CC-Ph-NH_2	CC-Ph-NH_2	CC-Ph-NO_2	CC-Ph-NO_2
e	NN-Ph-NH_2	NN-Ph-NH_2	NN-Ph-NO_2	NN-Ph-NO_2
f	NN-Py-NH_2	NN-Py-NH_2	NN-Tt-NO_2	NN-Tt-NO_2
g	C=C-Ph-NO_2	C=C-Ph-NO_2	C=C-Ph-NH_2	C=C-Ph-NH_2
h	CC-Ph-NO_2	CC-Ph-NO_2	CC-Ph-NH_2	CC-Ph-NH_2
i	NN-Ph-NO_2	NN-Ph-NO_2	NN-Ph-NH_2	NN-Ph-NH_2
j	NN-Py-NO_2	NN-Py-NO_2	NN-Tt-NH_2	NN-Tt-NH_2

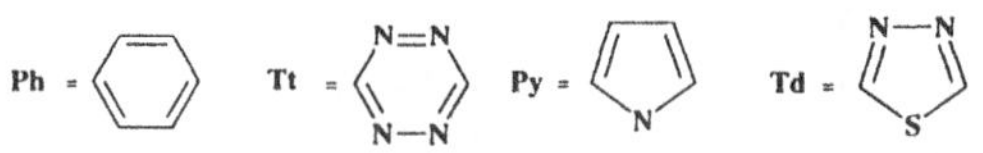

CC = C≡C NN = N=N

phenyls. An alternative possibility has been proposed for the smaller NLO response properties of polyynes: It has been speculated that[64a] an additional charge transfer in a perpendicular direction exists and that it opposes the usual charge transfer. That is, there are two charge transfer processes in push-pull polyynes in the two perpendicular π-frameworks and they tend to oppose one another, decreasing the response.

F. Organometallics: Pseudo-Centrosymmetry

The ZINDO model Hamiltonian has been parameterized to include organometallic and transition metal complex systems.[68,149] These are substantially more challenging than organics, since the valence manifold includes *s*, *p*, and *d* electrons. Nevertheless, there has been some activity in the study of organometallics for nonlinear response[150–171]: For the most part, organometallic chromophores have been slightly

Table 13 Linear and NLO Response Properties of SYBYL Force Field Optimized and Planar Geometries of Chromophore Set II at an Excitation Energy of 0.1 eV. Adapted from Ref: 146

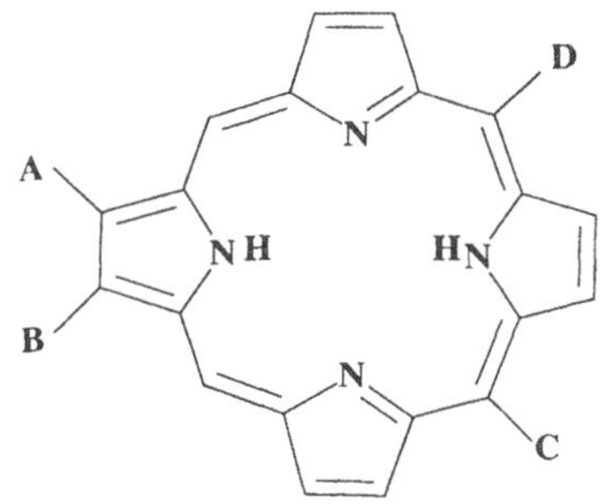

IV

Chromophore	λ_{max}(nm)		$\Delta\mu$ (Debye)		β_x (10^{-30} esu)		$\mu\beta$ (10^{-48} esu)	
	Optimized	Planar	Optimized	Planar	Optimized	Planar	Optimized	Planar
IIa	390.4	418.4	15.79	7.71	31.49	68.06	631.19	1511.70
IIb	414.2	701.3	9.02	26.58	202.42	253.51	3818.81	5372.12
IIc	377.7	439.2	18.38	7.76	57.32	125.12	967.45	2181.46
IId	403.5	405.5	13.04	12.94	157.22	156.61	2492.35	2395.24
IIe	650.1	650.6	6.71	6.71	259.05	259.71	4625.42	4633.60
IIf	489.7	492.6	12.77	13.11	345.75	493.50	3616.78	6443.74
IIg	380.5	439.6	21.50	7.80	68.50	139.69	1019.40	2215.47
IIh	402.1	403.6	9.09	8.99	99.96	100.29	2415.84	2525.47
IIi	523.2	528.9	8.49	5.79	130.32	162.35	3837.72	5215.87
IIj	537.6	537.6	10.02	10.02	305.22	305.21	8897.73	8897.70

disappointing, exhibiting smaller response than organics with a comparable number of electrons.

Some understanding of why organometallics are less effective, and some possible routes for making them better, can be gleaned from ZINDO/MECI calculations.[168–171] For example, Fig. 18 shows that the population change upon optical transition in a simple substituted ferrocene structure is very similar to that with methoxyphenyl substituted for the ferrocenyl donor. Here, the transition is essentially of donor/acceptor type, and the organometallic is only a moderate donor moiety.

Calculations on metal carbonyls generally show a relatively small nonlinear response. One possible explanation invokes effective electronic symmetry around the transition metal center: Although a structure such as cyclopentadienyl iron tricarbonyl formally has no center of symmetry, both the CO and cyclopentadienyl ligands act as π acceptors, so that, effectively, the electronic environment around the metal is almost centrosymmetric. Since centrosymmetric species cannot exhibit a $\chi^{(2)}$ response, these pseudo-centrosymmetric structures are expected to exhibit only small $\chi^{(2)}$.

Change in the effective local symmetry of the environment around the metal can theoretically lead to a large predicted NLO response. The hypothetical triam-

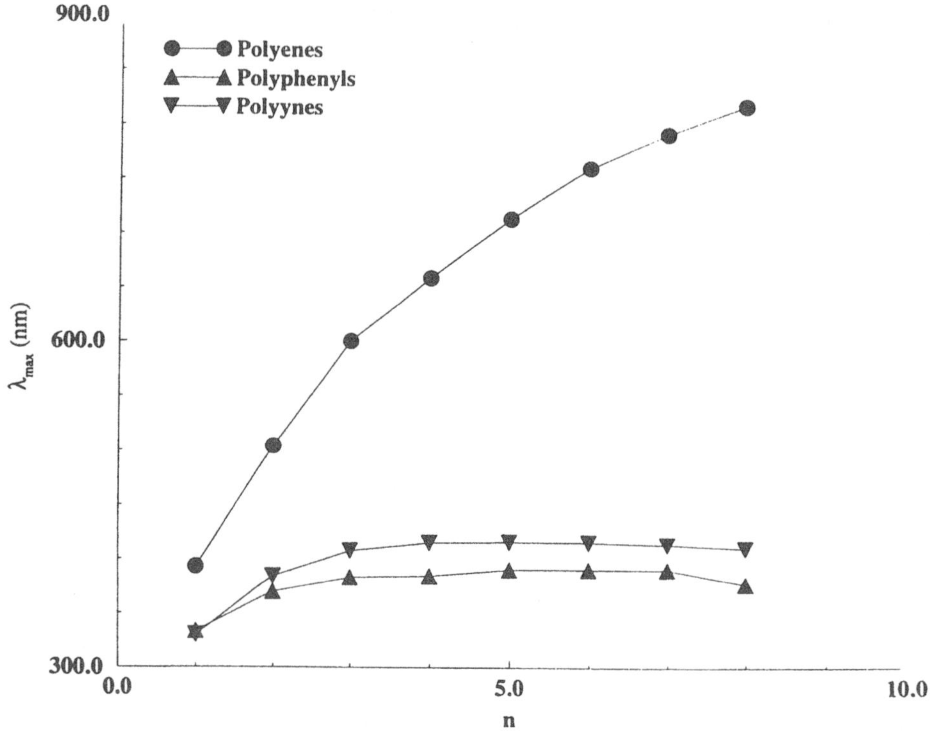

Figure 16 The computed wavelength of maximum absorption as a function of length of donor/acceptor polyenes, polyphenyls, and polyynes. The computations were done with the CNDO/VSB model, with all singles and doubles (from an active space of ten π-orbitals). The polyene maxima shift much more spectacularly with increasing size than do the polyphenyls or polyynes, both of which exhibit asymptotic behavior after only a few repeat units.

mine chromium structure shown in Table 14 has a highly acentric environment; ZINDO/MECI calculations suggest that this chromophore would exhibit a very large nonlinear response, as shown in Table 14.

Some of the organometallics that have been found to show reasonable frequency doubling responses are π-electron systems in which the metal acts to enforce planarity and to improve π-type mixing between donor and acceptor regions. More generally, as shown in Fig. 19, the theoretical frequency doubling responses calculated by ZINDO agree reasonably well with experiment.

Recent work on open-shell doublet organometallics shows that the ZINDO/S, MECI level of theory works well, but also suggests that open-shell organometallics, because of the availability of low-frequency electronic excitations, can exhibit a large β response, far larger than analogous closed shell organometallics. Examples, comparing open shell (Cu,Co) and closed shell (Ni) salophen complexes are given in Table 15.[171]

As we have shown in this section, for a wide variety of chemical design motifs ranging from conjugated systems to donor/bridge/acceptor substituted systems, including organometallics, frequency dependence, and isomerization, electronic

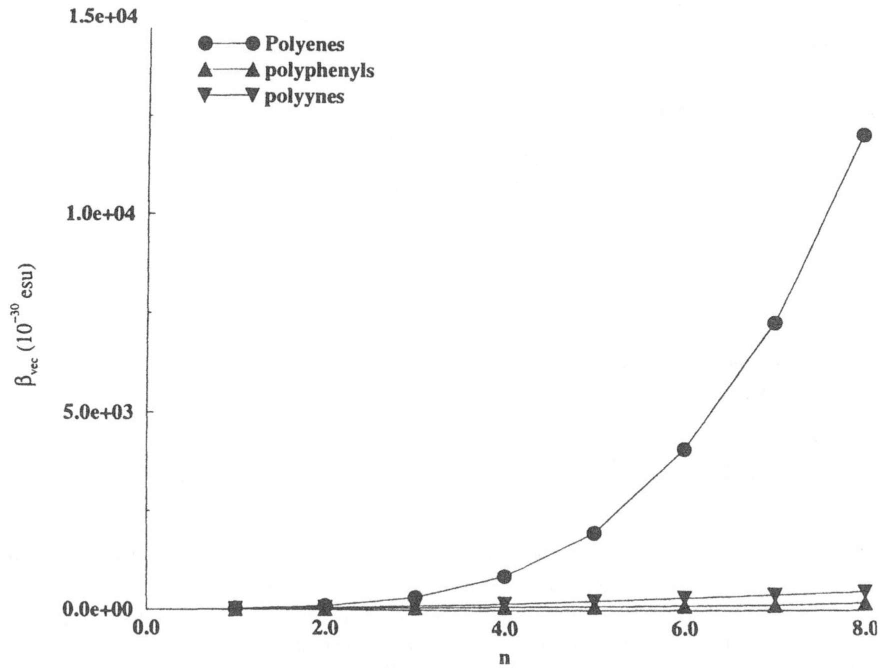

Figure 17 Calculated $\beta_{vec}(\omega = 0.1\text{eV})$ for the same chromophores as in Fig. 16. The polyenes, again, show substantial increase of nonlinear response with length, corresponding to more effective conjugation.

structure theory at the semiempirical ZINDO/MECI or (for pure-π electron systems), PPP/MECI is both accurate and helpful in suggesting modified chemical motifs to optimize nonlinear response.

VII. SOME SPECIAL TOPICS

The simplest nonlinear response calculation is for an isolated molecule in a fixed geometry in isolated vapor at zero temperature. Measurements, on the other hand, are generally carried out at finite temperatures in a material environment; the chromophore interacts with a host medium (a liquid, a glass, or a crystal), and possibly with other chromophores. Molecules can also be subject to geometric distortions, either from solvent interaction or from other chromophores. There has been extensive theoretical work examining these issues,[65,172–176] since they are critical in understanding and optimizing the hyperpolarizability of real materials. In this section, we sketch some of the approaches that have been taken and some of the progress that has been made in the study of the environmental effects on nonlinear response, and of the variation of nonlinear response with molecular geometry.

A. Environmental Effects

When the isolated vapor phase molecule at zero temperature is introduced into a real material environment, several effects on the nonlinear response occur. The

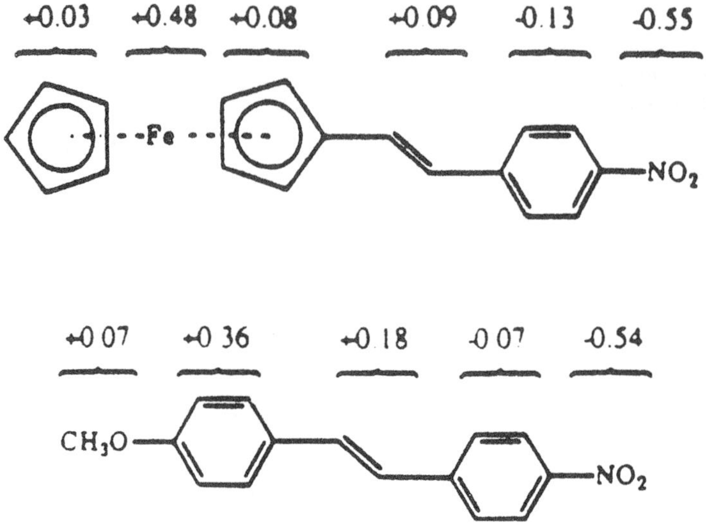

Figure 18 The difference in electronic population between groundstate and the crucial excited state for a prototypical ferrocenyl derivative and the traditional organic chromophore, as computed using MECI-INDO/S. The negative population indicates an increased electron density upon charge transfer. Notice that the ferrocene donor acts in very much the same way as the methoxy phenly substituent. (From Ref: 169.)

Table 14 ZINDO-Derived Molecular Hyperpolarizabilities of Two Hypothetical Organometallic Complexes at an Excitation Energy of 0.65 eV. The hyperpolarizability of 4-nitro-stilbene is reported for comparison. Adapted from Ref: 28

Compound	β_{vec} (10^{-30} esu)
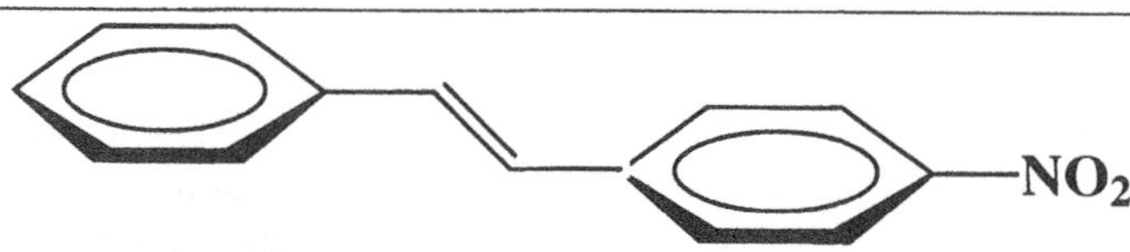	27.5
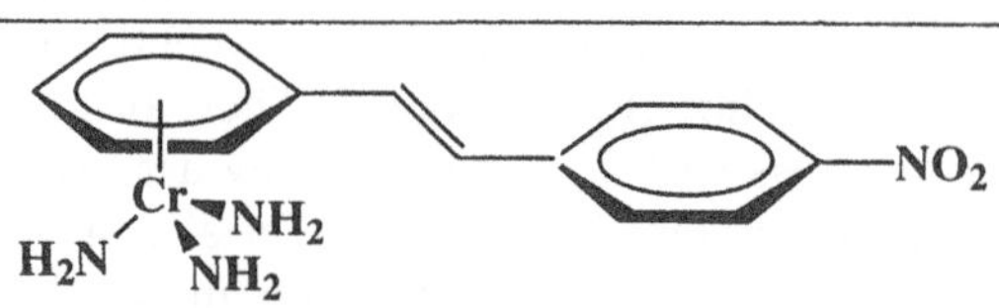	780
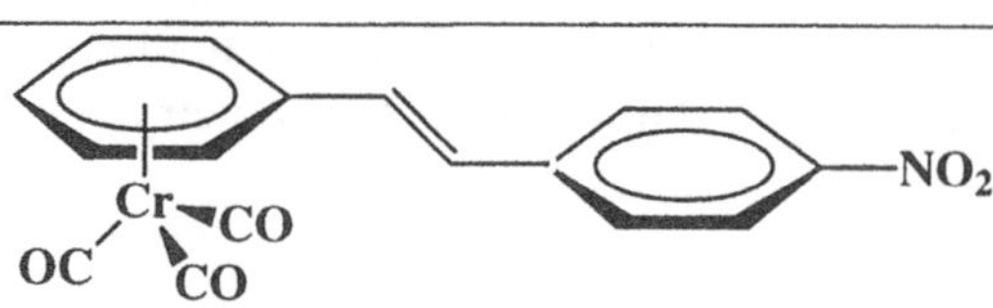	8.3

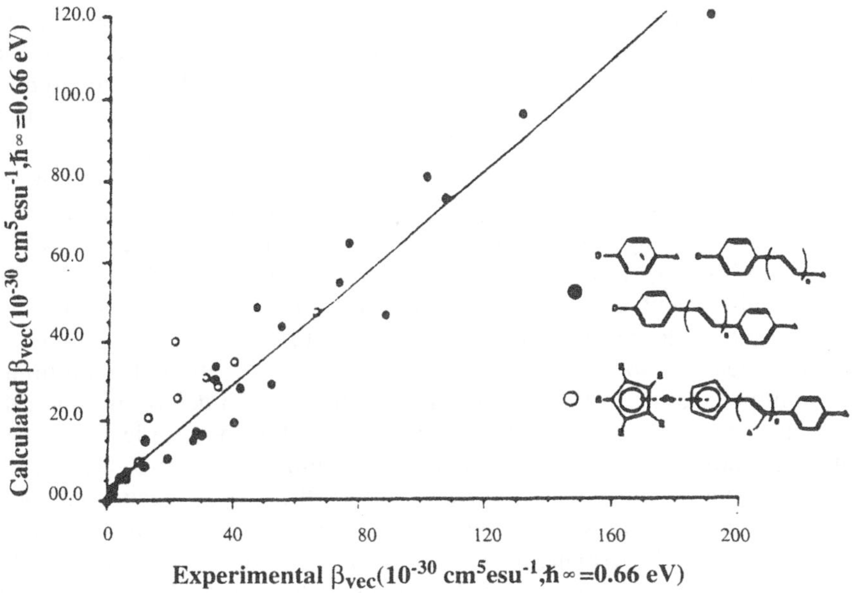

Figure 19 Plot of ZINDO-derived β_{vec} values for 25 π-electron organic and organometallic chromophores versus the experimental values. The slope is somewhat less than unity. (From Ref: 168.)

Table 15 INDO/SCI-SOS and Experimental EFISH-derived[a,b] Second-Order NLO Response Data (in units of 10^{-30} esu) for Coordination Complexes Having Open-Shell Doublet Ground State Electronic Structures. Data for the Homologous Closed-Shell Ni(II) Complexes are Reported for Comparison. Adapted from Ref: 171a

Compound	$\hbar\omega$ (eV)	β_{vec}^{calc}	β_{vec}^{expt}
Cu(salophen)	0.0	−37.2	
	0.92	−53.7	−50.0 ± 5[1]
Co(salophen)	0.0	−40.1	
	0.92	−105.1	−170 ± 40[1]
Ni(salophen)	0.0	−11.5	
	0.92	−17.3	−20.5 ± 2[1]
Co(4-OMe-sal-3-NO_2-ophen)	0.0	−90.4	
	0.92	−250	
Ni(4-OMe-sal-3-NO_2-ophen)	0.0	−25.6	
	0.92	−55.2	−55 ± 5[2]

[1]From Ref: 171b.
[2]From Ref: 171c.

simplest—increase of temperature causing internal excitation of the molecule resulting in differing vibrational envelopes for the nonlinear response—has not been extensively considered, and is probably of secondary importance. Far more significant, and more extensively analyzed in the literature, are effects due to other chromophores[176,177] and solvation effects[172,173] due to the environment.

In molecular crystals, such as NPP (N-(4-nitrophenyl)-L-prolinol), NSP (S-(-)-N[*P*-(4-nitrostyryl)phenyl)prolinol), POM(3-methyl-4-nitropysidine-1-oxide), MBANP (2(α-methyl benzylamino)-5-nitropyridine), MNA (2-methyl-4-nitroanibine), or NMBA (*p*-nitro-*p*-methylbenzylidine aniline), the local electromagnetic field environment at each site will differ from the environment of the bare vapor phase molecule, but will be the same from unit cell to unit cell. Under these conditions, one can rather accurately represent the state of affairs by summing the classical multipole terms in the electrostatic potential.[176] To do so, one should, in principle, self-consistently recalculate the polarization felt by a molecule in the presence of other molecules, at well-defined sites within the unit cell. Such calculations have not actually been carried out; instead, there have been published reports of the nonlinear response of a molecule in the electrostatic environment caused by the unperturbed, permanent multipoles of other identical molecules. In such a calculation, the aim is to include the so-called local field factors that are introduced in the relationship between the susceptibility and the molecular hyperpolarizability.[3,13] As discussed in Chapter 3, the second order susceptibility, $\chi^{(2)}$, is related to β according to:

$$\chi^{(2)}(-(\omega_1 + \omega_2); \omega_1, \omega_2) = NL(\omega_1)L(\omega_2)L(\omega_1 + \omega_2) \quad (18)$$
$$<\beta(-\omega_1 - \omega_2; \omega_1, \omega_2)>,$$

where N is the number of chromophores, and $L(\omega_1)$ is the local field factor, describing how the environment at a molecular site, responding at (ω_1), differs from that of the vapor phase. For isolated molecules, all the field factors are unity. In the simplest electrostatic picture of the molecular crystal, the field factors are implicitly included in a recalculation of β, where the local electromagnetic field felt by the molecule differs from the bare field by the multipole terms, as in

$$\vec{E}(\vec{r}) = \vec{F}(\vec{r}) + \sum_{n=1}^{\infty} \int \mathbf{d}\vec{r}' M_n(\vec{r}, \vec{r}')\, P_n(\vec{r}'), \quad (19)$$

where E, F, M_n and P_n are, respectively, the local field, the applied external field, the nth multipole interaction, and the nth multipole polarization at a point, $\vec{r}'$. Ewald techniques are used to perform the integral in Eq. (19). Munn has worked actively in the area, and has shown substantial effects in molecular crystals.[176] In most cases, the local field is larger than the applied field. For this to be true, the multipoles need to be aligned so that their fields are reinforced—this will be true in single domains of acentric point groups. Such alignment occurs in crystals such as NPP and NSP that exhibit large nonlinear response (as well as other characteristics such as birefringence, and ferroelectricity or piezoelectricity). Since the dominant materials of interest in organic nonlinear optical substances are polymeric or glassy systems, the crystalline analysis is generally less useful than it is in, say, inorganic species. Instead, studies have focused either on isolated molecular clusters or on environmental analysis using continuum treatments.

There have been only a few publications on intermolecular effects in molecular second order response.[175,176] This is somewhat surprising since molecules with a large molecular response should interact strongly with each other. Work in our laboratories has examined both dimeric and trimeric structures in which donor/acceptor molecules, such as *para*-nitroaniline interact with identical molecules or charge transfer partners.[175] Figure 20 shows the calculated nonlinear response as a function of measured response for a molecular pair of donor/acceptor species as a function at bridge length. We see that the nonlinear response, due largely to intramolecular excitation, increases substantially for particular geometries of the molecular pair. This is to be expected: These molecules have large dipole moments, and therefore the local field at one molecule due to the presence of the other can be substantially enhanced. This suggests that, for particularly favorable geometries, isolated molecular clusters will exhibit a substantially larger (or, in unfavorable cases, substantially smaller) response. A simple example of the latter is the situation in which the molecular cluster becomes effectively centrosymmetric, and the overall nonlinear response vanishes.

The second example that is important in understanding effects of electric field poling on molecular structures occurs when injection of an electron forms a doublet dimeric species.[177] Clear evidence of such structures has been seen in the EPR response of electric field poled thin films of the dimethyaminonitrostilbene molecule. Computationally, the frozen dimer structure corresponding to the doublet structure (that is, to an injected electron solvated by two molecules) has been

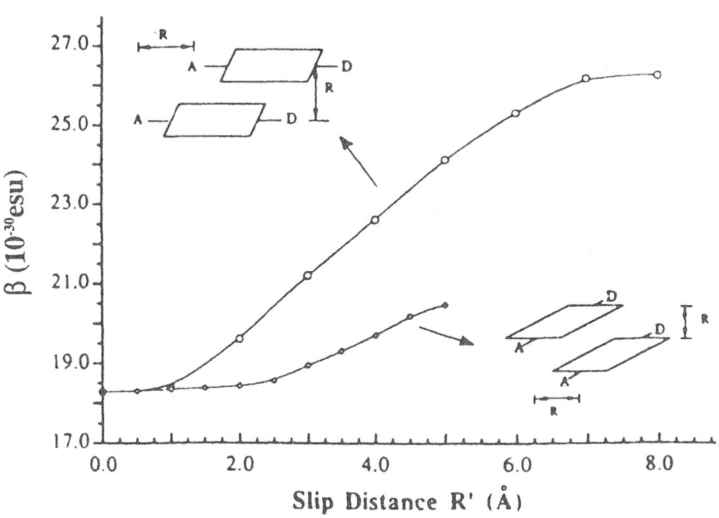

Figure 20 Calculations, at the MECI-INDO/S level of theory, on the interaction between donor/bridge/acceptor species held rigid with respect to one another. The substantial geometric dependence arises largely from the influence of the charge field, especially the dipole field, of these molecules upon one another. (From Ref: 178.)

studied.[178] The resulting nonlinear response is very strong, although effectively intramolecular. This is quite striking, since the structure is formally symmetric. The electronic structure studies (performed using small-basis ab-initio methods at the open shell Hartree-Fock level of theory; more sophisticated calculations are certainly desirable) show that the strong nonlinear response arises due to effective electron localization on one site, even in the absence of nuclear distortion. Essentially, this occurs because the polarization energy gained by localizing the excess electron is larger than the delocalization energy lost when the excess electron is no longer fully delocalized over the two sites. This results in asymmetric electron density, and a strong nonlinear response. Strikingly, this nonlinear response is almost entirely intramolecular, and largely perpendicular to the substantial dipole moment that occurs because of electron localization on the single site. This is an example in which an important experimental finding can be understood using a low-level electronic structure theory.

The great majority of electronic structure studies on environmental effects have focused on understanding the isolated molecule in a dielectric environment.[179–185] The usual method for performing such calculations is based on the Kirkwood-Onsager continuum picture.[184,185] In this approximation, the molecule is placed in a cavity in a dielectric continuum, characterized only by its dielectric function. Indeed, in almost all studies the frequency dependence of the dielectric function is ignored, and a simple static dielectric constant is used. A sketch of this picture is given in Fig. 21. In such a calculation, the only required empirical values are the shape and size of the cavity and the dielectric constant of the environment.

More sophisticated calculations employ a cavity shaped to the actual size of the molecule.[186] These calculations are complicated, because analytic forms for the self-consistent polarization are difficult to obtain. Numerically, however, several formulations have been given, and results are being reported. Far more common is the use of either elliptical or spherical cavities. In the spherical cavity picture, the interaction of the molecule with its environment is again represented by a multipole description (Eq. 19). In this case, however, one envisions the isolated molecule within the cavity as having an electrostatic charge distribution that induces electrostatic multipoles in the solvent environment. Those multipoles, in turn, react back on the molecule, causing a modified polarization. At the simplest Hartree-Fock level, this results in a modification of the effective Fock operator (that is, the one electron operator that defines the potential felt by each electron in the chromophore). The modified Fock operator in the presence of environmental perturbation is given by[179–183]

$$f = f^0 + \sum_{lm} <T_{lm}>^2 g_l(a). \tag{20}$$

Here, f, f^0, and T_{lm} are respectively the Fock operator in the cavity, the free Fock operator of the molecule, and the averaged self-consistent polarization moment (labeled by the usual spherical harmonic indexes l and m). The Onsager function g_l is given by

$$g_l(a) = \frac{1}{2} a^{-(2l+1)} \frac{(l+1)(\varepsilon - 1)}{l + \varepsilon(l+1)}, \tag{21}$$

where a is the cavity radius, and ε is the dielectric constant of the environment. In

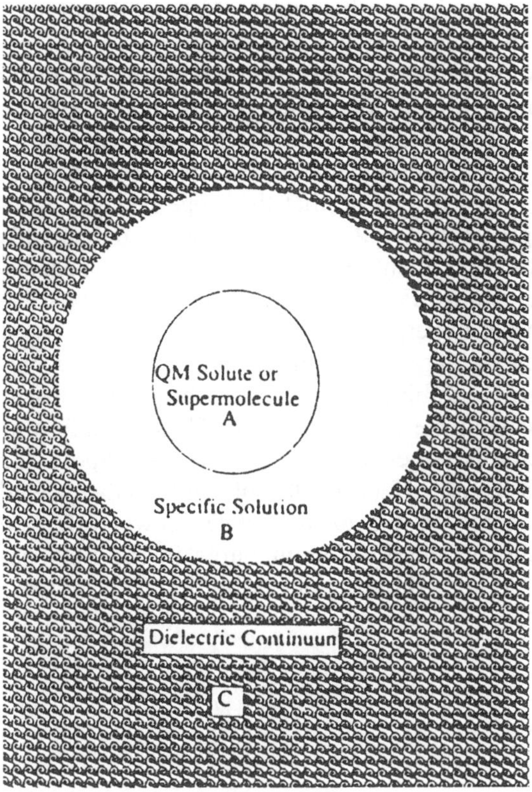

Figure 21 A schematic breakdown of solvation effects and solution. When the specific solution environment, B, vanishes, one obtains the standard Kirkwood-Onsager continuum structure.

the original Kirkwood-Onsager analysis, this self-consistent multipole picture was used to calculate solvation energies. In studies of the hyperpolarizability at the simplest MECI or CHF level, these self-consistently determined effects modify the Fock operators, and therefore the response of the molecule to the environment.

If one limits the sum over the spherical harmonics to the lowest (dipolar) term, and if one considers only an uncharged molecule, then Eq. (20) simplifies to

$$f_{dip} = f^0 + \sum_{m=-1}^{l} <T_{lm}>^2 \frac{1}{a^3} \cdot \frac{\varepsilon - 1}{2\varepsilon + 1}. \qquad (22)$$

If one further assumes that the dipole moment is dominated by a single Cartesian component, the m sum is replaced by the dominant m component. In this formulation, one simply needs to calculate the self-consistent dipole moment of the molecule in the environment, and multiply it by a factor depending on the inverse cube of the cavity radius, a.

An even greater simplification can be made for the approximate estimation of solvent effects: One can assume that only the dipolar term is important, that the cavity is spherical, and that self-consistency corrections to the isolated dipole moment are unimportant.[180,187] Further, in calculating the SOS expression of Eq. 5, one can assume that only the energy denominators are significantly modified by the environment.[187] This over-simplified picture describes the dominant effects, both of solvatochromism and of solvent modifications of the nonlinear response. Figure 22 shows the calculated versus these crude theoretical approximations in which the solvent is chloromethane (ε = 4.81). Note that the slope changes from roughly 0.65 to almost 0.9 (exact agreement would give a slope of unity); this substantially improved agreement with experiment arises, essentially, only because of energetic stabilization of the excited states. For systems in which β has positive sign, the dominant two-level terms in the hyperpolarizability expression show that the excited state dipole moment should be larger than the groundstate dipole moment, and such molecules will ordinarily exhibit positive solvatochromism (that is, bathochromic shifts in media of increasing dielectric constant). Conceptually, this means that since the excited state dipole is larger than the ground state dipole, the excited state is more stabilized by polarization effects of the solvent than the ground state, and this lowers the effective energy of the excited state. Lower effective energies of the excited state mean smaller energy denominators in the SOS expression of Eq. 5, and therefore larger computed hyperpolarizabilities. This increase

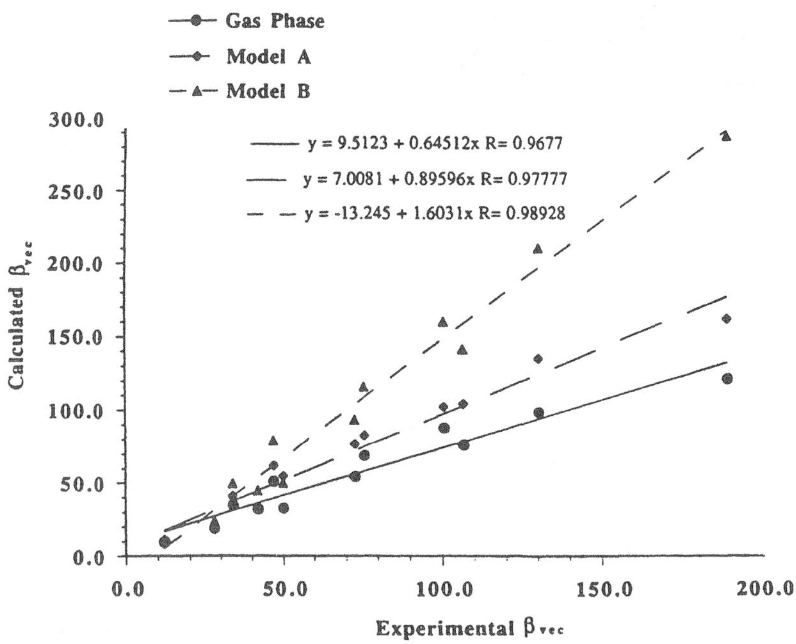

Figure 22 The effects of solvation environment on the first hyperpolarizability. The Model B calculation modifies the gas phase results using the full self-consistent environment as described by the factor of Eq. 22. Model A assumes that the only modification to the standard SOS formula of Eq. 5 is a change in the energy denominators due to a dielectric shift, to take into account the polarization stabilization of the excited state. This is adequate to substantially change the graphs, making the picture quantitatively reasonable. (From Ref: 118.)

in hyperpolarizability from solvent polarization is absent in the isolated molecule calculation, and its inclusion is the dominant effect in the modification of hyperpolarizability by solvent. Some numerical data are shown in Table 16; once again, we see substantially increased agreement between experiment and theory on making the simplest approximation for the solvent effect—that is, only modifying the energetics of the excited states.[118,187]

Far more sophisticated calculations are available; Scheme 1 is a conceptual road map of how one could improve the treatment of solvent, starting with the continuum picture of a dielectric environment acting on a single molecule; many such calculations, both semiempirical and ab-initio, have been reported in the literature.

Even the most sophisticated continuum treatment, however, will neglect certain important effects: For example, the specific interactions of two molecules with one another, ranging from self-consistent polarization at particular geometric orientations to specific interactions such as hydrogen-bonding or π stacking will be absent in any simple continuum picture. What is probably needed, therefore, is a sort of semi-continuum picture, such as that sketched in Fig. 23. Here, a small molecular cluster is treated directly, and the continuum treatment applied only to the environment of that molecular cluster. Such treatment is clearly necessary in situations where isolated clusters exist, for example in hydrogen bonding or strong local donor/acceptor pairs. While such calculations have been reported for electron transfer,[188] and have been discussed in the context of nonlinear applications, we know of no published work on this approach to β calculation.

Qualitatively, the effects of the environent on nonlinear molecular response can be understood using cavity models; the most difficult problem is determining the actual shape of the cavity, and its size. Indeed, one can see from the simple classical expression of Eq. 20 that the results will be sensitive to the choice of cavity size; by arbitrarily fitting the cavity size, one could come very close to exact agreement in theory and experiment. Usually, the cavity size is chosen in terms of the molecular volume, and several recipes are available.[179,181–184]

Calculations of environmental effects on molecular nonlinear optics are important not only for applications, but, more generally, for understanding how molecular properties are influenced by the solvent. Indeed, solvatochromic measurements involve precisely the same interactions, and the understanding of these optical and nonlinear optical responses, and their measurement is providing deeper insight into precisely how molecules interact with their solvent environment.

B. Variation of Molecular Response with Molecular Geometry

Whenever electronic structure calculations are performed using the Born-Oppenheimer approximation (that is to say, for essentially all electronic structure calculations), a nuclear geometry must be chosen before the electronic structure can be found. Choice of different nuclear geometries can affect the computed value of electronic properties such as dipole moment, total electronic energy, or electronic response properties including hyperpolarizabilities.[189,190] As an example, *cis*-dichloroethene will be non-centrosymmetric and exhibit a finite β, but the *trans* isomer will be centrosymmetric, with vanishing β. More recently, systematic study of the dependence of the nonlinear response in particular molecules as a function

Table 16 Comparison of the Experimental and the Calculated Linear Optical Properties and 1st Hyperpolarizabilities of the Two Series of Organic Chromophores from Models A and B in Chloroform Solution. The Numbering of the Molecules is Same in Chart 1 of Ref: 118. Adapted from Ref: 118

	Gas Phase		Model A			Model B			Experimental[1]	
Molecule	$\hbar\omega^2$	$\beta_{vec}{}^3$	$\mu_g{}^4$	$\hbar\omega^2$	$\beta_{vec}{}^3$	$\mu_g{}^4$	$\hbar\omega^2$	$\beta_{vec}{}^3$	$\hbar\omega^2$	$\beta_{vec}{}^3$
1	3.23	54.82	9.85	2.88	77.1	11.61	2.94	93.68	2.88	73.0
2	3.08	76.28	10.23	2.77	104.4	12.44	2.79	141.40	2.80	107.0
3	2.97	98.62	10.53	2.67	135.2	13.41	2.65	210.40	2.71	131.0
4	2.89	120.98	10.76	2.62	162.0	14.27	2.56	287.68	2.67	190.0
5	3.41	35.11	7.34	3.17	41.3	8.39	3.23	49.99	3.30	34.0
6	3.15	51.34	7.58	2.98	62.1	8.92	3.04	79.29	3.12	47.0
7	3.08	69.27	7.79	2.85	82.9	9.44	2.89	116.04	2.99	76.0
8	2.96	87.84	7.95	2.77	102.1	9.94	2.77	160.03	2.88	101.0
9	4.12	9.96	4.08	4.03	8.9	4.57	4.04	11.51	3.90	12.0
10	3.71	19.22	4.70	3.61	18.8	5.56	3.61	24.38	3.54	28.0
11	3.42	32.38	5.15	3.33	32.6	6.41	3.31	44.92	3.30	42.0
12	3.37	33.13	9.66	2.83	55.6	11.02	3.11	50.14	2.83	50.0

[1]The experimental data is taken from Ref: 135.
[2]The CT transition energy is in eV.
[3]The vector component of the 1st hyperpolarizability (in 10^{-30} cm^5 esu^{-1}) at an excitation energy of 0.65 eV.
[4]Groundstate dipole moment in Debyes. The groundstate dipole moments in the gas phase are the same as those in Model A.
These models are described in the legend of Fig. 22.

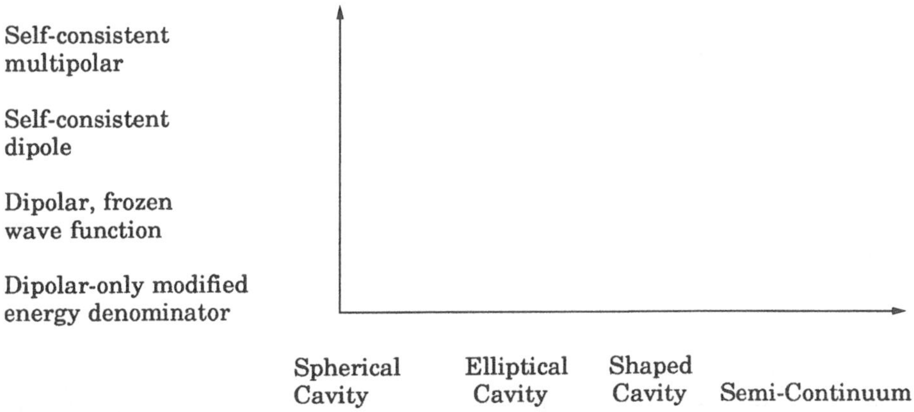

SCHEME 1 Increasing levels of sophistication for treating solvent (liquid or solid) environmental effects on dipolar response. The vertical line shows differing levels of treatment of the interaction effects on $\chi^{(2)}$ or $\chi^{(3)}$, while the treatment of the solvent is shown on the horizontal axis. The dielectric response of the solvent is usually treated assuming frequency independent ε, although some studies in which low-frequency and high-frequency response are differentiated are available.

of geometric distortions of those molecules has led to a deeper understanding of electronic structure effects on nonlinear optical response.

We give here three examples in which substantial modifications occur upon geometric variation.

1. Extensive Linearity

In classical electrostatics, the induced polarization in an ellipsoid is substantially larger than that in a sphere of identical volume. Similarly, one expects that for two electronic structures of roughly comparable size, the linear structure should show substantially larger nonlinear polarization than the two-dimensional structure. For

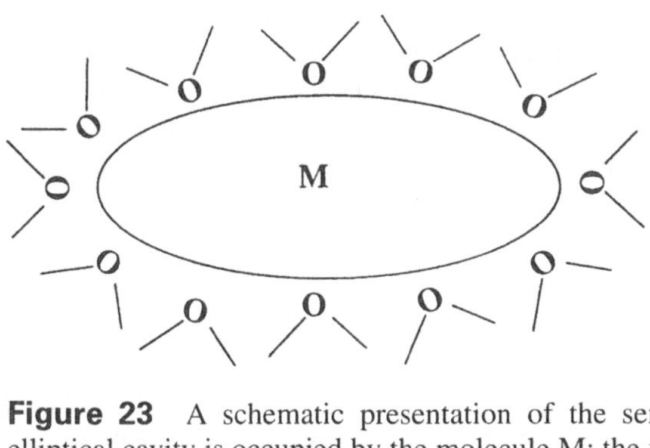

Figure 23 A schematic presentation of the semi-continuum model for solvation. The elliptical cavity is occupied by the molecule M; the first layer of water molecules surrounding that ellipsoid is treated by specific molecular quantum mechanics. Additional water molecules are treated as in the continuum model.

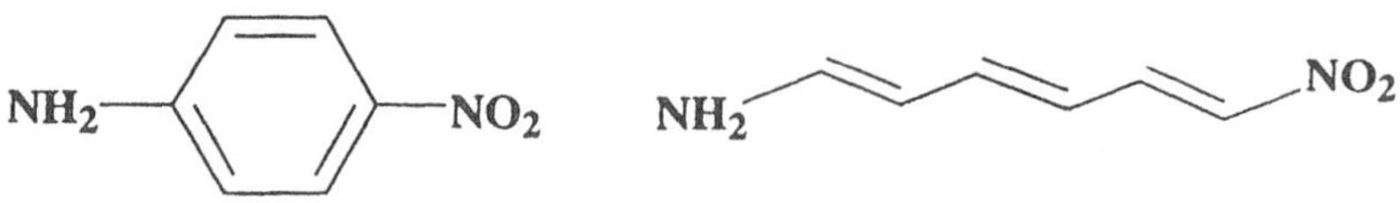

Ie (77.25 X 10^{-30} esu) **If (350.51 X 10^{-30} esu)**

Figure 24 β_{vec} calculated at an energy of 0.1eV using MECI-INDO/S level theory, for two six-member carbon chains: The more linear, hexatriene chain exhibits far larger response than the cyclic benzene structure.

example, Fig. 24 shows the computed hyperpolarizability for two six-π electron donor/acceptor species[135,142,143]; notice that the hyperpolarizability is larger for the linear structure than for the cyclic one. This is a general characteristic of molecular response, and for substantial polarization or hyperpolarizability, one therefore would tend to choose a linear extended structure. One difficulty is that such linear structures can exhibit, at finite temperatures, rotational isomerization; such isomerization reduces the effective π-type mixing between sub-units, and can result in localization of excitations. In polysilanes, such localization has been invoked, both experimentally and computationally, to explain the rapid onset, with increase in length, of essentially unchanging hyperpolarizability.[192]

2. Bond Length Alternation Effects

In an extended linear chain structure, the geometry of a molecule can become unstable with respect to particular distortions. For odd electron systems, such distortions result in the formation of so-called soliton defects[123,127,128,193]; these are also often referred to as "kinks," and are simply due to interrupted conjugation, as shown in Fig. 25. These defects are important in electrical transport of linear chain conductors,[193] as well as in optical response and $\chi^{(3)}$.[194] For symmetric structures, they are not generally important in the first hyperpolarizability.

A second set of distortions, however, does occur in the first hyperpolarizability and has provided a useful organizing principle for understanding effects of solvents and of donor/acceptor strength on linear structures.[195–198] For isolated polyenes, bond length alternation tends to occur: That is, as suggested by simple valence bond structures, all carbon-carbon bonds are not the same length, but the bond lengths tend to alternate from one C—C link to the next. In solid state systems, this is referred to as the "Peierls distortion," and is found whenever the band is not entirely filled. In the simple polyacetylene structure, the band is half-filled (that is, there is one π electron per site) and one expects dimerization; this is the basic

Figure 25 Sketch of a isolated carbon radical species, or soliton, in the odd-numbered chain. These structures are stable only for odd-numbered chains, and their conductivity and optical responses are unusual.

model for alternation. Simple examples occur in butadiene and hexatriene, in which the formally single and double bonds are indeed of substantially different length.

One can quantify the extent to which these distortions occur by the so-called bond length alternation (BLA); this is simply the average bond length difference between short and long bonds in an extended structure. Applying BLA ideas to understanding the variation of polarizabilities and hyperpolarizability with change in solvent or molecular substituent has been pioneered by Marder and collaborators[195]; these ideas are very useful for categorizing molecular response and its sensitivity to solvent polarization effects and to effects of donor and acceptor substituents.

As a simple example, consider the linear structure of an archetypical push-pull polyene of Fig. 26. The bridge in this chromophore could distort from a bond alternating polyene-like geometry to a uniform bond length symmetric cyanine-like geometry to a zwitterionic polymethine-like geometry, depending on the strength of the donor and acceptor groups, the solvent polarity or other environmental conditions.[195,196] The extent of distortion has been indexed by BLA. Marder and coworkers,[195] in a number of experimental and theoretical studies, demonstrated that this geometric distortion has many important consequences in NLO response properties of simple chromophores,such as the merocyanines. Important variations in the response properties of this class of chromophores with BLA are shown in Fig. 27.[196] These curves provide suggestions on the architectural nature of the bridge that might yield optimal NLO responses. For example, while a cyanine-like bridge architecture is not favored for second-order NLO applications, it yields the optimal bridge architecture to obtain the largest polarizability and hence refractive index. The most favorable structure for SHG applications, according to these studies, occurs when the bridge BLA lies in between the polyenic and cyanine-like structures. The curves also show the expected derivative relationships between the various NLO response properties. Studies have tried to understand the generality

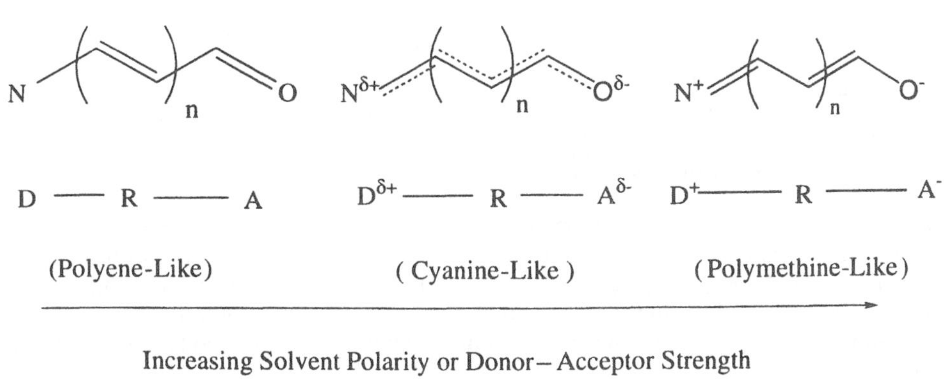

Figure 26 The chain geometry paradigm underlying the bond length alternation view of nonlinear response. With increasing solvent polarity or relative donor and acceptor strengths, the structure changes from polyene-like to polymethine-like. The intermediate, cyanine-like structure would be symmetric if D and A were identical, and therefore is expected to exhibit only very small hyperpolarizabilities. (From Ref: 196.)

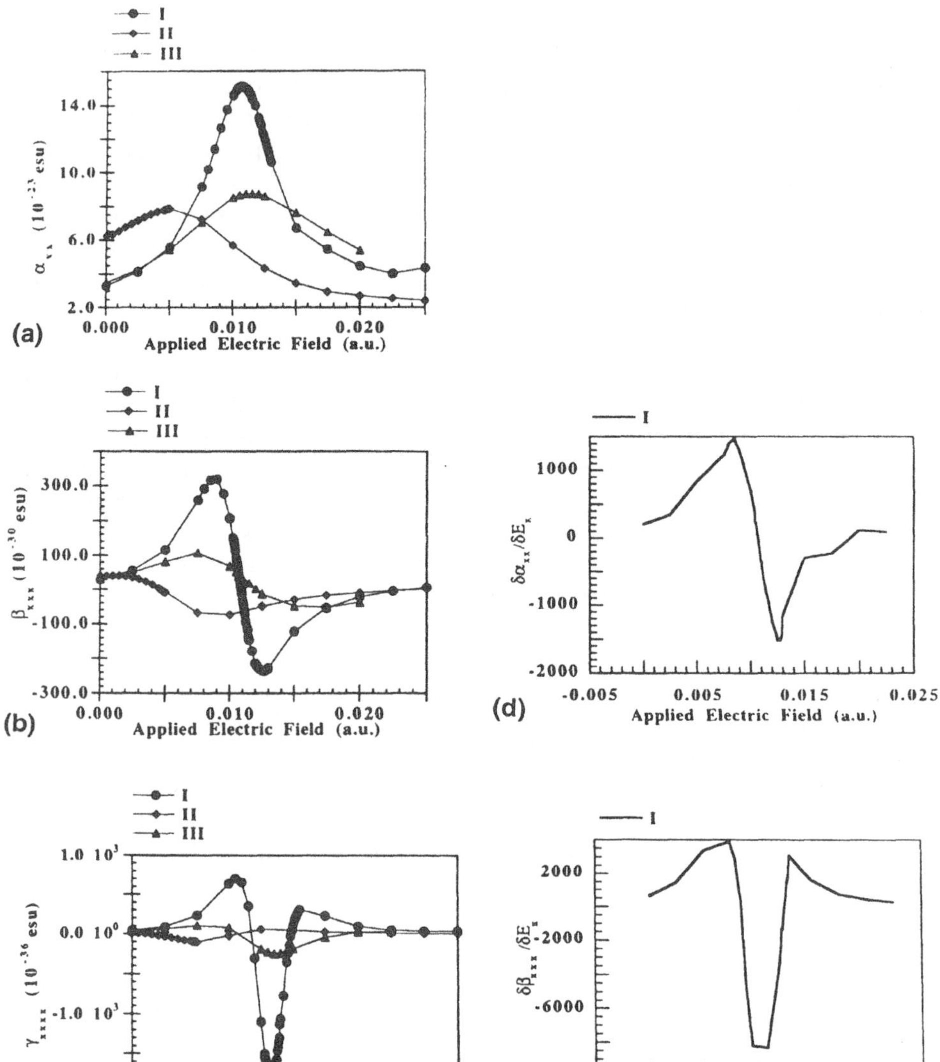

Figure 27 The polarizability, A, hyperpolarizability, B, and the second hyperpolarizability, γ, as a function of applied electric field, calculated at the MECI, INDO/S level for the three molecular chromophores of Fig. 26 with n = 4. The expected formal relationships (each higher order susceptibility being the derivative of the low order susceptibilities with respect to field, minimal β response near the cyanine limit) are numerically obeyed (D and E). (From Ref: 196.)

of such an approach in other chromophoric architectures. These studies show that large structural modifications and hence large variations in NLO response are possible only in structures that are stabilized by distortion.[196]

3. Tictoid Structures, and Bond Angle Effects

Even stronger variations of dipolar,[199–203] optical, and nonlinear optical[204] properties with molecular geometry can occur due to twists around certain conjugating double bonds. Consider, for example, structures shown in Fig. 28. The fully planar structure of Fig. 28a is quinoid and the dipole moment is relatively small. Upon twisting to 90 degrees, however, the inter-ring bond becomes effectively single, and the zwitterionic, strongly dipolar structure of Fig. 28b emerges. One would then expect much larger dipoles in the twisted structure, and differing nonlinear responses.

Very simple theory can be used to understand and predict the variation in the polarity and response properties of these molecules as a function of the twist angle. The simplest possible picture follows from Mulliken's analysis of charge transfer. Consider molecules such as those in Fig. 28 as comprising two sub-units, the left and the right, or donor and acceptor, linked by an interaction through the bond around which twisting occurs. One can then use a simple Hückel-like scheme, in which the donor sub-unit wave function is represented by ϕ_D, the acceptor sub-unit function by ϕ_A. These two units have characteristic one-electron energies which we will take equal to zero and α respectively (this α is reminiscent of the

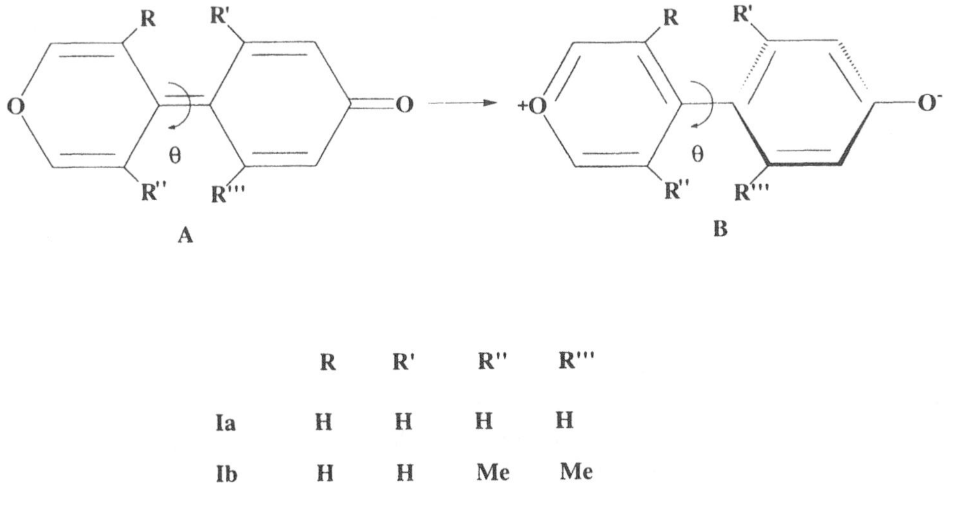

	R	R'	R''	R'''
Ia	H	H	H	H
Ib	H	H	Me	Me
Ic	Me	Me	Me	Me
Id	Me	t-But	Me	t-But
Ie	t-But	t-But	t-But	t-But

Figure 28 A characteristic structure demonstrating effects of tictoid distortions. With increasing bulkiness of R sustituent, the twist angle between the two rings will be pushed closer to 90 degrees, thus reducing the interring mixing and effective π conjugation. (From Ref: 204.)

Hückel α, and is the difference in effective electronegativity between the two halves of the molecule). The molecular sub-units are linked by a one-electron interaction that we call, again by analogy to simple Hückel theory, β. We expect β to scale approximately as the cosine of the twist angle between the two rings, so that it will have the largest absolute value (largest negative value) for co-planar rings, and will essentially vanish when the rings become perpendicular to each other. One can then represent the orthogonal wave functions for the ground and electronic excited states of a particular geometry in Mulliken terms as

$$\psi_g = C_D\phi_D + C_A\phi_A \tag{23a}$$

$$\psi_x = -C_A\phi_D + C_D\phi_A; \tag{23b}$$

orthogonality is assured by this choice of the coefficients C_A and C_D. Solving the simple two-site Hückel problem then, the ground state and excited state energies are

$$E_g = \frac{1}{2}(\alpha - \sqrt{\alpha^2 + 4\beta^2}) \tag{24a}$$

and

$$E_x = \frac{1}{2}(\alpha + \sqrt{\alpha^2 + 4\beta^2}). \tag{24b}$$

The coefficients follow by inserting these energies into the secular equation.

Given the coefficients and the energies, one can use the Pariser approximation to write the ground state and transition dipoles as

$$<g|\mu|g> = C_D^2R_D + C_A^2R_A \tag{25a}$$

$$<g|\mu|x> = C_DC_A(R_A - R_D), \tag{25b}$$

and the wavelength of the maximum absorption as

$$\lambda_{max} = \frac{hc}{E_x - E_g}, \tag{26}$$

where c and h are respectively the speed of light and Planck's constant.

These equations show that, given the values of α and β, one can immediately predict dipole moment, transition moment, and λ_{max} for the molecule. Most important, these will be controlled by only two electronic structure parameters, the effective electronegativity α, and the effective coupling strength β; both are negative energies. Twisting around the intramolecular double bond will change β substantially, therefore changing the dipole moment and the response properties.

Substitution of the solution to the Hückel expression into Eqs. 24, 25 and 26 yields

$$<g|\mu|g> = 1/2(R_A + R_D) - \frac{\alpha}{2(\alpha^2 + 4\beta^2)^{1/2}}(R_A + R_D), \tag{27a}$$

$$<g|\mu|x> = \frac{\beta}{(\alpha^2 + 4\beta^2)^{1/2}}(R_A + R_D), \tag{27b}$$

$$\Delta\mu_{ge} = \frac{2\alpha}{(\alpha^2 + 4\beta^2)^{1/2}} (R_A - R_D), \tag{27c}$$

and

$$\lambda_{\max} = \frac{hc}{\sqrt{\alpha^2 + 4\beta^2}}. \tag{27d}$$

This predicts that the groundstate dipole moment and λ_{max} are largest, and the transition dipole moment minimum when the bond angle twist is nearly 90 degrees, so that the π-electron systems are essentially orthogonal. (Remember that β is strongly dependent on twist angle, roughly as $\beta \propto \cos\theta$.)

The INDO/S model Hamiltonian and the MECI approximation give[204] the results in Figs. 29 and 30. Note that, as predicted, as the twist angle approaches 90 degrees, we see a maximum in the absorption wavelength, and a minimum in the transition dipole (Fig. 29). These results then lead to strong variations in both the polarizability and hyperpolarizability. For example, the hyperpolarizability variation, shown in Fig. 30, exhibits a very strong maximum for the twist angle near $\alpha = 90°$. This can be understood in terms of the standard two-level model of Eq. 6: the dipole moment difference, transition dipole, and energy difference that are required for this formula are given, approximately, in Eqs. 26 and 27; this indeed predicts that for twist angles very close to 90 degrees, a strong maximum in the β response should occur.

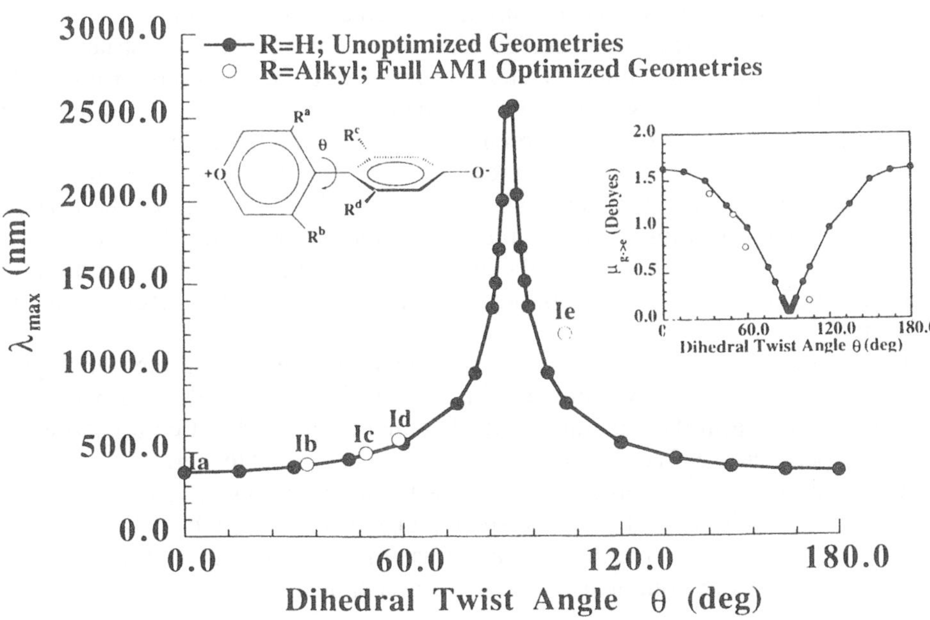

Figure 29 Variation of λ_{max} as a function of the imposed dihedral twist angle θ of the chromophore in Fig. 28. Open circles represent the corresponding values for geometry-optimized chromophore set **Ia-Ie**. The insert shows the corresponding transition moments of the first excited state as a function of θ for the same chromophores. (From Ref: 204.)

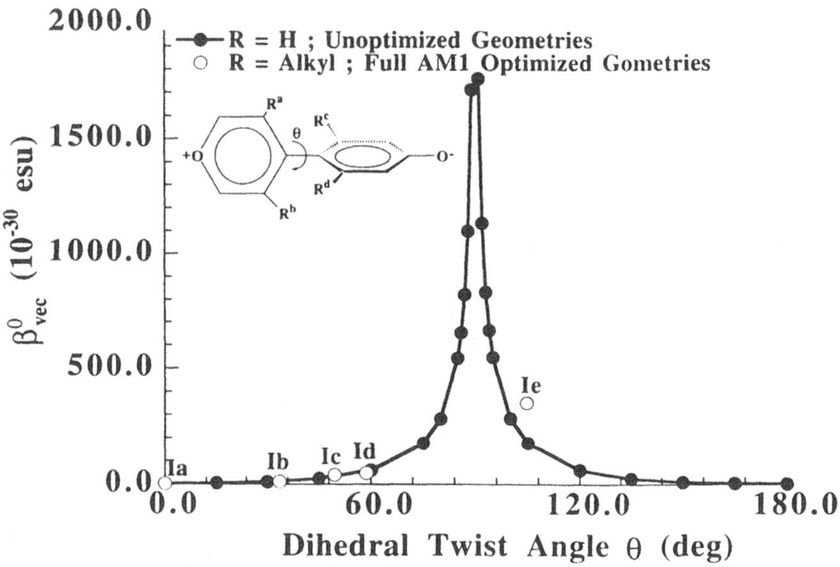

Figure 30 β_{vec}, at an excitation energy of 0.1eV (closed circles), as a function of imposed dihedral twist angle θ of chromophore in Fig. 28. Open circles represent the corresponding values for geometry-optimized chromophore set **Ia-Ie**. (From Ref: 204.)

This general class of structures can be referred to as tictoid structures, by analogy with a so-called twisted intramolecular charge transfer, or TICT, geometries of the excited states of molecules such as ethene. The tictoid structure geometry can be controlled either by steric crowding at the positions ortho to the linking bond between the two rings, or in structures where a particular orientation angle is demanded by the covalent bonding. In Figs. 29 and 30, we show, by open circles, the computed response for structures with differing ortho substitutents of bulkier alkyl groups; the reasonably close agreement between the standard parent structure and the substituted one is pleasing. More importantly, however, these suggest that by synthetically controllable variation of the twist angle, substantial differences in both linear and nonlinear optical properties of molecules can be obtained.

This section has shown that, using approximate semiempirical methodologies, important geometric and environmental effects can be fairly well-described. Such descriptions are within the general theme of semiempirical, MECI computations of the first hyperpolarizability: Such calculations can be used to correlate chemical structure and chemical bonding ideas with a nonlinear response, and to give semiquantitative predictions of actual response in molecules in which the response is relatively large. Thus it can be extremely useful in predicting nonlinear properties, and in correlating nonlinear response with chemical structure.

VIII. SOME REMARKS

Predicting the polarizability and hyperpolarizability of isolated or in condensed phase molecules is a challenging problem in electronic structure theory. Generally, electronic structure methods are more accurate for groundstate properties such as energies, geometries, and charge distributions than for response properties such as optical absorption, polarizability, or hyperpolarizability. Nevertheless, substantial progress has been made in theoretical models that vary in sophistication from simple Hückel theory to elaborate coupled cluster or multireference configuration interaction treatments using ab-initio model Hamiltonians with extended basis sets.

Perhaps the most striking overall result in the field of molecular nonlinear optical response is the utility and relative accuracy of simple semiempirical models. In a way this is not surprising: the spectral representations of Eqs. 4 and 5 show directly that the polarizability and first hyperpolarizability are functions of energy denominators, dipole moments, and transition dipole moments. These are the parameters that enter into the calculation of optical spectra. Since the 1950s it has been known that optical spectra of aromatic hydrocarbons and heterocycles, as well as conjugated hydrocarbon structures, could be very well-described using simple semiempirical models of PPP-type. Extension to more general molecular structures is possible using screened NDO model Hamiltonians.

These semiempirical models have been of substantial utility in understanding and predicting molecular nonlinear responses. As outlined above and detailed more extensively elsewhere, they can predict the hyperpolarizability as a function of molecular structure, substituents, distortion, solvent environment, intermolecular interaction, frequency, and geometric distortion. Moreover, these calculations are relatively easy and efficient to perform, and the results can be understood in terms of contributing structures (as seen in the SOS form of expression 4) and in terms of the electronic structure, as shown in Fig. 2. The semiempirical models use minimum valence basis sets, and therefore electronic population and electron assignment, using the Mulliken formulation, is both straightforward and unique.

Semiempirical models have the great disadvantage that their accuracy cannot be systematically improved. Therefore, extensive ab-initio studies with appropriately extended basis sets (at a minimum, using two sets of polarization functions and some diffuse orbitals) and correlation treatments (minimally, MP/2) represent an important contribution to the understanding of molecular and nonlinear optical response. One difficulty is that such calculations scale badly with respect to molecular size: The necessity for large basis sets, and for extensive treatments of correlation, if accurate results are to be obtained, enforces substantive resource requirements on the computation. Additionally, these elaborate wave functions are not so easily analyzed as the simple minimal basis associated with semiempirical models. For example, negative Mulliken populations can be obtained when extended basis sets are used. In terms of the excitation process, equation of motion techniques or response functions are used more often than the SOS expression because they provide increased computational efficiency. This occurs, however, at the cost of interpretability. Without the use of the SOS expression, it is not easy to understand which electronic configurations contribute to the nonlinear response.

Density functional theory (DFT) methods have become widely popular and very useful for predicting ground state properties of molecules.[205–215] These DFT methods scale attractively with molecular size (for small molecules, for which extensive correlated treatments vary roughly as the 7th power of the number of electrons, these DFT methods scale roughly as the third power). The difficulty is that, formally, density functional theory models are based on the Hohenberg-Kohn theorem,[209] which applies to the groundstate of the molecule. Therefore calculations of optical, nonlinear or any other response properties may become slightly problematic. The most obvious generalization is to compute, for example, the hyperpolarizabilities at zero-frequency with finite field calculations by taking the numerical derivatives of either the dipole moment or the total energy with respect to the applied electric field.[210–215] Very recent calculations using this method to study small molecules such as HF or NH_3 have appeared. While some publications have appeared in the area of DFT application to nonlinear response, this is still largely unexplored territory. On the basis of its extensive success in other areas of electronic structure, one might expect that DFT would be very useful for accurate calculations. Probably the most useful approach involves using the time-dependent generalization of DFT: The excitation energies and transition moments are then obtained directly. Work in this area, especially applied to absorption line shapes and to polarizabilities, is beginning. The importance of the diffuse regions of the electronic space, that require extended basis sets for proper treatment in an ab-initio context, place heavy demands on the exchange correlation functional to be used in DFT work.

From the viewpoint of the experimentalist, programs are now available to calculate hyperpolarizabilities. Commercial codes such as GAUSSIAN include zero-frequency response at the ab-initio level, as well as a treatment of solvent effects using the simple spherical cavity. Ab-initio codes such as HONDO or ACES II permit calculation of the hyperpolarizability using correlated approaches such as coupled cluster.

Despite these advances using the ab-initio approach, we feel that, for chemical interpretability, general applicability, ease of use, and efficiency, the semiempirical models have been, and will continue to be, of very substantial importance. Electronic structure methods are now at the point where they can help in understanding the nonlinear optical properties that have been measured, and in predicting nonlinear optical properties of molecular chromophores before their synthesis. This screening analysis makes such calculations of substantive utility in understanding, predicting, and optimizing first hyperpolarizabilities of molecules and molecular materials.

ACKNOWLEDGEMENTS

We are grateful to several research groups for sharing their insights with us. Particular thanks go to R. J. Bartlett, P. Jorgensen, K. Mikkelsen, K. Sylvester-Hvid, M. C. Zerner, S. Karna, D. Dudis, B. Kirtman, M. Dupuis, C. Dykstra, S. Marder, and P. Barbara. We are especially grateful to Professor C. W. Dirk for originally suggesting this project. Contributions to our work by D. R. Kanis, S. DiBella, and D. Li are gratefully acknowledged. Funding for this research was provided by the CAMP collaboration on optical electronic materials (ONR Grant #95-1-0717), and by the NSF Division of Materials Research through the Materials Research Center of Northwestern University (Grant #DMR-9632472).

REFERENCES

1. Prasad, P. N., and Williams, D. J., *Introduction to Nonlinear Optical Effects in Molecules and Polymers*, Wiley Interscience, New York, 1991.
2. Williams, D. J., ed., *Nonlinear Optical Properties of Organic and Polymeric Materials*, ACS Symp. Ser. 233, Washington DC, 1983.
3. Chemla, D. S., and Zyss, J., eds., *Nonlinear Optical Properties of Organic Molecules and Crystals*, Academic Press Inc. New York, 1987.
4. Lindsay, G. A., and Singer, K. D., eds., *Polymers for Second-Order Nonlinear Optics*, ACS Symp. Ser. 601, Washington DC, 1995.
5. Zyss, J., ed., *Quadratic Nonlinear Optics in Poled Polymer Films: From Physics to Device*, Academic Press Inc., San Diego, California, 1994.
6. Evans, M., and Keilich, S., eds., *Adv. Chem. Phys., 85* (1993).
7. Burland, D. M., ed., *Chem. Rev., 94* (1994).
8. Ratner, M. A., ed., *Int. J. Quantum Chem., 43* (1992).
9. Prasad, P. N., and Ulrich, D., eds., *Nonlinear Optical and Electroactive Polymers*, Plenum Press, New York, 1986.
10. Marks, T. J., and Ratner, M. A., *Angew. Chem. Int. Ed., 34*, 155 (1995).
11. Bloembergen, N., *Nonlinear Optics*, W. A. Benjamin, New York, 1965.
12. Boyd, R. W., *Nonlinear Optics*, Academic Press, New York, 1992.
13. Shen, Y. R., *The Principles of Nonlinear Optics*, Wiley, New York, 1984.
14. Zernike, F., and Midwinter, J., *Applied Nonlinear Optics*, Wiley, New York, 1973. Yariv, A., *Quantum Electronics*, Wiley, New York, 1975.
15. (a) Beljonne, B., Brèdas, J.-L., Chen, G., and Mukamel, S., *Chem. Phys. Lett., 210*, 353 (1996). (b) Wang, W. X., Mukamel, S., *Chem. Phys. Lett., 192*, 417 (1992). (c) *J. Chem. Phys., 97*, 8019 (1992). (d) Mukamel, S., *Principles of Nonlinear Optical Spectroscopy*, Oxford University Press, New York, 1995. (e) Treitiak, S., Chernyak, V., and Mukamel, S., *J. Chem. Phys., 105*, 8914 (1996).
16. (a) Krause, J. L., Schafer, K. J., and Kulander, K. C., *Phys. Rev. Lett., 68*, 3535 (1992). (b) *Phys. Rev. A., 45*, 4998 (1992). (c) L'Hullier, A., Schafer, K. J., and Kulander, K. C., *J. Phys. B., 24*, 3165 (1991). (d) Kulander, K. C., and Shore, B. W., *Phys. Rev. Lett., 62*, 524 (1989). (e) L'Hullier, A., Balcon, P., Candel, S., Schafer, K. J., and Kulander, K. C., *Phys. Rev. A., 46*, 2778 (1992).
17. Armstrong, J. A., Bloembergen, N., Ducuing, J., Pershan, P. S., *Phys. Rev., 127*, 1918 (1962).
18. (a) Ward, J. F., *Rev. Mod. Phys., 37*, 1 (1965). (b) Bloembergen, N., Lotem, H., and Lynch, R. T., *Ind. J. Pure and Appl. Phys., 16*, 151 (1978).
19. Orr, B. J., and Ward, J. F., *Mol. Phys., 20*, 513 (1971).
20. Flytzanis, C., in *Treatise of Quantum Electronics* (H. Rabin and C. L. Tang, eds.), Academic Press, New York, 1975.
21. Lalama, S. L., and Garito, A. F., *Phys., Rev. B., 20*, 1179 (1979).
22. Docherty, V. J., Pugh, D., and Morley, J. O., *J. Chem. Soc., Faraday Trans 2, 81*, 1179 (1985).
23. (a) Li, D., Marks, T. J., and Ratner, M. A., *Chem. Phys. Lett., 131*, 370 (1986). (b) Li, D., Marks, T. J., and Ratner, M. A., *J. Amer. Chem. Soc., 110*, 1707 (1988).
24. Dirk, C. W., Twieg, R. J., and Wagniere, G., *J. Amer. Chem. Soc., 108*, 5387 (1986).
25. Ulman, A., Willand, C. S., Kohler, W., Robello, D. R., Williams, D. J., and Handley, L., *J. Amer. Chem. Soc., 112*, 7083 (1990).
26. Kanis, D. R., Ratner, M. A., Marks, T. J., and Zerner, M. C., *Chem. Mater., 3*, 19 (1991).
27. Jain, M., and Chandrasekhar, J., *J. Phys. Chem., 97*, 4044 (1993).

28. Kanis, D. R., Ratner, M. A., and Marks, T. J., *Chem. Rev.*, *94*, 195 (1994) and references therein.
29. Svendsen, E. N., Willand, C. S., and Albrecht, A. C., *J. Chem. Phys.*, *83*, 5760 (1985).
30. Albert, I. D. L., Morley, J. O., and Pugh, D., *J. Chem. Phys.*, *99*, 5197 (1993).
31. Ramasesha, S., Shuai, Z., and Brèdas, J.-L., *Chem. Phys. Lett.*, *245*, 224 (1995).
32. (a) Dirk, C. W., Cheng, L.-T., and Kuzyk, M. G., *Int. J. Quantum Chem.*, *43*, 27 (1992). (b) Dirk, C. W., and Kuzyk, M. G., *Phys. Rev. A.*, *39*, 1219 (1989).
33. (a) Mazumdar, S., Guo, D. D., and Dixit, S. N., *Synth. Met.*, *57*, 3881 (1993). (b). Dixit, S. N., Guo, D. D., and Mazumdar, S., *Phys. Rev. B.*, *43*, 6781 (1991).
34. Linderberg, J., and Öhrn, Y., *Propogators in Quantum Chemistry*, Academic Press, London, 1974.
35. Yu, J., and Zerner, M. C., *J. Chem. Phys.*, *100*, 7487 (1994).
36. Rice, J. E., Amos, R. D., Colwell, S. M., Handy, N. C., and Sanz, J., *J. Chem. Phys.*, *93*, 8828 (1990).
37. Thouless, D. J., *Quantum Mechanics of Many Body Systems*, Academic Press, New York, 1961.
38. (a) Parkinson, W. A., and Zerner, M. C., *J. Chem. Phys.*, *94*, 478 (1991). (b) Luo, Y., Agren, H., Koch, H., Jorgensen, P., and Helgaker, T., *Phys. Rev. B*, *51*, 14949 (1995).
39. Oudar, J. L., and Chemla, D. S., *J. Chem. Phys.*, *66*, 2664 (1977).
40. Oudar, J. L., *J. Chem. Phys.*, *67*, 446 (1977).
41. (a) Ziolek, L., Palewska, K., Lipinski, J., Sworakowski, J., Nespurek, S., Bohm, S., and Meister, E. C., *Acta, Phys. Pol. A.*, *88*, 283 (1995). (b) Ziolek, L., Palewska, K., Lipinski, J., Sworakowski, J., Nespurek, S., *Mol. Cryst. Liq. Cryst. Sci. Tech. Sec. A-Mol. Cryst. Liq. Cryst.*, *283*, 125 (1996). (c) Sworakowski, J., Lipinski, J., Ziolek, L., Palewska, K., and Nespurek, S., *J. Phys. Chem.*, *100*, 12288 (1996).
42. Kim, O. K., and Lehn, J.-M., *Chem. Phys. Lett.*, *255*, 147 (1996).
43. Kolling, O. W., *J. Phys. Chem.*, *100*, 519 (1996).
44. Qin, J. G., Wu, X. D., Zhang, X. Y., Zhan, C. M., and Liu, D. Y., *Synth. Met.*, *71*, 1711 (1995).
45. Reichardt, C., *Chem. Rev.*, *94*, 2319 (1994) and references therein.
46. Heflin, J. R., Wong, K. Y., Zamini-Kamiri, O., and Garito, A. F., *Phys. Rev. B*, *38*, 1573 (1988).
47. Albert, I. D. L., Morley, J. O., and Pugh, D., *J. Chem. Phys.*, *102*, 237 (1995).
48. (a) Soos, Z. G., and Ramasesha, S., *J. Chem. Phys.*, *90*, 1067 (1988). (b) Ramasesha, S., and Soos, Z. G., *Chem. Phys. Lett.*, *153*, 171 (1988).
49. (a) Pierce, B., *J. Chem. Phys.*, *91*, 791 (1988). (b) Pierce, B. M., *Adv. Chem. Ser.*, *240*, 243 (1994). (c) Brèdas, J.-L., Adant, C., Tackx, P., Persoons, A., and Pierce, B. M., *Chem. Rev.*, *94*, 243 (1994). (d) Pierce, B. M., *Physica D*, *68*, 51 (1993).
50. Beljonne, D., Shuai, Z., and Brèdas, J.-L., *Int. J. Quantum Chem.*, *52*, 39 (1994).
51. Jorgensen, P., and Simons, J., *Second Quantization-based Methods in Quantum Chemistry*, Academic Press, New York, 1981.
52. Schatz, G. C., and Ratner, M. A., *Quantum Mechanics in Chemistry*, Prentice Hall, Engelwood Cliffs, New Jersey, 1993.
53. Salem, L., *Molecular Orbital Theory of Conjugated Systems*, Benjamin, New York, 1965.
54. Flurry, R. L., Jr., *Molecular Orbital Theories of Bonding in Organic Molecules*, Marcel Dekker, New York, 1968.
55. Ramasesha, S., and Albert, I. D. L., *Chem. Phys. Lett.*, *154*, 501 (1989). Albert, I. D. L., and Ramasesha, S., *Mol. Phys. Liq. Cryst.*, *168*, 95 (1989).
56. Marder, S. R., Gorman, C. B., Tiemann, B. G., and Cheng, L.-T., *J. Amer. Chem. Soc.*, *115*, 3006 (1993).

57. (a) Risser, S. M., Beratan, D. N., and Marder, S. R., *J. Amer. Chem. Soc. 115*, 7719 (1993). (b) Risser, S., Allender, D. W., and Lee, M. A., *Mol. Cryst. Liq. Cryst., 150b*, 631 (1987). (c) Hameka, H. F., *J. Chem. Phys., 67*, 2935 (1977). (d) Zamanikhamiri, O., and Hameka, H. F., *J. Chem. Phys.*, 73, 5693 (1980). (e) Zaminikhamiri, O., Hameka, H. F., and McIntyre, E. F., *J. Chem. Phys., 72*, 5906, 1280 (1980).
58. (a) Marder, S. R., Cheng, L.-T., and Beratan, D. N., *AIP Conf. Proc.* 1992. (b) Marder, S. R., Cheng, L.-T., Tiemann, B. G., and Beratan, D. N., *Proc. SPIE Int. Soc. Opt. Eng., 1560*, 86 (1991).
59. Instead of giving the original references, we have given the references of some excellent monographs on semiempirical methods. The original references can be found in these texts: (a) Dewar, M. J. S., *The Molecular Orbital Theory of Organic Chemistry*, McGraw Hill, New York, 1969. (b) Pople, J. A., and Beveridge, D. L., *Approximate Molecular Orbital Theory*, McGraw Hill, New York, 1970. (c) Murrell, J. N., and Harget, A. J., *Semi-Empirical Self-Consistent-Field Molecular Orbital Theory of Organic Molecules*, John Wiley, London, 1972. (d) Segal, G. A., *Semi-empirical Methods of Electronic Structure Calculation*, Plenum Press, New York, 1977. (e) Sadlej, J., *Semi-empirical Methods of Quantum Chemistry*, Halsted Press, New York, 1985.
60. Li, D., Ratner, M. A., and Marks, T. J., *J. Phys. Chem., 96*, 4325 (1992).
61. (a) Ramasesha, S., and Albert, I. D. L., *Phys. Rev. B.: Condens. Matter, 42*, 8587 (1990). (b) Albert, I. D. L., Das, P. K., and Ramasesha, S., *Phys. Rev. B: Condens. Matter, 43*, 7013 (1991).
62. (a) Morrell, J. A., and Albrecht, A. C., *Chem. Phys. Lett., 64*, 46 (1979). (b) Morrell, J. A., Albrecht, A. C., and Levin, K. H., *J. Chem. Phys., 71*, 5063 (1979). (c) Willand, C. S., and Albrecht, A. C., *Opt. Commun., 57*, 146 (1986).
63. (a) Teng, C. C., and Garito, A. F., *Phys. Rev. B., 28*, 6766 (1983). (b) *Phys. Rev. Lett., 50*, 350 (1983). (c) Lalama, S. J., Singer, K. D., Garito, A. F., and Desai, K. N., *Appl. Phys. Lett., 39*, 940 (1981). (d) Lipscomb, G. F., Narang, R. S., and Garito, A. F., *J. Chem. Phys., 75*, 1509 (1981).
64. (a) Morley, J. O., Docherty, V. J., and Pugh, D., *J. Chem. Soc. Perkin Trans., 2*, 1357, 1361 (1987). (b) *J. Mol. Electron., 5*, 117 (1989). (c) Morley, J. O., *J. Amer. Chem. Soc., 110*, 7660 (1988). (d) Morley, J. O., Pavlides, P., and Pugh, D., *Int. J. Quantum. Chem., 43*, 7 (1992).
65. Kanis, D. R., Ratner, M. A., and Marks, T. J., *J. Amer. Chem. Soc., 112*, 8203 (1990).
66. (a) Di Bella, S., Fragala, I. L., Ratner, M. A., and Marks, T. J., *J. Amer. Chem. Soc., 115*, 682 (1993). (b) Di Bella, S., Ratner, M. A., and Marks, T. J., *J. Amer. Chem. Soc., 114*, 5862 (1992).
67. (a) Kurtz, H. A., Stewart, J. J. P., Deiter, K. M., *J. Comput. Chem., 11*, 82 (1990). (b) Matsuzawa, N., and Dixon, D. A., *Int. J. Quantum Chem., 44*, 497 (1992). (c) *J. Phys. Chem., 96*, 6232 (1992). (d) Chandra, A. K., and Banerjee, M., *J. Mol. Struct. (THEOCHEM), 83*, 261 (1991). (e) Williams, G. R. J., *J. Mol. Struct. (THEOCHEM), 36*, 215 (1987).
68. (a) Ridley, J., and Zerner, M. C., *Theor. Chim. Acta (Berlin), 32*, 111 (1973). (b) Bacon, A. D., and Zerner, M. C., *Theor. Chim. Acta (Berlin), 53*, 21 (1979). (c) Anderson, W. P., Edwards, W. D., and Zerner, M. C., *Inorg. Chem., 25*, 2728 (1986).
69. (a) Waite, J., and Papadopoulos, M. G., *J. Chem. Phys., 82*, 1427 (1985). (b) *J. Phys. Chem., 94*, 1755 (1990). (c) *J. Mol. Struct. (THEOCHEM), 202*, 121 (1989). (d) Nicolaides, C. A., Papdopoulos, M. G., and Waite, J., *Theor. Chim. Acta, 61*, 427 (1982).
70. Mukhopadyay, D., and Soos, Z. G., *J. Chem. Phys., 101*, 5515 (1994).
71. (a) Ramasesha, S., and Soos, Z. G., *Int. J. Quantum Chem., 25*, 1004 (1984). (b) *J. Chem. Phys., 80*, 3278 (1984). (c) Soos, Z. G., and Ramasesha, S., *Phys. Rev. B., 32*, 5368 (1985).

72. Albert, I. D. L., Marks, T. J., and Ratner, M. A., in *Nonlinear Optical Materials. Theory and Modeling* (S. P. Karna and A. T. Yeates, eds.), ACS Symp. Ser. 628, ACS Washington DC, 1996.
73. Szabo, A., and Ostlund, N. S., *Modern Quantum Chemistry: An Introduction to Advanced Electronic Structure Theory*, McGraw Hill, New York, 1989.
74. Hehre, W. J., Radom, L., Schleyer, v. R. P., and Pople, J. A., *Ab-initio Molecular Orbital Theory*, Wiley, New York, 1986.
75. Shuai, Z., Ramasesha, S., and Brèdas, J.-L., *Chem. Phys. Lett.*, *250*, 14 (1996).
76. Lundquist, P. M., Yitzchaik, S., Zhang, T. G., Kanis, D. R., Ratner, M. A., Marks, T. J., and Wong, G. K., *Appl. Phys. Lett.*, *64*, 2194 (1994).
77. Karna, S. P., and Yeates, A. T., eds., *Nonlinear Optical Materials. Theory and Modeling*, ACS Symp. Ser. 628, ACS Washington DC, 1996.
78. Sadlej, A. J., *Theor. Chim. Acta*, *79*, 123 (1991).
79. Keshari, V., Wijekoon, M. K. P., Prasad, P. N., and Karna, S. P., *J. Phys. Chem.*, *99*, 9045 (1995).
80. (a) Bishop, D. M., *Rev. Mod. Phys.*, *62*, 343 (1990). (b) *J. Chem. Phys.*, *95*, 5489 (1991). (c) *Adv. Quantum Chem.*, *25*, 1 (1994). (d) Bishop, D. M., and Dalskov, E. K., *J. Chem. Phys.*, *104*, 1004 (1996). (e) Bishop, D. M., and Pipin, J., *J. Chem. Phys.*, *98*, 522 (1993).
81. (a) Kirtman, B., and Bishop, D. M., *Chem. Phys. Lett.*, *175*, 601 (1990). (b) D. M. Bishop, and Kirtman, B., *J. Chem. Phys.*, *95*, 2646 (1991). (c) *97*, 5255 (1992). (d) Bishop, D. M., Kirtman, B., Kurtz, H. A., and Rice, J. E., *J. Chem. Phys.*, *98*, 8024 (1993). (e) Papadopoulos, M. G., Willets, A., Handy, N. C., and Underhill, A. E., *Mol. Phys.*, *88*, 1063 (1996). (f) Cohen, M. J., Willets, A., Amos, R. D., and Handy, N. C., *J. Chem. Phys.*, *100*, 4467 (1994).
82. (a) Lowdin, P. O., *Adv. Chem. Phys.*, *2*, 207 (1959). (b) Nesbet, R. K., *Proc. Roy. Soc. (London)*, *A230*, 312 (1995). (c) Bender, C. F., and Davidson, E. R., *Phys. Rev. A.*, *183*, 21 (1969). (d) Raghavachari, K., and Anderson, J. B., *J. Phys. Chem.*, *100*, 12960 (1996).
83. (a) Sinanoglu, O., *Proc. US Natl. Acad. Sci.*, *47*, 1217 (1961). (b) Sinanoglu, O., *Three Approaches to Electron Correlations in Atoms*, Yale University Press, New Haven, 1970.
84. (a) Fulde, P., *Electron Correlations in Molecules and Solids*, Springer-Verlag, New York, 1993. (b) Wilson, S., *Electron Correlations in Molecules*, Clarendon Press, Oxford, 1984. (c) Hurley, A. C., *Electron Correlations in Small Molecules*, Academic Press, New York, 1976.
85. (a) Dykstra, C. E., *Ab initio Calculations of Structures and Properties of Molecules*, Elsevier, New York, 1988. (b) *Advanced Theories and Computational Approaches to the Electronic Structure of Molecules* (Dykstra, C.E., ed.) Dordrecht, Reidel, Boston, 1984. (c) Harris, F. E., Monkhorst, H. J., and Freeman, D. L., *Algebraic and Diagrammatic Methods in Many-Fermion Theory*, Oxford University Press, New York, 1992.
86. (a) Yeager, D. L., and Jorgensen, P., *Chem. Phys. Lett.*, *65*, 77 (1979). (b) Dalgaard, E., *J. Chem. Phys.*, *72*, 816 (1980).
87. (a) Olsen, J., and Jorgensen, P., *J. Chem. Phys.*, *82*, 3235 (1985). (b) Hettema, H., Jensen, H. J. Aa., Jorgensen, P., and Olsen, J., *J. Chem. Phys.*, *97*, 1174 (1992).
88. (a) Andres, L. S., Fulscher, M. P., Roos, B. O., and Merchan, M., *J. Phys. Chem.*, *100*, 6485 (1996). (b) Roos, B. O., Andersson, K., Fulscher, M. P., Malmqvist, P. A., Andres, L. S., Pierloot, K., and Merchan, M., *Adv. Chem. Phys.*, *93*, 219 (1996). (c) Andersson, K., Roos, B. O., *Int. J. Quantum Chem.*, *45*, 591 (1993). (d) Andres, L. S., Merchan, M., Nebotgil, I., Lindh, R., and Roos, B. O., *J. Chem. Phys.*, *98*, 3151

(1993). (e) Andersson, K., Malmqvist, P. A., and Roos, B. O., *J. Chem. Phys.*, *96*, 1218 (1992).

89. (a) Mikkelsen, K. V., and Sylvester Hvid, K. O., *J. Phys. Chem.*, *100*, 9116 (1996). (b) Luo, Y., Agren, H., Jorgensen, P., and Mikkelsen, K. V., *Adv. Quantum Chem.*, *26*, 165 (1995). (c) Mikkelsen, K. V., Cesar, A., Agren, H., and Jensen, H. J. Aa., *J. Chem. Phys.*, *100*, 6597 (1994).
90. (a) Mikkelsen, K. V., and Kmit, M., *Theor. Chim. Acta*, *90*, 307 (1995). (b) Mikkelsen, K. V., Luo, Y., Agren, J., and Jorgensen, P., *J. Chem. Phys.*, *100*, 8240 (1994).
91. (a) Møller, C., and Plesset, M. S., *Phys. Rev. A.*, *46*, 618 (1934). (b) Kelly, H. P., *Adv. Chem. Phys.*, *14*, 12 (1963). (c) Bartlett, R. J., and Silver, D. M., *Int. J. Quantum Chem. Symp.*, *8*, 271 (1974).
92. (a) Curtiss, L. A., Raghavachari, K., and Pople, J. A., *J. Chem. Phys.*, *103*, 4192 (1995). (b) *98*, 1293 (1993). (c) *Chem. Phys. Lett.*, *214*, 183 (1993). (d) Tozer, D. J., N. C., Handy, Amos, R. D., Pople, J. A., Nobes, R. H., Xie, Y. M., and Schaefer, H. F., *Mol. Phys.*, *79*, 777 (1993). (e) Knowles, P. J., Andrews, J. S., Amos, R. D., Handy, N. C., and Pople, J. A., *Chem. Phys. Lett.*, *186*, 130 (1991).
93. Frisch, M. J., Trucks, G. W., Head-Gordon, M., Gill, M. W., Wong, M. W., Foresman, J. B., Johnson, B. G., Schlegel, H. B., Robb, M. A., Replogle, E. S., Gomperts, D. J., Andres, J. L., Raghavachari, K., Binkley, J. S., Gonzalez, C., Martin, R. L., Fox, D. J. Defrees, D. J., Baker, J., Stewart, J. J. P., and Pople, J. A., *Gaussian 92*, Gaussian, Pittsburgh, Pennsylvania, 1992.
94. Schmidt, M. W., Balridge, K. K., Boatz, J. A., Elbert, S. T., Gordon, M. S., Jensen, J. H., Kosekia, S., Matsunaga, N., Nguyen, K. A., Su, S. J., Windus, T. L., Dupuis, M., and Montgomery, J. A., *J. Comput. Chem.*, *14*, 1347 (1993).
95. Hehre, W., *Spartan*, Wavefunction, Irvine, California, 1994.
96. Stanton, J. F., Gauss, J., Watts, J. Q., Lauderdale, W. J., and Bartlett, R. J., *ACES II: An ab-initio program*, Quantum Theory Project, University of Florida, 1993.
97. Sim, F., Chin, S., Dupuis, M., and Rice, J. E., *J. Phys. Chem.*, *97*, 1158 (1993). Rice, J. E., and Handy, N. C., *Int. J. Quantum Chem.*, *43*, 91 (1992); *J. Chem. Phys.*, *94*, 4959 (1991). Rice, J. E., Amos, R. D., Colwell, S. M., Handy, N. C., and Sanz, J., *J. Chem. Phys.*, *93*, 8828 (1990).
98. Bartlett, R. J., and Sekino, H., in ref. 73, p. 23.
99. (a) Cizek, J., *Adv. Chem. Phys.*, *14*, 35 (1969). (b) Bartlett, R. J., and Purvis III, G. D., *Int. J. Quantum Chem.*, *14*, 561 (1978). (c) Purvis III, G. D., and Bartlett, R. J., *J. Chem. Phys.*, *76*, 1910 (1982). (d) Watts, J. D., and Bartlett, R. J., *Chem. Phys. Lett.*, *258*, 581 (1996). (e) Watts, J. D., and Bartlett, R. J., *Int. J. Quantum Chem.*, *S27*, 51 (1993). (f) Kucharski, S. A., and Bartlett, R. J., *Chem. Phys. Lett.*, *206*, 574 (1993). (g) *J. Chem. Phys.*, *97*, 4282 (1992). (h) Bartlett, R. J., *Theor. Chim Acta*, *80*, 71 (1991).
100. (a) Sekino, H., Oliphant, N., and Bartlett, R. J., *J. Chem. Phys.*, *101*, 7788 (1994). (b) Oliphant, N., and Bartlett, R. J., *J. Chem. Phys.*, *100*, 6550 (1994). (c) Sekino, H., and Bartlett, R. J., *J. Chem. Phys.*, *94*, 3665 (1991). (d) Stanton, J. F., Gauss, J., and Bartlett, R. J., *Chem. Phys. Lett.*, *195*, 194 (1992). (e) Sekino, H., and Bartlett, R. J., *Int. J. Quantum Chem.*, *43*, 119 (1992). (f) Stanton, J. F., and Bartlett, R. J., *J. Chem. Phys.*, *99*, 5178 (1993). (g) Meisner, L., and Bartlett, R. J., *Int. J. Quantum Chem.*, *S27*, 67 (1993).
101. Sekino, H., and Bartlett, R. J., *J. Chem. Phys.*, *98*, 3022 (1993).
102. (a) Gwaltney, S. R., and Bartlett, R. J., *Chem. Phys. Lett.*, *241*, 26 (1995). (b) Nooijen, M., and Bartlett, R. J., *J. Chem. Phys.*, *102*, 3629 (1995). (c) Watts, J. D., and Bartlett, R. J., *Chem. Phys. Lett.*, *233*, 81 (1995). (d) *J. Chem. Phys.*, *101*, 3073 (1994). (e) Comeau, D. C., and Bartlett, R. J., *Chem. Phys. Lett.*, *207*, 414 (1993). (f) Stanton,

J. F., and Bartlett, R. J., *J. Chem. Phys.*, *98*, 7029 (1993). (g) Oddershede, J., Gruner, N. E., Diercksen, G. H. F., *Chem. Phys.*, *97*, 303 (1985).
103. Aiga, F., Sasagane, K., and Itoh, R., *Int. J. Quantum Chem.*, *51*, 87 (1994).
104. Albert, I. D. L., Marks, T. J., and Ratner, M. A., (Unpublished results).
105. (a) Bartlett, R. J., in *Geometrical Derivatives of Energy Surfaces and Molecular Properties* (P. Jorgensen, and J. Simons, eds.), Reidel, Dordrecht, The Netherlands, 1986, p. 35. (b) Pulay, P., *Mol. Phys.*, *17*, 197 (1969); *18*, 473 (1970). (c) Pulay, P., and Meyer, W., *Mol. Phys.*, *27*, 473 (1974).
106. Handy, N. C., Amos, R. D., Gaw, J. F., Rice, J. E., Simandrias, E. D., Lee, T. J., Harrison, R. J., Leidig, W. D., Fitzgerald, G. B., and Bartlett, R. J., in *Geometrical Derivatives of Energy Surfaces and Molecular Properties* (P. Jorgensen, and J. Simons, eds.), Reidel, Dordrecht, The Netherlands, 1986), p. 179.
107. Amos, R. D., *Adv. Chem. Phys.*, *67*, 99 (1987).
108. (a) Watts, J. D., Gauss, J., and Bartlett, R. J., *Chem. Phys. Lett.*, *200*, 1 (1992). (b) Salter, E. A., Trucks, G. W., and Bartlett, R. J., *J. Chem. Phys.*, *90*, 1752 (1989).
109. Bartlett, R. J., Stanton, J. F., and Watts, J. D., *Advances in Molecular Vibrations and Collision Dynamics*, Vol. 1B (J. Bowman and M. A. Ratner, eds.), JAI Press, Greenwich, Connecticut, p. 139.
110. (a) Sasagane, K., Aiga, F., and Itoh, R., *J. Chem. Phys.*, *99*, 3838 (1993). (b) Sasagane, K., Aiga, F., and Itoh, R., *J. Chem. Phys.*, *99*, 3738, 3779 (1993). (c) Aiga, F., Sasagane, K., and Itoh, R., *Chem. Phys.*, *167*, 277 (1992). (d) Aiga, F., and Itoh, R., *Chem. Phys. Lett.*, *251*, 372 (1996).
111. Sekino, H., and Bartlett, R. J., *Chem. Phys. Lett.*, *234*, 87 (1995).
112. Sekino, H., and Bartlett, R. J., *J. Chem. Phys.*, *94*, 3665 (1991).
113. Sekino, H., and Bartlett, R. J., *J. Chem. Phys.*, *98*, 3022 (1993).
114. Sekino, H., and Bartlett, R. J., in ref. 73, p. 79.
115. Bishop, D. M., *J. Chem. Phys.*, *90*, 3192 (1989).
116. Sekino, H., and Bartlett, R. J., (to be published).
117. Kanis, D. R., Griffith, F., O'Neal, P., Ratner, D., Shoemaker, B., Marks, T. J., and Ratner, M. A., *J. Phys. Chem.* (To be submitted).
118. Albert, I. D. L., Di Bella, S., Kanis, D. R., Marks, T. J., and Ratner, M. A., in ref. 4.
119. Morley, J. O., and Pugh, D., *Spec. Publ. Roy. Soc. Chem.*, *69*, 28 (1989).
120. Barzoukas, M., Blanchard-Desce, M., Josse, D., Lehn, J.-M., and Zyss, J., *Chem. Phys.*, *133*, 323 (1989).
121. Huijts, R. A., and Hesselink, G. L. J., *Chem. Phys. Lett.*, *156*, 209 (1989).
122. Dehu, C., Meyers, F., and Brèdas, J.-L., *J. Amer. Chem. Soc.*, *115*, 6198 (1993).
123. Baeriswyl, D., Campbell, D. K., and Mazumdar, S., in *Conjugated Conducting Polymers* (H. G. Keiss, ed.), Springer Series in Solid State Sciences 102, Springer, Berlin, 1986.
124. (a) Longuet-Higgins, H. C., and Salem, L., *Proc. Roy. Soc. London*, *A251*, 172 (1959). (b) Pople, J. A., and Walmsey, S. H., *Mol. Phys.*, *5*, 15 (1962).
125. Fincher, C. R., Jr., Chen, C. E., Heeger, A. J., McDiarmid, A. G., and Hastings, J. B., *Phys. Rev. Lett.*, *48*, 100 (1982).
126. Peierls, R. E., *Quantum Theory of Solids*, Clarendon Press, Oxford, 1955.
127. Brèdas, J.-L., and Silbey, R. J., eds., *Conjugated Polymeric Materials in Electronics, Optoelectronics, and Molecular Electronics*, Kluwer, Dordrecht, 1991.
128. Skotheim, T. A., ed., *Handbook of Conducting Polymers*, Marcel Dekker, New York, 1986.
129. de Melo, C. P., and Silbey, R., *J. Chem. Phys.*, *88*, 2558, 2567 (1988) and references therein.
130. Albert, I. D. L., Das, P. K., and Ramasesha, S., *Chem. Phys. Lett.*, *168*, 454 (1990 Ramasesha, S., and Das, P. K., *Chem. Phys.*, *145*, 343 (1989).

131. (a) Williams, D. J., *Thin Solid Films*, *216*, 117 (1992). *Angew Chem. Int. Ed. Engl.*, *23*, 690 (1984). (b) Williams, D. J., *Nonlinear Optical Properties of Organic Molecules V.*, *SPIE Proc.*, 1775 (1992). (c) Eaton, D. F., *Science*, *253*, 281 (1991).
132. (a) Marder, S. R., Beratan, D. N., and Cheng, L.-T., *Science*, *252*, 103 (1991). (b) Marder, S. R., Sohn, J. E., Stucky, G. D., eds., *Materials for Nonlinear Optics: Chemical Perspectives*, ACS Symp. Ser. 455, 1991.
133. Khanarian, G., ed. *Nonlinear Optical Properties of Organic Molecules*, *SPIE Proc.*, 977 (1989); *SPIE Proc. II*, 1147 (1990); *SPIE Proc. III*, 1337 (1990).
134. Hann, R. A., and Bloor, D., eds., *Organic Materials for Nonlinear Optics*, Roy. Soc. Chem., Monograph 69, Burlington House, London, 1988.
135. (a) Cheng, L.-T., Tam, W., Stevenson, S. H., Meridith, G. R., Rikken, G., and Marder, S. R., *J. Phys. Chem.*, *95*, 10631 (1991). (b) Cheng, L.-T., Tam, W., Marder, S. R., Stiegman, A. E., Rikken, G., and Spangler, C. W., *J. Phys. Chem.*, *95*, 10643 (1991).
136. Ulman, A., *J. Phys. Chem.*, *95*, 2385 (1988).
137. Bunsel, E., and Rajagopal, S., *Acc. Chem. Res.*, *23*, 226 (1990). *J. Org. Chem.*, *54*, 798 (1989).
138. Bensen, H. G., and Murrell, J. N., *J. Chem. Soc. Faraday Trans 2*, *68*, 137 (1972).
139. Yitzchaik, S., and Marks, T. J., *Acc. Chem. Res.*, *29*, 197 (1996).
140. Kanis, D. R., Marks, T. J., and Ratner, M. A., *Mol. Cryst. Liq. Cryst. Sci. Technol. Sec. B: Nonlinear Optics* (Submitted for publication).
141. Kanis, D. R., Marks, T. J., and Ratner, M. A., (Unpublished results).
142. (a) Jen, A.K.-Y., Rao, V. P., and Chandrasekhar, J., in ref. 4. (b) Rao, V. P., Jen, A. K.-Y., Wong, K. Y., and Drost, K. J., *Tetrhedron Lett.*, *34*, 1747 (1993). (c) Jen, A.K.-Y., Rao, V. P., Wong, K. Y., and Drost, K. J., *J. Chem. Soc. Chem. Commun.*, 90 (1993). (d) Wong, K. Y., Jen, A.K.-Y., and Rao, V. P., *Phys. Rev. A.*, *49*, 3077 (1994). (e) Wong, K. Y., Jen, A.K.-Y., Rao, V. P., and Drost, K. J., *J. Chem. Phys.*, *100*, 6818 (1994).
143. Albert, I. D. L., Marks, T. J., and Ratner, M. A., *J. Amer. Chem. Soc.* (Submitted for publication).
144. Albert, A., *Heterocyclic Chemistry*, Oxford University Press, New York, 1968.
145. (a) Reinhardt, B. A., *SPIE Proc. 1853*, 50 (1993). (b) Karna, S. P., Zhang, Y., Samoc, M., Prasad, P. N., Reinhardt, B. A., and Dillard, A. G., *J. Chem. Phys.*, *99*, 9984 (1993). (c) Reinhardt, B. A., Kannan, R., and Bhatt, J. C., *SPIE Proc. 2229*, 24 (1994). (d) Reinhardt, B. A., Kannan, R., Bhatt, J. C., Zieba, J., and Prasad, P. N., in ref. 4.
146. Albert, I. D. L., Marks, T. J., and Ratner, M. A., (Manuscript under preparation).
147. Albert, I. D. L., Pugh, D., and Morley, J. O., (Unpublished results).
148. Morley, J. O., *J. Phys. Chem.*, *99*, 10166 (1995).
149. Kotzian, M., Schroder, H., and Zerner, M. C., *J. Amer. Chem. Soc.*, *111*, 7687 (1989).
150. (a) Marder, S. R., Perry, J. W., Schaefer, W. P., Tiemann, B. G., Groves, P. C., and Perry, K., *Proc. SPIE-Int. Soc. Opt. Eng.*, *1147*, 108 (1990). (b) Green, M. L. H., Marder, S. R., Thompson, M. e., Bandy, J. A., Bloor, D., Kolinsky, P. V., and Jones, R. J., *Nature*, *330*, 360 (1987).
151. (a) Zyss, J., Dhenaut, C., Chauvan, T., and Ledoux, I., *Chem. Phys. Lett.*, *206*, 409 (1993). Laidlaw, W. M., Denning, R. G., Verbiest, T., Chauchard, E., and Persoons, A., *Nature*, *363*, 58 (1993). (b) Thami, T., Bassoul, P., Petit, M. A., Simon, J., Fort, A., Barzoukas, M., and Villaeys, A., *J. Amer. Chem. Soc.*, *114*, 915 (1992).
152. (a) Matsuo, T., Nakamura, H., Nakao, T., and Kawazu, M., *Chem. Lett.*, 2363 (1992). (b) Ledoux, I., *Synth. Met.*, *54*, 123 (1993). (c) Doisneau, G., Balavoine, G., Fillebeen-Khan, T., Client, J. C., Delaire, J., Ledoux, I., Loucif, R., and Pucceti, G., *J. Organomet. Chem.*, *421*, 299 (1991).
153. (a) Chiang, W., Thompson, M. E., Van Engen, D., *Spec. Publ. Roy. Soc. Chem.*, *91* (*Org. Mater. Nonlinear Opt.* 2), 210 (1991). (b) Lamberth, C., Murphy, D. M., and

Mingos, D. M. R., *Spec. Publ. Roy. Soc. Chem.*, *91* (*Org. Mater. Nonlinear Opt. 2*), 183 (1991). (c) Copley, R. C. B., Lamberth, C., Machell, J., Mingos, D. M. P., Murphy, D. M., and Powell, H., *J. Mater. Chem.*, 583 (1991).

154. Lesley, G., Yuan, Z., Stringer, G., Jobe, I. R., Taylor, N. J., Koch, L., Scott, K., Marder, T. B., Williams, I. D., and Kurtz, S. K., *Spec. Publ. Roy. Soc. Chem.*. *91* (*Org. Mater. Nonlinear Opt. 2*), 197 (1991).
155. Sakaguchi, H., Nagamura, T., and Matsuo, T., *Appl. Organomet. Chem.*, *5*, 257 (1991).
156. Chollet, P. A., Kajzar, F., Moigne, J. L., *Proc. SPIE-Int. Soc. Opt. Engg.*, *1273*, 87 (1990).
157. Zhang, N., Jiang, M. H., Yuan, D. R., Xu, D., Tao, X. T., and Shao, Z. S., *J. Cryst. Growth*, *102*, 581 (1990).
158. Coe, B. J., Jones, C. J., McCleverty, J. A., Bloor, D., Kolinsky, P. V., and Jones, R. J., *Chem. Soc., Chem. Commun.*, 1485 (1989).
159. (a) Tam, W., Eaton, D. F., Calabrese, J. C., Williams, I. D., Wang, Y., and Anderson, A. G., *Chem. Mater.*, *1*, 128 (1989). (b) Cheng, L.-T., Tam, W., and Eaton, D. F., *Organometallics*, *9*, 2856 (1990). (c) Tam, W., and Calabrese, J. C., *Chem. Phys. Lett.*, *144*, 79 (1988). (d) Eaton, D. F., Anderson, A. G., Tam, W., and Wang, Y., *J. Amer. Chem. Soc.*, *109*, 1886 (1987). (e) Calabrese, J. C., and Tam, W., *Chem. Phys. Lett.*, *133*, 244 (1987). (f) Anderson, A. G., Calabrese, J. C., Tam, W., and Williams, I. D., *Chem. Phys. Lett.*, *134*, 392 (1987). (g) Chang, L.-T., Tam, W., Meridith, G. R., and Marder, S. R., *Mol. Cryst. Liq. Cryst.*, *189*, 137 (1990).
160. (a) Bunting, H. E., Green, M. L. H., Marder, S. R., Thompson, M. E., Bloor, D., Kolinsky, P. V., and Jones, R. J., *Polyhedron*, *11*, 1489 (1992). (b) Bandy, J. A., Bunting, H. E., Garcia, M. H., Green, M. L. H., Marder, S. R., Thompson, M. E., Bloor, D., and Jones, R. J., *Polyhedron*, *11*, 1429 (1992). (c) Calabrese, J. C., Cheng, L.-T., Green, j. C., Marder, S. R., and Tam, W., *J. Amer. Chem. Soc.*, *113*, 7227 (1991). (d) Marder, S. R., *Inorg. Mater.*, 115 (1992).
161. Houlton, A., Jasim, N., Roberts, R. M. G., Silver, J., Cunningham, D., McArdle, P., and Higgins, T., *J. Chem. Soc. Dalton Trans.*, 2235 (1992).
162. Loucif-Saiebi, R., Delaire, J. A., Bonazzola, L., Doisneau, G., Balavoine, G., Khan-Fillebeen, T., Ledoux, I., Puccetti, G., Kimura, M., Abel-Halim, H., Robinson, D. W., and Cowan, D. O., *J. Organomet. Chem.*, *403*, 365 (1991).
163. Marder, S. R., Perry, J. W., Schaefer, W. P., and Tiemann, B. G., *Organometallics*, *10*, 1896 (1991).
164. Richardson, T., Roberts, G. G., Polywka, M. E. C., and Davies, S. G., *Thin Solid Films*, *179*, 405 (1989).
165. Bandy, J. A., Bunting, H. E., Garcia, M. H., Green, M. L. H., Marder, S. R., Thompson, M. E., Bloor, D., Kolinsky, P. V., and Jones, R. J., *Spec. Publ. Roy. Soc. Chem.*, *91* (*Org. Mater. Nonlinear Opt. 2*), 225 (1991).
166. Nalwa, H. S., *Appl. Organomet. Chem.*, *5*, 349 (1991).
167. (a) Whittal, I. R., and Humphrey, M. G., *Organometallics*, *14*, 5493 (1995). (b) Whittal, I. R., Humphrey, M. G., and Hockless, D. C. R., *Organometallics*, *14*, 3970 (1995). (c) Whittal, I. R., Humphrey, M. G., Persoons, A., and Houbrechts, S., *Organometallics*, *15*, 1935 (1996).
168. Kanis, D. R., Ratner, M. A., and Marks, T. J., *J. Amer. Chem. Soc.*, *114*, 10338 (1992).
169. Kanis, D. R., Ratner, M. A., and Marks, T. J., *J. Amer. Chem. Soc.*, *112*, 8203 (1990).
170. Kanis, D. R., Lacroix, P. A., Ratner, M. A., and Marks, T. J., *J. Amer. Chem. Soc.*, *116*, 10089 (1994).
171. (a) Di Bella, S., Fragala, I., Marks, T. J., and Ratner, M. A., *J. Amer. Chem. Soc.*, *118*, 12747 (1996). (b) Di Bella, S., Fragala, I., and Ledoux, I., *J. Amer. Chem. Soc.*, *117*, 9481 (1995). (c) Di Bella, S., Fragala, I., Ledoux, I., Diaz-Garcia, M. A., Lacroix, P. G., and Marks, T. J., *Chem. Mater.*, *6*, 881 (1994). (d) Lacroix, P. G., Di Bella, S.,

and Ledoux, I., *Chem. Mater.*, *8*, 541 (1996). Di Bella, S., Fragala, I., Ledoux, I., and Marks, T. J., *J. Amer. Chem. Soc.*, *117*, 9481 (1995).

172. (a) Zulani, P., Delzoppo, M., Castiglioni, C., Zerbi, G., Marder, S. R., and Perry, J. W., *J. Chem. Phys.*, *103*, 9935 (1995). (b) Dehu, C., Meyers, F., Hendrickx, E., Clays, K., Persoons, A., Marder, S. R., and Brèdas, L., *J. Amer. Chem. Soc.*, *117*, 10127 (1995).
173. Rice, J. E., and Willets, A., *J. Chem. Phys.*, *99*, 426 (1993).
174. Gorman, C. B., and Marder, S. R., *Chem. Mater.*, *99*, 426 (1995).
175. (a) Di Bella, S., Ratner, M. A., and Marks, T. J., *J. Amer. Chem. Soc.*, *114*, 5842 (1992). (b) Hamada, T., *J. Phys. Chem.*, *100*, 8777 (1996). (c) Ye, P., and Shen, Y. R., *Phys. Rev. B.*, *28*, 4288 (1983). Di Bella, S., Fragala, I. L., Ratner, M. A., and Marks, T. J., *Adv. Chem. Serv.*, *240*, 223 (1994).
176. (a) Munn, R. W., *J. Mol. Electron.*, *4*, 31 (1988) and references therein. (b) Hurst, M., Munn, R. W., and Morley, J. O., *J. Mol. Electron.*, *6*, 15 (1990). (c) Munn, R. W., *Mol. Phys.*, *64*, 1 (1988). (d) Hurst, M., Munn, R. W., and Aviram, A., eds., *Molecular Electronics*, Engineering Society, New York, 1989. (e) Hurst, M., and Munn, R. W., in *Organic Materials for Nonlinear Optics* (R. A. Hann, and D. Bloor, eds.), Roy. Soc. Chem., London, 1988, p. 3.
177. (a) Yam, R., Cohen, R., and Berkovic, G., *Nonlinear Optics*, *11*, 311 (1995). (b) Cohen, R., Berkovic, G., Yitzchaik, S., and Krongauz, V., *Mol. Cryst. Liq. Cryst.*, *240*, 169 (1994). (c) Yitzchaik, S., Krongauz, V., and Berkovic, G., *Nonlinear Optics*, *4*, 265 (1993). *J. Appl. Phys.*, *70*, 3949 (1991). (d) Lundquist, P., Yitzchaik, S., Marks, T. J., Di Bella, S., Cohen, R., and Berkovic, G., *Phys. Rev. Lett.* (submitted).
178. Di Bella, S., Fragala, I., Marks, T. J., and Ratner, M. A., *J. Amer. Chem. Soc.* (In press).
179. Karelson, M. M., and Zerner, M. C., *J. Phys. Chem.*, *96*, 6949 (1992). *J. Amer. Chem. Soc.*, *112*, 9405 (1990).
180. Amos, A. T., and Burrows, B. L., *Adv. Quantum Chem.*, *7*, 289 (1973); McRae, E. G., *J. Phys. Chem.*, *61*, 562 (1957).
181. (a) Tapia, O., Goscinskki, O., *Mol. Phys.*, *29*, 1653 (1975). (b) Tapia, O., in *Molecular Interactions*, vol. 3, Chap 2 (W. J. Orville-Thomas, ed.), Wiley, New York, 1982.
182. (a) Mikkelsen, K. V., Agren, H., Jensen, H. J. Aa., and Helgaker, T., *J. Chem. Phys.*, *89*, 308 (1988). (b) Mikkelsen, K. V., Dalgaard, E., and Swanstrom, P., *J. Phys. Chem.*, *91*, 3081 (1987). (c) Mikkelsen, K. V., and Ratner, M. A., *Int. J. Quantum Chem.*, *S21*, 341 (1987).
183. Cammi, R., Cossi, M., and Tomasi, J., *J. Chem. Phys.*, *104*, 4611 (1996).
184. Kirkwood, J. G., *J. Chem. Phys.*, *2*, 351 (1934). Kirkwood, J. G., and Westheimer, F. H., *J. Chem. Phys.*, *6*, 506 (1938).
185. Onsager, L., *J. Amer. Chem. Soc.*, *58*, 1486 (1936).
186. Tomasi, J., and Persico, M., *Chem. Rev.*, *94*, 2027 (1994).
187. Di Bella, S., Marks, T. J., and Ratner, M. A., *J. Amer. Chem. Soc.*, *116*, 4440 (1994).
188. Barbara, P. F., Meyer, T. J., and Ratner, M. A., *J. Phys. Chem.*, *100*, 13148 (1996) and references therein.
189. Kanis, D. R., Marks, T. J., and Ratner, M. A., *Int. J. Quantum Chem.*, *43*, 61 (1992).
190. Meyers, F., Brèdas, J.-L., and Zyss, J., *J. Amer. Chem. Soc.*, *114*, 2914 (1992).
191. (a) Hasegawa, T., Iwasa, Y., Koda, T., Kishida, H., Tokura, Y., Wada, S., Tashiro, H., Tachibana, H., Matsumoto, M., and Miller, R. D., *Synth. Met.*, *71*, 1679 (1995). (b) Tachibana, H., and Tokura, Y., *Synth. Met.*, *71*, 2005 (1995). (c) Kepler, R. G., and Soos, Z. G., *ACS Symp. Series*, *579*, 387 (1994) and references therein.
192. Andre, J. M., Delhalle, J., and Brèdas, J.-L., *Quantum Chemistry Aided Design of Organic Polymers: An Introduction to Quantum Chemistry of Polymers and its Applications*, World Scientific, Singapore, 1991.

193. (a) Rehwald, W., and Keiss, H. D., in *Conjugated Conducting Polymers* (H. G. Keiss, ed.), Springer Series in Solid State Sciences, 102, Springer, Berlin, 1992. (b) Chance, R. R., Boudreaux, D. S., Brèdas, J.-L., and Silbey, R., in ref. 123. (c) Brèdas, J.-L., and Street, G. B., *Acc. Chem. Res.*, *18*, 309 (1985).
194. (a) Torner, L., Menyuk, C. R., and Stegeman, G. I., *J. Opt. Soc. Amer. B.*, *12*, 889 (1995). (b) Cha, M., Torruellas, W. E., Yuan, S. H., Stegeman, G. I., and Lecrec, M., *J. Opt. Soc. Amer. B.*, *12*, 882 (1995). (c) Torner, L., Menyuk, C. R., Torruellas, W. E., and Stegeman, G. I., *Opt. Lett.*, *20*, 13 (1995). (d) Kivshar, Y. S., Chubykalo, O. A., Osatenko, O. V., and Grinyoff, D. V., *Int. J. Mod. Phys.*, *9*, 2963 (1995).
195. (a) Perry, W. A., Marder, S. R., Meyers, F., Lu, D., Chen, G., Goddard, W. A., Brèdas, J.-L., and Pierce, B. M., in ref. 4. Marder, S. R., *Adv. Chem. Ser.*, *245*, 189 (1995). (b) Meyers, F., Marder, S. R., Pierce, B. M., and Brèdas, J.-L., *Chem. Phys. Lett.*, *228*, 171 (1994). (c) Meyers, F., Marder, S. R., Pierce, B. M., and Brèdas, J.-L., *J. Amer. Chem. Soc.*, *116*, 10703 (1994). (d) Ortiz, R., Marder, S. R., Cheng, L.-T., Tiemann, B. G., Cavegnero, S., and Ziller, J. W., *J. Chem. Soc. Chem. Commun.*, 2263 (1994). (e) Gorman, C. B., Vandoremaele, G. H. J., and Marder, S. R., *Adv. Chem. Ser.*, *240*, 179 (1994). (f) Marder, S. R., Gorman, C. B., Meyers, F., Perry, J. W., Bourhill, G., Brèdas, J.-L., and Pierce, B. M., *Science*, *265*, 632 (1994). (g) Marder, S. R., Cheng, L.-T., Tiemann, B. G., Friedli, A. C., Blanchard-Desce, M., Perry, J. W., and Skindhoj, J., *Science*, *263*, 511 (1994). (h) Bourhill, G., Brèdas, J.-L., Cheng, L.-T., Marder, S. R., Meyers, F., Perry, J. W., and Tiemann, B. G., *J. Amer. Chem. Soc.*, *116*, 2619 (1994). (i) Gorman, C. B., and Marder, S. R., *Proc. Natl. Acad. Sci. USA*, *90*, 11297 (1993). (j) Tiemann, B. G., Cheng, L.-T., and Marder, S. R., *J. Chem. Soc. Chem. Commun.*, 735 (1993). (k) Marder, S. R., Perry, J. W., Bourhill, G., Gorman, C. B., Tiemann, B. G., and Mansour, K., *Science*, *261*, 186 (1993). (l) Marder, S. R., Perry, J. W., Tiemann, B. G., Gorman, C. B., Galmour, S., Biddle, S. L., and Bourhill, G., *J. Amer. Chem. Soc.*, *115*, 2524 (1993).
196. Albert, I. D. L., Marks, T. J., and Ratner, M. A., *J. Phys. Chem.*, *100*, 9714 (1996).
197. (a) Lu, D., Chen, G., Goddard, W. A., III, and Perry, W. A., *J. Amer. Chem. Soc.*, *116*, 10679 (1994). (b) Lu, D., Chen, G., and Goddard, W. A., III, *J. Chem. Phys.*, *101*, 5860 (1994).
198. Pierce, B. M., *Proc. SPIE-Int. Soc. Opt. Eng.*, *1560*, 148 (1991).
199. (a) Ramasesha, S., and Albert, I. D. L., *Chem. Phys.*, *142*, 395 (1990). (b) Albert, I. D. L., and Ramasesha, S., *J. Phys. Chem.*, *94*, 6540 (1990).
200. (a) Salem, L., *Acc. Chem. Res.*, *12*, 87 (1979). (b) Bonaic-Koutecky, V., Bruckmann, P., Hiberty, P. C., Koutecky, J., Laforestier, C., and Salem, L., *Agnew. Chem. Int. Ed. Engl.*, *14*, 575 (1978).
201. (a) Brunei, M. C., Duadey, J. P., Langlet, J., Malrieu, J. P., and Momcchioli, F., *J. Amer. Chem. Soc.*, *99*, 3587 (1977). (b) Malrieu, J. P., *Theor. Chim. Acta*, *54*, 59 (1979). (c) Nebot-Gil, I., and Malrieu, J. P., *J. Chem. Phys.*, *77*, 2475 (1982); *J. Amer. Chem. Soc.*, *104*, 3320 (1982). (d) Said, M., Mayanu, D., Malrieu, J. P., and Garcia-Bach, M. A., *J. Amer. Chem. Soc.*, *106*, 571 (1984).
202. Wulfmann, C. E., and Kumei, S. E., *Science*, *172*, 1061 (1978).
203. Brooks, B. R., and Schaefer, H. F., *J. Amer. Chem. Soc.*, *101*, 307 (1979).
204. Albert, I. D. L., Marks, T. J., and Ratner, M. A., *J. Amer. Chem. Soc.* (In press).
205. Ziegler, T., *Chem. Rev.*, *91*, 651 (1991). Laird, B. B., Ross, R. B., and Ziegler, T., *Chemical Applications of Density Functional Theory*, ACS Symposium Series, 629, ACS, Washington DC, 1996.
206. Ellis, D. E., *Density Functional Theory of Molecules, Clusters, and Solids*, Kluwer, 1995. Drezler, R. M., and Gross, E. K. U., *Density Functional Theory*, Springer, Berlin, 1990.
207. Jones, J. O., *Adv. Chem. Phys.*, *67*, 413 (1987).

208. Parr, R. G., and Yang, W., *Density Functional Theory of Atoms and Molecules*, Oxford University Press, New York, 1989.
209. Hohenberg, P., and Kohn, W., *Phys. Rev. A.*, *136*, 864 (1964). Kohn, W., and Sham, L. J., *Phys. Rev. A.*, *140*, 1133 (1964).
210. (a) Casida, M. E., Jamorki, C., Bohr, F., Guan, J., and Salahub, D. R., (in ref. 73). (b) Sim, F., Salahub, D. R., and Chin, S., *Int. J. Quantum Chem.*, *43*, 463 (1992).
211. Mahan, G. D., and Subbaswamy, K. R., *Local Density Theory of Polarizability*, Plenum Press, New York, 1996.
212. Guan, J., Duffy, P., Carter, J. T., Chong, D. P., Casida, K. C., Casida, M. E., and Wrinn, W., *J. Chem. Phys.*, *98*, 4753 (1993).
213. Colwell, S. M., Murray, C. W., Handy, N. C., and Amos, R. D., *Chem. Phys. Lett.*, *210*, 261 (1993). Colwell, S. M., Handy, N. C., Lee, A. M., *Phys. Rev. A.*, *53*, 1316 (1996). McDowell, S. A. C., Amos, R. D., and Handy, N. C., *Chem. Phys. Lett.*, *235*, 1 (1995).
214. (a) Matsuzawa, N., and Dixon, D. A., *J. Phys. Chem.*, *99*, 7698 (1995). (b) Dixon, D. A., Chase, B. E., Fitzgerald, G., and Matsuzawa, N., *J. Phys. Chem.*, *99*, 4486 (1995). (c) Matsuzawa, N., and Dixon, D. A., *Synth. Met.*, *71*, 1667 (1995).
215. Kohn, W., Becke, A. D., and Parr, R. G., *J. Phys. Chem.*, *100*, 12974 (1996).

3
Relationship Between the Molecular and Bulk Response

Mark G. Kuzyk
Washington State University
Pullman, Washington

I. INTRODUCTION

The purpose of this handbook is to tabulate the linear and nonlinear optical properties of the building blocks that make up a bulk material. Chapter 2 discusses how the vacuum molecular susceptibilities are calculated. This chapter, which shows how vacuum molecular susceptibilities are related to the bulk response of a material made of such units, serves as a bridge between Chapter 2 and the rest of the book.

It is rare for an experiment to directly measure the optical properties of a molecule without such properties being profoundly affected by the molecule's environment. Because most of the numbers reported in this volume are derived from bulk measurements, it is useful to understand how the local environment of a molecule effects optical measurements and how one can predict the optical properties of a molecule in a variety of host materials given that its response has been measured in only one bulk system.

Section II discusses how the environment of a molecule affects the local field experienced by that molecule. Both dipolar and non-dipolar cases are treated and quantities such as dressed energies, radiation fields, and dielectric properties are calculated. Section III discusses how local fields affect the nonlinear susceptibilities, while Section IV addresses how local field models depend on the concentration and how aggregating systems can be treated. The remainder of Section IV shows how the linear and nonlinear response of various material systems can be understood within the context of the dilute solute model. Section V introduces the Maxwell Garnett model and shows that it reduces to the dilute solute model in the low concentration limit. Section VI applies the dilute solute model to calculate the linear and nonlinear susceptibilities of a variety of mechanisms for both isotropic and orientationally ordered collections of molecules. Polymers, crystals, liquid crystals, gases and composites are treated. Section VII discusses novel local field effects that can have a profound influence on the observed bulk properties of a material. Specifically, aggregates of molecules, metals, and semiconductors and their affect on the linear and nonlinear optical properties of a material are treated.

II. LOCAL ELECTRIC FIELDS

As described in Chapter 1, the induced dipole moment of a molecule is related to the applied electric field through a power series in the field with the susceptibilities as expansion coefficients. The nth-order contribution to the induced dipole moment, $p^{(n)}$, is of the form:

$$p_i^{(n)}(\omega_\sigma) = \xi_{ijk\ldots l}^{(n)}(-\omega_\sigma;\omega_1,\omega_2,\ldots,\omega_n)F_j^{\omega_1}F_k^{\omega_2}\ldots F_l^{\omega_n}, \tag{1}$$

where $\vec{F}^\omega$ is the electric field at frequency ω at the molecule site, $\xi^{(n)}$ is the molecular nth-order nonlinear optical susceptibility tensor, ω_σ is the frequency of the dipole oscillation as excited by the n electric fields at frequencies (i.e., $\omega_\sigma = \omega_1 + \omega_2 + \ldots + \omega_n$) and where we use summation convention (double indices are summed over the three Cartesian coordinates). Note that we have omitted the degeneracy factor "K" that is discussed in Chapter 1 (depending on convention, the K-factor is sometimes not used). This degeneracy factor may precede each susceptibility but is dropped in this chapter to reduce clutter in the derivations that follow.

In a vacuum, $\vec{F}^\omega$ is the incident or emitted electric field. If the molecule is embedded in a material—either in a host material or a collection of similar molecules—the electric field at a particular molecular site is a superposition of the applied electric field and the electric field due to the nearby material. The electric field due to the nearby material that acts on a molecule can be grouped into four types:

1. The applied electric field can polarize the surrounding material, which then acts on the molecule.
2. The surrounding material can be the source of an electric field even in the absence of an applied field if there are local fluctuations in charge of an otherwise electrically neutral material.
3. If the molecule has a groundstate moment of any order (such as a dipole moment, quadruple moment, etc.), the electric field associated with this moment can polarize nearby material, which results in a reaction field, that is, a field that acts upon the molecule.
4. An applied electric field can induce a moment in a molecule which polarizes the material surrounding the molecule which results in a reaction field that changes the dipole moment of the molecule.

It is clear that calculations of the local electric field at a molecular site within the bulk material can be quite complicated and must involve the use of self-consistent methods (described later). In the following section, we summarize Lorentz's local field model and Onsager's local field model. Other important concepts, such as screening of embedded dipoles, reaction fields, the effect of an external electric field on a screened dipole, and radiating dipoles embedded in a dielectric, are also discussed.

A. The Continuum Approximation

Consider the local fields in a dipolar liquid or solid solution. Figure 1a represents a liquid in which the molecules sweep out a spherical volume. If the system is observed over a time scale that is long compared with the time it takes a molecule to collide many times with other molecules, the system will appear to be a continuous smear of molecules as represented in Fig. 1b. If a uniform and static electric field is applied to the isotropic and homogeneous liquid, the time-averaged polarization is given by:

$$\vec{P} = \chi^{(1)}\vec{E} = \frac{\epsilon - 1}{4\pi}\vec{E}, \tag{2}$$

where ϵ is the dielectric function of the liquid. Because it is understood that most measurements are done over time scales that are long compared with collision times in the liquid, we will refer to the time-averaged polarization as just the polarization. Under this assumption, the liquid is a continuous dielectric in which the substituent molecules can be considered spherical.

In the following two sections, we calculate the local electric field of a nondipolar liquid or solid. In a later section, we treat the more complicated case of a dipolar system.

B. Non-Dipolar Homogeneous Liquids and Solids

We are now ready to compute the local electric field experienced by a spherical molecule in a liquid (note that other shapes can be used to model the response). We will do so by removing a spherical part of the dielectric that represents the molecule, assuming that the charges do not redistribute as a result of the molecule being removed, and calculating the electric field that would be required to produce the observed polarization in the molecule. This electric field is the local electric field experienced by the molecule. Note that the following derivations draw on the well-known electrostatic boundary value problem of a dielectric sphere that is embedded in another dielectric and immersed in a uniform electric field. Such calculations are presented in texts on electrostatics.[1]

The dashed circle in Fig. 1b represents the spherical molecule (or chunk of dielectric) that is removed. Figure 1c shows a spherical "averaged" molecule (or dielectric chunk) after being removed from the liquid under the assumption that the polarization is "frozen in". Note that the polarization is uniform. Because the discontinuity of the polarization across a surface is related to the charge at that surface, it is clear that both the surface of the dielectric sphere and the remaining cavity are charged. Furthermore, because the surface normal unit vector in the cavity is in the opposite direction to the surface normal on the dielectric sphere for surface elements that were adjacent to each other before the sphere is removed (recall that the surface normal is defined to point away from the dielectric region under consideration), and the polarization is in the same direction on both sides of the interface, the surface charge on the sphere is of opposite sign to the surface charge on the cavity surface. This result is also a consequence of charge conservation: all positive and negative charge is balanced in an electrically neutral material.

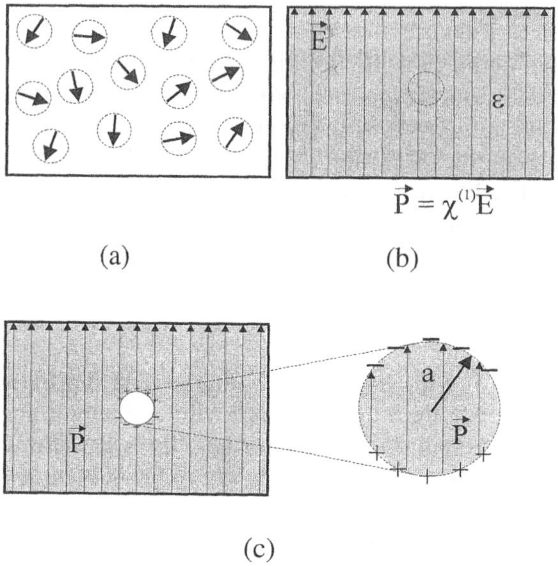

Figure 1 a) The collection of arrows represents a dipolar liquid in which the molecules sweep out a spherical volume, b) the shaded region represents the dielectric in an electric field (arrows) which is modeled as a continuum, and c) a spherical piece of the dielectric is removed without allowing the charges to move.

The local electric field experienced by the spherical dielectric is just that field which would induce a uniform polarization of magnitude P. It is well-known, however, that a dielectric sphere in a uniform electric field results in a uniform polarization. The induced dipole moment of such a dielectric sphere in vacuum, $\vec{p}$, in a uniform external field $\vec{F}$ is[1,2]:

$$\vec{p} = \alpha\vec{F} = \left(\frac{\epsilon - 1}{\epsilon + 2}\right) a^3\vec{F}, \tag{3}$$

where α is the polarizability of the sphere and a its radius. The polarization, $\vec{P}$, is the dipole moment per unit volume

$$\vec{P} = \frac{\vec{p}}{\frac{4}{3}\pi a^3} = \frac{3}{4\pi}\left(\frac{\epsilon - 1}{\epsilon + 2}\right)\vec{F}. \tag{4}$$

Setting Eq. 4 equal to Eq. 2 and solving for $\vec{F}$, we get:

$$\vec{F} = \left(\frac{\epsilon + 2}{3}\right)\vec{E}. \tag{5}$$

This is known as the Lorentz-Lorenz local field model. The local field factor is defined as:

$$L = \left(\frac{\epsilon + 2}{3}\right), \tag{6}$$

where $\vec{F} = L\vec{E}$.

C. Nondipolar Two Component System

Composite systems are made from two or more dissimilar components. It is therefore also useful to consider the local electric field at the site of an individual molecule (or chunk of dielectric) in a mixture. The simplest case is a low concentration solution in which the local electric field is calculated at the site of a solute molecule that is embedded in a sea of dielectric. Many important nonlinear optical materials, such as liquid solutions, dye-doped polymers, and colloids can be modeled as a low concentration solution.

Figure 2a shows the electric field lines of a dielectric sphere embedded in another dielectric under the influence of a uniform electric field. ϵ_1 and ϵ_2 are the dielectric constants of the surrounding medium (solvent or host) and sphere (solute or guest), respectively. The polarization in the sphere is different than the polarization of the surrounding dielectric. Relating the electric field inside the sphere to the external field allows us to calculate the polarization of the dielectric sphere in terms of the external field. Figure 2b shows the dielectric with the sphere removed. $\vec{P}_1$ and $\vec{P}_2$ are the polarizations of the surrounding medium far from the sphere and in the sphere, respectively.

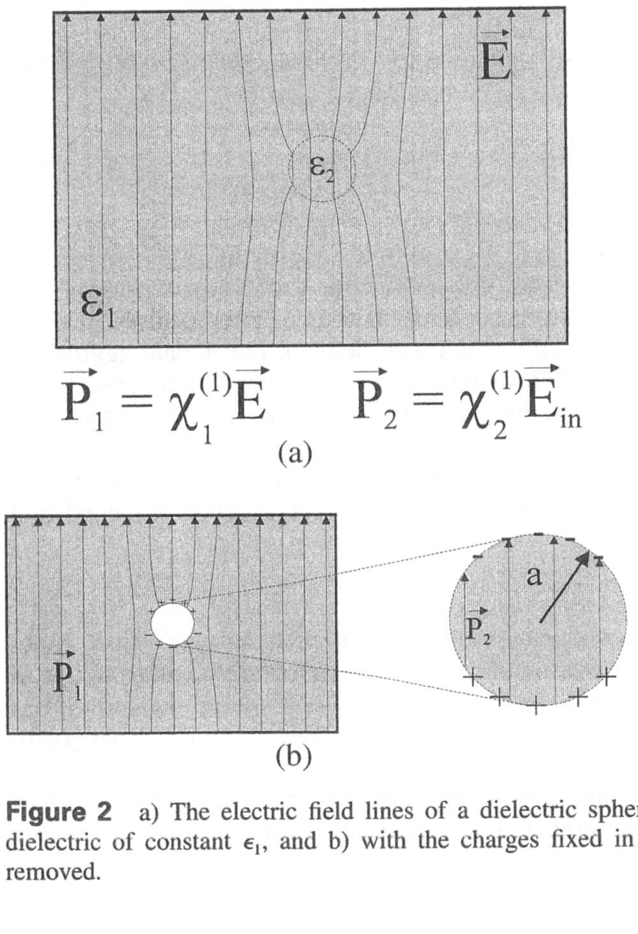

Figure 2 a) The electric field lines of a dielectric sphere of dielectric constant ϵ_2 in a dielectric of constant ϵ_1, and b) with the charges fixed in place and the dielectric sphere removed.

In analogy to Eq 2, the polarization inside the sphere is:

$$\vec{P}_2 = \left(\frac{\epsilon_2 - 1}{4\pi}\right) \vec{E}_{in}, \tag{7}$$

where $\vec{E}_{in}$ is the electric field inside the sphere. The field in the sphere is related to the uniform electric field far away from and outside the sphere, $\vec{E}$ [1]:

$$\vec{E}_{in} = \left(\frac{3\epsilon_1}{2\epsilon_1 + \epsilon_2}\right) \vec{E}. \tag{8}$$

Using Eqs. 7 and 8, the polarization inside the sphere can be expressed in terms of the applied electric field:

$$\vec{P}_2 = \left(\frac{\epsilon_2 - 1}{4\pi}\right) \left(\frac{3\epsilon_1}{2\epsilon_1 + \epsilon_2}\right) \vec{E}. \tag{9}$$

As in the previous calculation, the local electric field is the field that is required to induce a polarization P_2 in the sphere. Using the same argument that led to Eq. 4 with ϵ in Eq. 4 replaced with ϵ_2, the relationship between the local field F and the polarization of the sphere is:

$$\vec{P}_2 = \frac{3}{4\pi} \left(\frac{\epsilon_2 - 1}{\epsilon_2 + 2}\right) \vec{F}. \tag{10}$$

Setting Eqs. 9 and 10 equal, we can solve for the local electric field in terms of the applied electric field:

$$\vec{F} = \left(\frac{3\epsilon_1}{2\epsilon_1 + \epsilon_2}\right) \left(\frac{\epsilon_2 + 2}{3}\right) \vec{E}. \tag{11}$$

This expression looks similar to the Onsager local field formula. Note that when $\epsilon_1 = \epsilon_2$, we recover the Lorentz local field. While in principle, this expression should be used for any two component system such as a dye-doped polymer, the Lorentz local field model applied to the polymer host is usually used because the dielectric constant of a single molecule is not well defined. For dopant aggregates, however, Eq. 11 is the appropriate local field model. Section VII of this chapter shows how this local field model is applied to metal particles in a dielectric.

It is clear that models that account for nonspherical cavities and the tensor nature of the dielectric function of both the molecule and the surrounding material are far more complex. In such cases, the local field factor becomes a second rank tensor such that:

$$F_i = L_{ij} E_j, \tag{12}$$

where summation notation over repeated indices is implied. In many cases, however, the ensemble average of the principle axes associated with the dielectric properties of the molecule coincide with the principle axes of the bulk dielectric properties of the system so that the local field tensor is diagonal, a condition that greatly simplifies the problem.

Our prior derivation has considered only static electric fields. The local field tensors for an optical field are similar. If the guest molecule or solute particle is small compared to the wavelength of the illuminating source, the optical field is approximately spatially uniform in the vicinity of the sphere so that the same formalism applies. Because the dielectric function depends on the frequency of the applied field, the optical local field factors for an isotropic medium are generalized by evaluating them at the frequency of interest. For the two component system, the scalar local field factor at frequency ω, $L(\omega)$, is:

$$L(\omega) = \left(\frac{3\epsilon_1(\omega)}{2\epsilon_1(\omega) + \epsilon_2(\omega)}\right)\left(\frac{\epsilon_2(\omega) + 2}{3}\right). \tag{13}$$

D. The Clausius-Mossotti Equation

The Clausius-Mossotti Equation is a handy relationship between a material's dielectric constant, the polarizability of the substituent molecules, and their density. Such a relationship holds if the molecular polarizability is independent of molecule concentration.

The frequency of light oscillation in an optical electric field is so high that molecular dipoles cannot reorient in response to the light. The polarization of the material at optical frequencies is thus limited to electronic polarization, that is, the deformation of the electron cloud. If α_k is the electronic polarizability of molecule k (or dielectric chunk k), the polarization of the bulk material is:

$$\vec{P}(\omega) = \sum_k N_k \alpha_k(\omega) \vec{F}_k(\omega), \tag{14}$$

where N_k is the number density of species k and $\vec{F}_k$ is the local electric field at species k.

For a single species nondipolar material, the local electric field will be of the Lorentz form. Substituting Eq. 5 into Eq. 14, we get:

$$\vec{P}(\omega) = N\alpha(\omega)\left(\frac{n^2(\omega) + 2}{3}\right)\vec{E}(\omega), \tag{15}$$

where $n(\omega)$ is the refractive index which is related to the electronic part of ϵ, that is, $n^2(\omega) = \epsilon_{el}(\omega)$. The relationship between polarization and dielectric function (setting Eq. 2 equal to Eq. 15) yields:

$$\vec{P}(\omega) = \frac{n^2(\omega) - 1}{4\pi}\vec{E}(\omega) = N\alpha(\omega)\left(\frac{n^2(\omega) + 2}{3}\right)\vec{E}(\omega). \tag{16}$$

Rearranging the above yields the Clausius-Mossotti Equation:

$$\frac{n^2(\omega) - 1}{n^2(\omega) + 2} = \frac{4\pi}{3} N\alpha(\omega). \tag{17}$$

This Equation will be used later when the need for relating the molecular polarizability to the dielectric function arises.

The Clausius-Mossotti Equation can be generalized for lower frequencies where nonelectronic mechanisms become important. If we define α_{eff} to be the effective polarizability that includes both the electronic polarizability and the contribution to charge redistribution from molecular reorientation (α_{or}), that is:

$$\alpha_{eff} = \alpha + \alpha_{or}, \tag{18}$$

the Clausius-Mossotti Equation becomes:

$$\frac{\epsilon(\omega) - 1}{\epsilon(\omega) + 2} = \frac{4\pi}{3} N\alpha_{eff}(\omega). \tag{19}$$

This form applies at all frequencies. Equation 17, however, explicitly stresses that when frequencies are high, the electronic mechanisms dominate.

E. Screening and Dressed Dipoles

A dielectric surrounding a guest molecule can affect the molecule's static properties such as its static dipole moment. In this section, we show how the measured dipole moment of a guest molecule that is embedded inside a host medium is related to its vacuum dipole moment. In the absence of an applied electric field, a molecule is affected even in an electrically neutral host material. The molecule feels time varying electric fields due to local density fluctuations and a net electric field due to local polarization that is induced by the molecule's own electric moment. The electric field due to density fluctuations averages to zero over an appropriately long time interval, but, the local induced polarization can both screen the molecular moment and result in a reaction field that affects the molecule's electronic properties.

Consider a permanent dipole of vacuum moment μ in a cavity within a dielectric of dielectric constant ϵ_1. Figure 3a shows the dipole in the cavity while Fig. 3b shows the charge distribution of both the dipole and of the charge that is induced on the cavity wall. This surface charge will result in an electric field inside the cavity and will also contribute to the dipole field outside the cavity. It is straightforward to solve for the electric potential inside the cavity, ϕ_{in}, and outside of the cavity, ϕ_{out}:

$$\phi_{in} = \frac{\mu}{r^2} \cos\theta - \frac{2(\epsilon_1 - 1)}{2\epsilon_1 + 1} \frac{\mu}{a^3} r \cos\theta, \tag{20}$$

and

$$\phi_{out} = \frac{\mu}{r^2} \cos\theta - \frac{2(\epsilon_1 - 1)}{2\epsilon_1 + 1} \frac{\mu}{r^2} \cos\theta = \frac{3}{2\epsilon_1 + 1} \frac{\mu}{r^2} \cos\theta, \tag{21}$$

where θ is the angle between the field point and the z-axis, r the distance to the field point, and a the cavity radius. Note that the first term in Eq. 20 represents a dipole potential of moment $\vec{\mu}$ while the second term represents the uniform field in the cavity due to the induced surface charge. Similarly, the first term in Eq. 21 represents the dipole potential of the bare dipole $\vec{\mu}$ while the second term represents the dipole potential from the induced charge on the cavity surface. The second term in Eq. 21 is thus the electric potential due to a dipole moment:

$$-\frac{2(\epsilon_1 - 1)}{2\epsilon_1 + 1} \vec{\mu}. \tag{22}$$

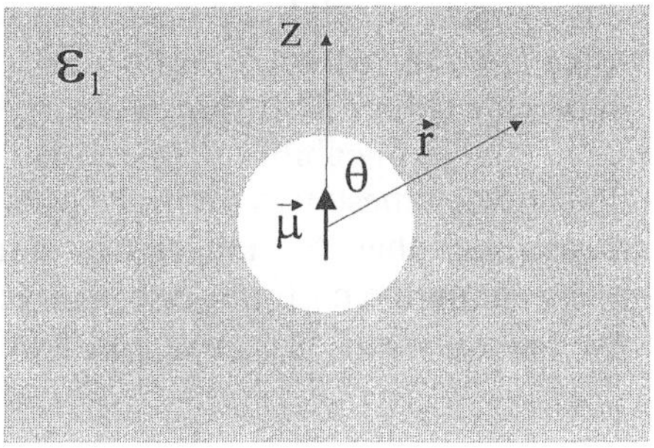

(a)

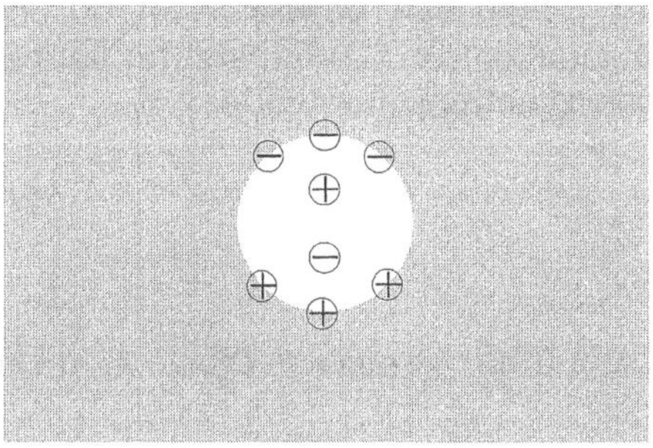

(b)

Figure 3 a) A dipole in a cavity within a dielectric, and b) a representation of the charges including the dipole-induced surface charge on the cavity wall and the dipole in the cavity.

The electric potential outside the cavity is related to the electric displacement field according to $\vec{D} = -\epsilon_1 \nabla \phi_1$. Because free charges are the source of $\vec{D}$, the electric dipole appears to have an effective free-charge dipole moment:

$$\vec{\mu}_e = \frac{3\epsilon_1}{2\epsilon_1 + 1} \vec{\mu}. \tag{23}$$

The electric field inside the cavity, however, is a superposition of a dipole field and a uniform electric field. The uniform electric field is due to the surface charge that is induced by the dipole and is called the reaction field, $\vec{R}$. Taking the gradient of the second term in Eq. 20 yields the reaction field:

$$\vec{R} = \frac{1}{a^3} \frac{2(\epsilon_1 - 1)}{2\epsilon_1 + 1} \vec{\mu}. \tag{24}$$

Note that the constant of proportionality, f, between the reaction field and the dipole moment is sometimes called the reaction field factor, that is:

$$\vec{R} = f\vec{\mu}. \tag{25}$$

In principle, it is difficult to estimate the cavity size, a. Bell assumed that a was the radius of the molecule [3] while Onsager assumed that a swept out a volume equal to the volume available to the molecule which he estimated to be $4\pi Na^3/3 = 1$, [4] where N is the number density. The applicability of these estimates depends on the material and generally falls between these two limiting cases.

The above calculation of the reaction field assumes that the molecule is not polarizable. If it is polarizable, the reaction field can act on the molecule and change its dipole moment which in turn can change the reaction field. The reaction field is then given by:

$$\vec{R} = f\vec{p} = f(\vec{\mu} + \alpha_2\vec{R}), \tag{26}$$

where $\vec{p}$ is the dipole moment in the cavity (spontaneous moment plus induced moment) and α_2 is the dipole's polarizability. (Note that because $\vec{R}$ in Eq. 26 depends on $\vec{R}$, it is said to be a self-consistent equation for $\vec{R}$.) Solving Eq. 26 for $\vec{R}$, and using Eqs. 24 and 25 for f, the self-consistent reaction field is:

$$\vec{R} = \frac{f}{1 - f\alpha_2}\vec{\mu} = \frac{1}{\dfrac{2\epsilon_1 + 1}{2(\epsilon_1 - 1)} - \dfrac{\alpha_2}{a^3}} \frac{1}{a^3}\vec{\mu}. \tag{27}$$

The resultant dipole moment, $\vec{\mu}^*$, is then the sum of the permanent dipole moment and the induced dipole moment and is given by

$$\vec{\mu}^* = \vec{\mu} + \alpha_2\vec{R} = \frac{\vec{\mu}}{1 - \dfrac{\alpha_2}{a^3}\dfrac{2(\epsilon_1 - 1)}{2\epsilon_1 + 1}}. \tag{28}$$

The dielectric surrounding the dipole has two effects: it changes the effective dipole moment as observed from outside the cavity and it results in a reaction field which changes the dipole moment even with no externally applied electric field. The dipole moment inside the cavity is called the *effective internal* dipole moment. Note that while the reaction field can polarize the molecule, it cannot cause it to reorient because the dipole moment and reaction field are always in the same direction. Thus, α_2 is the electronic polarizability of the molecule.

We apply the Onsager approximation for the molecular radius a ($4\pi Na^3/3 = 1$) and the Clausius-Mossotti Equation for the polarizability α_2 to Eq. 28. Before doing so, we stress that $\vec{\mu}^*$ is related to the static dielectric function of the host ($\epsilon_1(0)$), which includes both the zero frequency electronic polarization and zero frequency reorientational polarization, while the Clausius-Mossotti Equation for α_2 when applied to the optical dielectric function in the zero frequency limit of the guest ($n_2^2(0)$) includes only the zero frequency electronic polarization. Substituting Eq. 17 (with $\alpha(\omega) = \alpha_2$) into Eq. 28, and with the help of the Onsager approximation, we get the effective internal dipole moment:

$$\vec{\mu}^* = \frac{\vec{\mu}}{1 - \left(\dfrac{n_2^2(0) - 1}{n_2^2(0) + 2}\right)\left(\dfrac{2(\epsilon_1(0) - 1)}{2\epsilon_1(0) + 1}\right)} = \vec{\mu}\,\frac{(2\epsilon_1(0) + 1)(n_2^2(0) + 2)}{3(n_2^2(0) + 2\epsilon_1(0))}. \tag{29}$$

Just as the electric field from the vacuum dipole μ is screened by the dielectric, this effective internal dipole moment is also screened by the dielectric. In analogy to Eq. 23, the effective dipole moment measured by an observer whose perspective is external to the cavity is:

$$\vec{\mu}_e^* = \vec{\mu}^* \frac{3\epsilon_1(0)}{2\epsilon_1(0) + 1} = \frac{(n_2^2(0) + 2)\ \epsilon_1(0)}{n_2^2(0) + 2\epsilon_1(0)} \vec{\mu}. \tag{30}$$

F. Local Field Model of a Two Component Dipolar Composite

In this section, we compute the local electric field at a spherical molecule (of dielectric constant ϵ_2 and permanent vacuum dipole moment μ) that is embedded in a dielectric sea of constant ϵ_1. The dielectric sea can also be composed of a dipolar species.

The dipole moment of a molecule can be written as a series expansion in the electric field. The first two terms of such an expression for a dipole in a vacuum give:

$$p_i = \mu_i + \alpha_{ij} F_j, \tag{31}$$

where α is the electronic polarizability tensor. The above expression holds for a molecule whose orientation is fixed. If the molecule is allowed to freely reorient, the orientational average of the permanent dipole moment vanishes for non-zero temperature, leaving only the second term in Eq. 31. When an electric field is applied to such a molecule, the induced dipole moment will have two sources: the electronic polarization due to electron cloud deformation and the reorientational polarization due to partial alignment of the permanent dipole moment in response to the electric field. As discussed in Section VI of this chapter, the induced dipole moment of an orientationally unimpeded permanent vacuum dipole is of the form:

$$\vec{p} = \frac{\mu^2 \vec{F}}{3kT} + \alpha \vec{F}, \tag{32}$$

where α is the orientationally averaged polarizability. For a spherical molecule, α is independent of molecular orientation so that the tensor form is not required. We can rewrite this equation in the form:

$$\vec{p} = \alpha_{eff} \vec{F} = (\alpha_{or} + \alpha)\ \vec{F}, \tag{33}$$

where the effective polarizability α_{eff} is a sum of the orientational part and the electronic part.

When the dipole is placed inside a dielectric, the local electric field $\vec{F}$ can be viewed as arising from two sources:

$$\vec{F} = \vec{E}_c + \vec{R}, \tag{34}$$

the cavity field $\vec{E}_c$ and the reaction field $\vec{R}$. The cavity field is the sum of the applied electric field and the field due to the induced charge on the surface of the cavity, while the reaction field is due to the charge on the surface of the cavity that is induced by the dipole inside the cavity (as in Fig. 3b). Because the reaction field is *always* along the axis of the dipole, it can never reorient the molecule but it can polarize its electron could. The cavity field, on the other hand, will both reorient and polarize the electron cloud. The induced dipole moment of a molecule in a dielectric is thus of the form:

$$\vec{p} = \alpha_{or}\vec{E}_c + \alpha_2\vec{F}, \tag{35}$$

where α_2 is the electronic polarizability of the embedded molecule.

For a spherical cavity, the electric field inside is given by Eq. 8 with $\epsilon_2 = 1$:

$$\vec{E}_c = \left(\frac{3\epsilon_1}{2\epsilon_1 + 1}\right) E. \tag{36}$$

Because the reaction field is due to the total dipole moment (permanent plus induced) in the cavity, it is given by Eq. 24 with $\vec{\mu}$ replaced with $\vec{p}$:

$$\vec{R} = \frac{1}{a^3}\frac{2(\epsilon_1 - 1)}{2\epsilon_1 + 1}\vec{p}. \tag{37}$$

The induced dipole moment can thus be written in terms of the applied electric field by substituting Eqs. 36 and 37 into Eq. 34, followed by substitution into Eq. 35:

$$\vec{p} = \alpha_{or}\left(\frac{3\epsilon_1}{2\epsilon_1 + 1}\right)\vec{E} + \alpha_2\left[\frac{2(\epsilon_1 - 1)\vec{p}}{(2\epsilon_1 + 1)a^3} + \left(\frac{3\epsilon_1}{2\epsilon_1 + 1}\right)\vec{E}\right]. \tag{38}$$

This is a self-consistent expression for the dipole moment of the molecule in the cavity, $\vec{p}$. Solving for $\vec{p}$ we get:

$$\vec{p} = \left[\frac{\left(\frac{3\epsilon_1}{2\epsilon_1 + 1}\right)(\alpha_{or} + \alpha_2)}{1 - \frac{2(\epsilon_1 - 1)\alpha_2}{(2\epsilon_1 + 1)a^3}}\right]\vec{E}. \tag{39}$$

According to Eq. 3, the effective polarizability of the spherical molecule is related to its dielectric function:

$$\alpha_{or} + \alpha_2 = \alpha_{eff} = \left(\frac{\epsilon_2 - 1}{\epsilon_2 + 2}\right) a^3. \tag{40}$$

The electronic polarizability is related to the "fast" part of the response. The refractive index is related to the electronic part of the dielectric constant according to $n^2(\omega) = \epsilon_{el}(\omega)$. With the understanding that n^2 refers to the electronic part of ϵ:

$$\alpha_2 = \left(\frac{n_2^2 - 1}{n_2^2 + 2}\right) a^3. \tag{41}$$

Substituting Eqs. 40 and 41 into Eq. 39 yields:

$$\vec{p} = \frac{\epsilon_1(\epsilon_2 - 1)(n_2^2 + 2)}{(\epsilon_2 + 2)(n_2^2 + 2\epsilon_1)} a^3\vec{E}. \tag{42}$$

Equation 42 relates the applied electric field to the dipole moment. To get the relationship between the local field and applied field, we need to express the dipole moment of the dielectric sphere in terms of the local field in the sphere. This is most conveniently done by applying Eq. 4 to the sphere. Thus, using Eq. 4 with $\epsilon \rightarrow \epsilon_2$ and with $\vec{P} \rightarrow \vec{P}_2$, and substituting this into Eq. 42, we get:

$$\vec{F} = \frac{\epsilon_2(n_2^2 + 2)}{n_2^2 + 2\epsilon_1} \vec{E}. \tag{43}$$

This is the Onsager local field model. Note that our derivation, however, is for the two component system so that ϵ_1 refers to the dielectric constant of the *host*, while n_2 refers to the refractive index of the *guest*. The single component Onsager expression is obtained by removing the subscripts from Eq. 43.

G. Energy of an Embedded Dipole

In Section II.E. we calculated the effective dipole moment of a molecule that is embedded inside a cavity as determined from the dipole field outside the cavity. The dipole field is found to be a sum of the vacuum dipole field and the dipole field due to the induced charge distribution on the surface of the cavity. If a measurement of the field due to the polarized molecule is used to determine the dipole moment of the source, the effective dipole moment is what would be determined.

If we apply an electric field to a material, the molecules within reorient. The torque on the molecule is simply given in terms of its dipole moment and local field. From the experimental perspective, it is more useful to express the torque in terms of the *applied* electric field. In this section, we calculate the torque on an embedded dipole moment due to an *externally* applied electric field by calculating the molecule's energy as a function of its angle relative to the applied electric field. It is important to note that all sources of the local electric field do not result in a torque. For example, recall that the reaction field on a molecule within an isotropic material always points along the dipole moment. As the dipole is rotated, there is no change in its "reaction energy". We therefore would expect that the dipole energy will be proportional to an effective dipole moment, which we call $\vec{m}^*$, and to the applied field E. $\vec{m}^*$ thus represents that part of the screened dipole that feels a torque due to the external electric field. $\vec{m}^*$ will, in general, be different from $\vec{\mu}^*$.

We begin by considering a molecule in a cavity. Rather than consider a point dipole at the cavity center, we replace it with the mathematically equivalent dielectric sphere of polarizability α_2 and uniform zero field polarization $\vec{P}_0$. The vacuum dipole moment of such a sphere is simply:

$$\vec{\mu} = \frac{4}{3} \pi a^3 \vec{P}_0, \tag{44}$$

where a is the radius of the spherical molecule. With no external field applied, the electric field outside the dielectric sphere is dipolar for any distance from its center r as long as $r > a$. This is a convenient representation of the molecule because its radius is equal to the cavity size; that is, it can fully occupy the cavity. As such, it is then straightforward to apply the Onsager local field model to calculate the effective dielectric field inside the spherical dielectric that represents the molecule.

Figure 4 shows a molecule in a cavity and its representation as a spherical molecule (with dielectric function ϵ_2) in a dielectric represented by ϵ_1. The observed effective dipole moment of this embedded molecule is the same as calculated in

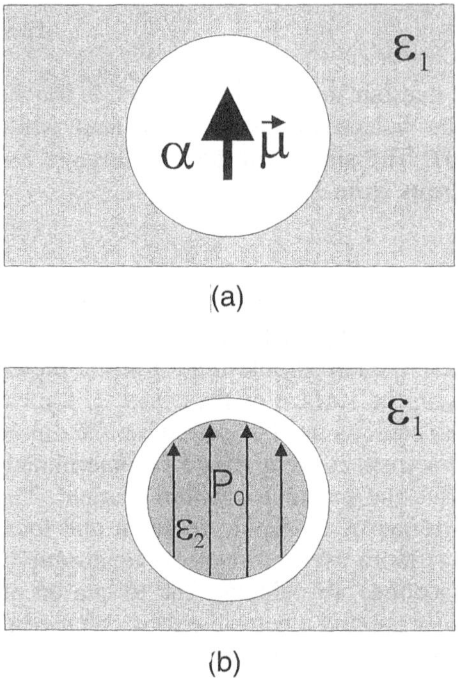

Figure 4 a) Molecule represented as a point dipole and in a dielectric of permittivity ϵ_1, b) represented as a dielectric sphere of permittivity ϵ_2 and polarization $\vec{P}_0$. No electric fields or induced polarization are shown.

the previous section. When an external electric field is applied to the material, the local field inside the cavity will be simply the sum of the cavity electric field due to the external field and the reaction field. Because the reaction field is always aligned along the dipole axes, the reaction field does not contribute to molecular reorientation. The orientational energy can thus be expressed as the dot product between the cavity electric field $\vec{E}_c$ (given by Eq. 36) and the effective internal dipole moment $\vec{\mu}^*$ (given by Eq. 29):

$$\begin{aligned} U &= -\vec{\mu}^* \cdot \vec{E}_c = -\mu^* \frac{3\epsilon_1}{2\epsilon_1 + 1} E \cos\theta \\ &= -\left(\frac{(2\epsilon_1(0) + 1)(n_2^2(0) + 2)}{3(n_2^2(0) + 2\epsilon_1(0))}\right) \mu \left(\frac{3\epsilon_1(0)}{2\epsilon_1(0) + 1}\right) E \cos\theta, \end{aligned} \tag{45}$$

where θ is the angle between the cavity electric field and the effective internal dipole. Note that the cavity field, $\vec{E}_c$, points in the same directions as the applied electric field, $\vec{E}$. Combining terms:

$$U = -\frac{\epsilon_1(0)(n_2^2(0) + 2)}{n_2^2(0) + 2\epsilon_1(0)} \vec{\mu} \cdot \vec{E}. \tag{46}$$

The energy of the dipole is:

$$U = -m^*E \cos \theta = -\vec{m}^* \cdot \vec{E}, \tag{47}$$

where $\vec{m}^*$ is called the dressed dipole moment:

$$\vec{m}^* = \left(\frac{3\epsilon_1(0)}{2\epsilon_1(0) + n_{2(0)}^2}\right)\left(\frac{n_2^2(0) + 2}{3}\right)\vec{\mu}. \tag{48}$$

Comparing Eq. 30 with Eq. 48, we find that $\vec{m}^* = \vec{\mu}_e^*$. For the isotropic models used in this chapter, these two dipole moments are the same.

H. Radiating Embedded Dipole

Materials are often studied with electromagnetic waves such as light. In light scattering experiments, for example, the sample is illuminated and the scattered light is detected. At the microscopic level, the light beam interacts with dipoles within the material which reradiate the light. This section describes the radiation from an embedded molecule that is illuminated with light. We only consider the first order response. Higher order terms will be formally introduced in Section III of this chapter.

Figure 5a shows a dipole in the cavity. A beam of light with electric field amplitude E is launched into a sample and the local electric field (F) at the molecule due to the light excites a dipole (p). The dipole oscillates and reradiates the light (E_{out}^*) which is screened to an external observer by the surrounding dielectric. The local electric field at the molecular site is most generally described with an Onsager local field because it accounts for molecules with *both* a permanent and induced dipole moment. The scattered light observed away from the dipole is simply a dipole radiation field of an effective dipole, p^*.

For a local electric field $\vec{F}$, the induced dipole moment $\vec{p}$ is:

$$\vec{p} = \alpha_2^{eff}\vec{F}, \tag{49}$$

where α_2^{eff} is the effective polarizability of the guest molecule and includes both the electronic and reorientational molecular polarizability. Using the Onsager local field model (Eq. 43), the induced dipole is of the form:

$$\vec{p}(\omega) = \alpha_2^{eff}(-\omega;\omega)\vec{E}(\omega)\left(\frac{3\epsilon_1(\omega)}{2\epsilon_1(\omega) + n_2^2(\omega)}\right)\left(\frac{n_2^2(\omega) + 2}{3}\right). \tag{50}$$

The dipole field observed outside the cavity will be screened, so that the effective dipole moment is of the form

$$\vec{p}^* = \vec{p}\,\frac{3\epsilon_1(\omega)}{2\epsilon_1(\omega) + 1}. \tag{51}$$

This expression is analogous to the effective dipole moment given by Eq. 23.

Combining Eqs. 50 and 51, we get:

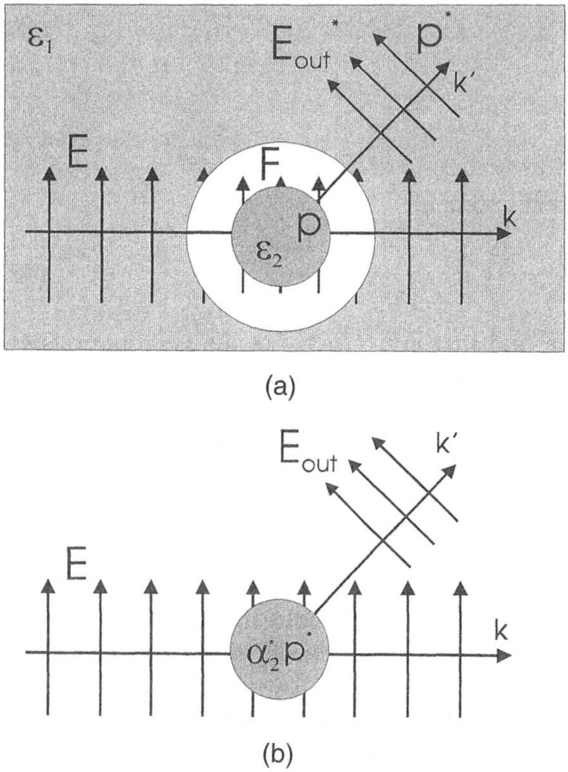

Figure 5 a) Molecule represented as a dielectric sphere in a dielectric of permittivity ϵ_1 with incident electric field E. The local electric field F induces a dipole moment. b) The same process represented as a "dressed" scattering event.

$$\vec{p}^{\,*} = \alpha_2^* E, \tag{52}$$

where

$$\alpha_2^* = \frac{3\epsilon_1(\omega)}{2\epsilon_1(\omega) + 1}\left(\frac{3\epsilon_1(\omega)}{2\epsilon_1(\omega) + n_2^2(\omega)}\right)\left(\frac{n_2^2(\omega) + 2}{3}\right)\alpha_2^{eff}, \tag{53}$$

and is called the dressed polarizability. Equations 52 and 53 summarize the light scattering process in the dressed polarizability picture. Figure 5b shows how this dressed process can be visualized as an "effective" vacuum process: Light with field E excites a dipole p^* (through polarizability α^*), which radiates light. At this point, it is important to emphasize that Eq. 53 holds only in the low concentration limit. If the dopant concentration is large, then the dielectric function of the host material ϵ_1 must be replaced by the dielectric function of the composite. Section V discusses the Maxwell Garnet theory, which is used to calculate the dielectric function of a composite material. In this case, then, the mean dielectric function ϵ in Eq. 101 replaces ϵ_1 in Eq. 53.

I. Review of Linear Dielectric Properties

In the process of deriving local field models in preceding sections, we defined several physical quantities such as the dressed dipole moment, cavity field, reaction field, etc. In this section, we briefly review the meaning of quantities and tabulate them to allow for quick reference.

Table 1 summarizes properties related to the dielectric constant. We use the notation that the dielectric constant $\epsilon(\omega)$ represents the total dielectric response at frequency ω. That is, all mechanisms are included. The refractive index, on the other hand, is calculated from the electronic part of the dielectric function ($n^2 = \epsilon_{el}$). While at optical frequencies the electronic mechanism may dominate so that refractive index can be equally well expressed in terms of the total dielectric function or the electronic part of the dielectric function (because $\epsilon(\omega) \approx \epsilon_{el}(\omega)$ for large ω), at low frequency both mechanisms can be comparable. The refractive index is thus a convenient way to express the electronic part of the response at any frequency.

The polarizability α is the induced dipole moment per unit of applied electric field to an atom, molecule, or small chunk of dielectric and refers to the electronic mechanism. (Note that we adopt the nonlinear susceptibility notation for α in which the first negative frequency argument represents the reradiated light, while the remaining arguments represent the frequencies of the incident light.) There are several other possible mechanisms that result in an induced dipole moment of which molecular reorientation is the most important. In addition to the free rotator model given by Eq. 32 and listed in the table, this chapter also discusses several impeded rotator models of reorientation and how they affect the linear and the nonlinear optical response. The effective polarizability, α_{eff}, includes electronic and reorien-

Table 1 Dielectric Constant, Refractive Index, and Polarizability Definitions and Models

Parameter	Description
$\epsilon(\omega)$	Dielectric constant at frequency ω. Includes all mechanisms.
$\epsilon_{el}(\omega)$	Electronic part of ϵ.
$n(\omega) = \sqrt{\epsilon_{el}(\omega)}$	Refractive index.
$\alpha(-\omega;\omega)$	Molecular polarizability. Electronic mechanism.
$\alpha^*(-\omega;\omega)$	Dressed molecular polarizability.
$\alpha(-\omega;\omega) = \dfrac{n^2(\omega) - 1}{n^2(\omega) + 2} a^3$	Electronic polarizability of sphere of radius a and refractive index $n^2(\omega)$.
$\alpha_{or} = \dfrac{\mu^2}{3kT}$ (free rotator model)	Reorientational molecular polarizability.
$\alpha_{eff}(-\omega;\omega) = \alpha(-\omega;\omega) + \alpha_{or}(-\omega;\omega)$	Effective molecular polarizability. Includes all mechanisms.
$\alpha_{eff}(-\omega;\omega) = \dfrac{\epsilon(\omega) - 1}{\epsilon(\omega) + 2} a^3$	Effective polarizability of sphere of radius a and dielectric function ϵ.

tational mechanisms. α is thus related to the refractive index of the material unit, while α_{eff} is related to the dielectric function of the material unit.

Table 2 summarizes the various electric fields and their relationship to each other. $\vec{E}$ is the applied electric field and $\vec{F}$ the local electric field acting on a microscopic unit within the material. The local field is comprised of the cavity field $\vec{E}_c$, which includes the applied field and external-field-induced charge on the surface of the cavity, and the reaction field, which is due to the cavity surface charge which is induced by the molecule inside the cavity. Relationships between the local electric field and the applied electric field are called local field models, which depend on assumptions pertaining to the geometry and composition of the material. The Lorentz local field model, for example, applies to nonpolar molecules, while the Onsager local field model applies to polarizable dipolar species.

Table 3 summarizes the various dipole moments. When a non-polarizable dipole is embedded inside *any* dielectric (ϵ_1), the dipole field in the dielectric is screened so that the dipole appears to have an effective moment $\vec{\mu}_e$. If a polarizable dipolar molecule is embedded in a dielectric, it induces polarization in the surrounding dielectric. The electric field from this polarization then induces a dipole moment back on the embedded molecule. The net dipole moment, $\vec{\mu}^*$, of the molecule is called the internal dipole moment. The effective dipole as measured by an external observer is called the screened internal effective dipole moment, $\vec{\mu}_e^*$. The energy of an embedded polarizable molecule is equal to the dot product between the dressed dipole moment, $\vec{m}^*$ and the external electric field, $\vec{E}$. Finally,

Table 2 Electric Fields

Parameter	Description
$\vec{E}$	Applied electric field.
$\vec{E}_c = \left(\frac{3\epsilon}{2\epsilon + 1}\right)\vec{E}$	Cavity field. Electric field inside a spherical cavity due to an applied field $\vec{E}$.
$\vec{R}$	Reaction field. Field at a dipole at the center of a cavity due to the dipole-induced surface charge.
$\vec{F} = \vec{E}_c + \vec{R}$	Local electric field at molecular site.
$\vec{R} = \frac{f}{1 - f\alpha_2}\vec{\mu} = \frac{1}{\frac{2\epsilon_1 + 1}{2(\epsilon_1 - 1)} - \frac{\alpha_2}{a^3}}\frac{1}{a^3}\vec{\mu}$	Reaction field due to a dipole of vacuum moment $\vec{\mu}$ and polarizability α_2 in a cavity of radius a in *any* dielectric medium of permittivity ϵ_1 and reaction field factor f.
$\vec{F} = \left(\frac{\epsilon + 2}{3}\right)\vec{E}$	Lorentz local field model for a nonpolar dielectric.
$\vec{F} = \left(\frac{3\epsilon_1}{2\epsilon_1 + \epsilon_2}\right)\left(\frac{\epsilon_2 + 2}{3}\right)\vec{E}$	Local field model for a nonpolar host dielectric of permittivity ϵ_1 and a dopant of permittivity ϵ_2.
$\vec{F} = \frac{\epsilon_1 (n_2^2 + 2)}{n_2^2 + 2\epsilon_1}\vec{E}$	Onsager local field model for species #2 in host of species #1 (both polar).

Table 3 Dipole Moments

Parameter	Description
$\vec{\mu}$	Vacuum dipole moment.
$\vec{\mu}_e = \dfrac{3\epsilon_1}{2\epsilon_1 + 1}\vec{\mu}$	Effective dipole moment. Moment observed by an outside observer of a dipole embedded in a dielectric of permittivity ϵ_1.
$\vec{\mu}^* = \vec{\mu} + \alpha_2\vec{R} = \vec{\mu}\,\dfrac{(2\epsilon_1(0) + 1)(n_2^2(0) + 2)}{3(n_2^2(0) + 2\epsilon_1(0))}$	Effective internal dipole moment. Includes induced dipole in response to reaction field.
$\vec{\mu}_e^* = \left(\dfrac{3\epsilon_1(0)}{2\epsilon_1(0) + n_2^2(0)}\right)\left(\dfrac{n_2^2(0) + 2}{3}\right)\vec{\mu}$	The screened internal effective dipole moment as seen by an external observer.
$\vec{m}^* = \left(\dfrac{3\epsilon_1(0)}{2\epsilon_1(0) + n_2^2(0)}\right)\left(\dfrac{n_2^2(0) + 2}{3}\right)\vec{\mu}$	Dressed dipole moment. Dipole moment that couples to external field to produce a torque.
$\vec{p}^* = \alpha^*\vec{E}$	Dressed dipole moment. The external dipole observed in response to an applied field $\vec{E}$.

when an embedded dipole radiates in response to an applied electric field, the dipole moment of the embedded molecule is the dressed dipole moment, $\vec{p}^*$.

III. EFFECT OF DIELECTRIC ON NONLINEAR MOLECULAR SUSCEPTIBILITIES

In the previous section, we discussed the effect of the surrounding medium on a molecule's dipole moment and polarizability. The nonlinear dielectric and optical properties of the molecule will also be affected. Furthermore, we previously ignored the fact that the polarizability of a molecule can change in the presence of a static electric field or a beam of light. Similarly, the nonlinear susceptibilities can also be influenced by electric fields. The linear and nonlinear optical susceptibilities are peaked functions of the optical frequency in which these peaks correspond to resonant excitations of the molecules. Fluctuations in the local electric field can result in peak broadening, while a static local electric field can effect both the shape and positions of these peaks. Broadening is commonly described by a phenomenological width parameter. [5] We will not discuss the broadening mechanisms in detail. The effects of the reaction field on the susceptibility through its effect on the structure of the molecule, however, warrants further attention.

To understand the effect of the reaction field on a molecule's response, we first need to understand how the eigenfunctions of a molecule are related to the susceptibility. While not explicitly stated, the molecular susceptibilities are functions of the transition frequencies, $\Omega_{nm} = \Omega_n - \Omega_m$, and the transition moments μ_{nm} of the molecule where m and n label the states of the molecule. (Chapter 2 gives a more detailed description of quantum calculations of susceptibilities.) When a molecule is embedded in a dielectric material such as a polymer, the energy levels shift and the transition moments are altered. The most pronounced effect on the linear absorption spectrum is a shift in the wavelength of maximum absorbance. Such a shift also appears in the nonlinear optical spectrum. When the transition moments are not strongly affected by the reaction field, the shifts can be formally introduced by an energy shift operator, $\hat{O}(\Omega_m, \Omega_m^0)$, that has the effect:

$$f(\Omega_m) = \hat{O}(\Omega_m, \Omega_m^0) f(\Omega_m^0), \tag{54}$$

where Ω_m^0 is a frequency so that $\hbar\Omega_m^0$ is the vacuum energy of level m and f is an arbitrary function. Note that the form of such an operator depends on the dielectric properties of the host matrix.

Taking into account the local fields through a local field tensor as described by Eq. 12, the nth-order induced dipole moment of a molecule in a dielectric as given by Eq. 1 can then be expressed in terms of the applied electric field and shift operators:

$$p_i^{(n)}(\omega_\sigma) = \left[\prod_m \hat{O}(\Omega_m, \Omega_m^0)\right] \xi_{ijk\ldots l}^{(n)}(-\omega_\sigma; \omega_1, \omega_2, \ldots, \omega_n) L_{jj'}(\omega_1) E_{j'}^{\omega_1} \times L_{kk'}(\omega_2) E_{k'}^{\omega_2} \times \ldots \times L_{ll'}(\omega_n) E_{l'}^{\omega_n}, \tag{55}$$

where $\xi_{ijk\ldots l}^{(n)}$ is the nth-order molecular susceptibility, $L_{jj'}$ the local field tensor, the index m spans all the states of the molecular system, and Π_m represents the product of shift operators over all states of the system.

By multiplying the above equation by $L_{i'i}(\omega_\sigma)$, we get:

$$p_i^{*(n)}(\omega_\sigma) = \xi_{ijk\ldots l}^{(n)*}(-\omega_\sigma; \omega_1, \omega_2, \ldots, \omega_n) E_j^{\omega_1} E_k^{\omega_2} \times \ldots \times E_l^{\omega_n}, \tag{56}$$

where the starred quantities are the "dressed" induced dipole moment and susceptibility:

$$p_i^{*(n)}(\omega_\sigma) = p_{i'}^{(n)}(\omega_\sigma) L_{i'i}(\omega_\sigma), \tag{57}$$

and

$$\xi_{i'j'k'\ldots l'}^{(n)*}(-\omega_\sigma; \omega_1, \omega_2, \ldots, \omega_n) = \left[\prod_m \hat{O}(\Omega_m, \Omega_m^0)\right] \xi_{ijk\ldots l}^{(n)}(-\omega_\sigma; \omega_1, \omega_2, \ldots, \omega_n) \times L_{ii'}(\omega_\sigma) L_{jj'}(\omega_1) L_{kk'}(\omega_2) \times \ldots \times L_{ll'}(\omega_n). \tag{58}$$

For a two component isotropic system, the local field factors are given by the Onsager local fields. Recall that according to Eq. 51, the radiating dipole will be screened by the dielectric. The local electric field for an isotropic medium associated with the radiation field at frequency ω_σ is thus of the form:

$$L_{ii'}(\omega_\sigma) = \delta_{ii'} \frac{3\epsilon_1(\omega_\sigma)}{2\epsilon_1(\omega_\sigma) + 1}, \tag{59}$$

where $\delta_{ii'}$ is the Kronecker delta. The remaining local field factors are of the Onsager form. For example, the local field factor at input frequency ω_1 is:

$$L_{jj'}(\omega_1) = \delta_{jj'}\left(\frac{3\epsilon_1(\omega_1)}{2\epsilon_1(\omega_1) + n_2^2(\omega_1)}\right)\left(\frac{n_2^2(\omega_1) + 2}{3}\right). \tag{60}$$

The beauty of the dressed susceptibility formalism is that the relationship between the dressed dipole moment and the applied electric field is identical in form to the vacuum relationship. Given the vacuum relationship $p^{(n)} = \xi^{(n)}F^n$ (Eq. 1), the dressed relationship is given by $p^{(n)} \rightarrow p^{*(n)}$, $\vec{F} \rightarrow \vec{E}$, and $\xi^{(n)} \rightarrow \xi^{(n)*}$, which yields $p^{*(n)} = \xi^{(n)*}E^n$.

IV. DILUTE SOLUTE SYSTEMS—ELECTRONIC RESPONSE

Over the past 10 years, work on composite nonlinear optical materials has accelerated. Such materials are attractive because various properties can be independently controlled to optimize the material's performance. For example, dye-doped polymers combine the high nonlinearity of the dopant with the low optical loss and fabricability of the bulk material. [6] Sol gel glasses doped with organic molecules combine the reliability of silica glass with the molecular engineering capabilities of organics. [7] Metal-doped polymers, on the other hand, have been demonstrated to enhance local electric fields so that a combination of dyes, metals, and a polymer host may potentially be tuned to make an optimum all-optical device material. [8] It is for this reason that it is useful to understand how the microscopic properties can be used as building blocks to make specialized optical materials. The description below outlines how nonlinear optical properties of a bulk material can be calculated from molecular properties. Such a formalism can be used to make educated guesses of the bulk optical properties of materials designed with the molecular building blocks tabulated in the remaining chapters.

A common trait of many composite materials is that they can be modeled as a dilute solution. A solute in a liquid solvent or a dilute gas are both obvious examples. In dye-doped polymers and dye-functionalized polymers, the polymer acts as a solid solvent whose rigidity allows the orientations of the dopant molecules to be controlled and held in place. The dilute solute model has been shown to apply in these materials up to almost 70 percent by weight solute where the solute molecules are known to strongly aggregate [9,10]. Dielectric measurements of an azo dye in PMMA shows that the dopant concentration reaches about 15 percent by weight before aggregation. [11] Third order nonlinear optical studies are also in agreement with dielectric studies in which the nonlinear response begins to saturate at 15 wt percent. [12] When the dye chromophores are covalently attached to the polymer, saturation occurs in the range between about 50 wt percent concentrations [9] and almost 70 wt percent [10]. As long as the molecules do not strongly associate with each other, the response of the material reflects on the molecular properties of the dopant. What is more surprising is that the rigid gas model also applies to many organic crystals owing to the fact that forces within an individual molecule are much larger than forces between the molecules in the crystal lattice. Because interactions between the molecular units are small in many organic material classes, it is possible to model the bulk response without the need for taking into account collective excitations.

Most models of the nonlinear optical response of composite organic materials have focused on the electronic mechanisms. In this section, we show how models of the electronic response are used to predict the nonlinear susceptibility of the material in terms of the material's structure or processing. Other mechanisms in organic composites, which have only recently piqued researcher's interests, are discussed in Section VI.

A. Dependence of Response on Concentration

Before presenting the dilute solute model of the linear and nonlinear-optical response, it is useful to paint a picture of how we can qualitatively use the concepts developed above to understand how a material's bulk properties change over a broad range of substituent concentration. We begin by considering the dilute gas limit and discuss when this approximation breaks down as the solute concentration increases.

1. Dilute Gas

For a dilute gas, in which the molecules do not interact with each other in any way, the bulk response of the system is identical to the molecular response. The number density of molecules, however, affects the strength of the response. For example, the linear absorption spectrum looks identical for both a single molecule and the gas except that the absorption strength of the gas is larger by a factor N over the molecular absorbance (N is the number density).

2. Dilute Liquid and Solid Solutions

If the molecules are incorporated into a transparent solid polymer or liquid solvent in *low concentration*, the reaction field causes a wavelength shift in the response function. Furthermore, because the local electric field at the molecular site is different from the applied field, the intensity of scattered light will be altered by an amount that can be calculated from local field models. As a result, peaks in the linear absorption spectrum will be shifted due to the reaction field and the absolute absorbance per molecule will be altered by local field factors. The strength of linear absorption, however, still scales linearly with the concentration as in the dilute gas, but the spectrum of the ensemble has the same properties as the dressed polarizability of a solute molecule rather than the vacuum polarizability.

3. High Concentration Without Aggregation

When the concentration of dopants is large but less than threshold for aggregation, the presence of the dopants alters the dielectric properties of the bulk system. Optical measurements then probe dressed susceptibilities as in the dilute limit; that is, the susceptibility depends on number density but the dependence is not necessarily linear. The bulk response is then determined from the dressed susceptibilities which include the dopant's contribution to the bulk dielectric constant. The response is thus dressed by both the host material and the other dopants.

4. Onset of Aggregation

When the concentration of dopants is high enough for aggregates to form, the material contains both aggregates and isolated molecules. In this regime, optical measurements probe a combination of species. It is difficult to compute the response of such a mixed system because the distribution of aggregate size is usually unknown. If the distribution is well defined, the response would be calculable from an ensemble average over the substituents.

5. *Aggregation Limit*

In the limit of high concentration, the number of aggregates greatly exceeds the number of unassociated molecules. If the optical response of the typical aggregate is known (i.e., its polarizability or higher order polarizability), the bulk response can be determined by considering the aggregate as the microscopic unit. As long as the aggregate is smaller than the wavelength of light and the spacing between the aggregates is large, the dilute solute model can once again be used. There are no particular restrictions on the material class of aggregates. They can include organic or inorganic microcrystals, metals, and/or semiconductors.

B. The Dilute Solute Model

In the dilute solute limit, we assume that a collection of microscopic units are embedded in a host material as shown in Fig. 6. The units must be sufficiently small so that a large ensemble (large enough for the ensemble to appear continuous) of such units fills a volume whose dimensions are small compared to the wavelength of light acting on the system. Under this long wavelength condition, the local field factors can be calculated as previously described. When the interaction between "dressed" guest units in a host material can be ignored, the bulk polarization is a vector sum over each dipole in the distribution. As we show below, the bulk nonlinear susceptibility is the tensor sum over the dressed nonlinear susceptibilities of the dopant molecules within a given volume of material. Because the dielectric function of the host polymer usually dominates the linear response of the

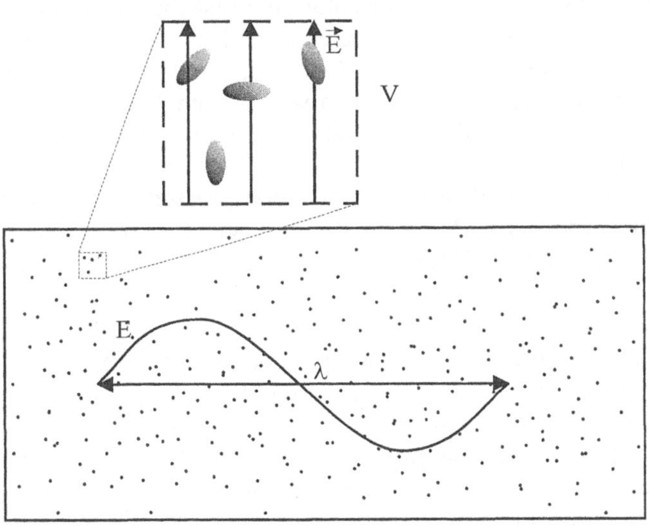

Figure 6 A schematic representation of a dilute solute in a host material. The wavelength of the light beam is large compared with the size of the microscopic unit and the size of a large ensemble.

composite system, the local field tensors of the doped and undoped polymer are almost equal unless the dopant concentration is large or the wavelength of the light or frequency of the electric field is in resonance with a transition of the dopant so that the dopant's contribution to the dielectric function is enhanced.

Within a volume V of material, the bulk polarization $\vec{P}$ is a simple vector sum over the induced dipole moments of the molecules within the volume:

$$\vec{P} = \frac{1}{V} \sum_{n=1}^{M} \vec{p}_n, \tag{61}$$

where M is the number of molecules in the volume, and the index n labels the individual molecules. If there are s different molecular species and M_i is the number of dipoles of species i (note $M = \Sigma_{i=1}^{s} M_i$), then the polarization is of the form:

$$\vec{P} = \frac{1}{V} \sum_{i=1}^{s} \sum_{n_i=1}^{M_i} \vec{p}^{\,*}_{n_i}. \tag{62}$$

If the dipole moment of species j is $\vec{p}^{(j)}$ in the principle coordinate frame of that molecule, the ith Cartesian component of the bulk polarization is then:

$$P_i = \sum_{j=1}^{s} N_j <p_{i'}^{(j)*}>_i, \tag{63}$$

where $N_j = M_j/V$ is the number density of species j and where the brackets denote an orientational average for each species:

$$<p_{i'}^{(j)*}>_i \equiv \frac{\int\int\int d^3\vec{\Omega} G_j(\vec{\Omega}) a_{ii'}(\vec{\Omega}) p_{i'}^{(j)*}}{\int\int\int d^3\vec{\Omega} G_j(\vec{\Omega})}. \tag{64}$$

$\vec{\Omega}$ represents the three Euler angles, $G_j(\vec{\Omega})$ is the orientational distribution function that describes the orientations of the dopants of species j, and $a_{ii'}$ is the rotation matrix that connects the dressed molecular frame to the laboratory frame. Because the generalization from one to s species is straightforward, from this point we consider a one species system only and will generalize the expression to multiple component systems as needed.

Because both the dipole moment and polarization are expressed as a power series in electric field, the nth-order term $P_i^{(n)} = N <p_{i'}^{*(n)}>_i$ yields a relationship between the molecular and bulk susceptibilities:

$$\chi_{ijk\ldots l}^{(n)} = N <\xi_{i'j'k'\ldots l'}^{(n)*}>_{ijk\ldots l}. \tag{65}$$

(Note that $p_{i'}^{*(n)}$ represents the nth-order dressed dipole moment. This should not be confused with $p_i^{(j)*}$ which is the dressed dipole moment of the jth species.) Equation 65 is the most general relationship between the bulk and microscopic nonlinear susceptibility of a dilute solute system when interactions between solutes is small and the host or solvent material is linear. Before applying this equation to specific examples, it is worthwhile to review essential ingredients to this relationship. First, because the molecular susceptibility is affected by the surrounding dielectric medium, it is clear that the bulk response is related to the dressed susceptibility. Secondly, the orientation of the microscopic units affects the relative magnitudes of each bulk tensor component and is taken into account by evaluating the orientational average with respect to a distribution function. Finally, if the molecules are noninteracting, the bulk response is linear in the concentration of molecular units.

Two important material classes are organic crystals and dye-doped polymers. Below we show how the dilute solute model can be applied to the calculation of the second harmonic coefficient in a crystal, and the second and third order electronic susceptibility of a dye-doped polymer.

C. Second Order Response

1. Organic Crystals

We first consider organic crystals. Motivated by the fact that the van der Waals interaction or hydrogen bond energies responsible for intermolecular cohesion of an organic lattice are several orders of magnitude smaller than intramolecular chemical bond energies, Zyss and Oudar proposed a relationship between the hyperpolarizability (second order molecular nonlinear optical susceptibility) and the second harmonic coefficient (proportional to the second order nonlinear optical susceptibility) of the form [13]:

$$d_{IJK}(-2\omega;\omega,\omega) = N f_I^{2\omega} f_J^{\omega} f_K^{\omega} \sum_{s=1}^{n} \cos(I, i(s)) \cos(J, j(s)) \cos(K, k(s))\, \beta_{ijk}^{(s)}(-2\omega;\omega,\omega), \tag{66}$$

where d_{IJK} is the second harmonic coefficient in the lattice frame, s is an index that labels the n molecules in the unit cell, $\cos(I, i(s))$ is the direction cosine between the lab frame (I) and the molecular frame (i) of molecule s, N is the number density of unit cells in the crystal, f_I^{ω} is the Lorentz local field factor in the lattice I direction and at frequency ω, and $\beta_{ijk}^{(s)}(-2\omega;\omega,\omega)$ is the second harmonic hyperpolarizability of molecule s. It is clear that this is a special case of Eq. 65 in which the orientational order is fixed by the crystal lattice. Zyss and Oudar applied this model to 2-methyl,4-nitro aniline (MNA) for which the crystal structure [14] and second harmonic coefficients are known [15]. They found good agreement between the model and measurements.

2. Isotropic Doped Polymers

The dye molecules in a dye-doped polymer can be similarly modeled. In crystalline materials, the orientations of the molecules in the unit cell and the structure of the crystal are well defined. In a doped polymer, on the other hand, the molecular orientations are continuously distributed. The distribution of molecules is usually represented by an orientational distribution function. The orienting forces acting on the molecules must therefore be known for the distribution function to be determined. The nonlinear optical properties of the doped polymer can then be calculated. As an example, we derive the second order susceptibility of an isotropic polymer that is poled in a uniform electric field and, in Section IV.D. the third order susceptibility of an isotropic polymer and liquid crystalline material. The general case for an anisotropic polymer (in which the dopants are axially ordered by either external stress or by a liquid crystalline state of the host) will also be presented.

We first discuss a thermodynamic model of the second order nonlinear optical susceptibility of a dye-doped isotropic polymer that is poled with an electric field above the glass transition temperature (T_g) of the system. [16,17] When cooling the system below T_g, the polymer hardens and the orientational order is frozen in place. In the thermodynamic model, the dye molecules are assumed to freely rotate in response to the applied electric field above T_g. Their orientational distribution is assumed to lock in place after the system is cooled below T_g so that the orientational distribution function below T_g is derivable from a Gibbs distribution function with $T = T_g$—the point at which the molecular orientations are frozen in place. Before deriving the relationship between the microscopic parameters and bulk response, it is worthwhile to review the conditions under which this model holds.

It is well-known that the orientational order of a dye-doped polymer decays over time when the poling field is turned off, and that the decay rate depends on the temperature of the system relative to the glass transition. [18–35] The thermodynamic model of the second-order response thus applies to materials just after poling; that is, on times scales that are short compared with the reciprocal decay rate. It should also be pointed out that the glass transition temperature is not well defined, so that the precise temperature at which the orientational order is locked in place is ambiguous. Nevertheless, it has been shown that the measured second harmonic coefficient agrees with the thermodynamic model of a poled polymer in which the orientational order is assumed to be locked in place at the standard T_g. [16,17,36–39] Models that account for limits of the thermodynamic model of poling [40] and aggregation have also been reported. [41,42]

The orientational distribution function of an ensemble of freely rotating dipoles under the influence of a static electric field is given by the Gibbs distribution:

$$G(\vec{\Omega}, E_p) = \frac{\exp\left[-\frac{1}{kT}(-\vec{m}^* \cdot \vec{E}_p)\right]}{\int_{-1}^{+1} d\vec{\Omega} \exp\left[-\frac{1}{kT}(-\vec{m}^* \cdot \vec{E}_p)\right]}, \tag{67}$$

where $\vec{m}^*$ is the dressed dipole moment of each molecule in the ensemble, $\vec{\Omega}$ is the triplet of three Euler angles (see Fig. 6 in Chapter 6 for the definition of Euler angles), $\vec{E}_p$ is the applied poling field, T is the temperature of the system, and k is Boltzmann's constant. (For convenience, the triple integral over the Euler angles is represented by a single integral.) Because the dipole orientation is assumed to freeze in place when the sample is cooled below T_g, the orientational order of the sample below T_g is given by the Gibbs distribution function with $T = T_g$.

Substituting Eq. 67 into Eq. 65 and evaluating the orientational average to first order in m^*E_p/kT yields:

$$\chi_{ijk}^{(2)} = \frac{N\beta^*_{I'J'K'}}{8\pi^2} \int d\vec{\Omega}\, \frac{\vec{m}^* \cdot \vec{E}}{kT}\, a_{iI'}(\vec{\Omega}) a_{jJ'}(\vec{\Omega}) a_{kK'}(\vec{\Omega}), \tag{68}$$

where $a_{iI'}(\vec{\Omega})$ is the rotation matrix and is defined in Eq. 52 of Chapter 6. Assuming that the molecule is one-dimensional, that is, the only nonvanishing component of $\vec{m}^*$ and β^* is m_z^* and β^*_{ZZZ}, respectively, the two independent tensor components of the bulk response are:

$$\chi_{zzz}^{(2)} = N\beta^* \frac{m^* E_p}{5kT}, \tag{69}$$

and

$$\chi_{xxz}^{(2)} = N\beta^* \frac{m^* E_p}{15kT}. \tag{70}$$

This thermodynamic model for poling was first shown to hold for a disperse red azo dye (DR1) in poly(methyl methacrylate) (PMMA). In particular, given the measured value of β^* for this dye in solution, the bulk second harmonic coefficient of DR1/PMMA was predicted within experimental uncertainties for a wide range of poling field and number density. [17] Figure 7 shows the concentration and poling field dependence of the second harmonic coefficient. Note that the shaded region corresponds to the theoretical prediction. The uncertainty in the theory stems from uncertainty in β^*. For low poling field strengths, this model has been used to guide the development of poled polymers with enhanced second order susceptibilities.

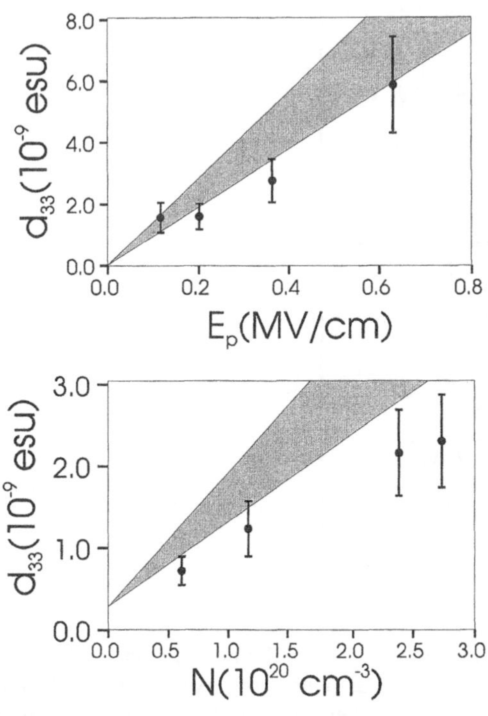

Figure 7 Dependence of second harmonic coefficient on poling field (upper) and number density (lower). Theory is represented by the shaded region.

3. Poled Anisotropic Polymer

While the above model is useful for a poled isotropic polymer, many materials are anisotropic. Examples of anisotropic materials include liquid crystals and stretched polymers. It is therefore useful to understand how the orientational order of a dopant dye molecule is affected when embedded in such a material. For stretched polymers and nonferroelectric materials, the orientational order is uniaxial. In other words, while the long axis of the molecules or chains are aligned along a single axis, the polar ends of the molecules are randomly oriented. Figure 8 shows a schematic representation of an isotropic material, axial order, and polar order.

When a dopant is incorporated into such a material, it will feel a local ordering potential even in the absence of an external electric field. Provided that the dye molecule concentration is sufficiently low so that it does not affect the local ordering forces, the energy of the system is lowest when the dye molecules take on the same axial order as the host. The axial ordering potential experienced by the dopants can be added to the Gibbs distribution function:

$$G(\Omega, E_p) = \frac{\exp\left[-\frac{1}{kT}(-\vec{m}^* \cdot \vec{E}_p + U(\cos(\theta)))\right]}{\int_{-1}^{+1} d\Omega \exp\left[-\frac{1}{kT}(-\vec{m}^* \cdot \vec{E}_p + U(\cos(\theta)))\right]}, \tag{71}$$

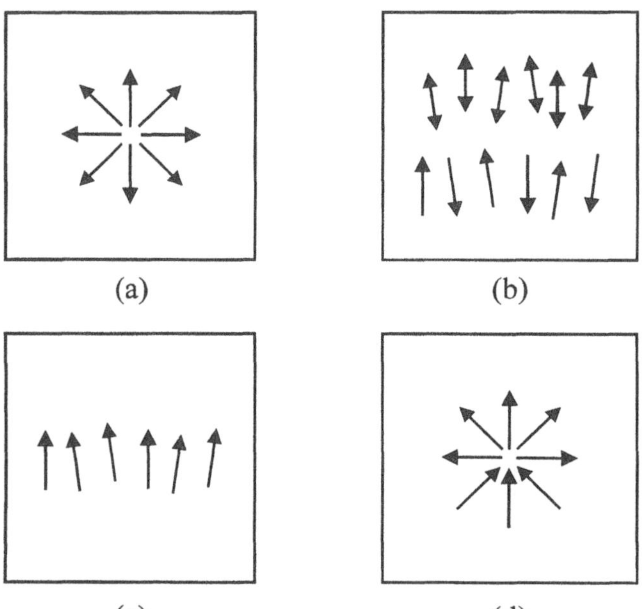

Figure 8 A schematic representation of a) an isotropic material where the odd-order poles such as the dipoles are randomly oriented, b) axial order, where the axes of the molecules are aligned with no net dipole moment, c) polar order where the dipoles are aligned, and d) polar order with no axial order, where there is a net dipole moment but the axes of the molecules are randomly oriented.

where $U(\cos(\theta))$ is the axial ordering potential. With no poling field applied, the order parameter of the dye molecule is given by:

$$<P_i> \equiv \int_{-1}^{+1} d(\cos\,\theta)P_i(\cos\,\theta) \frac{\exp\left[-\frac{1}{kT}\,U(\cos(\theta))\right]}{\int_{-1}^{+1} d\Omega\,\exp\left[-\frac{1}{kT}\,U(\cos(\theta))\right]}, \tag{72}$$

where P_i is the ith Legendre polynomial. (The Legendre polynomials are discussed in more detail in Section VI.) Because the axial forces as represented by $U(\cos\,\theta)$ are centrosymmetric, only even-order order parameters will be non-zero. Evaluating Eq. 65 for the second order distribution function with the generalized Gibbs distribution (Eq. 71) and using the definition of the order parameters (Eq. 72) yields:

$$\chi_{zzz}^{(2)} = N\beta^* \frac{m^*E_p}{kT}\left(\frac{1}{5} + \frac{4}{7}<P_2> + \frac{8}{35}<P_4>\right), \tag{73}$$

and

$$\chi_{xxz}^{(2)} = N\beta^* \frac{m^*E_p}{kT}\left(\frac{1}{15} + \frac{1}{21}<P_2> - \frac{4}{35}<P_4>\right). \tag{74}$$

We stress that the above equation relates the orientational order of the material before poling as quantified by $<P_2>$ and $<P_4>$ to the second order susceptibility after the material is poled. Without a poling field ($E_p = 0$), all tensor components vanish. When the applied electric field is large, however, it can affect the axial order parameters because a partial alignment of the material can induce strong interaction between molecules (as is the case for liquid crystals). Under these conditions, the calculations become more complicated for two reasons: The small field approximation given by Eq. 68 is no longer valid so that higher powers in E must be included when evaluating the integrals, and the electric field can force the material to undergo a phase transition which results in a drastic change in the order parameters. Under these conditions, a formalism which includes the axial interaction between molecules under a mean field approximation must be applied. This formalism is presented in Section IV.C.5.

4. Poling Materials with Axial Order

For any material in which the one particle potential is centrosymmetric and the even-order order parameters are not affected by the poling field, the order parameters can be calculated using Eq. 72. The second order susceptibility is then calculated using Eqs. 73 and 74 using these order parameters. Kuzyk and coworkers considered an axial potential of the form [43]:

$$U(\theta) = bP_2(\cos\,\theta), \tag{75}$$

where P_2 is the second Legendre polynomial and b the strength of the potential. When the axial potential arises from internal forces of a liquid crystal, the coefficient b is positive. The axial force thus acts to reinforce the poling field so that the orientational order is enhanced. In the limit when b is much larger than any other energy in the system, the order parameters approach unity and the two independent tensor components of $\chi^{(2)}$ (from Eqs. 73 and 74) become:

$$\chi^{(2)}_{zzz} = N\beta^* \frac{m^* E_p}{kT}, \tag{76}$$

and

$$\chi^{(2)}_{xxz} = 0. \tag{77}$$

The fact that $\chi^{(2)}_{xxz}$ vanishes can be easily understood. Recall that the derivation leading to Eqs. 73 and 74 assume that the dopant molecules are one-dimensional. If all the dipoles are perfectly aligned, the bulk second order nonlinear optical susceptibility reflects the microscopic hyperpolarizability tensor. Because β_{xxz} is zero in a one-dimensional molecules, the bulk response will also result in a vanishing $\chi^{(2)}_{xxz}$.

Such a vanishing component can have adverse consequences in certain nonlinear optical devices. In particular, if the information-carrying pulse of light is not polarized, an electro optic switch or modulator only acts on half of the light intensity. As a result, the modulating efficiency would never be over 50 percent and a switch would only divert half of the information to the proper port. While devices can be designed to act independent of light polarization in highly aligned materials, the design is not trivial. It would be much more convenient if all second order tensor components would be identical.

To evaluate a material's usefulness for polarization-independent device applications, it is convenient to define the tensor anisotropy, a:

$$a = \frac{\chi^{(2)}_{xxz}}{\chi^{(2)}_{zzz}}. \tag{78}$$

If $a = 1$, both tensor components are equal and a device built from such a material would act independent of the light polarization. Figure 9 shows a plot of the measured order parameters of several liquid crystalline materials. [43] The set of lines (calculated from Eqs. 73 and 74) represent constant tensor ratios as labeled on the figure. For an isotropic material, the ratio is $a = 1/3$. It is clear that the vast majority of liquid crystalline materials fall in the range $0 < a < 1/3$.

At least one of the order parameters must be negative if the tensor ratio a is to be unity. For an order parameter to be negative, the axial force should result in a depletion of molecules that are oriented along the poling field. Such behavior results when b is negative. Kuzyk and coworkers found that poling under uniaxial stress results in such a depletion and a tensor ratio near unity is found. [43] Section IV.C.6 discusses these findings.

5. *Poling Axial Materials with Strong Fields*

When the poling electric field is strong, it can induce a phase transition in the material. Calculations of the second order susceptibility in the high poling field case for a liquid crystalline material were first reported by van der Vorst and Picken. [44] They included in their potential function, U, the effect of poling to second order in the electric field and an effective single particle mean-field potential:

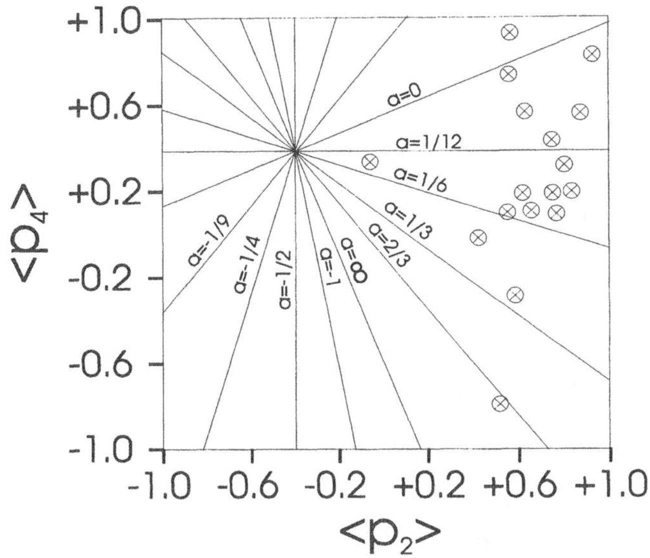

Figure 9 Order parameters of several liquid crystalline materials. Lines represent constant values of tensor ratio *a*.

$$U(\theta) = -\vec{m}^* \cdot \vec{E}_p - \frac{1}{2}\,\alpha E_p^2 - \frac{1}{3}\,\Delta\alpha E^2 P_2(\cos\,\theta) - \epsilon <P_2> P_2(\cos\,\theta), \qquad (79)$$

where the first term corresponds to dipolar poling, the second term corresponds to the energy shift of an isotropic material in response to the field (here α represents the isotropic average of the polarizability), the third term corresponds to poling of the induced dipole moment ($\Delta\alpha$ is the difference in the polarizability between the long and short axis of the cylindrical molecule and P_2 is the second order Legendre polynomial), and the fourth term corresponds to internal liquid crystalline forces as originally modeled by Maier and Saupe. [45] ϵ represents the strength of the mean field single particle potential and $<P_2>$ the second-order order parameter.

The second-order order parameter $<P_2>$ depends on the strength of the poling field and the single particle potential and must be calculated self-consistently. The upshot of such a self-consistent calculation is that the poling field affects the order parameter $<P_2>$ which can induce a phase transition between the isotropic and nematic phase of a liquid crystal.

Substituting Eq. 79 into Eq. 72 yields a self-consistent equation for the order parameters (the details of the calculation are too complex to present in this chapter). We can put the results of the self-consistent strong-field model in perspective by comparing it to the other models. The top portion of Fig. 10 shows the calculated order parameters [44] for a poled liquid crystal material with $m^* = 20.5D$, $\Delta\alpha = 47\text{Å}^3$, $T_g = 380$ K (glass transition), and $T_C = 340K$ (isotropic-nematic transition). Note that $kT_c \approx 0.22\epsilon$. The exact isotropic model (i.e., no weak field approximation and no liquid crystalline potential, $\epsilon = 0$) is given by the solid curve. The other

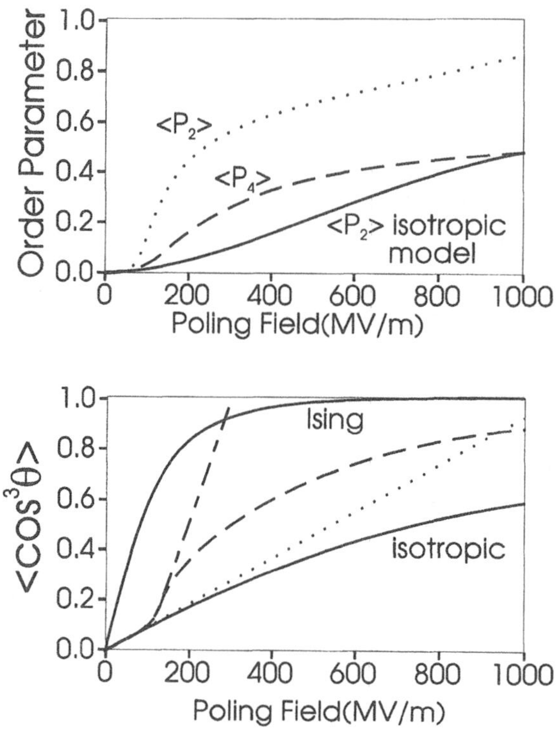

Figure 10 Top graph shows the self-consistent strong-field model (dotted and dashed curves) and exact isotropic model for axial order parameters. The bottom graph shows $<\cos^3 \theta>$ for the isotropic model (solid line), self-consistent field model (dashed line), Ising model (upper solid curve), and low-field model (dotted) that ignores the poling field's effect on the axial order parameters.

two curves represent a liquid crystal. For poling fields above 50 MV/m, it is clear that the liquid crystalline and isotropic models differ greatly.

The bottom portion of Fig. 10 shows $<\cos^3\theta>$ as a function of poling field strength. This parameter is directly proportional to $\chi^{(2)}_{zzz}$. The solid curve shows the exact isotropic model and the exact Ising model (the Ising model assumes that the dipole can point in only two discrete directions—up or down). The isotropic model assumes that the axial order parameters vanish while the Ising model assumes that they are all equal to unity. For low fields, the Ising model gives a value that is five times greater than the isotropic model. At infinite field, both curves approach unity. The dotted line represents the low-field approximation of the isotropic model. The two agree well below 200MV/m. The two dashed curves show the liquid crystalline system. The single-dashed curve is the self-consistent calculation of van der Vorst and Picken (VP model), [44] while the double-dashed curve represents the model of Singer, Kuzyk, and Sohn (SKS model) [17] in which the zero-field order parameters are used in Eqs. 73 and 74. Again, the two models agree for poling fields less than about 150MV/m but differ drastically at high poling fields.

6. Polar Ordered Materials

Finally, there may be instances in which the odd-order order parameters of a material are predetermined. An example of such a material is a ferroelectric liquid crystal or a Langmuir-Blodgett (LB) film. Because an even order response of order n depends only on the material's odd-order order parameters less than $n + 2$, the second order response will only depend on $<P_1>$ and $<P_3>$. All other order parameters are irrelevant. For the one-dimensional molecule, the two independent tensor components are:

$$\chi^{(2)}_{zzz} = N\beta^* \left(\frac{3}{5} <P_1> + \frac{2}{5} <P_3>\right), \tag{80}$$

and

$$\chi^{(2)}_{zzx} = N\beta^* \left(\frac{1}{5} <P_1> - \frac{1}{5} <P_3>\right). \tag{81}$$

To illustrate how Eqs. 80 and 81 are related to Eqs. 73 and 74, we consider the isotropic poled polymer. After a polymer is poled, its polar order can be described by odd-order order parameters. A straightforward calculation shows that the order parameter $<P_3>$ vanishes for small field strength ($m^*E/3kT \ll 1$) and $<P_1>$ (Eq. 72 with $U = 0$) is of the form,

$$<P_1> = \frac{m^*E}{3kT}. \tag{82}$$

Substituting Eq. 82 into Eq. 80 and 81 yields Eqs. 69 and 70, the thermodynamic model for the isotropic poled polymer.

Langmuir-Blodgett films form a material class in which polar order is induced in a monolayer by its interaction with a surface. [46] Figure 11 shows the measured polar order parameters for a few representative materials that were deposited with the LB technique. [47–51] The lines of constant tensor ratio were calculated using Eqs. 80 and 81. As in the case for liquid crystals, the aligning forces result in a tensor ratio in the range $0 < a < 1/3$. The shaded region, however, shows the range that was achieved by poling under uniaxial stress. [43] (The stress was applied by squeezing the sample with a clamp while the field is applied.) It is clear that poling with uniaxial stress results in materials that have a unique range of manipulable tensor components.

In summary, by choosing appropriate materials and poling them under well defined conditions, a wide range of nonlinear-optical properties result. When using the tabulated values of the molecular nonlinear optical properties that follow in the remaining chapters, one must keep in mind that the processing conditions can have a profound effect on the bulk response.

D. Third Order Response

We can evaluate Eq. 65 for the third order nonlinear optical susceptibility to relate the bulk response to the molecular susceptibility in terms of the even-order order parameters in analogy to the second order expression as given by Eqs. 80 and 81.

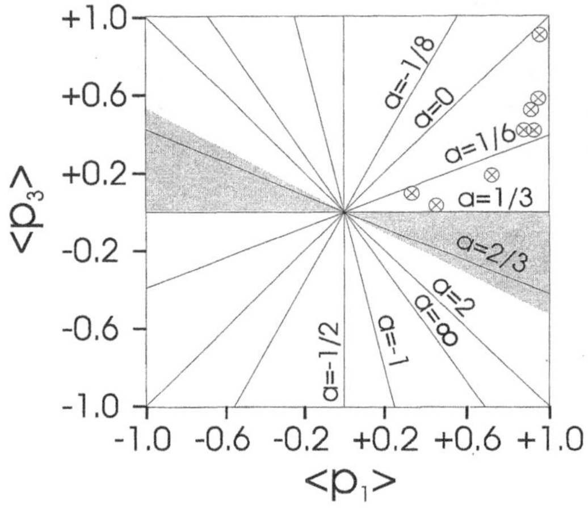

Figure 11 Order parameters of several Langmuir-Blodgett film materials. Lines represent constant values of tensor ratio a and shaded region represents the range of 'a's' that has been experimentally demonstrated with electric-field poling under uniaxial stress.

The result for an experiment that probes only along z or perpendicular to z (x-direction) is:

$$\chi^{(3)}_{zzzz} = N(\delta + 2\xi <P_2> + 8\eta <P_4>), \tag{83}$$

and

$$\chi^{(3)}_{xxxx} = N(\delta - \xi <P_2> + 3\eta <P_4>), \tag{84}$$

where

$$\delta = \frac{1}{5}(\gamma_{x'x'x'x'} + \gamma_{y'y'y'y'} + \gamma_{z'z'z'z'} + \gamma_{x'z'z'x'} + \gamma_{z'z'x'x'} + \gamma_{z'z'y'y'} + \gamma_{y'z'z'y'} + \gamma_{x'x'y'y'} + \gamma_{y'x'x'y'}), \tag{85}$$

$$\xi = \frac{1}{14}(4\gamma_{z'z'z'z'} - 2\gamma_{y'y'y'y'} - 2\gamma_{x'x'x'x'} + \gamma_{x'z'z'x'} + \gamma_{z'z'x'x'} + \gamma_{z'z'y'y'} + \gamma_{y'z'z'y'} - 2\gamma_{x'x'y'y'} + -2\gamma_{y'x'x'y'}), \tag{86}$$

and

$$\eta = \frac{1}{280}(8\gamma_{z'z'z'z'} + 3\gamma_{y'y'y'y'} + +3\gamma_{x'x'x'x'} - 12\gamma_{x'z'z'x'} - 12\gamma_{z'z'x'x'} - 12\gamma_{z'z'y'y'} - 12\gamma_{y'z'z'y'} + 3\gamma_{x'x'y'y'} + +3\gamma_{y'x'x'y'}), \tag{87}$$

and where the axial order parameters describe the degree of alignment along the z-direction and γ is the third order molecular susceptibility. Recall that this expression holds in the rigid gas approximation.

It is useful to consider Eqs. 83 and 84 for the case of a one-dimensional molecule in which the only non-zero component of the dressed molecular third order nonlinear susceptibility is $\gamma^*_{z'z'z'z'} = \gamma^*$:

$$\chi^{(3)}_{zzzz} = N\gamma^* \left(\frac{1}{5} + \frac{4}{7} <P_2> + \frac{8}{35} <P_4>\right), \tag{88}$$

and

$$\chi^{(2)}_{xxxx} = N\gamma^* \frac{m^* E_p}{kT} \left(\frac{1}{5} - \frac{2}{7} <P_2> + \frac{3}{35} <P_4>\right). \tag{89}$$

For an isotropic system (all order parameters set to zero), we find that the bulk response is one-fifth of the molecular susceptibility. When the molecules are fully aligned, all of the order parameters are unity, leading to $\chi^{(3)}_{zzzz} \rightarrow N\gamma^*_{z'z'z'z'}$ and $\chi^{(3)}_{xxxx} \rightarrow 0$.

1. Liquid Crystals

Wong and Garito applied third harmonic generation studies to determine the evolution of the order parameters in MBBA liquid crystals as a function of temperature. [52] Equations 83 and 84 can be used to relate $\chi^{(3)}_{zzzz}$ and $\chi^{(3)}_{xxxx}$ (both of which can be experimentally determined) to each axial order parameter:

$$<P_2> = \frac{\xi}{14\delta} \left(\frac{3\chi^{(3)}_{zzzz} - 8\chi^{(3)}_{xxxx}}{\chi^{(3)}_{iso}} + 5\right), \tag{90}$$

and

$$<P_4> = \frac{\xi}{14\eta} \left(\frac{\chi^{(3)}_{zzzz} + 2\chi^{(3)}_{xxxx}}{\chi^{(3)}_{iso}} - 3\right). \tag{91}$$

where $\chi^{(3)}_{iso}$ is the isotropic value of the third order susceptibility and is measured above the nematic-isotropic phase transition temperature.

To determine the order parameters, the coefficients δ, ξ, and η must be known. Recall that these parameters depend on the anisotropy of the dressed third order molecular susceptibility. As such, they should not depend on temperature. Wong and Garito determined the appropriate ratios of these parameters using a previously reported value of the order parameter $<P_2>$ at one temperature and their measured values of $\chi^{(3)}_{zzzz}$ and $\chi^{(3)}_{xxxx}$ at that same temperature. By substituting the known order parameters and measured susceptibilities into Eq. 90, they determined ξ/δ. Likewise, ξ/η can be determined.

Figure 12 shows the temperature dependence of the third harmonic susceptibility of MBBA as a function of temperature relative to the nematic-isotropic transition temperature. In the isotropic phase, the third harmonic susceptibility is found to be isotropic. Figure 13 shows $<P_2>$ as a function of temperature as determined from Fig. 12 and Eq. 90. The circles and crosses show $<P_2>$ as determined from diamagnetic studies [53] and NMR studies, [54] respectively.

The long range orientational order of a liquid crystal leads to a bulk birefringence. It is interesting to investigate the local electric fields for such a highly anisotropic system. Recall that the local electric fields are usually larger than the applied field so that the dressed susceptibility is larger than the vacuum suscepti-

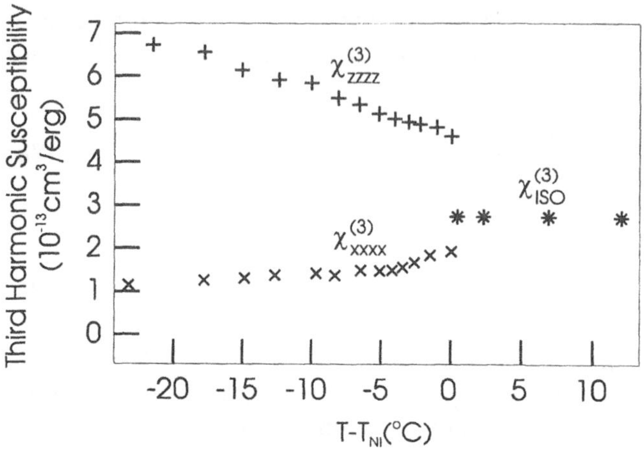

Figure 12 Third harmonic susceptibilities as a function of temperature (relative to the nematic-isotropic transition temperature) for a MBBA liquid crystal.

bility. If the molecular axis of each single molecule is on average aligned with the director (the axis in the bulk material corresponding to the largest refractive index), we would expect the local electric field to be larger along the molecular axis than perpendicular to it so that $F_{x'} < F_{z'}$. We would thus expect the dressed susceptibility to be more anisotropic than the vacuum susceptibility, that is:

$$\frac{\gamma_{z'z'z'z'}}{\gamma_{x'x'x'x'}} < \frac{\gamma^*_{z'z'z'z'}}{\gamma^*_{x'x'x'x'}}. \tag{92}$$

The dressed molecule thus appears to be more "one-dimensional" than the vacuum dipole if the local fields reflect the long range order of the liquid crystal.

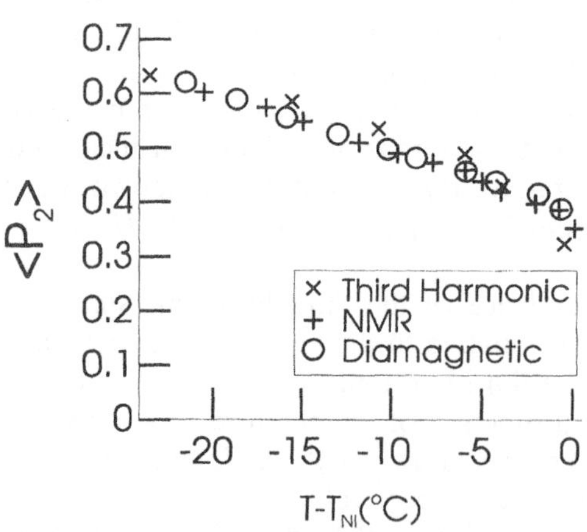

Figure 13 Order parameter $<P_2>$ as measured with third harmonic (×), NMR (+) and Diamagnetic studies (○) as a function of temperature (relative to the nematic-isotropic transition temperature) for a MBBA liquid crystal.

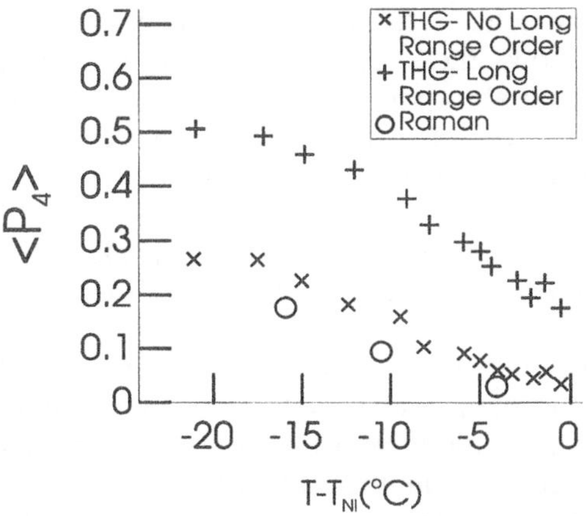

Figure 14 Order parameter $<P_4>$ as measured with third harmonic (×) without accounting for long range order, third harmonic accounting for long range order (+) and Raman scattering studies (○) as a function of temperature (relative to the nematic-isotropic transition temperature) for a MBBA liquid crystal.

Figure 14 shows a plot of the order parameter $<P_4>$ as determined from third harmonic generation with (+) and without (×) the long range order of the liquid crystal taken into account. The open circles show the order parameter as determined from Raman scattering. [55] It is clear that the third harmonic measurements are consistent with Raman scattering only if the long range order of the liquid crystal is not taken into account. This implies that the local field model appears isotropic in the vicinity of each molecule.

Durbin and Shen used two-photon absorption dichroism [56] to study the effects of long range order on the local fields. Two-photon absorption is described by the imaginary part of the third order nonlinear optical susceptibility. Figure 15

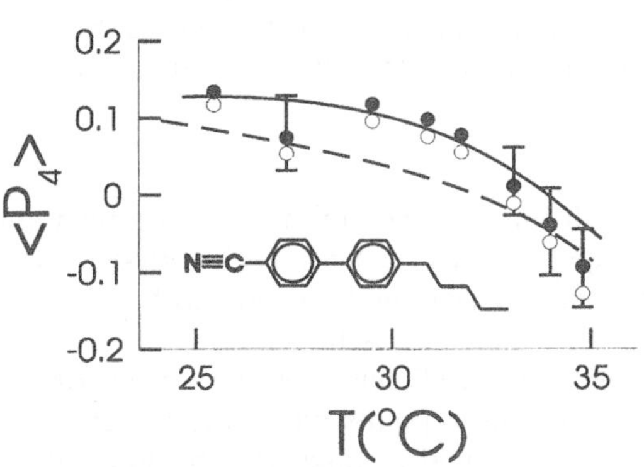

Figure 15 Order parameter $<P_4>$ as measured with two-photon dichroism (open circles assume a one-dimensional molecule and solid circles assume a small transverse two-photon cross-section). Solid curve includes effect of long range order in local field, while dashed curve assumes isotropic local field.

shows their determination of the order parameter $<P_4>$ of the liquid crystal 5CB as a function of temperature (the inset shows the structure of 5CB). The open circles are data assuming that the two-photon absorption tensor has only one non-zero component (i.e., a one-dimensional molecule), while the solid circles show the order parameter assuming that the two photon tensor has a perpendicular component of about 1.2 percent of the large component. The solid curve in Fig. 15 includes the effect of long range order and the dashed curve do not. It is clear that this data implies that the local electric field is affected by long range order. This contradicts the results of Wong and Garito.

These two separate experiments show that third order effects, such as third harmonic generation and two-photon dichroism, are useful tools for studying liquid crystalline order. Furthermore, they give information about the local field environment of the molecule. Such studies, however, illustrate the difficulty in understanding how local fields are included in models that relate molecular properties to the bulk response. While there may be questions as to how one might model highly anisotropic systems such as liquid crystals, nearly isotropic materials such as doped polymers are much more straightforward to understand and model.

V. EFFECTIVE DIELECTRIC THEORY

In the previous section, the dilute solute model is applied to several systems. An alternative method for relating the microscopic dielectric response of a composite material's substituents to the bulk response falls under the class of effective dielectric models. [2, 57] While there are many possible models, we concentrate on the simple case of a spherical inclusion in a host medium using the Maxwell Garnett formalism. [58] Both the linear and nonlinear optical response is treated. We conclude this section by showing the connection between the Maxwell Garnett result and the dilute solute model.

A. Maxwell Garnett Model

After considering the linear case in depth, we discuss the nonlinear problems as applied to the intensity-dependent refractive index and second harmonic generation. The reader is referred to the work of Sipe and Boyd who give rigorous derivation of the nonlinear case. [59]

1. Linear Case

Figure 16a shows a spherical inclusion with linear susceptibility $\chi_2^{(1)}$ in a medium of linear susceptibility $\chi_1^{(1)}$. A uniform applied electric field is distorted by the sphere. The average electric field in the host material can be shown to be uniform as follows. The total electric field outside the inclusion is the sum of the uniform applied electric field and a dipole field due to induced charge on the surface of the sphere. The average electric field due to the dipole points along the direction of the applied electric field. (Note that the average field is calculated by integrating the field per unit volume over the volume of interest.) If the inclusion concentration is low, the electric field contribution from the dipole to the average field is negligible.

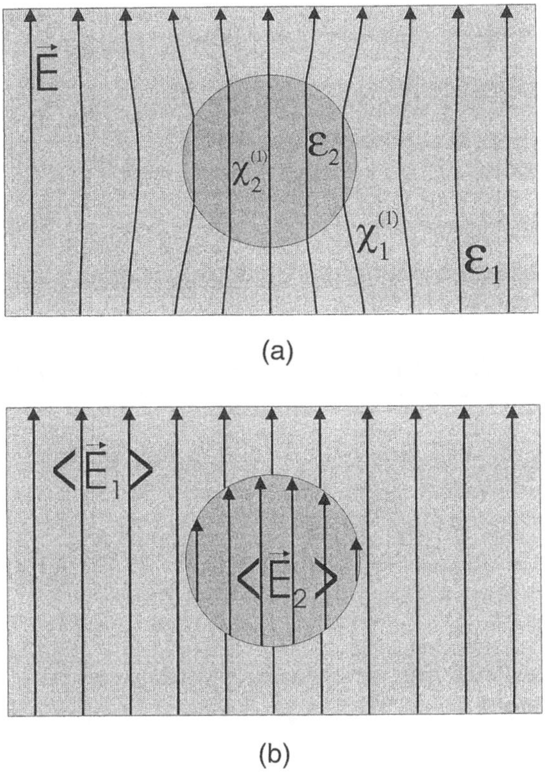

Figure 16 In the Maxwell Garnet treatment for a) a spherical inclusion, b) the average field in the sphere $<\vec{E}>$ and in the surrounding medium $<\vec{E}_2>$ are uniform.

If the volume fraction of the inclusion is f, and the average electric fields in the inclusion and surrounding medium are $<\vec{E}_2>$ and $<\vec{E}_1>$, respectively, the average field, $<\vec{E}_{av}>$, in the medium is:

$$<\vec{E}_{av}> = (1 - f) <\vec{E}_1> + f <\vec{E}_2>. \tag{93}$$

Because the average electric field is uniform in the material, we will express the average fields as scalar quantities. Furthermore, because it is understood that we are dealing with average fields, we drop the brackets and the "av" subscript so that Eq. 93 becomes:

$$E = (1 - f)E_1 + fE_2. \tag{94}$$

Maxwell Garnett starts with the relationship between the average electric field inside the sphere and outside the sphere as given by Eq. 8:

$$E_2 = \frac{3\epsilon_1}{2\epsilon_1 + \epsilon_2} E_1. \tag{95}$$

With the average polarizations given by:

$$P_1 = \chi_1^{(1)} E_1, \tag{96}$$

and

$$P_2 = \chi_2^{(1)} E_2, \tag{97}$$

the average polarization in the medium is:

$$P = \chi^{(1)} E = (1 - f) P_1 + f P_2, \tag{98}$$

where $\chi^{(1)}$ is the average or effective susceptibility of the composite medium. Substituting Eqs. 96 and 97 into Eq. 98 yields:

$$\chi^{(1)} E = (1 - f) \left(\frac{\epsilon_1 - 1}{4\pi} \right) E_1 + f \left(\frac{\epsilon_2 - 1}{4\pi} \right) E_2. \tag{99}$$

We can eliminate E_1 and E_2 from Eq. 99 with the help of Eqs. 94 and 95:

$$\chi^{(1)} = \frac{(1 - f) \left(\frac{\epsilon_1 - 1}{4\pi} \right) + f \left(\frac{\epsilon_2 - 1}{4\pi} \right) \left(\frac{3\epsilon_1}{2\epsilon_1 + \epsilon_2} \right)}{1 - f + f \left(\frac{3\epsilon_1}{2\epsilon_1 + \epsilon_2} \right)}. \tag{100}$$

After some algebraic manipulation, Eq. 100 can be expressed in terms of the effective dielectric function $\epsilon = 1 + 4\pi\chi^{(1)}$,

$$\epsilon = \epsilon_1 \left[\frac{\epsilon_2 + 2\epsilon_1 + 2f(\epsilon_2 - \epsilon_1)}{\epsilon_2 + 2\epsilon_1 - f(\epsilon_2 - \epsilon_1)} \right]. \tag{101}$$

This is the Maxwell Garnett result. For a small volume fraction, the Maxwell Garnett effective dielectric function reduces to:

$$\epsilon = \epsilon_1 \left(1 + 3f \left(\frac{\epsilon_2 - \epsilon_1}{\epsilon_2 + 2\epsilon_1} \right) \right). \tag{102}$$

The Maxwell Garnett calculation is based on considering an inclusion inside a host medium. It is often difficult to distinguish between the inclusion and the host, particularly when the components of a composite are of approximately equal concentration. For such a material, Bruggeman showed that the average dielectric function is of the form [60]:

$$f \frac{\epsilon_2 - \epsilon}{\epsilon_2 + 2\epsilon} + (1 - f) \frac{\epsilon_1 - \epsilon}{\epsilon_1 + 2\epsilon} = 0. \tag{103}$$

The applicability of each of these models depends on the system.

2. *Nonlinear Case*

In this subsection, we discuss the generalization of the Maxwell Garnett expression to the nonlinear case. The general problem is too complex to be rigorously treated here. Rather, we discuss two simple cases: The intensity-dependent refractive index, and second harmonic generation.

Intensity-dependent Refractive Index. There are several approaches to modeling the nonlinear response of a composite. We begin by discussing the intensity dependent refractive index. Through a third order nonlinear optical process, the refractive index, n, is a linear function of the intensity, I:

$$n = n_0 + n_2 I, \tag{104}$$

where n_0 is the low-intensity intensity-dependent refractive index, and n_2 is related to the third order nonlinear optical susceptibility. The dielectric function is then also a function of intensity. The simplest nonlinear optical composite material is one in which the inclusion is nonlinear and the surrounding material is linear. The dielectric function of the inclusion can be expanded in a power series of intensity and to lowest order of I is of the form:

$$\epsilon_2 = \epsilon_{20} + \epsilon_{22} I, \tag{105}$$

where ϵ_{02} is the zero-intensity dielectric function and ϵ_{22} is related to the third order susceptibility. ϵ_{22} is also related to n_{22} according to:

$$\epsilon_{22} = \frac{1}{2}\frac{n_2}{(n_0)^{1/2}}. \tag{106}$$

We can compute the nonlinear response for the composite material by recalling that:

$$n_2 = \left.\frac{\partial \epsilon^{1/2}}{\partial I}\right|_{I=0} = \left.\frac{1}{2\epsilon^{1/2}}\frac{\partial \epsilon}{\partial I}\right|_{I=0}. \tag{107}$$

In the low concentration limit (first-order in f), substituting Eq. 105 into Eq. 102, and performing the differentiation according to Eq. 107, we get:

$$n_2 = \frac{f\left(\dfrac{3\epsilon_{10}}{\epsilon_{20} + 2\epsilon_{10}}\right)^2 \epsilon_{22}}{2\epsilon_{10}^{1/2}}, \tag{108}$$

where ϵ_{10} is the linear dielectric functionof the host.

Second Harmonic Generation. Through the second order susceptibility, a light wave at frequency ω induces a polarization at 2ω. In the following, we derive a simplistic model that relates the effective material second order susceptibility $\chi^{(2)}$ to the susceptibilities of the host $\chi_1^{(2)}$ and the inclusion $\chi_2^{(2)}$. If an appreciable intensity of second harmonic light is generated in the material, it can interact with the fundamental leading to intractable complications. We therefore deal only with the low depletion limit, that is, we assume that the second harmonic intensity is always much smaller than the fundamental intensity at every point in the material. (This is true for either low intensities or small second harmonic coefficients).

As in the linear case, we define the average second order polarization in the form:

$$<P^{(2)}> = \chi^{(2)} <E^2> = (1 - f)P_1^{(2)} + fP_2^{(2)}, \tag{109}$$

where $\chi^{(2)}$ is the average or effective susceptibility of the composite medium, and

$$P_1^{(2)} = \chi_1^{(2)} <E_1^2>, \tag{110}$$

and

$$P_2^{(2)} = \chi_2^{(2)} <E_2^2>, \tag{111}$$

where the polarization in the dielectric is related to the average of the square of the fundamental field in that region. Because the fundamental electric field in the sphere is related to the external field according to Eq. 95, the relationship between the square of the fields is:

$$<E_2^2> = \left(\frac{3\epsilon_1}{2\epsilon_1 + \epsilon_2}\right)^2 <E_1^2>. \tag{112}$$

Similarly, the average of the square of the field is:

$$<E^2> = (1 - f) <E_1^2> + f <E_2^2>, \tag{113}$$

so we can combine Eqs. 109 through 113 to get the effective second order susceptibility:

$$\chi^{(2)} = \frac{(1 - f)\chi_1^{(2)} + f\chi_2^{(2)} \left(\frac{3\epsilon_1}{2\epsilon_1 + \epsilon_2}\right)^2}{1 - f + f\left(\frac{3\epsilon_1}{2\epsilon_1 + \epsilon_2}\right)^2}, \tag{114}$$

which for small f leads to:

$$\chi^{(2)} = \chi_1^{(2)} + f\left(\frac{3\epsilon_1}{2\epsilon_1 + \epsilon_2}\right)^2 (\chi_2^{(2)} - \chi_1^{(2)}). \tag{115}$$

B. Connection Between Maxwell Garnett and Dilute Solute Models

In the limit of low inclusion concentration, the Maxwell Garnett expression becomes the dilute solute model as given by Eq. 65. To illustrate, we calculate the linear response of a collection of molecules in a vacuum.

Figure 17 shows a collection of molecules that forms a cubic lattice, each occupying a space of volume V. In the dilute limit, the linear susceptibility is:

$$\epsilon = 1 + 4\pi\chi^{(1)} = 1 + 4\pi N\alpha. \tag{116}$$

According to Eq. 3, α for a spherical molecule is related to its diameter, a, and dielectric constant (ϵ_1), so that:

$$\epsilon = 1 + 4\pi N\left(\frac{\epsilon_1 - 1}{\epsilon_1 + 2}\right) a^3. \tag{117}$$

Because one molecule occupies a volume V, the number density is simply $N = 1/V$. The volume fraction can thus be expressed as:

$$f = \frac{\frac{4}{3}\pi a^3}{V} = \frac{4}{3} N\pi a^3. \tag{118}$$

Solving Eq. 118 for N and substituting into Eq. 117, we get:

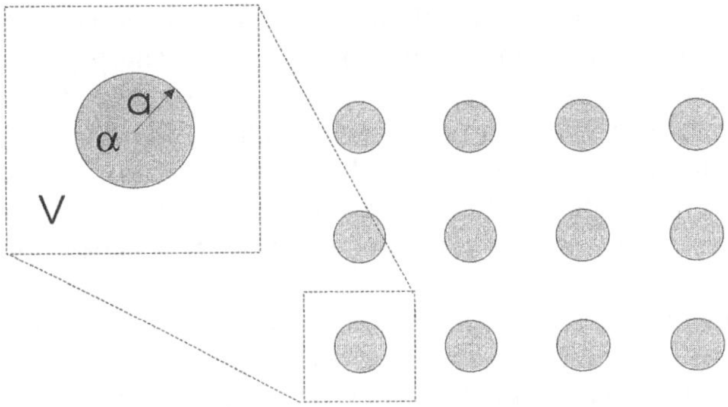

Figure 17 Lattice of molecules of radius a and polarizability α. On average, there is one molecule per volume V.

$$\epsilon = 1 + 3f\left(\frac{\epsilon_1 - 1}{\epsilon_1 + 2}\right). \tag{119}$$

This is just Eq. 102 with $\epsilon_2 = 1$. While we have chosen a cubic lattice, the final result applies to any material configuration.

VI. NONLINEAR MECHANISMS

Section IV concentrated on the electronic mechanism. There are several other mechanisms that lead to nonlinear optics, many of which have only recently been applied to the understanding of the interaction of light with composite nonlinear optical materials. In this section, we outline the underlying theory of these mechanisms and how they can be related to the bulk nonlinear optical response. As we have already seen in previous sections of this chapter, there are many ways in which an electric field can cause charges to redistribute in a material. The various mechanisms can be understood by considering Eq. 63. Recalling that the susceptibility is a derivative of the polarization with respect to the electric field, a particular mechanism is defined by the variation in one of the parameters in Eq. 63 with respect to field. Physical quantities that depend on the electric field include the number density, local field factor, molecular orientation, and molecular dipole moment. (The dependence of dipole moment of an immobile molecule whose nuclei are fixed—i.e., no vibrations or reorientation—when an electric field is applied is that electronic response.) Cross-terms can lead to all sorts of interesting coupled mechanisms, few of which would fit in this book if treated in detail. We are therefore forced to limit our discussion to those mechanisms that are most often encountered in the literature. It may very well be that those never-encountered mechanisms may be important in understanding the so called "artifacts" that are usually attributed to experimental quirks.

Before proceeding with a theoretical description of mechanisms, it is important to mention the role of nuclear vibrations. The electronic excited states of a molecule are generally separated by energies that are much larger than typical vibrational energies. In a linear absorption spectrum, these vibrational levels result in a broadening of the electronic peaks. If the temperature of the material is lowered, the vibrational states become more pronounced and appear as a series of sharp peaks that are superimposed on the electronic peak. This chapter does not treat the topic of vibrational states and their effect on the optical properties. Often, the effect of these states is taken into account by using a phenomenological broadening factor, which is the approach used in this chapter. While off-resonant electronic mechanisms are well approximated by using such a phenomenological parameter, the reader is cautioned that this may not always work.

To narrow the field, we begin this section by defining, in broad terms, two classes of nonlinear optical response and give the simplest examples of several mechanisms of both linear and nonlinear optical response. In the following subsections, we apply these concepts in detail to model the first, second, and third order responses. While we regrettably omit many possible mechanisms, it should be straightforward for the reader to dissect Eq. 63 and pull out the desired mechanism. This chapter is intended to provide the general armament to tackle all mechanisms that arise from Eq. 63.

We group nonlinear mechanisms into two broad classes: local mechanisms, which originate in the individual molecules, and nonlocal mechanisms, which originate solely from a collective response. Examples of local mechanisms include electric-field-induced electron cloud distortion and molecular reorientation. Nonlocal mechanisms include electric-field-induced density changes or changes in a sample's shape.

Formally, the mechanisms can be understood by considering Eq. 63. Because the nth-order nonlinear susceptibility is defined as the coefficient of the E^n term in the power series expansion of the polarization, any part of Eq. 63 that depends on the electric field will contribute to the susceptibility. Equations 63 and 64 can be combined into the form:

$$P_i = \sum_n N <p_{i'}^{*(n)}>_i = \sum_n N \frac{\int\int\int d^3\vec{\Omega} G(\vec{\Omega}) a_{ii'}(\vec{\Omega}) p_{i'}^{*(n)}}{\int\int\int d^3\vec{\Omega} G(\vec{\Omega})}, \tag{120}$$

where, for simplicity, we assume a single component system. Note also that the superscript, n, now refers to the order of the nonlinearity. The left-hand side of Eq. 120 is the polarization, which is related to a sum over all orders of the molecular induced dipole moment. Because many parameters in Eq. 120 implictly depend on the electric field (such as the orientational distribution function, which changes in the presence of an electric field and leads to the reorientational mechanism), the nth-order polarization will in the most general case be a function of all orders of the induced dipole moment. This is in contrast to the electronic mechanism in which the nth-order polarization is equal to the term containing the nth-order-induced dipole moment, which leads to the simple relationship given by Eq. 65.

To understand all possible mechanisms of the bulk nonlinearity of a material, we must carefully consider which parameters in Eq. 120 depend on electric field. In addition to the induced dipole moment due to the electron cloud, this includes the molecular number density, the volume of the sample, the local field factors, and the orientational distribution function. Because the nonlinear response arises from terms that are proportional to at least the second power of the electric field, second and higher order derivatives with respect to the electric field lead to a wonderfully rich array of mechanisms. Given the small number of mechanisms that have been carefully studied, it is highly probable that lots of interesting physics await discovery. Such new physics will certainly lead to practical applications beyond the horizon of imagination. We stress at this point that Eq. 120 does not include collective excitations (i.e., coherent superpositions of excited molecules, collision induced excitations, etc.) which, if acting in concert with the mechanisms described here, would further broaden the possibility of discovery.

The remainder of this chapter deals with the mechanism of the response of a composite material within the framework of the non-interacting model according to Eq. 120. While many of the mechanisms are discussed, only the better known ones are treated with a greater level of detail.

A. Example of Local Mechanisms—Molecular Reorientation

To illustrate local mechanisms, we focus on molecular reorientation and its contribution to the nonlinear response. In the subsections that follow, we apply the methodology from the molecular reorientation calculation to other mechanisms and to higher order response.

For simplicity, we consider a molecule whose orientation is described by a single angle. As such, this model holds only for molecules that have a rotational symmetry axis and for which the ensemble of molecules has azimuthal symmetry. Many materials, such as poled polymers, have this symmetry. Generalizing such a model to the case of a system that is described by three angles is straightforward and will not be discussed here.

Consider a single molecule inside a material with equilibrium orientation θ_0. When the molecule is reoriented by an external force to some angle θ, as measured in the laboratory frame, its total energy will increase. The potential energy of the molecule must therefore be a function of $\theta - \theta_0$. We call this potential energy function the internal energy $U_{int}(\theta - \theta_0)$ because it arises from interactions between the molecule and its surroundings.

If this molecule is placed in an external electric field $\vec{E}$ that is polarized along $\theta = 0$ (the laboratory z-axis) it will reorient toward the electric field. We define $U_E(\theta)$ to be the potential energy due to the action of the electric field. The Gibbs distribution function of this molecule is then of the form:

$$G(\theta, \theta_0) = \frac{\exp[-(U_E(\theta) + U_{int}(\theta - \theta_0))/kT]}{\int_{-1}^{+1} d(\cos\theta)\exp[-(U_E(\theta) + U_{int}(\theta - \theta_0))/kT]}. \tag{121}$$

The ith component of the dressed dipole moment of one molecule whose zero-field equilibrium orientation is θ_0 can then be determined from the distribution function:

$$<p_{i'}^*(\theta_0)>_i = \int_{-1}^{+1} d(\cos\theta) G(\theta, \theta_0) p_{i'}^* a_{i'i}(\theta). \tag{122}$$

This expression is similar to Eq. 64 except here we are dealing with a single molecule whose equilibrium orientation is θ_0.

To calculate the ensemble average of the dressed dipole moment, we need to know the distribution of equilibrium orientations. For an equilibrium distribution function $F(\theta_0)$, the bulk polarization can be written in the form:

$$P_i = N <p_{i'}^*>_i = N \int_{-1}^{+1} d(\cos\,\theta_0)F(\theta_0) <p_{i'}^*(\theta_0)>_i. \tag{123}$$

To evaluate the integrals in Eq. 122, we need to know the form of the internal potential. Furthermore, we need to understand how an optical field contributes to the potential energy. Let us first consider the electric field energy. In the time domain, the electric potential energy can be written in the form:

$$U_E(t) = -\int \vec{p}^*(\vec{E}(t)) \cdot d\vec{E}(t), \tag{124}$$

where the local field factor relating the dressed to vacuum dipole moments is given by Eq. 59. The dipole moment, $\vec{p}^*$, though, is a series in the electric field in which each term is of the form given by Eq. 56.

While most nonlinear optical experiments use a finite number of discrete monochromatic waves, for simplicity, we assume that there is a continuous distribution of frequencies. The induced dressed dipole moment $\vec{p}^*$ and applied electric field $\vec{E}$ can then be written in the form:

$$\vec{p}^*(t) = \sqrt{\frac{1}{2\pi}} \int_{-\infty}^{+\infty} d\omega\vec{p}^*(\omega) \exp[-i\omega t], \tag{125}$$

and

$$\vec{E}(t) = \sqrt{\frac{1}{2\pi}} \int_{-\infty}^{+\infty} d\omega\vec{E}(\omega) \exp[-i\omega t]. \tag{126}$$

(Note that in contrast to Chapter 1, we use the symmetric form of the Fourier transform in which all transforms have the $1/(2\pi)^{1/2}$ prefactor. The final results, however, are the same.) Later, we will limit the frequencies to a discrete set by setting $\vec{p}^*(\omega)$ and $\vec{E}(\omega)$ to a sum of Dirac delta functions. Note that because $\vec{p}^*(t)$ and $\vec{E}(t)$ are real, the complex conjugate of $\vec{p}^*(\omega)$ is $\vec{p}^*(-\omega)$ and the complex conjugate of $\vec{E}(\omega)$ is $\vec{E}(-\omega)$.

Substituting Eqs. 125 and 126 into Eq. 124, we get:

$$U_E(t) = -\frac{1}{2\pi} \int \left[\int_{-\infty}^{+\infty} d\omega\vec{p}^*(\omega) \exp(-i\omega t)\right] \times \left[\int_{-\infty}^{+\infty} d\omega' d\vec{E}(\omega') \exp(-i\omega' t)\right]. \tag{127}$$

We can Fourier transform Eq. 127 according to:

$$U(\omega'') = \sqrt{\frac{1}{2\pi}} \int_{-\infty}^{+\infty} dt U(t) \exp[+i\omega'' t]. \tag{128}$$

After some rearranging, this Fourier transform yields:

$$U(\omega'') = -\sqrt{\frac{1}{2\pi}}\int\int_{-\infty}^{+\infty} d\omega \int_{-\infty}^{+\infty} d\omega'$$
$$\times \left[\frac{1}{2\pi}\int_{-\infty}^{+\infty} dt\, \exp(-i(\omega + \omega' - \omega'')t)\right] \vec{p}^{\,*}(\omega)\cdot d\vec{E}(\omega'). \quad (129)$$

The bracketed term in Eq. 129 is the expression for the Dirac delta function of the frequency sum $(\omega + \omega' - \omega'')$. The frequency integrals thus collapse, yielding:

$$U(\omega'') = \sqrt{\frac{1}{2\pi}}\int\int_{-\infty}^{+\infty} d\omega \vec{p}^{\,*}(\omega)\cdot d\vec{E}(\omega'' - \omega), \quad (130)$$

where $U''(\omega)$ is the energy per unit frequency. Note that the energy density, averaged over many optical cycles, is just the zero frequency limit of $U(\omega)$, that is, $\bar{U} = U(\omega \rightarrow 0)$.

For a continuum of colors, the nth-order contribution to the induced dipole moment is of the form:

$$p_i^{*(n)}(\omega) = \left[\int_{-\infty}^{+\infty} d\omega_l \int_{-\infty}^{+\infty} d\omega_m \times \ldots\right] \xi_{ijk\ldots}^{*(n)}(-\omega; \omega_l, \omega_m, \ldots)$$
$$E_j(\omega_l)E_k(\omega_m)\ldots \times \delta(\omega_l + \omega_m + \ldots - \omega), \quad (131)$$

and its contribution to the Fourier transform of the internal energy is:

$$U^{(n)}(\omega'') = \sqrt{\frac{1}{2\pi}}\int\int_{-\infty}^{+\infty} d\omega dE_i(\omega'' - \omega)\left[\int_{-\infty}^{+\infty} d\omega_l \int_{-\infty}^{+\infty} d\omega_m \times \ldots\right]$$
$$\xi_{ijk\ldots}^{*(n)}(-\omega; \omega_l, \omega_m, \ldots)E_j(\omega_l)E_k(\omega_m)\ldots\delta(\omega_l + \omega_m + \ldots - \omega). \quad (132)$$

This is the most general expression for the *nth*-order contribution to the energy per unit frequency.

1. Example—Energy of a One-Dimensional Molecule

To illustrate how we can use Eq. 132 to calculate the electric potential energy, we will determine $U^{(1)}(\omega'')$ for the case of a one-dimensional molecule (only α_{zz}^{*} nonzero) exposed to a monochromatic beam (of frequency ω). The zero frequency limit of Eq. 132 gives the averaged energy:

$$U^{(1)}(\omega'') = \sqrt{\frac{1}{2\pi}}\int\int_{-\infty}^{\infty} d\omega \alpha_{zz}^{*}(-\omega; \omega)E_z(\omega)dE_z(\omega'' - \omega)$$
$$= \sqrt{\frac{1}{2\pi}}\int\int_{-\infty}^{\infty} d\omega \alpha_{zz}^{*}(-\omega; \omega)\left[\frac{\sqrt{2\pi}}{2}(E_z^{\omega}\delta(\omega' - \omega)\right.$$
$$\left. + E_z^{-\omega}\delta(\omega' - \omega))\right] \times \left[\frac{\sqrt{2\pi}}{2}(dE_z^{\omega}\delta(\omega' - \omega'' + \omega)\right.$$
$$\left. + dE_z^{-\omega}\delta(\omega' + \omega'' - \omega))\right], \quad (133)$$

where the bracketed terms in Eq. 133 are derived by applying a Fourier transform to a monochromatic field at frequency ω' of the form:

$$E_z(t) = \frac{E_z^{\omega'}}{2} \exp[-i\omega' t] + \frac{E_z^{-\omega'}}{2} \exp[+i\omega' t]. \tag{134}$$

The same Fourier transform applies to dE_z.

Applying intrinsic permutation symmetry ($\alpha_{ij}(-\omega;\omega) = \alpha_{ji}(\omega;-\omega)$) and integrating Eq. 133 over ω yields:

$$\begin{aligned} U^{(1)}(\omega'') = \frac{1}{2}\int & \alpha_{zz}^{*}(-\omega';\omega')\,[E_z^{\omega'}dE_z^{-\omega'}\delta(\omega'') \\ & + E_z^{-\omega'}dE_z^{\omega'}\delta(-\omega'') + E_z^{\omega'}dE_z^{\omega'}\delta(2\omega' - \omega'') \\ & + E_z^{-\omega'}dE_z^{-\omega'}\delta(2\omega' + \omega'')]. \end{aligned} \tag{135}$$

Analogous to the electric field's Fourier transform as given by Eq. 126, $U(\omega)$ can be expressed in terms of its energy amplitudes $U^{\pm\omega''}$. We are, however, only interested in the zero frequency limit of $U(\omega'')$. Considering only the amplitudes of the zero frequency terms:

$$U^{(1)}(\omega'') = \frac{\sqrt{2\pi}}{2}\,[U^0\delta(-\omega'') + U^0\delta(\omega'')]. \tag{136}$$

The average energy of the molecule is U^0. We can solve for U^0 by setting Eqs. 136 and 135 equal to each other and integrating over ω'' from $-\epsilon$ to ϵ in the limit $\epsilon \to 0$. Before doing so, we must consider two separate cases. When the applied electric field is time-independent, the integral over ω'' near $\omega'' = 0$ will pick up contributions from all four delta functions. If the applied field's frequency is nonzero, the integral only picks up the first two delta functions in Eq. 135. For a harmonic input ($|E^{\omega}| = |E^{-\omega}|$) at frequency ω, the average energy is:

$$U^0 = \frac{1}{4} D\alpha_{zz}^{*}(-\omega;\omega)|E_z^{\omega}|^2, \tag{137}$$

where $D = 2$ when $\omega = 0$ and $D = 1$ otherwise. The higher order contributions to the energy can be calculated using the same procedure with Eq. 132 as the starting point.

The nonlinear susceptibilities can be calculated by first evaluating Eq. 123. To do so, the orientational distribution function (Eq. 121) with $U_E = \bar{U}$ is substituted into Eq. 122 whch is used to evaluate Eq. 123. The nonlinear optical susceptibility is then calculated by differentiating Eq. 123 with respect to the electric fields:

$$\begin{aligned} &\chi_{ijk\ldots}^{(n)}(-\omega_1 - \omega_2 - \ldots;\omega_1,\omega_2,\ldots) \\ &= \frac{1}{K^{(n)}(-\omega_1 - \omega_2 - \ldots;\omega_1,\omega_2,\ldots)}\,\frac{1}{D_{ijk\ldots}^{(n)}(\omega_1,\omega_2,\ldots)} \\ &\quad \times \left[\frac{\partial^n}{\partial E_j^p(\omega_1)\partial E_k^r(\omega_2)\ldots} P_i\right]\Bigg|_{\vec{E}=0}, \end{aligned} \tag{138}$$

where p is the number of times that ω_1 appears with polarization j, r is the number of times that ω_2 appears with polarization k, and so on, and $p + r + \ldots = n$. Note that, because U_{int} is independent of the applied electric fields, we do not need to know its form to compute the derivative of Eq. 138. The factor $K^{(n)}(-\omega_1 - \omega_2 - \ldots;\omega_1,\omega_2, \ldots)$ is described in Chapter 1 and arises from the definition of the susceptibility in frequency space. The degeneracy factor, $D^{(n)}_{ijk\ldots}(\omega_1,\omega_2, \ldots)$, is given by:

$$D^{(n)}_{ijk\ldots}(\omega_1,\omega_2, \ldots) = \prod_{\alpha=1}^{3} \prod_{\beta=1}^{q} n^{\omega_\beta}_{x_\alpha}!, \tag{139}$$

where $n^{\omega_\beta}_{x_\alpha}$ is the number of distinct electric fields with frequency ω_β that are polarized along x_α, $(x_1,x_2,x_3) = (x,y,z)$ are the three Cartesian coordinates, β is an index that spans the q distinct frequencies, and α spans the three electric field polarizations.

2. Example—One-Dimensional Molecule in a Static Field

Evaluating Eq. 138 can be a horrendous task. We can, however, develop an intuitive perspective by calculating the static susceptibilities of a distribution of one-dimensional molecules whose equilibrium orientations are isotropic. In this case, the dressed induced dipole for a single molecule is:

$$p^*_i = \mu^*_i + \alpha^*_{ij} E_j + \beta^*_{ijk} E_j E_k + \ldots \tag{140}$$

If the applied electric field, $\vec{E}$, is in the z-direction and the molecule's long axis makes an angle θ with respect to the laboratory z-axis, the z component of the dressed induced dipole moment is:

$$p^*_z = \mu^* \cos\theta + \alpha^* E \cos^2\theta + \beta^* E^2 \cos^3\theta + \ldots, \tag{141}$$

and the averaged energy is calculated according to Eq. 104:

$$\bar{U} = \mu^* E \cos\theta + \frac{1}{2}\alpha^* E^2 \cos^2\theta + \frac{1}{3}\beta^* E^3 \cos^3\theta + \ldots \tag{142}$$

Because all the electric fields are of the same frequency and polarization, $K^{(1)} = 1$ and $D^{(1)} = 0!0!1! = 1$ (no fields polarized along x and y, and one field polarized along z). The dressed dipole moment of a single molecule is calculated according to Eq. 122:

$$<p^*_{z'}(\theta_0)>_z = \int_{-1}^{+1} d(\cos\theta) G(\theta, \theta_0) p^*_{z'} a_{z'z}(\theta). \tag{143}$$

Recall that z labels the laboratory axis corresponding to the direction of electric field polarization (z-axis) while z' labels the dressed molecular axis. Thus, $a_{z'z}(\theta) = \cos\theta$. Because there is only one unique axis in the laboratory and molecular frames, we drop all subscripts. The dressed dipole, then, is:

$$<p^*(\theta_0)> = \int_{-1}^{+1} d(\cos\theta) G(\theta, \theta_0) p^* \cos\theta. \tag{144}$$

By recasting the distribution function in the form:

$$G(\theta, \theta_0) = \frac{\exp(-\bar{U}/kT) \exp(-U_{int}/kT)}{\int_{-1}^{+1} d(\cos\theta) \exp(-\bar{U}/kT) \exp(-U_{int}/kT)}, \tag{145}$$

and defining the "internal" average of f according to:

$$\langle f \rangle_{int} \equiv \frac{\int_{-1}^{+1} d(\cos\theta) f(\theta) \exp(-U_{int}/kT)}{\int_{-1}^{+1} (d\cos\theta) \exp(-U_{int}/kT)}, \tag{146}$$

Eq. 144 substituted into Eq. 123 followed by substitution into Eq. 138 yields:

$$\chi^{(1)} = \frac{N}{2}\int_{-1}^{+1} d(\cos\theta_0) \left[\frac{\partial}{\partial E} \frac{\langle \exp[-\bar{U}/kT]\, p^* \cos\theta \rangle_{int}}{\langle \exp[-\bar{U}/kT] \rangle_{int}}\right]\Bigg|_{E=0}$$

$$= \frac{N}{2}\int_{-1}^{+1} d(\cos\theta_0) \left[\frac{\left\langle \left(\frac{\mu^*}{kT}\mu^* + \alpha^*\right)\cos^2\theta \right\rangle_{int}}{\langle 1 \rangle_{int}} + \frac{\langle \mu^* \cos\theta \rangle_{int}}{\langle 1 \rangle_{int}^2} \left\langle \frac{\mu^*}{kT}\cos\theta \right\rangle_{int}\right]. \tag{147}$$

Note that the above differentiation recognizes that three product terms are a function of the electric field: p^* and $\exp[-\bar{U}/kT]$ in the numerator of the upper part of Eq. 147 and $\exp[-\bar{U}/kT]$ in the denominator. Furthermore, we note that $\bar{U} \rightarrow 0$ as the electric field vanishes.

If, for the moment, we take the limiting case of a vanishing internal potential—the free rotator, in which the potential is independent of θ_0 so that the term in brackets in Eq. 147 is independent of θ_0 (and recall that we have assumed an isotropic zero-field distribution of orientations)—the average over the internal angle, θ_0, of any function $f(\theta)$ results in:

$$\frac{1}{2}\int_{-1}^{+1} d(\cos\theta_0) \frac{\langle f(\theta) \rangle_{int}}{\langle 1 \rangle_{int}} = \frac{1}{2}\int_{-1}^{+1} d(\cos\theta) f(\theta) \equiv \langle f(\theta) \rangle_{iso}, \tag{148}$$

where the brackets with *iso* subscripts denote an isotropic average. Because $\langle \cos\theta \rangle_{iso} = 0$, the last term in brackets in the second Eq. 147 vanishes. Applying Eq. 148 to Eq. 147, integration over $d(\cos\theta)$ yields:

$$\chi^{(1)} = \frac{N}{3}\left(\frac{\mu_z^{*2}}{kT} + \alpha_{zz}^*\right). \tag{149}$$

We have once again included the subscripts to remind the reader that the only non-vanishing response is along the molecular z-axis. Equation 149 can be used to understand the origin of Eq. 32 by noting that the isotropic average of the susceptibility $\alpha = \alpha_{zz}/3$.

There are two ways to account for the internal potential (by internal potential, we mean the potential due to a molecule's surroundings): direct calculation and phenomenology. For direct calculation, the functional form of the internal potential energy must be known if the orientational averages are to be calculated. (Generally, Eq. 148 will no longer hold.) Even for the simplest potentials, the integrals can only be evaluated numerically. In certain limiting cases (for example, when the internal energy is much smaller than all other energies), approximate analytical solutions exist. With the phenomenological approach, the internal potential can be characterized by a single parameter if we recognize that even for an isotropic material, an internal potential will lead to restricted molecular reorientation in response to an electric field (this is obviously not true for anisotropic materials such as liquid crystals). The susceptibility can then be written in the form:

$$\chi^{(1)} = \frac{N}{3}\left(b\,\frac{\mu_z^{*2}}{kT} + \alpha_{zz}^*\right), \tag{150}$$

where b is called the Kirkwood factor. [61] It is clear that for a free rotator $b = 1$, while for a restrained rotator $0 \leq b < 1$. If the molecules interact, the Kirkwood factor can exceed unity.

The computational complexity of deriving higher order nonlinearities grows exponentially with the order of nonlinearity. We will not describe such calculations in detail. Rather, we review certain key results later in this section. We can, however, guess the tensor form of the higher order nonlinearities by first considering the first order response. We begin by noting that $\chi^{(1)}$ is a second rank tensor so that all tensor products relating to $\chi^{(1)}$ must be second rank. μ^* is a first rank tensor, while α^* is a second-rank tensor. The only tensor product that yields a second rank tensor is $\mu^* \times \mu^*$, so that the response must be a linear combination of $\mu^* \times \mu^*$ and α^*. Furthermore, we can guess at the prefactor by using dimensional arguments. In Gaussian units, $\mu^* \times \mu^*$ has units of $erg \cdot cm^3$ and α^* has units of volume (cm^3). The prefactor to μ^{*2} must therefore have units of inverse energy. For a free rotator, the only parameter with units of energy is the thermal energy kT. Thus, aside from the geometric factor of 1/3, we could predict the form of the susceptibility as given by Eq. 149.

Next, let us consider the second order response. The possible tensor products leading to a third rank tensor are μ^{*3}, $\alpha^*\mu^*$ and β^*. Taking into account the units, the second order susceptibility of the free rotator must be of the form:

$$\chi^{(2)} = N\left(a_3\beta^* + a_2\left(\frac{1}{kT}\right)\mu^*\alpha^* + a_1\left(\frac{1}{kT}\right)^2\mu^{*3}\right), \tag{151}$$

where the a_i's are dimensionless constants that depend on the polar order of the material. Similarly, the third order response can be written in the form:

$$\begin{aligned}\chi^{(3)} = N\Bigg(a_4\gamma^* &+ a_3\left(\frac{1}{kT}\right)\beta^*\mu^* + a_2\left(\frac{1}{kT}\right)\alpha^{*2} \\ &+ a_1\left(\frac{1}{kT}\right)^2\alpha^*\mu^{*2} + a_0\left(\frac{1}{kT}\right)^3\mu^{*4}\Bigg)\end{aligned} \tag{152}$$

It is clear that the molecular reorientational mechanism terms are products of lower order susceptibilities (Eq. 150). The physical interpretation is as follows. First, consider the first order response, where α^* results from electron cloud deformation. The μ^{*2} term, on the other hand, is present even if the molecule is not polarizable and the induced dipole moment arises from a change in the molecule's orientation. The magnitude of the reoriented dipole should be proportional to the static dipole moment and to the degree of reorientation. The degree of reorientation, though, is proportional to the magnitude of the dipole moment and applied electric field and inversely proportional to the thermal energy which tends to randomize the molecule's orientation. The reorientational susceptibility is thus proportional to the square of the dipole moment and inversely proportional to the thermal energy.

The higher order nonlinearities can be similarly interpreted. Consider the second order response. The term $\alpha^*\mu^*$ has two physical interpretations. Both interpretations rely on electric-field-induced molecular reorientation. In the first interpretation, the molecule is reoriented by the electric field by its interaction with the permanent dipole moment. Subsequently, the molecule is polarized by the electric field. In the second interpretation, the electric field polarizes the molecule and reorients it through its interaction with the induced dipole moment. The reoriented molecule's permanent dipole moment is then the source of polarization.

As the order of the nonlinearity increases, there are many different ways in which the electric field can interact with a molecule to induce a polarization. In the remaining subsections, we discuss several limiting cases of higher order mechanisms.

B. Example of Nonlocal Mechanisms—Electrostriction

Nonlocal mechanisms arise from changes in the bulk properties of a material that rely on the presence of the collection of molecules. For example, even though the linear and nonlinear optical response of a molecule in a plane-wave electric field is translationally invariant, if the separation between the molecules in a collection of molecules changes, the bulk material response will change even when the molecules are noninteracting. An electric-field-induced strain is generally called electrostriction. This can be understood by once again considering the dilute solute model of the polarization as given by Eq. 120.

When evaluating Eq. 138 to compute the susceptibility as given by Eq. 147, we assumed that the number density is a constant—this was essentially the local mechanism approximation. for a purely nonlocal electrostrictive mechanism, the applied electric field effects only the number density. Then, the nth-order nonlinear optical response is calculated by differentiating the number density prefactor in Eq. 120 with respect to electric field n times and setting the field to zero.

Before writing a general expression for how we would differentiate Eq. 120, we first discuss how number density changes in response to an electric field. Let us consider a composite system made from dopant dipoles inside a polymer. The density changes can be divided into two broad mechanistic classes: uniform and differential strain. If the effect of the electric field is to exert a uniform force on all parts of the system, both the polymer host and the dopants will strain equally. An example of such a mechanism is electrode attraction, in which a film is pressed between two electrodes. When an electric field is applied perpendicular to the film plane, the plates attract and compress the film. Because this force is applied to the surface, the resulting uniaxial stress results in uniform strain. In such an experiment, however, there are also body forces, that is, forces inside the material due to the applied electric field. If the force on the embedded dipoles is different than the force on the polymer units, they will be displaced differently. The result is that the dopant molecules can move relative to the background polymer. Alternatively, if the applied electric field is non-uniform (as would be the case for finite electrodes due to edge effects), the molecules would be attracted to the region of higher electric field due to gradient forces. This is also an example of differential strain. The system need not be two component to observe differential strain because the inhomogeneity of the applied electric field can result in different regions of the polymer being strained differently.

If we consider a volume element within a material, there are two ways that the local density can change: The physical volume can change (uniform strain) or the material can redistribute itself within a given volume element (differential strain). Recall that the number density is defined as $N = n/V$, where n is the number of molecules with the volume V. If we consider a small volume element such that the number density is uniform in that volume, the change in number density can be expressed as:

$$\Delta N = \frac{\Delta n}{V} - \frac{n \Delta V}{V^2}. \tag{153}$$

The first term is associated with differential strain and the second term is associated with uniform strain. This can be understood by considering a dye-doped polymer in which the border of the volume element is "painted" with a marker that adheres to the polymer host. If an applied electric field exerts a body force on the molecules so that the dopant molecules move into the painted region while the border of that region remains fixed in space, the dopant molecules must certainly be moving relative to the host polymer. This corresponds to a constant volume number density change and is clearly associated with differential strain. The constant dye-number density change occurs when the volume within the painted border changes without a change in the number of dopant molecules within the border—the uniform stress limit.

In practice, the strain is not necessarily along the direction of applied force. For example, if we squeeze a material in the z-direction, it can deform in both the x- and y-directions. To account for such anisotropic deformations, we can write the number density of a small rectangular volume element whose dimensions are $x \times y \times z$ as:

$$N = \frac{n_x n_y n_z}{xyz}, \tag{154}$$

where n_i is the number of molecules along edge x_i. For simplicity, we assume that the molecules are arranged in linear arrays whose linear number density (n_i/x_i) is different along the three Cartesian axes. Furthermore, we assume that each surface of the volume element contains a two-dimensional array of molecules and that each edge of the volume element contains a linear array of molecules. While such assumptions may seem severely restrictive, it is straightforward to show that the following results hold for anisotropic systems provided that the volume element contains a large number of molecules.

Under the application of an external electric field, both the number of atoms in a particular array or the length of the array can change. Differentiating Eq. 154 with respect to the applied electric field polarized in the i direction E_i, we get:

$$\frac{dN}{dE_i} = N \sum_{j=1}^{3} \left[\frac{1}{n_j} \left(\frac{\partial n_j}{\partial E_i} \right) - \frac{1}{x_j} \left(\frac{\partial x_j}{\partial E_i} \right) \right]. \tag{155}$$

For the change in number density to be predicted, the physical quantities in parentheses in Eq. 155 must be known. The change in a material's strain with respect to applied static electric field is known as the piezoelectric tensor, d, and is defined by:

$$d_{ijj} = \frac{1}{x_j} \frac{\partial x_j}{\partial E_i}. \tag{156}$$

(Note that the tensor d_{ijk} in Eq. 156 should not be confused with the second harmonic coefficient.) There is no standard definition of the material property associated with the first term in parentheses in Eq. 155. We therefore define:

$$e_{ijj} = \frac{1}{n_j} \frac{\partial n_j}{\partial E_i}. \tag{157}$$

Equation 155 can then be written in terms of the known coefficients given by Eqs. 156 and 157.

C. Local-nonlocal Coupling

There are many processes in which both local and nonlocal mechanisms contribute to the material response. As our starting point, we again consider Eq. 120. The polarization is a function of both local and nonlocal variables. For example, the density N depends on the arrangement of the molecules, while the microscopic dipole moment is a local molecular property. When an external electric field is applied, both the number density and the molecular dipole moment changes relative to its zero-field state. Formally, the susceptibility is calculated by differentiating Eq. 120 with respect to the field. A local response is determined from its variation with respect to the electric field while the nonlocal response depends on the electric field dependence of the density. Upon taking second order or higher derivatives, cross-terms between these two types of response are common. We refer to this mechanism as local/nonlocal coupling.

Physically, local/nonlocal coupling can be understood by first considering reorientational mechanisms from the microscopic perspective: A photon from a beam of light or static field reorients a molecule and the subsequent photons measure the response of the reoriented molecule. Analogously, a photon can change the local density of molecules so that the subsequent photons measure the response of the newly arranged system. There are many possible scenarios of mixed processes in which the response originates from a combination of changes in molecular density, reorientation, and molecular dipole moment.

The mixed mechanisms are too complex to be further treated.

D. First Order Response

In this subsection, we describe the mechanisms that contribute to the first-order response.

1. Electronic Response

The purely electronic response arises from a collection of molecules whose orientations and positions are fixed. Under this assumption, the bulk response of a dilute collection of molecules will have a first order response:

$$\chi_{ij} = N <\alpha^*>_{ij}. \tag{158}$$

We illustrate how Eq. 158 is used to calculate the linear susceptibility for an ordered dilute ensemble of molecules. For one-dimensional molecules whose "long" axis we arbitrarily call the 3 axis, the dressed polarizability α^* has only one nonvanishing component, α^*_{33}. Assuming that the bulk material is azimuthally symmetric about the laboratory z-axis, a mathematically well-behaved orientational distribution function can be expanded in a series of orthogonal Legendre polynomials, $P_l(\cos\theta)$:

$$G(\cos\theta) = \sum_{l=0}^{\infty} \frac{2l+1}{2} <P_l> P_l(\cos\theta), \tag{159}$$

where the coefficients $<P_l>$ are called the order parameters. Because the Legendre polynomials are orthogonal, the coefficients can be determined by multiplying both sides of Eq. 159 by $P_m(\cos\theta)$ and integrating, yielding:

$$<P_m> = \int_{-1}^{+1} d(\cos\theta) G(\cos\theta) P_m(\cos\theta). \tag{160}$$

For reference, the first five Legendre polynomials are listed below:

$$\begin{aligned} P_0(x) &= 1, \\ P_1(x) &= x, \\ P_2(x) &= (3x^2 - 1)/2, \\ P_3(x) &= (5x^3 - 3x)/2, \\ P_4(x) &= (35x^4 - 30x^2 + 3)/8. \end{aligned} \tag{161}$$

Evaluating Eq. 65 for the first order susceptibility, and using the Legendre expansion of $G(\cos\theta)$, we get:

$$\chi_{ij}^{(1)} = \frac{N \int d\vec{\Omega} \alpha^*_{i'j'} a_{i'i}(\Omega) a_{j'j}(\Omega) \sum_{l=0}^{\infty} \frac{2l+1}{2} <P_l> P_l(\cos\theta)}{4\pi^2}. \tag{162}$$

We note that the θ-dependence of each tensor component of the rotation matrix $a(\Omega)$ can be expressed as the sum of $P_0(\cos\theta)$ and $P_1(\cos\theta)$. Because the first order response as given by Eq. 162 is proportional to the square of the rotation matrix, it can be written as an expansion over the first three Legendre polynomials. Evaluating the integral yields:

$$\chi_{zz}^{(1)} = \frac{N}{3} (1 + 2 <P_2>) \alpha_{33}^*, \tag{163}$$

and

$$\chi_{xx}^{(1)} = \chi_{yy}^{(1)} = \frac{N}{3} (1 - <P_2>) \alpha_{33}^*. \tag{164}$$

Note that in the limit of perfect alignment ($<P_2> = 1$), the bulk response is one-dimensional as is the molecular polarization. For an isotropic material ($<P_2> = 0$), the susceptibility is also isotropic, as expected. The refractive index is calculated from the above equations in the usual way:

$$n_{ii} = (1 + 4\pi\chi_{ii}^{(1)})^{1/2}. \tag{165}$$

As an example of applications of Eqs. 163 and 164, we calculate the birefringence of a poled dye-doped polymer. First, we need to evaluate the order parameter $<P_2>$ (according to Eq. 160) of the poling distribution function given by Eq. 145 with $U_{int} = 0$:

$$<P_2> = \int_{-1}^{1} d(\cos\theta) \left[\frac{\exp\left(\frac{m^*E}{kT}\cos\theta\right)}{\int_{-1}^{1} d(\cos\theta) \exp\left(\frac{m^*E}{kT}\cos\theta\right)} \right] \times \left(\frac{3}{2}\cos^2\theta - \frac{1}{2}\right), \tag{166}$$

where m^* is the dressed dipole moment that couples to the external field, and E is the applied poling field. Integrating Eq. 166 leads to:

$$<P_2> = 1 + \frac{3}{(m^*E/kT)^2} - \frac{3}{(m^*E/kT)} \coth\left(\frac{m^*E}{kT}\right). \tag{167}$$

As $E \to \infty$, $<P_2> \to 1$ as we would expect for a perfectly ordered material.

It is also useful to evaluate the order parameter in the limit as $E \to 0$. To do so, we expand $\coth(m^*E/kT)$. Recall that $\coth(\epsilon) = 1/\epsilon + \epsilon/3 - \epsilon^3/45 + \ldots$. Including the first three terms in the expansion, the order parameter is:

$$<P_2> \approx \frac{1}{15}\left(\frac{m^*E}{kT}\right)^2. \tag{168}$$

Next, we compute the linear susceptibility of an axially ordered dye-doped polymer. We note that this is a two component system that includes both the polymer host and the dopant dye. For low dye concentration, the bulk of the response is due to the polymer matrix. In the case of a poled polymer, the following derivation holds when the dipole moment of the dye molecule is much greater than the dipole moment of the polymer repeat units so that most of the electric-field-induced first order susceptibility is due to the dye. The linear susceptibility of a two component system can be written as:

$$\chi_{ij}^{(1)} = N_1 <\alpha_1^*>_{ij} + N_2 <\alpha_2^*>_{ij}, \tag{169}$$

where the 1 and 2 label each species. If the polymer contribution is given by the first term, for low dye concentrations, N_1 and the local field factors will be approximately independent of the amount of dye loading. Likewise, an applied electric field orients only the dyes, so it will not effect the polymer. Under these three conditions, the polymer's contribution to the first order susceptibility is independent of the dye concentration and poling fields. We therefore set the first term equal to the neat polymer's linear susceptibility and express the second term as a function of the order parameter $<P_2>$ according to Eqs. 163 and 164. The two independent diagonal components of the susceptibility become:

$$\chi_{zz}^{(1)} = \chi_{zz}^{(1)poly} + \frac{N}{3}(1 + 2 <P_2>)\alpha_{33}^*, \tag{170}$$

and

$$\chi_{xx}^{(1)} = \chi_{yy}^{(1)} = \chi_{xx}^{(1)poly} + \frac{N}{3}(1 - <P_2>)\alpha_{33}^*. \tag{171}$$

Defining $\epsilon_{ij}^0 = 1 + 4\pi\chi_{ij}^{(1)poly}$, and assuming that the dye contribution is small, the refractive index of the doped polymer can be expressed as:

$$\begin{aligned} n_{zz} &= \left[\epsilon_{zz}^0 + \frac{4\pi N}{3}(1 + 2 <P_2>)\alpha_{33}^*\right]^{1/2} \\ &\approx n_{zz}^0 + \frac{2\pi N}{3n_{zz}^0}(1 + 2 <P_2>)\alpha_{33}^*, \end{aligned} \tag{172}$$

and

$$\begin{aligned} n_{xx} &= \left[\epsilon_{xx}^0 + \frac{4\pi N}{3}(1 - <P_2>)\alpha_{33}^*\right]^{1/2} \\ &\approx n_{xx}^0 + \frac{2\pi N}{3n_{xx}^0}(1 - <P_2>)\alpha_{33}^*, \end{aligned} \tag{173}$$

where $n_{ij}^0 = (\epsilon_{ij}^0)^{1/2}$. For an isotropic host polymer, the birefringence of the material is:

$$n_{zz} - n_{xx} = \frac{2\pi N}{n_0} <P_2> \alpha_{33}^*. \tag{174}$$

Because the order parameter $<P_2>$ is proportional to the bulk birefringence, the dopant's orientational order can be determined by measuring the index ellipsoid. Equation 174 can be applied to a poled polymer directly by using Eq. 167 for $<P_2>$. The axial order parameter in the low-field limit is given by Eq. 168 yielding a birefringence:

$$n_{zz} - n_{xx} = \frac{2\pi N}{15n_0}\alpha_{33}^*\left(\frac{m^*E}{kT}\right)^2. \tag{175}$$

A birefringence measurement is thus one method for determining the degree of electric field poling.

2. Molecular Reorientation

In this subsection, we consider two different models of molecular reorientation: the low temperature elastic restraint model and the higher temperature thermodynamic model. Both models apply to materials in which a rigid, rod-like molecule is embedded in a viscoelastic host medium such as a crystal, liquid, or polymer.

Elastic Restraint Model—Low Temperature. We begin by assuming that the host medium acts as a rigid support and is not affected by the application of an electric field other than the effects caused by molecular reorientation of the guest species. Figure 18 shows a molecule of dressed dipole moment m^* which is oriented along θ_0 when there is no electric field applied. In the presence of an electric field, the molecule will partially align with the electric field until the elastic restraint permits it to reorient no further. The angle θ is measured with respect to the equilibrium angle. Given the elasticity of the host, when the field is turned off, the molecule will spring back to θ_0 provided that the field is on for a time that is shorter than the viscous response time. For a sinusoidally applied electric field, the time-averaged angle is θ_0 and the equilibrium angle will remain at θ_0 indefinitely.

The total torque due to the applied field, E and elastic restraint is:

$$\tau = m^*E \sin(\theta_0 + \theta) - k_\theta \theta, \tag{176}$$

where k_θ is the microscopic elasticity. Note that this model assumes that the elasticity is independent of frequency and that a single elasticity characterizes a polymer. While both of these assumptions are not generally valid for polymers, the qualitative results they yield are a useful paradigm for picturing an embedded molecule in a polymer. The shortcomings introduced by these assumptions will be discussed later. For small θ, the first order approximation for the torque is:

$$I\ddot{\theta} \equiv \tau = m^*E[\sin\theta_0 + \theta\cos\theta_0] - k_\theta \theta, \tag{177}$$

where I is the moment of inertia of the molecule, and $\ddot{\theta}$ is the second time derivative of θ. For a sinusoidal electric field of the form $E = E_0 \exp[i\omega t]$, the steady state solution for θ (when transients have decayed) is of the form $\theta = A \exp[i\omega t]$, where

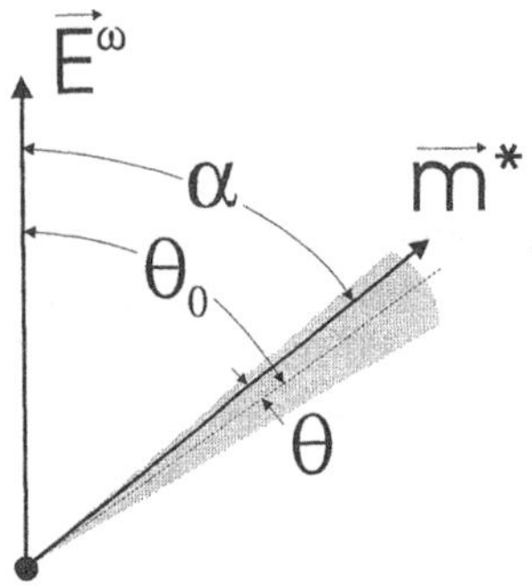

Figure 18 A sinusoidally time-dependent electric field will cause a molecule of dressed dipole moment m^* to oscillate about its equilibrium angle θ_0. The shaded region represents the angular sweep of the oscillating dipole. The angle θ is measured with respect to the equilibrium angle.

A is determined by substituting this solution into Eq. 177. Applying this procedure, the solution for θ is

$$\theta = \left[\frac{k_R - I\omega^2 - ik_I}{(k_R - I\omega^2)^2 + k_I^2}\right] m^* E_0 \sin\theta_0 \exp[i\omega t], \tag{178}$$

where $k_\theta = k_R + ik_I$ (the real part is the elasticity and the imaginary part is the viscosity).

To calculate the reorientational molecular susceptibility, α^{or}, we must calculate the induced dipole moment. The dipole moment along the direction of the applied electric field as seen by an observer outside the material is given by the projection of the screened internal effective dipole moment, μ_e^*, along E. As seen in Table 3, $\mu_e^* = m^*$. Because it is more convenient to express the results in terms of one dipole moment, we choose to write the expression in terms of m^*:

$$p_z^* = m^* \cos(\theta_0 + \theta) \approx m^* \cos\theta_0 - m^*\theta\sin\theta_0. \tag{179}$$

Note that this is the dressed dipole moment; that is, the one seen by an external observer. The first term is the static dipole moment and the second term is the induced dipole moment. Recalling that $p_z^\omega = \alpha_{zz}^{or}(-\omega;\omega)E_z^\omega$, and substituting Eq. 178 into Eq. 179, the reorientational polarizability is:

$$\alpha_{zz}^{or}(-\omega;\omega) = (m^* \sin\theta_0)^2 \left[\frac{k_R - I\omega^2 - ik_I}{(k_R - I\omega^2)^2 + k_I^2}\right]. \tag{180}$$

Up to this point, we have only considered a single molecule that is oriented along θ_0. For a distribution of molecules, we must average over the distribution of equilibrium orientations. The bulk linear susceptibility is related to the *averaged* susceptibility according to Eq. 65. For an isotropic material, the orientational average yields:

$$<\alpha_{zz}^{or}> = \frac{1}{2}\int_{-1}^{+1} d(\cos\theta_0)\alpha_{zz}^{or} = \frac{2}{3} m^{*2} \left[\frac{k_R - I\omega^2 - ik_I}{(k_R - I\omega^2)^2 + k_I^2}\right]. \tag{181}$$

When the material is not isotropic, we can expand the zero-field orientational distribution function in terms of Legendre polynomials (Eq. 159). Integrating over the distribution function and using Eq. 160, we get:

$$\begin{aligned} <\alpha_{zz}^{or}> &= \frac{1}{2}\int_{-1}^{+} G(\theta_0) d(\cos\theta_0)\alpha_{zz}^{or} \\ &= \frac{2}{3}(1 - <P_2>)m^{*2} \left[\frac{k_R - I\omega^2 - ik_I}{(k_R - I\omega^2)^2 + k_I^2}\right]. \end{aligned} \tag{182}$$

For an isotropic material, $<P_2> = 0$ and we recover Eq. 181. When $<P_2> = 1$, the electric field does not change the material alignment because it is already fully aligned in the direction of the field so that the reorientational response vanishes. Note that this is the dressed susceptibility so that the bulk response is given simply by $\chi^{(1)} = N <\alpha_{zz}^{or}>$.

It is instructive to apply the isotropic model to a well-known dye-doped polymer system. Before doing so, the reader is cautioned that the elasticity of a polymer depends on the frequency. The frequency dependence, however, is not large even over many decades in frequency so that the above expression can be used to get a qualitative picture of the frequency dependence at low temperature.

First, we comment on the temperature range over which this model holds. A thermodynamic model must be used when the thermal energy is comparable to or larger than the characteristic energies of the system. There are two energy scales in Eq. 182: The energy of oscillation at the natural elastic frequency ($(k_R/I)^{1/2}$) and the rotational kinetic energy at the driving frequency ($I\omega^2$). As long as these energies are large compared with kT, this model holds. Table 4 lists typical parameters for a disperse red azo dye as measured in the visible. Note that these are ballpark values intended solely for estimating the order of magnitude of the reorientational response.

We can use Eq. 181 to determine the reorientational polarizability as a function of frequency ω. In the limiting cases of high frequency ($\omega \to \infty$), resonant frequency ($\omega^2 = (k_R - k_I)/I$—the frequency at which the real part of the reorientational susceptibility is maximum), and low frequency ($\omega \to 0$):

$$<\alpha_{zz}^{or}>_{high} = \frac{2}{3} m^{*2} \frac{1}{I\omega^2} \to 0, \tag{183}$$

$$<\alpha_{zz}^{or}>_{resonance} = \frac{m^{*2}}{3k_I}(1 - i), \tag{184}$$

and

$$<\alpha_{zz}^{or}>_{low} = \frac{2}{3} m^{*2} \left(\frac{k_R - ik_I}{k_R^2 + k_I^2}\right). \tag{185}$$

Table 5 shows the real part of the reorientational susceptibility of DR1/PMMA at several frequencies. It is clear that at optical frequencies and higher, the response is dominated by the electronic mechanism. On resonance and at lower frequencies, the reorientational mechanism dominates the response. We stress that this model holds when the characteristic energies of the material are much larger than thermal energies. The response is therefore independent of temperature as long as this condition holds.

Table 4 Typical Values of Parameters of DR1 Dye in PMMA Polymer

Parameter	Value	Notes
m^*	$14 \times 10^{-18} esu \cdot cm$	
α^*	$9 \times 10^{-23} cm^3$	
β^*	$5 \times 10^{-28} cm^5/esu$	
γ^*	$1 \times 10^{-32} cm^7/esu^2$	
I	$7 \times 10^{-36} g \cdot cm^2$	
k_θ	$6 \times 10^{-13} esu^2/cm$	Room temperature, $\Omega \to 0$
kT	$4 \times 10^{-14} esu^2/cm$	Room temperature
n	1.48	Bulk doped system

Table 5 Reorientational Polarizability in Isotropic DR1/PMMA at Limiting Frequencies.

ω (*Hz*)	*Re* [$<\alpha_{zz}^{or}>$] (cm^3)
∞	0
10^{15}	1.9×10^{-29}
2.9×10^{11} (resonant)+	1.1×10^{-21}
0	2.2×10^{-22}

+assumes that $k_I \approx k_R/10$.

Free Gas Model—Low Temperature. The free gas model holds when the guest molecules are free to reorient and applies to dilute gases or guest/host systems in which the host voids are so large compared to the guest size that they appear to reorient without impediment from the host. This model is the zero-elasticity limit of the elastic restraint model. Including the anisotropy of the system, the reorientational response is:

$$<\alpha_{zz}^{or}> = \frac{2m^{*2}}{3I\omega^2}(1 - <P_2>). \tag{186}$$

This model is identical in form to the high frequency limit of the elastic restraint model, except that the free gas model holds at all frequencies (provided that θ remains small). The angular amplitude, given by Eq. 178 with $k_\theta = 0$, yields the small-angle condition:

$$\omega \gg \left(\frac{m^*E_0}{I}\right)^{1/2}. \tag{187}$$

For a $1W/cm^2$ light source, $E_0 \approx 0.09 statvolt/cm$. For unimpeded DR1, then, the free gas model would apply when $\omega \gg 4 \times 10^8 Hz$. At lower frequencies, the approximation given by Eq. 177 cannot be made.

Thermodynamic Model. The thermodynamic model holds when thermal energies are on the order of characteristic energies of the system, and when the thermal equilibration time is long compared with one cycle of the applied electric field.

The molecular reorientational thermodynamic model of a free gas is already presented in the beginning of this section. The result is given by Eq. 149. Equation 149 can be generalized to include a mean-field axial-orienting potential energy of the form:

$$U = <P_2>\, P_2(\cos\theta), \tag{188}$$

where $<P_2>$ is the axial order parameter. It is trivial to show that the reorientational part of Eq. 148, including the effects of the axial potential, is:

$$<\alpha_{zz}^{or}> = \frac{1}{3}\frac{m^{*2}}{kT}(1 - <P_2>). \tag{189}$$

It is useful to sketch the frequency dependence of the reorientational and electronic response. Figure 19 shows a schematic plot of several models of the frequency dependence of the magnitude of the real part of the polarizability for DR1 dye as a free gas or in PMMA polymer. Each model is briefly discussed.

Electronic dispersion. Chapter 2 discusses how the electronic response of a molecule is calculated. The dark solid curve in Figure 19 represents a two-level model of DR1 dye: It is assumed that only one excited state of the system dominates. The electronic response is flat until optical frequencies, at which point electronic excitations match the photon energy, resulting in a resonance.

Elastic reorientation. The dark broken curve is a plot of Eq. 181. The resonance shape is similar to the electronic case except the elastic response goes through a resonance at microwave frequencies, where the natural reorientational frequency matches the field frequency. Recall that the elastic model assumes that the elasticity is independent of frequency. In PMMA polymer, it is known that the elasticity increases with frequency. The net result is that the elastic model predicts the resonance to be at a lower frequency than the actual resonance. Recall also that polymers are not characterized by a single characteristic elasticity but by a broad range of elasticities. This results in a broadening in the resonance structure.

Free gas reorientation. The thin solid curve shows the free gas model at low temperature (Eq. 186). Recall that this model only applies beyond ($4 \times 10^8 Hz$). Note that in the limit of high frequency, both the free gas model and the elastic model become identical. This makes physical sense because at frequencies well beyond the natural reorientational frequency, the embedded molecule has a small reorientational oscillation amplitude so it is not influenced by the polymer and appears to be free.

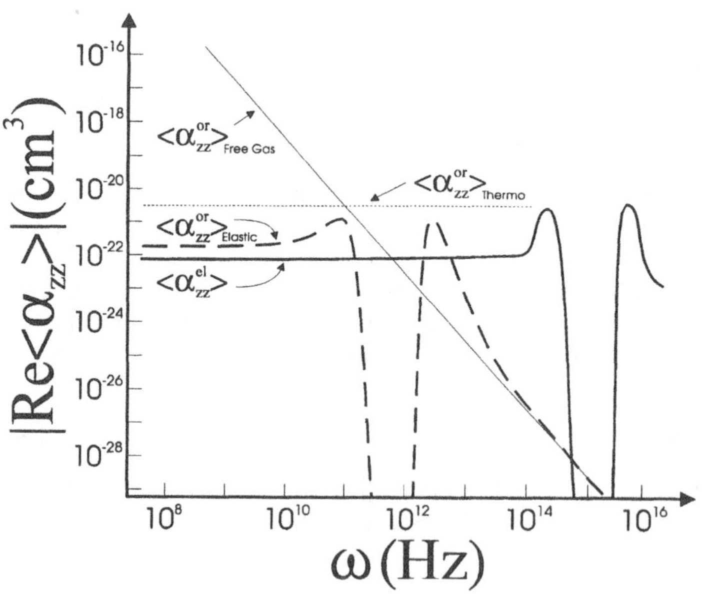

Figure 19 Schematic representation of several polarizability models of DR1 dye as a free gas or in PMMA polymer as a function of frequency.

Thermodynamic Model. The thin broken line represents the thermodynamic model (Eq. 189) at room temperature. Recall that this model only holds at frequencies that are smaller than the inverse of thermal equilibration time and can be applied to gases, liquids, and polymers above their glass transition temperatures.

It is clear that at optical frequencies the electronic response dominates, while at microwave frequencies the set of reorientational responses are all larger. Given the complexity of most material systems such as doped polymers and liquids, exact calculations of the various mechanisms become overwhelming.

3. Photoelasticity

This subsection discusses the effect of a mechanical load on a material's linear electronic susceptibility. As such, this can be viewed as a purely electronic response of a deformed material, whose linear optical properties can in principle be calculated if the exact state of the system is known. A more useful approach is to understand how the linear optical properties change under deformation in terms of applied stresses and material elasticity. Such an understanding has led to many applications including optical monitoring of material fatigue. An additional motivation for building such an understanding is to apply it to electrostriction: A third order mechanism that results in a change in the linear refractive index in response to optically induced stress. Electrostriction is discussed again in this section.

There are two broad mechanisms of linear susceptibility change: elastic and viscous. In the elastic case, the application of stress results in a redistribution of charges into a new equilibrium configuration. In the viscous case, a stress causes the material to flow. Often, the flow results in inhomogeneous shear that can reorient flowing molecules. Each case is separately treated below.

Elastic Materials. The photoelastic effect is the simplest example of a nonlocal mechanism and results from a redistribution of molecules or atoms in response to an applied stress. For an initially isotropic material of refractive index n, the refractive index of the stressed material is given by:

$$n_t = n + C_1\sigma_t + C_2(\sigma_r + \sigma_s), \tag{190}$$

where t, r, and s are the three orthogonal directions, σ_i is the stress along i, C_1 is the stress-optic coefficient that relates the stress along t to the refractive index along t, while C_2 is the stress-optic coefficient that relates the effect of a stress perpendicular to t to the refractive index along t.

The photoelastic effect can also be expressed in terms of strains. If ϵ_i is the strain along i it is related to stress according to:

$$\epsilon_i = \frac{1}{E}[\sigma_i - v(\sigma_j + \sigma_k)], \tag{191}$$

where i, j, and k are mutually orthogonal directions, v is Poisson's ratio, and E is the modulus of elasticity. The photoelastic effect can then be expressed as:

$$n_t = n + b_1\epsilon_t + b_2(\epsilon_r + \epsilon_s), \tag{192}$$

where it is straightforward to show:

$$C_1 = \frac{1}{E}(b_1 - 2\upsilon b_2), \tag{193}$$

and

$$C_2 = \frac{1}{E}(b_2 - \upsilon b_2 - \upsilon b_1). \tag{194}$$

We can understand the microscopic source of the photoelastic effect by considering the dilute solute model of the electronic refractive index. Expanding the susceptibility for small stress, the diagonal components of the dielectric constant, ϵ_{tt}, can be expressed as:

$$\epsilon_{tt} = 1 + 4\pi\chi_{tt}^{(1)}|_{\sigma=0} + \sum_{i=t,r,s} 4\pi \left.\frac{\partial\chi_{tt}^{(1)}}{\partial\sigma_i}\right|_{\sigma=0} \sigma_i. \tag{195}$$

Note that to differentiate the two ϵ's, the dielectric constant, ϵ_{tt}, is represented with two indices, while the strain, ϵ_t, is represented with a single index.

Using Eq. 165 for the refractive index, and defining n to be the non-stressed isotropic refractive index ($n = 1 + 4\pi\chi_{tt}^{(1)}|_{\sigma=0}$), the stressed refractive index for small stress is of the form:

$$n_{tt} = n + \frac{2\pi}{n} \sum_{i=t,r,s} \left.\frac{\partial\chi_{tt}^{(1)}}{\partial\sigma_i}\right|_{\sigma=0} \sigma_i. \tag{196}$$

Comparing Eq. 196 with Eq. 190, the stress-optic coefficients are of the form:

$$C_1 = \frac{2\pi}{n}\frac{\partial}{\partial\sigma_t}(N\langle\alpha^*\rangle_{tt}), \tag{197}$$

and

$$C_2 = \frac{2\pi}{n}\frac{\partial}{\partial\sigma_r}(N\langle\alpha^*\rangle_{tt}), \tag{198}$$

where we have substituted the dilute solute model for $\chi^{(1)}$ according to Eq. 158.

The stress-optic coefficients are clearly related to the change in the linear susceptibility with stress. Because the linear susceptibility is a function of concentration, local field factors, molecular orientation, and polarizability, the linear photoelastic effect results from stress-induced changes in each of these parameters. Therefore, several mechanisms contribute to the photoelastic effect.

Viscous Materials. Shear is exerted on a viscous material when it flows near a physical boundary. The substituent molecules or particles within the flowing host material will then experience aligning forces. Figure 20 shows two separate geometries in which shear results in alignment. Figure 20a shows a viscous fluid between two infinite planes. One plane is fixed, while the other plane moves parallel to the fixed surface. The arrows represent molecules within the fluid that are aligned in response to material flow. The alignment will be uniform as explained below. Note that such a shear can be used during crystallization to make thin-film surface-aligned crystals. Indeed, Thakur and coworkers have reported such a process that

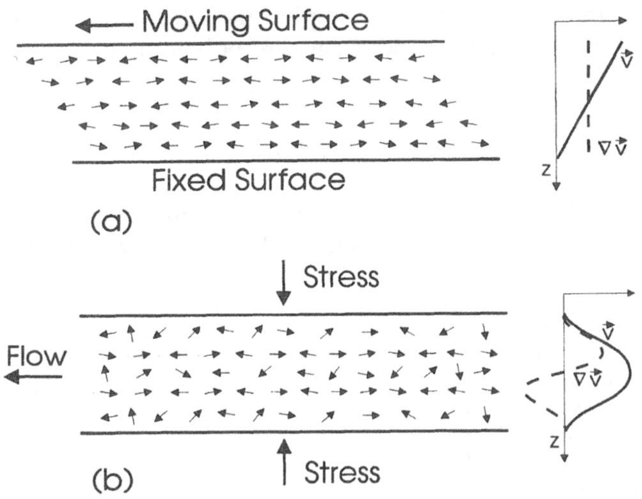

Figure 20 A viscous material between a fixed plate and a moving plate. Arrows represent substituent molecule orientation. Plot to right shows velocity field and velocity gradient as a function of position. b) Viscous liquid between two plates under uniaxial stress. Material flows radially away from center of stress point. Plot on right shows velocity and velocity gradient as a function of position.

has been applied to making high-quality nonlinear optical waveguides of PTS-PDA. [62–66]

The velocity of the fluid at the two physical boundaries is equal to the boundary velocity. These boundary conditions imply that the fluid will move with the moving surface and will be at rest at the fixed surface. The plot at the right portion of Fig. 20a shows the velocity distribution as a function of position between the plates. The magnitude of the shearing force at each molecular site is proportional to the magnitude of the gradient of the velocity field. For this geometry, then, the shear force is the same everywhere in the material and the molecules axially align in the direction of flow. Note that there is no net polar order.

Figure 20b shows the flow of a viscous material between two plates under uniaxial stress. The material will flow away from the stress points. Because the bounding surfaces have no velocity component along the material flow direction, the transverse velocity component of the material will vanish at the two boundary surfaces. This results in a transverse velocity distribution as shown in the right-hand portion of Fig. 20b. The dashed curve shows the velocity gradient. The shear will thus vanish at the two boundary surfaces and along the plane equidistant from the two plates. The material will thus remain isotropic in these three regions. In the two planes represented by the maximum magnitude of velocity gradient, the substituent molecules will align with their axes in the flow direction.

We next model the ordering forces on a molecule experiencing shear. Note that the shear force tends to align the molecules along the surface, which is perpendicular to the z-axis. Furthermore, the alignment is axial and the degree of alignment is proportional to the gradient of the flow rate. The ordering potential must therefore be at a minimum at 90 degrees from the z-axis. The simplest model of the ordering potential is thus approximated by the second order Legendre polynomial:

$$U(\theta) = b(\nabla v)P_2(\cos\theta), \tag{199}$$

where b is a constant that relates the gradient of the flow rate to the effective potential.

According to Eq. 170 and 171, the linear susceptibility of a collection of one-dimensional molecules is related to the order parameter $\langle P_2 \rangle$ which we evaluate using Eq. 199 substituted into Eq. 72:

$$\langle P_2 \rangle = \frac{\int_{-1}^{+1} dx P_2(x) \exp\left[-\frac{b\nabla v}{kT} P_2(x)\right]}{\int_{-1}^{+1} dx \exp\left[-\frac{b\nabla v}{kT} P_2(x)\right]}, \tag{200}$$

where we have made the convenient substitution $x = \cos x$. If the ordering potential is small, that is:

$$\frac{b\nabla v}{kT} \ll 1, \tag{201}$$

the order parameter $\langle P_2 \rangle$ to first order in $b\nabla v/kT$ is

$$\langle P_2 \rangle = \frac{\int_{-1}^{+1} dx P_2(x)\left[1 - \frac{b\nabla v}{kT} P_2(x)\right]}{\int_{-1}^{+1} dx \left[1 - \frac{b\nabla v}{kT} P_2(x)\right]} = -\frac{1}{5}\frac{b\nabla v}{kT}, \tag{202}$$

where we have used the orthogonality of Legendre polynomials:

$$\int_{-1}^{+1} dx P_n(x) P_m(x) = \frac{2}{2n+1}\delta_{n,m}. \tag{203}$$

We now apply the order parameter as given by Eq. 202 to calculate the linear susceptibility of several kinds of systems. First, consider a liquid composed of one-dimensional molecules. The first order susceptibility tensor is given by Eqs. 163 and 164, and with the help of Eq. 202, yields:

$$\chi_{zz}^{(1)} = \frac{N}{3}\left(1 - \frac{2}{5}\frac{b\nabla v}{kT}\right)\alpha_{33}^{*}, \tag{204}$$

and

$$\chi_{xx}^{(1)} = \frac{N}{3}\left(1 + \frac{1}{5}\frac{b\nabla v}{kT}\right)\alpha_{33}^{*}, \tag{205}$$

where x is the direction of flow. Note that the linear susceptibility in the direction of flow increases, while the susceptibility component perpendicular to the flow decreases. The birefringence can be calculated using Eq. 165, which, for the two independent tensor components gives:

$$n_{zz} = n_0 - \frac{4\pi}{15n_0}\left(\frac{b\nabla v}{kT}\right) N\alpha_{33}^*, \tag{206}$$

and

$$n_{xx} = n_0 + \frac{2\pi}{15n_0}\left(\frac{b\nabla v}{kT}\right) N\alpha_{33}^*, \tag{207}$$

where n_0 is the isotropic index given by:

$$n_0^2 = 1 + \frac{4\pi}{3} N\alpha_{33}^*. \tag{208}$$

The birefringence of the liquid is then of the form:

$$n_{zz} - n_{xx} = -\frac{2\pi}{5n_0}\left(\frac{b\nabla v}{kT}\right) N\alpha_{33}^*. \tag{209}$$

To lowest order, then, the birefringence is linear in the flow velocity gradient.

When a dye chromophore is embedded in a polymer matrix as a guest, flow of polymer will lead to an alignment of both the dye and the polymer. The above theory can be generalized by including the polymer's contribution to the susceptibility. The susceptibility of a doped polymer is given by the sum of the susceptibility of the polymer and the guest chromophore as expressed by Eqs. 170 and 171. It is straightforward to show that the refractive index is then of the form:

$$n_{zz} = n_{zz}^0 - \frac{4\pi}{15n_0}\left(\frac{b\nabla v}{kT}\right) N\alpha_{33}^*, \tag{210}$$

and

$$n_{xx} = n_{xx}^0 + \frac{2\pi}{15n_0}\left(\frac{b\nabla v}{kT}\right) N\alpha_{33}^*, \tag{211}$$

where n_{ii}^0 is the refractive index that excludes the contribution from oriented dyes and is given by:

$$n_{ii}^0 = \left(1 + 4\pi\left(\chi_{ii}^{(1)poly} + \frac{1}{3} N\alpha_{33}^*\right)\right)^{1/2}, \tag{212}$$

where $\chi_{ii}^{(1)poly}$ is the linear susceptibility of the stress-oriented polymer.

When the dye concentration is low enough so that its contribution to the linear susceptibility is small relative to the polymer's contribution, the refractive index n_{ii}^0 can be approximated:

$$n_{ii}^0 \approx n_{ii}^{poly} + \frac{2\pi}{3}\frac{N\alpha_{33}^*}{n_{ii}^{poly}}, \tag{213}$$

where

$$n_{ii}^{poly} = (1 + 4\pi\chi_{ii}^{(1)poly})^{1/2}. \tag{214}$$

This leads to a birefringence of the form:

$$n_{zz} - n_{xx} = n_{zz}^{poly} - n_{xx}^{poly} + \frac{2\pi}{3} N\alpha_{33}^{*} \left(\frac{1}{n_{zz}^{poly}} - \frac{1}{n_{xx}^{poly}} \right) - \frac{2\pi}{5n_0} \left(\frac{b\nabla v}{kT} \right) N\alpha_{33}^{*}. \tag{215}$$

The birefringence is thus a sum of three parts: one due to the polymer, one due to the dopant dye, and a "cross-term".

When the flow rate in a material results in inhomogeneous stress (as in Fig. 20b), the average orientational order or birefringence is calculated by integrating over the volume of the material. In the case of Fig. 20b, this would be an integration over z.

Electric Field Poling During Viscous Flow. A material can also be poled by an electric field while it is flowing. We consider the special case where the applied electric field is perpendicular to the direction of flow, that is, the field (E) is applied along the z-axis as defined by Fig. 20. The potential energy is then:

$$U = \frac{b\nabla v}{kT} P_2(\cos\theta) - \frac{m^*E}{kT} P_1(\cos\theta). \tag{216}$$

The order parameter $\langle P_2 \rangle$ can be evaluated using Eq. 72. To lowest order in both the electric field and flow rate, the integration yields:

$$\langle P_2 \rangle = -\frac{1}{5}\frac{b\nabla v}{kT} + \frac{1}{15}\left(\frac{m^*E}{kT}\right)^2. \tag{217}$$

The birefringence is then of the form:

$$n_{zz} - n_{xx} = n_{zz}^{poly} - n_{xx}^{poly} + \frac{2\pi}{3} N\alpha_{33}^{*} \left(\frac{1}{n_{zz}^{poly}} - \frac{1}{n_{xx}^{poly}} \right) + \frac{2\pi}{5n_0} \left[\frac{1}{3}\left(\frac{m^*E}{kT}\right)^2 - \frac{b\nabla v}{kT} \right] N\alpha_{33}^{*}. \tag{218}$$

It makes physical sense that the flow causes the order parameter to decrease, while the poling field causes the order parameter to increase. It is this combination of opposing effects that can be used to design a material whose nonlinear response tensor is anisotropic, while the refractive index tensor is isotropic. [43]

E. Second Order Response

In this section, we relate the microscopic second order nonlinear optical susceptibility to the bulk response of a collection of such units in the noninteracting approximation. This model applies to either ordered gases (as in the case of molecular crystals), or guest/host systems, in which a transparent host acts as a rigid support to the guest molecules.

Before proceeding with the calculation, it is worthwhile to outline the qualitative essentials of the problem from the microscopic perspective. First, consider a molecule in vacuum. Figure 21 shows a schematic representation of the electron cloud of the molecule whose principle axis is along z''. When this molecule is placed inside of a host material, its electron cloud changes shape. The new electron

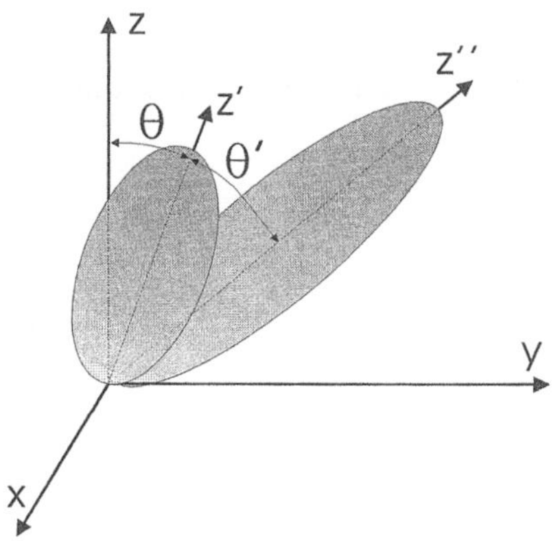

Figure 21 Molecule in vacuum (z″ frame), dressed molecule (z′ frame) and the laboratory frame (z frame). The x and y axes for the molecular frames are not shown to reduce clutter.

configuration essentially represents a different molecule whose principle axes are not necessarily along the principle axes of the molecule in vacuum. We define the "dressed" molecule as having its principle axes in the primed frame. The tensor properties of the dressed molecules are related to the tensor properties of the vacuum molecular properties through the local field tensors according to Eq. 58. When an electric field is applied to the molecule, its electron cloud can deform and the molecule can reorient. Molecular reorientation originates from dipole coupling to the permanent dipole moment of the molecule or the induced dipole moment of the molecule. In addition, changes in the molecule's electron cloud or orientation can result in a piezoelectric response in which the shape of the material changes as a result of its interaction with an electric field. This can therefore also act as one of the mechanisms of a second order response.

In the following subsection, we begin by calculating the second order electronic response of an ordered material. The subsequent subsections deal with the reorientational response and the piezo mechanism.

1. Electronic Response

In Section IV.C. the thermodynamic model of a poled polymer was presented and is embodied in Eq. 65. Here, we calculate the second order susceptibility of a polymer whose molecular dopants are aligned by an external electric field and an axial force (i.e., a force that results in axial dopant alignment). Such axial forces are found in liquid crystal polymers or stressed polymers. When the energy of the dipole in the electric field is small relative to thermal energies, $\vec{m}^* \cdot \vec{E}_p / kT \ll 1$, Eq. 71 can be expanded to first order in $\vec{E}_p$,

$$G \approx \frac{e^{-U/kT}\left(1 + \dfrac{\vec{m}^* \cdot \vec{E}_p}{kT}\right)}{\int e^{-U/kT}\, d\Omega}. \tag{219}$$

For a lab reference frame that is chosen so that poling field is along its z-axis,

$$\vec{m}^* \cdot \vec{E}_p = a_{3L} m_L^* E_p, \tag{220}$$

where m_L^* is the Lth-component of the dressed dipole moment in the molecular frame.

Using Eqs. 219 and 220, G reduces to:

$$G(\Omega, E_p) \approx \frac{e^{-U/kT}\left(1 + \dfrac{a_{3L} m_L^* E_p}{kT}\right)}{\displaystyle\int e^{-U/kT}\, d\Omega}. \tag{221}$$

For $n = 2$, Eq. 65 is:

$$\chi_{ijk} = N \langle \beta_{IJK}^* \rangle_{ijk}, \tag{222}$$

where $\chi_{ijk} \equiv \chi_{ijk}^{(2)}$. Because the laboratory frame axes, i, j, and $k = \{1, 2, 3\}$, and the molecular frame axes, i', j' and $k' = \{x, y, x\}$ are related through Euler angles, θ, φ, and ψ, [67] the orientation average is expressed in terms of the orientation distribution function, $G(\Omega,\xi)$, according to:

$$\langle \beta_{IJK}^* \rangle_{ijk} = \int \beta_{IJK}^*\, a_{iI}\, a_{jJ}\, a_{kK}\, G(\Omega,\xi)\, d\Omega, \tag{223}$$

where ξ represents all parameters that affect the orientations of the molecules.

For a film poled along the laboratory z-axis, the portion of G that depends on the poling field is a function of only the polar angle θ. If the mean-field orientation interaction potential, U, also depends only on θ, the independent terms are χ_{333}, χ_{113}, χ_{311}, and χ_{131} by virtue of ∞_{mm} point symmetry of the bulk. [68] Also, the 1 and 2 axes are equivalent so all indices 1 and 2 can be interchanged without affecting the susceptibility. Under the assumption that the even-order axial order parameters $\langle P_n \rangle$ [69] are independent of poling field, the second order susceptibility of the poled polymer can be expressed in terms of the axial (even-order) order parameters of the polymer prior to poling (as defined by Eq. 72).

$\chi^{(2)}$ can be written in the form:

$$\chi_{ijk} = \frac{NE_p}{kT}\,[u_{ijk}^{(0)} + u_{ijk}^{(2)} \langle P_2 \rangle + u_{ijk}^{(4)} \langle P_4 \rangle], \tag{224}$$

where the coefficients u_{ijk} are given in terms of the molecular polarizabilities, β_{ijk}^*, and the molecular dipole moment, m_I^*. Define:

$$\gamma_0 = \begin{pmatrix} \beta_{xxx}^* & 0 & 0 \\ 0 & \beta_{yyy}^* & 0 \\ 0 & 0 & \beta_{zzz}^* \end{pmatrix}, \tag{225}$$

$$\gamma_1 = \begin{pmatrix} 0 & \beta_{xyy}^* & \beta_{xzz}^* \\ \beta_{yxx}^* & 0 & \beta_{yzz}^* \\ \beta_{zxx}^* & \beta_{zyy}^* & 0 \end{pmatrix}, \tag{226}$$

$$\gamma_2 = \begin{pmatrix} 0 & \beta_{yxy}^* & \beta_{zxz}^* \\ \beta_{xyx}^* & 0 & \beta_{zyz}^* \\ \beta_{xzx}^* & \beta_{yzy}^* & 0 \end{pmatrix}, \tag{227}$$

$$\gamma_3 = \begin{pmatrix} 0 & \beta^*_{yyx} & \beta^*_{zzx} \\ \beta^*_{xxy} & 0 & \beta^*_{zzy} \\ \beta^*_{xxz} & \beta^*_{yyz} & 0 \end{pmatrix}, \tag{228}$$

$$M = \begin{pmatrix} m^*_x & m^*_y & m^*_z \\ m^*_x & m^*_y & m^*_z \\ m^*_x & m^*_y & m^*_z \end{pmatrix}, \tag{229}$$

and

$$\gamma = \gamma_1 + \gamma_2 + \gamma_3. \tag{230}$$

Also defining:

$$(AB)_{zz} = A_{zk}B_{kz}, \tag{231}$$

and

$$(AB)_{ii} = \text{Trace } \{AB\} = \sum_{i=1}^{3} \left(\sum_{k=1}^{3} A_{ik}B_{ki} \right), \tag{232}$$

the result of evaluating Eqs. 222 and 223 yields:

χ_{333} term;

$$u^{(0)}_{333} = \frac{1}{15} [M(3\gamma_0 + \gamma)]_{ii} \tag{233}$$

$$u^{(2)}_{333} = \frac{2}{7} [3(M\gamma_0)_{zz} - (M\gamma_0)_{ii}] - \frac{1}{21} [2(M\gamma)_{ii} - 3(M\gamma + \gamma M)_{zz}] \tag{234}$$

$$u^{(4)}_{333} = \frac{1}{35} [5(M\gamma_0)_{zz} + 3(M\gamma_0)_{ii} + (M\gamma)_{ii} - 5(M\gamma + \gamma M)_{zz}] \tag{235}$$

χ_{311} term;

$$u^{(0)}_{311} = \frac{1}{30} [2M\gamma_0 - \gamma M + 5\gamma_1 M]_{ii} \tag{236}$$

$$u^{(2)}_{311} = \frac{1}{42} [3(M\gamma_0)_{zz} - (M\gamma_0)_{ii} + 2(\gamma M)_{ii} - 3(M\gamma + \gamma M)_{zz} - 3(M\gamma_1)_{ii} - 4(\gamma_1 M)_{ii} + 21(\gamma_1 M)_{zz}] \tag{237}$$

$$u^{(4)}_{311} = -\frac{1}{70} [3(M\gamma_0)_{ii} + 5(M\gamma_0)_{zz} + (\gamma M)_{ii} - 5(M\gamma + \gamma M)_{zz}] \tag{238}$$

χ_{113} term;

$$u^{(0)}_{113} = \frac{1}{30} [2M\gamma_0 - \gamma M + 5\gamma_3 M]_{ii} \tag{239}$$

$$u^{(2)}_{113} = \frac{1}{42} [3(M\gamma_0)_{zz} - (M\gamma_0)_{ii} + 2(\gamma M)_{ii} - 3(M\gamma + \gamma M)_{zz} - 3(M\gamma_3)_{ii} - 4(\gamma_3 M)_{ii} + 21(\gamma_3 M)_{zz}] \tag{240}$$

$$u^{(4)}_{113} = -\frac{1}{70} [3(M\gamma_0)_{ii} + 5(M\gamma_0)_{zz} + (\gamma M)_{ii} - 5(M\gamma + \gamma M)_{zz}] \tag{241}$$

χ_{131} term;

$$u^{(0)}_{131} = \frac{1}{30} [2M\gamma_0 - \gamma M + 5\gamma_2 M]_{ii} \tag{242}$$

$$u^{(2)}_{131} = \frac{1}{42} [3(M\gamma_0)_{zz} - (M\gamma_0)_{ii} + 2(\gamma M)_{ii} - 3(M\gamma + \gamma M)_{zz} - 3(M\gamma_2)_{ii} - 4(\gamma_2 M)_{ii} + 21(\gamma_2 M)_{zz}] \tag{243}$$

$$u^{(4)}_{131} = -\frac{1}{70} [3(M\gamma_0)_{ii} + 5(M\gamma_0)_{zz} + (\gamma M)_{ii} - 5(M\gamma + \gamma M)_{zz}]. \tag{244}$$

The above expressions correct an error that appeared in the original SKS model. [70] The above calculation shows that the second order susceptibility is related to the magnitude of the static poling electric field.

Dye chromophores can also be polar aligned using light. This process is therefore called optical poling. Fiorini and coworkers have demonstrated the all-optical poling technique in polymers. [71–74] In these experiments, two beams of light at frequencies ω and 2ω are launched into the material. Through the third order process governed by $\chi^{(3)}(0;\omega,\omega,-2\omega)$, a static electric field is generated. Because of dispersion in the linear refractive index, the two optical beams travel at a different phase velocity. As a result, the static electric field is sinusoidally modulated with a period equal to the coherence length, resulting in a phase-matched material for those two colors (see Chapter 6 for a discussion of coherence length and phase-matching).

Zyss and coworkers have pointed out that any second order susceptibility can be represented as a combination of a polar and octupolar part (i.e., any third rank tensor can be represented as a combination of rank-1 and rank-3 tensor). While a uniform electric field induces a torque on a dipole moment, an all-optical poling method is required to align the octupoles. Such an optical poling process is called "Octupoling". Because the octupolar moment provides an additional degree of freedom in the design of molecules, novel second order nonlinear optical materials may result. Indeed, octupoling has already been demonstrated. [73, 74] While the topic of octupoling is scientifically interesting and potentially of practical importance, the field is new and little in the way of rigorous calculations are presently in print. Details of this novel poling process are therefore not presented here.

2. Molecular Reorientation

In this subsection, we begin with the general derivation that includes molecular reorientation and apply it the second order susceptibility. As before, we assume that the bulk material has azimuthal symmetry and that the molecules are randomly ordered along the axial angular coordinate in the dressed molecular frame. Under this assumption, the potential energy of a molecule depends only on its polar angle relative to the applied field, θ, and its polar angle relative to its equilibrium orientation. Defining the molecule's zero-field equilibrium polar angle as θ_0, the potential energy is:

$$U(\theta, \theta_0) = U_E(\theta) + U_{int}(\theta, \theta_0). \tag{245}$$

where $U_{int}(\theta, \theta_0)$ is the internal restraining potential which describes any forces in addition to the externally applied electric field, and $U_E(\theta)$ is the electric potential:

$$U_E = -\int_0^{\vec{E}} p^*_{q'}(\vec{E})dE_{q'}. \tag{246}$$

Note that the total electric field $\vec{E}$ includes both the optical and static fields that are applied to the sample. The electric field potential, then, depends only on the dressed molecule's polar coordinate in the laboratory frame while the interaction energy of the molecule with the host material is independent of the applied field. It is this separation of the potential that leads to a tractable solution of the nonlinear susceptibility.

The applied electric field has two effects. First, it induces a dipole moment in the molecule. Secondly, it causes the molecule to reorient through U_E. It is this combination of polarization and reorientation that leads to the reorientational mechanism of the nonlinear response of the molecule. Note that the theory of poling of the previous section can be considered the special case of a third order reorientational mechanism in which a static field and an optical field are applied to the material, and the second harmonic of the optical field results. The electronic dipole moment in the dressed frame, p^*, is specifically written as:

$$p^*_{i'} = m^*_{i'} + K'_1\alpha^*_{i'j'}E_{j'} + K'_2\beta^*_{i'j'k'}E_{j'}E_{k'} + K'_3\gamma^*_{i'j'k'l'}E_{j'}E_{k'}E_{l'} + ..., \tag{247}$$

where the K's are the convention-dependent degeneracy factors in the molecular frame.

When the material is in thermal equilibrium, the angle θ_0 is the average orientation of all molecules that experience the same zero-field potential. For noninteracting guest molecules, the bulk polarization in the laboratory frame (denoted by unprimed subscripts), $\vec{P}$, is related to the induced dipole moment through the vector sum of the individual induced molecular dipole moments. The orientational distribution function of the dipoles is determined by a Gibbs distribution as given by Eq. 63. For a continuous distribution of single-species of guest molecules, the average is given by an integral over the zero-field equilibrium orientation and the laboratory coordinate:

$$N <p^*_{i'}>_i = \frac{N\int d\Omega_0 \left[\int d\Omega\ [G(\theta, \theta_0)p^*_{i'}]\ a_{i'i}(\Omega)\right]}{\int d\Omega_0}, \tag{248}$$

where $G(\theta, \theta_0)$ is the orientational distribution function that gives the probability density of finding a molecule oriented at θ given that its zero-field orientation is θ_0, and a is the rotation matrix, which is a function of the three Euler angles $\vec{\Omega} = (\theta, \phi, \psi)$ that relate the dressed molecule's frame to the lab frame.

The bulk polarization is similarly expressed by a power series in the applied electric field, and in the laboratory frame, is given by:

$$P_i = \chi_i^{(0)} + K_1\chi_{ij}^{(1)}E_j + K_2\chi_{ijk}^{(2)}E_jE_k + K_3\chi_{ijkl}^{(3)}E_jE_kE_l + \ldots, \tag{249}$$

where the K's are the degeneracy factors in the lab frame. Note that while the above equations do not explicitly show that fields and susceptibilities depend on frequency, this is implied. The bulk n^{th}-order nonlinear susceptibility is then calculated from Eq. 240 by differentiating the polarization with respect to the electric field n times:

$$\chi_{ij\ldots}^{(n)}(-\omega;\omega_1,\omega_2,\ldots) = \frac{1}{K_n(-\omega;\omega_1,\omega_2,\ldots)D_n(E_j^{\omega_1},E_k^{\omega_2},\ldots)}\frac{\partial^n P_i^{(\omega)}}{\partial E_j^{\omega_1}\partial E_k^{\omega_2}\ldots}\bigg|_{\vec{E}=0}, \tag{250}$$

where D_n is the degeneracy factor that accounts for the number of distinct fields (fields are distinct if they have different frequencies or different vector components or both) and is given by $n_1!n_2!n_3\ldots!$, where $n! = n \times (n-1) \times (n-2)\ldots \times 1$. $n_1, n_2, n_3\ldots$ are the number of distinct fields at frequency $\omega_1, \omega_2, \omega_3, \ldots$ with $n_1 + n_2 + n_3 + \ldots = n$. Equation 250 is a generalization of Eq. 65. Equation 65 is clearly a special case of Eq. 250 when there is no molecular reorientation (i.e., when the induced dipole moment in the electric field potential given by Eq. 246 is replaced with the spontaneous dipole moment, m^*).

It is clear that the reorientational mechanism should be small for optical second harmonic generation because the response of the electron cloud is fast relative to molecular reorientation. When one or more of the applied electric fields are small or zero, the reorientational effects can dominate. We therefore apply the above theory to the electro optic susceptibility. Furthermore, Eq. 250 depends on the influence of the host material on the molecules by virtue of the dependence of Eq. 248 on the orientational distribution function. We therefore model the electro optic susceptibility in two limiting cases: the free gas limit and the strongly elastically impeded limit.

Electro Optic Susceptibility. In the following, we calculate the reorientational electro optic susceptibility for two special cases, the free gas and the elastically impeded rotator.

Free gas model. We consider a gas in which the molecules' orientations are influenced by some potential $U(\Omega)$. When the system is exposed to an electric field due to a beam of light and/or a static field, the equilibrium orientational order of the gas is determined by the competition between the forces acting on each molecule ($U(\Omega)$ and the electric field) and thermal jiggling. The susceptibility, then, is given by Eq. 250 where the bulk polarization is calculated according to the thermodynamic model given by Eq. 248. The orientational distribution function is determined from the total potential energy function which includes the electric field potential, U_E, and the potential $U(\Omega)$, which includes all other forces that cause molecules to align in the lab frame. Because the molecules are otherwise free to reorient, and $U(\Omega)$ is independent of internal angle, the electro optic susceptibility is of the form:

$$\frac{\chi^{(2)}_{ijk}(-\omega;0,\omega)}{N}$$

$$= \frac{\partial^2}{\partial E^{(0)}_j \partial E^{(\omega)}_k} \left[\left[\int d\Omega \left\{ \exp \left[\left(m^*_{i'} E^{(0)}_{i'} + \frac{1}{2}\, \alpha^*_{i'j'}(-0;\,0) E^{(0)}_{i'} E^{(0)}_{j'} + \ldots \right) \Big/ kT \right] \right.\right.\right.$$

$$\left. \times \exp\,[-U(\Omega)/kT] \left(m^*_{i'} + \frac{1}{2}\, \alpha^*_{i'j'}(-\omega;\omega) E^{(\omega)}_{j'} + \beta^*_{i'j'k'} E^{(0)}_{j'} E^{(\omega)}_{k'} \right) a_{i'i} \right\}$$

$$\left.\left. \times \frac{1}{\int d\Omega \exp \left[\left(m^*_{i'} E^{(0)}_{i'} + \frac{1}{2}\, \alpha^*_{i'j'}(-0;\,0) E^{(0)}_{i'} E^{(0)}_{j'} + \ldots \right) \Big/ kT \right] \exp \left[-\frac{U(\Omega)}{kT} \right]} \right]\right|_{\vec{E}=0} \tag{251}$$

Performing the differentiation with respect to electric field in the above expression, we get:

$$\frac{\chi^{(2)}_{ijk}(-\omega;0,\omega)}{N} = \left\langle \beta^*_{i'j'k'}(-\omega;0,\omega) \left(\frac{\partial E^{(0)}_{j'}}{\partial E^{(0)}_j} \right) \left(\frac{\partial E^{(\omega)}_{k'}}{\partial E^{(\omega)}_k} \right) a_{i'i} \right\rangle$$
$$+ \frac{1}{kT} \left\langle \left(m^*_{i'} \frac{\partial E^{(0)}_{i'}}{\partial E^{(0)}_j} \right) \alpha^*_{i'j'}(-\omega;\omega) \left(\frac{\partial E^{(\omega)}_{j'}}{\partial E^{(\omega)}_k} \right) \right\rangle$$
$$- \frac{1}{kT} \left\langle m^*_{i'} \frac{\partial E^{(0)}_{i'}}{\partial E^{(0)}_j} \right\rangle \left\langle \alpha^*_{i'j'}(-\omega;\omega) \left(\frac{\partial E^{(\omega)}_{j'}}{\partial E^{(\omega)}_k} \right) a_{i'i} \right\rangle, \tag{252}$$

where the brackets are defined as:

$$\langle x \rangle = \frac{\int d\Omega x \exp\,[-U(\Omega)/kT]}{\int d\Omega \exp\,[-U(\Omega)/kT]}. \tag{253}$$

Recognizing that:

$$\frac{\partial E^{(\omega)}_{j'}}{\partial E^{(\omega)}_k} = a_{j'k}, \tag{254}$$

Eq. 252 can be written as:

$$\frac{\chi^{(2)}_{ijk}(-\omega;0,\omega)}{N} = \langle \beta^*_{i'j'k'} \rangle_{ijk} + \frac{1}{kT} \langle m^*_{k'} \alpha^*_{i'j'} \rangle_{ijk} - \frac{1}{kT} \langle m^*_{k'} \rangle_k \langle \alpha^*_{i'j'} \rangle_{ij}, \tag{255}$$

where we have also assumed Kleinman symmetry and have adopted the tensor average definition:

$$\langle x_{i'j'k'\ldots}\rangle_{ijk\ldots} = \frac{\int d\Omega x_{i'j'k'\ldots} a_{i'i} a_{j'j} a_{k'k} \ldots \exp\left[-U(\Omega)/kT\right]}{\int d\Omega \exp\left[-U(\Omega)/kT\right]}. \tag{256}$$

It is instructive to apply Eq. 255 to a one-dimensional molecule; that is, for a molecule whose only non-zero tensor components are $\beta^* = \beta^*_{z'z'z'}(-\omega; 0,\omega)$, $\alpha^* = \alpha^*_{z'z'}(-\omega;\omega)$, and $m^* = m^*_{z'}$. The two independent tensor components that result are $\chi^{(2)}_{zzz}$ and $\chi^{(2)}_{xxz}$:

$$\chi^{(2)}_{zzz} = N\beta^* \left(\frac{3}{5}\langle P_1\rangle + \frac{2}{5}\langle P_3\rangle\right) + N\frac{m^*\alpha^*}{kT}\left[\left(\frac{3}{5}\langle P_1\rangle + \frac{2}{5}\langle P_3\rangle\right) - \langle P_1\rangle\left(\frac{1}{3} + \frac{2}{3}\langle P_2\rangle\right)\right], \tag{257}$$

and

$$\chi^{(2)}_{xxz} = N\beta^* \left(\frac{1}{5}\langle P_1\rangle - \frac{1}{5}\langle P_3\rangle\right) + N\frac{m^*\alpha^*}{kT}\left[\left(\frac{1}{5}\langle P_1\rangle - \frac{1}{5}\langle P_3\rangle\right) - \langle P_1\rangle\left(\frac{1}{3} - \frac{1}{3}\langle P_2\rangle\right)\right], \tag{258}$$

where the order parameters are defined according to Eqs. 159 and 160. It is instructive to observe that Eqs. 257 and 258 reduce to reasonable results in limiting cases. First, when the odd-order order parameters vanish, so does the susceptibility. When the material is fully aligned, the order parameters are all unity. In this limit, the reorientational response vanishes (if the molecules are already oriented, any applied electric field can not make them more oriented) but the electronic response remains. Note, however, that the *zzz* component is equal to $N\beta^*$ while the *xxz* component vanishes. This also makes sense because if all the one-dimensional molecules are aligned, the bulk response will also appear to be one-dimensional.

Given these expressions, we can estimate the relative importance of the reorientational mechanism to the electro optic response. In particular, ferroelectric liquid crystals are known to have non-zero even- and odd-order order parameters. Assuming reasonable numbers for the order parameters ($<P_1> = 0.2$, $<P_2> = 0.6$, and $<P_3> = 0.1$) and values of typical dressed susceptibilities and dipole moments, the ratio of reorientational response to electronic response is:

$$\frac{\chi^{(2)or}_{zzz}}{\chi^{(2)el}_{zzz}} = \frac{\frac{m^*\alpha^*}{kT}\left[\left(\frac{3}{5}\langle P_1\rangle + \frac{2}{5}\langle P_3\rangle\right) - \langle P_1\rangle\left(\frac{1}{3} + \frac{2}{3}\langle P_2\rangle\right)\right]}{\beta^*\left(\frac{3}{5}\langle P_1\rangle + \frac{2}{5}\langle P_3\rangle\right)} \approx 5, \tag{259}$$

and

$$\frac{\chi^{(2)or}_{xxz}}{\chi^{(2)el}_{xxz}} = \frac{\frac{m^*\alpha^*}{kT}\left[\left(\frac{1}{5}\langle P_1\rangle - \frac{1}{5}\langle P_3\rangle\right) - \langle P_1\rangle\left(\frac{1}{3} - \frac{1}{3}\langle P_2\rangle\right)\right]}{\beta^*\left(\frac{1}{5}\langle P_2\rangle - \frac{1}{5}\langle P_3\rangle\right)} \approx -21. \tag{260}$$

Note that the *xxz* component of the reorientational response and electronic response are of opposite sign. If the molecules are not constrained (as in a polymer above the glass transition temperature), the reorientational mechanism will dominate the response. Furthermore, in many studies of poled polymers, the ratio $\chi^{(2)}_{zzz}/\chi^{(2)}_{xxz} = 3$ is predicted when all other forces are absent. When the reorientational response dominates, this ratio can deviate significantly from 3. For example, if $<P_1>$ is the only non-zero order parameter, the ratio of the reorientational part of the tensor components is $\chi^{(2)}_{zzz}/\chi^{(2)}_{xxz} = -2$. Some unexplainable values of tensor ratios in the literature may originate in reorientational effects. This is yet an unexplored area and further research is required to understand when molecular reorientation may become important.

Elastically impeded model. When the molecules are held in place by a rigid constraint as modeled by the microscopic elasticity parameter of the polymer k_θ, the magnitude of the reorientational response is proportional to the inverse of k_θ (i.e., kT in Eqs. 259 and 260 are replaced by k_θ). For a polymer such as PMMA, $k_\theta/kT \approx 10$ so the reorientational response will be lower in the polymer than in the free gas model by that factor. While this greatly decreases the reorientational response, it may still be comparable to the electronic response. It is beyond the scope of this chapter to fully derive these results.

3. Piezoelectricity

The piezoelectric effect results in a change in the shape of a material in response to a static or slowly varying electric field. If the density of the material changes, so does its refractive index. The piezoelectric effect is thus a mechanism that contributes to the electro optic effect. There are two types of shape changes from piezoelectric mechanisms: fixed volume shape changes and nonconstant volume shape changes. In the fixed volume type, a compression along one direction is accompanied by an elongation in another direction so that the total volume stays fixed. An example of a nonconstant volume deformation is a compression along one axis with no accompanying strain along the other two axes. The former is often associated with a shear modulus while the latter is associated with a hydrostatic compression modulus.

Because the piezoelectric effect is a second order effect, it is governed by a third rank tensor. Consequently, the piezoelectric tensor is non-zero only when the bulk material has a net polar order. The most direct approach to calculating the shape change of a material is from the potential energy function. To first order in the applied electric field, the electric energy density of a polar-ordered material is:

$$U = \chi_i^{(0)} E_i, \tag{261}$$

where $\chi_i^{(0)}$ is the zeroth-order susceptibility. When the material is composed of an ensemble of dipoles, the dipole moment per unit volume can be calculated from the orientational order of those dipoles. Evaluating Eq. 65 for a collection of static one-dimensional dipoles in a field applied in the z-direction, we get:

$$U = Nm^*_{z'} \langle P_1 \rangle \, E_z(t), \tag{262}$$

where $\langle P_2 \rangle$ is the order parameter as defined by Eq. 160.

Consider a small rectangular-solid piece of material so that the applied electric field is perpendicular to the xy-face. The force in the z-direction on this face is calculated from the energy density according to:

$$F_z = -\frac{\partial}{\partial z}\int U(x, y, z)dxdydz = -\int U(x, y, z)dxdy. \tag{263}$$

When the electric field is homogeneous, the integrand in Eq. 263 is a constant. Substituting Eq. 262 into Eq. 263 and integrating yields:

$$\frac{F_z}{A} = -Nm^*_{z'}\langle P_1\rangle E_z(t), \tag{264}$$

where A is the area of the face in the xy-plane. The equation of motion is derived using Newton's second law for the resultant strain on that volume element in response to the applied electric field and the restraining elastic force. This yields:

$$\frac{F_z}{A} = \rho\frac{\partial^2 u_{zz}}{\partial t^2} = +Nm^*_{z'}\langle P_1\rangle E_z(t) - Ku_{zz}, \tag{265}$$

where $K = K_R + iK_I$ is the complex bulk modulus, ρ is the mass density, and u_{zz} is the strain in the z-direction. If the "static" applied field varies sinusoidally in time at frequency Ω (which is much smaller than optical frequencies), that is:

$$E_z(t) = E_z \exp[i\Omega t], \tag{266}$$

the driven harmonic oscillator solution of Eq. 265 is of the form:

$$u_{zz} = B \exp[i\Omega t], \tag{267}$$

where B is the complex amplitude which is found by substituting Eq. 266 and Eq. 267 into Eq. 265:

$$B = \frac{E_z N m^*_{z'}\langle P_1\rangle}{(\Omega_0^2 - \Omega^2)\rho + iK_I}, \tag{268}$$

where $\Omega_0^2 = K_R/\rho$ is the natural frequency of elastic oscillation.

There are two limiting cases of how the material strains: constant volume or pure compression. In the constant volume deformation, a stress in the z-direction causes the z-dimension to shrink while the xy-dimension expands. Such a deformation is described by the shear modulus. If the material is not permitted to strain perpendicular to the stress, it is a pure compression and the density change is proportional to the strain. This type of deformation is described by the bulk modulus. Most deformations fall between these two limits.

Electro optical measurements usually measure the electric-field-induced phase shift in a sample. Figure 22 shows a thin film sample through which a light beam passes with its k-vector along the surface normal $\hat{z}$ and polarization along $\hat{x}$. With no field applied, the phase difference over the distance d is $\Phi = nd$. If the applied static voltage results in a decrease in sample thickness, the phase difference over that same distance is the sum of the phase shift in air and the phase shift in the sample. When the sample thickness decreases, the refractive index also changes. The phase difference with the static field on is thus $\Phi = (n + \Delta n)(d - \Delta d) + \Delta d$. The difference in phase with field on and off, then, is:

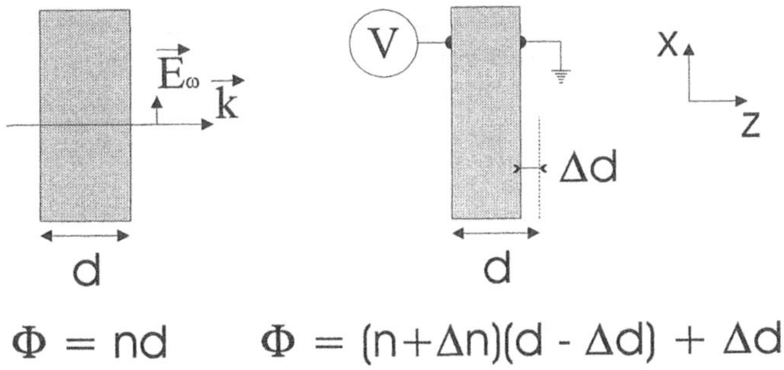

Figure 22 The phase difference between the two ends of a sample (left) and the phase difference over the length d when a voltage is applied to the sample.

$$\Delta\Phi = \Phi(E_0) - \Phi(E_0 = 0) = -n\Delta d + d\Delta n + \Delta d, \tag{269}$$

where we have only kept terms to lowest order in Δn and Δd.

The amplitude of a material's thickness oscillation in response to a quasistatic electric field is given by Eq. 268. The refractive index change associated with this thickness change can be modeled provided that certain conditions are met. As an example, consider a molecular crystal. Because the electric force between molecules are smaller than forces within a molecule, the refractive index can be calculated from the noninteracting rigid gas model. If such a material is slightly compressed, the density of molecules increases without affecting the optical properties of the individual molecules so that the refractive index is proportional to the material's density.

The noninteracting rigid gas model of the refractive index is:

$$n = (1 + 4\pi N \langle\alpha^*\rangle)^{1/2}. \tag{270}$$

If the number density, N, is the only parameter that varies with applied stress (i.e., the polarizablity is independent of N), then Eq. 270 yields

$$dn = \frac{n_0^2 - 1}{2n_0}\frac{dN}{N}, \tag{271}$$

where n_0 is the refractive index with no applied stress. But $dN \propto 1/V$, so:

$$\frac{dN}{N} = -\frac{\Delta d}{d} - \frac{\Delta x}{x} - \frac{\Delta y}{y} = -u_{zz} - u_{xx} - u_{yy} = (2\sigma - 1)u_{zz}, \tag{272}$$

where x, y, and d are the dimensions of the sample, u_{ii} is the strain along i, and σ is Poisson's ratio ($-1 \le \sigma \le 1/2$). Substituting Eq. 272 into Eq. 271 with the help of Eqs. 267 and 268, we get the change in refractive index:

$$dn(\Omega) = +\frac{(n_0^2 - 1)\,(2\sigma - 1)}{2n_0}\frac{Nm_z^*\langle P_1\rangle}{(\Omega_0^2 - \Omega^2)\,\rho + iK_l}E_z(\Omega), \tag{273}$$

where $E_z(\Omega)$ is the amplitude of the field at frequency Ω.

Equation 273 is the refractive index change of the material. Because the material is assumed to be isotropic, this refractive index change is independent of the light polarization. In a typical electro optic measurement, the phase change is measured, but it is often assumed to be due to only a refractive index change. Equation 269 is the total phase change experienced by a slab of material. We can calculate the "effective" refractive index change, Δn_{eff}, by dividing the phase by d:

$$\Delta n_{eff} = \frac{\Delta\Phi}{d} = n_0 u_{zz} - dn - u_{zz}, \tag{274}$$

where we have used $u_{zz} = \Delta d/d$ (because the strain is negative when $\Delta d > 0$, as defined in Fig. 22), dn is the change in material refractive index, and $(n_0 - 1)u_{zz}$ is due to the purely geometric effect of a change in material thickness. Substituting Eq. 273 into Eq. 274 and with the help of Eqs. 267 and 268, we get the change in the effective refractive index:

$$\Delta n_{eff}(\Omega) = +\frac{n_0^2 - 1}{2n_0}\left[\frac{2n_0}{n_0 + 1} + (1 - 2\sigma)\right]\frac{Nm^*_{z'}\langle P_1\rangle}{(\Omega_0^2 - \Omega^2)\,\rho + iK_I}E_z(\Omega). \tag{275}$$

It is useful to summarize the meaning of these results before proceeding to compute the contribution of electrostriction to the electro optic coefficient. Equation 273 expresses the intrinsic refractive index change of the material, while Eq. 275 assumes that the measured phase shift in a thin film measurement is due to a refractive index change and the extrinsic geometrical effect. There are certain measurements, such as waveguide electro optic measurements, that are not sensitive to geometric changes because the light is trapped inside the material. A waveguide Mach-Zehnder interferometer is an example of a geometry that is not affected by material shape changes. In such instances, Eq. 273 is the appropriate one to model the electrostrictive contribution. In each case, it is interesting to consider limiting values of σ. When $\sigma = 1/2$, the material density is unchanged with deformation. The refractive index is unchanged, but the effective refractive index changes by virtue of the change in sample thickness. When $\sigma = -1$, the sample is fully compressed and both the refractive index change and effective refractive index change is maximum. Under the condition that $\sigma = (3n_0 + 1)/(2(n_0 + 1))$, the effective refractive index doesn't change with applied field because the phase change due to the change in refractive index is balanced by the phase change due to the length change.

We are now ready to calculate the electro optic coefficient. At normal incidence (as shown in Fig. 22), the light is polarized perpendicular to the applied static voltage. The electro optic coefficient that is probed in this geometry is $r_{xxz}(-\omega;\omega,0) = r_{yyz}(-\omega;\omega,0)$. With the standard definition of the electro optic coefficient $r_{xz} \equiv r_{xxz}$,

$$dn_{xx}(\Omega) = \frac{n_0^3 r_{xz}(-\omega;\omega,0)}{2}E_z(\Omega), \tag{276}$$

where the quasi-static field oscillates at Ω. Comparing Eq. 275 and Eq. 276, the electro optic coefficient is:

$$r_{xz}^{eff}(-\omega;\omega,0) = +\frac{n_0^2 - 1}{n_0^4}\left[\frac{2n_0}{n_0 + 1} + (1 - 2\sigma)\right]\frac{Nm^*_{z'}\langle P_1\rangle}{(\Omega_0^2 - \Omega^2)\,\rho + iK_I}. \tag{277}$$

Note that Eq. 277 is in Gaussian units so the electro optic coefficient is in *cm* · *statvolt*$^{-1}$. The electro optic coefficient is usually quoted in *pm/V*. The conversion factor between the two is:

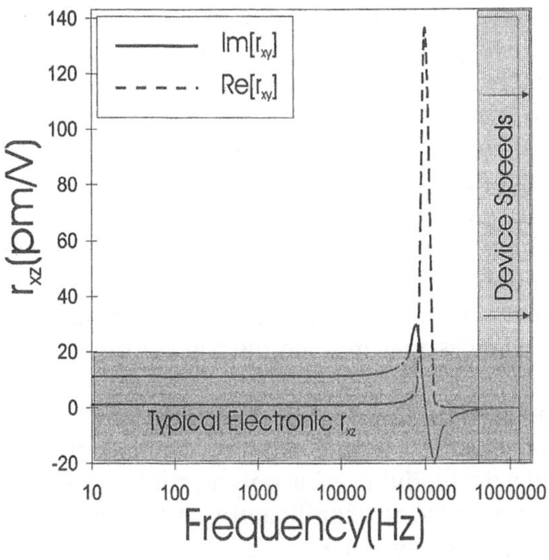

Figure 23 Electrostriction contribution to the electro optic coefficient (real part—solid curve; imaginary part—dashed curve) as a function of modulating frequency. Shaded region is typical electronic electro optic coefficient.

$$r(pm/V) = \frac{r(cm/statvolt)}{3 \times 10^{-8}}. \tag{278}$$

Figure 23 shows the theoretical piezo contribution to the electro optic coefficient given by Eq. 277 for DR1 dye in PMMA polymer. The parameters used for this calculation are listed in Table 6. There are several important conclusions from this plot. First, at low frequencies ($\Omega < 2 \times 10^5 Hz$) the electro optic coefficient

Table 6 Parameters of DR1 Dye in PMMA Polymer Used to Calculate Electrostrictive Contribution to the Electro Optic Coefficient (Plotted in Fig. 23)

Parameter	Value	Notes
m^*	$14 \times 10^{-18} esu \cdot cm$	
σ	3	
$<P_1>$	0.1	
N	$1.9 \times 10^{21} cm^{-3}$	
n_0	1.5	
ρ	$1.19 g/cm^3$	
K_R	$\approx 10^{-10} dyne/cm^2$	(varies with Ω)
K_I	$\approx 10^{-9} dyne/cm^2$	(varies with Ω)
$\Omega_0 = \left(\frac{K_R}{\rho}\right)^{1/2}$	$9.2 \times 10^4 Hz$	

due to the piezo contribution is large (about 10pm/V). If the electronic response ($\approx 20pm/V$) is comparable to the piezo response (which is most often the case), the low frequency measurement cannot be used to accurately determine the electronic response. The fact that thin film measurements often give the correct electronic response at low frequencies is a matter of chance. In particular, if the film sample is clamped in any way, the magnitude of thickness oscillations (Eq. 268) may be decreased and the resonance frequency shifted. Many thin film measurements use sandwich structures, in which the thin film is pressed between two massive glass plates. Under these conditions, the piezo contribution may be greatly reduced. One must be cautious, however, to avoid measurements at frequencies that coincide with new mechanical resonances that may be introduced by the substrate (s).

Note that at typical device frequencies exceeding 1MHz, the piezo contribution is negligible provided that the material elasticity is not strongly dependent upon frequency. It is well-known that the elastic modulus of a polymer increases with frequency so that the resonance structure may be significantly shifted to higher frequencies than predicted by our model as shown in Fig. 23. [75] It is therefore important that electrostrictive effects be understood in the particular geometry used in a measurement so that they can be correctly taken into account.

F. Third Order Response

1. Electronic

While the sophistication of second order measurements makes it possible to map out the second order susceptibility tensor, third order studies are generally more challenging so that most workers only measure a couple of the tensor components. In analogy to the second order case as given by Eq. 223, the orientational average over the second order hyperpolarizabilty γ^* is:

$$\langle \gamma^*_{IJKL} \rangle_{ijkl} = \int \gamma^*_{IJKL}\, a_{iI}\, a_{jJ}\, a_{kK}\, a_{lL}\, G(\Omega,\xi)\, d\Omega. \tag{279}$$

For noninteracting molecules, then, the bulk third order susceptibility is given by:

$$\chi^{(3)}_{ijkl} = N \langle \gamma^*_{IJKL} \rangle_{ijkl}. \tag{280}$$

The orientational average given by Eqs. 279 and 280 are stated in the beginning of Section IV.D.

2. Molecular Reorientation

Here we consider all possible sources of induced dipole moment to third order in the electric field from electronic polarization and molecular reorientation of a material in response to either a static voltage or optical field. We assume that the guest molecules are fixed with respect to translational motion, thereby effectively excluding such mechanisms as phonons and electrostriction—which are discussed in the following subsection. We also assume that the polymer doesn't contribute to the nonlinear polarization. Under these assumptions, we calculate both the D.C. Kerr effect susceptibility (quadratic electro optic effect) and the optical Kerr effect susceptibility. In both cases, we start with the model given by Eq. 250 which for the third order susceptibility is of the form:

$$\chi^{(3)}_{ijkl}(-\omega;\omega_1,\omega_2,\omega_3) = \frac{N}{K_3(-\omega;\omega_1,\omega_2,\omega_3)D_3(E_j^{\omega_1},E_k^{\omega_2},E_l^{\omega_3})} \times \left. \frac{\int d\Omega_0 \left[\dfrac{\partial^3 \left(\int d\Omega [G(\theta,\theta_0) p_{i'}^{*\omega}] a_{i'i}(\Omega) \right)}{\partial E_j^{\omega_1}\, \partial E_k^{\omega_2}\, \partial E_l^{\omega_3}} \right]}{\int d\Omega_0} \right|_{\vec{E}=0} . \quad (281)$$

Equation 281 can be evaluated for any model of internal energy by the choice of the orientational distribution function. The electric-field-derivative in Eq. 281 operates on two terms: the dressed dipole moment and the orientational distribution function. The electronic response is a result of the former, while the reorientational response results from a combination of field-induced changes in the distribution function and lower order electronic polarizations. The treatment below follows the calculations of Poga and coworkers. [76, 77]

D.C. Kerr Effect. A thermodynamic model of the orientational distribution function G can be computed from the Gibbs distribution by using the potential given by Eqs. 246 and 247:

$$G(\theta, \theta_0) = \frac{\exp\left[-\left(-\int_0^{\vec{E}} p_{i'}^{*}(\vec{E}) dE_{i'} + U_{int} \right) \Big/ kT \right]}{\int d\Omega \exp\,[-(-p_{i'}^{*}E_{i'} + U_{int})/kT]} . \quad (282)$$

The action of the third derivative with respect to electric field can be brought under the integral:

$$\left. \frac{\partial^3\, [G(\theta,\, \theta_0) p_{i'}^{*\omega}]}{\partial E_j^{\omega_1}\, \partial E_k^{\omega_2}\, \partial E_l^{\omega_3}} \right|_{\vec{E}=0} . \quad (283)$$

Setting the fields to zero, the integral over the orientational distribution function reduces to an integral over the zero-field orientational distribution function. We define this zero-field distribution function as $G_0(\theta, \theta_0)$:

$$G_0(\theta, \theta_0) = \frac{\exp\,[-U_{int}/kT]}{\int d\Omega \exp\,[-U_{int}/kT]} . \quad (284)$$

Zero frequency response. The zero frequency limit is the dielectric function. In this case, $D = 1$ if all the vector field components are different; otherwise, D is determined by considering the number of fields with the same vector component. The derivative of Eq. 283 becomes:

$$\left. \frac{\partial^3\, [G(\theta,\, \theta_0) p_{i'}^{*\omega}]}{\partial E_j^{\omega_1}\, \partial E_k^{\omega_2}\, \partial E_l^{\omega_3}} \right|_{\vec{E}=0} = \left. \frac{\partial^3\, [G(\theta,\, \theta_0) p_{i'}^{*}]}{\partial E_j \partial E_k \partial E_l} \right|_{\vec{E}=0} . \quad (285)$$

Substituting Eq. 282 into Eq. 283 and performing the derivative with respect to the static field (Eq. 285), we get:

$$\left.\frac{\partial^3 [G(\theta, \theta_0)p^*_{i'}}{\partial E_j \partial E_k \partial E_l}\right|_{\vec{E}=0} = G_0(\theta, \theta_0)\left[\frac{\partial^3 p^*_{i'}}{\partial E_j \partial E_k \partial E_l} - \frac{1}{kT}\frac{\partial^3 U_E}{\partial E_j \partial E_k \partial E_l} p^*_{i'}\right.$$

$$- \frac{1}{kT} P_{jkl}\left\{\frac{\partial^2 p^*_{i'}}{\partial E_j \partial E_k}\frac{\partial U_E}{\partial E_l} + \frac{\partial^2 U_E}{\partial E_j \partial E_k}\frac{\partial p^*_{i'}}{\partial E_l}\right\}$$

$$+ \left(\frac{1}{kT}\right)^2 P_{jkl}\left\{\frac{\partial U_E}{\partial E_j}\left(\frac{\partial U_E}{\partial E_k}\frac{\partial p^*_{i'}}{\partial E_l} + \frac{\partial^2 U_E}{\partial E_k \partial E_l} p^*_{i'}\right)\right\}$$

$$- \left(\frac{1}{kT}\right)^3 \frac{\partial U_E}{\partial E_j}\frac{\partial U_E}{\partial E_k}\frac{\partial U_E}{\partial E_l} p^*_{i'}$$

$$+ \left(\frac{1}{kT}\right) P_{jkl}\left\{\frac{\partial p^*_{i'}}{\partial E_j}\left\langle\frac{\partial^2 U_E}{\partial E_k \partial E_l}\right\rangle_U\right\}$$

$$- \left(\frac{1}{kT}\right)^2 P_{jkl}\left\{p^*_{i'}\frac{\partial U_E}{\partial E_j}\left\langle\frac{\partial^2 U_E}{\partial E_k \partial E_l}\right\rangle_U + \frac{\partial p^*_{i'}}{\partial E_j}\left\langle\frac{\partial U_E}{\partial E_k}\frac{\partial U_E}{\partial E_l}\right\rangle_U\right\}$$

$$+ \left(\frac{1}{kT}\right)^3 P_{jkl}\left\{p^*_{i'}\frac{\partial U_E}{\partial E_j}\left\langle\frac{\partial U_E}{\partial E_k}\frac{\partial U_E}{\partial E_l}\right\rangle_U\right\}$$

$$- \left(\frac{1}{kT}\right)^2 P_{jkl}\left\{p^*_{i'}\left\langle\frac{\partial U_E}{\partial E_j}\right\rangle_U \frac{\partial^2 U_E}{\partial E_k \partial E_l} + P_{jk}\left\{\frac{\partial p^*_{i'}}{\partial E_j}\frac{\partial U_E}{\partial E_k}\right\}\left\langle\frac{\partial U_E}{\partial E_l}\right\rangle_U\right\}$$

$$+ \left(\frac{1}{kT}\right)^3 P_{jkl}\left\{p^*_{i'}\frac{\partial U_E}{\partial E_j}\frac{\partial U_E}{\partial E_k}\left\langle\frac{\partial U_E}{\partial E_l}\right\rangle_U\right\}$$

$$+ \left(\frac{1}{kT}\right) P_{jkl}\left\{\left\langle\frac{\partial U_E}{\partial E_j}\right\rangle_U \frac{\partial^2 p^*_{i'}}{\partial E_k \partial E_l}\right\}$$

$$+ \left(\frac{1}{kT}\right) p^*_{i'}\left\langle\frac{\partial^3 U_E}{\partial E_j \partial E_k \partial E_l}\right\rangle_U$$

$$- \left(\frac{1}{kT}\right)^2 P_{jkl}\left\{p^*_{i'}\left\langle\frac{\partial^2 U_E}{\partial E_j \partial E_k}\frac{\partial U_E}{\partial E_l}\right\rangle_U\right\}$$

$$+ \left(\frac{1}{kT}\right)^3 p^*_{i'}\left\langle\frac{\partial U_E}{\partial E_j}\frac{\partial U_E}{\partial E_k}\frac{\partial U_E}{\partial E_l}\right\rangle_U$$

$$- 2\left(\frac{1}{kT}\right)^3 P_{jkl}\left\{p^*_{i'}\frac{\partial U_E}{\partial E_j}\left\langle\frac{\partial U_E}{\partial E_k}\right\rangle_U\left\langle\frac{\partial U_E}{\partial E_l}\right\rangle_U\right\}$$

$$+ 2\left(\frac{1}{kT}\right)^2 P_{jkl}\left\{\frac{\partial p^*_{i'}}{\partial E_j}\left\langle\frac{\partial U_E}{\partial E_k}\right\rangle_U\left\langle\frac{\partial U_E}{\partial E_l}\right\rangle_U\right\}$$

$$+ 2\left(\frac{1}{kT}\right)^2 P_{jkl}\left\{p_{i'}^* \left\langle\frac{\partial^2 U_E}{\partial E_j\,\partial E_k}\right\rangle_U \left\langle\frac{\partial U_E}{\partial E_l}\right\rangle_U\right\}$$

$$- 2\left(\frac{1}{kT}\right)^3 P_{jkl}\left\{p_{i'}^* \left\langle\frac{\partial U_E}{\partial E_j}\frac{\partial U_E}{\partial E_k}\right\rangle_U \left\langle\frac{\partial U_E}{\partial E_l}\right\rangle_U\right\}$$

$$+ 6\left(\frac{1}{kT}\right)^3 p_{i'}^* \left\langle\frac{\partial U_E}{\partial E_j}\right\rangle_U \left\langle\frac{\partial U_E}{\partial E_k}\right\rangle_U \left\langle\frac{\partial U_E}{\partial E_l}\right\rangle_U\Bigg], \tag{286}$$

where the permutation operator, P_{jkl}, includes only even permutations (i.e., $P_{jkl}\{f_{jkl}\} = f_{jkl} + f_{klj} + f_{ljk}$), and for two subscripts gives $P_{jk}\{f_{jk}\} = f_{jk} + f_{kj}$. The averaging denoted by the angled brackets is over the zero-field distribution function and is an integral over θ only. Note that summation notation over repeated indices does not apply for the permutation operator.

Substituting Eqs. 284 and 286 into Eq. 281 with the help of Eqs. 246 and 247 yields:

$$\chi_{ijkl}^{(3)} = \frac{N}{K_3 D_3}\Bigg\langle 6K_3'\,\langle\gamma_{ijkl}^*\rangle$$

$$+ \left(\frac{2}{kT}\right) P_{jkl}\,\{K_1'\eta_{jk}\,\langle\alpha_{il}^*\alpha_{jk}^*\rangle - \langle\alpha_{ij}^*\rangle\,\langle K_1'\eta_{kl}\alpha_{kl}^*\rangle\}$$

$$+ \left(\frac{6}{kT}\right) [K_2'\eta_{jkl}\langle m_i^*\beta_{jkl}^*\rangle - \langle m_i^*\rangle\,\langle K_2'\eta_{jkl}\beta_{jkl}^*\rangle]$$

$$+ \left(\frac{2}{kT}\right) P_{jkl}\,\{K_2'\,\langle m_l^*\beta_{ijk}^*\rangle - K_2'\,\langle m_j^*\rangle\,\langle\beta_{ikl}^*\rangle\}$$

$$+ \left(\frac{1}{kT}\right)^2 P_{jkl}\,\{\langle m_j^*m_k^*\alpha_{il}^*\rangle + 2K_1'\eta_{kl}\,\langle m_j^*m_i^*\alpha_{kl}^*\rangle + 2\,\langle\alpha_{ij}^*\rangle\,\langle m_k^*\rangle\,\langle m_l^*\rangle\}$$

$$+ \left(\frac{1}{kT}\right)^2 P_{jkl}\,\{4\,\langle m_i^*\rangle\,\langle K_1'\,\eta_{jk}\alpha_{jk}^*\rangle\,\langle m_l^*\rangle - 2\,\langle m_j^*\rangle\,\langle K_1'\eta_{kl}m_i^*\alpha_{kl}^*\rangle\}$$

$$- \left(\frac{1}{kT}\right)^2 P_{jkl}\,\{2\,\langle m_i^*m_j^*\rangle\,\langle K_1'\eta_{kl}\alpha_{kl}^*\rangle + 2\,\langle m_i^*\rangle\,\langle K_1'\eta_{jk}\alpha_{jk}^*m_l^*\rangle$$

$$+ \langle m_k^*m_l^*\rangle\,\langle\alpha_{ij}^*\rangle\}$$

$$- \left(\frac{1}{kT}\right)^2 P_{jkl}\,\{P_{jk}\,\{\langle m_l^*\rangle\,\langle\alpha_{ij}^*m_k^*\rangle\}\} - \left(\frac{1}{kT}\right)^3 [\langle m_i^*\rangle\,\langle m_j^*m_k^*m_l^*\rangle$$

$$+ 6\,\langle m_i^*\rangle\,\langle m_j^*\rangle\,\langle m_k^*\rangle\,\langle m_l^*\rangle]$$

$$+ \left(\frac{1}{kT}\right)^3 \{\langle m_i^*m_j^*m_k^*m_l^*\rangle - P_{jkl}\,\{\langle m_i^*m_j^*\rangle\,\langle m_k^*m_l^*\rangle - \langle m_i^*m_j^*m_k^*\rangle\,\langle m_l^*\rangle\}\}$$

$$+ 2\left(\frac{1}{kT}\right)^3 P_{jkl}\,\{\langle m_i^*m_j^*\rangle\,\langle m_k^*\rangle\,\langle m_l^*\rangle + \langle m_i^*\rangle\,\langle m_j^*m_k^*\rangle\,\langle m_l^*\rangle\}\Bigg\rangle_0, \tag{287}$$

where $\eta_{ijk...}^{(n)}$ is a factor that arises from evaluating the integral of Eq. 246 with p^* given by Eq. 247 and is given by $\eta_{ojk...}^{(n)} = \int_0^{\vec{E}} p_i^{*(n)} dE_i / \xi_{ijk...}^{(n)} E_i E_j E_k ...$, where $\xi^{(n)}$ is the nth-order molecular susceptibility and K_n' the degeneracy factor in the dressed molecular frame. Note that $\eta_{ijk...}^{(n)}$ is a dimensionless multiplicative factor which is not summed according to summation convention. Also, to conserve space, we have used the shorthand notation $\langle T_{ijk...} \rangle = \langle T_{i'j'k'...} \rangle_{ijk...}$. The brackets denote the average over the zero-field distribution function, i.e., with Eq. 284 as the distribution function, while the brackets $<>_0$ denote integration over the isotropic zero-field distribution of the guests, i.e., integration over θ_0.

Equation 287 is the static dielectric function. Any system can be modeled by substitituing the appropriate interaction potential between the substituent molecules and the surrounding material. In a gas, for example, the interaction potential vanishes and the results reduce to those of the standard dielectric response as given by Bottcher. [78] For an elastic restraint, on the other hand, a harmonic interaction potential is used if the elasticity of the surrounding material prevails (such as a dye molecule that is embedded in an elastic polymer). When the molecules are too small to fill completely the voids in a material, the free-volume model, where the molecules are allowed to freely rotate within a restricted range of angles, is more appropriate. Combinations of these potentials more accurately describe most real systems. Such calculations, however, will not be presented here but the generalized response for the quadratic electro optic effect is described below.

Electro optic response. Quadratic electro optic materials mediate the interaction between an optical field and a static field which results in a phase shift of the optical field, [79] where the phase shift is quadratic in the applied static field. The response of interest is then:

$$\chi^{(3)}_{ijkl}(-\omega; \omega, 0, 0). \tag{288}$$

Because observations time scales and thermodynamic reorientation time scales are slower than an optical cycle, it is appropriate to use the "average" potential energy in calculating the orientational distribution function. The average potential energy over one optical cycle is:

$$\bar{U} = \frac{1}{\tau} \int_0^{\tau} U dt, \tag{289}$$

where τ is the optical period. The interaction potential is explicitly time-independent ($\bar{U}_{int} = U_{int}$) so the time average of the total potential is:

$$\bar{U} = \bar{U}_E + U_{int}. \tag{290}$$

The applied field has a zero frequency component and an optical component ($\vec{E} = \vec{E}^{\omega} + \vec{E}^{0}$) so that the electric field potential energy as given by Eq. 246 and Eq. 247 will have terms of the form $(\vec{E}^0)^n(\vec{E}^{\omega})^m$. The time-averaged value of such a product vanishes for m-odd so the lowest order nonvanishing optical field term is $m = 2$. Because the third order polarization at the optical frequency is proportional to the square of magnitude of the static field and is directly proportional to the magnitude of the optical field, the $m = 2$ term will not contribute to the third order response. The potential energy in the exponent of the Gibbs distribution function will thus be a function of only the static field:

$$\begin{aligned} -\bar{U}_E = \int_0^{\vec{E}^0} p^*_{i'}(\vec{E}) dE^0_{i'} &= m^*_{i'} E^0_{i'} + \eta_{i'j'} \alpha^*_{i'j'}(0; 0) E^0_{i'} E^0_{j'} \\ &+ \eta_{i'j'k'} \beta^*_{i'j'k'}(0; 0, 0) E^0_{i'} E^0_{j'} E^0_{k'} \\ &+ \eta_{i'j'k'l'} \gamma^*_{i'j'k'l'}(0; 0, 0, 0) E^0_{i'} E^0_{j'} E^0_{k'} E^0_{l'} + \ldots, \end{aligned} \tag{291}$$

where $K_1' = 1$, $K_2' = 1$, and $K_3' = 1$ for the static Kerr susceptibility.

To evaluate the third order susceptibility, we evaluate Eq. 281 as we did for the static case. The induced dressed dipole moment now has both a static and optical component, $\vec{p} = \vec{p}^0 + \vec{p}^\omega$. Because the distribution function is independent of optical frequency, the induced dipole that multiplies the distribution function will contribute only at the optical frequency:

$$p_{i'} = p_{i'}^\omega = \alpha_{i'j'}^*(-\omega; \omega)E_{j'}^\omega + 2\beta_{i'j'k'}^*(-\omega; \omega, 0)E_{j'}^\omega E_{k'}^0 \\ + 3\gamma_{i'j'k'l'}^*(-\omega; \omega, 0, 0)E_{j'}^\omega E_{k'}^0 E_{l'}^0 + \ldots, \tag{292}$$

where for this case $K_1' = 1$, $K_2' = 2$, and $K_3' = 3$. Evaluating Eq. 281 for the quadratic electro optic susceptibility, we get:

$$\begin{aligned}\chi_{ijkl}^{(3)}(-\omega; \omega, 0, 0) = \frac{N}{6}\Bigg\langle & 3P_{kl}\ \{\langle\gamma_{ijkl}^*(-\omega; \omega, 0, 0)\rangle\} \\ & + \left(\frac{1}{kT}\right) P_{kl}\ \{2(\langle m_l^*\ \beta_{ijk}^*(-\omega; \omega, 0)\rangle \\ & - \langle m_l^*\rangle\ \langle\beta_{ijk}^*(-\omega; \omega, 0)\rangle) \\ & + (\langle\alpha_{ij}^*(0; 0)\alpha_{kl}^*(-\omega; \omega)\rangle - \langle\alpha_{lk}^*(0; 0)\rangle\ \langle\alpha_{ij}^*(-\omega; \omega)\rangle)\} \\ & + \left(\frac{1}{kT}\right)^2 [\langle m_l^* m_k^* \alpha_{ij}^*(-\omega; \omega)\rangle \\ & - \langle m_k^* m_l^*\rangle\ \langle\alpha_{ij}^*(-\omega; \omega)\rangle \\ & + 2\ \langle m_l^*\rangle\ \langle m_k^*\rangle\ \langle\alpha_{ij}^*(-\omega; \omega)\rangle \\ & - P_{kl}\ \{\langle m_l^*\rangle\ \langle m_k^*\alpha_{ij}^*(-\omega; \omega)\rangle\}]\Bigg\rangle_0,\end{aligned} \tag{293}$$

where we have set $D_3 = 2$ owing to the degeneracy of the two zero frequency fields. Note that Eq. 293 is of the form predicted by Eq. 152.

Using the above general equation, the quadratic electro optic response of a given system can be modeled by choice of appropriate interaction potential. The following section considers the elastic and free volume model.

Model potentials. In this section, we evaluate the quadratic electro optic susceptibility for two different model potentials under the assumption that the molecule is one-dimensional (the only non-zero tensor components of the dressed susceptibilities are $m_{z'}^*$, $\alpha_{z'z'}^*$, $\beta_{z'z'z'}^*$, $\gamma_{z'z'z'z'}^*$, etc.). Many molecules with large nonlinearities have been shown to behave as nearly one-dimensional—predictions of thermodynamic models of their nonlinear optical susceptibilities agree with experiments. [43] For convenience, we choose the external electric field along the laboratory 3-direction so that the bulk material has azimuthal symmetry in the 1-2 plane. Typical quadratic electro optic modulation experiments use a thin film sample in which the modulating field is applied perpendicular to the film plane, so that the laboratory 3-axis is defined to be along the film normal.

In these electro optic experiments, the optical field propagates through the sample at some arbitrary angle to the film normal. There are therefore two possible independent polarizations: s-polarization, where the optical field is polarized in the film plane, and p-polarization, where the optical field polarization is in the plane of incidence so that it has both a component in the film plane and perpendicular to it. There are, therefore, two tensor components that are probed under these conditions; that is, $\chi^{(3)}_{1133}(-\omega; \omega, 0, 0)$, $\chi^{(3)}_{3333}(-\omega; \omega, 0, 0)$ and any permutation of these tensor subscripts. We therefore specifically evaluate these two components for the model potentials.

The two model potentials that we choose to evaluate are the harmonic restraint:

$$U_{int} = \frac{1}{2} k_\theta (\theta - \theta_0)^2, \tag{294}$$

where k_θ is the microscopic polymer elasticity, and the restricted free rotator:

$$U_{int} = \begin{cases} 0 & \text{if } \theta_0 - \dfrac{\Delta\theta}{2} < \theta < \theta_0 + \dfrac{\Delta\theta}{2} \\ \infty & \text{otherwise} \end{cases}, \tag{295}$$

where $\Delta\theta$ is the limit of angular motion, and θ_0 is the average zero-field orientation. In the following subsections, we evaluate the quadratic electro optic susceptibility for the elastic restraint and the impeded rotator potentials.

Elastic potential. In this subsection, we evaluate the third order susceptibility as given by Eq. 293 for the elastic potential (Eq. 294). Instead of presenting these calculations in detail, we summarize the results below. Note that the thermodynamic averages of Eq. 294 for the elastic potential gives rise to integrals of the form:

$$\int_0^\pi d\theta \sin^n \theta \cos^m \theta \exp\left[-\frac{pk_\theta}{kT}(\theta - \theta_0)^2\right], \tag{296}$$

where n and m are integers and p a rational number. An integral of this form cannot be solved in closed form. For integration limits that span from $-\infty$ to ∞, however, the integral can be evaluated exactly. [80] Our integrals from 0 to π can be approximated with the integral of infinite range when the microscopic elasticity, k_θ, is much greater than thermal energies, kT (i.e., $kT/k_\theta \ll 1$). In the opposite limit, where the elasticity is small relative to thermal energies, the exponential can be approximated by unity and the integrals become trivial to evaluate. Both limiting cases are evaluated below.

High elasticity limit. In the high elasticity limit, the $\chi^{(3)}_{1133}$-component of a one-dimensional molecule as calculated from Eqs. 293 and 294 is:

$$\begin{aligned}
&\chi^{(3)}_{1133}(-\omega; \omega, 0, 0) \\
&\quad = \frac{N}{6}\left[\frac{3}{16}\gamma^*(-\omega; \omega, 0, 0)\left(2 + \frac{1}{3}\exp\left[-4\frac{kT}{k_\theta}\right] - \frac{1}{5}\exp\left[-12\frac{kT}{k_\theta}\right]\right)\right. \\
&\qquad + \frac{1}{8kT}\beta^*(-\omega; \omega, 0)m^*\left(2 - \frac{8}{3}\exp\left[-3\frac{kT}{k_\theta}\right] + \frac{1}{3}\exp\left[-4\frac{kT}{k_\theta}\right]\right. \\
&\qquad \left. + \frac{8}{15}\exp\left[-9\frac{kT}{k_\theta}\right] - \frac{1}{5}\exp\left[-12\frac{kT}{k_\theta}\right]\right) \\
&\qquad - \frac{1}{16kT}\alpha^*(-\omega; \omega)\alpha^*(0; 0)\left(1 + \frac{1}{3}\exp\left[-4\frac{kT}{k_\theta}\right]\right.
\end{aligned}$$

$$
\begin{aligned}
&- \frac{23}{15} \exp\left[-8\frac{kT}{k_\theta}\right] + \frac{1}{5}\exp\left[-12\frac{kT}{k_\theta}\right]\Bigg) \\
&- \frac{1}{32}\left(\frac{1}{kT}\right)^2 m^{*2}\alpha^*(-\omega;\omega)\left(1 - \frac{8}{3}\exp\left[-3\frac{kT}{k_\theta}\right] + \frac{1}{3}\exp\left[-4\frac{kT}{k_\theta}\right]\right. \\
&+ \frac{56}{15}\exp\left[-7\frac{kT}{k_\theta}\right] - \frac{23}{15}\exp\left[-8\frac{kT}{k_\theta}\right] - \frac{16}{15}\exp\left[-9\frac{kT}{k_\theta}\right] \\
&\left.+ \frac{1}{5}\exp\left[-12\frac{kT}{k_\theta}\right]\right)\Bigg], \qquad (297)
\end{aligned}
$$

where we have set $D_3 = 2$ owing to the degeneracy of the two zero-frequency fields. Given that the molecules are one-dimensional, we have dropped the tensor indices of the molecular polarizabilities ($m^* = m^*_{z'}$, $\alpha^* = \alpha^*_{z'z'}$, ...).

Similarly, the $\chi^{(3)}_{3333}$-component is:

$$
\begin{aligned}
&\chi^{(3)}_{3333}(-\omega;\omega, 0, 0) \\
&= \frac{N}{6}\Bigg[\frac{3}{8}\gamma^*(-\omega;\omega, 0, 0)\left(2 + \exp\left[-4\frac{kT}{k_\theta}\right] + \frac{1}{5}\exp\left[-12\frac{kT}{k_\theta}\right]\right) \\
&+ \frac{1}{4kT}\beta^*(-\omega;\omega, 0)m^*\left(2 - \frac{8}{3}\exp\left[-3\frac{kT}{k_\theta}\right] + \exp\left[-4\frac{kT}{k_\theta}\right]\right. \\
&\left.- \frac{8}{15}\exp\left[-9\frac{kT}{k_\theta}\right] + \frac{1}{5}\exp\left[-12\frac{kT}{k_\theta}\right]\right) \\
&+ \frac{1}{8kT}\alpha^*(-\omega;\omega)\alpha^*(0;0)\left(1 + \frac{1}{3}\exp\left[-4\frac{kT}{k_\theta}\right]\right. \\
&\left.- \frac{23}{15}\exp\left[-8\frac{kT}{k_\theta}\right] + \frac{1}{5}\exp\left[-12\frac{kT}{k_\theta}\right]\right) \\
&+ \frac{1}{16}\left(\frac{1}{kT}\right)^2 m^{*2}\alpha^*(-\omega;\omega)\left(1 - \frac{8}{3}\exp\left[-3\frac{kT}{k_\theta}\right] + \frac{1}{3}\exp\left[-4\frac{kT}{k_\theta}\right]\right. \\
&+ \frac{56}{15}\exp\left[-7\frac{kT}{k_\theta}\right] - \frac{23}{15}\exp\left[-8\frac{kT}{k_\theta}\right] - \frac{16}{15}\exp\left[-9\frac{kT}{k_\theta}\right] \\
&\left.+ \frac{1}{5}\exp\left[-12\frac{kT}{k_\theta}\right]\right)\Bigg]. \qquad (298)
\end{aligned}
$$

For a highly rigid material, the elasticity is large so only the lowest order terms in $1/k_\theta$ contribute to the susceptibility. Equations 297 and 298 to lowest order in $1/k_\theta$ *in each term* reduce to:

$$
\begin{aligned}
\chi^{(3)}_{1133}(-\omega;\omega, 0, 0) = \frac{N}{15}\Bigg[&\gamma^*(-\omega;\omega, 0, 0) + \frac{4}{3}\frac{\beta^*(-\omega;\omega, 0)m^*}{k_\theta} \\
&- \frac{4}{3}\frac{\alpha^*(-\omega;\omega)\alpha^*(0;0)}{k_\theta} - \frac{1}{3}\frac{m^{*2}\alpha^*(-\omega,\omega)}{k_\theta^2}\Bigg], \qquad (299)
\end{aligned}
$$

and

$$\chi^{(3)}_{3333}(-\omega;\omega,0,0) = \frac{N}{15}\left[3\gamma^*(-\omega;\omega,0,0) + 4\frac{\beta^*(-\omega;\omega,0)m^*}{k_\theta} + \frac{8}{3}\frac{\alpha^*(-\omega;\omega)\alpha^*(0;0)}{k_\theta} + \frac{2}{3}\frac{m^{*2}\alpha^*(-\omega,\omega)}{k_\theta^2}\right]. \quad (300)$$

Low elasticity limit. In the low elasticity limit ($k_\theta \ll kT$) for a one-dimensional molecule, Eqs. 293 and 294 give:

$$\begin{aligned}\chi^{(3)}_{1133}(-\omega;\omega,0,0) = \frac{N}{6}\Bigg\{&\left[6\gamma^*(-\omega;\omega,0,0) + \left(\frac{4}{kT}\right)\beta^*(-\omega;\omega,0)m^*\right]\\ &\times\left[\frac{1}{15} - \frac{11}{3375}\left(\frac{k_\theta}{kT}\right)\right]\\ &+ \frac{\alpha^*(-\omega;\omega)}{kT}\left[2\alpha^*(0;0) + \left(\frac{1}{kT}\right)m^{*2}\right]\\ &\times\left[-\frac{2}{45} + \frac{92}{10125}\left(\frac{k_\theta}{kT}\right)\right]\Bigg\},\end{aligned} \quad (301)$$

and

$$\begin{aligned}\chi^{(3)}_{3333}(-\omega;\omega,0,0) = \frac{N}{6}\Bigg\{&\left[6\gamma^*(-\omega;\omega,0,0) + \left(\frac{4}{kT}\right)\beta^*(-\omega;\omega,0)m^*\right]\\ &\times\left[\frac{1}{5} - \frac{76}{1125}\left(\frac{k_\theta}{kT}\right)\right]\\ &+ \frac{\alpha^*(-\omega;\omega)}{kT}\left[2\alpha^*(0;0) + \left(\frac{1}{kT}\right)m^{*2}\right]\\ &\times\left[\frac{4}{45} - \frac{184}{10125}\left(\frac{k_\theta}{kT}\right)\right]\Bigg\}.\end{aligned} \quad (302)$$

Impeded rotor. In the case of the impeded rotator, and, for Δ—the allowed amplitude of angular oscillation—small, we get:

$$\begin{aligned}\chi^{(3)}_{1133}(-\omega;\omega,0,0) = \frac{N}{6}\Bigg\{&6\gamma^*(-\omega;\omega,0,0)\\ &\times\left(\frac{1}{15} + \frac{1}{90}\sin^2\Delta - \frac{1}{50}\sin^4\Delta\right)\\ &+ \left(\frac{4}{kT}\right)\beta^*(-\omega;\omega,0)m^*\\ &\times\left(\frac{2}{45}\sin^2\Delta + \frac{1}{75}\sin^4\Delta\right)\\ &+ \left(\frac{2}{kT}\right)\alpha^*(-\omega;\omega)\alpha^*(0;0)\\ &\times\left(-\frac{4}{45}\sin^2\Delta + \frac{44}{675}\sin^4\Delta\right)\\ &- \frac{16}{675}\left(\frac{1}{kT}\right)^2\alpha^*(-\omega;\omega)m^{*2}\sin^4\Delta\Bigg\},\end{aligned} \quad (303)$$

Table 7 High Elasticity Terms for DR1/PMMA

$\frac{\beta^* m^*}{k_\theta}$ $(10^{-32} cm^7/esu)$	$\frac{(\alpha^*)^2}{k_\theta}$ $(10^{-32} cm^7/esu)$	$\frac{(m^*)^2}{k_\theta^2}$ $(10^{-32} cm^7/esu)$	$\gamma^*(-\omega; \omega, 0, 0)$ $(10^{-32} cm^7/esu^2)$
1.2	1.4	5.2	1.1

and

$$\chi^{(3)}_{3333}(-\omega; \omega, 0, 0) = \frac{N}{6}\left\{6\gamma^*(-\omega; \omega, 0, 0) \times \left(\frac{1}{5} - \frac{2}{15}\sin^2\Delta + \frac{1}{25}\sin^4\Delta\right) + \left(\frac{4}{kT}\right)\beta^*(-\omega; \omega, 0)m^* \times \left(\frac{2}{15}\sin^2\Delta - \frac{2}{75}\sin^4\Delta\right) + \left(\frac{2}{kT}\right)\alpha^*(-\omega; \omega)\alpha^*(0; 0) \times \left(+\frac{8}{45}\sin^2\Delta - \frac{88}{675}\sin^4\Delta\right) + \frac{32}{675}\left(\frac{1}{kT}\right)^2 \alpha^*(-\omega; \omega)m^{*2}\sin^4\Delta\right\}, \tag{304}$$

where $\Delta = \Delta\theta/2$.

Tables 7 and 8 show the various reorientational terms of DR1 dye in PMMA for which the measured microscopic elasticity is $k_\theta = 6 \times 10^{-13} esu^2/cm$. It is evident that the reorientational terms and electronic response are comparable below the glass transition temperature but are much larger than the electronic response at elevated temperatures. For ISQ dye, $\gamma(-\omega; \omega, 0, 0) \approx 10 \times 10^{-32} cm^7/esu$ with $\beta^* = m^* = 0$. In this case, the electronic response is expected to dominate below the glass transition. Clearly, the relative contribution of each mechanism depends on the wavelength of measurement. Nevertheless, the electronic response should generally be dominant below the glass transition and small relative to the reorientational term above the glass transition.

Table 8 Low Elasticity Terms for Room Temperature DR1/PMMA

$\frac{\beta^* m^*}{kt}$ $(10^{-32} cm^7/esu)$	$\frac{(\alpha^*)^2}{kT}$ $(10^{-32} cm^7/esu)$	$\frac{(m^*)^2}{(kT)^2}$ $(10^{-32} cm^7/esu)$	$\gamma^*(-\omega; \omega, 0, 0)$ $(10^{-32} cm^7/esu^2)$
18	22	78	1.1

Experimental evidence of molecular reorientation. In polymers, the reorientational mobility increases with temperature. This can be modeled by assuming that at high temperature (i.e., above the glass transition temperature) the molecules freely rotate, while at low temperature they are elastically impeded. For a centrosymmetric molecule where the dipole moment and second order molecular susceptibility vanish ($u^* = \beta^* = 0$), the two independent tensor components in the two corresponding limits are [77]:

1. In the high elasticity limit (from Eqs. 299 and 300):

$$\chi^{(3)}_{1133}(-\omega;\omega,0,0) = \frac{N}{15}\left[\gamma^*_{z'z'z'z'}(-\omega;\omega,0,0) - \frac{4}{3}\frac{\alpha^*_{z'z'}(-\omega;\omega)\alpha^*_{z'z'}(0;0)}{k_\theta}\right], \tag{305}$$

and

$$\chi^{(3)}_{3333}(-\omega;\omega,0,0) = \frac{N}{15}\left[3\gamma^*_{z'z'z'z'}(-\omega;\omega,0,0) + \frac{8}{3}\frac{\alpha^*_{z'z'}(-\omega;\omega)\alpha^*_{z'z'}(0,0)}{k_\theta}\right]. \tag{306}$$

2. In the low elasticity limit (from Eqs. 301 and 302):

$$\chi^{(3)}_{1133}(-\omega;\omega,0,0) = \frac{N}{6}\left\{6\gamma^*_{z'z'z'z'}(-\omega;\omega,0,0)\left[\frac{1}{15} - \frac{11}{3375}\left(\frac{k_\theta}{kT}\right)\right] + \frac{2\alpha^*_{z'z'}(-\omega;\omega)\alpha^*_{z'z'}(0;0)}{kT}\left[-\frac{2}{45} + \left(\frac{92}{10125}\right)\left(\frac{k_\theta}{kT}\right)\right]\right\}, \tag{307}$$

and

$$\chi^{(3)}_{3333}(-\omega;\omega,0,0) = \frac{N}{6}\left\{6\gamma^*_{z'z'z'z'}(-\omega;\omega,0,0)\left[\frac{1}{5} - \frac{76}{1125}\left(\frac{k_\theta}{kT}\right)\right] + \frac{2\alpha^*_{z'z'}(-\omega;\omega)\alpha^*_{z'z'}(0;0)}{kT}\left[\frac{4}{45} - \frac{184}{10125}\left(\frac{k_\theta}{kT}\right)\right]\right\}, \tag{308}$$

where k_θ is the microscopic elasticity, α^* is the dressed polarizability, and γ^* is the molecular third order susceptibility.

If the electronic response dominates, then the tensor ratio is $a = \chi^{(3)}_{3333}/\chi^{(3)}_{1133} = 3$ while a reorientionally dominated response would yield $a = -2$. The response mechanisms can then be studied by measuring the third order susceptibility tensor. Figure 24 shows the measured tensor ratio of the third order susceptibility as a function of temperature for a centrosymmetric squarylium dye, ISQ in PMMA polymer. [76, 77] It is clear that at the measurement wavelength of 633nm, the electronic mechanism dominates below the glass transition temperature, while the reorientational mechanism dominates above the glass transition.

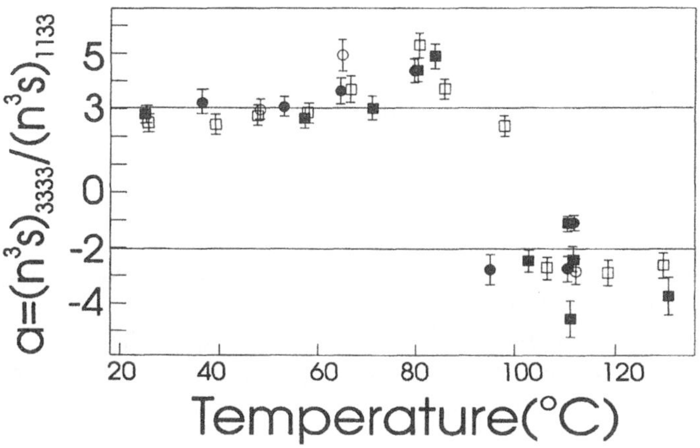

Figure 24 Tensor ratio of third order susceptibility $\chi^{(3)}_{3333}/\chi^{(3)}_{1133}$ as a function of temperature.

More recently, Poga and coworkers have shown that at other wavelengths, the reorientational response is large at all temperatures. Such studies show the importance of understanding the response mechanisms when designing materials. [76, 77]

Optical Kerr Effect. When a collection of molecules is reoriented by an optical field, the refractive index tensor of the ensemble is altered. This change in refractive index is the reorientational mechanism of the optical Kerr effect. The ordering force on a molecule can be understood by considering a single molecule in an alternating (optical) electric field. The interaction energy between the molecule and field is given by Eq. 246 and the time-averaged energy is given by Eq. 289. For a monochromatic beam, the time-averaged potential is:

$$\begin{aligned} -\bar{U}_E &= \int_0^{\vec{E}^\omega} p_{i'}^*(-\omega)dE_{i'}^\omega \\ &= \eta_{i'j'}(-\omega;\omega)E_{i'}^{-\omega}E_{j'}^{\omega} + \eta_{i'j'k'l'}\gamma^*_{i'j'k'l'}(-\omega;\omega,-\omega,\omega)E_{i'}^{-\omega}E_{j'}^{\omega}E_{k'}^{-\omega}E_{l'}^{\omega} + \ldots \end{aligned} \tag{309}$$

This time-averaged electric field potential can be used to calculate the thermodynamic model of the orientational distribution function via Eq. 282 from which the Kerr susceptibility can then be calculated from Eq. 281 with the three field components given by E_i^ω, $E_f^{-\omega}$, and E_k^ω. In analogy to the derivation of the DC Kerr susceptibility that leads to Eq. 293, the general thermodynamic model for the optical Kerr effect yields:

$$\chi^{(3)}_{ijkl}(-\omega;\omega,-\omega,\omega) = \frac{N}{6}\Big\langle 3P_{kl}\ \{\langle\gamma^*_{ijkl}(-\omega;\omega,-\omega,\omega)\rangle\} + \left(\frac{1}{kT}\right) P_{kl}\ \{\langle\alpha^*_{ij}(-\omega;\omega)\alpha^*_{kl}(-\omega;\omega)\rangle - \langle\alpha^*_{lk}(-\omega;\omega)\rangle\,\langle\alpha^*_{ij}(-\omega;\omega)\rangle\}\Big\rangle_0. \quad (310)$$

As with the quadratic electro optic case, Eq. 310 can be evaluated for any interaction potential between the nonlinear molecular molecule and its surroundings. As an example, for an elastic potential in the limit of high elasticity (in analogy to the results for the quadratic electro optic response given by Eq. 299), the 1133 component of the Kerr susceptibility is:

$$\chi^{(3)}_{1133}(-\omega;\omega,-\omega,\omega) = \frac{N}{15}\left[\gamma^*(-\omega;\omega,-\omega,\omega) - \frac{4}{3}\frac{\alpha^*(-\omega;\omega)\alpha^*(-\omega;\omega)}{k_\theta}\right]. \quad (311)$$

It is straightforward to calculate the susceptibility of other tensor components and other interaction potentials so the results will not be presented here. The computational procedure follows those described in the previous section and many of the optical Kerr expressions can be written down by inspection due to their similarity to the analogous quadratic electro optic response.

3. Light-induced Density Change Mechanisms

The refractive index of a material depends on its density. There are many mechanisms that result in light-induced density changes. For example, electrostriction is the process by which matter is attracted (repelled) to (from) regions of bright light. A high intensity laser pulse can induce a change in its optical path by locally stressing the material. This stress can result in density and length changes, both of which affect the phase of the light that exits the material. Indeed, electrostrictive effects have been shown to contribute to the Kerr effect in glass fibers [81] and has been observed to result in fast length changes in a polymer optical fiber. [82] Electrostrictive response times are about 1ps, so that measurements of the Kerr effect on these time scales need to account for the electrostrictive contribution if the underlying mechanisms are to be understood.

Many other mechanisms can lead to density and length changes, including photothermal heating, changes in lattice structure due to electron cloud deformation, molecular reorientation, photoisomerization, photostriction, [82–85] and many others. Because the study of light-induced mechanics is relatively new and activity in the area sparse, many of these mechanisms have not yet been studied. This offers both an opportunity for researchers to study new phenomena in materials, as well as a concern for existing measurements that may be inadvertently "contaminated" by a large background from unknown mechanisms. In this section, we will treat these mechanisms briefly.

Photothermal Heating. When a material absorbs energy from a beam of light, the associated thermal heating results in material expansion. the heating effects are largest in materials that strongly absorb light. The thermal mechanism response time is on the order of 1s to 1*ms* and is certainly negligible in experiments that employ subnanosecond laser pulses. Because the magnitude of the effect is large, though, the contribution to Kerr effect measurements may be substantial in the nanosecond to microsecond regime.

To show how photothermal heating results in a Kerr effect, we calculate this mechanism for an isotropic guest-host system. If the guest molecule is one dimensional and has a dressed polarizability α^*, the first order susceptibility of the system is:

$$\chi^{(1)} = \chi^{(1)}_{host} + \frac{N}{3}\alpha^*, \tag{312}$$

where $\chi^{(1)}_{host}$ is the first order susceptibility of the host and N is the number density of guest molecules. The dielectric function is then of the form:

$$\epsilon = \epsilon_{host} + \frac{4\pi}{3}N\alpha^* = \epsilon_R^{host} + \frac{4\pi}{3}N\alpha_R^* + i\left(\epsilon_I^{host} + \frac{4\pi}{3}N\alpha_I^*\right), \tag{313}$$

where we have expressed the dielectric function and polarizability in terms of their real (subscript R) and imaginary (subscript I) parts.

When the imaginary part of the dielectric function is small relative to the real part, the refractive index of the system is approximately:

$$n = \epsilon^{1/2} \approx n_R + \frac{i}{2n_R}\left(\epsilon_I^{host} + \frac{4\pi}{3}N\alpha_I^*\right), \tag{314}$$

where n_R is the real part of the refractive index of the system. Note that the real part of the refractive index typically is dominated by the host so that it does not vary much as a function of guest concentration in the low doping limit. This is generally true at all wavelengths.

If the intensity of light incident on the material is I_0, the intensity after propagating a distance z into the material is:

$$I(z) = I_0 \exp\left[-2n_I\frac{\omega}{c}z\right], \tag{315}$$

where ω is the frequency of the optical field, c is the speed of light in vacuum, and n_I is given by:

$$n_I = \frac{1}{2n_R}\left(\epsilon_I^{host} + \frac{4\pi}{3}N\alpha_I^*\right). \tag{316}$$

For a sample of length L and illuminated area A, the energy absorbed by the sample over a time interval Δt is:

$$\Delta E = A\Delta t I_0\left(1 - \exp\left[-2n_I\frac{\omega}{c}L\right]\right). \tag{317}$$

The absorbed energy, after thermalization, leads to an increase in sample temperature. If the inverse of the cooling rate of the sample is long compared with the time of photothermal heating, all of the absorbed energy contributes to the temperature increase. For a sample of mass density ρ, the mass being heated is $m = \rho AL$ so that the temperature increase is:

$$\Delta T = \frac{\Delta E}{AL\rho c_v}, \tag{318}$$

where c_v is the specific heat of the material. The phase of the light is affected by changes in the refractive index and length of the material with temperature:

$$\Delta n = \left(\frac{1}{n}\frac{\partial n}{\partial T}\right)\Delta T, \tag{319}$$

and

$$\Delta L = \left(\frac{1}{L}\frac{\partial L}{\partial T}\right)\Delta T, \tag{320}$$

where $1/n(\partial n/\partial T)$ is the thermal refractive index coefficient and $1/L(\partial L/\partial T)$ is the linear thermal expansion coefficient. The calculation of the intensity-dependent phase shift through the sample, then, must account for both the length and refractive index change.

In the following, we derive the thermal Kerr coefficient and evaluate it for typical material parameters. Substituting Eq. 317 into Eq. 318 and using the result in Eq. 319, we get:

$$\Delta n = \left(\frac{1}{n}\frac{\partial n}{\partial T}\right)\frac{\Delta t}{\rho c_v}\frac{\left(1 - \exp\left[-2n_I\frac{\omega}{c}L\right]\right)}{L} I_0. \tag{321}$$

Because the Kerr coefficient n_2 is defined according to $\Delta n = n_2 I_0$, it is simply the coefficient of the intensity in Eq. 321. When the sample is highly absorbing, the argument of the exponential is large and the Kerr coefficient takes the form:

$$n_2 = \left(\frac{1}{n}\frac{\partial n}{\partial T}\right)\frac{\Delta t}{\rho c_v}\frac{1}{L}. \tag{322}$$

When the material is transparent and not highly absorbing, the Kerr coefficient becomes:

$$n_2 = 2n_I\left(\frac{1}{n}\frac{\partial n}{\partial T}\right)\frac{\Delta t}{\rho c_v}\frac{\omega}{c}. \tag{323}$$

To get a rough estimate of the magnitude of the photothermal contribution to the Kerr effect, we evaluate Eqs. 322 and 323 for a polymeric material. Table 9 shows typical order-of-magnitude values of the relevant parameters.

Table 9 Typical Order-of-Magnitude Parameters for a Polymer

$\frac{1}{n}\frac{\partial n}{\partial T}$ (K^{-1})	c_v $(J/kg \cdot K)$	ρ (Kg/m^3)
1×10^{-4}	1×10^3	1×10^3

In the high absorption limit, we get:

$$n_2 = 10^{-10} \frac{\Delta t}{L}, \tag{324}$$

and in the low absorption limit we get:

$$n_2 = 10^{-10} \frac{\Delta t}{L} \left(\frac{2n_I \omega L}{c} \right). \tag{325}$$

Note that the above equations are in MKS units so that n_2 has units of m^2/W. In the low absorption limit, the term in parentheses is much smaller than unity. A value of 0.1 or lower for this factor is consistent with the low absorption approximation. We therefore only consider the high absorption limit. (The low absorption numbers, then, are 10 percent or less of the numbers that we calculate for the high absorption limit.)

For a sample of thickness $L = 1mm$, for a $\Delta t = 1s$ heating time, we get a Kerr response of $n_2 = 10^{-7}m^2/W = 10^{-3}cm^2/W$. For a 1ns heating time, the Kerr coefficient is $n_2 = 10^{-12}cm^2/W$. This is several orders of magnitude larger than the electronic response. Even for a 1ps pulse, the heating effect is an order of magnitude larger than typical electronic Kerr coefficients. If the material is highly transparent, though, thermal effects may be small enough over picosecond time scales for a measurement to detect the electronic and reorientational mechanisms alone.

Electrostriction. There are many ways that an electric field can cause changes in the density of a material. For illustration, we treat the electrostrictive force on the surface of an optical fiber due to electric field gradients and show how these forces can result in a change in the fiber length and refractive index. Figure 25a shows a dark fiber. When light is coupled into the fiber so that it is brightest at the center and monotonically decreases away from the center in the radial direction, the largest intensity gradient is at the surface of the fiber, $r = a$. As we show formally below, the stress on the fiber's surface is related to the field gradient, which results in a decrease in the fiber diameter and an increase in the fiber length. Figure 25b shows the uniformly illuminated fiber. If the fiber is illuminated with a light pulse, the electrostrictive force will be maximum at the brightest part of the pulse. The electrostrictive stress, then, travels with the pulse as shown in Fig. 25c. In the following discussion, we calculate the degree of length change due to a uniformly illuminated fiber, or when the light intensity changes slowly with time. Consider a slice of the fiber of thickness dz in an electric field $E(r)$. If the optical pulse intensity varies slowly along the length of the fiber, there is no applied stress in the z-direction. The electrostrictive stress, then, is along the radial direction. For a cylindrical shell at radius ρ, the compressive force per unit area along the radial direction, $\sigma_{\rho\rho}$, is:

$$\sigma_{\rho\rho} = \frac{1}{2} \chi^{(1)} E^2(\rho). \tag{326}$$

The change in radius induced by the radial stress results in a strain along the fiber axis, m_{zz}, and is given by:

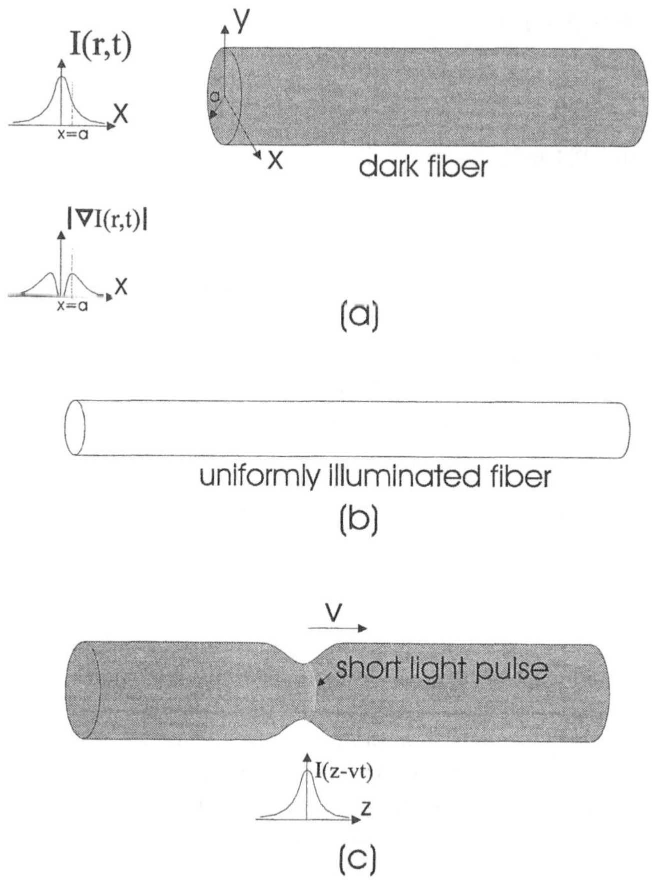

Figure 25 a) A dark fiber, b) uniformly illuminated fiber, and c) fiber with pulsed light source.

$$m_{zz} = -\chi^{(1)} \frac{2\sigma}{Y} \sigma_{\rho\rho} = -\chi^{(1)} \frac{\sigma}{Y} E^2(\rho), \tag{327}$$

where we have used Eq. 326. σ is Poisson's ratio and Y is Young's modulus. Note that Eq. 327 is the quasi-static approximation and applies when the duration of the pulse is long compared with the response time of the material. Because the electric field depends on ρ, the length change is calculated by integrating the strain both over ρ and z.

In the case of a short pulse, light-induced stress acts as a source in the mechanical wave equation. This problem is too complex to solve here. [86] The result of such a calculation is qualitatively described by Fig. 25c. The stress wave travels with the light pulse, which results in a local decrease in the waist and an accompanying strain along the fiber axis. Note that the complicated temporal evolution of the fiber's shape after the pulse passes (i.e., behind the light pulse) is not shown.

VII. NOVEL LOCAL FIELD EFFECTS

A. Local Fields

A method for increasing the nonlinearity of a material is to increase the local electric field at the molecular sites. As described in the beginning of this chapter, there are many local field models. The simplest model that adequately predicts local fields in a homogeneous and isotropic dielectric is that of Lorentz and Lorenz. Applying this to a doped polymer, the dopant molecule, or inclusion when the wavelength of the light is small compared with the size of the dopant, the local field at the site is:

$$F = \frac{\epsilon + 2}{3} E. \tag{328}$$

For a typical polymer of refractive index $n = 1.5$, $\epsilon = n^2$ yields an enhancement factor of 1.42. Another approach is motivated by the long history of metal particle doping in silica glass. Red stained glass windows, for example, are made by heating gold and glass together to form gold-doped silica. The source of the red color is an ensemble of gold spheres that are smaller than the wavelength of light but large enough to have a band structure. (Such small particles are sometimes called quantum dots.) The intense color of such materials implies that the local electric fields must be much higher than the applied field.

The use of quantum dots to enhance local electric fields with the goal of making highly nonlinear optical materials was pioneered by Flytzanis, Hache, and coworkers. [87–90] The effects of the local field on the nonlinearity of gold and silver colloids was first established in 1985 by Ricard and coworkers with optical phase-conjugation experiments. [87] The colloidal system was shown by electron microscopy to be free of aggregates and the gold and silver particles were found to be spherical with 10nm diameter. Within the duration of the 28ps laser source, the enhanced nonlinear optical response was found to be instantaneous. Such a fast response rules out slower effects such as direct heating of the bulk sample.

The electric field at the metal sphere is the central theme that is common to all response mechanisms and is calculated by using the two component local field model given by Eq. 13. (We note that for particles of other shapes, the results are qualitatively similar to the sphere results.) When the sphere radius is much smaller than the wavelength of the optical source and if the skin depth of the metal is large compared with the radius, the quasi-static approximation can be used. Under the influence of an externally applied static field, $\vec{E}$, the electric field inside the sphere, $\vec{E}_{in}$ is:

$$\vec{E}_{in} = \vec{E}\left[\frac{3\epsilon_d(\omega)}{\epsilon_m(\omega) + 2\epsilon_d(\omega)}\right]. \tag{329}$$

The local electric field on the metal sphere ($\vec{F}$) of dielectric constant ϵ_m that is imbedded in a host material of dielectric constant ϵ_d can be calculated directly from Eq. 13 and is:

$$\vec{F} = \vec{E}\left[\frac{3\epsilon_d(\omega)}{\epsilon_m(\omega) + 2\epsilon_d(\omega)}\right] \times \left(\frac{\epsilon_d(\omega) + 2}{3}\right), \tag{330}$$

where ω is the frequency of the applied electric field.

In a transparent material, the dielectric constant is mostly real and varies negligibly with wavelength over the window of transparency. Metallic materials, on the other hand, typically have an absorption edge in the near ultraviolet. When the real part of the metal inclusion's dielectric function becomes negative at $Re|\epsilon_m| = -2\epsilon_d$, the local field factor is maximum and is limited only by the magnitude of the imaginary part of the metal's dielectric constant. The large local field in the metal sphere, then, can strongly excite the electrons in the metal, thereby enhancing its intrinsic nonlinearity.

There are several mechanisms that lead to a nonlinearity in a metal. These include intraband electronic transitions, interband electronic transitions, and hot electrons. In the earlier measurements by Flytzanis and coworkers, it was believed that intraband transitions were dominant, [88] where quantum confinement of conduction electrons leads to the nonlinearity. The measured frequency dependence of gold colloid solutions and gold-doped glasses is consistent with the predicted behavior. This same theory also predicts a particle size effect, which was not observed in later experiments. [90, 91] It was therefore concluded that the hot electron contribution and interband mechanism are dominant. In the hot electron contribution, a nonequilibrium distribution of high temperature conduction electrons is generated by the laser beam when the beam frequency matches an electronic transition. This state of disequilibrium results in a modification of the dielectric response of the sphere which manifests itself as a third order response. A similar hot electron mechanism was proposed earlier by Heilweil and Hochstrasser to explain the temporal third order response of an aqueous gold colloid as probed by picosecond transient grating experiments. [92].

Another intriguing possibility was proposed by Neeves and Birnboim. [93] They calculated the field enhancements of a metallic sphere inclusion of gold or aluminum that is coated with a nonlinear optical shell and embedded in a nonlinear optical medium. The theoretical enhancements in phase conjugate reflectivities were 10^8 times the neat material's response. While this is an interesting approach, no materials of that type—having largely enhanced nonlinearities—have been reported.

B. Fractal Clusters

In the metal colloid studies reviewed above, the samples were carefully prepared to prevent aggregation. An opposite approach is to use aggregates to enhance local fields. The general argument for such an effect is that, in areas of high particle concentration (as found in aggregates), the local field at one inclusion site is produced by the cumulative effect of many nearby particles. At each individual particle site, then, the local field is the sum of the applied field and the polarization field of the other particles. Because each sphere in the cluster is individually polarized by such an enhanced field, the self-consistent local field produced by the cluster acts on each sphere and leads to even a larger field than is possible with noninteracting particles. More significantly, the cluster local field enhancement is found to be only a slowly varying function of the light's frequency.

In most inclusion systems, the metal particle density is low. The criteria for a large nonlinearity in a low density material, however, is for the local density to be as high as possible. Having the property of high local density with low overall density, fractal clusters may fulfill this requirement. Indeed, recent work using the nonlinear optical technique of degenerate four-wave mixing in a colloidal fractal cluster material suggests that signal enhancements are as large as 10^6. [95]

A fractal is an object with fractional dimension that is quantified by the dependence of the local particle density on scale. If the number of particles contained within a sphere of radius r varies as the cube of the radius, the dimension is $D = 3$. A fractal, on the other hand, shows a fractional exponential dependence. While a fractal is a complex object, its dimension is a parameter that quantifies how it affects light propagation. Below we outline a qualitative theoretical argument for large field enhancements within a fractal cluster.

There are two physical properties of a cluster that endows it with a large local field near the surface plasmon resonance frequency. First, if the cluster is composed of noble metal spheres, the local field inside or near an individual particle is enhanced at the surface plasmon frequency as described above. First, consider an isolated sphere. In the quasi-static approximation, the field just outside the sphere is a superposition of the applied uniform electric field and the dipole field of the polarized sphere, where the dipole field is nearly as large as the field within the sphere. Next, consider a second sphere that is placed near the first one. The effective applied field on the second sphere is then the sum of the applied external field and the surface-plasmon-enhanced field of the first sphere. The electric field inside the second sphere is thus larger than what was found for the isolated sphere. The presence of the second sphere, however, increases the local field at the first sphere thereby enhancing the field even more in the first sphere. The total electric field inside a cluster is thus best calculated by summing over all the particles in the cluster in a self-consistent way. As seen in the two-sphere argument, the self-consistent field in the region between the spheres and within a cluster will be greatly enhanced.

While the electric field at a sphere is calculated self-consistently from the sum of the applied field and the induced dipole fields from the other spheres, the fractal nature of the cluster complicates the procedure. Owing to strong sphere density fluctuations, the local fields are far from uniform and simplifying assumptions need to be made in order to estimate the degree of field enhancement. Typically, this type of approach yields a lower and upper limit on enhancement factors that are attainable for an object of given fractal dimension. An important consideration in determining the enhancement factor is whether the cluster size is small enough for the nonlinear optical process to remain coherent while the light passes through the fractal. An interesting theoretical result suggests that processes with degenerate field components such as the intensity dependent refractive index (quantified by n_2) are much more strongly enhanced than sum frequency and parametric processes. [94–96] This result bodes well for all-optical logic and computing devices which all depend on n_2. In the limit of $D \rightarrow 3$, however, the fractal local field enhancements vanish (excluding, of course, the individual surface plasmon contributions) for all processes.

While the arguments presented above describe enhancement of the electronic nonlinearity in a noble metal sphere, other deliberately engineered material systems should also show an enhanced nonlinearity. Butenko and coworkers calculate the enhanced susceptibility of a material that incorporates both noble metal fractal clusters and what they call impurities. [96] The impurities are low concentrations of nonlinear optical units (atoms, molecules, or microscopic aggregates that are small compared to the metal spheres). In these materials, if the nonlinear optical impurities are of low concentration, they will not interfere with the local field enhancement mechanism. If these units of high nonlinearity reside within a cluster, their nonlinearity contribution may dominate the material's response.

1. Local Fields Near Quantum Dots

In this section, we consider the local electric fields at a molecule near or within a quantum dot. Figure 26 shows two possible positions at which a molecule can be placed. The applied electric field, $\vec{E}$, is far away from the sphere. This electric field induces a polarization in the sphere which results in a total electric field outside the sphere, $\vec{E}_{out}$, which is a sum of the dipole field and the uniform applied electric field, and a uniform electric field inside the sphere, $\vec{E}_{in}$.

The local electric field at a molecule inside the sphere can be calculated by applying the Lorentz-Lorenz local field model to a void in the metal. Given that the electric field in the metal is given by Eq. 329, the local electric field at the molecule is:

$$\vec{F} = \vec{E}_{in}\left(\frac{\epsilon_m(\omega) + 2}{3}\right) = \vec{E}\left[\frac{3\epsilon_d(\omega)}{\epsilon_m(\omega) + 2\epsilon_d(\omega)}\right] \times \left(\frac{\epsilon_m(\omega) + 2}{3}\right). \tag{331}$$

Next, we calculate the local electric field just outside the surface of the sphere. To do so, we first calculate the electric field in the dielectric at the point where the electric field is normal to the surface of the sphere. Using continuity in $\vec{D}$, the electric displacement at this point, we get:

$$\epsilon_d E_{out} = \epsilon_m E_{in}. \tag{332}$$

Using Eqs. 332 and 329, the electric field at the surface of the sphere in the dielectric is:

$$\vec{E}_{out} = \vec{E}\left[\frac{3\epsilon_m(\omega)}{\epsilon_m(\omega) + 2\epsilon_d(\omega)}\right]. \tag{333}$$

Applying the Lorentz-Lorenz local field model in the dielectric, we get:

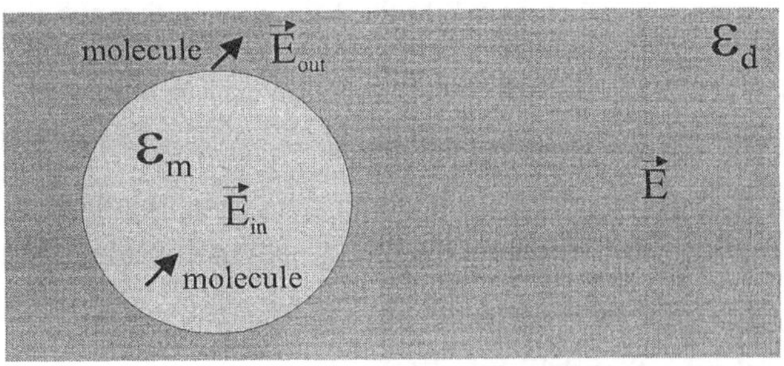

Figure 26 The high local electric field inside and near a metal inclusion can be used to enhance the nonlinearity of a molecule.

$$\vec{F} = \vec{E}_{out}\left(\frac{\epsilon_d(\omega) + 2}{3}\right) = \vec{E}\left[\frac{3\epsilon_m(\omega)}{\epsilon_m(\omega) + 2\epsilon_d(\omega)}\right] \times \left(\frac{\epsilon_d(\omega) + 2}{3}\right). \tag{334}$$

It is interesting to evaluate the point at which the local electric field is largest. The ratio of the local electric field just outside the sphere to the local electric field just inside the sphere (Eq. 334 divided by Eq. 331) is:

$$\frac{F_{out}}{F_{in}} = \frac{\epsilon_d(\omega) + 2}{\epsilon_m(\omega) + 2} \times \frac{\epsilon_m(\omega)}{\epsilon_d(\omega)}. \tag{335}$$

At the surface plasma resonance, when $\epsilon_m(\omega) = -2\epsilon_d(\omega)$, the ratio of local electric fields is:

$$\frac{F_{out}}{F_{in}} = \frac{\epsilon_d(\omega) + 2}{\epsilon_d(\omega) - 1}. \tag{336}$$

For a typical polymer of refractive index $\epsilon_d = n^2 = 2.25$, the ratio if 3.4. The largest local fields are thus for the case where the molecule is just outside of a metal sphere in the dielectric.

C. Nonlinear Local Fields and Cascading

1. *Introduction*

When an electric field is applied to a material, it induces a polarization. At the microscopic level, polarization originates at the atoms or molecules that collectively make the material. Each of these dipoles, however, radiates an electric field which can interact with the other molecules. Indeed, this idea was used in Section II to calculate the local field models. Section II, however, assumes that the induced dipole moment is linear in the applied electric field. In a nonlinear optical material, the induced dipole moment is a nonlinear function of the electric field so that the dipole field of each molecule can radiate at any color even when the applied field is due to a monochromatic light source. The problem is further complicated by the fact that these "multiple colors" that are generated by a given molecule interact with the other molecules in the material which subsequently reradiate a field that acts back on the first molecule. It is clear that such a process requires a self-consistent calculation of the local electric field.

Figure 27 shows a simple schematic of the process to lowest order. An electric field, E, is applied to the material. At molecule κ, the electric field is the sum of the applied electric field and the field due to the neighboring molecules. The electric field due to molecular λ is a function of its dipole moment and position relative to the molecule κ. The tensor $g_{ii'}^{\kappa\lambda}$ accounts for the geometric factor. The total polarization field is then the sum over all molecules in the material (i.e., over λ). To calculate a self-consistent field, this procedure must be done for each molecule in the system. The process is then repeated to get successive approximations to the self-consistent local field.

2. *Iterative Calculation of the Self-Consistent Local Field*

In this subsection, we use an iterative approach to calculate the local electric field. [97–99] The induced dipole moment of molecule κ is related to the local electric

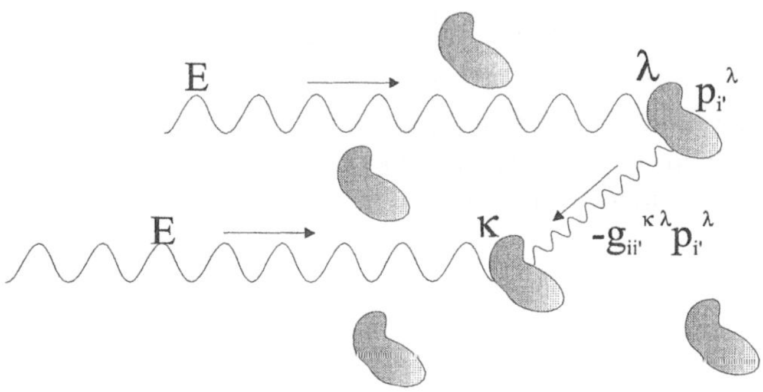

Figure 27 The local field at molecule κ is due to that applied field and the polarization field due to other molecules (λ).

field F (which is yet to be calculated) and the polarizabilities. For the purposes of illustration, we consider the first two terms:

$$p_i^\kappa = \alpha_{ij}^\kappa F_j^\kappa + \beta_{ijk}^\kappa F_j^\kappa F_k^\kappa, \tag{337}$$

where summation notation applies for all indices except for κ. To first approximation, the local electric field is the sum of the applied field and the polarization field due to the other molecules in the material:

$$F_i^\kappa = E_i^\kappa - g_{ii'}^{\kappa\lambda} p_{i'}^\lambda, \tag{338}$$

where the index λ is summed over *all* molecules in the material (excluding molecule κ) and the g tensor depends on the position and orientation of each dipole in the sum. Substituting Eq. 338 into the first term of Eq. 337 and bringing the term with the dipole moment to the left-hand side of the equation, we get:

$$(\delta_{ij'}^{\lambda\kappa} + \alpha_{ij}^\kappa g_{jj'}^{\kappa\lambda})\, p_{j'}^\lambda = \alpha_{ij}^\kappa E_j^\kappa + \beta_{ijk}^\kappa F_j^\kappa F_k^\kappa, \tag{339}$$

where $\delta_{ij'}^{\lambda\kappa}$ is the Kronecker delta and is non-zero only when $i = j'$ and $\lambda = \kappa$.

To simplify the expressions, we define the operator:

$$R_{ij'}^{\lambda\kappa} = (\delta_{ij'}^{\lambda\kappa} + \alpha_{ij}^\kappa g_{jj'}^{\kappa\lambda})^{-1}, \tag{340}$$

and adopt the tensor product notation:

$$A \cdot B = A_{ij\ldots k} B_{kl\ldots m}, \tag{341}$$

and

$$A : EE = A_{ijk} E_j E_k. \tag{342}$$

These definitions will allow us to avoid the cumbersome subscripts.

Using this notation and multiplying both sides of Eq. 339 by $R_{ij'}^{\lambda\kappa}$, we get

$$p = R \cdot \alpha \cdot E + R \cdot \beta : FF. \tag{343}$$

Note that the first term in Eq. 343 is now expressed as a function of E, the applied electric field, so that the coefficient $R \cdot \alpha$ is the dressed polarizability of the molecule. Defining α^* to be the dressed polarizability:

$$\alpha^* = R \cdot \alpha, \tag{344}$$

and

$$\beta' = R \cdot \beta, \tag{345}$$

with the help of Eqs. 344 and 345, Eq. 343 becomes:

$$p = \alpha^* \cdot E + \beta' : FF. \tag{346}$$

For the next iteration, we substitute $F = E - g \cdot p$ into Eq. 346. Completing the square and grouping together the terms that depend only on the applied field, E, we get

$$p = \{\alpha^* \cdot E + \beta' : EE\} - 2\beta' : E\,(g \cdot p) + \beta' : (g \cdot p)\,(g \cdot p). \tag{347}$$

It is clear that the term in braces is the first approximation to the induced dipole moment where β' is the dressed hyperpolarizability to first approximation. We can thus get the second order correction by substituting $p = \alpha^* \cdot E + \beta' : EE$ into the second and the third terms in Eq. 347. To third order in the electric field, this yields:

$$\begin{aligned} p = {} & \{\alpha^* \cdot E + \beta' : EE\} + \beta' : (g \cdot \alpha^* \cdot E)^2 - 2\beta' : E\,(g \cdot \alpha^* \cdot E) \\ & - 2\beta' : E\,(g \cdot \beta' : E^2) \\ & + \beta' : [(g \cdot \alpha^* \cdot E)\,(g \cdot \beta' : E^2) \\ & + (g \cdot \beta' : E^2)\,(g \cdot \alpha^* \cdot E)]. \end{aligned} \tag{348}$$

Recall that by Klienman symmetry, the two terms in the brackets are equal. We then get

$$\begin{aligned} p = {} & \alpha^* \cdot E + \beta^* : EE + 2\beta' : [(g \cdot \alpha^* \cdot E)\,(g \cdot \beta' : E^2) \\ & - E\,(g \cdot \beta' : E^2)], \end{aligned} \tag{349}$$

where β^* is the dressed hyperpolarizability and is defined as:

$$\beta^* = \beta'[(1)(1) + (g \cdot \alpha^*)(g \cdot \alpha^*) - 2(1)(g \cdot \alpha^*)]. \tag{350}$$

Note that 1 in parentheses is the unit tensor of rank 2.

3. Cascading

The last term in Eq. 349 (which results in cascading) is third order in the applied electric field so it contributes to the third order response. We can rewrite this term so that Eq. 349 takes the form:

$$p = \alpha^* \cdot E + \beta^* : EE + \delta\gamma \vdots E^3, \tag{351}$$

where

$$\delta\gamma = 2\beta' : [(g \cdot \alpha^*)\,(g \cdot \beta') - (1)\,(g \cdot \beta')] \tag{352}$$

is a fourth rank tensor that is proportional to the square of the second order nonlinear optical susceptibility and is therefore a correction term to the third order nonlinear optical susceptibility. Figure 28 shows a pictorial representation of the process described by Eq. 352 for the case of three distinct input colors. In general,

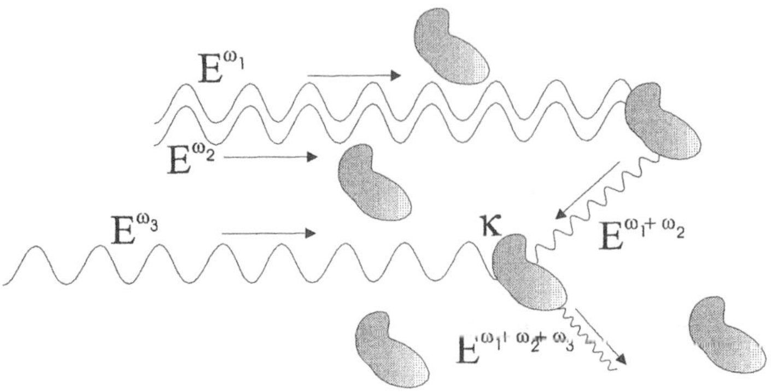

Figure 28 Two molecules interact through a second order process to yield a third order susceptibility.

a third order process describes the mixing of three fields. We therefore choose the most general case of three incident fields at frequency ω_1, ω_2, and ω_3. ω_1 and ω_2 interact together at one molecular site through a second order process yielding sum frequency generation; that is, a field at $\omega_1 + \omega_2$ is radiated. This field, then, interacts with the field at ω_3 at the second molecule to generate a field at frequency $\omega_1 + \omega_2 + \omega_3$. This process is called cascading because two sequential second order processes yield a higher order effect. Given that each pair of molecules can interact through a second order molecular susceptibility to yield a third order response, we need to sum over all pairs of interactions; and, because each molecule radiates a dipole field pattern, generally only nearby neighbors of molecules will contribute significantly to the cascading process. Note that the cascading calculation is somewhat complicated by the fact that a molecule can be found in the near field zone of another. A rigorous cascading calculation can be found in the literature. [100] Finally, it is instructive to represent the process given by Fig. 28 in the frequency domain. The induced dipole moment to third order in the electric field is:

$$\begin{aligned} p^{(\omega_1+\omega_2+\omega_3)} &= \delta\gamma(-\omega_1 - \omega_2 - \omega_3; \omega_1,\omega_2,\omega_3)E^{\omega_1}E^{\omega_2}E^{\omega_3} \\ &= 2[\beta'(-\omega_1 - \omega_2 - \omega_3; \omega_1 + \omega_2,\omega_3) \\ &\quad : (g \cdot \alpha^*(-\omega_1 - \omega_2; \omega_1 + \omega_2)) \, (g \cdot \beta'(-\omega_1 \\ &\quad - \omega_2; \omega_1,\omega_2))] \vdots E^{\omega_1}E^{\omega_2}E^{\omega_3}. \end{aligned} \tag{353}$$

The order of photon absorption/conversion is clear. In general, then, the local field models must include nonlinear terms. There are cases in which macroscopic cascading of two second order effects yields an effective third order response that greatly exceeds the third order susceptibility of the material. [101, 102] Recently, cascading has been used to make devices. [103–105] These nonlinear local fields, then, can be large and used for enhancing nonlinear effects.

VIII. CONCLUSION

In this chapter, we have related the molecular susceptibilities and the interaction between a molecule and its surroundings to the bulk optical properties of a material.

The tabulated values of properties in the following chapters, when used in conjunction with the models presented in this chapter, can be used to understand how composite materials can be formed and their bulk properties controlled.

IX. ACKNOWLEDGMENTS

This work was supported by the Air Force Office of Scientific Research; Army Research Office; Sentel Technologies; and the National Science Foundation, Lightwave Technology and Polymer Chemistry Division. I thank Brian Canfield, Dennis Garvey, and Robert Kruhlak for a thorough reading of this chapter.

REFERENCES

[1] Jackson, J. D., *Classical Electrodynamics*, 2nd ed., Wiley, New York, 1975.
[2] *Theory of Electric Polarization*, Vol. I, Elsevier, New York, 1973, p. 205; Bottcher and Bordewijk, *Theory of Electric Polarization*, Vol. II, Elsevier, New York, 1978.
[3] Bell, R. P., *Trans. Faraday. Soc.*, *27*, 797 (1931).
[4] Onsager, L., *J. Am. Chem. Soc.*, *58*, 1486 (1936).
[5] Dirk, C. W., and Kuzyk, M. G., *Phys. Rev. B*, *41*, 1636 (1990).
[6] *Molecular Nonlinear Optics* (J. Zyss, ed.) Academic Press, New York, 1994.
[7] Pang, Y., Samoc, M., and Prasad, P. N., *J. Chem. Phys.*, *94*, 5282 (1991).
[8] See, for example, Kuzyk, M. G., All-Optical Materials and Devices, in *Organic Thin Films for Waveguiding Nonlinear Optics* (F. Kajzar and J. D. Swalen, eds.), Gordon and Breach, 1996, p. 759.
[9] Matsumoto, S., and Kaino, T., *Proc. SPIE*, *1337*, 144 (1990).
[10] Li, L., Lee, J., Zhu, X., Kumar, J., and Tripathy, S. K., *Jpn. J. Appl. Phys.*, *31*, L89 (1992).
[11] Lei, D., Runt, J., Safari, A., and Newnham, R. E., *Macromol*, *20*, 1797 (1987).
[12] Kuzyk, M. G., Dirk, C. W., and Sohn, J. E., *J. Opt. Soc. Am. B*, *5*, 842 (1990).
[13] Zyss, J., and Oudar, J. L., *Phys. Rev. A*, *26*, 2028 (1982).
[14] Lipscomb, G. F., Garito, A. F., and Narang, R. S., *J. Chem. Phys.*, *75*, 1509 (1981).
[15] Levine, B. F., Bethea, C. G., Thurmond, C. D., Lynch, R. T., and Bernstein, J. L., *J. Appl. Phys.*, *50*, 2523 (1979).
[16] Singer, K. D., Sohn, J. E., and Lalama, S. J., *Appl. Phys. Lett.*, *49*, 248 (1986).
[17] Singer, K. D., Kuzyk, M. G., and Sohn, J. E., *J. Opt. Soc. Am. B*, *4*, 968 (1987).
[18] Singer, K. D., Kuzyk, M. G., Holland, W. R., Sohn, J. E., Lalama, S. J., Comizzoli, R. B., Katz, H. E., and Schilling, M. L., *Appl. Phys. Lett.*, *53*, 1800 (1988).
[19] Hampsch, H. L., Yang, J., Wong, G. K., and Torkelson, J. M., *Macromolecules*, *21*, 526 (1988).
[20] Singer, K. D., and King, L. A., *J. Appl. Phys.*, *6*, 744 (1989).
[21] Kohler, W., Robello, D. R., Dao, P. T., Willand, C. S., and Williams, D. J., *J. Chem. Phys.*, *93*, 9157 (1990).
[22] Mortazavi, M. A., Knoesen, A., Kowel, S. T., Higgens, B. G., and Dienes, A., *J. Opt. Soc. Am. B*, *6*, 733 (1989).
[23] Eich, M., Sen, A., Looser, H., Bjorklund, G. C., Swalen, J. D., Twieg, R., and Yoon, D. Y., *J. Appl. Phys.*, *66*, 2559 (1989).
[24] Eich, M., Reck, B., Yoon, D. Y., Wilson, C. G., and Bjorklund, G. C., *J. Appl. Phys.*, *66*, 3241 (1989).
[25] Hampsch, H. L., Yang, J., Wong, G. K., and Torkelson, J. M., *Polym. Commun.*, *30*, 40 (1989).

[26] Hampsch, H. L., Yang, J., Wong, G. K., and Torkelson, J. M., *Macromol*, *23*, 3640 (1990).
[27] Hampsch, H. L., Yang, J., Wong, G. K., and Torkelson, J. M., *Macromol.*, *23*, 3648 (1990).
[28] Hampsch, H. L., Torkelson, J. M., Bethke, S. J., and Grubb, S. G., *J. Appl. Phys.*, *67*, 1037 (1990).
[29] Boyd, G. T., Francis, C. V., Trend, J. E., and Ender, D. A., *J. Opt. Soc. Am. B*, *8*, 887 (1991).
[30] Ghebremichael, F., Kuzyk, M. G., and Dirk, C. W., *Nonlinear Optics*, *6*, 123 (1993).
[31] Ghebremichael, F., and Kuzyk, M. G., *J. Appl. Phys.*, *77*, 2896 (1995).
[32] Burland, D. M., Miller, R. D., and Walsh, C. A., *Chem. Rev.*, *94*, 31 (1994).
[33] Zimmerman, K., Ghebremichael, F., Kuzyk, M. G., and Dirk, C. W., *J. Appl. Phys.*, *75*, 1267 (1994).
[34] Dhinojwala, A., Wong, G. K., and Torkelson, J. M., *Macromol.*, *25*, 7395 (1992); Dhinojwala, A., Wong, G. K., and Torkelson, J. M., *Macromol.*, *26*, 5943 (1993).
[35] Wang, C. H., Gu, S. H., and Guan, H. W., *J. Chem. Phys.*, *99*, 5597 (1993).
[36] Kim, Y. W., Jin, J. I., Jin, M. Y., Choi, K. Y., Kim, J. J., and Zyung, T., *Polymer*, *38*, 2269 (1997).
[37] Kubo, Y., Takaba, T., and Aramaki, S., *Chem. Lett.*, *3*, 255 (1997).
[38] Norwood, R. N., Kuzyk, M. G., and Keosian, R. A., *J. Appl. Phys.*, *75*, 1869 (1994).
[39] Fenimore, D. L., Schepler, K. L., Zelmon, D., Kuck, S., Ramabadran, U. B., Von Richter, P., and Small, D., *J. Opt. Soc. Am. B*, *13*, 1935 (1996).
[40] Dalton, L. R., Harper, A. W., and Robinson, B. H., *Proceedings of the National Academy of Sciences of the United States of America*, *94*, 4842–4847 (1997).
[41] Lundquist, P. M., Yitzchiak, S., Marks, T. J., Wong, G. K., DiBella, S., Cohen, R., and Berkovic, G., *Phys. Rev. B*, *55*, 14055 (1997).
[42] DiBella, S., Lanza, G., Fragala, I., Yitzchaik, S., Ratner, M. A., and Marks, T. J., *J. Am. Chem. Soc.*, *119*, 3003 (1997).
[43] Kuzyk, M. G., Singer, K. D., Zahn, H. E., and King, L. A., *J. Opt. Soc. Am. B*, *6*, 742 (1989).
[44] van der Vorst, C. P. J. M., and Picken, S. J., *J. Opt. Soc. Am. B*, *7*, 320 (1990).
[45] Maier, W., and Saupe, A., *Z. Naturforsch*, *13A*, 564 (1958); *114A*, 882 (1959); *15A*, 287 (1960).
[46] Girling, I. R., Kolinsky, P. V., Cade, N. A., Earls, J. D., and Peterson, I. R., *Opt. Commun.*, *55*, 4 (1985).
[47] Neal, D. B., Petty, M. C., Roberts, G. G., Amhed, M. M., Feast, W. J., Girling, I. R., Cade, N. A., Kolinsky, P. V., and Peterson, I. R., *Electron. Lett.*, *22*, 461 (1986).
[48] Blinov, L. M., Dubinin, N. V., Mikhnev, L. V., and Yudin, S. G., *Thin Solid Films*, *120*, 161 (1984).
[49] Blinov, L. M., Dubinin, N. V., Rumyantsev, V. G., and Yudin, S. G., *Opt. Spectrosc. (U.S.S.R.)*, *55*, 403 (1984).
[50] Blinov, L. M., Dubinin, N. V., Yudin, S. G., *Opt. Spectrosc. (U.S.S.R.)*, *56*, 173 (1984).
[51] Dubinin, N. V., Yudin, S. G., and Blinov, L. M., *Opt. Spectrosc. (U.S.S.R.)*, *59*, 53 (1985).
[52] Garito, A. F., and Wong, K. Y., *Polymer Journal*, *19*, 51 (1987); Wong, K. Y., and Garito, A. F., *Phys. Rev. A*, *34*, 5051 (1986).
[53] Lee, Y. S., Hsu, Y. Y., and Dolphin, D. H., presentation at the ACS Symposium on Ordered Fluids and Liquid Crystals, Chicago, 1973 (unpublished).
[54] Rose, P. I., presentation at the Fourth International Conference on Liquid Crystals, Kent, Ohio, 1972 (unpublished).

[55] Jen, S., Clark, N. A., Pershen, P. S., and Priestly, E. B., *J. Chem. Phys.*, *66*, 4635 (1977).
[56] Durbin, S. D., and Shen, Y. R., *Phys. Rev. A*, *30*, 1419 (1984).
[57] Bohren, C. F., and Huffman, D. R., *Absorption and Scattering of Light by Small Particles*, Wiley, New York, 1983.
[58] Maxwell Garnet, J. C., *Philos. Trans. R. Soc. A*, *203*, 385 (1904).
[59] Sipe, J. E., and Boyd, R. W., *Phys. Rev. A*, *46*, 1614 (1992).
[60] Bruggeman, D. A. G., *Ann. Phys. (Leipzig)*, *24*, 636 (1935).
[61] Kirkwood, J. G., *J. Chem. Phys.*, *7*, 911 (1939).
[62] Greene, B. I., Orenstein, J., Thakur, M., and Rapkine, D. H., in *Nonlinear Optical Properties of Polymers*, Materials Research Society Symposium Proceedings (A. J. Heeger, J. Orenstein, and S. R. Ulrich, eds.), Pittsburgh, 1988, *109*, 159.
[63] Gabriel, M. C., Whitaker, Jr., N. A., Dirk, C. W., Kuzyk, M. G., and Thakur, M., *Opt. Lett.*, *16*, 1334 (1991).
[64] Krol, D. M., and Thakur, M., *Appl. Phys. Lett.*, *56*, 1406 (1990).
[65] Thakur, M., Verbeek, B., Chi, G. C., and O'Brien, K. J., in *Nonlinear Optical Properties of Polymers*, Materials Research Society Symposium Proceedings (A. J. Heeger, J. Orenstein, and S. R. Ulrich eds.), Pittsburgh, 1988, *109*, 41.
[66] Thakur, M., and Meyler, S., *Macromol.*, *18*, 2341 (1985).
[67] Arfkin, G., *Mathematical Methods for Physicists*, Academic Press, New York, 1970.
[68] Nye, J. F., *Physical Properties of Crystals*, Oxford Press, Oxford, 1976.
[69] Legrange, J. D., Kuzyk, M. G., and Singer, K. D., *Mol. Cryst. Liq. Cryst.*, *150b*, 567 (1987).
[70] Ghebremichael, F., Kuzyk, M. G., Andrews, J. H., and Singer, K. D., Manuscript in preparation.
[71] Fiorini, C., Nunzi, J. M., Charra, F., Kajzar, F., Lequan, M., Lequan, R. M., and ChaneChing, K., *Chem. Phys. Lett.*, *271*, 335 (1997).
[72] Chalupczak, W., Fiorini, C., Charra, F., Nunzi, J. M., and Raimond, P., *Opt. Commun.*, *126*, 103 (1996).
[73] Fiorini, C., Charra, F., Nunzi, J. M., Samuel, I. D. W., and Zyss, J., *Opt. Lett.*, *20*, 2469 (1995).
[74] Fiorini, C., Nunzi, J. M., Charra, F., Samuel, I. D. W., and Zyss, J., *J. Nonlinear Opt. Phys. & Mat.*, *5*, 653 (1996).
[75] Kuzyk, M. G., Moore, R. C., and King, L. A., *J. Opt. Soc. Am. B*, *7*, 64 (1990).
[76] Kuzyk, M. G., and Poga, C., in *Molecular Nonlinear Optics* (J. Zyss, ed.) Academic Press, New York, 1994, p. 299.
[77] Poga, C., Kuzyk, M. G., and Dirk, C. W., *J. Opt. Soc. Am. B*, *11*, 80 (1994).
[78] See, for example, Bottcher, *Theory of Electric Polarization*, Vol. I, Elsevier, New York, 1973, p. 205; Bottcher and Bordewijk, *Theory of Electric Polarization*, Vol. II, Elesvier, New York, 1978, p. 404.
[79] Kaminow, I. P., *An Introduction to Electrooptic Devices*, Academic Press, New York, 1974.
[80] Gradshteyn, I. S., and Ryzhik, I. M., *Tables of Integrals, Series, and Products*, Academic Press, New York, 1980, p. 480.
[81] Buckland, E. L., and Boyd, R. W., *Opt. Lett.*, *15*, 1117 (1996).
[82] Kuzyk, M. G., Welker, D. J., and Zhou, S., *Nonlinear Optics*, *10*, 409 (1995).
[83] Welker, D. S., and Kuzyk, M. G., *Appl. Phys. Lett.*, *66), 809 (1994).*
[84] Welker, D. S., and Kuzyk, M. G., *Appl. Phys. Lett.*, *64*, 2792 (1995).
[85] Welker, D. S., and Kuzyk, M. G., *Appl. Phys. Lett.*, *69*, 1835 (1996).
[86] Zhou, S., and Kuzyk, M. G., *SPIE Proc*, *2285*, 125 (1994).
[87] Ricard, D., Roussignol, Ph., and Flytzanis, Chr., *Opt. Lett.*, *10*, 511 (1985).
[88] Hache, F., Ricard, D., and Flytzanis, Chr., *J. Opt. Soc. Am. B*, *3*, 1647 (1986).

[89] Hache, F., Ricard, D., and Girard, C., *Phys. Rev. B*, *38*, 7990 (1988).
[90] Hache, F., Ricard, D., Flytzanis, C., and Kreibig, U., *Appl. Phys. A*, *47*, 347 (1988).
[91] Bloemer, M. J., Haus, J. W., and Ashley, P. R., *J. Opt. Soc. Am. B*, *7*, 790 (1990).
[92] Heilweil, E. J., and Hochstrasser, R. M., *J. Chem. Phys.*, *82*, 4672 (1985).
[93] Neeves, A. E., and Birnboim, M. H., *Opt. Lett.*, *13*, 1087 (1988).
[94] Markel, V. A., Muratov, L. S., Stockman, M. I., and George, T. F., *Phys. Rev. B*, *43*, 8183 (1991).
[95] Butenko, A. V., Chubakov, P. A., Danilova, Yu. E., Korpov, S. V., Popov, A. K., Rautian, S. G., Safonov, V. P., Slabko, V. V., Shalaev, V. M., and Stockman, M. I., *Z. Phys. D.*, *17*, 283 (1990).
[96] Butenko, A. V., Shalaev, V. M., and Stockman, M. I., *Z. Phys. D.*, *10*, 81 (1988).
[97] Bedeaux, D., and Mazur, P., *Physica*, *67*, 23 (1973).
[98] Bedeaux, D., and Vlieger, J., *Physica*, *67*, 55 (1973).
[99] Bedeaux, D., and Bloembergen, N., *Physica*, *69*, 57 (1973).
[100] Andrews, J. H., Kowalski, K. L., and Singer, K. D., *Phys. Rev. A*, *46*, 4172 (1992).
[101] Ueno, Y., Stegeman, G. I., and Tajima, K., *Japanese J. of Appl. Phys. Part 2 Lett.*, *36*, 613 (1997).
[102] Stegeman, G. I., *Quantum and Semiclasical Optics*, *9*, 139 (1997).
[103] Stegeman, G. I., Hagan, D. J., and Torner, L., *Opt. and Quant. Elec.*, *29*, 532 (1997).
[104] Stegeman, G. I., Hagan, D. J., and Torner, L., *Opt. and Quant. Elec.*, *28*, 1691 (1996).
[105] Schiek, R., Baek, Y., Krijnen, G., Stegeman, G. I., Baumann, I., and Sohler, W., *Opt. Lett.*, *21*, 940 (1996).

4
Linear Optical Properties

J. D. Swalen
University of California at Santa Cruz
Santa Cruz, California

C. R. Moylan
Lightwave Microsystems Corporation
Santa Clara, California

I. INTRODUCTION

In investigations and applications of nonlinear optical (NLO) polymers, besides obtaining a thorough understanding of their nonlinear behavior, it is necessary to characterize their linear optical properties. These are the real and imaginary parts of the index of refraction as a function of wavelength and the concomitant dielectric function in the optical region. The real part (n) is the standard index of refraction, which can be anisotropic; the imaginary part (κ) is directly related to the optical absorbance and it can be dichroic. In this Chapter, we will explain and discuss the various methods for determining these optical constants of a NLO polymer in thin film form, the common configuration for device applications. Although we will be addressing exclusively $\chi^{(2)}$ materials, many of the techniques also apply to $\chi^{(3)}$ materials.

There are many NLO polymers and materials in the literature whose properties have been tabulated. These involve many different host polymeric systems with many different incorporated NLO chromophores. A chromophore can be present as a guest mixed with a polymer, or can be chemically attached to the polymer as a side chain or incorporated into the main chain. In addition, the concentration of the chromophore can vary from sample to sample depending on the method of preparation or by design. Even the generic and ubiquitous host polymer, poly(methyl methacrylate) (PMMA), exhibits indices of refraction that vary with the degree of polymerization, i.e., the molecular weight and the polydispersity. Finally, as one searches the literature, the reported values are not all measured at the same wavelengths. It has been standard procedure to report values at the sodium

D lines, a doublet at 589 nm.[1–3] We also see results at 633 nm, 830 nm, 1.32 μm, and 1.55 μm, but rarely do we find data over a sufficient number of wavelengths to give a good representation of a dispersion curve, an exception being reference [4] where a number of models are investigated. Hence we concluded that a random collection of uncorrelated data would serve little purpose. A recent review by Singh was published in the *Chemical Rubber Handbook on Laser Science and Technology*.[5] It contains a potpourri of spectra, indices of refraction, and other data on published NLO materials. Although it is most useful, there is no reported trend or main theme. The reader is directed to this review for some recent data and references.

If one has a thin film of an NLO polymer, then the index of refraction and absorption can be measured directly by either ellipsometry[6–8] or optical waveguiding.[9] Further, the optical absorption spectrum over a wide spectral range can be transformed from κ, the imaginary part of the index of refraction, to n, the real part, by either the Kramers-Krönig relations[10] or the Lorentz equation[11] (see below).

When a thin film is not available for measurement, it is useful to have a procedure to predict the index of refraction at a specified wavelength for a particular polymer and an incorporated chromophore at a specified concentration. We describe below a computational scheme to accomplish this from a set of simple measurements. Data on a number of selected chromophores have been published.[12–16] A number of the most widely used chromophores are given in Table 1.[17] Here, the wavelength of the maximum absorption (λ_{max}), the extinction coefficient at this wavelength (ε_{max}), and the full linewidth at half maximum (FWHM) in a given solvent are listed, as well as the dipole moment (μ) and the molecular hyperpolarizability at zero frequency (β_0). This information, along with the dispersion in the index of refractive of the host polymer, can be used to calculate the index of refraction at some other non-absorbing wavelength. This is rigorously true only in the absence of absorption, but the method seems to be applicable even with a moderate amount of absorption. In this chapter we will give the procedure and illustrate it with a sample calculation on the dye Disperse Red #1 in PMMA. The known dispersion in the index of refraction for PMMA will be used.[8,9,18]

II. PROCEDURE

The procedure for determining the linear susceptibility is outlined below. The remainder of this section discusses each of these steps.

The method is as follows:

1. The molar concentration of the chromophore in the polymer must be specified for a desired film or determined for an existing film.
2. From this molar concentration, the weight fraction of each component can be derived.
3. The molar concentration of the chromophore as a neat solid must be estimated.
4. The hypothetical absorption of this neat condensed-phase chromophore at λ_{max} can then be estimated.
5. From either a Kramers-Krönig[10] or Lorentz[11] analysis, the refractive index of the neat chromophore can be calculated at any other wavelength.

Table 1

	Chromophore	Solvent	λ_{max}	ε_{max}	FWHM	μ	β_0
1	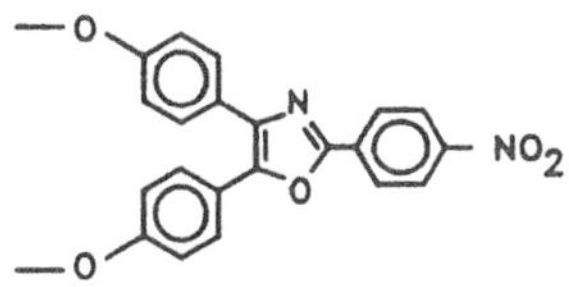	D	390	19,300	5,510	6.3	34.0
2	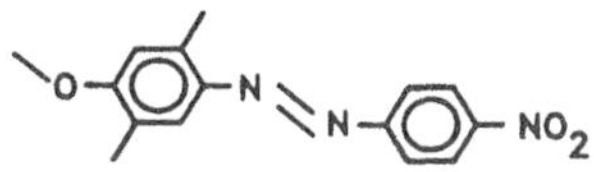	C	398	23,500	4,890	5.3	34.0
3	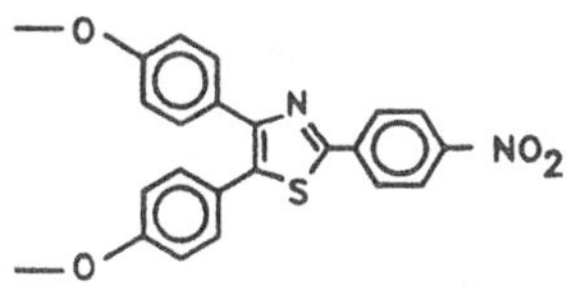	C	400	16,200	4,950	8.1	40.7
4	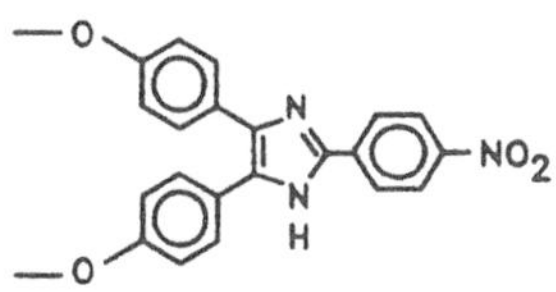	C	412	17,900	5,200	6.4	29.7
5	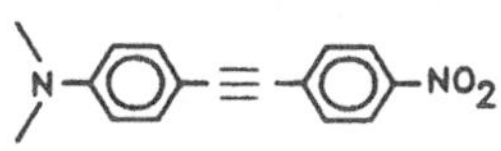	C	416	21,000	5,430	6.6	63.3
6	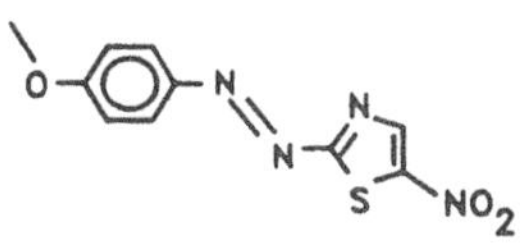	D	424	26,200	4,790	6.2	43.8
7	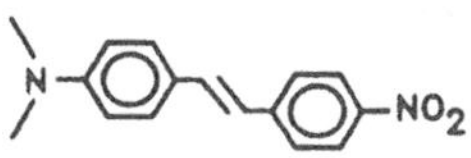	C	438	30,600	5,150	6.7	81.3
8	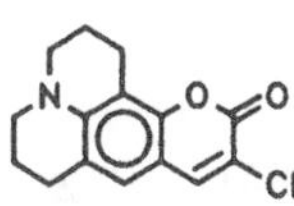	C	446	45,500	2,460	10.2	32.4
9	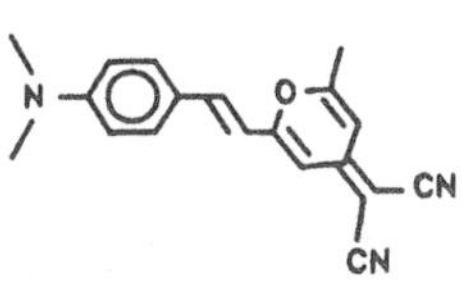	C	468	38,500	4,260	10.2	63.3
10	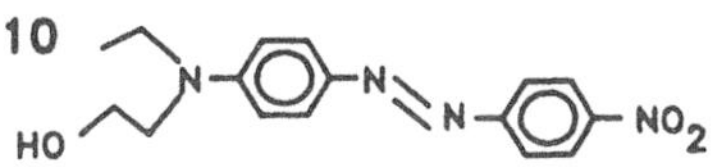	C	478	31,000	4,790	7.2	94.7
11	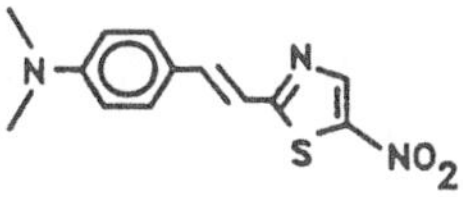	D	478	35,500	4,430	8.2	127

Table 1 Continued

Chromophore		Solvent	λ_{max}	ε_{max}	FWHM	μ	β_0
12	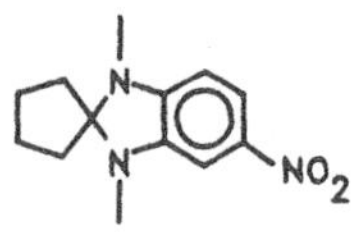	C	480	11,100	4,370	8.2	17.8
13	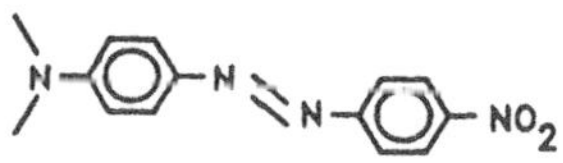	C	480	36,000	4,770	7.7	77.0
14	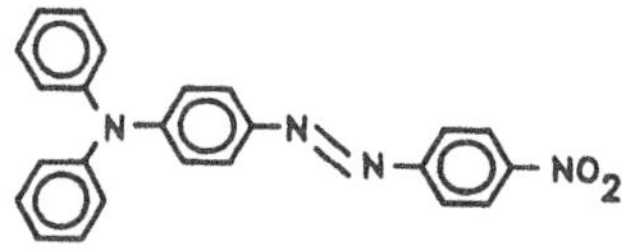	C	486	31,200	4,660	5.9	105
15	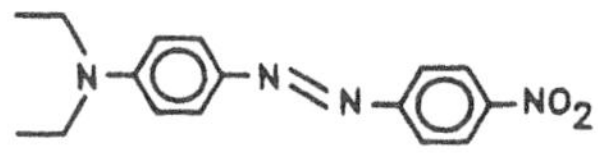	C	494	35,400	4,490	8.0	95.2
16	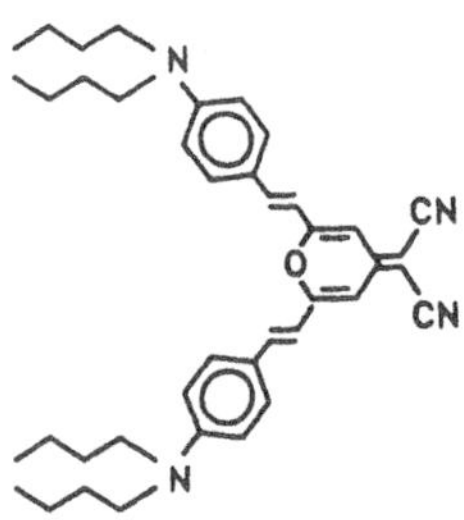	C	496	77,000	4,040	12.6	134
17	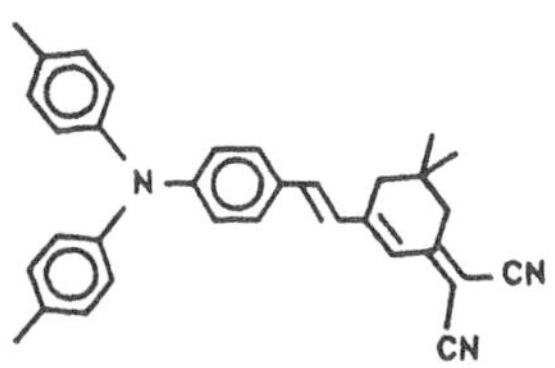	C	520	43,600	3,920	7.3	229
18	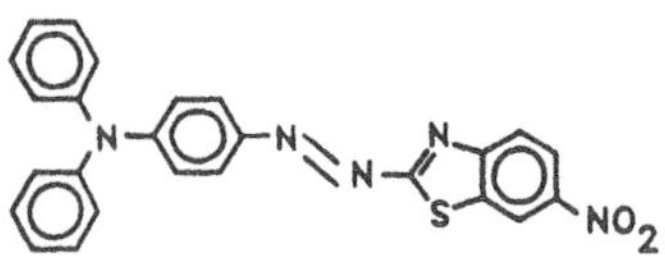	C	550	52,400	3,770	7.2	139
19	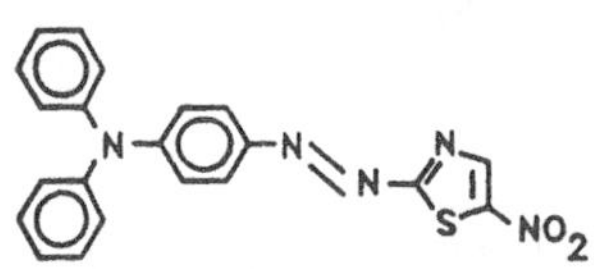	C	582	41,700	3,560	6.9	132
20	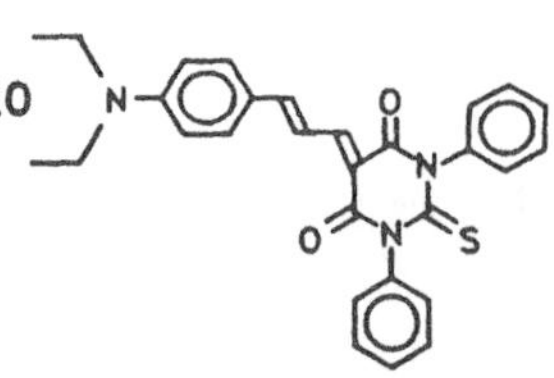	C	588	114,000	1,700	8.3	181

Table 1 Continued

Chromophore		Solvent	λ_{max}	ε_{max}	FWHM	μ	β_0
21	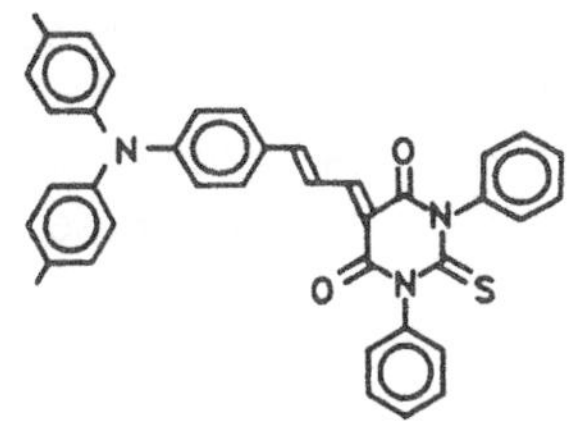	C	594	76,400	2,800	6.5	230
22	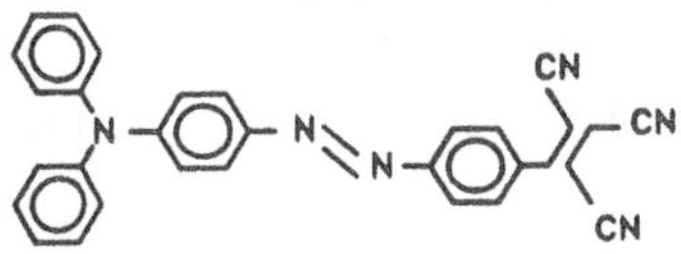	C	602	30,300	4,040	7.1	259
23	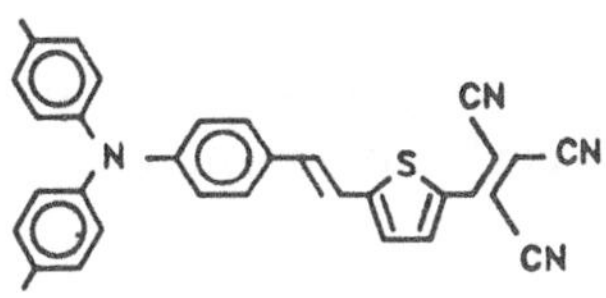	C	680	49,700	3,200	8.3	479
24	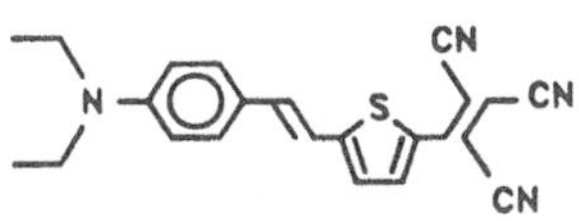	C	698	70,200	2,880	10.4	359

Chromophores listed in order of increasing absorption maximum. Absorption maxima are given in nm; extinction coefficients in $M^{-1}cm^{-1}$; full widths at half maxima in cm^{-1}; dipole moments in debye; and zero-frequency hyperpolarizabilities in 10^{-30} esu. C = chloroform; D = 1,4-dioxane. Molecular hyperpolarizabilities are referenced to the currently preferred value for the nonlinearity of quartz[a,b] and use the Willets et al. B convention.[c]

a. Jerphagnon, J., and Kurtz, S. K., *Phys. Rev. B*, *1*, 1739–1744 (1970).
b. Roberts, D. A., *IEEE J. Quantum Electron.* *28*, 2057–2074 (1992).
c. Willetts, A., Rice, J. E., Burland, D. M., and Shelton, D. P., *J. Chem. Phys.*, *97*, 7590–7599 (1992).

6. From a Sellmeier fit to several indices of refraction of the polymer as a function of wavelength, the index of the polymer film can be determined at the same desired wavelength.
7. Given the dielectric functions of the polymer and the chromophore, the square of the indices at optical frequencies, the weight fraction, the composite dielectric function and index can be estimated quite accurately at the desired wavelength. There are three methods: (1) use of the Lorenz-Lorentz equation, that is, the Clausius-Mossotti equation with $\varepsilon = n^2$; (2) the application of the Maxwell Garnett theory (see Chapter 3 for more details) based on a low concentration of spherical particles in an isotropic host matrix; or (3) a weighted average. These methods are discussed below.

A. Concentration

If the weight fraction is known or specified, then one proceeds to step (3). Otherwise, there are several ways to determine the concentration of the NLO chromophore in a prepared NLO polymer film. The first one, the chemical method, is applicable if the mole fraction of the chromophore and oligomer is specified or is known from the details of how the material was synthesized. The second and third methods use optical measurements to determine the concentration. These three methods are discussed below.

1. Chemical Method

Let M be the molecular weight and A be Avogadro's number. Then the molecule's mass is given by M/A and the molecular volume by $M/A\rho$, where ρ is the density.

The number of molecules/1000cc = $1000A\rho/M$ and the molar concentration, c, is:

$$c = 1000\rho/M \text{ moles/liter.} \tag{1}$$

For the composite dye/polymer material, the molecular weight, M, is the sum of the molecular weight of the oligomer associated with each chromophore, plus the chromophore's molecular weight.

The weight fraction of the chromophore is:

$$f = M_{\text{chromophore}}/M_{\text{chromophore+polymer}}. \tag{2}$$

The volume fraction is:

$$\nu = cM_{\text{chromophore}}/1000\rho, \tag{3}$$

and the weight fraction is:

$$f = \frac{\nu\rho_{\text{chromophore}}}{\nu\rho_{\text{chromophore}} + (1 - \nu)\rho_{\text{polymer}}}. \tag{4}$$

Hence, because the composition of a material is known when it is made by mixing polymer and chromophore or from the starting materials used in the chemical reaction, we can determine the concentration of the chromophore in the NLO polymer system using Eqs. 1–4.

2. Spectroscopic Correlation Method

Alternatively, from absorbance measurements at several concentrations in solution, one can produce a curve of absorbance as function of concentration for a fixed-thickness sample cell. An absorbance measurement of a sample film of unknown chromophore concentration can be used to calculate the concentration provided that the thickness, t, of the sample is known.

3. Spectroscopic Absorption Method

From a measurement of the absorption spectrum in solution, the extinction coefficient, ε_{max}, can be calculated (see Table 1 for the values of selected chromophores). From the absorbance, A, of a film of known thickness, t, the concentration, c, can be determined by:

$$c = A/\varepsilon_{max}t \tag{5}$$

Alternatively, from ellipsometric data at λ_{max}—which measures the refractive index—the absorbance can be calculated from:

$$A = 4\pi\kappa t/(\lambda_{max} \times \ln 10), \tag{6}$$

where κ is the imaginary part of the refractive index. Again, this absorbance, A, can be substituted into Eq. 5 to get the concentration.

The concentration of a neat dye can be estimated by assuming no aggregation and no intermolecular interaction. While this is a fictitious situation, it is useful for our analyses. Using Eq. 1 with the molecular weight of the chromophore and its crystalline density, the concentration of the neat dye can be determined. Using this concentration, the absorbance, A, can then be calculated with Eq. 5.

B. Dispersion of Chromophores

Although the Kramers-Krönig[10] method of analysis could be used to get the refractive index from the absorption spectrum, we will illustrate the above technique with the Lorentz equation which gives dispersion in both the absorbance and refractive index.[11] This equation is applicable to many compounds because they usually have only one or two main absorption bands in the visible region of the spectrum that can be easily fit to a single Lorentzian or a sum of two Lorentzians. On the other hand, for semiconductors and many inorganic solids, the absorbance is large over broad regions of the spectrum so a Lorentz model is not applicable. Here the Kramers-Krönig method should be used to obtain the refractive index from the absorbance.

The Lorentz equation for the dielectric function ε is

$$\varepsilon = \varepsilon_b + \frac{\omega_p^2}{\omega_0^2 - \omega^2 - i\omega\gamma}. \tag{7}$$

In molecules where multiple states contribute, the dielectric function is a sum of Lorentzians over several resonance frequencies ω_0.

In this equation, ε_b is the background dielectric function arising from the absorption bands in the ultraviolet. This background term is generally not a constant and has some small amount of dispersion. We have, however, not found dispersion in ε_b to significantly contribute to the measured dispersion of ε.[11,22] The parameter ω_0 is the angular frequency of the absorption band at λ_{max}. The parameter γ is the full width at half maximum (FWHM) of the absorption band in units of s^{-1} (i.e., γ in cm^{-1} multiplied by $2\pi c$) ω is the angular frequency ($2\pi\nu$) of the applied electric field and ω_p is the plasma frequency. For optical frequencies, $\omega \sim 10^{15}$ Hz.

Equation 7 can be resolved into real and imaginary parts:

$$\varepsilon' = \varepsilon_b + \frac{\omega_p^2(\omega_0^2 - \omega^2)}{(\omega_0^2 - \omega^2)^2 + \omega^2\gamma^2}, = n^2 - \kappa^2 \tag{8}$$

and

$$\varepsilon'' = \frac{\omega_p^2\omega\gamma}{(\omega_0^2 - \omega^2)^2 + \omega^2\gamma^2}, = 2n\kappa \tag{9}$$

respectively.

At $\omega = \omega_0$, Eqs. 8 and 9 reduce to

$$\varepsilon_0' = \varepsilon_b \tag{10}$$

and

$$\varepsilon_0'' = \omega_p^2/\omega_0\gamma. \tag{11}$$

Ramo, Whinnery and van Duzer[19] define the propagation vector as:

$$k = k_0(n - i\kappa) = \omega\sqrt{\mu(\varepsilon' - i\varepsilon'')} = -i\alpha + \beta, \tag{12}$$

where k_0 is the propagation vector in free space, n and κ the real and imaginary parts of the refractive index, and μ the magnetic permeability. α is the absorption and upon resolving Eq. 12, is given by:

$$\alpha^2 = \frac{k_0^2\varepsilon'}{2}\left(\sqrt{1 + \left(\frac{\varepsilon''}{\varepsilon'}\right)^2} - 1\right). \tag{13}$$

Substituting Eqs. 10 and 11 into Eq. 13, we obtain after some rearrangement:

$$\omega_p^2 = \omega_0\varepsilon_b\gamma\sqrt{\eta + \frac{\eta^2}{4}}, \tag{14}$$

where

$$\eta = \frac{4\alpha^2}{k_0^2\varepsilon_b} = \left(\frac{(A/t)\ \ln 10}{2\pi\bar{\nu}_0 n_b}\right)^2. \tag{15}$$

Note that for a material with one resonance peak, $\omega_p \approx \omega_0$.

A is the measured absorbance at λ_{max} of a sample of thickness t, n_b the background refractive index ($= \sqrt{\varepsilon_b}$) and $\bar{\nu}_0$, the inverse wavelength, in units of inverse centimeters (called wave numbers) at λ_{max}. We used a value of 1.53 for n_b, based on the dispersion curves in the index of refraction for organic NLO crystals.[22] These experimentally determined values are used to determine η which is used to determine the plasma frequency, ω_p, with Eq. 14. This parameter can then be used to determine ε' and ε'' as a function of wavelength from Eqs. 8 and 9 and hence the complex index of refraction.

C. Dispersion of Polymers

As mentioned above, the indices of refraction of some well-known polymers are found in other handbooks.[1–3,20] Unfortunately, they usually give the index at only one wavelength, with the exception of PMMA, poly(styrene) and maybe a few others. What is most useful is the dispersion curve for a given polymer, but the index and dispersion curve for a polymer can depend on how the polymer was made. For example, the method of polymerization can affect the linear optical properties so that two polymer samples formed by different methods of polymerization could exhibit slightly different values of n. Hence, if a researcher needs the precise value for the refractive index, the given polymer should be measured rather than relying on numbers in a tabulation. A measurement can be most conveniently and accurately made by using a 1 to 2 μm film and measuring the ellipsometric parameters, ψ, the phase difference, and Δ, the amplitude difference between the two polarizations[6–8] or the waveguide modes at the desired wavelength.[9] The eigenvalue equation for each polarization (s & p) is

$$k_{z2}t = \tan^{-1} \beta_{12} + \tan^{-1} \beta_{32} + m\pi, \tag{16}$$

where for s-polarization $\beta^{s}_{i2} = k_{zi}/k_{z2}$ and for p-polarization $\beta^{p}_{i2} = (k_{zi}/\varepsilon_{xi})/(k_{z2}/\varepsilon_{x2})$.

The propagation vector, β, depends on the polarization of the light in the guide: s (=TE) or p (=TM), that is, transverse electric or transverse magnetic with respect to the plane of incidence. The axes are oriented with the z axis perpendicular to the film, the x axis along the direction of propagation, and the y axis perpendicular to these two axes and to the plane of incidence. The thickness is t and k_{zi} is the effective propagation vector in the z direction for the ith layer. For s polarization $k_z^2 = \varepsilon_y k_0^2 - k_x^2$, and for p polarization $k_z^2 = \varepsilon_x k_0^2 - \varepsilon_x k_x^2/\varepsilon_z$. See references [9,11] for details on the solution of this eigenvalue Eq. 16 which yields the refractive index. Optical constants of a thin film can also be obtained by transmittance measurements.[10]

Once the refractive index is measured at several wavelengths, the dispersion can then be fit to the Sellmeier equation:

$$n^2 = A + \frac{B}{1 - C/\lambda^2}, \tag{17}$$

where the parameters A, B, and C are determined by the fit. For materials that have several absorption bands in the visible, a sum of Sellmeier terms as given by Eq. 17 may be needed to describe the data.

D. Models of Composite Dielectrics

1. Lorenz-Lorentz Equation

The Lorenz-Lorentz equation is the optical analogue of the Clausius-Mossotti Equation, i.e., $\varepsilon \rightarrow n^2$:

$$\frac{n^2 - 1}{n^2 + 2} = \frac{4\pi}{3} N\alpha, \tag{18}$$

where the microscopic polarizability, α, of the molecule is related to the macroscopic index of refraction, n. N is the number of molecules per unit volume and is equal to $A\rho/M$, that is Avogadro's number times the density, divided by the molecular weight. Our goal is to determine the composite index of a polymer-chromophore system. The effective polarizability is

$$N_t\alpha_{\text{eff}} = N_t\alpha_c + N_t\alpha_p, \tag{19}$$

where $N_t = A\rho_t/(M_p + M_c)$ is the total number of NLO oligomers per unit volume and ρ_t the mass density of the composite material. M_p is the molecular weight of the polymer oligomer associated with one chromophore, M_c the molecular weight of the chromophore, α_c the polarizability of chromophore, and α_p the polarizability of the oligomer that is associated with one chromophore.

Depending on the relative concentration of chromophores to polymer repeat units in the oligomer, we assign the proper number of repeat units of the polymer to one molecule of the chromophore. As a result, a unit volume contains the same number of chromophores as oligomers. Hence, the coefficients multiplying the different polarizabilities (α's) in Eq. 19 are all the same and equal; that is, the number of molecules per unit volume and the number of oligomers associated with each molecule are both N_t.

For the neat oligomer material, the number density is

$$N_p = A\rho_p/M_p. \tag{20}$$

where ρ_p is the mass density of the neat oligomer. The number density of the fictitious neat chromophore material is:

$$N_c = A\rho_c/M_c, \tag{21}$$

where ρ_c is the mass density of the neat chromophore.

Substituting Eqs. 18, 20 and 21 into Eq. 19, we obtain

$$\frac{n_{\text{eff}}^2 - 1}{n_{\text{eff}}^2 + 2} = \frac{f\rho_t}{\rho_c}\frac{n_c^2 - 1}{n_c^2 + 2} + \frac{(1 - f)\rho_t}{\rho_p}\frac{n_p^2 - 1}{n_p^2 + 2}, \tag{22}$$

where n_c is the index of refraction at a given wavelength of the chromophore as a hypothetical neat solid with no intermolecular interactions, n_p is the index of refraction of the neat oligomer at the same wavelength, $f(f = M_c/(M_c + M_p))$ is the weight fraction of the chromophore, and n_{eff} is the effective index of the two component system that is, the oligomer and chromophore. Remember that M_p is the molecular weight of the oligomer and not that of the whole polymeric chain. Hence, for example, if we have an NLO polymer with 10 mole percent of chromophore within it, then we have one chromopore molecule per one oligomer with nine repeat units.

To determine the effective index, we must know the densities, the indices, and the weight fraction. In regard to the densities,[1–3,21] most of the common polymers and chromophores have densities around 1.2, polyimides being a slight exception with a density of ~1.4. Also, the density of the NLO polymer will be close to that of the neat polymer without chromophore added. The density of the composite is the weighted average of the inverse of each component:

$$\rho_t^{-1} = f\rho_c^{-1} + (1 - f)\rho_p^{-1} \tag{23}$$

Here we assume that the volume occupied by the NLO polymer is the sum of the volumes occupied by the chromophore and by the polymer, that is, these two components are noninteracting, there is no volume reduction in combining the two, and the chromophore does not fill any voids in the polymer. If we now assume that the densities are all very similar, as is normally the case, then Eq. 22 reduces to

$$\frac{n_{\text{eff}}^2 - n_p^2}{n_{\text{eff}}^2 + 2} = f\,\frac{n_c^2 - n_p^2}{n_c^2 + 2}. \tag{24}$$

2. Maxwell Garnett Equation

As derived in Chapter 3, the effective dielectric function, ε, of a composite material made of a host polymer with a dielectric function ε_p and guest or chemically attached chromophore with a dielectric function ε_c is

$$\frac{\varepsilon - \varepsilon_p}{\varepsilon + 2\varepsilon_p} = f\,\frac{\varepsilon_c - \varepsilon_p}{\varepsilon_c + 2\varepsilon_p}, \tag{25}$$

where f is the weight fraction. Note that the derivation of Eq. 25 assumes spherical particles in an isotropic host matrix. Although the chromophores are not spherical and the host matrix after poling will not be isotropic, the equation gives reasonable results. Note also the similarity between Eq. 24 and 25 if one makes the substitution $\varepsilon = n^2$.

3. *Weighted Average*

The simplest model of the refractive index is the weighted average. The sum of each component is weighted by its fractional weight:

$$n_{\mathrm{eff}} = f \times n_c + (1 - f) \times n_p. \tag{26}$$

III. SAMPLE CALCULATION OF DR1/PMMA

For illustration, we will apply the procedure to each model discussed in subsection II.D. For the chromophore Disperse Red No. 1, Figure 1 shows the structure of a side chain polymer schematically and Figure 2 shows the optical absorption spectrum of DR1 in chloroform. The main band in the visible region of the spectrum peaks at 478 nm and a weaker band is centered at 290 nm.

The linear optical properties of the PMMA were measured[9] by optical waveguiding at five different wavelengths. The refractive index values reported in the AIP Handbook[20] are slightly higher (~0.002) at comparable wavelengths than those

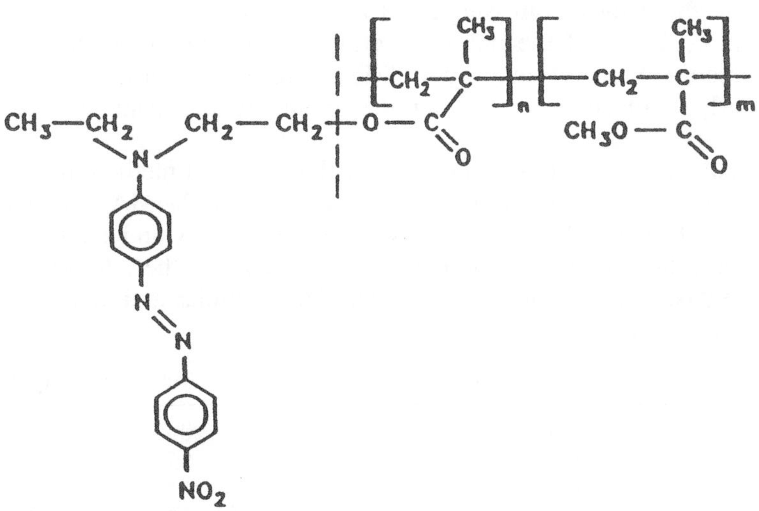

Figure 1 A structural schematic of the polymer poly(methyl methcrylate) (PMMA) with the chromophore Disperse Red #1 (DR1) chemically attached. For our materials, $n = 1$ and $m = 9$. The dashed line shows our separation between the chromophore and the polymer in our calculations.

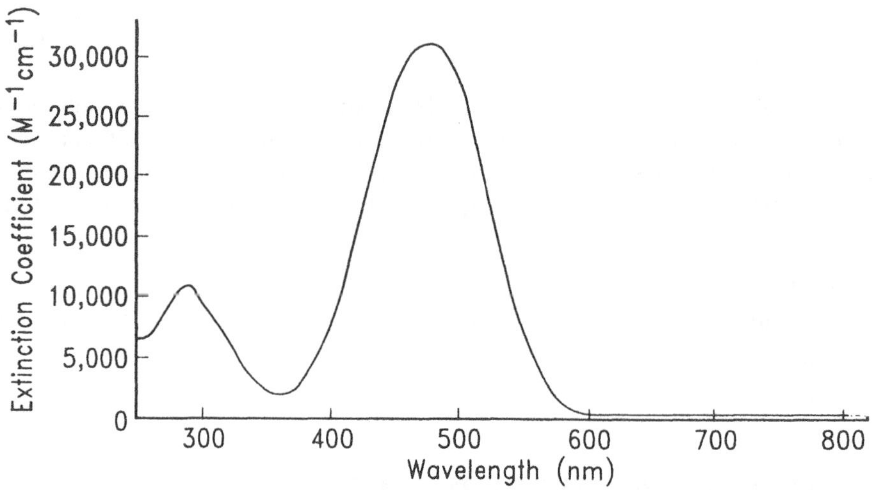

Figure 2 The absorption spectrum of DR1 in a chloroform solution, measured at a concentration of 2.57×10^{-5} moles/liter in a one centimeter absorption cell. The main absorption band at 478 nm has an oscillator strength of 0.64.

reported by the IBM group. Hence, one can appreciate the point that indices of refraction of polymers can vary, depending on the source of the material. Recently, Ishigure et al.[18] have fit the Sellmeier equation with a sum of three wavelength terms to their ellipsometric values. Wu et al.[22] have confirmed these results with their ellipsometric measurements. Because the two reported values are in very good agreement, we present the values from Ishigure et al. in Table 2.

For our sample calculation of the index of refraction of a polymer/chromophore film, we will use the dye DR1 attached to PMMA (10mer%, that is 10 repeat units of PMMA per dye molecule). This material is commercially available from IBM.[23]

We will first calculate the molar concentration by the chemical method involving the molecular weights. Here we have nine repeat units of PMMA and one repeat unit with DR1 attached. The dividing line between PMMA and DR1 was arbitrarily taken between the CH_2 group and the oxygen attached in the ester group. (See Fig. 1 where the dashed line shows the division.) The formulas and molecular weights of the three parts are:

Material	Formula	Molecular weight
PMMA	$C_5H_8O_2$	100
PMMA (part)	$C_4H_5O_2$	85
DR1	$C_{16}H_{17}O_2N_2$	297
Total		985 + 297 = 1282

For the density[1,21] $\rho = 1.18$, the molar concentration, $c = 1000 \times 1.18/1282 = 0.92$ moles/liter and $f = 0.23$.

For comparison, we calculated the concentration from ellipsometric data[8,22] at $\lambda_{max} = 476$ nm and for a film with thickness of $t = 592$ nm. The real and imaginary

Table 2

λ (nm)	n_{PMMA}	$n_{chromophore}$	$n_{ellipsometry}$	$n_{Lorentz}$	$n_{Maxwell\ Garnet}$	$n_{average}$
600	1.4921	2.0900	1.6301	1.5961	1.5995	1.6027
610	1.4917	2.0732	1.6178	1.5934	1.5964	1.5993
620	1.4913	2.0574	1.6100	1.5908	1.5935	1.5960
630	1.4909	2.0427	1.6065	1.5884	1.5908	1.5930
640	1.4906	2.0289	1.6007	1.5861	1.5882	1.5902
650	1.4902	2.0162	1.5960	1.5839	1.5858	1.5875
660	1.4899	2.0044	1.5919	1.5819	1.5836	1.5851
670	1.4896	1.9934	1.5883	1.5800	1.5815	1.5828
680	1.4893	1.9832	1.5851	1.5782	1.5796	1.5807
690	1.4890	1.9738	1.5825	1.5765	1.5777	1.5787
700	1.4888	1.9651	1.5803	1.5750	1.5760	1.5769
710	1.4885	1.9569	1.5777	1.5735	1.5744	1.5752
720	1.4882	1.9493	1.5755	1.5721	1.5729	1.5735
730	1.4880	1.9423	1.5740	1.5708	1.5715	1.5720
740	1.4878	1.9356	1.5722	1.5696	1.5702	1.5706
750	1.4875	1.9295	1.5704	1.5684	1.5690	1.5693
760	1.4873	1.9237	1.5689	1.5673	1.5678	1.5680
770	1.4871	1.9182	1.5676	1.5663	1.5667	1.5669
780	1.4869	1.9131	1.5664	1.5653	1.5656	1.5657
790	1.4867	1.9083	1.5653	1.5644	1.5646	1.5647
800	1.4865	1.9037	1.5640	1.5635	1.5637	1.5637
810	1.4863	1.8994	1.5629	1.5626	1.5628	1.5628
820	1.4862	1.8954	1.5619	1.5618	1.5619	1.5619
830	1.4860	1.8916	1.5610	1.5611	1.5611	1.5610
840	1.4858	1.8879	1.5603	1.5604	1.5604	1.5602
850	1.4857	1.8845	1.5593	1.5597	1.5596	1.5594
860	1.4855	1.8812	1.5583	1.5590	1.5589	1.5587
870	1.4853	1.8781	1.5575	1.5584	1.5583	1.5580
880	1.4852	1.8752	1.5567	1.5578	1.5576	1.5573
890	1.4851	1.8723	1.5561	1.5572	1.5570	1.5567
900	1.4849	1.8697	1.5556	1.5566	1.5564	1.5561
910	1.4848	1.8671	1.5547	1.5561	1.5559	1.5555
920	1.4846	1.8647	1.5543	1.5556	1.5553	1.5549
930	1.4845	1.8623	1.5540	1.5551	1.5548	1.5544
940	1.4844	1.8601	1.5536	1.5546	1.5543	1.5539
950	1.4843	1.8580	1.5532	1.5542	1.5538	1.5534
960	1.4841	1.8559	1.5526	1.5537	1.5534	1.5529
970	1.4840	1.8540	1.5521	1.5533	1.5529	1.5525
980	1.4839	1.8521	1.5516	1.5529	1.5520	1.5520
990	1.4838	1.8503	1.5509	1.5525	1.5521	1.5516
1000	1.4837	1.8485	1.5509	1.5521	1.5517	1.5512

The indices of refraction in Column 2 are for PMMA and these were calculated from the Sellmeier equation constants given in Ref. [18]. The indices in Column 3 were calculated with Eqs. 7, 14 and 15 for a hypothetical neat crystalline chromophore sample. The parameters used were 21008 cm^{-1} for λ_0, 21056 cm^{-1} for the plasma wavelength, and 5267 cm^{-1} for the full line width at half maximum. Note that the plasma wavelength is close to that of λ_0 because ε' goes through zero in its dispersion near λ_0, the slight difference comes from the shift produced by the background index, n_b.

The indices of refraction follow from the determined epsilon values: $\varepsilon' = n^2 - \kappa^2$ and $\varepsilon'' = 2n\kappa$. The indices in Column 4 are the ellipsometric results from Refs. [8] and [22] on a 592 nm film of PMMA with DR1 chemically attached at 10 mole percent.

The indices in Column 5 were calculated with the Lorentz Eq. 24 with the indices of PMMA (Column 2) and of the chromophore (Column 3), and an optimized weight fraction of 0.21 to "best fit" the ellipsometric values.

The indices in Column 6 were calculated in a similar manner with the Maxwell Garnett Eq. 25 and a weight fraction of 0.195.

The indices in Column 7 were calculated in a similar manner with the weighted average Eq. 26 and a weight fraction of 0.18. If we had an independent and accurate determination of the weight fraction, we could decide which method is most consistant with the data.

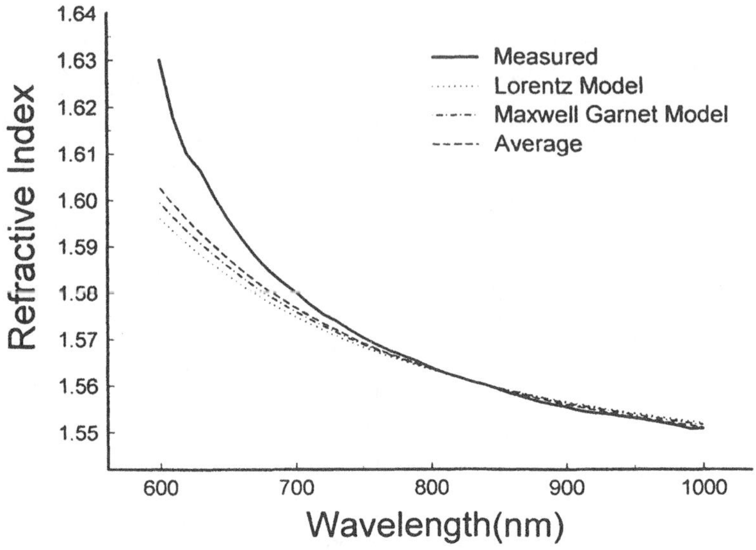

Figure 3 Data and theory values from Table 2 of the refractive index as a function of wavelength.

parts of the index of refraction were found to be $n = 1.5309$ and $\kappa = 0.1978$. This gives an absorbance of

$$A = 4\pi\kappa t/(\lambda_{max} \times \ln 10)$$
$$= 4\pi\, 0.1978 \times 592/476 \times \ln 10 = 1.3425 = \varepsilon_{max}ct.$$

With the extinction coefficient given in Table 1, ($\varepsilon_{max} = 31000$), the concentration c was calculated to be 0.73 and $f = 0.18$. Although the two methods differ slightly, $c \sim 0.8$ moles/liter and $f \sim 0.20$.

For the second step, we calculated that a neat (100 percent) sample of DR1 with a molecular weight of 297 and a density $\rho \sim 1.2$ should have a concentration of 3.97 moles/liter and this gives an absorbance, A, equal to 7.286. Using Eqs. 7 to 15, we can estimate n and κ at any other wavelength. The indices of refraction, which were determined with this procedure, are given in Table 2 and plotted in Figure 3.

All three equations 24, 25 and 26 were fit to the ellipsometric measurements[8,22] by adjusting the weight fraction. Sasaki[24] had similar results for the indices of refraction of DR1 in PMMA at various concentrations as a guest-host system. The Lorenz-Lorentz equation required a weight fraction of 0.21 for a "best fit", the Maxwell Garnett equation required a value of $f \cong 0.195$, and the weighted average method required $f \cong 0.18$. Hence $f = 0.20 \pm 0.02$.

IV. CONCLUSIONS

By combining the index of refraction of a neat polymer (determined either by an ellipsometric or waveguide method or by a fit to the Sellmeier equation and calculated at the desired wavelength) with the index of the chromophore (determined by use of the Lorentz equation at the desired wavelength), one can estimate the index of refraction of an NLO polymer quite well for a given weight fraction of the chromophore. Although the Lorentz model seems to have a more logical basis, the other two give comparable curves. Only with more detailed experiments where the two indices and the concentration are known accurately can one decide which method is to be preferred.

As a final note, NLO polymers become anisotropic when poled. Up to this point, we have considered all the films to be isotropic. Hence, we cannot verify whether this method will work as well for anisotropic films. On the other hand, it should be correct for each independent direction.

In conclusion, we believe that with all the variations in the linear optical properties of polymers, it was not practical to compile long tables of values of indices of refraction versus wavelength for many NLO polymers. Instead, we have described a formalism by which such values may be calculated for a given chromophore-polymer system. The methods described here can be successfully used to give good estimates of the refractive index away from absorption.

ACKNOWLEDGMENTS

We wish to thank discussions and comments by Robert Boyd and Wolfgang Knoll. The unpublished experimental results by André Knoesen, Scott Hamilton, Lui Ming Wu and Philippe Pretre at UC Davis and by K. Sasaki and his graduate student at Keio University are most appreciated to illustrate the agreement between theory and experiment.

REFERENCES

1. van Krevelen, D. W., and Hoftyzer, P. J., *Properties of Polymers, Their Estimation and Correlation with Chemical Structures*, Elsevier Scientific Publishing, Amsterdam, 1976.
2. Wunderlich, W., *Polymer Handbook* (J. Brandrup and E. H. Immergut, eds.), John Wiley & Sons, New York, 1989.
3. Raine, H. C., Plastic Glasses, *Proc. London Conf. Opt. Instruments*, 243 (1950), published in the *American Institute of Physics Handbook* (D. W. Gray, ed.), McGraw-Hill, New York, 1963, pp. 6-93 and 6-94.
4. Otoma, A., Ph.D. Dissertation, University of Central Florida, 1995.
5. Singh, S., *CRC Handbook of Laser Science and Technology, Supplement 2: Optical Materials* (M. J. Weber, ed.), Chemical Rubber, Boca Raton, Florida, 1995, Section 7, pp. 147–249.
6. Azzam, R., and Bashara, W. M., *Ellipsometry and Polarized Light*, North Holland, Amsterdam, 1977.
7. Toussaere, E., and Zyss, J., *Thin Solid Films 234*, 432 (1993); ibid., 454 (1993).
8. Hamilton, S., and Knoesen, A., (Unpublished results).
9. Swalen, J. D., Santo, R., Tacke, M., and Fischer, J., *IBM Journal of Research and Development 21*, 168 (1977).
10. Palmer, K. F., and Williams, M. Z., *Applied Optics, 24*, 1788 (1985).
11. Swalen, J. D., *J. of Molecular Electronics 2*, 155 (1986).
12. Twieg, R. J., and Dirk, C. W., Design, Properties and Applications of Nonlinear Optical Chromophores, in *Organic Thin Films for Waveguiding Nonlinear Optics* (F. Kajzar and J. D. Swalen, eds.), Gordon & Breach, Amsterdam, 1996, Chapter 2, pp. 45–136.

13. Miller, R. D., Poled Polymers for $\chi^{(2)}$ Applications, in *Organic Thin Films for Waveguiding Nonlinear Optics*, (F. Kajzar and J. D. Swalen, eds.) Gordon & Breach, Amsterdam, 1996, Chapter 8, pp. 329–456.
14. Cheng, T.-L., Tam, W., Stevenson, S. H., Meredith, G. R., Rikken, G., and Marder, S. R., *Phys. Chem. 95*, 10631 (1991).
15. Burland, D. M., Miller, R. D., Reiser, O., Twieg, R. J., and Walsh, C. A., *J. App. Phys. 71*, 410 (1992).
16. Moylan, C. R., Swanson, S. A., Walsh, C. A., Thackara, J. I., Twieg, R. J., Miller, R. D., and Lee, V. Y., Nonlinear Optical Properties of Organic Materials VI, *Proceedings of the SPIE—The International Society for Optical Engineering, 2025*, 192 (1993).
17. Moylan, C. R., measurements made at the IBM Almaden Research Center.
18. Ishigure, T., Nihei, E., and Koike, Y., *App. Optics 35*, 2048 (1996).
19. Ramo, S., Whinnery, J. R., and van Duzer, T., *Fields and Waves in Communications Electronics*, John Wiley & Sons, New York, 1965, p. 335.
20. *American Institute of Physics Handbook* McGraw-Hill, New York, 1963, p. 6-94.
21. Welsh, J., Densities of Amorphous and Crystalline Polymers, in *Physical Properties of Polymers Handbook* (J. E. Mark, ed.), American Institute of Physics, Woodbury, New York, 1996, pp. 401–407.
22. Prêtre, P., Wu, L.-M., Knoesen, A., and Swalen, J. D., *J. Opt. Soc. Am., B.*, 15, No. 1 (1998).
23. IBM synthesized NLO material (Contact Dr. Dan Dawson, IBM Almaden Research Center, 650 Harry Road, San Jose, California 95120).
24. Sasaki, K., (Private communication).

5
Electro-Optic Effect

George F. Lipscomb, John I. Thackara, and Rick Lytel*
Akzo Nobel Electronic Products, Inc.
Sunnyvale, California

I. INTRODUCTION

Electro-optic effects are macroscopic materials properties that are defined as the change in the optical characteristics of a material under the influence of an applied electric field. The electro-optic coefficients are conventionally defined in terms of the change in the impermeability tensor B_{ij}[1]:

$$B_{ij}(E) - B_{ij}(0) = \Delta\left(\frac{1}{n^2}\right)_{ij} = \sum_k r_{ij,k}E_k + \sum_{k,l} s_{ij,kl}E_kE_l. \tag{1}$$

E_k is the applied electric field, $r_{ij,k}$ is the linear electro-optic or Pockel's coefficient, and $s_{ij,kl}$ is the quadratic or D. C. Kerr coefficient. The names of the coefficients give a testament to the colorful history of the original discoveries of these effects. This chapter is concerned with techniques for the measurement of electro-optic coefficients as macroscopic materials parameters. In the following two subsections, we give a brief introduction to the linear and quadratic electro-optic effects as realized in organic and polymeric materials. Experimental techniques for determining those electro-optic coefficients are then described in Section II. The measurement methods covered in that section range from those suitable for rapid, coarse screening to those intended for precise characterization. Differences between techniques for measuring crystals and those for polymer thin films are delineated. Finally, the results of electro-optic measurements on organic polymeric and crystalline materials are presented in tables in Section III.

**Current affiliation*: Amp Inc., Palo Alto, California

A. Linear Electro-Optic Effects

The electro-optic (EO) coefficients can be related to the nonlinear susceptibility $\chi^{(2)}$ defined in Chapter 1 of this book under the condition that one of the applied fields has a frequency far below optical, $\omega \rightarrow 0$. For the linear electro-optic coefficient, this relationship is given by[2]:

$$\chi^{(2)}_{ij,k}(-\omega;\omega,0) = -\tfrac{1}{2}\varepsilon_{ii}(\omega)\varepsilon_{jj}(\omega)r_{ij,k}(-\omega;\omega,0). \tag{2}$$

$\varepsilon_{ii}(\omega)$ is the component of the dielectric constant of the material at the optical frequency. The nonlinear susceptibility can then be related to the microscopic molecular susceptibilities, as described in Chapters 1, 2, and 3 of this book. In this chapter, we will be concerned with the measurement of electro-optic coefficients as macroscopic materials parameters.

For convenience, the coefficients are usually written in matrix notation with contracted indices, where[1]:

$$1 \rightarrow (11);\ 2 \rightarrow (22);\ 3 \rightarrow (33);\ 4 \rightarrow (23,32);\ 5 \rightarrow (13,31);\ 6 \rightarrow (12,21).$$

In this case, the linear electro-optic coefficient $r_{ij',k'}$ is represented by r_{ij} where i ranges from 1 to 6 and j ranges from 1 to 3.

In general, for useful applications, the frequency of the modulating electric field is, of course, finite and non-zero, but is far less than optical frequencies. Depending on this frequency, the linear electro-optic effect can therefore—unlike second harmonic generation—contain terms that depend on a coupling of the acoustic and optic phonon modes to the applied electric field.[1] A conceptual plot of the dispersion in the linear electro-optic coefficient r is shown in Fig. 1 from Singer et al.[3] r^{el} is the electronic contribution from the molecular hyperpolarizabilities described in previous chapters. r^R is the contribution from the optic phonon modes and r^B is the contribution from the acoustic phonon modes. It is a striking feature of organic electro-optic materials that r^{el} dominates, whereas for inorganic electro-optic materials, such as lithium niobate, r^R dominates. The origin and dispersion of the electro-optic effect can have a significant affect on the performance of a device. Therefore, we will make a brief digression to illustrate one of the potential advantages of organic electro-optic materials and one of the prime reasons that so much research effort has been directed toward developing these materials.

The same optic phonon interactions that enhance r^R in inorganic EO materials also contribute to the linear dielectric constant, leading to a similar dispersion curve. Thus, in inorganic EO materials a large electro-optic coefficient is accompanied by a large dielectric constant, which causes increased device power consumption and reduced speed of operation. In lithium niobate, for example, $r_{33} = 30.8\ 10^{-12}$ m/V and $\varepsilon_3 = 28$, while $n_3 = 2.2$.[1] By contrast, there is very little dispersion in the dielectric constant of organic electro-optic materials where ε is usually between 3 and 4 over large frequency ranges.[3]

The low dielectric constant can significantly improve device performance. The frequency response of a standard integrated optic Mach-Zehnder modulator with a traveling wave electrode is limited by the velocity mismatch between the optical wave, which travels at c/n, and the electrical wave, which travels at $c/\sqrt{\varepsilon}$, where ε is the effective dielectric constant of the microwave electrode at the rf frequency.[4] The length-bandwidth product is given by:

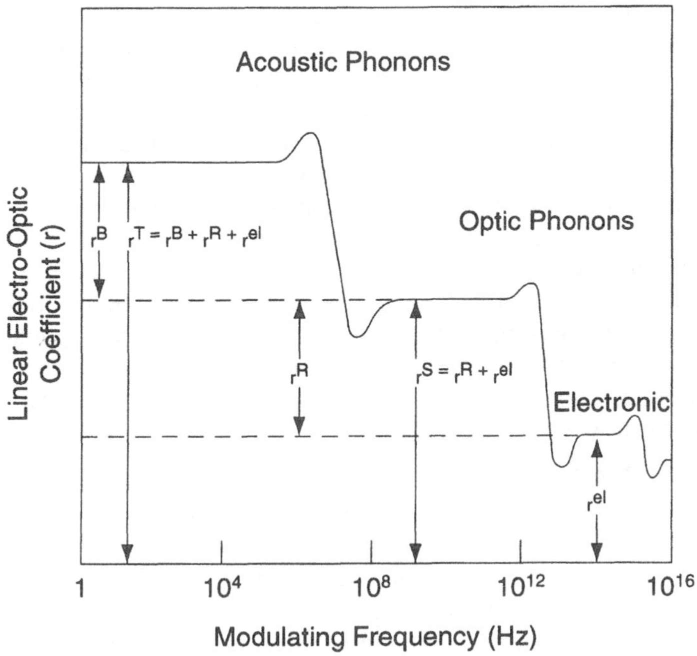

Figure 1 Dispersion of the linear electro-optic effect. (From K. D. Singer, S. J. Lalama, J. E. Sohn, and R. D. Small, Electro-Optic Organic Materials, in *Nonlinear Optical Properties of Organic Molecules and Crystals—Vol. 1* (D. S. Chemla and J. Zyss, eds.), Academic Press, New York, 1987, Fig. 2.)

$$\Delta f L = \left| \frac{c}{n - \sqrt{\varepsilon}} \right|, \tag{3}$$

where Δf is the bandwidth of the device, L is the interaction length of the light with the modulating electric field, n is the index of refraction, and ε is the dielectric constant of the microwave structure. For lithium niobate in a standard Mach-Zehnder configuration, $\Delta f L = 10$ GHz-cm,[4] while the number for a good EO polymer configuration would be $\Delta f L = 120$ GHz-cm. All other things being equal, EO polymers would have a significant advantage in modulation bandwidth over conventional inorganic materials. However, all other things have not been equal due to the relative immaturity of EO polymer technology, but rapid progress is being made in polymer devices. Researchers have exploited the potential advantage of EO polymers to build high-speed organic-polymer-based modulators moving from 20 GHz,[5] to wideband modulators at 40 GHz,[6] 50 GHz,[7] and now 60 GHz.[8,9]

The number of independent and non-zero elements in the electro-optic tensor depends on the symmetry of the material to be measured. For example, the linear electro-optic coefficient will be identically zero in any crystal or polymer material which is centrosymmetric.[1] This fact is central to any discussion of linear electro-optic effects in organic materials. The vast majority of organic molecular crystals possess point groups with a center of inversion symmetry and amorphous polymers are intrinsically centrosymmetric. A major portion of the research effort in organic electro-optics over the last 20 years has focused on how to utilize the extremely large electronic nonlinearities of organic molecular chromophores (described in the first chapters of this book) to make macroscopic materials with usefully large electro-optic coefficients. Unless a non-centrosymmetric point symmetry can be achieved, the large responses of individual molecules will cancel out in the bulk material. There are, in general, two routes to inducing and optimizing non-centrosymmetry. First, large numbers of organic molecular crystals can be tested and minute changes can be made in molecular structure in the hope of discovering naturally occurring non-centrosymmetric organic crystals. Secondly, the non-centrosymmetry can be induced by brute force, as it were, through the application of a large electric field, and then the asymmetry can be locked in by an amorphous polymer matrix. This second process is called electric field poling. Both of these two approaches have been used with considerable success by numerous researchers, and will be discussed in the next two subsections.

1. Organic Molecular Crystals

From a measurement point of view, organic single crystals are not much different from inorganic crystals. Once a high optical quality single crystal of a highly nonlinear organic molecule in a non-centrosymmetric structure has been grown, the electro-optic coefficients can be identified and measured using standard techniques, such as those described in Yariv.[10] Growth of such high quality, organic molecular crystals is extremely difficult because nonlinear chromophores with large hyperpolarizabilities, β, usually exhibit very large groundstate dipole moments. The preferred crystal structure of highly dipolar molecules is centrosymmetric, and when a non-centrosymmetric structure is attained, the crystals tend to be of poor quality or twinned. In the cases where good quality crystals can be grown, the resulting electro optic coefficients are sometimes spectacular due to the large molecular nonlinearities and high density of aligned molecules.

2-methyl-4-nitroaniline (MNA) presents a good example of the study of the linear electro-optic effect in an organic material. A diagram of the molecular structure of MNA is shown in Fig. 2.[11] MNA is a relatively small molecule, and exhibits, certainly by present day standards, a relatively small molecular hyperpolarizability, $\beta_{zzz} = 20 \pm 6 \times 10^{-30}$ cm^5/esu at $\lambda = 1.06$ μm.[3] MNA, however, crystallizes in the space group *Cc* with the non-centrosymmetric point group *m*. As can be seen in the diagram of the unit cell of MNA shown in Fig. 3, the polar axes of the individual molecules are nearly aligned with each other along the crystallographic *a* axis.[11] This means that the nonlinearities of the individual molecules add together to produce a macroscopic electro-optic effect. The relationship between microscopic molecular properties and macroscopic materials properties is described in Chapter 3 of this book.

MNA crystallizes with the point group *m*, leading to 10 independent electro-optic coefficients. In this point group, with the optic *y* axis defined to be along the *b* crystallographic axis, the effect on the index ellipsoid through the electro-optic effect from an applied electric field *E* is given by[10]:

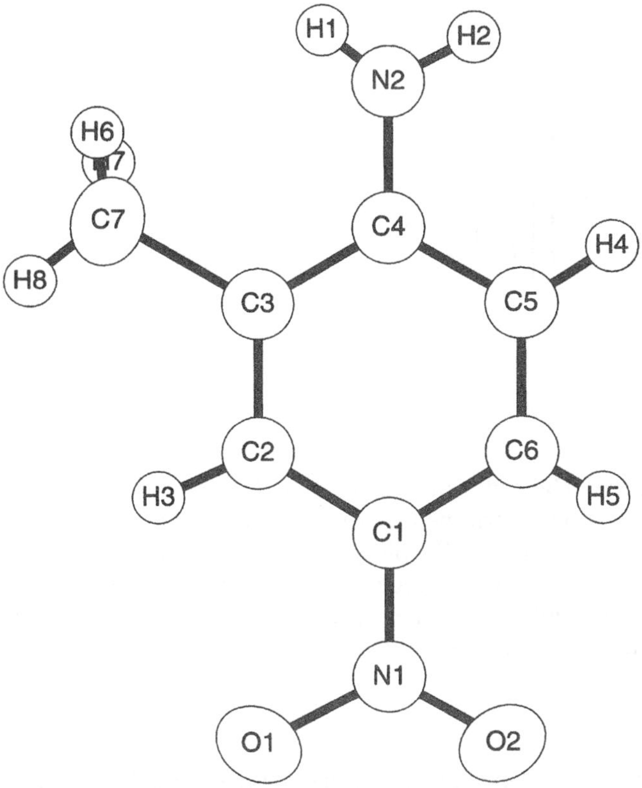

Figure 2 Molecular structure of 2-methyl-4-nitroaniline (MNA). (From G.F. Lipscomb, A. F. Garito, and R. S. Narang, A large linear electro-optic effect in a polar organic crystal 2-methyl-4-nitroaniline, *J. Chem. Phys.*, *75*, 1509 (1981), Fig. 1.)

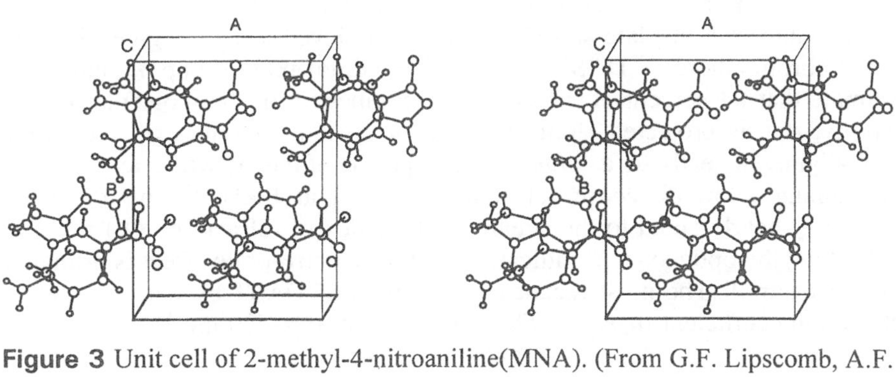

Figure 3 Unit cell of 2-methyl-4-nitroaniline(MNA). (From G.F. Lipscomb, A.F. Garito, and R. S. Narang, A large linear electro-optic effect in a polar organic crystal 2-methyl-4-nitroaniline, *J. Chem. Phys.*, *75*, 1509 (1981), Fig. 2.)

$$\begin{bmatrix} \Delta\left(\frac{1}{n^2}\right)_1 \\ \Delta\left(\frac{1}{n^2}\right)_2 \\ \Delta\left(\frac{1}{n^2}\right)_3 \\ \Delta\left(\frac{1}{n^2}\right)_4 \\ \Delta\left(\frac{1}{n^2}\right)_5 \\ \Delta\left(\frac{1}{n^2}\right)_6 \end{bmatrix} = \begin{bmatrix} r_{11} & & r_{13} \\ r_{21} & & r_{23} \\ r_{31} & & r_{33} \\ & r_{42} & \\ r_{51} & & r_{53} \\ & r_{62} & \end{bmatrix} \begin{bmatrix} E_1 \\ E_2 \\ E_3 \end{bmatrix}, \tag{4}$$

where a blank represents a zero. The optic main axes X and Z are not determined from crystal symmetry, and must be determined experimentally. (When relating the optic main axes of the crystal to the index ellipsoid, $(X,Y,Z) = (1,2,3)$.) The X optic main axis is found to lie at an angle of 8 degrees to the crystallographic a axis. For the case of an electric field applied only along the X axis, the change in the index of refraction in the X direction is given by:

$$\Delta n_1(E_1) = -\frac{1}{2} n_1^3 r_{11} E_1. \tag{5}$$

The different elements of the electro-optic tensor can then be measured by applying fields along different directions, with different polarizations and propagation directions of light, and determining the change in index by various common-path and non-common-path interferometric techniques. For the case of MNA, r_{11} is found to be the dominant component of the electro-optic tensor, with a value at $\lambda = 633$ nm of $r_{11} = 60 \times 10^{-12}$ m/V.[12,13] Electro-optic coefficients are usually given in units of picometers/Volt, pm/V $= 10^{-12}$ m/V. The electro-optic coefficient of MNA is almost entirely electronic in origin, and Wah et al. have used crystals of MNA to modulate light at 94 GHz.[14]

Much larger electro-optic coefficients are possible with more highly nonlinear molecules. For example, Perry et al. crystallized an ionic chromophore, dimethylaminostilbazolium in a tosylate salt (DAST) and achieved a high degree of alignment of the chromophores.[15] DAST crystals also exhibit point group m, and Perry et al. estimate that 83 percent of the dominant component of the molecular hyperpolarizability, β_{zzz}, is projected along the polar a crystallographic axis. This high degree of alignment leads to a large electro-optic coefficient, while at the same time maintaining a low dielectric constant, $\varepsilon_{11} = 5.1$ at 1 kHz.[15] This dielectric constant is very nearly equal to the square of the index of refraction $n(X) = 2.216$ ($n^2 = 4.9$) along the optic axis, X, indicating that the electro-optic effect is primarily electronic in nature.[15] The most recent measurement of DAST by Pan et al. gives an electro-optic coefficient of $r_{11} = 92 \pm 9$ pm/V at $\lambda = 720$ nm.[16]

2. Poled Polymer Films

In spite of the successes that have been achieved with organic single crystals, most molecules with large nonlinearities cannot be used in a crystal form. Chromophores with a microscopic nonlinearity more than 100 times larger than that of MNA have now been synthesized and tested.[17] Electric field poling has now become nearly ubiquitous in the field of organic electro-optics because it enables the use of chromophores which do not lend themselves to crystal growth. The concept of poling, as described below, was employed by Havings et al.[18] to freeze in an anisotropic distribution of solute molecules in a host polymer matrix for electrochromic studies. The technique was subsequently used by Singer et al.[19] to create nonlinear optical polymer films. Poled polymer films are thin films of polymer materials containing electro-optically active chromophores, into which a polar order has been induced through the application of a large electric field.

A diagram of the poling process is shown in Fig. 4.[3] First, the nonlinear chromophores are placed in a polymer matrix. Although not completely accurate, the molecular chromophores are highly polar and can be viewed as tiny one-dimensional vectors. The chromophores can be mixed with the polymer matrix as

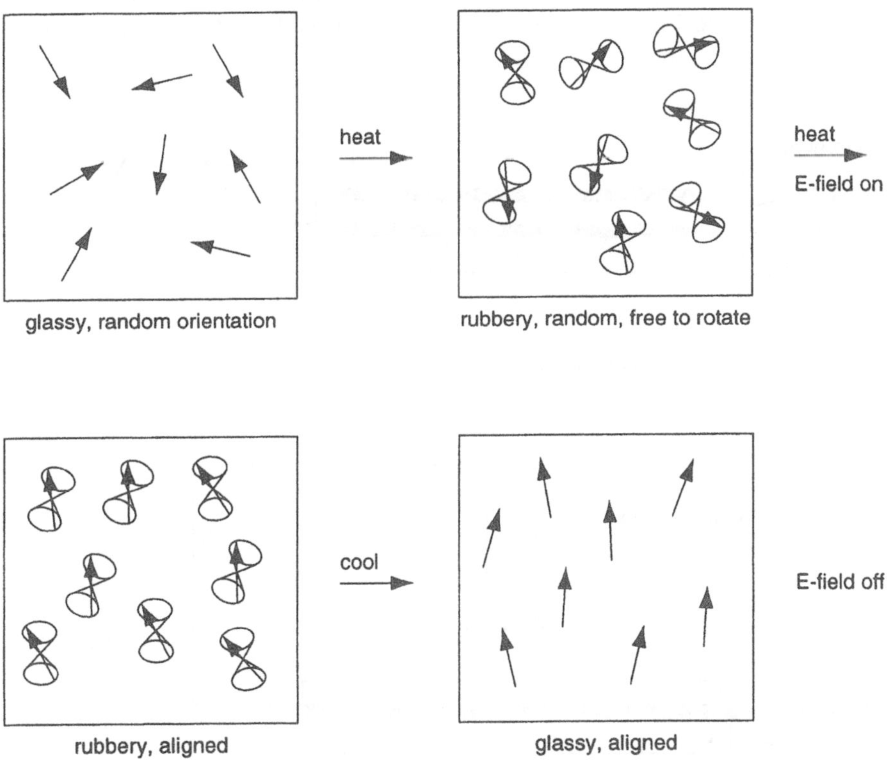

Figure 4 Electric field poling of an electro-optic polymer film. (From K. D. Singer, S. J. Lalama, J. E. Sohn, and R. D. Small, Electro-Optic Organic Materials, in *Nonlinear Optical Properties of Organic Molecules and Crystals—Vol. 1* (D. S. Chemla, and J. Zyss, eds.), Academic Press, New York, 1987, Fig. 10.)

a guest-host, they can be covalently attached to the matrix polymer as a side chain, they can form part of the main chain or they can be part of a more complex cross-linking system. Thin films are then produced by a variety of techniques such as spin coating, casting, doctor-blading or spray coating. As produced, the films are centrosymmetric, with the chromophores pointed in random directions. The polymer film is then heated to near or above its glass transition temperature, a large electric field is applied to align the chromophores that are in thermodynamic equilibrium with the polymer and the film is then cooled back to room temperature with the electric field still on. The alignment of the nonlinear chromophores are presumedly locked in place when the electric field is removed if the final temperature is well below the glass transition temperature of the polymer. Of course, the alignment is thermodynamically unstable and will eventually decay back to a random state. Engineering materials to maintain sufficient alignment stability over the practical lifetime of a device is a major focus of the research on these materials.

The large poling field can be applied using either metallic electrodes or through a corona discharge, as shown in Fig. 5. In the first instance, metallic electrodes are fabricated near the polymer in a variety of geometries, so that the polymer essentially forms the dielectric medium in a capacitor. A voltage, V, applied across the electrodes produces an electric field in the polymer layer. For the case shown in Fig. 5, the sample forms a parallel plate capacitor completely filled by the polymer. The field, defined to be in the $z = 3$ direction normal to the plane of the film, is

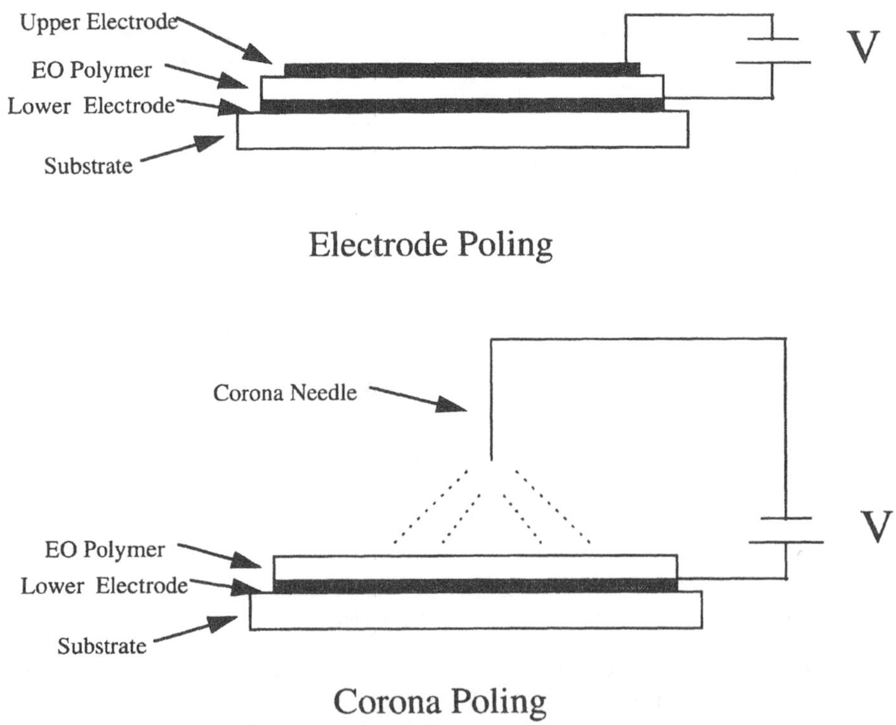

Figure 5 Poling geometries for thin film electro-optic polymers.

$E_3 = V/d$, where d is the film thickness. Electrode poling can also be performed in the plane of the film using co-planar electrodes, as demonstrated by Berkovic et al.[20] and by Wu et al.,[21] although with increased field non-uniformity.

Also shown in Fig. 5 is a diagram of the corona poling geometry. In this case, the polymer thin film is placed in the corona discharge from a needle placed at several thousand volts and positioned just above the film. In electrode poling, the metallic electrodes are placed with a well-defined geometry, so the magnitude of the poling field is known precisely. However, breakdown and shorting between the electrodes at any point in the film completely collapses the poling field and is more likely to result in catastrophic breakdown due to run away effects of increasing current at the weakest point in the material. This limits the achievable poling field in electrode poling to several hundred volts per micron. In corona poling, there are no metal electrodes, and the corona itself has fairly high impedance, so breakdown in the film is less likely to happen. When breakdown occurs at one point, the rest of the film remains exposed to a high electric field because of the high impedance of the surface of the film. Therefore, in general, higher poling voltages can be achieved with corona poling than with electrode poling. However, with corona poling, the poling field is not precisely known and may take days to decay, since charge has been sprayed onto the highly resistive surface of the polymer with no path to ground. Because electrodes are eventually required to apply a static field for purposes of electroptic modulation, electrode poling is often more convenient. For second harmonic measurements, the presence of electrodes is a detriment, and corona poling is therefore often used.

A poled polymer film is uniaxial and has one unique direction determined by the poling field and belongs to the point group ∞ mm and the electro-optic tensor has five non-zero components, r_{33}, $r_{13} = r_{23}$, and $r_{42} = r_{51}$.[3] The change in the index ellipsoid due to an applied electric field through the electro-optic effect is then given by[3]:

$$\begin{bmatrix} \Delta\left(\frac{1}{n^2}\right)_1 \\ \Delta\left(\frac{1}{n^2}\right)_2 \\ \Delta\left(\frac{1}{n^2}\right)_3 \\ \Delta\left(\frac{1}{n^2}\right)_4 \\ \Delta\left(\frac{1}{n^2}\right)_5 \\ \Delta\left(\frac{1}{n^2}\right)_6 \end{bmatrix} = \begin{bmatrix} & & r_{13} \\ & & r_{13} \\ & & r_{33} \\ & r_{42} & \\ r_{51} & & \\ & & \end{bmatrix} \begin{bmatrix} E_1 \\ E_2 \\ E_3 \end{bmatrix}. \tag{6}$$

For the common case, where the electric field drive signal is applied along the same direction as the poling field ($z = 3$), the changes in the extraordinary index (n_3) and the ordinary index ($n_1 = n_2$) are then given by:

$$\Delta n_3 = -\frac{1}{2} n_3^3 r_{33} E_3,$$

and (7)

$$\Delta n_i = -\frac{1}{2} n_3^3 r_{12} E_3. \qquad (i = 1,2)$$

The index changes induced are actually quite small. For example, $r_{33} = 30$ pm/V is usually considered a high value. Coupled with a standard drive field of 1 V/μm and a polymer index of $n_3 = 1.6$, $\Delta n_3 = 10^{-4}$. This small change in index, however, can be easily measured using interferometric techniques and, conversely, can be used to build practical optical modulators when integrated optic waveguides are used.

After ideal thermodynamic poling and ignoring any subsequent decay of the alignment, the electro-optic nonlinear susceptibility along the polar axis is given by Singer[3]:

$$\chi^{(2)}_{333}(-\omega;\omega,0) = NL^{\omega}L^{\omega}L^{0}\beta_{zzz}(-\omega;\omega,0)L_s(p),$$

$$\text{where } L^{\omega} = \frac{n^2 + 2}{3}, \tag{8}$$

$$\text{and } L^{0} = \frac{\varepsilon(n^2 + 2)}{n^2 + 2\varepsilon}.$$

N is the number density of chromophores in the material, the L's are the local field factors at the designated frequency as given by Singer[3] (and described in Chapter 3), β is the molecular hyperpolarizability and L_3 is the third order Langevin function. $\chi^{(2)}$ is linearly related to the electro-optic coefficient r through Eq. 2 above. p is given by[3]:

$$p = \left[\frac{\varepsilon(n^2 + 2)}{n^2 + 2\varepsilon}\right] \frac{\mu E_p}{kT} = L^0 \frac{\mu E_p}{kT}, \tag{9}$$

where μ is the vacuum dipole moment of the chromophore, E_p is the poling field, k is the Boltzman constant, and T is the poling temperature (usually near the glass transition temperature) in Kelvin. To first order in the applied field, which is usually the case for practical poling conditions, $L_3(p) = p/5$ and the electro-optic coefficient is given by:

$$r_{33}(-\omega;\omega,0) = \frac{2NL^{\omega}L^{\omega}L^{0}}{n_3^4(\omega)} \beta_{zzz}(-\omega;\omega,0) \frac{L^0}{5} \frac{\mu E_p}{kT}. \tag{10}$$

In most practical cases, the electro-optic coefficient of a poled polymer therefore depends linearly on the number density, N, and the poling field, E_p. In a crystal, nearly 100 percent of the molecular chromophores can be aligned to nearly 100 percent, leading to the maximum electro-optic response. In poled polymers, the number density of chromophores is reduced to less than 50 percent of what it would be in a pure crystal, due to the presence of the binding polymer matrix. For practical poling fields of 200 V/μm and typical poling temperatures of 150–250 degrees C, the thermodynamic alignment is limited to 30 percent or so of that in a single crystal. With the large molecular nonlinearities that are now available, these reductions in alignment and concentration still produce materials with large r_{33} coefficients. For example, Ahlheim et al. have reported a poled polymer with $r_{33} = 55$ pm/V at 1.3 μm for poling at 80 degrees C and 150 V/μm.[17]

This chapter is concerned with the measurement of the electro-optic coefficient of organic materials. Of course, much more than just a large electro-optic coefficient is required for a material to be technologically interesting. The material must be processable, form waveguides with low optical propagation loss at commercially important wavelengths, be chemically, optically, and thermally stable, and must meet dozens of other secondary property requirements.

B. Quadratic Effects

Organic materials potentially also exhibit a relatively large change in the index of refraction that is proportional to the square of the applied electric field. As shown in Eq. 1, the quadratic electro-optic effect is designated by the D. C. Kerr coefficient, s_{ijkl}, named after Kerr, who first discovered the effect in liquids in 1875.[22] Unlike the linear electro-optic, or Pockels effect, the D. C. Kerr effect does not require a non-centrosymmetric symmetry, and is non-zero in any point group. Quadratic effects are therefore easier to find, since there is no need to pole a polymer or to search for a rare crystal structure. Understanding quadratic effects, however, is usually quite complicated, since the lack of symmetry limitations means that almost any mechanism can, and usually does, contribute to the observed signal. Quadratic effects in organic materials can arise from ultra-fast electronic process, from reasonably fast small molecular segment reorientation, from slow chain segment reorientation, or from even slower thermal heating, as well as other mechanisms such as space charge, electrostriction, electrode attraction, or dimensional changes in the sample. Care must therefore be exercised in the planning and execution of experiments to measure quadratic electro-optic effects.

The recent interest in quadratic effects in organic materials has centered on the use of quadratic electro-optic measurements as a probe of the electronic structure of organic molecules and to help elucidate the all-optical third order effects for direct device applications.[23,24] In particular, quadratic electroabsorption spectroscopy has emerged as a powerful tool for assessing the two-photon all-optical device figure of merit for organic systems.[25] Quadratic electroabsorption arises from changes in absorption described by the imaginary part of the quadratic electro-optic tensor, $(s_{ij,kl})_I$. The real part of the quadratic electro-optic tensor, $(s_{ij,kl})_R$, results in a change in the real part of the index of refraction and a corresponding phase change in the propagating light. Both of these effects are useful probes of materials parameters. This is in contrast to the case of the linear electro-optic effect, where the focus has been on direct device applications, and very little attention has been paid to the imaginary part of $r_{ij,k}$.[26] Linear electro-optic devices depend on there being a long, low loss path length, and most work has centered away from resonances, where the absorption is low. For studies of the potential all-optical response in organic materials, resonant behavior is critical and absorption must be taken into account.

The measurement approaches and techniques for quadratic electro-optic effects are very similar to those for linear electro-optic effects. Many of the experimental setups are, in fact, identical. The primary difference is that the response is now proportional to the square of the applied electric field, leading to a frequency doubling in modulation experiments; that is, the light is modulated at twice the frequency of the applied voltage. Experimental techniques for quadratic electro-optic measurements are described in Section II.C.1, and techniques for quadratic electroabsorption measurements are described in Section II.C.2.

II. MEASUREMENT TECHNIQUES

The measurement of the linear and quadratic electro-optic coefficients involves the detection of relatively small changes in the index of refraction of a material under the influence of an electric field. Fortunately, interferometry provides an excellent tool for measuring very small changes in refractive index. A wide variety of interferometric techniques, both common path and non-common path, have been applied to the measurement of these effects in organic and polymeric materials. We first describe the techniques that have been employed to measure the linear electro-optic coefficients in poled polymer films and in molecular crystals and discuss the relative advantages and disadvantages of each method. Then measurement techniques for the quadratic electro-optic effect and electroabsorption are described.

A. Pockel's Effect in Poled Polymer Films

This section reviews the measurement techniques for determining the electro-optic coefficients (Pockel's effect) in poled polymer films.

Although there are many different techniques for determining the electro-optic coefficients, r_{ij}, only two basic sample geometries are generally employed. Since poled polymers are thin films, light can either propagate through the narrow dimension of film at some angle, or it can propagate some distance in the long direction of the film as a guided mode. We will refer to the first case as the "through-plane" geometry and the second case as the "waveguide" geometry. The through-plane geometry is usually used for material evaluation studies, while the waveguide geometry is used in device applications. As such, the through-plane measurements provide more useful information on the materials parameters, while the waveguide measurements provide more practical information on how the material performs in a waveguide device.

1. Through-Plane Measurement Techniques

Through-plane sample geometries are often used in materials evaluation studies. The sample geometry is quite simple and is shown in Fig. 6 from Norwood et al.[27] Usually, a transparent electrode, such as ITO, is deposited on a transparent substrate, such as glass, and the polymer material is spun onto the substrate. A second electrode, transparent or reflective, is deposited on top of the thin polymer layer. The material is then poled using the electrode sandwich, and the electro-optic effects are measured by transmitting light through the transparent electrode, through the thin film and either reflecting off of the second electrode or transmitting through to the other side and measuring the voltage-induced phase shift due to the sample using an interferometer. No complex layer structure is required, as is the case in a true waveguide structure, and the electric fields are known precisely. Since the light only passes through a few microns of the polymer material, measurements can be

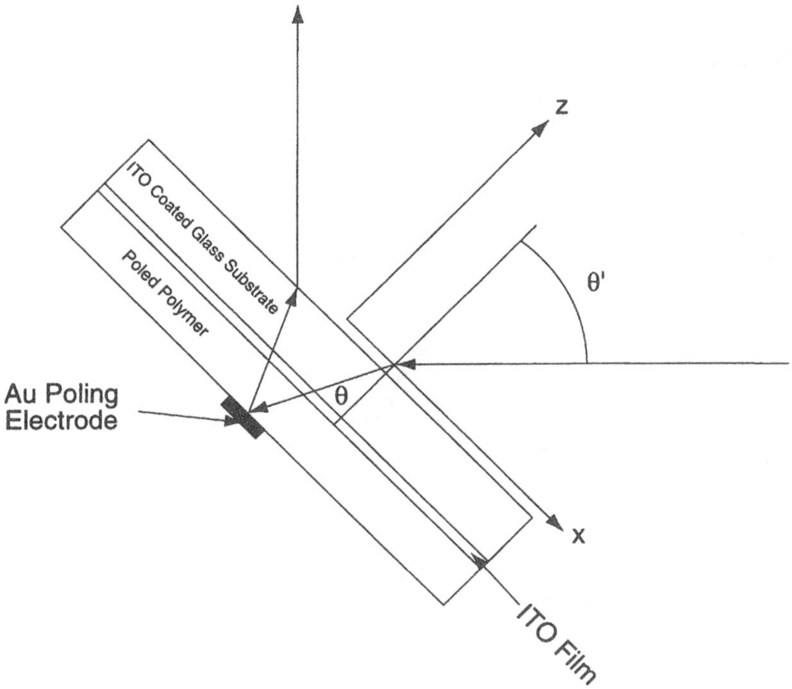

Figure 6 Through-plane sample geometry. (From R.A. Norwood, M. G. Kuzyk, and R. A. Keosian, Electro-optic tensor ratio determination of side-chain copolymers with electro-optic interferometry, *J. Appl. Phys.*, *75*, 1869 (1994), Fig. 4.)

made on materials that exhibit considerable absorption. All of these factors make the through-plane geometry ideal for materials evaluation. As will be seen in the next few subsections, a great deal can be learned about the material parameters, depending on which experimental techniques are employed. The actual modulation of light, however, is small, and, in general, through-plane geometries are not practical for actual devices.

Thin Film Reflection Techniques. The most commonly used technique to measure the electro-optic coefficients in a poled polymer film is the through-plane reflection technique first proposed by Teng and Mann,[28] which is shown in Fig. 7.[29] This technique has the advantage that sample preparation is simple, measurement is quick, and the interpretation is straight forward, and has, therefore, become widely used for evaluation measurements. One disadvantage, however, is that this simple reflection technique does not give an independent measurement of r_{33} and r_{13}, but assumes a relationship from symmetry considerations. Independent measurements of r_{33} and r_{13} can be made with a slightly more complex version of the experiment shown in Fig. 7.

The measurement is performed as follows: A laser beam polarized at 45 degrees to the plane of incidence enters the back surface of the sample at an angle θ. The light travels through the glass substrate, the lower ITO electrode and the electro-optic polymer layer, before reflecting off of the upper gold electrode and

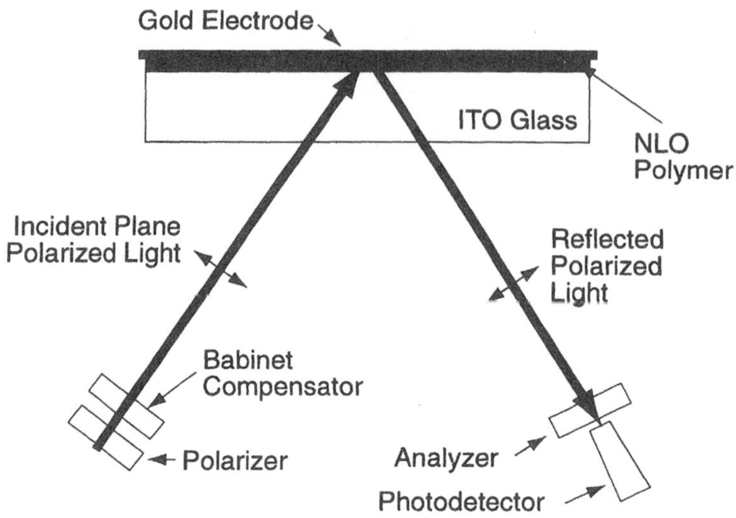

Figure 7 Through-plane reflection technique. (From R. N. DeMartino, D. E. Allen, R. Keosian, G. Khanarian, and D. R. Haas, Side Chain Polymers for Electro-Optic Applications, *Mat. Res. Soc. Symp. Proc.*, *228*, 39 (1992), Fig. 6.)

exciting the sample through the same layers. Once outside the sample, the laser beam travels through a Soleil-Babinet compensator, an analyzer set at −45 degrees and into a detector. The sample is effectively placed between crossed polarizers, with equal amounts of the *s* and *p* polarizations launched. The intensity at the detector, I_{out}, is therefore:

$$I_{out} = I_{max} \sin^2(\Gamma_{sp}/2), \tag{11}$$

where I_{max} is the intensity just before the analyser, and Γ_{sp} is the total phase shift between the *s* and *p* polarizations. A modulating voltage, $V_m \sin(\Omega t)$, is then applied across the electrodes on the sample, and the modulated portion of the optical signal at the detector is measured with a lock-in amplifier. The total phase shift then has a static portion, Γ_o, due to the difference in refractive index and path length between the two polarizations, a static portion due to the Soleil-Babinet compensator, Γ_{SB}, and a modulated part due to the applied voltage, $\Gamma_m(V_m)\sin(\Omega t)$:

$$\Gamma_{sp} = \Gamma_o + \Gamma_{SB} + \Gamma_m(V_m)\sin(\Omega t). \tag{12}$$

The phase retardation from the compensator is adjusted until $\Gamma_o + \Gamma_{SB} = \pi/2$, so that the intensity at the detector for zero applied voltage ($\Gamma_m = 0$) is $I_{out} = I_{max}/2$. Substituting Eq. 12 into Eq. 11 and assuming the modulation from the applied voltage is small, the output intensity is approximated by:

$$\frac{I_{out}}{I_{max}} = \frac{1}{2} + \frac{\Gamma_m}{2} \sin(\Omega t). \tag{13}$$

When an electric field, E_3, is applied in the *z* direction, due to the voltage on the parallel plate electrodes, the ordinary, $n_o = n_1 = n_2$, and extra-ordinary, $n_e = n_3$, indices of refraction change differently, as given in Eq. 7 above. The *s* polarization is perpendicular to the plane of incidence of the light and, therefore, encounters only the ordinary index. The p polarization is parallel to the plane of incidence, and therefore encounters a combination of the ordinary and the extraordinary index. The ratio of p- to s-polarized components depends on the angle of incidence inside the sample. At 45 degrees, there is an equal amount of s- and p-polarized light. The field-induced changes in index result in a differential phase shift between the s and p polarizations directly from the index difference, and, also, from the change in the optical path length through the electro-optic material. A light wave with an external angle of incidence θ to the plane-normal, will propagate inside the film at an angle θ' to the plane normal, as determined by Snell's law, $\sin(\theta) = n \sin(\theta')$. With the assumptions $n_o = n_e = n$ and $r_{33} = 3r_{13}$, and assuming that there are no reflections from the different layers, Teng and Mann solved for r_{33} in terms of experimentally measurable values.[28] As restated by De Martino et al.,[29] Mortazavi et al.,[30] and by Levy et al.,[31] the expression for r_{33} becomes:

$$r_{33} = \frac{3\lambda}{2\pi V_m} \frac{(n^2 - \sin^2 \theta)^{1/2}}{n^2 \sin^2 \theta} \frac{I_m}{I_{max}}, \tag{14}$$

where I_m is the amplitude of the modulated portion of the detected optical signal, that is, from Eq. 13, $I_m/I_{max} = (\Gamma_m/2)$. This expression assumes that the metal layer is a perfect reflector. The value for r_{33} depends on the ratio of two optical intensities, and the detector responsivity drops out of the equation, provided the two quantities are measured in the same manner. For example, it is common for I_{max} to be measured by inserting a chopper in the optical path and using the lock-in amplifier to measure a voltage signal equivalent to I_{max}. The chopper produces a square wave signal, and the rms value measured by the lock-in will be different than the rms value measured for the sine wave modulated signal I_m. For the case where a chopper is used, $I_m = 2\sqrt{2}I_m^{\text{rms}}$. similarly, I_m and V_m must be measured in the same way, either peak to peak or rms.

As was noted by Teng and Mann[28] in their original paper and subsequently by Schildkraut,[32] the assumption that there are no other reflections other than the one at the intended mirror is not valid in practice. In fact, the reflections off of the different layers are significant and can cause substantial errors in the electro-optic coefficients calculated with the above formula. This is not surprising, since the ITO transparent electrode has a significant enough index difference to cause multiple reflections within the film. These multiple reflections cause interference effects and intensity modulation when only one of the s or p polarizations is present. These intensity modulation effects arising from a single polarization are superimposed on the desired intensity modulation signal caused by interference between the two polarizations, causing the results to be distorted. Teng and Mann note, however, that the contributions to the modulation from each polarization should be independent of the setting of the compensator, while the desired interference signal between the two polarizations will change sign when the bias point is changed from $\pi/2$ to $3\pi/2$ or $-\pi/2$. The change of sign is illustrated in Fig. 8 after Teng and Mann.[28] To first order, the average of the signal at the two quadrature points will give the correct value for r_{33}.

The assumption that $r_{33} = 3r_{13}$ is more problematical, and places limits on the interpretation of measurements using this simple reflection technique. It can be shown that under certain ideal conditions, r_{33} exactly equals $3r_{13}$.[2] These conditions

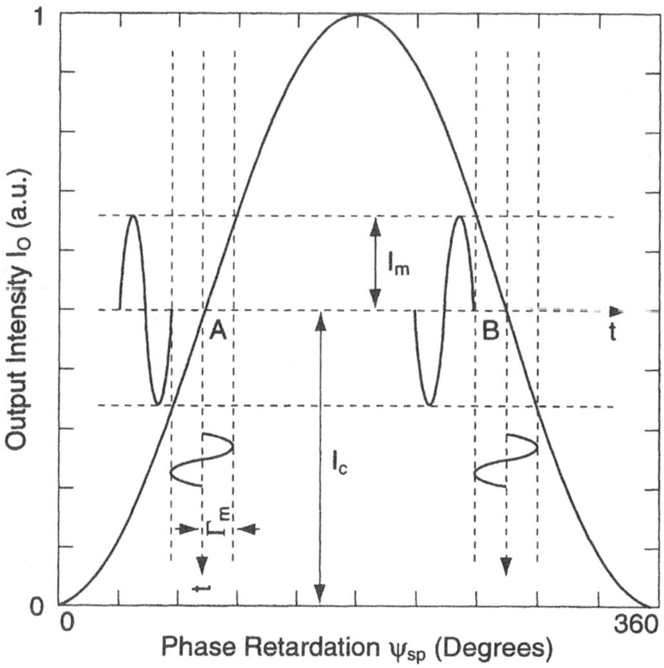

Figure 8 Modulated output intensity versus static phase retardation between the *s* and *p* polarizations. (From C. C. Teng and H. T. Mann, Simple reflection technique for measuring the electro-optic coefficient of poled polymers, *Appl. Phys. Lett.*, *56*, 1734 (1990), Fig. 2.)

are that the molecule is essentially one-dimensional, i.e., only β_{zzz} and μ_z are significant, and that the poling orientation distribution is that of an ideal gas of dipoles in an electric field at the poling temperature. These conditions are often not met in practice. Norwood et al. have shown using an external interferometer, described in the subsection entitled "External Mach-Zehnder Interferometer," below, which can measure r_{33} and r_{13} independently, that the ratio can range from 3 to 6.[27] They ascribe the change in the ratio to hindered molecular rotation during alignment, to the fact that the molecules are not one-dimensional, and to piezoelectric contributions to the electro-optic effect.

The reflection technique is very useful for material screening measurements, for quickly determining the relative size of the electro-optic coefficient, and for studies of poling conditions and stability. If precise and independent values of r_{33} and r_{13} are required, then different, more time-consuming techniques are required.

Fabry-Perot Etalon Techniques. An independent measurement of the electro-optic coefficient r_{13} can also be made using an extension of the technique described in the previous section. As developed by Eldering et al.,[33] van der Vorst et al.,[34] and Kobayashi et al.,[35] this technique makes use of the partial reflections at the electrodes, which were an unwanted complication in the reflection technique, and treats the sample as a Fabry-Perot etalon of relatively low finesse. The sample geometry is similar to that for the crossed polarizer technique described above, and the sample can be measured in reflection or in transmission. For this example we will consider the transmission geometry, as shown in Fig. 9.[33] The experimental set-up is identical to the previous technique, except that the Soleil-Babinet compensator and the analyzer are not present. The intensity modulation that is measured arises in the Fabry-Perot sample itself due to interference between exiting waves that had bounced back and forth in the cavity a different number of times.

Because the sensitivity of intensity modulation that results from an index change depends on the phase difference between the two faces of the interferometer, one additional parameter must be varied to determine the transmission function of the Fabry-Perot sample. This additional parameter can either be the wavelength[35] or the angle of incidence of the incoming beam,[33] both of which change the optical path length through the etalon. In order to measure r_{13}, the input polarization is set perpendicular to the plane of incidence (s), so that no contribution is present from r_{33}. r_{13} is then given by[34]:

$$r_{13} = \frac{2d\Delta I \sin\theta \cos\theta}{n_0^4 V \, dI/d\theta} \tag{15}$$

where d is the spacing between the two electrodes, θ is the external angle of incidence, ΔI is the change in transmitted intensity due to the applied field, V is the applied voltage, n_o is the ordinary index of refraction of the film, and $dI/d\theta$ is

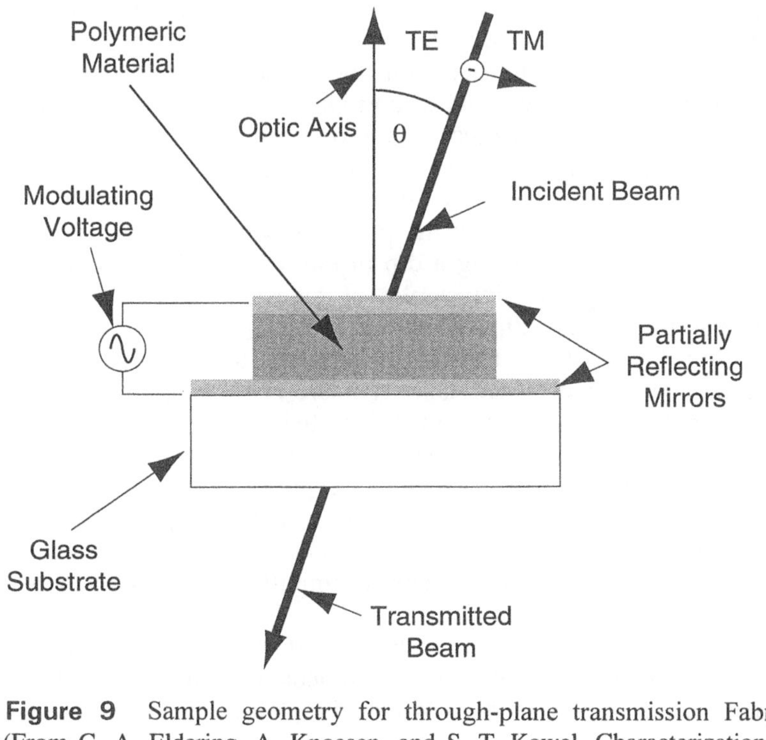

Figure 9 Sample geometry for through-plane transmission Fabry-Perot technique. (From C. A. Eldering, A. Knoesen, and S. T. Kowel, Characterization of polymeric films using metal mirror/electrode Fabry-Perot etalons, *Proc. SPIE*, *1337*, 348 (1990), Fig. 1a.)

the slope of the Fabry-Perot transmission curve at the point where the measurement is made. Shown in Fig. 10 from Eldering et al.[33] is a plot of the transmittance of a Fabry-Perot etalon containing a poled polymer film versus angle of incidence, from which $dI/d\theta$ can be obtained. Also shown is a plot of the change in transmittance versus angle of incidence, from which ΔI can be obtained. The curves for the TE polarization are of interest for determining r_{13}.

By using a combination of the crossed polarizer reflection technique and the Fabry-Perot interferometric modulator technique, r_{13} and r_{33} can be measured independently. The sample preparation and the experimental set-up are relatively simple, and the measurement and analysis are straight forward.

Attenuated Total Reflectance (ATR) Spectroscopy. The somewhat different technique of attenuated total reflectance (ATR) spectroscopy was used by Dentan et al. to measure electro-optic effects in ferroelectric polymers.[36] ATR was also employed by Herminghaus, Smith, and Swalen to independently measure r_{13} and r_{33} of poled polymer film samples.[37,38] A diagram of the experimental set-up is shown in Fig. 11 from Smith et al.[38] The sample consists of a thin polymer film sandwiched between two thin gold electrodes and brought into contact with a prism. The sample can be prepared directly on the prism or, more frequently, prepared on a substrate and contacted to the prism with an index matching fluid. Similarly, the sample can be directly poled using the two gold electrodes, or can be corona-poled before the second gold electrode is deposited. A thin buffer layer between the electro-optic polymer and the poling electrode can be used to increase the poling field and to improve film quality.[37]

The measurement is made by reflecting a laser beam off of the gold electrode in contact with the prism, as shown, and varying the angle of incidence and the polarization. If the gold electrode is thin enough, the evanescent field of the reflected light will penetrate the gold layer and extend into the electro-optic polymer layer. Smith et al. used a gold layer thickness of approximately 50 nm.[38] At angles of incidence where the longitudinal component of the propagation vector of the incident light in the plane of the film matches that of the propagation vector of a waveguide mode, light will be coupled into the guided mode in the electro-optic film and out of the reflected beam, causing a dip in the reflected light. If several modes are present in the waveguide film, a plot of reflectance versus angle of incidence will show a series of sharp dips, as shown in the upper curve in Fig. 12 from Smith et al.[38] From the position of the sharp dips, the thickness of the film and the index of refraction can be obtained using standard prism coupling analysis. By using two different polarizations to excite the film, the ordinary n_{TE} and extraordinary n_{TM} indices of refraction can be measured independently.

The electro-optic coefficients are measured by applying a modulating field across the two gold electrodes. The applied field causes a small change in the index of refraction of the film, which in turn causes a small change in the optimum coupling angle. The signal detected with a lock-in amplifier at the fundamental frequency from a modulating field is then proportional to the first derivative of the reflectance versus angle curve, as shown in the lower half of Fig. 12. The ratio of the modulated signal to the first derivative of the reflectance curve is the field-induced change in the coupling angle from which the change in index due the

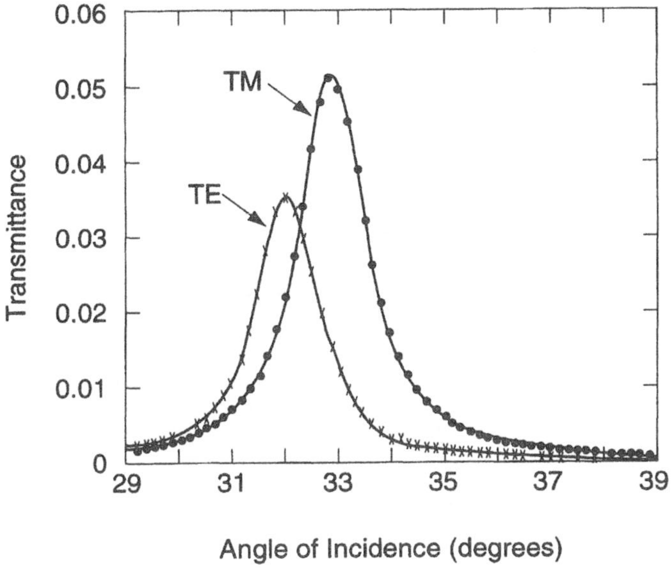

(a)

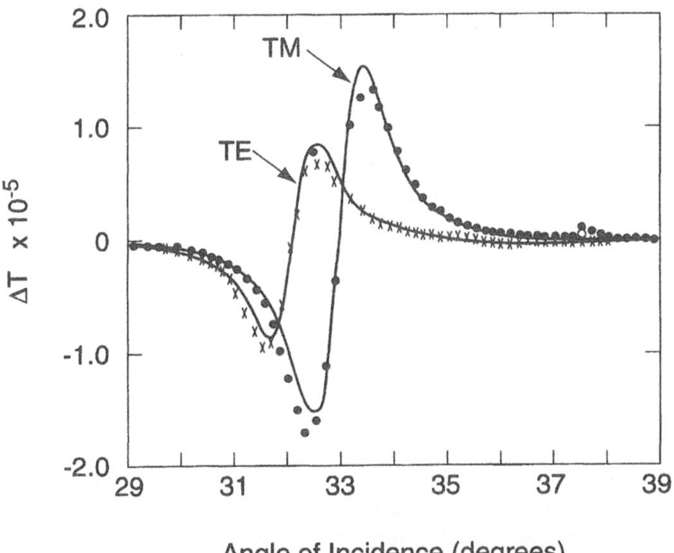

(b)

Figure 10 a) Transmission vs. angle of incidence for an etalon containing a poled polymer film. b) Change in transmittance vs. angle of incidence for an applied modulating field. (From C. A. Eldering, A. Knoesen, and S. T. Kowel, Characterization of polymeric films using metal mirror/electroe Fabry-Perot etalons, *Proc. SPIE*, *1337*, 348 (1990), Fig. 4.)

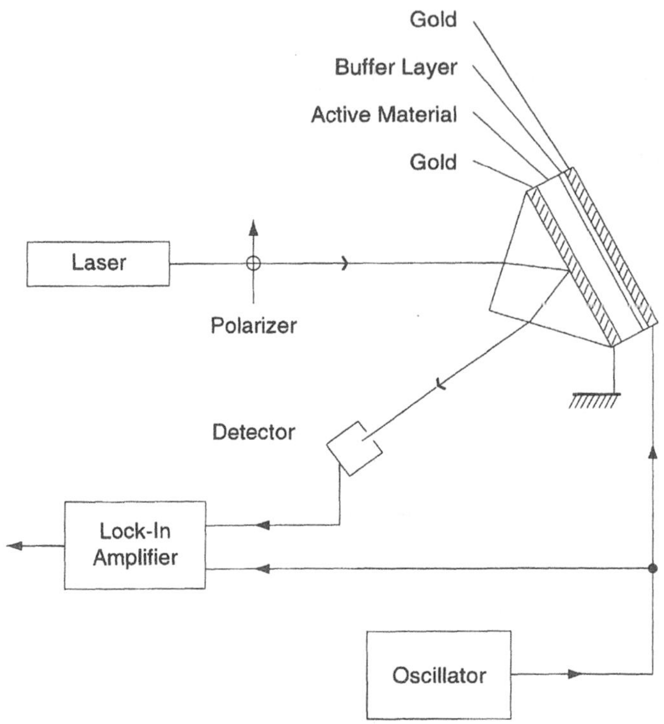

Figure 11 Attenuated total reflectance management experiment. (From B. A. Smith, S. Herminghaus, and J. D. Swalen, Electrooptic coefficients in electric field poled polymer waveguides, *Proc. SPIE*, 1560, 400 (1991), Fig. 1.)

applied field, $\delta n/\delta E$, can be determined.[38] The electro-optic coefficients for a poled polymer film can then be determined[38] from Eq. 7:

$$r_{13} = \frac{2}{n_{TE}^3} \frac{\partial n_{TE}}{\partial E}, \tag{16}$$

and

$$r_{33} = \frac{2}{n_{TM}^3} \frac{\partial n_{TM}}{\partial E}, \tag{17}$$

where E is the modulating electric field. In an ATR measurement, the mode coupling angle changes in response to an electric field due to the electro optic effect (changes in refractive index) and piezoelectric effect (changes in sample thickness). To determine the electro optic coefficient from ATR, the thickness change must be independently determined. However, Smith et al.[38] report that piezoelectric thickness changes did not noticeably affect their measurements.

ATR techniques provide a useful method for independently determining r_{13} and r_{33}. Furthermore, since several modes can be measured—and the electrostriction contribution varies with mode—ATR also provides information about film thickness changes. A disadvantage of ATR is that the sample preparation and measure-

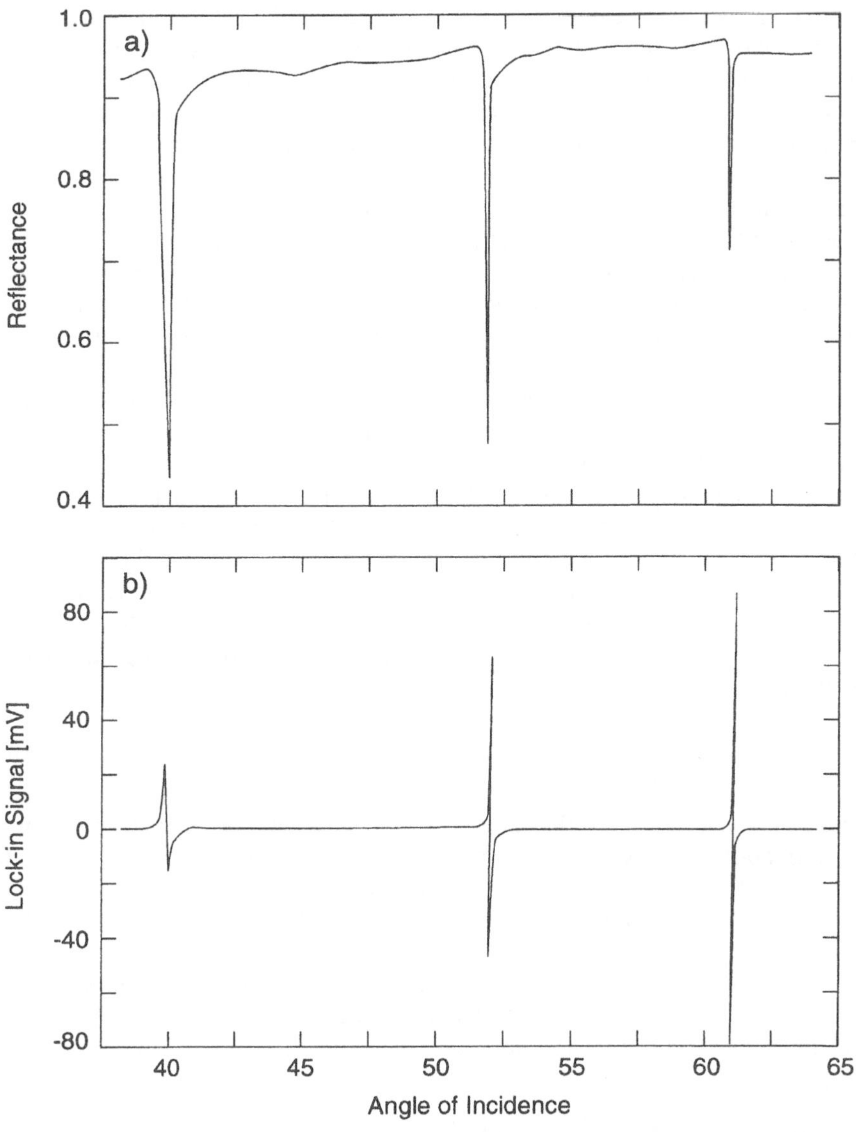

Figure 12 Attenuated reflectance (ATR) spectroscopy data: a) reflected intensity, b) lock-in signal obtained by electric field modulation. Experiment. (From B. A. Smith, S. Herminghaus, and J. D. Swalen, Electrooptic coefficients in electric field poled polymer waveguides, *Proc. SPIE*, *1560*, 400 (1991), Fig. 2.)

ment are slightly more complicated than in the crossed polarizer technique or in the Fabry-Perot interferometer technique, and the data analysis is significantly more complicated.

External Mach-Zehnder Interferometer. An independent measurement of the electro-optic components of a poled polymer film can also be made by placing the poled polymer sample in an external arm Mach-Zehnder interferometer, as developed by Singer et al.[39] A diagram of the experimental configuration is shown in Fig. 13, after Norwood et al.[27] The sample geometry is identical to the simple through-plane measurement described in the subsection entitled "Thin Film Reflection Techniques," above as was shown in Fig. 6. Once again, the sample can be measured in transmission or in reflection. For illustration, we will consider the case where the sample is measured in reflection, as shown in Fig. 13. Unlike the simple measurement, however, the reference beam does not travel on a common path with the modulated signal beam, but rather follows a separate path in the other arm of a Mach-Zehnder modulator. This approach has the advantage that each polarization can be measured separately, r_{13} and r_{33} can be determined independently, and the data analysis is fairly straight forward. As in any external arm interferometer, however, a laser with a substantial coherence length must be used and the reference arm must be isolated from any vibration or noise that is not also identically present in the signal arm, leading to a somewhat more complicated and expensive experimental apparatus.

A voltage at frequency Ω is applied to the sample, leading to a small phase modulation in the signal arm. At the same time, the phase in the reference arm, ϕ, is swept through a full fringe by rotating a glass plate intersecting the reference beam. A plot of the reference signal, i.e., the dc light intensity of the interferometer, and the modulated signal at Ω from the electro-optic effect in the sample, is shown in Fig. 14 after Norwood et al.[27] The electro-optic coefficient for each polarization is then given by[27]:

$$r^i = \frac{\lambda|I_1 + I_2|^i}{2\sqrt{2}(n^i)^3(V_{rms})^i(I_r)^i}\frac{\cos\theta'^i}{2}, \tag{18}$$

where i represents the s or p polarization, n^i is the index of refraction for the indicated polarization, V_{rms} is the rms applied modulating voltage at frequency Ω, θ' is the internal reflection angle within the film, I_1 and I_2 are the intensities of the two peaks in the signal modulated at frequency Ω, and I_r is the intensity of the reference signal.

r_{13} is given directly by the s polarization measurement. Once r_{13} is known, the ratio between r_{33} and r_{13} can be determined by a measurement with the p polarization at an internal angle θ'^p in the sample. The ratio r_{33}/r_{13} is related to r^s and r^p according to [27]:

$$\frac{r_{33}}{r_{13}} = \left[\left(\frac{n_1}{n_3}\right)^4\left(\frac{n^p}{n_1}\frac{(n^p)^3 r^p}{(n^s)^3 r^s} - \cos^2\theta'^p\right)\right]\Big/(\sin^2\theta'^p), \tag{19}$$

where θ'^p is the internal angle of the beam in the p polarization. n^p is the index of the beam for the p polarization, and is given by[27]:

$$\frac{1}{n^p(\theta')^2} = \frac{\cos^2\theta'^p}{n_1^2} + \frac{\sin^2\theta'^p}{n_3^2}, \tag{20}$$

where n_1 and n_3 are the ordinary and extraordinary indices of the film. Norwood et al.[27] used this electro-optic interferometry technique to measure several samples, and demonstrated that, contrary to the simple model, the tensor ratio can range from $r_{33}/r_{13} = 3$ to 6. They ascribe the variation to piezoelectric contributions to the electro-optic measurement, restricted motion of the chromophores during pol-

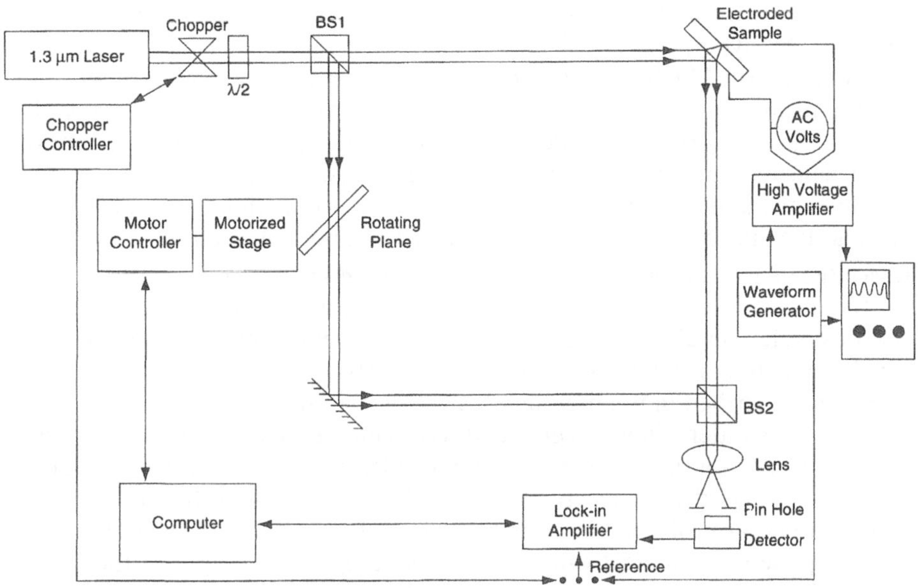

Figure 13 External arm Mach-Zehnder interferometer configuration. (From R. A. Norwood, M. G.Kuzyk, and R. A. Keosian, Electro-optic tensor ratio determination of side-chain copolymers with electro-optic interferometry, *J. Appl. Phys.*, *75*, 1869 (1994), Fig. 2.)

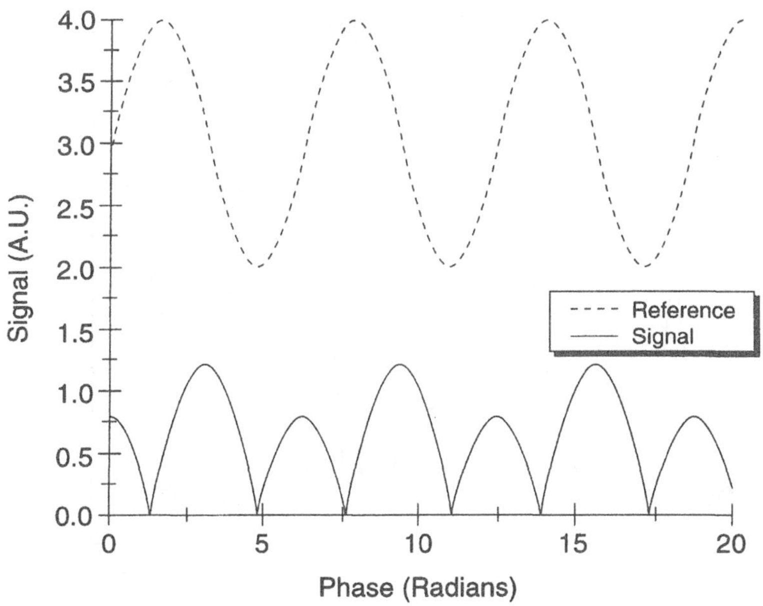

Figure 14 Reference and signal intensities from an external arm Mach-Zehnder electro-optic measurement. (From R. A. Norwood, M. G. Kuzyk, and R. A. Keosian, Electro-optic tensor ratio determination of side-chain copolymers with electro-optic interferometry, *J. Appl. Phys.*, *75,* 1869 (1994), Fig. 3.)

ing, and deviations from the assumption that the molecule has only one significant component of the hyperpolarizability.

The external arm Mach-Zehnder interferometer provides a powerful and flexible tool for measuring the electro-optic coefficients of poled polymer films. The tensor components can be independently measured and the contributions from different mechanisms can be separated. It has the disadvantage, however, that the experimental apparatus is more complex and the data acquisition is somewhat more difficult than the simple measurements described in previous sections.

Heterodyne External Mach-Zehnder Interferometer. An interesting variation of the external arm Mach-Zehnder interferometer measurement technique has been developed by Valley et al.[40] A diagram of the experimental set-up is shown in Fig. 15 after Valley and Wu.[41] The configuration is nearly identical to the external arm Mach-Zehnder described in the previous section, except that the rotating glass plate (and one beam splitter) have been replaced with an acousto-optic modulator (AOM). The electro-optic polymer sample is identical to that for the standard interferometer, and is designated by *S* in Fig. 15. The AOM shifts the frequency of the light in one arm of the Mach-Zehnder by a frequency ω. In the absence of any voltage applied to the electro-optic polymer sample, the output of the interferometer oscillates at frequency ω. When a voltage at a second frequency, Ω, is applied to the sample, the beat frequencies $\omega \pm \Omega$ also appear in the output spectrum of the interferometer. An example from Valley et al., where the heterodyne

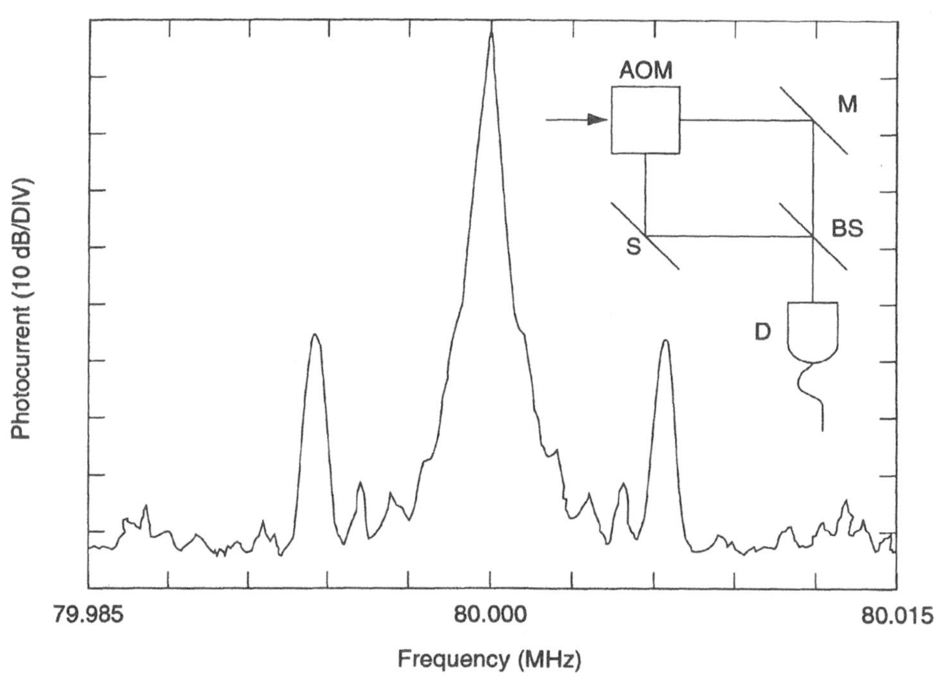

Figure 15 Experimental configuration and data from a heterodyne Mach-Zehnder interferometer measurement. (From J. F. Valley and J. W. Wu, Measurement of electro-optic poling transient effects in polymer thin films, *Proc. SPIE, 1337,* 226 (1990) Fig. 1.)

frequency is $\omega = 80$ Mhz and the signal frequency is $\Omega = 6$ kHz, is shown in Fig. 15.[41]

The heterodyne technique has the advantage that the measurement is shifted up to 80 Mhz, above most mechanical vibration and where low noise electronics eliminate dc drift and other instabilities. The technique has the obvious disadvantage that it requires a stable rf spectrum analyzer. In effect, the heterodyne technique simultaneously performs the reference measurement and the signal measurement of the standard Mach-Zehnder measurement. Valley et al. showed that the electro-optic coefficient can be determined from the ratio of the intensity of the zeroth order peak at ω and the electro-optically induced side-bands at $\omega \pm \Omega$[40]:

$$\left| \frac{I(\omega \pm \Omega)}{I(\omega)} \right| = \frac{J_1(\delta)}{J_o(\delta)}, \tag{21}$$

where I is the intensity of the signal on the rf spectrum analyzer, and J_i is the Bessel function of order i. For the s polarization, δ is given directly in terms of r_{13}:

$$\delta = 2\pi r_{13} n^3 d E_\Omega / \lambda, \tag{22}$$

where d is the thickness of the EO polymer, n is the index for the s polarization, and E_Ω is the applied modulating field. As in the case of the standard Mach-Zehnder, a combination of contributions from r_{13} and r_{33} can be deconvolved from the p polarization data.[40] The primary advantage of the heterodyne technique, however, is the speed of measurement combined with insensitivity to environmental factors, making it ideal for dynamic measurements. Valley et al. used this technique to measure the dynamics of poling as a function of temperature.[41]

Phase Grating Electro-optic Modulator. Another technique that is well-suited for transient studies is the phase grating electro-optic modulation method developed by Shi et al.[42] A diagram of the experimental set-up is shown in Fig. 16. A static grating is etched directly in the electro-optic sample itself either by

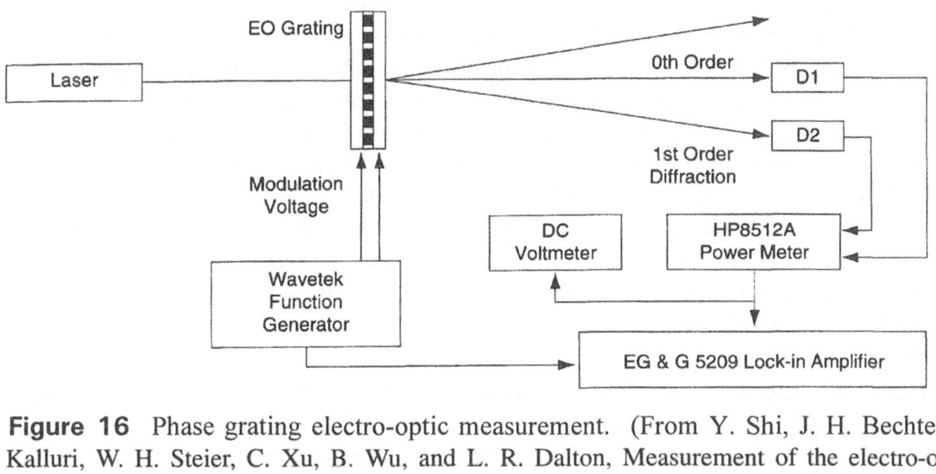

Figure 16 Phase grating electro-optic measurement. (From Y. Shi, J. H. Bechtel, S. Kalluri, W. H. Steier, C. Xu, B. Wu, and L. R. Dalton, Measurement of the electro-optic coefficient dispersion in poled polymer materials, *Proc. SPIE, 2285*, 131 (1994), Fig. 3.)

etching the polymer layer directly or by producing the grating pattern in the transparent electrode. In the configuration used by Shi et al., the sample is measured in transmission and the laser beam enters normal to the surface.[42] As shown in Fig. 16, the zeroth order beam travels straight through the sample, while the first order beam is diffracted at an angle determined by the grating spacing. Both the zeroth and the first order beams are detected with photodetectors and a lock-in amplifier. The amplitude of the diffracted beam will be determined by the magnitude of the refractive index modulation depth of the grating, so changes in the refractive profile through electro optic modulation result in a change of diffraction efficiency. Shi et al. show that, for the case of normal incidence, the r_{13} component of the electro-optic tensor can be determined by[42]:

$$r_{13} = \frac{\lambda d}{2n^3 E} \frac{\delta I_1}{\sqrt{I_0 I_1}}, \tag{23}$$

where d is the thickness of the film, and E is the applied electric field. δI_1 is the change in intensity of the first order diffracted beam due to the applied voltage, and I_0 and I_1 are the intensities of the first and second order diffracted beams.

The phase grating approach provides a rapid measurement of the electro-optic coefficient in a poled polymer film through the simple comparison of the intensities of the two diffracted beams. Shi et al.[42] used this technique to measure the dispersion of the electro-optic coefficient in a poled polymer film and to monitor the stability of a poled polymer film. Unlike the simple techniques described above, a grating with a spacing near the wavelength of light must be formed in sample. This requires some capability in photolithography and significantly complicates sample preparation. For these reasons, the phase grating technique is best suited for extended studies on a relatively small number of samples, rather than for general screening of many materials.

2. Waveguide Measurement Techniques

The through-the-plane electro-optic measurement techniques described in Section II.A.1 are very useful in evaluating poled polymer films for electro-optic activity. However, they all involve the detection of a small electro-optically modulated signal in a much larger unmodulated background signal, usually through phase sensitive detection. The small EO signal is due to the short interaction length of a few microns achievable when propagating through the plane of a poled polymer film. With the currently achievable electro-optic coefficients in poled polymer films, the performance of a useful device function, such as complete switching or modulation, requires the propagation of light within the plane of the film as a guided wave mode. The guided mode can be a slab mode confined only in one dimension, or a channel mode confined in both transverse dimensions. After a material has been evaluated with one of the through-plane techniques and found to be potentially useful for device applications, it must then be tested in a waveguide device configuration. Waveguide electro-optic measurement techniques are described in the following sections.

Phase Modulator in an External Interferometer. The slab waveguide phase modulator is perhaps the simplest device geometry that allows electro-optic measurements of a poled film to be made with light propagating in the plane of the film. In this geometry, no effort is made to confine the propagating mode in the lateral direction so a simple three layer structure, composed of the poled film sandwiched between two buffer layers, is all that is required to support the guided mode. A cross-section of a slab waveguide phase modulator utilized by Thackara et al. is shown in Fig. 17.[43] Any waveguide geometry, however, is more complicated than the simple through-plane geometries described in previous sections. The poled active layer is typically 1 to 2 microns thick. To prevent excessive attenuation of

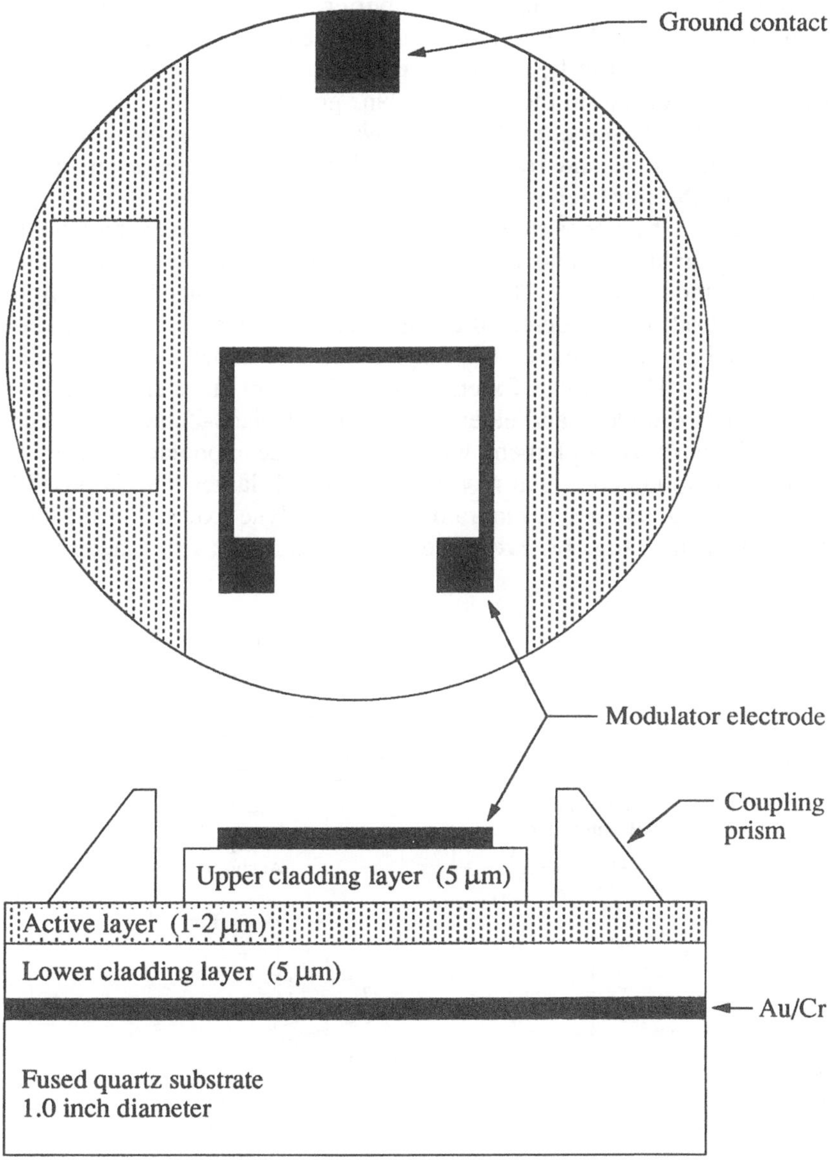

Figure 17 Slab waveguide phase modulator geometry. (From J. I. Thackara, M. Jurich, and J. D. Swalen, Electro-optic measurements of tolane-based polymeric phase modulators, *J. Opt. Soc. Am. B*, *11*, 835 (1994), Fig. 3.)

the mode due to contact with the device electrodes, the buffer layers must have refractive indices lower than the active layer and must be thick enough to confine the mode within the waveguide layers. The core layer is poled between the ground electrode and a temporary electrode deposited onto either the core or the top buffer layer. In either case, to efficiently pole the core layer, the conductivity of the buffer material must be close to, or preferably greater than, the conductivity of the core material. This assures that a significant fraction of the applied poling field drops across the core layer. After poling, the temporary electrode is removed, and the modulator is completed by depositing the top buffer layer (if it is not present already) and the drive electrode. For electro-optic measurement purposes, this means that both the drive electric field and the poling electric field are not known as precisely as in the case of the simple through-plane geometries, leading to greater measurement uncertainty. On the other hand, the waveguide geometry is the most useful for device applications, and from a practical standpoint, measurements of r_{33} in a waveguide are the most relevant.

As shown in Fig. 17, by depositing the top buffer layer over only part of the active layer, coupling light into and out of the modulator can be accomplished easily by using high index prisms. The phase modulation produced by a signal applied to the top electrode can be detected using a number of techniques, including either of the approaches described in the subsections "External Mach-Zehnder Interferometer"[27,39] and "Heterodyne External Mach-Zehnder Interferometer."[40,41] Both the external Mach-Zehnder[44] and heterodyne external Mach-Zehnder[43,45] techniques are especially easy to implement with a waveguide modulator, since the amplitude of the phase modulation can now be made much larger than is possible in the through-the-plane geometry. A diagram of a heterodyne external arm Mach-Zehnder testbed is shown in Fig. 18. From measurements made with a heterodyne

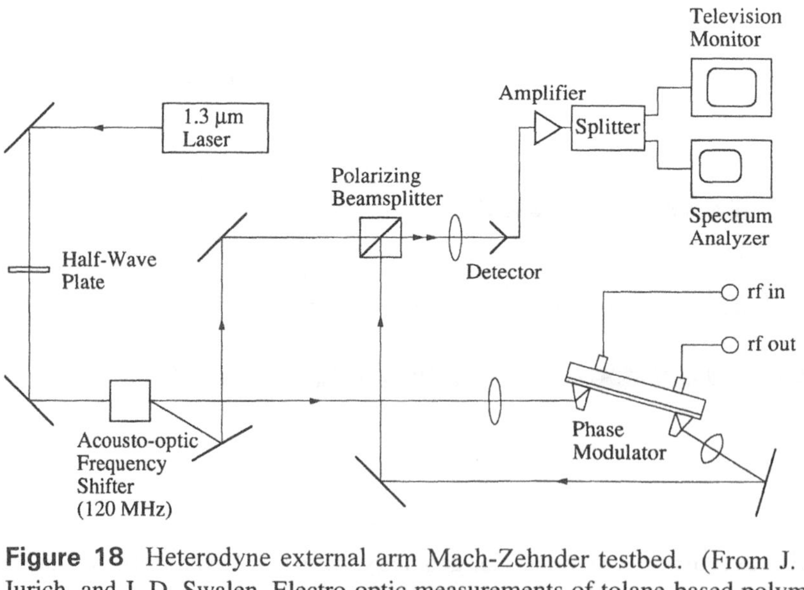

Figure 18 Heterodyne external arm Mach-Zehnder testbed. (From J. I. Thackara, M. Jurich, and J. D. Swalen, Electro-optic measurements of tolane-based polymeric phase modulators, *J. Opt. Soc. Am. B*, *11*, 835 (1994), Fig. 4.)

system, the EO coefficient of the active layer can be calculated by using the following equation[43]:

$$r_{33} = \frac{\sqrt{2}\lambda d_c 10^{-\Delta/20}}{\pi L V_s (n_c^{TM})^3}\left(1 + \frac{2\varepsilon_c d_b}{\varepsilon_b d_c}\right)\Theta_m, \tag{24}$$

where λ is the carrier wavelength, d_c and ε_c are the thickness and dielectric constant of the core layer in the poled region, d_b and ε_b are the thickness and dielectric constant of each buffer (assumed to be the same) in the poled region, Δ is the difference between the electrical powers in the central beat and the side peaks measured in decibels, L is the length of the active region, V_s is the rms applied voltage and n_c^{TM} is the TM refractive index of the poled core layer. Θ_m is a correction factor accounting for the difference between the change in the effective TM mode index and the change in the actual core TM index induced by the electro-optic effect. The capacitive voltage division of the drive signal between the core and buffer layers is accounted for by the factor[43]:

$$\left(1 + \frac{2\varepsilon_c d_b}{\varepsilon_b d_c}\right). \tag{25}$$

Just as in the case of the trough-plane measurements, the external arm Mach-Zehnder experiment provides independent measurement of r_{33} and r_{13}. It is assumed in Eq. 24 that optical coupling has been made to the zeroth order TM slab waveguide mode. The r_{13} coefficient can be determined by coupling to the zeroth order TE mode and then replacing n_c^{TM} with the corresponding n_c^{TE} in Eq. 24 and by using the appropriate Θ_m correction factor.

The accuracy of this method obviously depends not only on the measurements of Δ and V_s and the calculation of Θ_m, but also on how accurately the device parameters d_c, ε_c, d_b, ε_b, L, and n_c^{TM} or n_c^{TE} are known. Thackara et al.[43] estimate the error associated with the calculation of the EO coefficients to be approximately ± 10 percent.

Integrated Mach-Zehnder Modulator. One of the principal device applications for electro-optic polymer materials is an integrated Mach-Zehnder modulator. Unlike the external arm test structure described in the previous section, both the reference arm and the signal arm are waveguides integrated onto the same substrate. Integrated Mach-Zehnder modulators based on electro-optic polymer materials have been demonstrated with multi-GHz bandwidths.[5,6] These integrated optic Mach-Zehnder modulators can, of course, also be used to measure the electro-optic coefficient of the core waveguide material. In fact, this type of measurement provides the most useful information for device design, but at the cost of greatly increased sample fabrication complexity. For the through-plane measurements, a thin layer of EO polymer was coated on an ITO glass substrate and a thin layer of gold was evaporated on top, with only gross patterning. In contrast, to build an integrated Mach-Zehnder modulator, channel waveguides must be defined in a three layer slab structure, and electrodes must be patterned and aligned with the waveguides. This process requires multiple coating, metallization and lithographic steps, and can only be carried out successfully with reasonable yield in a clean room. Furthermore, materials interaction issues make fabrication much more difficult, and in the end it is often difficult to separate out a measured parameter related only to a single material. Integrated Mach-Zehnder modulators are certainly not suited for screening new materials, but provide the most useful information on successful candidate materials. After a material has been successfully tested in an integrated Mach-Zehnder, the greater expenditure of effort to design and build a high frequency device can be justified.

A diagram of an integrated Y-branch Mach-Zehnder modulator is shown in Fig. 19 from Lipscomb et al.[46] The Mach-Zehnder modulator consisted of a straight channel input region followed by a Y-branch to divide the light into two separate channels. The light then travels in two straight parallel segments for some distance, and is then combined back into one output channel by a second Y-branch splitter/combiner. For the Mach-Zehnder pictured in Fig. 19, the channel waveguides were formed by photobleaching, the Y-branches had a 1 degree splitting angle, and the straight segments were 17 mm long.[46] Waveguides can also be formed by reactive ion etching, micromachining, poling, or a variety of other techniques. As shown in Fig. 19, electrodes are positioned to apply an electric field to one or both of the arms of the Mach-Zehnder. Due to the long interaction length that can be achieved in an integrated waveguide modulator, the device can be modulated through its entire range (i.e., the intensity turned off from fully on) with only a few volts, and V_π can be determined directly. V_π is the voltage required to turn the modulator completely on from the off state. Shown in Fig. 20 is the output of a polymer based Mach-Zehnder modulator. The triangular trace in the center is the applied voltage, and the curved trace showing the classic $\cos^2$ behavior is the output of the Mach-Zehnder. The cusps in the Mach-Zehnder output correspond to the turning points in the applied voltage. Cursors placed on the maximum and minimum of the modulator output give a direct reading of V_π. The electro-optic coefficient can then be calculated from V_π[47]:

$$r = \frac{\lambda d}{n^3 V_\pi \Gamma L}. \tag{26}$$

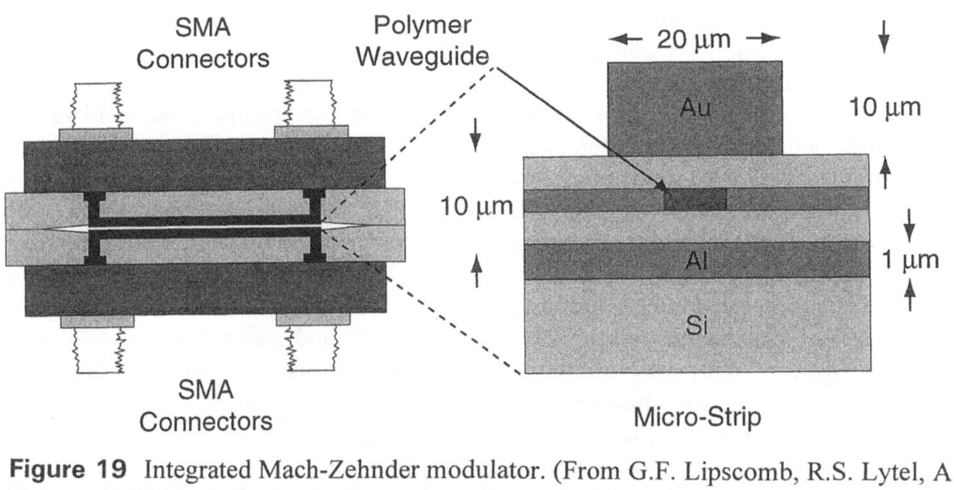

Figure 19 Integrated Mach-Zehnder modulator. (From G.F. Lipscomb, R.S. Lytel, A. J. Ticknor, T. E. Van Eck, D. G. Girton, S. P. Ermer, J. F. Valley, J. Kenney, and E. Binkley, Organic Electro-Optic Devices for Optical Interconnections, *Proc. SPIE*, *1560*, 388 (1991), Fig. 8.)

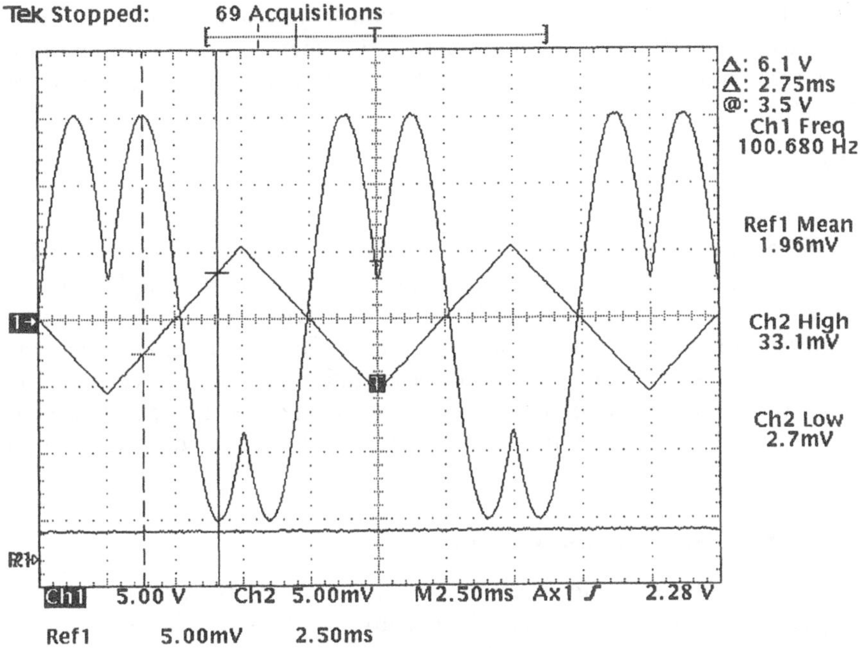

Figure 20 Measurement of V_π for an integrated Mach-Zehnder modulator.

Here it is assumed that only one arm of the Mach-Zehnder is driven with a voltage. r_{33} is measured for TM polarization and r_{13} is measured for TE polarization. L is the interaction length of the electrode over the arm of the Mach-Zehnder, d is the gap between the lower electrode and the upper drive electrode, and Γ is the overlap between the electric field and the optical field. For most test structures with large drive electrodes, Γ is very close to 1. For impedance-matched high-speed electrodes, where fringing fields are important, Γ can be somewhat smaller. For a stack height of $d = 8\ \mu\text{m}$ and a 50 Ω microstrip configuration, $\Gamma = 0.95$.[47] By contrast, lithium niobate with a coplanar electrode typically has an overlap integral of $\Gamma = 0.45$.[47]

Electro-optic coefficients can also be measured in other types of integrated optic devices. Among those that have been demonstrated in poled polymer materials are phase modulators,[48] directional couplers,[48] rail-taps,[49] fiber gyroscope modulators,[50] and Mach-Zehnder switches.[51]

TE/TM Interference Modulator. Both the external arm and integrated Mach-Zehnder measurements described in the previous two sections provide independent measurements of r_{33} and r_{13} for the TM and TE polarizations respectively. A combination of the two coefficients can be measured using a single straight planar slab or channel waveguide.[52] This experimental set-up is very similar to the through-plane measurement of the subsection "Thin Film Reflection Techniques," above, with the incidence angle set to 90 degrees, or to the crystal measurement described in Section II.B.1. The incident light is polarized at an angle of 45 degrees to the poling direction, so that half of the incident light is in the TM polarization and half is in the TE polarization. Due to the long path length achievable in a waveguide, leading to a favorable geometrical factor d/L, where d is the sample thickness and L is the length of the waveguide, a full V_π modulation can be achieved with a reasonable voltage. V_π is then given by:

$$V_\pi = \frac{\lambda d}{L}(n_3^3 r_{33} - n_1^3 r_{13})^{-1}. \tag{27}$$

If we make the assumption that $r_{33} = 3\, r_{13}$, then the electro-optic coefficient can be measured with this modulator structure, just as in the case of the simple through-plane measurement described in "Thin Film Reflection Techniques," above. Unlike the crystal measurements described below in Section II.B.1, the tensor elements cannot be separated easily by using different propagation and field directions. The poling geometry fixes the accessible measurement geometries. Independent measurements of r_{33} and r_{13} require non-common path geometries, as described in the previous two sections.

B. Pockel's Effect in Organic Crystals

The measurement of electro-optic effects in organic crystals is similar to that of inorganic crystals and has been described in many standard text books.[1,10] Measurements on crystals are both simpler and more complex than measurements on poled polymer films. They are simpler in that, once macroscopic crystals are grown (no easy feat in itself), the crystals can be cut so that light can propagate in any desired direction and an electric field can also be applied in any arbitrary direction. Furthermore, long interaction lengths are possible in every direction, leading to large signals and low noise. Crystals are not limited by the thin film geometry of poled polymer films. By using different crystal cuts, different propagation directions and different applied electric fields, each electro-optic tensor element can be measured independently.

Electro-optic measurements in organic crystals are more difficult, however, in that materials can crystallize in any point group with any number of independent and significant tensor elements. Most organic crystals, in fact, crystallize in centrosymmetric point groups where the Pockel's effect is identically zero. Of the organic crystals that exhibit non-centrosymmetric point groups, however, most are monoclinic, and the tensorial character of the electro-optic response can be quite complex. The first task in crystal measurements is to determine the point group of the crystal, look up which electro-optic tensor components are non-zero and determine the change in index associated with a given applied field. The electro-optic coefficients can be determined from the change in index as a function of applied field. Two principle geometries are generally employed. The first measures the phase retardation between the two polarizations traveling along a common path through the crystal, in a manner similar to the TE/TM waveguide modulator described in "Integrated Mach-Zehnder Modulator," above. The second approach utilizes an external arm Mach-Zehnder modulator, and is very similar to the external arm Mach-Zehnder used with poled polymers, as described in "External Mach-Zehnder Interferometer" and "Phase Modulator in an External Interferometer," above. These two approaches are described in the following two sections.

1. Phase Retardation Measurements in Crystals

The phase retardation measurement technique uses a basic electro-optic crystal modulation technique that has been described in standard reference works.[1,10] A diagram of the experimental set-up is shown in Fig. 21.[53] Unlike poled polymer films, the geometry of each organic crystal is different, and therefore the orientation of the crystallographic axes, the propagation direction and the direction of the applied field will all be different. The most advantageous geometry must be worked out for each different crystal. As an example, we will consider the organic molecular crystal 2-methyl-4-nitroaniline (MNA), as was described briefly in the Introduction.

MNA crystallizes in the point group *m*, which possesses one symmetry element, a mirror plane, which is conventionally taken to be perpendicular to the *Y* axis. MNA crystals grown from solution exhibit a monoclinic platelet habit. A diagram of the crystal morphology is shown in Fig. 22.[53] Also shown are the relationship of the crystallographic axes (*a*, *b*, *c*) and the optical axes (*X*, *Y*, *Z*) to the crystal morphology. The *b* and *Y* axes are fixed by crystal symmetry and are perpendicular to the crystal platelet. In the point group *m*, however, the optical axes *X* and *Z* are not determined by symmetry, and must be determined experimentally. The optical axes are the principal axes of the index ellipsoid, and can be determined by observing the crystal between crossed polarizers. The directions for which extinction is observed are the optical axes. In MNA, the *X* axis makes an 8 ± 1 degree angle with the cleavage plane parallel to the long platelet edge and cleavage plane. The *Z* axis intersects the crystal edge an angle of 82 ± 1 degrees.[53]

The effect of an applied voltage on the propagation of light through a crystal due to the electro-optic effect can be determined by examining changes in the index ellipsoid. It is convenient to use the optical axes coordinate system where the index tensor is diagonal. In this case, the index ellipsoid is given by[1]:

$$\frac{X^2}{n_{11}^2} + \frac{Y^2}{n_{22}^2} + \frac{Z^2}{n_{33}^2} = 1. \tag{28}$$

A new index ellipsoid is then found for the case where the electric field is applied. The equation for the change in the index from the electro-optic effect due to an applied electric field is given in Eq. 4. For the case where the electric field is applied only along the *X* axis ($E_2 = E_3 = 0$), the index ellipsoid becomes:

$$\left(\frac{1}{n_1^2} + r_{11}E_1\right) X^2 + \left(\frac{1}{n_2^2} + r_{21}E_1\right) Y^2 + \left(\frac{1}{n_1^2} + r_{31}E_1\right) Z^2 + 2r_{51}E_1XZ = 1. \tag{29}$$

The elliptical cross-section of the index ellipsoid through the origin and perpendicular to the wave vector determines the ordinary and extra-ordinary polarizations for any direction of propagation. Light linearly polarized along one of the semiprincipal axes of the ellipse will remain linearly polarized as it propagates. Light polarized in other directions will become elliptically polarized with propagation. The magnitude of a semiprincipal axis is equal to the index of refraction for that polarization.[54] For the case of MNA, if we choose the *Y* axis as the propagation direction, the Y^2 term drops out of Eq. 29. r_{11} and r_{31} contribute to the change in

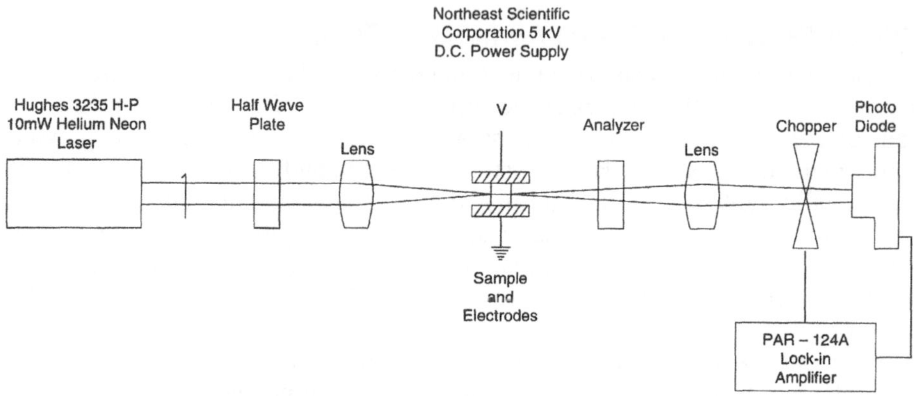

Figure 21 Schematic diagram of phase retardation electro-optic measurement. (From G. F. Lipscomb, Experimental Studies for Electro-Optic and Ferroelectric Phenomena in Organic Solids, University Microfilms, Ann Arbor, 1980, Fig. 3.7.)

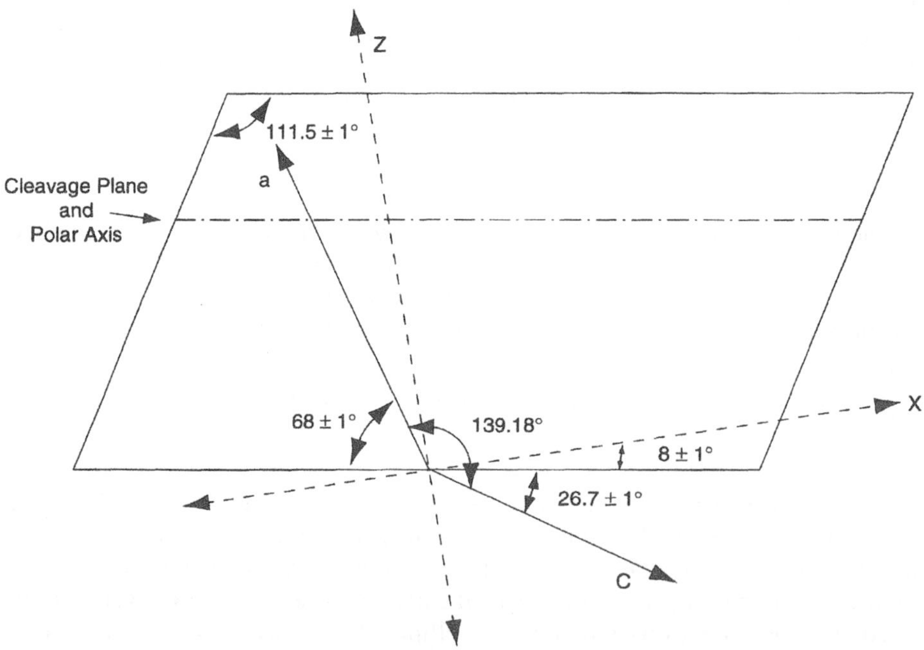

Figure 22 Diagram of the crystal habit of MNA. (From Experimental Studies for Electro-Optic and Ferroelectric Phenomena in Organic Solids, University Microfilms, Ann Arbor, 1980, Fig. 3.2.)

the magnitude of the index to first order, but do not change the direction of the optical axes. r_{51} causes a very small rotation in the direction of the principal axes and only contributes to the change in the magnitude of the indices to second order in the applied electric field.[53] To first order, then, the equation of the intersection ellipse becomes:

$$\left(\frac{1}{n_1^2} + r_{11}E_1\right) X^2 + \left(\frac{1}{n_3^2} + r_{31}E_1\right) Z^2 = \left(\frac{1}{(n_1 + \Delta n_1)^2}\right) X^2 + \left(\frac{1}{(n_3 + \Delta n_3)^2}\right) Z^2 = 1. \quad (30)$$

To first order in a Taylor series expansion, Δn_1 and Δn_3 become:

$$\Delta n_1 = -\frac{1}{2} n_1^3 r_{11} E_1,$$

and (31)

$$\Delta n_3 = -\frac{1}{2} n_3^3 r_{31} E_1.$$

In a similar manner, the change in indices for different directions of the applied field and different propagation directions can be determined. For the case of MNA, when the crystal is cut so that the field is applied along X and propagation is along Y, polarized light launched at 45 degrees to the principal axes will undergo a relative phase change of Γ:

$$\Gamma = \frac{\omega}{c}(n_1 - n_3)L + \frac{1}{2}\frac{\omega}{c}(n_1^3 r_{11} - n_3^3 r_{31})E_1 L, \quad (32)$$

where ω is the radial frequency of the light, c is the speed of light, and L is the length of the crystal. As in the case of the thin film measurements described in "Thin Film Reflection Techniques," above, for the experimental configuration shown in Fig. 21, the un-modulated phase difference is adjusted with the half wave plate to bias the signal to the linear region at half the maximum intensity. The small signal modulated at the fundamental frequency is then measured with a lock-in amplifier, and $|n_1^3 r_{11} - n_3^3 r_{31}|$ is determined from the slope of the signal with applied field strength. Similarly, when the field is applied along the Z direction and the propagation direction is still along Y, $|n_1^3 r_{13} - n_3^3 r_{33}|$ can be determined. From these measurements, the dominant electro-optic coefficient of MNA was determined to be $r_{11} = 67 \pm 25 \times 10^{-12}$ m/V at $\lambda = 632.8$ nm.[53]

Ho et al.[12] performed a similar measurement on MNA crystals utilizing a different geometry. In this case, the electric field was still applied along X, but the propagation direction was along Z to increase the path length. Using this geometry, Ho et al. measured $r_{11} = 60 \pm 10 \times 10^{-12}$ m/V at $\lambda = 632.8$ nm.[12]

2. *Interferometric Measurements in Crystals*

Electro-optic measurements can also be performed in crystals using the standard external arm Mach-Zehnder interferometer configuration described in "External Mach-Zehnder Interferometer" and "Phase Modulator in an External Interferometer," above, for poled polymer films. As in those cases, the interferometric approach has the advantage that through the selection of crystal cuts, applied field direction, polarization direction, and propagation direction, each tensor component can be measured separately. Sigelle et al. employ a variation on this technique, which is shown in Fig. 23.[55] The layout is identical to the poled polymer film measurement shown in Fig. 13, except that a second electro-optic sample containing a known crystal is placed in the reference arm. Instead of calculating the electro-optic coefficient from a modulated signal, Sigelle et al. adjust the voltage applied to the reference crystal until it exactly cancels the signal from the sample to be measured. After correcting for geometrical factors, the ratio of the voltages applied to the two crystals is equal to the ratio of the electro-optic coefficients, and the unknown sample can be calculated with respect to the known reference.

Sigelle et al. used this technique to measure three different electro-optic coefficients for the organic crystal 3-methyl 4-nitropyridine 1-oxide (POM).[55] The technique has been widely used to study both organic and inorganic electro-optic crystals.[56] As with other external arm interferometric techniques, vibration and noise in the reference arm must be controlled, increasing the experimental complexity and expense.

C. Quadratic Measurements

Away from resonances where the materials are non-absorbing, quadratic electro-optic effects in organic materials are measured with similar apparatus and sample geometries to those used for poled polymer films described in Section II.A and for crystals described in Section II.B. Unlike electro-optic measurements, however, care must be taken in quadratic measurements to separate out the different contributions to the signal, each of which has a different frequency dependence. In the case of linear electro-optic measurements in poled polymers, symmetry considerations usually eliminate slow response mechanisms. The dominant contribution is nearly always from the electronic molecular response, and coefficients determined at low frequency can give a reliable indication of final device performance at high frequencies (Chapter 3 describes piezo and reorientational contributions to r.) For quadratic electro-optic effects, where non-centrosymmetric symmetries are not required, many different mechanisms can contribute. Low frequency interferometric measurements performed in the audio frequency range will contain many contributions that will not be present in ultra-high frequency all optical switching devices. Nonetheless, due to the small signals and relatively large drive voltages required, most of the quadratic electro-optic measurements are made at low frequency. The results are then compared with faster all optical measurements, such as short pulse degenerate four-wave mixing, optical frequency tripling, or pump-probe depolarization, to separate out the contributing mechanisms.[23] Non-resonant, non-absorbing quadratic electro-optic measurement techniques are described in Section II.C.1.

Near resonance, valuable information can be learned from measurements of electro-absorption which is described by the imaginary part of the quadratic electro-optic coefficient.[24,25] Again, different mechanisms can contribute to the quadratic electro-absorption effect on different time scales. Poga et al. have used measurements of quadratic electro-absorption in dye-doped polymers as a function of wavelength and temperature to separate molecular reorientational and electronic contributions to the third order nonlinear susceptibility.[57] Techniques for measuring

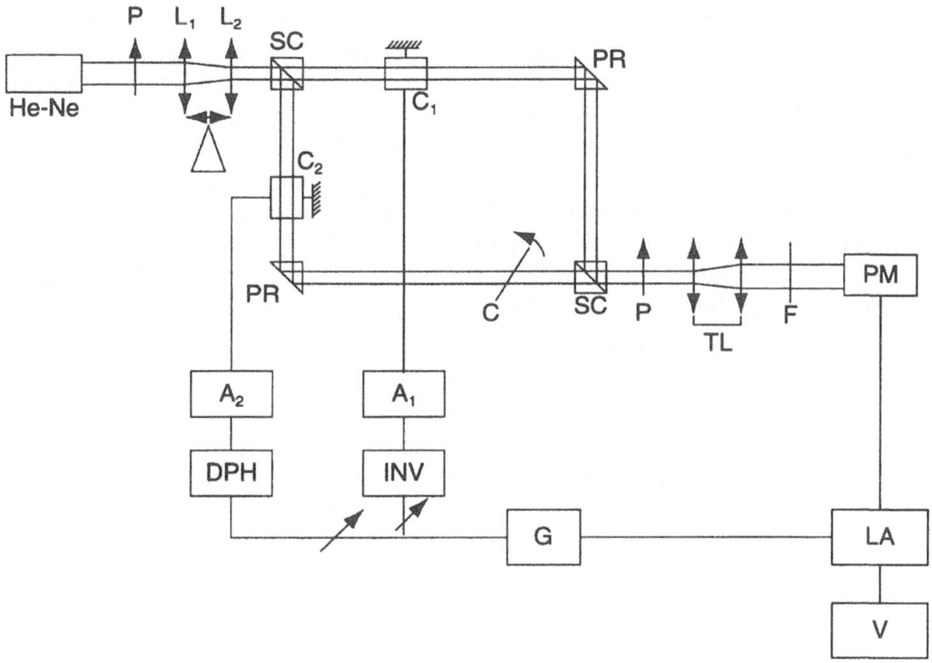

Figure 23 Schematic diagram of a Mach-Zehnder interferometer measurement of the electro-optic coefficients in organic crystals. C_1 and C_2 are the reference and sample crystals. (From M. Sigelle, and R. Hierle, Determination of the electrooptic coefficients of 3-methyl 4-nitropyridine 1-oxide by an interferometric phase modulation technique, *J. Appl. Phys.*, *52*, 4199 (1981), Fig. 1.)

the imaginary part of the quadratic electro-optic tensor are discussed in Section II.C.2.

1. Quadratic Electro-optic Measurements

Measurements of the real part of the quadratic electro-optic (QEO) coefficient, $s_{ij,kl}$, are very similar to measurements of the linear electro-optic effect. The only difference is that the change in index is proportional to the square of the electric field which for the isotropic case gives[57]:

$$\Delta n = \frac{n^3}{2} sE^2. \tag{33}$$

Any of the techniques used to measure Δn for linear electro-optic samples can also be applied to measure the Δn caused by the quadratic effect. Since the induced change in index is proportional to the square of the applied field, the signal will be at twice the fundamental modulating frequency, Ω. Kuzyk et al.[23] used the external arm Mach-Zehnder interferometer shown in Fig. 13 to measure the quadratic electro-optic tensor element $S_{11,33}$ in a dye-doped polymer. Unlike Norwood et al.,[27] who measured the linear electro-optic coefficient with this set-up, they measured the frequency doubled component of the signal by setting the lock-in amplifier to twice the modulation frequency. The sample geometry is identical to that for the linear electro-optic measurements; a thin film of polymer between transparent and/or reflective electrodes. As in the case of linear electro-optic measurements, a voltage at frequency Ω is applied to the sample, which for QEO leads to a small phase modulation in the signal arm at frequency 2Ω. At the same time, the phase in the reference arm, ϕ, is swept through a full fringe by rotating a glass plate in the reference beam. A plot of the reference signal, i.e., the dc light intensity of the interferometer, and the modulated signal at 2Ω from the quadratic electro-optic effect in the sample, is shown in Fig. 24 from Kuzyk et al.[23] For a centrosymmetric system, where r_{ij} is identically zero, Kuzyk et al. show that the quadratic electro-optic tensor element is given by[23]:

$$s_{11,33} = \frac{\lambda d}{2\pi n^3 V_{rms}^2} \frac{(I_{max}^{2\Omega} + I_{min}^{2\Omega})}{I_{sig}/\sqrt{8}}, \tag{34}$$

where I_{sig} is the magnitude of the unmodulated signal from the interferometer, $I_{max}^{2\Omega}$ and $I_{min}^{2\Omega}$ are the maxima and minima of the signal modulated at 2Ω, d is the thickness of the sample, and V_{rms} is the applied voltage. It can be seen from the

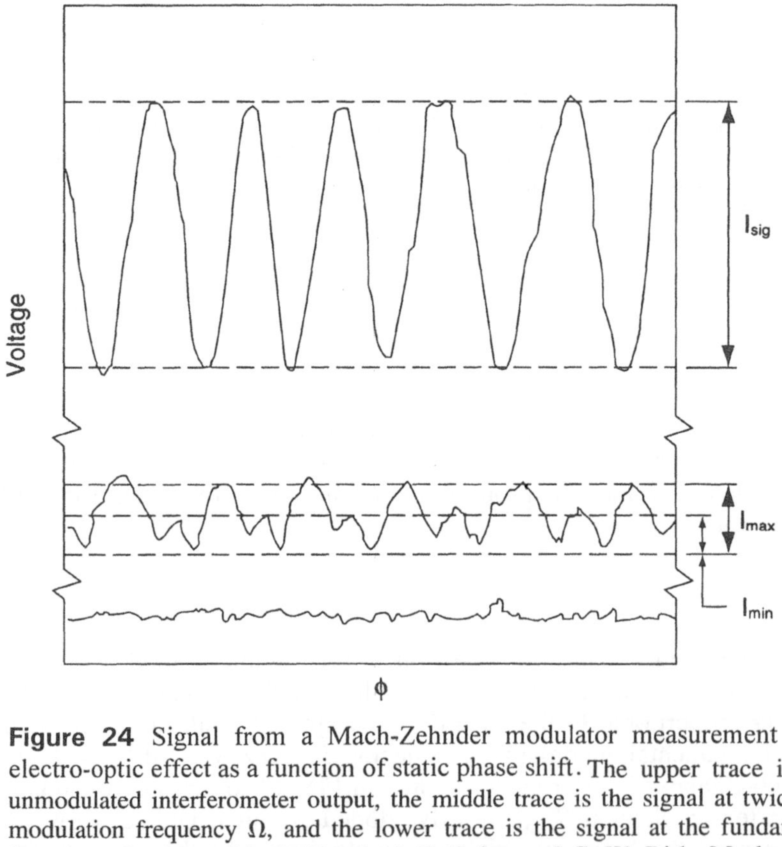

Figure 24 Signal from a Mach-Zehnder modulator measurement of the quadratic electro-optic effect as a function of static phase shift. The upper trace is the fundamental unmodulated interferometer output, the middle trace is the signal at twice the fundamental modulation frequency Ω, and the lower trace is the signal at the fundamental modulation frequency Ω. (From M. G. Kuzyk, J. E. Sohn, and C. W. Dirk, Mechanisms of quadratic electro-optic modulation of dye doped polymer films, *J. Opt. Soc. Am. B*, *7*, 842 (1990), Fig. 1.)

bottom trace in Fig. 24 that the signal at the fundamental modulation frequency in a centrosymmetric sample is negligible, as expected from symmetry considerations. Of course quadratic effects also occur in noncentrosymmetric samples with strong linear electro-optic effects. In this case, the measurement of the quadratic coefficient s must be corrected to account for terms at 2Ω that arise from the linear effect, r.[23]

2. Electro-absorption Measurements

Quadratic electroabsorption measurements can also be performed with similar experimental geometries to standard linear electro-optic measurements. Shown in Fig. 25 is a diagram of the quadratic electro-absorption experiment developed by Poga et al.[57] The sample geometry is identical to that used for the simple through-the-

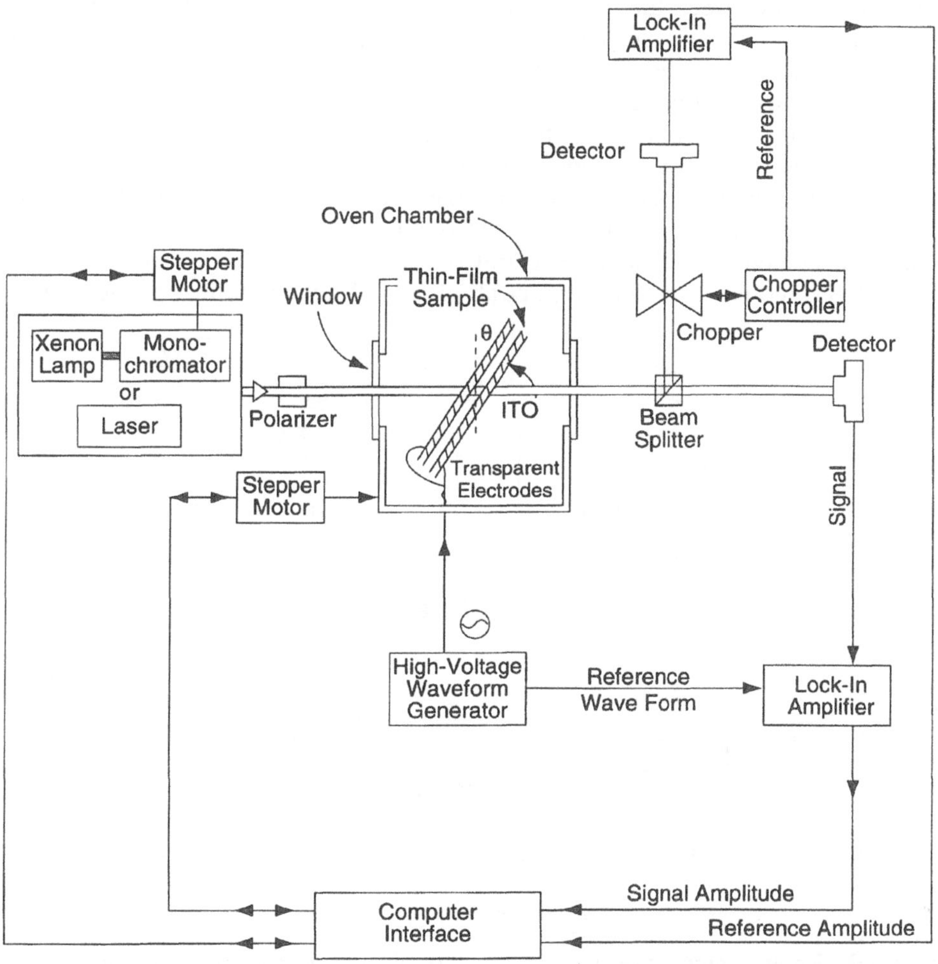

Figure 25 Quadratic electro-absorption experiment. (From C. Poga, M.G. Kuzyk, and C. W. Dirk, Quadratic electroabsorption studies of third-order susceptibility mechanisms in dye doped polymers, *J. Opt. Soc. Am. B*, *11*, 80 (1994), Fig. 2.)

plane linear electro-optic measurement described in "Thin Film Reflection Techniques," above, for the case of transmission. A thin film of polymer is placed between two glass substrates with transparent ITO electrodes. The sample is set at an angle θ to the incoming beam. Since the electric field is always applied along the 3 direction perpendicular to the film, for s polarized light, the electric field is polarized in the plane of the film and only the $s_{11,33}$ component of the quadratic electro-optic coefficient is accessed. Poga et al. show that the imaginary part of the index change is then given by[57]:

$$(n_o^3 s_{11,33})_I = \frac{\lambda d}{4\pi V_{RMS}^2} \cos(\theta') \frac{I^{2\Omega}}{I_{sig}}, \tag{35}$$

where d is the thickness of the sample, θ' is the internal angle of the beam in the sample, $I^{2\Omega}$ is the signal at twice the fundamental modulating frequency Ω, I_{sig} is the unmodulated signal, and V_{rms} is the modulating voltage. As in the case of the linear electro-optic effect, the p polarization will contain terms from both the $s_{11,33}$ and $s_{33,33}$ tensor elements, and the ratio between the two coefficients can be determined by measuring the signal as a function of incidence angle θ.[57]

As also shown in Fig. 25, Poga et al. replaced the laser beam with a white light source and a monochrometer to measure the wavelength dependence of the electro-absorption coefficient. Shown in Fig. 26 is a measurement from Poga et al.[25] of the wavelength dependence of the quadratic electro-absorption of a PMMA film doped with a squaraine dye (ISQ) through resonance. They used this measurement to demonstrate that the spectral dependence of $(s_{11,33})_I$ was closely explained by a three level model. Poga et al. also used temperature dependent quad-

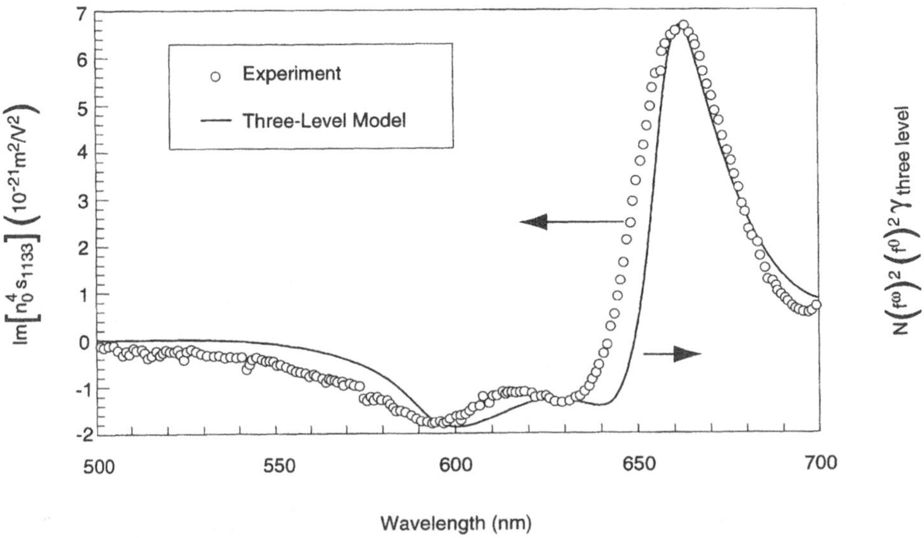

Figure 26 Room temperature quadratic electro-absorption spectrum. (From C. Poga, T. M. Brown, M. G. Kuzyk, and C. W. Dirk, Characterization of the excited states of a squaraine molecule with quadratic electroabsorption spectroscopy, *J. Opt. Soc. Am. B*, *12*, 531 (1995), Fig. 5.)

ratic electro-absorption spectroscopy to show that the third order nonlinear susceptibility in the PMMA-ISQ system was dominated by purely electronic contributions below T_g and by molecular reorientational mechanisms above T_g.[57]

Measurements of the real and imaginary parts of the quadratic electro-optic tensor have proven to be very useful in investigations of the origins of the third order nonlinear optical response in organic materials.[58–63] They are particularly important in elucidating the mechanisms that contribute to the all-optical switching response of organic materials. Quadratic electro-optic effects in and of themselves, however, do not have the immediate technological interest that centers around linear electro-optic effects in organic and polymeric materials and are therefore not tabulated.

III. TABLES OF ELECTRO-OPTIC MEASUREMENTS IN ORGANIC MATERIALS

Over the last two decades extensive work has been done on the development and measurement of organic electro-optic materials. In this section we present summary tables of measurements of electro-optic effects in organic and polymeric materials.

A. Through-plane Measurements on Poled Polymer Films

A summary of measurements of the linear electro-optic coefficients of poled polymer films using through-plane techniques is presented in Table 1. These measurements were made using, in some form or other, the simple sample geometry pictured in Fig. 6. A code name is given for each material. For the cases where the chemical structure has been published, the code name corresponding to a given structure is given in the appendix. For each material, the material type specifies whether the material is a guest-host, side-chain polymer or main-chain polymer. Where possible, the active nonlinear molecular element and the polymer main chain are listed. Then the poling method, concentration of the nonlinear chromophore, molecular $\mu\beta$, poling temperature, and poling field are listed. Finally, the measured electro-optic coefficient r_{33} and the measurement wavelength are given, as well as r_{13} where it is available. A code then relates the measurement to the literature reference where it was published. A similar list of parameters is given in Table 2 for electro-optic polymers measured using attenuated total reflection (ATR) techniques. In this case, the sample geometry is different from that of Table 1 and requires a coupling prism, as shown in Fig. 11.

B. Waveguide Measurements in Poled Polymer Films

A summary of measurements of linear electro-optic coefficients in planar slab waveguides is presented in Table 3. In these measurements, the light is confined in the vertical dimension, but not in the horizontal dimension. As before, the material code name, material type, active nonlinear molecular unit, polymer backbone, and poling conditions are listed. For the few cases where such information is available, the propagation loss in the waveguide is also listed. Table 4 presents a summary of similar information for measurements of the linear electro-optic coefficient in channel waveguides. In this case, the light is confined in both the vertical and

(text continues on page 287)

Table 1 Through-plane Linear Electro-optic Measurements on Poled Polymer Films

Code Name	Material Type	Active Molecule	Polymer	Poling Method	Concentration	$\mu\beta$ 10^{-48} esu	Poling Temp. °C	E_p (V/μm)	λ (μm)	r_{33} (pm/V)	r_{13} (pm/V)	Ref. Code
3RDCVXY	Side Chain	4-(4-(dicyanovinyl)phenylazo)-3,5,-xylidine	3RDCVXY-Diazo dye-methacrylate backbone	Electrode	15 Mol%		140		1.31	30		Amano94
Anlheim-5	Guest Host	3-(dicyanomethylidenyl)-2,3-dihydrob enzothiophen-2-ylidenyl-1,1-dioxide polyene	Polycarbonate	Electrode	20 wt%	13500	80	150	1.3	55		Anlheim95
NQP2	Side Chain	naphthoquinone methide	Polyurethane side chain		23% wt.	6100		.5	1.3	11		Aramaki94
P(MAP)-DR1	Side Chain	DR1	poly(maleic anhydride-co-propylene)-DR 1 side chain	Electrode			185	135	.780	6		Bauer93
PMMA/DR1	Guest Host	DR1	PMMA/DR1 guest-host	Electrode	12% wt.		90	135	.780	2		Bauer93
MMA-MA1	Side Chain	MA1	MMA-MA1 side chain		16.5% mol.			50	1.3	2		Beeson92
MMA-DR1	Side Chain	DR-1	MMA-DR1 side chain	Electrode					.8	10		Chollet92
Durham-1	Main Chain	nitroaniline	Diethylene diphenylsilicon	Corona			85	12	.633	.38		Cross92
PC/DAN	Main Chain	2-(N,N-dimethylamino)-5-nitroacetanilide (DAN)	Bis-phenol-A polycarbonate DAN guest host		6.0 E20		82	31	.633	.96		Cross92
PC/NMBA	Main Chain	(4-methyl-N-(4-nitro-phenyl) methylene benzamine) (NMBA)	Bis-phenol-A polycarbonate NMBA guest host		5.3E20		82	27	.633	.16		Cross92
PMMA/DR1	Guest Host	DR-1	PMMA/DR1	Electrode	1.6E20		127	80	.633	1.65	.52	Eldering90
MMA-DR1	Side Chain	DR-1	MMA-DR1		35%				.6	30		Florin94
PMMA/JPL6	Guest Host	#6 3-phenyl-5-isoxazolone acceptor-Thiophene	PMMA		2 Mol%	1060		100	1.3	2.7		Gilmour94

PMMA/JPL7	Guest Host	#7 3-phenyl-5-isoxazolone acceptor-Thiophene	PMMA		2 Mol%	1550		100	1.3	3.3		Gilmour94
PMMA/JPL8	Guest Host	#8 3-phenyl-5-isoxazolone acceptor-Thiophene	PMMA		2 Mol%	1250		100	1.3	4.9		Gilmour94
PMMA/JPL9	Guest Host	#9 3-phenyl-5-isoxazolone acceptor-Thiophene	PMMA		2 Mol%	1970		100	1.3	5.3		Gilmour94
MO3ONS	Side Chain	OxyNitroStilbene	Methacrylate-ONS Side Chain				114		.633	11		Goldberg90
ENIChem-1	Side Chain	N,N-diethylamino-4-tricyanovinylbenzene	Polyimide sidechain	Electrode	27% wt.	650	200	50	.83	15		Jen94
NNDN-NAN	Cross-linked	NNDN+NAN	NNDN-NAN crosslinked	Corona	63% wt.		120		.531		6.5	Jungbauer90
PIQ2200/DASS-62	Guest Host	DAS-62	PIQ2200		50% wt.	113	160	130	.633	21		Kim95
PαMS/DHANS	Guest Host	N,N-dihexyl-4-amino-4′-nitrostilbene (DHANS)	PαMS/DHANS poly (α methyl styrene)guest host		14		160	.5	.633	.5		Knabe89
PQI2200/DASS-6	Guest Host	DASS-6	Hitachi PIQ2200 polyimide		40% wt.		170	135	.6328	13		Lee94
PMMA-CNET-S	Guest Host	CNET-S	PMMA/CNET-S	Corona	15% wt.		100		.830	1.39		Levenson91
PMMA/CNET-S	Guest Host	CNET-S	PMMA/CNET-S	Corona	15% wt.		100		1.32	.47		Levenson91
PMMA/CNET-TP	Guest Host	CNET-TP	PMMA/CNET-TP	Corona	15% wt.		100		1.32	1.0		Levenson91
PMMA/CNET-TP	Guest Host	CNET-TP	PMMA/CNET-TP	Corona	15% wt.		100		.830	2.7		Levenson91
MMA-DR1	Side Chain	DR-1	MMA-DR1 Side chain	Corona	29%		130		1.32	12.4		Levenson94
MMA-DR1	Cross-linked	Red Acxid MAGLY	MMA-Red Acid MAGLY Side chain	Corona	30%		130		1.32	12.6		Levenson94

Table 1 Continued

Code Name	Material Type	Active Molecule	Polymer	Poling Method	Concentration	$\mu\beta$ 10^{-48} esu	Poling Temp. °C	E_p (V/μm)	λ (μm)	r_{33} (pm/V)	r_{13} (pm/V)	Ref. Code
9000/DADC	Guest Host	DADC	Ultradel 9000D Polyimide	Electrode	9% wt.		275	30	.8		.26	Melnhard94
9000D/DCM	Guest Host	DCM	Ultradel 9000D Polyimide	Electrode	17% wt.		220	30	.8		.75	Melnhardt94
PU11	Side Chain	DR17	Polyurethane DR17 side chain	Electrode	20% wt.	381	140	120	.830	23	4.5	Meyruelx93
BC17	Side Chain	DR17	Maleimide DR17 side chain cross-linked network	Electrode	20% wt.	381	200	140	.830	8		Meyruelx93
BPE 2.5	Main Chain	BISMAN 9,9-bis-4-hyroxy-phynel-2-dimethylaminoo-7-nitrofluorene	BPE 2.5 cardo polyester	Electrode	1.44E21	276	205	50	.81	2.6		Nahata94
MMA-DANS	Side Chain	DANS	MMA-DANS side chain		35% mol			100	1.3	4.08	1.22	Norwood94
MMA-DCVHT	Side Chain	icyanovinyl-hexatriene (DCVHT)	MMA-DCVHT side chain		50% mol			100	1.3	14	2.4	Norwood94
PMMA/DR1	Guest Host	DR-1	PMMA/DR1 Guest Host	Corona			70	480	.633	11	2.9	Page90
PPNA	Guest Host	para nitroaniline	poly(N-(4-nitro-phenyl(allylamine)	Corona			100	400	.532	26	7.5	Page90
A-095.11	Guest Host		styrene-maleic-anhydride side chain	Electrode			137	150	1.3		6.7	Pretre94
A-097.07	Guest Host		styrene-maleic-anhydride side chain	Electrode			149	120	1.3		2.3	Pretre94
A-148.02	Guest Host		styrene-maleic-anhydride side chain	Electrode			172	150	1.3		6.5	Pretre94
ROI-4	Side Chain	tricyanovinyl	polyimide	Electrode	17% m		215	100	1.3	16		Rao95
ROI-1	Guest Host	tricyanovinyl	polyquinoline	Electrode	25% wt.	6200	180	80		30		Rao95

AM3-0095.11	Guest Host	DR-1	Styrene-maleic anhydride DR1 side chain (Sandoz AM3-0095.11)	Corona			155	100	1.55	18		Ren93
PMMA/DR1	Guest Host	DR-1	PMMA/DR1	Corona	12% wt.		90	90	1.55	2.6		Ren94
PUR-DR19	Side Chain	DR-19	Polyurethane-DR19 side chain	Corona			160		.633		13	Shi92
PUR-DR19	Side Chain	DR-19	Polyurethane-DR19 side chain	Corona			160		.800		5	Shi92
PUR-DR19	Side Chain	DR19	Polyurethane-DR19 Side chain	Corona					.633		8	Shi94
PMMA/DR1	Guest Host	DR-1	PMMA Guest Host with DR1	Electrode	2% mol		120	100	.82	.8		Skindhoj94
PMMA/JPL-A	Guest Host	JPLA	PMMA Guest Host with JPLA	Electrode	2% mol	1200	120	100	.82	5		Skindhoj94
PMMA/JPL-B	Guest Host	JPLB	PMMA Guest Host with JPLB	Electrode	2% mol	1460	120	100	.82	6		Skindhoj94
Fujitsu-1	Side Chain	Azobenzene Dye	Side Chain Polyimide	Electrode			180	170	1.3	10.8		Sotoyama94
TS-PUR-DR19	Side Chain	DR-19	Thermosetting polyurethane DR19 side chain	Corona					.633		13.5	Steier92
PE-DR19	Side Chain	DR-19	Polyester DR19 side chain	Corona					.633		17.2	Steier92
Akzo-DANS	Side Chain	DANS	DANS Side Chain	Electrode	1.26 E21		139	12.5	.633	2.7	.9	V.d.Vorst90
Akzo-DANS	Side Chain	DANS	Akzo DANS side chain	Electrode			140	50	.633	9.5	3.1	V.d.Vorst92
PE-MONS	Side Chain	methoxy nitro stilbene	polyester-MONS side chain	Electrode			95	10	.633	5.4		V.d.Vorst93-2
LQ2200/DR1	Guest Host	DR-1	LQ2200/DR1 guest host	Corona	11% wt.		190	50	.8		1.5	Valley92

Table 1 Continued

Code Name	Material Type	Active Molecule	Polymer	Poling Method	Concentration	$\mu\beta$ 10^{-48} esu	Poling Temp. °C	E_p (V/μm)	λ (μm)	r_{33} (pm/V)	r_{13} (pm/V)	Ref. Code
TR	Side Chain	thiazole TR	PMMA	Electrode			140	40	1.3	12		Watanabe94
BIN2-HDT	Cross-linked	Azo Benzimidazole dye	polyurea-polyurethane Cross-Linked Polymer	Electrode			122	1	1.3	7		White94
PIQ-2200/ENIChem-1	Guest Host	N,N′-diethyl-amino-tricyanovinyl substituted thiophene stilbene	PIQ-2200 Polyimide/ENIChem-2 guest host	Electrode	12% wt.	6200	220	100	1.52	10.8		Wong94
AM3-0095.11	Side Chain		Sandoz side chain polymer	Electrode			155	83	1.55	15		Yilmaz93
Chicago-PI-A	Side Chain		Polyimide	Corona	44.6% mol		230		780	4.3		Yu95
Chicago-PI-B	Side Chain		Polyimide	Corona	51% mol		209		780	4.0		Yu95

Table 2 ATR Linear Electro-optic Measurements on Poled Polymer Films

Code Name	Material Type	Active Molecule	Polymer	Poling Method	Concentration	$\mu\beta$ 10^{-48} esu	Poling Temp. °C	E_p (V/μm)	λ (μm)	r_{33} (pm/V)	r_{13} (pm/V)	Ref. Code
RIKEN-LBK1	Lang-muir-Blodge								.633	5.3	3.5	Aust93
GT-P1	Side Chain	DR-1	MMA side chain	Electrode	39% wt		180	100	1.541	9.5		Eich94
GT-P2	Side Chain	thiophene	PMMA side chain	Electrode	39% wt		180	100	1.541	12		Eich94
GT-P3	Side Chain	DR-1	Polyimide side chain	Electrode	62% wt		180	100	1.541	12		Eich94
GT-P4	Side Chain	DR-1	Polyimie side chain	Electrode	20% wt		180	100	1.541	5		Eich94
NPP-PEI	Side Chain		Polyimide				225		1.3	2		Moylan94
DR-PEI	Side Chain		Polyimie				170		1.3	7		Moylan94
ODPA-NPP	Side Chain		Polyimide				257		1.3	2		Moylan94
TMEG-DR	Side Chain		Polyimide				185		1.3	7		Moylan94
ODPA-DAPDP	Side Chain		Polyimide		47% wt	620	227		1.3	10.5		Moylan94
IBM-2	Side Chain		Polyimide		57% wt	620	227		1.3	13		Moylan94
Bis-Λ-ANT	Cross-linked	poly(bisphenol-Λ-amino-nitro-tolane)	poly(bisphenol-Λ-amino-nitro-tolane)	Corona			115		.633	4.5	1.2	Smith91

Table 3 Linear Electro-optic Measurements on Slab Waveguide Poled Polymer Films

Code Name	Material Type	Active Molecule	Polymer	Poling Temp. °C	E_p (V/μm)	λ (μm)	r_{33} (pm/V)	r_{13} (pm/V)	Loss dB/cm	Ref. Code
MMA-NA	Side Chain	Nitroaniline	MMA-Nitroaniline side chain	111		.633	7			Dubois93
MMA-DR1	Side Chain	DR1	MMA-DR1 side chain	126		1.32	10			Dubois93
Ultem/IBM-A	Guest Host	IBM-A	Ultem Polyimide	160	300	1.3	2.5			Moylan93
Ultem/IBM-B	Guest Host	IBM-B	Ultem Polyimide	155	300	1.3	4.0			Moylan93
Ultem/IBM-C	Guest Host	IBM-C	Ultem Polyimide	167	300	1.3	6.8			Moylan93
Ultem/IBM-D	Guest Host	IBM-D	Ultem Polyimide	190	300	1.3	2.5			Moylan93
3RDCVXY	Side Chain	diethylamino	diazo-dye substituted meth-acrylate polymer	140	200	.633	40			Shuto91
E828/DR1	Guest Host	DR1	Epoxy Guest Host with DR1	150	15	.82	.184			Skindhoj94
Epofix/DR1	Guest Host	DR1	Epoxy Guest Host with DR1	80	13	.82	.091			Skindhoj94
E828/JPL-A	Guest Host	JPLA	Epoxy Guest Host with JPLA	150	6.7	.82	.044			Skindhoj94
PMMA-NAT	Side Chain	nitroamino tolane (NAT)	PMMA-NAT side chain with nitroamino tolane (NAT)	115		1.32	26		3–5	Thack-ara94

Table 4 Linear Electro-optic Measurements on Channel Waveguides

Code Name	Material Type	Active Molecule	Polymer	Poling Temp. °C	E_p (V/μm)	Waveguide Fab. Method	λ (μm)	r_{33} (pm/V)	Loss dB/cm	Ref. Code
3RDCVXY	Side Chain	4-(4-dicyanovinyl)phenylazo)-3,5,-xylidine	3RDCVXY	140		RIE	1.31	9	1.1	Amano94
MMA-MA1	Side Chain	MA1	MMA-MA1 side chain		125	Photobleaching	.838	13	4	Beeson92
MA1	Side Chain	DR-1	DR1 side chain		125	Photobleaching	.838	13		Beeson93
FI:MMA-DR1	Side Chain	DR-1	MMA-DR1 side chain	90	20	Inverted Rib	1.3	4.7		Brauer93
MMA-DR1	Side Chain	DR-1	MMA-DR1 Side chain	113	204	Laser Etching	1.3	10.5	.33	Chon94
U-PI/DCM	Guest Host	4-(dicyanomethylene)-2-methyl-I-6-(p-dimethylami-nostyryl4H-pyran DCM	Ultradel Polyimide U-PI/ DCM guest host	190	312		.830	3.4		Ermer92
HCC-DANS	Side Chain	DANS	HCC-DANS side chain		165	Photobleaching	1.3	21		Findakly93
Akzo-DANS	Side Chain	DANS	Akzo-DANS side chain	120	70	Photobleaching	1.3	18		Girton91
4212/DCM	Guest Host	DCM	Ultradel 4212/DCM guest Host		200	Photobleaching	1.3	12		Girton94
4212/DCM	Guest Host	DCM	Ultradel 4212/DCM guest Host		200	Photobleaching	1.3	2		Girton94
4212/DCM	Guest Host	DCM	Amoco 4212		230	Photobleaching	1.3	13		Girton95
PMMA/CNET-TP	Guest Host	CNET-TP	PMMA/CNET-TP	100			.830	1.76		Levenson91
MMA-DR1	Cross-linked	Red Acxid MAGLY	MMA-Red Acid MAGLY Side chain	130		RIE	1.06	4		Levenson94
Akzo-DANS	Side Chain	DANS	DANS Side Chain	140	70		1.3	18		Lipscomb91
Akzo-DANS	Side Chain	DANS	DANS Side Chain	130	170	Photobleaching	1.3	34		Mohlman90

Table 4 Continued

Code Name	Material Type	Active Molecule	Polymer	Poling Temp. °C	E_p (V/μm)	Waveguide Fab. Method	λ (μm)	r_{33} (pm/V)	Loss dB/cm	Ref. Code
Akzo-DANS	Side Chain	DANS	Akzo-DANS	131	105	Photobleaching	1.3	16		Mohlman91
Akzo-DANS	Side Chain	DANS	DANS Side Chain	140		Photobleaching	1.33	34		Mohlman94
Hercules-1			Hercules polymer	160			1.3	32.5		Paul94
3RDCVXY	Side Chain	Diazo Dye	3RDCVXY		70	RIE	1.3	26		Shuto95
PUR-DR19	Side Chain	DR-19	PUR-DR19 Polyurethane DR19 side chain	155		RIE	1.06	12		Steier93
BE-AEANP	Side Chain	2(2-animoethylamino)-5-nitropyridine	tetramethyl-biphenylepoxy BE-AEANP side chain		20		.633	.1		Tatsuura92
PMMA-DR1	Side Chain	DR1	MMA-DR1 side chain	112	200		1.32	15		Thackara93
6FDA-DPDR1	Side Chain	diphenyl DR1	6FDA-DPDR1 side chain	220		Photobleaching	1.32	6		Thackara95
DOW-TP7	Side Chain		Dow Chemical TP7	140	70	Inverted Rib	1.3	8		Tumolillo93
DOW-TP7	Side Chain		Dow Chemical TP7	140	70	Inverted Rib	1.3	10		Tumolillo93
Akzo-DANS	Side Chain	DANS	Akzo DANS Side Chain	130	50	Photobleaching	1.3	14		VanEck91
RA-HDT	Cross-linked		Cross linked polymer		250	Photobleaching	1.30	7.8	1	Watson94
BIN2-HDT	Cross-linked		Cross linked polymer		250	Photobleaching	1.30	15	1.2	Watson94

horizontal dimensions. The method used to fabricate the waveguides is also reported.

C. Electro-optic Measurements in Organic Crystals

A summary of measurements of the electro-optic coefficients in organic molecular crystals is presented in Table 5. The crystal point group and the molecular hyperpolarizability, β, are given, along with the relevant elements of the electro-optic tensor. Finally, the references used in all of the tables in this section are correlated with the appropriate reference code in Table 6.

IV. CONCLUSIONS

Great progress has been made over the last two decades in engineering organic electro-optic materials. The great efforts expended have been rewarded with r_{11} = 92 pm/V in a molecular crystal,[16] r_{33} = 55 pm/V[17] in a poled polymer, and polymer-based integrated optic devices with a bandwidth of 60 GHz.[8] These advances have been made possible not only through the ingenuity of organic chemists, but also through the development of appropriate measurement techniques for initial rough screening, optimization, and device evaluation. The measurement approaches described in this chapter will play a major role in the future advances necessary before organic electro-optic materials achieve their technological potential.

(References begin on page 307)

Table 5 Linear Electro-optic Measurements in Organic Crystals

Code Name	Active Molecule	Point Group	β 10^{-30} esu	λ (μm)	r_{11} (pm/V)	r_{33} (pm/V)	r_{13} (pm/V)	r_{63} (pm/V)	r_{52} (pm/V)	Ref. Code
COANP	2-cyclooctylamino-5-nitropyridine (COANP)	mm2	22	.633		15	3.4			Bosshard92
PNP	2-(N-prolinol)-5-nitropyridine (PNP)	2	21	.633		12.8	13.1			Bosshard92
NMBA	4-nitro-4′-methylbenzylidine aniline	m		.633		25.2				Halfpenny93
NMBA	4-nitro-4′-methylbenzylidine aniline	m		.488	37	25.2				Halfpenny93
MNA	2-methyl-4-nitroaniline	m		.633	60					Ho91
MNA	2-methyl-4-nitroaniline	m	17	.633	67					Lipscomb81
DAST	DAST	m		1.535	47		5			Pan96
DAST	DAST	m		1.313	53		6			Pan96
DAST	DAST	m		.800	77		15			Pan96
DAST	DAST	m		.720	92		16			Pan96
DAST	4′-dimethylamino-N-methyl-4-stilbaxolium tosylate	m		.820		400				Penny91
MNBA	4′-Nitrobenzylidene-3-Acetamino-4-Methoxyaniline	m	30	.633	29	2.4	1.5			Schlesser95
MNBA	4′-Nitrobenzylidene-3-Acetamino-4-Methoxyaniline	m	30	.532	50	4	2			Schlesser95
POM	3-methyl 4-nitropyridine 1-oxide	222		.633				2.6	3.6	Sigelle81
m-NA	meta-nitroaniline	mm2				16.7	7.4			Stevenson73
MNA	2-methyl-4-nitroaniline	m		.633	30					Wah93

Table 6 Reference List for Tables of Electro-optic Measurements

Amano94	M. Amano, M. Hikita, Y. Shuto, T. Watanabe, A. Tomaru, M. Yaita, and T. Nagatsuma **Diazo Dye Attached Electro-optical Polymer and its Applications to Waveguide Devices** *Proceedings SPIE*, Los Angeles, 1994, Vol. 2143, p. 68.
Anlheim95	M. Anlheim, J.-L. Bredas, M. Barzoukas, P. V. Bedworth, Y. Cai, C. Dehu, M. Blanchard-Desce, A. Fort, Z. Hu, A. K.-Y. Jen, S. Marder, F. Meyers, J. Perry, C. Runser, M. Staehelin, and B. Zysset **Experimental and Theoretical Studies of Nonlinear Optical Chromophores** OSA 1995 Technical Digest Series, Portland, Oregon, 1995, Vol. 21, p. 244.
Aramaki94	S. Aramaki, Y. Okamoto, T. Murayama, and Y. Kubo **Nobel Naphthoquinone Methide Dyes for Poled Polymers** *Proc. SPIE*, San Diego, 1994, Vol. 2285, p. 58.
Aust93	E. Aust, W. Knoll, W. Hickel, H. Knobloch, and H. Orendi **Electro-optic Characterization of Polymeric Materials for Integrated Optics** *Proc. SPIE*, San Diego, 1993, Vol. 2025, p. 255.
Bauer93	S. Bauer, W. Ren, S. Yilmaz, W. Wirges, W. Molzow, R. Gerhard-Multhaupt, U. Oertel, B. Hanel, L. Haussler, H. Kombr, and K. Lunkwitz **Nonlinear Optical Side-chain Polymer With High Thermal Stability and its Pyroelectric Thermal Analysis** *Appl. Phys. Lett.*, 63(15), p. 2018 (1993).
Beeson92	K. Beeson, P. Ferm, K. Horn, M. Mcfarland, A. Nahata, J. Shan, C. Wu, and J. Yardley **Loss Measurements in Electro-optic Polymer Waveguides** *Proc. SPIE*, San Diego, 1992, Vol. 1775, p. 133.
Beeson93	K. Beeson, P. Ferm, K. Horn, C. Knapp, M. Mcfarland, A. Nahata, J. Shan, C. Wu, and J. Yardley **Polymeric Electro-optic Materials and Devices: Meeting The Challenges of Practical Applications** *Proc. SPIE*, San Diego, 1993, Vol. 2025, p. 488.
Beeson94	K. Beeson, K. Horn, M. Mcfarland, A. Nahata, C. Wu, and J. Yardley **Polymeric Materials For Guided-wave Devices** *Proc. SPIE*, San Diego, 1990, Vol. 1337, p. 195.
Bosshard92	C. Bosshard, K. Sutter, R. Schlesser, and P. Gunter **Electro-optic Effects in Organic Single Crystals** *Proc. SPIE*, San Diego, 1992, Vol. 1775, p. 271.
Brauer93	A. Brauer, L. Erdmann, T. Gase, P. Dannberg, W. Karthe, and C. Wachter **Electro-optic Effects in Polymer Waveguide Devices for Optical Communication** *Proc. SPIE*, San Diego, 1993, Vol. 2025, p. 547.
Chollet92	P. Chollet, G. Gadret, F. Kajzar, P. Raimond, and M. Zagorska **Electrooptic Organic Thin Films for Applications in Waveguided Devices** *Proc. SPIE*, San Diego, 1992, Vol. 1775, p. 121.
Chon94	J. Chon, J. Thackara, P. Comita, M. Jurich, and J. Swalen **Characterization of Low Loss Polymer Modulators Fabricated by Laser Ablation** *Proc. SPIE*, San Diego, 1994, Vol. 2285, p. 340.

Table 6 Continued

Cross92	G. Cross, Y. Karakus, D. Gray, D. Bloor, W. Douglas, R. Corriu, and Z. Yang **Cooperative Processes in Polar Ordered Polymers** *Proc. SPIE*, San Diego, 1992, Vol. 1775, p. 144.
Dubois93	J. Dubois, P. Robin, and V. Dentan **Properties and Applications of Polymers in Optics and Electrooptics** *Proc. SPIE*, San Diego, 1993, Vol. 2025, p. 467.
Eich94	M. Eich, H. Beisinghoff, B. Knodler, M. Ohl, M. Sprave, J. Vydra, M. Eckl, P. Strohriegl, M. Dorr, R. Zentel, M. Anlheim, M. Stahelin, B. Zysset, J. Liang, R. Levenson, and J. Zyss **Electrooptical Properties and Poling Stability of High Glass Transition Polymers** *Proc. SPIE*, San Diego, 1994, Vol. 2285, p. 104.
Eldering90	C. Eldering, A. Knoesen, and S. Kowel **Characterization of Polymeric Electro-optic Films Using Metal Mirror/Electrode Fabry-Perot Etalons** *Proc. SPIE*, San Diego, 1990, Vol. 1337, p. 348.
Ermer92	S. Ermer, J. Valley, R. Lytel, G. Lipscomb, T. Van Eck, and D. Girton **Dcm-polyimide System for Triplepstack Poled Polymer Electro-optic Devices** *Appl. Phys. Lett.*, 61(19), 2272 (1992).
Findakly93	T. Findakly, and C. C. Teng **Wideband NLO Polymer Modulators** *Proc. SPIE*, San Diego, 1993, Vol. 2025, p. 526.
Fiorini94	C. Fiorini, P. Chollet, F. Charra, F. Kajzar, and J. Nunzi **Quadratic Optical Polarizabilities in Polymer Film Obtained by DC-Electric Field and Pure Optical Poling** *Proc. SPIE*, San Diego, 1994, Vol. 2285, p. 92.
Gilmour94	S. Gilmour, S. R. Marder, J. W. Perry, and L.-T. Cheng **Enhanced Second-order Nonlinearities in Extended Thiophene Containing Compounds** *Proc. SPIE*, Los Angeles, 1994, Vol. 2143, p. 117.
Girton91	D. Girton, S. Kwiatkowski, G. Lipscomb, and R. Lytel **20 Ghz Electro-optic Polymer Mach-Zehnder Modulator** *Appl. Phys. Lett.*, 58(16), 1730 (1991).
Girton94	D. Girton, S. Ermer, J. Valley, T. E. Van Eck, S. Lovejoy, D. Leung, and J. Marley **Considerations For Dcm-polyimide Systems used in Electro-optic Devices** *ACS Ntl Mtg Abstracts*, Washington, D.C., 1994, p. 219.
Girton95	D. G. Girton, W. W. Anderson, J. A. Marley, T. E. Van Eck, and S. Ermer **Current Flow in Doped and Undoped Electro-optic Polymer Films During Poling** *Tech. Digest Series*, Portland, Oregon, 1995, Vol. 21.
Goldberg90	H. Goldberg, A. East, R. Johnson, G. Khanarian, R. Norwood, M. Sansone, I. Kalnin, D. Haas, and R. Keosian **Electro-optic and Nonlinear Optical Polymers and Devices** *Proc. SPIE*, San Diego, 1990, Vol. 1337, p. 326.
Halfpenny93	P. Halfpenny, E. Shepherd, J. Sherwood, and G. Simpson **Organic Crystals for Optoelectronic Devices** *Proc. SPIE*, San Diego, 1993, Vol. 2025, p. 171.

Table 6 Continued

Ho91	E. S. S. Ho, C. Wah, K. Iizuka, A. Freundorfer, and C. Wah **An Organic MNA Crystal Phase Modulator** *J. Light. Tech.*, 9, 3110 (1993).
Ho92	A. Ho, R. Chen, and R. Shih **Electro-optic Phenomena in Gelatin-based Poled Polymer** *Appl. Phys. Lett.*, 61(1), 4 (1992).
Horsthuis92	W. Horsthuis, F. Van Veggel, B. Hams, C. Van Der Vorst, J. Heideman, H. Mertens, M. Rheede, and G. Mohlmann **Photonic Polymers in Passive and Active Waveguide Devices** *Proc. SPIE*, San Diego, 1992, Vol. 1775, p. 360.
Jen94	A. Jen, V. Rao, K. Drost, Y. Cai, R. Mininni, J. Kenney, E. Binkley, and L. Dalton **The Development of Highly Active, Thermally Stable Chromophores and Polymers for Electro-optic Apps.** *Proc. SPIE*, San Diego, 1994, Vol. 2285, p. 49.
Jungbauer90	D. Jungbauer, B. Reck, R. Twieg, D. Yoon, C. Willson, and J. Swalen **Highly Efficient and Stable Nonlinear Optical Polymers Via Chemical Cross-linking Under Electric Field** *Appl. Phys. Lett.*, 56(26), 2610 (1990).
Kim95	H. K. Kim, H. J. Lee, M. H. Lee, S. K. Han, H.-Y. Kim, K. H. Kang, Y. H. Min, and Y. H. Won **Thermally Stable Nonlinear Optical Moiety-doped Polymides for Photonic Devices** *Polymers for 2nd-Order Nonlin. Opt.*, ACS Symp. Series 601, p. 111.
Knabe89	G. Knabke, H. Franke, and W. Frank **Electro-optical Properties of a Nonlinear Dye in Poly-methyl Methacrylate) And Poly (α-methyl Styrene)** *J. Opt. Soc. Am. B*, 6, No. 4, 761 (1989).
Lee94	M. H. Lee, H. K. Kim, H.-Y. Kim, H. J. Lee, K. H. Kang, Y. H. Won, E. S. Jeon, and J. W. Wu **NIo Moiety Doped Polymers With Improved Optical Properties for Photonic Devices** *Proc. SPIE*, Los Angeles, 1994, Vol. 2143, p. 209.
Levenson91	R. Levenson, J. Liang, E. Toussaere, A. Carenco, and J. Zyss **A Quasi-relaxation Free Guest-host Poled Polymer Waveguide Modulator** *Proc. SPIE*, San Diego, 1991, Vol. 1560, p. 251.
Levenson94	R. Levenson, J. Liang, and J. Zyss **Nonlinear Optical Cross-linked Polymers for Electro-optic Applications** *ACS Ntl Mtg Abstracts*, Washington, D.C., 1994, p. 162.
Levy92	Y. Levy, P. Chollet, G. Gadret, and F. Kajzar **Ellipsometry Techniques in Thin Film Electrooptic Coefficients Measurements** *Proc. SPIE*, San Diego, 1992, Vol. 1775, p. 299.
Lipscomb81	G. F. Lipscomb, A. F. Garito, and R. S. Narang **A Large Linear Electro-optic Effect in a Polar Organic Crystal 2-methyl-4-nitroaniline** *Appl. Phys. Lett*, 38, 663 (1981).

Table 6 Continued

Lipscomb91	G. Lipscomb, R. Lytel, A. Ticknor, T. E. Van Eck, D. Girton, S. Ermer, J. Valley, J. Kenney, and E. Binkley **Organic Electro-optic Devices for Optical Interconnection** *Proc. SPIE*, San Diego, 1991, Vol. 1560, p. 388.
Meinhardt94	M. B. Meinhardt, P. A. Cahill, C. J. Seager, A. J. Beuhler, D. A. Wargowski, K. D. Singer, T. C. Kowalczyk, T. Z. Kose, and S. Ermer **Characterization of Crosslinked Electrooptic Polyimides** *Proc. SPIE*, Los Angeles, 1994, Vol. 2143, p. 110.
Meyrueix93	R. Meyrueix, G. Tapolsky, M. Dickens, and J. Lecomte **NLO Polyurethane and Crosslinked Bismaleimide Network With Exceptional Thermal Stability** *Proc. SPIE*, San Diego, 1993, Vol. 2025, p. 117.
Mohlman90	G. Mohlmann, and W. Horsthuis **Optically Nonlinear Polymeric Switches and Modulators** *Proc. SPIE*, San Diego, 1990, Vol. 1337, p. 215.
Mohlman91	G. Mohlmann, W. Horsthuis, J. Mertens, M. Diemeer, F. Suyten, B. Hendriksen, C. Duchet, P. Fabre, C. Brot, J. Copeland, J. Mellor, E. Van Tomme, P. Van Daele, and R. Baets **Optically Nonlinear Polymeric Devices** *Proc. SPIE*, San Diego, 1991, Vol. 1560, p. 426.
Mohlman94	G. Mohlmann, M. Kleinkoerkamp, J. Heideman, T. Tumolillo, M. Van Rheede, D. Ramsamoedj, C. Van Der Vorst, and R. Van Gassel **Optical Polymers and Related Photonic Devices** *Proc. SPIE*, San Diego, 1994, Vol. 2285, p. 366.
Moylan93	C. Moylan, S. Swanson, C. Walsh, J. Thackara, R. Twieg, R. Miller, and V. Lee **From Efish to Electro-optic Measurements of Nonlinear Chromophores** *Proc. SPIE*, San Diego, 1993, Vol. 2025, p. 192.
Moylan94	C. Moylan, R. Twieg, V. Lee, R. Miller, W. Voksen, J. Thackara, and C. Walsh **Synthesis and Characterization of Thermally Robust Electro-optic Polymers** *Proc. SPIE*, San Diego, 1994, Vol. 2285, p. 17.
Nahata94	A. Nahata, C. Wu, C. Knapp, V. Lu, J. Shan, and J. Yardley **Thermally Stable Polyester Polymers for Second Order Nonlinear Optics** *Appl. Phys. Lett.*, 64(25), 3371 (1994).
Norwood94	R. Norwood, M. Kuzyk, and R. Keosian **Electro-optic Tensor Ratio Determination of Side-chain Copolymers With Electro-optic Interferometry** *J. Appl. Phys.*, 75(4), 1869 (1994).
Page90	R. Page, M. Jurich, B. Reck, A. Sen, R. Twieg, J. Swalen, G. Bjorklund, and C. Wilson **Electrochromic and Optical Waveguide Studies of Corona-poled Electro-optic Polymer Films** *J. Opt. Soc. Am. B*, 7, 1239 (1990).
Pan96	F. Pan, G. Knopfle, Ch. Bosshard, S. Follonier, R. Spreiter, M. S. Wong, and P. Gunther **Electro-optic Properties of the Organic Salt 4-n,n-dimethylamino-4′-n′-methyl-stilbazolium Tosylate** *Appl. Phys. Lett.*, 69, 13 (1996).

Table 6 Continued

Paul94	D. Paul **High-yield, High-performance NIo Polymer Integrated Optical Channel Waveguide Devices** *ACS Ntl Mtg Abstracts*, Washington, D.C., 1994, p. 217.
Perry91	J. Perry, S. Marder, K. Perry, and E. Sleva **Organic Salts With Large Electro-optic Coefficients** *Proc. SPIE*, San Diego, 1991, Vol. 1560, p. 302.
Pretre94	P. Pretre, P. Kaatz, U. Meier, P. Gunter, B. Zysset, M. Ahliheim, M. Stahelin, and F. Lehr **Modified Polyimide Side-chain Polymers for Electro-optical Applications** *ACS Natl Mtg Abstracts*, Washington, D.C., 1994, p. 136.
Rao95	V. P. Rao, Y. M. Cai, and A. K. Y. Jen **New Developments in Thermally Stable Second Order Nonlinear Optical Chromophores and E-O Polymers** *Proc. SPIE*, San Diego, Vol. 2527, p. 84 (1995).
Ren94	W. Ren, S. Bauer, S. Yilmaz, W. Wirges, and R. Gerhard-Multhaupt **Optimized Poling of Nonlinear Optical Polymers Based on Diapole-orientation and Dipole-relaxation Studies** *J. Appl. Phys.*, 75(11), 7211 (1994).
Schlesser95	R. Schlesser, G. Knopfle, K. Sutter, Ch. Bosshard, and P. Gunter **Electro-optic and Photorefractive Properties of 4′-Nitrobenzylidene-3-Acetamino-4-Methoxyaniline (MNBA) Single** *Nonlinear Optics*, Vol. 9, No. 1–4, pp. 171–179 (1995).
Shi92	Y. Shi, W. Steier, M. Chen, L. Yu, and L. Dalton **Thermosetting Nonlinear Optical Polymer: Polyurethane With Disperse Red 19 Side Groups** *Appl. Phys. Lett.*, 60(21), 2577 (1992).
Shi94	Y. Shi, J. Bechtel, S. Kalluri, W. Steier, C. Xu, B. Wu, and L. Dalton **Measurement of the Electro-optic Coefficient Dispersion in Poled Polymer Materials** *Proc. SPIE*, San Diego, 1994, Vol. 2285, p. 131.
Shuto91	Y. Shuto, M. Amano, and T. Kaino **Electrooptical Light Modulation in Novel Azo-dye-substituted Poled Polymers** *Proc. SPIE*, San Diego, 1991, Vol. 1560, p. 184.
Shuto95	Y. Shuto, S. Tomaru, M. Hikita, and M. Amano **Optical Intensity Modulation in Diazo-dye-substituted Polymer Channel Waveguides** ECIO'95 (Delft, The Netherlands, April 3–6, 1995).
Sigelle81	M. Sigelle, and R. Hierle **Determination of the Electro-optic Coefficients of 3-methyl 4-nitropyridine 1-oxide** *J. Appl. Phys.*, 52, 4199 (1981).
Singer87	K. Singer, M. Kuzyk, and J. Sohn **Secon-order Nonlinear-optical Processes in Orientationally Ordered Materials** *J. Opt. Soc. Am. B*, 4, No. 6, 968 (1987).
Skindhoj94	J. Skindhoj, J. Perry, and S. Marder **Development of Electrooptic Polymers for High Voltage Instrument Transformers** *Proc. SPIE*, San Diego, 1994, Vol. 2285, p. 116.

Table 6 Continued

Smith91	B. Smith, S. Herminghaus, and J. Swalen **Electrooptic Coefficients in Electric Field Poled Polymer Waveguides** *Proc. SPIE*, San Diego, 1991, Vol. 1560, p. 400.
Sotoyama94	W. Sotoyama, S. Tatsuura, and T. Yoshimura **Electro-optic Side-chain Polyimide System With Large Optical Nonlinearity and High Thermal Stability** *Appl. Phys. Lett.*, 64(17), 2197 (1994).
Steier92	W. Steier, Y. Shi, L. Yu, M. Chen, and L. Dalton **Nonlinear Optics and Optical Micro-patterning in Polymers With Disperse Red 19 Sidegroups** *Proc. SPIE*, San Diego, 1992, Vol. 1775, p. 379.
Steier93	W. Steier, Y. Shi, P. Ranon, C. Xu, B. Wu, L. Dalton, W. Wang, D. Chen, and H. Fetterman **Waveguide Photonic Devices Made From Thermally Crosslinked Second-order Nonlinear Optical Polymers** *Proc. SPIE*, San Diego, 1993, Vol. 2025, p. 535.
Stevenson73	J. L. Stevenson *J. Phys. D.: Appl. Phys.*, 6 L13 (1973).
Tatsuura92	S. Tatsuura, W. Sotoyama, and T. Yoshimura **Electro-optic Polymer Waveguide Fabricated Using Electric-field-assisted Chemical Vapor Deposition** *Appl. Phys. Lett.*, 60(14), 1661 (1992).
Thackara93	J. Thackara, G. Bjorklund, W. Fleming, M. Jurich, B. Smith, and J. Swalen **A Polymeric Electro-optic Phase Modulator for Broadband Data Transmissions** *Proc. SPIE*, San Diego, 1993, Vol. 2025, p. 564.
Thackara94	J. I. Thackara, M. Jurich, and J. D. Swalen **Electro-optic Measurements of Tolane-based Polymeric Phase Modulators** *J. Opt. Soc. Am. B*, 11, 835 (1994).
Thackara95	J. Thackara, J. C. Chon, G. Bjorklund, W. Volksen, and D. M. Burland **Polymeric Electro-optic Mach-Zehnder Devices** *Appl. Phys. Lett.*, 67, 3874 (1995).
Tumolillo93	T. Tumolillo, and P. Ashley **Fabrication and Design Considerations for Multilevel Active Polymeric Devices** *Proc. SPIE*, San Diego, 1993, Vol. 2025, p. 507.
Tumolillo93	T. Tumolillo, and P. Ashley **Multilevel Registered Polymeric Mach-Zehnder Intensity Modulator Array** *Appl. Phys. Lett.*, 62(24), 3068 (1993).
V.d.Vorst90	C. Van Der Vorst, and C. Van Weerdenburg **Electro-optic Measurements in Poled Polymer Films** *Proc. SPIE*, San Diego, 1990, Vol. 1337, p. 246.
V.d.Vorst92	C. Van Der Vorst, and M. Rheede **Electro-optic Measurements in Poled Polymers** *Proc. SPIE*, San Diego, 1992, Vol. 1775, p. 186.
V.d.Vorst93-2	C. Van Der Vorst, and S. Picken **Electro-optic Measurements in a Nematic Sidechain Homopolymer During Poling** *Proc. SPIE*, San Diego, 1993, Vol. 2025, p. 243.

Table 6 Continued

Valley92	J. Valley, J. Wu, S. Ermer, M. Stiller, E. Binkley, J. Kennry, G. Lipscomb, and R. Lytel **Thermoplasticity and Parallel-plate Poling of Electro-optic Polyimide Host Thin Films** *Appl. Phys. Lett.*, 60(2), 160 (1992).
VanEck91	T. Van Eck, A. Ticknor, R. Lytel, and G. Lipscomb **Complementary Optical Tap Fabricated in an Electro-optic Polymer Waveguide** *Appl. Phys. Lett.*, 58(15), 1588 (1991).
Wah93	C. Wah, K. Iizuka, and A. Freundorfer **94-ghz Phase Modulation of Light by Organic Mna (2-methyl-4nitroaniline) Crystal** *Appl. Phys. Lett*, 63(23), 3110 (1993).
Watanabe94	T. Watanabe, M. Amano, and S. Tomaru **Second-order Optical Nonlinearities of an Electrooptic Polymer Containing a Thiazole Ring** *Jpn. J. Appl. Phys.*, Vol. 33 (1994) pp. L 1683–L 1685, Part 2, No. 12A
Watson94	J. Watson, C. Francis, R. Moshrefzadeh, K. White, P. Kitipichai, E. Cross, G. Boyd, and P. Pedersen **Progress in NLO Polymeric Waveguide Devices At 3 M** *Proc. SPIE*, San Diego, 1994, Vol. 2285, p. 328.
White94	K. White, E. Cross, and C. Francis **Bifunctional Dyes for Cross-linked Nonlinear Optical Polymers** *ACS Ntl Mtg Abstracts*, Washington, D.C., 1994, p. 172.
Wong94	K. Wong, and A. Jen **Thermally Stable Poled Polyimides Using Heteroaromatic Chromophores** *J. Appl. Phys.*, 75(71), 3308 (1994).
Wu91	J. Wu, J. Valley, M. Stiller, S. Ermer, E. Binkley, J. Kenney, G. Lipscomb, and R. Lytel **Poled Polyimides as Thermally Stable Electro-optic Polymer** *Proc. SPIE*, San Diego, 1991, Vol. 1560, p. 196.
Yilmaz93	S. Yilmaz, S. Bauer, W. Wirges, and R. Gerhard-Multhaupt **Scanning Electro-optical and Pyroelectrical Microscopy for the Investigation of Polarization Patterns** *Appl. Phys. Lett.*, 63(13), 1724 (1993).
Yu95	D. Yu, A. Gharavi, and L. Yu **Novel Second-order Nonlinear Optical, Aromatic, and Aliphatic Polyimides Exhibiting High Temp. Stability** *Appl. Phys. Lett.*, 66(9), 1050 (1995).

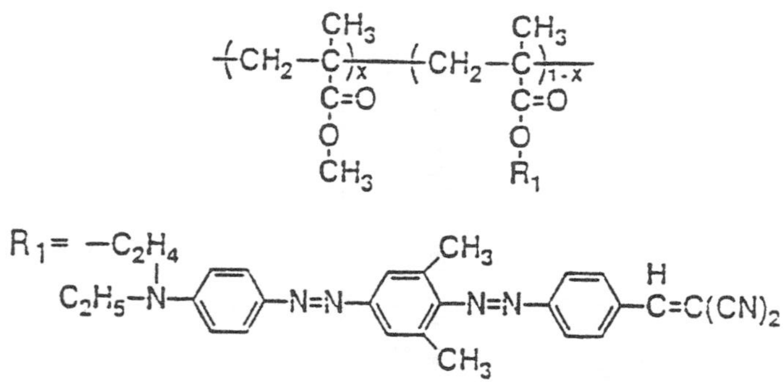

3RDCVXY

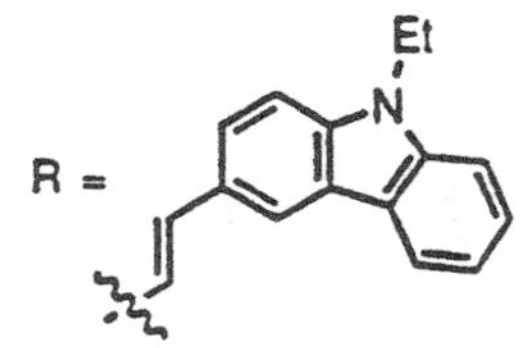

DADC

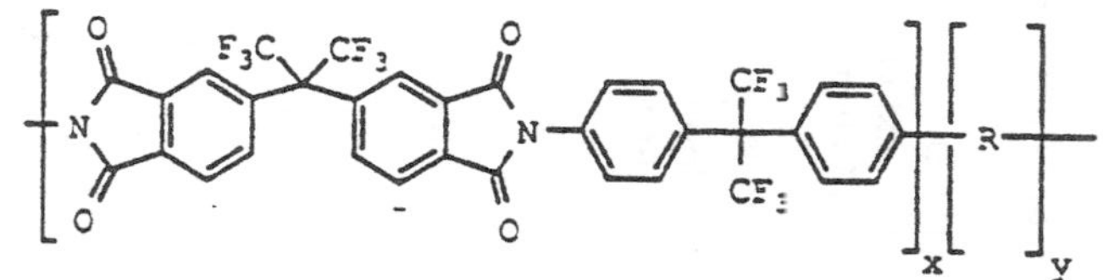

9000D/DADC

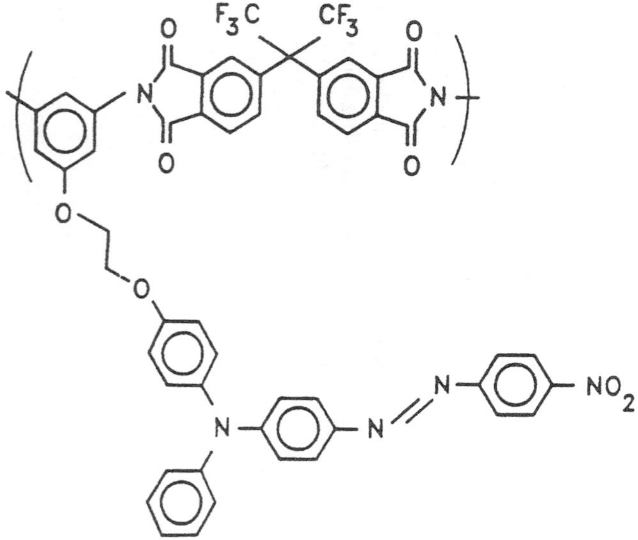

6FDA-DPDR1

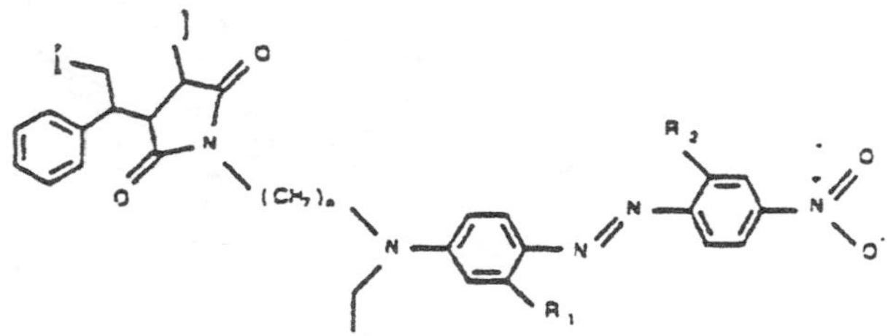

Polymer	n	R_1	R_2	λ_{max}	T_g (°C)
A-095.11	3	CH_3	H	490	137
A-097.07	3	CH_3	Cl	510	149
A-148.02	2	H	H	470	172

A-095.11

A-097.07

A-148.02

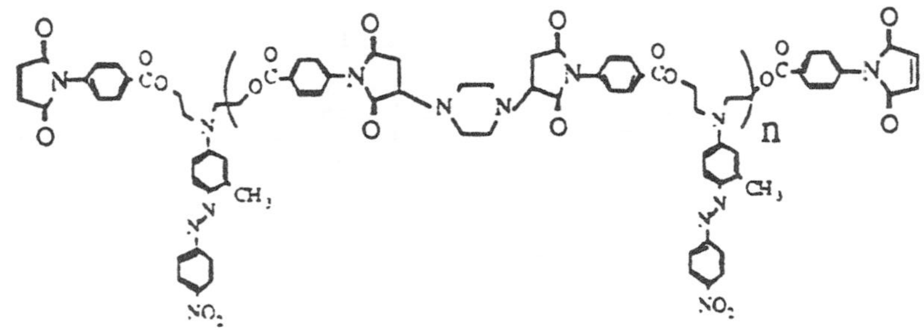

BC17

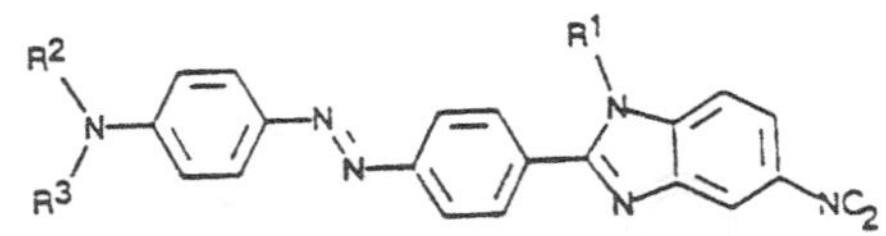

BIN2-HDT

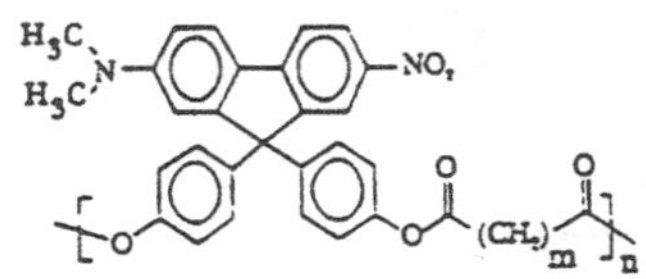

BPE 2.5

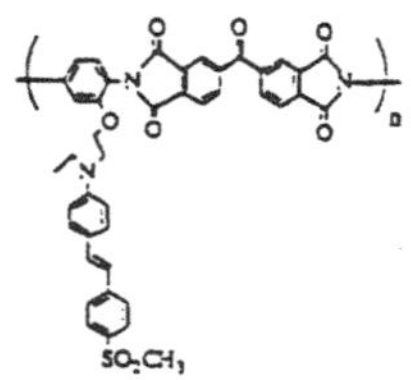

Polyimide A

Chicago—PI-A

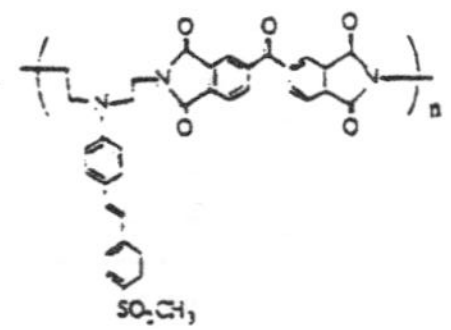

Polyimide B

Chicago-PI-B

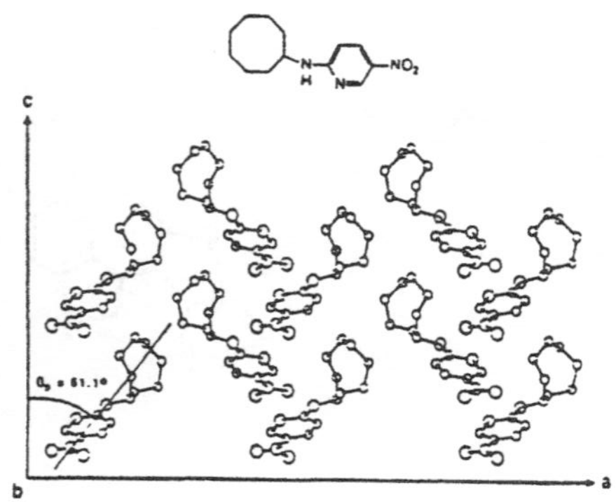

COANP

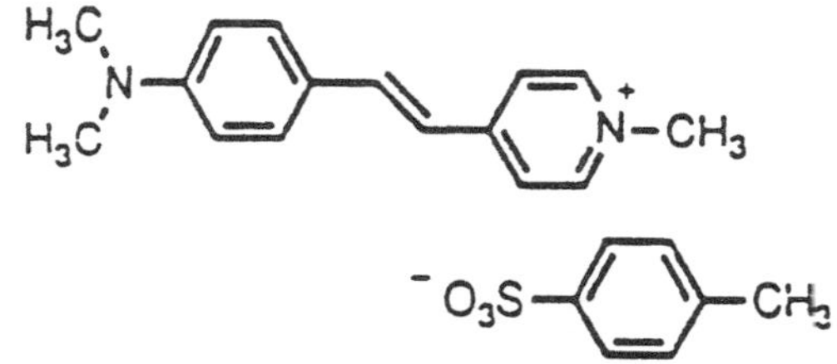

DAST

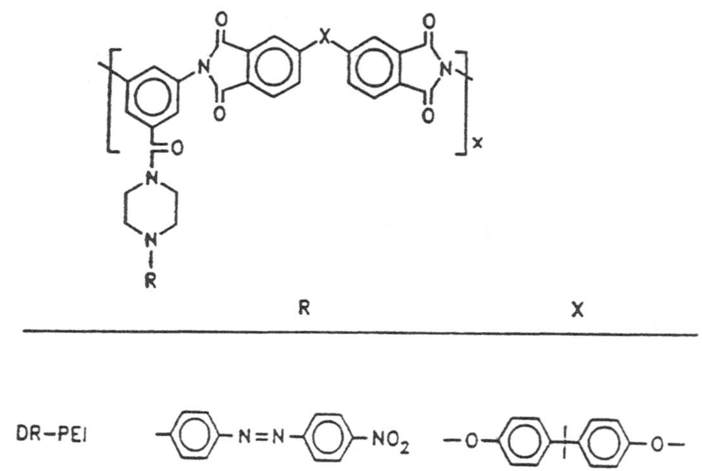

DR-PEI

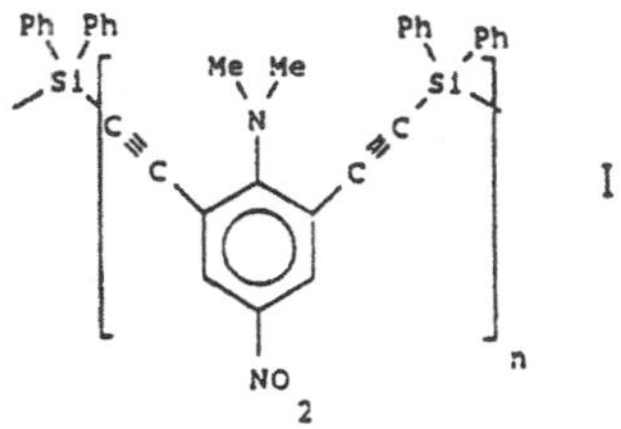

Durham-1

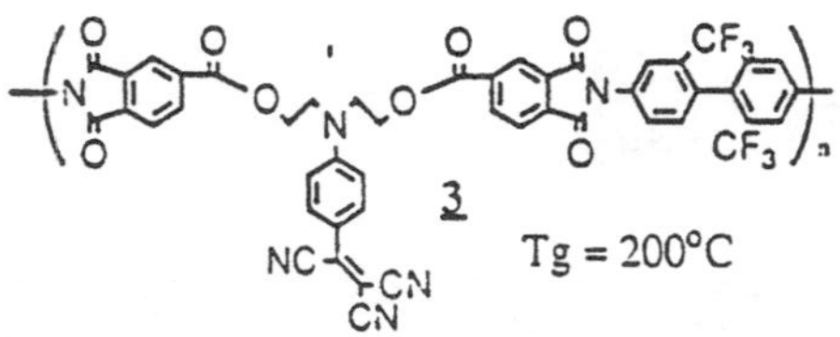

ENIChem-1

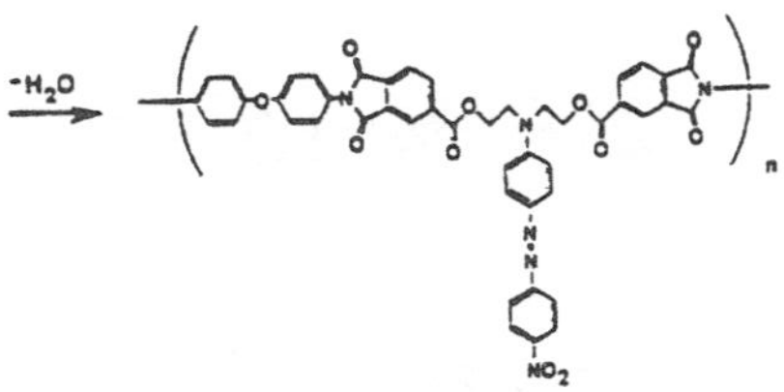

Fujitsu-1

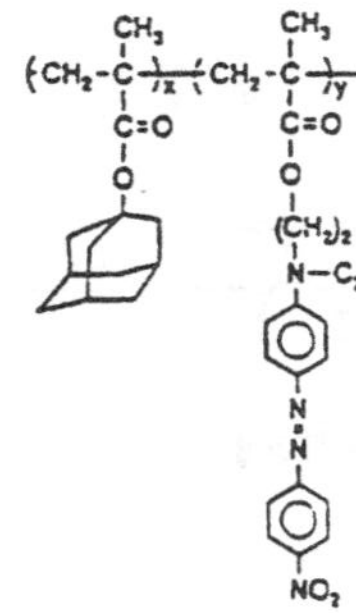

GT-P1

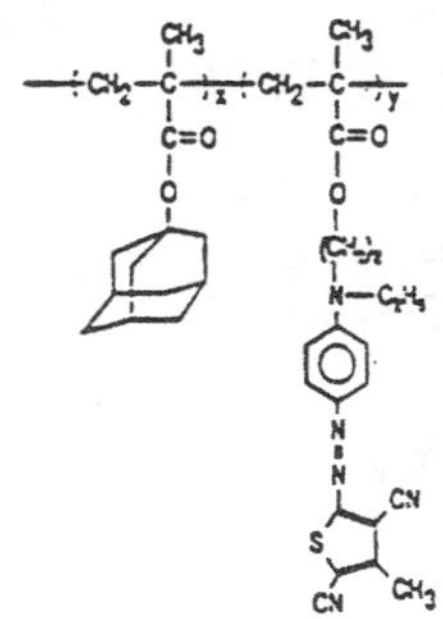

GT-P2

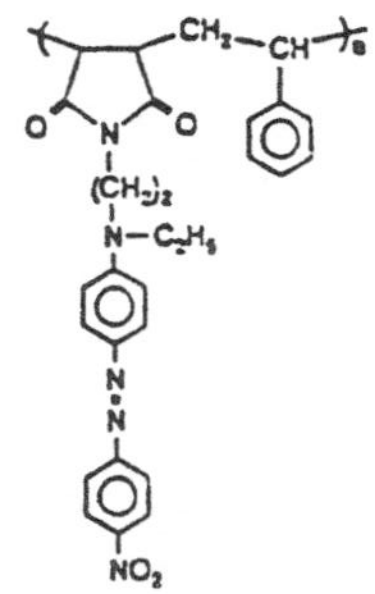

GT-P3

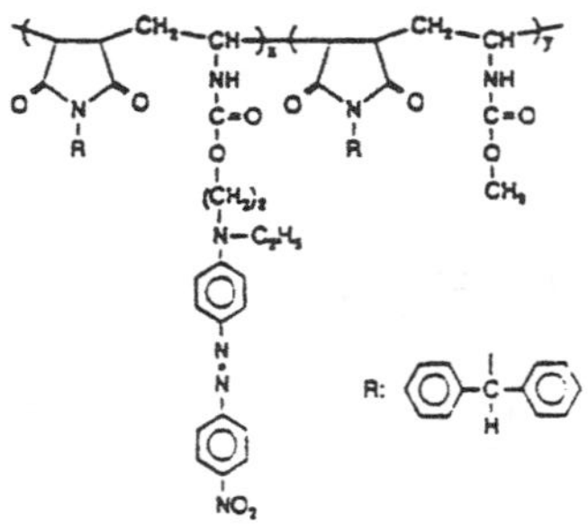

GT-P4

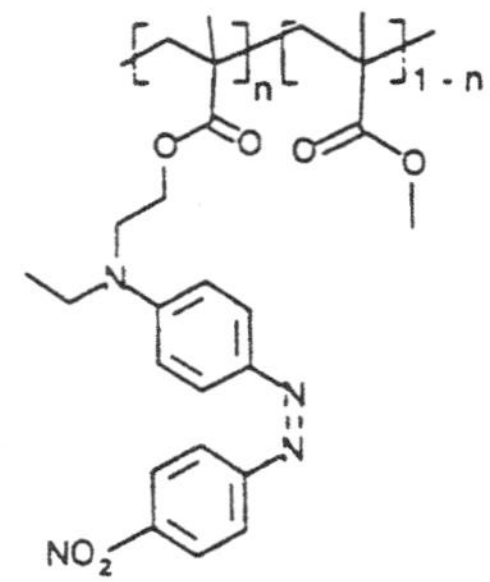

MA1

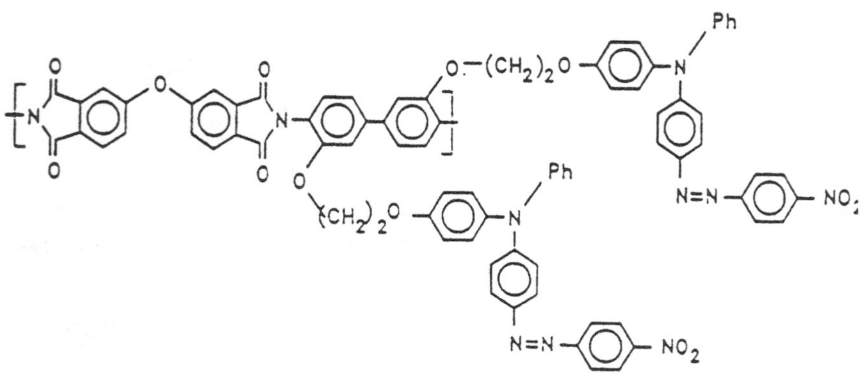

IBM-2

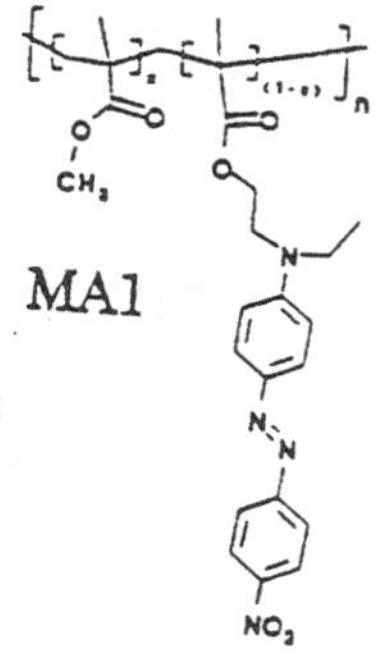

MA1-MMA

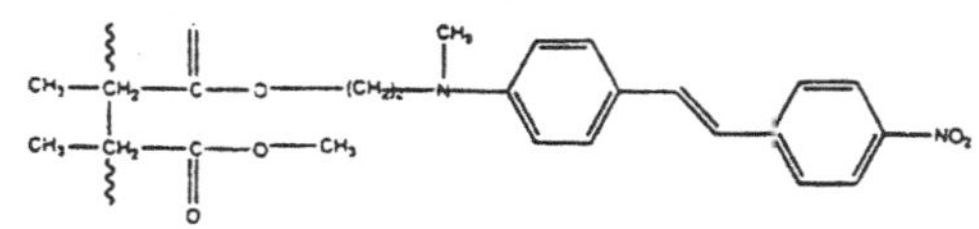

MMA-DANS

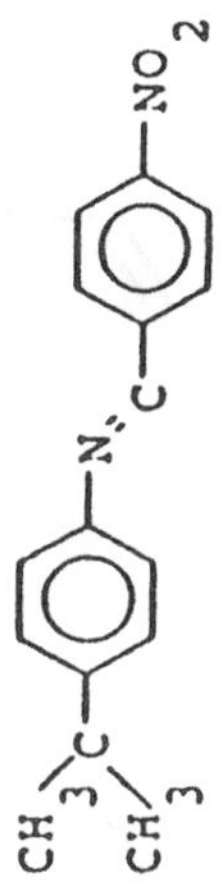

NMBA

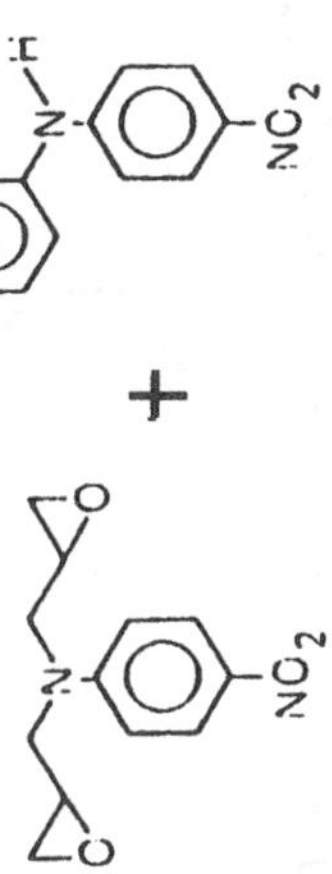

NNDN NAN

NNDN-NAN

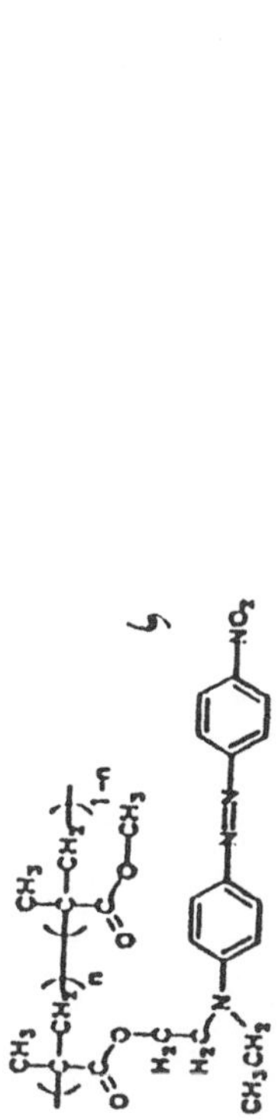

MMA-DR1

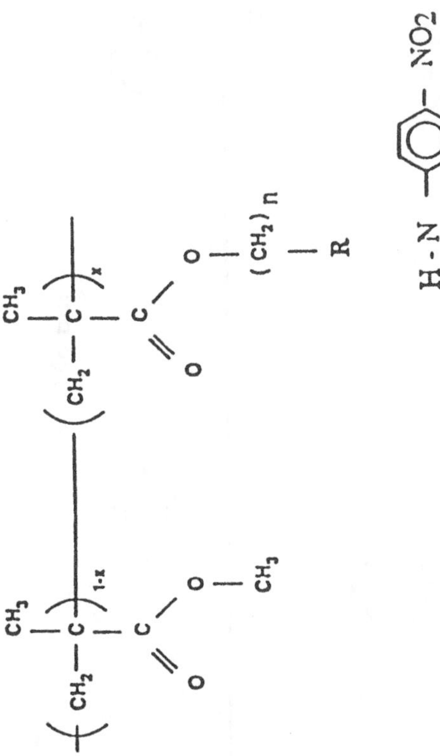

MMA-NA

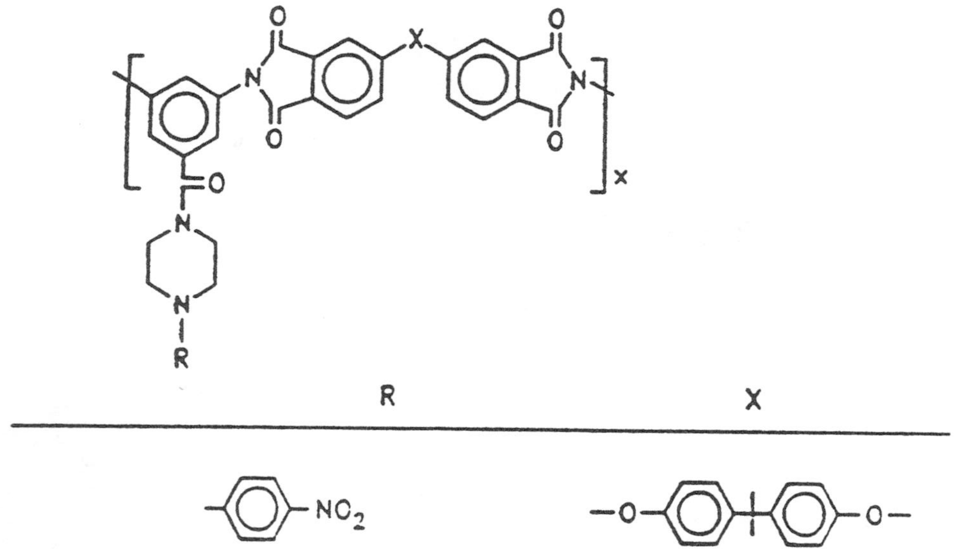

NNP-PEI

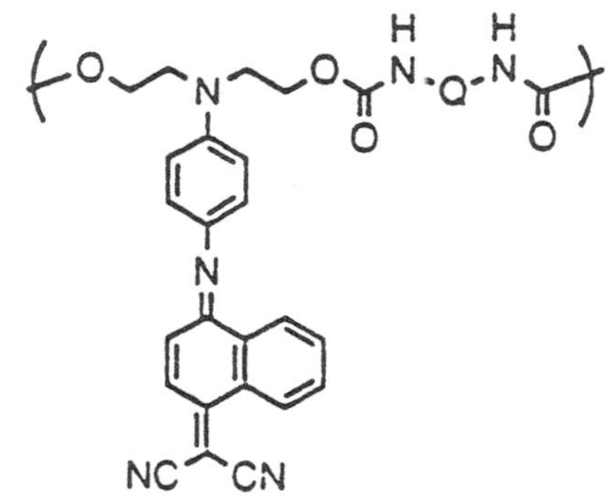

NQP2

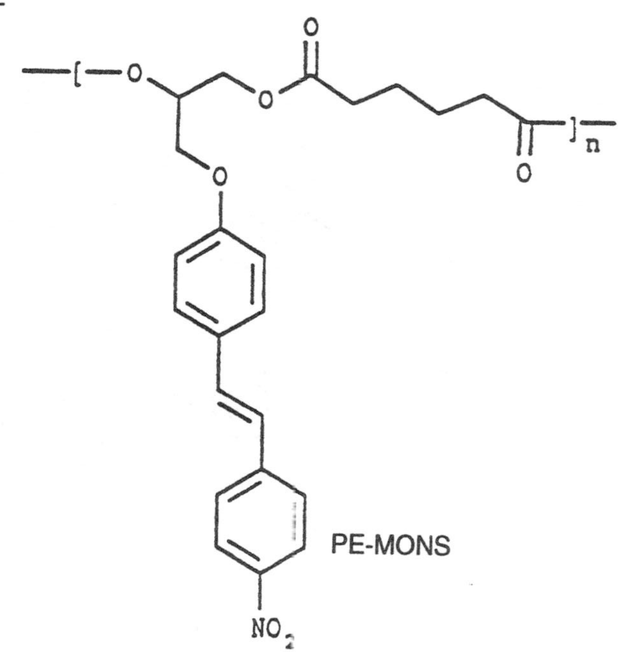

PE-MONS

DASS-6

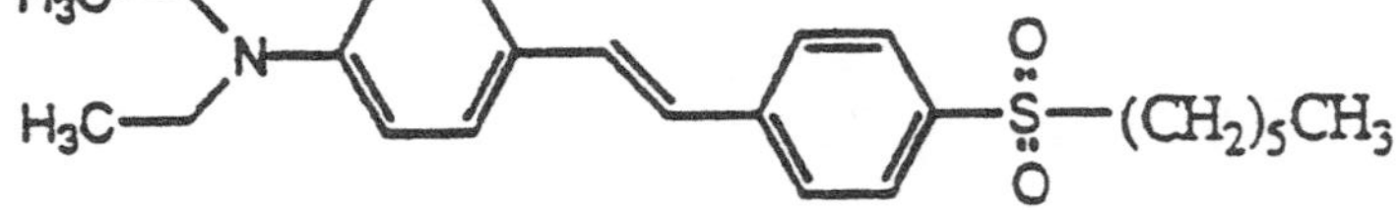

PQ12200/DASS

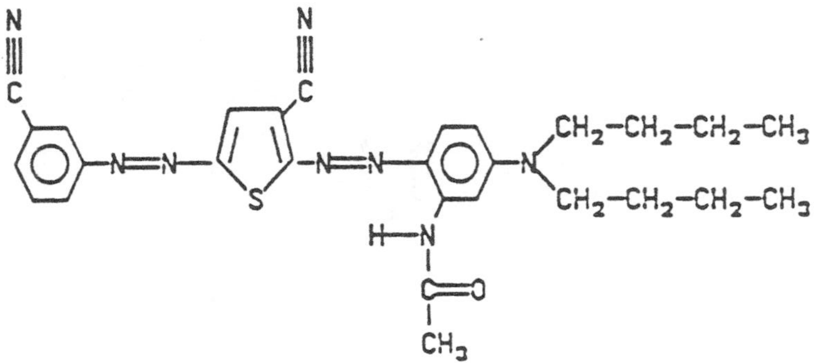

PMMA/CNET-S

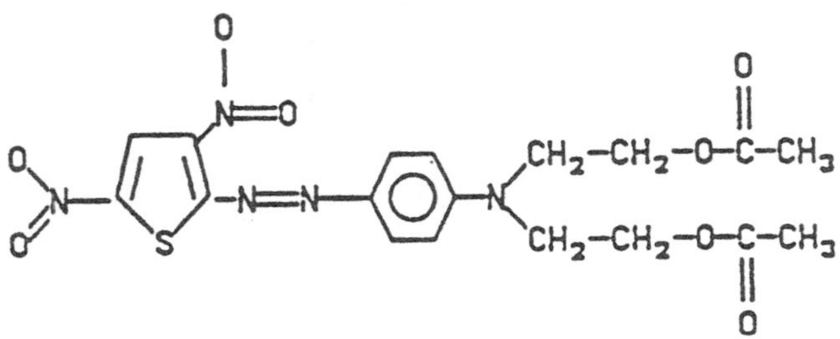

PMMA/CNET-TP

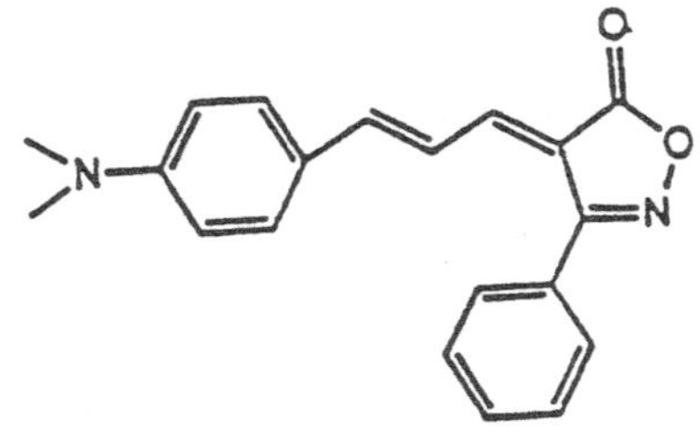

PMMA/JPL-A

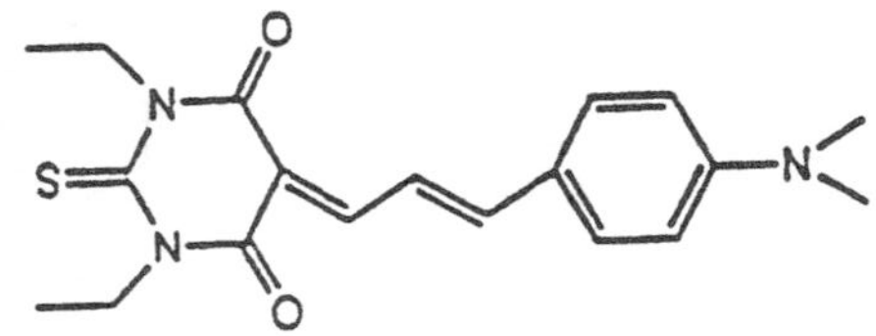

PMMA/JPL-B

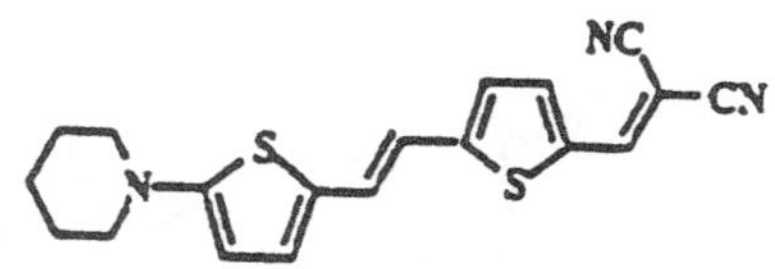

PMMA/JPL6

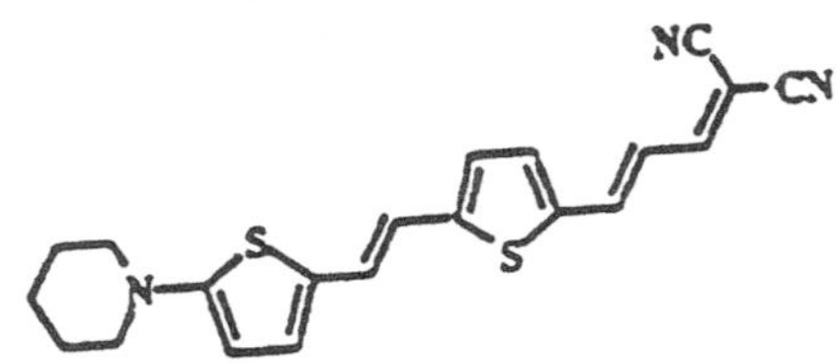

PMMA/JPL7

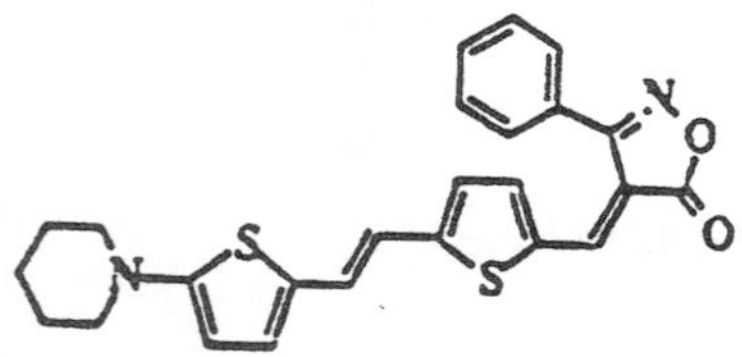

PMMA/JPL8

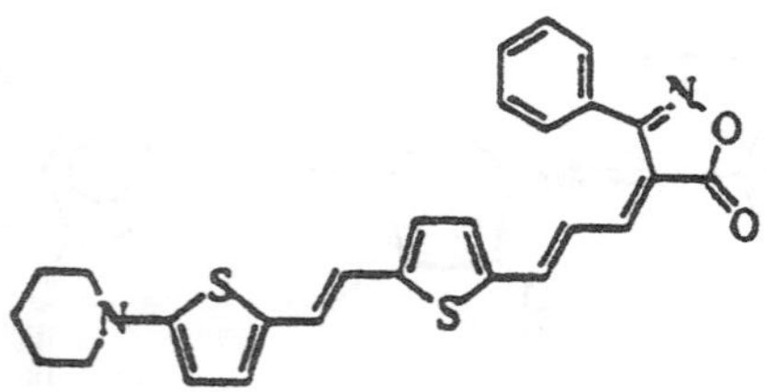

PMMA/JPL9

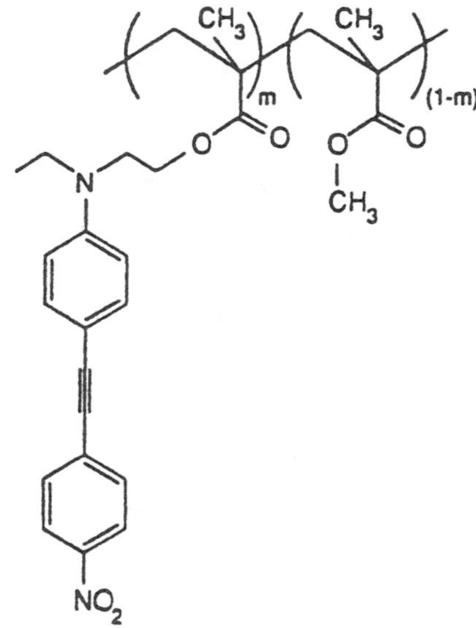

PMMA-NAT

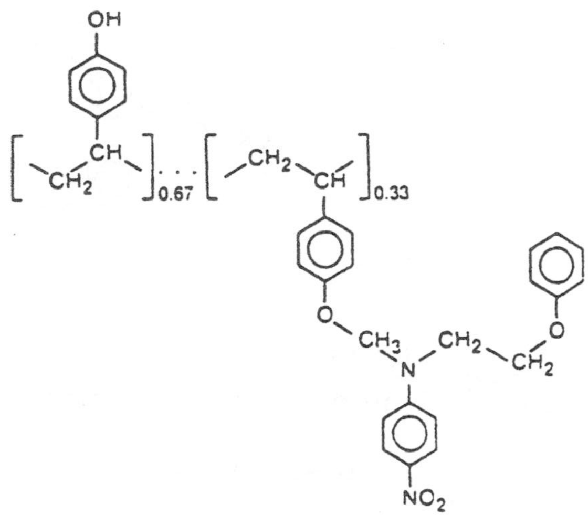

PPNA

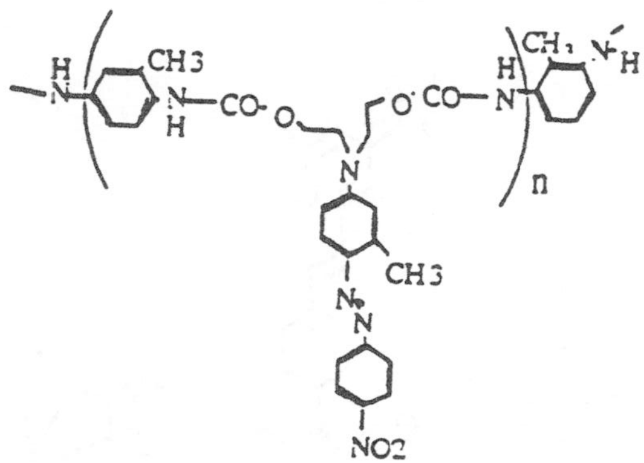

PU11

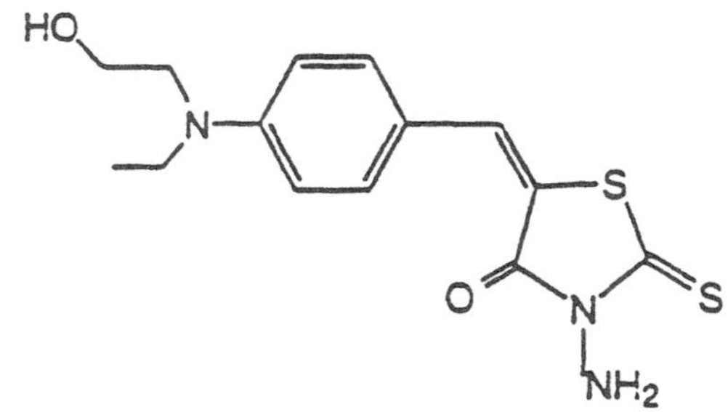

RA-HDT

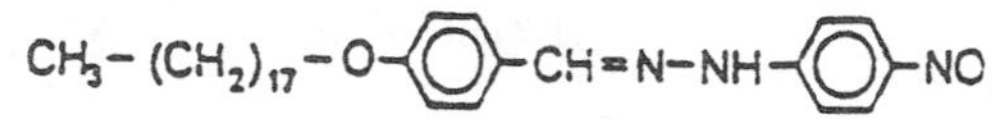

A

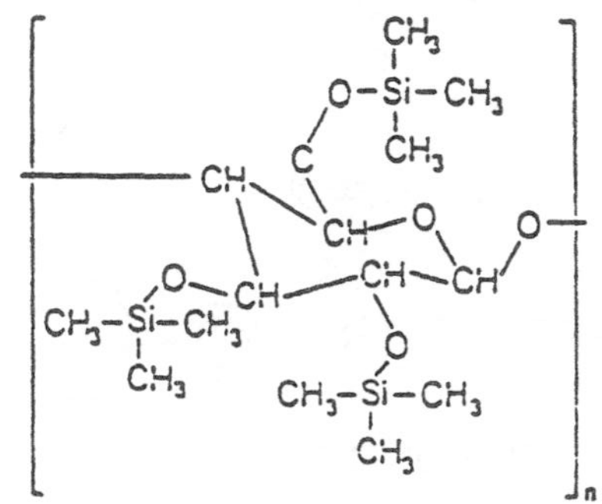

B

RIKEN-LBK1

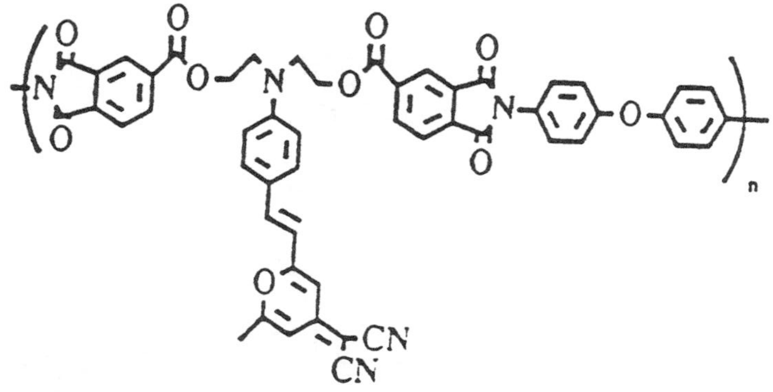

ROI-4

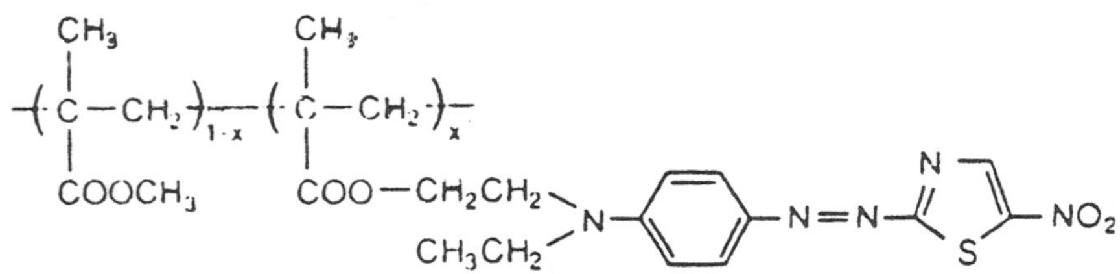

TR

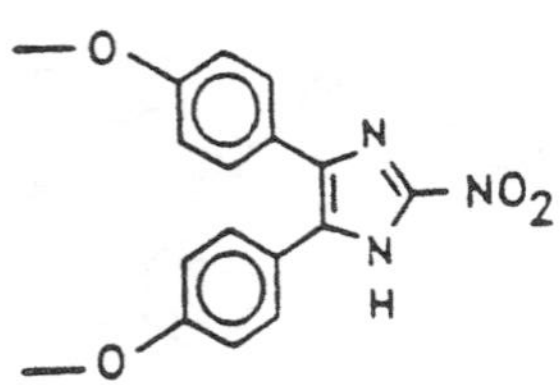

Ultem/IBM-A

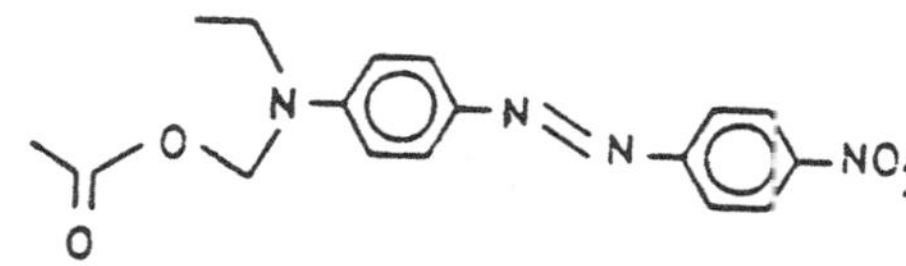

Ultem/IBM-B

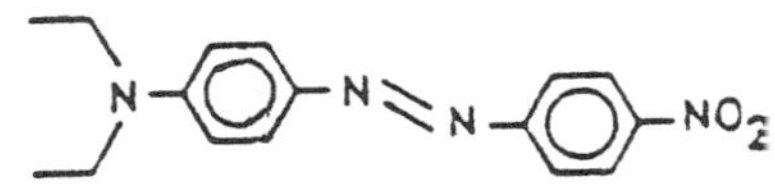

Ultem/IBM-C

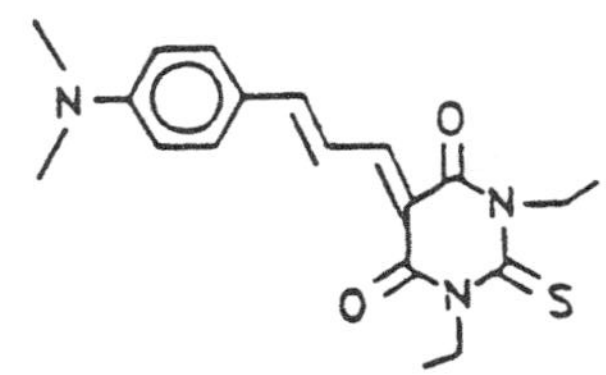

Ultem/IBM-D

REFERENCES

1. Kaminow, I. P., *An Introduction to Electro-Optic Devices*, Academic Press, New York, 1974, pp. 40–53.
2. Singer, K. D., Kuzyk, M. G., and Sohn, J. E., Second-order nonlinear-optical processes in orientationally ordered materials: relationship between molecular and macroscopic properties, *J. Opt. Soc. Am. B*, *4*, 968 (1987).
3. Singer, K. D., Lalama, S. J., Sohn, J. E., and Small, R. D., Electro-Optic Organic Materials, in *Nonlinear Optical Properties of Organic Molecules and Crystals Vol. 1*, (D. S., Chemla, and J. Zyss, eds.), Academic Press, New York, 1987.
4. Korotky, K. K., and Alferness, R. C., $TiLiNbO_3$ Integrated Optics Technology, in *Integrated Optical Circuits and Components* (L. D. Hutcheson, ed.) Marcel Dekker, New York, 1987, p. 169.
5. Girton, D. G., Kwiatkowski, S. L., Lipscomb, G. F., and Lytel, R. S., 20 GHz electro-optic polymer Mach-Zehnder modulator, *Appl. Phys. Lett.*, *58*, 1730 (1991).
6. Teng, C. C., Traveling-wave polymeric optical intensity modulator with more than 40 GHz of 3-dB electrical bandwidth, *Appl. Phys. Lett.*, *60*, 1538 (1992).
7. Ermer, S. E., Anderson, W. W., van Eck, T. E., Girton, D. G., Marley, J. A., Harwit, A., Lovejoy, S. M., and Leung, D. S., Progress Toward Practical Polymer Electro-Optic Devices, in *Abstracts of Organic Thin Films for Photonic Applications*, OSA 1995 Technical Digest Series, Portland, Oregon, 1995, *Vol. 21*, p. 285.
8. Wang, W. W., Chen, D., Fetterman, H. R., Shi, Y., Bechtel, J. H., Kalluri, S., Steier, W. H., and Dalton, L. R., 60 Ghz Electro-optic Modulation by Polymer Waveguide Phase Modulators, in *Abstracts of Organic Thin Films for Photonic Applications*, OSA 1995 Technical Digest Series, Portland, Oregon, 1995, *Vol. 21*, p. 293.
9. Wang, W. W., Chen, D., and Fetterman, H. R., Optical heterodyne detection of 60 GHz electro-optic modulation from polymer waveguide modulators, *Appl. Phys. Lett.*, *67*, 1806 (1995).
10. Yariv, A., *Quantum Electronics*, John Wiley & Sons, New York, 1975, Chapter 14.
11. Lipscomb, G. F., Garito, A. F., and Narang, R. S., An exceptionally large linear electro-optic effect in the organic solid MNA, *J. Chem. Phys.*, *75*, 1509 (1981).
12. Ho, E. S. S., Iizukp, K., Freundorfer, A. P., and Wah, C. K. L., An Organic MNA Crystal Phase Modulator, *J. Lightwave Tech.*, *9*, 101 (1991).
13. Lipscomb, G. F., Garito, A. F., and Narang, R. S., A large linear electro-optic effect in a polar organic crystal 2-methyl-4-nitroaniline, *Appl. Phys. Lett.*, *38*, 633 (1981).
14. Wah, C. K. L., Iizuka, K., and Freundorfer, A. P., 94-Ghz phase modulation of light by organic MNA (2-methyl-4-nitroaniline) crystal, *Appl. Phys. Lett.*, *63*, 3110 (1993).
15. Perry, J. W., Marder, S. R., Perry, K. J., Sleva, E. T., Yakymyshyn, C., Stewart, K. R., and Boden, E. P., Organic Salts With Large Electro-Optic Coefficients, *Proc. SPIE*, *1560*, 302 (1991).
16. Pan, F., Knopfle, G., Bosshard, Ch., Follonier, S., Spreiter, R., Wong, M. S., and Gunther, P., Electro-optic properties of the organic salt 4-N,N-dimethylamino-4′-N′-methyl-stilbazolium tosylate," *Appl. Phys. Lett.*, *69*, 13 (1996).
17. Anlheim, M., Brédas, J.-L., Barzoukas, M., Bedworth, P. V., Cai, Y., Dehu, C., Blanchard-Desce, M., Fort, A., Hu, Z., Jen, A. K.-Y., Marder, S. R., Meyers, F., Perry, J. W., Runser, C., Staehelin, M., and Zysset, B., Experimental and Theoretical Studies of Nonlinear Optical Chromophores, OSA 1995 Technical Digest Series, Portland, Oregon, 1995, *Vol. 21*, p. 244.
18. Havings, E. E., and van Pelt, P., Electrochromism of Substituted Polyalkenes in Polymer Matrices; Influence of Chain Length on Charge Transfer, *Ber. Bunsenges. Phys. Chem.*, *83*, 816 (1979).

19. Singer, K. D., Sohn, J. E., and Lalama, S. J., Second harmonic generation in poled polymer films, *Appl. Phys. Lett.*, *49*, 248 (1986).
20. Berkovic, G., Yitzchaik, S., and Krongauz, V., In-plane poling of doped polymer films: Creation of two asymmetric directions for second order optical nonlinearity, *Proc. SPIE*, *1560*, 238 (1991).
21. Wu, J. W., Valley, J. F., Stiller, M., Ermer, S., Binkley, E. S., Kennery, J. T., Lipscomb, G. F., and Lytel, R., Poled polyimides as thermally stable electro-optic polymer, *Proc. SPIE*, *1560*, 196 (1991).
22. Kerr, J., *Phil. Mag*, *50*, 337 (1875).
23. Kuzyk, M. G., Sohn, J. E., and Dirk, C. W., Mechanisms, of quadratic electro-optic modulation of dye doped polymer films, *J. Opt. Soc. Am. B*, *7*, 842 (1990).
24. Goodson III, T., and Wang, C. H., Third Order Nonlinear Optical Effects of a Guest-Host System Investigated by Quadratic Electro-Optic and Quadratic Electro-Absorption Measurements, *Proc. SPIE*, *2527*, 286 (1995).
25. Poga, C., Brown, T. M., Kuzyk, M. G., and Dirk, C. W., Characterization of the excited states of a squaraine molecule with quadratic electroabsorption spectroscopy, *J. Opt. Soc. Am. B*, *12*, 531 (1995).
26. O'Brien, N., Dominic, V., Caracci, S., Electro-refraction and electro-absorption in poled-polymer films, OSA Technical Digest Series, Portland, Oregon, 1995 *Vol. 21*.
27. Norwood, R. A., Kuzyk, M. G., and Keosian, R. A., Electro-optic tensor ratio determination of side-chain copolymers with electro-optic interferometry, *J. Appl. Phys.*, *75*, 1869 (1994).
28. Teng, C. C., and Mann, H. T., Simple reflection technique for measuring the electro-optic coefficient of poled polymers, *Appl. Phys. Lett.*, *56*, 1734 (1990).
29. DeMartino, R. N., Allen, D. E., Keosian, R., Khanarian, G., and Haas, D. R., Side-Chain Polymers for Electro-Optic Applications, *Mat. Res. Soc. Symp. Proc.*, 1992, *Vol. 228*, p. 39.
30. Mortazavi, M. A., Knoesen, A., Kowel, S. T., Henry, R. A., Hoover, J. M., and Lindsay, G. A., Second-Order Nonlinear Optical Properties of Poled Coumaromethacrylate Co-polymers, *Appl. Phys. B*, *53*, 287 (1991).
31. Levy, Y., Dumont, M., Chastaing, E., Robin, P., Chollet, P. A., Gadret, G., and Kajzar, Reflection Method for Electro-Optic Coefficient Determination in Stratified Thin Film Structures, *Nonlinear Optics*, *4*, 1 (1993).
32. Schildkraut, J. S., Determination of the electrooptic coefficient of a poled polymer film, *Applied Optics*, *29*, 2839 (1990).
33. Eldering, C. A., Knoesen, A., and Kowel, S. T., Characterization of polymeric films using metal mirror/electrode Fabry-Perot etalons, *Proc. SPIE*, *1337*, 348 (1990).
34. van der Vorst, C. P. J. M., and van Weerdenburg, C. J. M., Electro-optic Measurements in Poled Polymer Films Between Plane Parallel, Semitransparent Metal Electrodes, *Proc. SPIE*, *1337*, 246 (1990).
35. Kobayashi, T., Uchiki, H., and Minoshima, K., Electro-Optic Constand of Doped Polymers by a Novel Method, in *Springer Proceedings in Physics*, *Vol. 36*, *Nonlinear Optics of Organics and Semiconductors* (T. Kobayashi ed.), Springer Verlag, Berlin, 1989, p. 140.
36. Dentan, V., Levy, Y., Dumont, M., Robin, P., and Chastaing, E., Electrooptic Properties of a Ferroelectric Polymer Studied by Attenuated Total Reflection, *Optics Communications*, *69*, 379 (1988).
37. Herminghaus, S., Smith, B. A., and Swalen, J. D., Electro-optic coefficients in electric-field-poled polymer waveguides, *J. Opt. Soc. Am. B*, *8*, 2311 (1991).
38. Smith, B. A., Herminghaus, S., and Swalen, J. D., Electrooptic coefficients in electric field poled polymer waveguides, *Proc. SPIE*, *1560*, 400 (1991).

39. Singer, K. D., Kuzyk, M. G., Holland, W. R., Sohn, J. E., and Lalama, S. J., Electro-optic phase modulation and optical second harmonic generation in corona-poled polymer films, *Appl. Phys. Lett.*, *53*, 1800 (1988).
40. Valley, J. F., Wu, J. W., and Valencia, C. L., Heterodyne measurement and poling transient effects in electro-optic polymer thin films, *Appl. Phys. Lett.*, *57*, 1084 (1990).
41. Valley, J. F., and Wu, J. W., Measurement of electro-optic poling transient effects in polymer thin films, *Proc. SPIE*, *1337*, 226 (1990).
42. Shi, Y., Bechtel, J. H., Kalluri, S., Steier, W. H., Xu, C., Wu, B., and Dalton, L. R., Measurement of the electro-optic coefficient dispersion in poled polymer materials, *Proc. SPIE*, *2285*, 131 (1994).
43. Thackara, J. I., Jurich, M., and Swalen, J. D., Electro-optic measurements of tolane-based polymeric phase modulators, *J. Opt. Soc. Am. B*, *11*, 835 (1994).
44. Kalluri, S., Chen, A., Chuyanov, V., Ziari, M., and Steier, W. H., Integration of Polymer Electrooptic Devices on Non-Planar Silicon Integrated Circuits, *Proc. SPIE*, *2527*, 375 (1995).
45. Smith, B. A., Jurich, M., Moerner, W. E., Volksen, W., Best, M. E., Fleming, W., Swalen, J. D., and Bjorklund, G. C., Lightwave transmission of multiple television signals using an organic polymer electro-optic phase modulator, *Proc. SPIE*, *2025*, 499 (1993).
46. Lipscomb, G. F., Lytel, R. S., Ticknor, A. J., Van Eck, T. E., Girton, D. G., Ermer, S. P., Valley, J. F., Kenney, J., and Binkley, E., Organic Electro-Optic Devices for Optical Interconnections, *Proc. SPIE*, *1560*, 388 (1991).
47. Findakly, T., and Teng, C. C., Wideband NLO Polymer Modulators, *Proc. SPIE*, *2025*, 526 (1993).
48. Mohlman, G. R., Horsthuis, W. H. G., McDonach, A., Copeland, M. J., Diemeer, M. B. J., Trommel, E. S., and Suyten, F. M. M., Optically nonlinear polymeric switches and modulators, *Proc. SPIE*, *1337*, 215 (1990).
49. Van Eck, T. E., Ticknor, A. J., Lytel, R. S., and Lipscomb, G. F., Complementary optical tap fabricated in an electro-optic polymer waveguide, *Appl. Phys. Lett.*, *58*, 1588 (1991).
50. Ashley, P. R., and Gulotty, R. J., Polymeric Devices for Fiber Optical Gyroscope Applications, ACS Ntl Mtg Abstracts, Washington, D.C., 1994, p. 215.
51. Thackara, J. I., Chon, J. C., Bjorklund, G. C., Volksen, W., and Burland, D. M., Polymeric electro-optic Mach-Zehnder switches, *Appl. Phys. Lett.*, *67*, 3874 (1995).
52. Khanariuan, G., Che, T., DeMartino, R. N., Haas, D., Leslie, T., Man, H. T., Sansone, M., Stamatoff, J. B., Teng, C. C., and Yoon, H. N., Characterization of polymeric nonlinear optical materials, *Proc. SPIE*, *824*, 72 (1987).
53. Lipscomb, G. F., Experimental Studies of Electro-Optic and Ferroelectric Phenomena in Organic Solids, University Microfilms, Ann Arbor, 1980.
54. Nye, J. F., Physical Properties of Crystals, Clarendon Press, Oxford, 1957.
55. Sigelle, M., and Hierle, R., Determination of the electrooptic coefficients of 3-methyl 4-nitropyridine 1-oxide by an interferometric phase modulation technique, *J. Appl. Phys.*, *52*, 4199 (1981).
56. Bosshard, Ch., Sutter, K., Schlesser, R., and Gunther, P., Electro-optic effects in organic single crystals, *Proc. SPIE*, *1775*, 271 (1992).
57. Poga, C., Kuzyk, M. G., and Dirk, C. W., Quadratic electroabsorption studies of third-order susceptibility mechanisms in dye doped polymers, *J. Opt. Soc. Am. B*, *11*, 80 (1994).
58. Kobayashi, T., Minoshima, K., Nomura, S., Fukaya, S., and Ueki, A., *Proc. SPIE*, *1147*, 182 (1989).
59. Gelsen, O. M., Halliday, D. A., Bradley, D. D. C., Burn, P. L., Holmes, A. B., Murata, H., Tsutsui, T., and Saito, S., *Mol. Cryst. Liq. Cryst.*, *216*, 117 (1992).

60. Mortazavi, M. A., Knoeson, A., Kowel, S. T., Higgens, B. G., and Dienes, A., *J. Opt. Soc. Am. B*, *6*, 733 (1989).
61. Page, R. H., Jurich, M. C., Reck, B., Sen, A., Twieg, R. J., Swalen, J. D., Bjorklund, G. C., and Willson, C. G., *J. Opt. Soc. Am. B*, *7*, 1239 (1990).
62. Kobayashi, T., Uchiki, H., and Minoshima, K., *Syn. Met.*, *28*, D 699 (1989).
63. Uchiki, H., and Kobayashi, T., *J. Appl. Phys.*, *64*, 2625 (1988).

6
Second Harmonic Generation

Kenneth D. Singer, Steven F. Hubbard, and Andrew Schober
Case Western Reserve University, Cleveland, Ohio

L. Michael Hayden and Kennita Johnson
University of Maryland Baltimore County, Baltimore, Maryland

I. INTRODUCTION

Interest in organic materials as second-order nonlinear optical materials has centered on molecular materials in which the nonlinearity arises mainly from an appropriately arranged collection of relatively independent molecular sources of optical nonlinearity. Until recently, most work has focused on the dipolar nonlinearity which, in model systems, arises from donor and acceptor substituted π-electron compounds possessing a substantial molecular dipole moment.[1] This type of molecular source is most effectively exploited in a polar macroscopic material. A number of approaches for generating polar macroscopic structures have been investigated. The earliest type which was vigorously studied was molecular crystals.[2] In this case, the molecular units are grown into a crystal usually by solution or melt techniques with the resulting crystal formed by dipolar (van der Waals) or hydrogen bonding interactions. Due to the high molecular density and the highly polar structure, in some cases a very high optical nonlinearity can result. These high nonlinearities, even based on relatively inefficient compounds by the most modern standards, surpassed previously studied inorganic crystals, and accounted for early interest in organic materials. This line of research has matured, and high-quality crystals for parametric interactions have been developed. Manufacturing and robustness of these crystals have always been at issue, but the current state-of-the-art includes crystals that are of commercial quality. In addition, some organic crystals provide functions unattainable in other crystals.[3]

Other approaches to forming polar materials for the potentially larger market for electrooptic devices have been favored due to their more natural ability to be processed into optical waveguides. These include monomolecular growth or deposition,[4,5] and poled polymer films. Poled polymers are the most mature of these approaches.[6,7,8,9] In this case, nonlinear optical molecules are dissolved or covalently attached in amorphous polymers. In a state of high molecular mobility, a strong electric field is applied to break the center of symmetry and impart polar order. With the field applied, the polymer is put into an immobile state by cooling or chemical reaction, thus locking in the polar order. In this approach, high-quality waveguides can be fabricated. The statistical nature of the alignment and the reduced density present is a challenge in attaining high nonlinearities. This has stimulated a great deal of work in increasing the nonlinearity per molecule, which accounts for much of the data presented in the Tables. Thus, the Tables are mainly divided between smaller molecules (and nonlinearities) for crystalline materials, and larger molecules (and nonlinearities) for polymeric materials.

In all cases—crystals, polymers, and monomolecular films—the origin of the observed macroscopic nonlinear optical susceptibility is molecular. That is, the bulk is considered to be a simple sum of the molecular nonlinearities with corrections taking into account the effects of local fields and simple interactions. With this in mind, most of the studies involving the origin of optical nonlinearity have focused on the molecular units, with oriented gas models applied to collections of molecules to determine the bulk nonlinearity. One can quantify this discussion by considering the molecular polarization oscillating at the second harmonic frequency, $p(2\omega) = \beta(2\omega;\ \omega,\ \omega)F^{\omega}F^{\omega}$, where β is the relevant molecular hyperpolarizability that is the subject of the Tables in this Chapter, and F^{ω} is the applied electric optical local field. Measurement of this second harmonic hyperpolarizability has been used to assess potential molecular compounds for use in second harmonic, electrooptic, and other second order nonlinear optical devices. This hyperpolarizability can be calculated using quantum chemical techniques. One approach is a sum over states perturbation technique, in which case the hyperpolarizability is proportional to a sum of quantum mechanical transition moments and energy denominators. Frequently, a two-level approximation is taken for highly dipolar molecules with a dominant transition to the lowest excited state. Also assuming a one-dimensional molecule (see Chapter 2), we have:

$$\beta_{xxx}(2\omega;\ \omega,\ \omega) = \frac{e^3}{\hbar^2}\,|\mu_{01}|^2\,\Delta\mu\,\frac{3\omega_0^2}{(\omega_0^2 - \omega^2)(\omega_0^2 - 4\omega^2)}, \qquad (1)$$

where e is the electronic charge, μ_{01} the transition moment, $\Delta\mu$ the difference between the excited state and groundstate dipole moments, ω_0 the frequency of the first excited state, and ω the fundamental measurement frequency. This equation has guided much of the development of new nonlinear optical chromophores. Once the hyperpolarizability is known, the macroscopic second harmonic susceptibility, in the weakly interacting approximation, is given by an appropriate sum over the molecular units taking local fields into account (see Chapter 3). Thus:

$$\chi^{(2)}(2\omega;\ \omega,\ \omega) \propto Nf\beta(2\omega;\ \omega,\ \omega) \qquad (2)$$

where N is the number density, and f the local field factors. The addition of the individual molecules must take into account their overall orientation or position.

In this Chapter, we focus on hyperpolarizability measurements using second harmonic generation relating to macroscopic material measurements in the weakly interacting approximation. We begin with a general description of coherent second harmonic generation from a bulk material. This theory of Maker fringes lays the foundation for a variety of measurements including crystal second harmonic generation, thin film second harmonic generation, and electric field induced second harmonic generation (EFISH). Four important techniques for measuring molecular nonlinearities—surface second harmonic generation; second harmonic generation in poled polymer films; EFISH; and Hyper-Rayleigh scattering (HRS)—are then described. The nature of the susceptibilities, the method for deducing the molecular susceptibility tensor components, measurement conventions, the role of local fields, and experimental techniques are described in the limit of weakly interacting molecules. Also included is a description of the relationship between molecular and bulk nonlinear optical susceptibilities. The Chapter ends with extensive tables of experimental results on second order nonlinear properties as deduced from second harmonic generation.

II. MACROSCOPIC MEASUREMENTS

The second harmonic (SH) coefficient d_{ijk} $(= \frac{1}{2}\chi^{(2)}_{ijk})$ can be determined through absolute and relative methods. Two absolute methods,[10] SH phase matching and parametric fluorescence, are the most precise, while a third method, Raman scattering, can give the relative sign of the coefficients (see Chapter 1). These absolute methods are difficult to implement since an accurate determination of the fundamental and SH beam powers is required. As a result, these experiments are rarely performed, and considerable controversy remains concerning the measured values, as described in Chapter 1. Two relative techniques, Maker fringes and the wedge technique, circumvent this problem and hence are typically favored.

A. Relative Measurements

For the relative techniques discussed here, a high-powered laser is made incident on the sample while the intensity of the transmitted frequency-doubled light is monitored as a function of the sample translation distance (wedge) or the angle of incidence (Maker fringe). In each case, the SH intensity generated by the unknown sample is compared to that generated from a known reference, typically quartz or some other crystal that has been carefully characterized by one of the absolute techniques mentioned above. (See Chapter 1.) The wedge and Maker fringe techniques are essentially the same technique except that in the wedge method the sample is translated perpendicular to the incident beam, while in the Maker fringe technique the sample is rotated about an axis that is perpendicular to the incident beam direction.

1. Wedge Technique

The wedge technique, with separate considerations for birefringence and absorption, has been fully detailed previously by Chemla and Kupecek.[11] As Kurtz[10] points out, the wedge technique is useful in the infrared where the long coherence lengths prevent the observation of the fringes necessary for the Maker fringe technique. A disadvantage of the wedge method is that the unknown sample must be made into the shape of a wedge. This requirement is sometimes not achievable with thin films or some crystals. A major advantage of the wedge method is that multiple reflections of the fundamental and second harmonic waves are eliminated. In their derivations, both Kurtz, and Chemla and Kupecek, use the boundary conditions as given by Jerphagnon and Kurtz (next section) for their description of Maker fringes. These boundary conditions are incomplete and give rise to errors in the determination of $\chi_{ijk}^{(2)}$ for dispersive materials. We will discuss the correct boundary conditions and apply them to the Maker fringe case in the next section, but note that the consequences apply to the wedge method also.

2. Maker Fringe Technique

Since 1970, the determination of the nonlinear optical (NLO) coefficients d_{ij} (i = 1-3, j = 1-6) associated with second harmonic generation (SHG) has been performed almost exclusively by the Maker fringe technique[12] as described in detail by Jerphagnon and Kurtz (JK).[13] Over the years, this technique has been applied to inorganic and organic crystals and organic polymers. However, some of the original assumptions made in the derivations in JK, while applicable to the transparent crystals of the day, are not uniformly applicable to all material classes currently under investigation. Also, the theory presented by JK did not consider absorbing materials nor the general case of anisotropic media. Many of the second order NLO organic materials under investigation today, such as liquid crystals, Langmuir-Blodgett films, poled polymers, and organic crystals, are absorbing at the wavelength of the typical SHG experiment and exhibit a fair degree of anisotropy. Unfortunately, during the past 24 years, the results of JK have been applied by many workers to systems that were either anisotropic or absorbing, resulting in errors in the determination of d_{ij}.

In solving the boundary value problem for SHG, JK used approximate boundary conditions for the SH waves by ignoring the SH wave reflected from the second interface when considering the boundary conditions at the first interface, and assumed, in portions of their calculations, that the nonlinear material had the same index of refraction at the fundamental frequency ω as it had at the second harmonic frequency 2ω. As a result, their expression for $p(\theta)$, the angular projection factor for the nonlinearity, is incorrect for materials where there is dispersion. These problems are not widely realized in the literature and continue to be propagated, even recently.[14,15,16] Some authors have realized the problem with $p(\theta)$ and have replaced it in the theory of JK, ad hoc, with the correct term.[17] As a result, a combination of JK with an ad hoc correction for the projection factor is sometimes used to analyze SHG experiments.

Ignoring reflections, expressions for the sum-frequency fields generated by an isotropic transparent nonlinear parallel slab were given three decades ago by Bloembergen and Pershan.[18] Subsequent to JK, Chemla and Kupecek[11] obtained Poynting vectors for nonlinear wedges in air for anisotropic nonabsorbing materials with 32-point symmetry and for absorbing isotropic materials, but, like JK, considered the boundary conditions for the input and output faces of the nonlinear material separately. Further, Okamoto et al.[19] obtained expressions for Maker fringes in anisotropic parallel slabs with $C_{\infty v}$ space symmetry, but, again, they used the JK boundary condition approach leading to complicated expressions that are not only difficult to interpret physically, but are also incorrect if reflections are kept and no assumptions are made about the dispersion. Recently, a new derivation of the Maker fringes using complete boundary conditions for the SH waves was presented by Herman and Hayden.[20] They provide a general solution for birefringent media and a separate solution for materials with non-zero absorption of the fundamental and SH wave. The theory includes the reflections of the SH wave in the nonlinear material and predicts the same results as JK for transparent, isotropic materials, but exhibits significant differences for materials with non-zero dispersion. The differences have implications for experimentalists who are investigating the order in poled polymers by the technique of measuring the ratio d_{31}/d_{33}. They also show that the effect of neglecting the birefringence in the analysis of Maker fringe data can result in large errors in the determination of d_{ij}. Since that theory is generally applicable to crystals, polymers, films, and bulk media, and includes the effects of multiple reflections of the harmonic waves, absorption, and material birefringence, we will briefly summarize it here. The effects of multiple reflections of the fundamental are not described here, although Neher[21] has included them for the case of third harmonic generation. They found that for the case where the nonlinear film is on the side of the substrate facing the incident beam (as in Fig. 1), the effect of neglecting the multiple reflections of the fundamental is small.

We take as an example structure one that is frequently found in the field of nonlinear optical polymers: a nonlinear, non-absorbing, birefringement film supported by a linear, nonabsorbing isotropic substrate as shown in Fig. 1. We assume that an electromagnetic wave of frequency ω is incident upon the structure from the bottom at an angle θ (the x-z plane is the plane of incidence). For a frequency $m\omega$ (m = 1 for the fundamental and m = 2 for the second harmonic) the nonlinear layer of thickness L has an index n_m, and the substrate has an index n_{ms}. For free-standing films or crystal plates, setting the substrate index n_{2s} equal to one will suffice.

B. Birefringent Case

In a uniaxial material, the dielectric permittivity is a second rank tensor and can be represented as:

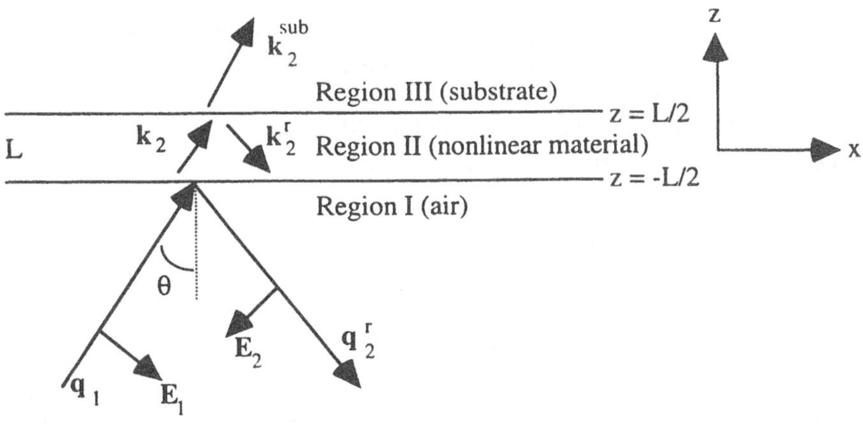

Figure 1 Three-layer slab geometry for SHG due to a nonlinear layer in Region II with the origin at the center of the nonlinear film.

$$\overleftrightarrow{\varepsilon}_i = \begin{pmatrix} n_{io}^2 & 0 & 0 \\ 0 & n_{io}^2 & 0 \\ 0 & 0 & n_{ie}^2 \end{pmatrix}, \quad i = 1,2, \tag{3}$$

where n_{io} and n_{ie} are the ordinary and extraordinary indices of refraction at the fundamental ($i = 1$) and SH ($i = 2$) frequencies respectively. If we write the electric field of the incident fundamental wave as:

$$\mathbf{E}_{1v} = \hat{\mathbf{e}}_v E_1 \exp[i(\mathbf{q}_1 \cdot \mathbf{r} - \omega t)], \tag{4}$$

with $\hat{\mathbf{e}}_v = (0,1,0)$ or $(\cos\theta,0,-\sin\theta)$ for $v = s$ or p respectively and $\mathbf{q}_1 = (\theta/c)(\sin\theta,0,\cos\theta)$, then for a p-polarized input, the SH fields in the three regions shown in Fig. 1 are given in Region I by:

$$\mathbf{E}_2 = R\hat{\mathbf{e}}_p^r \exp\left[i\mathbf{q}_2^r \cdot \left(\mathbf{r} + \frac{L}{2}\hat{\mathbf{z}}\right)\right], \tag{5a}$$

$$\mathbf{H}_2 = -\frac{ic}{2\omega}\nabla \times \mathbf{E} = R\hat{\mathbf{e}}_s^r \exp\left[i\mathbf{q}_2^r \cdot \left(\mathbf{r} + \frac{L}{2}\hat{\mathbf{z}}\right)\right], \tag{5b}$$

in Region II by:

$$\mathbf{E}_2 = \mathbf{e}_b \exp[i2\mathbf{k}_1 \cdot \mathbf{r}] + A\hat{\mathbf{e}}_{2p} \exp[i\mathbf{k}_2 \cdot \mathbf{r}] + B\hat{\mathbf{e}}_{2p}^r \exp[i\mathbf{k}_2^r \cdot \mathbf{r}], \tag{6a}$$

and

$$\mathbf{H}_2 = \mathbf{h}_b \exp[i2\mathbf{k}_1 \cdot \mathbf{r}] + n_2(\theta_2) \times \cos\gamma_2(A\hat{\mathbf{e}}_{2s} \exp[i\mathbf{k}_2 \cdot \mathbf{r}] + B\hat{\mathbf{e}}_{2s}^r \exp[i\mathbf{k}_{2r} \cdot \mathbf{r}]), \tag{6b}$$

and in Region III by:

$$\mathbf{E}_2 = T\hat{\mathbf{e}}_{2p}^{sub} \exp\left[i\mathbf{k}_2^{sub} \cdot \left(\mathbf{r} - \frac{L}{2}\hat{\mathbf{z}}\right)\right], \tag{7a}$$

and

$$\mathbf{H}_2 = n_{2s}T\hat{\mathbf{e}}_{2s}^{sub} \exp\left[i\mathbf{k}_2^{sub} \cdot \left(\mathbf{r} - \frac{L}{2}\hat{\mathbf{z}}\right)\right], \tag{7b}$$

where $\hat{\mathbf{e}}_v^r = (0,1,0)$ or $(-\cos\theta,0,-\sin\theta)$, for $v = s$ or p respectively, is the polarization unit vector for the reflected SH field in the air. A and B are the complex amplitudes for the forward and backward SH free waves in the nonlinear media, respectively, while T and R are the corresponding SH amplitudes in the substrate and air, respectively. The polarization unit vectors for the forward and backward SH waves in the nonlinear media are:

$$\hat{\mathbf{e}}_m = (\cos(\theta_m - \gamma_m),0,-\sin(\theta_m - \gamma_m)), \qquad m = 1,2 \tag{8a}$$

and

$$\hat{\mathbf{e}}_2^r = (-\cos(\theta_2 - \gamma_2),0,-\sin(\theta_2 - \gamma_2)). \tag{8b}$$

The walkoff angle γ_m is defined in Fig. 2. For the substrate, $\hat{\mathbf{e}}_{2s}^{sub} = (0,1,0)$ and $\hat{\mathbf{e}}_{2p}^{sub} = (c_{2s},0,-s_{2s})$. The wave vectors are (outside the slab) $\mathbf{q}_2^r = (2\omega/c)(\sin\theta,0,-\cos\theta)$ and (in the substrate) $\mathbf{k}_2^{sub} = (2\omega n_{2s}/c)(s_{2s},0,c_{2s})$, where $s_m = (1/n_m)\sin\theta$ and $c_m = \sqrt{1 - s_m^2}$.

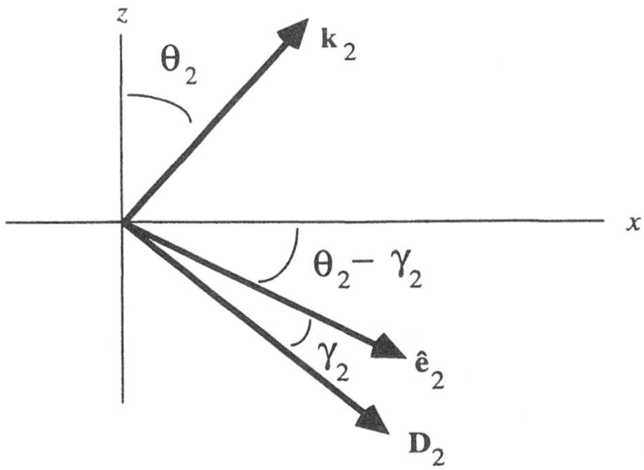

Figure 2 Definition of the walk-off angle γ in a birefringent medium.

Maxwell's Equations require that the SH field $\mathbf{E}_2$ satisfy:

$$\nabla \times \nabla \times \mathbf{E}_2 - \left(\frac{2\omega}{c}\right)^2 \overleftrightarrow{\varepsilon}_2 \cdot \mathbf{E}_2 = \left(\frac{2\omega}{c}\right)^2 4\pi \mathbf{P}^{NL} \exp[2i\mathbf{k}_1 \cdot \mathbf{r}], \tag{9}$$

where use is made of the electric displacement vector $\mathbf{D}_2 = \overleftrightarrow{\varepsilon}_2 \cdot \mathbf{E}_2 + 4\pi\mathbf{P}^{NL} \times \exp[2i\mathbf{k}_1 \cdot \mathbf{r}]$, $\overleftrightarrow{\varepsilon}$ is the permittivity tensor given in Eq. 3, and $\mathbf{P}^{NL}$ is the nonlinear polarization. In the linear media (air, substrate), $\mathbf{P}^{NL}$ is zero, and the analogous form of Eq. 9 is homogeneous.

In the nonlinear film (Region II of Fig. 1), the propagation vectors are:

$$\mathbf{k}_m = \frac{m\omega n_m(\theta_m)}{c}(s_m, 0, c_m), \qquad m = 1,2 \tag{10a}$$

and

$$\mathbf{k}_2^r = \frac{2\omega n_2(\theta_2)}{c}(s_2, 0, -c_2), \tag{10b}$$

with the angle-dependent refractive indices $n_m(\theta_m)$ given by:

$$n_i(\theta_i) = \left(\frac{\cos^2\theta_i}{n_{io}^2} + \frac{\sin^2\theta_i}{n_{ie}^2}\right)^{-1/2} \qquad i = 1,2 \tag{11a}$$

and

$$n_2(\theta_1) = \left(\frac{\cos^2\theta_1}{n_{2o}^2} + \frac{\sin^2\theta_1}{n_{2e}^2}\right)^{-1/2}. \tag{11b}$$

While $\mathbf{D}_2$ is still perpendicular to $\mathbf{k}_2$, the electric field deviates from the perpendicular by the walk-off angle[22] γ, as shown in Fig. 2.

The bound wave, $\mathbf{E}_{2b}$, is a particular solution to the inhomogeneous Eq. 9 in the nonlinear material.[23] Inserting $\mathbf{E}_{2b} = \mathbf{e}_b \exp[i\mathbf{k}_b \cdot \mathbf{r}]$, with $\mathbf{k}_b = 2\mathbf{k}_1$, into Eq. 9 gives the following equations for the bound wave amplitude $\mathbf{e}_b$:

$$\mathbf{e}_b = \frac{4\pi n_2^2(\theta_1)}{n_1^2(\theta_1) - n_2^2(\theta_1)} \left[(\overleftrightarrow{\varepsilon}_2)^{-1} \cdot \mathbf{P}^{NL} - \frac{\mathbf{k}_b(\mathbf{k}_b \cdot \mathbf{P}^{NL})}{((^{2\omega}/_c) n_{2o} n_{2e})^2} \right], \tag{12a}$$

where

$$\mathbf{h}_b = n_1(\theta_1)\hat{\mathbf{k}}_1 \times \mathbf{e}_b, \tag{12b}$$

and

$$\mathbf{P}^{NL} \equiv |E_1|^2 \overleftrightarrow{\mathbf{d}}:\hat{\mathbf{e}}_1\hat{\mathbf{e}}_1. \tag{12c}$$

Using $\mathbf{D} = \overleftrightarrow{\varepsilon} \cdot \mathbf{E}$ and $\mathbf{D} = [n(\theta)/c]^2\, \hat{\mathbf{k}} \times (\mathbf{E} \times \hat{\mathbf{k}})$ for the linear electric displacement field, we can derive the following useful relations:

$$\tan \gamma_m = \tfrac{1}{2} n_m(\theta_m)^2 \sin 2\theta_m \left(\frac{1}{n_{mo}^2} - \frac{1}{n_{me}^2} \right), \tag{13}$$

$$\cos \gamma_m = \frac{n_{mo} n_{me}}{n_m(\theta_m)\sqrt{n_{mo}^2 + n_{me}^2 - n_m(\theta_m)^2}}, \tag{14}$$

$$\sin \theta_m = \frac{n_{me} \sin \theta}{\sqrt{(n_{mo} n_{me})^2 + (n_{me}^2 - n_{mo}^2) \sin^2 \theta}}. \tag{15}$$

For the $s \rightarrow p$ case, the fundamental beam is s-polarized, so that $\gamma_1 = 0$ and:

$$\mathbf{k}_1 = (^{\omega}/_c) n_{1o}(s_1, 0, c_1) \tag{16a}$$

$$\hat{\mathbf{e}}_1 = (0,1,0), \tag{16b}$$

and

$$\sin \theta_1 = \sin \theta / n_{1o}, \tag{16c}$$

while Eqs. 8, 10, and 13–15 remain valid for the SH wave ($m = 2$).

The boundary conditions at the I-II interface ($z = -L/2$) requiring the continuity of E_{2x} and H_{2y} result in the equations:

$$-\cos(\theta)R = e_{bx} \exp(-i\phi_1) + \cos(\theta_2 - \gamma_2)[A \exp(-i\phi_2) - B \exp(i\phi_2)], \tag{17}$$

and

$$R = h_{by} \exp(-i\phi_1) + n_2(\theta_2) \cos \gamma_2 [A \exp(-i\phi_2) + B \exp(i\phi_2)], \tag{18}$$

respectively. Continuity of these field components at the II-III interface ($z = L/2$) yields:

$$c_{2s} T = e_{bx} \exp(i\phi_1) + \cos(\theta_2 - \gamma_2)[A \exp(i\phi_2) - B \exp(-i\phi_2)], \tag{19}$$

and

$$n_{2s} T = h_{by} \exp(i\phi_1) + n_2(\theta_2) \cos \gamma_2 [A \exp(i\phi_2) + B \exp(-i\phi_2)], \tag{20}$$

where $\phi_1 = k_1 c_1 L$ and $\phi_2 = k_2 c_2 L/2$. Note that in the JK approach, $B = 0$ in Eqs. 17–20. Solving Eqs. 17–20 for T we obtain the transmitted second harmonic field:

$$T = \frac{4\pi i t_{fs}^{(2p)}}{n_2(\theta_2)\cos\gamma_2\cos(\theta_2-\gamma_2)}\left(\frac{2\pi L}{\lambda}\right)\left[\frac{n_2(\theta_1)}{n_{2o}}\right]^2\left[\frac{n_1(\theta_1)^2-n_2(\theta_2)^2}{n_1(\theta_1)^2-n_2(\theta_1)^2}\right]$$
$$\times\frac{\left[\hat{\mathbf{e}}_2\cdot\mathbf{P}^{NL}\dfrac{\sin\Psi_{bi}}{\Psi_{bi}}\exp(-i\phi_2)-r_{af}^{(2p)}\hat{\mathbf{e}}_2^r\cdot\mathbf{P}^{NL}\dfrac{\sin\Phi_{bi}}{\Phi_{bi}}\exp(i\phi_2)\right]}{[\exp(-2i\phi_2)+r_{af}^{(2p)}r_{fs}^{(2p)}\exp(2i\phi_2)]}, \tag{22}$$

where

$$\Psi_{bi} = \phi_1 - \phi_2 = \frac{2\pi L}{\lambda}[n_1(\theta_1)c_1 - n_2(\theta_2)c_2], \tag{23a}$$

and

$$\Phi_{bi} = \phi_1 + \phi_2 = \frac{2\pi L}{\lambda}[n_1(\theta_1)c_1 + n_2(\theta_2)c_2]. \tag{23b}$$

The components of $\mathbf{P}^{NL}$ along the SH **E**-fields are given by:

$$\hat{\mathbf{e}}_2\cdot\mathbf{P}_{NL} = -\frac{8\pi}{c}d_{eff}I_1[t_{af}^{(1\gamma)}]^2, \tag{24a}$$

and

$$\hat{\mathbf{e}}_2^r\cdot\mathbf{P}^{NL} = -\frac{8\pi}{c}d_{eff}^rI_1[t_{af}^{(1\gamma)}]^2, \tag{24b}$$

where I_1 is the intensity of the incident fundamental wave, and $t_{af}^{(1\gamma)}$ is the transmission coefficient for the fundamental at the air-film interface:

$$t_{af}^{(1\gamma)} = \begin{cases}\dfrac{2\cos\theta}{\cos(\theta_1-\gamma_1)+n_1(\theta_1)\cos\gamma_1\cos\theta} & \text{for } \gamma = p\\[2ex] \dfrac{2\cos\theta}{\cos\theta_1+n_{1o}\cos\theta} & \text{for } \gamma = s.\end{cases} \tag{25}$$

The transmission coefficient for the SH wave at the film-substrate interface is:

$$t_{fs}^{(2p)} = \frac{2n_2(\theta_2)\cos\gamma_2\cos(\theta_2-\gamma_2)}{n_{2s}\cos(\theta_2-\gamma_2)+n_2(\theta_2)c_{2s}\cos\gamma_2}, \tag{26}$$

and the p-polarized reflection coefficients for the SH wave at the air-film and film-substrate interfaces for the SH wave are:

$$r_{af}^{(2p)} = -\frac{\cos(\theta_2-\gamma_2)-n_2(\theta_2)\cos\gamma_2\cos\theta}{\cos(\theta_2-\gamma_2)+n_2(\theta_2)\cos\gamma_2\cos\theta}, \tag{27}$$

and

$$r_{fs}^{(2p)} = \frac{n_{2s}\cos(\theta_2-\gamma_2)-n_2(\theta_2)c_{2s}\cos\gamma_2}{n_{2s}\cos(\theta_2-\gamma_2)+n_2(\theta_2)c_{2s}\cos\gamma_2}. \tag{28}$$

The effective d-coefficients for $C_{\infty v}$ space symmetry are:

$$d_{eff} = -\hat{\mathbf{e}}_2 \cdot \vec{\mathbf{d}}:\hat{\mathbf{e}}_1\hat{\mathbf{e}}_1 \tag{29a}$$
$$= \begin{cases} d_{15}\cos(\theta_2 - \gamma_2)\sin 2(\theta_1 - \gamma_1) + \sin(\theta_2 - \gamma_2) & p \to p \\ \quad \cdot [d_{31}\cos^2(\theta_1 - \gamma_1) + d_{33}\sin^2(\theta_1 - \gamma_1)], & \\ d_{31}\sin(\theta_2 - \gamma_2), & s \to p \end{cases}$$

and

$$d^r_{ef} = -\hat{\mathbf{e}}^r_2 \cdot \vec{\mathbf{d}}:\hat{\mathbf{e}}_1\hat{\mathbf{e}}_1 \tag{29b}$$
$$= \begin{cases} -d_{15}\cos(\theta_2 - \gamma_2)\sin 2(\theta_1 - \gamma_1) + \sin(\theta_2 - \gamma_2) & p \to p \\ \quad \cdot [d_{31}\cos^2(\theta_1 - \gamma_1) + d_{33}\sin^2(\theta_1 - \gamma_1)], & \\ d_{31}\sin(\theta_2 - \gamma_2), & s \to p \end{cases}$$

where the notation $s \to p$ refers to an incident s-polarized fundamental and a p-polarized second harmonic field.

The SH power transmitted through the substrate into air is then given by $(c/8\pi)[t^{(2p)}_{sa}]^2|T|^2 A$, where A is the cross-sectional area of the fundamental beam, and the transmission coefficient for the p-polarized SH wave from the substrate into air is:

$$t^{(2p)}_{sa} = \frac{2n_{2s}c_{2s}}{n_{2s}\cos\theta + c_{2s}}. \tag{30}$$

Thus, we get:

$$P^{\gamma \to p}_{2\omega} = \frac{128\pi^3}{cA} \frac{[t^{(2p)}_{sa}]^2[t^{(2p)}_{fs}]^2[t^{(1\gamma)}_{af}]^4 P^2_\omega}{n_2(\theta_2)^2\cos^2\gamma_2\cos^2(\theta_2 - \gamma_2)}\left(\frac{2\pi L}{\lambda}\right)^2$$
$$\cdot \left[\frac{n_2(\theta_1)}{n_{2o}}\right]^4 \left[\frac{n_1(\theta_1)^2 - n_2(\theta_2)^2}{n_1(\theta_1)^2 - n_2(\theta_1)^2}\right]^2$$
$$\times d^2_{eff} \frac{\left\{\dfrac{\sin^2\Psi_{bi}}{\Psi^2_{bi}} + [r^{(2p)}_{af}]^2R^2\dfrac{\sin^2\Phi_{bi}}{\Phi^2_{bi}} - 2r^{(2p)}_{af}R\dfrac{\sin\Psi_{bi}}{\Psi_{bi}}\dfrac{\sin\Phi_{bi}}{\Phi_{bi}}\cos 2\phi_2\right\}}{\{1 + [r^{(2p)}_{af}r^{(2p)}_{fs}]^2 + 2r^{(2p)}_{af}r^{(2p)}_{fs}\cos 4\phi_2\}}, \tag{31}$$

where R is the ratio d^r_{eff}/d_{eff}. For the case of $s \to p$ generation, the bound waves $\mathbf{e}_b$ and $\mathbf{h}_b$ are given by Eq. 12 with the substitution $n_1(\theta_1) \to n_{1o}$. This substitution in Eq. 31 gives the corresponding SH power. The isotropic result is obtained from Eq. 31 by letting $n_{1e} = n_{1o} \to n_1$ and $n_{2e} = n_{2o} \to n_2$.

A comparison of the Maker fringe pattern for X-cut quartz predicted by JK and Herman and Hayden (HH) given by the isotropic version of Eq. 31 is shown in Figure 3. The high frequency oscillations on the HH curve are the result of the reflections of the SH wave in the nonlinear media. The magnitude of the oscillations is related to the index difference between the nonlinear media and Regions I and III. In order to carefully map out the high frequency oscillations, an angular resolution far beyond that which is normally attempted is required.

In terms of the ability to measure d_{ij} accurately, it is usually true that the difference between coefficients determined by considering reflections and neglecting them (<2 percent) is smaller than the typical experimental error in those mea-

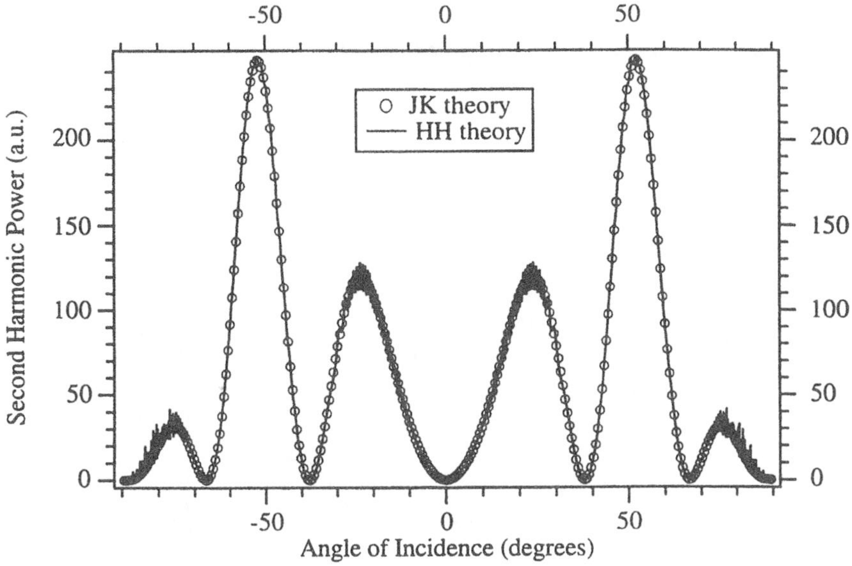

Figure 3 Maker fringe pattern predicted by JK and HH for a 265 μm thick piece of X-cut quartz. Rotation is about the z-axis.

surements ($\approx$10–20 percent). Because of the small difference predicted in the calculated d_{ij}, it is usually safe to neglect the reflections by setting $r_{af}^{(2p)} \approx 0$ in Eq. 31. This action is different from the method used by JK to neglect reflections. In JK, their approximation of no reflections in the boundary conditions results in the necessity to later assume no dispersion ($n_1 \approx n_2$) and incident angles $\theta_i \leq \pi/4$ in order to reduce the complexity of their result [see Eqs. A36 and A38 in JK]. The consequences of these assumptions can be significant for materials with even modest dispersion.[20] As the dispersion increases, the ratio of d_{31}/d_{33} calculated with JK decreases. Since the standard rigid-gas model[24] describing the electric field induced orientation of dipoles predicts $d_{31}/d_{33} = 1/3$ for low poling fields and ratios smaller than 1/3 for high poling fields, measured SHG ratios of $d_{31}/d_{33} < 1/3$ are generally attributed to more order. In these cases, however, if JK is used to analyze SHG experiments in dispersive media, the implied high degree of order may be an artifact caused by the inability of JK to account properly for the effects due to dispersion.

From the standpoint of an experimentalist's ability to accurately determine d_{ij} using the Maker fringe technique, what is the effect of neglecting birefringence when trying to determine d_{ij}? Because birefringence is nearly always present when studying NLO organic films and because of the current interest in these materials, we will use representative values found in the current literature to illustrate the magnitude of the effects. The theory, of course, is not limited to these materials.

We start our discussion on the effect of birefringence by examining the Maker fringe plots for each case for X-cut quartz. Figure 4 shows only a slight difference between the two cases. Calculation of d_{11} and the crystal thickness for each case yields a difference of <2 percent and 2 μm, respectively, for a 265 μm thick

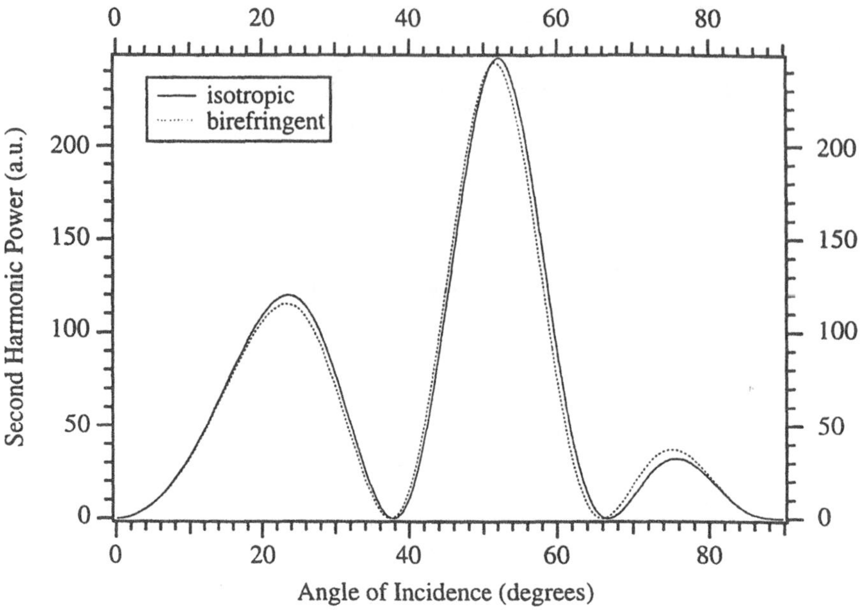

Figure 4 Maker fringe plots for 265 μm thick X-cut quartz.

crystal. The close agreement between the isotropic and birefringent treatments is due to the small birefringence of quartz. Quartz has a birefringence ($\Delta n_i \equiv n_{ie} - n_{io} = 3\sigma_i$) of 0.0088 and 0.0091 at 1.064 μm ad 0.532 μm, respectively.

The standard definition of birefringence, Δn, can be related to the poling induced birefringence in polymers, σ_i, through the following relations:

$$n_{io} = n_{iu} - \sigma_i \tag{32}$$

and

$$n_{ie} = n_{iu} + 2\sigma_i, \tag{33}$$

where $i = 1{,}2$ and n_{iu} is the unpoled refractive index at the frequency $i\omega$. Equations 32 and 33 are generally valid for amorphous poled polymers. Recent data for poling-induced birefringence in a polymer (BISA-ANT, $\sigma_i = 0.020 \rightarrow 0.025$) is given by Jungbauer et al.[26] In Fig. 5, we have calculated, using Eq. 31, the ratio of d_{33} (birefringent)/d_{33} (isotropic) as a function of film thickness and degree of birefringence for an incident angle of 50 degrees, a fundamental wavelength of 1.064 μm, and assuming $d_{31} = d_{15}$ and $d_{33} = 3d_{31}$. It can be seen that large errors occur as the thickness approaches an integer multiple of twice the coherence length. The use of the ordinary indices gives similar qualitative results as in Fig. 5 but the magnitude of the error is worse than that obtained for extraordinary indices in a $p \rightarrow p$ experiment. Figure 5 also shows us that experiments that attempt to determine d_{33}, while making comparisons at only a single angle and ignoring birefringence, are subject to very large errors depending on the thickness of the film, regardless of the degree of birefringence. A complete angular scan and subsequent fit of the data to the theory is the only way to avoid large errors.

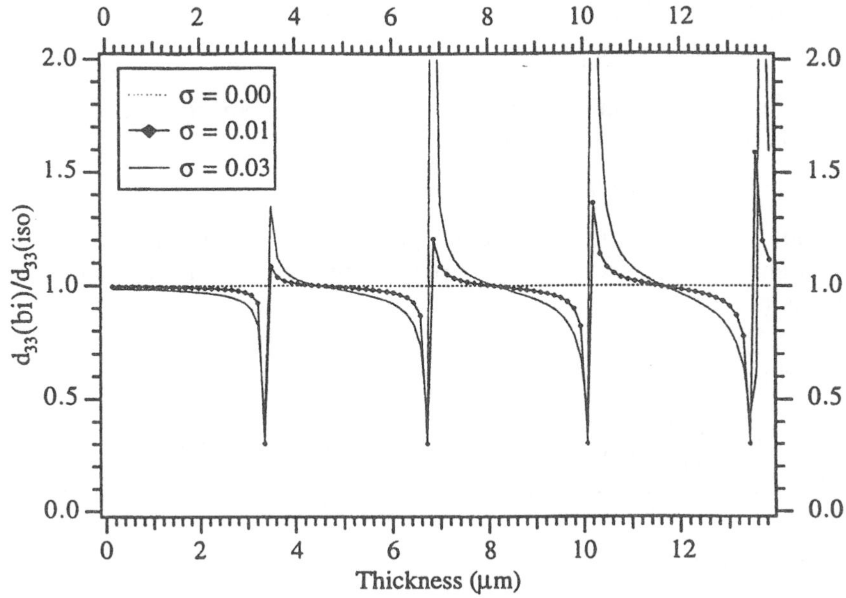

Figure 5 Effect of birefringence on the calculation of d_{33} in a poled polymer at a fixed angle of incidence = 50°. Extraordinary indices were used for the isotropic case. The unpoled indices were $n_1 = 1.653$ and $n_2 = 1.795$.

C. Absorbing Case

If the nonlinear material has non-zero absorption for either the fundamental or SH wave, Eq. 31 no longer holds. In order to introduce the effects of absorption, the index of refraction for the nonlinear film can be written as a complex quantity and the same boundary value problem solved. The effects of absorption have been considered to some extent for EFISH[26] and Maker fringes from wedges[11] and for poled polymer films.[27] A complete description of absorption for use in Maker fringe experiments, including reflections, has also been given.[20] To date, there is no published closed form expression describing the results of a Maker fringe experiment that incorporates both birefringence and absorption simultaneously. The expression given below is valid only for the case of no birefringence. It may be applied successfully to films with small birefringences provided the experimenter has taken care to check the magnitude of the possible error by using the birefringent theory given above.

For an isotropic, absorbing nonlinear material, the fields in the three regions are given by Eqs. 4–7 with $n_{1e} = n_{1o} \rightarrow n_1$ and $n_{2e} = n_{2o} \rightarrow n_2$, and by writing the complex refractive indices[28] as $\tilde{n}_m = n_m(1 + i\kappa_m)$. Snell's Law for a transparent medium (n) to an absorbing medium ($\tilde{n}$), $\tilde{n} \sin \tilde{\theta} = n \sin \theta$, then implies complex angles of refraction, $\tilde{\theta} = \theta_r + i\theta_i$. Snell's Law can be separated into real and imaginary parts in order to solve for θ_r and θ_i giving:

$$\tanh \theta_i = -\kappa \tan \theta_r, \tag{34}$$

and

$$\cos^2 \theta_r = \frac{(\cos^2 \theta_R + \kappa^2) + \sqrt{(\cos^2 \theta_R - \kappa^2)^2 + 4\kappa^2}}{2(1 + \kappa^2)}, \tag{35}$$

where θ_R is the angle of refraction in the limit $\kappa \to 0$, viz., $\sin \theta_R = n \sin \theta / n_m$. For small κ, θ_r is only slightly larger than θ_R giving:

$$\tilde{n} \cos \tilde{\theta} \approx n \cos \theta_R + in\kappa / \cos \theta_R. \tag{36}$$

This approximation is used in terms having a phase, and real angles are assumed in transmission and reflection coefficients. Applying the same boundary conditions as earlier results in:

$$T = \frac{4\pi i t_{fs}^{(2p)}}{n_2 c_2}\left(\frac{2\pi L}{\lambda}\right) \cdot \frac{\left[\hat{\mathbf{e}}_2 \cdot \mathbf{P}^{NL} \frac{\sin \tilde{\Psi}}{\tilde{\Psi}} \exp(-i\tilde{\phi}_2) - r_{af}^{(2p)} \hat{\mathbf{e}}_2^r \cdot \mathbf{P}^{NL} \frac{\sin \tilde{\Phi}}{\tilde{\Phi}} \exp(i\tilde{\phi}_2)\right]}{[\exp(-2i\tilde{\phi}_2) + r_{af}^{(2p)} r_{fs}^{(2p)} \exp(2i\tilde{\phi}^2)]}, \tag{37}$$

where, with the aid of Eq. 36:

$$\tilde{\phi}_m = \frac{2\pi L}{\lambda} \tilde{n}_m \tilde{n}_m = \phi_m + i\delta_m, \quad \delta_m = \frac{2\pi L}{\lambda} \frac{n_m \kappa_m}{c_m}, \tag{38}$$

$$\tilde{\Psi} = \tilde{\phi}_1 - \tilde{\phi}_2 = \Psi + i\chi, \quad \chi = \frac{2\pi L}{\lambda}\left(\frac{n_1 \kappa_1}{c_2} - \frac{n_2 \kappa_2}{c_2}\right), \tag{39}$$

and

$$\tilde{\Phi} = \tilde{\phi}_1 + \tilde{\phi}_2 = \Phi + i\Gamma, \quad \Gamma = \frac{2\pi L}{\lambda}\left(\frac{n_1 \kappa_1}{c_1} + \frac{n_2 \kappa_2}{c_2}\right), \tag{40}$$

and the transmission and reflection coefficients are:

$$t_{fs}^{(2p)} = \frac{2n_2 c_2}{n_{2s} c_2 + n_2 c_{2s}}, \tag{41a}$$

$$r_{af}^{(2p)} = \frac{n_2 \cos \theta - c_2}{n_2 \cos \theta + c_2}, \tag{41b}$$

and

$$r_{fs}^{(2p)} = \frac{n_{2s} c_2 - n_2 c_{2s}}{n_{2s} c_2 + n_2 c_{2s}}. \tag{41c}$$

For the case of absorption, $\mathbf{e}_b$ in Eq. 6 is the value of the bound wave amplitude at the middle of the film. Consequently, $\mathbf{P}^{NL}$ in Eq. 12 depends on the fundamental wave amplitude at the center of the film, viz.:

$$\mathbf{P}^{NL} = |E_1|^2 \overleftrightarrow{\mathbf{d}} : (\hat{\mathbf{e}}_1 \exp[-\delta_1/2])(\hat{\mathbf{e}}_1 \exp[-\delta_1/2]), \tag{42}$$

where E_1 is the magnitude of the fundamental wave in air. Using Eq. 37 in $P_{2\omega} = (c/8\pi)(t_{sa}^{(2p)})^2 |T|^2 A$, we obtain:

$$P_{2\omega}^{(\gamma\to p)} = \frac{\frac{128\pi^3}{cA}[t_{sa}^{(2p)}]^2[t_{fs}^{(2p)}]^2[t_{af}^{(1\gamma)}]^4 P_\omega^2}{n_2^2 c_2^2}\left(\frac{2\pi L}{\lambda}\right)^2 d_{eff}^2 \exp[-2(\delta_1 + \delta_2)]$$
$$\times \frac{\left\{\begin{array}{l}|\sin\tilde{\Psi}/\tilde{\Psi}|^2 + [r_{af}^{(2p)}]^2 R^2 \exp[-4\delta_2]|\sin\tilde{\Phi}/\tilde{\Phi}|^2 \\ \quad -2r_{af}^{(2p)} R \exp[-2\delta_2]|\sin\tilde{\Psi}/\tilde{\Psi}||\sin\tilde{\Phi}/\tilde{\Phi}|\cos(2\phi_2 - \theta_x + \theta_y)\end{array}\right\}}{\{1 + [r_{af}^{(2p)} r_{fs}^{(2p)}]^2 \exp[-8\delta_2] + 2r_{af}^{(2p)} r_{fs}^{(2p)} \cos 4\phi_2 \exp[-4\delta_2]\}}, \tag{43}$$

where

$$|\sin\tilde{\Psi}/\tilde{\Psi}|^2 = (\sin^2\Psi + \sinh^2\chi)/(\Psi^2 + \chi^2), \tag{44a}$$

$$|\sin\tilde{\Phi}/\tilde{\Phi}|^2 = (\sin^2\Phi + \sinh^2\Gamma)/(\Phi^2 + \Gamma^2), \tag{44b}$$

$$\tan\theta_x = \frac{\Psi\tanh\chi - \chi\tan\Psi}{\Psi\tan\Psi + \chi\tanh\chi}, \tag{44c}$$

and

$$\tan\theta_y = \frac{\Phi\tanh\Gamma - \Gamma\tan\Phi}{\Phi\tan\Phi + \Gamma\tanh\Gamma}, \tag{44c}$$

where $R = d_{eff}^r/d_{eff}$ with d_{eff} and d_{eff}^r given as in Eq. 29, but with $\gamma_1 = \gamma_2 = 0$.

III. RELATIONSHIP BETWEEN SECOND HARMONIC GENERATION AND THE LINEAR ELECTROOPTIC EFFECT

A method for predicting the electronic contribution to the electrooptic (EO) coefficient based on SHG measurements was presented by Singer et al.[29] This method is based on a two-level model[30] for β_{ijk}. For the chromophores listed in this book, this model gives reasonable predictions far from resonance,[31] but it is known to overcorrect for the dispersion in β_{ijk} when the harmonic beam is near resonance. Nevertheless, such a model is still useful and the equation below can be used to calculate the electronic component of the EO coefficient, r_{ijk}:

$$r_{ijk}^{el} = \frac{4d_{kij}}{n_i^2(\omega)n_j^2(\omega)} \frac{L_{ii}(\omega)L_{jj}(\omega)L_{kk}(0)}{L_{kk}(2\omega')L_{ii}(\omega')L_{jj}(\omega')} \frac{(3\omega_0^2 - \omega^2)(\omega_0^2 - \omega'^2)(\omega_0^2 - 4\omega'^2)}{3\omega_0^2(\omega_0^2 - \omega^2)^2}, \tag{45}$$

where ω' is the frequency of the fundamental used in measuring d_{kij}, the EO coefficient is evaluated at frequency ω, $n_{ii}(\omega) \sim n_{jj}(\omega)$, and $L_{ii}(\omega) \sim L_{jj}(\omega)$ and $L(0)$ are local field factors of the Onsager type.[32] (See Chapter 3 for a derivation of the local field factors.)

IV. THIN-FILM SECOND HARMONIC GENERATION

Second harmonic generation in very thin films of organic molecules has been studied extensively to evaluate both the films and the constituent molecules as nonlinear optical materials. A common example is for ordered monomolecular films. Films are produced by depositing one molecular layer at a time, either through mechanical deposition (Langmuir-Blodgett technique) or chemical self-assembly.[4,5] Usually, a well-defined order exists in the layer direction, but the in-plane order is generally multidomain, and may exhibit a strong in-plane anisotropy. Like the case of poled polymers, the relationship between molecular order and optical nonlinearity depends on a distribution of molecular orientation and an order parameter. The description in Chapter 3 of the role of order parameters in the nonlinear optical response would apply to these materials. However, in monomolecular films, a number of assumptions are commonly made which simplify the analysis as proposed by Shen and coworkers.[33] Firstly, many dipolar molecules are nearly one dimensional; secondly, local field corrections are often ignored; and thirdly, the orientational distribution is assumed to be highly peaked and isotropic in the plane of the film. In this case, the average angle of the molecular axis with respect to the film's normal, θ, fully describes the orientation of the molecule, and the point group $C_{\infty v}$ applies. As such, two independent second harmonic measurements can be carried out: (1) both fundamental and second harmonic *p*-polarized; and (2) fundamental *s*- and second harmonic *p*-polarized. From these two measurements, the two unknowns, θ and β_{zzz}, can be determined from,

$$\chi^{(2)}_{\perp\perp\perp} = N\beta^{*}_{zzz}\langle\cos^3\theta\rangle, \tag{46}$$

and

$$\chi^{(2)}_{\perp 11} = \frac{1}{2}N\beta^{*}_{zzz}\langle\cos\theta\sin^2\theta\rangle, \tag{47}$$

where N is the number density of molecules. The analysis presented in the earlier section on Maker fringes in this Chapter is used to deduce the two macroscopic susceptibilities above and, from them, the value of β^{*}_{zzz} where * denotes a non-gas phase dressed hyperpolarizability as described in Chapter 3.

V. POLED POLYMER FILMS

As described above, poled polymer films incorporate nonlinear optical chromophores in a glassy polymer matrix, either by guest-host doping or covalent attachment. Following this, the polymer is put into a state of high mobility, usually by heating. An electric field is applied and the material then put in a state of low mobility by chemical reaction and/or cooling. This results in a polar alignment of the nonlinear optical chromophores and thus second order nonlinear optical effects are allowed. Because these materials are readily formed into thin films, they have received a great deal of attention as potential materials for electrooptic devices. Some of the entries for molecular hyperpolarizability results in this Chapter's tabulations have been obtained from measurements in poled polymer films. In order to obtain a value for the hyperpolarizability, the relationship between the measured second harmonic coefficient and the molecular quantity must be derived. This was carried out in detail in Chapter 3, but we reproduce the main result here:[7]

$$d_{33}(-2\omega;\omega,\omega) = L_{zz}(2\omega)L_{zz}(\omega)L_{zz}(\omega)\beta_{zzz}(-2\omega;\omega,\omega)\left(\frac{\varepsilon(n^2+2)}{n^2+2\varepsilon}\right)\frac{\mu E_p}{5kT}. \tag{48}$$

The coefficient, d_{33}, is measured using second harmonic generation in thin films and is obtained as described earlier in this Chapter. In Eq. 48, the quantity N is the number density of chromophores in the polymer, the L's are appropriate Lorentz local field factors (see Chapter 3), ε the static dielectric constant, n the refractive index, μ the molecular dipole moment, E_p the amplitude of the poling field, and kT the Boltzmann factor. For most chromophores of interest, one-dimensionality is a reasonable assumption. Thus, the measured component of the molecular hyperpolarizability is β_{zzz}. Generalizations of Eq. 48 are described in Chapter 3.

VI. ELECTRIC FIELD INDUCED SECOND HARMONIC GENERATION (EFISH)

Electric field induced second harmonic generation is a well-established technique to determine the second order nonlinear optical response of molecules. By applying an electric field, the center of inversion of a gas or liquid is broken allowing bulk second harmonic generation.[34,35,36,37,38] This is applicable only in the case of dipolar molecules where orientational coupling to the static electric field is allowed. In addition, the presence of a strong static field limits the technique to non-charged species. This third order process also contains contributions from molecular third order responses. Original experiments were carried out on molecules in the gas phase.[37,38] Though the contributions of local fields and interactions are minimized in the gas phase, many compounds of interest cannot develop the necessary vapor pressure for such measurements. The basic technique was later adapted for use in measurements in solution.[39,40] Such measurements can give reliable results, but care must be taken to minimize the effects of local fields and interactions. An infinite dilution technique was developed to address this issue.[41] Later improvements included dispersion measurements,[42] determination of the real and imaginary parts of the response,[43] and simultaneous measurement of the third harmonic coefficient.[44]

A. EFISH Susceptibility

In this subsection, we apply the developments of Chapter 3 to the case of electric field induced second generation. The macroscopic susceptibility that is measured is related to the molecular quantities of interest.[45] The macroscopic nonlinearity is third order in nature since it depends on two optical fields and a static field, thus we will look for polarization dependencies that are third order in the applied field. Since EFISH measurements involve dilute liquid solutions, isotropic averaging of the molecular quantities is performed. A number of issues peculiar to EFISH arise, most especially regarding conventions, and will be described. We begin with an expression for the molecular dipole moment in terms of a time-dependent applied electric field, assuming a spatially local response (independent of $\mathbf{k}$). Thus:

$$\begin{aligned} p_i^{(2)}(t) &= \varepsilon_0 \int_0^\infty d\tau_1 \int_0^\infty d\tau_2 \beta_{ijk}(t;\tau_1,\tau_2)E_j(t-\tau_1)E_k(t-\tau_2) \\ &+ \varepsilon_0 \int_0^\infty d\tau_1 \int_0^\infty d\tau_2 \int_0^\infty d\tau_3 \gamma_{ijk}(t;\tau_1,\tau_2,\tau_3)E_j(t-\tau_1)E_k(t-\tau_2)E_l(t-\tau_3), \end{aligned} \tag{49}$$

where the superscript (2) reminds us that we will eventually take the Fourier component of the dipole at the second harmonic frequency. The quantities $\beta_{ijk}(t;\tau_1,\tau_2)$, and $\gamma_{ijk}(t;\tau_1,\tau_2,\tau_3)$, are the second and third order molecular nonlinear optical susceptibilities. We start in the time domain since it is more convenient to carry out the derivation with an unspecified time-dependent field, and take the Fourier transform at the end. This makes dealing with conventions simpler. From now on, we suppress the integrals. Integration is understood when independent variables are unmatched between the left and right hand side of the equation. The dipole above refers to the individual molecules in solution, and the tensor and vector indices correspond to the molecular reference frame. We now apply the oriented gas model which relates the macroscopic polarization density in the lab frame to the sum over individual molecular dipoles. Thus, the macroscopic polarization density is given by:

$$P_I(t) = N\langle p_i(t)\rangle_I, \tag{50}$$

where N is the molecular number density, and the brackets denote molecular ensemble averaging. The ensemble average assumes a Boltzmann distribution, and is given by:

$$\begin{aligned}\langle p_i(t)\rangle_I = \Bigg[& \frac{1}{\int \exp\left(-\frac{\mathbf{p}\cdot\mathbf{E}}{kT}\right) d\mathbf{E}} \int \exp\left(-\frac{\mathbf{p}\cdot\mathbf{E}}{kT}\right) \\ & \cdot \Big\{ \beta_{ijk}(t;\tau_1,\tau_2)E_J(t-\tau_1)E_K(t-\tau_2) \\ & + \gamma_{ijlk}(t;\tau_1,\tau_2,\tau_3)E_J(t-\tau_1)E_K(t-\tau_2)E_L(t-\tau_3)\Big\}\, d\mathbf{E}\Bigg]_I.\end{aligned} \tag{51}$$

As discussed in Chapter 3, we now consider the limit where the dipolar orienting energy (as we will see later, we refer now to the static field) is less than kT so that $\exp[(\mathbf{p}\cdot\mathbf{E})/kT] \approx 1 + (\mu\cdot\mathbf{E})/kT$ where, for a dipolar molecule, we have also taken $\mathbf{p}$ to lowest order, namely the molecular static dipole moment vector μ. The electric fields have been written in the laboratory frame IJKL since they are applied in that frame. As previously discussed in Chapter 3, to carry out the average we take the ensemble average of the direction cosines relating the molecular to the laboratory frame. These direction cosines are described by the rotation matrix containing the Euler angles which connect the molecular dipole frame to the laboratory frame as shown in Fig. 6. The rotation matrix is given by:

$$\mathbf{a} = \begin{bmatrix} \cos\theta\cos\varphi\cos\psi & \cos\theta\sin\varphi\cos\psi & -\sin\theta\cos\psi \\ \quad -\sin\varphi\sin\psi & \quad +\cos\varphi\sin\psi & \\ -\cos\theta\cos\varphi\sin\psi & -\cos\theta\sin\varphi\sin\psi & \sin\theta\sin\psi \\ \quad -\sin\varphi\cos\psi & \quad +\cos\varphi\cos\psi & \\ \sin\theta\cos\varphi & \sin\theta\sin\varphi & \cos\theta \end{bmatrix}. \tag{52}$$

We can now rewrite the polarization as:

$$\begin{aligned}\langle p_i(t)\rangle_I = \frac{1}{\int d\Omega}\int & (1 + \mu_l a_{lL}E_L^0)\{\beta_{ijk}(t;\tau_1,\tau_2)a_{iI}a_{jJ}a_{kK}E_J(t-\tau_1)E_K(t-\tau_2) \\ & + \gamma_{ijlk}(t;\tau_1,\tau_2,\tau_3)a_{iI}a_{jJ}a_{kK}a_{lL}E_J(t-\tau_1)E_K(t-\tau_2)E_L(t-\tau_3)\}\, d\Omega,\end{aligned} \tag{53}$$

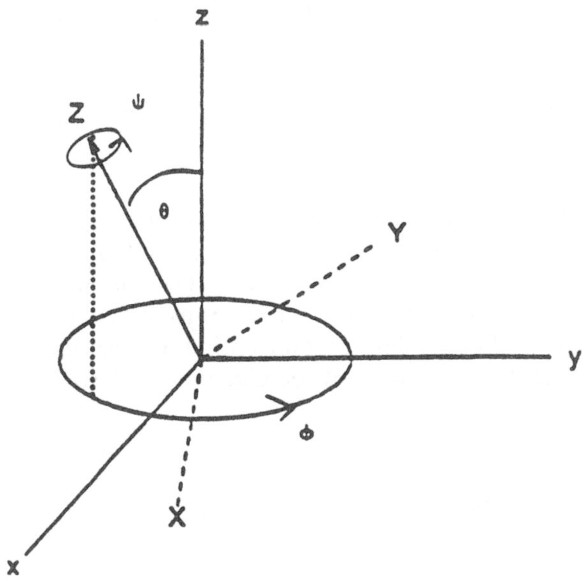

Figure 6 Euler angles relating the molecular to the laboratory reference frames.

where Ω denotes the solid Euler angles, and where we have taken the static field, E_L^0, which orients the molecules in solution. The orientation average then becomes the average of products of the direction cosines. These have been tabulated by Cyvin, Rauch and Decius.[46] It is found that the terms containing three factors of a_{uU} do not contribute to the integrals. We can now define a macroscopic susceptibility, Γ_{IJKL}:

$$\Gamma_{IJKL}E_L = N(\chi_1\delta_{IJ}E_K + \chi_2\delta_{IK}E_J + \chi_3\delta_{JK}E_I), \tag{54}$$

where

$$\begin{aligned}
\chi_1 &= \frac{1}{30}(4\gamma_{uuvv} - \gamma_{uvuv} - \gamma_{uvvu}) + \frac{1}{30kT}(4\beta_{uuv}\mu_v - \beta_{uvu}\mu_v - \beta_{uvv}\mu_u) \\
\chi_2 &= \frac{1}{30}(4\gamma_{uvuv} - \gamma_{uvvu} - \gamma_{uuvv}) + \frac{1}{30kT}(4\beta_{uvu}\mu_v - \beta_{uvv}\mu_u - \beta_{uuv}\mu_v) \\
\chi_3 &= \frac{1}{30}(4\gamma_{uvvu} - \gamma_{uuvv} - \gamma_{uvuv}) + \frac{1}{30kT}(4\beta_{uvv}\mu_u - \beta_{uuv}\mu_v - \beta_{uvu}\mu_v).
\end{aligned} \tag{55}$$

We now insert the local field factors as described in detail in Chapter 3, define the time-dependence of the electric field, take the Fourier transform of the polarization, and pick off the component of the polarization density oscillating at the optical second harmonic. We further assume a quasi-monochromatic optical field and an instantaneous and local material response with the electric field defined by $E(t) = E_0^{\omega}\cos(\omega t) + E^0$ where both a static and optical field are present. For the $\beta\mu$ terms, integrals over τ_1 and τ_2 are carried out, whereas for the γ term, integrals over τ_1, τ_2, and τ_3 are carried out as specified in Eq. 49. Experiments generally measure only the Γ_{1111} component of the EFISH susceptibility. We can now write this susceptibility as:

$$\Gamma_{1111} = \frac{NL(0)L(2\omega)L(\omega)L(\omega)}{15}\{K(2\gamma_{uuvv} + \gamma_{uvvu}) + \frac{K'}{kT}(2\beta_{uuv}\mu_v + \beta_{uvv}\mu_u)\}, \tag{56}$$

where u and v are summed over the molecular coordinates. The factors K and K' are numerical factors resulting from the Fourier transform and, in the convention given by the definition of $E(t)$ above, are $K = 3/2$ and $K' = 1/2$. They are unequal due to the differing number of integrals over time in these two terms. These factors K and K' may or may not be suppressed depending on convention as described in the next section. The L's are local field factors and will be discussed below. Thus, we have:

$$P_1^{2\omega} = \Gamma_{1111}E_1^{\omega}E_1^{\omega}E_1^{0}, \tag{57}$$

where the cos $2\omega t$ has been eliminated. We can now define the scalar part of γ_{uvwx} as:

$$\gamma = \frac{1}{15}\sum_{\substack{u,v\\=x,y,z}}(2\gamma_{uuvv} + \gamma_{uvvu}), \tag{58}$$

and the inner product of the vector part of β_{uvw} and μ_u as:

$$\beta\mu = \frac{1}{3}\sum_{\substack{u,v\\=x,y,z}}(2\beta_{uuv}\mu_v + \beta_{uvv}\mu_u). \tag{59}$$

With these definitions:

$$\Gamma_{1111} = NL(0)L(2\omega)L(\omega)L(\omega)\left(K\gamma + \frac{K'\beta\mu}{5kT}\right). \tag{60}$$

B. Conventions

Though Chapter 1 discussed conventions in detail, we will elaborate here in order to account for issues peculiar to EFISH and the experimental determination of EFISH susceptibilities. We carried through the development in the time domain in the previous section to suppress the convention issue until the end of the calculation when the conversion to the frequency domain took place. This clearly connects the factors K and K' to the definition of the field. A complication in discussing EFISH conventions for determination of β involves the fact that one is measuring a second order molecular quantity by using a third order process which is normalized to a second order process in a reference standard.

As mentioned in Chapter 1, the most commonly used convention for early EFISH measurements on organics involved a power series expansion of the field, and the I convention (of Chapter 1) where the numerical factors K and K' are suppressed, $K = K' = 1$.[39,40,41] This early work grew out of a desire to understand molecular mechanisms in organic crystals where an analogous convention is used.

Convention II of Chapter 1 has $K = 3/2$ and $K' = 1/2$. In analogy with the development above, one can also describe two conventions for second harmonic generation, one with the factor ½ explicit, and one where it is suppressed. We will call the one with the factor of ½ explicit the second harmonic susceptibility, $\chi^{(2)}_{ijk}(-2\omega;\omega,\omega)$, and the one where it is suppressed the second harmonic coefficient, $d_{ijk}(-2\omega,\omega,\omega)$ so that $\chi^{(2)} = 2d$. In keeping with the definitions for EFISH and the development in Chapter 1, we denote the d_{ijk} convention as I and the $\chi^{(2)}_{ijk}$ convention as II. Note that we use the term coefficient when referring to the macroscopic measured quantity in the I convention, and the term susceptibility when referring to the II convention where the factors K and K' are explicitly written. This is in keeping with the literature on second harmonic generation. The crystal susceptibilities in both conventions are related to the molecular ones by[47]:

$$d_{ijk}(-2\omega;\omega,\omega) = Nf_i^{2\omega}f_j^{\omega}f_k^{\omega} \sum_{I,J,K} \sum_{s=1}^{n} \cos[i,I(s)] \cos[j,J(s)] \cos[k,K(s)]\beta^{I}_{IJK}(s), \tag{61a}$$

and

$$\chi^{(2)}_{ijk}(-2\omega;\omega,\omega) = Nf_i^{2\omega}f_j^{\omega}f_k^{\omega} \sum_{I,J,K} \sum_{s=1}^{n} \cos[i,I(s)] \cos[j,J(s)] \cos[k,K(s)]\beta^{II}_{IJK}(s), \tag{61b}$$

where s refers to molecules in the unit cell, and the direction cosines relate the crystal frame to the molecular frame. Similar expressions obtain for noncrystalline materials using oriented gas model, for example, as described in Chapter 3. Note that in both conventions, the second harmonic coefficient, d_{ijk} and the second harmonic susceptibility, $\chi^{(2)}_{ijk}$, are simply related to their molecular analogs, β^{I}_{IJK} and β^{II}_{IJK}, respectively. These β's are the ones deduced from EFISH as in Eq. 60 in either convention. Other conventions have been proposed that do not result in such simple relationships.[48]

As we show below, the measured EFISH response is related to the measured ratio of sample and reference standard intensities, and the reference response by,

$$K\gamma + \frac{K'\beta\mu}{5kT} \propto \sqrt{\frac{A^L_m}{A^R_m}} M\chi^{(2)}_{ref}, \tag{62}$$

where the A's are the measured intensities, and $M = 1$ or ½ depending on convention consistent with K and K'. Normally, the measured susceptibility and the reference standard are the same order so that once one chooses a convention the reference value takes on its appropriate magnitude. This is not the case, in general, for EFISH, since the third order response, Γ^M_L, is compared to a second order reference standard. If one uses convention I for EFISH, $K = K' = 1$, and $M = 1$ for the reference, then by Eq. 62 the factors K' and M cancel for the $\beta\mu$ term, and the β^I in Eq. 61 is measured. Similarly, when using convention II for EFISH, $K = 3/2$, $K' = 1/2$, and $M = 1/2$, then by Eq. 62 the factors K' and M cancel for the $\beta\mu$ term, and the β^{II} in Eq. 61 is measured. For the γ term in Eq. 62, a factor of three remains in that equation in the II convention. Though the γ term is often ignored in the data analysis, it is also possible to measure that term separately using four-wave mixing or third harmonic generation.[26,44,49] Care must be taken to use the proper conventions when converting the measured γ from another process to the EFISH γ.

The polarization can be expressed as a Taylor series (see Chapter 1) with a factorial denominator preceding each term rather than a power series as has been done in this Chapter.[48] This convention results in an additional overall multiplicative factor. This convention is not in general use for EFISH in organic chromophores.

Table 2 in Chapter 1 assigns conventions to a number of works in the literature. As discussed in that Chapter, although a number of works are assigned, the relevant literature did not explicitly state which convention is used. In those works, and many others, it is sometimes unclear what convention is used. Though one can often deduce the convention used if not explicitly stated by looking for factors K and K' in the definition of the polarization, some authors define the convention one way in the polarization definition, but then use a value for the reference nonlinearity appropriate for another convention. However, the convention is presented in the definition of polarization, the cancellations in Eq. 62 provide that the value of the reference nonlinearity is the sole determining factor of the convention used. Thus, by examining the magnitude of the reference nonlinear optical susceptibility or coefficient, a convention can be assigned. The references values by convention differ by a factor of two (convention II being twice that of convention I). In assembling the Tables at the end of this Chapter, we examined the value of the reference nonlinearity in order to assign conventions. However, we were unable to assign conventions in many cases, and thus we left many of those out of the Tables. It does seem, however, than most measurements of β have used convention I. The ambiguity in reported conventions obviously increases the uncertainty when comparing numbers form different workers; however, comparing results for one worker is still reasonable. For this reason, we have grouped the compounds measured according to referenced work. Thus, measurements by several workers on a single compound will appear in various places in the table, so that results from different workers may be compared.

Another source of inaccuracy associated with the Tables involves the reference standard number used. Although most workers refer their intensities to quartz d_{11}, there is some controversy as to the best value to use. Most have used the quartz value obtained by comparison to parametric fluorescence in $LiIO_3$, namely $d_{11} = 1.13 \pm 0.11 \times 10^{-9}$ *esu*, or twice this value when convention II is used. Recent results described in Chapter 1 indicate the value of d_{11} for quartz may be quite different. This may introduce additional uncertainty when considering the absolute magnitude of results in the Tables and comparison to quantum chemical calculations, but does not affect comparison of experimental values when properly compensating for different values.

C. Local Fields and Interactions

We can now apply the development of Chapter 3 regarding local fields to the particular case of EFISH measurements.[41] We provide a technique that most accurately accounts for the effects of local fields by carrying out concentration-dependent measurements on dilute solutions of the molecule of interest. A related and equivalent technique has recently been introduced.[50] To measure β accurately, we will require concentration-dependent measurements of not only the nonlinear optical susceptibility but also the static dielectric constant and linear polarizability, and develop a concentration extrapolation procedure. To begin, we consider the EFISH susceptibility corresponding to a two-component solution in convention I:

$$\Gamma_{1111} = N_0 L_0(0) L_0(\omega) L_0(\omega) L_0(2\omega)\gamma_0' + N_1 L_1(0) L_1(\omega) L_1(\omega) L_1(2\omega)\gamma_1' \qquad (63)$$

where γ' contains the two terms of the microscopic susceptibility of Eq. 60, the subscripts 0 and 1 correspond to the solvent and solute respectively. We know that the concentration dependence appears in Eq. 63 not only explicitly through N, but is also implicitly contained in the local field factors. In order to apply a concentration-dependent procedure, we take the derivative of Eq. 63 with respect to concentration, and we consider how to evaluate the concentration dependence of the local fields. To do this, we begin with a corresponding expression for the static dielectric constant, namely:

$$\frac{\varepsilon - 1}{4\pi} = N_0 \alpha_0^0 L_0(0) + \frac{N_0 \mu_0^2 L_0^*}{3kT} + N_1 \alpha_1^0 L_1(0) + \frac{N_1 \mu_1^2 L_1^*}{3kT}, \qquad (64)$$

where ε is the static dielectric constant, α the linear polarizability, and μ the static dipole moment. The 0 and * superscript denote static local fields appropriate for each term. The * local field also "dresses" the dipole. Thus, a concentration-dependent measurement of the static dielectric constant must be carried out. To estimate the linear polarizability, we also consider the concentration dependence of the index of refraction of the solution:

$$\frac{n^2 - 1}{4\pi} = N_0 \alpha_0^e L_0(\infty) + N_1 \alpha_1^e L_1(\infty), \qquad (65)$$

where α^e is the linear polarizability at optical frequencies, and the superscript ∞ denotes the optical limit. We can write $\alpha_i^0 = \alpha_i^e + \alpha_i^e$, where the a superscript refers to the distortional contribution which we will neglect at here optical frequencies. For experimental convenience, we rewrite the number density N in terms of the weight fraction w, specific volume (inverse density) v, and Avagadro's number N_A. We then take the derivatives of Eqs. 63–65, and the infinite dilution (zero-concentration limit) yielding:

$$\frac{N_A}{M_1} L_1 \gamma_1' = \left.\frac{\partial \Gamma_{1111}}{\partial w}\right)_0 + \left.\frac{\Gamma_0}{v_0}\frac{\partial v}{\partial w}\right)_0 + \Gamma_0 - \left.\frac{\Gamma_0}{L_0}\frac{\partial L_0}{\partial_w}\right)_0,$$

$$\frac{4\pi N_A}{M_1 v_0}\left(\alpha_1^e L_1(0) + \frac{\mu_1^2 L_1^*}{3kT}\right) = \left.\frac{\partial \varepsilon}{\partial w}\right)_0 + \left.\frac{\varepsilon_0 - 1}{v_0}\frac{\partial v}{\partial w}\right)_0$$
$$+ \varepsilon_0 - 1 - \left.\frac{n_0^2 - 1}{L_0(0)}\frac{\partial L_0(0)}{\partial w}\right)_0 - \left.\frac{\varepsilon_0 - n_0^2}{L_0^*}\frac{\partial L_0^*}{\partial w}\right)_0,$$

and

$$\frac{4\pi N_A}{M_1 v_0}\alpha_1^e L_1(\infty) = \left.\frac{\partial n^2}{\partial w}\right)_0 + \left.\frac{n_0^2 - 1}{v_0}\frac{\partial v}{\partial w}\right)_0 + n_0^2 - 1 - \left.\frac{n_0^2 - 1}{L_0(\infty)}\frac{\partial L_0(\infty)}{\partial w}\right)_0, \qquad (66)$$

where M_1 is the solute molecular weight and the 0 subscript refers to the infinite dilution limit.

In order to evaluate the expressions given in Eq. 66, we must choose a local field model. It has been shown that good results are obtained when one chooses Debye type local field factors at optical frequencies and Onsager type in the static limit (see Chapter 3 and references therein).[41] Thus:

$$L_i(0) = \frac{\varepsilon(n_i^2 + 2)}{(2\varepsilon + n_i^2)}, \qquad L_i^* = \frac{\varepsilon(2\varepsilon + 1)(n_i^2 + 2)^2}{3(2\varepsilon + n_i^2)^2} \tag{67}$$

for the Onsager type. Note that n_i refers to a "index" of refraction for the species. Normally, this will apply to the solute. One need not use the crystal index of refraction; this number can be determined by a procedure explained below. For the Debye type local fields,

$$L_i(\infty) = \frac{n^2 + 2}{3}, \tag{68}$$

where this applies to optical frequencies. Inserting the local field definitions into Eq. 66 and ignoring concentration-dependence of the specific volume we obtain:

$$\frac{N_A}{M_1}\gamma_1' = \frac{(2\varepsilon_0 + n_1^2)(2n_0^2 + n_1^2)^3}{(n_1^2 + 2)^4 n_0^6 \varepsilon_0}\left\{ v_0 \left.\frac{\partial \Gamma_{1111}}{\partial w}\right)_0 + v_0\Gamma_0 - v_0\Gamma_0\left[\left.\frac{1}{n_0^2}\frac{\partial n^2}{\partial w}\right)_0 + \left.\left(\frac{1}{\varepsilon_0} - \frac{2}{2\varepsilon_0 + n_0^2}\right)\frac{\partial \varepsilon}{\partial w}\right)_0\right]\right\} \tag{69}$$

$$\frac{4\pi N_A}{M_1}\left\{\alpha_1^e\left[\frac{\varepsilon_0(n_1^2 + 2)}{2\varepsilon_0 + n_1^2}\right] + \frac{\mu_1^2}{3kT}\left[\frac{\varepsilon_0(2\varepsilon_0 + 1)(n_1^2 + 2)^2}{3(2\varepsilon_0 + n_1^2)^2}\right]\right\} = v_0 \left.\frac{\partial \varepsilon}{\partial w}\right)_0 + (\varepsilon_0 - 1)v_0 - \left.\frac{\partial \varepsilon}{\partial w}\right)_0\left[(n_0^2 - 1)v_0\left(\frac{1}{\varepsilon_0} - \frac{2}{2\varepsilon_0 + n_0^2}\right) + (\varepsilon_0 - n_0^2)v_0\left(\frac{1}{\varepsilon_0} + \frac{2}{2\varepsilon_0 + 1} - \frac{4}{2\varepsilon_0 + n_0^2}\right)\right] \tag{70}$$

$$\frac{4\pi N_A}{3M_1}\alpha_1^e = \frac{3v_0}{(n_0^2 + 2)^2}\left.\frac{\partial n^2}{\partial w}\right)_0 + v_0\frac{n_0^2 - 1}{n_0^2 + 2}. \tag{71}$$

Equations 69–71 constitute the extrapolation equations. By measuring the concentration-dependence of the EFISH susceptibility, the static dielectric constant, and the index of refraction, one obtains linear dependence at low concentration. From the slopes and intercepts of the measured plots, one obtains $\partial\Gamma_{1111}/\partial w)_0$, $\partial\varepsilon/\partial w)_0$, $\partial n^2/\partial w)_0$, Γ_0, ε_0, n_0^2. The quantity n_1^2 is calculated from $N_A\alpha_1^e$. From these quantities, the quantities μ_1, α_1^e, and γ_1' can be determined.

Recall that $\gamma_1' = \gamma + \beta\mu/5kT$. The quantity γ can be measured independently using four-wave mixing, third harmonic generation, or some other third order process. Though care must be used to keep the conventions straight, this will give the most accurate results. Many workers, however, assume that the γ term is negligible which generally gives results to within 20 percent. We now describe the experimental techniques used to carry out the required measurements.

D. Experiment

Three experiments must be carried out to properly determine β: EFISH, dielectric measurements, and refractive index measurements. All should be performed for a series of dilute concentrations of solutions containing the compound of interest. EFISH is generally carried out using the wedge Maker fringer technique. A typical experimental arrangement is shown in Fig. 7.[51] Because much recent interest has focused on compounds for poled polymer materials for electrooptics, a source in the 1–2 μm wavelength range is needed to measure these visible-absorbing dyes. *It should be stressed that the data analysis described here will be applicable only to weakly absorbing and weakly dispersive solutions.* The wedge Maker fringe analysis relies on these assumptions. For this reason, results based on resonant measurements are suspect. An earlier section in this Chapter dealt directly with pitfalls of ignoring these assumptions.

The laser sources are generally based on 1.06 μm wavelength Nd:YAG or closely related solid-state lasers. A number of techniques are available to tune these lasers further into the infrared. Stimulated Raman scattering (SRS),[51] optical parametric oscillation (OPO),[52] and optical parametric generation and amplification (OPG-OPA) are the most common techniques.[53] Stimulated Raman scattering yields optical frequencies shifted by integral multiples of the vibrational frequency of the scattering gas. A commonly used technique is SRS using hydrogen gas to yield a

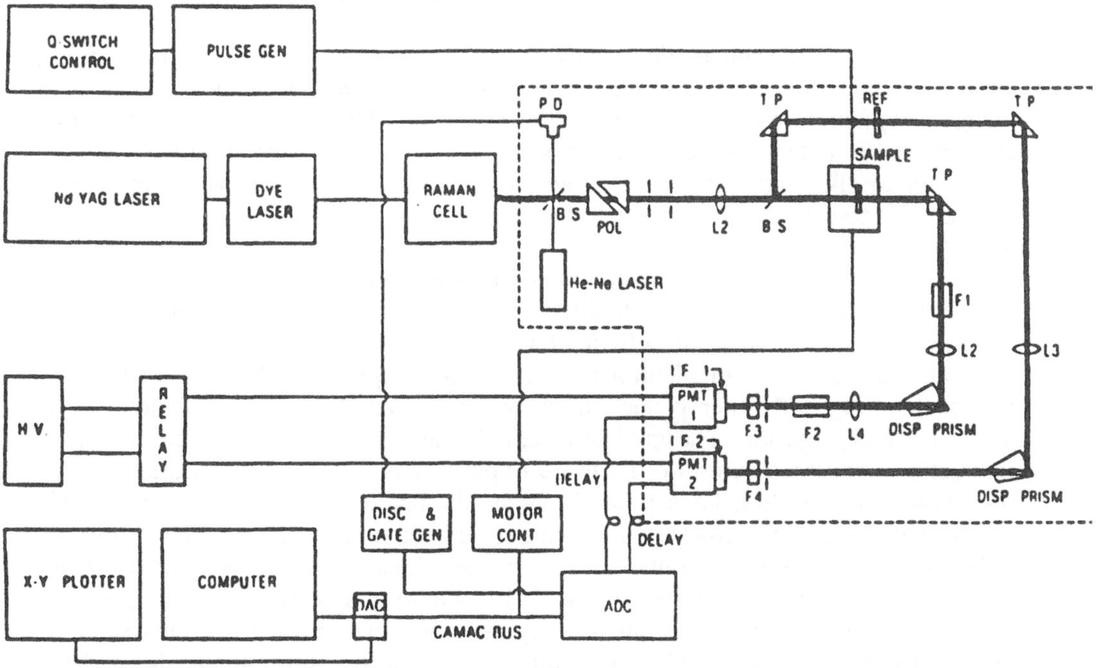

Figure 7 Experimental arrangement for EFISH. P.D., photodiode; B.S., beamsplitter; L, lens; T.P., turning prism; F, filter; DISP.PRISM, dispersing prism; I.F., interference filter; PMT, photomultiplier tube; REF, reference; ADC, analog-to-digital converter; DAC, digital-to-analog converter; H.V., high voltage power supply; POL, polarizer.

source at 1.907 μm.[44] Frequency doubling the Nd:YAG laser which pumps a dye laser which in turn pumps an SRS cell yields a continuously tunable source[51] as do the OPO, OPG-OPA techniques.

The laser source can then be directed at the two-arm system shown in Fig. 7. In one arm is the sample, and in the other a reference crystal. This reference crystal provides a photomultiplier signal quadratic in the incident intensity in order to reduce the effects of source fluctuations on the data. The exact arrangement of the sample/reference arms can be adjusted for optimum signal to noise for the particular system used.[54] The incident light should be properly filtered and polarized to eliminate spurious signals. A pulsed high voltage source must be synchronized with the arrival of the laser pulse in the sample. A pulsed source of duration in the micro to millisecond range is necessary in order to reduce the problems associated with the finite conductivity of the solutions. By appropriate timing, the high voltage pulse approximates a DC source as compared to the nanosecond or picosecond duration optical pulses.

The signal must then be filtered and its polarization analyzed before detection by a photomultiplier tube or avalanche photodiode. One must realize that the conversion efficiency for EFISH is very low: the signal is many orders of magnitude smaller than the pump laser's intensity. This makes the filtering of signal from background challenging. Filtering is done in two stages. First, absorbing solutions, water, or colored absorbing glass filters are used to differentially reduce the pump to signal by many orders of magnitude. This broad filter can then be followed by interference filters or a monochromator to select out only the second harmonic wavelength.

The sample cell (schematic is shown in Fig. 8[51]) consists of two glass slips of about two mm thickness sandwiched between two metal electrodes such that a wedged gap runs down the center of the electrode. The outer edge of the glass slips must be at least three electrode spacings from the electrode edge. This is necessary to eliminate the electric field from this surface which would introduce additional interference in the signal. The solution must be introduced into the gap without bubbles.

Some care must be used in the choice of solvents. Concentrations up to 0.01 M are generally used, so the solubility up to that level is a prime consideration. The next consideration is the effect of particular solvents on the measurements. Solvents such as water or alcohol are usually not appropriate because of the hydrogen bonding and other protic effects which make the local field models we described inappropriate. The goal is usually to seek solvents with a minimum of interaction with the solute in order to best approximate "gas phase" values for the nonlinear susceptibility. Less polar is generally better, but is usually traded off against solubility. Chloroform and dioxane have been often used.[41,50] The *p*-dioxane molecule, though globally nonpolar, does possess local moments which may interact with the solute, but has been found to be generally a reasonable choice. Chloroform is also a good choice and is only slightly polar. Strong solvents such as DMSO have also been used, but results may disagree substantially from "gas phase" values.[41] Care must be taken to use pure solvents, and to ensure that chemical and photochemical processes do not degrade the solution.

A series of concentrations can be produced by preparing in advance or by consecutive dilution or concentration during the experimental runs. Data is accu-

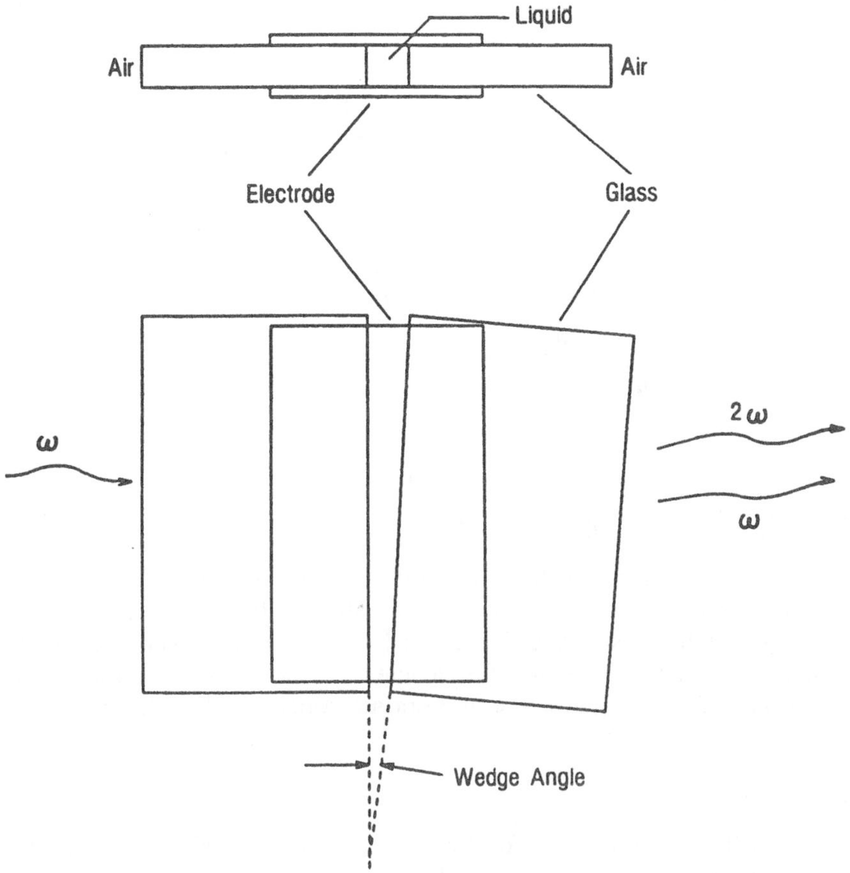

Figure 8 Schematic of sample cell.

mulated in each solution run by driving the wedge laterally across the laser beam while acquiring the data using averaging techniques. A gated integrator or boxcar averager can be used. The driven wedge maps out the Maker fringe function which we will describe below. Following this run, the intensity must be calibrated. This is done by placing a wedged reference standard such as quartz in the exact position as the sample in the previous run. The technique here is crucial to accurate results. Maker fringes from the wedge reference standard are acquired. The signal here is usually several orders of magnitude more intense than the EFISH signal, so carefully calibrated neutral density filters are required to decrease the signal so that the detectors or electronics do not saturate. This is also crucial to accurate results.

These measurements constitute the necessary data to perform the analysis. The first step consists of fitting the data to the wedge Maker fringe function. The wedge Maker fringe technique was first discussed by Oudar and Levine and their collaborators.[39,40] We present a version developed later by Teng and Garito.[42] The intensity of the second harmonic from the sample solution is given by:

$$I_{2\omega} = AE_0^2 I_\omega^2 (T_1 \Gamma_G l_c^G - T_2 \Gamma_L l_c^L)^2 f(l) \tag{72}$$

with

$$f(l) = 2 \exp\left[-\left(a_\omega + \frac{a_{2\omega}}{2}\right) l\right]\left\{\cosh\left[a_\omega - \frac{a_{2\omega}}{2}\right] - \cos\left(\frac{\pi l}{l_c^L}\right)\right\} \tag{73}$$

and

$$A - \frac{8\pi}{c}\left(\frac{8\omega}{c}\right)^2, \quad T_1 = t_{2\omega}^G t_\omega^2 \frac{1}{n_{2\omega}^G + n_\omega^G} \frac{n_\omega^G + n_{2\omega}^L}{n_{2\omega}^G + n_{2\omega}^L},$$
$$T_2 = t_{2\omega}^G t_\omega^2 t_\omega'^2 \frac{1}{n_{2\omega}^L + n_\omega^L} \frac{n_\omega^L + n_{2\omega}^L}{n_{2\omega}^G + n_{2\omega}^L}, \tag{74}$$

and where

$$t_{2\omega}^G = \frac{2n_{2\omega}^G}{1 + n_{2\omega}^G}, \quad t_\omega = \frac{2}{1 + n_\omega^G}, \quad t'_\omega = \frac{2n_\omega^G}{n_\omega^G + n_\omega^L}. \tag{75}$$

In addition, E_0 is the magnitude of the static field, I_ω is the incident laser intensity, the n's are the appropriate refractive indices at the indicated frequencies, and where G refers to the glass windows and L the liquid being measured. The a's are the absorption coefficients at the indicated frequencies for the liquid; l is the propagation length through the gap, and l_c the appropriate coherence length defined by,

$$l_c = \frac{\pi c}{2\omega(n_{2\omega} - n_\omega)}. \tag{76}$$

The Γ's are the EFISH susceptibilities for the glass and liquid. A similar expression is applicable to the reference standard data. Both Maker fringe sets are fit to:

$$y = A_1 \sin^2\left[\frac{\pi l}{2A_3} + \frac{A_4}{2}\right] + A_2. \tag{77}$$

The mean intensity is calculated from:

$$A_m = \frac{A_1}{2} + A_2. \tag{78}$$

Taking the ratio of the mean intensities from the liquid and reference standard data, we obtain:

$$\frac{A_m^L}{A_m^R} = \frac{E_0^2(T_1\Gamma_G l_c^G - T_2\Gamma_L l_c^L)^2[\exp(-2a_\omega \bar{l}) + \exp(-a_{2\omega}\bar{l})]}{(Q_1^2 + Q_2^2)d_{11}^2(l_c^R)^2}, \tag{79}$$

where

$$Q_1 = \frac{2n_{2\omega}^R}{n_{2\omega}^R + 1} \frac{n_\omega^R + 1}{n_{2\omega}^R + 1} \frac{1}{n_{2\omega}^R + n_\omega^R} \left(\frac{2}{n_\omega^R + 1}\right)^2 \tag{80}$$

and

$$Q_2 = \frac{n_{2\omega}^R + n_\omega^R}{n_{2\omega}^R + 1} \frac{1}{n_{2\omega}^R + n_\omega^R} \left(\frac{2}{n_\omega^R + 1}\right)^2. \tag{81}$$

The R superscript refers to the reference standard, and $\bar{l}$ is the mean gap length. The coherence lengths are obtained from A_3, and the T's from index of refraction measurements. The values for d_{11}, Γ_G, and l_c^G are taken from independent measurements or the literature. Their values are discussed in detail in Chapter 1. Clearly the level of accuracy in the absolute measurements of the standard affects the overall accuracy of measurements.

Dipole moment measurements can be carried out with an appropriate liquid dielectric cell and a ratio arm transformer capacitance bridge.[45] Such a bridge has sufficient precision to measure the capacitance differences in dilute solutions. Low polarity solvents and careful selection of electrodes are necessary to reduce the effects of electrode polarization and ionic drift. Measurements may be carried out at a number of frequencies and should approximate the response at the EFISH field conditions.

Index of refraction measurements of these dilute solutions generally require that a differential refractometer be used. Such a setup is shown in Fig. 9.[55] It consists of a two-part liquid cell with an angled divider. The cell is placed in an interferometer, and the cell driven laterally across the laser beam in one arm of the interferometer. In one section of the cell is placed pure solvent, and in the other the solution of the molecules of interest. The resultant interferogram measures the difference in index of refraction between the two liquids directly. The index difference is related to the lateral position of subsequent fringes, maxima, or minima. Thus, the index difference is given by:

$$\Delta n = \frac{\lambda}{2[x(\varphi = 2\pi) - x(0)]\tan\alpha}, \tag{82}$$

where α is the angle of the cell separator. Practical cells can measure index differences to nearly one part in 10^5.

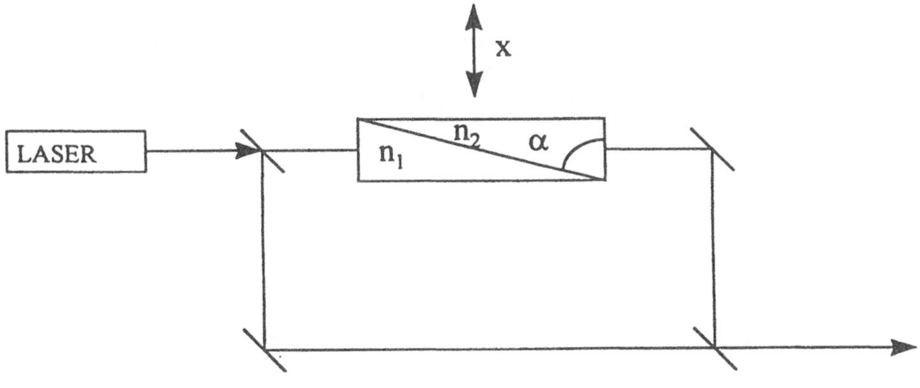

Figure 9 Differential refractometer consisting of a two-part liquid cell inserted into one arm of a Mach-Zehnder interferometer.

VII. HYPER-RAYLEIGH SCATTERING

Hyper-Rayleigh scattering (HRS) is an alternative to EFISH as a method of measuring the second order molecular susceptibility of nonlinear optical chromophores. The HRS experimental procedure involves measuring the amount of light incoherently scattered at exactly twice the frequency of an intense incident laser beam. Figure 10 shows a basic representation of HRS. The technique was first demonstrated in 1965[56] soon after the development of high peak-power lasers. EFISH became the standard method of measuring molecular nonlinear coefficients, and HRS was seldom used until the early 1990s.[57]

HRS has important distinctions from EFISH, most notable that HRS does not require an externally applied electric field. With EFISH, the external field aligns the dipole moments of the nonlinear optical chromophores to remove the inversion symmetry from the solution being studied. The measured quantity is the scalar product of the dipole moment and the vector portion of the susceptibility tensor, which are assumed to be parallel. This limits the EFISH technique to studying only compounds with a permanent electric dipole whose dipole moment must be measured in an independent experiment. In addition, the presence of a high voltage DC field in EFISH prohibits measurements of materials containing ionic species.

HRS measures the second order molecular susceptibility without an externally applied field. As such, it can be used to measure molecules that lack a permanent dipole moment but have an octupolar moment.[58,59,60,61] It also allows experiments with ionic or conductive species. HRS can measure several of the tensor components of the susceptibility while EFISH can measure only the component parallel to the dipole moment. The two techniques combined can determine the angle between the dipole moment and the vector portion of the nonlinear susceptibility.

A. Theory

In the presence of an optical field **E**, a material will become polarized according to

$$P_I = P_I^o + A_{IJ}E_J + B_{IJK}E_JE_K + \dots, \tag{83}$$

where P is the polarization, A_{IJ} the uncorrelated macroscopic linear susceptibility, B_{IJK} the uncorrelated macroscopic second order susceptibility, and summing over repeated indices is implied. The indices I, J, and K refer to fixed axes in the laboratory frame of reference. The third term on the right of Eq. 83 corresponds

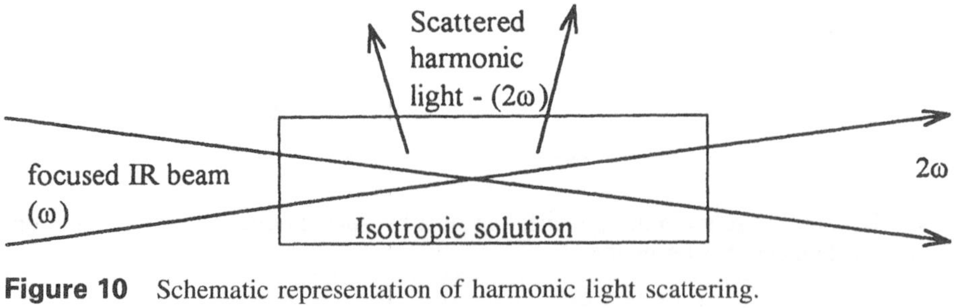

Figure 10 Schematic representation of harmonic light scattering.

to a macroscopic polarization that oscillates at the second harmonic frequency of the incident light:

$$P_I^{2\omega} = B_{IJK} E_J^{\omega} E_K^{\omega}. \tag{84}$$

In isotropic media, the average value of B_{IJK} vanishes because the scattering units are randomly oriented, and fields emitted from each individual source (the NLO molecules in solution) interfere destructively. However, fluctuations in orientation can produce small regions of alignment that generate a measurable amount of incoherently scattered second harmonic light.[62] Assuming that values of B_{IJK} are correlated only over distances much smaller than the wavelength, the second harmonic intensity is proportional to the average value of the square of B_{IJK} as follows:

$$I_{IL}^{2\omega} \propto \int d^3x_1 \, \langle B_{IJK,1} B_{LMN,2}^* \rangle \int d^3x_2 E_J E_K E_M^* E_N^*, \tag{85}$$

where the brackets indicate averaging over all orientations. The subscripts 1 and 2 on B_{IJK} refer to different positions within the scattering media. The assumption that B_{IJK} is uncorrelated over long distances allows those subscripts to be dropped. If the scatterers are randomly oriented molecules, the total intensity, $I^{2\omega}$, is the sum of the contributions from individual molecules and can be written as[56]:

$$I^{2\omega} = GN \sum_{IJKLMN} \langle \beta_{IJK} \beta_{LMN}^* \rangle \, (I^{\omega})^2, \tag{86}$$

where N is the number density of scattering units, β the molecular second harmonic coefficient (first hyperpolarizability), I^{ω} the incident intensity, and G is a factor that incorporates corrections for local fields and experimental geometry. According to Zyss and Ledoux[61]:

$$G = \frac{2^{11}\pi^5}{cR^2\lambda^4} \left(\frac{n_{2\omega}+2}{3}\right) \left(\frac{n_{\omega}+2}{3}\right)^2, \tag{87}$$

where the n's are the refractive indices at the indicated frequencies, λ the wavelength, and R the distance from the scattering center to the detector. Internal referencing to the solvent or to a solution of known nonlinearity determines G.

In the limit of low absorption, Eq. 86 can be extended to include scattering from two species:

$$I^{2\omega} = G(N_{\text{solvent}} \langle \beta_{\text{solvent}}^2 \rangle + N_{\text{dye}} \langle \beta_{\text{dye}}^2 \rangle)(I^{\omega})^2. \tag{88}$$

Equation 88 predicts a linear relationship between $I^{2\omega}$ and N_{dye} in the limit of a dilute solution ($N_{\text{dye}} << N_{\text{solvent}}$). It is important to note that the values of $\langle \beta^2 \rangle$ in Eq. 88 are defined with respect to laboratory-fixed axes. The transformation to molecule-fixed axes, i, j, and k, as well as the method for determining β_{ijk}, will be reviewed later.

B. Experiment

A typical experimental setup is shown in Fig. 11. The HRS process is extremely inefficient. Typically, the number of scattered photons is about 10^{-14} times the number of incident photons. Intense input radiation is required to produce measurable output. For this reason, Q-switched Nd:YAG-based sources are popular be-

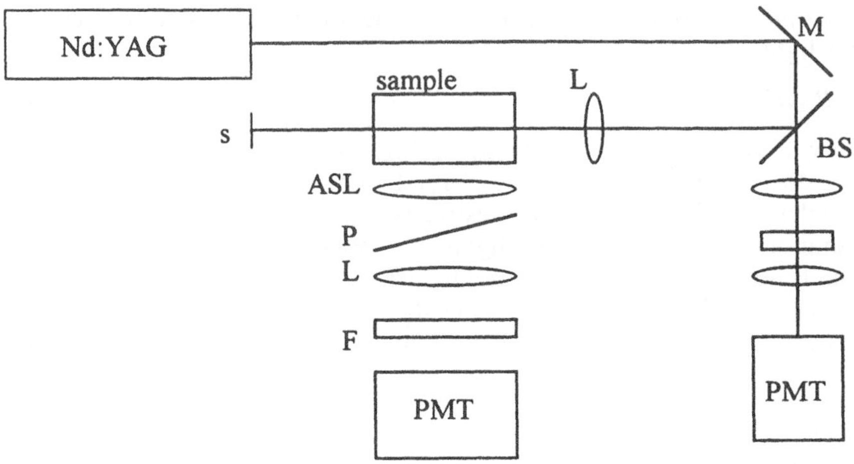

Figure 11 Experimental set-up for Hyper-Rayleigh Scattering. Nd:YAG is the optical source. L—lens, ASL—aspheric lens, M—mirror, BS—beam splitter, Q—quartz reference crystal. P—polarizer, F—filters—beamstop, PMT—photomultiplier tube.

cause they provide very high peak power.[63] It is also possible to use mode-locked CW sources such as Ti:Sapphire lasers.[64] Tunable optical sources are very desirable to conduct experiments such that the second harmonic light is not in the linear absorption band of the chromophore[65] or to probe resonant behavior of nonlinear optical chromophores.

The setup in Fig. 11 uses alternating half-wave plates and linear polarizers to set the polarization and continuously vary the incident intensity. It is necessary to vary the intensity to verify the $(I^{\omega})^2$ dependence of the scattered light. At very high fundamental intensities, additional nonlinear optical processes such as self-focusing lead to subquadratic behavior in the intensity of the scattered light.[61]

Figure 11 shows the 90 degree scattering geometry and uses an efficient optical system for the collection of photons at the harmonic wavelength. The first lens must be of a very short focal length (large numerical apperature) in order to collect and collimate as much light as possible. However, a trade-off exists for the selection of the entrance pupil, namely that too small a pupil will limit usable signal whereas too large a pupil affects assumptions concerning the scattering angle and G.[66] Aspheric lenses are ideal for this purpose. The sample cell can be a rectangular glass cuvette and should be long enough (~40 mm) so that the sample can be viewed while avoiding the second harmonic light produced at the air-glass interface. Once collected and collimated, the scattered light passes through an IR filter to remove scattered fundamental radiation, and a linear polarizer or waveplate can analyze the polarization state of the scattered light. A final lens then focuses the beam into the detector. A small portion of the incident beam is split off before it reaches the sample for referencing. Figure 11 references HRS to second harmonic generation in quartz which has the same $(I^{\omega})^2$ dependence as HRS.

The detector consists of a narrow-band filter and a photomultiplier tube. Fixed narrow bandpass (3–10 nm) filters or a grating monochromator can be used as filters. A boxcar integrator averages the outputs of the photomultiplier tubes.

Multiphoton fluorescence occurs in many dyes across the visible spectrum and has an $(I^{\omega})^2$ dependence which competes with HRS.[67,68,69,70] Fluorescence can sometimes be separated from HRS either temporally[70] or spectrally.[69,71] However, when the fluorescence is largest near the two-photon absorption maximum of the molecule, the Hyper-Rayleigh signal can be completely extinguished due to dephasing of the molecule's virtual excited state.[69] A tunable band-pass filter, such as a grating monochromator in front of the detector, allows spectral isolation of the narrow Hyper-Rayleigh line from broad-band fluorescence when dephasing is not too severe. High-intensity tunable sources such as OPOs[52,53] can be used to minimize fluorescence by conducting experiments in far off-resonant regimes. Though some resonant results for HRS are given in the Tables, these results are highly suspect due to dephasing and fluorescence. In fact, dephasing of HRS can occur far from the two-photon resonance peak.[69]

Consider the experimental setup in Fig. 11. With incident light polarized perpendicular to the plane of incidence, an analyzer with the same orientation will measure

$$I_Z^{2\omega} = GN\langle \beta_{ZZZ}^2 \rangle (I^{\omega})^2, \tag{89}$$

where Z indicates the laboratory axis perpendicular to the plane of incidence in Fig. 11. With the same setup, an analyzer parallel to the direction of travel of the incident beam will measure:

$$I_X^{2\omega} = GN\langle \beta_{ZXX}^2 \rangle (I^{\omega})^2. \tag{90}$$

With right-angle scattering and linear polarization, Eqs. 89 and 90 represent the maximum amount of unique information available with off-resonant HRS. Scattering at other angles and with circular and elliptically polarized input and output radiation will provide more information about the susceptibility.[72,73] Parametric light scattering (incoherently scattered light at the sum frequency of two input beams) is required to obtain the maximum amount of information on the complete molecular susceptibility tensor.[74]

C. Rotational Invariants

As mentioned earlier, nonlinear light scattering can measure several components of the susceptibility tensor. The susceptibility tensor β is a rank-3 tensor and in the most general case has 27 independent components. However, there are considerations in nonlinear light scattering that reduce the number of measurable components. Consider the decomposition of β in the most general case as a tensor product of three vectors:

$$\begin{aligned} \beta &\sim 1 \otimes 1 \otimes 1 \\ &\sim (0 \oplus 1 \oplus 2) \otimes 1 \\ &\sim 0 \oplus (1 \oplus 1 \oplus 1) \oplus (2 \oplus 2) \oplus 3, \end{aligned} \tag{91}$$

where 0, 1, 2, and 3 correspond to the rank, J, of various components. The tensor decomposition and multiplication depicted in Eq. 91 follow the rules of the addition of angular momenta.[75] The $\oplus$ symbols in Eq. 91 show that the susceptibility is a tensorial sum of one pseudo-scalar, three vectors, two second-rank pseudo-tensors, and one rank-3 tensor (all embedded in a rank-3 tensor space using appropriate unit tensors[75]). Since nonlinear scattering measures the average square of the susceptibility ($\langle\beta \otimes \beta\rangle$), interference will produce three complex terms among the vectors and two complex terms between the pseudo-tensors. There are a total of 15 observable quantities—10 real and five imaginary. The rotational invariants of the tensor components may be determined with polarization studies of parametric light scattering.[74]

In the case of HRS, the incident fields are indistinguishable so β is symmetric with respect to the interchange of two indices. The decomposition now becomes:

$$\begin{aligned}\beta &\sim 1 \otimes (0 \oplus 2)\\ &\sim (1 \oplus 1) \oplus 2 \oplus 3,\end{aligned} \tag{92}$$

so in this case, a measurement of $\langle\beta\beta^*\rangle$ as in Eq. 86 would yield six components: Two vectors, one deviator, one rank-3 tensor, and a complex term describing the interference between the vectors.

In cases where Kleinman symmetry is valid, the tensor must be symmetric under interchange of all indices. In that case, the fully symmetric rank-3 tensor reduces to

$$\beta \sim 1 \oplus 3 \tag{93}$$

which indicates that one vector ($J = 1$) and one rank-3 tensor ($J = 3$) are measurable with HRS if Kleinman symmetry is present.[76] The $J = 1$ component is a dipolar contribution measurable with EFISH or HRS while the $J = 3$ part has octupolar symmetry measurable only with HRS.

D. Data Analysis

For linear polarizations and right-angle scattering, Eqs. 88–90 provide the ratio of β_{dye} to β_{solvent} in the reference frame of the laboratory. Molecular symmetry of the solvent and dye molecules must be considered (and the susceptibility of the solvent must be known) to calculate the molecule-fixed frame susceptibility of the chromophore from $\langle\beta^2_{ZXX}\rangle$ and $\langle\beta^2_{ZZZ}\rangle$. Cyvin et al. proposed a method to calculate $\langle\beta^2_{ZXX}\rangle$ and $\langle\beta^2_{ZZZ}\rangle$ in linear combinations of the following five expressions quadratic in β_{ijk}, where ijk are molecular frame fixed axes[46]:

$$\sum_i \beta^2_{iii}, \quad \sum_{i\neq j} \beta_{iii}\beta_{ijj}, \quad \sum_{i\neq j} \beta^2_{iij}, \quad \sum_{ijk,\text{cyclic}} \beta_{iij}\beta_{jkk}, \quad \beta^2_{ijk}.$$

As an example, consider planar molecules such as p-nitroaniline (pNA) where only 2 indices occur in the susceptibility because out-of-plane nonlinearities are assumed to be negligible. In this case, both $\langle\beta^2_{ZXX}\rangle$ and $\langle\beta^2_{ZZZ}\rangle$ are measurable in the lab frame. According to Ref. [46], these values correspond to the following linear combinations:

$$\langle\beta^2_{ZZZ}\rangle_{\text{pNA}} = \frac{1}{7}\beta^2_{zzz} + \frac{9}{35}\beta^2_{zxx} + \frac{6}{35}\beta_{zzz}\beta_{zxx} \tag{94}$$

and

$$\langle\beta^2_{ZXX}\rangle_{\text{pNA}} = \frac{1}{35}\beta^2_{zzz} + \frac{11}{105}\beta^2_{zxx} - \frac{2}{105}\beta_{zzz}\beta_{zxx}, \tag{95}$$

where z and x are defined in the molecule-fixed frame with z corresponding to the symmetry axis of the molecule. Note that the ratio of Eq. 94 to Eq. 95 gives the depolarization ratio R:

$$R = \frac{3 + 11u^2 - 2u}{15 + 27u^2 + 18u}, \tag{96}$$

where $u = \beta_{zxx}/\beta_{zzz}$.

For example, in the case of pNA, HRS using long wavelength (>1064 nm) fundamental radiation, only two components can be measured according to Eq. 93. The quantity u relates to the ratio of the vector to octupolar contributions to the nonlinearity[60] as follows:

$$\frac{\beta(J = 1)}{\beta(J = 3)} = \frac{1}{\sqrt{3}} \left(\frac{1 - 3u}{1 + u} \right), \tag{97}$$

where $\beta(J = 1)$ is the vector contribution which is exactly that measured by EFISH while $\beta(J = 3)$ is the octupolar (rank-3 tensor) contribution. For pNA-like compounds in a Kleinman valid regime[76]:

$$\begin{aligned} \beta(J = 1) &= (\beta_{zzz} + \beta_{zxx}) \\ \beta(J = 3) &= (3\beta_{zxx}^2 + \beta_{zzz}^2 - \tfrac{3}{5}\beta(J = 1)^2)^{1/2}. \end{aligned} \tag{98}$$

As an example of another molecular symmetry, consider the T_d compound, carbon tetrachloride (CCl_4). This was among the first compounds measured with nonlinear light scattering.[56] CCl_4 has only one susceptibility component, β_{xyz}. In this case, linearly polarized input light with no analyzer in place will measure the sum of Eqs. 89 and 90. The numerical coefficient relating $\langle\beta^2\rangle$ to β_{xyz}^2 is given Refs. [46] and [61] as:

$$\langle\beta^2\rangle_{T_d} = \frac{4}{7}\beta_{xyz}^2. \tag{99}$$

E. Reference Methods

In order to calculate the susceptibility from Eqs. 88–90, it is necessary to know the nonlinearity and the point group of the solvent being used. When $I^{2\omega}$ is plotted as a function of N_{dye}, the ratio of the slope to the intercept is proportional to the ratio $\langle\beta_{\text{dye}}^2\rangle/\langle\beta_{\text{solvent}}^2\rangle$. However, solvent signals can be extremely small, so it is often advantageous to reference measurements to a solution of known nonlinearity.

Although referencing schemes can be worked out for the general case, it is much simpler when the reference and unknown belong to the same symmetry group and especially if they have similar values of u from Eq. 96. If $I^{2\omega}$ is plotted as a function of N_{dye}, Eq. 90 yields a straight line of slope m and intercept b. These relate to $\langle\beta_{\text{dye}}^2\rangle$ by:

$$\frac{m}{R} = \frac{\langle\beta_{\text{dye}}^2\rangle}{\langle\beta_{\text{reference}}^2\rangle}, \tag{100}$$

where $R = b/N_{\text{solvent}}$ if the solvent is used as an internal reference. If the experiment is referenced to another material of known nonlinearity, R is the slope of $I^{2\omega}$ vs. N for that material in the same solvent.

If the unknown and reference have the same value of u and the same symmetry (such as pNA and acetone), Eq. 97 implies that $\beta(J = 1)$ and $\beta(J = 3)$ have the same fixed ratio between the two species. In that case, Eq. 96 can directly yield the molecule-fixed frame hyperpolarizability according to:

$$\left(\frac{m}{R}\right) \beta^2_{\text{reference},J} = \beta^2_{\text{dye},J}, \tag{101}$$

where $J = 1$ or $J = 3$.

If, as will often be the case, the unknown dye is not of the same symmetry as the reference material, group theory can be used to derive equations similar to Eqs. 94 and 95 which can be used to calculate the molecule-fixed frame nonlinearity. It is reasonable to assume that pNA-like compounds have only one dominant tensor component; β_{zzz} (β_{zxx} is only $\sim$10% of β_{zzz}). With that assumption, a dominantly one-dimensional molecule like disperse red 1 can be referenced with a pNA-like compound using Eq. 100.

VIII. GUIDE TO THE TABLES

The Tables that follow contain many of the reported measurements of molecular second order nonlinear optical responses. Reproducibility is generally reported in the range of 10–20 percent. Based on the discussion in the above sections, the overall accuracy may be considerably worse. The factors that contribute to accuracy include a lack of a clearly defined convention by the authors and the quartz standard value used, which has recently been shown to be questionable (see Chapter 1). Other factors affecting accuracy are the assumptions regarding whether γ can be ignored. This may lead to errors over 20 percent. Other assumptions commonly made which affect comparison with theory include the one-dimensionality of the β tensor. Dispersive and absorptive factors contributing to assumptions in the Maker fringe derivation may be a significant source of inaccuracy when near-resonant measurements are made. In addition, solvent shifts of gas-phase absorptions will also affect comparison with theory. Thus, the best use of the Tables is for comparing values for various compounds by a single worker. When different workers measure the same compound, a scale for comparison between workers can be developed by comparing the magnitudes of compounds that are common to both workers.

A number of the entries in the Tables, for all techniques, are highly resonant (absorption peak near the two-photon resonance). For a number of reasons, results of such measurements can be completely inaccurate (e.g., no contribution from the second harmonic coefficient). Even comparisons between such measurements can be inaccurate due to differing amounts of fluorescence between compounds, for example.

A measure of the nonlinear response that reduces the effect of dispersion is useful in comparing various compounds of differing excitation energy determined at a single wavelength. The quantity β_0 in the Tables is such a measure. Here, β_0 is based on a two-level model, and attempts to eliminate the dispersion in β due to the lowest excited state. While the two-level model is not adequate to precisely describe the dispersion in many compounds, it is adequate for this purpose, thus allowing an assessment of compounds on a common scale. It is applicable only to dipolar nonlinearities. The quantity is defined through[51]

$$\beta_0 = \beta \cdot \frac{(\omega_0^2 - \omega^2)(\omega_0^2 - 4\omega^2)}{\omega_0^4}, \tag{102}$$

where ω_0 is the frequency of the first excited molecular electronic state, and ω the fundamental laser frequency of the measurement.

Note that Ref. [2] in the Table refers to estimates of β taken from solvatochromic measurements. Since this Chapter deals with second harmonic measurements, a description of the technique is beyond the scope of the Chapter. The reader is referred to that reference and references therein. These results are included for the convenience of the reader.

All β values listed in the Tables are vacuum values. Because of ambiguities in the literature, some of the tabulations may contain dressed β values.

IX. LEGEND

Technique:	HRS = Hyper-Rayleigh Scattering
	EFISH = Electric Field Induced Second Harmonic Generation
	LB = Thin Film (Langmuir-Blodgett)
	PLS = Parametric Light Scattering
	PP = Poled polymer
λ_{meas}:	Measurement wavelength (nm)
λ_{max}:	Peak absorption wavelength (nm)
μ:	Ground state molecular dipole moment (D = 10^{-18} esu)
β:	Molecular nonlinear response (usually projection along μ) (10^{-30} esu)
	Conversion to SI: 1×10^{-30} *esu* = 2.386×10^{-39} *SI*
β_0:	Defined above (10^{-30} esu)

ACKNOWLEDGMENTS

CWRU gratefully acknowledges support from the National Science Foundation under the Science and Technology Center ALCOM (DMR89-20147). UMBC gratefully acknowledges support from the National Science Foundation (DMR-9318948) and a Cottrell Scholars Award from the Research Corporation.

References begin on page 504

Hyperpolarizability tabulations. Shaded entries are calculated from values given in the cited reference.

	MOLECULE NAME, OR STRUCTURE	tech	solvent
	Conv. I: d_{11} (quartz) ≅ 1×10^{-9}esu II: $X^{(2)}$(quartz) ≅ 2×10^{-9}esu.		
1	n=1 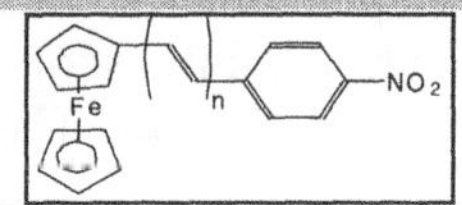	EFISH	dichloromethane
1	n=2 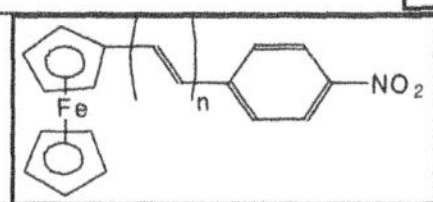	EFISH	dichloromethane
1	n=0 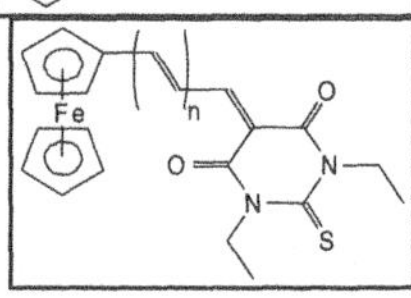	EFISH	dichloromethane
1	n=1 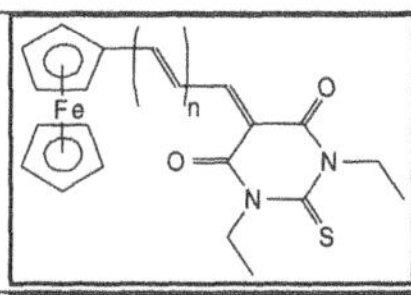	EFISH	dichloromethane
1	n=2 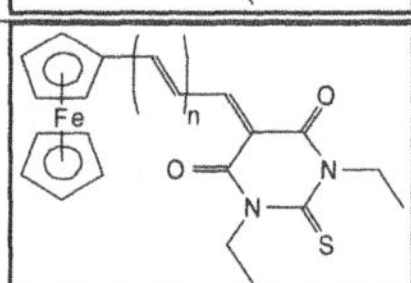	EFISH	dichloromethane
1	n=0 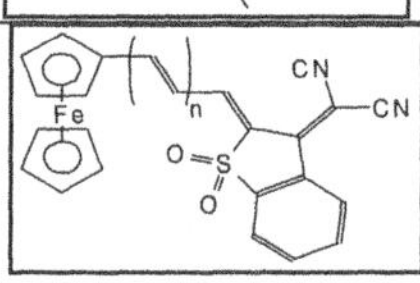	EFISH	dichloromethane

λ_{meas}	λ_{max}	μ	β	β_0	$\mu\beta$	$\mu\beta_0$	footnote	convention
					$\times 10^{-30}$			
(nm)	(nm)	(D)	($\times 10^{-30}$esu)		esu $\times$ D			
		Debye	$\times 10^{-40}$ m^4/V ( )=ERROR					
1907					140.0			I
1907					300.0			I
1907					100.0			I
1907					850.0			I
1907					1900.0			I
1907					160.0			I

Footnote entries on page 503.

Hyperpolarizability tabulations. Shaded entries are calculated from values given in the cited reference.

	MOLECULE NAME, OR STRUCTURE	tech	solvent
1	n=1 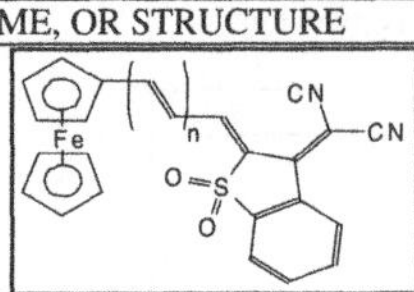	EFISH	dichloromethane
1	n=2 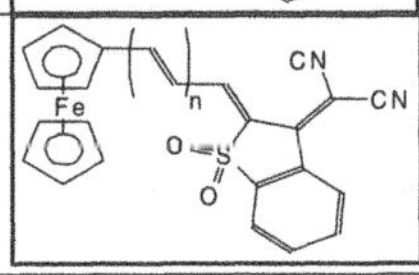	EFISH	dichloromethane
2	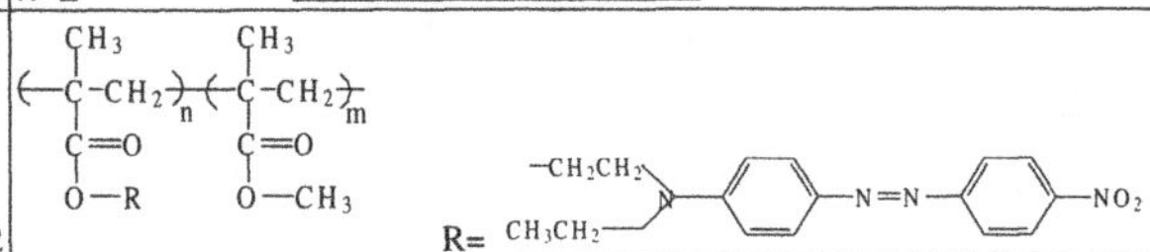	solvatochromism	
2	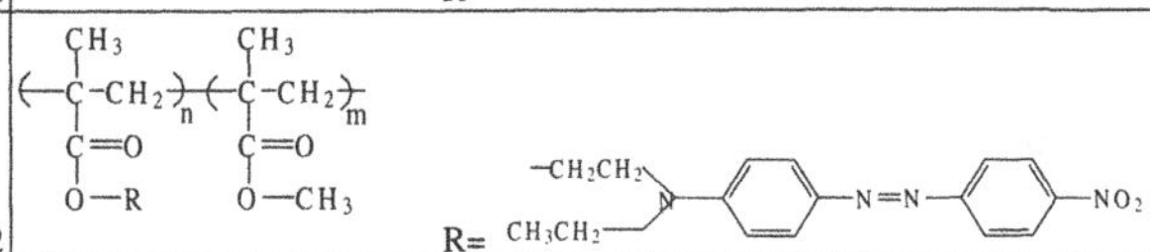	solvatochromism	
2	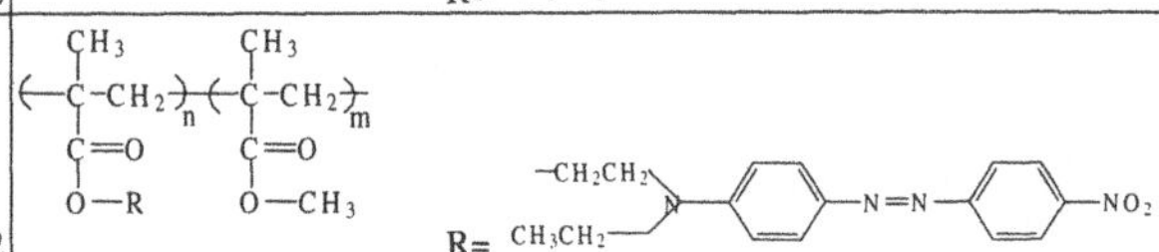	solvatochromism	
2	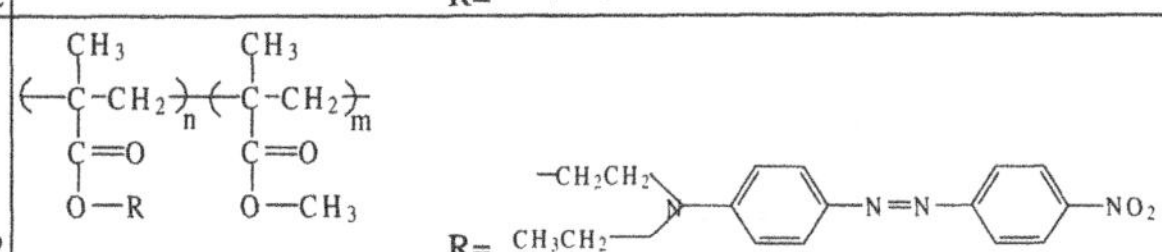	solvatochromism	
2	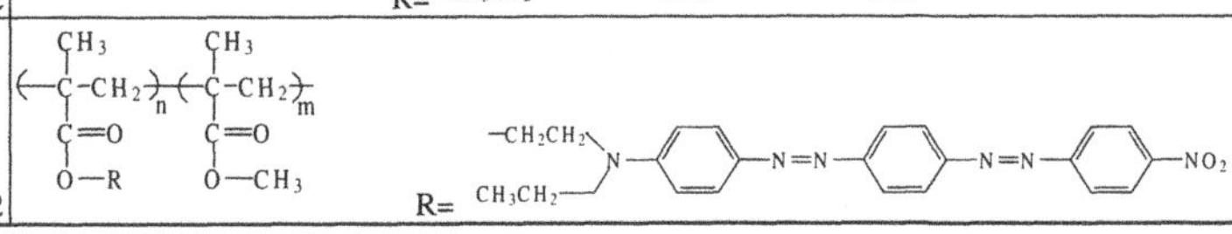	solvatochromism	

λ_{meas}	λ_{max}	μ	β	β_0	$\mu\beta$	$\mu\beta_0$	footnote	convention
1907					1400.0			I
1907					3000.0			I
1064					2000.0			
1300					830.0			
1500					630.0			
1700					540.0			
1064					19000.0			

Footnote entries on page 503.

Hyperpolarizability tabulations. Shaded entries are calculated from values given in the cited reference.

	MOLECULE NAME, OR STRUCTURE	tech	solvent
2	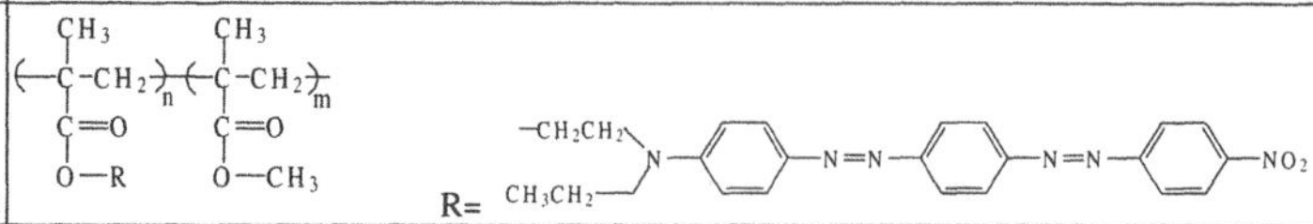	solvatochromism	
2	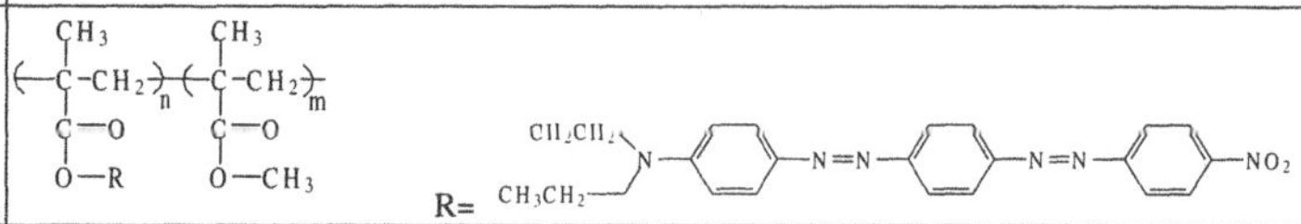	solvatochromism	
2	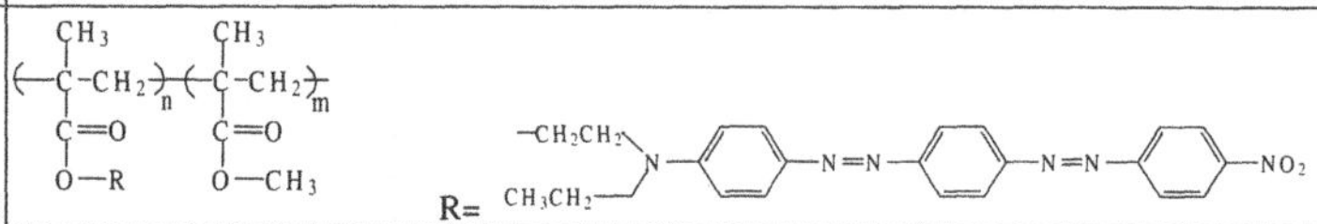	solvatochromism	
3	pentacyano(4-octadecylaminopyridine)-ferrate(III)	LB	monolayer
4	stilbene derivatives: D=Me, A=NO_2	EFISH	chloroform
4	stilbene derivative: D=Me, A=CN	EFISH	chloroform
4	stilbene derivative: D=MeO, A=NO_2	EFISH	chloroform
4	stilbene derivative: D=MeO, A=CN	EFISH	chloroform
4	stilbene derivative: D=MeS, A=NO_2	EFISH	chloroform
4	stilbene derivative: D=Me2N, A=NO_2	EFISH	chloroform
4	stilbene derivative: D=Me2N, A=CN	EFISH	chloroform
4	stilbene derivative: D=Br, A=NO_2	EFISH	chloroform
4	stilbene derivative: D=Br, A=CN	EFISH	chloroform
4	stilbene derivative: D=MeS, A=CN	EFISH	chloroform
5	NPAN N-(4-nitrophenyl)-N-methylaminoacetonitrile	EFISH	acetone
5	NPP N-(4-nitrophenyl)-L-prolinol	EFISH	acetone
6	N,N-dimethethyl-4-nitroaniline	EFISH	acetone
6	CH_3-S-Φ-NO (MNS)	EFISH	acetone
6	Φ-S-Φ-NO_2 (NDS)	EFISH	acetone
6	Φ-S-Φ-NH (ADS)	EFISH	acetone
6	NO_2-Φ-S-Φ-NH (ANDS)	EFISH	acetone

λ_{meas}	λ_{max}	μ	β	β_0	μβ	$\mu\beta_0$	footnote	convention
1300					4600.0			
1500					3200.0			
1700					2700.0			
1064	560		172	-13.4				
1064	371	5.30	51	23.0	270.3	121.9		
1064	330	5.00	28	16.0	140.0	80.0		
1064	379	5.60	85	36.0	476.0	201.6		
1064	340	5.50	36	19.0	198.0	104.5		
1064	382	5.60	104	43.0	582.4	240.8		
1910	436	7.10	90	70.0	639.0	497.0		
1064	388	7.00	120	47.0	840.0	329.0		
1064	356	3.10	63	31.0	195.3	96.1		
1064	326	3.30	23	13.0	75.9	42.9		
1064	353	5.60	36	18.0	201.6	100.8		
1064	360	7.20	23	11.0	165.6	79.5		I
1064	397	7.30	42	16.0	306.6	116.9		I
1064	388	6.90	26	12	179.4	82.8		
1064	335	4.40	20	10	88.0	44.0		
1064	334	4.30	19	10	81.7	43.0		
1064	320	3.00	14	8	42.0	24.0		
1064	341	5.80	28	15	162.4	87.0		

Footnote entries on page 503.

Hyperpolarizability tabulations. Shaded entries are calculated from values given in the cited reference.

	MOLECULE NAME, OR STRUCTURE	tech	solvent
6	NO_2-Φ-SO_2-Φ-NH_2 (ANPS)	EFISH	acetone
6	NH_2-Φ-SO_2-Φ-NH_2 (APS)	EFISH	acetone
7	4-(N,N-dimethylamino)-3- acetemidonitrobenzene	EFISH	dioxane
8	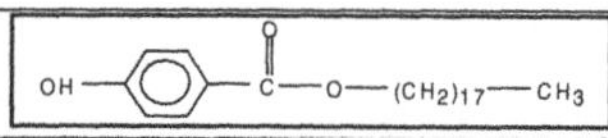	LB	monolayer
8	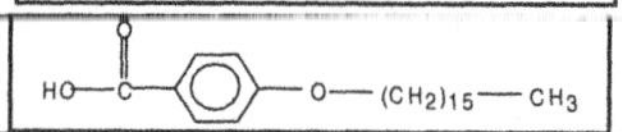	LB	monolayer
8	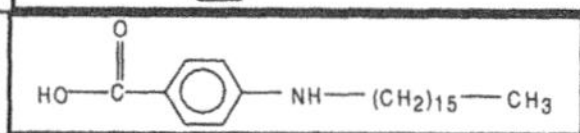	LB	monolayer
9	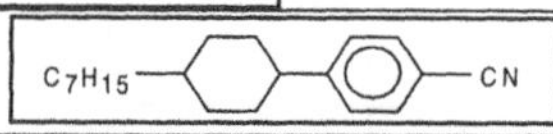	LB	monolayer
9	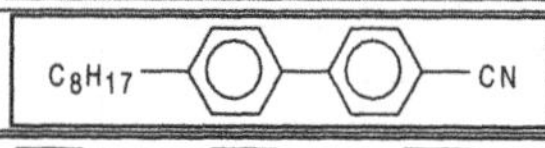	LB	monolayer
9	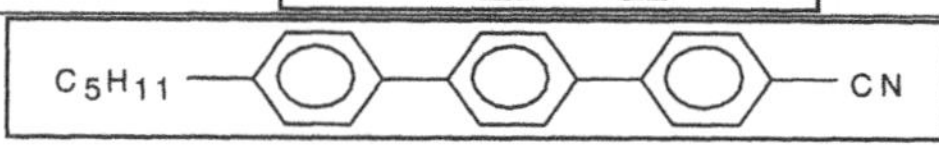	LB	monolayer
9	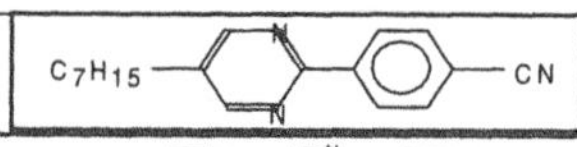	LB	monolayer
9	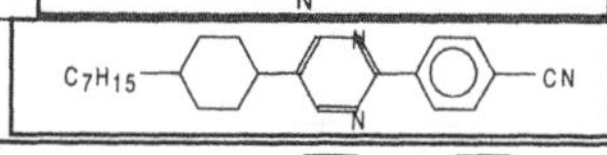	LB	monolayer
9	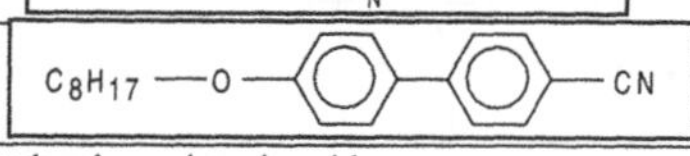	LB	monolayer
10	ODCA 4-Octadecyloxy cinamic acid	LB	monolayer
10	ODCAM 4-Octadecyloxy cinnamamide	LB	monolayer
11	$C_8H_{17}(C_6H_4)_2CN$	LB	monolayer
11	$C_{12}H_{25}(C_6H_4)_2CN$	LB	monolayer
11	$C_8H_{17}(C_6H_4)_2COOH$	LB	monolayer
11	$C_7H_{15}(C_4N_2H_2)C_6H_4CN$	LB	monolayer
11	$C_5H_{11}(C_6H_4)_3CN$	LB	monolayer
11	$C_5H_{11}(C_6H_4)_3CN$	LB	monolayer

λ_{meas}	λ_{max}	μ	β	β_0	$\mu\beta$	$\mu\beta_0$	footnote	convention
1064	332	8.00	19	13	152.0	104.0		
1064	295	7.80	8	5	62.4	39.0		
1064		8.10	30		243.0			
1224			3					
1224			2					
1224			7					
1064			4					
1064			25					
1064			13					
1064			8					
1064			8					
1064			40					
1224			5					
1224			5					
1064			25					
1064			25					
1064			6					
1064			8					
1064			7.5					
1172			13					

Footnote entries on page 503.

Hyperpolarizability tabulations. Shaded entries are calculated from values given in the cited reference.

	MOLECULE NAME, OR STRUCTURE	tech	solvent
12	2-cyclooctylamino-5-nitropyridine (COANP)	EFISH	p-dioxane
12	2-cyclooctylamino-5-nitropyridine (COANP)	EFISH	p-dioxane
12	2-cyclooctylamino-5-nitropyridine (COANP)	EFISH	p-dioxane
12	2-docosylamino-5-nitropyridine (DCANP)	EFISH	p-dioxane
12	2-docosylamino-5-nitropyridine (DCANP)	EFISH	p-dioxane
12	2-docosylamino-5-nitropyridine (DCANP)	EFISH	p-dioxane
12	2-(N-prolinol)-5-nitropyridine (PNP)	EFISH	p-dioxane
12	2-(N-prolinol)-5-nitropyridine (PNP)	EFISH	p-dioxane
12	2-(N-prolinol)-5-nitropyridine (PNP)	EFISH	p-dioxane
12	2-methyl-4-nitroaniline (MNA)	EFISH	p-dioxane
12	2-methyl-4-nitroaniline (MNA)	EFISH	p-dioxane
12	2-methyl-4-nitroaniline (MNA)	EFISH	p-dioxane
13	2-octadecylamino-5-nitropyridine (ODANP)	LB	multilayer
13	2-eicosylamino-5-nitropyridine (ECANP)	LB	multilayer
13	2-docosylamino-5-nitropyridine(DCANP)	LB	multilayer
13	2-tetracosylamino-5-nitropyridine (TCANP)	LB	multilayer
13	2-hexacosylamino-5-nitropyridine (HCANP)	LB	multilayer
14	calix[4]arenes donor / conformation		
14	OCH_3 / cone	EFISH	
14	OCH_3 / cone	HRS	
14	OCH_3 / partial cone	EFISH	
14	OCH_3 / 1,3 -alt	EFISH	
14	OCH_3 / 1,3-alt	EFISH	
14	paranitrophenol	EFISH	
15	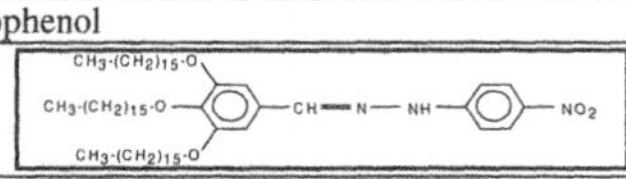	LB	monolayer
15	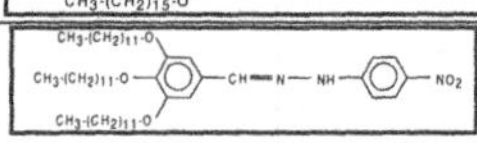	LB	monolayer
15	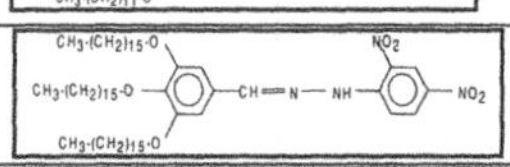	LB	monolayer
15	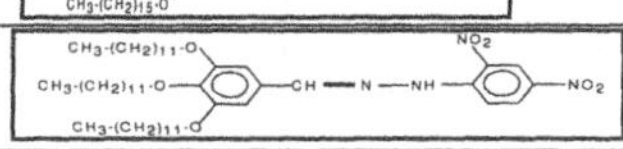	LB	monolayer
15	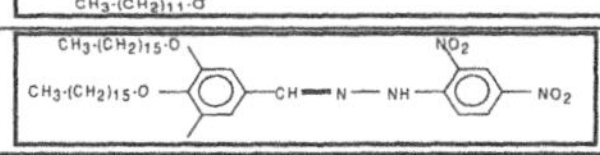	LB	monolayer

λ_{meas}	λ_{max}	μ	β	β_0	μβ	$\mu\beta_0$	footnote	convention
1064		6.80	21.95		149.3			I
1318		6.80	11.93		81.1			I
1907		6.80	10.98		74.6			I
1064		6.70	12.88		86.3			I
1318		6.70	9.31		62.3			I
1907		6.70	5.97		40.0			I
1064		7.20	18.13		130.6			I
1318		7.20	11.21		80.7			I
1907		7.20	11.21		80.7			I
1064		7.40	18.37		136.0			I
1318		7.40	9.07		67.1			I
1907		7.40	7.64		56.5			I
1064			26.01					I
1064			22.91					I
1064			25.53					I
1064			14.08					I
1064			9.78					I
1064	291	13.80	30	19.5	414.0	268.4		
1064	291	13.80	24	15.6	331.2	214.7		
1064	291	6.70	27	17.5	180.9	117.3		
1064	291	0.00	0	0.0				
1064	291	0.00	0	0.0				
1064	302	4.60	12	7.5	55.2	34.4		
1064			44					
1064			21					
1064			64					
1064			13					
1064			60					

Footnote entries on page 503.

Hyperpolarizability tabulations. Shaded entries are calculated from values given in the cited reference.

	MOLECULE NAME, OR STRUCTURE	tech	solvent
15	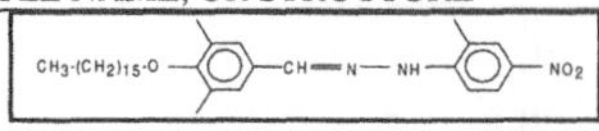	LB	monolayer
15	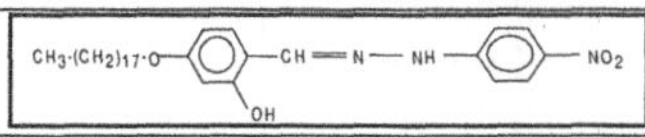	LB	monolayer
15	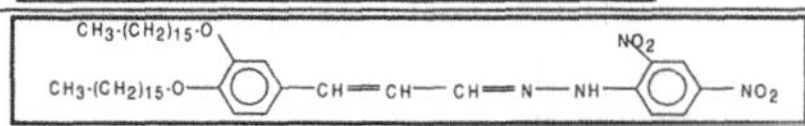	LB	monolayer
15	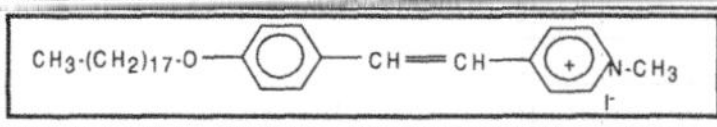	LB	monolayer
15	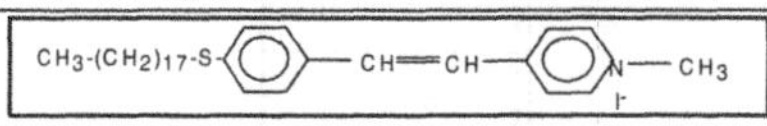	LB	monolayer
15	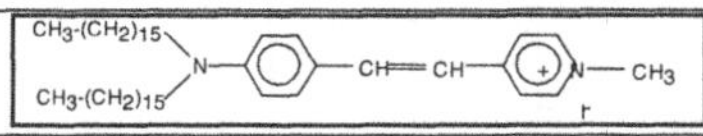	LB	monolayer
15	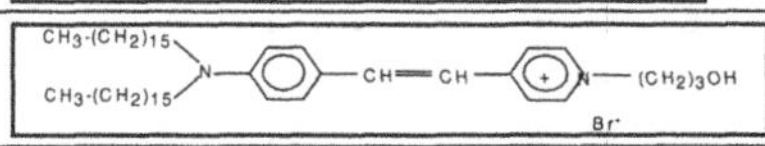	LB	monolayer
15	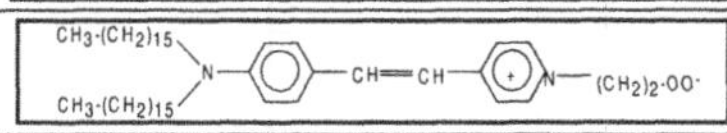	LB	monolayer
15	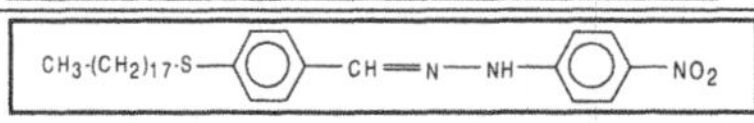	LB	monolayer
16	1)Stilbenes D/A 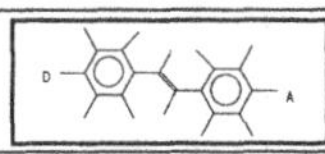		
16		EFISH	p-dioxane
16	CH_3O/SO_2CH_3	EFISH	p-dioxane
16	CH_3O/SO_2CF_3	EFISH	p-dioxane
16	$(CH_3)_2N/SO_2CF_3$	EFISH	p-dioxane
16	$(CH_3)_2N/SO_2CF_3$	EFISH	p-dioxane
16	Φ-S/SO_2CH_3	EFISH	p-dioxane
16	Φ-S/NO_2	EFISH	p-dioxane
16	2) Tolanes D/A 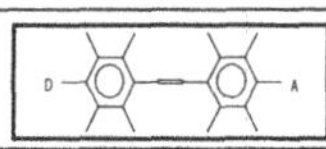		

λ_{meas}	λ_{max}	μ	β	β_0	$\mu\beta$	$\mu\beta_0$	footnote	convention
1064			57					
1064			66					
1064			62					
1064			42					
1064			65					
1064			304					
1064			273					
1064			197					
1064			81					
1064	364	4.50	59.6	27.0	268.2	121.5		
1064	335	6.10	9.1	5.0	55.5	30.5		
1064	350	6.60	34.1	17.3	225.1	114.2		
1064	376	6.90	65.6	28.7	452.6	198.0		
1906	370	5.10	68	29.4	346.8	149.9		
1064	344	4.40	19.4	10.1	85.4	44.4		
1064	364	4.40	57.6	27.0	253.4	118.8		

Footnote entries on page 503.

Hyperpolarizability tabulations. Shaded entries are calculated from values given in the cited reference.

	MOLECULE NAME, OR STRUCTURE	tech	solvent
16	CH_3O/SO_2CH_3	EFISH	p-dioxane
16	CH_3O/SO_2CF_3	EFISH	p-dioxane
16	$(CH_3)_2N/NO_2$	EFISH	p-dioxane
16	$(CH_3)_2N/SO_2CH_3$	EFISH	p-dioxane
16	$(CH_3)_2N/SO_2CF_3$	EFISH	p-dioxane
16	CH_3S/SO_2CH_3	EFISH	p-dioxane
16	H/NO_2	EFISH	p-dioxane
16	H/SO_2CH_3	EFISH	p dioxane
17	Foron Brilliant Blue S-R	EFISH	acetonitrile
17	Foron Brilliant Blue S-R	EFISH	acetonitrile
18	Urea	EFISH	
19	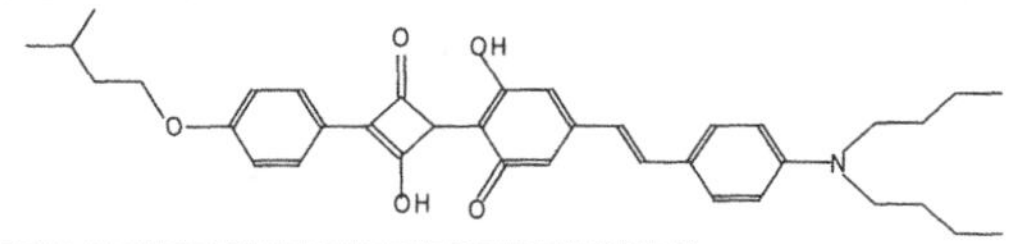	EFISH	chloroform
19	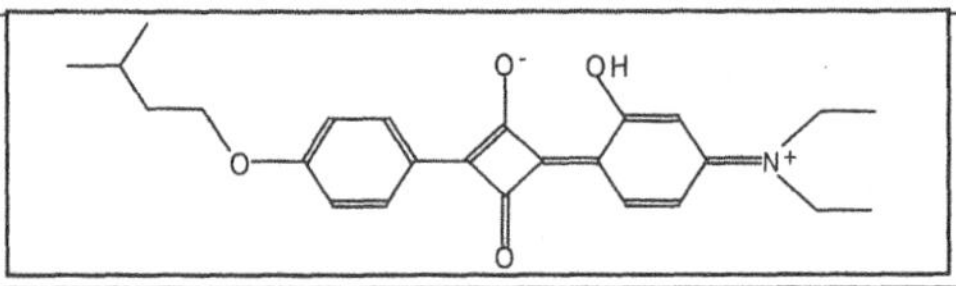	EFISH	chloroform
19	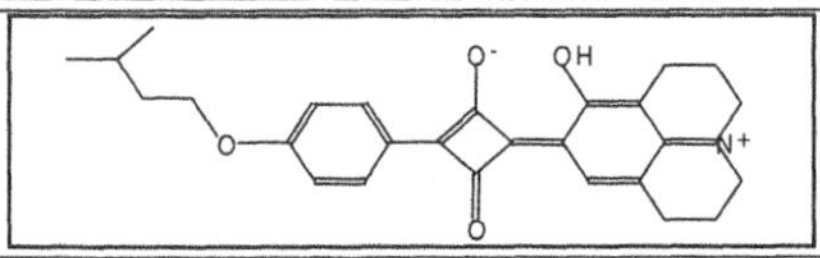	EFISH	chloroform
	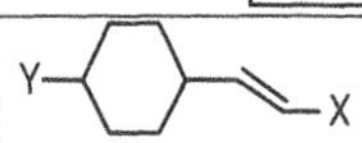 X Y		
20	CN OMe	EFISH	chloroform
20	CN NMe_2	EFISH	chloroform
20	CHO Br	EFISH	chloroform
20	CHO OMe	EFISH	chloroform
20	CHO NMe_2	EFISH	chloroform

λ_{meas}	λ_{max}	μ	β	β_0	μβ	$\mu\beta_0$	footnote	convention
1064	310	5.90	10.8	7.1	63.7	41.9		
1064	327	6.20	20.8	11.7	129.0	72.5		
1064	402	7.10	102	37.5	724.2	266.3		
1064	358	7.50	39.8	19.3	298.5	144.8		
1064	388	8.40	5.6	34.8	47.0	292.3		
1064	320	5.20	15.9	9.2	82.7	47.8		
1064	326	4.60	16	9.1	73.6	41.9		
1064	310	5.30	3.75	2.3	19.9	12.0		
1580		6.80	163	50.0	1108.4	340.0		
954		6.80	-118	50.0	-802.4	340.0		
1064	200		2.3	1.9				
1907	732	8.10	481	170.0	3896.1	1377.0		
1907	570	6.50	-53	-31.0	-344.5	-201.5		
1907	572	7.40	-62	-36.0	-458.8	-266.4		
1064	304	4.20	7.0	4.3	29.4	18.2		II
1064	364	6.00	23	10.8	138.0	64.8		II
1064	298	2.00	6.5	4.1	13.0	8.2		II
1064	318	4.20	11	6.4	46.2	27.0		II
1064	384	5.60	30	12.5	168.0	70.0		II

Footnote entries on page 503.

Hyperpolarizability tabulations. Shaded entries are calculated from values given in the cited reference.

	MOLECULE NAME, OR STRUCTURE	tech	solvent
20	COMe OMe	EFISH	chloroform
20	NO_2 H	EFISH	chloroform
20	NO_2 OH	EFISH	chloroform
20	NO_2 OMe	EFISH	chloroform
20	NO_2 NMe_2	EFISH	chloroform
20	NMe_2 NO_2	EFISH	chloroform
	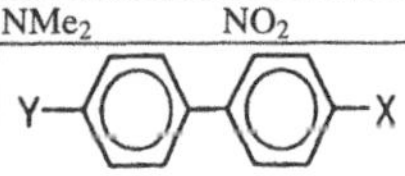		
20	X Y		
20	H H	EFISH	p-dioxane
20	CN H	EFISH	p-dioxane
20	$COCH_3$ H	EFISH	p-dioxane
20	NO_2 H	EFISH	p-dioxane
20	$SO_2C_6H_{12}OH$ NMe_2	EFISH	chloroform
20	CN OH	EFISH	p-dioxane
20	$COCH_3$ OMe	EFISH	p-dioxane
20	NO_2 Br	EFISH	p-dioxane
20	NO_2 OH	EFISH	p-dioxane
20	NO_2 OMe	EFISH	p-dioxane
20	NO_2 NH_2	EFISH	chloroform
20	NO_2 NH_2	EFISH	N-methyl-pyrollidone
20	NO_2 NMe_2	EFISH	chloroform
	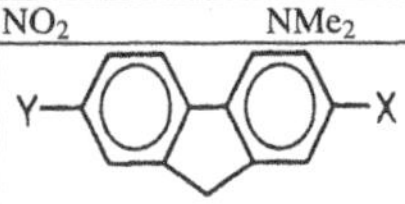		
20	X Y		
20	H H	EFISH	p-dioxane
20	CN H	EFISH	p-dioxane
20	NO_2 H	EFISH	p-dioxane
20	NO_2 Br	EFISH	p-dioxane
20	NO_2 OMe	EFISH	p-dioxane
20	NO_2 NMe_2	EFISH	p-dioxane
20	NO_2 NMe_2	EFISH	chloroform

λ_{meas}	λ_{max}	μ	β	β_0	$\mu\beta$	$\mu\beta_0$	footnote	convention
1064	316	4.00	8.9	5.3	35.6	21.0		II
1064	312	3.80	8.0	4.8	30.4	18.2		II
1064	352	5.10	18	9.0	91.8	46.0		II
1064	352	4.60	17	8.5	78.2	39.2		II
1064	438	6.50	50	13.4	325.0	87.0		II
1064	438	5.90	35	9.4	206.5	55.3		II
								II
1064	252	0.00	0.0	0.0				II
1064	272	4.00	1.9	1.3	7.6	5.2		II
1064	280	3.10	2.0	1.3	6.2	4.2		II
1064	304	3.80	4.1	2.5	15.6	9.6		II
1064	340	6.00	13	6.9	78.0	41.4		II
1064	292	4.80	6.3	4.1	30.2	19.5		II
1064	304	3.40	4.9	3.0	16.7	10.3		II
1064	306	2.70	4.4	2.7	11.9	7.3		II
1064	334	4.90	7.7	4.2	37.7	20.6		II
1064	332	4.50	9.2	5.1	41.4	22.8		II
1064	372	5.00	24	10.8	120.0	53.8		II
		7.80	24		187.2	0.0		II
1064	390	5.50	50	20.0	275.0	110.1		II
								II
1064	262	0.00	0.0	0.0				II
1064	284	3.90	3.0	2.0	11.7	7.8		II
1064	328	4.10	5.1	2.9	20.9	11.7		II
1064	330	2.80	6.0	3.3	16.8	9.3		II
1064	356	4.70	11	5.4	51.7	25.4		II
1064	410	5.60	40	13.8	224.0	77.5		II
1064	417	6.00	55	18.0	330.0	107.7		II

Footnote entries on page 503.

Hyperpolarizability tabulations. Shaded entries are calculated from values given in the cited reference.

	MOLECULE NAME, OR STRUCTURE		tech	solvent
	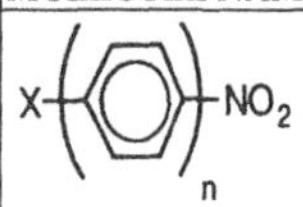			
20	n	X		
20	1	NH_2	EFISH	NMP
20	2	NH_2	EFISH	NMP
20	3	NH_2	EFISH	NMP
20	4	NH_2	EFISH	NMP
20	1	OMe	EFISH	p-dioxane
20	2	OMe	EFISH	p-dioxane
20	3	OMe	EFISH	p-dioxane

20	X	Y		
20	SO_2Me	NH_2	EFISH	chloroform
20	CO_2Me	SMe	EFISH	chloroform
20	CO_2Me	NH_2	EFISH	chloroform
20	COMe	SMe	EFISH	chloroform
20	COMe	NH_2	EFISH	chloroform
20	COPh	NH_2	EFISH	chloroform
20	CN	SMe	EFISH	chloroform
20	CN	NH_2	EFISH	chloroform
20	CN	NHMe	EFISH	chloroform
20	CN	NMe_2	EFISH	chloroform
20	NO_2	Br	EFISH	chloroform
20	NO_2	OMe	EFISH	p-dioxane
20	NO_2	SMe	EFISH	chloroform
20	NO_2	NH_2	EFISH	chloroform
20	NO_2	NH_2	EFISH	NMP
20	NO_2	NHMe	EFISH	chloroform
20	NO_2	NMe_2	EFISH	chloroform

λ_{meas}	λ_{max}	μ	β	β_0	$\mu\beta$	$\mu\beta_0$	footnote	convention
								II
1064	370	7.80	10	4.5	78.0	35.4		II
1064	372	7.80	24	10.8	187.2	84.0		II
1064	360	7.60	16	7.7	121.6	58.4		II
1064	344	10.00	11	5.7	110.0	57.3		II
1064	302	4.60	5.1	3.2	23.5	14.6		II
1064	332	4.50	9.2	5.1	41.4	22.8		II
1064	340	5.00	11	5.8	55.0	29.2		II
								II
1064	338	6.50	13	7.0	84.5	45.3		II
1064	328	2.90	8	4.5	23.2	13.0		II
1064	332	3.80	15	8.3	57.0	31.4		II
1064	336	3.70	9.8	5.3	36.3	19.6		II
1064	334	3.30	12	6.6	39.6	21.6		II
1064	352	3.70	19	9.5	70.3	35.2		II
1064	333	4.00	15	8.2	60.0	32.9		II
1064	342	5.20	20	10.5	104.0	54.7		II
1064	358	5.70	27	13.1	153.9	74.7		II
1064	372	6.10	29	13.0	176.9	79.4		II
1064	335	3.00	10	5.4	30.0	16.3		II
1064	356	4.40	14	6.9	61.6	30.2		II
1064	362	4.00	20	9.5	80.0	38.0		II
1064	380	5.50	24	10.3	132.0	56.4		II
1064	410	5.50	40	13.8	220.0	76.1		II
1064	400	5.70	46	17.2	262.2	97.9		II
1064	415	6.10	46	15.3	280.6	93.1		II

Footnote entries on page 503.

Hyperpolarizability tabulations. Shaded entries are calculated from values given in the cited reference.

	MOLECULE NAME, OR STRUCTURE			tech	solvent
	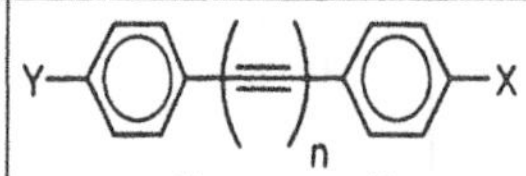				
20	n	X	Y		
20	1	CN	SMe	EFISH	chloroform
20	2	CN	SMe	EFISH	chloroform
20	1	CN	NH_2	EFISH	chloroform
20	2	CN	NH_2	EFISH	NMP
20	1	NO_2	SMe	EFISH	chloroform
20	2	NO_2	SMe	EFISH	chloroform
20	1	NO_2	NH_2	EFISH	NMP
20	2	NO_2	NH_2	EFISH	chloroform
20	2	NO_2	NH_2	EFISH	NMP
20	3	NO_2	NH_2	EFISH	NMP
	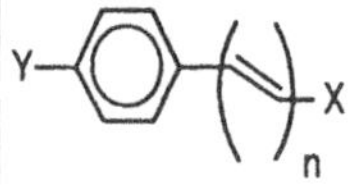				
20	n	X	Y		
20	0	CHO	MeO	EFISH	neat
20	1	CHO	MeO	EFISH	chloroform
20	2	CHO	MeO	EFISH	chloroform
20	3	CHO	MeO	EFISH	chloroform
20	0	CHO	NMe_2	EFISH	p-dioxane
20	1	CHO	NMe_2	EFISH	chloroform
20	2	CHO	NMe_2	EFISH	chloroform
20	3	CHO	NMe_2	EFISH	chloroform
20	0	$CHC(CN)_2$	MeO	EFISH	chloroform
20	1	$CHC(CN)_2$	MeO	EFISH	chloroform
20	0	$CHC(CN)_2$	NMe_2	EFISH	chloroform
20	1	$CHC(CN)_2$	NMe_2	EFISH	chloroform
20	2	$CHC(CN)_2$	NMe_2	EFISH	chloroform

λ_{meas}	λ_{max}	μ	β	β_0	μβ	$\mu\beta_0$	footnote	convention
								II
1064	333	4.00	15	8.2	60.0	32.9		II
1064	330	3.70	17	9.5	62.9	35.0		II
1064	342	5.20	20	10.5	104.0	54.7		II
1064	388				110.0	44.6		II
1064	362	4.00	20	9.5	80.0	38.0		II
1064	338	3.90	17	9.1	66.3	35.5		II
1064	410				220.0	76.1		II
1064	334	6.30	28	15.3	176.4	96.3		II
1064	416				240.0	79.0		II
1064	440				280.0	73.3		II
								II
1064	280	3.50	2.2	1.5	7.7	5.2		II
1064	318	4.00	12	7.0	48.0	28.1		II
1064	350	4.30	28	14.2	120.4	60.9		II
1064	376	4.60	42	18.4	193.2	84.6		II
1064	326	5.10	6.3	3.6	32.1	18.2		II
1064	384	5.60	30	12.5	168.0	70.0		II
1064	412	6.00	52	17.7	312.0	106.2		II
1064	434	6.30	88	24.5	554.4	154.6		II
1064	350	5.50	9.8	5.0	53.9	27.3		II
1064	392	5.80	32	12.6	185.6	73.3		II
1064	420	7.80	32	10.2	249.6	79.4		II
1064	486	8.40	82	10.7	688.8	90.2		II
1064	520	9.00	163	5.5	1467.0	49.8		II

Footnote entries on page 503.

Hyperpolarizability tabulations. Shaded entries are calculated from values given in the cited reference.

	MOLECULE NAME, OR STRUCTURE	tech	solvent
	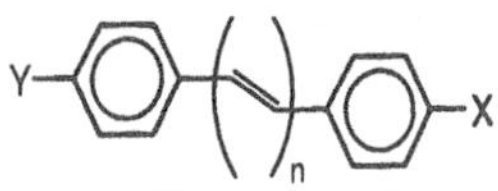		
20	n X Y		
20	1 CN OMe	EFISH	chloroform
20	2 CN OMe	EFISH	chloroform
20	3 CN OMe	EFISH	chloroform
20	1 NO_2 Br	EFISH	chloroform
20	2 NO_2 Br	EFISH	chloroform
20	3 NO_2 Br	EFISH	chloroform
20	1 NO_2 MeO	EFISH	chloroform
20	2 NO_2 MeO	EFISH	chloroform
20	3 NO_2 MeO	EFISH	chloroform
20	4 NO_2 MeO	EFISH	chloroform
20	1 NO_2 MeS	EFISH	chloroform
20	2 NO_2 MeS	EFISH	chloroform
20	1 NO_2 NMe_2	EFISH	chloroform
20	2 NO_2 NMe_2	EFISH	chloroform
20	3 NO_2 NMe_2	EFISH	chloroform
20	4 NO_2 NMe_2	EFISH	chloroform
	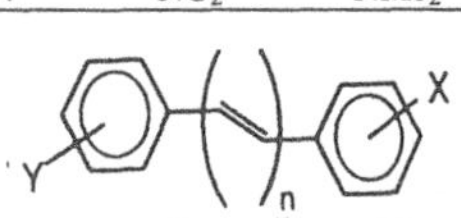		
20	n X Y		
20	2 2'-CN 2-OMe	EFISH	chloroform
20	3 2'-CN 2-OMe	EFISH	chloroform
20	2 2'-CN 4-OMe	EFISH	chloroform
20	3 2'-CN 4-OMe	EFISH	chloroform
20	2 4'-CN 2-OMe	EFISH	chloroform
20	1 2'-NO_2 2-OMe	EFISH	chloroform
20	2 2'-NO_2 2-OMe	EFISH	chloroform
20	3 2'-NO_2 2-OMe	EFISH	chloroform
20	1 2'-NO_2 4-OMe	EFISH	chloroform
20	2 2'-NO_2 4-OMe	EFISH	chloroform

λ_{meas}	λ_{max}	μ	β	β_0	$\mu\beta$	$\mu\beta_0$	footnote	convention
								II
1064	340	3.80	19	10.1	72.2	38.3		II
1064	360	4.50	27	13.0	121.5	58.3		II
1064	380	4.40	40	17.1	176.0	75.2		II
1064	356	3.40	18	8.8	61.2	30.0		II
1064	378	3.50	21	9.1	73.5	31.8		II
1064	400	3.80	35	13.1	133.0	49.6		II
1064	376	4.50	34	14.9	153.0	67.0		II
1064	397	4.80	47	17.9	225.6	86.1		II
1064	414	5.10	76	25.4	387.6	129.7		II
1064	430	5.80	101	29.3	585.8	169.9		II
1064	380	4.30	34	14.5	146.2	62.5		II
1064	398	4.50	55	20.8	247.5	93.7		II
1064	430	6.60	73	21.2	481.8	139.8		II
1064	442	7.60	107	27.4	813.2	208.4		II
1064	458	8.20	131	27.6	1074.2	226.5		II
1064	464	9.00	190	36.8	1710.0	331.4		II
								II
1064	354	3.80	4.5	2.2	17.1	8.5		II
1064	376	3.80	7.1	3.1	27.0	11.8		II
1064	356	3.90	2.6	1.3	10.1	5.0		II
1064	378	3.90	4.3	1.9	16.8	7.3		II
1064	358	4.90	16	7.8	78.4	38.0		II
1064	360	3.80	4.4	2.1	16.7	8.0		II
1064	376	3.70	6.4	2.8	23.7	10.4		II
1064	392	4.10	11	4.3	45.1	17.8		II
1064	390	3.50	3.8	1.5	13.3	5.3		II
1064	370	3.80	4.9	2.2	18.6	8.5		II

Footnote entries on page 503.

Hyperpolarizability tabulations. Shaded entries are calculated from values given in the cited reference.

	MOLECULE NAME, OR STRUCTURE	tech	solvent
20	3 2'-NO_2 4-OMe	EFISH	chloroform
20	1 4'-NO_2 2-OMe	EFISH	chloroform
20	2 4'-NO_2 2-OMe	EFISH	chloroform
20	3 4'-NO_2 2-OMe	EFISH	chloroform
	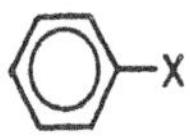 X		
21	H	EFISH	neat
21	p-dioxane	EFISH	neat
21	p-xylene	EFISH	neat
21	CH_3	EFISH	neat
21	OH	EFISH	p-dioxane
21	SH	EFISH	neat
21	OMe	EFISH	neat
21	SMe	EFISH	neat
21	NH_2	EFISH	neat
21	NMe_2	EFISH	neat
21	julolidinamine	EFISH	p-dioxane
21	F	EFISH	neat
21	Cl	EFISH	neat
21	Br	EFISH	neat
21	I	EFISH	neat
21	SO_2CH_3	EFISH	p-dioxane
21	SO_2F	EFISH	neat
21	CN	EFISH	neat
21	COH	EFISH	neat
21	$COCF_3$	EFISH	p-dioxane
21	NO	EFISH	p-dioxane
21	NO_2	EFISH	neat
21	$C_2H(CN)_2$	EFISH	p-dioxane
21	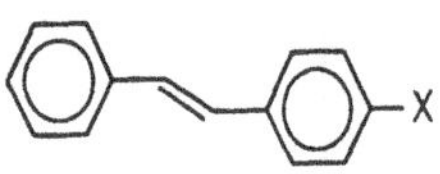 X		
21	H (trans)	EFISH	chloroform

λ_{meas}	λ_{max}	μ	β	β_0	μβ	$\mu\beta_0$	footnote	convention
1064	392	3.80	11	4.3	41.8	16.5		II
1064	362	5.00	22	10.4	110.0	52.2		II
1064	380	4.30	17	7.3	73.1	31.2		II
1064	412	4.80	56	19.1	268.8	91.5		II
								II
1064		0.00	0.0		0.0			II
1064		0.00	0.0		0.0			II
1064		0.00	0.0		0.0			II
1064		0.38	0.2		0.1			II
1064		1.50	0.2		0.3			II
1064		1.40	0.2		0.3			II
1064		1.40	0.2		0.3			II
1064		1.30	0.2		0.3			II
1064		1.50	0.55		0.8			II
1064		1.60	1.1		1.8			II
1064		1.60	1.3		2.1			II
1064		1.50	0.2		0.3			II
1064		1.60	0.2		0.3			II
1064		1.40	0.2		0.3			II
1064		1.30	0.2		0.3			II
1064		3.90	0.2		0.8			II
1064		4.60	0.3		1.4			II
1064		3.90	0.36		1.4			II
1064		2.80	0.80		2.2			II
1064		3.30	1.3		4.3			II
1064		3.10	1.7		5.3			II
1064		4.00	1.9		7.6			II
1064		4.80	3.1		14.9			II
								II
1064		0.00	0.0		0.0			II

Footnote entries on page 503.

Hyperpolarizability tabulations. Shaded entries are calculated from values given in the cited reference.

	MOLECULE NAME, OR STRUCTURE	tech	solvent
21	H(cis)	EFISH	chloroform
21	NH_2	EFISH	p-dioxane
21	NMe_2	EFISH	p-dioxane
21	NO_2	EFISH	p-dioxane
	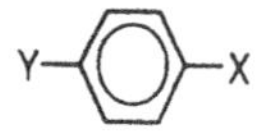		
21	X Y		
21	SO_2Me OH	EFISH	p-dioxane
21	CN Me	EFISH	neat
21	CN Cl	EFISH	p-dioxane
21	CN Br	EFISH	p-dioxane
21	CN OPh	EFISH	p-dioxane
21	CN OMe	EFISH	p-dioxane
21	CN SMe	EFISH	p-dioxane
21	CN NH_2	EFISH	p-dioxane
21	CN NMe_2	EFISH	p-dioxane
21	CHO Me	EFISH	neat
21	CHO OPh	EFISH	neat
21	CHO OMe	EFISH	neat
21	CHO SMe	EFISH	neat
21	CHO NMe_2	EFISH	p-dioxane
21	i-$SO_2C_3F_7$ 0Me	EFISH	chloroform
21	$COCF_3$ 0Me	EFISH	p-dioxane
21	$COCF_3$ 0Ph	EFISH	p-dioxane
21	$COCF_3$ NMe_2	EFISH	p-dioxane
21	NO NMe_2	EFISH	p-dioxane
21	NO_2 Me	EFISH	p-dioxane
21	NO_2 Br	EFISH	p-dioxane
21	NO_2 OH	EFISH	p-dioxane
21	NO_2 OPh	EFISH	p-dioxane
21	NO_2 OMe	EFISH	p-dioxane
21	NO_2 SMe	EFISH	p-dioxane
21	NO_2 N_2H_3	EFISH	p-dioxane
21	NO_2 NH_2	EFISH	acetone
21	NO_2 NMe_2	EFISH	acetone
21	NO_2 CN	EFISH	p-dioxane

λ_{meas}	λ_{max}	μ	β	β_0	$\mu\beta$	$\mu\beta_0$	footnote	convention
1064		0.00	0.0		0.0			II
1064		2.20	7.4		16.3			II
1064		2.10	10		21.0			II
1064		4.20	11		46.2			II
								II
1064	290	3.40	1.3	0.8	4.4	2.9		II
1064		4.40	0.7		3.1			II
1064		2.30	0.8		1.8			II
1064		2.40	1.1		2.6			II
1064		4.10	1.2		4.9			II
1064	248	4.80	1.9	1.4	9.1	6.8		II
1064		4.40	2.8		12.3			II
1064	270	5.00	3.1	2.2	15.5	10.8		II
1064	290	5.60	5.0	3.3	28.0	18.2		II
1064		3.00	1.7		5.1			II
1064	269	2.80	1.9	1.3	5.3	3.7		II
1064	269	3.50	2.2	1.5	7.7	5.4		II
1064	310	3.10	2.6	1.6	8.1	4.9		II
1064	326	5.10	6.3	3.6	32.1	18.2		II
1064	290	5.40	3.3	2.1	17.8	11.6		II
1064	292	3.50	3.6	2.3	12.6	8.1		II
1064	292	4.00	3.6	2.3	14.4	9.3		II
1064	356	5.90	10	4.9	59.0	28.9		II
1064	407	6.20	12	4.2	74.4	26.3		II
1064	272	4.20	2.1	1.4	8.8	6.1		II
1064	274	3.00	3.3	2.3	9.9	6.8		II
1064	304	5.00	3.0	1.9	15.0	9.3		II
1064	294	4.20	4.0	2.6	16.8	10.8		II
1064	302	4.60	5.1	3.2	23.5	14.6		II
1064	322	4.40	6.1	3.5	26.8	15.4		II
1064	366	6.30	7.6	3.5	47.9	22.2		II
1064	365	6.20	9.2	4.3	57.0	26.6		II
1064	376	6.40	12	5.3	76.8	33.6		II
1064		0.90	0.6		0.5	0.0		II

Footnote entries on page 503.

Hyperpolarizability tabulations. Shaded entries are calculated from values given in the cited reference.

	MOLECULE NAME, OR STRUCTURE	tech	solvent
21	NO_2 CHO	EFISH	p-dioxane
21	$CHC(CN)_2$ OMe	EFISH	p-dioxane
21	$CHC(CN)_2$ NMe_2	EFISH	chloroform
21	$CHC(CN)_2$ julolidine	EFISH	dichloromethane
21	$C_2(CN)_3$ NH_2	EFISH	dichloromethane
21	$C_2(CN)_3$ NMe_2	EFISH	dichloromethane
21	$C_2(CN)_3$ julolidine	EFISH	dichloromethane
	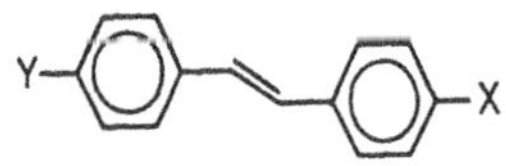		
21	X Y		
21	$SO_2C_5H_{11}$ HOHexO	EFISH	chloroform
21	SO_2C6F_{13} OMe	EFISH	p-dioxane
21	$COCF_3$ 0Me	EFISH	p-dioxane
21	CN OH	EFISH	p-dioxane
21	CN OMe	EFISH	chloroform
21	CN $N(Me)_2$	EFISH	chloroform
21	NO_2 H	EFISH	p-dioxane
21	NO_2 Me	EFISH	p-dioxane
21	NO_2 Br	EFISH	p-dioxane
21	NO_2 Br	EFISH	chloroform
21	NO_2 OH	EFISH	p-dioxane
21	NO_2 OPh	EFISH	p-dioxane
21	NO_2 OMe	EFISH	p-dioxane
21	NO_2 OMe	EFISH	chloroform
21	NO_2 SMe	EFISH	p-dioxane
21	NO_2 SMe	EFISH	chloroform
21	NO_2 NH_2	EFISH	chloroform
21	NO_2 $N(Me)_2$	EFISH	chloroform
21	NO_2 $N(Me)_2$	EFISH	NMP
21	NO_2 julolidinamine	EFISH	chloroform
21	NO_2 COOMe	EFISH	dichloromethane
21	NO_2 CHO	EFISH	p-dioxane
21	Br OMe	EFISH	p-dioxane

λ_{meas}	λ_{max}	μ	β	β_0	$\mu\beta$	$\mu\beta_0$	footnote	convention
1064	376	2.50	0.2	0.1	0.5	0.2		II
1064	345	5.50	9.8	5.1	53.9	27.9		II
1064	420	7.80	32	10.2	249.6	79.4		II
1064	458	8.00	44	9.3	352.0	74.2		II
1064	498	7.80	39	3.8	304.2	29.4		II
1064	516	8.20	50	2.3	410.0	18.6		II
1064	556	8.50	60	-4.0	510.0	-34.2		II
								II
1064	336	6.50	10	5.4	65.0	35.2		II
1064	347	7.80	14	7.2	109.2	56.1		II
1064	368	4.20	16.4	7.5	68.9	31.6		II
1064	344	4.50	13	6.8	58.5	30.5		II
1064	340	3.80	19	10.1	72.2	38.3		II
1064	382	5.70	36	15.2	205.2	86.6		II
1064	345	4.20	11	5.7	46.2	24.0		II
1064	351	4.70	15	7.5	70.5	35.5		II
1064	344	3.20	14	7.3	44.8	23.3		II
1064	356	3.40	18	8.8	61.2	30.0		II
1064	370	5.50	17	7.7	93.5	42.4		II
1064	350	4.60	18	9.1	82.8	41.9		II
1064	364	4.50	28	13.1	126.0	59.2		II
1064	370	4.50	34	15.4	153.0	69.4		II
1064	374	4.30	26	11.5	111.8	49.6		II
1064	380	4.30	34	14.5	146.2	62.5		II
1064	402	5.10	40	14.7	204.0	75.0		II
1064	427	6.60	73	21.8	481.8	143.8		II
1064		7.20	70		504.0			II
1064	438	7.00	96	25.7	672.0	179.8		II
1064	350	4.00	4.0	2.0	16.0	8.1		II
1064	352	4.10	6.0	3.0	24.6	12.3		II
1064	325	4.00	2.5	1.4	10.0	5.7		II

Footnote entries on page 503.

Hyperpolarizability tabulations. Shaded entries are calculated from values given in the cited reference.

	MOLECULE NAME, OR STRUCTURE	tech	solvent
21	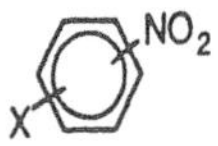 meta X		
21	Me	EFISH	neat
21	Br	EFISH	p-dioxane
21	OH	EFISH	p-dioxane
21	OMe	EFISH	neat
21	NH_2	EFISH	p-dioxane
21	CN	EFISH	p-dioxane
21	CHO	EFISH	p-dioxane
21	X Ortho		
21	Me	EFISH	neat
21	Br	EFISH	p-dioxane
21	OH	EFISH	p-dioxane
21	OMe	EFISH	p-dioxane
21	NH_2	EFISH	p-dioxane
21	CN	EFISH	p-dioxane
21	CHO	EFISH	p-dioxane
21	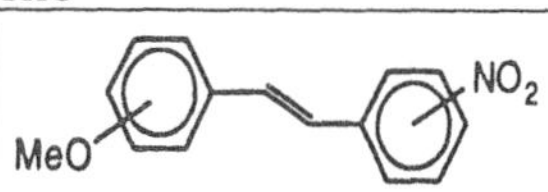 MeO NO_2		
21	2 2'	EFISH	chloroform
21	3 2'	EFISH	chloroform
21	4 2'	EFISH	chloroform
21	2 3'	EFISH	chloroform
21	3 3'	EFISH	chloroform
21	4 3'	EFISH	p-dioxane
21	2 4'	EFISH	chloroform
21	3 4'	EFISH	chloroform
21	4 4'	EFISH	chloroform

λ_{meas}	λ_{max}	μ	β	β_0	$\mu\beta$	$\mu\beta_0$	footnote	convention
								II
1064		3.90	1.0		3.9			II
1064		4.00	0.4		1.6			II
1064	384	3.40	1.2	0.5	4.1	1.7		II
1064	318	3.80	1.4	0.8	5.3	3.1		II
1064		4.10	2.5		10.3			II
1064		5.50	1.2		6.6			II
1064		4.00	0.8		3.2			II
								II
1064		3.90	1.5		5.9			II
1064		3.40	1.0		3.4			II
1064		3.60	0.8		2.9			II
1064	326	3.90	1.6	0.9	6.2	3.5		II
1064	396	4.70	1.9	0.7	8.9	3.4		II
1064		3.70	0.8		3.0			II
1064		2.80	1.7		4.8			II
								II
1064	360	3.80	4.4	2.1	16.7	8.0		II
1064	370	3.70	1.6	0.7	5.9	2.7		II
1064	390	3.50	3.8	1.5	13.3	5.3		II
1064	320	4.40	5.5	3.2	24.2	14.0		II
1064	292	3.90	4.5	2.9	17.6	11.3		II
1064	318	3.90	5.3	3.1	20.7	12.1		II
1064	362	5.00	22	10.4	110.0	52.2		II
1064	352	4.00	21	10.5	84.0	42.1		II
1064	370	4.50	34	15.4	153.0	69.4		II

Footnote entries on page 503.

Hyperpolarizability tabulations. Shaded entries are calculated from values given in the cited reference.

	MOLECULE NAME, OR STRUCTURE	tech	solvent
	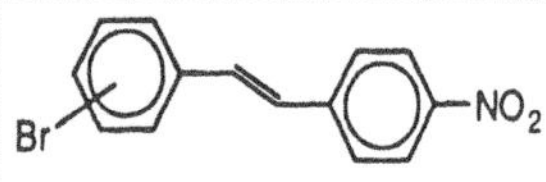		
21	Br NO$_2$		
21	2 4'	EFISH	chloroform
21	3 4'	EFISH	chloroform
21	4 4'	EFISH	chloroform
	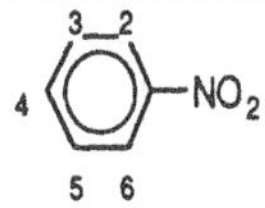		
21	substitution		
21	3-aza; 4-NH$_2$	EFISH	acetone
21	2-aza; 4-OMe	EFISH	p-dioxane
21	3-Me; 4-NH$_2$	EFISH	p-dioxane
21	3-Cl; 4-NH$_2$	EFISH	p-dioxane
21	3-OMe; 4-NH$_2$	EFISH	p-dioxane
21	2-F; 4-OMe	EFISH	p-dioxane
21	2,5-F; 4-OMe	EFISH	p-dioxane
21	2,3,5,6-F; 4-OMe	EFISH	p-dioxane
	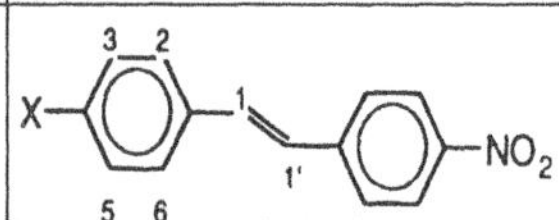		
21	X substitution		
21	H 1-aza	EFISH	p-dioxane
21	Me 1-aza	EFISH	p-dioxane
21	OMe 1-aza	EFISH	p-dioxane
21	OMe 1'-aza	EFISH	p-dioxane
21	NH$_2$ 1,1'-azo	EFISH	p-dioxane
21	NEtEtOH 1,1'-azo	EFISH	p-dioxane
21	OMe 3-Me	EFISH	p-dioxane
21	OMe 3-OMe	EFISH	p-dioxane
21	OMe 3-F	EFISH	p-dioxane
21	OMe 2-OMe	EFISH	p-dioxane
21	OMe 1'-CN	EFISH	p-dioxane

λ_{meas}	λ_{max}	μ	β	β_0	$\mu\beta$	$\mu\beta_0$	footnote	convention
								II
1064	346	4.60	12	6.2	55.2	28.5		II
1064	346	3.40	14	7.2	47.6	24.6		II
1064	356	3.40	18	8.8	61.2	30.0		II
								II
1064		6.50	3.7		24.1			II
1064		3.50	2.2		7.7			II
1064		6.20	8.7		53.9			II
1064	350	5.90	6.8	3.4	40.1	20.3		II
1064		6.00	8.7		52.2			II
1064	304	4.40	2.5	1.5	11.0	6.8		II
1064	304	4.90	2.6	1.6	12.7	7.9		II
1064	270	4.20	1.7	1.2	7.1	5.0		II
								II
1064	346	4.40	4.9	2.5	21.6	11.1		II
1064	351	4.70	15	7.5	70.5	35.5		II
1064	376	4.40	14	6.1	61.6	27.0		II
1064	349	5.40	6.6	3.4	35.6	18.1		II
1064	420	5.80	29	9.2	168.2	53.5		II
1064	455	7.00	49	10.8	343.0	75.3		II
1064	366	5.20	26	12.1	135.2	62.8		II
1064	380	4.70	23	9.8	108.1	46.2		II
1064	363	4.10	18	8.5	73.8	34.8		II
1064	395	4.80	32	12.4	153.6	59.4		II
1064	361	5.30	21	10.0	111.3	53.1		II

Footnote entries on page 503.

Hyperpolarizability tabulations. Shaded entries are calculated from values given in the cited reference.

	MOLECULE NAME, OR STRUCTURE	tech	solvent
21	Br 1'-CN	EFISH	p-dioxane
21	Br 1-aza,1'-CN	EFISH	p-dioxane
21	OMe 2'-NO_2	EFISH	p-dioxane
22	PNA	HRS	chloroform
22	4-methoxy-4'-nitrostilbene (MONS)	HRS	chloroform
22	4-hydroxy-4'-nitrostilbene (HONS)	HRS	chloroform
23	Crystal violet(CV+)	HRS	methanol
24	para-nitroniline		chloroform
24	4-methoxy-4'-nitrostilbene	HRS	Dioxane
24	4-Hydroxy-4'-nitrostilbene	HRS	chloroform
25	retinal	HRS	dioxane
25	retinal	HRS	chloroform
25	retinal	HRS	dichloroethane
25	retinal	HRS	methanol
25	retinal	HRS	nitromethane
26	Z-(beta)-[(N,N-diethylmethylammonium)-(alpha)-cyano-4-styryl dicyanomethanide]	EFISH	dimethylformamide
27	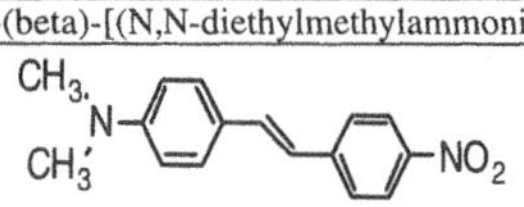	EFISH	
27	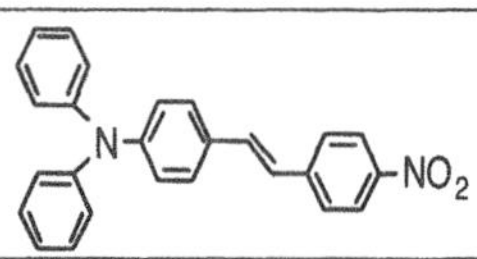	EFISH	
27	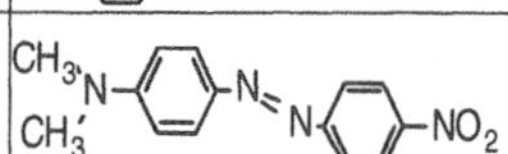	EFISH	
27	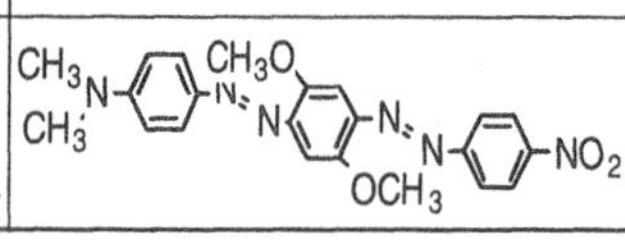	EFISH	

λ_{meas}	λ_{max}	μ	β	β_0	$\mu\beta$	$\mu\beta_0$	footnote	convention
1064	340	4.60	8.0	4.2	36.8	19.5		II
1064	382	4.10	2.1	0.9	8.6	3.6		II
1064	390	4.70	22	8.8	103.4	41.4		II
1064			23					
1064			105					
1064			93					
800			450					
1064			23					
1064			81					
1064			93					
1064	385	5.30	80	33.1	424.0	175.5		
1064	385	5.30	270	111.8	1431.0	592.3		
1064	385	5.30	380	157.3	2014.0	833.6		
1064	385	5.30	730	302.2	3869.0	1601.5		
1064	385	5.30	1100	455.3	5830.0	2413.2		
1907	694	45.00	-190	-85.0	-8550.0	-3825.0	m	
1907					580.0			
1907					788.0			
1580					800.0			
1064					1040.0			

Footnote entries on page 503.

Hyperpolarizability tabulations. Shaded entries are calculated from values given in the cited reference.

	MOLECULE NAME, OR STRUCTURE	tech	solvent
27	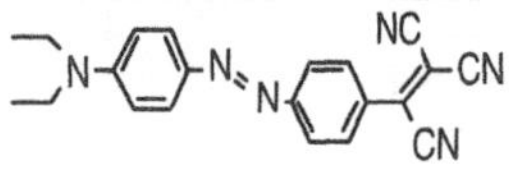	EFISH	
27	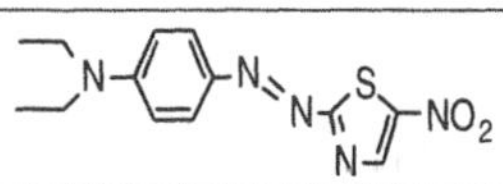	EFISH	
27	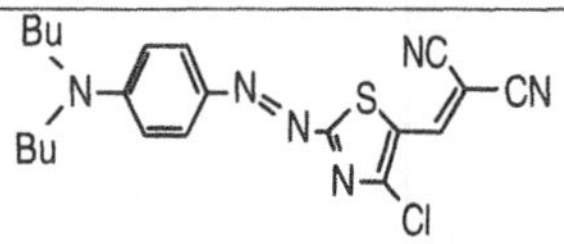	EFISH	
27	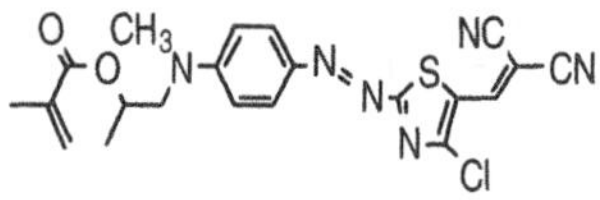	EFISH	
27	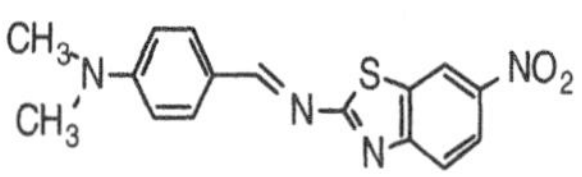	EFISH	
27	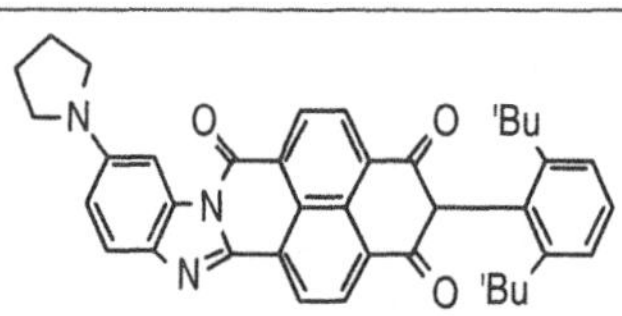	EFISH	
27	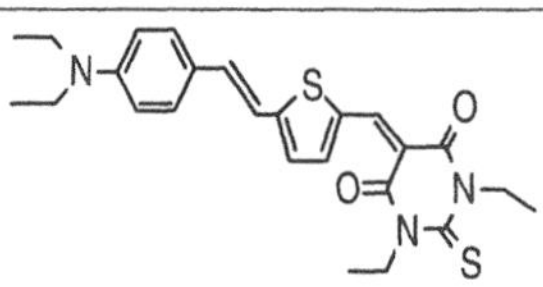	EFISH	

λ_{meas}	λ_{max}	μ	β	β_0	$\mu\beta$	$\mu\beta_0$	footnote	convention
1580					4100.0			
1580					2190.0			
1580					5320.0			
1907					1658.0			
1064					640.0			
1907					1500.0			
1907					2400.0			

Footnote entries on page 503.

Hyperpolarizability tabulations. Shaded entries are calculated from values given in the cited reference.

	MOLECULE NAME, OR STRUCTURE	tech	solvent
28	X=O 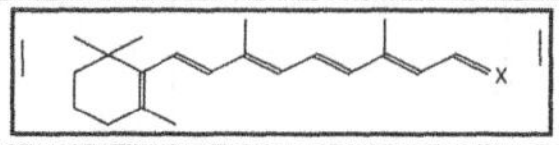	HRS	methanol
28	X=O 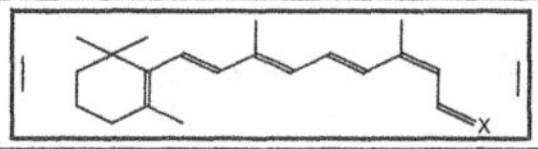	HRS	methanol
28	X=N-Bu 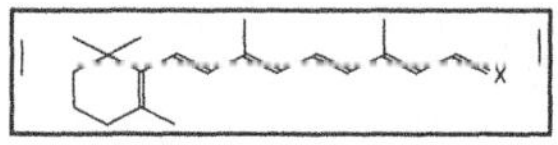	HRS	methanol
28	X=N-Bu 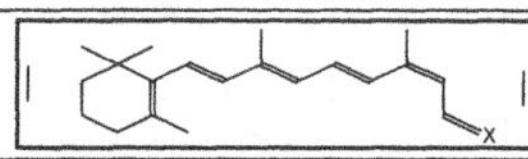	HRS	methanol
28	X=(NH)+-Bu 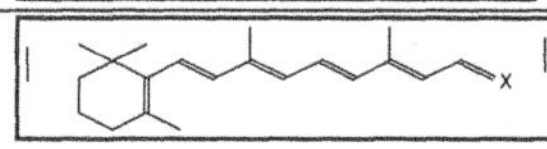	HRS	methanol
28	X=(NH)+-Bu 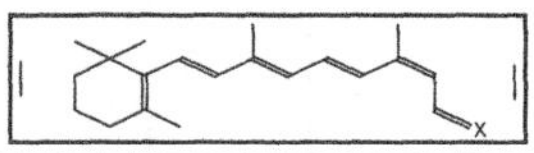	HRS	methanol
29	1-4[4-(2-cyano-2-methoxycarbonyl)ethenyl]fenyl]-4-piperidinyl acetate	EFISH	chloroform
29	1-[4-[2,2-dicyanoethenyl]fenyl]-4-piperidinyl acetate	EFISH	chloroform
29	N-[4-[(4[cyanofenyl)azo]fenyl]methylaminoethyl acetate	EFISH	chloroform
29	4-methoxy,4'nitro-stilbene	EFISH	chloroform
29	2-[2-(4-methoxyfenyl)ethenyl]-5-nitrothiophene	EFISH	chloroform
30	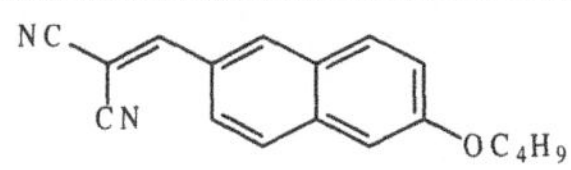	HRS	chloroform
30	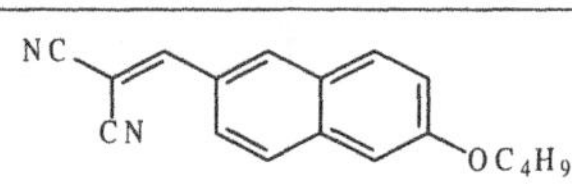	EFISH	chloroform

λ_{meas}	λ_{max}	μ	β	β_0	$\mu\beta$	$\mu\beta_0$	footnote	convention
1064	385		730	300.0				
1064	385		480	198.7				
1064	365		470	220.0				
1064	365		260	121.4				
1064	445		3600	900.0				
1064	445		2440	604.6				
1064		5.00	64		320.0			I
1064		7.00	84		588.0			I
1064		6.00	177		1062.0			I
1064		4.50	100		450.0			I
1064		6.20	139		861.8			I
1064	394		79	30.8				
1064	394		87	33.9			n	

Footnote entries on page 503.

Hyperpolarizability tabulations. Shaded entries are calculated from values given in the cited reference.

	MOLECULE NAME, OR STRUCTURE	tech	solvent
30	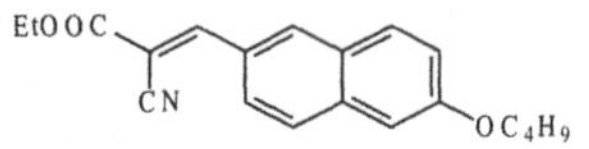	HRS	chloroform
30	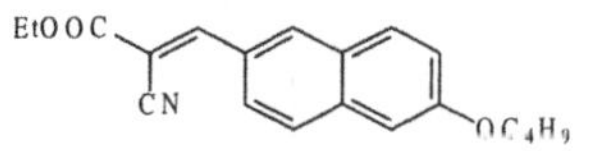	EFISH	chloroform
30	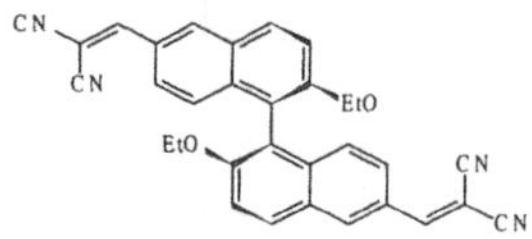	HRS	chloroform
30	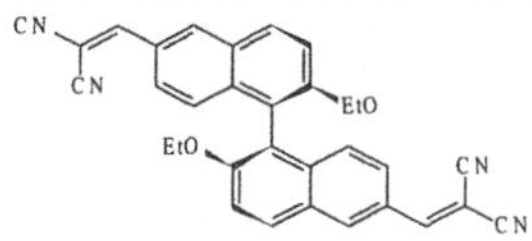	EFISH	chloroform
30	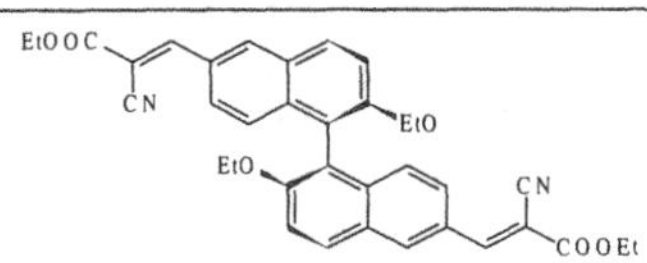	HRS	chloroform
30	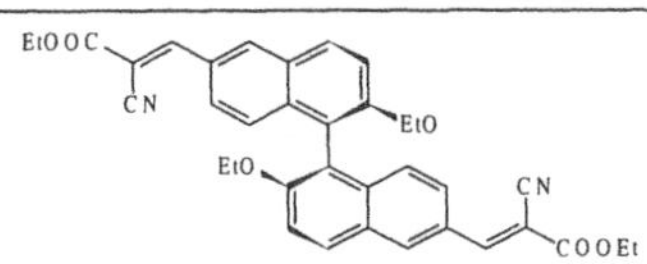	EFISH	chloroform

λ_{meas}	λ_{max}	μ	β	β_0	$\mu\beta$	$\mu\beta_0$	footnote	convention
1064	377		64	27.9			n	
1064	377		64	27.9			n	
1064	407		136	48.1			n	
1064	407		135	47.8			n	
1064	389		118	47.6			n	
1064	389		103	41.5			n	

Footnote entries on page 503.

Hyperpolarizability tabulations. Shaded entries are calculated from values given in the cited reference.

	MOLECULE NAME, OR STRUCTURE	tech	solvent
30	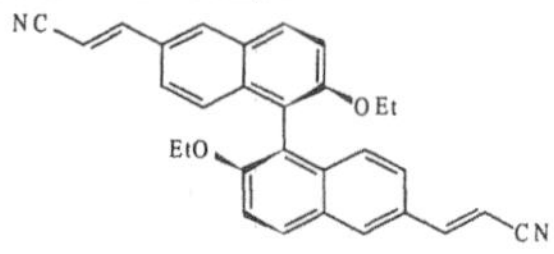	HRS	chloroform
30	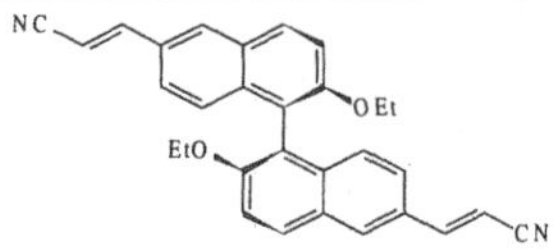	EFISH	chloroform
30	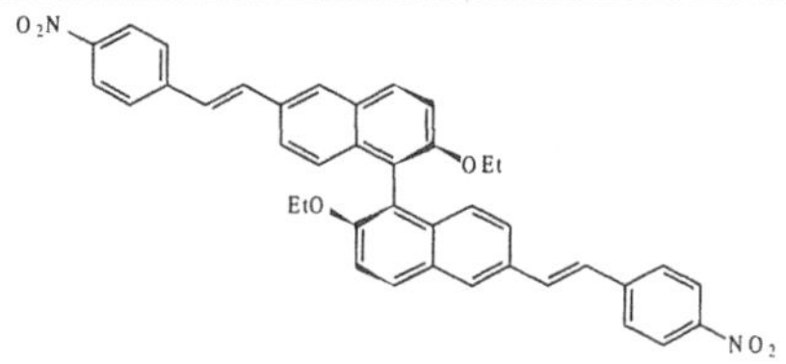	HRS	chloroform
30	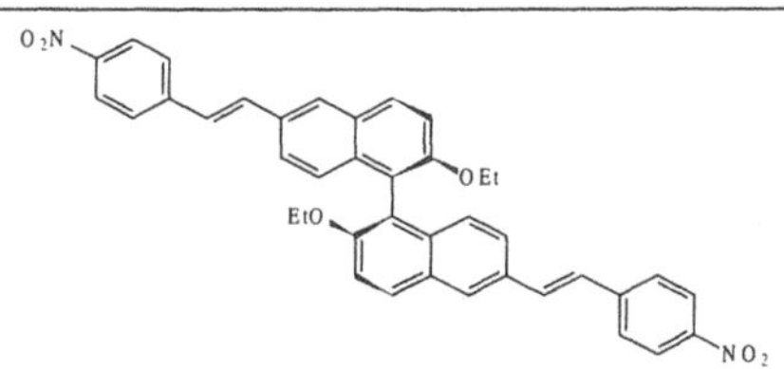	EFISH	chloroform
30	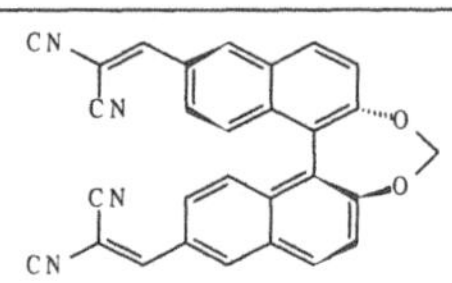	HRS	chloroform

λ_{meas}	λ_{max}	μ	β	β_0	$\mu\beta$	$\mu\beta_0$	footnote	convention
1064			94					
1064			102					
1064			364					
1064			344					
1064	385		68	28.1			n	

Footnote entries on page 503.

Hyperpolarizability tabulations. Shaded entries are calculated from values given in the cited reference.

	MOLECULE NAME, OR STRUCTURE	tech	solvent
30	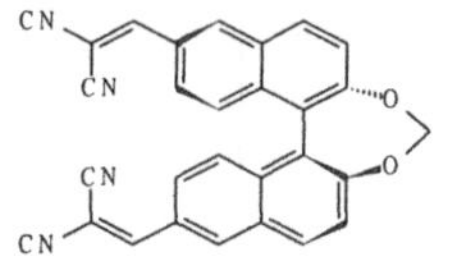	EFISH	chloroform
30	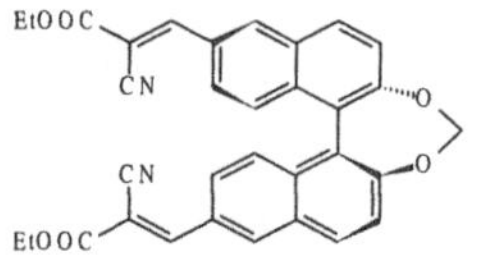	HRS	chloroform
30	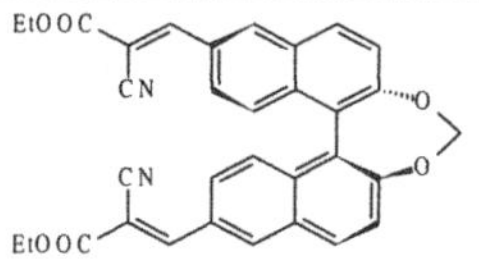	EFISH	chloroform
30	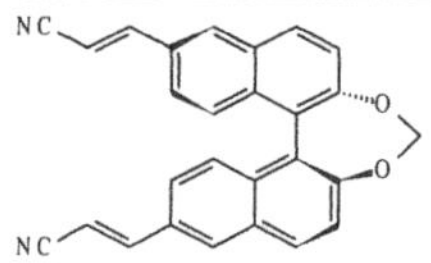	HRS	chloroform
30	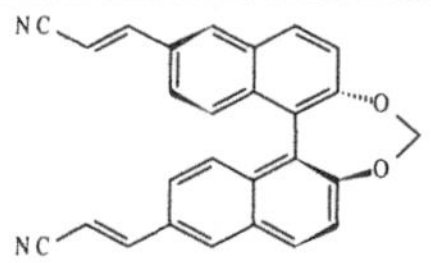	EFISH	chloroform

λ_{meas}	λ_{max}	μ	β	β_0	$\mu\beta$	$\mu\beta_0$	footnote	convention
1064	385		65	26.9			n	
1064	378		58	25.1			n	
1064	378		95	41.1			n	
1064			34					
1064			34					

Footnote entries on page 503.

Hyperpolarizability tabulations. Shaded entries are calculated from values given in the cited reference.

	MOLECULE NAME, OR STRUCTURE	tech	solvent
30	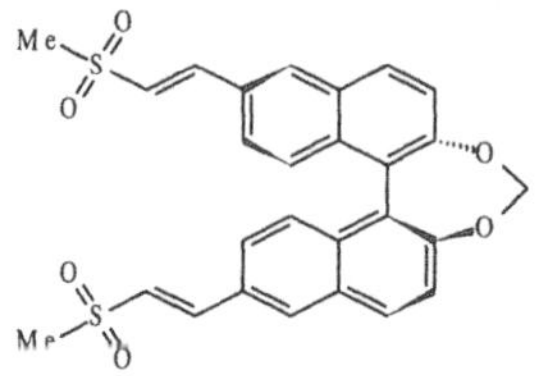	HRS	chloroform
30	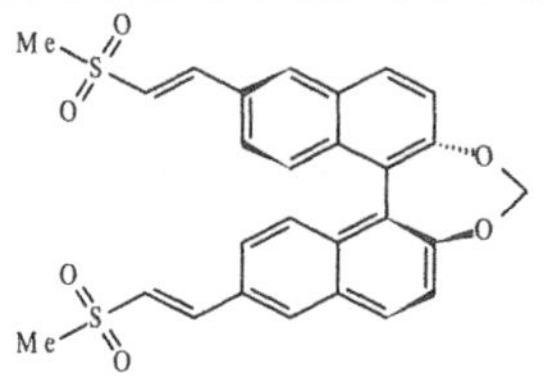	EFISH	chloroform
30	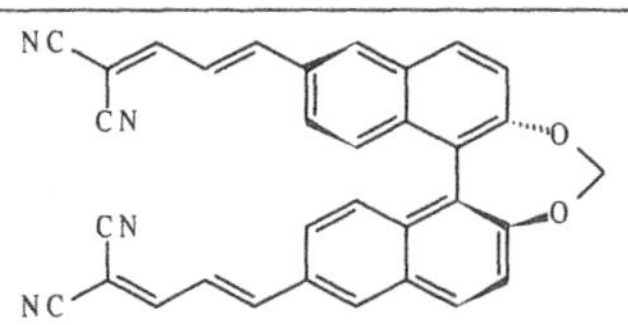	HRS	chloroform
30	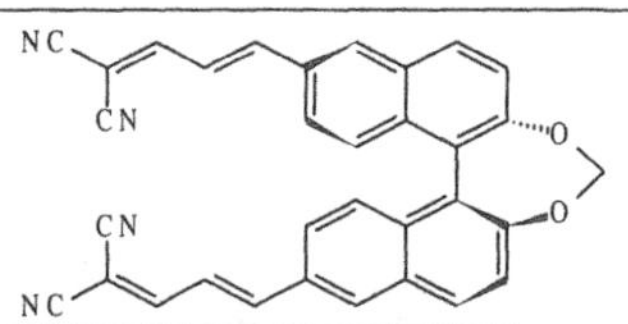	EFISH	chloroform
31	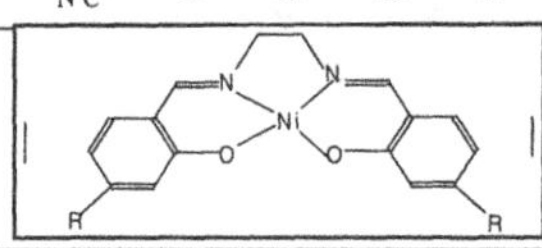	EFISH	chloroform

λ_{meas}	λ_{max}	μ	β	β_0	$\mu\beta$	$\mu\beta_0$	footnote	convention
1064			39					
1064			45					
1064			243					
1064			219					
	414	8.90	-9.3		-85			

Footnote entries on page 503.

Hyperpolarizability tabulations. Shaded entries are calculated from values given in the cited reference.

	MOLECULE NAME, OR STRUCTURE	tech	solvent
31	R=OMe 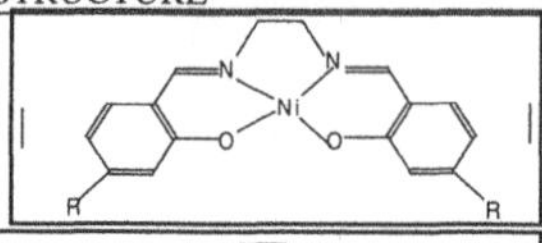	EFISH	chloroform
31	R=H R1=H 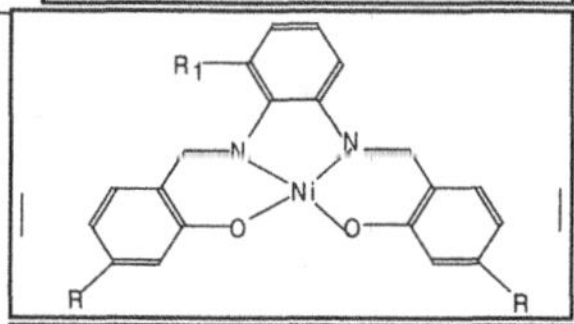	EFISH	chloroform
31	R=H R1=NO_2 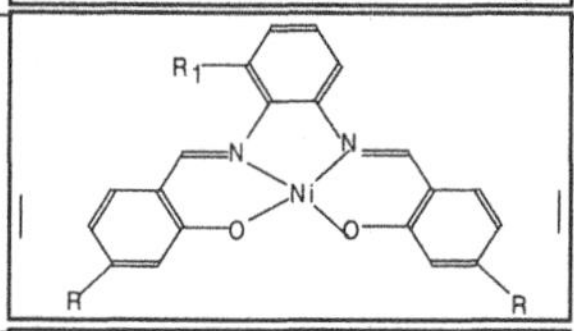	EFISH	acetone
31	R=OMe R1=NO_2 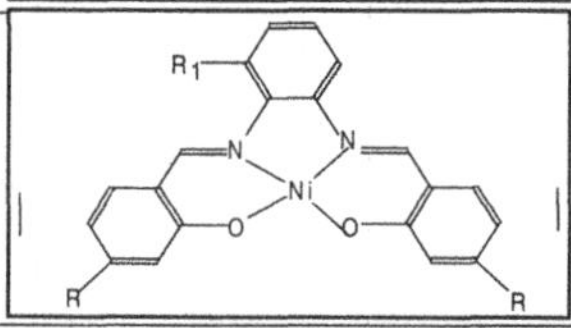	EFISH	acetone
32	stilbazolium p-toluenesulfonates 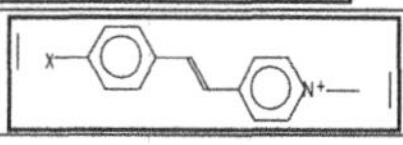		
32	X=CN	HRS	methanol
32	X=H	HRS	methanol
32	X=OH	HRS	methanol
32	X=Cl	HRS	methanol
32	X=CH_3	HRS	methanol
32	X=Br	HRS	methanol
32	X=OCH_3	HRS	methanol
32	methoxystilbazoliums w/ p-toluenesulfonate counter ion	HRS	methanol
32	methoxystilbazoliums w/ iodide counter ion	HRS	methanol
32	sodium p-toluenesulfonate	HRS	methanol

λ_{meas}	λ_{max}	μ	β	β_0	$\mu\beta$	$\mu\beta_0$	footnote	convention
	398	8.60	-9.6		-108			
	480	7.00	-20.5		-147			
	570	7.70	-43		-298		b	
	570	9.90	-55		-546		b	
1064	335		105	57.1				
1064	347		190	97.6				
1064	480		390	57.8				
1064	346		290	149.6				
1064	361		280	133.7				
1064	348		305	155.8				
1064	381		360	152.9				
1064	383		360	150.9				
1064	381		330	140.1				
1064	219		22	17.5				

Footnote entries on page 503.

Hyperpolarizability tabulations. Shaded entries are calculated from values given in the cited reference.

	MOLECULE NAME, OR STRUCTURE	tech	solvent
33	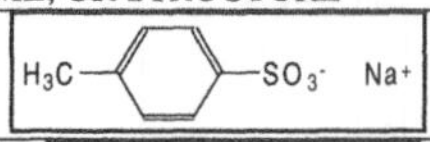	HRS	methanol
33	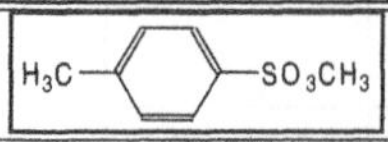	HRS	methanol
34	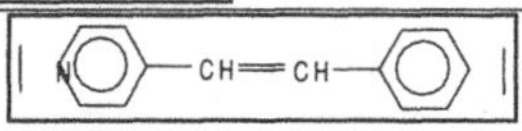	EFISH	ethanol
34	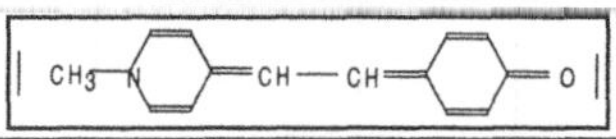	EFISH	dimethylsulfoxide
34	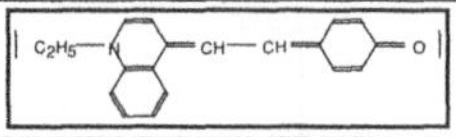	EFISH	dimethylsulfoxide
34	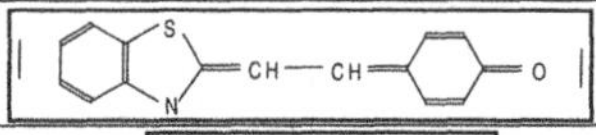	EFISH	dimethylsulfoxide
34	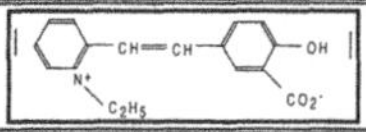	EFISH	dimethylsulfoxide
34	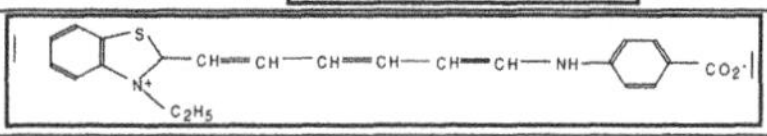	EFISH	dimethylsulfoxide
35	H_3-C-NC_5H_4-CH-CH-C_6H_4-O	EFISH	dimethylsulfoxide
36	para-disubstituted benzenes D/A	EFISH	dimethylsulfoxide
36	CH_3/CN	EFISH	dimethylsulfoxide
36	OCH_3/CN	EFISH	dimethylsulfoxide
36	NH_2/CN	EFISH	dimethylsulfoxide
36	$N(CH_3)_2$/CN	EFISH	dimethylsulfoxide
36	OCH_3/CHO	EFISH	dimethylsulfoxide
36	$N(CH_3)_2$/CHO	EFISH	dimethylsulfoxide
36	NH_2/$COCH_3$	EFISH	dimethylsulfoxide
36	CH_3/NO_2	EFISH	dimethylsulfoxide
36	OCH_3/NO_2	EFISH	dimethylsulfoxide
36	NH_2/NO_2	EFISH	dimethylsulfoxide
36	$N(CH_3)_2$/NO_2	EFISH	dimethylsulfoxide

λ_{meas}	λ_{max}	μ	β	β_0	$\mu\beta$	$\mu\beta_0$	footnote	convention
1064	222		22	17.4				
1064	225		12	9.4				
1890	307				36.0			I
1890	570				7600.0			I
1890	660				5100.0			I
1890	560				5000.0		a	I
1890	408				3700.0			I
1890	498				1500.0			I
1890	570	8.00	1000	578.3	7600.0	4395.2		
1890								I
1890	232	4.4	3.0	2.7	13	12.0	s	I
1890	247	4.8	4.8	4.4	23	21.1	s	I
1890	269	5.9	13.4	12.1	79	71.1	s	I
1890	297	6.6	14.2	12.5	94	82.6	s	I
1890					27			I
1890	352			0.0	129			I
1890	310	4.5	24.2	21.0	109	94.7	s	I
1890	280	4.0	8.3	7.4	33	29.4	s	I
1890	314	4.5	15.8	13.6	71	61.4	s	I
1890	378	6.1	46.9	37.8	286	230.6	s	I
1890	418	6.8	52.2	39.9	355	271.6	s	I

Footnote entries on page 503.

Hyperpolarizability tabulations. Shaded entries are calculated from values given in the cited reference.

	MOLECULE NAME, OR STRUCTURE	tech	solvent
37	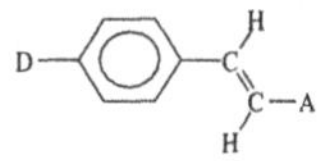 D/A		
37	OCH_3/CHO	EFISH	dimethylsulfoxide
37	$N(CH_3)_2$/CHO	EFISH	dimethylsulfoxide
37	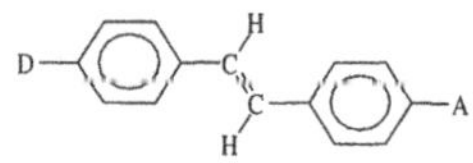 D/A		
37	OCH_3/CN	EFISH	dimethylsulfoxide
37	$N(CH_3)_2$/CN	EFISH	dimethylsulfoxide
37	CH_3/NO_2	EFISH	dimethylsulfoxide
37	$N(CH_3)_2/NO_2$	EFISH	dimethylsulfoxide
37	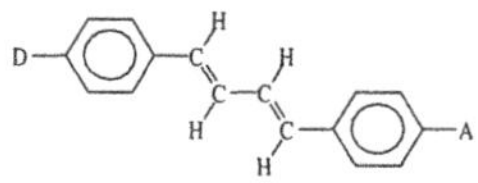 D/A		
37	$N(CH_3)_2$/CN	EFISH	dimethylsulfoxide
	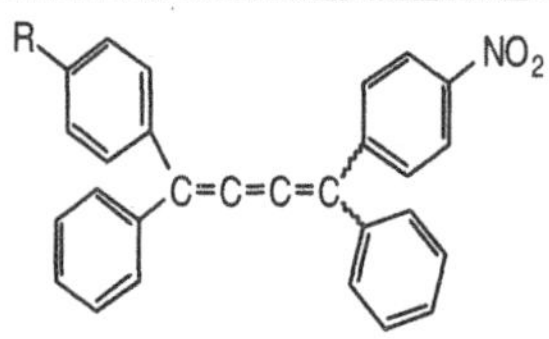		
38	R=H	EFISH	p-dioxane
38	R=H	EFISH	p-dioxane
38	$R=CH_3$	EFISH	p-dioxane
38	$R=OCH_3$	EFISH	p-dioxane
38	$R=O(CH_2)_{11}CH_3$	EFISH	p-dioxane
39	dimethyl-amino-nitro-tolan	EFISH	chloroform
39	dimethyl-amino-nitro-stilbene	EFISH	chloroform

λ_{meas}	λ_{max}	μ	β	β_0	$\mu\beta$	$\mu\beta_0$	footnote	convention
1890	320				80			
1890	387				370			
1890	342				98			
1890	390				820			
1890	368				200			
1890	447				4200			
1890	410				1700			
1064	442				675.0	173.0		
1900	442				145.0	107.5		
1064	448				600.0	143.6		
1064	459				850.0	176.8		
1907	461				280.0	202.0		
	415		42					
	436		70					

Footnote entries on page 503.

Hyperpolarizability tabulations. Shaded entries are calculated from values given in the cited reference.

	MOLECULE NAME, OR STRUCTURE	tech	solvent
40	RCHO 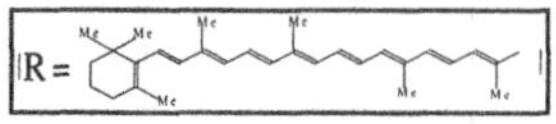	EFISH	chloroform
40	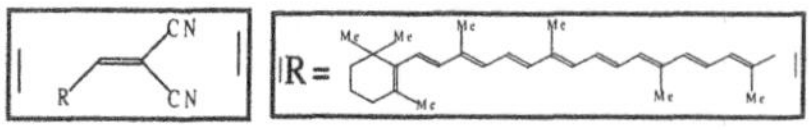	EFISH	chloroform
40	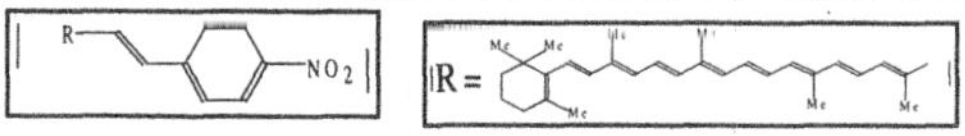	EFISH	chloroform
41	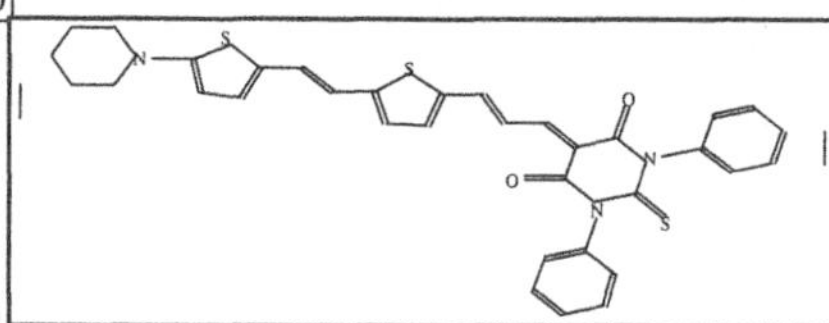	EFISH	chloroform
41	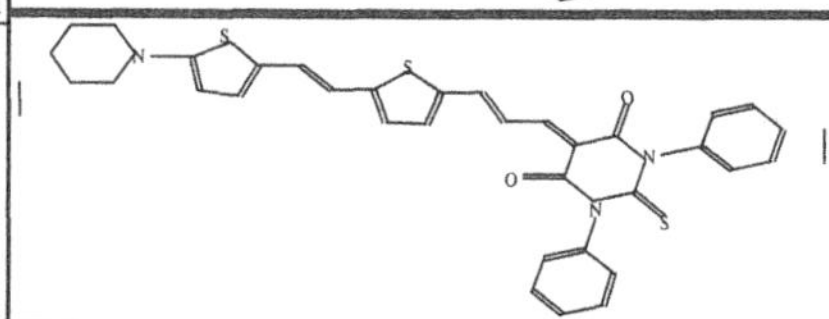	EFISH	dichloromethane
42	PNA	EFISH	p-dioxane
42	PNA	EFISH	p-dioxane
42	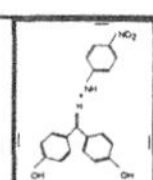	EFISH	tetrahydrofuran
42	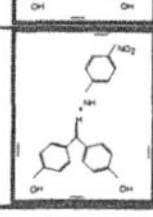	EFISH	tetrahydrofuran

λ_{meas}	λ_{max}	μ	β	β_0	$\mu\beta$	$\mu\beta_0$	footnote	convention
1907	476	6.50	147	104.0	956.0	676.0		
1907	566	7.70	570	337.0	4389.0	2594.9		
1907	502	7.00	220	148.0	1540.0	1036.0		
1907	710				8600.0	3300.5		
1907	710				8600.0	3300.5		
1064	354	6.20		14.4	172.0	89.3		I
1579	354	6.20		14.4	118.0	89.3		I
1064	421	6.20		48.8	827.0	302.6		I
1579	421	6.20		48.8	455.0	302.6		I

Footnote entries on page 503.

Hyperpolarizability tabulations. Shaded entries are calculated from values given in the cited reference.

	MOLECULE NAME, OR STRUCTURE	tech	solvent
42	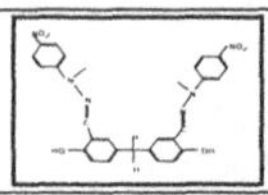	EFISH	tetrahydrofuran
42	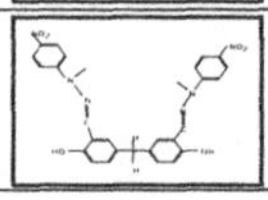	EFISH	tetrahydrofuran
42	DANS	EFISH	tetrahydrofuran
42	DANS	EFISH	tetrahydrofuran
43	X=NH_2 Y=NO_2 R=H_4 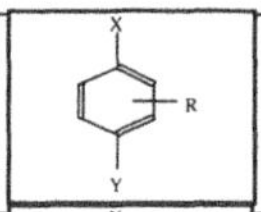	EFISH	p-dioxane
43	X=NH_2 Y=NO_2 R=F_4 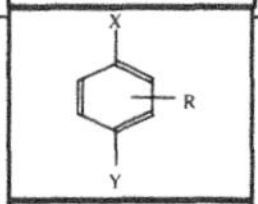	EFISH	p-dioxane
43	X=NH_2 Y=CN R=H_4 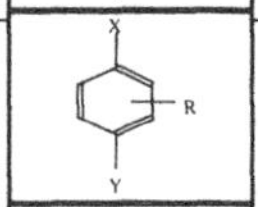	EFISH	p-dioxane
43	X=NH_2 Y=CN R=F_4 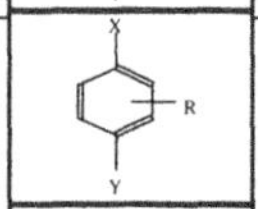	EFISH	p-dioxane
43	X=Br Y=CN R=H_4 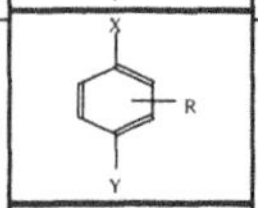	EFISH	p-dioxane
43	X=Br Y=CN R=F_4 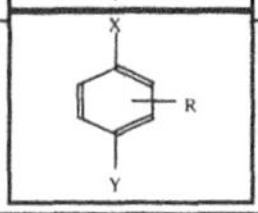	EFISH	p-dioxane
44	Retinoic acid	HRS	
44	Vitamin A acetate	HRS	

λ_{meas}	λ_{max}	μ	β	β_0	$\mu\beta$	$\mu\beta_0$	footnote	convention
1064	418	6.20		72.5	2570.0	449.5		I
1579	418	6.20		72.5	672.0	449.5		I
1064	434	6.20		132.0	2760.0	818.4		I
1579	434	6.20		132.0	1270.0	818.4		I
1590	314	6.50	11.8	9.6	76.7	62.2		
1590	288	6.00	5.5	4.6	33.0	27.7		
1590	258	6.40	2.4	2.1	15.4	13.4		
1590	271	6.50	2.6	2.2	16.9	14.5		
1590	241	2.50	2.2	2.0	5.5	4.9		
1590	250	3.50	1.35	1.2	4.7	4.2		
1064	351		310	156.0				
1064	326		140	79.2				

Footnote entries on page 503.

Hyperpolarizability tabulations. Shaded entries are calculated from values given in the cited reference.

	MOLECULE NAME, OR STRUCTURE	tech	solvent
45	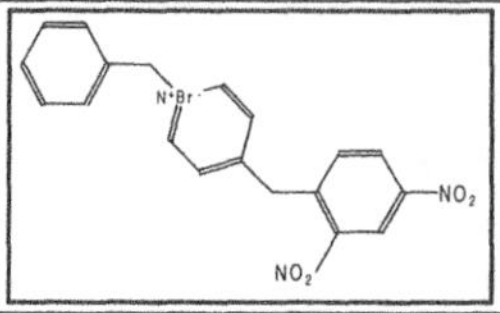	HRS	acetonitrile
45	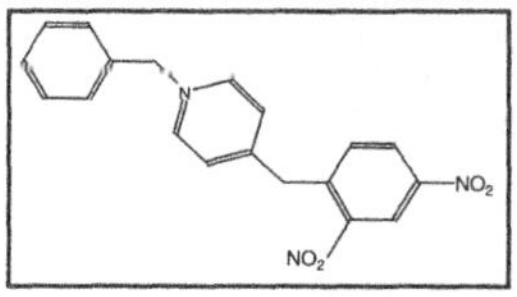	HRS	acetonitrile
45	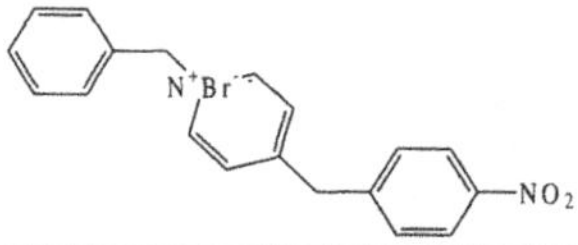	HRS	acetonitrile
45	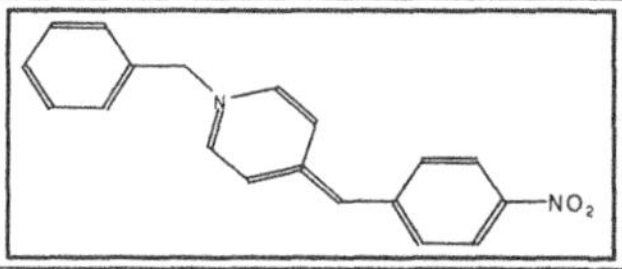	HRS	acetonitrile
45	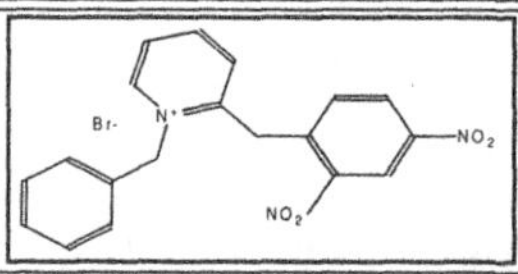	HRS	acetonitrile
45	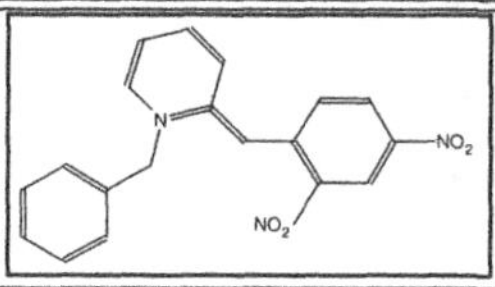	HRS	acetonitrile

λ_{meas}	λ_{max}	μ	β	β_0	$\mu\beta$	$\mu\beta_0$	footnote	convention
1064	254		2	1.0				
1064	582		450	62.0				
1064	262		5	3.0				
1064	548		1170	52.0				
1064	266		7	5.0				
1064	567		84	8.0				

Footnote entries on page 503.

Hyperpolarizability tabulations. Shaded entries are calculated from values given in the cited reference.

	MOLECULE NAME, OR STRUCTURE	tech	solvent
45	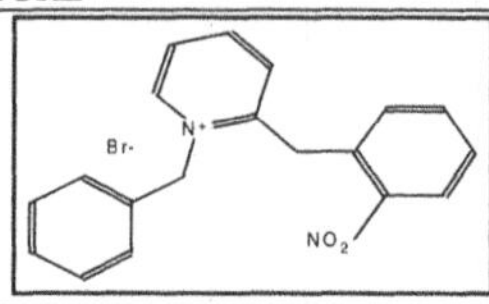	HRS	acetonitrile
45	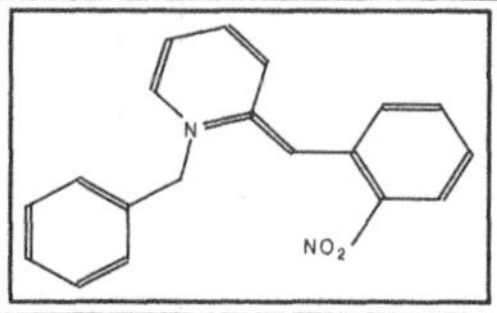	HRS	acetonitrile
46	pNA	HRS	acetone
46	pNA	HRS	acetone
46	pNA	HRS	acetone
46	pNA	HRS	acetone
46	Disperse Red 1	HRS	acetone
46	Disperse Red 1	HRS	acetone
47	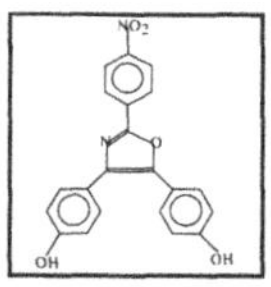 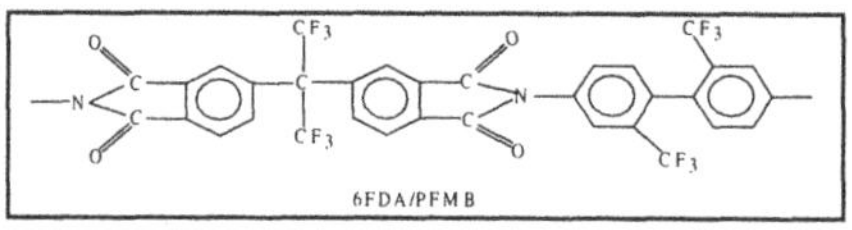 6FDA/PFMB—TNON guest—host system	EFISH	p-dioxane
48	p-methoxynitrobenzene	EFISH	p-dioxane
48	4-methoxy-β-nitrostyrene	EFISH	p-dioxane
48	α-p-methoxyphenyl-ω-p-nitrophenylpolyene 1	EFISH	p-dioxane
48	α-p-methoxyphenyl-ω-p-nitrophenylpolyene 2	EFISH	p-dioxane
48	α-p-methoxyphenyl-ω-p-nitrophenylpolyene 3	EFISH	p-dioxane
48	α-p-methoxyphenyl-ω-p-nitrophenylpolyene 4	EFISH	p-dioxane
48	α-p-methoxyphenyl-ω-p-nitrophenylpolyene 5	EFISH	p-dioxane

λ_{meas}	λ_{max}	μ	β	β_0	$\mu\beta$	$\mu\beta_0$	footnote	convention
1064	266		9	6.0				
1064	545		105	4.0				
1064	385		14.8	6.0				
1111	385		12.1	6.0				
1229	385		11.8	6.0				
1323	385		10.2	6.0				
1350	509		125	50.0				
1294	509		148	50.0				
1344		7.70	36		277.2			
1064		4.90	6	8.0	29.4	39.2		I
1064		5.50	12	12.0	66.0	66.0		I
1064		5.70	81	78.0	461.7	444.6		I
1064		6.00	135	106.0	810.0	636.0		I
1064		6.60	274	188.0	1808.4	1240.8		I
1064		6.70	367	211.0	2458.9	1413.7		I
1064		7.00	623	340.0	4361.0	2380.0		I

Footnote entries on page 503.

Hyperpolarizability tabulations. Shaded entries are calculated from values given in the cited reference.

	MOLECULE NAME, OR STRUCTURE	tech	solvent
	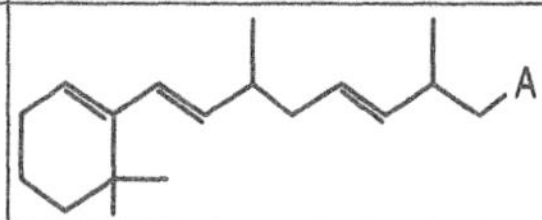		
49	A=-CHO	EFISH	dimethylsulfoxide
49	A= 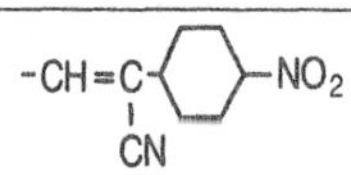	EFISH	dimethylsulfoxide
49	A= 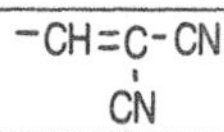	EFISH	dimethylsulfoxide
49	A= 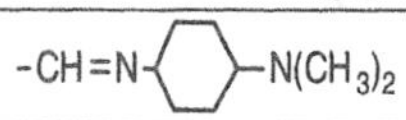	EFISH	dimethylsulfoxide
49	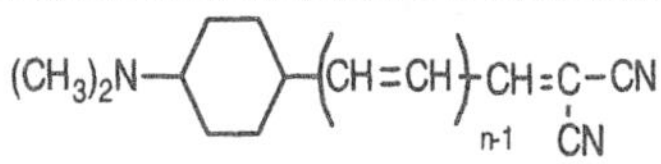		
49	n=1	EFISH	dimethylsulfoxide
49	n=2	EFISH	dimethylsulfoxide
49	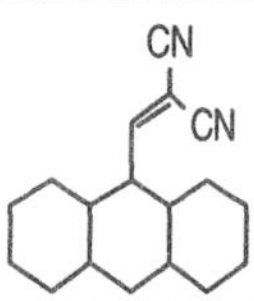	EFISH	dimethylsulfoxide
	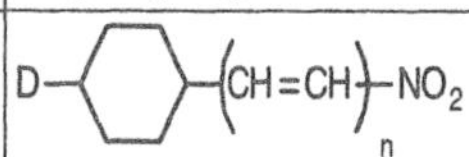		
50	D=CH_3O n=1	EFISH	dimethylsufloxide
50	D=CH_3O n=2	EFISH	dimethylsufloxide
50	D=$(CH_3)_2N$ n=1	EFISH	dimethylsufloxide
50	D=$(CH_3)_2N$ n=2	EFISH	dimethylsufloxide

λ_{meas}	λ_{max}	μ	β	β_0	$\mu\beta$	$\mu\beta_0$	footnote	convention
1064	380				230.0	98.3		
1064	480				1400.0	207.3		
1064	470				1300.0	229.7		
1064	440				1500.0	392.9		
1064	440				620.0	162.4		
1064	500				3400.0	309.1		
1064	430				100.0	29.0		
1064	370				81.0	36.8		
1064	400				300.0	112.0		
1064	460				1740.0	357.0		
1064	490				5540.0	662.0		

Footnote entries on page 503.

Hyperpolarizability tabulations. Shaded entries are calculated from values given in the cited reference.

	MOLECULE NAME, OR STRUCTURE	tech	solvent
50	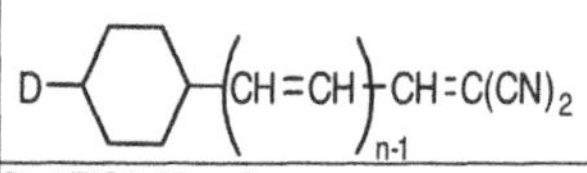		
50	D=$(CH_3)_2$N n=1	EFISH	dimethylsufloxide
50	D=$(CH_3)_2$N n=2	EFISH	dimethylsufloxide
50	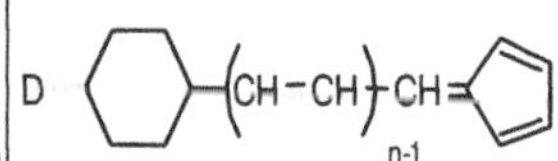		
50	D=CH_3O n=2	EFISH	dimethylsufloxide
50	D=$(CH_3)_2$N n=1	EFISH	dimethylsufloxide
50	D=$(CH_3)_2$N n=2	EFISH	dimethylsufloxide
	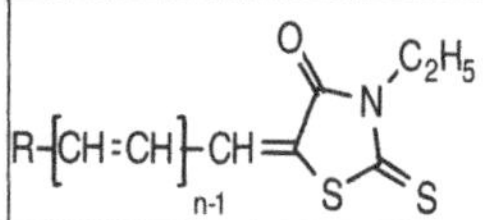		
51	R=dimethylamino, n=2	EFISH	dimethylsulfoxide
51	R= morpholino, n=3	EFISH	dimethylsulfoxide
51	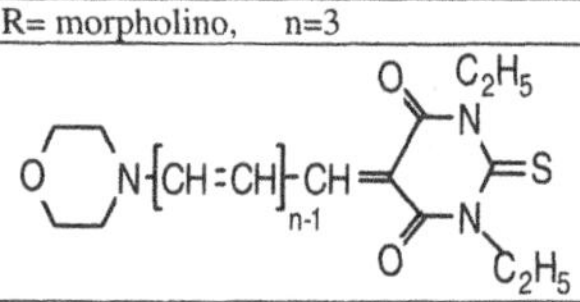		
51	n=3	EFISH	dimethylsulfoxide
51	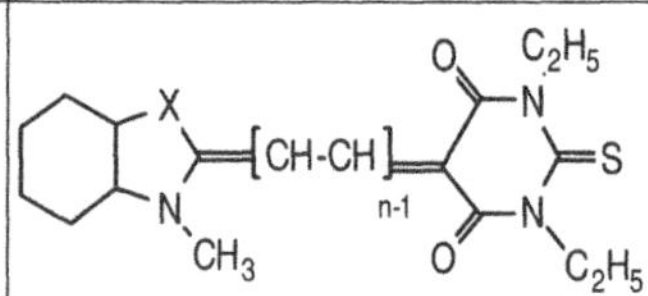		
51	x=$C(CH_3)_2$ n=3	EFISH	dimethylsulfoxide
51	X=$C(CH_3)_2$ n=4	EFISH	dimethylsulfoxide
51	X=S n=4	EFISH	dimethylsulfoxide
51	X=-CH=CH- n=4	EFISH	dimethylsulfoxide

λ_{meas}	λ_{max}	μ	β	β_0	$\mu\beta$	$\mu\beta_0$	footnote	convention
1064	440				610.0	159.8		
1064	500				3600.0	327.3		
1064	370				70.0	31.8		
1064	420				440.0	139.9		
1064	450				2000.0	467.2		
1064	470				740.0	130.7		
1064	550				48000.0	-2420.5		
1064	530				21000.0	118.5		
1064	600				17000.0	-3153.3		
1064	700				5200.0	-2156.8		
1064	700				13000.0	-5392.1		
1064	700				20000.0	-8295.5		

Footnote entries on page 503.

Hyperpolarizability tabulations. Shaded entries are calculated from values given in the cited reference.

	MOLECULE NAME, OR STRUCTURE	tech	solvent
51	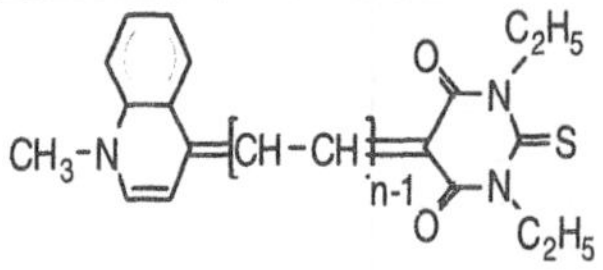		
51	n=2	EFISH	dimethylsulfoxide
51	n=4	EFISH	dimethylsulfoxide
51	n=4 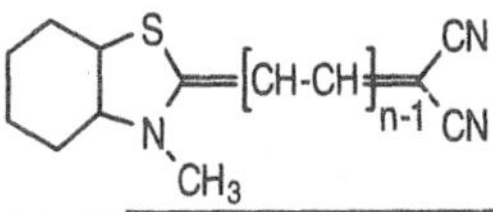	EFISH	dimethylsulfoxide
52	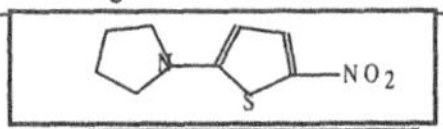	EFISH	p-dioxane
52	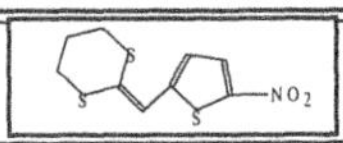	EFISH	p-dioxane
52	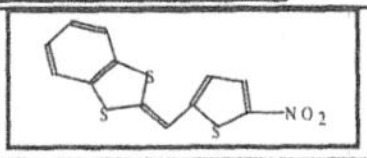	EFISH	p-dioxane
52	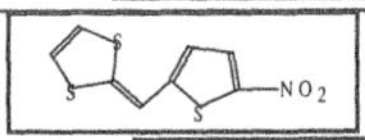	EFISH	p-dioxane
52	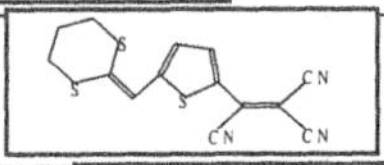	EFISH	p-dioxane
52	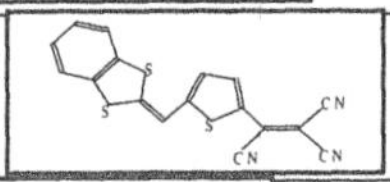	EFISH	p-dioxane
52	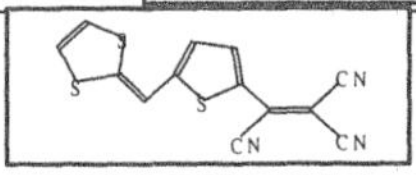	EFISH	p-dioxane

λ_{meas}	λ_{max}	μ	β	β_0	$\mu\beta$	$\mu\beta_0$	footnote	convention
1064	560				13000.0	-1015.4		
1064	720				27000.0	-12172.3		
1064	650				10000.0	-3088.9		
1907	430				125.0	95.0		
1907	438				200.0	150.0		
1907	478				253.0	178.0		
1907	493				350.0	239.0		
1907	558				940.0	565.0		
1907	604				1350.0	727.0		
1907	625				1600.0	815.0		

Footnote entries on page 503.

Hyperpolarizability tabulations. Shaded entries are calculated from values given in the cited reference.

	MOLECULE NAME, OR STRUCTURE	tech	solvent
52	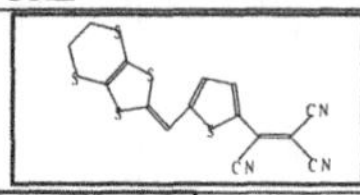	EFISH	p-dioxane
53	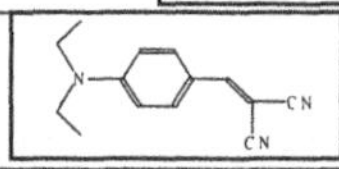	EFISH	p-dioxane
53	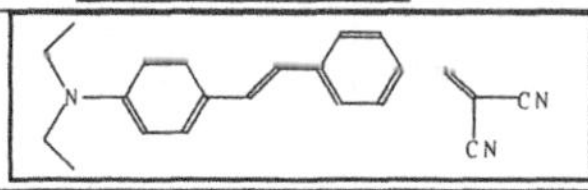	EFISH	p-dioxane
53	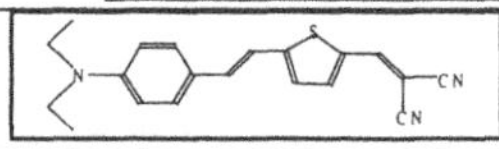	EFISH	p-dioxane
53	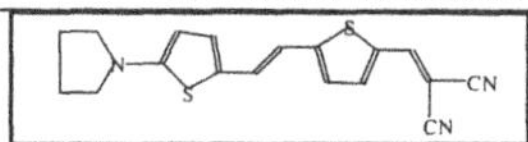	EFISH	p-dioxane
53	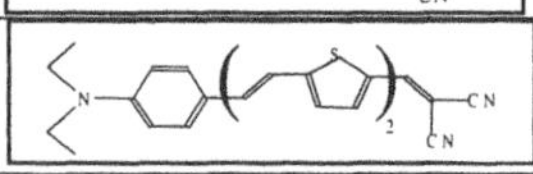	EFISH	p-dioxane
53	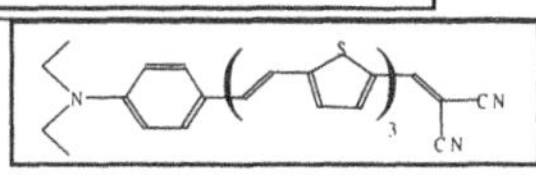	EFISH	p-dioxane
54	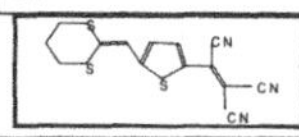	EFISH	p-dioxane
54	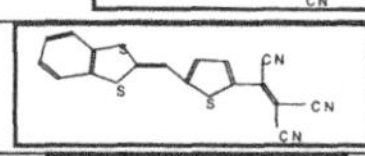	EFISH	p-dioxane
54	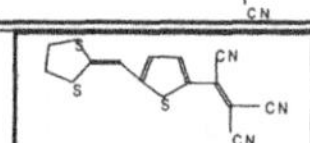	EFISH	p-dioxane
54	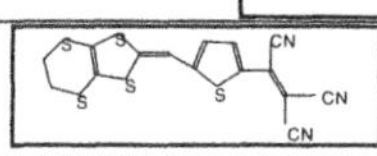	EFISH	p-dioxane

λ_{meas}	λ_{max}	μ	β	β_0	μβ	$\mu\beta_0$	footnote	convention
1907	640				2400.0	1170.0		
1907	419				300.0	230.4		
1907	468				1100.0	784.7		
1907	513				1300.0	856.9		
1907	584				2600.0	1472.3		
1907	547				2300.0	155.0		
1907	556				3800.0	2294.7		
1907	556				940.0	567.6		
1907	604				1350.0	727.2		
1907	625				1600.0	814.5		
1907	638				2400.0	1177.1		

Footnote entries on page 503.

Hyperpolarizability tabulations. Shaded entries are calculated from values given in the cited reference.

	MOLECULE NAME, OR STRUCTURE	tech	solvent
54	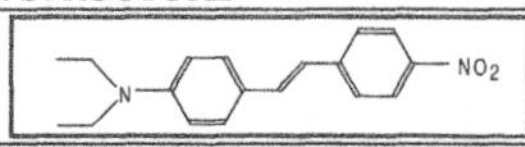	EFISH	p-dioxane
54	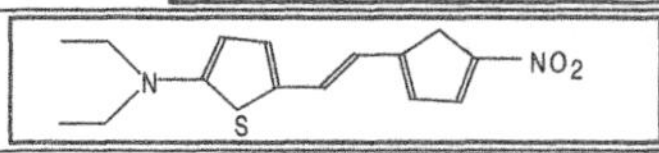	EFISH	p-dioxane
54	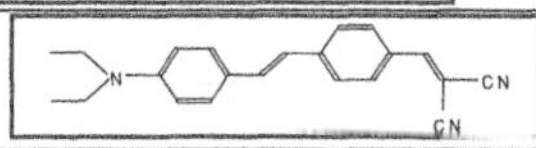	EFISH	p-dioxane
54	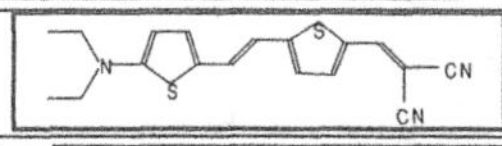	EFISH	p-dioxane
54	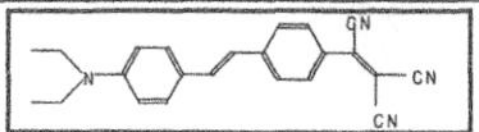	EFISH	p-dioxane
54	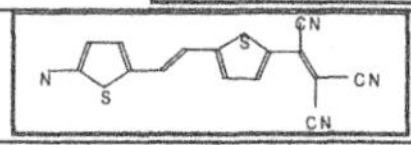	EFISH	p-dioxane
54	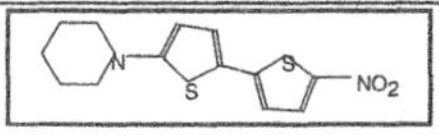	EFISH	p-dioxane
54	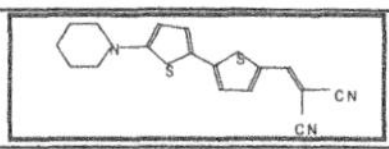	EFISH	p-dioxane
54	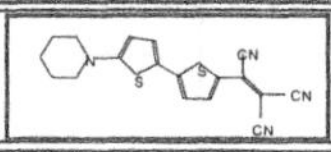	EFISH	p-dioxane
54	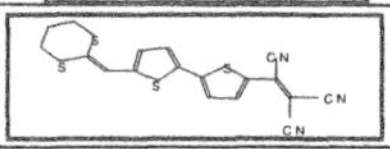	EFISH	p-dioxane
54	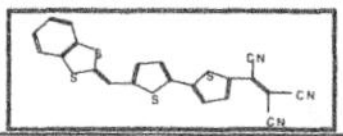	EFISH	p-dioxane
54	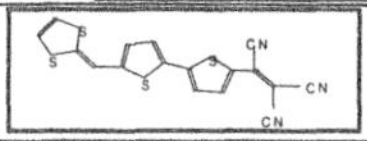	EFISH	p-dioxane

λ_{meas}	λ_{max}	μ	β	β_0	$\mu\beta$	$\mu\beta_0$	footnote	convention
1907	424				580.0	442.3		
1907	516				1040.0	681.6		
1907	468				1100.0	784.7		
1907	584				2600.0	1472.3		
1907	594				2700.0	1491.9		
1907	718				6900.0	2564.0		
1907	500				563.0	380.1		
1907	558				1250.0	751.5		
1907	650				2850.0	1348.3		
1907	594				1500.0	828.8		
1907	635				3300.0	1632.8		
1907	660				3800.0	1742.2		

Footnote entries on page 503.

Hyperpolarizability tabulations. Shaded entries are calculated from values given in the cited reference.

	MOLECULE NAME, OR STRUCTURE	tech	solvent
54	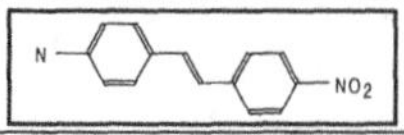	EFISH	p-dioxane
54	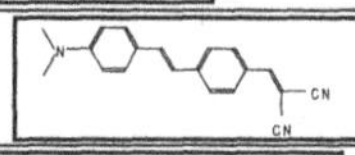	EFISH	p-dioxane
54	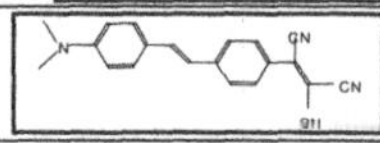	EFISH	p-dioxane
55	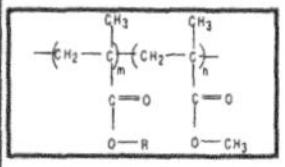 R= 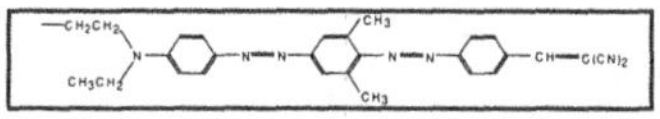	thin film	poled polymer
55	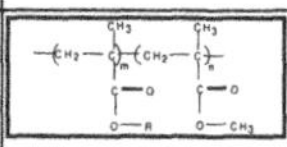 R= 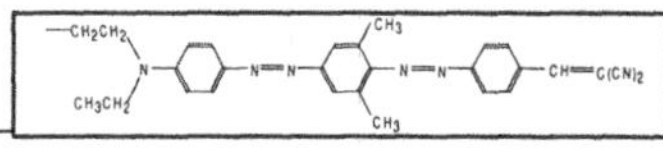	thin film	poled polymer
55	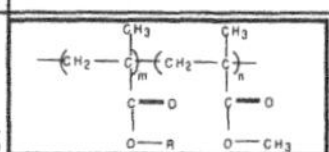 R= 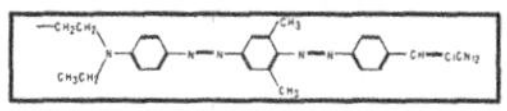	thin film	poled polymer
55	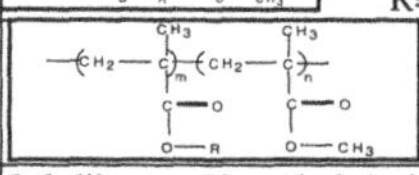 R= 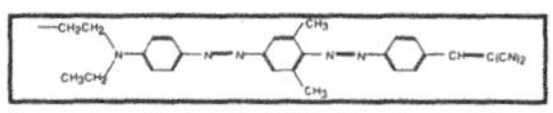	thin film	poled polymer
56	2,6-dibromo-N-methyl-4-nitroaniline(DBMNA)	crystal SHG	
56	2,6-dibromo-N-methyl-4-nitroaniline(DBMNA)	crystal SHG	
56	2,6-dibromo-N-methyl-4-nitroaniline(DBMNA)	EFISH	dioxane
57	2-nitro-1-(2-thienyl)ethene (NTE)	EFISH	chloroform
57	4-nitro-1-(2-thienyl)-1,3-butadine (NTB)	EFISH	chloroform
57	PNA	EFISH	
58	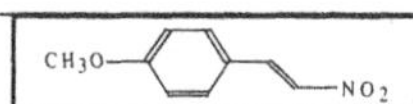	EFISH	dimethylsulfoxide
58	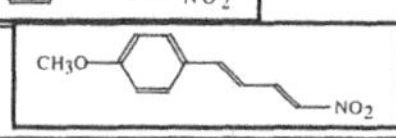	EFISH	dimethylsulfoxide
58	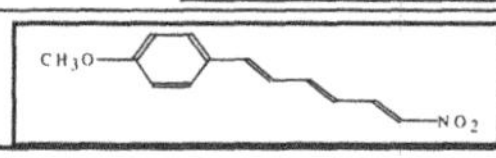	EFISH	dimethylsulfoxide

λ_{meas}	λ_{max}	μ	β	β_0	$\mu\beta$	$\mu\beta_0$	footnote	convention
1907	424				580.0	442.3		
1907	468				1100.0	784.7		
1907	594				2700.0	1491.9		
1064	510				24000.0	1497.3		
1300	510				4200.0	1365.9		
1500	510				2900.0	1378.8		
1700	510				2400.0	1397.8		
1064	453	5.50	50	3.1	275.0	17.1	b	I
1620		5.50	7.2		39.6			I
1064		5.50	14.8		81.4			I
1064	350.5	5.20	20.2	10.2	105.0	53.0		
1064	381.5	5.40	66.6	28.2	359.6	152.2		
1064	347	6.20	20.2	10.4	125.2	64.3		
1064	370				70.0	31.8		
1064	400				200.0	74.6		
1064	420				240.0	76.3		

Footnote entries on page 503.

Hyperpolarizability tabulations. Shaded entries are calculated from values given in the cited reference.

	MOLECULE NAME, OR STRUCTURE	tech	solvent
58	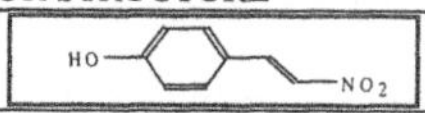	EFISH	dimethylsulfoxide
58	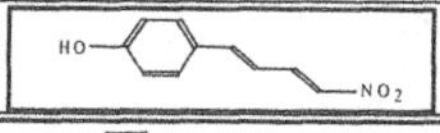	EFISH	dimethylsulfoxide
58	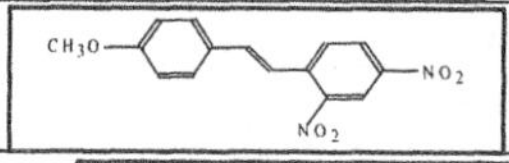	EFISH	dimethylsulfoxide
58	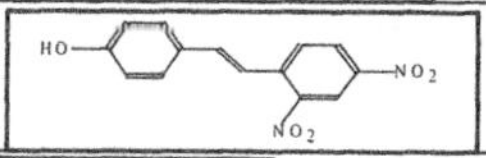	EFISH	dimethylsulfoxide
58	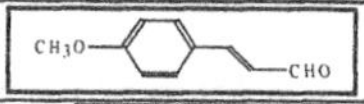	EFISH	dimethylsulfoxide
58	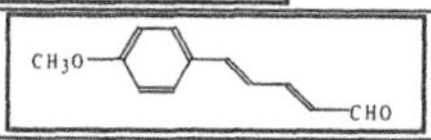	EFISH	dimethylsulfoxide
58	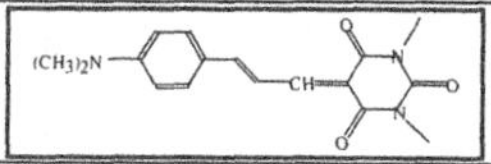	EFISH	dimethylsulfoxide
58	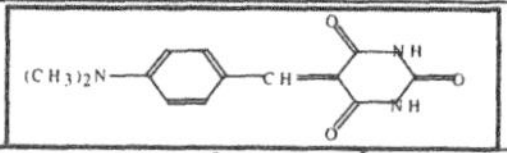	EFISH	dimethylsulfoxide
58	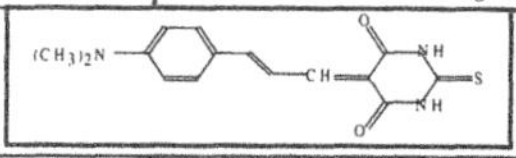	EFISH	dimethylsulfoxide
58	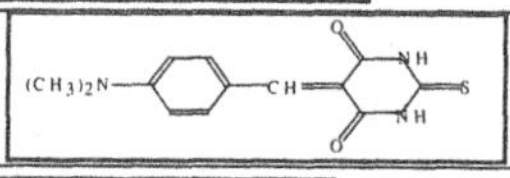	EFISH	dimethylsulfoxide
58	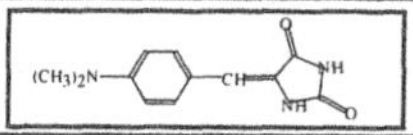	EFISH	dimethylsulfoxide
58	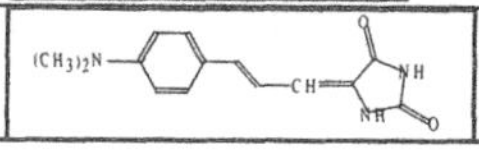	EFISH	dimethylsulfoxide

λ_{meas}	λ_{max}	μ	β	β_0	$\mu\beta$	$\mu\beta_0$	footnote	convention
1064	370				130.0	59.0		
1064	400				470.0	175.4		
1064	400				300.0	112.0		
1064	410				470.0	259.0		
1064	324				40.0	855.0		
1064	350				110.0	2151.0		
1064	539				1330.0	5095.0		
1064	464				630.0	394.0		
1064	574				140.0	1159.0		
1064	500				1150.0	3069.0		
1064	389				120.0	8019.0		
1064	414				340.0	14920.0		

Footnote entries on page 503.

Hyperpolarizability tabulations. Shaded entries are calculated from values given in the cited reference.

	MOLECULE NAME, OR STRUCTURE	tech	solvent
58	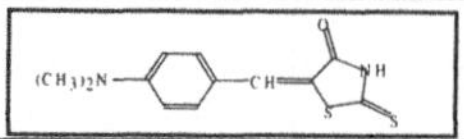	EFISH	dimethylsulfoxide
58	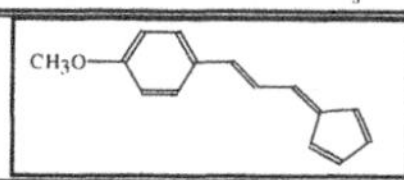	EFISH	dimethylsulfoxide
58	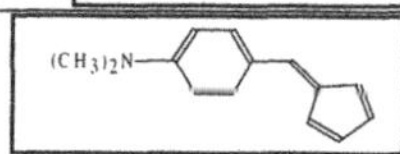	EFISH	dimethylsulfoxide
58	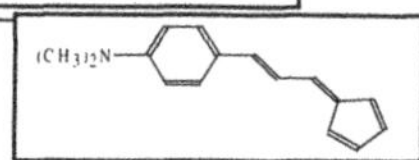	EFISH	dimethylsulfoxide
58	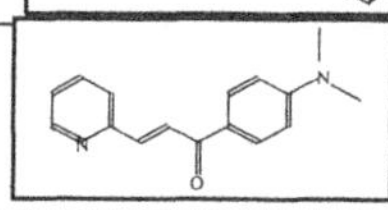	EFISH	dimethylsulfoxide
58	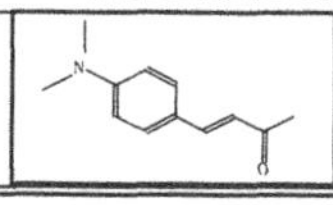	EFISH	dimethylsulfoxide
58	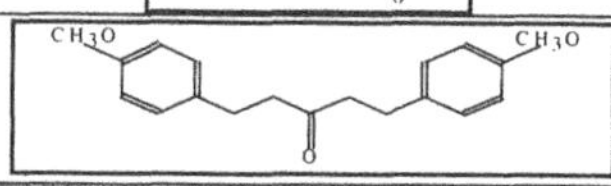	EFISH	dimethylsulfoxide
58	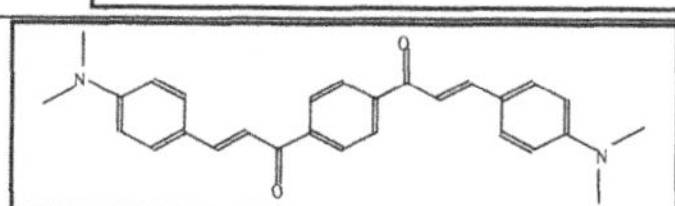	EFISH	dimethylsulfoxide
58	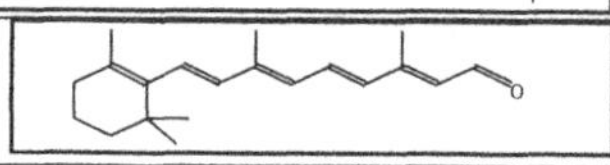	EFISH	dimethylsulfoxide
58	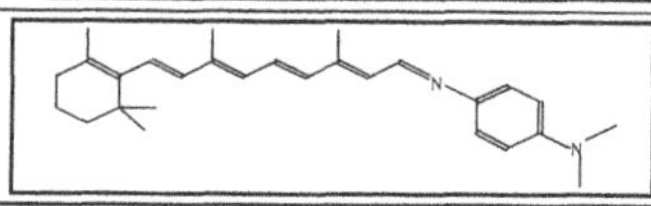	EFISH	dimethylsulfoxide

λ_{meas}	λ_{max}	μ	β	β_0	$\mu\beta$	$\mu\beta_0$	footnote	convention
1064	473				1050.0	221.0		
1064	360				40.0	771.0		
1064	420				300.0	1895.0		
1064	449				1300.0	4696.0		
1064	370				70.0	328.0		
1064	380				180.0	919.0		
1064	360				70.0	3000.0		
1064	370				70.0	4753.0		
1064	381				230.0	13600.0		
1064	430				1500.0	363.0		

Footnote entries on page 503.

Hyperpolarizability tabulations. Shaded entries are calculated from values given in the cited reference.

	MOLECULE NAME, OR STRUCTURE	tech	solvent
58	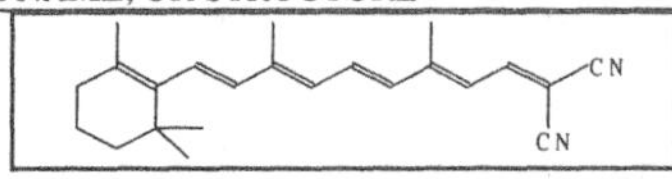	EFISH	dimethylsulfoxide
58	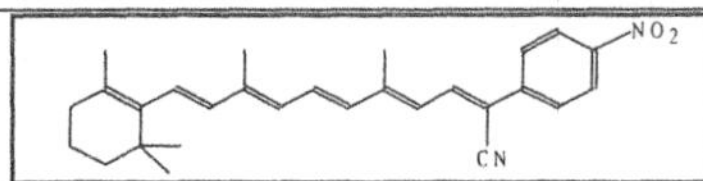	EFISH	dimethylsulfoxide
58	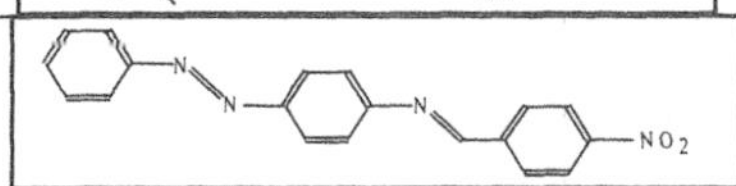	EFISH	dimethylsulfoxide
58	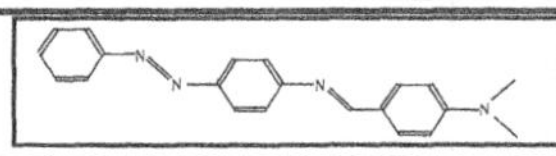	EFISH	dimethylsulfoxide
58	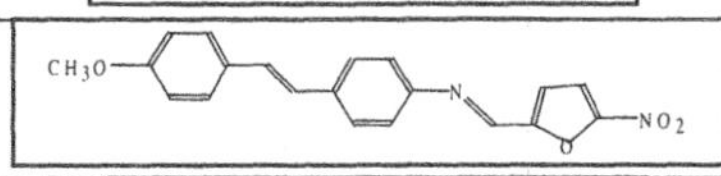	EFISH	dimethylsulfoxide
58	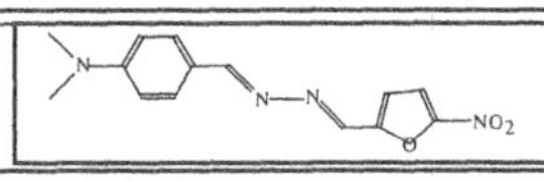	EFISH	dimethylsulfoxide
58	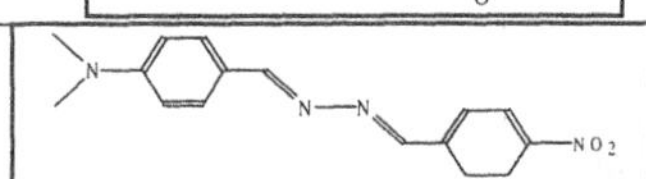	EFISH	dimethylsulfoxide
58	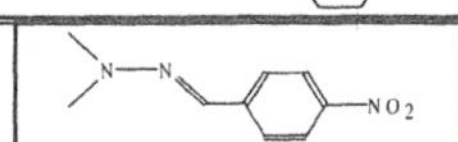	EFISH	dimethylsulfoxide
58	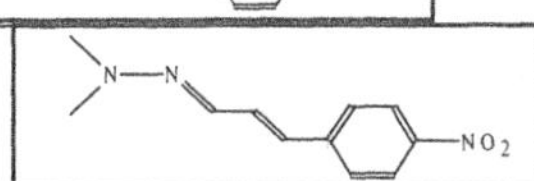	EFISH	dimethylsulfoxide
58	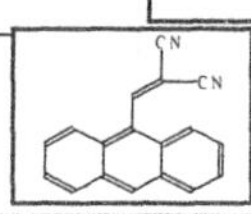	EFISH	dimethylsulfoxide

λ_{meas}	λ_{max}	μ	β	β_0	μβ	$\mu\beta_0$	footnote	convention
1064	459				1300.0	608.0		
1064	480				1400.0	779.0		
1064	380				230.0	1197.0		
1064	410				450.0	70.0		
1064	430				640.0	90.0		
1064	454				1400.0	102.0		
1064	400				370.0	110.0		
1064	405				440.0	150.0		
1064	420				1100.0	138.0		
1064	430				100.0	528.0		

Footnote entries on page 503.

Hyperpolarizability tabulations. Shaded entries are calculated from values given in the cited reference.

	MOLECULE NAME, OR STRUCTURE	tech	solvent
58	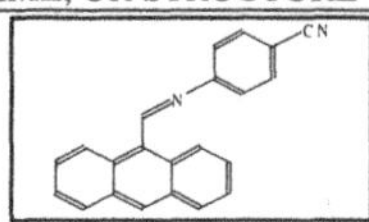	EFISH	dimethylsulfoxide
58	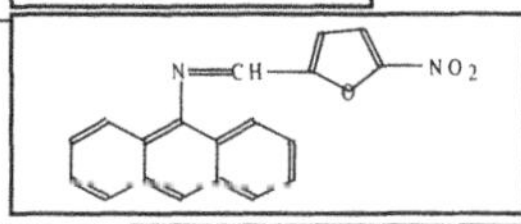	EFISH	dimethylsulfoxide
58	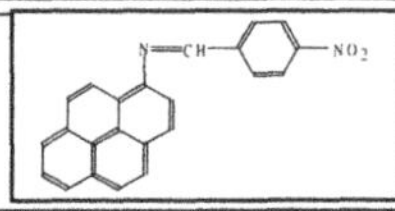	EFISH	dimethylsulfoxide
58	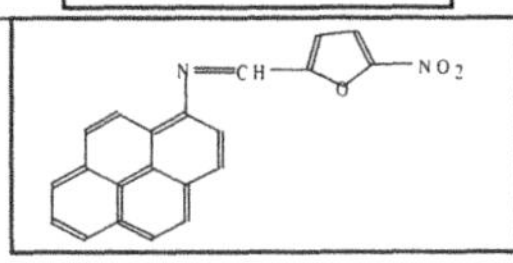	EFISH	dimethylsulfoxide
58	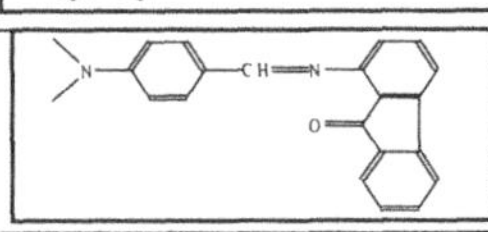	EFISH	dimethylsulfoxide
58	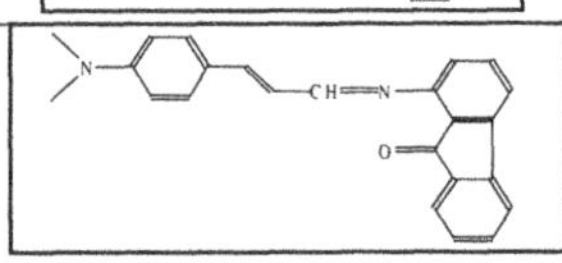	EFISH	dimethylsulfoxide
59	tetranitro-tetrapropoxycalix[4]arene	HRS	chloroform
59	tetranitro-tetrapropoxycalix[4]arene	EFISH	chloroform
59	4-nitroanisole	HRS	chloroform
59	4-nitroanisole	EFISH	chloroform
59	tetrakis(nitrostilbene)-tetrapropoxycalix[4]arene	HRS	chloroform
59	tetrakis(nitrostilbene)-tetrepropoxycalix[4]arene	EFISH	chloroform
59	4-methoxy-4'-nitrostilbene	HRS	chloroform
59	4-methoxy-4'-nitrostilbene	EFISH	chloroform
59	5,17-dihydroxy-tetrakis(nitrophenylazo)-11.23-dipropoxy calix[4]arene	HRS	chloroform

λ_{meas}	λ_{max}	μ	β	β_0	$\mu\beta$	$\mu\beta_0$	footnote	convention
1064	414				95.0	565.0		
1064	444				260.0	1803.0		
1064	439				740.0	829.0		
1064	469				370.0	150.0		
1064	380				470.0	280.0		
1064	410				820.0	565.0		
1064		13.80	27		372.6			
1064		13.80	30		414.0			
1064		4.60	12		55.2			
1064		4.60	12		55.2			
1064		15.30	278		4253.4			
1064		15.30	280		4284.0			
1064		4.50	105		472.5			
1064		4.50	100		450.0			
1064		13.00	231		3003.0			

Footnote entries on page 503.

Hyperpolarizability tabulations. Shaded entries are calculated from values given in the cited reference.

	MOLECULE NAME, OR STRUCTURE	tech	solvent
59	5,17-dihydroxy-tetrakis(nitrophenylazo)-11,23-dipropoxy calix[4]arene	EFISH	chloroform
59	4-hydroxy-3,5-dimethyl-4'-nitroazostilbene	EFISH	chloroform
59	4-methoxy-3,5-dimethyl-4'-nitroazostilbene	EFISH	chloroform
60	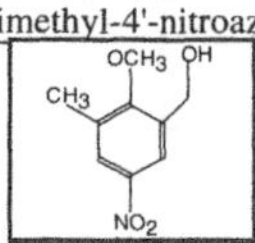	EFISH	chloroform
60	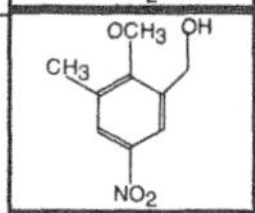	HRS	chloroform
60	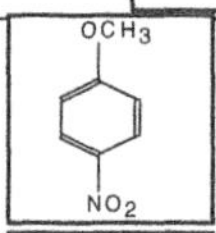	EFISH	chloroform
60	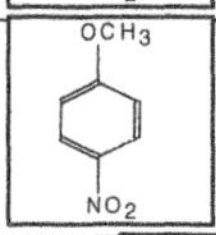	HRS	chloroform
60	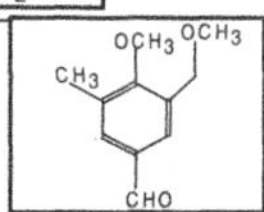	EFISH	chloroform
60	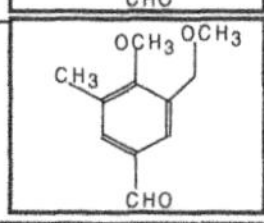	HRS	chloroform
60	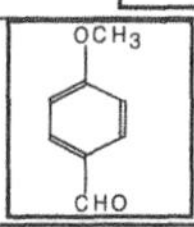	EFISH	chloroform

λ_{meas}	λ_{max}	μ	β	β_0	$\mu\beta$	$\mu\beta_0$	footnote	convention
1064		13.00	161		2093.0			
1064		5.50	92		506.0			
1064		4.30	42		180.6			
1064	290	6.00	6	3.9	36.0	23.4		
1064	290	6.00	15	9.8	90.0	58.6		
1064	302	4.60	12	7.5	55.2	34.4		
1064	302	4.60	12	7.5	55.2	34.4		
1064	272	4.30	10	6.9	43.0	29.7		
1064	272	4.30	14	9.7	60.2	41.6		
1064	269	3.20	8	5.6	25.6	17.8		

Footnote entries on page 503.

Hyperpolarizability tabulations. Shaded entries are calculated from values given in the cited reference.

	MOLECULE NAME, OR STRUCTURE	tech	solvent
60	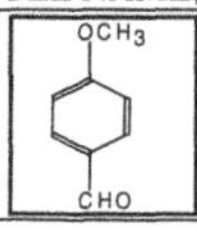	HRS	chloroform
60	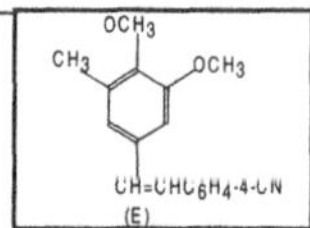	EFISH	chloroform
60	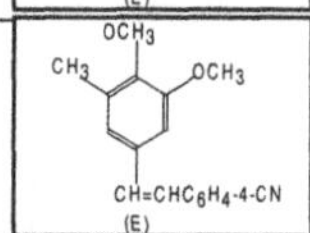	HRS	chloroform
60	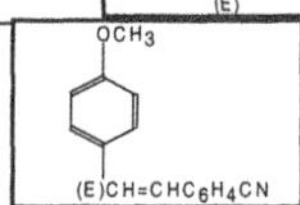	EFISH	chloroform
60	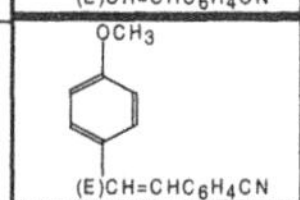	HRS	chloroform
60	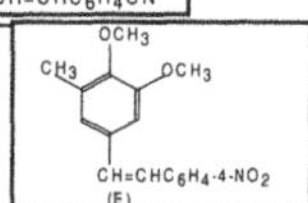	EFISH	chloroform
60	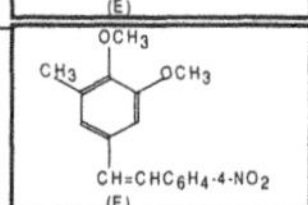	HRS	chloroform
60	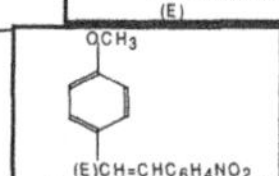	EFISH	chloroform
60	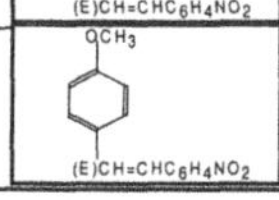	HRS	chloroform

λ_{meas}	λ_{max}	μ	β	β_0	$\mu\beta$	$\mu\beta_0$	footnote	convention
1064	269	3.20	9	6.3	28.8	20.1		
1064	340	3.30	48	25.5	158.4	84.1		
1064	340	3.30	80	42.5	264.0	140.2		
1064	340	5.40	46	24.4	248.4	131.9		
1064	340	5.40	47	25.0	253.8	134.8		
1064	376	4.60	179	78.4	823.4	360.6		
1064	376	4.60	338	148.0	1554.8	681.0		
1064	377	4.50	100	43.5	450.0	195.9		
1064	377	4.50	105	45.7	472.5	205.7		

Footnote entries on page 503.

Hyperpolarizability tabulations. Shaded entries are calculated from values given in the cited reference.

	MOLECULE NAME, OR STRUCTURE	tech	solvent
61	p-dimethylaminobenzylidene-1,3-indanodine (DABI)	EFISH	p-dioxane
61	p-dimethylaminobenzylidene-1,3-indanodine (DABI)	EFISH	p-dioxane
62	4'-nitrobenzylidene-3acetamino-4-methoxyaniline	EFISH	p-dioxane
62	4'-nitrobenzylidene-3acetamino-4-methoxyaniline	EFISH	p-dioxane
63	N-(2,4-dinitrophenyl)-1-chloro-3-phenyl-2-propamine	EFISH	p-dioxane
63	nitrobenzene	EFISH	p-dioxane
64	MBANP (-)2-(α–methylbenzylamino)-5-nitropyridine	EFISH	p-dioxane
64	nitrobenzene	EFISH	p-dioxane
64	p-nitroaniline	EFISH	p-dioxane
64	MNA	EFISH	p-dioxane
65	2-amino-5-nitro-pyridine	EFISH	acetone
65	2-amino-5-nitro-pyridine	EFISH	acetone
66	R=NH_2 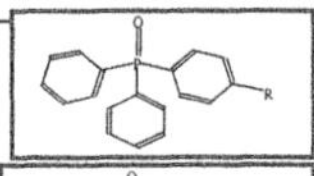	EFISH	p-dioxane
66	R=NH_2 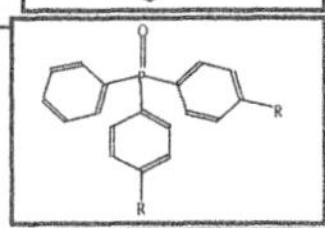	EFISH	p-dioxane
66	R=OH 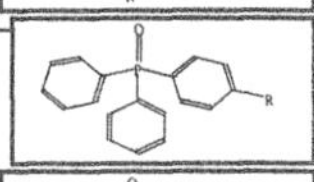	EFISH	chloroform
66	R=OH 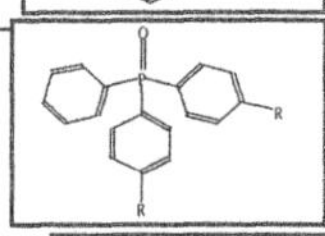	EFISH	p-dioxane
66	R=OH 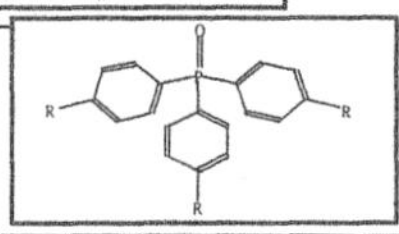	EFISH	p-dioxane
66	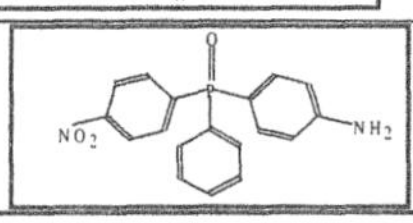	EFISH	p-dioxane

λ_{meas}	λ_{max}	μ	β	β_0	$\mu\beta$	$\mu\beta_0$	footnote	convention
1064	472	3.58	253.9	43.4	908.9	155.4		I
1907	472	3.58	253.9	179.9	908.9	644.1		I
1064	530	8.70	30.1		261.6			I
1907	530	8.70	15.3		132.9			I
1064		6.42	7.26		46.6			I
1064		4.82	2.41		11.6			I
1064	359	6.07	15.0	7.2	91.1	43.9		I
1064	255	4.07	2.0	1.5	8.1	5.9		I
1064	368	5.50	12.6	5.8	69.3	31.8		I
1064	373	5.63	14.2	6.3	79.9	35.7		I
1340	347	6.30	20.3	13.9	127.9	87.3		I
1064	347	6.30	27	13.9	170.1	87.3		I
1064	264	5.50	4.3	3.0	23.7	158.0		
1064	266		4	2.8				
1064	238	7.60	4.6	3.5	35.0	26.6		
1064	240	5.90	1.5	1.1	8.9	6.7		
1064	242	5.70	1	0.8	5.7	4.3		
1064	340	8.30	9	4.8	74.7	39.7	g	

Footnote entries on page 503.

Hyperpolarizability tabulations. Shaded entries are calculated from values given in the cited reference.

	MOLECULE NAME, OR STRUCTURE	tech	solvent
66	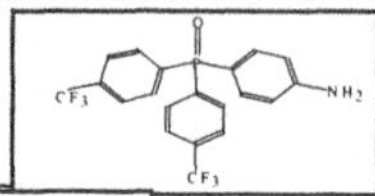	EFISH	p-dioxane
66	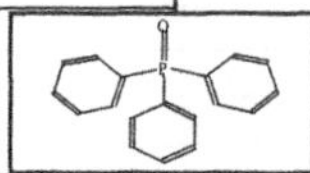	EFISH	p-dioxane
66	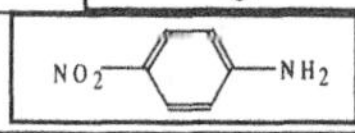	EFISH	p-dioxane
67	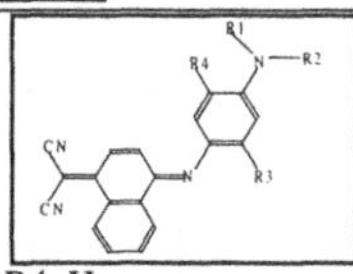 R1=Et, R2=CH_2CH_2OH, R3=Me, R4=H	poled polymer	PMMA
67	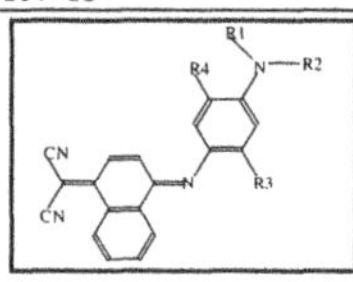 R1=R2=Et, R3=R4=H	poled polymer	PMMA
67	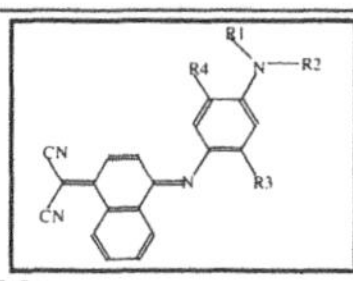 R1=R2=Et, R3=NHCOMe, R4=OMe	poled polymer	PMMA
68	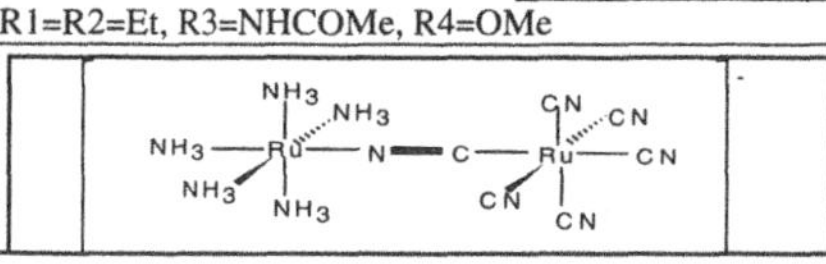	HRS	H_2O

λ_{meas}	λ_{max}	μ	β	β_0	$\mu\beta$	$\mu\beta_0$	footnote	convention
1064	298	4.70	10.1	6.4	47.5	30.0	g	
1064	272	4.80	0.8	0.6	3.8	2.7	h	
1064	355	7.40	19.2	9.5	142.1	70.0		
1310	752				6100.0	3080.0	i	
1310	728				5700.0	3040.0	i	
1310	802				7400.0	3300.0	i	
1064			1440					

Footnote entries on page 503.

Hyperpolarizability tabulations. Shaded entries are calculated from values given in the cited reference.

	MOLECULE NAME, OR STRUCTURE	tech	solvent
68	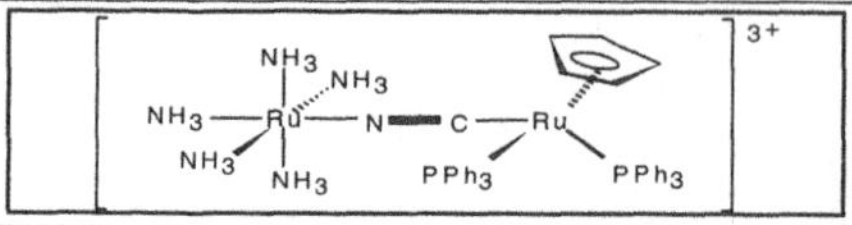	HRS	nitromethane
69	2-(4-dicyanomethylenecyclohexa-2.5-dienylidine)-inidozolidine	EFISH	dimethylsulfoxide
70	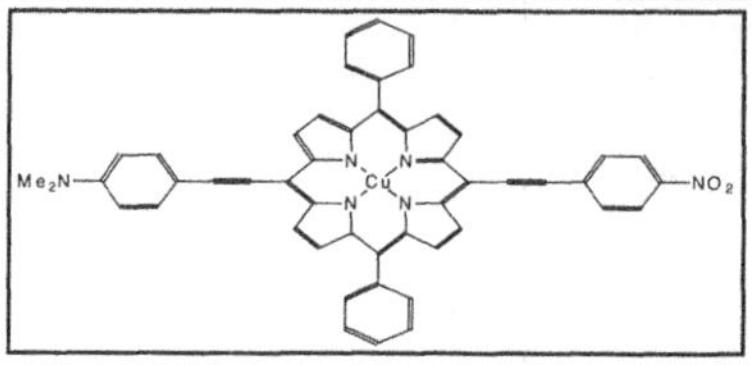	HRS	chloroform
70	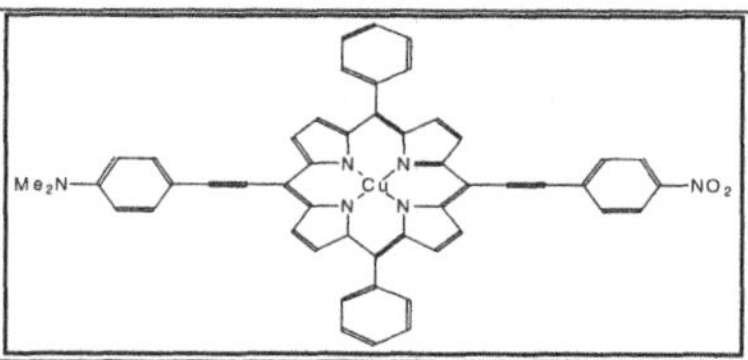	HRS	chloroform
70	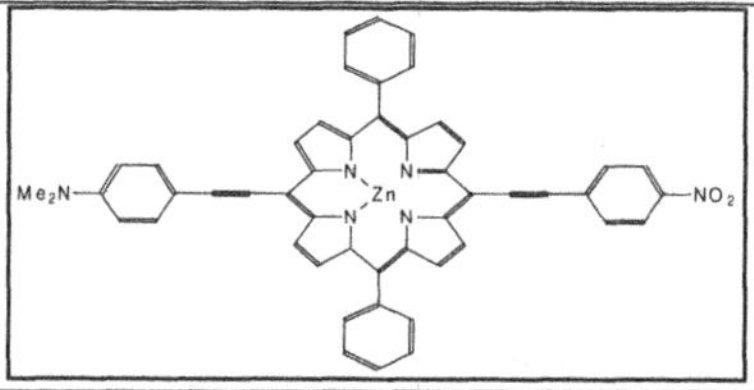	HRS	chloroform

λ_{meas}	λ_{max}	μ	β	β_0	$\mu\beta$	$\mu\beta_0$	footnote	convention
1064			1080					
1064		3.50	-240		-840.0			
830			4374					
1064			1501					
830			5142					

Footnote entries on page 503.

Hyperpolarizability tabulations. Shaded entries are calculated from values given in the cited reference.

	MOLECULE NAME, OR STRUCTURE	tech	solvent
70	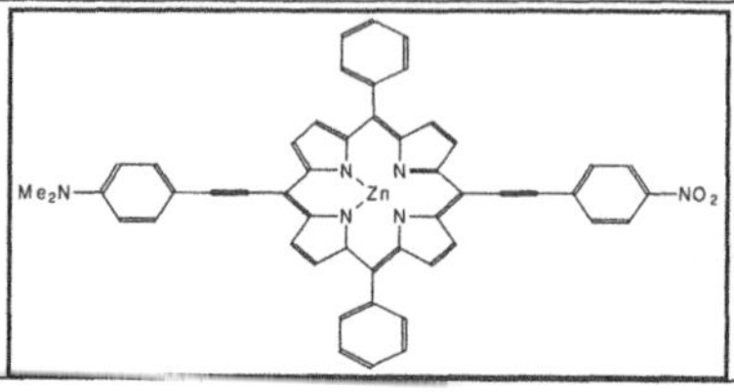	HRS	chloroform
71	urea	EFISH	H_2O
71	urea	EFISH	dimethylformamide
71	urea	EFISH	dimethylsulfoxide
71	monomethylurea	EFISH	H_2O
71	monomethylurea	EFISH	dimethylformamide
71	monomethylurea	EFISH	dimethylsulfoxide
71	dimethylurea	EFISH	H_2O
71	dimethylurea	EFISH	dimethylformamide
71	dimethylurea	EFISH	dimethylsulfoxide
71	tetramethylurea	EFISH	neat
72	DPNA prepared with Cd+2	LB	
72	DPNA prepared without Cd+2	LB	
73	$Sn(C_6H_4NMe_2)_4$	HRS	dichloromethane
73	$Sn(C_6H_4C_6H_4NMe_2)_4$	HRS	dichloromethane
74	*o*-nitroaniline	EFISH	neat
74	*m*-nitroaniline	EFISH	neat
74	*p*-nitroaniline	EFISH	neat
74	nitrobenzene	EFISH	neat
74	aniline	EFISH	neat
75	nitrobenzene	EFISH	benzene
76	p-xylene	EFISH	neat
76	toluene	EFISH	neat
76	fluorobenzene	EFISH	neat
76	chlorobenzene	EFISH	neat
76	bromobenzene	EFISH	neat
76	iodobenzene	EFISH	neat
76	*o*-diiodobenzene	EFISH	neat
76	aniline	EFISH	neat
76	nitrobenzene	EFISH	neat

λ_{meas}	λ_{max}	μ	β	β_0	$\mu\beta$	$\mu\beta_0$	footnote	convention
1064			4933	800.0		9840.0		
1064		4.56	0.45		2.1			I
1064		4.56	0.3		1.4			I
1064		4.56	0.29		1.3			I
1064		4.34	0.45		2.0			I
1064		4.34	0.65		2.8			I
1064		4.34	0.43		1.9			I
1064		4.66	0.6		2.8			I
1064		4.66	0.28		1.3			I
1064		4.66	0.18		0.8			I
1064		3.92	0.41		1.6			I
1064	464		1.0737	0.2				
1064	504		0.78738	0.1				
1064	274		13	8.9				
1064	319		24	14.0				
1318		5.00	6.4		32.0			I
1318		5.50	4.2		23.1			I
1318		7.20	21.1		151.9			I
1318		4.10	1.97		8.1			I
1318		1.50	7.9		11.9			I
1064		4.10	20		82.0			I
1064		0.00	0					I
1064		0.34	0.18		0.1			I
1064		-1.45	0.44		-0.6			I
1064		-1.45	0.281		-0.4			I
1064		-1.38	0.042		-0.1			I
1064		-1.14	-0.279		0.3			I
1064		-1.34	-0.291		0.4			I
1064		1.50	0.89		1.3			I
1064		-4.10	-2.04		8.4			I

Footnote entries on page 503.

Hyperpolarizability tabulations. Shaded entries are calculated from values given in the cited reference.

	MOLECULE NAME, OR STRUCTURE	tech	solvent
76	*o*-fluoronitrobenzene	EFISH	neat
76	*m*-fluoronitrobenzene	EFISH	neat
76	*p*-fluoronitrobenzene	EFISH	neat
77	Pyridine +I_2	EFISH	p-dioxane
77	4-Aminopyridine +I_2	EFISH	p-dioxane
77	Pyridine +ICl	EFISH	p-dioxane
78	Pyridine-I_2 CT complex	EFISH	p-dioxane
79	poly-(gamma)-benzyl-L-glutamate	EFISH	p-dioxane
80	trans-stilbene	EFISH	neat
80	p-nitroaniline	EFISH	trans-stilbene
80	DANS	EFISH	trans-stilbene
81	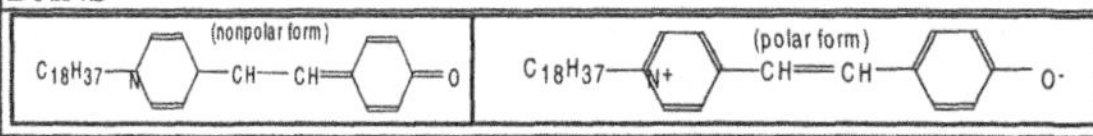	EFISH	methanol
81	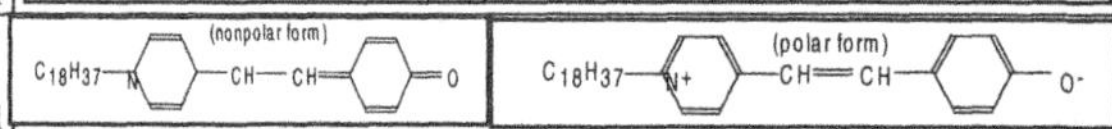	EFISH	50methanol/ 50pyridine
81	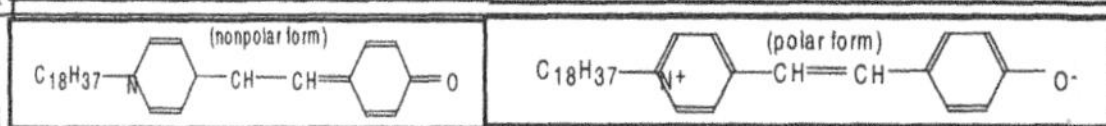	EFISH	pyridine
82	n=0 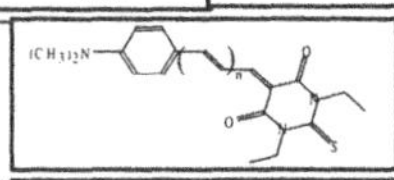	EFISH	chloroform
82	n=1 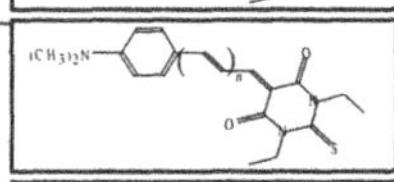	EFISH	chloroform
82	n=2 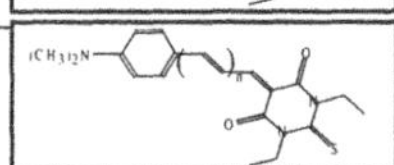	EFISH	chloroform
82	n=3 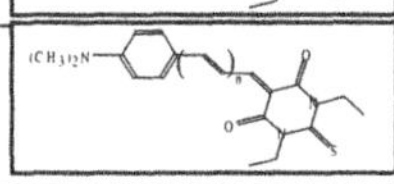	EFISH	chloroform
82	n=0 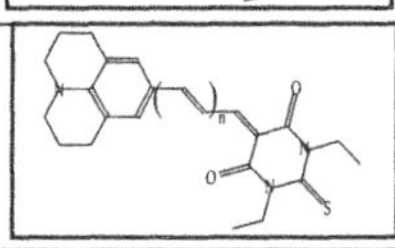	EFISH	chloroform

λ_{meas}	λ_{max}	μ	β	β_0	$\mu\beta$	$\mu\beta_0$	footnote	convention
1064		-5.00	-1.75		8.8			I
1064		-3.60	-1.64		5.9			I
1064		-2.70	-2.14		5.8			I
1318		4.20	9.6		40.3			I
1318		7.50	4.5		33.8			I
1318		8.20	1.4		11.5			I
1318			9.6					I
1064		8000.00	500		4000000.0			I
1064		0.00	0					I
1318		9.20	6.4		58.9			I
1318		7.40	39		288.6			I
1318			-100	-51.0				
1318			-130	-37.0				
1318			-210	-34.0				
1907	484	5.40	68	48.0	370.0	259.2		
1907	572	5.70	256	150.0	1457.0	855.0		
1907	604	6.20	636	347.0	3945.0	2151.4		
1907	624	6.60	1490	772.0	9831.0	5095.2		
1907	522	7.00	87	56.0	609.0	392.0		

Footnote entries on page 503.

Hyperpolarizability tabulations. Shaded entries are calculated from values given in the cited reference.

	MOLECULE NAME, OR STRUCTURE	tech	solvent
82	n=1 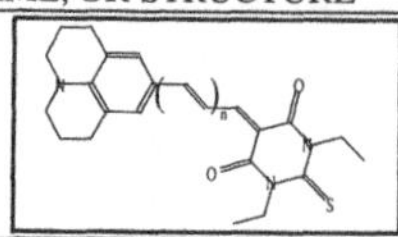	EFISH	chloroform
82	n=2 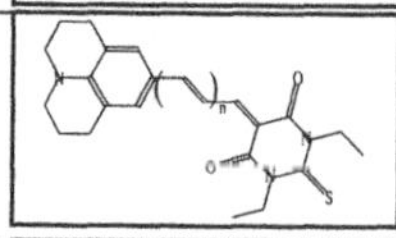	EFISH	chloroform
82	n=3 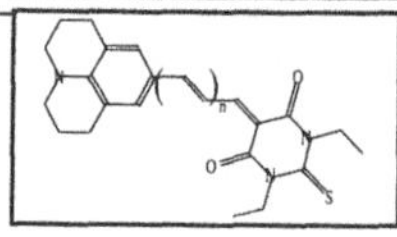	EFISH	chloroform
82	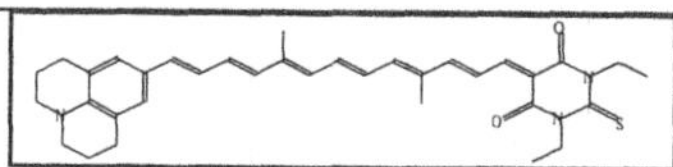	EFISH	chloroform
82	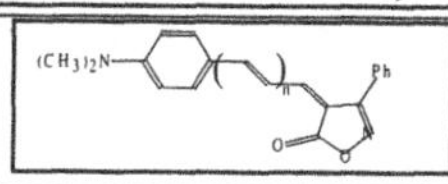	EFISH	chloroform
82	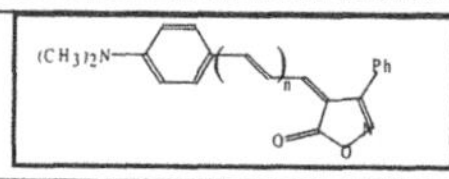	EFISH	chloroform
82	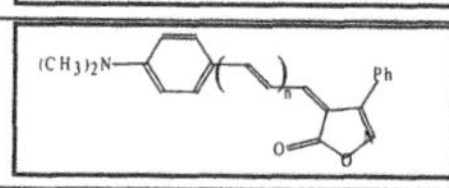	EFISH	chloroform
82	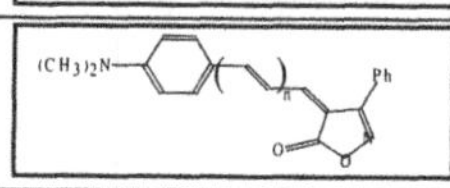	EFISH	chloroform
82	n=0 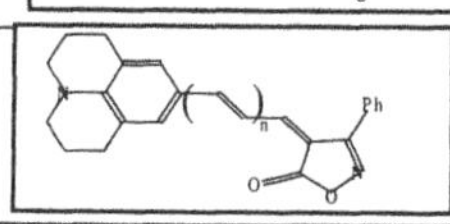	EFISH	chloroform

λ_{meas}	λ_{max}	μ	β	β_0	$\mu\beta$	$\mu\beta_0$	footnote	convention
1907	614	6.60	355	186.0	2210.0	1227.6		
1907	680	6.30	1141	490.0	7152.0	3087.0		
1907	686	8.80	2169	911.0	19086.0	8016.8		
1907	680				34770.0	14913.5		
1907	478	8.30	37	27.0	312.0	224.1		
1907	530	8.60	140	90.0	1202.0	774.0		
1907	562	8.70	362	218.0	3156.0	1896.6		
1907	582	8.90	918	528.0	8171.0	4699.2		
1907	504	9.50	51	34.0	488.0	323.0		

Footnote entries on page 503.

Hyperpolarizability tabulations. Shaded entries are calculated from values given in the cited reference.

	MOLECULE NAME, OR STRUCTURE	tech	solvent
82	n=1 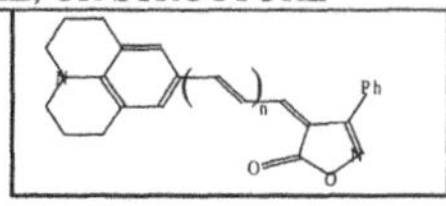	EFISH	chloroform
82	n=2 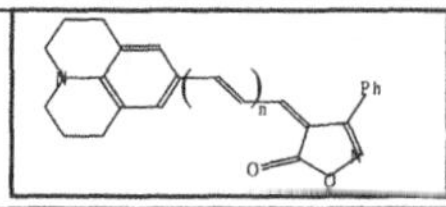	EFISH	chloroform
82	n=3 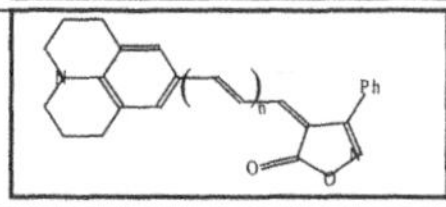	EFISH	chloroform
82	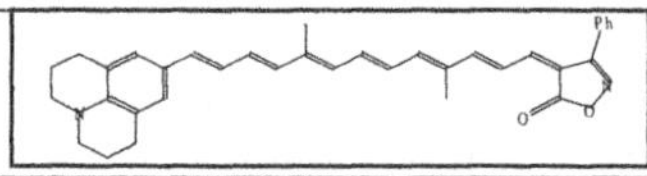	EFISH	chloroform
82	n=1 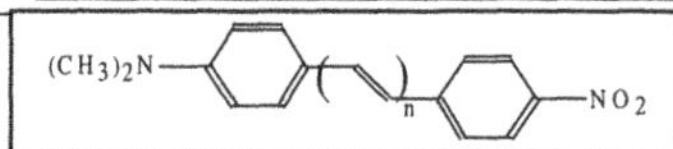	EFISH	chloroform
82	n=2 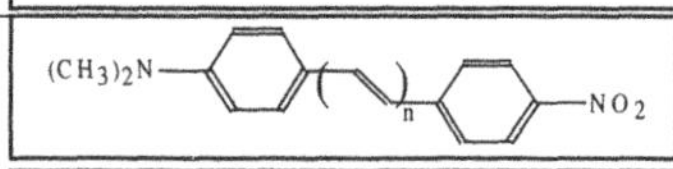	EFISH	chloroform
82	n=3 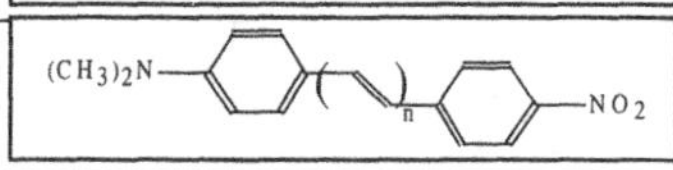	EFISH	chloroform
82	n=4 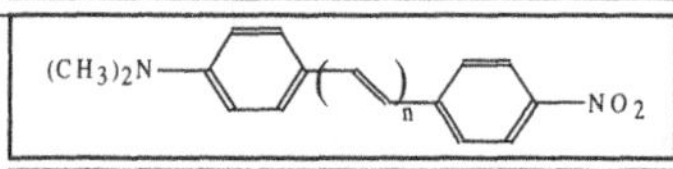	EFISH	chloroform
82	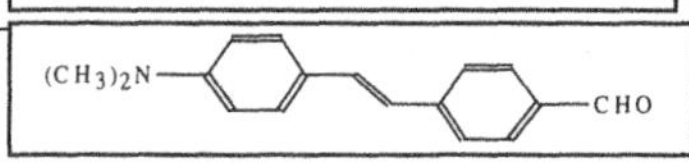	EFISH	chloroform
83	D A n Donor-Acceptor Polyenes: D(CH=CH)nA		
83	Me_2N CHO 0	EFISH	chloroform
83	Me_2N CHO 1	EFISH	chloroform
83	Et_2N CHO 2	EFISH	chloroform

λ_{meas}	λ_{max}	μ	β	β_0	$\mu\beta$	$\mu\beta_0$	footnote	convention
1907	586	9.10	180	100.0	1638.0	910.0		
1907	620	9.00	656	339.0	5812.0	3051.0		
1907	640	9.80	995	485.0	9750.0	4753.0		
1907	647	16.00	1781	849.0	28500.0	13584.0		
1907	430	6.60	73	55.0	482.0	363.0		
1907	442	7.60	107	80.0	813.0	608.0		
1907	458	8.20	131	95.0	1074.0	779.0		
1907	464	9.00	190	133.0	1700.0	1197.0		
1907	360	3.50	24	20.0	84.0	70.0		
1907	240	3.50	0.5	0.5	1.8	1.8		
1907	284	6.30	3.3	2.9	20.8	18.3		
1907	363	6.50	20	16.5	130.0	107.3		

Footnote entries on page 503.

Hyperpolarizability tabulations. Shaded entries are calculated from values given in the cited reference.

	MOLECULE NAME, OR STRUCTURE			tech	solvent
83	Me_2N CHO 3			EFISH	chloroform
83	Me_2N $CH{=}C(CN)_2$ 0			EFISH	chloroform
83	Me_2N $CH{=}C(CN)_2$ 1			EFISH	chloroform
83	Et_2N $CH{=}C(CN)_2$ 2			EFISH	chloroform
83	Me_2N $CH{=}C(CN)_2$ 3			EFISH	chloroform
84	D=c-$(CH_2)_6$N A=NO_2	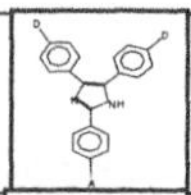		EFISH	chloroform
84	D=c-$(CH_2)_5$N NO_2	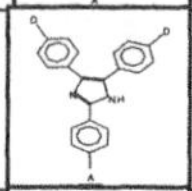		EFISH	chloroform
84	D= $CH_3OPhC{\equiv}C$— A=NO_2	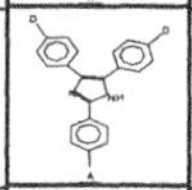		EFISH	p-dioxane
84	D=CH_3O A= $O_2NPhC{\equiv}C$—	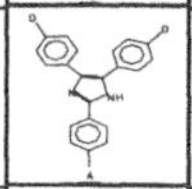		EFISH	p-dioxane
84	D=CH_3OPhS A=NO_2	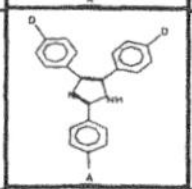		EFISH	p-dioxane
84	D=$C_4H_9CH(C_2H_5)CH_2O$ A=NO_2	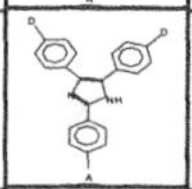		EFISH	chloroform
84	D=CH_3O A=NO_2	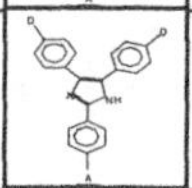		EFISH	p-dioxane
84	D=CH_3O A=NO_2	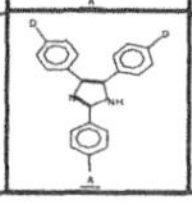		EFISH	chloroform

λ_{meas}	λ_{max}	μ	β	β_0	$\mu\beta$	$\mu\beta_0$	footnote	convention
1907	422	6.90	53	41.0	365.7	282.9		
1907	352	7.60	1	0.9	7.6	6.8		
1907	374	8.90	6.1	5.0	54.3	44.5		
1907	476	10.70	45	32.0	481.5	342.4		
1907	550	9.90	211	129.0	2088.9	1277.1		
1907	467	8.30	78.7	55.4	653.2	459.8		
1907	438	7.20	45.5	34.0	327.6	244.8		
1064	344	8.00	53.2	27.7	425.6	221.6		
1064	400	8.10	69.1	25.8	559.7	209.0		
1064	398	6.60	55.9	21.2	368.9	139.9		
1907	416	6.30	24.5	18.9	154.4	119.1		
1064	410	7.00	52.3	18.1	366.1	126.7		
1907	412	6.40	19.9	15.4	127.4	98.6		

Footnote entries on page 503.

Hyperpolarizability tabulations. Shaded entries are calculated from values given in the cited reference.

	MOLECULE NAME, OR STRUCTURE	tech	solvent
84	D=$C_4H_9CH(C_2H_5)CH_2O$ A=$SO_2C_4F_9$ 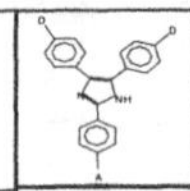	EFISH	chloroform
84	D=CH_3O A=SO_2Ph 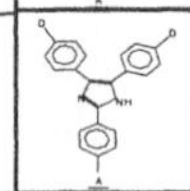	EFISH	chloroform
84	D=CH_3O A=NO_2 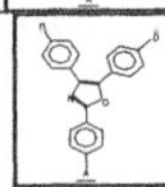	EFISH	p-dioxane
84	D=CH_3O A=NO_2 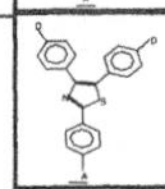	EFISH	p-dioxane
84	D=CH_3O A= $O_2NPhC{\equiv}C$— 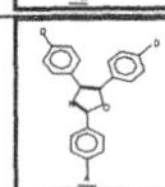	EFISH	p-dioxane
84	D=CH_3O A=SO_2Ph 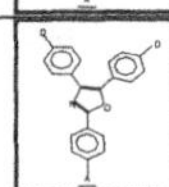	EFISH	p-dioxane
84	D=$C_4H_9CH(C_2H_5)CH_2O$ A=NO_2 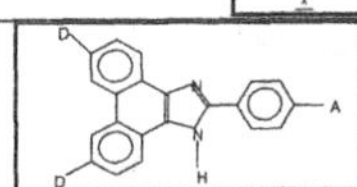	EFISH	chloroform
84	D=$C_4H_9CH(C_2H5)CH_2O$ A=$SO_2C_4F_9$ 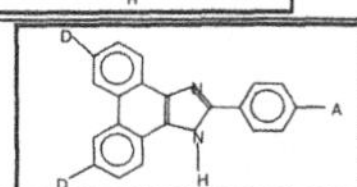	EFISH	chloroform
84	D=$C_4H_9CH(C_2H_5)CH_2O$ A=NO_2 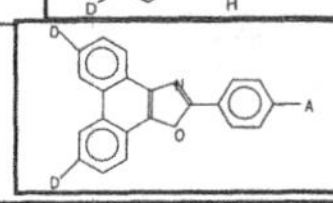	EFISH	chloroform

λ_{meas}	λ_{max}	μ	β	β_0	$\mu\beta$	$\mu\beta_0$	footnote	convention
1907	384	6.50	13.3	11.2	86.5	72.8		
1907	362	8.00	10.1	4.8	80.8	38.4		
1064	390	6.30	46.35	18.1	292.0	114.0		
1064	400	8.10	48.15	21.1	390.0	170.9		
1064	378	7.10	61.83	27.6	439.0	196.0		
1064	358	7.00	18.29	8.3	128.0	58.1		
1907	436	5.50	60.18	25.5	331.0	140.3		
1907	416	5.90	54.75	23.6	323.0	139.2		
1907	330	5.20	52.31	22.4	272.0	116.5		

Footnote entries on page 503.

Hyperpolarizability tabulations. Shaded entries are calculated from values given in the cited reference.

	MOLECULE NAME, OR STRUCTURE	tech	solvent
84	D= c-$(CH_2)_5$N A=NO_2 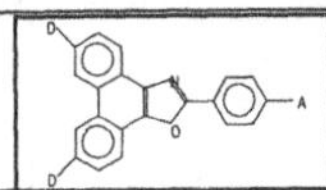	EFISH	chloroform
85	A= $SO_2C_4F_9$ D=MeOPh 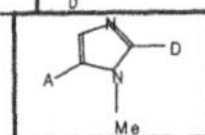	EFISH	p-dioxane
85	A=$SO_2C_4F_9$ D= MeOPhC≡C— 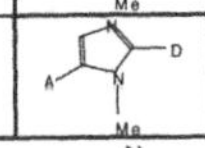	EFISH	p-dioxane
85	A=$SO_2C_4F_9$ D=MeOPh 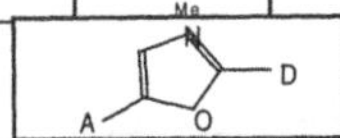	EFISH	p-dixoane
85	A=$SO_2C_4F_9$ D= MeOPhC≡C— 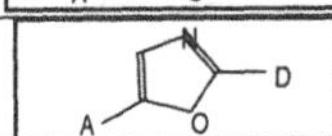	EFISH	p-dioxane
85	A=$SO_2C_4F_9$ D=MeOPh 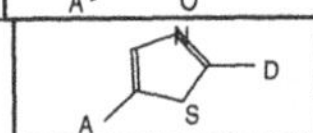	EFISH	p-dioxane
85	A=$SO_2C_4F_9$ D= MeOPhC≡C— 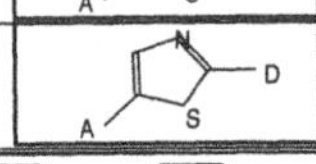	EFISH	p-dioxane
85	A=$SO_2C_4F_9$ D=MeO 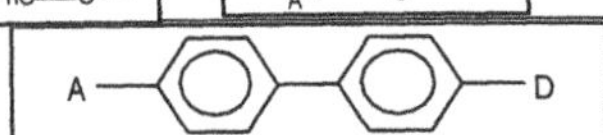	EFISH	p-dioxane
85	A=$SO_2C_4F_9$ D=MeO 	EFISH	p-dioxane
85	A=NO_2 D= 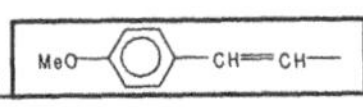 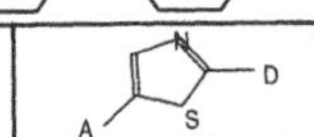	EFISH	p-dioxane
85	A=NO_2 D= 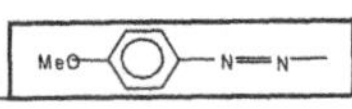 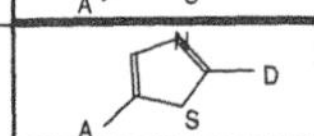	EFISH	p-dioxane
85	A=NO_2 D= 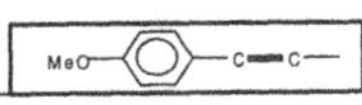 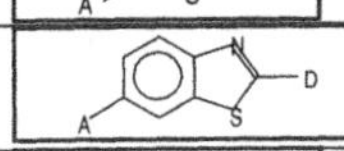	EFISH	p-dioxane
85	A=NO_2 D= 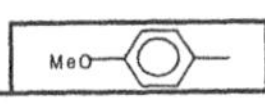 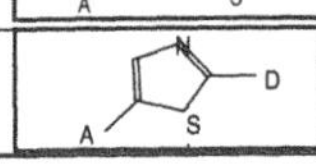	EFISH	p-dioxane

λ_{meas}	λ_{max}	μ	β	β_0	$\mu\beta$	$\mu\beta_0$	footnote	convention
1907	462	6.70	96.27	39.7	645.0	266.0		
1064	282	6.00	2.7	1.8	16.2	10.8		
1064	312	7.60	9.9	5.9	75.2	44.8		
1064	312	5.10	11.8	7.1	60.2	36.2		
1064	328	6.20	21.4	12.0	132.7	74.4		
1064	338	6.00	15	8.0	90.0	48.0		
1064	358	7.40	26.9	13.1	199.1	96.9		
1064	310	6.30	13.6	8.2	85.7	51.7		
1064	334	6.80	19.6	10.7	133.3	72.8		
1064	406	6.00	67.7	24.2	406.2	145.2		
1907	424	6.20	29.7	22.7	184.1	140.7		
1064	362	6.00	46.9	22.3	281.4	133.8		
1064	378	5.40	41.2	17.8	222.5	96.1		

Footnote entries on page 503.

Hyperpolarizability tabulations. Shaded entries are calculated from values given in the cited reference.

	MOLECULE NAME, OR STRUCTURE	tech	solvent
85	A=NO_2 D= 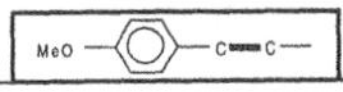 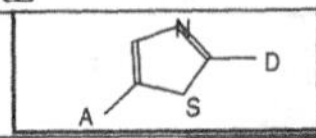	EFISH	p-dioxane
85	A=$SO_2C_4F_9$ D= 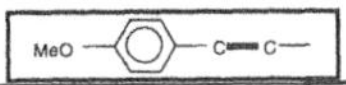 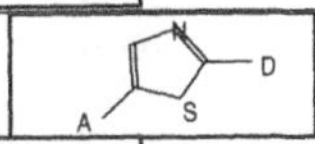	EFISH	p-dioxane
85	A=NO_2 D= 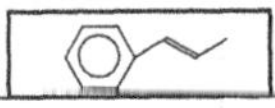 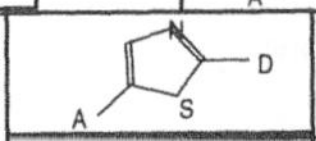	EFISH	p-dioxane
85	A=$SO_2C_4F_9$ D= 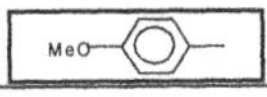 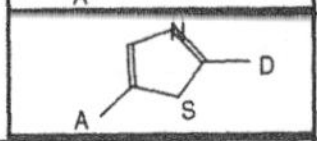	EFISH	p-dioxane
85	A=SO_2Me D= 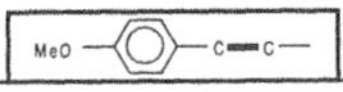 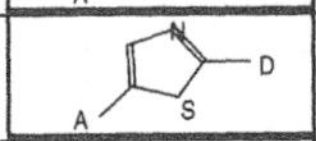	EFISH	p-dioxane
85	A=NO_2 D= 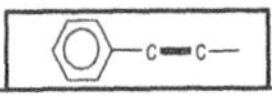 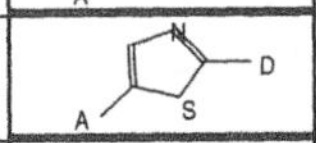	EFISH	p-dioxane
85	A=NO_2 D= 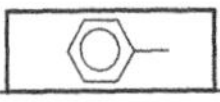 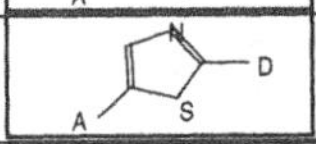	EFISH	p-dioxane
85	D= 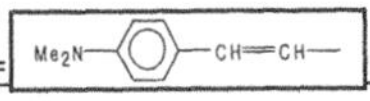 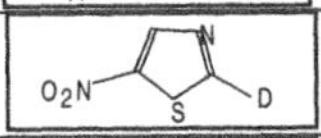	EFISH	p-dioxane
85	D= 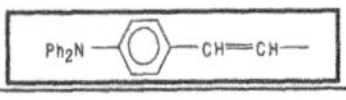 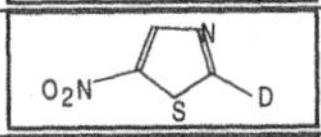	EFISH	p-dioxane
85	D= 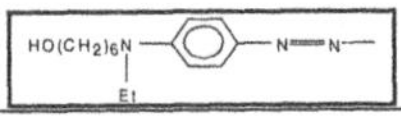 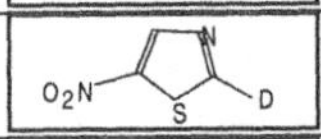	EFISH	p-dioxane
85	D= 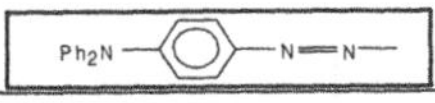 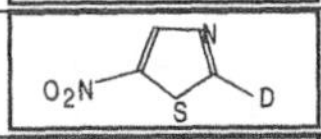	EFISH	p-dioxane
85	D= 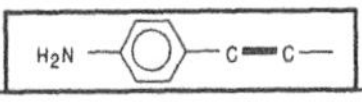 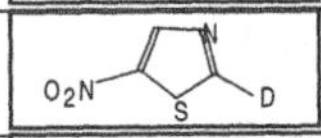	EFISH	p-dioxane
85	D= 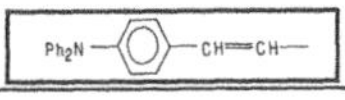 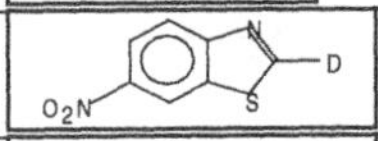	EFISH	p-dioxane
85	D= 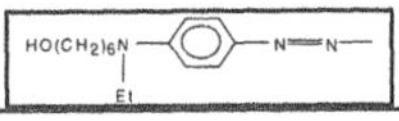 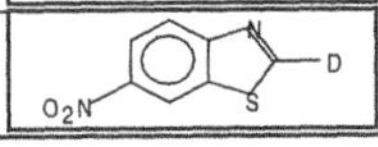	EFISH	p-dioxane

λ_{meas}	λ_{max}	μ	β	β_0	$\mu\beta$	$\mu\beta_0$	footnote	convention
1064	382	5.90	33.7	14.2	198.8	83.8		
1064	358	7.40	26.9	13.1	199.1	96.9		
1064	372	4.90	26	11.7	127.4	57.3		
1064	338	6.00	15	8.0	90.0	48.0		
1064	334	6.00	13.2	7.2	79.2	43.2		
1064	354	5.50	10.9	5.4	60.0	29.7		
1064	346	4.10	8.1	4.2	33.2	17.2		
1907	478	8.20	93.8	65.8	769.2	539.6		
1907	492	5.20	83	56.8	431.6	295.4		
1907	582	9.00	91.7	52.2	825.3	469.8		
1907	582	6.90	120	68.2	828.0	470.6		
1064	426	7.10	71.9	21.7	510.5	154.1		
1907	458	5.50	78.8	57.2	433.4	314.6		
1907	548	9.50	97.8	60.1	929.1	571.0		

Footnote entries on page 503.

Hyperpolarizability tabulations. Shaded entries are calculated from values given in the cited reference.

	MOLECULE NAME, OR STRUCTURE	tech	solvent
85	D= 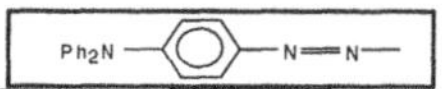 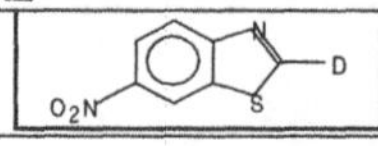	EFISH	p-dioxane
86	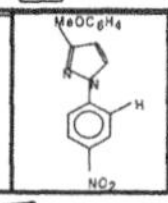	EFISH	p-dioxane
86	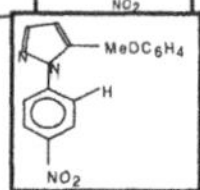	EFISH	p-dioxane
86	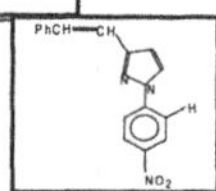	EFISH	p-dixoane
86	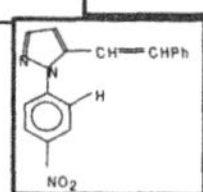	EFISH	p-dioxane
86	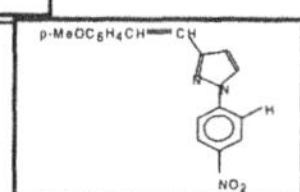	EFISH	p-dioxane
86	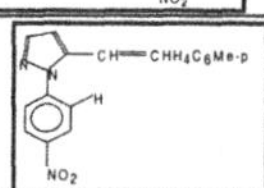	EFISH	p-dioxane
86	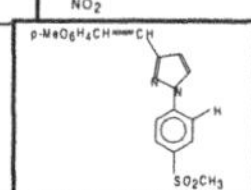	EFISH	p-dioxane
86	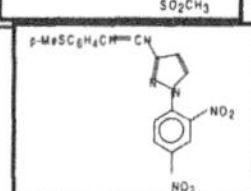	EFISH	p-dioxane
86	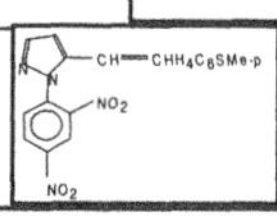	EFISH	p-dioxane

λ_{meas}	λ_{max}	μ	β	β_0	$\mu\beta$	$\mu\beta_0$	footnote	convention
1907	550	7.20	117	71.8	842.4	517.0		
1064	354	6.60	31.4	15.6	207.2	103.0		
1064	306	6.30	12.7	7.8	80.0	49.1		
1064	346	6.10	32.2	16.6	196.4	101.3		
1064	298	4.60	7.3	4.6	33.6	21.2		
1064	368	5.70	40.1	18.4	228.6	104.9		
1064	302	5.50	13.6	8.5	74.8	46.8		
1064	330	5.60	15.6	8.7	87.4	48.7		
1064	354	4.10	41.6	20.6	170.6	84.5		
1064	312	3.30	11.5	6.9	38.0	22.8		

Footnote entries on page 503.

Hyperpolarizability tabulations. Shaded entries are calculated from values given in the cited reference.

	MOLECULE NAME, OR STRUCTURE	tech	solvent
86	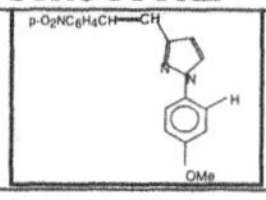	EFISH	p-dioxane
86	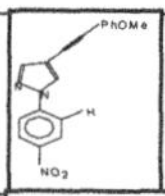	EFISH	p-dioxane
86	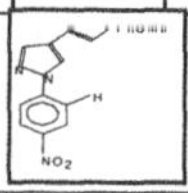	EFISH	p-dioxane
87	methanol	HRS	
87	CH_3NO_2	HRS	methanol
87	MNA	HRS	methanol
87	4-hydroxystilbazolium p toluenesulfonate	HRS	methanol
87	Cyclopentadienyl-bistriphenylphosphine-ruthenium(II)-(mu-cyano)-pentammine-ruthenium(III) triflate	HRS	methanol
87	4-hydroxy-4'-nitrostilbene	HRS	methanol
88	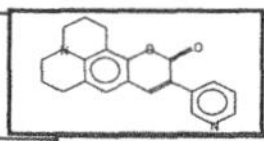	EFISH	chloroform
88	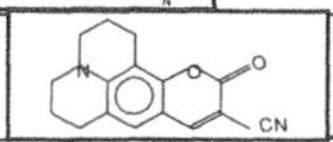	EFISH	chloroform
88	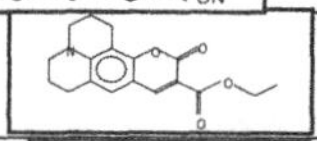	EFISH	chloroform
88	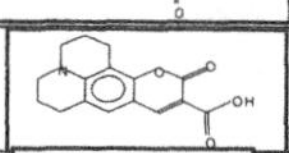	EFISH	chloroform
88	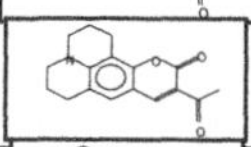	EFISH	chloroform
88	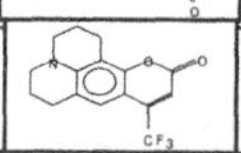	EFISH	chloroform
88	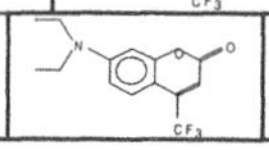	EFISH	chloroform

λ_{meas}	λ_{max}	μ	β	β_0	$\mu\beta$	$\mu\beta_0$	footnote	convention
1064	370	7.00	48.2	21.9	337.4	153.3		
1064	340	5.60	20	10.6	112.0	59.4		
1064	358	5.10	63.3	30.7	322.8	156.6		
1064			0.14					
1064			0.3					
1064			40					
1064			70					
1064			78					
1064			120					
1064	430	7.44	114	32.9	848.2	244.8		
1064	446	10.20	132	32.4	1346.4	330.5		
1064	436	8.20	106	29.0	869.2	237.8		
1064	448	9.86	120	28.6	1183.2	282.0		
1064	452	7.56	111	25.4	839.2	192.0		
1064	418	6.55	49.1	15.8	321.6	103.5		
1064	400	6.07	41.8	15.7	253.7	95.3		

Footnote entries on page 503.

Hyperpolarizability tabulations. Shaded entries are calculated from values given in the cited reference.

	MOLECULE NAME, OR STRUCTURE	tech	solvent
88	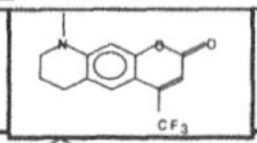	EFISH	chloroform
88	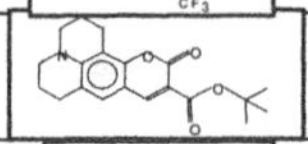	EFISH	chloroform
88	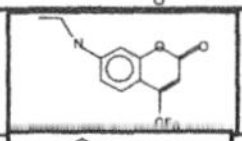	EFISH	chloroform
88	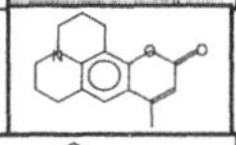	EFISH	chloroform
88	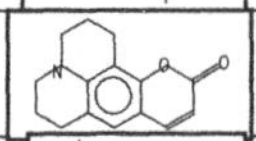	EFISH	chloroform
88	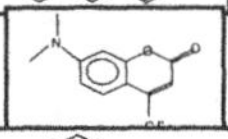	EFISH	chloroform
88	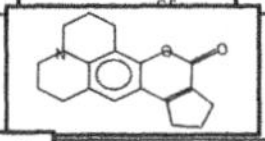	EFISH	chloroform
88	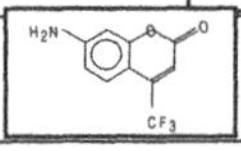	EFISH	chloroform
89	fluorinated 4'-methoxystilbenes (derivative given) 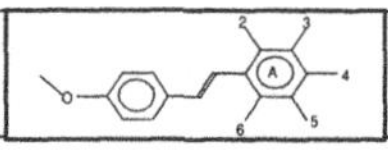		
89	perfluorophenyl-4-vinyl	EFISH	
89	perfluorohexyl-4-vinyl	EFISH	
89	4-trifluoromethyl tetrafluoro	EFISH	
89	4-trifluoromethyl	EFISH	
89	pentafluoro	EFISH	
89	3,5-di-trifluoromethyl	EFISH	
89	2,3,4,5-tetrafluoro	EFISH	
89	2,3,5-trifluoro	EFISH	
89	3,4,5-trifluoro	EFISH	
89	3,5-difluoro	EFISH	
89	parent compound	EFISH	
89	thiazoles D / A 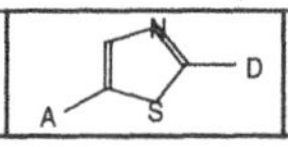		

λ_{meas}	λ_{max}	μ	β	β_0	$\mu\beta$	$\mu\beta_0$	footnote	convention
1064	406	6.14	42.8	15.3	262.8	93.9		
1064	434	8.12	49.5	13.9	401.9	112.9		
1064	378	5.59	29	12.6	162.1	70.4		
1064	388	6.98	30.6	12.4	213.6	86.6		
1064	392	6.76	27.9	11.0	188.6	74.4		
1064	392	5.71	27	10.6	154.2	60.5		
1064	382	6.39	22.1	9.4	141.2	60.1		
1064	358	4.59	18.7	9.0	85.8	41.3		
1064	366	3.90	36	16.7	140.4	65.1		
1064	346	4.10	26	13.4	106.6	54.9		
1064	334	4.00	23	12.6	92.0	50.4		
1064	326	4.20	16.4	9.3	68.9	39.1		
1064	322	3.30	15.7	9.0	51.8	29.7		
1064	326	4.20	10.6	6.0	44.5	25.2		
1064	324	3.50	11.8	6.7	41.3	23.5		
1064	322	3.40	10.4	6.0	35.4	20.4		
1064	320	4.60	6.8	3.9	31.3	17.9		
1064	322	3.10	9.2	5.3	28.5	16.4		
1064	320	2.80	6.1	3.5	17.1	9.8		

Footnote entries on page 503.

Hyperpolarizability tabulations. Shaded entries are calculated from values given in the cited reference.

	MOLECULE NAME, OR STRUCTURE	tech	solvent
89	$H_3COPhCC$ / Ph(fused)NO_2	EFISH	
89	H_2NPhCC / NO_2	EFISH	
89	$H_3COPhCH$=CH / NO_2	EFISH	
89	$H_3COPhCC$ / $SO_2C_4F_9$	EFISH	
89	$H_3COPhCC$ / NO_2	EFISH	
89	Pyrazoles Donor (position) / Acceptor (position) 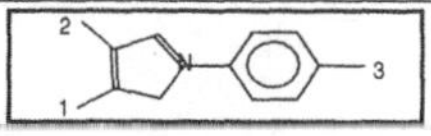		
89	H_9C_4OPhCH=CH (2) / $PhNO_2$ (3)	EFISH	
89	H_3COPh (3) / CH=$CHPhNO_2$ (1)	EFISH	
89	PhCH=CH (1) / $PhNO_2$(3)	EFISH	
89	H_3COPh (1) / $PhNO_2$ (3)	EFISH	
89	$H_3COPhCH$=CH (1) / $PhNO_2$ (3)	EFISH	
90	D/A 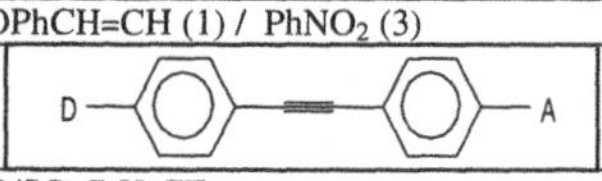		
90	$CH_3O/SO_2C_6H_4CF_3$	EFISH	p-dioxane
90	$CH_3O/SO_2C_6F1_3$	EFISH	p-dioxane
90	$CH_3O/SO_2C_6H_5$	EFISH	chloroform
90	$CH_3O/SO_2C_6H_5$	EFISH	p-dioxane
90	CH_3O/SO_2CH_3	EFISH	p-dioxane
90	CH_3/NO_2	EFISH	p-dioxane
90	CH_3O/CF_3	EFISH	p-dioxane
90	D=$(CH_3)_2N$ A=NO_2 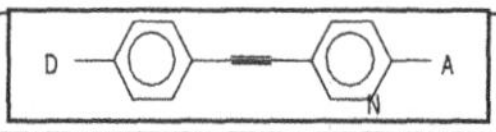	EFISH	p-dioxane
90	D=$(CH_3)_2N$ A=NO_2 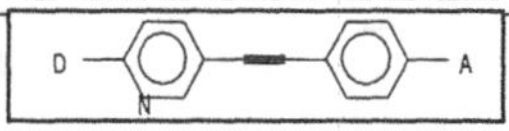	EFISH	p-dioxane
90	D=$(CH_3)_2N$ A=NO_2 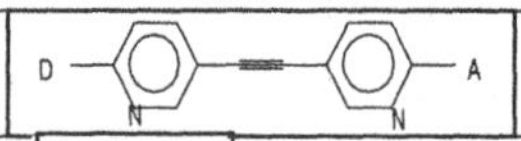	EFISH	chloroform
91	D / A 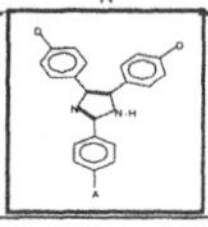		
91	c-$(CH_2)_6N$ / NO_2	EFISH	chloroform

λ_{meas}	λ_{max}	μ	β	β_0	$\mu\beta$	$\mu\beta_0$	footnote	convention
1064	362	6.00	46.9	22.3	281.4	133.8		
1064	426	7.10	71.9	21.7	510.5	154.1		
1064	406	6.00	67.7	24.2	406.2	145.2		
1064	358	7.40	26.9	13.1	199.1	96.9		
1064	382	5.90	33.7	14.2	198.8	83.8		
1064	358	5.10	63.3	30.7	322.8	156.6		
1064	370	7.00	48.2	21.9	337.4	153.3		
1064	346	6.10	32.2	16.6	196.4	101.3		
1064	354	6.60	31.4	15.6	207.2	103.0		
1064	368	5.70	40.1	18.4	228.6	104.9		
								I
1064	330	5.20	29.9	16.6	155.5	86.3		I
1064	334	6.80	19.6	10.7	133.3	72.8		I
1064	322	5.60	18.2	10.5	101.9	58.8		I
1064	330	4.70	20.6	11.4	96.8	53.6		I
1064	310	5.90	10.8	6.5	63.7	38.4		I
1064	334	5.80	12.3	6.7	71.3	38.9		I
1064	316	3.90	7.6	4.5	29.6	17.6		I
1064	418	7.20	101	32.8	727.2	236.2		I
1064	388	7.20	64.9	26.3	467.3	189.4		I
1064	416	6.10	88	29.0	536.8	176.9		I
1907	476	8.30	78.7	55.4	653.2	459.8		

Footnote entries on page 503.

Hyperpolarizability tabulations. Shaded entries are calculated from values given in the cited reference.

	MOLECULE NAME, OR STRUCTURE	tech	solvent
91	$CH_3OPhC \equiv C$— / NO2	EFISH	p-dioxane
91	c-$(CH_2)_5N$ / NO_2	EFISH	chloroform
91	CH_3O / $O_2NPhC \equiv C$—	EFISH	p-dioxane
91	CH_3OPhS / NO_2	EFISH	p-dioxane
91	CH_3O / NO_2	EFISH	p-dioxane
91	$C_4H_9CH(C_2H_5)CH_2O$ / NO_2	EFISH	chloroform
91	CH_3O / NO_2	EFISH	chloroform
91	$C_4H_9CH(C_2H_5)CH_2O$ / $SO_2C_4F_9$	EFISH	chloroform
91	CH_3O / SO_2Ph	EFISH	p-dioxane
91	D / A 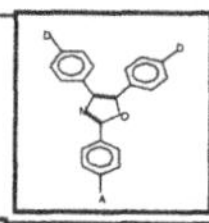		
91	CH_3O / $O_2NPhC \equiv C$—	EFISH	p-dioxane
91	CH_3O / SO_2PhCF_3	EFISH	chloroform
91	CH_3O/NO_2	EFISH	p-dioxane
91	CH_3O / $SO_2C_4F_9$	EFISH	p-dioxane
91	CH_3O/ SO_2Ph	EFISH	p-dioxane
91	CH_3O/SO_2CH_3	EFISH	p-dioxane
91	CH_3O/C_6F_{13}	EFISH	p-dioxane
91	CH_3O/CF_3	EFISH	p-dioxane
91	D / A 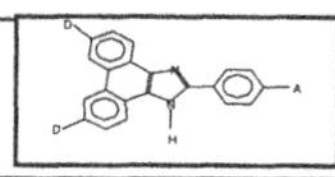		
91	$C_4H_9CH(C_2H_5)CH_2O$ / NO_2	EFISH	chloroform
91	$C_4H_9CH(C_2H_5)CH_2O$ / $SO_2C_4F_9$	EFISH	chloroform
92	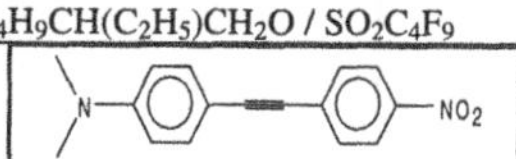	EFISH	chloroform
92	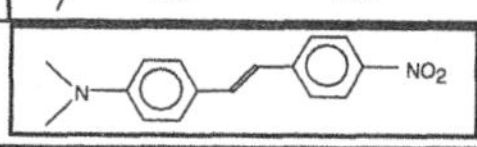	EFISH	chloroform

λ_{meas}	λ_{max}	μ	β	β_0	$\mu\beta$	$\mu\beta_0$	footnote	convention
1064	344	8.00	53.2	27.7	425.6	221.6		
1907	438	7.20	45.5	34.0	327.6	244.8		
1064	400	8.10	69.1	25.8	559.7	209.0		
1064	398	6.60	55.9	21.2	368.9	139.9		
1064	410	7.00	52.3	18.1	366.1	126.7		
1907	416	6.30	24.5	18.9	154.4	119.1		
1907	412	6.40	19.9	15.4	127.4	98.6		
1907	384	6.50	13.9	11.2	90.4	72.8		
1064	362	8.00	10.1	4.8	80.8	38.4		
1064	378	7.10	63.9	27.6	453.7	196.0		
1064	370	8.00	31.5	14.3	252.0	114.4		
1064	390	6.30	47.3	18.9	298.0	119.1		
1064	378	7.90	31.1	13.4	245.7	105.9		
1064	358	7.00	17.1	8.3	119.7	58.1		
1064	352	7.00	15.6	7.8	109.2	54.6		
1064	344	5.70	13.7	7.2	78.1	41.0		
1064	338	5.30	9.2	4.2	48.8	22.3		
1907	436	5.50	34.02	25.5	187.1	140.3		
1907	416	5.90	30.60	23.6	180.6	139.2		
1907	416	6.58	42.5	32.8	279.7	215.8		I
1907	438	6.69	56.4	42.1	377.3	281.6		I

Footnote entries on page 503.

Hyperpolarizability tabulations. Shaded entries are calculated from values given in the cited reference.

	MOLECULE NAME, OR STRUCTURE	tech	solvent
92	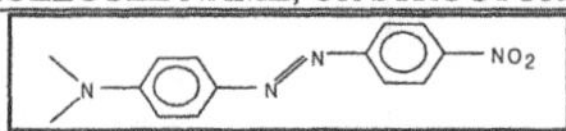	EFISH	chloroform
92	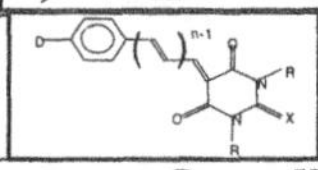		
92	D n R X		
92	Et_2N 1 Me O	EFISH	chloroform
92	Me_2N 1 Et S	EFISH	chloroform
92	Et_2N 1 Et S	EFISH	chloroform
92	Me_2N 1 Et S	EFISH	chloroform
92	MeO 2 Et S	EFISH	chloroform
92	Me_2N 2 Me O	EFISH	chloroform
92	Me_2N 2 Et S	EFISH	chloroform
93	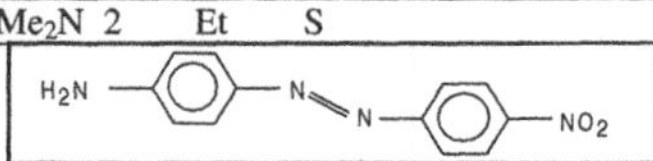	EFISH	chloroform
93	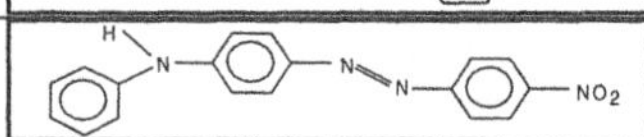	EFISH	chloroform
93	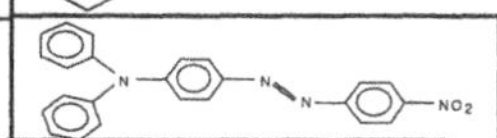	EFISH	chloroform
93	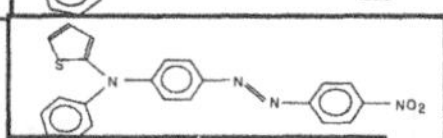	EFISH	chloroform
93	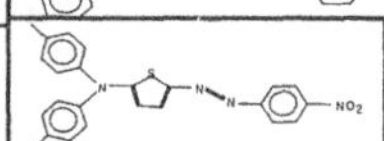	EFISH	chloroform
93	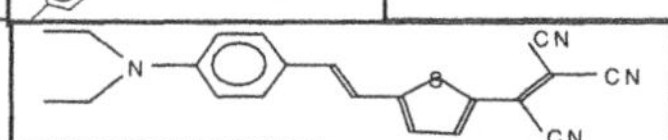	EFISH	chloroform
93	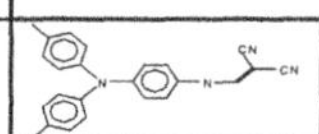	EFISH	chloroform
93	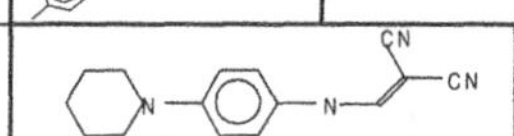	EFISH	chloroform
93	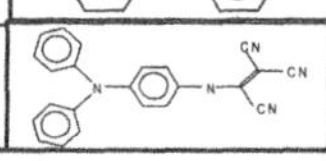	EFISH	chloroform

λ_{meas}	λ_{max}	μ	β	β_0	$\mu\beta$	$\mu\beta_0$	footnote	convention
1907	480	7.66	57	39.9	436.6	305.6		I
								I
								I
1907	470	6.21	27.8	19.8	172.6	123.0		I
1907	494	6.40	29.2	19.9	186.9	127.4		I
1907	500	7.10	30	20.3	213.0	144.1		I
1907	498	6.22	30.8	20.9	191.6	130.0	c	I
1907	454	3.54	47.4	34.6	167.8	122.5		I
1907	534	6.23	104	65.8	647.9	409.9		I
1907	570	7.18	114	66.6	818.5	478.2		I
1907	410	6.60	25.76	52.3	170.0	345.2	e	
1907	452	6.00	26.67	68.5	160.0	411.0	e	
1907	486	5.90	34.07	105.0	201.0	619.5	e	
1907	468	5.50	23.09	72.4	127.0	398.2	e	
1907	562	8.70	32.41	97.9	282.0	851.7	e	
1907	698	10.40	179.81	361.0	1870.0	751.6	e	
1907	452	7.60	172.37	48.6	1310.0	369.4	e	
1907	440	8.50	16.35	32.0	139.0	272.0	e	
1907	534	7.80	119.36	75.5	931.0	588.9	e	

Footnote entries on page 503.

Hyperpolarizability tabulations. Shaded entries are calculated from values given in the cited reference.

	MOLECULE NAME, OR STRUCTURE	tech	solvent
93	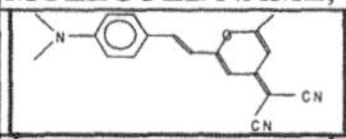	EFISH	choroform
93	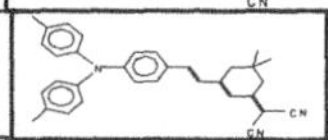	EFISH	chloroform
93	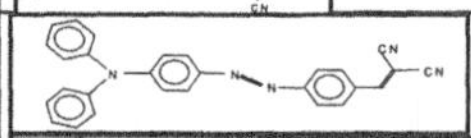	EFISH	chloroform
93	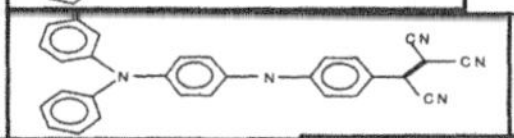	EFISH	chloroform
94	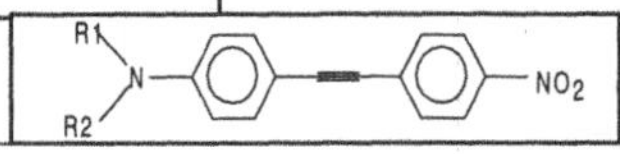		
94	R1=R2=Me	EFISH	chloroform
94	R1=R2=Ph	EFISH	chloroform
94	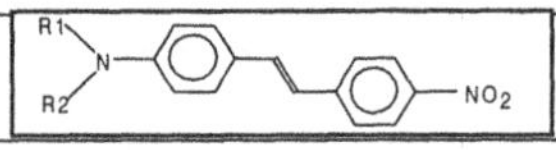		
94	R1=R2=Me	EFISH	chloroform
94	R1=R2=Ph	EFISH	chloroform
94	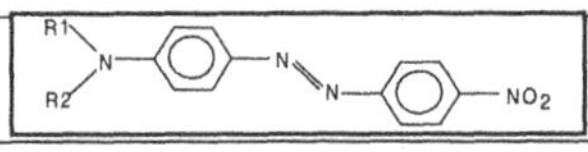		
94	R1=R2=Me	EFISH	chloroform
94	R1=R2=Et	EFISH	chloroform
94	R1=R2=Ph	EFISH	chloroform
94	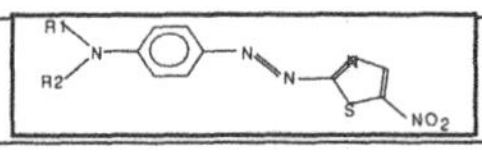		
94	R1=Et R2=$HO(CH_2)_6$	EFISH	chloroform
94	R1=R2=Ph	EFISH	chloroform
94	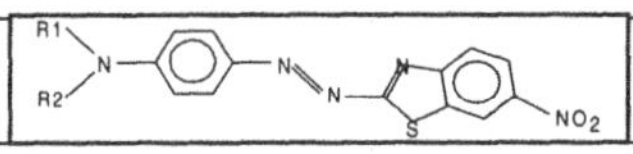		
94	R1=Et R2=$HO(CH_2)_6$	EFISH	chloroform
94	R1=R2=Ph	EFISH	chloroform
94	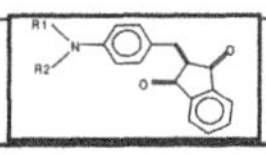		

λ_{meas}	λ_{max}	μ	β	β_0	$\mu\beta$	$\mu\beta_0$	footnote	convention
1907	468	10.20	26.37	63.3	269.0	645.7	e	
1907	518	7.80	35.00	123.0	273.0	959.4	e	
1907	526	6.70	47.76	149.0	320.0	998.3	e	
1907	602	7.10	86.90	259.0	617.0	334.1	e	
1907	416	6.58	75.5	32.8	497.0	215.8	d	
1907	418	4.84	65.1	3.2	315.0	15.2	d	
1907	436	6.69	99.0	42.1	662.0	281.6	d	
1907	436	4.75	87.6	37.3	416.0	177.2	d	
1907	480	7.66	98.0	39.9	751.0	305.6	d	
1907	494	8.04	123.9	49.6	996.0	398.8	d	
1907	486	5.87	134.2	54.3	788.0	318.7	d	
1907	582	8.96	146.2	52.2	1310.0	467.7	d	
1907	582	6.89	191.6	68.2	1320.0	469.9	d	
1907	548	9.47	160.5	60.1	1520.0	569.1	d	
1907	550	7.21	192.8	71.8	1390.0	517.7	d	

Footnote entries on page 503.

Hyperpolarizability tabulations. Shaded entries are calculated from values given in the cited reference.

	MOLECULE NAME, OR STRUCTURE	tech	solvent
94	R1=R2=Et	EFISH	chloroform
94	R1=R2=4-MeC_6H_4	EFISH	chloroform
94	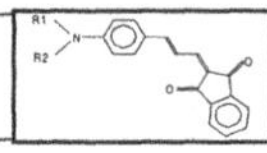		
94	R1=R2=Et	EFISH	chloroform
94	R1=R2=4-MeC_6H_4	EFISH	chloroform
94	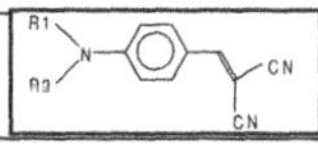		
94	R1=R2=4-MeC_6H_4	EFISH	chloroform
94	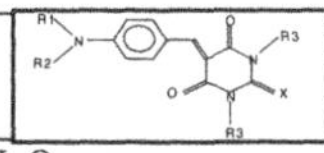		
94	R1=R2=Et R3=Me X=O	EFISH	chloroform
94	R1=R2=4-MeC_6H_4 R3=Me X=O	EFISH	chloroform
94	R1=R2=Me R3=Et X=S	EFISH	chloroform
94	R1=R2=Et R3=Et X=S	EFISH	chloroform
94	R1=R2=4-MeC_6H_4 R3=Et X=S	EFISH	chloroform
94	R1=R2=Et R3=Ph X=S	EFISH	chloroform
94	R1=R2=4-MeC_6H_4 R3=Ph X=S	EFISH	chloroform
94	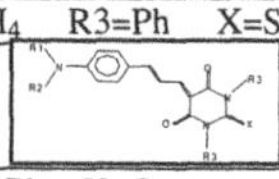		
94	R1=R2=Et R3=Ph X=O	EFISH	chloroform
94	R1=R2=4-MeC_6H_4 R3=Ph X=O	EFISH	chloroform
94	R1=R2=Et R3=Ph X=S	EFISH	chloroform
94	R1=R2=4-MeC_6H_4	EFISH	chloroform
95	fluorinated 4-methoxystilbenes (derivative given)		
95	2,3,4,5,6-F	EFISH	p-dioxane
95	2,3,4,5-F	EFISH	p-dioxane
95	2,3,5-F	EFISH	p-dioxane
95	3,4,5-F	EFISH	p-dioxane
95	3,5-F	EFISH	p-dioxane
95	2,3,5,6-F	EFISH	p-dioxane
95	2,4,5-F	EFISH	p-dioxane
95	4-F	EFISH	p-dioxane
95	2,3,4-F	EFISH	p-dioxane
95	2,4-F	EFISH	p-dioxane

λ_{meas}	λ_{max}	μ	β	β_0	μβ	$\mu\beta_0$	footnote	convention
1907	490	3.74	94.1	37.9	352.0	141.7	d	
1907	498	2.02	159.9	63.7	323.0	128.7	d	
1907	552	4.35	246.0	91.8	1070.0	399.3	d	
1907	544	2.02	247.0	93.0	499.0	187.9	d	
1907	452	7.60	60.5	25.2	460.0	191.5	d	
1907	470	6.21	48.1	19.8	299.0	123.0	d	
1907	486	4.47	99.6	40.2	445.0	179.7	d	
1907	494	6.40	49.7	19.9	318.0	127.4	d	
1907	500	7.10	51.0	20.3	362.0	144.1	d	
1907	518	5.18	115.6	45.0	599.0	233.1	d	
1907	502	7.59	64.6	25.6	490.0	194.3	d	
1907	520	5.66	146.6	56.9	830.0	322.1	d	
1907	558	7.06	246.5	91.1	1740.0	643.2	d	
1907	560	4.00	262.5	97.1	1050.0	388.4	d	
1907	588	8.29	265.4	93.7	2200.0	776.8	d	
1907	594	6.48	339.5	119.0	2200.0	771.1	d	
1064	322	3.30	15.7	9.0	51.8	29.7		
1064	324	3.50	11.8	6.7	41.3	23.5		
1064	322	3.40	10.4	6.0	35.4	20.4		
1064	320	4.60	6.8	3.9	31.3	17.9		
1064	322	3.10	9.2	5.3	28.5	16.4		
1064	324	2.40	11.6	6.6	27.8	15.8		
1064	322	3.10	6.7	3.8	20.8	11.8		
1064	320	2.90	6.8	4.0	19.7	11.6		
1064	318	4.00	4.4	2.6	17.6	10.4		
1064	316	3.30	5.3	3.1	17.5	10.2		

Footnote entries on page 503.

Hyperpolarizability tabulations. Shaded entries are calculated from values given in the cited reference.

	MOLECULE NAME, OR STRUCTURE	tech	solvent
95	parent	EFISH	dioxane
95	3-F	EFISH	dioxane
95	2,4,6-F	EFISH	dioxane
95	2-F	EFISH	dioxane
95	related 4'-methoxystilbenes (derivative given)		
95	4-CF_3	EFISH	dioxane
95	3,5-CF_3	EFISH	dioxane
95	4-CF_3, 2,3,4,6-F	EFISH	dioxane
95	4-CH=CH-C_6F_{13}	EFISH	dioxane
95	4-CH=CH-C_6F_5	EFISH	dioxane
96	ethyl violet	HRS	propane-1,3-diol
97	n=1 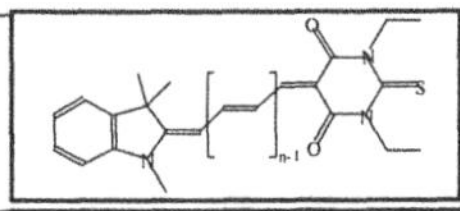	EFISH	chloroform
97	n=2 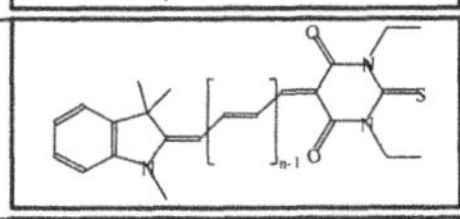	EFISH	chloroform
97	n=3 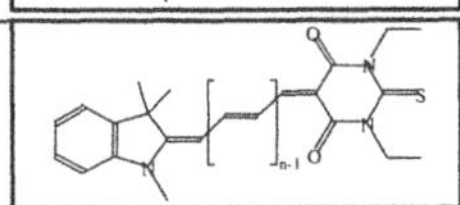	EFISH	chloroform
97	n=3 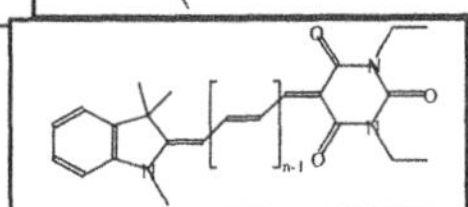	EFISH	chloroform
97	n=3 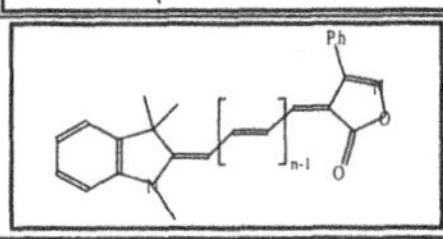	EFISH	chloroform
97	n =3 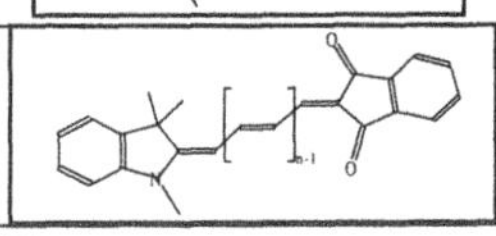	EFISH	chloroform
98	4-Nitrostilbene	EFISH	benzene

λ_{meas}	λ_{max}	μ	β	β_0	$\mu\beta$	$\mu\beta_0$	footnote	convention
1064	320	2.80	6.1	3.5	17.1	9.8		
1064	322	3.10	4.2	3.0	13.0	9.3		
1064	318	2.20	6.8	4.0	15.0	8.8		
1064	320	1.80	3.8	2.2	6.8	4.0		
1064	326	4.20	16.4	9.3	68.9	39.1		
1064	326	4.20	10.6	6.0	44.5	25.2		
1064	334	4.00	23	12.6	92.0	50.4		
1064	346	4.10	26	13.4	106.6	54.9		
1064	366	3.90	36	16.7	140.4	65.1		
1064			510					
1907	494	7.40	-46	-31.0	-340.0	-229.4		
1907	592	8.90	52	28.0	462.0	249.2		
1907	686	9.10	300	120.0	2730.0	1092.0		
1907	664	11.00	280	126.0	3080.0	1386.0		
1907	648	6.90	360	171.0	2484.0	1179.9		
1907	658	4.40	500	230.0	2200.0	1012.0		
1064		4.56	29		132.2			I

Footnote entries on page 503.

Hyperpolarizability tabulations. Shaded entries are calculated from values given in the cited reference.

	MOLECULE NAME, OR STRUCTURE	tech	solvent
98	4-Aminostilbene	EFISH	benzene
98	4-Dimethylaminostilbene	EFISH	benzene
98	4-Chlorostilbene	EFISH	benzene
98	4-Chloro-4'-nitrostilbene	EFISH	chloroform
98	4-Chloro-4'-dimethylaminostilbene	EFISH	chloroform
98	4-Nitro-4'-aminostilbene	EFISH	acetone
98	4-Nitro-4'-dimethylaminostilbene	EFISH	acetone
98	DMA-NS	EFISH	chloroform
98	DMA-PNB	EFISH	acetone
99	Benzene	EFISH	neat
99	Nitrobenzene	EFISH	neat
99	Aniline	EFISH	neat
99	Phenol	EFISH	neat
99	Fluorobenzene	EFISH	neat
99	Chlorobenzene	EFISH	neat
99	Bromobenzene	EFISH	neat
99	Iodobenzene	EFISH	neat
99	Benzonitrile	EFISH	neat
99	N-N-dimethylaniline	EFISH	neat
99	P-nitroaniline	EFISH	methanol
99	4-dimethylamino-4' nitrostilbene	EFISH	acetone
100	p-NO_2-C_6H_4-NO_2	EFISH	benzene
100	m-NO_2-C_6H_4-NO_2	EFISH	benzene
100	m-NO_2-C_6H_4-NH_2	EFISH	acetone
100	m-NO_2-C_6H_4-Cl	EFISH	benzene
100	m-NO_2-C_6H_4-Br	EFISH	benzene
100	m-OH-C_6H_4-NH_2	EFISH	acetone
100	C_6H_6	EFISH	neat
100	F-C_6H_5	EFISH	neat
100	Cl-C_6H_5	EFISH	neat
100	Br-C_6H_5	EFISH	neat
100	I-C_6H_5	EFISH	neat
100	NO_2-C_6H_5	EFISH	neat
100	CN-C_6H_5	EFISH	neat
100	NH_2-C_6H_5	EFISH	neat
100	$N(CH_3)_2$-C_6H_5	EFISH	neat
100	OH-C_6H_5	EFISH	neat
101	DANS	HRS	chloroform

λ_{meas}	λ_{max}	μ	β	β_0	$\mu\beta$	$\mu\beta_0$	footnote	convention
1064		2.07	12		24.8			I
1064		2.40	29		69.6			I
1064		1.50	-3.6		-5.4			I
1064		3.10	39		120.9			I
1064		4.02	42		168.8			I
1064		7.54	260		1960.4			I
1064		7.42	450		3339.0			I
1064		7.90	220		1738.0			I
1064		8.80	630		5544.0			I
1064			0					I
1064			-2.4					I
1064			1.2					I
1064			-0.2					I
1064			0.5					I
1064			0.2					I
1064			0					I
1064			0.6					I
1064			-0.5					I
1064			1.7					I
1064			32					I
1064			340					I
		0.00	0					
		3.89	1.2		4.7			
		4.90	6.1		29.9			
		3.40	1.6		5.4			
		3.40	1.2		4.1			
		1.83	1.2		2.2			
		0.00	0					
		-1.60	-0.7		1.1			
		-1.70	-0.33		0.6			
		-1.70	-0.2		0.3			
		-1.70	0.46		-0.8			
		-4.00	2.27		-9.1			
		-4.10	0.48		-2.0			
		1.53	1.17		1.8			
		1.75	1.61		2.8			
		1.45	0.17		0.2			
1064			1360				p	

Footnote entries on page 503.

Hyperpolarizability tabulations. Shaded entries are calculated from values given in the cited reference.

	MOLECULE NAME, OR STRUCTURE	tech	solvent
101	DANS	HRS	chloroform
102	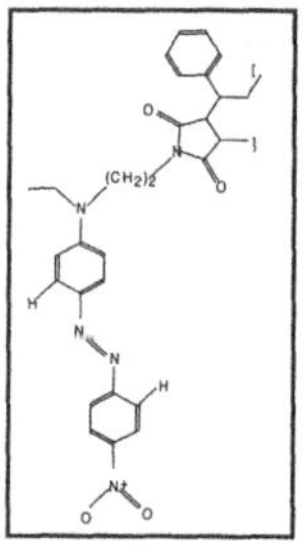	poled po	thin film
103	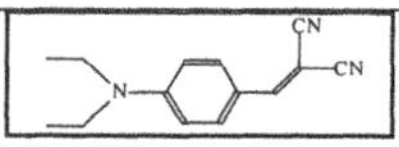	EFISH	p-dioxane
103	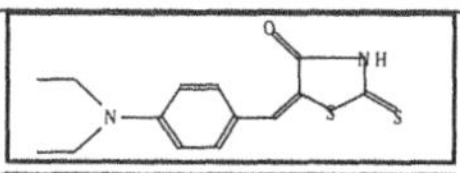	EFISH	p-dioxane
103	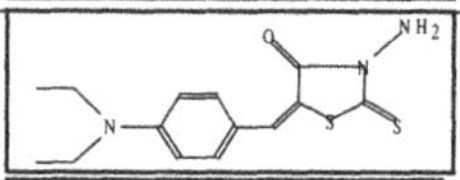	EFISH	p-dioxane
103	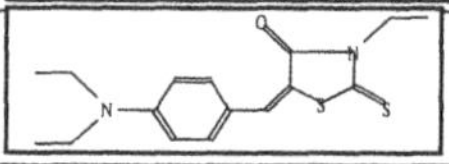	EFISH	p-dioxane
103	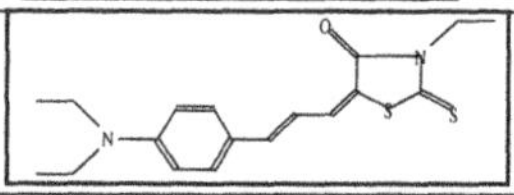	EFISH	p-dioxane
103	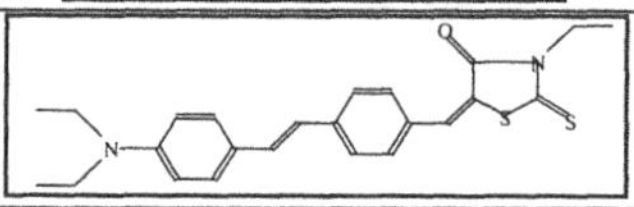	EFISH	p-dioxane

λ_{meas}	λ_{max}	μ	β	β_0	$\mu\beta$	$\mu\beta_0$	footnote	convention
1064			1610				q	
1338			75.16					I
1907	419				300.0	230.4		
1907	452				300.0	219.5		
1907	473				410.0	290.1		
1907	465				420.0	301.1		
1907	493				908.0	620.8		
1907	466				885.0	633.4		

Footnote entries on page 503.

Hyperpolarizability tabulations. Shaded entries are calculated from values given in the cited reference.

	MOLECULE NAME, OR STRUCTURE	tech	solvent
103	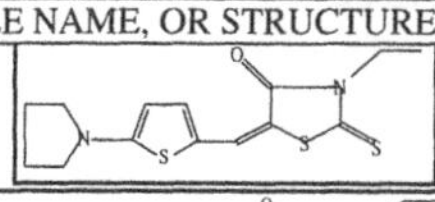	EFISH	p-dioxane
103	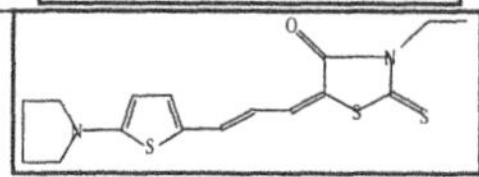	EFISH	p-dioxane
103	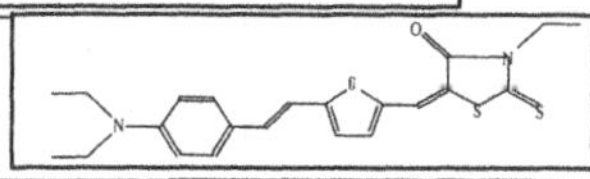	EFISH	p-dioxane
104	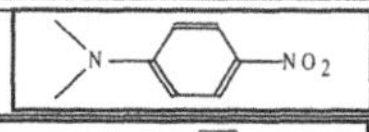	EFISH	p-dioxane
104	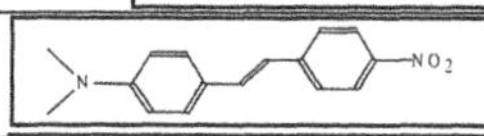	EFISH	p-dioxane
104	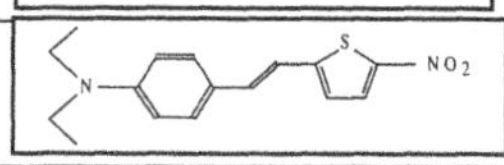	EFISH	p-dioxane
104	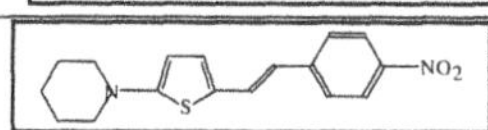	EFISH	p-dioxane
104	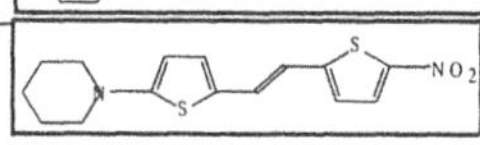	EFISH	p-dioxane
104	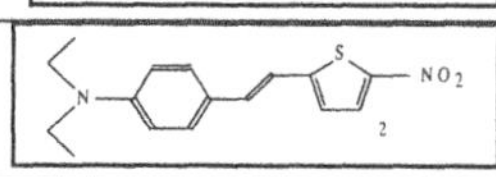	EFISH	p-dioxane
104	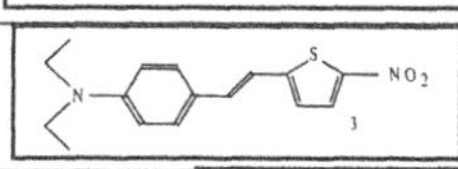	EFISH	p-dioxane
105	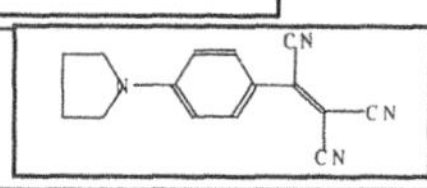	EFISH	p-dioxane
105	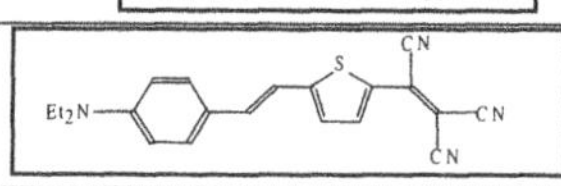	EFISH	p-dioxane

λ_{meas}	λ_{max}	μ	β	β_0	$\mu\beta$	$\mu\beta_0$	footnote	convention
1907	501				383.0	258.1		
1907	525				980.0	631.1		
1907	527				1165.0	747.3		
1907	370				110.0	89.9		
1907	424				580.0	442.3		
1907	478				600.0	421.0		
1907	460				660.0	476.9		
1907	516				1040.0	681.6		
1907	506				1400.0	934.9		
1907	514				2100.0	1381.5		
1907	505				710.0	475.0		
1907	640				6200.0	3023.0		

Footnote entries on page 503.

Hyperpolarizability tabulations. Shaded entries are calculated from values given in the cited reference.

	MOLECULE NAME, OR STRUCTURE	tech	solvent
105	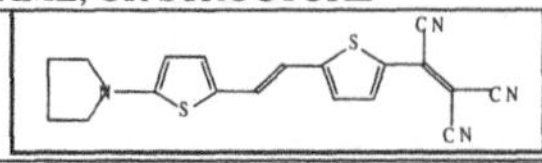	EFISH	p-dioxane
105	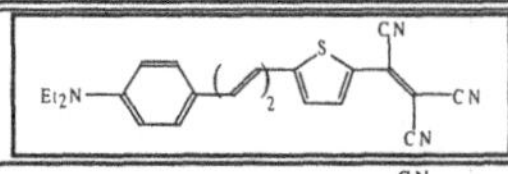	EFISH	p-dioxane
105	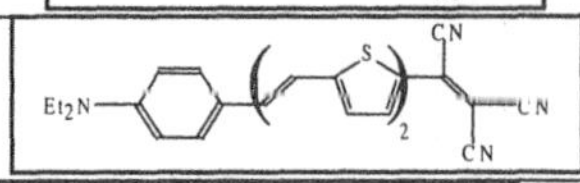	EFISH	p-dioxane
105	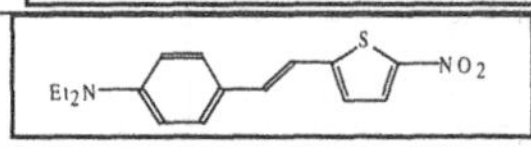	EFISH	p-dioxane
105	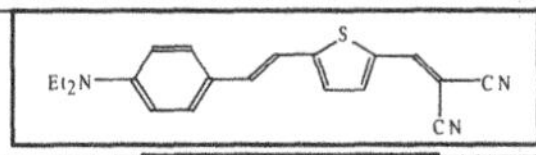	EFISH	p-dioxane
106	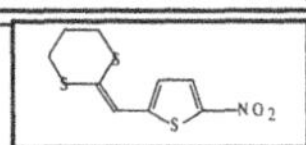	EFISH	p-dioxane
106	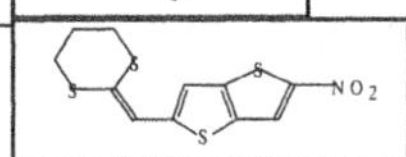	EFISH	p-dioxane
106	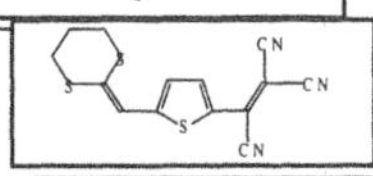	EFISH	p-dioxane
106	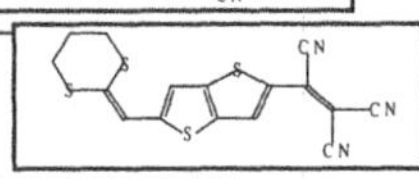	EFISH	p-dioxane
107	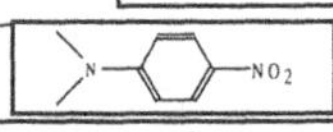	EFISH	p-dioxane
107	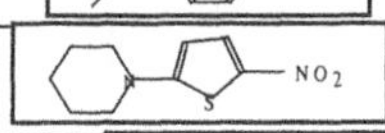	EFISH	p-dioxane
107	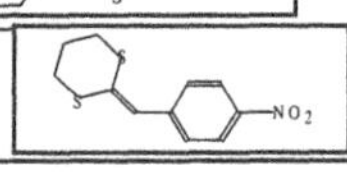	EFISH	p-dioxane

λ_{meas}	λ_{max}	μ	β	β_0	$\mu\beta$	$\mu\beta_0$	footnote	convention
1907	718				6900.0	2564.0		
1907	662				9100.0	4145.5		
1907	653				7400.0	3468.6		
1907	478				600.0	421.0		
1907	513				1300.0	856.9		
1907					200.0	150.0		
1907					380.0	280.0		
1907					940.0	565.0		
1907					2200.0	1287.0		
1907	370				110.0	89.9		
1907	430				125.0	94.5		
1907	380				130.0	105.0		

Footnote entries on page 503.

Hyperpolarizability tabulations. Shaded entries are calculated from values given in the cited reference.

	MOLECULE NAME, OR STRUCTURE	tech	solvent
107	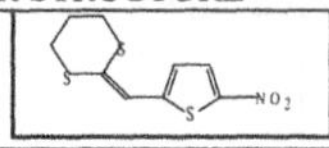	EFISH	p-dioxane
107	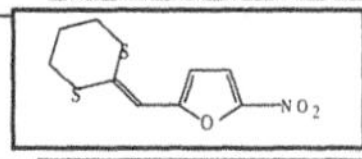	EFISH	p-dioxane
107	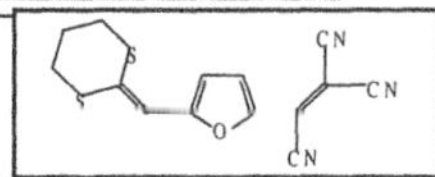	EFISH	p-dioxane
107	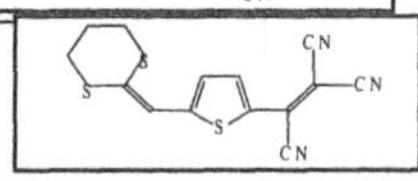	EFISH	p-dioxane
107	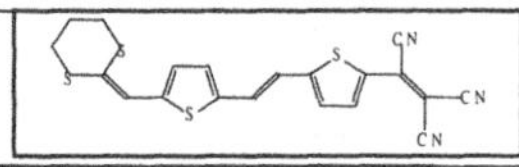	EFISH	p-dioxane
107	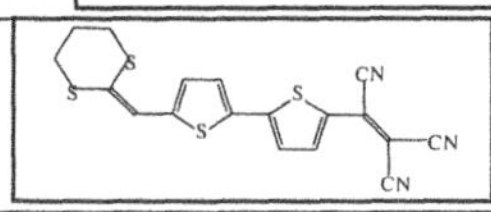	EFISH	p-dioxane
108	$C_8H_{17}(C_6H_4)_2CN$	LB	monolayer
108	$C_9H_{19}(C_6H_4)_2CN$	LB	monolayer
108	$C_{10}H_{21}(C_6H_4)_2CN$	LB	monolayer
108	$C_{12}H_{21}(C_6H_4)_2CN$	LB	monolayer
108	$C_{14}H_{29}COOH$	LB	monolayer
108	$C_{17}H_{35}COOH$	LB	monolayer
108	$C_{22}H_{45}COOH$	LB	monolayer
108	$C_{17}H_{35}CH_2OH$	LB	monolayer
108	$C_{12}H_{25}C_{10}H_6SO_3Na$	LB	monolayer
109	Y 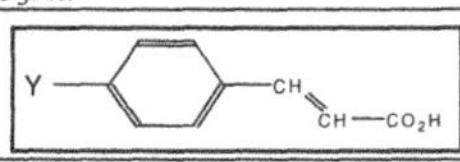		
109	NMe_2	HRS	chloroform
109	NH_2	HRS	chloroform
109	OMe	HRS	chloroform
109	OH	HRS	chloroform

λ_{meas}	λ_{max}	μ	β	β_0	$\mu\beta$	$\mu\beta_0$	footnote	convention
1907	438				200.0	149.5		
1907	418				180.0	138.4		
1907	542				850.0	528.9		
1907	558				940.0	565.2		
1907	612				3420.0	1804.0		
1907	595				1500.0	826.7		
532			25					II
532			25					II
532			25					II
532			25					II
532			0.08					II
532			0.07					II
532			0.07					II
532			0.11					II
532			0.76					II
1064	320	8.30	50	29.4	415.0	244.0		
1064	326	8.10	46	26.3	372.6	213.0		
1064	212	7.80	42	25.3	327.6	197.3		
1064	308	7.60	36	21.8	273.6	165.7		

Footnote entries on page 503.

Hyperpolarizability tabulations. Shaded entries are calculated from values given in the cited reference.

	MOLECULE NAME, OR STRUCTURE	tech	solvent
109	Cl	HRS	chloroform
109	Br	HRS	chloroform
109	Me	HRS	chloroform
110	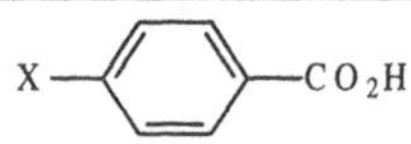 X		
110	OMe	HRS	ethanol
110	OH	HRS	ethanol
110	Me	HRS	ethanol
110	I	HRS	ethanol
110	Cl	HRS	ethanol
110	Br	HRS	ethanol
110	NO_2	HRS	ethanol
110	NMe_2	HRS	ethanol
110	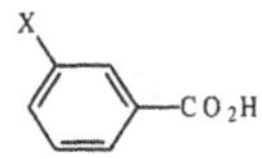 X	HRS	ethanol
110	OMe	HRS	ethanol
110	OH	HRS	ethanol
110	Me	HRS	ethanol
110	I	HRS	ethanol
110	Cl	HRS	ethanol
110	Br	HRS	ethanol
110	NO_2	HRS	ethanol
110	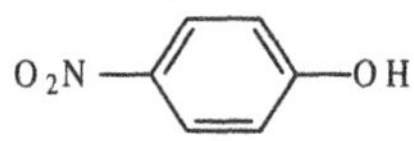	HRS	ethanol
110	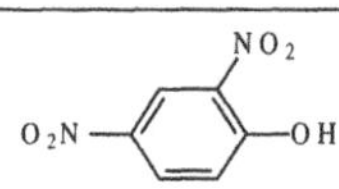	HRS	ethanol
110	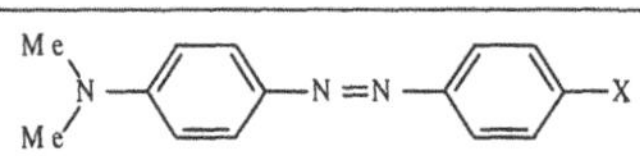 X		

λ_{meas}	λ_{max}	μ	β	β_0	$\mu\beta$	$\mu\beta_0$	footnote	convention
1064	296	7.40	32	20.2	236.8	149.5		
1064	310	7.30	30	18.2	219.0	132.9		
1064	301	7.00	27	16.9	189.0	118.3		
1064	298.6	9.40	36	22.7	338.4	213.5	o	
1064	279.6	9.20	30	20.2	276.0	186.0	o	
1064	276.8	9.00	29	19.7	261.0	177.5	o	
1064	289.4	8.80	26	17.0	228.8	149.2	o	
1064	284.1	8.70	24	15.9	208.8	138.6	o	
1064	280.6	8.50	22	14.8	187.0	125.6	o	
1064	267.8	8.20	20	14.0	164.0	114.7	o	
1064	294.6	10.20	20	12.8	204.0	130.6	o	
1064							o	
1064	287.6	8.20	24	15.7	196.8	129.1	o	
1064	272.8	8.00	18	12.4	144.0	99.2	o	
1064	269.1	7.60	18	12.5	136.8	95.3	o	
1064	284.6	7.40	20	13.3	148.0	98.1	o	
1064	279.2	7.30	16	10.8	116.8	78.8	o	
1064	271.6	7.10	13	9.0	92.3	63.8	o	
1064	261.8	9.00	12	8.5	108.0	76.9	o	
1064	311	7.40	22	13.2	162.8	98.0	o	
1064	319	9.20	25	14.6	230.0	134.1	o	

Footnote entries on page 503.

Hyperpolarizability tabulations. Shaded entries are calculated from values given in the cited reference.

	MOLECULE NAME, OR STRUCTURE	tech	solvent
110	CO_2H	HRS	ethanol
110	SO_3H	HRS	ethanol
110	water	HRS	
110	ethylene glycol	HRS	
110	methanol	HRS	
110	ethanol	HRS	
110	propanol	HRS	
111	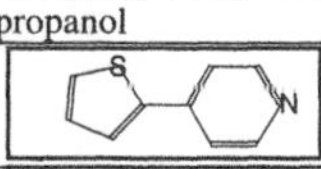	EFISH	
111	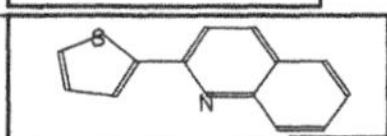	EFISH	
111	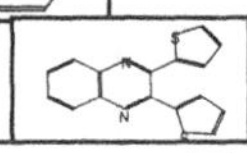	EFISH	
111	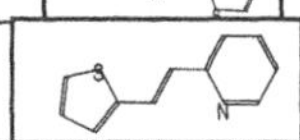	EFISH	
111	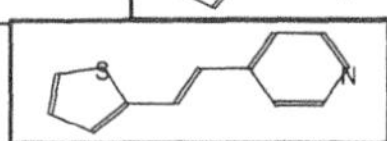	EFISH	
111	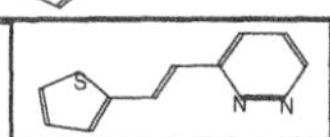	EFISH	
111	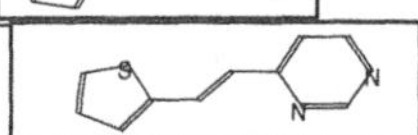	EFISH	
111	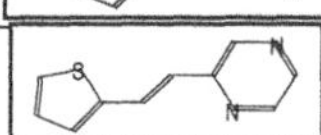	EFISH	
111	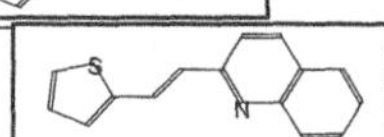	EFISH	
111	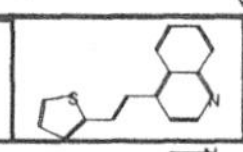	EFISH	
111	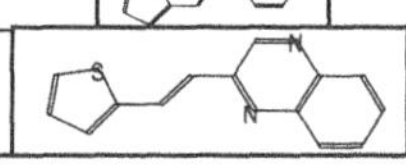	EFISH	

λ_{meas}	λ_{max}	μ	β	β_0	$\mu\beta$	$\mu\beta_0$	footnote	convention
1064	441	15.00	135	35.0	2025.0	524.7	o	
1064	458	16.20	164	34.6	2656.8	560.3	o	
1064			0.05					
1064			0.41					
1064			0.52					
1064			0.61					
1064			0.48					
1063	304	2.20	3	1.9	6.6	4.1		
1063	350	1.90	3	1.5	5.7	2.9		
1063	381	1.37	13	5.5	17.8	7.5		
1063	333	1.50	3	1.6	4.5	2.5		
1063	328	2.80	33	18.5	92.4	51.8		
1063	329	3.70	11	6.1	40.7	22.7		
1063	343	2.65	35	18.3	92.8	48.5		
1063	351	1.10	32	16.1	35.2	17.7		
1063	363	1.91	12	5.7	22.9	10.8		
1063	351	2.62	38	19.1	99.6	50.0		
1063	382	0.92	78	32.8	71.8	30.2		

Footnote entries on page 503.

Hyperpolarizability tabulations. Shaded entries are calculated from values given in the cited reference.

	MOLECULE NAME, OR STRUCTURE	tech	solvent
111	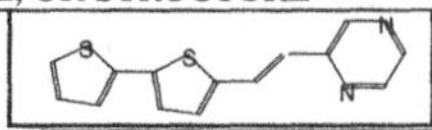	EFISH	
111	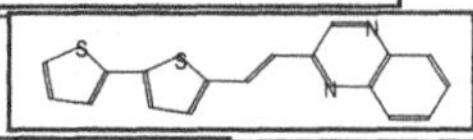	EFISH	
111	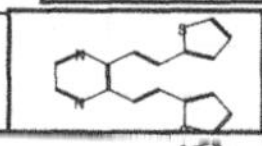	EFISH	
111	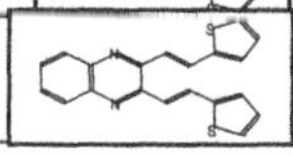	EFISH	
112	PNA	EFISH	chloroform
112	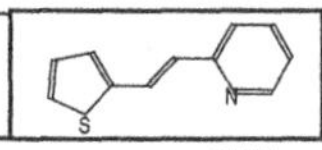	EFISH	chloroform
112	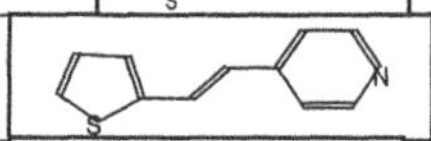	EFISH	chloroform
112	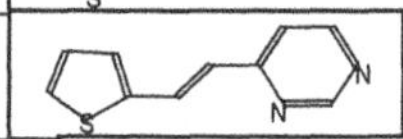	EFISH	chloroform
112	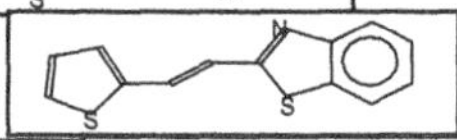	EFISH	chloroform
112	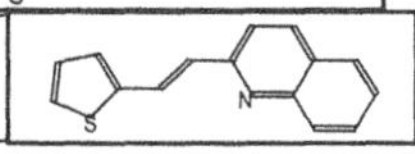	EFISH	chloroform
113	Nitrobenzene	EFISH	hexane
113	Nitrobenzene	EFISH	dimethylsulfoxide
113	Nitrobenzene	EFISH	neat
114	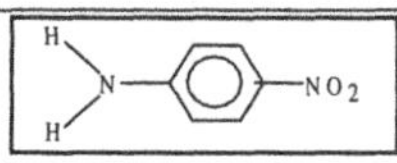	EFISH	dimethylsulfoxide
114	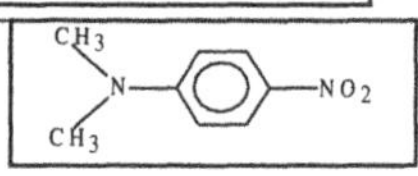	EFISH	dimethylsulfoxide

λ_{meas}	λ_{max}	μ	β	β_0	$\mu\beta$	$\mu\beta_0$	footnote	convention
1063	395	1.19	121	46.7	144.0	55.6		
1063	420	1.09	224	71.0	244.2	77.4		
1063	365	1.09	68	31.7	74.1	34.5		
1063	410	0.60	167	57.6	100.2	34.5		
1064	347	6.20	20	10.3	124.0	63.7		
1064	333	1.40	4.6	2.5	6.4	3.5		
1064	328	2.60	21	11.8	54.6	30.6		
1064	341	2.70	19	10.0	51.3	27.1		
1064	358	1.50	25	12.1	37.5	18.2		
1064	363	1.70	11	5.2	18.7	8.8		
1064		4.06	1.11		4.5			I
1064		4.31	1.39		6.0			I
1064		4.10	2.04		8.4			I
1580		6.30	12	9.0	75.0	56.7		I
1356		7.10	21	12.0	138.0	85.2		I

Footnote entries on page 503.

Hyperpolarizability tabulations. Shaded entries are calculated from values given in the cited reference.

	MOLECULE NAME, OR STRUCTURE	tech	solvent
114	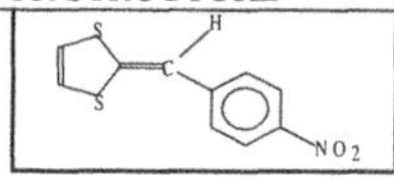	EFISH	dimethylsulfoxide
114	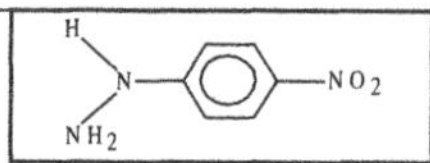	EFISH	dimethylsulfoxide
114	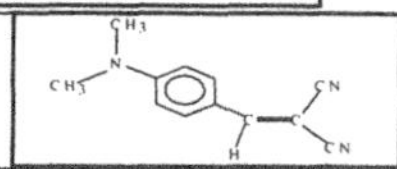	EFISH	dimethylsulfoxide
114	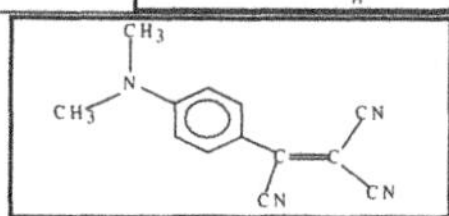	EFISH	dimethylsulfoxide
114	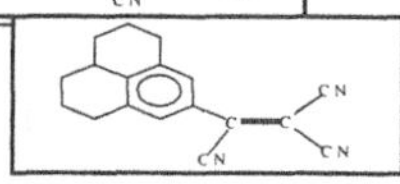	EFISH	dimethylsulfoxide
114	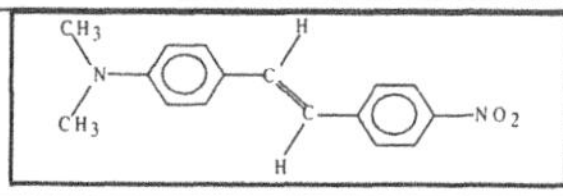	EFISH	dimethylsulfoxide
114	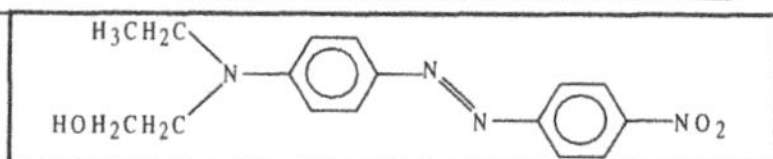	EFISH	dimethylsulfoxide
114	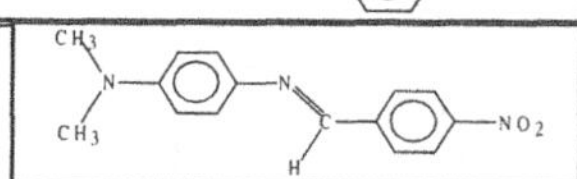	EFISH	dimethylsulfoxide
114	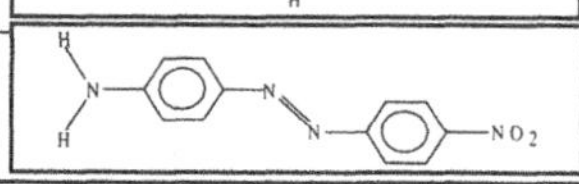	EFISH	dimethylsulfoxide
114	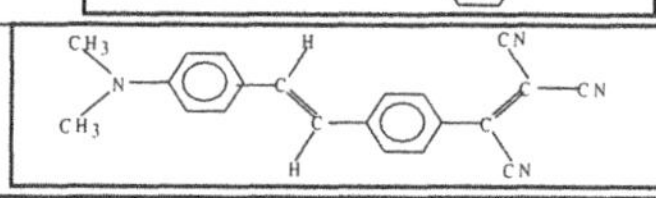	EFISH	dimethylsulfoxide

λ_{meas}	λ_{max}	μ	β	β_0	$\mu\beta$	$\mu\beta_0$	footnote	convention
1356		6.90	52	25.0	358.0	172.5		I
1356		6.50	15	9.5	102.0	61.8		I
1356		8.70	31	16.0	271.0	139.2		I
1356		10.90	78	26.0	846.0	283.4		I
1356		10.20	72	19.0	730.0	193.8		I
1356		7.20	105	52.0	760.0	374.4		I
1356		8.70	125	47.0	1090.0	408.9		I
1356		6.60	77	37.0	500.0	244.2		I
1356		7.50	103	47.0	770.0	352.5		I
1356		8.20	323	133.0	2650.0	1090.6		I

Footnote entries on page 503.

Hyperpolarizability tabulations. Shaded entries are calculated from values given in the cited reference.

	MOLECULE NAME, OR STRUCTURE	tech	solvent
114	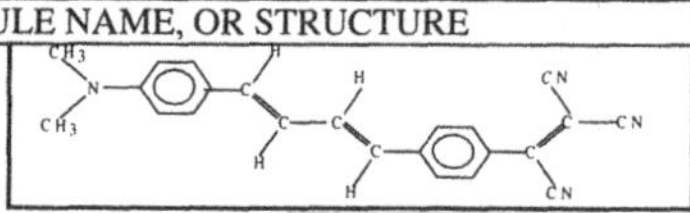	EFISH	dimethylsulfoxide
114	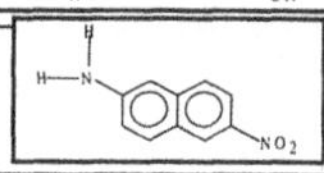	EFISH	dimethylsulfoxide
114	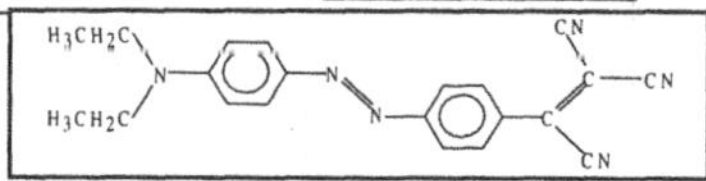	EFISH	dimethylsulfoxide
114	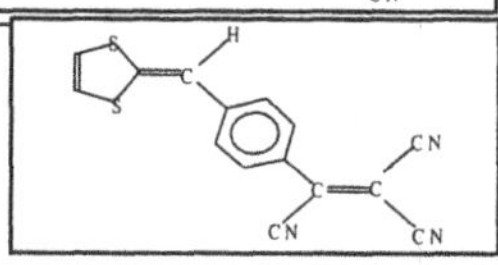	EFISH	dimethylsulfoxide
115	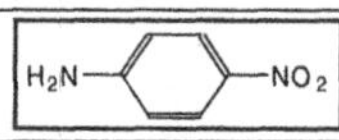	HRS	chloroform
115	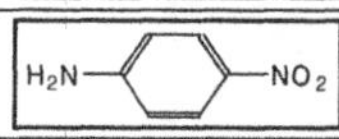	HRS	chloroform
115	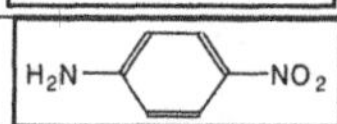	EFISH	chloroform
115	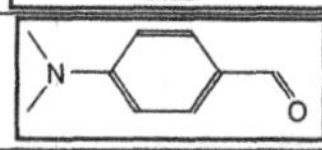	HRS	chloroform
115	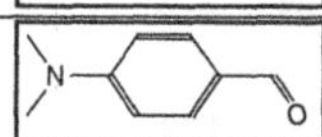	HRS	chloroform
115	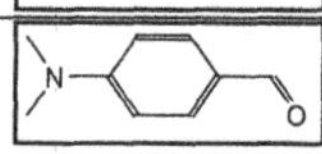	HRS	chloroform
115	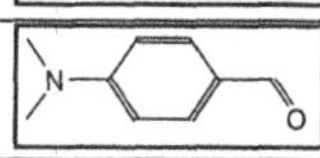	EFISH	chloroform

λ_{meas}	λ_{max}	μ	β	β_0	$\mu\beta$	$\mu\beta_0$	footnote	convention
1356		7.90	167	73.0	1320.0	576.7		I
1356		5.20	45		233.0			I
1580		10.50	390	154.0	4110.0	1617.0		I
1580					1200.0			I
1064	348		17	8.7			p	
1064	348		19	9.7			r	
1064	348		15	7.7				
1064	326		10	5.7			p	
1064	326		12	6.8			q	
1064	326		13	7.4			r	
1064	326		10	5.7				

Footnote entries on page 503.

Hyperpolarizability tabulations. Shaded entries are calculated from values given in the cited reference.

	MOLECULE NAME, OR STRUCTURE	tech	solvent
115	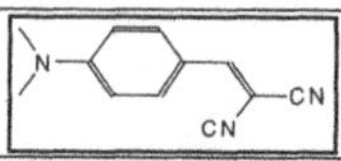	HRS	chloroform
115	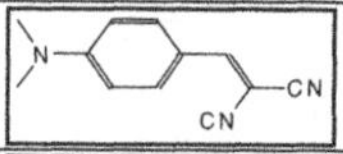	HRS	chloroform
115	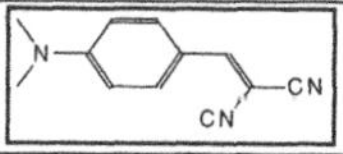	HRS	chloroform
115	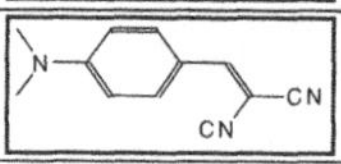	EFISH	chloroform
116	Crystal Violet	HRS	acetonitrile
117	5-cyclohexyl-2-(4'-N,N-dimethylaminophenylimino)-2,3-dihydro-benzo[b]thiophen-3-on3	HRS	chloroform
118	PNA	EFISH	chloroform
118	PNA	EFISH	dichloromethane
118	PNA	EFISH	p-dioxane
118	PNA	EFISH	ethylacetate
118	PNA	EFISH	tetrahydropyran
118	PNA	EFISH	tetrahydrofuran
118	PNA	EFISH	dimethoxyethane
118	PNA	EFISH	acetonitrile
118	PNA	EFISH	acetone
118	PNA	EFISH	2-methoxyethyl ether
118	PNA	EFISH	methanol
118	PNA	EFISH	dimethylformamide
118	PNA	EFISH	N-methyl-pyrollidone
119	hemicyanine	LB	monolayer
120	PNA	EFISH	p-dioxane
120	PNA	EFISH	p-dioxane
120	PNA	EFISH	p-dioxane
120	PNA	EFISH	p-dioxane
120	PNA	EFISH	p-dioxane
120	MNA	EFISH	p-dioxane
120	MNA	EFISH	p-dioxane
120	MNA	EFISH	p-dioxane
120	MNA	EFISH	p-dioxane

λ_{meas}	λ_{max}	μ	β	β_0	μβ	$\mu\beta_0$	footnote	convention
1064	430		150	43.5			p	
1064	430		120	34.8			q	
1064	430		140	40.6			r	
1064	430		87	25.2				
1450			117	34.0				
1500	515		78	36.0				
1064	348	6.40	16.8	8.6	107.5	55.0		I
1064	350	6.20	16.9	8.5	104.8	52.7		I
1064	354	7.00	16.3	8.1	114.1	56.7		I
1064	358	6.70	22.6	11.0	151.4	73.7		I
1064	360	7.10	22.5	10.8	159.8	76.7		I
1064	362	7.10	21.4	10.2	151.9	72.4		I
1064	366	7.70	19.5	9.1	150.2	70.1		I
1064	366	6.20	29.2	13.6	181.0	84.3		I
1064	368	7.30	25.9	11.9	189.1	86.9		I
1064	368	7.80	23.8	11.0	185.6	85.8		I
1064	370	6.10	32	14.5	195.2	88.5		I
1064	382	7.30	20	12.6	146.0	92.0		I
1064	386	6.80	38.4	15.8	261.1	107.4		I
1064			63					I
1907			9.6					I
1370			11.8					I
1060			16.9					I
909			25					I
830			40					I
1907			9.5					I
1370			12.8					I
1060			16.7					I
909			27					I

Footnote entries on page 503.

Hyperpolarizability tabulations. Shaded entries are calculated from values given in the cited reference.

	MOLECULE NAME, OR STRUCTURE	tech	solvent
120	MNA	EFISH	p-dioxane
121	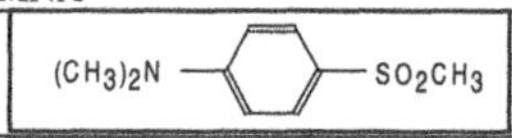	EFISH	chloroform
121	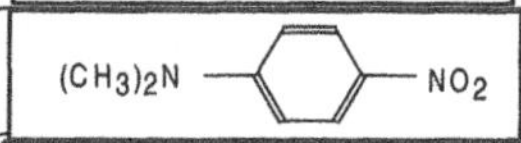	EFISH	chloroform
121	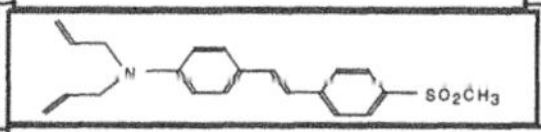	EFISH	chloroform
121	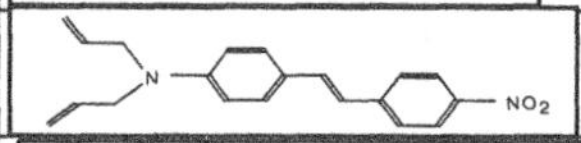	EFISH	chloroform
121	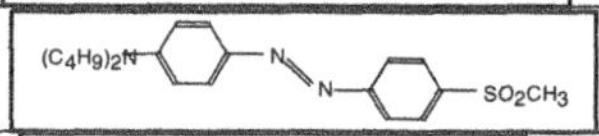	EFISH	chloroform
121	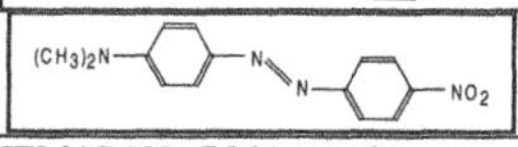	EFISH	chloroform
122	HEMAPAN - DMA copolymer	LB	monolayer
123	4-[4-[(2-hydroxylethyl)methylamino]-phenylazo]benzonitrile (HEMAPAN) -MMA copolymer	LB	monolayer
123	4-[4-(N,N-dimethyl-amino)phenylazo]benzonitrile (DIMAPAN)	EFISH	chloroform
124	4-[4-[(2-hydroxyethyl)methylamino-]-phenylazo]-benzonitrile (HEMAPAN) -DMA copolymer	LB	monolayer
125	DMACS 4-dimethylamino-4'-cyanostilbene	HRS	
125	OMACS 4-octadecylmethylamino-4'-cyanostilbene	HRS	
125	PMMA-MACS	HRS	
125	bacteriorhodopsin	HRS	Triton X-100 in acetate
126	tricyanomethanide ion	HRS	H_2O
126	tricyanomethanide ion	HRS	MeOH
126	tricyanomethanide ion	HRS	EtOH
126	tricyanobenzene	HRS	p-dioxane
126	trinitrobenzene	HRS	chloroform
126	benzonitrile	HRS	p-dioxane
126	nitrobenzene	HRS	chloroform
127	triisopropylaminotrinitrobenzene	HRS	chloroform
127	triisopropylaminotrinitrobenzene	HRS	nitromethane
127	trihydroxytrinitrobenzene	HRS	acetone
127	trihydroxytrinitrobenzene	HRS	methanol

λ_{meas}	λ_{max}	μ	β	β_0	$\mu\beta$	$\mu\beta_0$	footnote	convention
830			45					I
1907			26					
1907			143					
1907			573					
1907			1150					
1907			512					
1907			1270					
1064	442		10.1	2.6			k	
1064	445		5	1.2			j	
1064	448	8.17	360	86.1	2941.2	703.8		
1064			8.55					
			240					
			240					
			230					
1064			2100					
1064			7					
1064			7					
1064			7					
1064			1.8					
1064			4.6					
1064			0.7					
1064			2.2					
1064	375		16	7.1				
1064	383		19	8.0				
1064	395		27	10.4				
1064	385		25	10.3				

Footnote entries on page 503.

Hyperpolarizability tabulations. Shaded entries are calculated from values given in the cited reference.

	MOLECULE NAME, OR STRUCTURE	tech	solvent
127	crystal violet	HRS	methanol
127	crystal violet	HRS	methanol
127	tris-(4-methoxyphenyl)cyclopropenylium bromide	HRS	methanol
127	nitrophenol	HRS	methanol
128	4-octadecylmethylamino-4'-cyanostilbene	HRS	chloroform
128	4-hydroxy-4'-nitrostilbene	HRS	methanol
128	4-tetrabutylammonium-oxy-4' nitrostilbene	HRS	methanol
128	benzodithiopropenal	HRS	
128	crystal violet	HRS	methanol
129	para-nitroanisole	PLS	methanol
130	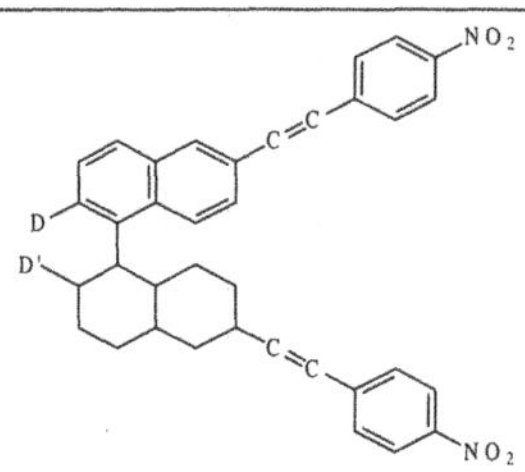 D=D'=OH	EFISH	dioxane
130	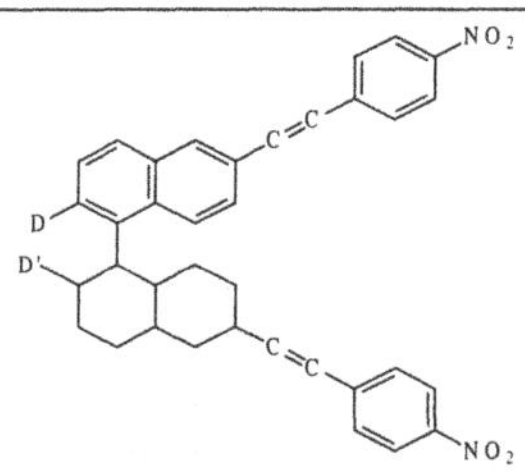 D=D'=OH	EFISH	chloroform

λ_{meas}	λ_{max}	μ	β	β_0	$\mu\beta$	$\mu\beta_0$	footnote	convention
872	588		866	-386.7				
1064	588		6120	-942.0				
1064	359		90	43.4				
1064	313		25	14.9				
1064			258					
1064			287					
1064			847					
1064			30					
1064			3060					
1064/954			12					
1910		6.80		28.0	230.0	190.4		
1910		6.80		21.0	150.0	142.8		

Footnote entries on page 503.

Hyperpolarizability tabulations. Shaded entries are calculated from values given in the cited reference.

	MOLECULE NAME, OR STRUCTURE	tech	solvent
130	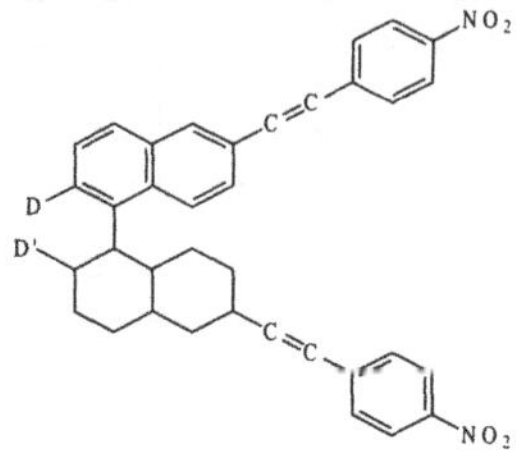 D=D'=OH	EFISH	dimethylsulfoxide
130	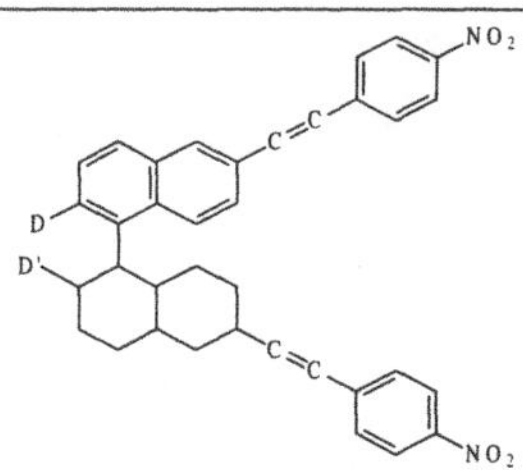 D-D' = $OCH_2CH_2CH_2O$	EFISH	dioxane
130	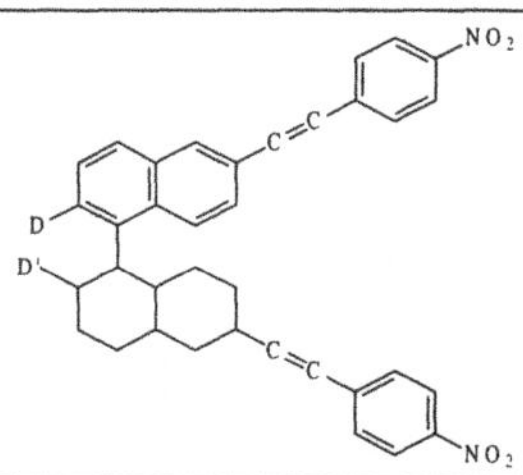 D-D' = $OCH_2CH_2CH_2O$	EFISH	chloroform

λ_{meas}	λ_{max}	μ	β	β_0	$\mu\beta$	$\mu\beta_0$	footnote	convention
1910		6.80			480.0			
1910		8.40		25.0	250.0	210.0		
1910		8.40		31.0	340.0	260.4		

Footnote entries on page 503.

Hyperpolarizability tabulations. Shaded entries are calculated from values given in the cited reference.

	MOLECULE NAME, OR STRUCTURE	tech	solvent
130	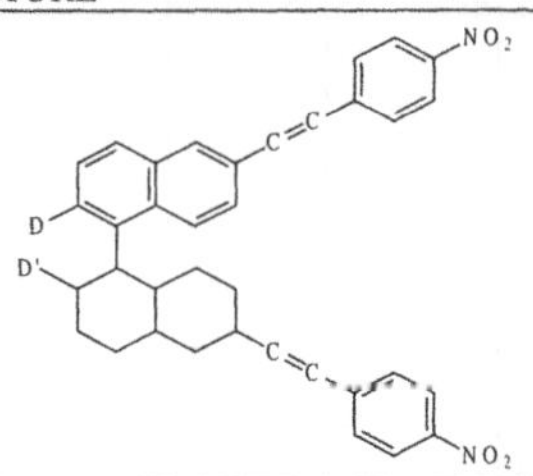 D-D' = $OCH_2CH_2CH_2O$	EFISH	dimethylsulfoxide
130	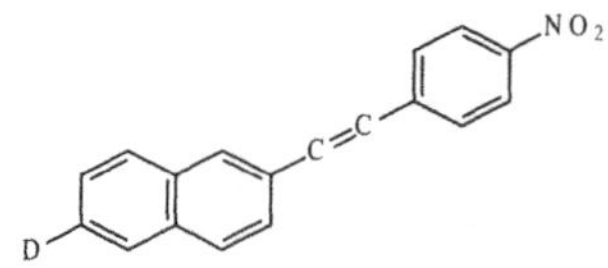 D=OH	EFISH	dioxane
130	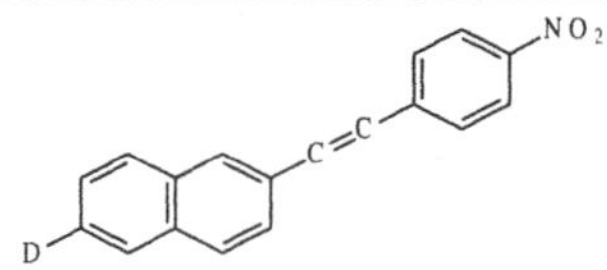 D=OH	EFISH	chloroform
130	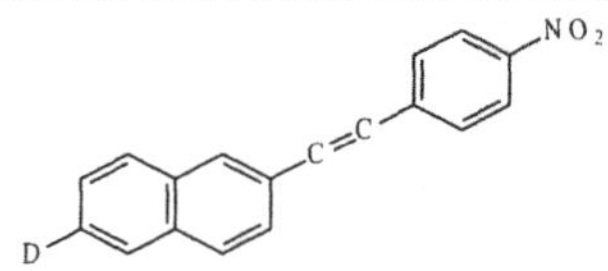 D=OH	EFISH	dimethylsulfoxide
130	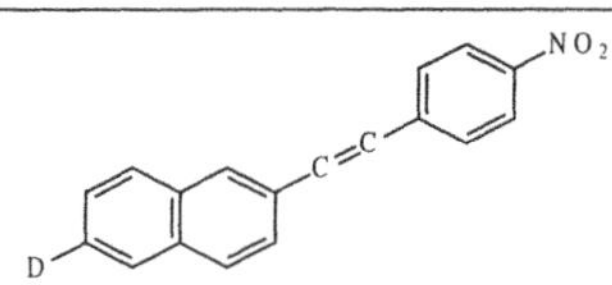 D=OC_6H_{13}	EFISH	dioxane

λ_{meas}	λ_{max}	μ	β	β_0	$\mu\beta$	$\mu\beta_0$	footnote	convention
1910		8.40			420.0			
1910		6.10		23.0	170.0	140.3		
1910		6.10		22.0	160.0	134.2		
1910		6.10			215.0			
1910		5.80		25.0	175.0	145.0		

Footnote entries on page 503.

Hyperpolarizability tabulations. Shaded entries are calculated from values given in the cited reference.

	MOLECULE NAME, OR STRUCTURE	tech	solvent
130	$D=OC_6H_{13}$ 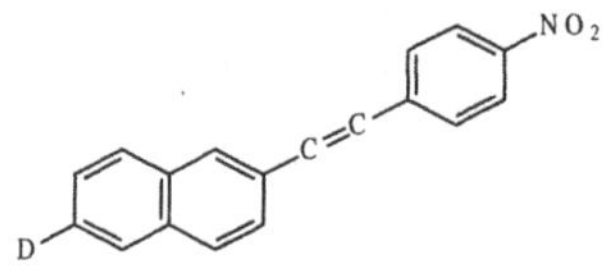	EFISH	chloroform
130	$D=OC_6H_{13}$ 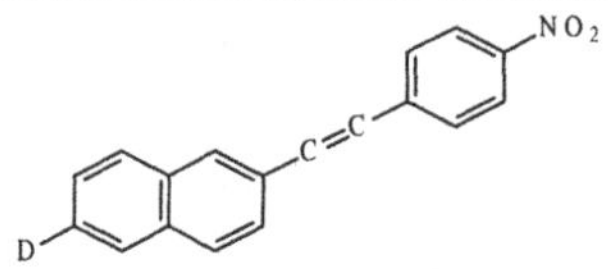	EFISH	dimethylsulfoxide
131	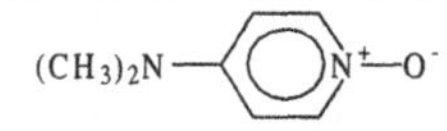	EFISH	chloroform
131	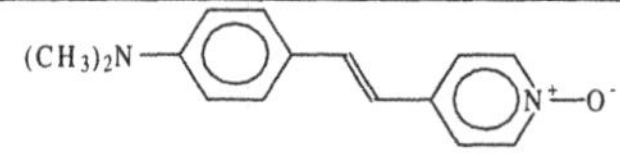	EFISH	chloroform

λ_{meas}	λ_{max}	μ	β	β_0	$\mu\beta$	$\mu\beta_0$	footnote	convention
1910		5.80		25.0	200.0	145.0		
1910		5.80			175.0			
1064		6.76	5		33.8			
1064		6.90	500		3450.0			

a) $\mu\beta$ value reported is an upper bound.
b) μ is a calculated value, not experimentally determined.
c) Molecule contains a chlorine substituent meta to the dimethylamino group.
d) Experiment was performed at 1907 nm, but $\mu\beta$ was extrapolated to 1300 nm.
e) Reported value for $\mu\beta$ is $\mu\beta$ divided by the molecular weight.
f) β is the mean value taken from measurements with several bond angles.
g) β determined using the infinite dilution method.
h) The reported beta value is an upper bound.
i) Determined using Pockel's coefficients.
j) 11.9% functionalized copolymer, β decreased with increasing functionalization.
k) 10% functionalized copolymer, β decreased with increasing functionalization.
l) λ_{max} not measured in this reference.
m) Sign was obtained from solvatochromic techniques.
n) λ_{max} was measured in $CHCl_2$.
o) λ_{max} was reported for the acidic form.
p) Using the internal reference method.
q) Using the external reference method.
r) Using the normalized internal reference method.
s) μ is the average of reported values.

REFERENCES

1. Zyss, J., *Molecular Nonlinear Optics: Materials, Physics, and Devices*, Academic Press, Boston, 1994.
2. Chemla, D. S., and Zyss, J., *Nonlinear Optical Properties of Organic Molecules and Crystals*, Academic Press, Orlando, 1987.
3. Ledoux, I., Badan, J., Zyss, J., Migus, A., Hulin, D., Etchepare, J., Grillon, G., and Antonelli, A., *J. Opt. Soc. Am. B.*, *4*, 987 (1987).
4. Ulman, A., *An Introduction to Ultrathin Organic Films*, Academic Press, New York, 1991.
5. Marks, T. J., and Ratner, M. A., *Angew. Chem. Int. Ed. Engl.*, *34*, 155 (1995).
6. Meredith, G., VanDusen, J., and Williams, D., *Macromol.*, *15*, 1385 (1982).
7. Singer, K. D., Sohn, J. E., and Lalama, S. J., *Appl. Phys. Lett.*, *49*, 248 (1986).
8. Burland, D. M., Miller, R. D., and Walsh, C. A., *Chem. Rev.*, *94*, 31 (1994).
9. *Polymers for Second-Order Nonlinear Optics* (G. A. Lindsay and K. D. Singer, eds.), ACS Symp. Ser. 601, Washington DC, 1995.
10. Kurtz, S., in *Quantum Electronics* (H. Rabin and C. Tang, ed.), Academic Press, New York, (1975).
11. Chemla, D., and Kupecek, P., *Rev. Phys. Appl.*, *6*, 31 (1971).
12. Maker, P. D., Terhune, R. W., Nisenhoff, M., and Savage, C. M., *Phys. Rev. Lett.*, *8*, 21 (1962).
13. Jerphagnon, J., and Kurtz, S. K., *J. Appl. Phys.*, *41*, 1667 (1970).
14. Okamoto, N., Hirano, Y., and Sugihara, O., *J. Opt. Soc. Am. B.*, *9*, 2083 (1992).
15. Eich, M., Reck, B., Yoon, D. Y., Willson, C. G., and Bjorklund, G. C., *J. Appl. Phys.*, *66*, 3241 (1989).
16. Wang, C. H., and Guan, H. W., *J. Polymer Science B., Poly. Phys.*, *31*, 1983 (1993).
17. Kuzyk, M. G., Singer, K. D., Zahn, H. E., and King, L. A., *J. Opt. Soc. Am. B.*, *6*, 742 (1989).
18. Bloembergen, N., and Pershan, P. S., *Phys. Rev.*, *128*, 606 (1962).
19. Okamoto, N., Hirano, Y., and Sugihara, O., *J. Opt. Soc. Am. B.*, *9*, 2083 (1992).
20. Herman, W. N., and Hayden, L. M., *J. Opt. Soc. Am. B.*, *12*, 416 (1995).
21. Neher, D., Wolf, A., Bubek, C., and Wegner, G., *Chem. Phys. Lett.*, *163*, 116 (1989).
22. See, for example, M. Born and W. Wolf, *Principles of Optics*, Pergammon Press, New York, 1975, Section 14.2.
23. Bloembergen, N., *Nonlinear Optics*, W. A. Benjamin, New York, 1965, Chapter 4.
24. Williams, D. J., Nonlinear optical properties of guest-host polymer structures, in *Nonlinear Optical Properties of Organic Molecules and Crystals* (D. S. Chemla and J. Zyss, eds.), Academic Press, New York, 1987, p. 405.
25. Jungbauer, D., Teraoka, I., Yoon, D. Y., Reck, B., Swalen, J. D., Tweig, R., and Willson, C. G., *J. Appl. Phys.*, *69*, 8011 (1991).
26. Oudar, J. L., Optical nonlinearities of conjugated molecules. Stilbene derivatives and highly polar aromatic compounds, *J. Chem. Phys.*, *67*, 446 (1977).
27. Hayden, L. M., Sauter, G. F., Ore, F. R., Pasillas, P. L., Hoover, J. M., Lindsay, G. A., and Henry, R. A., *J. Appl. Phys.*, *68*, 456 (1990).
28. See Section 14.6 of Ref: 22.
29. Singer, K. D., Kuzyk, M., and Sohn, J. E., *J. Opt. Soc. Am. B.*, *4*, 968 (1987).
30. Oudar, J. L., and Chemla, D. S., *J. Chem. Phys.*, *66*, 2664 (1977).
31. Singer, K. D., Sohn, J. E., King, L. A., Gordon, H. M., Katz, H. E., and Dirk, C. W., *J. Opt. Soc. Am. B.*, *6*, 1339 (1989).
32. Onsager, L., *J. Am. Chem. Soc.*, *58*, 1486 (1936).
33. Heinz, T. F., Tom, H. W. K., and Shen, Y. R., *Phys. Rev. A.*, *28*, 1883 (1983).
34. Mayer, G., *C.R. Acad. Sci. Ser. B.*, *267*, 54 (1968).

35. Kielich, S., *IEEE J. Quant. Electron.*, *QE-5*, 562 (1969).
36. Hauchcorne, G., Kerherve, F., and Mayer, G., *J. Phys. (Paris)*, *32*, 47 (1971).
37. Finn, R. S., and Ward, J. F., *J. Chem. Phys.*, *60*, 454 (1974).
38. Ward, J. F., and Bigio, I. J., *Phys. Rev. A.*, *11*, 60 (1975).
39. Levine, B. F., and Bethea, C. G., *J. Chem. Phys.*, *63*, 2666 (1975).
40. Oudar, J. L., *J. Chem. Phys.*, *67*, 446 (1977).
41. Singer, K. D., and Garito, A. F., *J. Chem. Phys.*, *75*, 3572 (1981).
42. Teng, C. C., and Garito, A. F., *Phys. Rev. B.*, *28*, 6766 (1983).
43. Kajzar, F., Ledoux, I., and Zyss, J., *Phys. Rev. A.*, *36*, 2210 (1987).
44. Cheng, L.-T., Tam, W., Stevenson, S. H., Meredith, G. R., Rikken, G., and Marder, S. R., *J. Phys. Chem.*, *95*, 10631 (1991).
45. Singer, K. D., Ph.D. Dissertation, University of Pennsylvania, 1981.
46. Cyvin, S. J., Rauch, J. E., and Decius, J. C., *J. Chem. Phys.*, *43*, 4083 (1965).
47. Zyss, J., and Oudar, J. L., *Phys. Rev. A.*, *26*, 2028 (1982).
48. Willetts, A., Rice, J. E., Burland, D. M., and Shelton, D. P., *J. Chem. Phys.*, *93*, 7120 (1992).
49. Oudar, J. L., Chemla, D. S., and Batifol, E., *J. Chem. Phys.*, *67*, 1626 (1977).
50. Burland, D. M., Walsh, C. A., Kajzar, F., and Sentein, C., *J. Opt. Soc. Am. B.*, *8*, 2269 (1991).
51. Singer, K. D., Sohn, J. E., King, L. A., Gordon, H. M., Katz, H. E., and Dirk, C. W., *J. Opt. Soc. Am. B.*, *6*, 1339 (1989).
52. Tang, C. L., Bosenberg, W. R., Ukachi, T., Lane, R. J., and Cheng, L. K., *Proc. IEEE*, *80*, 365 (1992).
53. Khaydarov, J. D. V., Andrews, J. H., and Singer, K. D., *J. Opt. Soc. Am. B.*, *12*, 2199 (1995).
54. Andrews, J. H., Khaydarov, J. D. V., Singer, K. D., Hull, D. L., Chuang, K. C., *J. Opt. Soc. Am. B.*, *12*, 2360 (1995).
55. Singer, K. D., Merlin, M. S., Lalama, S. J., and Garito, A. F., *Rev. Sci. Instrum.*, *53*, 202 (1982).
56. Terhune, R. W., Maker, P. D., and Savage, C. M., *Phys. Rev. Lett.*, *14*, 681 (1965).
57. Clays, K., and Persoons, A., *Phys. Rev. Lett.*, *66*, 2980 (1991).
58. Verbiest, T., Clays, K., Persoons, A., Meyers, F., and Brèdas, J.-L., *Opt. Lett.*, *18*, 525 (1993).
59. Zyss, J., *Nonlinear Optics*, *1*, 3 (1991).
60. Zyss, J., *J. Chem. Phys.*, *98*, 6583 (1993).
61. Zyss, J., and Ledoux, I., *Chem. Rev.*, *94*, 77 (1994).
62. Clays, K., Persoons, A., and De Maeyer, L., *Modern Nonlinear Optics*, *Part 3*, *Advances in Chemical Physics*, J. Wiley & Sons, New York, 1993.
63. Clays, K., and Persoons, A., *Rev. Sci. Instrum.*, *63*, 3285 (1992).
64. Clays, K., and Persoons, A., *Rev. Sci. Instrum.*, *65*, 2190 (1994).
65. Stadler, S., Dietrich, R., Bourhill, G., and Bräuchle, Ch., *Optics Lett.*, *21*, 251 (1995).
66. Wong, S. W., and Wong, K. Y., *Optics Communications* (In press).
67. Flipse, M. C., de Jonge, R., Woudenberg, R. H., Marsman, A. W., van Walree, C. A., and Jenneskens, L. W., *Chem. Phys. Lett.*, *245*, 297 (1995).
68. Morrison, I. D., Denning, R. G., Laidlaw, W. M., and Stammers, M. A., *Rev. Sci. Instrum.*, *67*, 1445 (1996).
69. Hubbard, S. F., Petschek, R. G., and Singer, K. D., *Opt. Lett.*, *21*, 1774 (1996).
70. Noordman, O. F. J., and can Hulst, N. F., *Chem. Phys. Lett.*, *253*, 145 (1996).
71. Kaatz, P., and Shelton, D. P., *Rev. Sci. Instrum.*, *67*, 1438 (1996).
72. McClain, W. M., *J. Chem. Phys.*, *57*, 2264 (1972).
73. Kaatz, P., and Shelton, D. P., *J. Chem. Phys.*, *105*, 3918 (1996).
74. Kauranen, M., and Persoons, A., *J. Chem. Phys.*, *104*, 3445 (1996).

75. Jerphagnon, J., Chemla, D., and Bonneville, R., *Advances in Physics*, *27*, 609 (1978).
76. Jerphagnon, J., *Phys. Rev. B.*, *2*, 1091 (1970).

TABLE REFERENCES

1. Alain, V., Fort, A., Barzoukas, M., Chen, C. T., BlanchardDesce, M., Marder, S. R., The Linear and Non-Linear Optical Properties of Some Conjugated Ferrocene Compounds, *Inorganica Chimica Acta.*, *242(1-2)*, 43–49 (1996).
2. Amano, M., and Kaino, T., Second-order Nonlinearity of a Novel Diazo-dye Attached Polymer, *J. Appl. Phys.*, *68(12)*, 6024 (1990).
3. Armand, F., Sakuragi, H., Tokumaru, K., Tamada, T., Kajikawa, K., Takezoe, H., Okada, S., Matsuda, H., and Nakanishi, H., Second Harmonic Generation in a Langmuir Monolayer of the New Amphiphilic Coordination Compound Pentacyano(4-octadecylaminopyridine)-ferrate(III), *Mol. Eng.* *2*, 315–324 (1993).
4. Barzoukas, M., Fort, A., Klein, G., Serbutoviez, C., Oswald, L., and Nicoud, J. F., Conformational Dependence of the Quadratic Hyperpolarizabilities of a Series of Push-Pull Stilbenes-Characterization and Investigation of Empirical Correlations, *Chem. Phys.*, *164(3)*, 395–406 (1992).
5. Barzoukas, M., Josse, D., Fremaux, P., Zyss, J., Nicoud, J. F., and Morley, J. O., Quadratic Nonlinear Properties of N-(4-nitrophenyl)-L-prolinol and of a Newly Engineered Molecular Compound N-(4-nitrophenyl)-N-methylaminoacetonitrile: A Comparative Study, *J. Opt. Soc. Am. B.*, *4(6)*, 977 (1987).
6. Barzoukas, M., Josse, D., Zyss, J., Gordon, P., and Morley, J. O., Quadratic Hyperpolarizability of Sulfur-Containing Systems, *Chem. Phys.* *139*, 359–367 (1989).
7. Baumert, J. C., Twieg, R. J., Bjorklund, G. C., Logan, J. A., and Dirk, C. W., Crystal Growth and Characterization of 4-(N,N-dimethylamino)-3-acetamidonitrobenzene, A New Organic Material for Nonlinear Optics, *Appl. Phys. Lett.*, *51(19)*, 1484 (1987).
8. Berkovic, G., New Studies of Liquid and Solid Surfaces Using Second Harmonic Generation, *Phys. A.*, *168*, 140–148 (1990).
9. Berkovic, G., and Shen, Y., The Effect of Conjugation Length and Electron Donor Groups on the Second Order Nonlinear Polarizability of Cyano Substituted Aromatic Molecules, *Mol. Cryst. Liq. Cryst.*, *150b*, 607–616 (1987).
10. Berkovic, G., and Weissbuch, I., Second Harmonic Generation Studies of Monolayers: Orientations, Surface Coverages and Interactions, *Mol. Cryst. Liq. Cryst.*, *211*, 183–188 (1991).
11. Berkovic, G., Rasing, Th., and Shen, Y. R., Second-Order Nonlinear Polarizability of Various biphenyl Derivatives, *J. Opt. Soc. Am. B.*, *4*, 945–949 (1987).
12. Bosshard, C., Knopfle, G., Pretre, P., and Gunter, P., Second-order Polarizabilities of Nitropyridine Derivatives Determined with Electric-field-induced Second-harmonic Generation and a Solvatochromic Method: A Comparative Study, *J. Appl. Phys.*, *71(4)*, 1594–1605 (1992).
13. Bosshard, Ch., Kupfer, M., Florsheimer, M., Borer, T., Gunter, P., Tang, Q., and Zahir, S., Investigation of Chromophore Orientation of 2-docosylamino-5-nitropyridine and Derivatives by Nonlinear Optical Techniques, *Thin Solid Films*, *210/211*, 198–201 (1992).
14. Brouyѐre, E., and Brѐdas, J. L., Theoretical Investigation of the Second-order Nonlinear Optical Properties of Calix[4]arene Molecules, *Synthetic Metals*, *71(1–3)*, 1699–1700 (1995).
15. Bubeck, C., Laschewsky, A., Lupo, D., Neher, D., Ottenbreit, P., Paulus, W., Pras, W., Ringsdorf, H., and Wegner, G., Amphiphilic Dyes for Nonlinear Optics: Dependence of Second Harmonic Generation on Functional Group Substitution, *Adv. Mater*, *3(1)*, 54–58 (1991).

16. Burland, D. M., Miller, R. D., Reiser, O., Twieg, R. J., and Walsh, C. A., The Design, Syntheses, and Evaluation of Chromophores for Second-harmonic Generation in a Polymer Waveguide, *J. Appl. Phys.*, *71(1)*, 410–417 (1992).
17. Cahill, P. A., Singer, K. D., and King, L. A., Anomalous Dispersion Phase-matched Second-harmonic Generation, *Opt. Lett.*, *14(20)*, 1137–1139 (1989).
18. Cassidy, C., Halbout, J. M., Donaldson, W., and Tang, C. L., Nonlinear Optical Properties of Urea, *Opt. Commun.*, *29(2)*, 243–246 (1979).
19. Chen, C. T., Marder, S. R., Cheng, L.-T., Syntheses and Linear and Nonlinear Optical Properties of Unsymmetrical Squaraines with Extended Conjugation, *J. Am. Chem. Soc.*, *116(7)*, 3117–3118 (1994).
20. Cheng, L.-T., Tam, W., Marder, S. R., Stiegman, A. E., Rikken, G., Spangler, C. W., Experimental Investigations of Organic Molecular Nonlinear Optical Polarizabilities, 2: A Study of Conjugation Dependences, *J. Phys. Chem.*, *95(26)*, 10643–10652 (1991).
21. Cheng, L.-T., Tam, W. Stevenson, S. H., Meredith, G. R., Rikken, G., and Marder, S. R., Experimental Investigations of Organic Molecular Nonlinear Optical Polarizabilities, 1: Methods and Results on Benzene and Stilbene Derivatives, *J. Phys. Chem.*, *95(26)*, 10631–10643 (1991).
22. Clays, K., and Persoons, A., Hypr-Rayleigh Scattering in Solution, *Phys. Rev. Lett.*, *66(23)*, 2980 (1991).
23. Clays, K., and Persoons, A., Hyper-Rayliegh Scattering in Solution with Tunable Femtosecond Continuous-wave Laser Source, *Rev. Sci. Instrum.*, *65(7)*, 2190–2194 (1994).
24. Clays, K., and Persoons, A., Hyper-Rayliegh Scattering in Solution, *Rev. Sci. Instrum.*, *63(6)*, 3285–3289 (1992).
25. Clays, K., Hendrickx, E., Triest, M., Verbeist, T., Persoons, A., Dehu, C., and Brèdas, J.-L., Nonlinear Optical Properties of Proteins Measured by Hyper-Rayleigh Scattering in Solution, *Science*, *262*, 1418–1422 (1993).
26. Cross, G. H., Bloor, D., Axon, T. L., Farsari, M., Gray, D., Healy, D., Swann, M., and Szablewski, M., High Dipole, High β Molecules with Blue Window Transparency, *SPIE Proc.*, *2285*, 11–16.
27. Dalton, L. R., Harper, A. W., Ghosn, R., Steier, W. H., Ziari, M., Fetterman, H., Shi, Y., Mustacich, R. V., Jen, A. K. Y., and Shea, K. J., Synthesis and Processing of Improved Organic Second-order Nonlinear Optical Materials for Applications in Photonics, *Chemistry of Mater.*, *7*, 1060–1081 (1995).
28. Dehu, C., Hendrickx, E., Clays, K., Persoons, A., and Brèdas, J. L., Theoretical and Experimental Investigation of the Second-order Nonlinear Optical Properties of Retinal and Derivatives, *Synthetic Metals*, *71(1–3)*, 1697–1698 (1995).
29. Derhaeg, L., Samyn, C., and Persoons, A., EFISH-studies of Stilbene- and Thiophene-derivatives, (J. Messier et al., eds.), *Organic Molecules for Nonlinear Optics and Photonics*, Kluwer Academic Publishers, The Netherlands (1991), pp. 177–183.
30. Deussen, H. J., Hendrickx, E., Boutton, C., Krog, D., Clays, K., Bechgaard, K., and Persoons, A., Novel Chiral Bis-dipolar 6,6′-disubstituted Binaphthol Derivatives for Second-order Nonlinear Optics: Synthesis and Linear and Nonlinear Optical Properties, *J. Am. Chem. Soc.*, *118*, 6841 (1996).
31. Dibella, S., Fragala, I., Ledoux, I., Diazgarcia, M. A., Lacroix, P. G., Marks, T. J., Sizable 2nd-Order Nonlinear-optical Response of Donor-acceptor Bis-(Salicylaldiminato)Nickel(II) Schiff-base Complexes, *Chem. of Mater.*, *6(7)*, 881–883 (1994).
32. Duan, X.-M., Okada, S., Nakanishi, H., Wantanabe, A., Matsuda, M., Clays, K., Persoons, A., and Matsuda, H., Evaluation of $|\beta|$ of Stilbazolium P-toluenesulfonates by the Hyper/Rayleigh Scattering Method, *SPIE Proc.*, *2143*, 41–51 (1994).

33. Duan, X.-M., Okada, S., Oikawa, H., Matsuda, H., and Nakanishi, H. Comparatively Large Second-order Hyperpolarizability of Aromatic Sulfonate Anion with Short Cut-off Wavelength, *Jpn. J. Appl. Phys.*, *33(part 2, 11A)*, 1559 (1994).
34. Dulcic, A., Some Aspects of Optical Nonlinearity in a New Class of Conjugated Molecules, *Chem. Phys.*, *37*, 57–61 (1979).
35. Dulcic, A., and Flytzanis, C., A New Class of Conjugated Molecules with Large Second Order Polarizability, *Optics Comm.*, *25(3)*, 402 (1978).
36. Dulcic, A., and Sauteret, C., The Regularities Observed in the Second Order Hyper-polarizabilities of Variously disubstituted Benzenes, *J. Chem. Phys.*, *69(8)*, 3453 (1978).
37. Dulcic, A., Flytzanis, C., Tang, C. L., Pepin, D., Fetizon, M., and Hoppilliard, Y., Length Dependence of the Second-order Optical Nonlinearity in Conjugated Hydro-carbons, *J. Chem. Phys.*, *74(3)*, 1559–1563 (1981).
38. Ermer, S., Lovejoy, S., Leung, D., Sptizer, R., Hansen, G., and Stone, R., Synthesis and Nonlinear Optical Activity of Cumulenes, *SPIE Proc.*, *1560*, 120–129 (1991).
39. Fort, A., Barzoukas, M., Boeglin, A., Serbutoviez, C., Oswald, L., and Nicoud, J. F., Conformational Dependence of the Quadratic Hyperpolarizabilities of Two Series of Push-pull Diarylacetylenes and Stilbenes, *Nonlinear Optics*, *5*, 321–327 (1993).
40. Gilmour, S., Marder, S. R., Tiemann, B. G., Cheng, L. T., Synthesis and 1st Hyper-polarizabilities of Acceptor-substituted Beta-apo-8′-carotenal Derived Compounds, *J. Chem. Soc.—Chem. Comm.*, *(5)*, 432–433 (1993).
41. Gilmour, S., Montogomery, R. A., Marder, S. R., Cheng, L.-T., Jen, A. K. Y., Cai, Y. M., Perry, J. W., Dalton, L. R., Synthesis of Diarylthiobarbituric Acid Chromo-phores with Enhanced 2nd-Order Optical Nonlinearities and Thermal-stability, *Chem. of Mater.*, *6(10)*, 1603–1604 (1994).
42. Gulotty, R. J., Brennan, D. J., Chartier, M. A., Gillie, J. K., Haag, A. P., Hazard, K. A., and Inbasekaran, M. N., Nitrophenylhydrazone Based Phenoxy Thermoplastic Electro-optic Polymers, *Tech. Digest Series*, *1*, 6–9 (1993).
43. Henderson, C. C., Cahill, P. A., Kowalczyk, T. C., and Singer, K. D., Thermal, Optical, and Nonlinear-optical Properties of Tetrafluorinated Donor-acceptor Benzenes, *Chem. of Mater.*, *5(8)*, 1059–1060 (1993).
44. Hendrickx, E., Verbiest, T., Clays, K., Samyn, C., and Persoons, A., Hyper-Rayleigh Scattering (HRS) in Isotropic Media, *Mat. Res. Soc. Symp. Proc.*, *328*, 565–575 (1994).
45. Houbrechts, S., Clays, K., Persoons, A., Pikramenou, Z., Lehn, J. M., Hyper-Rayleigh Scattering Investigation of Nitrobenzyl Pyridine Model Compounds, *Chem. Phys. Lett.*, *258(3–4)*, 485–489 (1996).
46. Hubbard, S. F., Petschek, R. G., and Singer, K. D., Spectral Content and Dispersion of Hyper-Rayleigh Scattering, *Optics Lett.*, *21(21)*, 1774 (1996).
47. Hubbard, S. F., Singer, K. D., Li, F., Cheng, S. Z. D., and Harris, F. W., Nonlinear Optical Studies of a Fluorinated Poled Polyimide Guest-host System, *Appl. Phys. Lett.*, *65*, 265 (1994).
48. Huijts, R. A., and Hesselink, G. L. J., Length Dependence of the Second-order Po-larizability in Conjugated Organic Molecules, *Chem. Phys. Lett.*, *156(2,3)* 209–212 (1989).
49. Ikeda, H., Kawabe, Y., Sakai, T., and Kawasaki, K., Nonlinear Optical Properties of Retinal Derivatives, (Chemical Society of Japan) *Chem. Lett.*, 1285–1288 (1989).
50. Ikeda, H., Kawabe, Y., Sakai, T., and Kawasaki, K., Second Harmonic Generation in Nonbenzoid Aromatics, *Chem. Phys. Lett.*, *157(6)*, 576–578 (1989).
51. Ikeda, H., Sakai, T., and Kawasaki, K., Nonlinear Optical Properties of Cyanine Dyes, *Chem. Phys. Lett.*, *179(5,6)*, 551–554 (1991).

52. Jen, A. K. Y., Rao, V. P., Drost, K. J., Wong, K. Y., Cava, M. P., Optimization of Thermal-stability and 2nd-Order Nonlinear-optical Properties of Thiophene Derived Chromophores, *J. Chem. Soc.—Chem. Comm.*, (*18*), 2057–2058 (1994).
53. Jen, A. K. Y., Rao, V. P., Wong, K. Y., and Drost, K. J., Functionalized Thiophenes—2nd-Order Nonlinear Optical Materials, *J. Chem. Soc.—Chem. Comm.*, (*1*), 90–92 (1993).
54. Jen, A. K. -Y., Rao, V. P., Drost, K. J., Cai, Y. M., Mininni, R. M., Kenney, J. T., Binkley, E. S., Dalton, L. R., and Marder, S. R., Progress on Heteroaromatic Chromophores in High Temperature Polymers for Electro-optic Applications, *SPIE Proc.*, *2143*, 30–40 (1994).
55. Kaino, T., Amano, M., and Shuto, Y., Nonlinear Optical Properties of Chromophore-functionalized Polymeric Materials, *Nonlinear Optics*, 163–178 (1992).
56. Kajzar, F., Horn, K., Nahata, A., and Yardley, J. T., Nonlinear Optical Properties of an Organic Single Crystal: Relation between Macroscopic and Microscopic Hyperpolarizabilities, *Nonl. Opt. 8*, 205 (1994).
57. Karna, S. P., Zhang, Y., Samoc, M., Prasad, P. N., Reinhardt, B. A., and Dillard, A. G., Nonlinear Optical Properties of Novel Thiophene Derivatives: Experimental and Ab Initio Time-dependent Coupled Perturbed Hartree-Fock Studies, *J. Chem. Phys.*, *99(12)*, 9984–9993 (1993).
58. Kawabe, Y., Ikeda, H., Sakai, T., and Kawasaki, K., 2nd-Order Nonlinear Optical-properties of New Organic Conjugated Molecules, *J. Mater. Chem.*, *2(10)*, 1025–1031 (1992).
59. Kelderman, E., Heesink, G. J. T., Derhaeg, L., Verbiest, T., Klaase, P. T. A., Verboom, W., Engbersen, F. J., vanHulst, N. F., Clays, K., Persoons, A., and Reinhoudt, D. N., Highly Ordered Films of Neat Calix[4]arenes for Second Order Nonlinear Optics, *Adv. Mater.*, *5(12)*, 925–930 (1993).
60. Kelderman, E., Starmans, W. A. J., Vanduynhoven, J. P. M., Verboom, W., Engbersen, J. F. J., Reinhoudt, D. N., Derhaeg, L., Verbiest, T., Clays, K., Persoons, A., Triphenylcarbinol Derviatives as Molecules for 2nd-Order Nonlinear Optics, *Chem. of Mater.*, *6(4)*, 412–417 (1994).
61. Keller, B. A., Knopfle, G., Pretre, Ph., Bosshard, C. and Gunter, P., Second-order Nonlinear Optical Properties of P-dimethylaminobenzylidene-1,3-indandione (DABI) Measured by EFISH N Dioxane, *Molecular Eng.*, *2*, 325–338 (1993).
62. Knopfle, G., Bosshard, C., Schlesser, R., and Gunter, P., Optical, Nonlinear Optical, and Electro-Optical Properties of 4′-nitrobenzylidene-3-acetamino-4-methoxyaniline (MNBA) Crystals, *IEEE J. Quantum Electron.*, *30(5)*, 1303–1312 (1994).
63. Komatsu, H., Sugiyama, T., Shigemoto, T., Ukachi, T., and Niinomi, N., Linear and Nonlinear-optical Properties of Molecule and Crystal of N-(2,4-dinitrophynyl)-1-chloro-3-phenyl-2-propamine, *Jpn. J. Appl. Phys. Part 1—Regular Papers, Short Notes and Review Papers*, *33(11)*, 6185–6189 (1994).
64. Kondo, T., Morita, R., Ogasawara, N., Umegaki, S., and Ito, R., A Nonlinear Optical Organic Crystal for Waveguiding SHG Devices: (−2)-(a-methylbenzylamino)-5-nitropyridine (MBANP). *Jpn. J. Appl. Phys.*, *28(9)*, 1622 (1989).
65. Kotler, Z., Hierle, R., Josse, D., Zyss, J., and Masse, R., Quadratic Nonlinear-optical Properties of a New Transparent and Highly Efficient Organic-Inorganic Crystal: 2-amino-5-nitropyridinium-dihydrogen Phosphate (2A5NPDP), *J. Opt. Soc. Am. B.*, *9(4)*, 534 (1992).
66. Kott, K. L., Whitaker, C. M., and McMahon, R. J., 2nd-Order Nonlinear-optical Properties of Substituted Arylphosphine Oxides, *Chem. of Mater.*, *7(2)*, 426–439 (1995).
67. Kubo, Y., Aramaki, S., Okamoto, Y., and Murayama, T., Novel Naphthoquinone Methide Dyes for 2nd-Order Nonlinear-optical Materials, *J. Chem. Soc.—Chem. Comm.*, (*9*), 969–970 (1995).

68. Laidlaw, W. M., Denning, R. G., Verbiest, T., Chauchard, E., and Persoons, A., Large Second-order Optical Polarizabilities in Mixed Valency Metal Complexes, *Nature*, *363*, 58–59 (1993).
69. Lalama, S. J., Singer, K. D., Garito, A. F., and Desai, K. N., Exceptional Second-order Nonlinear Optical Susceptibilities of Quinoid Systems, *Appl. Phys. Lett.*, *39(12)*, 940–942 (1981).
70. LeCours, S. M., Guan, H. W., DiMagno, S. G., Wang, C. H., and Therien, M. J., Push-pull Arylethynyl Porphyrins: New Chromophores that Exhibit Large Molecular Hyperpolarizabilities, *J. Am. Chem. Soc.*, 118(6) (1996).
71. Ledoux, I., and Zyss, J., Influence of the Molecular Environment in Solution Measurements of the Second-order Optical Susceptibility for Urea and Derivatives, *Chem. Phys.*, *73*, 203–213 (1982).
72. Ledoux, I., Josse, D., Vidakovic, P., Zyss, J., Hann, R. A., Gordon, P. F., Bothwell, B. D., Gupta, S. K., Allen, S., Robin, P., Chastaing, E., and Dubois, J. C., Second Harmonic Generation by Langmuir-Blodgett Multilayers of an Organic Azo Dye, *Europhys. Lett.*, *3(7)*, 803–809 (1987).
73. Lequan, M., Branger, C., Simon, J., Thami, T., Chauchard, E., and Persoons, A., Hyperpolarizability of Tetraorganotin Compounds Determined by the Hyper-Rayleigh Scattering Technique, *Chem. Phys. Lett.*, *229*, 101–104 (1994).
74. Levine, B. F., Donor-acceptor Charge Transfer Contributions to the Second Order Hyperpolarizability, *Chem. Phys. Lett.*, *37(3)*, 516–520 (1996).
75. Levine, B. F., and Bethea, C. G., Molecular Hyperpolarizabilities Determined from Conjugated and Nonconjugated Organic Liquids, *Appl. Phys. Lett.*, *24(9)*, 445 (1974).
76. Levine, B. F., and Bethea, C. G., Second and Third Order Hyperpolarizabilities of Organic Molecules, *J. Chem. Phys.*, *63(6)*, 2666 (1975).
77. Levine, B. F., and Bethea, C. G., Charge Transfer Complexes and Hyperpolarizabilities, *J. Chem. Phys.*, *66(3)*, 1070–1074 (1976).
78. Levine, B. F., and Bethea, C. G., Hyperpolarizability of the Pyridine-iodine Charge Transfer Complex, *J. Chem. Phys.*, *65(6)*, 2439 (1976).
79. Levine, B. F., and Bethea, C. G., Second Order Hyperpolarizability of a Polypeptide α-helix: Poly-γ-benzyl-L-glutamate, *J. Chem. Phys.*, *65(5)*, 1989–1993 (1976).
80. Levine, B. F., and Bethea, C. G., Ultraviolet Disperson of the Donor-acceptor Charge Transfer Contribution to the Second Order Hyperpolarizability, *J. Chem. Phys.*, *69(12)*, 5240–5245 (1978).
81. Levine, B. F., Bethea, C. G., Wasserman, E., and Leenders, L., Solvent Dependent Hyperpolarizability of a Merocyanine Dye, *J. Chem. Phys.*, *68(11)*, 5042–5045 (1978).
82. Marder, S. R., Cheng, L. T., Tiemann, B. G., Friedli, A. C., Blancharddesce, M., Perry, J. W., Skindhoj, J., Large 1st Hyperpolarizabilities in Push-pull Polyenes by Tuning of the Bond-length Althernation and Aromaticity, *Science*, *263(5146)*, 511–514 (1994).
83. Marder, S. R., Gorman, C. B., Tiemann, B. G., Cheng, L. T., Stronger Acceptors Can Diminish Nonlinear Optical-response in Simple Donor-acceptor Polyenes, *J. Am. Chem. Soc.*, *115(7)*, 3006–3007 (19xx).
84. Miller, R. D., Betterton, K. M., Burland, D. M., Lee, V. Y., Moylan, C. R., Twieg, R. J., Walsh, C. A., and Volksen, W., Thermally Stable Chromophores and Polymers for $\chi^{(2)}$ Applications, *Proc. SPIE 2042*, 354 (1993).
85. Miller, R. D., Lee, V. Y., and Moylan, C. R., Substituted Azole Derivatives as Nonlinear Optical Chromophores, *Chem. Mater.*, *6*, 1023–1032 (1994).
86. Miller, R. D., Moylan, C. R., Reiser, O., and Walsh, C. A., Heterocyclic Azole Nonlinear Optical Chromophores, Donor-acceptor substituted Pyrazole Derivatives, *Chem. Mater*, *5(5)*, 625–632 (1993).

87. Morrison, I. D., Denning, R. G., Laidlaw, W. M., Stammers, M. A., Measurement of First Hyperpolarizabilities by Hyper-Rayleigh Scattering, *Rev. Sci. Instrum.*, *67*(*4*), 1445–1453 (1996).
88. Moylan, C. R., Molecular Hyperpolarizabilities of Coumarin Dyes, *J. Phys. Chem.*, *98*, 13513–13516 (1994).
89. Moylan, C. R., Advances in Characterization of Molecular Hyperpolarizabilities, *SPIE Proc.*, *1775*, 198–205 (1992).
90. Moylan, C. R., and Walsh, C. A., The Hyperpolarizabilities of Tolanes and Pyridyl Analogues, *Nonlinear Optics*, *6*, 113–121 (1993).
91. Moylan, C. R., Miller, R. D., Twieg, R. J., Betterton, K. M., Lee, V. Y., Matray, T. J., and Nuygen, C., Synthesis and Nonlinear Optical Properties of Donor-acceptor Substituted Triaryl Azole Derivatives, *Chem. Mater.*, *5*(*10*), 1499–1508 (1993).
92. Moylan, C. R., Swanson, S. A., Walsh, C. A., Thackara, J. I., Twieg, R. J., Miller, R. D., and Lee, V. Y., From EFISH to Electro-optic Measurements of Nonlinear Chromophores, *SPIE*, *2025*, 192–201 (1993).
93. Moylan, C. R., Twieg, R. J., Lee, V. Y., Miller, R. D., Volksen, W., Thachara, J. I., and Walsh, C. A., Synthesis and Characterization of Thermally Robust Electro-optic Polymers, *SPIE Proc.*, *2285*, 17–30 (1994).
94. Moylan, C. R., Twieg, R. J., Lee, V. Y., Swanson, S. A., Betterton, K. M., and Miller, R. D., Nonlinear Optical Chromophores with Large Hyperpolarizabilities and Enhanced Thermal Stabilities, *J. Am. Chem. Soc.*, *115*(*26*), 12599–12600 (1993).
95. Moylan, C. R., Walsh, C. A., Betterton, K. M., and Twieg, R. J., The Hyperpolarizabilities of Fluorinated 4-methoxystilbenes, *Nonlinear Optics*, *8*, 69–78 (1994).
96. Nunzi, J.-M., Charra, F., Fiorini, C., Zyss, J., Transient Optically Induced Noncentrosymmetry in a Solution of Octupolar Molecules, *Chem. Phys. Lett.*, *219*, 349–354 (1994).
97. Ortiz, R., Marder, S. R., Cheng, L. T., Tiemann, B. G., Cavagnero, S., Ziller, J. W., The Dependence of the Molecular First Hyperpolarizabilities of Merocyanines on Ground-state Polarization and Length, *J. Chem. Soc.—Chem. Comm.*, (*19*), 2263–2264 (1994).
98. Oudar, J.-L., Optical Nonlinearities of Conjugated Molecules: Stilbene Derivatives and Highly Polar Aromatic Compounds, *J. Chem. Phys.*, *67*(*2*), 446–457 (1977).
99. Oudar, J. L., and LePerson, H., Second-order Polarizabilities of Some Aromatic Molecules, *Opt. Commun.*, *15*(*2*), 258–262 (1975).
100. Oudar, J. L., Chemla, D. S., and Batifol, E., Optical Nonlinearities of Various Substituted Benzene Molecules in the Liquid State, and Comparison with Solid State Nonlinear Susceptibilities, *J. Chem. Phys.*, *67*(*4*), 1626–1635 (1977).
101. Pauley, M. A., Guan, H. W., Wang, C. H., Jen, A. K. Y., Determination of First Hyperpolarizability of Nonlinear Optical Chromophores, *J. Chem. Phys.*, *104*(*20*), 7821 (1996).
102 Pretre, P., Kaatz, P., Bohren, A., Gunter, P., Zysset, B., Ahlheim, M., Stahelin, M., and Lehr, F., Modified Polymide Side-chain Polymers for Electro-optics, *Macromolecules*, *27*, 5476–5486 (1994).
103. Rao, V. P., Jen, A. K. Y., Caldwell, J. B., Rhodanine-methine as Pi-electron Acceptor in 2nd-Order Nonlinear-optical Chromophores, *Tetrahedron Lett.*, *35*(*23*), 3849–3852 (1994).
104. Rao, V. P., Jen, A. K. Y., Wong, K. Y., and Drost, K. J., Novel Push-pull Thiophenes for 2nd-Order Nonlinear Optical Applications, *Tetrahedron Lett.*, *34*(*11*), 1747–1750 (1993).
105. Rao, V. P., Jen, A. K. Y., Wong, K. Y., Drost, K. J., Dramatically Enhanced 2nd-Order Nonlinear-optical Susceptibilities in Tricyanovinylthiophene Derivatives, *J. Chem. Soc.—Chem. Comm.*, (*14*), 1118–1120 (1993).

106. Rao, V. P., Wong, K. Y., Jen, A. K. Y., and Drost, K. J., Functionalized Fused Thiophenes—A New Class of Thermally Stable and Efficient 2nd-Order Nonlinear Optical Chromophores, *Chem. of Mater.*, *6(12)*, 2210–2212 (1994).
107. Rao, V. P., Cai, Y. M., Jen, A. K. Y., Ketene Dithioacetal as a Pi-electron Donor in 2nd-Order Nonlinear-optical Chromophores, *J. Chem. Soc.—Chem. Comm.*, *(14)*, 1689–1690 (1994).
108. Rasing, Th., Berkovic, G., Shen, Y. R., Grubb, S. G., and Kim, M. W., A Novel Method for Measurements of Second-order Non-linearities of Organic Molecules, *Chem. Phys. Lett.*, *130(1,2)*, 1–5 (1986).
109. Ray, P. C., Second-order Polarizability of P-substituted Cinnamic Acids, *Chem. Phys. Lett.*, *248(1–2)*, 27–30 (1996).
110. Ray, P. C., Das, P. K., Dissociation Constants of Weak Organic Acids in Protic Solvents Obtained From the First Hyperpolarizabilities in Solution, *J. Phys. Chem.*, *99(51)*, 17891 (1995).
111. Reinhardt, B. A., Kannan, R., and Bhatt, J. C., The Expanding Role of Aromatic Heterocyclic Rings as Functional Groups in the Design of New Nonlinear Optical Materials, *SPIE 2229*, 24–32 (1994).
112. Reinhardt, B. A., Kannan, R., and Dillard, A. G., Synthetic Approaches for Novel Optoelectronic and Polymeric Materials, *SPIE*, *1853*, 51–59 (1993).
113. Singer, K. D., and Garito, A. F., Measurements of Molecular Second-order Optical Susceptibilities Using dc Induced Second Harmonic Generation, *J. Chem. Phys.*, *75*, 3572 (1981).
114. Singer, K. D., Katz, H. E., Sohn, J. E., Dirk, C. W., King, L. A., and Gordon, H. M., Second-order Nonlinear Optical Properties of Donor and Acceptor Substituted Aromatic Compounds, *J. Opt. Soc. Am. B.*, *6*, 1339 (1989).
115. Stadler, S., Bourhill, G., Brauchle, C., Problems Associated with Hyper-Rayleigh Scattering as a Means to Determine the Second Polarizability of Organic Chromophores, *J. Phys. Chem.*, *100(17)*, 6927 (1996).
116. Stadler, S., Dietrich, R., Bourhill, G., Brauchle, C., Long-wavelength First Hyperpolarizability Measurements by Hyper-Rayleigh Scattering, *Opt. Lett.*, *21(4)*, 251–253 (1996).
117. Stadler, S., Dietrich, R., Bourhill, G., Brauchle, C., Pawlik, A., Grahn, W., First Hyperpolarizability Measurements via Hyper-Rayleigh Scattering at 1500 nm, *Chem. Phys. Lett.*, *247(3)*, 271–276 (1995).
118. Stahelin, M., Burland, D. M., and Rice, J. E., Solvent Dependence of the Second Order Hyperpolarizability in P-nitroaniline, *Chem. Phys. Lett.*, *191(3,4)*, 245–250 (1992).
119. Su, W.-F. A., Kurata, T., Nobutoki, H., and Koezuka, H., Quantitative Study of Molecular Orientation in Hemicyanine Langmuir-Blodgett Films by Fourier Transform Infrared Spectroscopy and Second Harmonic Generation, *Langmuir*, *8(3)*, 915–918 (1992).
120. Teng, C. C., and Garito, A. F., Dispersion of the Nonlinear Second-order Optical Susceptibility of Organic Systems, *Phys. Rev. B.*, *28(12)*, 6766–6773 (1983).
121. Ulman, A., Willand, C. S., Kohler, W., Robello, D. R., Williams, D. J., and Handley, L., New Sulfonyl-containing Materials for Nonlinear Optics: Semiemperical Calculations, Synthesis, and Properties, *J. Am. Chem. Soc.*, *112(20)*, 7083–7090 (1990).
122. Verbeist, T., Persoons, A., and Samyn, C., Synthesis and Nonlinear Optical Properties of Preformed Polymers Forming Langmuir-Blodgett Films, *SPIE*, *1560*, 353–361 (1991).
123. Verbeist, T., Samyn, C., and Persoons, A., Second-order Hyperpolarizabilities and Conformations of Functionalized Copolymers, *Org. Mol. for Nonlinear Optics and Photonics*, 201–206 (1991).

124. Verbeist, T., Samyn, C., and Persoons, A., Second Harmonic Generation in Langmuir-Blodgett Films of Preformed Polymers, *Thin Solid Films*, *210/211*, 188–190 (1992).
125. Verbiest, T., Clays, K., Persoons, A., Meyers, F., and Brèdas, J. L., Determination of the Hyperpolarizability of an Octopolar Molecular Ion by Hyper-Rayleigh Scattering, *Opt. Lett.*, *18(7)*, 525–527 (1993).
126. Verbiest, T., Clays, K., Persoons, A., Samyn, C., Meyers, F., and Brèdas, J.-L., Investigations of the β-response in Organic Molecules From Dipolar to Octopolar Systems, *SPIE*, *2025*, 31–36 (1993).
127. Verbiest, T., Clays, K., Samyn, C., Wolff, J., Reinhoudt, D., and Persoons, A., Investigations of the Hyperpolarizability in Organic Molecules From Dipolar to Octopolar Systems, *J. Am. Chem. Soc.*, *116*(*20*), 9320–9323 (1994).
128. Verbiest, T., Hendrickx, E., Persoons, A., and Clays, K., Measurements of Molecular Hyperpolarizabilities using Hyper-Rayleigh Scattering, *SPIE*, *1775*, 206–211 (1992).
129. Verbiest, T., Kauranen, M., and Persoons, A., Parametric Light Scattering, *J. Chem. Phys.*, *101*(*2*), 1745–1747 (1992).
130. Wong, M. S., Nicoud, J. F., Runser, C., Fort, A., Barzoukas, M., and Marchal, E., *Chem. Phys. Lett.*, *253*(*1–2*), 141–144 (1996).
131. Zyss, J., Chemla, D. S., and Nicoud, J. F., Demonstration of Efficient Nonlinear Optical Crystals with Vanishing Molecular Dipole Moment: Second-harmonic Generation in 3-methyl-4-nitropyridine-1-oxide, *J. Chem. Phys.*, *74*(*9*), 4800 (1981).

7
Two-Photon Absorption

Stephen Kershaw
BT Laboratories
Martlesham Heath, Suffolk, England

I. INTRODUCTION

Two-photon absorption was one of the first nonlinear optical phenomena to be experimentally observed, shortly after the advent of intense pulsed laser sources.[1] Although Goeppert-Mayer had established the basis of the quantum mechanical process for the simultaneous absorption of two quanta many years earlier,[2] it remained in the theoretical realms (as far as optical radiation was concerned) until the early 1960s when Kaiser and Garrett first observed blue fluorescence from CaF_2:Eu^{2+} crystals pumped at 694.3 nm by a focused, pulsed ruby laser beam.[3] Ever-increasing peak pulse powers afforded by Q-switching and mode locking laser cavities, and steady improvements in temporal and spatial profiles of the output beams over the following decades has led to a wide range of experimental information on two-photon absorption coefficients and the absorption dynamics in a wide range of materials. These include, for example, compound semiconductors, quantum well structures, inorganic crystals, glasses, as well as what may broadly be classed as molecular materials specifically covered in this handbook, i.e., organic molecules (solvents and dyes), biological materials, organometallics, polymers, and liquid crystals. Data for the former groups of materials have been collated in previous handbooks[4] and subsequent updated editions. Improvements in the sensitivity and flexibility of the various experimental techniques now allows a wide variety of material configurations ranging from bulk solid/crystalline form through to ultra thin films, incorporated into supporting host materials such as polymers[5] or sol-gel glasses,[6] and taken a stage further, as prototype all-optical switching devices.[7]

Two-photon absorption is of great importance in the performance of switching devices for optical computing and telecommunications applications. For all-optical switches, two-photon absorption must be minimized relative to the nonlinear refractive index change. A figure of merit for such, B, defined as:

$$B = \frac{n_2}{2\alpha^{(2)}\lambda}, \tag{1}$$

must exceed unity. Here n_2 is the nonlinear refractive index and $\alpha^{(2)}$ is the two-photon absorption coefficient. For the sake of consistency with other chapters in this handbook, the two-photon absorption coefficient will be written as, $\alpha^{(2)}$ though the reader will often encounter it represented by the symbol β. Note that the two-photon figure of merit B, used here is simply the inverse of the figure of merit proposed by Mizrahi et al.[8]

There are other applications where two-photon absorption is a help rather than a hindrance, for example, in the design of laser dyes for wavelength tuneable systems where the absorption process is used to populate excited states directly.[9] Nonlinear absorptions, including two-photon processes, are of potential interest for ultra-fast optical limiters[10] for applications such as laser pulse protection devices. The presence of two-photon absorptions in spectral regions where there may be no linear absorption bands to take advantage of has led to materials processing applications where two-photon absorption has been usefully applied for surface machining and finishing by laser ablation,[11] and to drive photolytic chemical reactions.[12] Wavelength and spatial selectivity (by controlling the location of beam overlap) in materials addressed with two photons of differing energies to populate a reactive excited state has proved useful for applications in volume optical storage devices[12] and ultra-lithography (i.e., high resolution) photoresists. Perhaps the most widespread use for two-photon absorption to date remains that of a spectroscopic tool. In molecules with spectroscopic states possessing definite parity as restricted by symmetry, linear absorption spectroscopy can only access electric dipole operator connected states, i.e., transitions between states with opposite parity. Note, however, that in the case of molecules such as nitrobenzene, for example, parity is not a good quantum number and the parity selection rule is relaxed. Two-photon absorption spectroscopy selection rules have an even instead of an opposite (as for odd number photon absorption; 1, 3, 5...) parity restriction since the transition strength is controlled by the square of the dipole operator[14] allowing transitions between states of like symmetry to be observed. McClain[15] also established that the symmetries of two-photon allowed transitions could be determined from the polarization sensitivity of transitions accessed by a pair of photons with differing energies even where the material concerned was in a randomly oriented gas or liquid state. In addition to opening up a wide range of transitions in many molecules, two-photon absorption spectroscopy also has the advantage that it may be used to probe bulk properties in the vicinity of strong linear absorption bands where one-photon absorption is limited to surface layers due to high beam attenuation.

After a brief review of the basics of two-photon absorption and the interrelationship with other (often competing) nonlinear absorption and emission processes, this chapter describes a variety of two-photon characterization techniques, discussing both the applicability of the measurement method, and the likely caveats that should be considered in interpreting the results from a given experiment. In addition, a couple of experimental techniques are discussed where measurements of the refractive nonlinearity may be substantially distorted by contributions due to two-photon absorption. While the techniques in question are not methods as such for measuring the two-photon absorption, the data analysis described allows the nonlinear absorption contribution to be separated, and may be used to check for consistency with independent two-photon measurements.

II. DEFINITIONS AND CONVENTIONS

The general equation for the transmitted intensity, I_1, at wavelength λ_1 with propagation in the z direction in a system possessing multiple-photon absorption processes may be written in the form:

$$\frac{dI_1}{dz} = -\alpha^{(1)}I_1 - \alpha^{(2)}I_1I_2 - \alpha^{(3)}I_1I_2I_3 - \ldots\ldots \tag{2}$$

The wavelengths λ_2 and λ_3 corresponding to intensities I_2 and I_3, etc., may in general differ from λ_1. The expansion in terms of a power series is analogous to the expansion of the polarization field in terms of the electric field with complex susceptibility coefficients and the two-photon absorption coefficient, $\alpha^{(2)}$, may thus be related to the complex part of the third order nonlinear susceptibility, Im $[\chi^{(3)}]$. In the case of an isotropic system, treating the susceptibility as a scalar quantity:

$$\alpha^{(2)} = \frac{3\omega}{2\varepsilon_0 c^2 n_0^2}\,\mathrm{Im}[\chi^{(3)}], \tag{3}$$

where n_0 is the linear refractive index, ω is the light's frequency, ε_0 is the permittivity of free space, and c is the speed of light. Similar expressions hold in crystalline or otherwise ordered systems though specific components of the susceptibility tensor will be involved and thus nonlinear dichroism may be observable.[16] It is often the case that where the refractive part of the nonlinearity is measured, it will be quoted in terms of a nonlinear refractive index n_2, related to the real part of the susceptibility, Re$[\chi^{(3)}]$ thus (in *SI* units):

$$n_2 = \frac{1}{\varepsilon_0 c n_0^2}\,\mathrm{Re}[\chi^{(3)}], \tag{4}$$

where n_2 and Re$[\chi^{(3)}]$ have units of m^2W^{-1} and m^2V^{-2}, respectively. The optically induced refractive index change due to excitation at intensity, I (Wm^{-2}) is then simply, n_2I. In the presence of nonlinear absorption, n_2 may be treated as a complex quantity, i.e.:

$$n_2 = n_2' + in_2'', \tag{5}$$

with results often quoted in terms of n_2' and the ratio $\frac{n_2'}{n_2''}$. In experiments such as electro-absorption[17] and transient grating experiments[18] where a Kramers-Kronig analysis is appropriate, relationships between the imaginary susceptibility component, $\chi'^{(3)} \equiv \mathrm{Im}[\chi^{(3)}]$, and the field or intensity induced change in refractive index, $\Delta n''$, are useful, e.g.:

$$\chi'^{(3)} = \frac{n^2\,\Delta n''}{Z_0 I}, \tag{6}$$

where $Z_0 = 120\pi\ \Omega$ is the impedance of the light wave in free space, and the intensity-dependent change in the imaginary refractive index is:

$$\Delta n'' = \frac{\Delta\alpha}{2k} = \frac{\alpha^{(2)}Ic}{2\omega}, \tag{7}$$

where k is the wavevector and $\Delta\alpha$ the change in absorbance. Where measurements have been made on nonlinear polymers or on materials dispersed in a solid matrix to allow film formation, generally with switching device applications ultimately in mind, it is the usual convention to quote the nonlinear absorption in terms of $\alpha^{(2)}$, qualified with an indication of the number of repeat units or molecular concentration per unit volume. Unfortunately, the concentration of active species is not always cited, making direct comparisons between materials difficult. Spectroscopic studies on the other hand, usually express the two-photon absorption in terms of the cross section, $\sigma^{(2)}$, expressed in units of $cm^4s\ photon^{-1}\ molecule^{-1}$. The two quantities are simply related as:

$$\sigma^{(2)} = \frac{\hbar\omega}{N}\alpha^{(2)}, \tag{8}$$

where N is the number density of molecules. $\alpha^{(2)}$ and $\sigma^{(2)}$ values are typically expressed in multiples of cm GW^{-1} and $10^{-50}cm^4s\ photon^{-1}\ molecule^{-1}$, respectively. As the units suggest, the cross-section may be defined by the simple expression for the decay in an incident light flux, F, along a propagation direction, z, due solely to two-photon absorption, namely:

$$\frac{-dF}{dz} = N\sigma^{(2)}F^2, \tag{9}$$

where F has units of photon $cm^{-2}sec^{-1}$, z is in cm, and N in molecules cm^{-3}.

It should be pointed out that the two-photon cross-section defined by Eq. 8 is a "dressed" quantity and will in general be dependent upon the choice of solvent or host matrix to a degree dictated by the polarity and polarizability of the latter. As described in Chapter 3, the host molecules will modify the externally applied field, in this case that due to the incident light wave(s). In local field models such as the Onsager model, this field modification can be encapsulated as an interaction between the vacuum fields via a modified or "dressed" total susceptibility. The latter is simply the vacuum susceptibility of the guest or solute molecule scaled by a local field tensor, e.g., $L_{i'i}(\omega)$.

III. ENERGY LEVEL TRANSITIONS

In many cases, two-photon absorption is not the only excitation process present. In general, there may be several excitation and emission pathways and these will generally interconnect far more than two or three electronic energy levels. This may make assignment of an observed nonlinear absorption ambiguous without taking steps to discriminate amongst the various mechanisms. A simple example is shown in Fig. 1, where an excited state, (3), may be reached either by simultaneous absorption of two photons (and not necessarily any change in symmetry from ground to excited level), or by promotion from the groundstate to excited state (2) by a single-photon transition, and then a further single-photon excitation to level (3). However, each of the single-photon excitations must satisfy the symmetry selection rule. Contributions to the measured nonlinear absorption from the

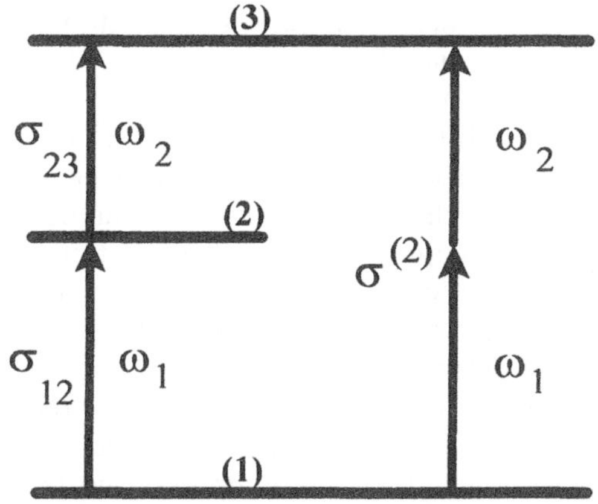

Figure 1 Two-photon and single-photon excited state absorption processes. Both populate excited state (3) but in the latter case the symmetry of the intermediate state, (2), must be opposite to both upper and lower levels. While two-photon absorption is an instantaneous process, the stepwise excitations may be separated by a period determined by the lifetime of level (2). The two photons involved in either process are not necessarily of equal frequency (ω_1, ω_2). Cross-sections $\sigma^{(2)}$, σ_{12}, σ_{23}, correspond to two-photon, single-photon and excited state absorption processes respectively. While σ_{12} andσ_{23} together combine to give an effective cross-section for population of state (3), a two-photon absorption spectrum would distinguish between this *step-wise* route and the *instantaneous* two-photon absorption path.

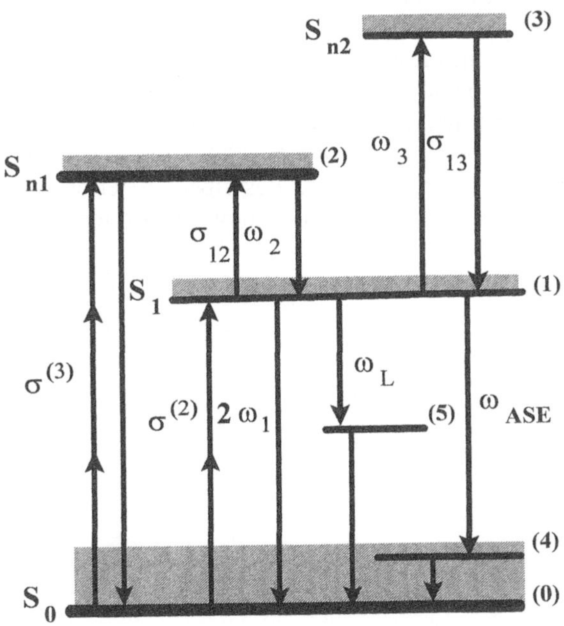

Figure 2 Processes affecting the population of a two-photon excited state, S_1, of importance in the design of molecules for dye laser applications. Cross-sections $\sigma^{(3)}$ and $\sigma^{(2)}$ denote direct three-photon and two-photon absorption, respectively, while σ_{1n} corresponds to excited state absorption from the two-photon state to a singlet state S_n. Where strong ASE or lasing is present at a frequency, ω_{ASE} or ω_L, matching such an excited state absorption (ω_3 or ω_2 in this case) depopulation of the two-photon state may occur.

two processes may be separated using multiple beam techniques if the lifetime of the intermediate single-photon state is long compared with the duration of the measuring pulses, by delaying one pulse relative to the other(s).

The above scheme is of course grossly simplified in that vibrational structure has been neglected as have possible depopulation routes from the intermediate and final states. Figure 2 shows a system of singlet levels such as those encountered in dye molecules as considered by Penzkofer and Leupacher.[9] Here, two-photon absorption is the only significant route to the first excited singlet state S_1 (level 1). Again, further excited state absorption may populate a higher singlet state such as S_2 (level 2). This could also conceivably occur directly by three-photon absorption, although the cross-section, $\sigma^{(3)}$, is small, typically of the order of $10^{-81} cm^6 s^2$ molecule^{-1} photon^{-2}. Three-photon excitation followed by a decay from level 2 to level 1 would populate the same excited state as the two-photon process and both would result in fluorescence in the same wavelength region on decay to the groundstate.

The magnitude of the fluorescence from the two-photon state (1) will also be strongly influenced by any excited state absorption, e.g., to level 3. This may be particularly apparent in dye systems used for laser applications where population inversion is in favor of a two-photon state. In such materials, there may be a significant cross-section for amplified spontaneous emission (ASE) in addition to the emission at the lasing frequency, ω_L. With such intense emissions there may be significant reabsorption by the two-photon excited state to further populate higher excited singlet states as shown in the transition from S_1 to S_{n1}. The involvement of such additional processes may particularly complicate the measurement of two-photon absorption cross-sections by luminescence techniques.

In addition to simultaneous two-photon absorption, there are other nonlinear absorption processes which may be encountered in the course of measurements and which may complicate the determination of $\alpha^{(2)}$ or $\sigma^{(2)}$. Saturated absorption occurs in many metallophthalocyanines,[18] which may be described in terms of a three- or five-level model (Fig. 3). Two *sequential* single-photon absorptions may

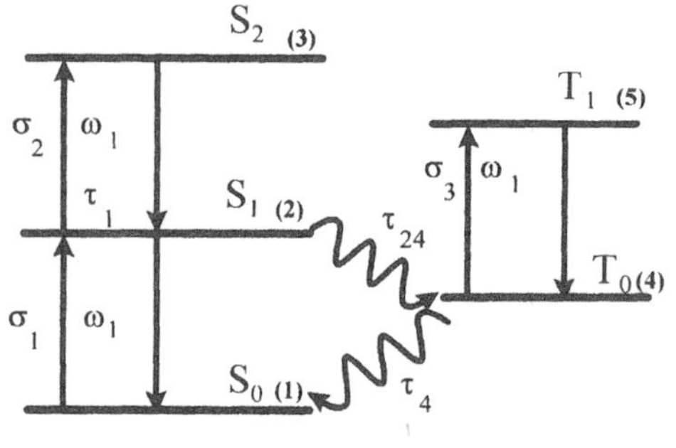

Figure 3 Three and five level model applicable in cases involving saturable and reverse saturable absorption. For measurements with fast pulses a three level model incorporating only singlet states may be necessary to account for RSA, while a five level model including intersystem crossing to a set of triplet states may explain RSA when using long measurement pulses.

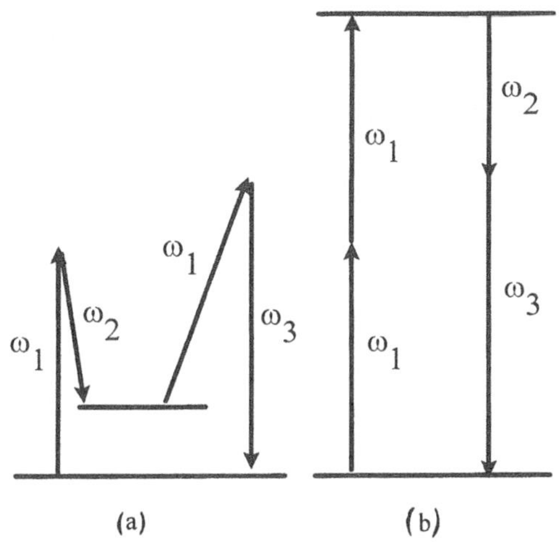

Figure 4 Energy levels involved in CARS; (a) Raman scattering and (b) two-photon absorption contributions in the resonance case. In general, both processes may be present in the signal measured in a CARS experiment and must be separated to extract $\chi^{(3)}$ and $\alpha^{(2)}$ values.

populate an excited singlet state via a lower intermediate level, S_1. Where the second absorption cross-section is small compared with that for the initial transition, i.e., $\sigma_2 < \sigma_1$, and where the lifetime τ_1 is sufficiently long, then the absorption process at ω_1 eventually becomes bleached out as more of the groundstate population becomes significantly depleted, hence the term saturable absorption. The nonlinear transmission of saturable absorbers will increase with increasing power, while two-photon absorption shows the opposite trend.

Where excited state absorption from S_1 dominates over the initial excitation, reverse saturable absorption (RSA) at frequency ω_1 may occur as the total absorption cross-section increases, while the population of the excited states rises. For singlet to singlet transitions, the depopulation of excited states is rapid and so the extra absorption from the excited state may not be all that great for long pulses. If there is substantial intersystem crossing to a set of triplet states, however, a substantial fraction of the excited population is prevented from rapidly decaying back to the groundstate and so may absorb further quanta. Again this increases the total absorption cross-section and is likewise classed as RSA, though here the nonlinear transmission may remain significant for longer durations (determined by the lowest triplet state lifetime). Once more, transitions involving real intermediate excited states may be probed by the use of multiple pulse experiments where the arrival of one of the pulses at the sample may be delayed.

Where spectroscopic measurements have identified the major energy level locations for a given molecule, it is often useful to make a rate equation analysis to determine the primary excitation and energy level depopulation pathways in order to fully understand the significance of two-photon absorption for that material. Taking the energy level diagram of Fig. 3 as an example, Tutt and Boggess[10] obtain the following set of equations for the population rates of each level:

$$\frac{\delta N_1}{\delta t} = -\sigma_1(N_T - N_2 - N_3 - N_4 - N_5)\frac{I}{h\upsilon} + \frac{N_2}{\tau_2} + \frac{N_4}{\tau_4} \tag{10}$$

$$\frac{\delta N_2}{\delta t} = \sigma_1(N_T - N_2 - N_3 - N_4 - N_5)\frac{I}{h\upsilon} - \frac{N_2}{\tau_2} - \frac{N_4}{\tau_{24}} - \frac{\sigma_2 N_2 I}{h\upsilon} \tag{11}$$

$$\frac{\delta N_3}{\delta t} = \frac{\sigma_2 N_2 I}{h\upsilon} - \frac{N_3}{\tau_3} \tag{12}$$

$$\frac{\delta N_4}{\delta t} = \frac{-\sigma_4 N_4 I}{h\upsilon} - \frac{N_4}{\tau_4} + \frac{N_2}{\tau_{24}} + \frac{N_5}{\tau_5} \tag{13}$$

$$\frac{\delta N_5}{\delta t} = \frac{\sigma_4 N_4 I}{h\upsilon} - \frac{N_5}{\tau_5} \tag{14}$$

$$N_T = N_1 + N_2 + N_3 + N_4 + N_5, \tag{15}$$

where $h\upsilon$ is the single-photon energy at frequency ω_1 and N_i the population of state i. These simultaneous equations may be solved to give the steady state populations, N_i from which the overall absorption coefficient may be found using:

$$\frac{\delta I}{\delta z} = -\sigma_1(N_T - N_2 - N_3 - N_4 - N_5)I - \sigma_2 N_2 I - \sigma_4 N_4 I. \tag{16}$$

It may often be safe to assume that there is little population of the triplet states leading to a much simpler set of expressions. A five-state analysis of this type has been applied to several metal phthalocyanines by Mansour et al.[19] where the materials were of interest for fast all-optical power limiting devices.

A full analysis of the absorption dynamics of an energy level system similar to that shown in Fig. 2 has been given by Penzkofer and Leupacher.[9]

Three wave mixing techniques, such as Coherent Anti-Stokes Raman Spectroscopy (CARS), are often used to measure the dispersion of the two-photon absorption by determining the real and complex $\chi^{(3)}$ spectra. The relevant energy level transitions are shown in Fig. 4, together with the competing two-photon process in the case of a resonant transition. Interpretation of CARS spectra require the separation of the two-photon response from the Raman signal.[20]

It may be appreciated that for an unequivocal measurement of $\alpha^{(2)}$ or $\sigma^{(2)}$, it is strictly necessary to use several techniques or at the very least make cross-checks to ensure that the response obtained is solely due to straight two-photon absorption. In some work reported in the literature, this appears not to have been done, either because it has become widely accepted that two-photon absorption is the most likely and predominant process occurring for a particular material or group of materials at a given wavelength, or perhaps that the focus of the papers in question concerned the refractive part of the nonlinearity rather than the nonlinear absorption, leading to the tacit assumption that only direct two-photon absorption was present. In some instances these assumptions have proved to be reasonable, with several independent groups obtaining broadly similar results. However, there are certain cases, such as cited by Chen and McCann[21] for data on transitions in liquid benzene, where a very broad range of results can be found (see note for benzene in data tables). This effectively forms the caveat for the data tables listed later in the chapter: Take into account the precise experimental conditions—measurement wavelength relative to resonances, choice of pulse widths, any delays between arrival of excitation pulses, the technique used (and its validity)—and finally look how closely it fits in with the data of other groups.

IV. EXPERIMENTAL TECHNIQUES

In this section, several common experimental techniques for measuring $\alpha^{(2)}$ or $\sigma^{(2)}$ are described and their subtleties discussed. Some of the four wave mixing experiments described are, strictly speaking, more suitable for measurement of the refractive nonlinearity than two-photon absorption, but have been modified to take account of the presence of an imaginary susceptibility component. The methods in such cases are intended for use in separating the real and imaginary quantities given an independent measurement of the nonlinear absorption. As such they are useful as a test of internal measurement consistency. The third harmonic Maker fringe technique is included in the wave mixing category and is an increasingly popular method for measuring both real and imaginary susceptibility spectra provided three-photon resonances do not make significant contributions. In most cases, the issues influencing measurement accuracy and sensitivity are discussed and any necessary secondary measurements or relevant calibration procedures are included in the descriptions. It must be emphasized that the most potentially reliable and currently popular measurement techniques for quantitative two-photon absorption coefficients are single beam nonlinear absorption, two-photon luminescence, and Z-scan. Other techniques such as four wave mixing are included since two-photon effects are often encountered in such measurements and values for $\alpha^{(2)}$ estimated by the departure of the response from that of an ideal purely refracting material. The discussion of such experiments merely puts such cited values in context and is not meant in general to endorse them as measurement techniques for two-photon absorption. Older and less direct measurement techniques such as thermal lensing, photoionization spectroscopy, and photoacoustic spectroscopy are mentioned to place some of the more historical literature values in context.

A. Single Beam, Single Wavelength Techniques

1. Nonlinear Transmission Measurements

The transmission, T, of a material of thickness, L, and with linear and nonlinear transmissions, α and $\alpha^{(2)}$ respectively, follows from the solution of Eq. 2, where terms higher than $\alpha^{(2)}$ are neglected and $I_1 = I_2$. In the case of a CW beam of uniform spatial intensity profile, the transmission is:

$$T = T_l T_{nl}, \tag{17}$$

where the linear transmission, T_l is given by:

$$T_l = (1 - R)^2 e^{-\alpha L},$$

and the nonlinear factor T_{nl} is:

$$T_{nl} = \frac{1}{1 + f},$$

with

$$f = \frac{\alpha^{(2)}(1 - R)I_1(1 - e^{-\alpha L})}{\alpha}. \tag{18}$$

The light reflected from the rear surface of the sample is linearly accounted for by the reflectivity factor R, though it is assumed that $R << 1$ so that the assumption of negligible two-photon absorption of the internally reflected light is valid. Note that because $R << 1$, etalon, effects can be ignored. In the more general case of a time varying and spatially nonuniform beam, the full expression for the total transmission is:

$$T = \frac{\int_{-\infty}^{\infty} dt \int_0^{\infty} 2\pi\, r dr I(r, L, t)}{\int_{-\infty}^{\infty} dt \int_0^{\infty} 2\pi\, r dr I(r, 0, t)} \tag{19}$$

Solution of Eq. 19 for a uniform beam with a Gaussian time profile, i.e., $I_1 \propto e^{-(t/\tau)^2}$, rectangular uniform spatial profile, and no appreciable spreading leads to a modified T_{nl} contribution,

$$T_{nl} = \frac{1}{\sqrt{\pi}} \int_{-\infty}^{\infty} \frac{e^{-\xi^2}}{1 + f\, e^{-\xi^2}}\, dz, \tag{20}$$

where ξ is the normalized time $\xi \equiv t/\tau$.

For a CW beam with a Gaussian cross-section T_{nl} is given by:

$$T_{nl} = \frac{\ln(1 + f)}{f}, \tag{21}$$

while in perhaps the most frequently encountered case, where both time and spatial profiles are Gaussian:

$$T_{nl} = \frac{2}{f\sqrt{\pi}} \int_0^{\infty} \ln(1 + f\, e^{-\xi^2})\, dz. \tag{22}$$

For computational convenience, when $f < 1$ the previous expression may be closely approximated by the series:

$$T_{nl} = \sum_{n=1}^{\infty} (-f)^{n-1} n^{-3/2}. \tag{23}$$

With only 10 terms, the summation differs from the numerically integrated form by only up to 3 percent or so over the range $f = 0$ to 1. In this form, $\alpha^{(2)}$ may easily be determined from the variation of the transmitted intensity against incident intensity with the experimental layout shown in Fig. 5.

A primary consideration in the applicability of the above expressions is the temporal and spatial beam quality of the laser source used. While these may be quite stable and approximate reasonably to a Gaussian form for many commercially available Nd/YAG and YLF lasers, sources such as dye lasers may suffer from poorer beam quality. With such systems, it is generally desirable to calibrate measurements against a suitable reference material. Typically, ZnSe is used for mea-

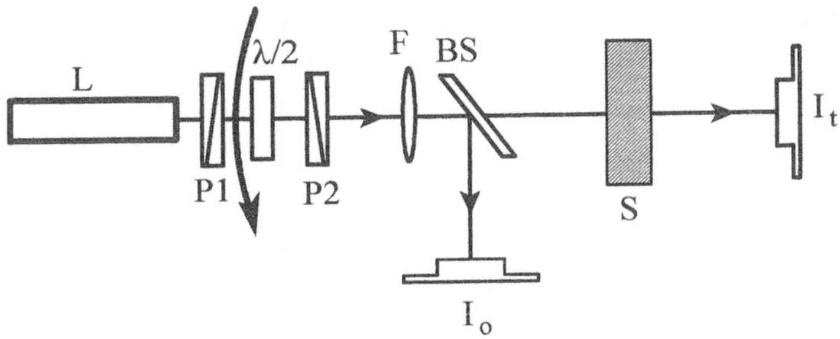

Figure 5 Typical single beam nonlinear absorption experiment. The intensity of the pulses from laser, L, is controlled with a polarizer/half-wave plate combination, P1, $\lambda/2$, P2. The intensity of each pulse, I_o, is monitored via a beam splitter BS and detector prior to the sample, S. The beam is typically focused to a spot several tens of microns in diameter by lens, F, chosen to have a depth of field (confocal length) greater than the optical path length of the sample. Under these circumstances, with the sample positioned at the beam waist, the thin sample approximation may be applied. The nonlinear transmission is measured as I_t/I_o after the sample. The low intensity (linear) transmission is determined separately by standard spectroscopic measurement.

surements in the visible having a $\alpha^{(2)}$ of ~5.5 cm/GW for incident radiation at 532 nm.[22] In some instances, cross-referencing is done on a shot-by-shot basis with the laser source split between a sample and reference material simultaneously (often referred to as two-channel normalization). The spatial beam quality is determined by the choice of focusing lens. Furthermore, the above analyses assume a thin sample such that within the sample volume the beam is effectively parallel. The choice of lens focal length therefore limits the maximum sample thickness to the diffraction length within the sample,

$$\frac{n_0 \pi w_0^2}{\lambda},$$

where w_0 is the beam waist at the lens focus, n_0 is the refractive index and λ is the wavelength.

By way of a numerical example, assuming first a material with negligible linear absorption, a two-photon absorption coefficient of 10 cm/GW and ignoring reflections at the interfaces, with a Gaussian profile beam with a beam waist of 200 μm and peak irradiance of 0.8 GW/cm^2 (of the order of 0.1 mJ in a 100 ps pulse), a T_{nl} of 86 percent would be expected for a 5 mm path length. At a measurement wavelength of 1.5 μm, the confocal length of a material with a refractive index of 1.5 would be approximately π cms, i.e., more than suitable for solution measurements in glass cuvettes. More realistically, with a linear absorption coefficient of 2 cm^{-1}, T_{nl} drops to only 95.9 percent, while T_l becomes 36.8 percent. Typically, nonlinear transmission measurements are made at somewhat higher sample concentrations than the linear absorption coefficient measurements and a simple Beer law relationship between linear transmission and concentration assumed. This may well be valid in many cases but may not be where there are significant intermolecular interactions, e.g., aggregation effects, dimer formation, etc.

The analysis of nonlinear transmission by this technique may be complicated by the possible presence of other multi-photon absorption processes, particularly excited state absorption. In this case, Eq. 2 becomes:

$$\frac{dI_1}{dz} = -\alpha I_1 - \alpha^{(2)} I_1^2 - \sigma_{ex} N I_1, \tag{24}$$

where σ_{ex} is the excited state absorption cross-section, and N is the excited state population density due to the absorption. The latter is governed by the rate equation:

$$\frac{dN}{dt} = \frac{\alpha I}{h\upsilon} - \frac{N}{\tau}, \tag{25}$$

where $h\upsilon$ is the energy of incident photons, and τ is the excited state decay time. Contributions to the excited state population due to two-photon absorption are neglected here. Plotting the observed nonlinear transmission data as $T_{nl}^{-1} - 1$ on a log-log plot against peak irradiance should then lead to a curve with gradient between the two limiting cases of 1 for ESA alone, and 2 for two-photon absorption in isolation.[23]

2. Open Aperture Z-scan

The Z-scan measurement technique and the more sensitive EZ-scan variation[24] have become increasingly popular for the determination of refractive nonlinearities. (See Chapter 8.) The presence of two-photon absorption in the restricted aperture (or field stop) scan is usually accounted for in a simultaneous open aperture scan where the Z-position dependent signal is determined only by absorptive effects. Provided that the laser wavelength is well-removed from any single photon resonance or significant vibrational overtones, and also that the pulse rep. rate is sufficiently low (<10 Hz typically), linear thermal effects can be neglected and the signal is predominantly due to the nonlinear absorption, $\alpha^{(2)}$. Figure 6 shows a typical Z-scan experimental arrangement.

For a temporally Gaussian pulse with an incident Gaussian spatial profile, it can be shown[25] that the on axis transmission as a function of sample position relative to the lens focal point, $T(z)$, is given:

$$T(z) = \sum_{m=0}^{\infty} \frac{(-q(z))^m}{(m+1)^{3/2}}, \tag{26}$$

provided $|q| < 1$, which is essentially identical in form to Eq. 23 though with f replaced by $q(z)$ where,

$$q(z) = \frac{I_1 L_{eff} \alpha^{(2)}}{1 + \frac{z^2}{z_0^2}}. \tag{27}$$

As before, should q be greater than 1, the approximation leading to Eq. 26 is not valid so the full integral expression corresponding to Eq. 22 should be evaluated numerically in place of the approximation in Eq. 26.

The confocal parameter, z_0 is simply $\pi w_0^2/\lambda$, where w_0 is the beam waist at the

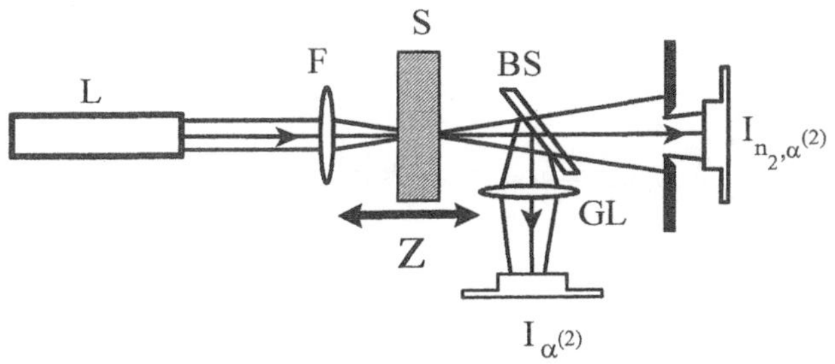

Figure 6 Two-photon absorption measurement by Z-scan method. The beam from laser, L, is focused to a few tens of microns diameter spot by lens F. In the thin sample limit the beam waist is effectively parallel over a length exceeding the optical path through the sample. Self-focusing (or defocusing) *and* nonlinear absorption give rise to transmission changes at the apertured detector, $I_{n2,\alpha^{(2)}}$. By inserting a beam splitter, BS, after the sample the nonlinear absorption may be separately measured at a second detector, $I_{\alpha^{(2)}}$. Including a large diameter collecting lens, GL, ensures that nonlinear focusing effects are excluded. In practice, a third normalizing beam splitter and detector combination prior to lens F is also used (not shown) to allow for shot-to-shot intensity variations in the laser. The sample S is translated through the focal position of lens, F, along axis Z.

lens focus, and the effective sample length, L_{eff}, takes into account the linear absorption, α:

$$L_{eff} = \frac{(1 - e^{-\alpha L})}{\alpha}. \tag{28}$$

As in the case of nonlinear absorption measurements, the maximum sample thickness for the above analysis is limited to the diffraction length in the sample, $n_0 z_0$.

Curve fitting experimental data to Eq. 26 with terms up to $m = 10$ is usually sufficient to determine the factor $I_1 L_{eff} \alpha^{(2)}$, from which $\alpha^{(2)}$ can be extracted provided the linear absorption is known and the peak on-axis irradiance, I_1, is also known. Again, it is useful to have a reference material against which to cross-calibrate this type of experiment so that errors due to uncertainties in either the spatial or temporal beam profiles can be avoided. A typical open aperture Z-scan trace is shown in Fig. 7. The incident power levels used in this case were comparatively low (10 kW peak power pulses, 100 ps pulse widths) so relatively small peak intensity induced changes in transmission were observed (~3 percent in transmission at focus).

In the nonlinear transmission experiment described above, excited state absorption can also lead to a signal trace similar to the two-photon response. Van Stryland and co-workers at CREOL[26,27] have obtained approximate expressions for the change in transmission. ΔT, in the open aperture scan, for both two-photon and ESA processes, respectively:

$$\Delta T \approx \frac{\alpha^{(2)}}{2\sqrt{2}} I_1 (1 - r) L_{eff}, \tag{29}$$

(two-photon).

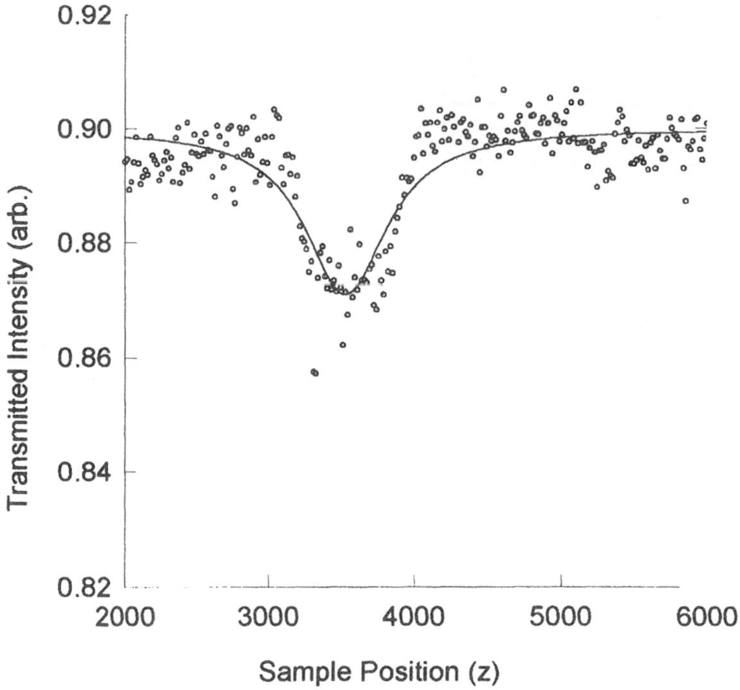

Figure 7 Typical Open Aperature Z-Scan data for a Nickel dithiolene compound. Measurement wavelength, 1.32 μm, pulse widths 100 ps, path length 1 cm, $\alpha^{(2)}$ = 1.3 GW/cm, I_1 ~0.7 GW/cm^2. The curve is fitted using Eqs. 26 and 27.

$$\Delta T \approx \frac{\sigma_{ex}\alpha}{4\hbar\omega} F_1(1 - r)L_{eff}, \tag{30}$$

(ESA)
both provided $\Delta T < 0.1$.

Since Eq. 29 is *irradiance*-dependent, while Eq. 30 follows the on-axis *fluence*, F_1, it is possible to distinguish between the two processes by making measurements with equal fluence but different duration pulses. With picosecond pulsed lasers this may often be easily accomplished by interchanging a cavity etalon and adjusting the output power level.

With picosecond laser sources it is also possible to identify slow, non-electronic contributions to the open aperture signal by using fast detectors. This, however, makes optical alignment somewhat more critical because fast photodiode detectors tend to have small junction diameters to limit the device capacitance and therefore minimize rise and decay times. Photomultipliers are sometimes used but have restricted wavelength coverage only up to the near-infrared. Time-resolved Z-scan may usefully discriminate against thermal effects and in many cases may be able to pick up transient response times associated with reorientational effects in large molecules (rotational dichroism). Measurement sensitivity is largely determined by the available laser power and shot-to-shot stability. With a moderate level of signal averaging, two-photon induced changes in transmission of greater than 1 percent can usually be resolved.

As reasonably high power tuneable laser sources are becoming more economically viable and particularly where the eclipsing Z-scan technique can be applied, it is becoming increasingly more favorable to use the open aperture Z-scan method to record two-photon absorption *spectra* rather than covering only one or two fixed wavelengths.

B. Thermal Lensing

Although this technique[28] strictly employs two beams, it is essentially related to the above techniques and may be considered as a forerunner of the Z-scan type of experiment. Z-scan is, however, the preferred technique by far. A sample is placed either just before or just after the focal point of a lens and an intense pump signal is used to populate excited states (see Fig. 8). Relaxation of these states via vibrational transitions and intermolecular collisions causes a proportion of the absorbed energy to be released as heat in the sample which gives rise to a thermally induced time-dependent change in refractive index, hence the term "thermal lensing." The index change is probed by a weaker collinear probe laser beam at a differing wavelength, preferably well-removed from any significant absorption band. The distortion in the far field profile of the probe beam is monitored by a detector placed behind a pinhole. The distortion is monitored as a function of time where

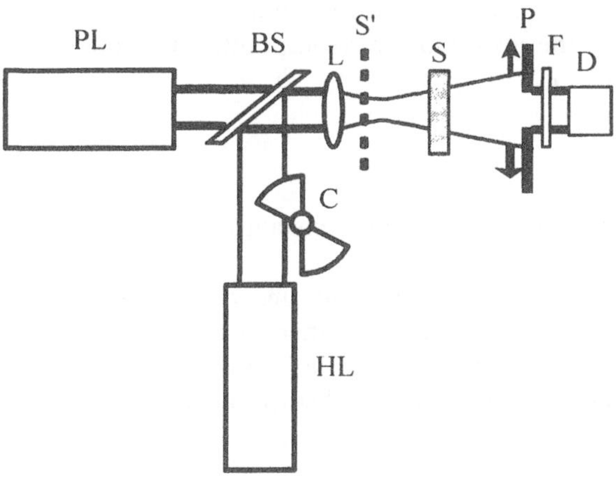

Figure 8 A typical thermal lensing experiment. HL, heating laser; PL, probe laser; C, chopper wheel (if HL not pulsed); BS, dichroic beam splitter, L, focusing lens; S, sample; P, pinhole at far field; F, blocking filter to remove HL beam; D, detector. With the arrangement shown, a heating pulse will result in a defocusing of the probe beam profile in the far field, represented by the vertical arrows. If the sample is placed at S′, i.e., before the focus, the probe beam will undergo focusing upon laser induced heating. Other arrangements are also possible, e.g., HL and PL beams may be counter propagating, or may be non-collinear. The two beams may be orthogonally polarized to permit separation with polarizing beam splitters rather than wavelength filters.

the pump signal is either a CW beam chopped on a time-scale long compared with the thermal response time of the material under study, or a fast pulsed source. The fraction of light absorbed per unit length for a multi-photon absorption process is:

$$\alpha = mN\sigma^{(m)}I^{(m-1)}, \tag{31}$$

where m is the number of photons absorbed in each absorption event, N is the number of molecules per cm^3 in the sample, I is the incident photon flux at the absorption wavelength in photons cm^2/S, and $\sigma^{(m)}$ is the cross-section for the multi-photon absorption process in cm^{2m} s$^{(m-1)}$ molecules^{-1} photon$^{(1-m)}$. For moderate single-photon absorption cross-sections, $\sigma^{(1)}$ is typically of the order 10^{-17} cm^2 molecule^{-1} and may readily be observed with CW pump lasers, while two-photon absorptions, typically in the range of $\sigma^{(2)} \sim 10^{-50}$ cm^{4}s molecule^{-1} photon^{-1}, require the use of moderately powered pulsed lasers delivering better than, say, 100 mJ in 10 ns. Three-photon absorptions in the range $\sigma^{(3)} \sim 10^{-81}$ cm^{6}s^2 molecule^{-1} photon^{-2} would be observable with power levels an order of magnitude greater. With currently commercially available sources, the technique is generally used in the pulsed pump configuration for the evaluation of one- and two-photon absorptions.

The approximate magnitude of the signal response immediately following the heating pulse is:

$$\Delta I(0)_{\text{probe}} \propto \frac{2Z}{f_0}, \tag{32}$$

where f_0 is the lens focal length, and Z is the offset of the sample from the focal position. The latter is positive when the sample is placed after the focal point, or negative when the sample is between the lens and the focal point. The modulus of the signal transient decays according to:

$$\Delta I(t)_{\text{probe}} \approx \left(1 - \frac{2mt}{t_c}\right)^{-2}, \tag{33}$$

with t_c being a material dependent thermal decay time constant and m the number of photons absorbed in the excitation.

A more detailed analysis of the magnitude of the one- and two-photon contributions to the transient thermal lensing signal has been made by Takatori et al.[29] Referring to the geometry given in Fig. 9, the radial index distortions within the sample ($z_1 \leq z \leq z_1 + 1$) due to one- and two-photon absorptions are:

$$\delta n_1(r) = \frac{E}{V_1 C_p \rho} \frac{\partial n}{\partial T} I(x, y)\, e^{-\alpha z}, \tag{34}$$

and

$$\delta n_2(r) = \frac{E}{V_2 C_p \rho} \frac{\partial n}{\partial T} I(x, y)^2\, e^{-\alpha^{(2)} I_0 z}, \tag{35}$$

where E is the heat dissipated in the sample, C_p is the heat capacity at constant pressure, I_0 is the incident intensity, ρ is the density of the material, $I(x, y)$ is the dimensionless transverse intensity profile of the input beam normalized to $I(0,0) = 1$, and $\partial n/\partial T$ is the thermo-optic coefficient. The latter quantity is almost invariably

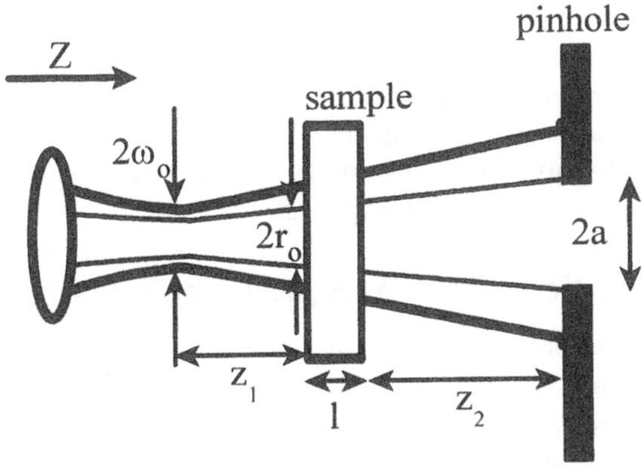

Figure 9 Thermal lensing geometry referred to in Eqs. 34 to 38.

negative owing to the usual reduction in density of a medium upon heating. (See Chapter 3.) Volumes V_1 and V_2 are defined by:

$$V_1 = \int_{-\infty}^{\infty}\int_{-\infty}^{\infty}\int_{z^1}^{z_1+l} I(x, y)\, e^{-\alpha z}\, dx\, dy\, dz, \tag{36}$$

and

$$V_2 = \int_{-\infty}^{\infty}\int_{-\infty}^{\infty}\int_{z_1}^{z_1+l} I^2(x, y)\, e^{-2\alpha^{(2)} I_0 z}\, dx\, dy\, dz. \tag{37}$$

The one- and two-photon contributions to radial probe beam profile, $P(x, y)$, can then be calculated by substitution of Eq. 34 or 35 into the general ray equation:

$$\frac{d}{ds}\left[n(r)\frac{dr}{ds}\right] = \nabla n(r), \tag{38}$$

where s is the coordinate in the z-direction in the sample and the path of light along z is given by $r(s)$. After spatial integration over the pinhole diameter, the thermal lensing signal may be obtained.

While this technique has been widely used for spectroscopic analysis of weak absorption features, especially in highly dilute systems, there may be several difficulties in interpreting results. Experimentally, as with all multiple beam techniques, precise beam overlap and good spatial beam quality are essential if measurements are to be reasonably quantitative. Also there may be many absorption processes, both single- and multi-photon, and additionally excited state transitions can contribute to the absorption of heat by the sample. In the case of materials in solution, the solvent may also contribute to the signal. An elegant solution to this problem may be found by including a reference cell filled with solvent at an almost equal but opposite offset from the focus relative to that of the sample cell which effectively nulls the solvent's contribution.[30] Finally, contributions to the signal from refractive nonlinearities may also in general be present. Where the nonlinearity is sufficiently strong, and the dimensions of the pinhole, etc., are of appropriate dimensions, the experiment will effectively correspond to that used in the Z-scan method with a small aperture to measure the real component of the optical nonlinearity. These considerations effectively limit applicability of the technique to materials with weak refractive nonlinearities, e.g., for weak spectroscopic absorption features, well-removed from resonant enhancement of n_2, or as a measurement linked with other characterization techniques to assist in the separation of signal contributions from several different types of absorption process.[29,31]

C. Dual Beam Nonlinear Absorption Spectroscopic Techniques

Two-photon absorption spectra may be obtained by scanning the wavelength of one (or both) of the incident beams in a nonlinear absorption experiment where the two beams are collinear. As mentioned in the previous section, it is of paramount importance that the beam overlap and beam profiles are well-optimized for quantitative spectra. Before the advent of tuneable laser sources, the only intense wideband light sources were CW incandescent or arc lamps, and so these were used in conjunction with a monochromator (after the sample) to provide the scanned beam, while the second quanta were provided by a fixed wavelength pulsed laser.[32] The nonlinear absorption at the scanned frequency ω_1 then depends upon the product of the intensities of the two beams at ω_1 and ω_2 and the two-photon absorption coefficient, $\alpha^{(2)}$, i.e.:

$$\frac{dI_1}{dz} = -\alpha I_1 - \alpha^{(2)} I_1 I_2, \tag{39}$$

in the absence of any other competing excitation process. ω_1 is generally termed the probe (CW), and ω_2 the pump (pulse) frequency, respectively.

Present day tuneable pulsed laser sources allow near continuous coverage of the wavelength range typically from 600 nm with combinations of dye lasers, Raman shifted lasers, and optical parametric oscillators. However, it is still somewhat expensive to cover such a large tuning range, with some of these systems being troublesome to operate and often leaving plenty to be desired in terms of shot-to-shot stability and spatial beam quality. For a completely rigorous study, it is desirable to be able to independently scan both pump and probe lasers so that excited state absorptions can be fully identified. For pulse durations faster than the expected lifetime of any singly resonant intermediate states, it is also possible to identify two-step sequential absorption processes by temporally delaying one pulse relative to the other. In the case of a true instantaneous two-photon excitation, the absorption signal should follow the cross-correlation of pump and probe pulses against delay time.[33] Of course it is always necessary to separate the pump and probe pulses to monitor changes in transmission at the probe wavelength only. This may be done by selecting orthogonal polarizations and using broadband polarizing prisms to separate the beams, or by separation in the wavelength domain using a monochromator linked to the scan laser.

An increasingly popular approach covering probe energies in the range 1–2 eV[33–35] involves the use of a pulsed white light continuum probe generated by focusing a portion of the pulsed pump beam into a flow cell containing liquids such as ethylene glycol (resulting in a 1.4-1.8 eV continuum), heavy water (resulting in a 1.3–2.2 eV continuum) or water. Probe pulse durations are broadly similar to those of the pump when generated in this way, and pulse widths as short as 100 fs have been reported.[33] Nonlinear transmission spectra may be recorded by dispersing the broadband probe through a spectrograph and then recording the distribution using a CCD array or other imaging device. A rectangular array where image lines are clocked sequentially across the device are useful in that a sequence of complete spectra may be gathered rapidly at the clock rate, typically several kHz, and then processed afterwards on readout of the entire diode array. Probe-pulse delay with fast pulses can again be used to give useful information on the dynamics of the transitions involved.[35] Measurements of this type are often accompanied by observation of the transient excited state fluorescence spectra since the spectrometer is already effectively part of the experiment, and this provides additional help in assigning the excitation processes.

Pfeffer et al.[36] have recently reported measurements of two-photon absorption in polythiophenes using an extension of the pump-continuum probe technique which they term transient Kerr ellipsometry. Rather than probe the nonlinear transmission directly, they place the sample cell between (initially) crossed polarizers. The pump-induced nonlinear dichroism rotates the transmitted probe beam by a small angle which may be determined by rotating the analyzer through an angle, ϕ, to minimize the probe at a given wavelength. The transmitted intensity between crossed polarizers will be determined by a combination of the nonlinear dichroism and the nonlinear refractive index. The nonlinear refractive index alone will not cause a continuous rotation of the elliptically polarized light, but rather will cause the ellipticity to change. Consider, for example, the case where the incident probe beam is linearly polarized at 45 degrees to the pump polarization. As a function of increasing nonlinearity, the output polarization is first elliptical with major axis aligned with the input polarization. For higher nonlinearity, the output state is circularly polarized then elliptical with the minor axis parallel to the input polarization. For high enough nonlinearity, the output polarization is 90 degrees from the input. The transmitted intensity through the crossed polarizers, I, is $I = I_0\ (1 + \sin(2\alpha)\cos\delta)$, where α is the angle between the pump and input polarization and δ the phase retardation due to the nonlinearity. Clearly, I is minimum for $\alpha = \pm 45$ degrees depending on the sign of $\cos\delta$. When nonlinear dichroism is present, the fast axis of the input ellipse may be continuously rotated. Resolving the input polarization state onto orthogonal principle axes of the anisotropic absorption, it is clear that one of the components will be more greatly attenuated than the other. This results in a rotation of the ellipse by an angle ϕ.[200] The angle is therefore related to the imaginary component of the susceptibility:

$$\phi = \frac{3\pi l I_{\text{pump}}}{2c\varepsilon_0 \lambda_{\text{probe}} n_{\text{pump}} n_{\text{probe}}} \text{Im}[\chi^{(3)}_{xxxx} - \chi^{(3)}_{xyyx}], \tag{40}$$

where l is the sample thickness, and n_{pump}, n_{probe}, I_{pump} are indices at pump and probe wavelengths, and the pump intensity, respectively. In the case reported, it was possible to make the simplifying quasi-one-dimensional assumption that $\chi^{(3)}_{xxxx} = 3\chi^{(3)}_{xyyx}$ (see Chapter 3), so that the two-photon absorption is simply given by the expression:

$$\alpha^{(2)} = \frac{3\phi}{l I_{\text{pump}}}. \tag{41}$$

D. Two-photon Luminescence Techniques

The use of nonlinear transmission techniques is limited to materials with sufficiently large two-photon cross-sections, and measurements with two-photon excitation intensities high enough to achieve induced transmission changes of the order of >0.1–1 percent in order to extract meaningful data above the typical power level fluctuations present in high power lasers. For materials with smaller cross-sections, two-photon excited fluorescence may be a more sensitive alternative. This is in fact the method first used by Kaiser and Garret[3] to detect two-photon absorption in CaF_2:Eu^{2+}. A sample is excited at half the nonlinear absorption wavelength and fluorescence monitored at a single wavelength within the two-photon emission peak. To record spectra, the excitation energy is scanned while keeping the fluorescence detection wavelength fixed. In practice, it is also necessary to measure linear absorption spectra and the fluorescence emission spectra in order to derive absolute $\sigma^{(2)}$ values, and so both source and detector system are usually wavelength tuneable as shown in the typical experimental geometry in Fig. 10.

The fluorescence, in units of power, due to a single-photon absorption process may be written as:

$$F_1 = \eta_F N\sigma^{(1)} \int_{-l/2}^{l/2} dz \int_0^{\infty} 2\pi r\, e^{-\alpha(z+l/2)}\, I_1(r, z)\, dr. \tag{42}$$

Here, l is the length of the cell, η_F ($\eta_F = 0 \rightarrow 1$) is the fluorescence yield for either one- or two-photon excited processes, N is the number of molecules per cm^3, and $\sigma^{(1)}$ the linear absorption cross-section. Since $N\sigma^{(1)}$ amounts to the linear ab-

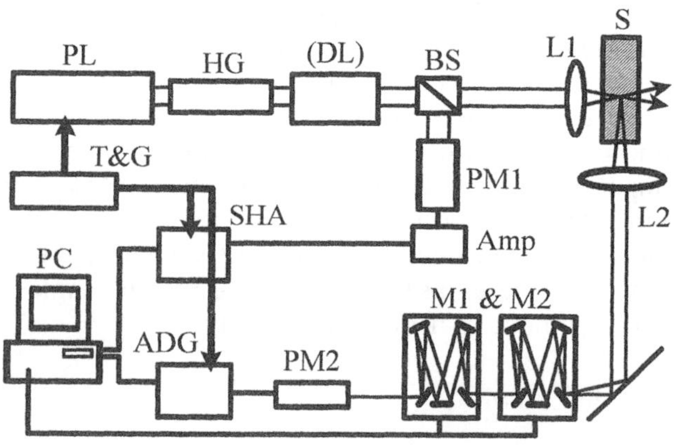

Figure 10 Typical experimental arrangement for two-photon absorption spectra by fluorescence measurement. PL, pulsed laser, e.g., Nd: YAG; HG, harmonic generator; (DL), tuneable dye laser pumped by Nd harmonic which may be removed to allow use of harmonic beam: BS, beam splitter; PM1, photodetector or power meter; Amp, pre-amplifier, L1, focusing lens to define excitation volume; S, sample position; L2, relay lens to define sampling volume imaged at monochromator slit; M1 & M2, double monochromator; PM2, photon counting photomultiplier; T&G, Laser system trigger and gating electronics for measuring system; SHA, gated sample and hold and analogue to digital converter; ADG, amplifier-discriminator and gated pulse counter; PC, control and data collection computer.

sorption coefficient α, and carrying out the spatial integration over the beam intensity profile, $I_1(r, x)$, the above expression may be more simply expressed as:

$$F_1 = \eta_F E_1(1 - e^{(-\alpha l)}), \tag{43}$$

with E_1 being the incident power at a given wavelength.

For the same fluorescence process excited at twice the wavelength by two-photon absorption, the fluorescence power level is:

$$F_2 = \eta_F N\sigma^{(2)} \int_{-l/2}^{l/2} dz \int_0^{\infty} 2\pi r I_2^2(r, z)\, dr, \tag{44}$$

where $I_2(r, z)$ is the incident intensity at the two-photon wavelength. The latter expression assumes negligible linear absorption at this wavelength. Measurement of the linear absorption spectrum and single-photon fluorescence may therefore be used to determine the fluorescence yield which can then be substituted in Eq. 44 to determine $\sigma^{(2)}$ from the two-photon fluorescence spectrum, provided that the beam profile is known. In practice, other factors are also required such as the optical collection efficiency of the spectrometers, overall detector gains, and the influence of any processes such as Rayleigh scattering which may also contribute to the signal. The problems encountered in determining the instrumental factors often lead to the publication of relative spectra with perhaps a fixed wavelength value measured by alternative means quoted for comparison. An alternative approach used by several groups is to calibrate the whole experimental apparatus against a material with a known fluorescence yield. Chen and McCann[37] have used dilute Rhodamine 6G perchlorate as a reference at a concentration guaranteed to permit complete population of the excited level in all molecules in the probe volume. The instrument calibration then allows more accurate determination of the fluorescence yield of the unknown samples from the linear fluorescence spectra. However the spatial beam profile is still required in order to calculate the two-photon coefficient from the two-photon fluorescence, Eq. 44. Where the fluorescence yield is low, phase-sensitive detection techniques or photon counting detectors may be used. For CW detection, luminescence techniques are generally limited to materials with $\eta_F > 10^{-3}$.[31]

Where more than one excited state is being probed, there may be difficulties in calculating absolute or even relative spectral intensities due to the presence of non-radiative decay channels. In general, these may be different for each excited state. The tacit assumption made in the above expressions for the powers F_1 and F_2, that the fluorescence yield was the same for one- and two-photon excited fluorescence, may also be invalid if non-radiative depopulation of the excited level may occur. Absolute calculations may also be hampered by the occurrence of self-focusing or thermal lensing problems[38] since the excitation beam is invariably focused to a small spot size (typ. 50–100 μm diameter) to increase the two-photon fluorescence. Under such circumstances the material may exhibit a significant intensity-dependent refractive index, distorting the spot size and so making an accurate estimation of the laser profile within the sample volume problematic. Self-absorption may also pose a problem. If part of the fluorescence is absorbed by the sample, that portion of absorbed light must be taken into account. Note that self-absorption can lead to shifted peaks in the fluorescence spectrum.

As with the two-beam nonlinear absorption experiments, two-beam, two-photon fluorescence measurements with one beam delayed relative to the other can give useful information on the nature of the excitation process often allowing direct two-photon excitation to be distinguished from stepwise excitation via an intermediate level with a finite lifetime.[33]

E. Wave Mixing Techniques

1. Coherent Anti-Stokes Raman Spectroscopy

Coherent anti-stokes Raman spectroscopy (CARS) is often termed three wave mixing though it is strictly a four wave mixing process where two of the photons involved are derived from the same beam (see Chapter 9). In a typical CARS experiment, shown in Fig. 11, intense beams from two tuneable lasers intersect at a phase-matching angle governed by the wave vector condition:

$$k_3 = 2k_1 - k_2,$$

where suffixes 1 and 2 refer to the two incoming beams. The wave vector k_3 denotes

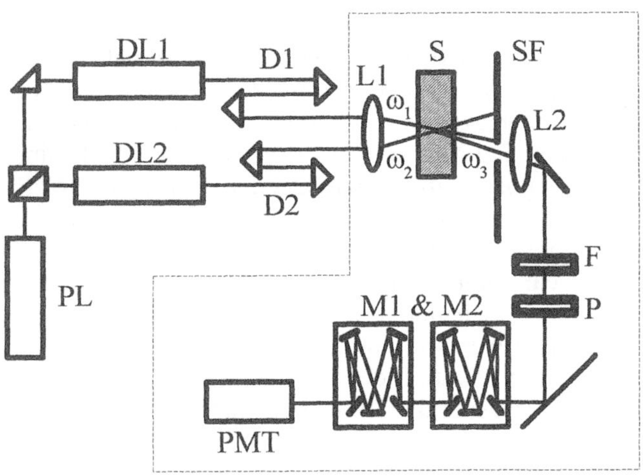

Figure 11 Typical CARS experiment. A high power pump laser PL such as a frequency doubled Nd:YAG laser, etc., is used to simultaneously pump a pair of independently tuneable dye lasers, DL1 and DL2. The output pulse from each of the tuneable lasers may be individually synchronized by separate delay lines, D1 and D2, respectively. This also allows for the pulse arrival times at the sample to be offset if required. The two beams at frequencies ω_1, and ω_2 are brought to a focus at the phase matching angle within the sample by an achromatic lens system represented by L1.[39] The CARS signal at ω_3 is separated from the input beams by both spatial filtering with pinhole SF and passage through a double monochromator system, M1 & M2. A photomultiplier is the usual type of detector. The inclusion of a polarizer, P, may allow a variant of CARS to be used to remove the non-resonant contribution to $\chi^{(3)}$ from measurements. A reference block consisting of a duplicate of the region bounded by the dashed box may be used to null out contributions due to solvent or container nonlinearities. The inclusion of an adjustable filter, F, then allows the signals from sample and reference arms to be properly balanced.

a third beam generated by a combination of two-photon absorption and Raman scattering wave mixing processes as shown earlier in the energy level diagram, Fig. 4. Typically the phase-matching angle amounts to a few degrees and the CARS signal at frequency ω_3 can be separated from the two excitation beams by a combination of spatial filtering and the use of a double monochromator system. Spectra are recorded by scanning ω_1, ω_2 and the difference $\omega_3 = 2\omega_1 - \omega_2$ is recorded. Note that the wavelength resolution of the technique is not limited by the spectrometer performance but by the linewidths of the two source lasers.

Frequencies ω_1, ω_2, and ω_3 are connected via the susceptibility term:

$$\chi^{(3)}(-\omega_3;\ \omega_1,\ \omega_1,\ -\omega_2).$$

Where one-photon dispersion may be neglected, Lotem and Lynch[20] express this in the form:

$$\chi^{(3)}(-\omega_3;\omega_1,\omega_1,-\omega_2) = \sum_T \left[\frac{A_T}{\omega_T - 2\omega_1 - i\Gamma_T} + \frac{A_T}{\omega_T + 2\omega_1 + i\Gamma_T}\right] + \sum_R \left[\frac{A_R}{\omega_R - (\omega_1 - \omega_2) - i\Gamma_R} + \frac{A_R}{\omega_R + (\omega_1 - \omega_2) + i\Gamma_R}\right], \tag{45}$$

where A_i, ω_i, Γ_i (with i = T or R) are the strength, frequency and half-width of the two-photon and Raman transitions, respectively. With $\omega_1 > \omega_2$ this may be simplified to:

$$\chi^{(3)}(-\omega_3,\ \omega_1,\omega_1,\ -\omega_2) = \eta_{NR} + \sum_T \left[\frac{A_T}{\omega_T - 2\omega_1 - i\Gamma_T} + \frac{A_T}{\omega_T + 2\omega_1 + i\Gamma_T}\right] + \sum_R \left[\frac{A_R}{\omega_R - (\omega_1 - \omega_2) - i\Gamma_R}\right], \tag{46}$$

where η_{NR} contains all Raman anti-resonances, and resonances far removed from $\omega_1 - \omega_2$ which will be small and not strongly frequency dependent. There are thus three principle contributions to the CARS susceptibility close to $\omega_1 + \omega_2$: a background non-resonant term, a summation of terms over the Raman resonances close to the difference frequency, and a summation over all two-photon resonances which is identical to that in the definition of the non-linear susceptibility for the direct two-photon absorption effect with identical photons:

$$\chi^{(3)}(-\omega_1;\ \omega_1,\ \omega_1,\ -\omega_1).$$

In the simple case of a single Raman resonance, the intensity of the CARS signal can be written as,[31]

$$I = [\chi^{(3)}_{NR}]^2 + [\chi^{(3)}_T]^2 + \frac{2A_R(\omega_R\chi^{(3)}_{NR} - \Delta\chi^{(3)}_{NR} + \Gamma_R\chi^{(3)}_T)}{(\omega_R - \Delta)^2 + \Gamma_R^2} + \frac{A_R^2}{(\omega_R - \Delta)^2 + \Gamma_R^2}, \tag{47}$$

where $\Delta = \omega_1 - \omega_2$. Here, $\chi^{(3)}_{NR}$ covers the non-resonant Raman and the dispersive (real) part of the two-photon contribution to the susceptibility, while $\chi^{(3)}_T$ is the imaginary part of the two-photon susceptibility. The latter is usually extracted from the CARS spectrum by fitting the experimental linewidth to the above expression.

For example, Foggi et al.,[31] measuring two-photon absorption in liquid pyridine, observe that in that particular case $\chi_{NR}^{(3)}$ is largely independent of ω_1. Thus shifts in the characteristic minima seen in CARS spectra (see Fig. 12) are attributable to $\chi_T^{(3)}$. Lynch and Lotem[19] also point out that the width of the CARS minima are sensitive to $\chi_T^{(3)}$ and use linewidth fitting techniques to extract values. The two-photon absorption coefficient (for an isotropic medium) can be calculated according to the expression:

$$\sigma^{(2)} = 3\chi_I^{(3)} \frac{4^3\pi^3 h\omega_1^2}{n^2 N}, \tag{48}$$

where N is the number of molecules per cm^3. Note that the CARS technique in this form requires a convenient Raman resonance with which the two-photon absorption can interfere. This may be inherent in the material to be studied or may perhaps arise from a judicious choice of solvent for solution measurements. Raman parameters such as transition strength, resonant frequency, and linewidth for all on-resonance transitions must all be independently measured by conventional Raman scattering methods for a fully quantitative analysis of the spectra. Single or double Raman resonance may also further complicate the interpretation of the spectra. The above analysis assumes that there are no excitation processes that will significantly alter the real populations of the energy levels involved, so perturbing $\chi^{(3)}$. This is usually a reasonable assumption for transparent materials at moderate pumping levels.

Other variations on this technique include coherent Stokes Raman scattering (CSRS) where $\omega_2 > \omega_1 > \omega_3$ (reverse ranking in CARS mode), which has been used by Shand et al.[40] to extend the wavelength coverage in measurements on polydiacetylenes for a given set of source lasers. Polarization CARS has also been used to reduce the effect of a large background non-resonant nonlinearity.[14]

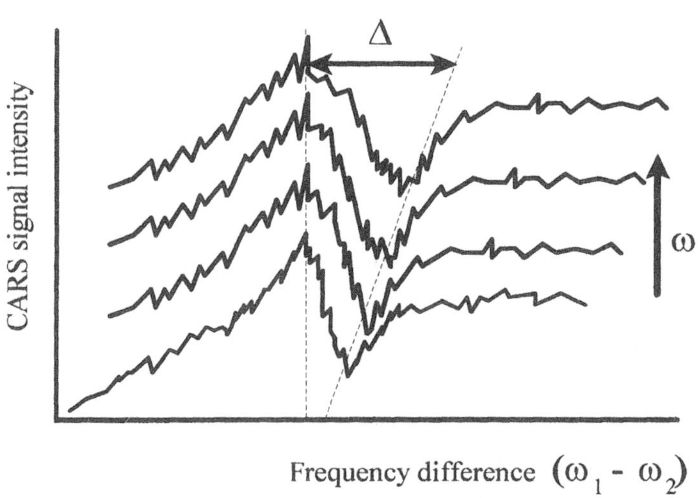

Figure 12 CARS signal vs. frequency difference between incident beams, for several ω_1 in the vicinity of a two-photon resonance, after Foggi et al.[31] As $2\omega_1$ approaches the two-photon absorption, the shift in the minimum, Δ, increases.

2. Four Wave Mixing Measurements

Four wave mixing techniques, and degenerate four wave mixing (DFWM) experiments in particular, are often used to measure the refractive nonlinearity, $\chi'^{(3)}$ (see Chapter 9). In a typical DFWM experiment with all beams of the same wavelength and the geometry given in Fig. 13, interference between the forward and probe beams creates a coarse phase grating which scatters the backward propagating beam. Also, the interference of the probe and backward beams creates a finer pitched phase grating which scatters the forward propagating beam. The scattered radiation is phase conjugate to the incident probe beam and the grating reflectivity, ρ, is given by:

$$\rho = \frac{I_{pc}}{I_{\text{pump}}} \propto (\chi'^{(3)})^2 \, I_{\text{pump}}^2 \frac{A}{n^4}, \tag{49}$$

where only real third order nonlinear processes are present. It is assumed that the forward, backward, and probe beams are all identically proportional to I_{pump} and that there is negligible pump depletion. I_{pc} is the intensity of the phase conjugate beam. The A factor accounts for attenuation by linear absorption of the beams.

If the third order nonlinearity is complex, the imaginary component, $\chi''^{(3)}$, will lead to the formation of further fine and coarse gratings. For each type there are contributions due to population changes driven by two-photon absorption, and also a thermal grating component driven by both one- and two-photon absorption.[41–44] The real and imaginary contributions to the measured $\chi''^{(3)}$ may be separated in several ways, as discussed below.

Time Resolved Nanosecond DFWM Measurements. The temporal response of each of the grating contributions will in general be quite different. This has been used by Sutherland et al.[41] to extract two-photon absorption values from DFWM data. The phase conjugate electric field may be written as:

$$E_c(t) = iL[K(t)E_p^* + K_{\text{TPA}}^c(t)E_b(t) + K_{\text{TPA}}^f(t)E_f(t)], \tag{50}$$

where L is the sample thickness. The field subscripts refer to Fig. 13 and the superscripts denote fine and coarse grating factors. The coupling coefficient, $K(t)$, is identical to that in the case of a purely refractive nonlinearity. The K_{TPA}^i terms are related to an index modulation δn_p by:

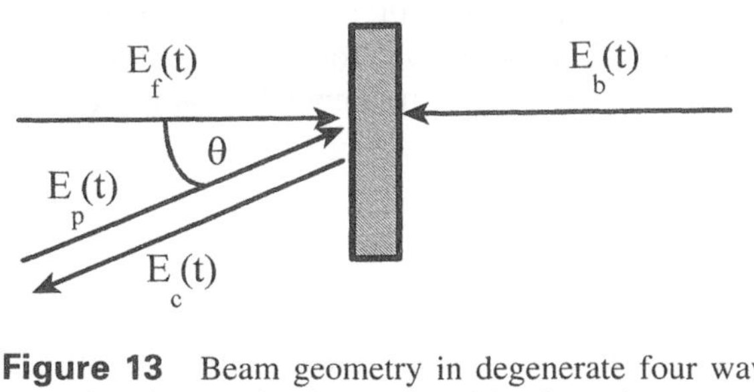

Figure 13 Beam geometry in degenerate four wave mixing experiment. Subscripts f, b, p, and c denote the forward and backward pump, and probe and phase conjugate fields, respectively. With all fields parallel polarized, the $\chi_{1111}^{(3)}$ component of the nonlinear susceptibility tensor is measured, while with the probe polarization crossed with respect to both the pump beams, the $\chi_{1221}^{(3)}$ component is measured.

$$K^{i}_{\text{TPA}} = \frac{\pi \, \delta n_p}{\lambda}. \tag{51}$$

Apart from the pitch, both fine and coarse gratings can be treated identically and so are not distinguished in the following.

The two-photon-induced population grating will have a spatial concentration profile, $\Delta N(x, t)$ determined by the rate equation:

$$\frac{d(\Delta N(x,t)}{dt} + \frac{\Delta N(x,t)}{T_1} = \frac{\sigma^{(2)} N_0}{(\hbar\omega)^2} I^2 (x,t). \tag{52}$$

Here, T_1 is the decay time from the two-photon excited state to the ground level. It is assumed that the original groundstate population, N_0, is not substantially altered by the excitation process. The time-dependent population grating amplitude, $N(t)$ is given by:

$$N(t) = \frac{1}{\hbar\omega} \int_{-\infty}^{t} q(t') \exp\left[\frac{(t - t')}{T_1}\right] dt', \tag{53}$$

where

$$q(t) = 2\alpha^{(2)}(2\varepsilon_0 nc)^2 \, |E_f(t)|^2 \, E_f(t) E_p^*(t), \tag{54}$$

assuming $|E_p| << |E_f|$. From the population grating, the time-dependent index grating amplitude, $\delta n_p(t)$, can be found:

$$\delta n_p(t) = \Delta\chi^{(1)} \frac{N(t)}{nN_0}, \tag{55}$$

with $\Delta\chi^{(1)}$ being the difference in the linear susceptibilities of groundstate and excited state.

The index modulation for the thermally driven time-dependent grating is derived starting from the heat diffusion equation for a temperature distribution, $\Delta T(x,y)$:

$$\rho C_p \frac{\partial(\Delta T(x,t))}{\partial t} + k_T \nabla^2(\Delta T) = \phi_{\text{int}} \alpha^{(2)} I^2(x,t), \tag{56}$$

where ρ, C_p, k_T are the density, specific heat and thermal conductivity, respectively, and ϕ_{int} is the fraction of energy absorbed by the two-photon process that is converted into heat by radiationless transfer. A further source term proportional to $\alpha I(x, t)$ could be added to the above expression to account for one-photon thermal effects but will be neglected for the time being. The time-dependent amplitude of the temperature grating, $T(t)$ is given by:

$$T(t) = \frac{\phi_{\text{int}}}{\rho C_p} \int_{-\infty}^{t} q(t') \exp\left[\frac{(t - t')}{\tau_D}\right] dt', \tag{57}$$

where $q(t)$ has already been given (Eq. 54). The thermal index grating amplitude $\delta n_t(t)$ is then given by:

$$\delta n_t(t) = 2T(t) \frac{dn}{dT}, \tag{58}$$

where the thermo-optic coefficient, dn/dT, is usually negative. The factor 2 arises because $T(t)$ is half the *peak-to-peak* time-dependent amplitude of the temperature grating.

Given the time-dependent grating amplitudes, the phase conjugate reflectivity response to the laser pulse can then be calculated. Sutherland et al. assume a Gaussian temporal profile, i.e.:

$$I_{\text{pump}} = I_0 \exp\left[-\left(\frac{t}{\tau}\right)^2\right]. \tag{59}$$

Provided that the two-photon excited state lifetime and the thermal diffusion time constant are much greater than the laser pulse width, τ, the time-dependent phase conjugate reflectivity, $\rho(t)$, may exhibit a double peak as shown in Fig. 14. This can be viewed as an interference effect between the instantaneous grating created by the real nonlinearity, with a linear dependence of index modulation on the intensity, $I(x, t)$ and the slower two-photon-created gratings where the dependence is quadratic. Note that all beams are temporally coincident, i.e., there is no delay of probe and pump beams. The extent to which the two peaks are resolved depends upon the ratio of the real and complex components of the nonlinearity. If a minimum in-between the two peaks can be observed, the reflectivity at that point, ρ_{min}, together with the peak pump intensity, I_{pump}, and time position within the pump envelope, t_{min}, can be used to determine the imaginary susceptibility component. According to Sutherland[41]:

$$\rho_{\text{min}} = B(0.5\pi^{3/2} w_0 \tau\, I_{\text{pump}})^2 \exp\left[-3\left(\frac{t_{\text{min}}}{\tau}\right)^2\right], \tag{60}$$

where w_0 is the $1/e^2$ beam radius within the sample. The factor B is related to $\chi''^{(3)}$ via

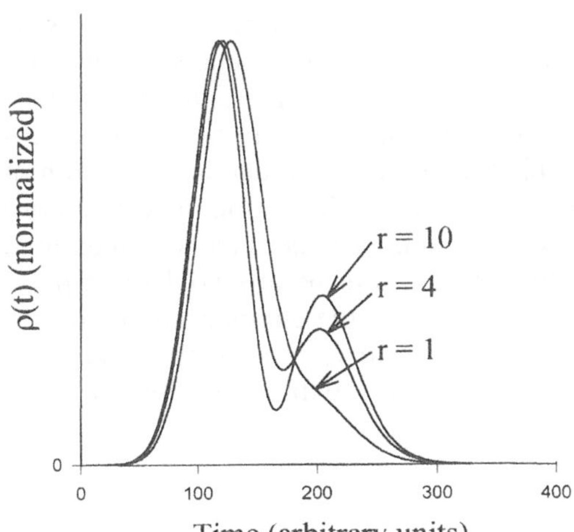

Figure 14 Phase conjugate signal in time resolved degenerate four wave mixing experiment after Sutherland et al.[41] The pulse profiles show the dependence on the ratio of the real and imaginary susceptibility terms through the factor $r = (\chi'^{(3)}/\chi''^{(3)})^2$.

$$B = \left(\frac{3L\omega\, \chi''^{(3)}}{\pi^{3/2} \varepsilon_0 n^2 c^2 w_0^2 \tau} \right)^2, \tag{61}$$

and the two-photon absorption coefficient then follows from Eq. 3. B is determined from a fit of ρ_{min} against I_{pump} with t_{min} also determined for each point.

As mentioned above, single-photon absorption may also contribute to the thermal grating. If the thermal grating is largely dominated by single-photon absorption, the same double peak in the phase conjugate signal is expected but, in this case, since the grating will have the same intensity-dependence as the D.C. Kerr (real n_2) grating, the interference (ρ_{min}) will occur at a fixed point in time for all pump intensities and the ratios of the two phase conjugate intensities remains fixed.[43]

The above analysis has some limitations. For example, the population grating could equally well have contributions from single photon and step wise multiphoton excited states. The simple two-level model likewise does not include excited states which could influence the rate at which excited species are returned to the groundstate population. The analysis and measurements were originally applied to experiments on the nanosecond (~10 ns) time scale. If measurements are carried out with ps pulses, it should be borne in mind that the approximate rise-time for a thermal lens is of the order of magnitude, w_0/v where v is the velocity of sound in the medium. For typical beam sizes, this is of the order of 10 ns so measurements with much shorter pulses are unlikely to pick up much of the thermal grating contribution. Also, the rate at which energy is converted into heat by radiationless transfer may become significant on shorter time scales. In addition, the possibility of contributions from amplitude gratings as well as phase gratings has not been taken into account. Finally, Sutherland et al. point out that the quadratic dependence on the intensity of the two-photon created gratings and the need to measure the simultaneous temporal response of both the phase conjugate and pump beams necessitates the use of a streak camera so that the on-axis portion of each beam is sampled. The use of a spatially integrating detector would tend to smear out the interference of the various gratings in the time domain.

Gass et al.[45] have demonstrated the effect of nonlinear absorption in a DFWM experiment using the arrangement shown in Fig. 15. Since independent parametric generators were used to derive the pump (p) and test (t) beams, the two beams, although spatially and temporally coincident at the sample, were not mutually coherent. This considerably reduces the two-photon contribution due to absorption of one photon from the pump and one from the test beam. Absorption of two photons from either beam will still occur however. The authors experimentally measured the ratio of the self and induced two-photon processes to be around 10:1 in polydiacetylene samples with 20 ps pulses. Taking $\chi^{(3)}$ to be the complex susceptibility, the following set of coupled equations can be used to a good approximation to represent the pump, test, and signal optical fields (with subscripts p, t, s, respectively):

$$\frac{dE_p}{dz} = \frac{i\omega\, \chi^{(3)}}{8nc} 3|E_p|^2 E_p, \tag{62}$$

$$\frac{dE_t}{dz} = \frac{i\omega\, \chi^{(3)}}{8nc} 3|E_t|^2 E_t, \tag{63}$$

and

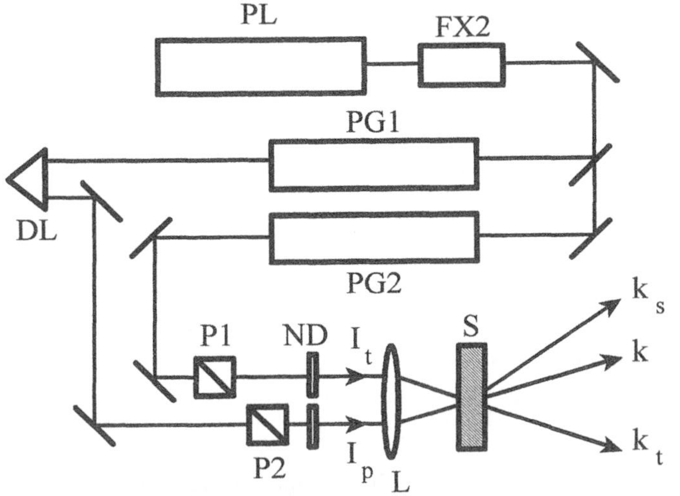

Figure 15 DFWM experiment with non-mutually coherent pump beams. PL, pump laser (e.g., Nd:YAG); FX2, frequency doubler; PG1, PG2, parametric generators; P1, P2, polarizers; ND, neutral density filter; L, lens; S, sample. Pump beams have incident intensities, I_p and I_t, k_p, k_t, k_s are the two pump and phase conjugate wave vectors, respectively.

$$\frac{dE_s}{dz} = A\,\frac{i\omega\,\chi^{(3)}}{8nc}\,[(|E_p|^2 + |E_t|^2)6E_s + 3E_pE_pE_t^*\,\exp(i\Delta kz)]. \tag{64}$$

The coherence factor, A, in the above reference was approximately 0.1, and the phase mismatch, Δk, is given by the wave vector equation:

$$\Delta k = 2k_p - k_t - k_s. \tag{65}$$

Note that A accounts for the parametric generation process not being coherent. In particular, the two pump waves have an arbitrary relative phase.

In their paper, Gass et al. solve the set of differential Eqs. 62–64 numerically having independently measured the nonlinear absorption, and in this way retrieve linear reflectivity plots (Fig. 16) with two-photon absorption effects removed. If the coherence factor A could be determined without measuring the two-photon absorption directly, the method could be used in an iterative fashion to calculate the imaginary susceptibility component required to linearize the plots. The coherence factor could be calculated by using a reference material with known complex susceptibility to calibrate the measuring system. Apart from this consideration, the technique has the disadvantage that two generators must be used, though the flexibility in wavelength tuning is useful for spectral measurements. It should also be borne in mind that the output of the parametric generators is not too close an approximation to the usually assumed Gaussian profile and is often rather less stable shot-to-shot than many other commonly used sources. Also, the geometry used requires careful spatial filtering of the diffracted signal beam from the pump and test beams to maintain a reasonable signal to background noise ratio.

By using a differential delay on one of the input beams, Gass et al. tentatively attributed the nonlinear absorption that they observed in the experiment to a direct

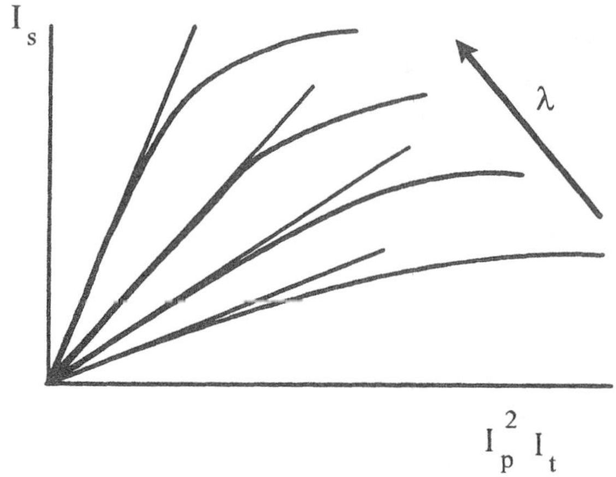

Figure 16 Influence of two-photon absorption on phase conjugate intensity in DFWM experiment. As λ approaches a resonance, the two-photon absorption increases, progressively distorting the plot at high pump powers. If $\alpha^{(2)}$ is measured independently by monitoring the nonlinear transmission of a pump beam, the value may be used to extract the linear plots from the raw data.

triplet state excitation rather than an instantaneous two-photon absorption process though the precise details of the mechanism remained unexplained.

Picosecond DFWM With Coherent Pump and Probe Beams. Zhao et al.[42] have considered the effect of the two-photon induced population grating on the observed phase conjugate beam in the conventional DFWM geometry of Fig. 13. Due to the short experimental time scale, thermal grating effects are neglected since these will not be detected by the probe pulse, provided that the repetition rate is sufficiently slow for such gratings to have decayed between pulses. Considering only the cascaded effect of two-photon-absorption-induced population changes on the linear refractive index, Zhao et al. obtain a modified expression for the peak amplitude of the phase conjugate field (i.e., time-dependence neglected):

$$E_c(z) = i\,\frac{2\pi\omega^2}{kc^2}\,\chi'^{(3)}E_fE_bE_p^* \times \frac{\ln\left[\dfrac{(1+bl)}{(1+bz)}\right]}{b}, \tag{66}$$

with the subscripts c, f, b, and p referring to the phase conjugate, forward and backward pump, and probe beams, respectively. The incident fields are also taken as the peak values, i.e., time-dependence likewise neglected. The factor b contains the imaginary third order susceptibility:

$$b = \frac{4\pi\omega^2}{kc^2}\,3\chi''^{(3)}E_f^2. \tag{67}$$

Comparison of Eq. 66 with the corresponding expression in the absence of two-photon population gratings leads to the correction factor, F:

$$F = \frac{bL}{\ln[1 + bL]}. \tag{68}$$

The third order susceptibility is calculated by multiplying the square root of the signal intensity F. This removes the effects of TPA on $\chi^{(3)}$. Since the correction factor also has a quadratic field dependence, materials with a substantial two-photon absorption will exhibit an apparent fifth order nonlinearity where the population shift is sufficient to markedly alter the groundstate linear susceptibility. Although it may be possible to use a data fitting procedure to determine an experimental value for F in order to retrieve a cubic dependence for the phase conjugate signal, it is not likely that such an approach would be all that useful or particularly accurate. In many instances, DFWM in moderately two-photon absorbing materials will give phase conjugate signals with exponents intermediate between 3 and 5, indicating a mixture of coherent and cascaded processes. In favorable cases, the technique may be useful in confirming or cross-checking the magnitude of $\alpha^{(2)}$, measured by nonlinear transmission for example. It will often be more straightforward to make the latter type of measurements and calculate F in order to extract the real susceptibility, since the DFWM experiment needs little modification to do so.

DFWM Measurements of Solutes in Nonlinear Solvents. If two-photon absorption is present but is not too great it may not be possible to observe a strong fifth order response by DFWM. This may be especially true in dilute solution measurements. DFWM measurements effectively probe the modulus of the complex nonlinearity, i.e.:

$$|\chi'^{(3)} + i\chi''^{(3)}|$$

For a dilute solution in a solvent with a purely refractive nonlinearity, γ solvent, the ratio of the measured solution and solvent "effective" $\chi^{(3)}$ values can be expressed in terms of the real and complex hyperpolarizabilities γ' and γ'', and the number densities of solute and solvent units[46,47]:

$$\frac{|\chi^{(3)}_{\text{solution}}|^2}{|\chi^{(3)}_{\text{solvent}}|^2} = 1 + \frac{2N_{\text{solvent}}N_{\text{solute}}\gamma'}{N^2_{\text{neat}}\gamma_{\text{solvent}}} + \frac{N^2_{\text{solute}}}{N^2_{\text{neat}}\gamma^2_{\text{solvent}}}(\gamma'^2 + \gamma''^2). \tag{69}$$

N_{solvent} and N_{solute} are the number densities of solvent and solute in the solution, while N_{neat} is the number density of solvent molecules in the pure solvent. This expression assumes that the solution is sufficiently dilute that the Lorentz field factors of solution and solvent are effectively equal and can be neglected in the ratio of the hyperpolarizabilities. The hyperpolarizabilities can then be found by fitting the concentration dependence of the DFWM data to the above expression with the solute and solvent concentrations as the variables.

Again, this is an approximate method of obtaining the two-photon absorption coefficient, relying on a sensible balance between the real nonlinearity of the solvent, and the real and imaginary components of the solute nonlinearity if reasonably accurate values are to be obtained. As before, it is not readily verified if the apparent imaginary nonlinearity is due to two-photon absorption or which two-photon gratings are involved in the mixing process.

3. THG Maker Fringe Measurements

The experimental layout used for the observation of Maker fringes due to third harmonic generation (see Chapter 10) is shown in Fig. 17. From the variation of the third harmonic intensity with the incident angle, the modulus of the complex third order susceptibility $|\chi^{(3)}|$ can be determined. Obviously, if there is any crystalline or otherwise ordered structure in the sample then $\chi^{(3)}$ will be a combination of tensor coefficients. The sample is usually in the form of a thin solid film, typically of the order of 0.1 μm thick. In practice, the thin film sample is usually supported on a substrate such as a fused silica plate which itself has a third order nonlinearity comparable with that of the film. The standard procedure is therefore to make a Maker fringe measurement of substrate and film together and then to dissolve away the film with a solvent wash while the plate is still mounted on the rotation table. In this way the Maker fringes due to the silica plate alone can be measured at exactly the same point on the sample and the effect of any variations in intensity due to changes in substrate thickness can be reduced.

Aramaki,[48] Kajzar,[49] and more recently Hasegawa et al.[50,51] have shown how the technique can be extended to measure not only the modulus but also the phase of the complex third order susceptibility (see Chapter 10). This is by virtue of the fact that the third harmonic signals from the sample (generally complex $\chi^{(3)}$) interferes with that from the substrate (at most measurement wavelengths usually a

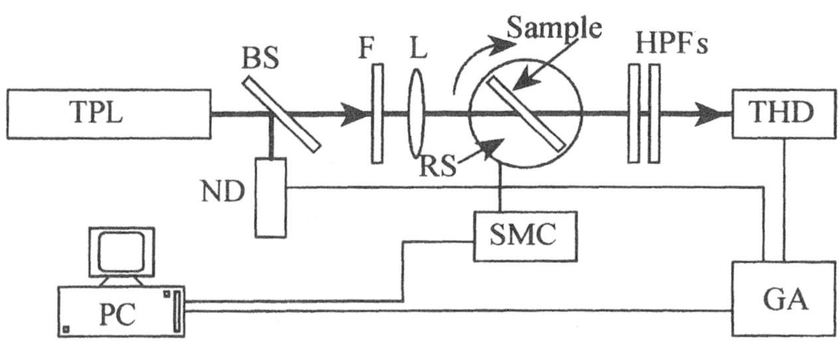

Figure 17 Maker fringe technique for third harmonic measurement. The sample substrate is usually fused silica which acts as an in situ reference material. Two scans are made, one with the film present on the silica substrate and a second scan with the film removed (without disturbing the silica plate). For complete spectral measurements a tuneable pulsed laser source is required (TPL). If there is significant shot-to-shot intensity variation it is best to include a monitoring detector to normalize measurements (ND) or reject spurious values. This signal along with that from the third harmonic detector (THD) are simultaneously monitored by a gated analyzer (e.g., box car), to reduce noise contributions. The sample is mounted on a temperature controlled rotation stage (RS) controlled by a stepper or servo motor controller (SMC). Normalized TH intensity vs. angle scans are collected by a controlling PC. Filter F removes any short wave radiation from the tuneable source (e.g., pump harmonics, flashlamp background, etc.). An optional lens, L, may be used to increase the local intensity. The focal length would normally be chosen to ensure that beam is reasonably parallel within both sample *and* substrate, i.e., the confocal parameter, $z_0 > n_\omega l$, where n_ω is the refractive index and l the overall sample thickness. Filters HPF pass only the third harmonic through to the detector.

purely real $\chi^{(3)}$). To illustrate, taking a plane wave approximation, the third harmonic field arising from the substrate. $E_{3\omega,s}$ is given by:

$$E_{3\omega,s} = \frac{3\omega}{2\pi n_s c} \chi_s^{(3)} l_{c,s} \left\{1 - \exp\left(-i\frac{\pi l_s}{l_{c,s}}\right)\right\} E_\omega^3, \tag{70}$$

while that from the film can be written:

$$E_{3\omega,f} = \frac{3i\omega}{2n_f c} |\chi^{(3)}| \, e^{-i\phi} l E_\omega^3. \tag{71}$$

Here, $l_{c,s}$ is the coherence length of the substrate, l_s and l are the optical path lengths in the substrate and film, and n_s and n_f are the refractive indices of the substrate and film. $\chi_s^{(3)}$ is the substrate nonlinearity, while $|\chi^{(3)}| \, e^{-i\phi}$ is the complex nonlinearity of the sample. The total third harmonic intensity detected after the film is proportional to $|E_{3\omega,f} + E_{3\omega,s}|^2$. Expressions such as these may be used to fit the experimental data and both ϕ and $|\chi^{(3)}|$ can be extracted. Hasegawa et al. also point out that the transmitted intensity through the film + substrate at angles where a minimum appears in the silica only scan gives the locus of the third harmonic intensity due to the film only (proportional to the modulus of the film susceptibility). Figure 18 shows examples of calculated Maker fringe plots neglecting Fresnel intensity factors, multiple reflections at film and substrate interfaces, and assuming only an incident plane wave according to Eq. 70 and 71. Traces are shown for the film and substrate alone (F and S) and for complex susceptibility phase factors (ϕ) of 0, $\pi/2$, π, $3\pi/2$.

Hasegawa et al. have successfully applied this technique to polysilane[51] and polydiacetylene[50] films and have shown the location and *relative* strengths of two-photon absorption features. However, it is also shown that three-photon resonances may be detected and may in certain cases give rise to interference with the two-photon signal making interpretation of the $\chi^{(3)}$ spectrum difficult. The presence of long-lived single-photon excited states which may then be further excited bring similar difficulties. While THG is often used to identify multi-photon absorption features, it is rare that absolute absorption cross-sections are quoted by this technique. In order to make spectral measurements, the Maker fringe measurements must be made at a range of wavelengths and in order to have sufficiently high power in the fundamental beam a ns pump laser is often required. This may restrict the amount of information on the excitation dynamics that can be extracted particularly whether a two-photon excitation is instantaneous or via an intermediate state with a finite lifetime. Then, only if parity is a good quantum number for the material in question, it may be possible to infer the dynamics from the symmetry of the states involved. For such reasons, it may not always be totally clear whether a particular feature in the $\chi''^{(3)}(\omega)$ spectrum should be attributed to a $\sigma^{(2)}$, $\sigma^{(3)}$ or perhaps $\sigma^{(01)}$ type of cross-section.

F. Two-photon Ionization Spectroscopy

If a two-photon absorption process accesses an excited state above the ionization limit, then an electron-ion pair may be created which, if it exists for sufficient time to separate and migrate, can be detected by electronic rather than optical means. In the gas phase, a biased electrode may be used to collect photogenerated species.

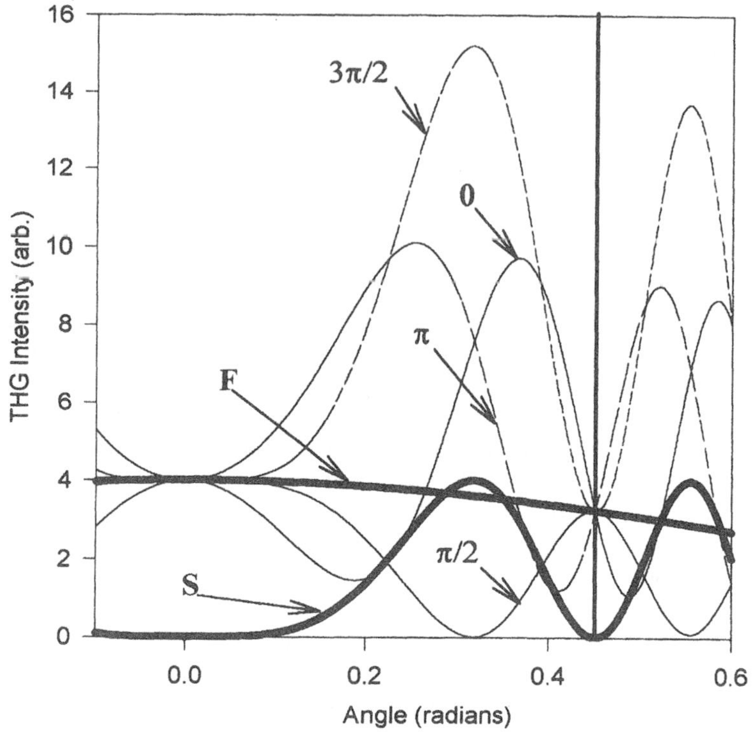

Figure 18 Maker fringe plots of third harmonic vs. angle of incidence relative to normal. **S**—THG due to substrate alone (usually wholly real, e.g., silica), **F**—locus of THG due to film only, 0, $\pi/2$, π, $3\pi/2$,—THG due to film on substrate. Given curve **S**, $|\chi^{(3)}|$ and ϕ can be obtained by fitting the experimental curves to expressions such as Eqs. 70 and 71.

Similarly, photoconductivity may be used in solid materials between metallized electrodes to monitor the excitation mechanism. The population of the direct two-photon excited state is again given by Eq. 52. As shown in Fig. 19, it is also possible to excite molecules into ionization states by sequential multi-photon absorption, and, if one of the pump wavelengths is close to resonance with any of the excitation steps, the overall ionization process will be significantly enhanced. The stepwise excitation processes involved may include two-photon absorption as one of the transition sequences and so provide a means of observing the spectral dependence of $\sigma^{(2)}$ even though the two-photon state does not itself result directly in ionization. Again, the two-photon or other absorption processes may be resonantly enhanced so improving the inherent sensitivity of the technique. In addition, use of excitation to auto-ionization states improves the probability of separation of the charged species further enhancing the sensitivity. Ionization probability may also be increased by including a large DC bias field in the photo-ionization region.[14] The probability of electron tunnelling from the distorted bound state increases, leading to greater charge separation. It should be borne in mind that strong DC fields will also lead to superposition of the DC Stark effect on the photo-ionization spectrum.

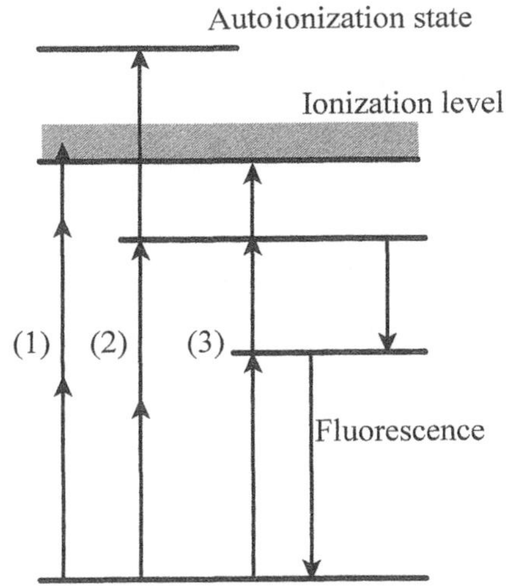

Figure 19 Examples of possible multi-photon ionization processes in general. (1) Direct three-photon transition to the ionization level; (2) two-photon transition followed by a further excited state transition to an autoionization level; (3) three sequential single-photon transitions to the ionization level. The transitions may also be accompanied by fluorescence from any intermediate excited states. In practice, measurements are usually performed with pump photons of a single energy. The autoionization states are discrete states that coexist with the continuum states (see Y.R. Shen, *Principles of Nonlinear Optics*, John Wiley, New York, 1984, Chapter 18).

The basic experimental arrangement for gas phase multi-photon ionization spectroscopy is shown in Fig. 20. Ionization is detected by a negatively biased wire while the chamber walls are grounded. If the probe wire is heated, thermionic emission leads to a zone of space charge around the wire. Photogenerated ions present in the chamber around the wire attract the electronic charge leading to a far larger current flow than would be detected simply from the flow of ions towards the wire, typically by several orders of magnitude. To reduce the DC Stark effect due to the operating bias, a shielding grid is often incorporated around the ionization zone, at the same bias as the probe wire. Further more sophisticated detection schemes may make use of electron multiplication in the same manner as the gain stages of photomultipliers, again giving several orders of magnitude of current gain. With these techniques it is easily possible to carry out experiments down to a detection limit imposed by background noise of a count rate around one electron per minute. The instrumental sensitivity can exceed even that of two-photon-induced fluorescence measurements. Applicability of the technique is limited to materials where the decay of the two-photon excited state is low enough to permit significant further promotion to an ionized level if two-photon absorption does not itself exceed the ionization limit. Also, detection schemes involving charged wire ionization detectors are limited to gas phase materials. Vapors may often be studied in the presence of inert buffer gases, e.g., Argon, to allow the total pressure to be varied and to control the collision-induced ionization processes.

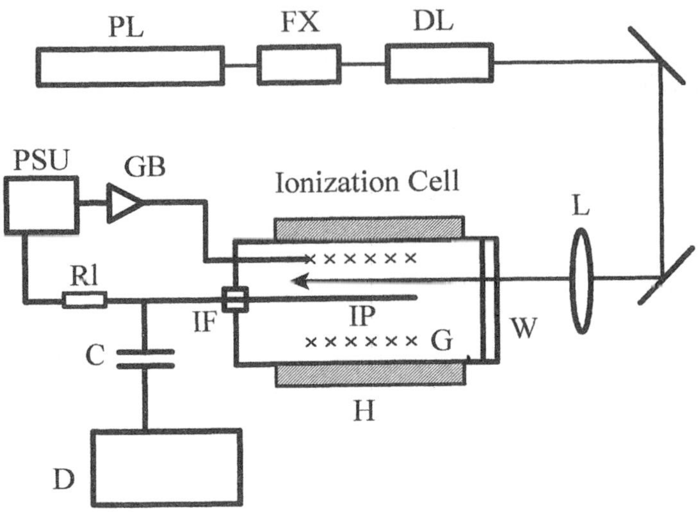

Figure 20 Multi-photon ionization spectroscopy experiment. A pulsed pump laser, PL, is typically frequency doubled or tripled (FX) and the output used to pump a tuneable dye laser, DL. The output is focused by a lens, L, onto an ionization cell through a window, W. The photoinization region is typically in a region between the negatively biased ionization probe, IP, and a shielding grid, G. The latter permits the formation of space charge around the wire ionization probe due to thermionic emission when the latter is radiatively heated by the external heater, H. The grid also prevents the ionization process from being perturbed by the bias between the ionization probe and the grounded cell walls. The probe is insulated from the latter by a dielectric feedthrough, IF. The remaining components make up the bias and detection sections; GB, grid bias; PSU, probe bias supply; Rl, load resistor; C, output signal coupling capacitor; D, data recording system.

G. Acousto-optic Measurements

De-excitation of materials excited by two-photon absorption may also be detected by the photoacoustic technique.[52] Non-radiative decay of the excited state under pulsed excitation results in localized heating of the sample which sets up an acoustic wave via a thermoelastic expansion. The time-dependent pressure wave is monitored with a piezoelectric microphone, typically made from the ceramic PZT. The technique, as shown in Fig. 21, is best suited to the study of fluid samples since coupling to the acoustic transducer is usually quite efficient and simple to arrange. In highly absorbing materials or at very high pulse energies, other mechanisms may give rise to a photoacoustic signal, i.e., localized vaporization in liquid samples and dielectric breakdown. The temporal profiles of each type of process are used to distinguish the required thermoelastic signal from those of the destructive processes.[53] Although absorption cross-sections have been measured by photoacoustic methods,[52] the technique is most often used to record relative spectra, particularly where the sample concentration is varied through a series of measurements. This is partly due to the difficulties in analyzing the response in terms of the effective acoustic source geometry. Depending on the effective penetration depth of the sample, the "source" may be spherical, cylindrical, or even planar.

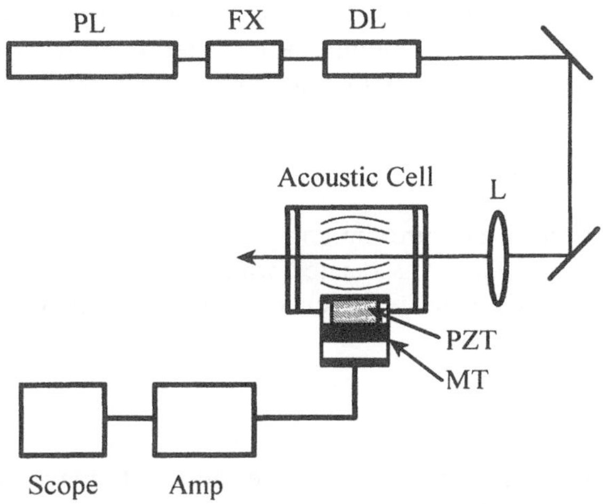

Figure 21 Photoacoustic experiment for recording two-photon absorption spectra. A tuneable pulsed source consisting of a pulsed Nd:YAG laser, PL, a frequency doubler, FX and a Dye Laser, DL, is used to excite ultrasonic waves in an acoustic cell. The beam is focused by lens, L, through the cell windows and into the sample liquid. The acoustic signal is monitored by a piezoceramic transducer, PZT, mounted in a mechanically damped holder, MT, usually consisting of layers of lead and copper to allow electrical conduction while reducing acoustic reflections back onto the transducer. The response is amplified (Amp) and monitored on an oscilloscope.

All non-radiative de-excitation processes contribute to the thermoelastic signal. These include principally relaxations from one-photon and two-photon populated states, but may also include decay from states populated by excited state absorption or multi-photon absorption with $n > 2$. Experimentally, the amplitude of the photoacoustic relaxation signal is measured (Fig. 22) as a function of the laser pulse intensity. For a one-photon process the intensity dependence will be linear, while two-photon absorption results in quadratic behavior. In practice, the exponent is often between 1 and 2 with a mix of excitation processes being present. Despite

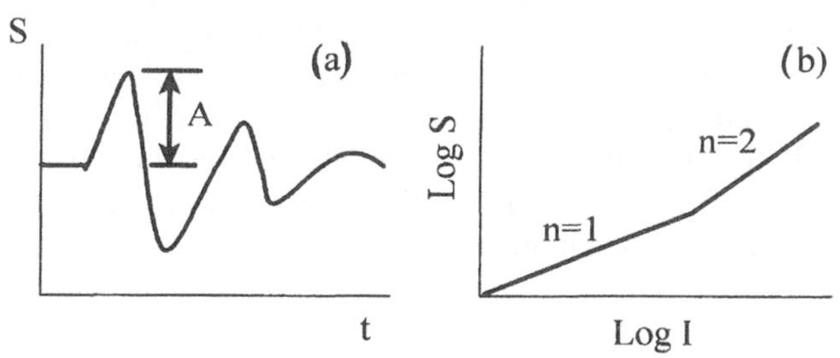

Figure 22 Photoacoustic measurements. a) Time-dependence of photoacoustic signal, S, showing signal amplitude, A. b) Photoacoustic signal dependence on laser intensity. At low powers single-photon absorption dominates (exponent $n = 1$), while at higher powers the quadratic dependence ($n = 2$) signifies a two-photon absorption mechanism.

the limitations in being able to diagnose the excitation processes present, the technique may be a sensitive and useful alternative to luminescence experiments where the fluorescence yield of the material in question is low and only relative spectral features are required.

H. Electro-absorption

Electro-absorption (EA) has been used to measure the complex DC field induced (Kerr) third order susceptibility, $\chi^{(3)}(-\omega;0,0,\omega)$, associated with the Stark shift of absorption levels (see Chapter 5). The nonlinearity is related to the field-induced change in the complex refractive index $\overline{\Delta n}$ by:

$$\chi^{(3)}(-\omega;0,0,\omega) = \frac{\overline{n\Delta n}}{2\pi F^2}, \tag{72}$$

where F is the applied electric field. The complex refractive index difference is calculated from the field-induced change in the absorption spectrum via a Kramers-Kronig relationship.[54] Typically, field strengths of the order of 100 kV/cm are applied via interdigitated grating electrodes deposited on the free surface of the sample. This of course requires that the sample is a solid and that other field-induced effects such as reorientation of dichroic materials or electrostrictive effects can be neglected. Polymeric samples are usually studied well below the glass transition temperature and so field-induced reorientation is not generally a problem. The contribution from field-induced effects that are linear in the applied field may be removed by using a low frequency (typically 10 kHz) electric field and using phase-sensitive detection at twice this frequency. Phase-sensitive detection improves the accuracy of the technique since sample transmission changes of 10^{-3} to 10^{-4} are typically measured.

In addition to the field-induced Stark shift, the intense DC electric field may induce a distortion of the symmetry of the excited states of the molecule. States with A_g symmetry will acquire a measure of B_u character. Some of the oscillator strength associated with one-photon allowed transitions will be shifted to forbidden transitions associated with two-photon absorption processes, i.e., they will become weakly one-photon allowed. These field-induced transitions will be at the expense of the regular excitation processes, i.e., the latter will exhibit a field-strength-dependent bleaching as observed in polysilanes by Tachibana and Kawabata.[55] The bleaching effect is also known to markedly distort the profile of the Stark effect band and this is useful in correlating induced absorption and bleaching features.

Apart from the limitations on sample format, the electro-absorption technique has been shown to be a useful technique in identifying the excitation processes present in nonlinear materials, though it is not readily applicable to absolute cross-section calculations. Also, the features of electro-absorption spectra are themselves field-dependent and so are not quite representative of the zero-field two-photon absorption spectrum. However, EA spectra are a useful guide to the likely location of two-photon transitions.

I. Waveguide Device Measurements

Nonlinear absorption effects in polymer waveguides and guiding planar films have been observed by several groups[7,56] and have been used to determine the complex part of the third order nonlinearity. Townsend et al.[7] have modeled the nonlinear transmission, T_{nl}, of a polydiacetylene directional coupler as an approximation to a single channel waveguide where:

$$\frac{T_0}{T_{nl}} = 1 + \frac{\alpha^{(2)} L_{eff} P_0}{A_{eff}}. \tag{73}$$

L_{eff} and A_{eff} are the effective device length accounting for linear absorption, and the effective guided mode cross-section. T_0 and P_0 are the low transmittance and peak incident power in the applied laser pulse. This simple analysis tends to overestimate the nonlinear absorption coefficient, $\alpha^{(2)}$, since in the directional coupler geometry the power in the mode will be distributed over the two guides and the evanescent coupling region between. In practice, a complete analysis will include the effect of the nonlinear refraction. Townsend et al. quote the expression for the field amplitude, E_1, in arm 1 of the directional coupler, as:

$$\frac{\partial E_1}{\partial z} = \frac{i\pi E_2}{2L_c} - \frac{\alpha E_1}{2} - \frac{\alpha^{(2)} E_1 |\mathrm{E}_1|^2}{2} + ik_0 \Delta n E_1 |E_1|^2, \tag{74}$$

where Δn is the intensity-dependent refractive index change, k_0 is the free space wavenumber, and L_c is the length of the coupling region. Similarly, the field amplitude in arm 2 is described by an expression analogous to the above but with subscripts 1 and 2 interchanged. The two coupled equations may then be solved numerically to give the output transmission in terms of the optical coefficients. Unfortunately, this approach is not readily inverted to give $\alpha^{(2)}$ in terms of the transmission against power dependence unless both the linear absorption and the refractive nonlinearity are known to be effectively zero.

Although only approximate values for $\alpha^{(2)}$ may be obtained from this type of measurement, it may give useful information if the material is anisotropic. Either TE or TM polarized light may be launched into the guide and the relative magnitudes of nonlinear absorption in either plane measured. This is especially important for polymeric devices fabricated by spin coating processes where the material may form oriented films during manufacture, either by flow alignment or due to anisotropic consolidation on loss of solvent from the surface.

McBranch et al.[56] have used a technique involving the measurement of film reflectivity in the angular range close to the point at which a guided mode is launched into a polymer film. The mode launch angle gives the total real refractive index, i.e., the sum of linear and nonlinear dispersion. In the presence of an intense pump beam, a weaker probe beam will therefore show a mode profile slightly shifted with respect to the linear refractive index profile. The difference in the reflectivity induced by the pump was typically of the order of 0.01 percent, corresponding to a real refractive index change of around -10^{-7} for polydiacetylene films. If nonlinear absorption is present, the shape of the mode profile is altered in the presence of the pump beam as shown in Fig. 23. McBranch et al. derive the following expression valid in the symmetric central region of a mode reflectivity profile:

$$\Delta R = A\left(\frac{\partial R}{\partial \theta}\right) \Delta n' + B\left(\frac{\partial^2 R}{\partial \theta^2}\right) \Delta k, \tag{75}$$

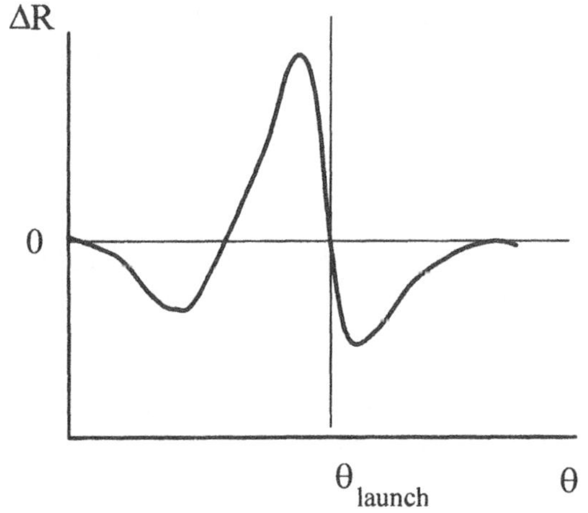

Figure 23 Example of modulated reflectivity profile, ΔR vs. θ. An optically induced index change of approximately 10^{-4} typically results in a change of 0.1% in the reflectivity close to the mode launch angle and the latter may shift by as much as 0.1–0.2°.

where ΔR is the pump induced change in reflectivity, A and B are fitting parameters, while $\Delta n'$ and Δk are the intensity-dependent refractive index and absorption coefficient, respectively. The latter are then related to the real and imaginary parts of the complex nonlinear refractive index:

$$n_2 = n_2' + ik_2. \tag{76}$$

The planar waveguide technique has the advantage that the real and imaginary components of the refractive index may be measured simultaneously in a film geometry with the same characteristics as would be encountered in an integrated device. TE and TM pump beams may be used to probe any anisotropy in the nonlinear properties and the inclusion of a differential time delay in the pump and probe beams allows the dynamics of any contributing processes to be resolved, i.e., thermal (slow) vs. electronic (fast) processes, etc. Limitations in the data analysis arise from the fact that Eq. 75 is only strictly valid in the vicinity of the mode peak. Film quality is important since surface imperfections and inhomogeneities will give rise to interference effects, distorting the mode profile. In the geometry used by McBranch et al. the pump beam is reflected from the rear surface of the film and so a nonlinear phase grating is established due to interference of the forward and backward pump beams. The resulting refractive index modulation should be included in the derivation of the nonlinear refractive index from the change in the mode index. McBranch et al. make the simplifying assumption that the mode index modulation, ΔN, can be approximated as, $\Delta N = 2n_2'I$, i.e., the modulated index profile is replaced by an average nonlinear refractive index over the cross-sectional area of the launched mode. Finally, the technique is not easily adapted to spectral measurements particularly to cover wide spectral ranges, since the film will tend to propagate additional higher order modes with decreasing wavelength, and the beam parameters will change for each launch angle.

V. MODELING TWO-PHOTON SPECTRA

After obtaining a two-photon absorption coefficient value or a good clean two-photon absorption spectrum with all competing excitation processes removed, the results must be interpreted in terms of the participating molecular energy levels. It should be made clear that what follows is not an attempt to go into this subject in great detail, which is beyond the scope of this chapter. Many highly refined models have emerged in recent years which take into account many of the subtleties of the known electronic structure of the material in question.[57] The main difficulties (at least from an experimentalist's point of view) are, as always, in obtaining a tractable expression where most of the necessary parameters can be measured by independent means or otherwise reliably inferred. In the following subsections, a few examples are shown where it has been possible to fit experimental results to rather *simple* models, though usually over a restricted frequency range. A few examples are also given where rather more detailed sum-over-states quantum mechanical calculations have been used to gain insight into the implications of the involvement of particular excited states on the two-photon absorption spectrum. In these cases, only an overview is given, and the reader is referred to the original articles for further details.

A. Polymers

1. Polydiacetylenes

Chance et al.[40] have obtained both real and imaginary components of the complex two-photon hyperpolarizability spectra, $\gamma_T(2\omega_1)$ for various colored forms of the polydiacetylene 4BCMU in chloroform solution using CARS and CSRS techniques. They obtain an approximate form for the imaginary two-photon hyperpolarizability:

$$\mathrm{Im}(\gamma_T) = \mathrm{Im}\left(\frac{A}{(\langle\omega_{gf}\rangle - 2\omega_1 - i\Gamma_{gf})}\right), \tag{77}$$

with

$$A = \frac{3e^4\phi\, f_{gi} f_{if}}{16\hbar m^2\langle\omega_{gi}\rangle\langle\omega_{if}\rangle}\left(\frac{1}{\langle\omega_{gi}\rangle - \omega_1}\right)\left(\frac{1}{\langle\omega_{gi}\rangle - \omega_2} + \frac{1}{\langle\omega_{gi}\rangle - \omega_3}\right), \tag{78}$$

where the subscripts g, i and f represent the ground, intermediate, and two-photon states, respectively, while the frequency terms, ω_1, ω_2, ω_3 are the two incident beams and the CARS signal beam, respectively. Orientational averaging of the four wave mixing signal is included in the factor ϕ which ranges from 1 (isotropic) to 0.2 (1D chromophore). The single-photon oscillator strength, f_{gi}, is obtained from the linear absorption spectrum, leaving $\langle\omega_{gf}\rangle$, Γ_{gf} and f_{if} as data fitting parameters. These terms represent the average two-photon excitation frequency, inhomogeneous line width, and the oscillator strength for excitation from the intermediate state to the two-photon state. An example of the spectral dependence for $\mathrm{Im}(\gamma_T)$ and the corresponding real component of the hyperpolarizability, $\mathrm{Re}(\gamma_T)$, using experimentally fitted parameters[40] appropriate to the yellow form of 4 BCMU in chloroform is shown in Fig. 24.

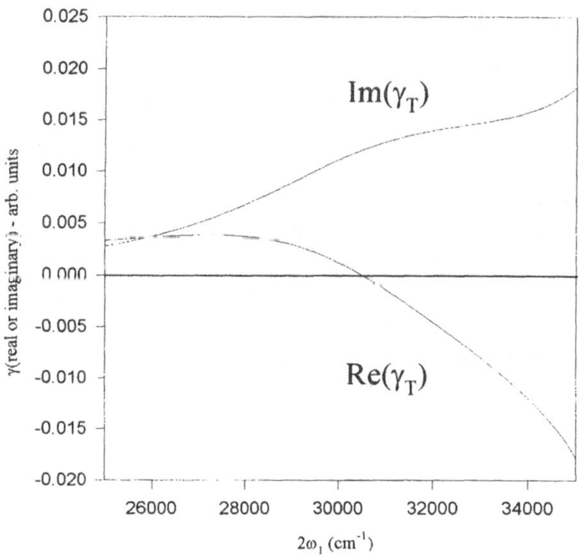

Figure 24 Real and imaginary parts of the susceptibility of the yellow form of 4BCMU in chloroform after Chance et al.[40] The imaginary component may be viewed as a combination of a single-photon term, increasing with two-photon energy, and a much weaker two-photon resonance centered around 30500 cm^{-1}.

2. Polysilanes

Soos and Kepler[58] have measured the two-photon absorption of poly(di-n-hexylsilane) films at cryogenic temperatures using fluorescence techniques and have successfully fitted their data using an exact Pariser-Parr-Pople model. Their analysis accounts for the correlated behavior of the σ delocalized bonding along the silicon chain. The calculated dispersion is shown in Fig. 25, and may be compared with the experimental spectrum for PDHS listed in the data table section (Fig. 30). The graph is scaled in terms of the energy gap E_g, the energy of the odd parity 1^1B_u state. In this particular example it was shown that reasonable agreement could be obtained taking the 10 lowest excited states. In the vicinity of the two-photon absorption peak, the dispersion can simply be approximated as:

$$\alpha^{(2)}(\omega) \propto \frac{|M(Y)|^2}{\{[E(Y) - 2\hbar\omega]^2 + \Gamma^2(Y)\}}, \tag{79}$$

where Y is the even-parity two-photon excited state index, i.e., $Y = 2^1A_g$, $E(Y)$ is the energy of the two-photon excited state, $1/\Gamma(Y)$ is the excited state lifetime, and $M(Y)$ is the two-photon transition moment.

B. Isolated Chromophores

In small molecules, as opposed to extended polymers, it is in principal simpler to model the nonlinear dispersion, due to the smaller number of sites involved in the localized electronic structure. In many cases, an adequate description may be ob-

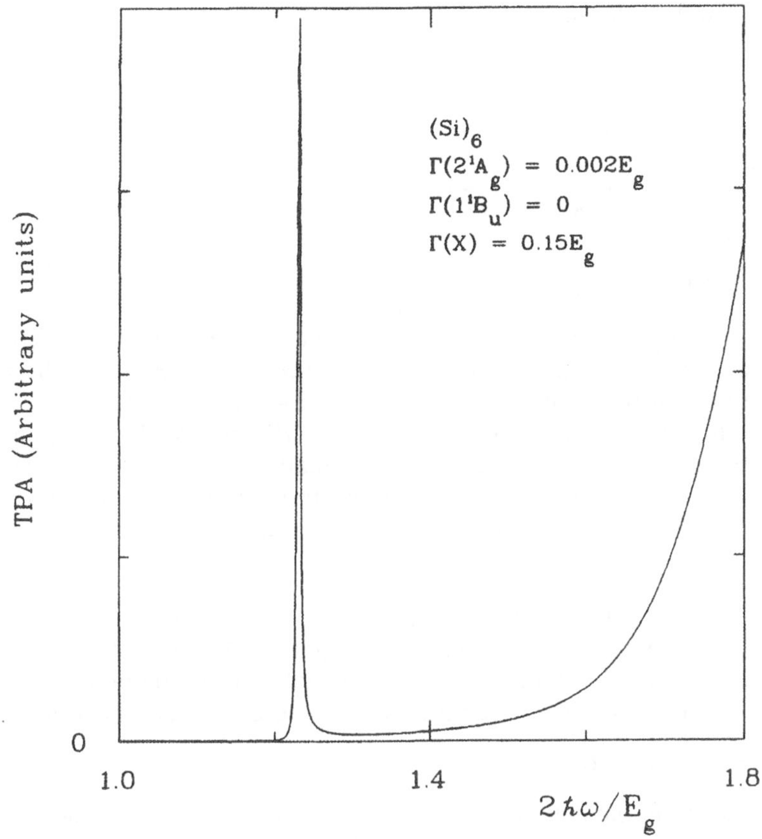

Figure 25 Exact PPP calculation of PDHS two-photon absorption by Soos and Kepler.[58] [Reproduced from: Phys. Rev. B, *43(14)*, 11 908–11 912, (1991).]

tained with fewer excited states though the extent to which this is true depends strongly on the molecule in question. Using a standard perturbation approach it is possible to obtain an expression for the third order nonlinearity in terms of a sum-over-states which may be further rearranged[59,60] to give[61]:

$$\begin{aligned}
&\gamma(-\omega;\omega_1,\omega_2,\omega_3) \\
&\quad = \frac{e^4}{\hbar^3} I_{1,2,3} \left[\sum_{l,m,n} \left\{ \frac{\langle x_i \rangle_{gl} \langle \bar{x}_j \rangle_{lm} \langle \bar{x}_k \rangle_{mn} \langle x_l \rangle_{ng}}{(\Omega_{lg} - \omega)(\Omega_{mg} - \omega_1 - \omega_2)(\Omega_{ng} - \omega_1)} \right. \right. \\
&\qquad + \frac{\langle x_i \rangle_{gl} \langle \bar{x}_j \rangle_{lm} \langle \bar{x}_k \rangle_{mn} \langle x_l \rangle_{ng}}{(\Omega^*_{lg} + \omega)(\Omega_{mg} - \omega_1 - \omega_2)(\Omega_{ng} - \omega_1)} \\
&\qquad + \frac{\langle x_i \rangle_{gl} \langle \bar{x}_j \rangle_{lm} \langle \bar{x}_k \rangle_{mn} \langle x_l \rangle_{ng}}{(\Omega^*_{lg} + \omega)(\Omega^*_{mg} + \omega_1 + \omega_2)(\Omega_{ng} - \omega_1)} \\
&\qquad \left. + \frac{\langle x_i \rangle_{gl} \langle \bar{x}_j \rangle_{lm} \langle \bar{x}_k \rangle_{mn} \langle x_l \rangle_{ng}}{(\Omega^*_{lg} + \omega)(\Omega^*_{mg} + \omega_1 + \omega_2)(\Omega^*_{ng} + \omega_1)} \right\}
\end{aligned}$$

$$- \sum_{m,n} \left\{ \frac{\langle x_i \rangle_{gm} \langle x_j \rangle_{mg} \langle x_k \rangle_{gn} \langle x_l \rangle_{ng}}{(\Omega_{mg} - \omega)(\Omega_{mg} - \omega_3)(\Omega_{ng} - \omega_1)} + \frac{\langle x_i \rangle_{gm} \langle x_j \rangle_{mg} \langle x_k \rangle_{gn} \langle x_l \rangle_{ng}}{(\Omega_{mg} - \omega_3)(\Omega^*_{ng} + \omega_2)(\Omega_{ng} - \omega_1)} + \frac{\langle x_i \rangle_{gm} \langle x_j \rangle_{mg} \langle x_k \rangle_{gn} \langle x_l \rangle_{ng}}{(\Omega^*_{mg} + \omega)(\Omega^*_{mg} + \omega_3)(\Omega^*_{ng} - \omega_1)} + \frac{\langle x_i \rangle_{gm} \langle x_j \rangle_{mg} \langle x_k \rangle_{gn} \langle x_l \rangle_{ng}}{(\Omega^*_{mg} + \omega_3)(\Omega_{ng} - \omega_2)(\Omega^*_{ng} + \omega_1)} \right\} \Bigg]. \quad (80)$$

In this equation, $I_{1,2,3}$ is a permutation operator such that the terms following are averaged over frequencies ω_1, ω_2, ω_3, and coordinate indices j, k, and l. In each of the summations, the groundstate term is excluded. Quantities of the form Ω_{ng}, etc., denote the complex transition frequency, $\Omega_{ng} = \omega_{ng} - i\Gamma_n$, where the damping factor Γ_n accounts for any broadening mechanisms. Finally, $\langle x_i \rangle_{gm} = \langle g|x_i|m \rangle$ and $\bar{x} \equiv x - \langle x \rangle_{gg}$, where x is the position vector of the electron, are used as shorthand notation in the description of the states involved in the summation.

1. *Two-level Models*

Where the nonlinear interaction involves absorption of beams of only one wavelength $\gamma(-\omega;\omega,\omega,-\omega)$ is computed. For a system adequately described by a one-dimensional molecular model involving only a groundstate and a one-photon excited state, the previous expression can be reduced to a more compact notation:

$$\gamma = -[\mu_{01}^4 D_{11} - \mu_{01}^2 (\Delta\mu_{01})^2 D_{111}], \quad (81)$$

where D_{11} and D_{111} represent the denominators of the double and triple summations in Eq. 80, μ_{01} is the dipole transition moment for the excitation, and $\Delta\mu_{01}$ is the difference in dipole moments of the groundstate and excited state. An example of a two level calculation based on transition moment data for disperse red 1 is shown in Fig. 26.

2. *Three-level Models*

A more elaborate model can be used while the molecule in question can still be considered centrosymmetric:

$$\gamma = -\left[\mu_{01}^4 D_{11} - \mu_{12}^2 \mu_{01}^2 D_{121} \right]. \quad (82)$$

This model incorporates a further (gerade) two-photon excited state in addition to the previous one-photon excited state. The expression is an approximate form of the more general equation given by Kuzyk and Dirk.[60] Figure 27 shows an example of a three level calculation for the squarine, ISQ, invoking the 2^1Ag and 1^1Bu excited states.

3. *Four-level Models*

Further refinements to the above models may be achieved by introducing further excited states. For example, an extra two-photon transition may be induced:

$$\gamma = -\mu_{01}^2 [\mu_{01}^2 D_{11} - \mu_{12}^2 D_{121} - \mu_{13}^2 D_{131}], \quad (83)$$

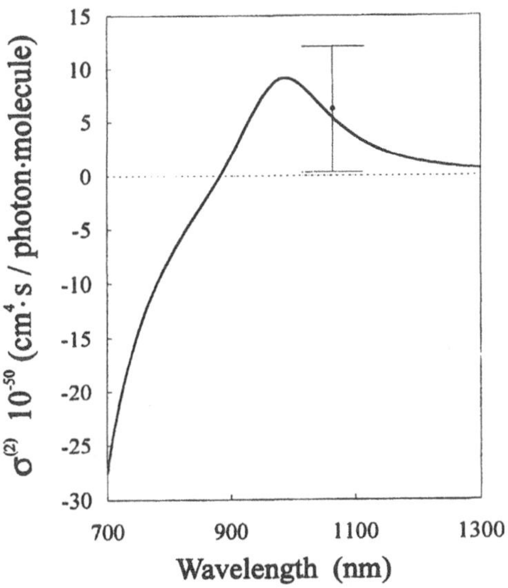

Figure 26 Two-level calculated two-photon absorption spectrum of Disperse Red 1.[61] Also shown for comparison is the result of a nonlinear absorption measurement at 1064 nm. [Reproduced from: D.M. Cohen, MSc. Thesis, Washington State University, 1994.]

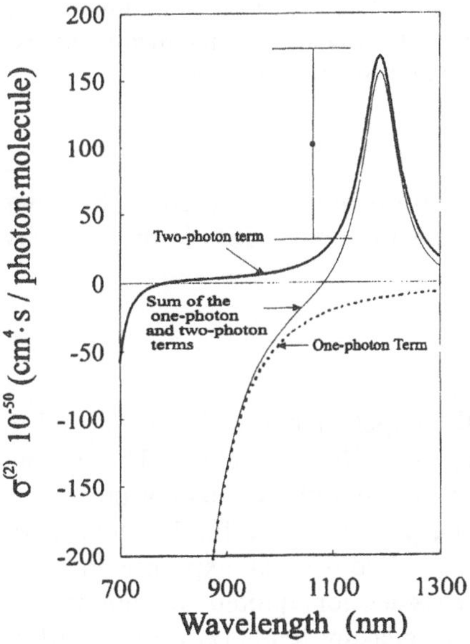

Figure 27 Three-level model for the ISQ molecule showing relatively poor agreement with experimentally measured value.[61] The sum of the one- and two-photon terms has opposite sign to the measured $\sigma^{(2)}$ at 1064 nm. [Reproduced from: D.M. Cohen, MSc. Thesis, Washington State University, 1994.]

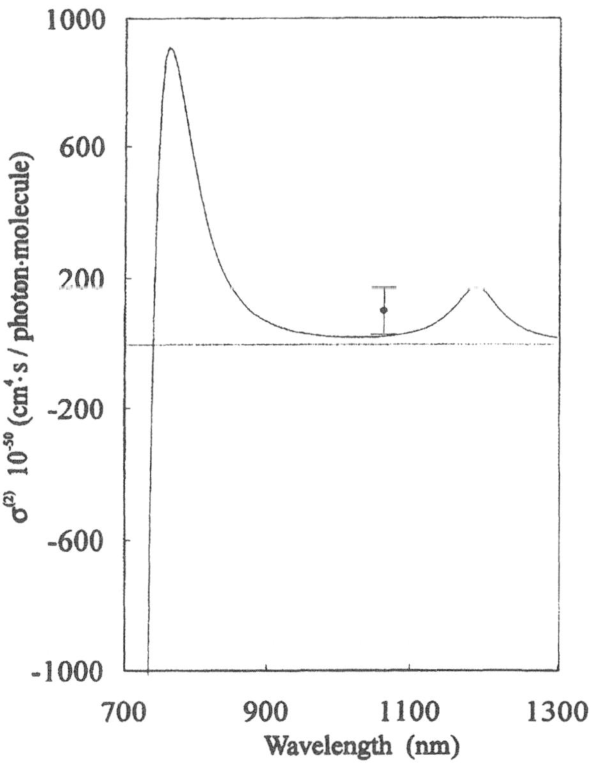

Figure 28 Four-level model for ISQ showing improved agreement by including an additional two-photon excited state[61] at shorter wavelengths than in the previous figure. The overall sum now has the correct sign at 1064 nm compared with experimental measurements. [Reproduced from: D.M. Cohen, MSc. Thesis, Washington State University, 1994.]

where again it is assumed that the molecule is centrosymmetric. Figure 28 shows how the predicted dispersion of the two-photon absorption for ISQ is modified by including the further 3^1Ag excited state.

VI. DATA TABLES

Table 1 lists the abbreviations used to classify the experimental techniques used to obtain the two-photon absorption coefficients in the following tables. Throughout the tables, $\alpha^{(2)}$ and $\sigma^{(2)}$ are quoted as a function of the two-photon excitation energy in eV. Where the conversion from $\alpha^{(2)}$, etc., to the perhaps more fundamental two-photon absorption cross-section, $\sigma^{(2)}$, has been made, the original experimental data are generally quoted first and the calculated cross-section quoted in parenthesis. This was felt preferable since, for example, most measurements on polymer films will be for assessing device performance where $\alpha^{(2)}$ is the most frequently used quantity, while the performance of dyes, for example, is more conveniently classified in terms of $\sigma^{(2)}$. In the original articles, it is usually possible to determine the chromophore number densities fairly readily and some authors even take the

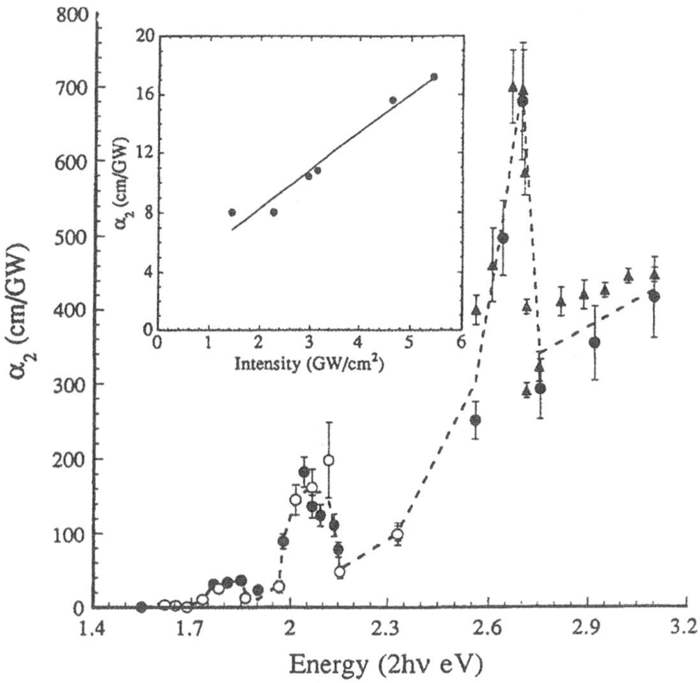

Figure 29 Two-photon absorption (α_2) spectrum for PTS measured by Lawrence et al.[114] using the open aperture Z-scan technique. Empty and filled circles are for different PTS samples, measured with 65 ps pulses, while the triangles correspond to measurements with 4 ps pulses. The inset shows intensity dependence of nonlinear absorption, indicating the presence of nonlinear terms of higher order than three. [Reproduced from: Phys. Rev. Letts., *73(4)*, 597–600 (1994).]

trouble to quote results in either form. In some instances, insufficient information was available to make a reasonably accurate conversion. In such cases, order of magnitude calculations may be made bearing in mind that pure materials generally have a number density in the range 10^{21}–10^{22} molec./cm^3, while most dilute solution measurements tend to be in the region of 10^{18} molec./cm^3. Where concentrations were explicitly stated in the original publications, number densities have been listed. Where a reasonable estimate has been made, e.g., assuming a material density of 1 g/cm^3, then this is indicated with the symbol ~.

A note of caution is necessary since it is not always wise to assume that the measured nonlinear absorption ($\alpha^{(2)}$) is solely due to an instantaneous two-photon absorption. Excited state absorption (step-wise two-photon absorption) may lead to rather inflated apparent two-photon cross-sections especially where longer excitation pulses are used and close to resonance. The reader is urged to consult the original texts for further details where this is suspected. In many cases, this issue will have been discussed and steps to compensate for such effects may or may not have been taken.

The materials tables are grouped into the following broad categories: molecules of biological interest (Table 2), materials commonly used as dyes (Table 3), liquid

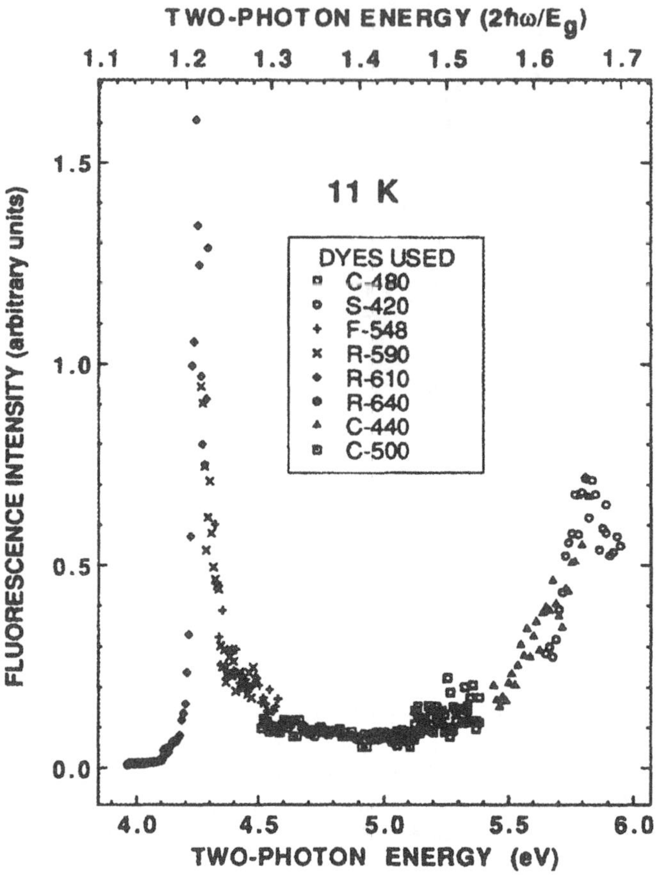

Figure 30 Two-photon excited luminescence for a poly(di-n-hexylsilane) film at 11 K by Soos and Kepler.[58] The excitation source was a dye laser followed by a monochromator to remove dye amplified spontaneous emission. The symbols correspond to measurements made with the dyes listed in the inset. [Reproduced from: Phys. Rev. B, *43*(*14*), 11 908–11 912 (1991).]

crystals (Table 4), and polymers (Table 5). Organo-metallic materials (Table 6) are grouped separately from the dyes even though it is recognized that some were, and still are, used as laser dyes in passive mode-locking devices. Organo-metallic polymers are listed with the rest of the polymers in Table 5. Finally, Table 7 is headed simply "Organic Materials" and covers the remaining types of molecules such as solvents, pure liquids, crystalline materials, vapors and gases, etc.

In addition to the tabulated data, several two-photon absorption spectra are also given. These include spectra for polyacetylenes (Figs. 29 and 31), polysilanes (Figs. 30, 32, and 33), pyridine (Fig. 35), and anthracene (Fig. 36). Figure 34 shows the two-photon absorption *cross section* versus concentration for a selection of essentially planar metal dithiolenes, indicating that in some instances intermolecular effects can substantially enhance the nonlinear absorption coefficient.

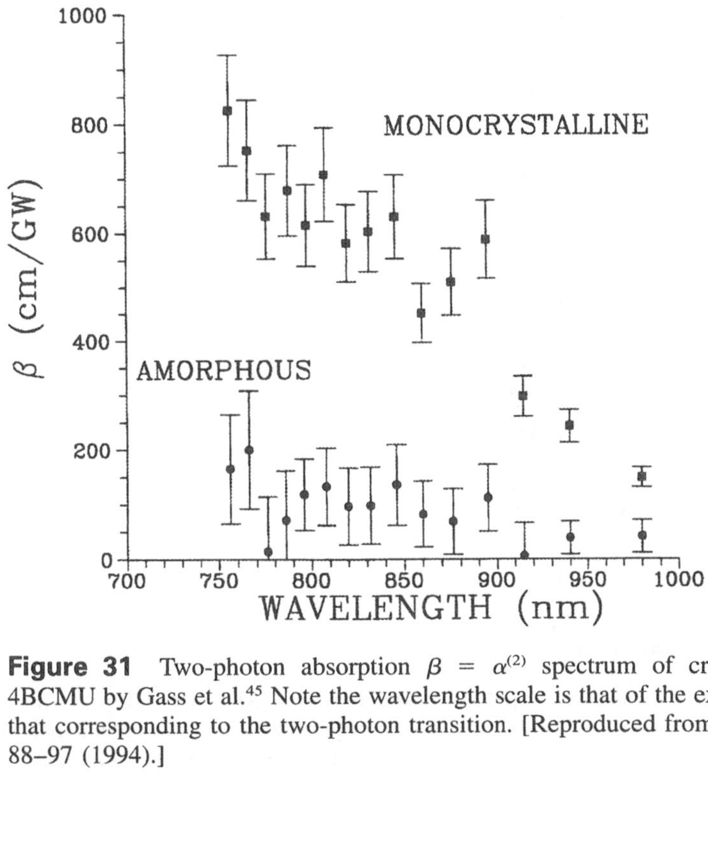

Figure 31 Two-photon absorption $\beta = \alpha^{(2)}$ spectrum of crystalline and amorphous 4BCMU by Gass et al.[45] Note the wavelength scale is that of the excitation beam rather than that corresponding to the two-photon transition. [Reproduced from: J. Chem. Phys., *100(1)*, 88–97 (1994).]

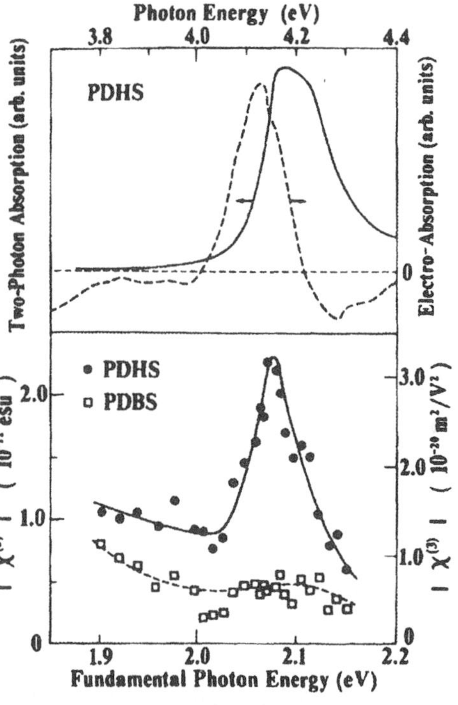

Figure 32 Third order nonlinearity of polydihexylsilane (PDHS) and polydibutylsilane (PDBS) films (lower part) and the corresponding two-photon and electro-absorption spectra for PDHS (upper part). The two-photon data are derived from complex THG measurements.[51] The lower abscissa is the incident photon-energy while for the absorption spectra the upper energy scale refers to the *two-photon* energy. [Reproduced from: Phys. Rev. B., *45(11)*, 6317–6320 (1992).]

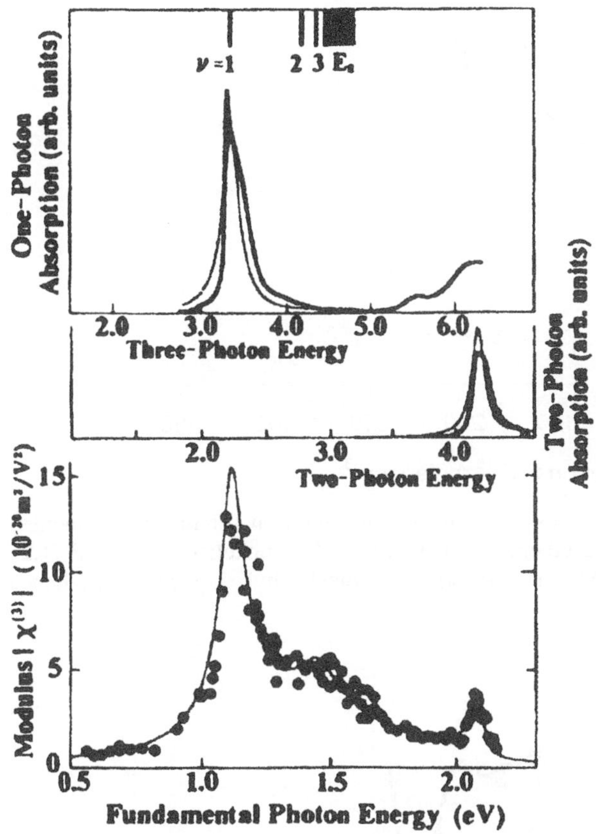

Figure 33 One- and two-photon absorption spectra and NLO spectra for PDHS measured using the THG technique of Hasegawa et al.[121] Top curve—linear absorption spectrum scaled to demonstrate the multiphoton resonant enhancement in the NLO spectrum. Middle curve—two photon absorption spectrum vs. two-photon energy. Lower curve—NLO spectrum with corresponding enhancement features. [Reproduced from: Phys. Rev. Letts., *69*(*4*), 668–71, (1992).]

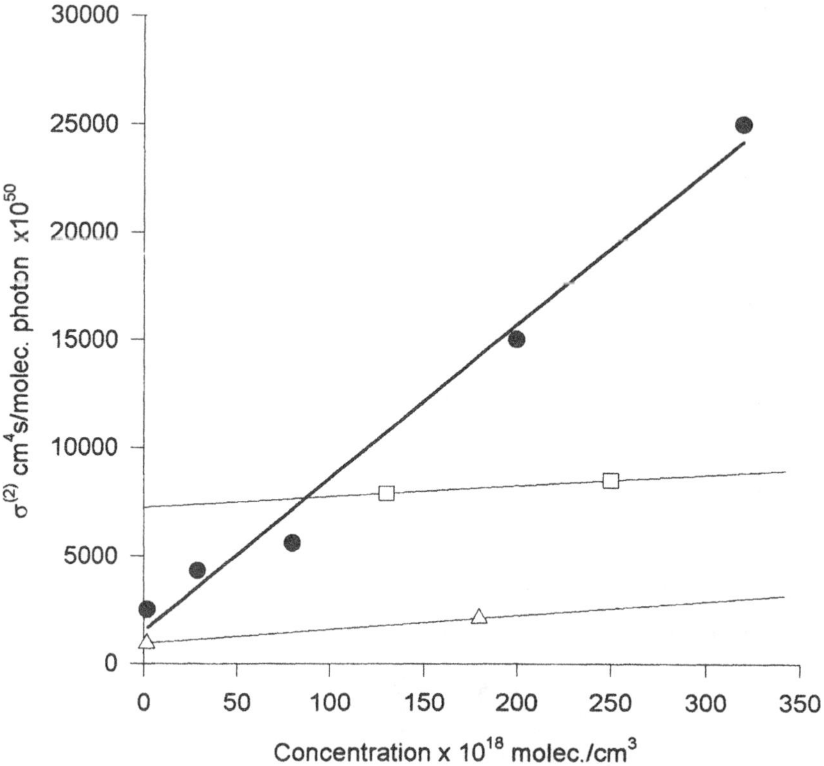

Figure 34 Two-photon absorption cross-section vs. concentration for metal dithiloenes; filled circles, bis[1-butyl.2-phenyl-1,2-ethenedithiolato (2-) S,S′] nickel; triangles, bis[1,2-diethyl-1,2-ethene dithiolato (2-)S,S′] nickel; squares, bis[1-ethyl 2-phenyl-1,2-ethene-dithiolato (2-)S,S′] nickel.

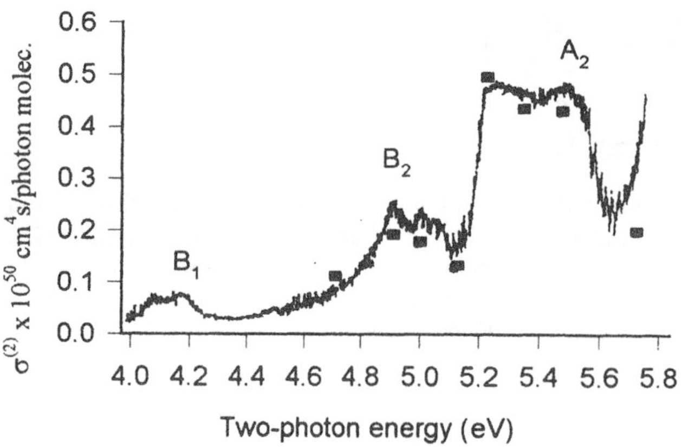

Figure 35 Comparison of thermal lensing two-photon absorption spectrum and calculated cross-sections, $\sigma^{(2)}$, derived from CARS measurements for liquid pyridine, by Foggi et al.[31] The symmetries of the transitions corresponding to the maxima are as indicated. A further n to π^* transition with A_2 symmetry occurs to higher energy (~6.17 eV). [Reproduced from: J. Molecular Structure, *175*, 147–152 (1988).]

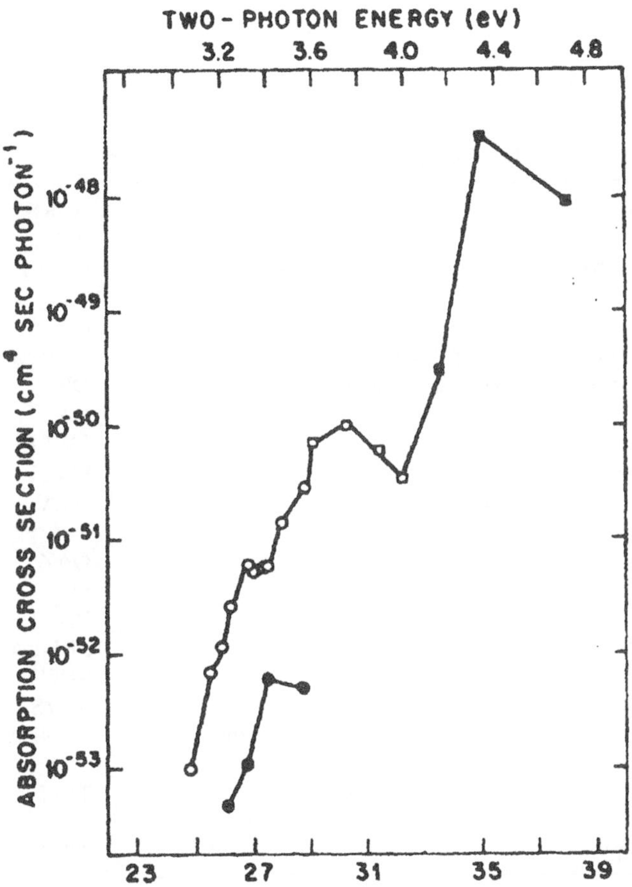

Figure 36 Two-photon absorption spectrum of anthracene according to Warlock[199] from the data in references [133], [138], and [139]. [Reproduced from: Chapter 3, p. 1364, in *Laser Handbook*, Vol. 2 (F.T. Arecchi and E.O. Schulz-Dubois, eds.), North Holland, New York, 1972.]

Table 1 Abbreviations for Measurement Techniques Found in Following Tables

Anharmonic forced Rayleigh scattering	AFRS
Acousto-optic measurements	AO
Coherent anti-Stokes' Raman spectroscopy and coherent Stokes Raman scattering	CARS
Comparison of two-photon loss and Raman gain	CR
Fitting parameter in degenerate four wave mixing measurements	DFWM
Directional coupler overlay device or other type of waveguide device	DRC
Electro-absorption	EA
Four wave mixing in solvent with real n_2	FWMS
Non-degenerate four wave mixing	FWM
Gain measurement in a two-photon pumped amplifier	GTA
Luminescence measurements	L
Open aperture Z-scan	OAZ
Photoconductivity	PC
Raman cross-section data	RX
Single beam nonlinear absorption	SBNLA
Fitting parameter in SHG analysis	SHG
Stimulated thermal Rayleigh scattering	STRS
Two beam nonlinear absorption, one beam scanned in frequency, or intense pump beam probed with pulsed continuum derived from pump	TBNLA
Two channel normalization	TCN
Thermal lensing	TL
Two-photon ionization spectroscopy	TPIS

Table 2 Biological Materials. All Values of $\sigma^{(2)}$ are Dressed Values of the Two-Photon Absorption Cross-Section

Material	Technique	Applied two-photon energy (eV)	$\sigma^{(2)}$ (10^{-50} cm^{4}s/ photon-molecule)	Pulse duration (ns)	Ref	Solvent	Concentration molec/cm^3	Other
Bacteriorhodopsin	TBNLA	2.07	169	6	62	D_2O		
Bacteriorhodopsin	TBNLA	2.12	207	6	62	D_2O		
Bacteriorhodopsin	TBNLA	2.16	247	6	62	D_2O		
Bacteriorhodopsin	TBNLA	2.21	289	6	62	D_2O		
Bacteriorhodopsin	TBNLA	2.30	288	6	62	D_2O		
Bacteriorhodopsin	TBNLA	2.36	244	6	62	D_2O		
Bacteriorhodopsin	TBNLA	2.56	201	6	62	D_2O		
Bacteriorhodopsin	TBNLA	2.70	167	6	62	D_2O		
Bacteriorhodopsin	TBNLA	2.78	127	6	62	D_2O		
Bacteriorhodopsin	TBNLA	2.92	174	6	62	D_2O		
Bacteriorhodopsin	TBNLA	3.02	199	6	62	D_2O		
β-carotene	SBNLA	2.33	1.3×10^4 [$\alpha^{(2)} = 4.0$ cm/GW]	~50–100	63	DCM/DMF/THF	2.9×10^{18}	
β-carotene	SBNLA	2.33	8.5×10^4 [$\alpha^{(2)} = 26$ cm/GW]	~50–100	63	DCM/DMF/THF	2.9×10^{18}	
β-carotene	SBNLA	2.33	720 [$\alpha^{(2)} = 0.22$ cm/GW]	0.035	63	DCM/DMF/THF	2.9×10^{18}	
Chlorophyll-a	L	2.33	2.0	30	64	Ethanol		
Chlorophyll-a	SBNLA	3.6	—	2	65, 66			
H_2O	PC	9.32	0.081	0.03	67	Neat	3.3×10^{22}	
NADH (enzyme)	SBNLA, TCN	3.56	340	20	68			
Retinal	L	3.57	27	40	69	Ethanol		
11-cis-locked Retinal (locked Rhodopsin analogue)	TL	2.48–3.41	Relative spectrum	Not stated (pulsed N_2 or SH-YAG pumped dye laser excitation)	70	D_2O	Not stated	Room temp. Bound to Opsin. Non-photoble achable form
Retinyl acetate	L	3.57	26	40	69	n-hexane		

Table 2 Continued

Material	Technique	Applied two-photon energy (eV)	$\sigma^{(2)}$ (10^{-50} cm^{4}s/photon-molecule)	Pulse duration (ns)	Ref	Solvent	Concentration molec/cm^3	Other
Retinyl acetate	L	3.57	29	40	69	5:5:2 mixture petroleum ether/diethyl ether/isopropanol		
Trans-β-carotene	SBNLA	1.86	300 [$\alpha^{(2)}$ = 0.8 cm/GW]	QS*	48	Chloroform	7.9×10^{18} (5%)	
Trans-β-carotene	SBNLA	3.22	1.5×10^3 [$\alpha^{(2)}$ = 3.3 cm/GW]	QS*	48	Chloroform	7.9×10^{18} (5%)	
Trans-retinal	L	2.8–3.0	Relative spectrum		71			77K
Trans-retinal	L	2.61–3.60	Relative spectrum	Not stated (see 11-cis-locked retinal)	70	Ethyl-ether/iso-pentane/ethanol, 5:5:2 (v/v)	Not stated	77K
Trans-retinol (vitamin A)	L	3.4–4.0	Relative spectrum 20 @ 3.52 eV	1200	72	Ethyl-ether/iso-pentane/ethanol, 5:5:2 (v/v)	4.8×10^{17}	77K

Trans-retinol	L	3.29–3.84	Relative spectrum	Not stated (see 11-cis-locked retinal)	70	Ethyl-ether/iso-pentane/ethanol, 5:5:2 (v/v)	Not stated	77K
Trans-N-retinylidene-n-butylamine (Schiff base)	L	2.73–3.60	Relative spectrum	Not stated (see 11-cis-locked retinal)	70	Ethyl-ether/iso-pentane/ethanol, 5:5:2 (v/v)	Not stated	77K
Trans-N-retinylidene-n-butylamine: HCl (protonated Schiff base)	TL	2.54–3.48	Relative spectrum	Not stated (see 11-cis-locked retinal)	70	Carbon tetrachloride	Not stated	Room temp.

*QS-Q-switched laser, pulse width not specified
Note: Ref. 70 also gives CNDO calculation of excited state energy levels for a free-base porphyrin and the dianion for comparison with Retinol derivatives.

Table 3 Dressed Two-Photon Absorption Cross-Section for Dyes and Related Materials

Material	Technique	Applied two-photon energy (eV)	$\sigma^{(2)}$ (10^{-50} cm^{4}s/ photon-molecule)	Pulse duration (ns)	Ref	Solvent	Concentration molec/cm^3	Other
Acridine	L	3.56	2.0		73	Ethanol		
Acridine orange	L	2.33	Rel. values	~200, ~40	74–77			
Acridine red	L	2.33	Rel. values	~200, ~40	74–77			
Acridine red	L	2.33	2.8	60	78	Ethanol		
Acridine yellow	L	2.33	Rel. values	~200, ~40	74–77			
BBOT	L	3.56	4	40	79	Methylcyclo-hexane		
BBQ	L	4.66	81	<<2000	37	Toluene/ethanol	Not stated	
BSQ	SBNLA	2.33	<107	0.035	61	PGMEA solution	$\sim 1 \times 10^{18}$	
Coumarin 450	L	4.66	0.08	<<2000	37	Methanol	Not stated	
Coumarin 480	L	4.66	0.12	<<2000	37	Methanol	Not stated	
Cyanine I, II, III, IV	L	2.33	Rel. values	~200, ~40	74–77			
DCM	L	2.33	9.0	110	80			
Dimethyl-POPOP	L	3.56	11.5	40	79	Methyl-cyclohexane		
Dimethyl-POPOP	GTA	3.56	600	0.005	81	Methyl-cyclohexane		
Disodium fluorescein	L	2.33	0.18	60	78	Ethanol		
Disperse Red 1	SBNLA	2.33	6.2	0.035	61	PGMEA solution	$\sim 4 \times 10^{19}$	
DNFA	SBNLA	2.33	<25	0.035	61	PGMEA solution	$\sim 2 \times 10^{19}$	
DODCI	L	2.33	22	60	78	Methanol		
DODCI I	L	2.33	38	110	80			
DPS390	L	4.66	5		37	Methanol	Not stated	
Eosin	L	2.33	18	30	64			
Eosine B, Y	L	2.33	Rel. values	~200, ~40	74–77			
Fuschin	SBNLA	2.35	10	0.005	9	Methanol	1.2×10^{20}	
HMICI	SBNLA	2.35	20	0.005	9	Hexalfluoro-isopropanol	4.8×10^{19}	

HSQ	SBNLA	2.33	<286	0.035	61	PGMEA solution	$\sim 1 \times 10^{18}$	
HITC/PMMA (1,1′,3,3,3′,3′,-heaxamethylindo-tricarbocyanine iodide)	OAZ	2.33	5600 [$\alpha^{(2)}$ = 30 cm/GW]	0.1	82	PMMA host	$\sim 10^{20}$ (10%w/w)	Film
Hydroxycoumarin	L	3.8–5.0	Spectrum 7.9 @ 4.1 eV	4	83			
Imidocarbo-cyanine	L	2.23–3.29	Relative spectrum		84	Diethylene-glycol		
Indo-1	L	3.42–3.97	Relative spectrum	0.0001–0.00015 @ 78 MHz	85	Cell stain preparation	$\sim 6 \times 10^{16}$ (100 μM)	Relative to p-bis (o-methyl-styryl) benzene ref.
Indo-1 bound to Ca^{2+} ion	L	3.42–3.97	Relative spectrum	0.0001–0.00015 @ 78 MHz	85	Cell stain preparation	$\sim 6 \times 10^{16}$ (100 μM)	Relative to p-bis (o-methyl-styryl) benzene ref.
ISQ	SBNLA	2.33	102	0.035	61	PGMEA	$\sim 1 \times 10^{18}$	
LD490	L	4.66	0.05		37	Methanol	Not stated	
Methyldiethyl-aminocoumarin	L, SBNLA	3.56	14.5, 50	40	79	Ethanol		
Methylene blue	SBNLA	2.35	7	0.005	9	Methanol	1.2×10^{20}	
NCAE	SBNLA	2.33	<38.2	0.035	61	PGMEA	$\sim 3.5 \times 10^{19}$	
Oxazine 725	L	2.33	0.41	110	80			
Phloxine B	L	2.33	Relative values	~200, ~40	74–77			
PSQ	SBNLA	2.33	<102	0.035	61	PGMEA	$\sim 1 \times 10^{18}$	
PYC	SBNLA	2.35	18	0.005	9, 86	Hexafluoro-isopropanol	6×10^{19}	

Table 3 Continued

Material	Technique	Applied two-photon energy (eV)	$\sigma^{(2)}$ (10^{-50} cm^4s/ photon-molecule)	Pulse duration (ns)	Ref	Solvent	Concentration molec/cm^3	Other
Rhodamine 101	L	2.33	20	110	80			
Rhodamine 110	L	2.33	0.19	110	80			
Rhodamine 6G	L	2.33	26	30	64			
Rhodamine 6G	SBNLA	2.33	12.9		87			
Rhodamine 6G	SBNLA	3.56	355		87			
Rhodamine 6G	L	2.1–3.5	Relative spectrum	0.025	88			
Rhodamine 6G	L	2.33	11	110	80			
Rhodamine 6G	L	2.33	5.5	60	78	Ethanol		
Rhodamine 6G	TBNLA	4.83–5.45	Spectrum	8	89	Ethanol	Stated as 6×10^{-8} M ($=3.6 \times 10^{13}$!)	

Rhodamine 6G	SBNLA	2.35	10	0.005	9	Methanol	1.2×10^{20}
Rhodamine 6G	SBNLA	3.57	180	0.015	90	Acidic methanol	
Rhodamine B	L	2.33	13	CW	91		
Rhodamine B	SBNLA	2.33	14.3		87		
Rhodamine B	SBNLA	3.56	148		87		
Rhodamine B	L	2.33	12	110	80		
Rhodamine B	L	2.33	14	60	78	Dichloroethane	
Rhodamine B	SBNLA	3.57	120	0.015	90	Acidic methanol	
Rhodamine B, 6G	L	2.33	Relative values	~200, ~40	74–77		
Rhodamine C	L	2.33	26	30	64		
Rubicene	L	2.33	Relative values	~200, ~40	74–77		
Safranine O	L	2.33	Relative values	~200, ~40	74–77		
Safranine T	SBNLA	2.35	5	0.005	9	Hexafluoro-isopropanol	2×10^{20}
Stilbene	L	4.66	8		37	Methanol	6×10^{13}
TSQ	SBNLA	2.33	−105	0.035	61	PGMEA	1×10^{18}
Uranine	L	2.33	Relative values	~200, ~40	74–77		

Table 4 Two-Photon Absorption Coefficient, $\alpha^{(2)}$, for Liquid Crystals

Material	Technique	Applied two-photon energy (eV)	$\alpha^{(2)}$ (cm/GW)	Pulse duration (ns)	Ref	Solvent	Concentration molec/cm^3	Other
BBIPA	SBNLA	4.66	0.5	0.030	92	Neat	$\sim 2 \times 10^{21}$	Isotropic phase
ETBBA	SBNLA	4.66	0.5	0.030	92	Neat	$\sim 2 \times 10^{21}$	Isotropic phase
IPBBA	SBNLA	4.66	0.5	0.030	92	Neat	$\sim 2 \times 10^{21}$	Isotropic phase
MEBBA	SBNLA	4.66	0.55	0.030	92	Neat	$\sim 2 \times 10^{21}$	Isotropic phase
PEBBA	SBNLA	4.66	0.55	0.030	92	Neat	$\sim 2 \times 10^{21}$	Isotropic phase
Pentyl cyanobiphenyl	SBNLA	4.31	0.031* [$\sigma^{(2)} \sim 0.45 \times 10^{-50}$ cm^{4}s/photon-molecule]	.0003	93	Neat	$\sim 2 \times 10^{21}$	Isotropic phase, 42°C
Pentyl cyanobiphenyl	OAZ	4.66	320 parallel to director	7	15	Neat	$\sim 2 \times 10^{21}$	25 micron LC cells, planar alignment, 25°C-nematic phase
Pentyl cyanobiphenyl	OAZ	4.66	<32 (estimated) perpendicular to director	7	15	Neat	$\sim 2 \times 10^{21}$	25 micron LC cells, planar alignment, 25 deg. C-nematic phase

*Authors quote value as $3.4 \pm .4$ (GW.cm)$^{-1}$ from intercept of plot of reciprocal transmitted pulse energy against reciprocal incident pulse energy. The conversion to $\alpha^{(2)}$ assumes a beam waist of 120 μm diameter and pulse widths of 300 fs.

Table 5 Two-Photon Absorption Coeficients, $\alpha^{(2)}$, and Dressed TPA Cross-Sections, $\sigma^{(2)}$, for Polymers

Material	Technique	Applied two-photon energy (eV)	$\alpha^{(2)}$ (cm/GW) [or cross section $\sigma^{(2)}$ (10^{-50} cm^{4}s/photon-molecule)*]	Pulse duration (ns)	Ref	Solvent	Concentration molec/cm^3	Other
BBTDOT	OAZ	4.66	0.084 [$\sigma^{(2)} = 260$]	0.032	94	THF	1.2×10^{19}	
3BCMU (3 butoxy-carbonylmethyl-urethane)	CR	4.0	[$\sigma^{(2)} = 6000$]	~5	95	$CHCl_3$		
3BCMU	SBNLA	2.33	0.52	0.033	96	Chlorobenzene	Probably $\sim 10^{19}$	2.3% (blue gel)
Poly[3BCMU]	DRC	3.10	24	0.0012	97		$\sim 10^{21}$	Highly anisotropic film. TE polarized measurement.
Poly[3BCMU]	DRC	2.33	1.1 (Intensity-dependent, $\alpha^{(3)} = -0.24$ cm^3/GW^2)	0.035	97		$\sim 10^{21}$	Highly anisotropic film. TE polarized measurement.

Table 5 Continued

Material	Technique	Applied two-photon energy (eV)	$\alpha^{(2)}$ (cm/GW) [or cross section $\sigma^{(2)}$ (10^{-50} cm^{4}s/photon-molecule)*]	Pulse duration (ns)	Ref	Solvent	Ccncentration molec/cm^3	Other
Poly[3BCMU]	DRC	1.63	<0.02	0.007	97		$\sim 10^{21}$	Highly anisotropic film. TE polarized measurement.
4BCMU (4 butoxy-carbonylmethly-urethane)	CARS	3.2–4.2	Relative spectrum	~5	40	$CHCl_3$		
4BCMU	SBNLA	1.88	<0.25	0.06	98		$\sim 10^{21}$	Polymer waveguide.
4BCMU	SBNLA	1.55–3.1	Spectrum (5 @ 2.8 eV)		99			
4BCMU	DRC	2.33	~1 (TE polarization)	0.030	7		$\sim 10^{21}$	Directional coupler device (thin film) with partial inplane alignment.
4BCMU	DRC	2.33	~0.05 (TM polarization)	0.030	7		$\sim 10^{21}$	Directional coupler device (thin film) with partial inplane alignment.

4BCMU	DRC	1.8–3.2	Relative spectrum	2	100		~10^{21}	TE spectrum in thin film waveguide geometry
4BCMU	TBNLA	3.77–4.48	Spectrum and 0.06 @ 3.8 eV	<0.0001	101	DMF	~1.4×10^{20}	12% wt solution in DMF
4BCMU	SBNLA	2.33	0.76	0.033	96	Chlorobenzene	~10^{19}	1.7% (red) gel
4BCMU	SBNLA	2.33	<0.1	0.033	96	DMF	~1.6×10^{20}	14% (yellow) solution
4BCMU	SBNLA	Spectrum	4500 @ ~2.78 eV	10	102		~10^{21}	Amorphous film, large esa contribution probably included[45,100]
4BCMU	SBNLA	2.53–3.31	Spectrum, ~100 @ 3.02 eV	0.020	45		~10^{21}	Amorphous film
4BCMU	SBNLA	2.53–3.31	Spectrum, ~600 @ 3.02 eV	0.020	45		~10^{21}	Monocrystalline film
DANS/side chain polymer	OAZ	2.70	5.5	0.003	103		~10^{21}	50% DANS side-chains by volume. Films

Table 5 Continued

Material	Technique	Applied two-photon energy (eV)	$\alpha^{(2)}$ (cm/GW) [or cross section $\sigma^{(2)}$ (10^{-50} cm^4s/photon-molecule)*]	Pulse duration (ns)	Ref	Solvent	Concentration molec/cm^3	Other
DANS/side chain polymer	OAZ	3.10	1.5	0.003	103		$\sim 10^{21}$	50% DANS side-chains by volume. Films
DANS/side chain polymer	OAZ	2.92	3.5	0.003	103		$\sim 10^{21}$	50% DANS side-chains by volume. Films
DANS/side chain polymer	OAZ	2.33	2.1	0.035	103		$\sim 10^{21}$	50% DANS side-chains by volume. Films
DANS/side chain polymer	OAZ	1.55–2.07	<0.1 (across range)	0.065	103		$\sim 10^{21}$	50% DANS side-chains by volume. Films
DANS/side chain polymer	DRC	2.35	0.4	0.180–0.200	104		$\sim 10^{21}$	50% DANS side-chains by volume. Films

DANS/MMA copolymer 50/50	SBNLA/L	1.93	<3.5	~6	105			2mm thick pressed disks. Measurement limited by one-photon induced damage.
DANS/MMA copolymer 50/50 (mole/mole)	SBNLA/L	1.93	<0.3	~6	105	Cyclohexanone	Not given—presumably in region $10^{19} - 10^{20}$	
Didecyl-hexathiophene oligomer	OAZ	4.66	0.23 [$\sigma^{(2)} = 2.4 \times 10^4$]	0.030	106	Dioxane	3.6×10^{17}	TPA *saturated* at 35GW/cm^2
Didecyloxy para-polyphenyl oligomer (n = 7)	DFWM	4.12	0.73 [$\sigma^{(2)} \sim 4 \times 10^3$]	0.0004	42	THF	$\sim 6 \times 10^{18}$	
DDOS	OAZ	4.66	0.048 [$\sigma^{(2)} = 600$]	0.032	94	THF	3×10^{18}	
PBTAB Poly(α-{5,5′-bithiophenediyl}p-acetoxy-benzylidene)	SBNLA	2.33	100 [$\sigma^{(2)} \sim 2.1 \times 10^6$]	~50–100	63	DCM/DMF THF	9×10^{17} (1.5×10^{-3} M)	
PBTAB	SBNLA	2.33	6.9 [$\sigma^{(2)} \sim 1.4 \times 10^5$]	~50–100	63	DCM/DMF THF	9×10^{17} (1.5×10^{-3} M)	

Table 5 Continued

Material	Technique	Applied two-photon energy (eV)	$\alpha^{(2)}$ (cm/GW) [or cross section $\sigma^{(2)}$ (10^{-50} cm^{4}s/photon-molecule)*]	Pulse duration (ns)	Ref	Solvent	Concentration molec/cm^3	Other
PBTAB	SBNLA	2.33	0.019 [$\sigma^{(2)} \sim 400$]	0.035	63	DCM/DMF THF	9×10^{17} (1.5 $\times 10^{-3}$ M)	
PBTB Poly(α-{5,5′-bithiophenediyl} benzylidene)	SBNLA	2.33	320 [$\sigma^{(2)} \sim 4.9 \times 10^6$]	~50–100	63	DCM/DMF/THF	1.2×10^{18} (2 $\times 10^{-3}$ M)	
PBTB	SBNLA	2.33	580 [$\sigma^{(2)} \sim 9.0 \times 10^6$]	~50–100	63	DCM/DMF/THF	1.2×10^{18} (2 $\times 10^{-3}$ M)	
PBTB	SBNLA	2.33	0.094 [$\sigma^{(2)} \sim 1500$]	0.035	63	DCM/DMF/THF	1.2×10^{18} (2 $\times 10^{-3}$ M)	
PDTDS (poly(di-n-tetradecylsilane)	L	3.5–6.0	Relative spectra		95			77K thin films
PMMPS (poly methyl-4-methylpentyl-silane)	L	3.5–6.0	Relative spectra		107			77K thin films
PMPS (poly (methyl-phenylsilane)	L	4–6	Relative spectra	5	41			77K & 294K thin films
Poly (3-octylthiophene)	TBNLA	2–4.2	Spectrum, peak ~0.27 @ 3.5 eV	0.033	36	THF	7.7×10^{19}	(25 g/l)
Poly(di-n-butylsilane)	EA	3–6	Relative spectra		55			77K thin film
Poly(di-n-butylsilane)	L	4.2–5.0	Relative spectra		108			

Poly(di-n-hexylsilane)	L	1.15–1.75	Relative spectra		58			High mol. wt. Thin films 295K and 11K
Poly(di-n-hexylsilane)	EA	3–6	Relative spectra		55, 109			77K thin film
Poly(di-n-hexylsilane)	L	3.7–4.5	Relative spectra		108			
Poly-[Pd(PBu$_3$)$_2$DEB] (Palladium poly-yne)	FWMS	3.76	$\gamma'' \sim 380 \times 10^{-36}$ esu/ru** {equiv. to approx. $\sigma^{(2)} \sim$ 600}	8	46	THF	Typ. $\sim 7 \times 10^{19}$	
Poly-[Pt(PBu$_3$)$_2$ (DEX)$_2$] (Platinum poly-yne)	SBNLA	4.66	3.5 [$\sigma^{(2)} \sim 1900$]	6, 0.021	23	THF	6.7×10^{19}	
Poly-[Pt(PBu)$_3$)$_2$ (DEB)$_2$] (Platinum poly-yne)	SBNLA	4.66	5 [$\sigma^{(2)} \sim 7.8 \times 10^3$]	0.023	110	THF	2.4×10^{19}	
Poly-[Pt(PBu)$_3$)$_2$DEB] (Platinum poly-yne)	SBNLA	4.66	6.7 [$\sigma^{(2)} \sim 7.4 \times 10^3$]	0.023	110	THF	3.4×10^{19}	
Poly-[Pt(PBu)$_3$)$_2$DEB] (Platinum poly-yne)	FWMS	3.76	$\gamma'' \sim 1300 \times 10^{-36}$ esu/ru** {equiv. to approx. $\sigma^{(2)} \sim$ 2300}	8	46	THF	Typ. $\sim 7 \times 10^{19}$	
Poly-[Pt(PBu)$_3$)$_2$DEB] (Platinum poly-yne)	FWMS	3.76	$\gamma'' \sim 949 \times 10^{-36}$ esu/ru** {equiv. to approx. $\sigma^{(2)} \sim$ 1600}	8	46	Benzene	Typ. $\sim 7 \times 10^{19}$	

Table 5 Continued

Material	Technique	Applied two-photon energy (eV)	$\alpha^{(2)}$ (cm/GW) [or cross section $\sigma^{(2)}$ (10^{-50} cm^{4}s/photon-molecule)*]	Pulse duration (ns)	Ref	Solvent	Concentration molec/cm^3	Other
Poly-[Pt(PBu)$_3$)$_2$DEX] (Platinum poly-yne)	SBNLA	4.66	5 [$\sigma^{(2)} \sim 2.6 \times 10^3$]	0.023	110	THF	7.2×10^{19}	
Poly-[Pt(PBu)$_3$)$_2$DEX] (Platinum poly-yne)	FWMS	3.76	$\gamma'' \sim 464 \times 10^{-36}$ esu/ru** {equiv. to approx. $\sigma^{(2)} \sim$ 800}	8	46	Benzene	Typ. $\sim 7 \times 10^{19}$	
Poly-[Pt(PBu)$_3$)$_2$DEX$_2$] (Platinum poly-yne)	SBNLA	4.66	15.6 [$\sigma^{(2)} \sim 9.5 \times 10^3$]	0.023	110	THF	6.1×10^{19}	
Poly-[Pt(PBu)$_3$)$_2$DEX$_2$] (Platinum poly-yne)	FWMS	3.76	$\gamma'' \sim 344 \times 10^{-36}$ esu/ru** {equiv. to approx. $\sigma^{(2)} \sim$ 600}	8	46	Benzene	Typ. $\sim 7 \times 10^{19}$	
Poly-[Pt(PBu)$_3$)$_2$(C C-C C]$_n$ (Platinum poly-yne)	FWMS	3.76	$\gamma'' \sim 1296 \times 10^{-36}$ esu/ru** {equiv. to approx. $\sigma^{(2)} \sim$ 2300}	8	46	THF	Typ. $\sim 7 \times 10^{19}$	
Poly-chloro-trifluoroethylene	SBNLA	7.07	0.012	1.1, 7	4			
Polymethylphenyl silane	TBNLA	4.0–4.4	Relative spectrum	~0.002	111			Amorphous film
PPB poly(2,5-pyrrole-diyl benzylidene)	SBNLA	2.33	26 [$\sigma^{(2)} \sim 2.5 \times 10^5$]	~50–100	63	DCM/DMF/THF	1.9×10^{18} (3.2×10^{-3} M)	

PPB	SBNLA	2.33	26 [$\sigma^{(2)} \sim 2.5 \times 10^5$]	~50–100	63	DCM/DMF/THF	1.9×10^{18} (3.2×10^{-3} M)	
PPB	SBNLA	2.33	0.031 [$\sigma^{(2)} \sim 300$]	0.035	63	DCM/DMF/THF	1.9×10^{18} (3.2×10^{-3} M)	
PPPV (poly(phenyl p-phenylene-vinylene))	TBNLA	4.03	<0.5 {equiv. to approx. $\sigma^{(2)} \sim 24$}	0.0001	33	Polycarbonate	$\sim 6.7 \times 10^{20}$	Thin film, 20 wt.% blend with poly-carbonate 1.77 eV probe 2.25 eV pump
PPV (poly(p-phenyl-enevinylene))	SBNLA	4	6.8	0.00006	6			PPV in silica sol-gel film
PPV	L	1.8–4.2	Relative spectrum	3–4	112			77K, thin film
PPV	TBNLA	4.03	5 {equiv. to approx. $\sigma^{(2)} \sim 27$}	0.0001	33		$\sim 5.9 \times 10^{21}$	Thin film 1.77 eV probe 2.25 eV pump
PTB poly(2,5-thiophenediyl benzylidene	SBNLA	2.33	1.5 [$\sigma^{(2)} \sim 1.1 \times 10^4$]	~50–100	63	DMF/DCM/THF	2.6×10^{18} (4.4×10^{-3} M)	
PTB	SBNLA	2.33	12 [$\sigma^{(2)} \sim 8.5 \times 10^4$]	~50–100	63	DMF/DCM/THF	2.6×10^{18} (4.4×10^{-3} M)	

Table 5 Continued

Material	Technique	Applied two-photon energy (eV)	$\alpha^{(2)}$ (cm/GW) [or cross section $\sigma^{(2)}$ (10^{-50} cm^4s/photon-molecule)*]	Pulse duration (ns)	Ref	Solvent	Concentration molec/cm^3	Other
PTB	SBNLA	2.33	0.017 [$\sigma^{(2)} \sim 120$]	0.035	63	DMF/DCM/THF	2.6×10^{18} (4.4×10^{-3} M)	
PTFE poly(tetrafluoro-ethylene)		10.0	18 [$\sigma^{(2)} \sim 240$]	0.0005	11		$\sim 6 \times 10^{21}$	Solid. Pulsed laser ablation modeling
PTS Polydiacetelyne-bis(toluene sulphonate)		2.33	[$\sigma^{(2)} = 500$]		113			Crystal
PTS	OAZ	1.54–2.22	Spectrum, 200 @ 2.1 eV, 30 @ 1.8 eV, 5 @ 1.6 eV	0.065	114			Single crystal
PTS	OAZ	2.56–3.1	Spectrum, 700 @ 2.7 eV	0.065	114			Single crystal
PTS	OAZ	2.34	Spectrum	0.035	114			Single crystal
PTS	OAZ	2.60–3.1	Spectrum, 700 @ 2.7 eV	0.004	114			Single crystal
PTS	TBNLA	2.3–3	28 [$\sigma^{(2)} \sim 300$] @ 2.28 eV	0.015	115		1.7×10^{21}	Crystal (*see [40] also)
PTS	TBNLA	2.3–3	~36 [$\sigma^{(2)} \sim 400$] @ 2.33 eV	0.015	115		1.7×10^{21}	Crystal (*see [40] also)
PTS	TBNLA	2.3–3	100 [$\sigma^{(2)} \sim 1200$] @ 2.55 eV	0.015	115		1.7×10^{21}	Crystal (*see [40] also)
PTS	TBNLA	2.3–3	283 [$\sigma^{(2)} \sim 3400$] @ 2.78 eV	0.015	115		1.7×10^{21}	Crystal (*see [40] also)

PTS	TBNLA	2.3–3	510 [$\sigma^{(2)} \sim 7100$] @ 2.96 eV	0.015	115		1.7×10^{21}	Crystal (*see [40] also)
PTS	OAZ	2.29	65	0.010	116		1.7×10^{21}	Crystal (Polarization parallel to polymer chain axis, negligible signal perpendicular to chain)
PTS	OAZ	1.91	20 (Intensity-dependent, $\alpha^{(3)} = 5.2\ cm^3/GW^2$)	~ps (not specified)	117		1.7×10^{21}	Single crystal
PTS	OAZ	1.55	<0.5	0.065	118		1.7×10^{21}	Single crystal
PTS	OAZ	2.33	100, $\alpha^{(3)} = -5\ cm^3/GW^2$	0.035	119			Single crystal
PTTB Poly(α-{5,5″-terthiophenediyl} benzylidene)	SBNLA	2.33	120 [$\sigma^{(2)} \sim 2.5 \times 10^6$]	~50–100	63	DCM/DMF/THF	9×10^{17} (1.5×10^{-3} M)	
α-sexithieneyl (polythiophene oligomer)	L	2.73–2.10	Relative spectrum	5	120			Thin films 4.2K
Silicon phthalocyanine/MMA copolymer 0.2/99.8	SBNLA/L	1.93	75	~6	105		1.4×10^{20}	Solid pressed disks (1.23 mm)
TCDU Polydiacetelyne-bis(phenyl-urethane)	TBNLA	2.3–3	1 @ 2.2 eV	0.015	115		1.4×10^{21}	Crystal (pump 0.81 eV), (*see [40] also)

Table 5 Continued

Material	Technique	Applied two-photon energy (eV)	$\alpha^{(2)}$ (cm/GW) [or cross section $\sigma^{(2)}$ (10^{-50} cm^{4}s/photon-molecule)*]	Pulse duration (ns)	Ref	Solvent	Concentration molec/cm^3	Other
TCDU	TBNLA	2.3–3	7.6 [$\sigma^{(2)} \sim 100$] @ 2.3 eV	0.015	115		1.4×10^{21}	Crystal (pump 1.17 eV), (*see [40] also)
TCDU	TBNLA	2.3–3	3 @ 2.4 eV	0.015	115		1.4×10^{21}	Crystal (pump 0.81 eV), (*see [40] also)
TCDU	TBNLA	2.3–3	6.4 @ 2.6 eV	0.015	115		1.4×10^{21}	Crystal (pump 0.81 eV), (*see [40] also)
TCDU	TBNLA	2.3–3	13 [$\sigma^{(2)} \sim 200$] @ 2.6 eV	0.015	115		1.4×10^{21}	Crystal (pump 1.17 eV)
TCDU	TBNLA	2.3–3	39 [$\sigma^{(2)} \sim 700$] @ 2.9 eV	0.015	115		1.4×10^{21}	Crystal (pump 1.17 eV), (*see [40] also)
Tetrathiophene oligomer	OAZ	4.66	0.11 [$\sigma^{(2)} = 1.7 \times 10^4$]	0.030	106	Dioxane	2.4×10^{17}	

*Conversions from $\sigma^{(2)}$ to cross sections assume that two-photon absorption is the only mechanism contributing to the nonlinear absorption. This may not necessarily be valid, especially where the conversion leads to large apparent $\sigma^{(2)}$ values. It may be readily apparent that excited state absorption is a contributory factor, for example, a short pulse measurement may give a low cross-section, while a longer pulse duration may give much larger apparent values. The values in square brackets should therefore be relied on for guidance only and for making relative comparisons weighted by sample concentration and excitation energy. For PTS and TCDU, the $\sigma^{(2)}$ values shown alongside Lequime and Herman's data[115] are derived from fitted curves from reference[40] by Chance et al.

**γ'' ~ esu/ru, imaginary component of complex hyperpolarizability per repeat unit of polymer.

Table 6 Two-Photon Absorption Coefficients, $\alpha^{(2)}$, and Dressed TPA Cross-Sections, $\sigma^{(2)}$, for Organo-Metallic Materials

Material	Technique	Applied two-photon energy (eV)	$\alpha^{(2)}$ (cm/GW) [or $\sigma^{(2)}$ (10^{-50} cm^{4}s/photon-molecule)]	Pulse duration (ns)	Ref	Solvent	Concentration molec/cm^3	Other
Benzoporphyrin 1	SBNLA	4.66	2.5 [$\sigma^{(2)} \sim 3.3 \times 10^5$]	6, 0.021	23	THF	2.8×10^{17}	
Benzoporphyrin 2	SBNLA	4.66	<0.5 [$\sigma^{(2)} \sim 5.8 \times 10^4$]	6, 0.021	23	THF	3.2×10^{17}	
Bis[1,2-diethyl-1,2-ethene dithiolato(2-)S,S′] Nickel	OAZ	2.35	20* [$\sigma^{(2)} \sim 2100$]	0.1	122	PMMA	1.8×10^{20}	PMMA-doped film
Bis[1,2-diethyl-1,2-ethene dithiolato(2-)S,S′] Nickel	Estimated from FWM	2.33	<0.05 [$\sigma^{(2)} \sim 930$]	0.1	123	DCM	1×10^{18}	
Bis[1,2-dimethyl-1,2-ethenedithiolato (2-)-S,S′] Nickel	SBNLA	2.33	<0.01 [$\sigma^{(2)} \sim 250$]	10	124	DCM	7.3×10^{17}	
Bis[1,2-dimethyl-phenyl-1,2-ethenedithiolato (2-)-S,S′] Nickel	SBNLA	2.33	0.36 [$\sigma^{(2)} \sim 3400$]	0.1	5	DCM	2×10^{18}	
Bis[1,2-dimethyl-phenyl-1,2-ethenedithiolato (2-)-S,S′] Nickel	OAZ	1.88	<0.1 [$\sigma^{(2)} \sim 1500$]	0.1	125	DCM	1×10^{18}	
Bis[1,2-diphenyl-1,2-ethene-dithiolato(2-)-S,S′] Nickel	SBNLA	2.33	2.9 [$\sigma^{(2)} \sim 7.8 \times 10^4$]	10	124	DCM	6.9×10^{17}	

Table 6 Continued

Material	Technique	Applied two-photon energy (eV)	$\alpha^{(2)}$ (cm/GW) [or $\sigma^{(2)}$ (10^{-50} cm^{4}s/ photon-molecule)]	Pulse duration (ns)	Ref	Solvent	Concentration molec/cm^3	Other
Bis[1-butyl 2-phenyl-1,2-ethenedithiolato (2-)-S,S′] Nickel	OAZ	2.33	0.28* [$\sigma^{(2)} \sim 2500$]	0.1	122	DCM	2.1×10^{18}	
Bis[1-butyl 2-phenyl-1,2-ethenedithiolato (2-)S,S′] Nickel	OAZ	2.33	6.7* [$\sigma^{(2)} \sim 4300$]	0.1	122	PMMA	2.9×10^{19}	PMMA-doped film
Bis[1-butyl 2-phenyl-1,2-ethenedithiolato (2-)S,S′] Nickel	OAZ	2.33	24* [$\sigma^{(2)} \sim 5600$]	0.1	122	PMMA	8.0×10^{19}	
Bis[1-butyl 2-phenyl-1,2-ethenedithiolato (2-)S,S′] Nickel	OAZ	2.33	158* [$\sigma^{(2)} \sim 1.5 \times 10^4$]	0.1	122	PMMA	2.0×10^{20}	
Bis[1-butyl 2-phenyl-1,2-ethenedithiolato (2-)S,S′] Nickel	OAZ	2.33	423* [$\sigma^{(2)} \sim 2.5 \times 10^4$]	0.1	122	PMMA	3.2×10^{20}	

Bis[1-decyl 2-phenyl-1,2-ethenedithiolato (2-),S,S′] Nickel	Estimated from FWM	2.33	<0.05 [$\sigma^{(2)} \sim 930$]	0.1	123	DCM	1×10^{18}	
Bis[1-ethyl 2-phenyl-1,2-ethenedithiolato (2-)S,S′] Nickel	OAZ	2.33	55* [$\sigma^{(2)} \sim 7900$]	0.1	122	PMMA	1.3×10^{20}	PMMA-doped film
Bis[1-ethyl 2-phenyl-1,2-ethenedithiolato (2-)S,S′] Nickel	OAZ	2.33	114* [$\sigma^{(2)} \sim 8500$]	0.1	122	PMMA	2.5×10^{20}	PMMA-doped film
C11S4NiDT	OAZ	1.88	0.5 [$\sigma^{(2)} \sim 7500$]	0.1	125	DCM	1×10^{18}	
C7S4NiDT	OAZ	1.88	1.3 [$\sigma^{(2)} \sim 1.9 \times 10^4$]	0.1	125	DCM	1×10^{18}	
Palladium polyyne model system $Pd(PBu_3)_2Cl_2$	FWMS	3.76	$\gamma'' \sim 23 \times 10^{-36}$ esu/molec.**	8	126	THF		
Platinum poly-yne model system $(Pt(PBu_3)_2)_2DEBCl_2$	FWMS	3.76	$\gamma'' \sim 580 \times 10^{-36}$ esu/molec.**	8	126	THF		

Table 6 Continued

Material	Technique	Applied two-photon energy (eV)	$\alpha^{(2)}$ (cm/GW) [or $\sigma^{(2)}$ (10^{-50} cm^{4}s/ photon-molecule)]	Pulse duration (ns)	Ref	Solvent	Concentration molec/cm^3	Other
Cis-Pt(PBu_3)$_2$(DEB)$_2$	FWMS	3.76	$\gamma'' \sim 256 \times 10^{-36}$ esu/molec.**	8	46	THF		
Cis-Pt(PBu_3)$_2$$Cl_2$	FWMS	3.76	$\gamma'' \sim 95 \times 10^{-36}$ esu/molec.**	8	46	THF		
Pt(PBu_3)$_2$(C C-C C-H)$_2$	FWMS	3.76	$\gamma'' \sim 108 \times 10^{-36}$ esu/molec.**	8	46	THF		
Trans-Pt(PBu_3)$_2$(DEB)$_2$	FWMS	3.76	$\gamma'' \sim 221 \times 10^{-36}$ esu/molec.**	8	46	Benzene		
Trans-Pt(PBu_3)$_2$(DEB)$_2$	FWMS	3.76	$\gamma'' \sim 246 \times 10^{-36}$ esu/molec.**	8	46	THF		
Trans-Pt(PBu)$_3$)$_2$(DEX)$_2$	FWMS	3.76	$\gamma'' \sim 173 \times 10^{-36}$ esu/molec.**	8	46	Benzene		
Tris-cyclopentadienyl ytterbium (Yb(cp)$_3$)	SBNLA	2.33	4.1 [$\sigma^{(2)} \sim 2500$]	10	124	THF	3×10^{19}	

*Values obtained from two-photon absorption measurements quoted as ratio of Re(χ^3)/Im(χ^3) and refractive nonlinearity, n_2.
**γ'' ~ hyperpolarizability in esu/molecule.

Table 7 Dressed Two-Photon Absorption Cross-Sections for Other Organic Molecules

Material	Technique	Applied two-photon energy (eV)	$\sigma^{(2)}$ (10^{-50} cm^{4}s/photon-molecule)	Pulse duration (ns)	Ref	Solvent	Concentration molec/cm^3	Other
Acetone		4.66	~0.073 [$\alpha^{(2)} = 1.6 \times 10^{-2}$ cm/GW]	0.025	127, 128	Neat		Referenced to CS_2[129]
Acetone		Rydberg states: 3s $1B_2$; 3p $2A_1$, $2A_2$, & 2 B_2	Values rel. to $2B_2$ state and relative spectrum.		130	Neat		
Actone	SBNLA	3.57	~0.062 [$\alpha^{(2)} = 1.77 \times 10^{-2}$ cm/GW]	20	128	Neat		
Aminophthalimide	L	3.56	2		67	Ethanol		
Aniline	TL	3.96–5.69	Relative spectrum (8.8 × benzene at 4.10 eV)		131	Neat		Relative to benzene @ 4.65 eV
Anthracene	L	3.56	25		73			Crystal
Anthracene	L	3.56	3.5		73	Ethanol		
Anthracene	L	3.24–3.42	Relative spectrum		132	Hexane		
Anthracene	L, PC	4.14–4.70	Relative spectrum	40	133			Crystal (see fig. 36)

Table 7 Continued

Material	Technique	Applied two-photon energy (eV)	$\sigma^{(2)}$ (10^{-50} cm^4s/photon-molecule)	Pulse duration (ns)	Ref	Solvent	Concentration molec/cm^3	Other
Anthracene	L	3.8–4.6	Relative spectrum	8	134			
Anthracene	L	3.60–5.1	Spectrum 70 @ 4.6 eV	20	135			Crystal
Anthracene	TBNLA	4.4	70	20	136	Chloroform		
Anthracene	L	3.60–5.1	Spectrum ~1 @ 4.6 eV	25	135	Cyclohexane, benzene		
Anthracene	L	3.56	Relative values		137			
Anthracene	L	3.10–3.56	Spectrum 0.01 @ 3.2 eV	20–25	138			Crystal (see fig. 36)
Anthracene	L	3.10–3.56	Spectrum 0.0005 @ 3.2 eV	20–25	138	Benzene		(See fig. 36)
Anthracene	TBNLA	3.45–4.25	Spectrum	60	139			10K, 77K (see fig. 36)
Anthracene	L	3.56	0.16		140	Cyclohexane		
Anthracene	L	3.56	0.23	40	79	Cyclohexane		
Anthracene	L	3.9–5.0	Spectrum ~2 @ 4 eV	5	141			Crystal
Anthracene	L	4.66 eV	150		142			
Anthracene	L	4.41 eV	300		142			

Anthracene	L	3.92 eV	0.5		142			
Anthracene	L	3.57	0.09	30	143		~4.6 × 10^{18}	Gas (503K, 1.7 Torr)
Anthracene	L	3.57	14	40	69	n-hexane		
Azulene	L	4.0–5.7	Relative spectrum	20	144			
Azulene	AFRS	4.66	1070	42–67	145	CCl_4		
BBP	SBNLA	2.33	~5.0 × 10^4 [$\alpha^{(2)}$ = 16 cm/GW]	~50–100	63	DCM/DMF/THF	6 × 10^{18} (10^{-2} M)	
Benzanthracene	L	3.56	Relative values		137			
Benzene	CR	6.7	2.5	15	146			
Benzene	L	4.6–5.2	Relative spectrum	5	147			
Benzene	TL	4.6–5.3	Relative spectrum	~5	148			
Benzene	L	5.8–6.8	Relative spectrum	6	149			4.2K, Crystal
Benzene	L	4.4–6.0	Relative spectrum	20	150			
Benzene	L	5.4–6.8	Relative spectrum	~5	151			
Benzene	AO	4.8–5.44	Spectrum 10 @ 4.7 eV	1000	52			
Benzene	CR	4.34	8.7	~5	152			2K
Benzene	L	3.4–4.4	Relative spectrum		153			2K, Crystal
Benzene	L	4.6–5.1	Relative spectrum	5	154			2K, Crystal
Benzene	CR	4.72	<0.003	4	155			
Benzene		4.66	~0.26 [$\alpha^{(2)}$ = 4.8 × 10^{-2} cm/GW]	0.025	127, 128, 129	Neat		

Table 7 Continued

Material	Technique	Applied two-photon energy (eV)	$\sigma^{(2)}$ (10^{-50} cm^{4}s/photon-molecule)	Pulse duration (ns)	Ref	Solvent	Concentration molec/cm^3	Other
Benzene	TPIS	4.92	0.0126	10	156		$\sim 8 \times 10^{23}$	30 Torr vapor + 7 Torr Ar buffer, valence state transition
Benzene	TPIS	8.43	0.0016	10	156		1.3×10^{23}	Vapor, Rydberg state
Benzene	TPIS	8.53	0.00146	10	156		1.3×10^{23}	Vapor, Rydberg state
Benzene	TPIS	8.44	0.00075	10	156		1.3×10^{23}	Rydberg state
Benzene	TPIS	8.87	0.00001	10	156		1.3×10^{23}	Rydberg state

Benzene	STRS	7.2	4		157	Neat	*See note regarding spread of data for benzene
Benzene	RX	6.7	3.4		146		*
Benzene		6.2	<0.1		158	Neat	*
Benzene	L	5.2	0.034	5	21	Neat	*
Benzene	L	4.96	0.082	5	21	Neat	*
Benzene	L	4.92	0.2		159	Neat	*
Benzene	FWM/RX	4.81	8.7		160		*1.6K crystal
Benzene		4.86	0.17		161	Neat	*
Benzene	AO	4.81	0.1		52	Neat	*
Benzene	L	4.86	0.06	5	21	Neat	*
Benzene	L	4.77	0.025	5	21	Neat	*
Benzene	L	7.0	12.4	5	21	Neat	*
Benzene	L	4.66	2.5×10^{-4}	5	21		*
Benzene	TL	4.46–5.69	Relative spectrum		131	Neat	Liquid

Table 7 Continued

Material	Technique	Applied two-photon energy (eV)	$\sigma^{(2)}$ (10^{-50} cm^4s/photon-molecule)	Pulse duration (ns)	Ref	Solvent	Concentration molec/cm^3	Other
Benzene, deuterated	L	4.7–5.1	Relative spectrum	5	154			2K, Crystal
Benzopyrene	SBNLA	3.31–4.68	Spectrum 10 @ 3.6 eV	10	162			
Benzopyrene	L	3.56	Relative values		137			
Binaphthyl	L	4.1–5.6	Relative spectrum		163			
Biphenyl	TBNLA, L	4.1–5.8	Spectrum 1 @ 4.8 eV	20	164			
Biphenyl-d_{10}	L	4.1–4.3	Relative spectrum		164			2K
Biphenyl-h_{10}	L	4.1–4.3	Relative spectrum		165			2K
Bipyridine	L	3.7–4.7	Relative spectrum	~5	166	CCl_4		
Bipyridine	TBNLA, L	4.1–4.6	Relative spectrum	20	164			
p-Bis(o-methylstyryl) benzene	L	3.57–4.62	Spectrum 690 @ 2.18 eV	0.002	167	Cyclohexane	~6 × 10^{17} (~1 mM)	Both circular and linear polarization spectra. Proposed as a TPA reference material.

Carbon disulphide	CR	4.72	<0.057	4	155	Neat	
Carbon disulphide	CR	4.34–5.58	Spectrum 1.1 @ 5.58 eV		168	Neat	
Carbon disulphide	SBNLA	3.56	0.5	40	129	Neat	Used as reference material in[127,128]
Carbon disulphide	SBNLA	3.56	0.26	~30	169	Neat	
Carbon tetrachloride	CR	4.72	<0.014	4	155	Neat	
Chloroform		4.66	0.11 [$\alpha^{(2)}$ = 2.1 cm/GW]	0.025	127, 128, 129	Neat	
Chloronaphthalene	TBNLA	3.97–4.34	Relative spectrum	0.020	170		
Chloronaphthalene	TBNLA	4.2–5.3	Spectrum 1 @ 4.8 eV		158	Neat	
Chloronaphthalene	TBNLA	3.7	Spectrum 1 @ 5.2 eV		171	Neat	

Table 7 Continued

Material	Technique	Applied two-photon energy (eV)	$\sigma^{(2)}$ (10^{-50} cm^4s/photon-molecule)	Pulse duration (ns)	Ref	Solvent	Concentration molec/cm^3	Other
Chrysene	L	3.56	Relative values		137			
Chrysene	L	3.35–3.84	Relative spectrum		172			Crystal
Cyclohexane	CR	3.92	1.9		173			
Dibenzanthracene	L	3.56	Relative values		137			
o-dichlorobenzene	TPIS	4.59–5.11	Relative spectrum	7	174			Vapor
Dichlorobiphenyl	TBNLA, L	4.1–5.6	Relative spectrum	20	164			
Difluorenyl	L	4.1–4.7	Relative spectrum		163			
Difluorobenzene	L	4.4–4.5	Relative spectrum	5	147			
Difluorobiphenyl	TBNLA, L	4.1–5.6	Relative spectrum	20	164			
9,10-dihydro-anthracene	L	3.49–3.70	Relative spectrum		175			Crystal with *low* (<1.1 M/M) anthracene impurity levels. 4.2K
Dioxane	OAZ	4.66	0.14 [$\alpha^{(2)} = 0.026$ cm/GW]	0.030	106	Neat		
Diphenyl	TBNLA	3.97–4.34	Relative spectrum	0.020	170			
Diphenylbutadiene	CR	4.08	14.4	CW	39	Benzene	8.49×10^{19}	
Diphenylbutadiene	L	3.4–4.0	Relative spectrum		176	EPA glass		77K
Diphenylbutadiene	TBNLA	3.50	1.0	~30	177			
Diphenylbutadiene	SBNLA	4.66	34 [$\alpha^{(2)} = 0.11$ cm/GW]	7	41	Chloroform	1.2×10^{20}	
Diphenylbutadiene	DFWM	4.66	40	10	41	Chloroform	1.2×10^{20}	

Diphenylbutadiene	L	3.72	36.6		178			
Diphenylbutadiene	AO	4.13	30		179			
Diphenylhexatriene	CR	4.08	43.4	CW	39	Benzene	1.32×10^{19}	
Diphenylhexatriene	L	2.9–3.5	Relative spectrum		180	EPA glass		77K
Diphenyloctatetraene	CR	4.08	61.0	CW	39	Benzene	0.25×10^{19}	
Diphenyloxazole	CR	4.72	0.41	4	155			
DPP 5,10-diphenyl-phenazine	SBNLA	2.33	$\sim 3.7 \times 10^5$ [$\alpha^{(2)}$ = 120 cm/GW]	~50–100	63	DCM/DMF/THF	6×10^{18} (10^{-2} M)	
DPP	SBNLA	2.33	$\sim 2.3 \times 10^5$ [$\alpha^{(2)}$ = 74 cm/GW]	~50–100	63	DCM/DMF/THF	6×10^{18} (10^{-2} M)	
Fluoranthene	L	3.56	Relative values		137			
Fluorene	L	4.1–4.7	Relative spectrum		163			
Fluorene	L	3.56	Relative values		137			
Fluorobenzene	L	4.5–5.1	Relative spectrum	5	181			Vapor
Fluorobenzene	L	4.6–5.0	Relative spectrum	5	147			
Fluorobenzene	TL	4.46–5.69	Relative spectrum 1.5 × benzene @ 4.64 eV, 5.5 × benzene @ 5.69 eV		131			Liquid
Formladehyde	AO	4.28–4.43	Relative spectrum	QSW	182			Vapor phase

Table 7 Continued

Material	Technique	Applied two-photon energy (eV)	$\sigma^{(2)}$ (10^{-50} cm^{4}s/photon-molecule)	Pulse duration (ns)	Ref	Solvent	Concentration molec/cm^3	Other
Hexatriene	TL	4.1–6.0	Spectrum 2 @ 5.2 eV		183			
MBA-NP	SHG	2.33	0.013 [$\alpha^{(2)} = 2.35 \times 10^{-3}$ cm/GW]	0.032	184		$\sim 3.4 \times 10^{21}$	Crystal
MBA-NP	SHG	4.66	3.1 [$\alpha^{(2)} = 0.28$ cm/GW]	0.023	184		$\sim 3.4 \times 10^{21}$	Crystal
Mesitylene	L	4.66	0.096	5	21	Neat		*
Methylene iodide	SBNLA	4.66	2.1	0.004	185			
2-methyl-4-nitroaniline (MNA)	FWMS	3.76	$\gamma'' = 42 \times 10^{-36}$ esu/molecule	8	126	THF		
Naphthacene	L	3.56	Relative values		137			
Naphthalene	L	4.1–5.8	Relative spectrum		163	Cyclohexane		
Naphthalene	TBNLA	3.8–3.95	Spectrum 0.2 @ 3.9 eV	20	186			
Naphthalene	CR	5.46	2.0		187	CS_2		
Naphthalene	L	3.84–6.19	Relative spectrum	6	188	Cyclohexane		

Naphthalene	L	3.84–6.19	Relative spectrum	6	188			Crystal
Naphthalene	CARS	4.34–5.70	Spectrum 50 @ 4.96 eV		20	CS_2	7.2×10^{20}	
Naphthalene	CARS	4.34–5.70	Spectrum 1 @ 4.96 eV		20	$CHCl_3$	5.4×10^{20}	
Naphthalene	CARS	4.34–5.70	Spectrum 1 @ 4.96 eV		20	Benzene	5.4×10^{20}	
Naphthalene-d_8	L	3.84–6.19	Relative spectrum	6	188	Cyclohexane		
Naphthalene-d_8	L	3.84–6.19	Relative spectrum	6	188			Crystal
2-(1-naphthyl)5-phenyloxazole	L	3.57–4.62	Relative spectrum	0.002	167	Cyclohexane	$\sim 6 \times 10^{17}$ ($\sim$1 mM)	Relative to bis-MSB
4-nitroaniline (PNA)	FWMS	3.76	$\gamma'' = 49 \times 10^{-36}$ esu/molecule	8	126	THF		
Nitrobenzene	CR	4.72	0.11	4	155			
Nitrobenzene	TBNLA	4.1–4.5	Relative spectrum		189			
α-picoline		4.66	0.39 [$\alpha^{(2)}$ = 6.4 cm/GW]	0.025	127, 128, 129			

Table 7 Continued

Material	Technique	Applied two-photon energy (eV)	$\sigma^{(2)}$ (10^{-50} $cm^4 s$/photon-molecule)	Pulse duration (ns)	Ref	Solvent	Concentration molec/cm^3	Other
Paracyclophane	L	3.87–6.03	Relative spectrum	~5	190			Crystal
Perylene	L	3.56	Relative values		137			
Perylene	L	3.57	1.2	30	143			Gas (583K, 0.6 Torr)
Phenanthrene	L	3.56	Relative values		137			
Phenol	TL	4.21–5.69	Relative spectrum 0.8 × benzene @ 4.39 eV, 8.6 × benzene @ 5.45 eV		131			Relative to benzene @ 4.65 eV
POPOP	L	3.57	0.05	30	143			Gas (603K, 1.23 Torr)
Pyrene	L	3.56	Relative values		137			
Pyrene	SBNLA	3.56	0.22	~30	169			

Pyridine	CARS & TL	3.97–5.83	Spectra 0.1 @ 4.09 eV, 0.2 @ 4.72 eV, 0.5 @ 5.08 eV		31	Neat		Detailed band symmetry and assignments given also
Pyridine	CARS & TL	3.97–5.83	>3 @ 6.08 eV		31	Neat		
Pyrimidine	L	3.8–6.0	Relative spectrum	~5	191	Hexane		
Stilbene	TBNLA	4.4	80	20	136	Chloroform		
Stilbene	L	4.3–5.5	Relative spectrum		192	Methylcyclohexane		
Stilbene	TBNLA	3.56	80	200	193	Chloroform		
Stilbene	L, TL	4.6–5.4	Relative spectrum	~5	194	Hexane		
Stilbene	CR	4.82	12.1	CW	39	Cyclohexane		
Stilbene	L	3.57	~2.7 [$\alpha^{(2)}$ = 1.5 cm/GW]	20	195		Assumed ~10^{22}	Crystal
Stilbene	CR	4.08	12.1	CW	39	Cyclohexane	3.31×10^{19}	
Terphenyl	TBNLA, L	4.1–5.6	Relative spectrum	20	164			
p-terphenyl	L & PC	5.3–6.5	Relative spectrum	8	196			Crystal

Table 7 Continued

Material	Technique	Applied two-photon energy (eV)	$\sigma^{(2)}$ (10^{-50} cm^{4}s/photon-molecule)	Pulse duration (ns)	Ref	Solvent	Concentration molec/cm^3	Other
1,2,3,4 tetramethyl benzene	L	4.66	0.076	5	21			*
1,2,3,4 tetramethyl benzene	L	4.66	0.18	5	21			*
Thianthrene	PC	3.88–5.52	Relative spectrum	10	197			Crystal
Toluene	L	4.66	0.036 [$\alpha^{(2)} = 5.33 \times 10^{-3}$ cm/GW]		198	Neat		
Toluene	L	5.00	0.0054	5	21	Neat		*
Toluene	L	4.88	0.012	5	21	Neat		*
Toluene	L	4.79	0.011	5	21	Neat		*
Toluene	L	4.70	0.036	5	21	Neat		*
Toluene	L	7.01	10	5	21	Neat		
Toluene	L	4.4–6.0	Relative spectrum	20	150	Neat		
Toluene	L	4.5–4.7	Relative spectrum	5	147			

Toluene	TL	4.46–5.69	Relative spectrum 2.1 × benzene @ 4.59 eV, 3.3 × benzene @ 5.62 eV		131		Liquid. Relative to benzene @ 4.65 eV
1,2,3 trimethyl benzene	L	4.66	0.0021	5	21		*
1,2,4 trimethyl benzene	L	4.66	0.075	5	21		*
Triphenylene	L	3.56	Relative values		137		
m-Xylene	L	4.86	0.12	5	21	Neat	*
m-Xylene	L	4.73	0.12	5	21	Neat	*
m-Xylene	L	4.48	0.031	5	21	Neat	*
m-Xylene	L	7.01	30	5	21	Neat	*

Table 7 Continued

Material	Technique	Applied two-photon energy (eV)	$\sigma^{(2)}$ (10^{-50} cm^{4}s/ photon-molecule)	Pulse duration (ns)	Ref	Solvent	Concentration molec/cm^3	Other
m-Xylene	L	4.66	0.028	5	21	Neat		*
o-Xylene	L	4.86	0.21	5	21	Neat		*
o-Xylene	L	4.77	0.21	5	21	Neat		*
o-Xylene	L	4.48	0.050	5	21	Neat		*
o-Xylene	L	4.66	0.035	5	21	Neat		*
p-Xylene	L	4.79	0.16	5	21	Neat		*
p-Xylene	L	4.70	0.080	5	21	Neat		*
p-Xylene	L	4.49	0.020	5	21	Neat		*
p-Xylene	L	4.66	0.052	5	21	Neat		*

*Chen and McCann discuss the spread of the literature values for benzene two-photon absorption cross-sections.[21] As shown in their Table 2, values in the vicinity of 4.8–5 eV range from 0.082×10^{-50} to 8.7×10^{-50} cm^{4}s molec^{-1} photon^{-1}. In the latter case, the result was deduced from four wave mixing measurements and Raman scattering cross-sections[152] and is probably an order of magnitude overestimated. Opto-acoustic[157] and two-photon fluorescence[21] measurements tend to give results in the range 0.05×10^{-50} to 0.2×10^{-50} cm^4 s molec^{-1} photon^{-1} at these two-photon energies. Boissel et al.[52] and Prior and Voigt[146] have cited values of 4×10^{-50} and 3.4×10^{-50} cm^4 s molec^{-1} photon^{-1} at slightly higher excitation energies (7.2 and 6.7 eV, respectively) using stimulated thermal Rayleigh and Raman scattering techniques, respectively.

Material	**Structure**
Bacteriorhodopsin	In dark adapted state 1:1 ratio of all-trans-retinal and 13-cis-retinal chromophores bound to Lysine 216 site of a protein chain. In light adapted state only all-trans-isomer present e.g. 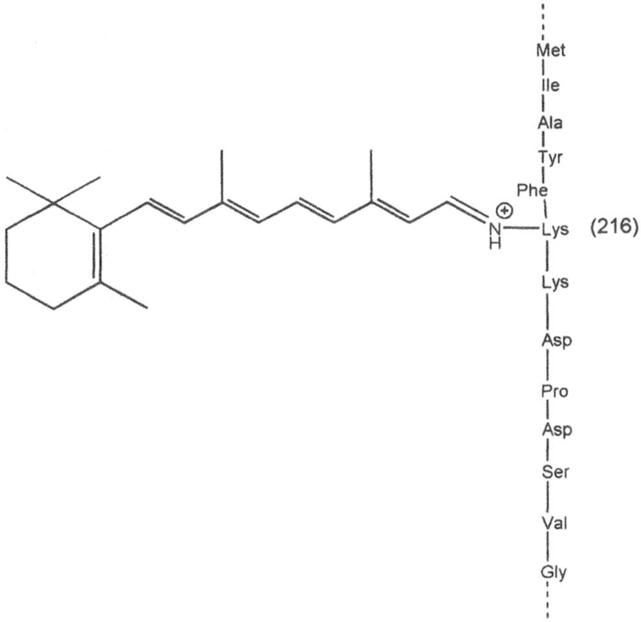
β-carotene	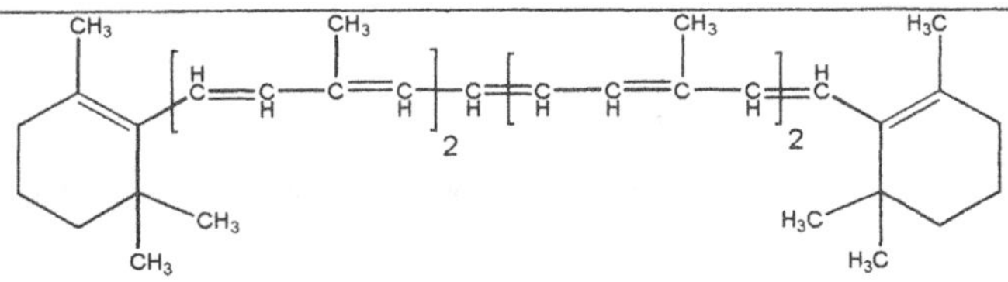

Chlorophyll-a	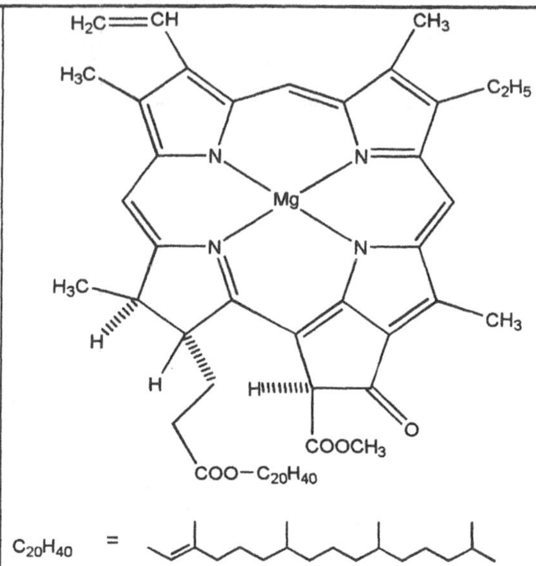
NADH (enzyme)	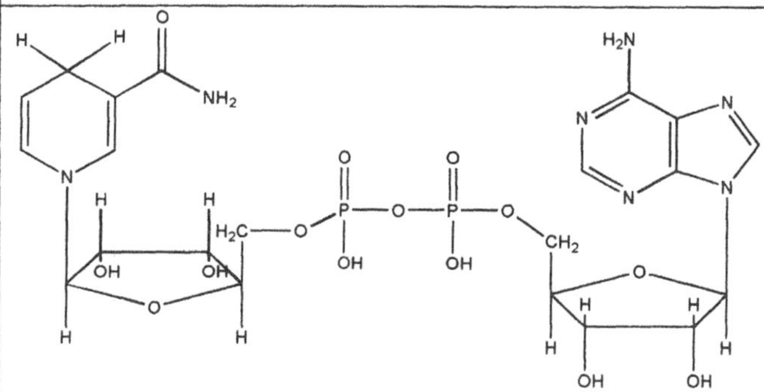 di-Hydro derivative of Nicotinamide Adenine Dinucleotide
Retinal	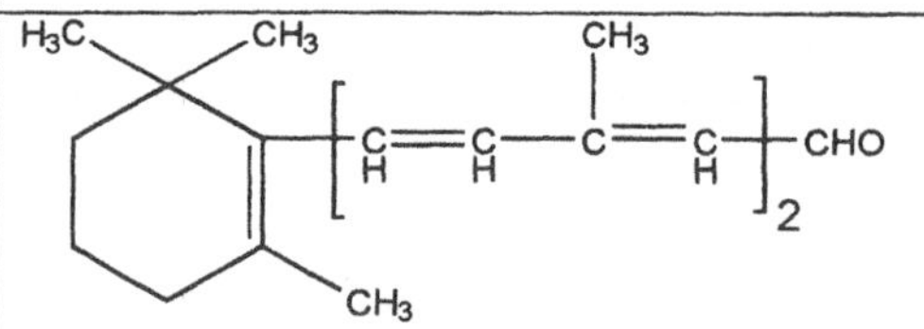

Retinyl acetate	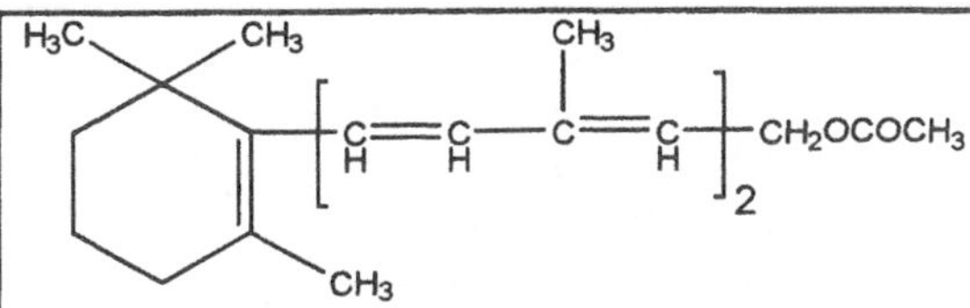
Trans-retinol (vitamin A alcohol)	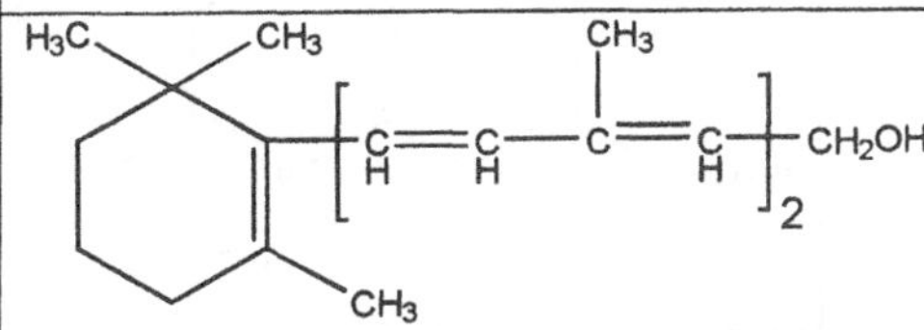
11-cis-locked Retinal : present in 11-cis-locked Rhodopsin bound to Opsin (See Bacteriorhodopsin for details of binding site)	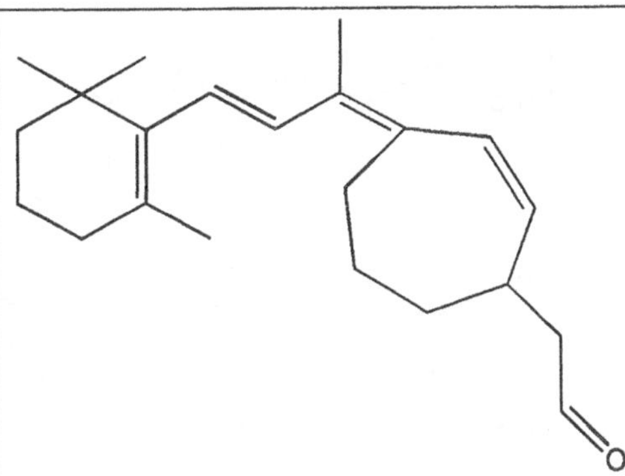
all-trans-N-Retinylidene-n-butylamine (Schiff base)	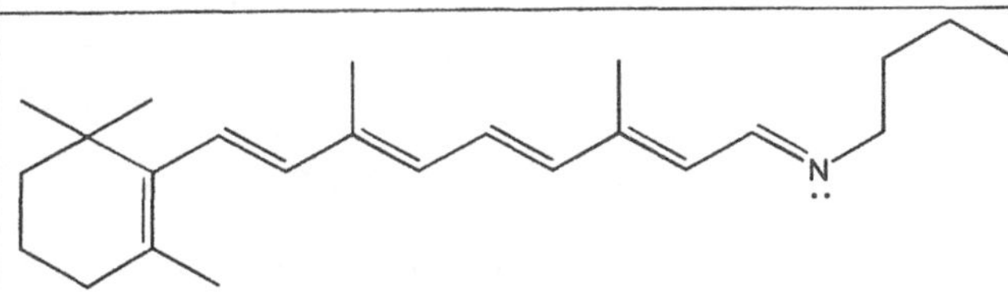
all-trans-N-Retinylidene-n-butylamine:HCl (protonated Schiff base)	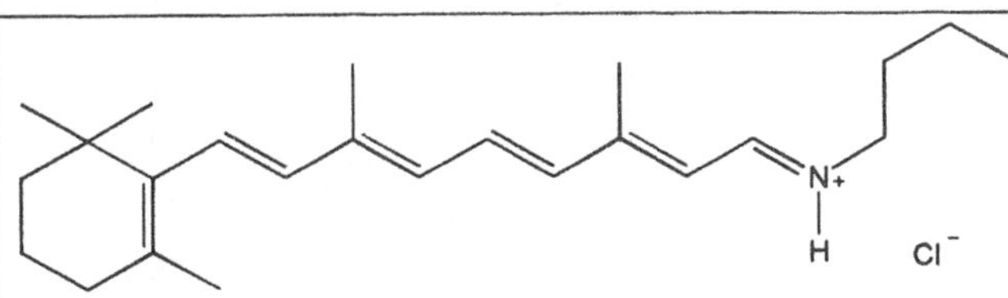

Acridine	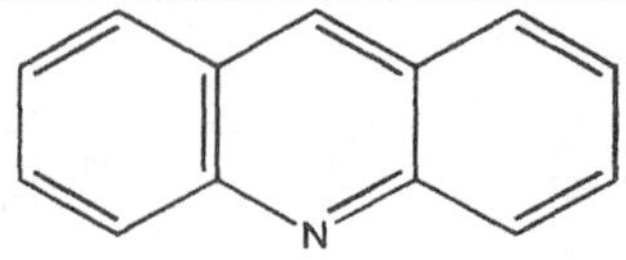
Acridine orange	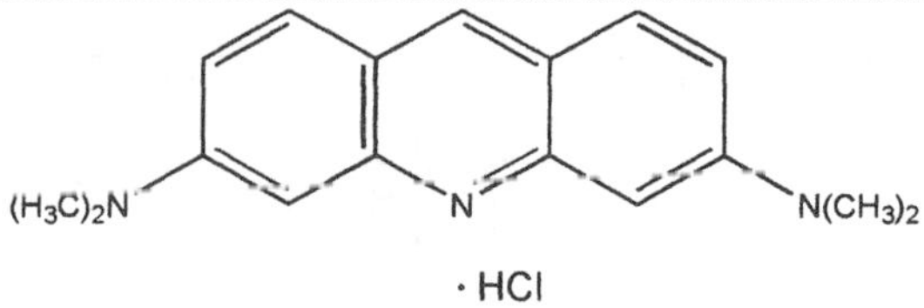
Acridine red	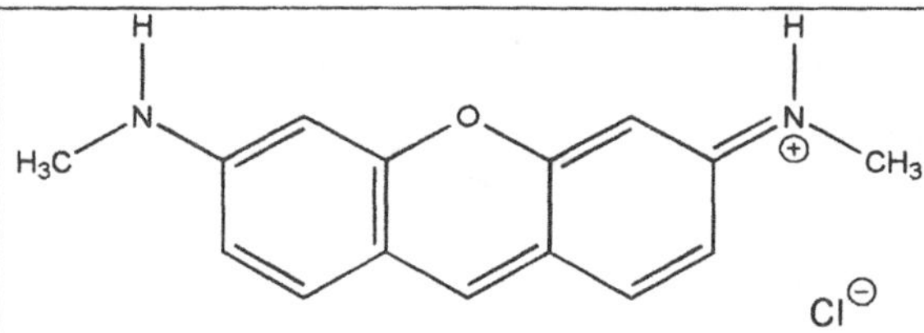
Acridine yellow	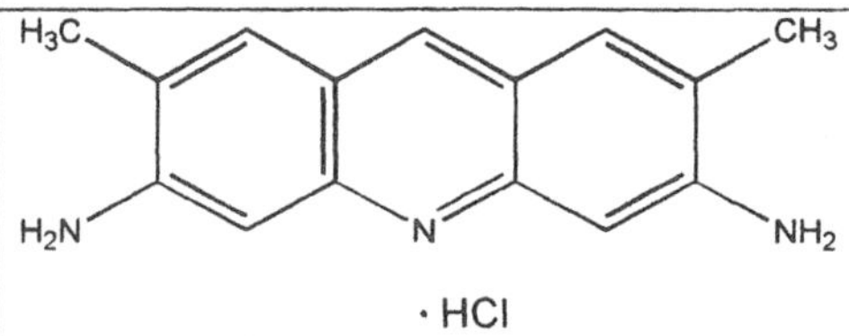
BBOT	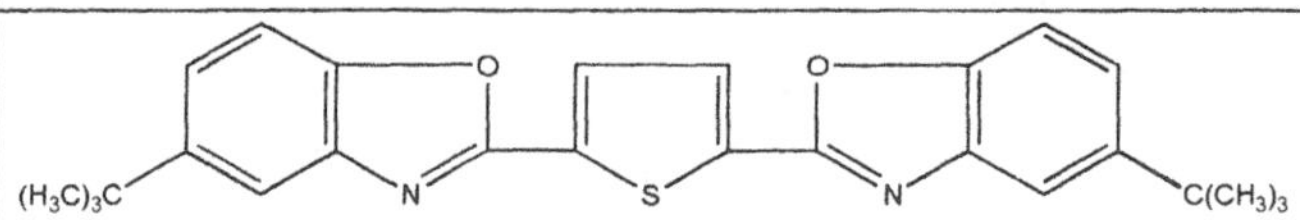 2,5-Bis(5'-*tert*-butyl-2-benzoxazolyl)thiophene
BBQ	**
BSQ	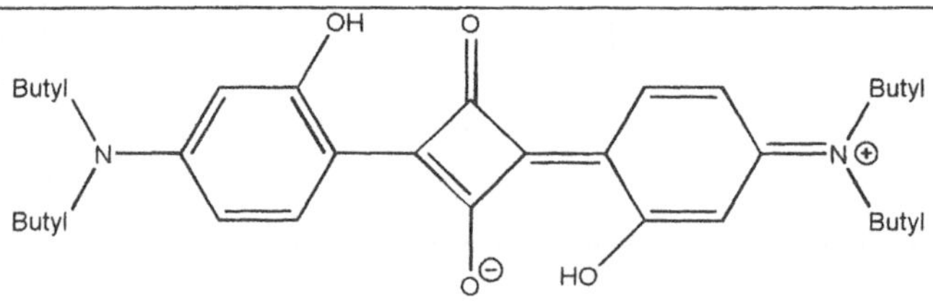

Coumarin 450 (Coumarin 2)	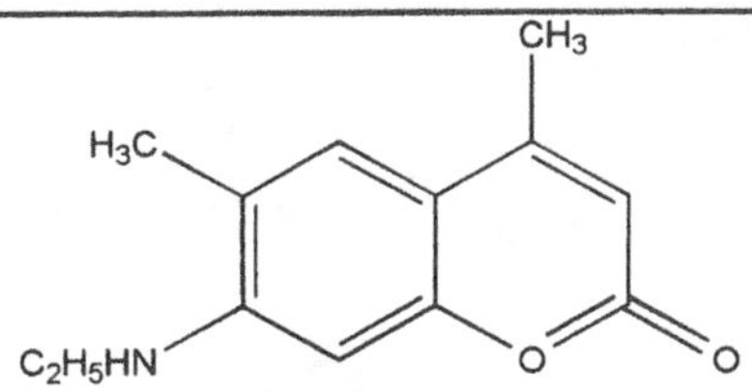 4,6-Dimethyl-7-ethylaminocoumarin
Coumarin 480 (Coumarin 102)	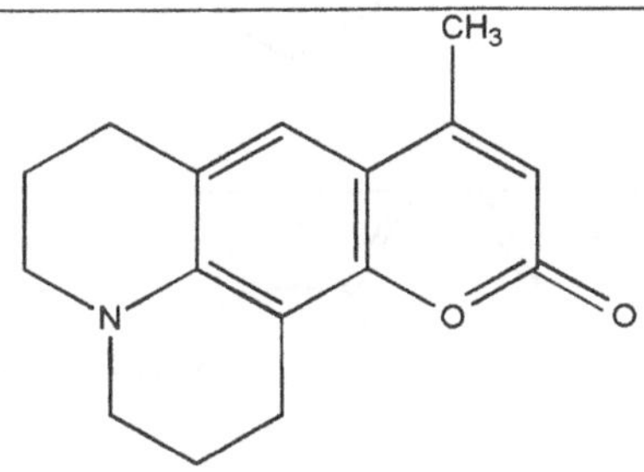
Cyanine I, II,III,IV	*************************************
DCM (dye)	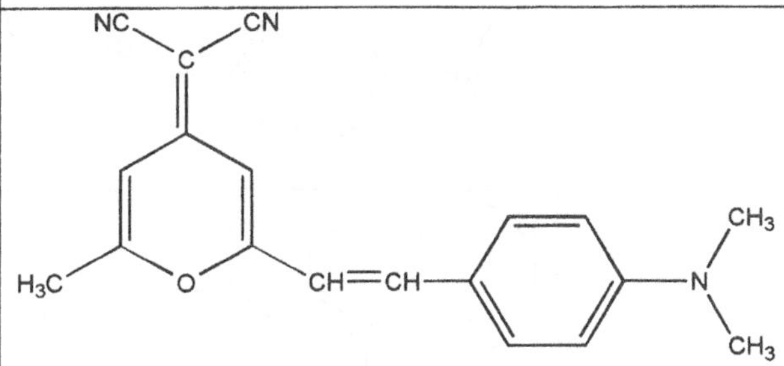 4-(Dicyanomethylene)-2-methyl-6-(4-dimethylaminostyryl)-4H-pyran
Dimethyl-POPOP	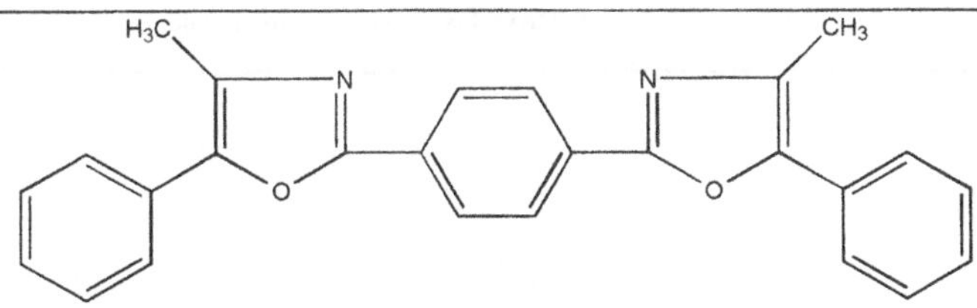

Disodium fluorescein	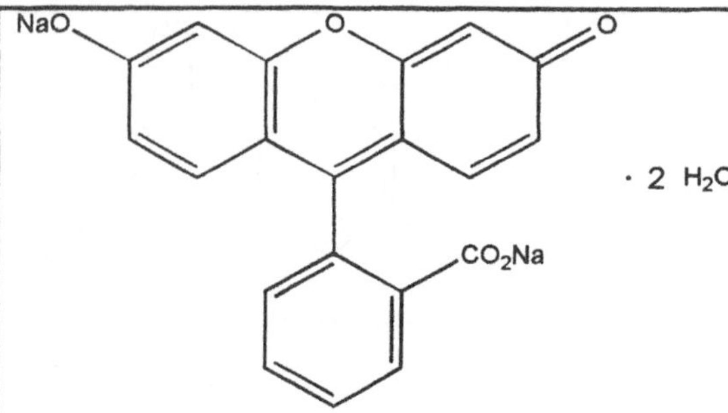 Uranin
Disperse Red 1	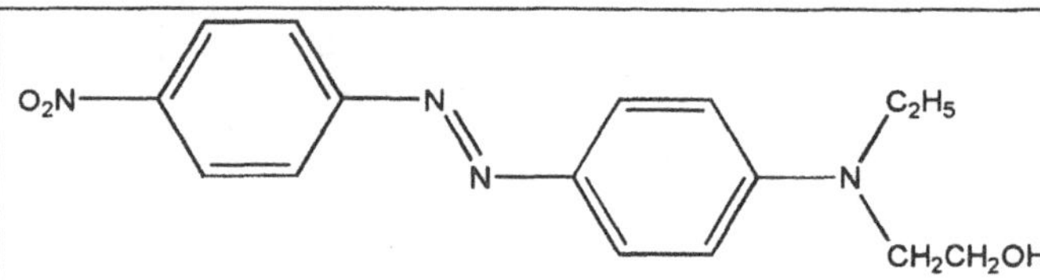
DNFA	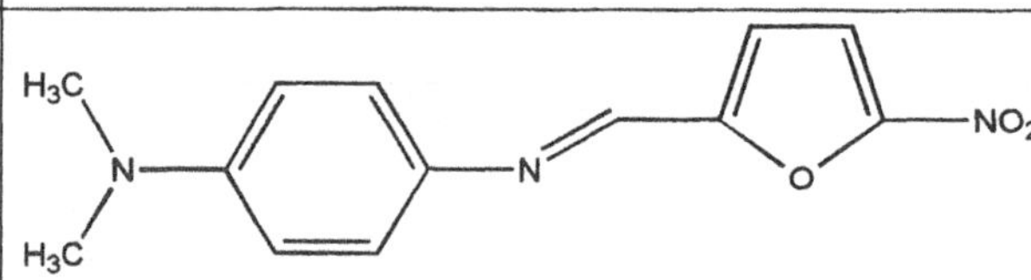
DODCI	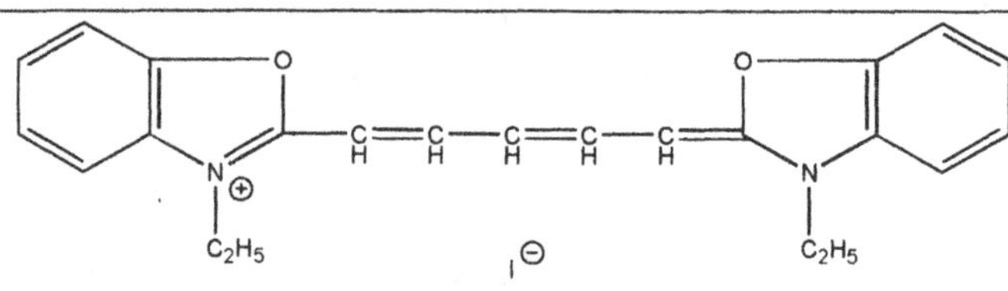 3,3'-Diethyloxadicarbocyanine Iodide

DPS390	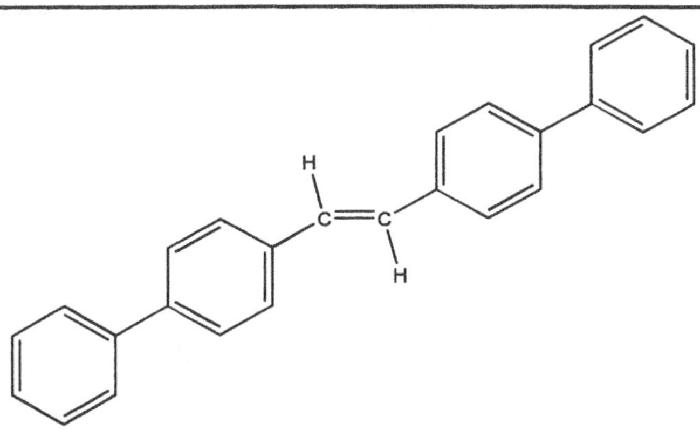
Eosin (Eosin Y)	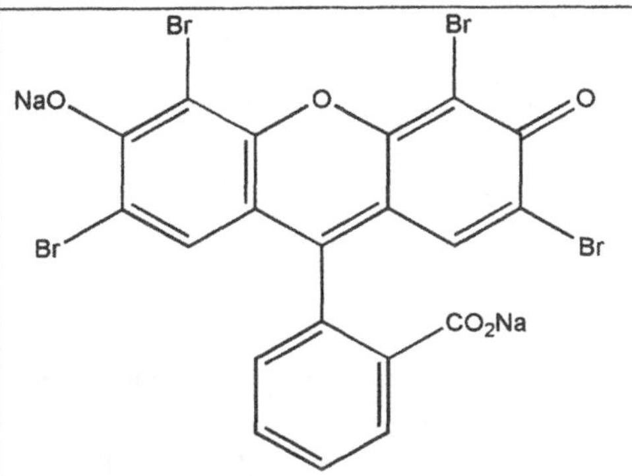
Eosine B	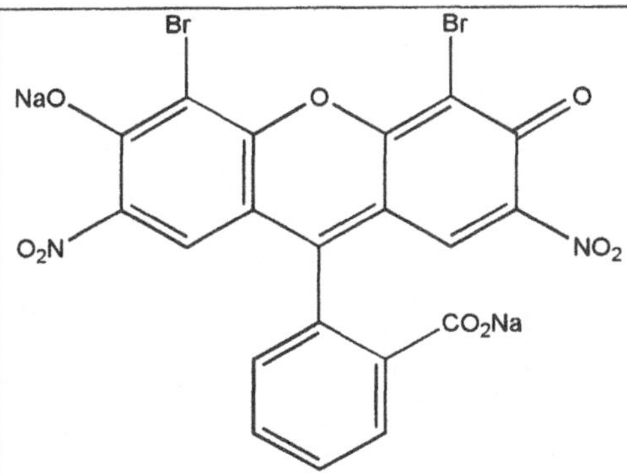

Fuchsin	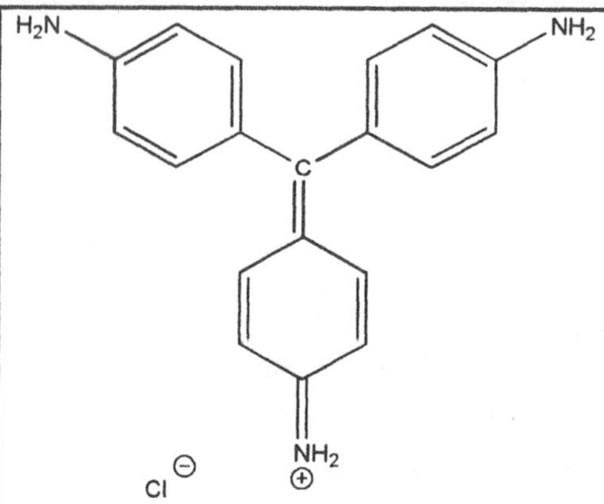
HMICI	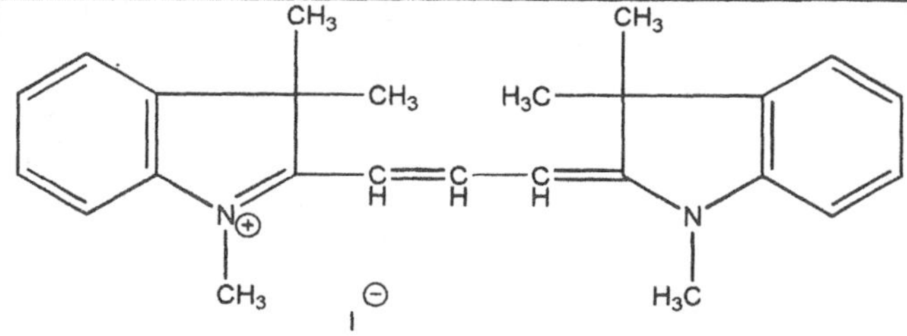 1,3,3,1',3',3'-hexamethylindocarbocyanine iodide
HSQ	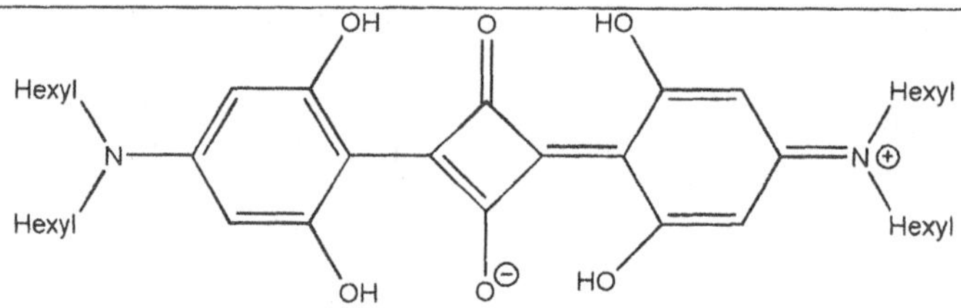
HITC	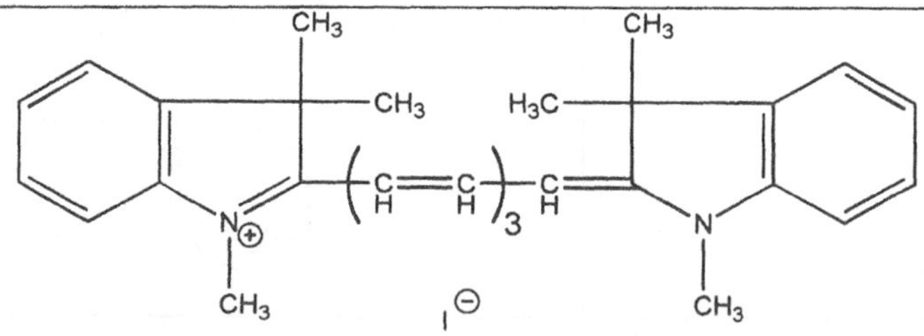 (1,1',3,3,3',3',-heaxamethylindotricarbocyanine iodide)

Hydroxycoumarin	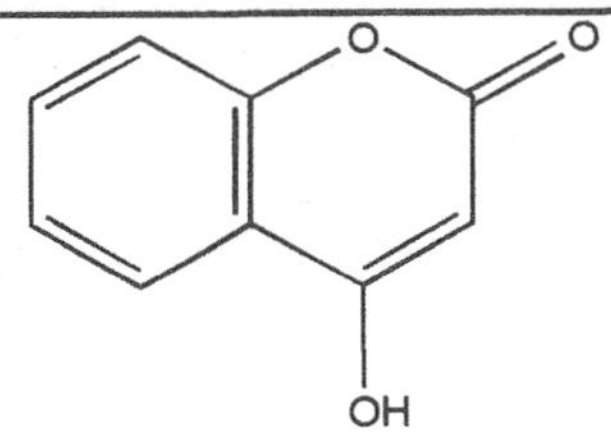
Imidocarbocyanine	********************************
Indo-1	[1-[2-amino-5-(6-carboxyindol-2-yl)phenoxy]-2-(2-amino-5-methylphenoxy)ethane-N,N,N',N'-tetraacetic acid]
ISQ	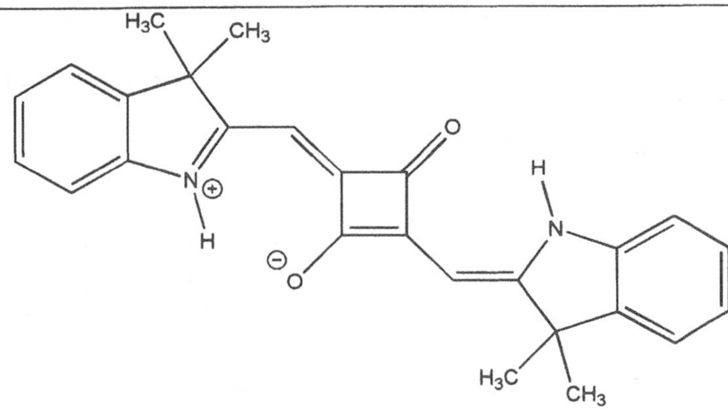 1,3-bis(3', 3'-dimethyl-2'-indoleninylidenyl)-cyclobutene-2,4-dione
LD490 (Coumarin 6H)	******************************
Methyldiethylaminocoumarin (Coumarin 1)	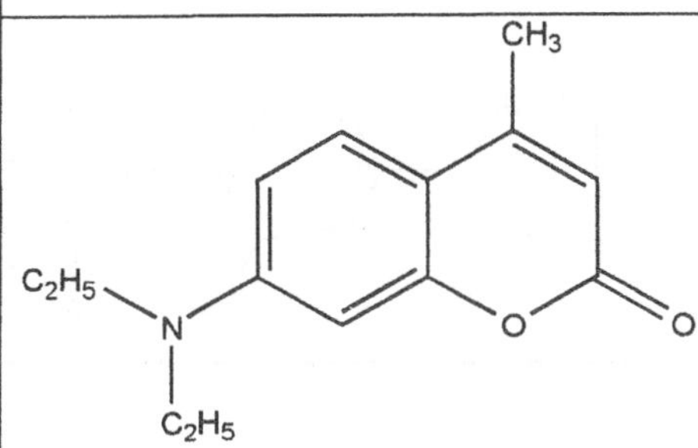 7-diethylamino-4-methyl coumarin

Methylene blue	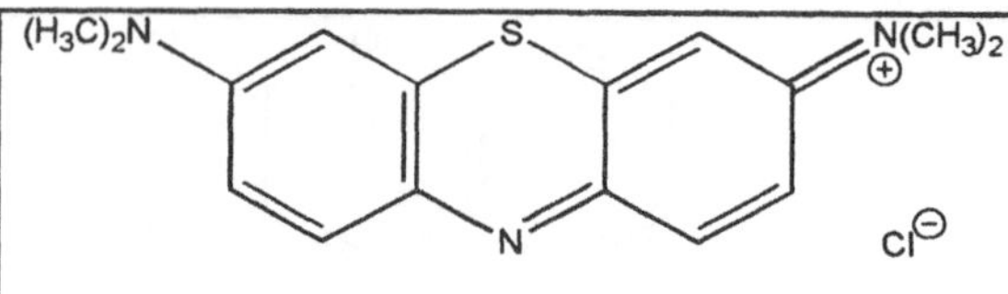
NCAE	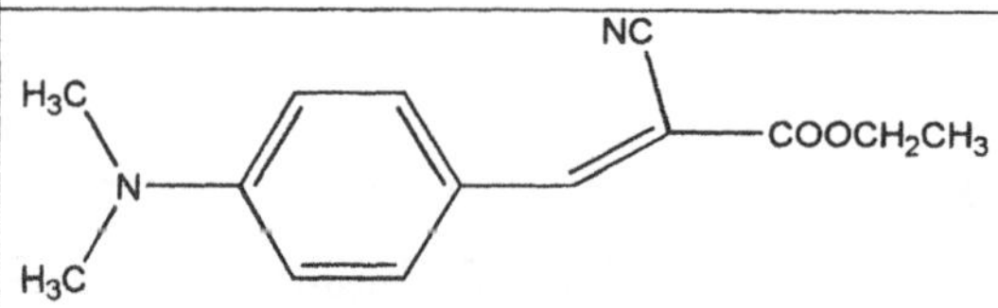
Oxazine 725 (Oxazine 1 Perchlorate)	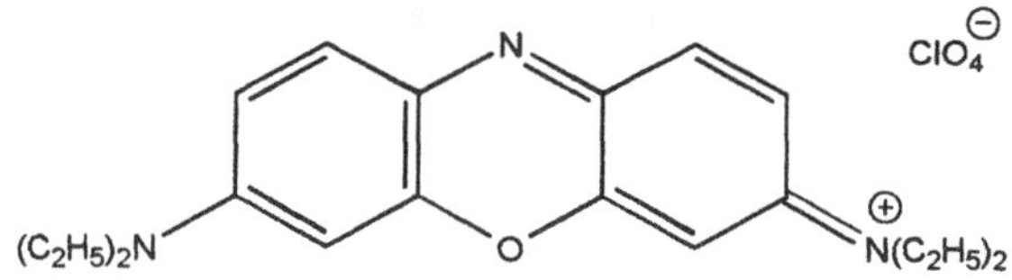
Phloxine B	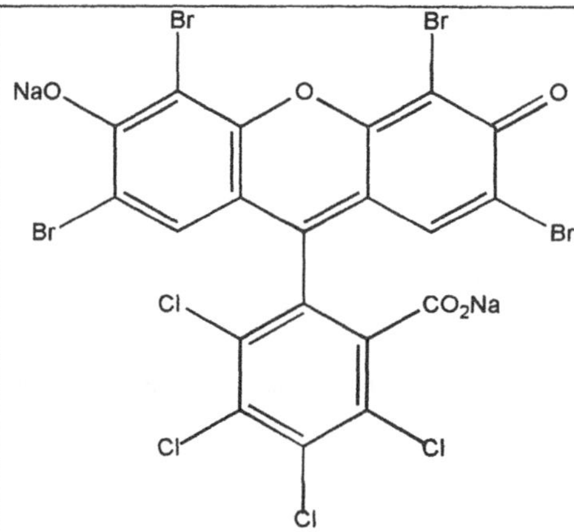
PSQ	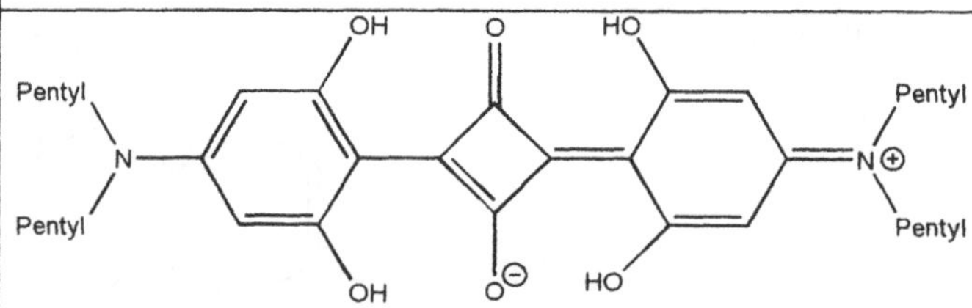

PYC	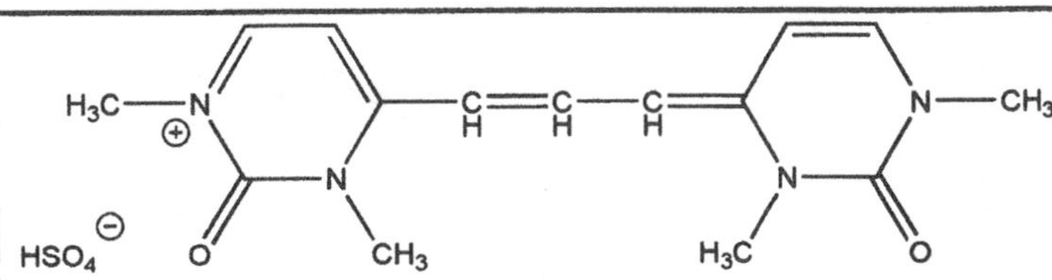 1,3,1',3'-tetramethyl-2,2'-dioxopyrimido-6,6'-carbocanine hydrogen sulphate
Rhodamine 101	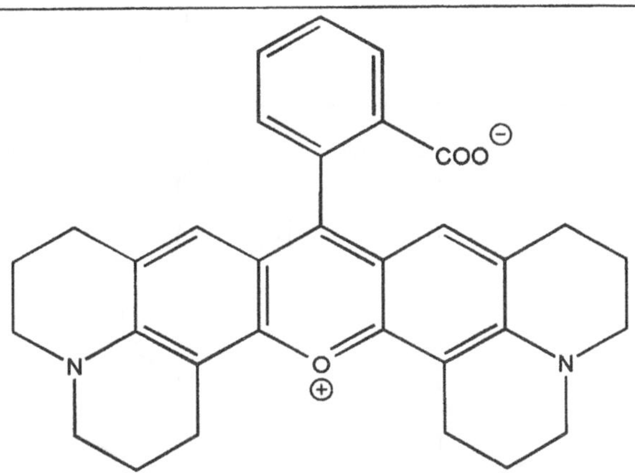
Rhodamine 110	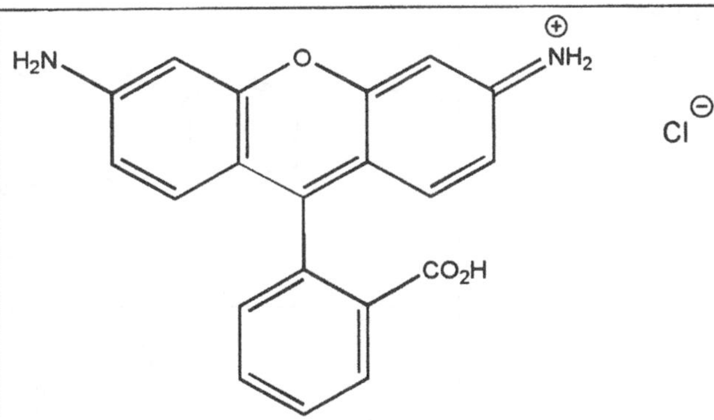

Rhodamine 6G	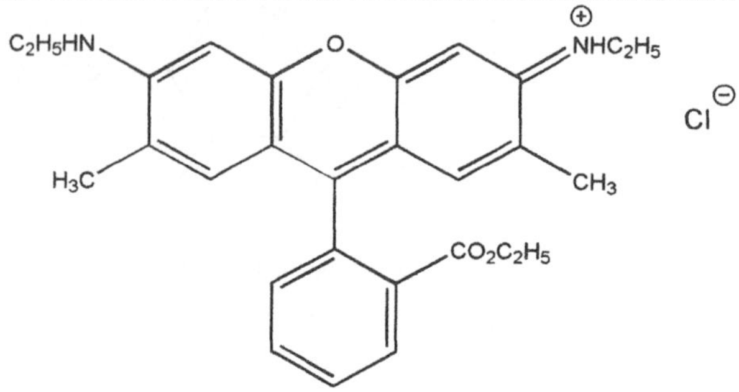
Rhodamine B	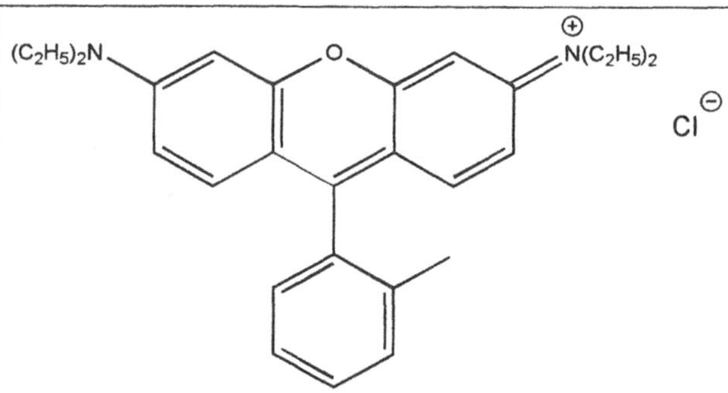
Rhodamine C (Rhodamine S)	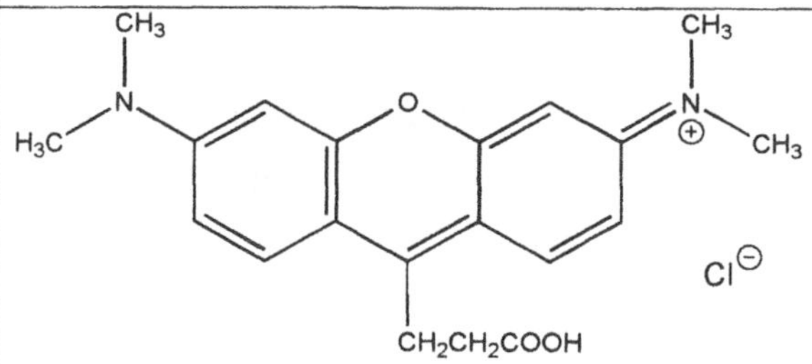
Rubicene	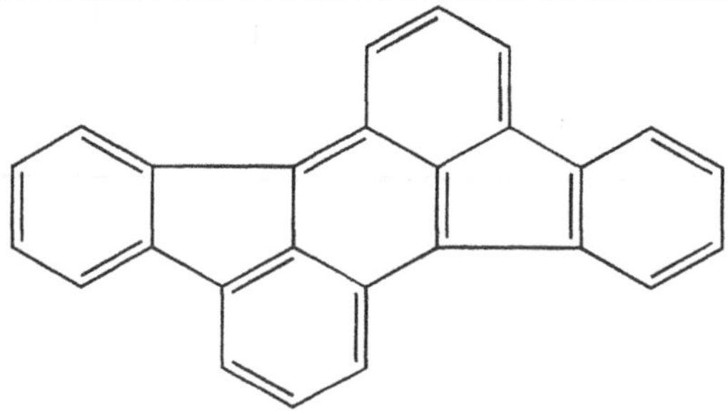

Safranine O (Safranine T)	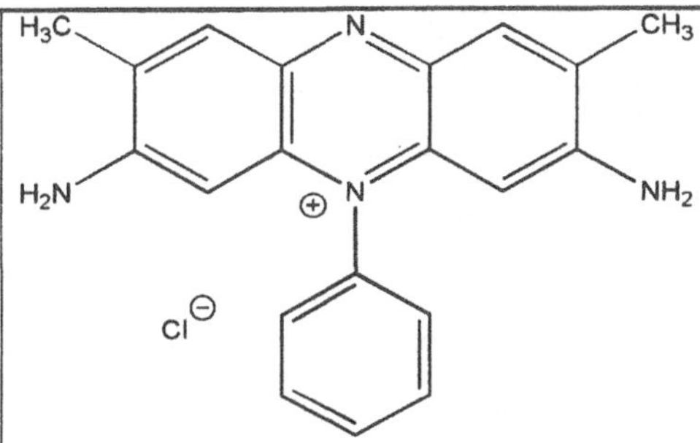
Trans-Stilbene	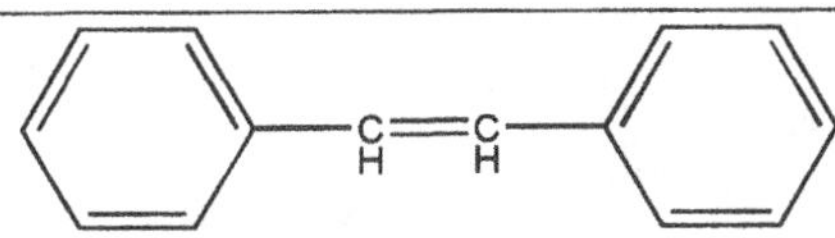
TSQ	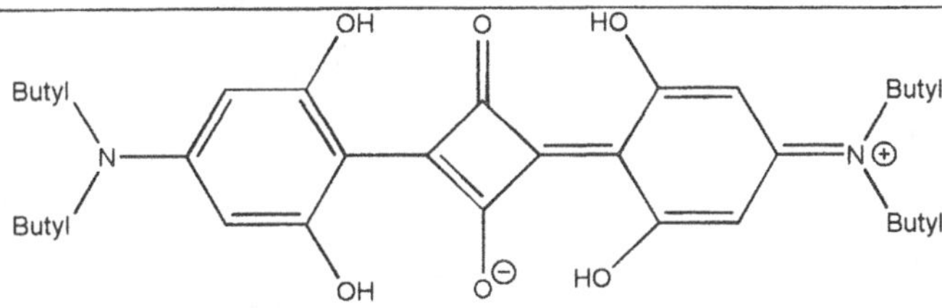
Uranine	See *Disodium Fluorescein*
BBIPA	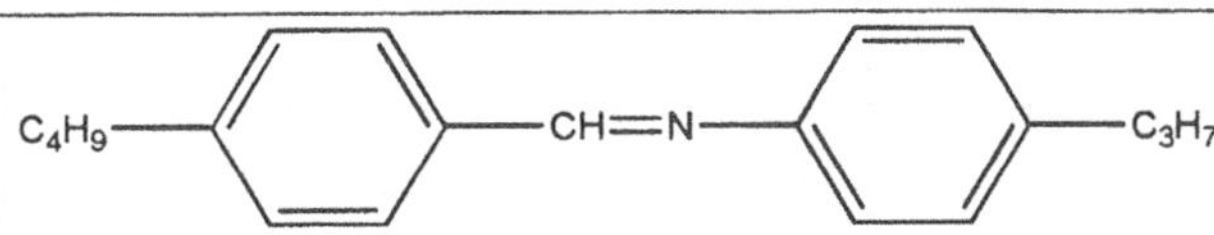
ETBBA	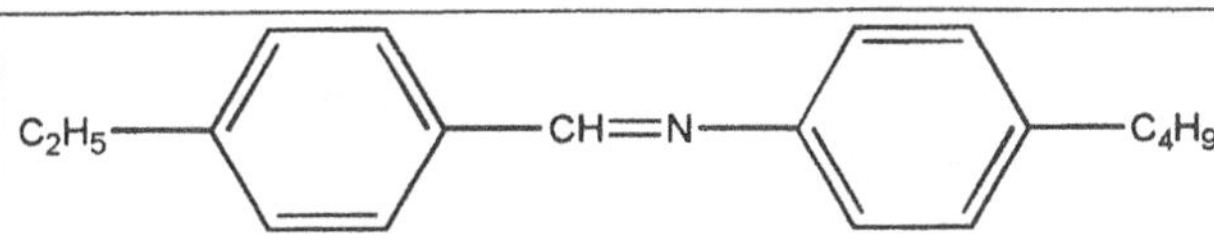
IPBBA	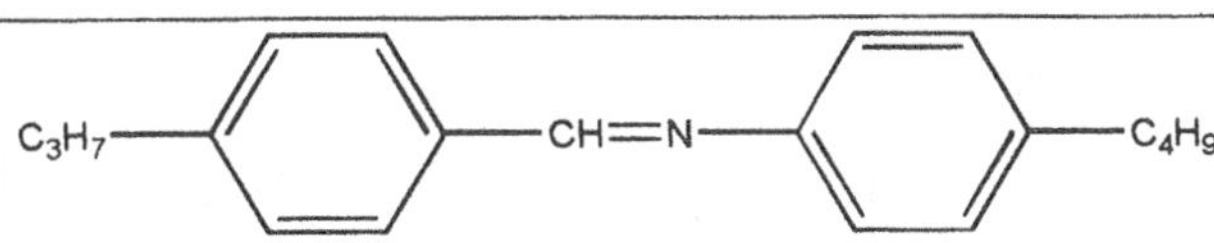
MEBBA	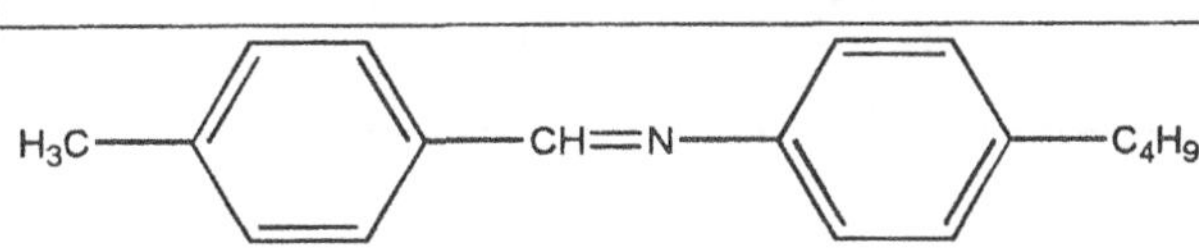

PEBBA	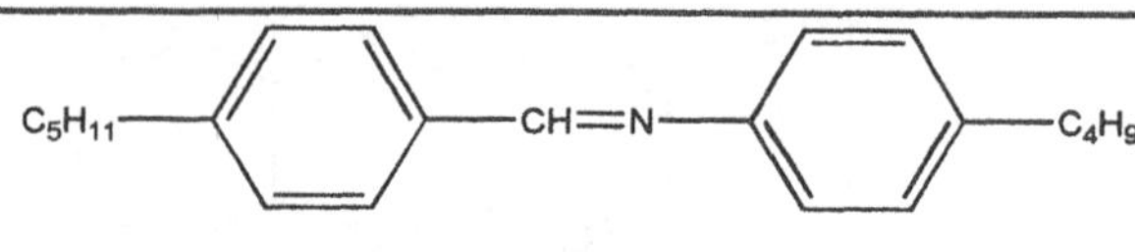
Pentyl cyanobiphenyl	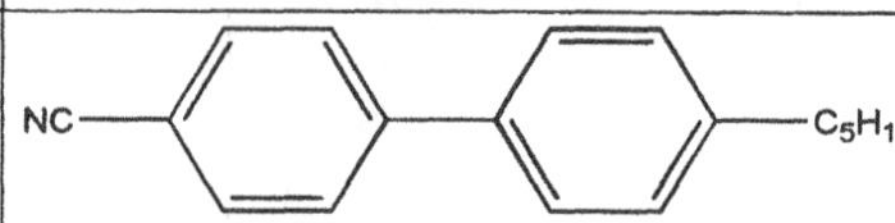 5CB, K15 (Merck)
BBTDOT (2,5-bis(2-benzothiazoyl)-3,4-didecyloxy-thiophene	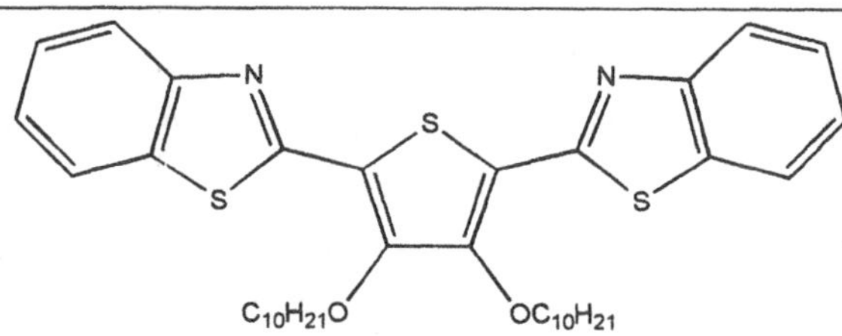
3BCMU polymer (3-butoxycarbonylmethyl-urethane)	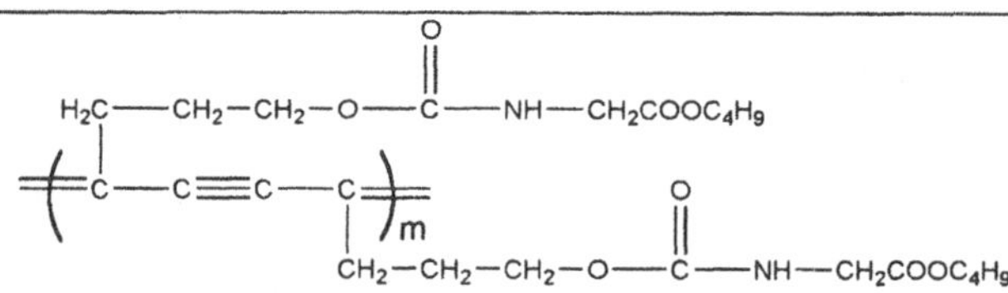
4BCMU (4-butoxycarbonylmethyl-urethane)	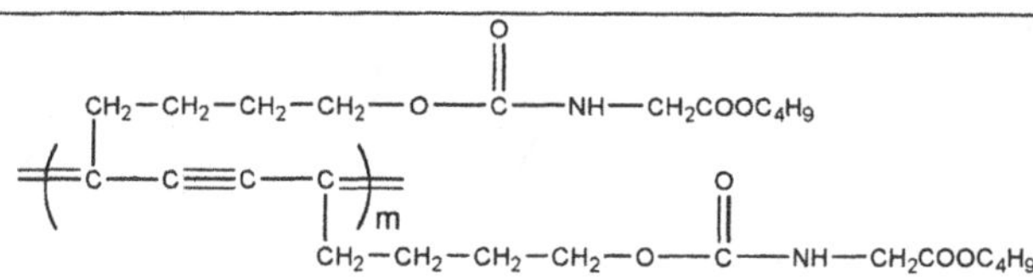 poly[5,7-dodecadyine-1,12-diolbis(n-butoxycarbonyl methylurethane)]
DANS/ side chain polymer (~ 50% by volume di-alkyl-amino-nitro-stilbene side groups)	Chromophore side group, DANS: 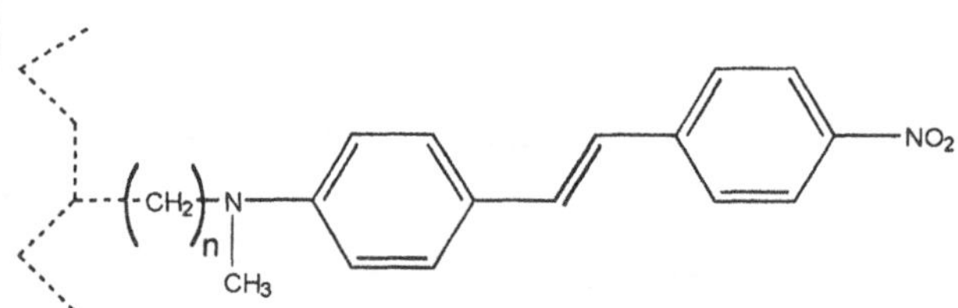

Didecyl-hexathiophene oligomer	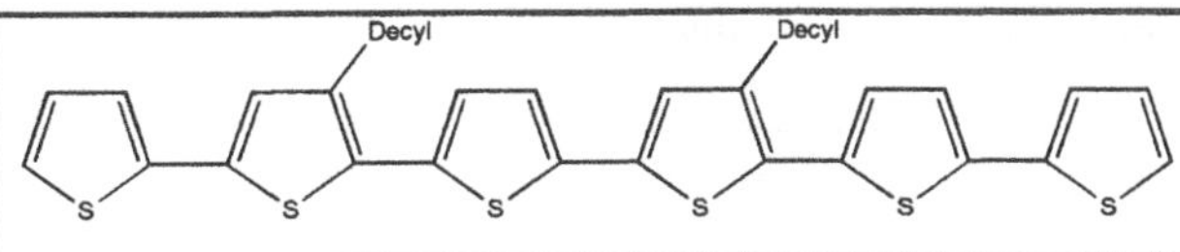
Didecyloxy para-polyphenyl heptamer	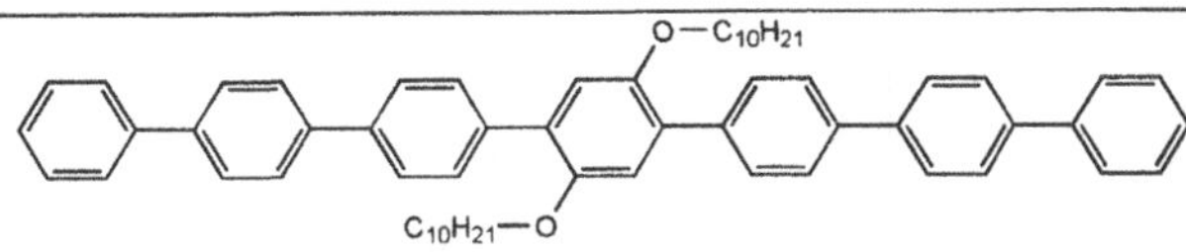
DDOS 2''',5'''-didecyloxy-1,1':4',1'':4'',1''':4''',1'''':4'''',1''''':4''''',1''''''-septiphenyl	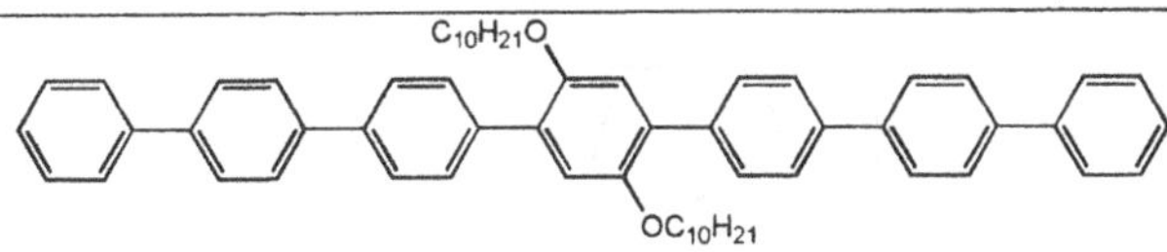
PBTAB Poly (α-{5,5'-bithiophenediyl} p-acetoxy-benzylidene)	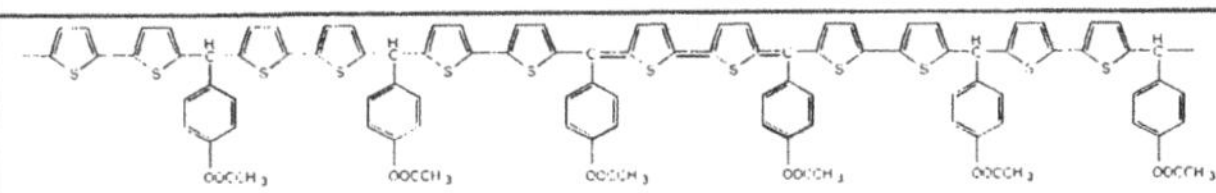
PBTB Poly (α-{5,5'-bithiophenediyl} benzylidene)	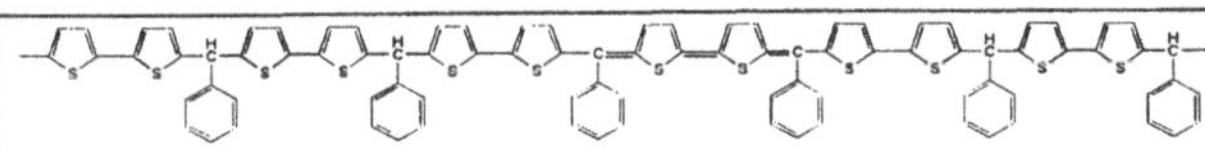
PDTDS (poly(di-n-tetradecylsilane)	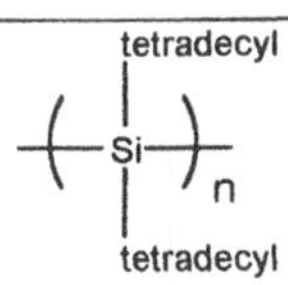
PMMPS (poly (methyl-4-methylpentyl-silane)	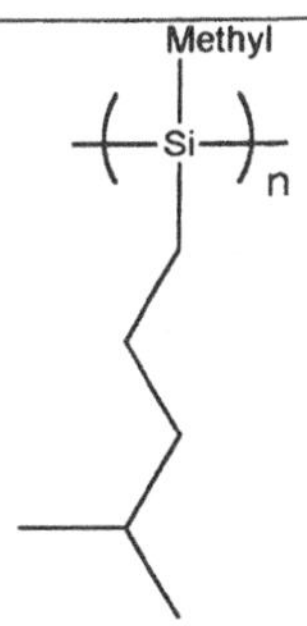

PMPS (poly (methyl phenyl-silane)	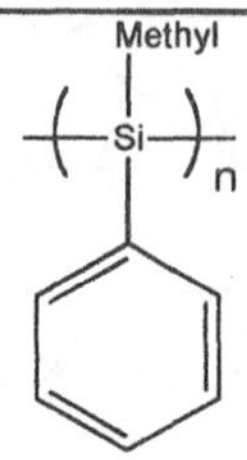
Poly (3-octylthiophene)	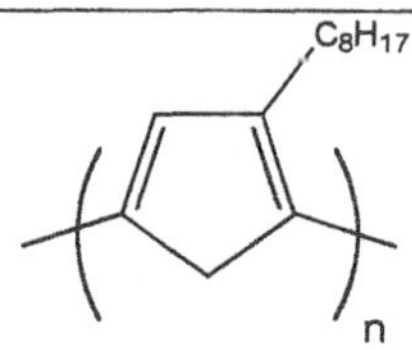
Poly(di-n-butylsilane)	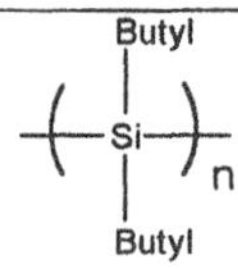
Poly(di-n-hexylsilane)	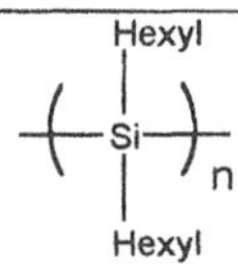
Poly-[Pd(PBu_3)$_2$DEB] (Palladium poly-yne)	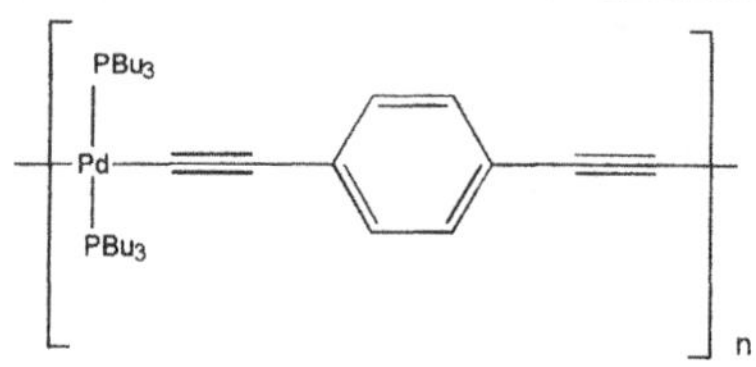
Poly-[Pt(PBu_3)$_2$(DEX)$_2$] (Platinum poly-yne)	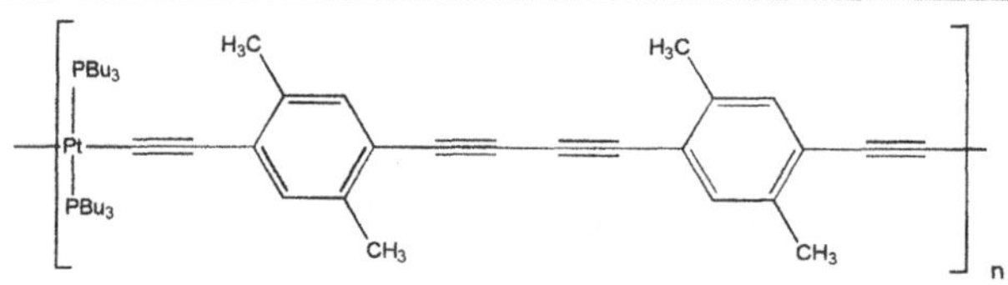

Poly-[Pt(PBu)$_3$)$_2$ (DEB)$_2$] (Platinum poly-yne)	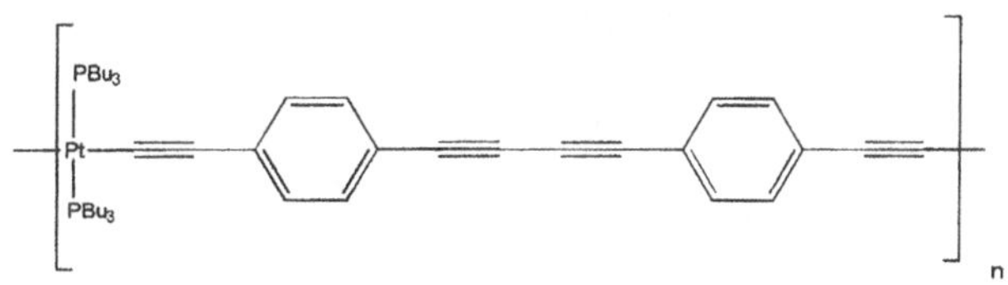
Poly-[Pt(PBu)$_3$)$_2$DEB] (Platinum poly-yne)	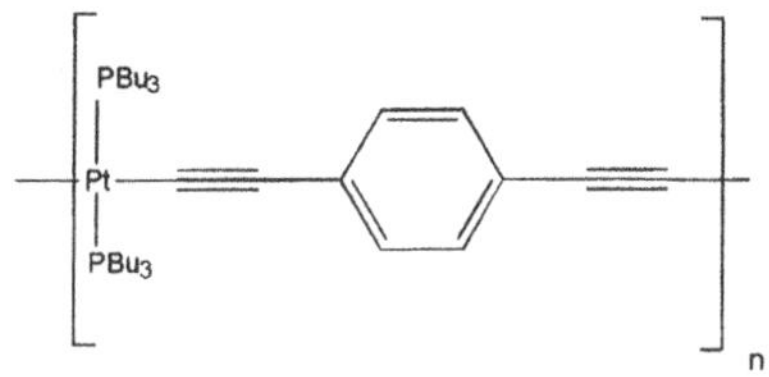
Poly-[Pt(PBu)$_3$)$_2$DEX] (Platinum poly-yne)	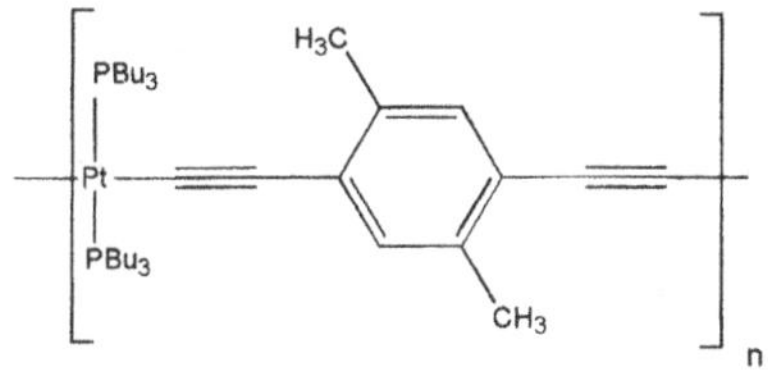
Poly-[Pt(PBu$_3$)$_2$(C C-C C]$_n$ (Platinum poly-yne)	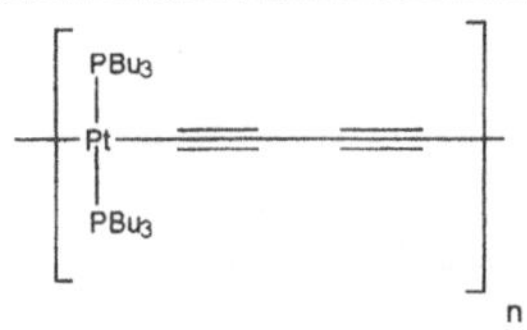
Poly (chloro-trifluoroethylene)	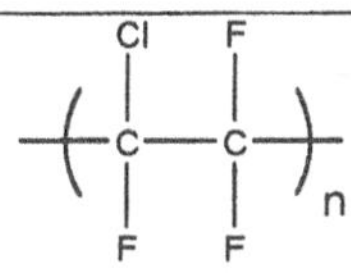
PPB poly (2,5-pyrrole-diyl benzylidene)	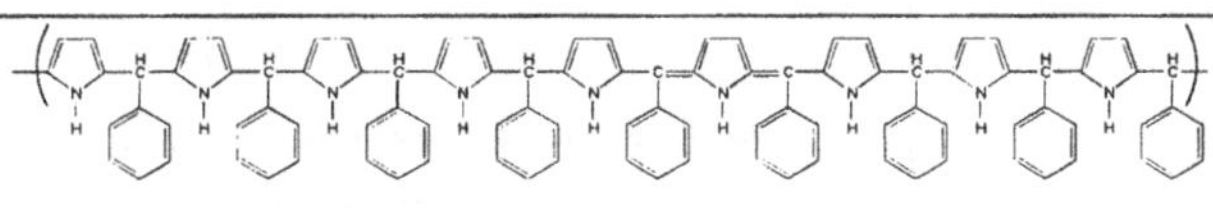

PPPV (poly (phenyl p-phenylene vinylene))	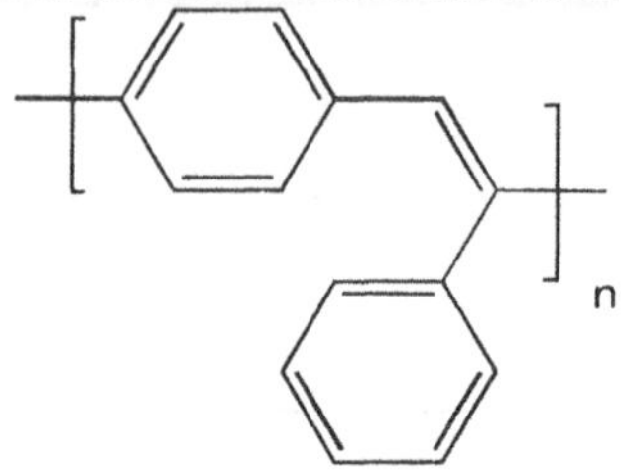
PPV (poly (p-phenylene vinylene))	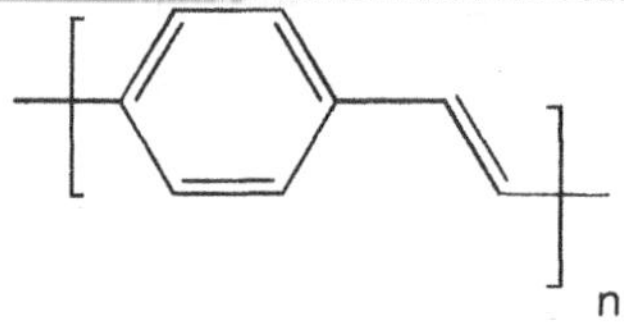
PTB poly(2,5-thiophenediyl benzylidene	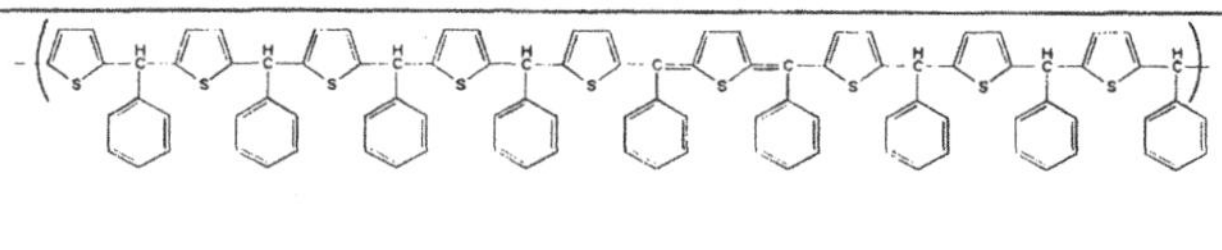
PTFE poly (tetrafluoro-ethylene)	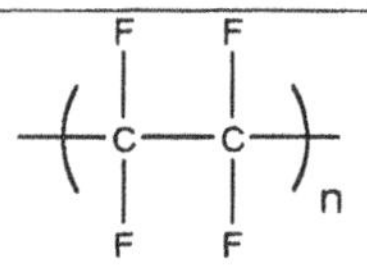
PTS Polydiacetelyne-bis(toluene sulphonate)	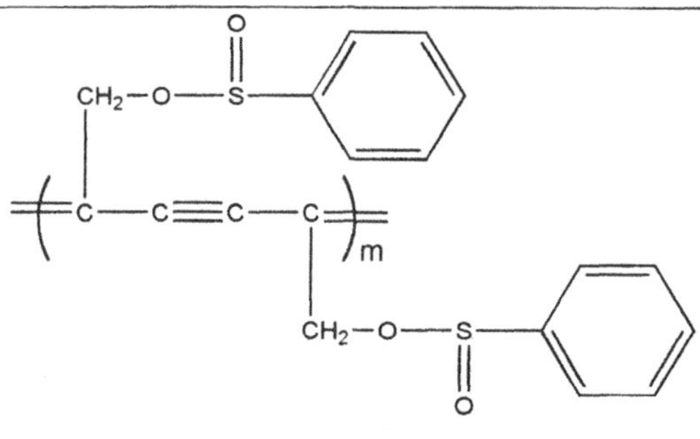 poly[2,4-hexdiyne-1,6-diol-bis-p-toluene-sulphonate]
PTTB Poly (α-{5,5"-terthiophenediyl} benzylidene)	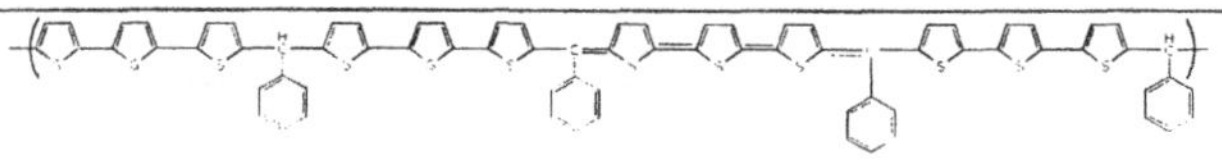

α-sexithieneyl (polythiophene oligomer)	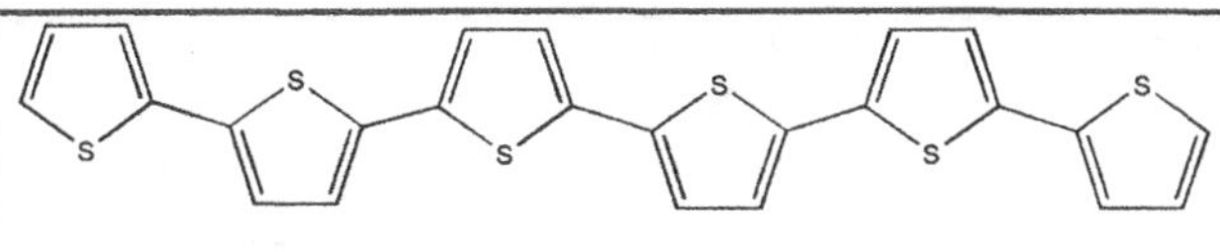
TCDU Polydiacetelyne-bis(phenyl-urethane)	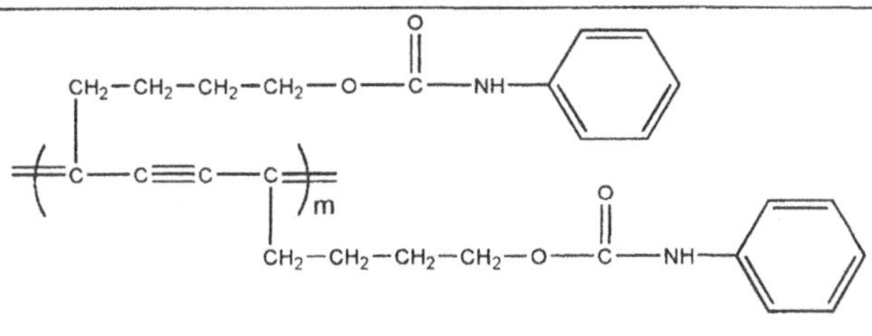 Bis-(phenylurethane) of 5,7-dodecadiyne-1,2-diol
Tetrathiophene oligomer	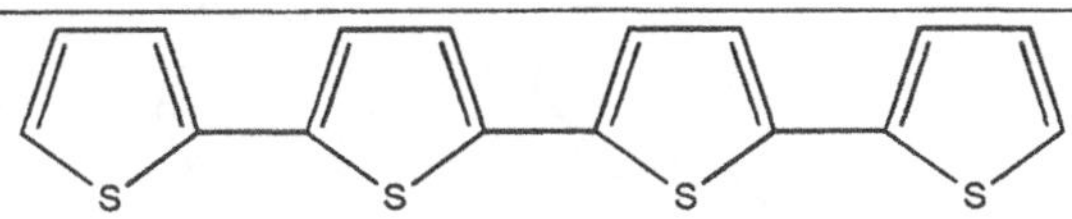
Benzoporphyrin 1	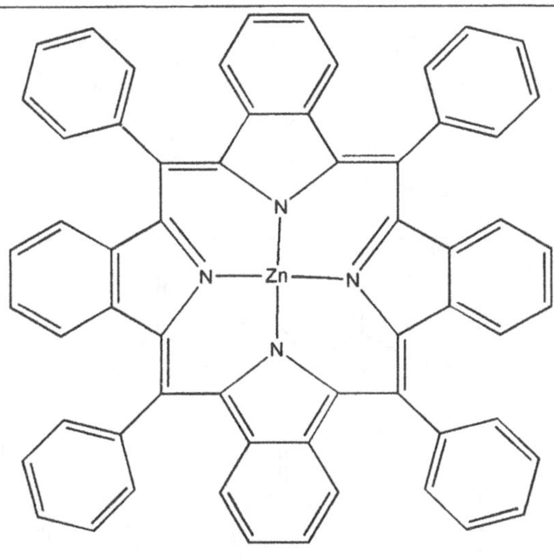 Zinc meso-tetraphenyltetrabenzporphyrin

Benzoporphyrin 2	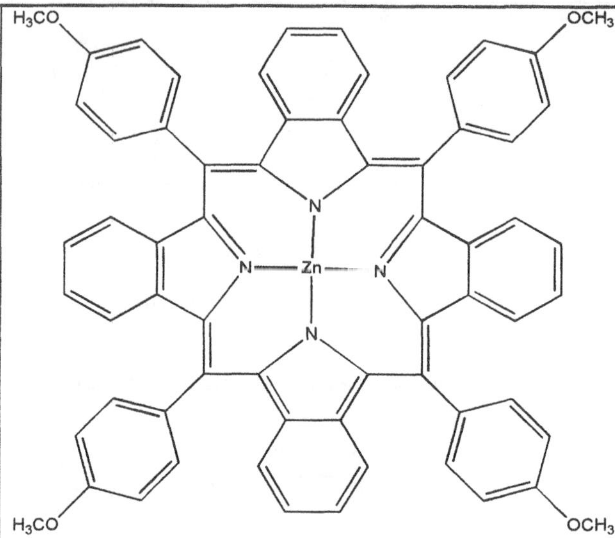 Zinc meso-tetra-(p-methoxpheny) tetrabenzporphyrin
Bis[1,2-diethyl-1,2-ethene dithiolato (2-)S, S'] Nickel	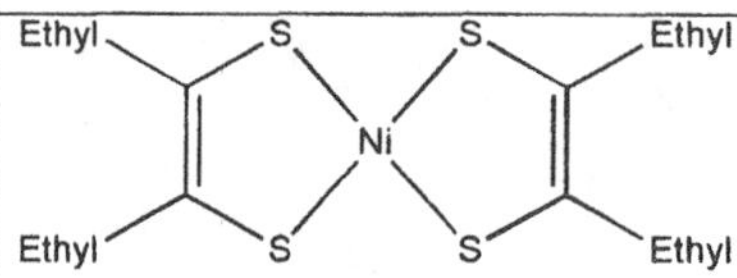
Bis[1,2-dimethyl-1,2ethenedithiolato (2-)-S,S']Nickel	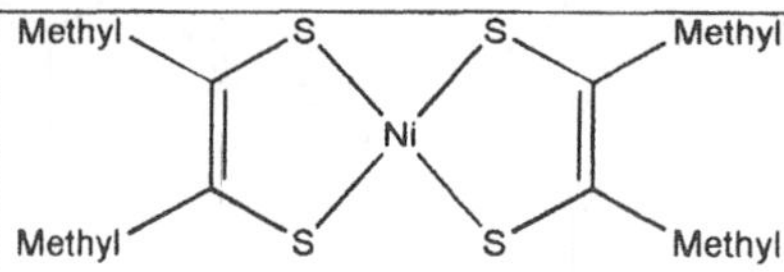
Bis[1,2-methylphenyl-1,2-ethenedithiolato (2-)-S,S']Nickel	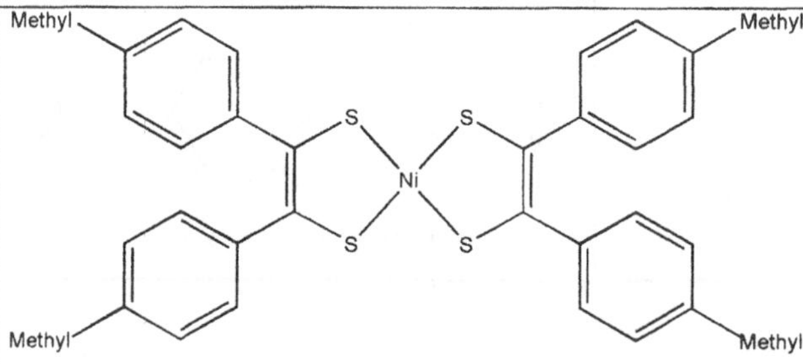

Bis[1,2-diphenyl-1,2-ethene-dithiolato(2-)-S,S']Nickel	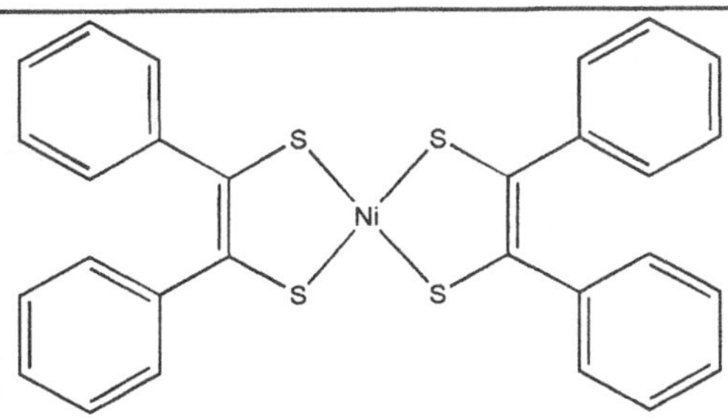
Bis[1-butyl 2-phenyl-1,2-ethenedithiolato (2-)S,S']Nickel	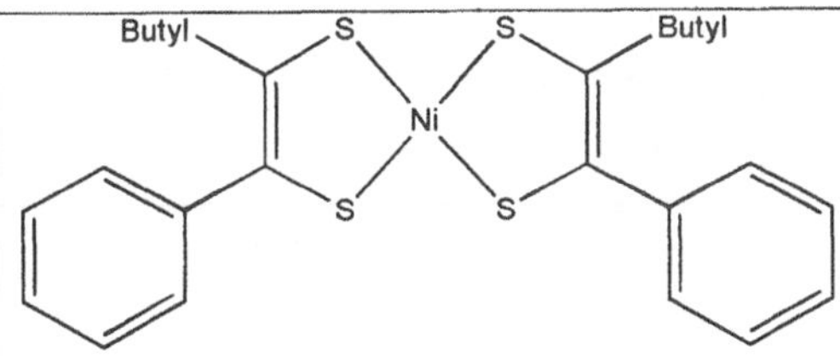
Bis[1-decyl 2-phenyl-1,2-ethenedithiolato (2-)S,S']Nickel	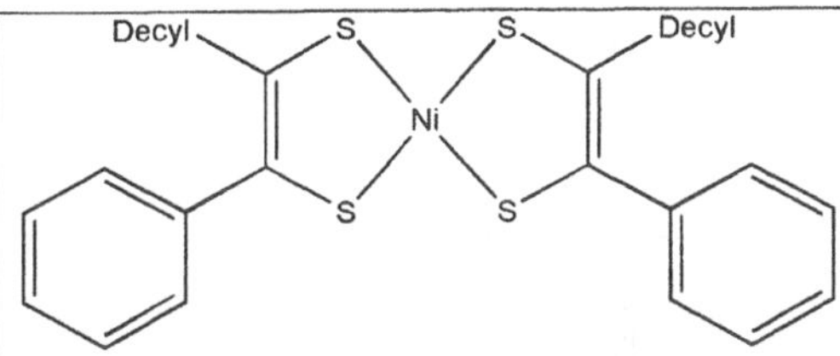
Bis[1-ethyl 2-phenyl-1,2-ethenedithiolato (2-)S,S']Nickel	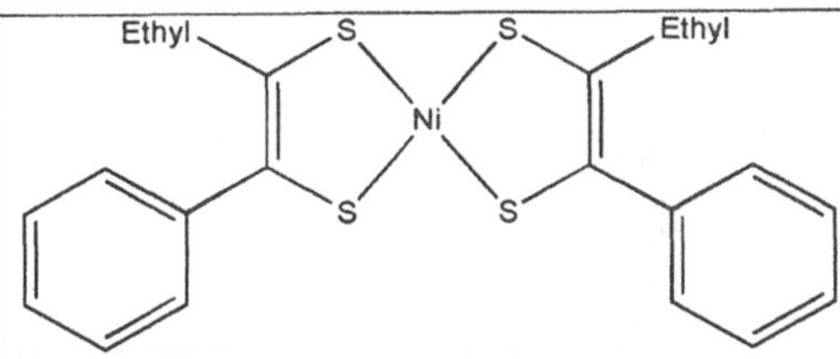
C11S4NiDT	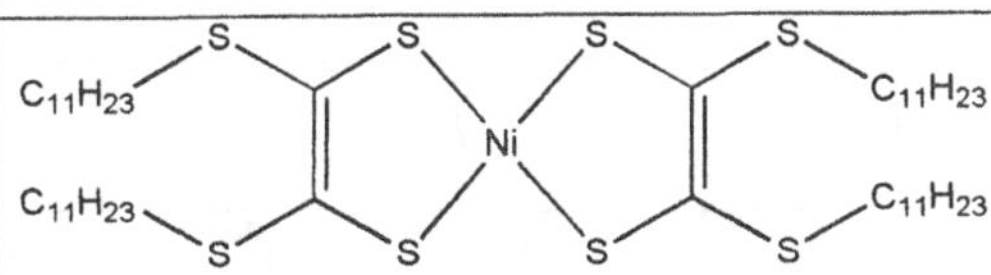

C7S4NiDT	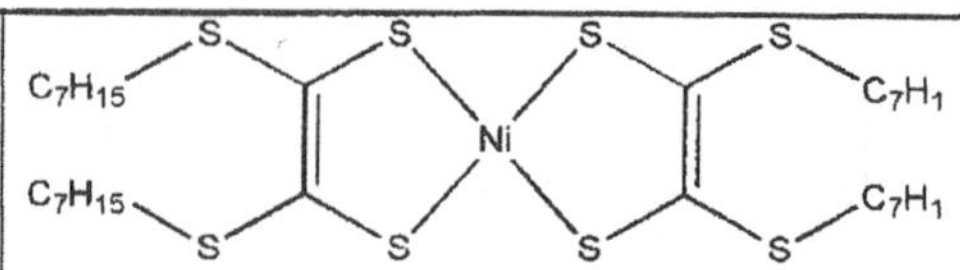
Palladium poly-yne model system $Pd(PBu_3)_2Cl_2$	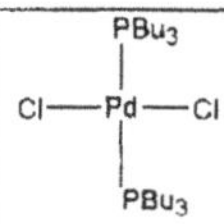
Platinum poly-yne model system $(Pt(PBu_3)_2)_2$DEB Cl_2	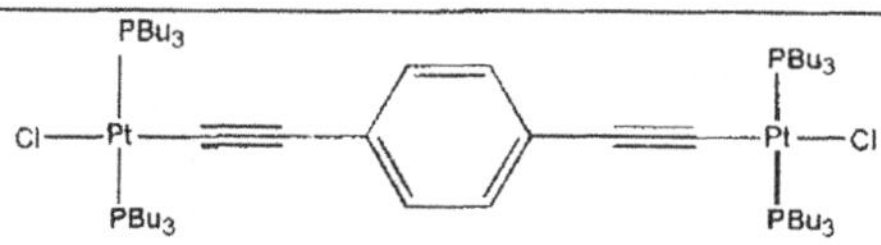
Cis-$Pt(PBu_3)_2(DEB)_2$	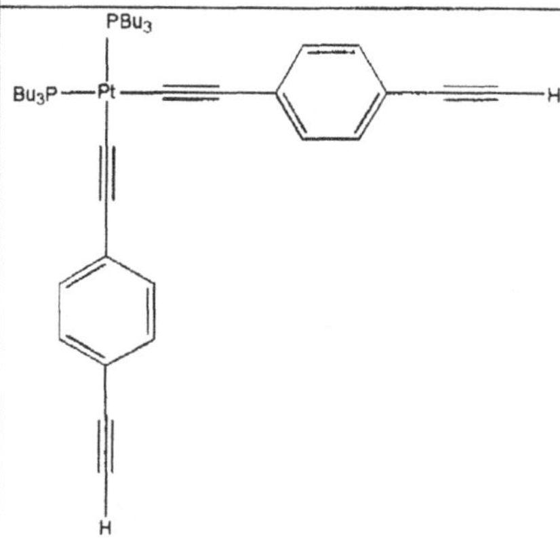
Cis-$Pt(PBu_3)_2Cl_2$	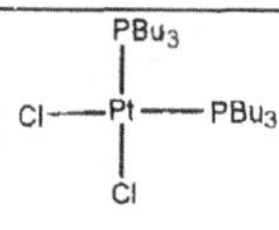
$Pt(PBu_3)_2$(C C-C C-H$)_2$	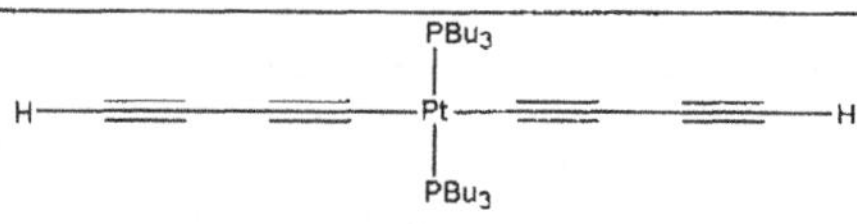
Trans-$Pt(PBu_3)_2(DEB)_2$	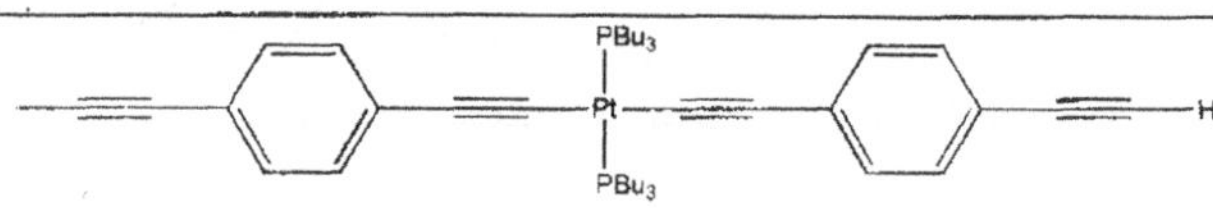

Trans-Pt(PBu$_3$)$_2$(DEX)$_2$	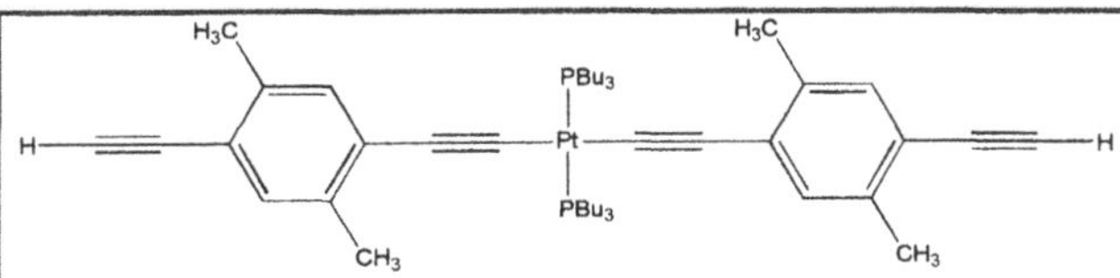
Tris-cyclopentadienyl ytterbium (Yb(cp)$_3$)	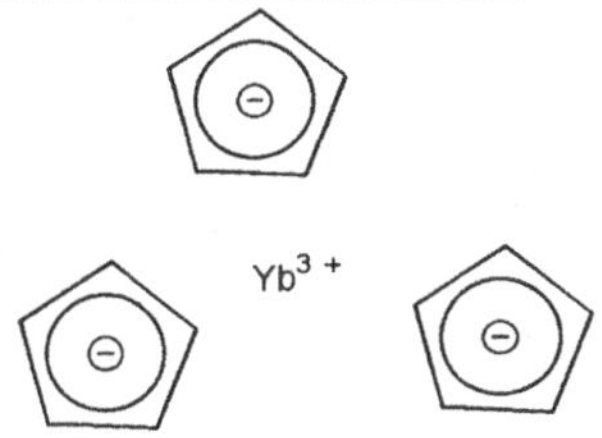
Acetone	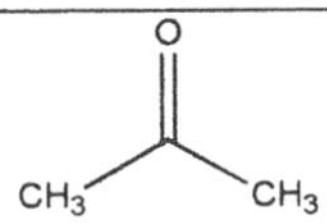
Aminophthalimide	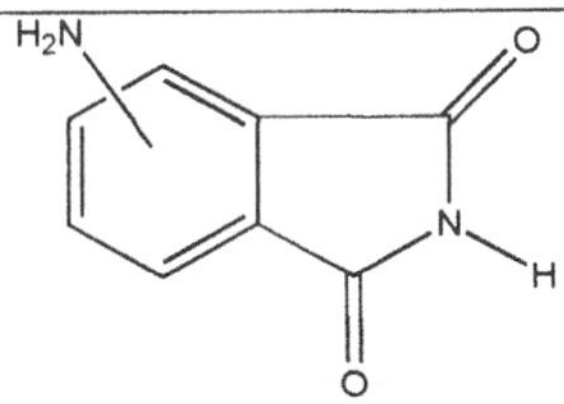
Aniline	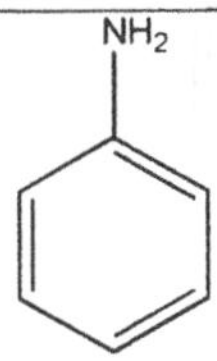
Anthracene	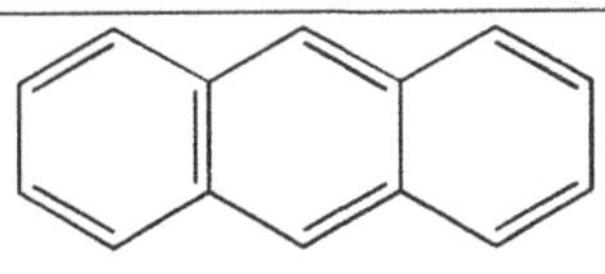

Azulene	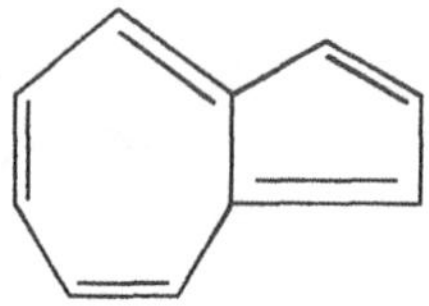
BBP	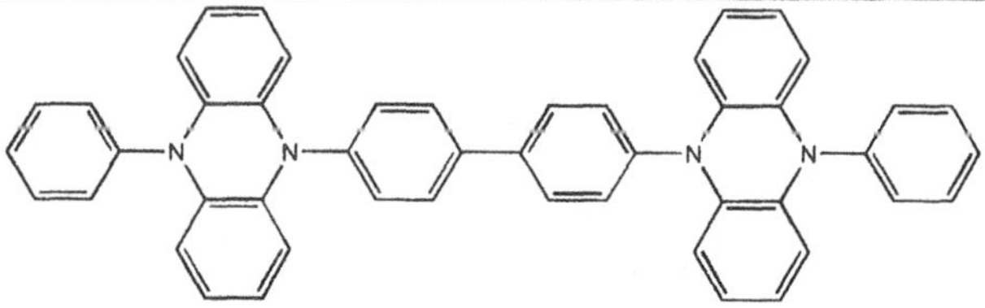 (Biphenylene bridged phenazine)
Benzanthracene	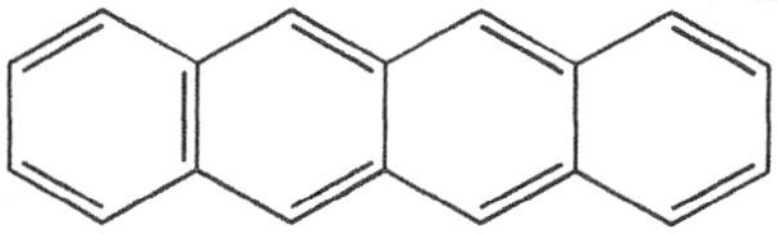
Benzene	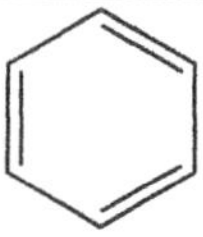
Benzopyrene	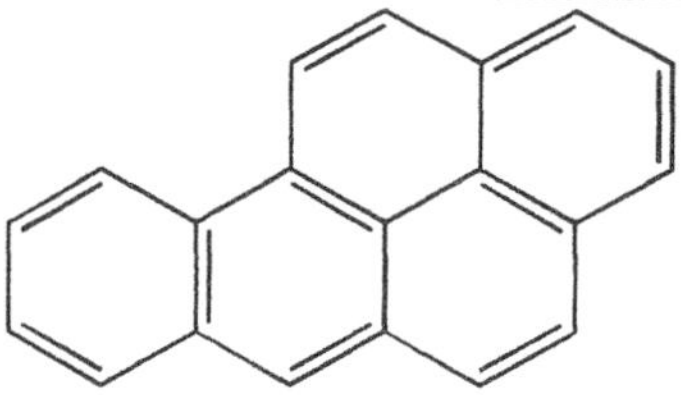

Binaphthyl	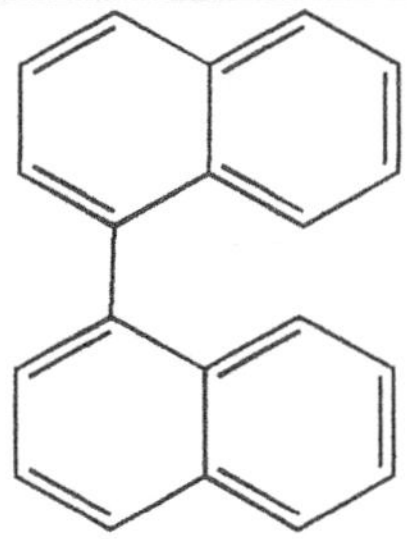
Biphenyl	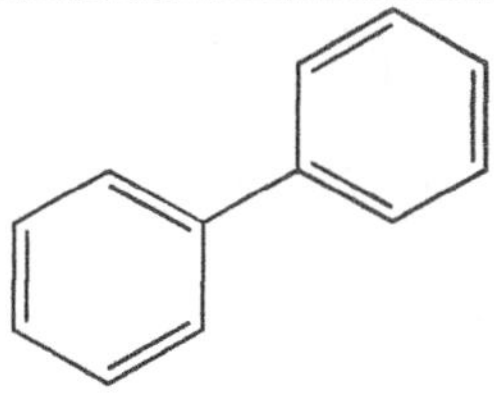
Bipyridine	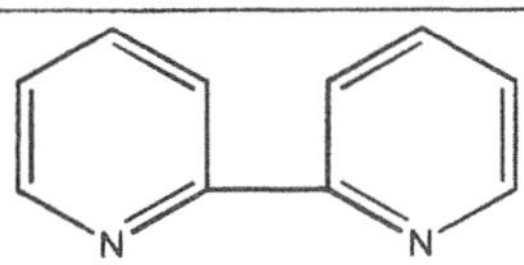
p-Bis(o-methylstyryl) benzene	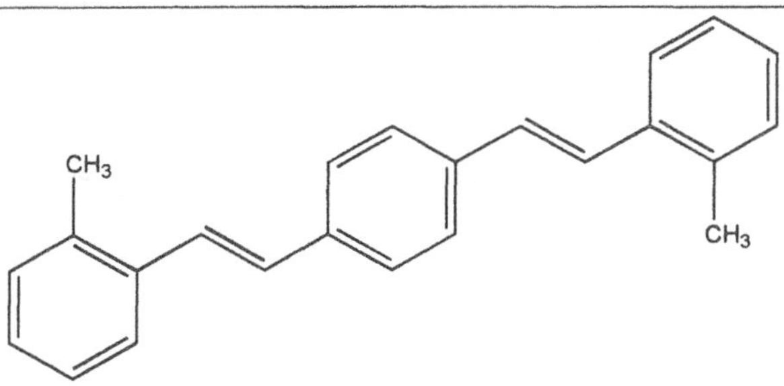 1,4-Bis(2-methylstyryl)benzene
Carbon disulphide	S=C=S
Carbon tetrachloride	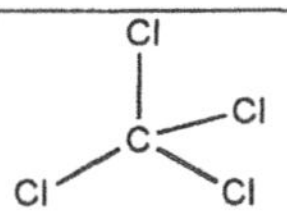

Chloroform	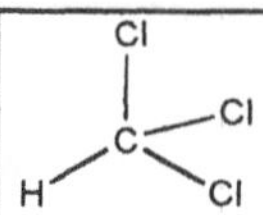
Chloronaphthalene	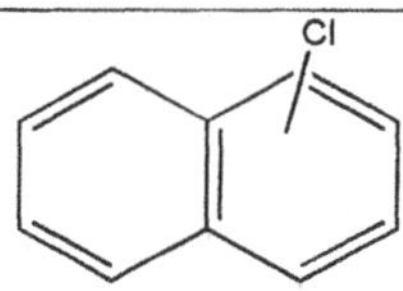
Chrysene	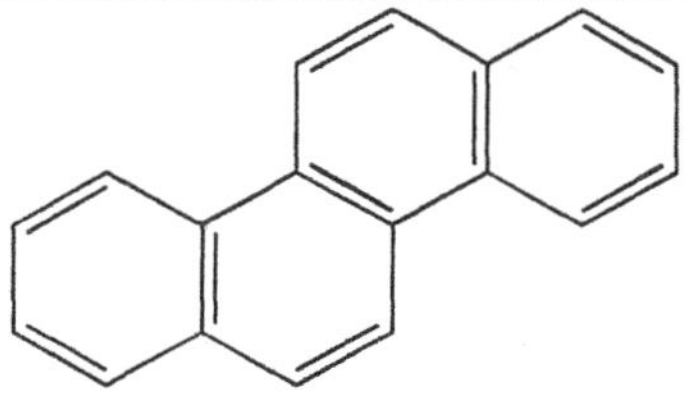 (1,2-Benzphenanthrene)
Cyclohexane	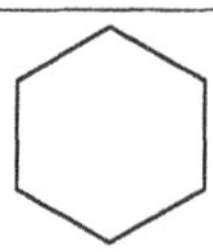
Dibenzanthracene	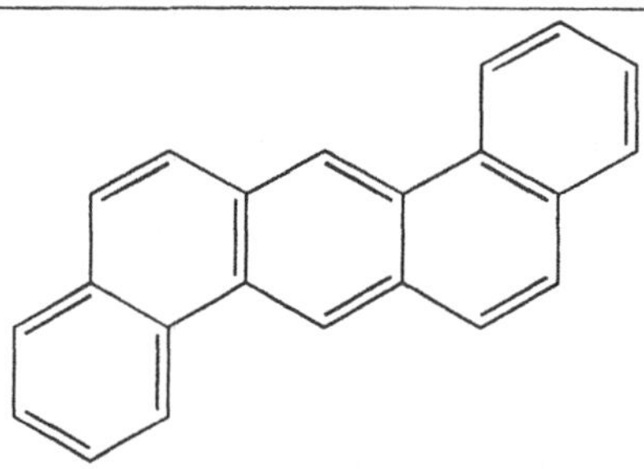
o-dichlorobenzene	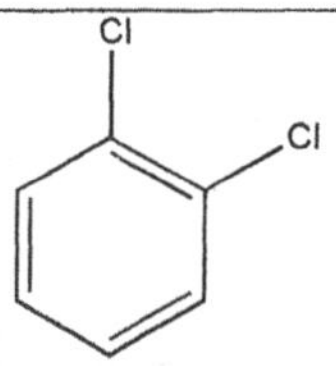

Dichlorobiphenyl	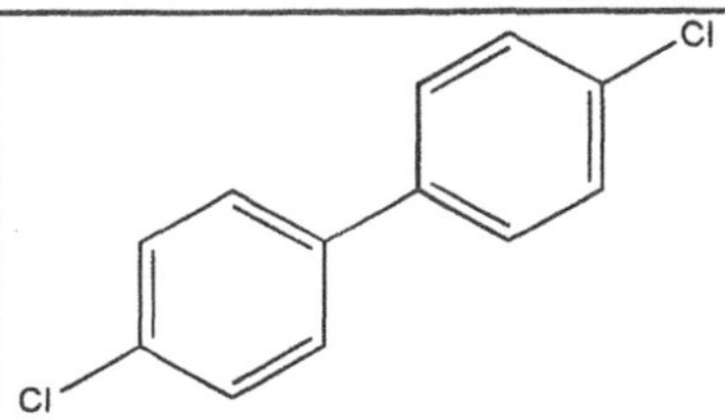
Difluorenyl	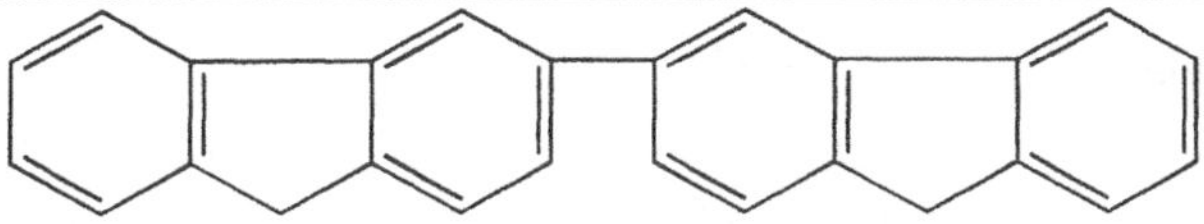
Difluorobenzene	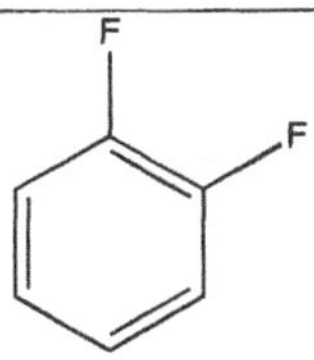
Difluorobiphenyl	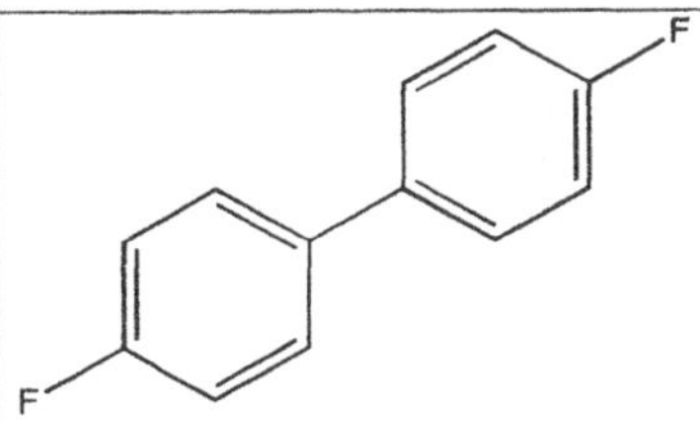
9,10-dihydroanthracene	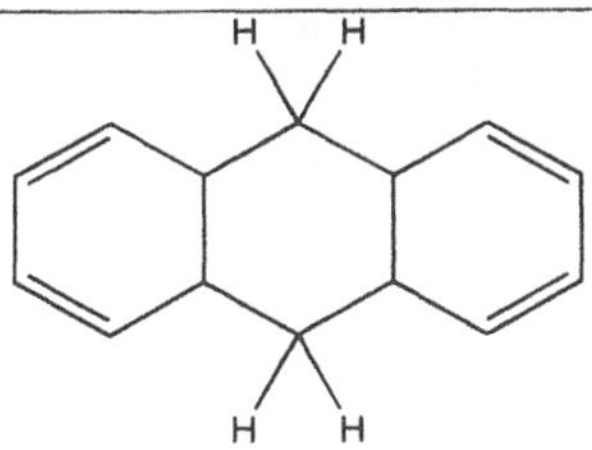
Dioxane	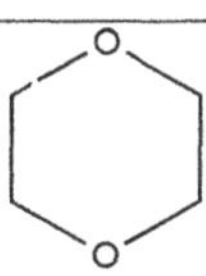
Diphenyl	see *biphenyl*

Diphenylbutadiene	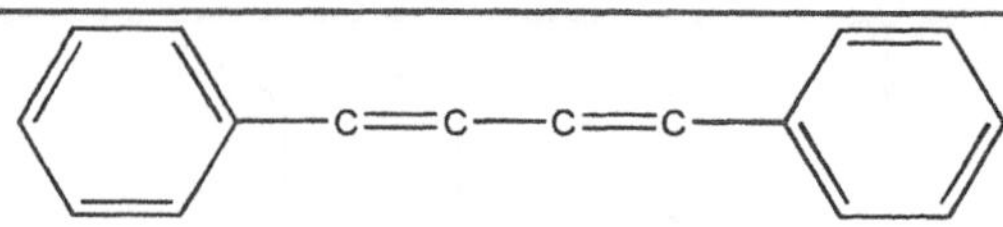
Diphenylhexatriene	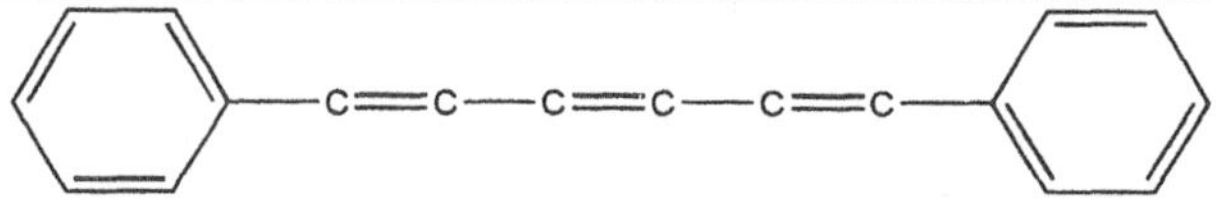
Diphenyloctatetraene	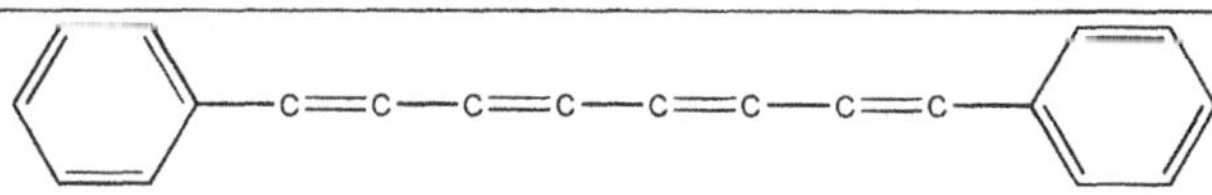
2,5-Diphenyloxazole	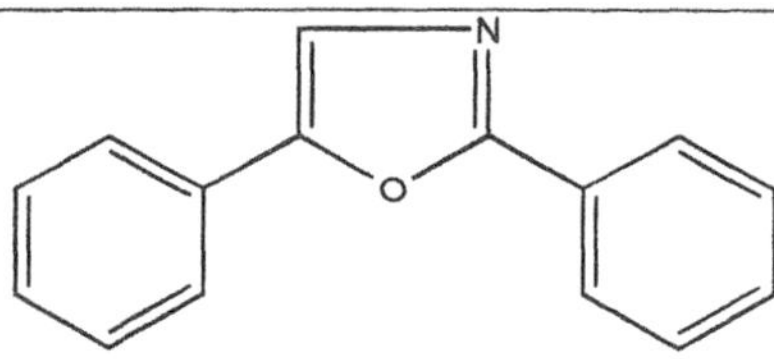 (PPO)
DPP (5,10-diphenyl-phenazine)	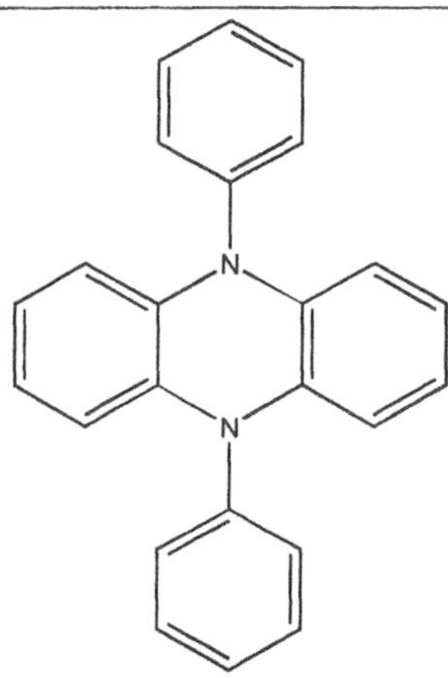

Fluoranthene	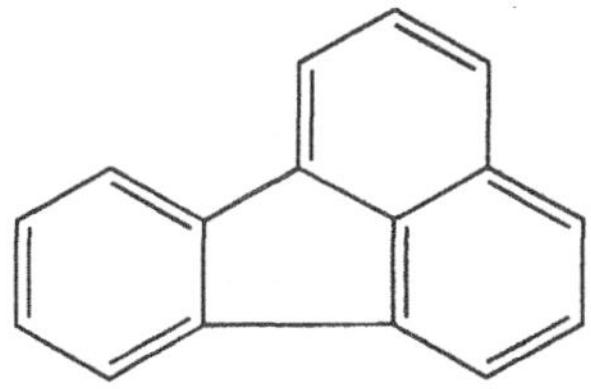
Fluorene	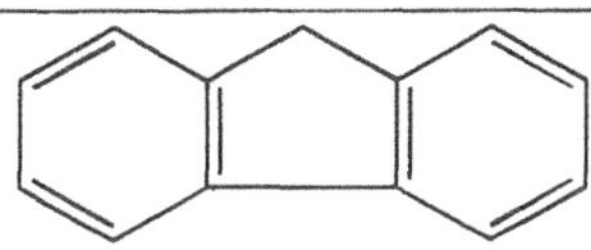
Fluorobenzene	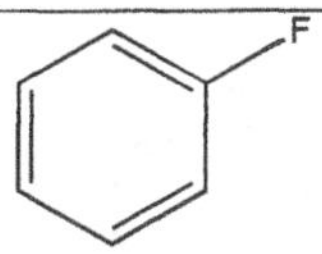
Formaldehyde	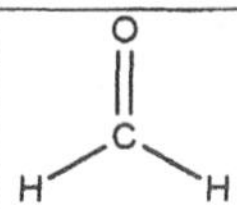 (Formalin)
Hexatriene	$H_2C{=}CH{-}CH{=}CH{-}CH{=}CH_2$
MBA-NP	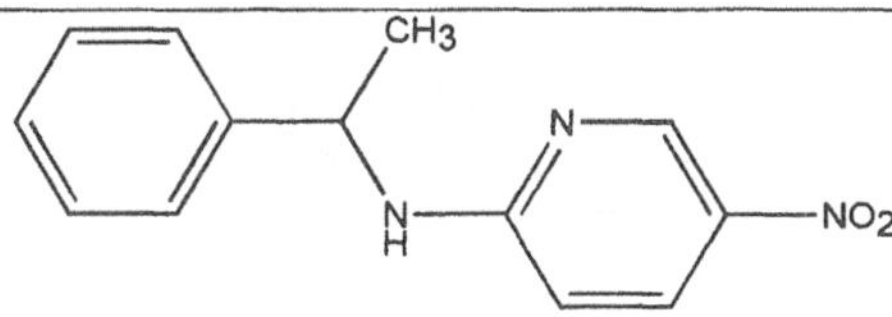
Mesitylene	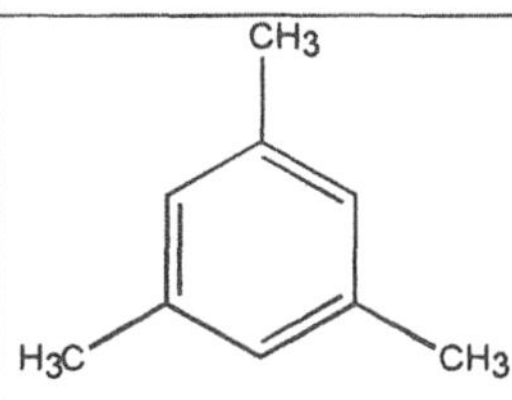 1,3,5-trimethylbenzene

Methylene iodide	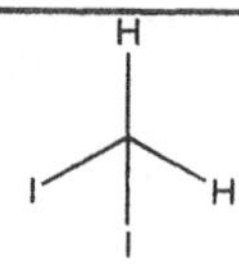
2-methyl-4-nitroaniline (MNA)	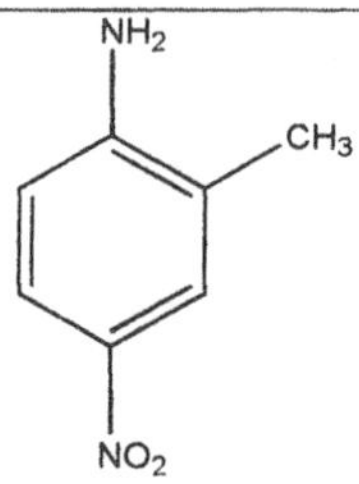
Naphthacene	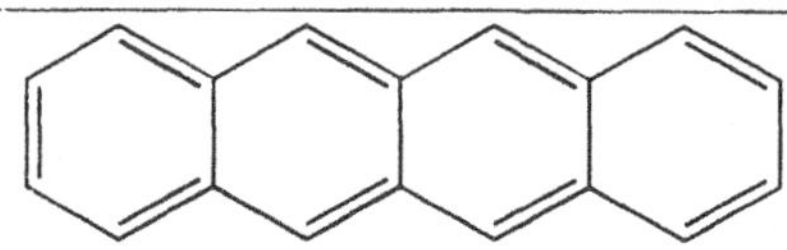
Naphthalene	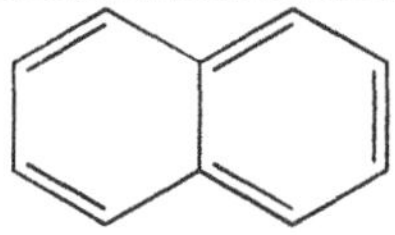
2-(1-naphthyl)5-phenyloxazole	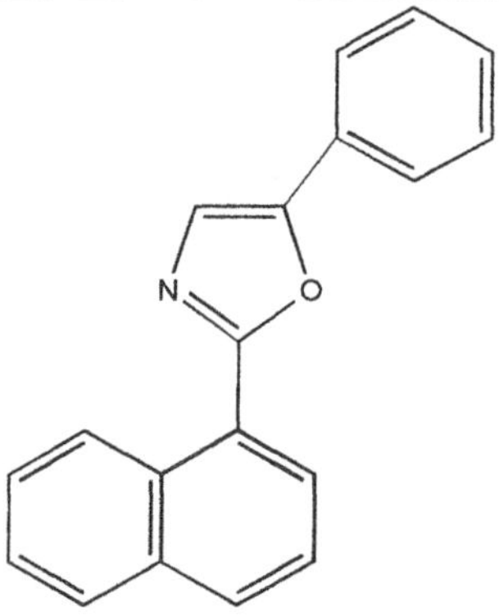 α-NPO

4-nitroaniline (pNA)	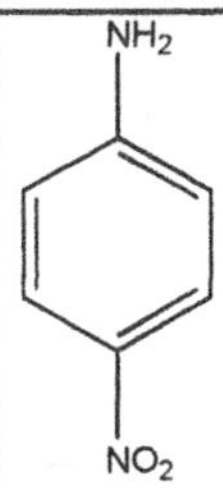
Nitrobenzene	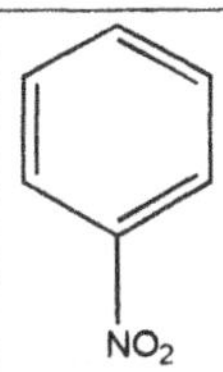
α–picoline	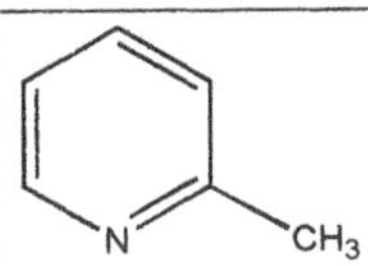
Paracyclophane	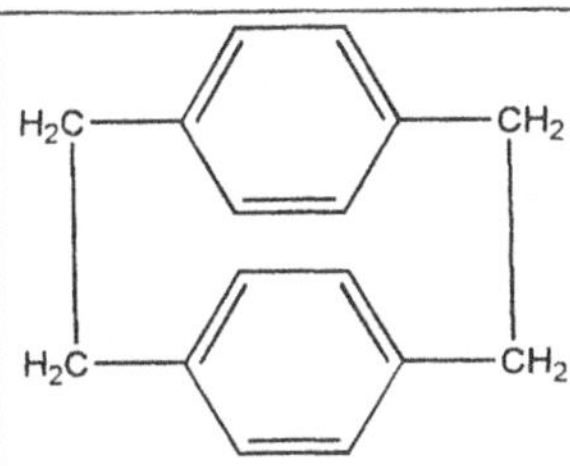
Perylene	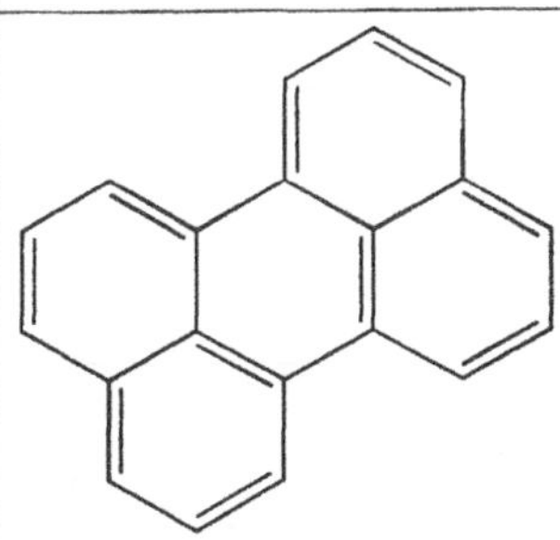

Phenanthrene	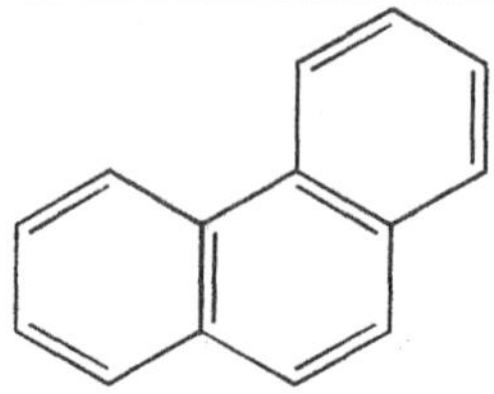
Phenol	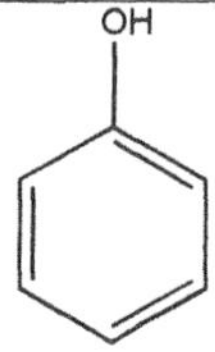
POPOP	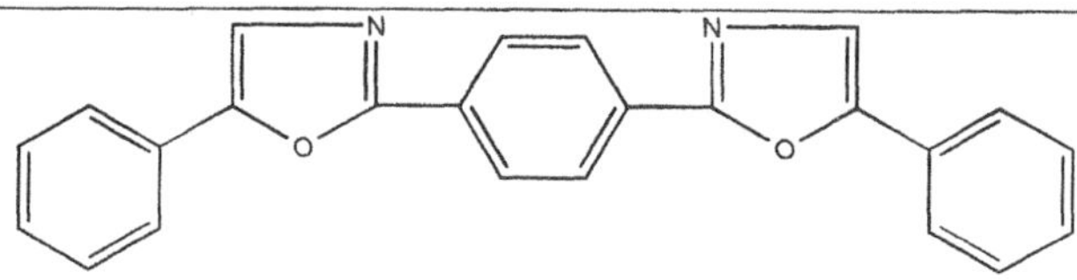
Pyrene	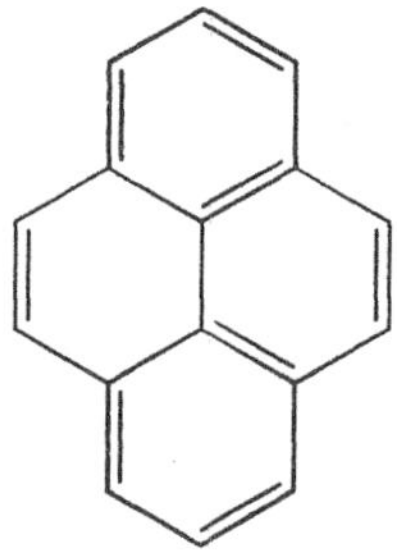
Pyridine	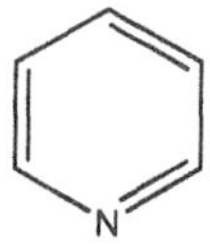
Pyrimidine	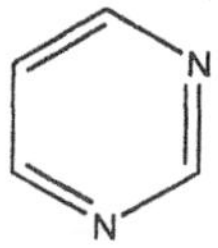

	1,3-diazine
p-Terphenyl	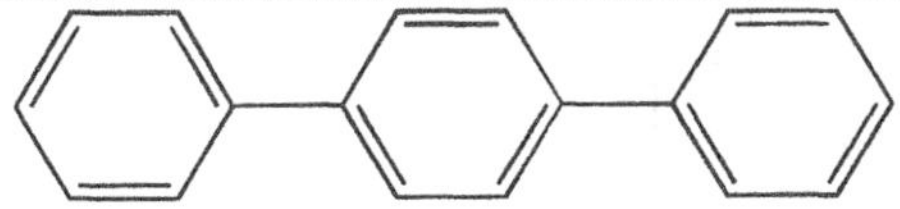
1,2,3,4 tetramethyl benzene	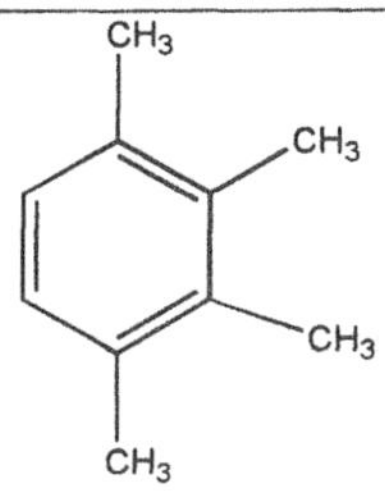
Thianthrene	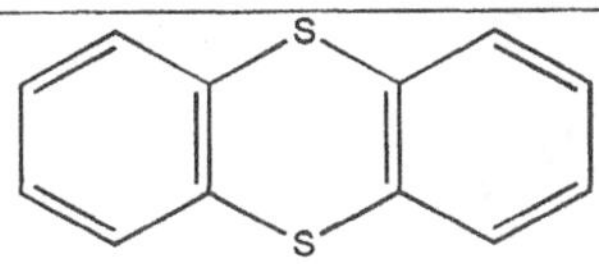
Toluene	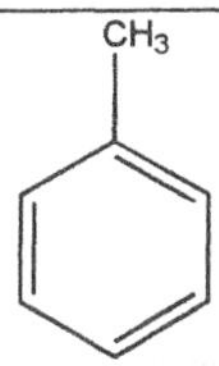
1,2,3 trimethyl benzene	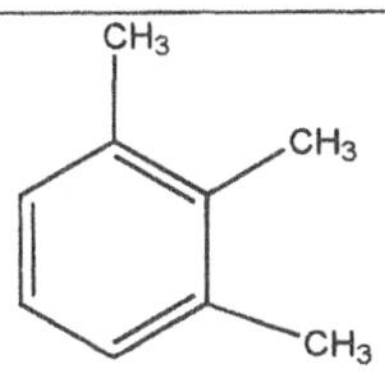

1,2,4 trimethyl benzene	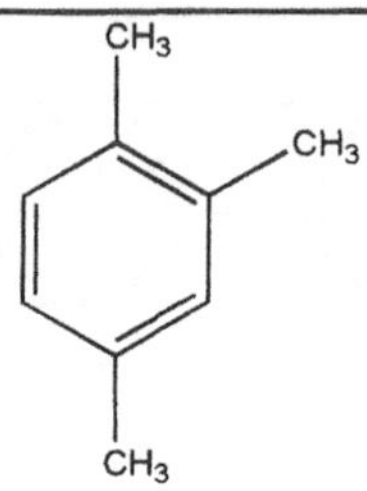
Triphenylene	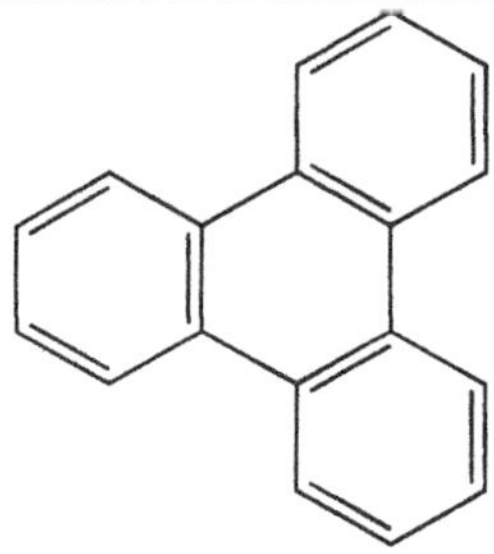
Xylene	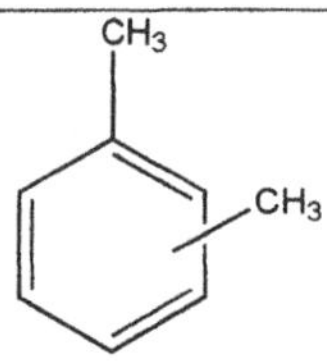

Silicon Phthalocyanine / MMA copolymer	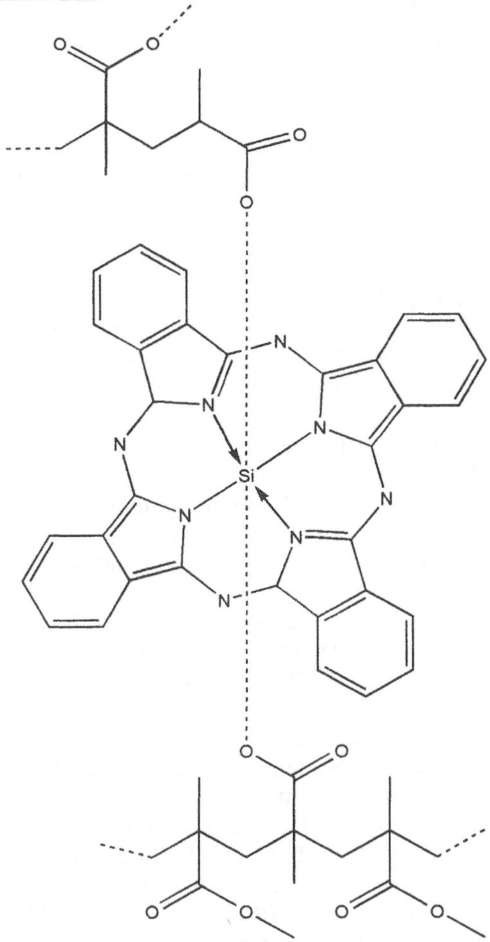

VII. ACKNOWLEDGMENTS

The author would particularly like to thank Drs. Steve Oliver and Clive Day for useful discussions and for proofreading this chapter, and BT Labs. for supporting the work. Thanks are also due to the several authors who contributed some of the figures and their most recent results and to Prof. John Griffiths for supplying the structures of some of the more venerable molecules listed in the tables.

REFERENCES

1. Collins, R. J., Nelson, D. F., Schawlow, A. L., Bond, W. L., Garrett, C. G. B., and Kaiser, W. K., Coherence, narrowing, directionality and relaxation oscillations in the light emissions from ruby, *Phys. Rev. Lett. 5*, 303 (1960).
2. Goeppert-Mayer, M., *Ann. Physik*, Uber elementarakte mit zwei quantensprungen, *9*, 273 (1931).
3. Kaiser, W., Garrett, C. G. B., Two-photon excitation in CaF_2:Eu^{2+}, *Phys. Rev. Lett.*, *7(6)*, 229–231 (1961).
4. Lee Smith, W., Two-Photon Absorption in Condensed Media, in *CRC Handbook of Laser Science and Technology* (M. J. Weber, ed.), CRC Press, Inc., Boca Raton, Florida, 1986, Volume III, Optical Materials Part 1: Nonlinear Optical Properties/Radiation Damage, Section 1.2.
5. Kershaw, S. V., Oliver, S. N., Manning, R. H. J., Rush, J. D., Hill, C. A. S., Underhill, A. E., and Charlton, A., Large refractive nonlinearities and two-photon absorption in aryl substituted dithiolenes, *SPIE*, *2025*, Nonlinear Optical Properties of Organic Materials VI, 388–399 (1993).
6. Pang, Y., Samoc, M., and Prasad, P. N., Third order nonlinearity and two-photon induced molecular dynamics: femtosecond time resolved transient absorption, Kerr gate, and degenerate four-wave mixing studies in poly (p-phenylene vinylene)/sol-gel silica films, *J. Chem. Phys.*, *94(8)*, 5282 (1991).
7. Townsend, P. D., Jackel, J. L., Baker, G. L., Shelburne, J. A., and Etemad, S., Observation of nonlinear optical transmission and switching phenomena in polydiacetylene-based directional couplers, *Appl. Phys. Lett.*, *55(18)*, 1829–1831 (1989).
8. Mizrahi, V., DeLong, K. W., Stegman, G. I., Saifi, M. A., and M. J., Andrejco, Two-photon absorption as a limitation to all-optical switching, *Opt. Letts.*, *14(20)*, 1140–1142 (1989).
9. Penzkofer, A., and Leupacher, W., S_0-S_1 two-photon absorption dynamics of organic dye solutions, *Opt. Quantum Electron.*, *19*, 327–349 (1987).
10. Tutt, L. W., and Boggess, T. F., A review of optical limiting mechanisms and devices using organics, fullerenes, semiconductors and other materials, *Prog. Quantum Electron.*, *17*, 299–338 (1993).
11. Preuss, S., Spath, M., Zhang, Y., and Stuke, M., Time resolved dynamics of subpicosecond laser ablation, *Appl. Phys. Lett.*, *62(23)*, 3049–3051 (1993).
12. Chen, C. H., and McCann, M. P., Two-photon induced chemical reactions in liquids, *AIP Conf. Proc.*, *191*, 354–356 (1989).
13. Dvornikov, A. S., and Rentzepis, P. M., Studies on 3D volume memory, *SPIE*, *1662*, Image storage and retrieval systems, 197–204 (1992).
14. Shen, Y. R., The principles of nonlinear optics, John Wiley, New York, 1984, Chapter 12.
15. Yuan, H. J., Li, L., and Palffy-Muhoray, P., Nonlinear birefringence of nematic liquid crystals, *SPIE*, *1307*, Electro-Optical Materials for Switches, Coatings, Sensor Optics and Detectors, 363–373 (1990).
16. McClain, W. M., *J. Chem. Phys.*, *55*, 2789 (1971).

17. Gelsen, O. M., Bradley, D. D. C., Murata, H., Takada, N., Tsutsui, T., and Saito, S., Spectroscopic investigation of the electro-optic nonlinearity in poly(2,5-thienylene vinylene), *J. Appl. Phys.*, *71*(*2*), 1064–1066 (1992).
18. Wiese, N., Eichler, H. J., and Salk, J., Nonlinear complex susceptibility of Cresyl Violet solution measured with a dynamic grating method, *IEEE J. Quantum Electron.*, *25*(*3*), 403–407 (1989).
19. Mansoor, K., Alvarez, D., Perry, K. J., Choong, I., Marder, S. R., and Perry, J. W., Dynamics of optical limiting in heavy-atom substituted phthalocyanines, *SPIE*, *1853*, Organic and Biological Optoelectronics, 132–141 (1993).
20. Lynch, R. T., and Lotem, H., Two-photon absorption measurements in organic liquids via nonlinear light mixing spectroscopy, *J. Chem. Phys.*, *66*(*5*), 1905–1913 (1977).
21. Chen, C. H., and McCann, M. P., Measurements of two-photon absorption cross sections for liquid benzene and methyl benzenes, *J. Chem. Phys.*, *88*(*8*), 4671–4677 (1988).
22. Van Stryland, E. W., Woodall, M. A., Vanherzeele, H., and Soileau, M. J., Energy band-gap dependence of two-photon absorption, *Opt. Lett.*, *10*, 490–492 (1985).
23. Guha, S., Kang, K., Porter, P., Roach, J. F., Remy, D. E., Aranda, F. J., and Rao, D. V. G. L. N., Third order optical nonlinearities of metallobenzoporphyrins and a platinum poly-yne, *Opt. Lett.*, *17*(*4*), 264–266 (1992).
24. Kershaw, S. V., Analysis of the E-Z Scan measurement technique, *J. Mod. Opt.*, *42*(*7*), 1361–1366 (1995).
25. Sheik-bahe, M., Said, A. A., Wei, T. H., Hagan, D. J., and Van Stryland, E. W., Sensitive Measurement of Optical Nonlinearities using a Single Beam, *IEEE J. Quantum Electronics*, *24*(*4*), 760–769 (1990).
26. Wei, T. H., Hagan, D. J., Spence, M. J., Van Stryland, E. W., Perry, J. W., and Coulter, D. R., Direct measurements of nonlinear absorption and refraction in solutions of phthalocyanines, *Appl. Phys. B.*, *B54*, 46–51 (1992).
27. Soileau, M. J., Wei, T. H., Sheik-bahae, M., Hagan, D. J., Spence, M., and Van Stryland, E. W., Nonlinear optical characterisation of organic materials, *Mol. Cryst. Liq. Cryst.*, *207*, 97–101 (1991).
28. Kliger, D. S., Thermal lensing: A new spectroscopic tool, *Acc. Chem. Res.*, *13*(*5*), 129–134 (1980).
29. Takatori, Y., Kajii, Y., Shibuya, K., and Obi, K., Dissociation of highly excited triplet benzophenone into phenyl radicals and carbon monoxide: determination of the reaction quantum yield and the heat of reaction by time resolved thermal lensing technique, *Chem. Phys.*, *180*, 99–107 (1994).
30. Dovichi, N. J., and Harris, J. M., presented at the ACS Analytical Division Summer Symposium, Purdue University, June 1979.
31. Foggi, P., Salvi, P. R., Bini, R., and Castellucci, E., Two-Photon Absorption of Liquid Pyridine: A study using Thermal Lensing and CARS Spectroscopy, *J. Molecular Structure*, *175*, 147–152 (1988).
32. Hopfield, J. J., Wolock, J. M., and Park, K., Two-quantum absorption spectrum of KI, *Phys. Rev. Lett.*, *11*(*9*), 414–417 (1963).
33. Lemmer, U., Fischer, R., Feldmann, J., Mahrt, R. F., Yang, J., Greiner, A., Bassler, H., Gobel, E. O., Heesel, H., and Kurz, H., Time-resolved studies of two-photon absorption processes in poly(p-phenylenevinylene)s, *Chem. Phys. Lett. 203*(*1*), 28–32 (1993).
34. Sinclair, M. B., McBranch, D., Hagler, T. W., and Heeger, A. J., Subpicosecond photoinduced absorption in poly[2,5-thienylene vinylen] and poly[3-methoxy-6-(2-ethylhexyloxy) phenylene vinylene], *Synthetic Metals*, *49–50*, 593–602 (1992).

35. Halle, S. D., Yoshizawa, M., Murata, H., Tsutsui, T., Saito, S., and Kobayashi, T., Ultrafast non-linear optical properties and photoinduced relaxation dynamics of poly(2,5-thienylenevinylene), *Synthetic Metals*, *49–50*, 429–438 (1992).
36. Pfeffer, N., Raimond, P., Charra, F., and Nunzi, J. M., Determination of the two-photon absorption spectrum of a soluble polythiophene, *Chem. Phys. Lett.*, *201*, 357–363 (1993). See also Beljonne, D., and Brèdas, J. L., Theoretical investigation of two-photon absorption and third harmonic generation in a long thiophene oligomer: α-heptathiophene, *J. Opt. Soc. Am. B. 11(8)*, 1380–1387 (1994), for a theoretical discussion of the two- and three-photon excited states of polythiophenes.
37. Chen, C. H., and McCann, M. P., Measurements of two-photon absorption cross sections of common blue dyes, *Opt. Commun.*, *63(5)*, 335–338 (1987).
38. Drucker, R. D., and McClain, W. M., *J. Chem. Phys.*, *61*, 2609 (1974).
39. Anderson, R. J. M., Holtom, G. R., and McClain, W. M., Two-photon absorptivities of the all trans α,ω-diphenylpolyenes from stilbene to diphenyloctatetraene via three wave mixing, *J. Chem. Phys.*, *70(9)*, 4310–4315, (1979).
40. Chance, R. R., Shand, M. L., Hogg, C., and Silbey, R., Three-wave mixing in conjugated polymer solutions: two-photon absorption in polydiacetylenes, *Phys. Rev. B.*, *22(8)*, 3540–3550 (1980).
41. Sutherland, R. L., Rea, E., Natarajan, V., Pottenger, T., and Fleitz, P. A., Two-photon absorption and second hyperpolarizability measurements in diphenylbutadiene by degenerate four wave mixing, *J. Chem. Phys.*, *98(4)*, 2593–2603, (1993).
42. Zhao, M., Cui, Y., Samoc, M., Prasad, P. N., Unroe, M. R., and Reinhardt, B. A., Influence of two-photon absorption on third order nonlinear optical processes as studied by degenerate four wave mixing: The study of soluble didecyloxy substituted polyphenyls, *J. Chem. Phys.*, *95(6)*, 3991–4001 (1991).
43. Fleitz, P. A., Sutherland, R. L., Natarajan, L. V., Pottenger, T., Fernelius, N. C., Effects of two-photon absorption on degenerate four-wave mixing in solutions of diphenyl polyenes, *Opt. Lett.*, *17(10)*, 716–718 (1992).
44. Maruani, A., *IEEE J. Quantum Electron.*, *QE-16*, 558 (1980).
45. Gass, P. A., Abram, I., Raj, R., and Schott, M., The nonlinear optical spectrum of polydiacetylene in the near infrared, *J. Chem. Phys.*, *100(1)*, 88–97 (1994).
46. Frazier, C. C., Guha, S., Porter, P. L., Cockerham, P. M., and Chauchard, E. A., Nonlinear optical properties of transition metal poly-ynes, *SPIE*, *971*, Nonlinear Optical Properties of Organic Materials, 186–196 (1988).
47. Shand, M. L., and Chance, R. R., Nonlinear optical properties of polydiacetylenes, in *Nonlinear optical properties of organic and polymeric materials* (D. J. Williams, ed.), ACS Symposium, *233*, 187–212 (1983).
48. Aramaki, S., Torruellas, W., Zanoni, R., and Stegeman, G. I., Tunable third harmonic generation of trans-β-carotene, *Optics Commun.*, *85*, 527–535 (1991).
49. Kajzar, F., Messier, J., and Rosilio, R., *J. Appl. Phys.*, *60*, 3040 (1986).
50. Hasegawa, T., Ishikawa, K., Koda, T., Takeda, K., Kobayashi, H., and Kubodera, K., Nonlinear spectroscopy on one-dimensional excitons in conjugated polymer diacetylenes, proceedings of International Topical Conference: "Optical Probes of Conjugated Polymers", 1991, published in *Synth. Metals*.
51. Hasegawa, T., Iwasa, Y., Kishida, H., Koda, T., Tokura, Y., Tachibana, H., and Kawabata, Y., Two-photon resonant third harmonic generation in polysilanes, *Phys. Rev. B.*, *45(11)*, 6317–6320 (1992); Hasegawa, T., Iwasa, Y., Kishida, H., Koda, T., Tokura, Y., Tachibana, H., and Kawabata, Y., Nonlinear optical properties of polysilanes, *Mol. Cryst. Liq. Cryst.*, *217*, 25–30 (1992).
52. Tam, A. C., and Patel, C. K. N., Two-photon absorption spectra and cross section measurements in liquids, *Nature*, *280*, 304 (1979).

53. Philip, R., Sathy, P., Nampoori, V. P. N., Philip, J., and Vallabhan, C. P. G., Characteristics of two-photon absorption in methanol solutions of Rhodamine 6G using laser pulsed photoacoustics, *J. Phys. B.*, *25*, 155–161 (1992).
54. Gelsen, O. M., Halliday, D. A., Bradley, D. D. C., Burn, P. L., Holmes, A. B., Murata, H., Tsutsui, T., and Saito, S., Electro-Absorption spectroscopy on poly(arylene vinylenes), *Mol. Cryst. Liq. Cryst.*, *216*, 117–121 (1992).
55. Tachibana, H., and Kawabata, Y., Electro-absorption spectroscopy of electronic structures in polysilanes, *Synth. Metals*, *41–43*, 1385–1388 (1991).
56. McBranch, D., Sinclair, M., Hays, A., Moses, D., Heeger, A. J., Picosecond waveguide modulation in conjugated polymers, *Synth. Metals*, *49–50*, 147–157 (1992).
57. See for examples not referred to elsewhere in Section V: Beljononne, D., and Brèdas, J. L., Theoretical investigation of two-photon absorption and third-harmonic generation in a long thiophene oligomer: α-heptathiophene, *J. Opt. Soc. Am. B*, *11(8)*, 1380–1387 (1994); Hasegawa, T., Ishikawa, K., Koda, T., Takeda, K., Kobayashi, H., and Kubodera, K., Analysis of excitonic resonant effect in the third-order nonlinear optical susceptibility of polydiacetylene films, *Synth. Met.*, *41–43*, 3151–3156 (1991); Kajzar, F., and Messier, J., Third harmonic generation and two-photon absorption in a polydiacetylene solution, *Phys. Rev. A*, *32*, 2352 (1985); Townsend, P. D., Fann, W. S., Etemad, S., Baker, G. L., Soos, Z. G., and McWilliams, P. C. M., Non-linear optical spectroscopy of correlated electrons in a one-dimensional semiconductor, *Chem. Phys. Letts.*, *180(5)*, 485–489 (1991).
58. Soos, Z. G., and Kepler, R. G., Two-photon absorption spectrum of poly(di-n-hexylsilane) films, *Phys. Rev. B*, *43(14)*, 11 908–11 912 (1991). See also Kepler, R. G., and Zoos, Z. G., Electronic excitations of poly(methylphenylsilane) films, *Phys. Rev. B*, *43(15)*, 12 530–12 537 (1991).
59. Orr, B. J., and Ward, J. F., *Mol. Phys.*, *20*, 513 (1971).
60. Kuzyk, M. G., and Dirk, C. W., *Phys. Rev. A*, *41*, 5098 (1990).
61. Cohen, D. M., Two-photon absorption studies in organic molecules, MSc. thesis, Washington State University, 1994.
62. Birge, R. R., Zhang, C. F., Two-photon double resonance spectroscopy of bacteriorhodopsin. Assignment of the electronic and dipolar properties of the low lying ${}^1A_g{}^{*-}$-like and ${}^1B_g{}^{*+}$-like π, π^* states, *J. Chem. Phys.*, *92(12)*, 7178 (1990).
63. Lo, S., Jenekhe, S. A., and Wellinghoff, S. T., Optical power limiting behaviour of novel nonlinear optical polymers, *SPIE*, *824*, Advances in Nonlinear Polymers and Inorganic Crystals, Liquid Crystals, and Laser Media, 162–170 (1987).
64. Vsevolodov, N. N., Kostikov, L. P., Kayushin, L. P., and Corbatenkov, V. I., Two-photon absorption of laser radiation by chlorophyll-a and certain organic dyes, *Biofizika* (*USSR*), *18*, 755 (1973); *Biophysics* (*GB*), *18*, 807 (1973).
65. Leupold, D., Mory, S., and Hoffmann, O., Nonlinear absorption of chlorophyll-a, *Acta Phys. Chem.*, *23(1)*, 33 (1977).
66. Leupold, D., Voight, B., Hoxtermann, E., Ehlert, J., Hoffmann, P., and Hieke, B., Nonlinear absorption of chlorophyll-a oligomers in dry n-hexane in comparison with chlorophyll in vivo, *Stud. Biophys.*, *83*, 41 (1981).
67. Nikogosyan, D. N., and Angelov, D. A., Production of free radicals in water by high-power laser UV radiation, *Doklady Akademii Nauk SSSR*, *253(3)*, 733 (1980).
68. Catalano, I. M., and Cingolani, A., Absolute determination of the two-photon cross section in NADH, *Opt. Commun.*, *32(1)*, 156 (1980).
69. Bachilo, S. M., and Bondarev, S. L., Spectral and polarisation features of two-photon absorption in retinal and retinyl acetate, *J. Appl. Spectrosc.*, *45(4)*, 1078 (1986).
70. Birge, R. L., Two-photon spectroscopy of protein-bound chromophores, *Acc. Chem. Res.* *19*, 138–146 (1986).

71. Birge, R. R., Bennett, J. A., Fang, H. L. -B., and Leroi, G. E., The two-photon spectroscopy of all trans retinal and related polyenes, *Advances in Laser Chemistry*, Springer-Verlag, Berlin 1978.
72. Birge, R. R., Bennett, J. A., Pierce, B. M., and Thomas, T. M., Two-photon spectroscopy of the visual chromophores. Evidence for a lowest excited $^1A_g^-$-like $\pi\pi^*$ state in all trans retinol (Vitamin A), *J. Am. Chem. Soc.*, *100(5)*, 1533 (1978).
73. Galanin, M. D., and Chizhikova, Z. A., Effective cross sections of two-photon absorption in organic molecules, *Sov. Phys. JETP Lett.*, *4*(2), 27 (1966).
74. Schafer, F. P., and Schmidt, W., Geometrical model and experiment verification of two-photon absorption in organic dye solutions, *IEEE J. Quantum Electron.*, *QE 2(9)*, 357 (1966).
75. Schafer, F. P., and Schmidt, W., Comments on two-photon absorption in organic dyes—relation with symmetry of the levels, *Opt. Commun.*, *17(1)*, 11–13 (1976).
76. Foucault, B., and Hermann, J. P., Two-photon absorption in organic dyes—relation with symmetry of the levels, *Opt. Commun.*, *15(3)*, 412 (1975).
77. Ducuing, J., Foucault, B., and Hermann, J. P., Two-photon absorption in organic dyes—reply to a comment, *Opt. Commun.*, *17(3)*, 267 (1976).
78. Bradley, D. J., Hutchinson, M. H. R., and Koetser, H., Interactions of picosecond laser pulses with organic molecules. II. Two-photon absorption cross-sections, *Proc. Roy. Soc. London A*, *329*, 105 (1972).
79. Hermann, J. P., and Ducuing, J., Absolute measurement of two-photon cross sections, *Phys. Rev. A*, *5(6)*, 2557 (1972).
80. Li, S., and Shi, C. Y., Two-photon absorption cross-section measurements in common laser dyes at 1.06 μm, *Opt. Acta*, *29(3)*, 281 (1982).
81. Topp, M. R., and Rentzepis, P. M., Picosecond stimulated emission in a fluorescent solution following two-photon absorption, *Phys. Rev. A*, *3(1)*, 358 (1971).
82. Planas, S. A., Duarte, A. S., Mazzali, C., Palange, E., Fragnito, H. L., Cardoso, V. L., Sanches, M. P. R., Villani, M. F., and Reggiani, A. E., Nonlinear refractive index of dye doped organic polymers, *SPIE*, *2025*, 381–387 (1993).
83. Parma, L., and Omenetto, N., Two-photon absorption of 7-hydroxycoumarine, *Chem. Phys. Lett.*, *54(3)*, 541 (1978).
84. Aslanidi, E. B., and Tikhonov, E. A., Two-photon absorption spectra of organic dye molecules, *Opt. Spect.*, *37(4)*, 446 (1974).
85. Piston, D. W., Kirby, M. S., Cheng, H., Lederer, W. J., and Webb, W. W., Two-photon-excitation fluorescence imaging of three-dimensional calcium-ion activity, *App. Opt.*, *33(4)*, 662–669 (1994).
86. Leupacher, W., and Penzkofer, A., Efficient phase-matched third harmonic light generation in hexafluoroisopropanol solutions of a pyrimidonecarbocyanine dye, *Appl. Phys. B*, *44*, 133–140 (1987).
87. Hermann, J. P., and Ducuing, J., Dispersion of the two-photon cross section in rhodamine dyes, *Opt. Commun.*, *6*(2), 101 (1972).
88. Rulliere, C., and Kottis, P., Two-photon spectroscopy of organic molecules recorded by the fluorescence induced by a picosecond light continuum generated in D_2O, *Chem. Phys. Lett.*, *75(3)*, 478 (1980).
89. Gaisenok, V. A., Damm, T., Slobodyanyuk, A. I., and Rench, S., Two-photon absorption of rhodamine 6G in the range of double resonance, *Opt. Spect.* (*USSR*), *64*(*1*), 33–36 (1988).
90. Sperber, P., and Penzkofer, A., S_0-S_n two-photon absorption dynamics of rhodamine dyes, *Opt. Quantum Electron.*, *18*, 281 (1986).
91. Catalano, I. M., and Cingolani, A., Absolute two-photon fluorescence with low power cw lasers, *Appl. Phys. Lett.*, *38(10)*, 745 (1981).

92. Soileau, M. J., Van Stryland, E. W., and Guha, S., Two-photon absorption and nonlinear refraction in isotropic liquid crystals, *SPIE*, *682*, Molecular and Polymeric Optoelectronic Materials: Fundamentals and Applications, 135–137 (1986).
93. Deeg, F. W., and Fayer, M. D., Analysis of complex molecular dynamics in an organic liquid by polarisation selective subpicosecond transient grating experiments, *J. Chem. Phys.*, *91(4)*, 2269–2279 (1989).
94. Said, A. A., Wamsley, C., Hagan, D. J., Van Stryland, E. W., Reinhardt, B. A., Roderer, P., and Dillard, A. G., *Chem. Phys. Letts. 228*, 646–650 (1994).
95. Shand, M. L., and Chance, R. R., Third order nonlinear mixing in polydiacetylene solutions, *J. Chem. Phys.*, *69(10)*, 4482 (1978).
96. Nunzi, J. M., and Grec, D., Picosecond phase conjugation in polydiacetylene gels, *J. Appl. Phys.*, *62(6)*, 2198–2202 (1987).
97. Pender, W. A., Boyle, A. J., Lambkin, P., Mazaheri, K., Westland, D. J., Skarda, V., and Sparpaglione, M., Nonlinear absorption in polydiacetylene waveguides, *Appl. Phys. Letts.*, *66(7)*, 786–788 (1995).
98. Rockford K., Zanoni, R., Stegeman, G. I., Krug, W., Miao, E., and Beranek, M. W., Measurement of nonlinear refractive index and transmission in polydiacetylene waveguides, *Appl. Phys. Lett. 58(1)*, 13 (1991).
99. Stegeman, G. I., Materials requirements for nonlinear third order phenomena in waveguides, in *Nonlinear Optics. Fundamentals, Materials and Devices*, (S. Miyata, ed.), Elsevier Science, North-Holland 1992, p. 337 et seq. See also Torruellas, W. E., Rochford, K. B., Zanoni, R., and Stegeman, G. I., *Opt. Commun.*, *82*, 94 (1991), and Stegeman, G. I., Torruellas, W., Rochford, K. B., Zanoni, R., Krug, W., Miao, E., and Beranek, M. W., Materials requirements for all-optical switching devices: Nonlinear properties of poly4-BCMU, Proc. MRS Spring Meeting, Anaheim, California, 1991.
100. Townsend, P. D., Fann, W. S., Etemad, S., Baker, G. L., Soos, Z. G., and McWilliams, P. C. M., Non-linear optical spectroscopy of correlated π-electrons in a one-dimensional semiconductor, *Chem. Phys. Lett. 180(5)*, 485–489 (1991).
101. Philppart, V., Dumont, M., Nunzi, J. M., and Charra, F., Femtosecond two-photon absorption and Raman resonances in polydiacetylene yellow solutions, *Synth. Metals*, *49–50*, 187–194 (1992).
102. Torruellas, W. E., Rochford, K. B., Zanoni, R., Aramaki, S., Stegeman, G. I., *Opt. Commun. 82*, 94 (1991).
103. Cha, M., Torruellas, W. E., Stegeman, G. I., Horsthuis, W. H. G., Mohlmann, G. R., and Meth, J., Two-photon absorption of di-alkyl-amino-nitro-stilbene side chain polymer, *Appl. Phys. Letts. 65(21)*, 2648–2650 (1994).
104. van Schoot, J. B. P., Characterisation and application of organic materials for all-optical integrated optic devices, Ph.D Thesis, Twente University, Netherlands, 1994.
105. Norwood, R. A., Holcomb, D. R., and So, F. F., Polymers for Nonlinear Optics: Absorption, Two-Photon Absorption and Photodegradation, *Nonlinear Optics*, *6*, 193–204 (1993).
106. Hein, J., Bergner, H., Lenzner, M., and Rentsch, S., Determination of real and imaginary part of $\chi^{(3)}$ of thiophene oligomers using the z-scan technique, *Chem. Phys.*, *179*, 543–548 (1994).
107. Tachibana, H., Matsumoto, M., Tokura, Y., Moritomo, Y., Yamaguchi, A., Koshihara, S., Miller, R. D., and Abe, S., Spectra of one-dimensional excitons in polysilanes with various backbone conformations, *Phys. Rev. B*, *47(8)*, 4363–4371 (1993).
108. Hasegawa, T., Iwasa, Y., Kishida, H., Koda, T., Tokura, Y., Tachibana, H., and Kawabata, Y., Two-photon resonant third harmonic generation in polysilane films, *Phys. Rev. B*, *45(11)*, 6317–6320 (1992).
109. Tachibana, H., Kawabata, Y., Koshihara, S., and Tokura, Y., Exciton states of polysilanes by electro-absorption spectra, *Solid. State Commun.*, *75(1)*, 5–9 (1990).

110. Guha, S., Frazier, C. C., Porter, P. L., Kang, K., and Finberg, S. E., Measurement of the third-order hyperpolarizability of platinum poly-ynes, *Opt. Lett.*, *14(17)*, 952–954 (1989).
111. Nakano, H., Ishida, Y., Yanagawa, T., and Matsumoto, N., Permanent grating induced by nonlinear absorption in polysilane films, *Appl. Phys. Lett.*, *57(27)*, 2876–2878 (1990).
112. Baker, C. J., Two-photon absorption in poly(p-phenylene vinylene), MSc. Thesis, University of Cambridge, U.K. 1993.
113. Reimer, B., Bassler, H., *Chem. Phys. Lett.*, *55*, 315 (1978).
114. Lawrence, B., Torruellas, W. E., Cha, M., Sundheimer, M. L., Stegeman, G. I., Meth, J., Etemad, S., and Baker, G., Identification and role of two-photon excited states in a π-conjugated polymer, *Phys. Rev. Letts.*, *73(4)*, 597–600 (1994).
115. Lequime, M., Herman, J. P., Reversible creation of defects by light in one dimensional conjugated polymers, *Chem. Phys. 26*, 431 (1977).
116. Thakur, M., and Quintero-Torres, R., The sign and magnitude of the off-resonant nonlinearities of polydiacetylene crystal measured by Z-scan, *SPIE*, *2025*, 446–449 (1993).
117. Kim, D. Y., Lawrence, B. L., Torruellas, W. E., Stegeman, G. I., Baker, G., and Meth, J., Assessment of single crystal p-toluene sulfonate as an all-optical switching material at 1.3 μm, *Appl. Phys. Letts. 65(14)*, 1742–1744 (1994).
118. Lawrence, B. L., Cha, M., Kang, J. U., Torruellas, W., Stegeman, G., Baker, G., Meth, J., and Etemad, S., Large purely refractive nonlinear index of single crystal p-toluene sulphonate (PTS) at 1600 nm, *Electron. Letts. 30(5)*, 447–448 (1994).
119. Lawrence, B. L., Cha, M., Torruellas, W. E., Stegeman, G. I., Etemad, S., Baker, G., and Kajzar, F., Measurement of the complex nonlinear refractive index of single crystal p-toluene sulphonate at. 1064 nm, *Appl. Phys. Letts.*, *64(21)*, 2773–2775 (1994).
120. Periasamy, N., Danieli, R., Ruani, G., Zamboni, R., and Taliani, C., Location of the low-energy ^1A_g state in a polythiophene oligomer by two-photon absorption spectroscopy: α-sexithienyl, *Phys. Rev. Lett.*, *68(7)*, 919–922 (1992).
121. Hasegawa, T., Iwasa, Y., Sunamura, H., Koda, T., Tokura, Y., Tachibana, H., Matsumoto, M., and Abe, S., Nonlinear optical spectroscopy on one-dimensional excitons in silicon polymer, polysilane, *Phys. Rev. Letts.*, *69(4)*, 668–671 (1992).
122. Winter, C. S., Manning, R. J., Oliver, S. N., and Hill, C. A. S., Measurement of the large optical nonlinearity of nickel dithiolene doped polymers, *Opt. Commun. 90*, 139–143, (1992).
123. Winter, C. S., Hill, C. A. S., and Underhill, A. E., Near resonance optical nonlinearities in nickel dithiolene complexes, *Appl. Phys. Lett. 58(2)*, 107–109 (1991).
124. Winter, C. S., Oliver, S. N., Rush, J. D., Manning, R. J., Hill, C., and Underhill, A., Third order near-resonance nonlinearities in dithiolenes and rare earth metallocenes, ACS Symposium Series, Chapter 41, *455*, "Materials for Nonlinear Optics", ACS, Washington, DC, 1991.
125. Kershaw, S. V., author's unpublished data.
126. Frazier, C. C., Chauchard, E. A., Cockerham, M. P., and Porter, P. L., Four wave mixing in metal poly-ynes, *Mater. Res. Soc. Symp. Proc.*, *109*, 323–328 (1988).
127. Xuan, N. G., Ferrier, J. L., Gazengel, J., Rivoire, G., Brekhovskikh, G. L., Kudriavtseva, A. D., Sokolovskaya, A. I., Tcherniega, N. V., Changes in the space structures of light beams induced by nonlinear optical phenomena: Application to phase contrast and image processing, *Opt. Commun.*, *68*, 244 (1988).
128. Chernega, N. V., Brekhovskikh, G. L., Kudryavtseva, A. D., Kirsanov, B. P., and Sokolovskaya, A. I., Transformation of amplitude-phase characteristics of focused laser radiation with the aid of nonlinear optical effects in acetone and liquid nitrogen, *Sov. J. Quantum Electron.*, *19(12)*, 1626–1631 (1989).

129. Giordmaine, J. A., and Howe, J. A., Intensity-induced optical absorption cross section in CS_2, *Phys. Rev. Lett. 11(5)*, 207 (1963).
130. Gedanken, A., and McDiarmid, R., Experimental one- and two-photon absorption properties of acetone, *J. Chem. Phys. 92(5)*, 3237–3238 (1990).
131. Rice, J. K., and Anderson, R. W., Two-photon thermal lensing spectroscopy of mono-substituted benzenes in $^1B_{2u}(^1L_b)$ – $^1A_{1g}(^1A)$ and $^1B_{1u}(^1L_a)$ – $^1A_{1g}(^1A)$ transition regions, *J. Chem. Phys., 90(26)*, 6793 (1986).
132. Aleksandrov, A. P., Bredikhin, V. I., and Genkin, V. N., Electron-vibrational nature of two-photon absorption in centrally-symmetrical organic molecules, *Sov. Phys. JETP Lett., 10*, 117 (1969).
133. Strome, F. C., Direct, two-photon photocarrier generation in anthracene, *Phys. Rev. Lett., 20(1)*, 3 (1968).
134. Nikumb, S. K., Hari, V., Inamdar, A. S., and Mehendale, N., Two-photon absorption in anthracene crystal with a nitrogen laser pumped dye laser, *Ind. J. Phys., 54B*, 100 (1980).
135. Bergman, A., and Jortner, J., Two-photon spectroscopy utilizing dye lasers, *Chem. Phys. Lett. 15(3)*, 309 (1972).
136. Kleinschmidt, J., Tottleben, W., and Rentsch, S., Measurements of two-photon absorption in the cavity of an organic dye laser, *Exp. Tech. Phys., 22*, 191 (1974).
137. McMahon, D. H., Soref, R. A., and Franklin, A. R., Quantitative measurements of double-photon absorption in the polycyclic benzene ring compounds, *Phys. Rev. Lett., 14(26)*, 1060 (1965).
138. Webman, I., Jortner, J., Energy dependence of two-photon absorption cross sections in anthracene, *J. Chem. Phys., 50(6)*, 2706 (1969).
139. Frohlich, D., and Mahr, H., Two-photon spectroscopy in anthracene, *Phys. Rev. Lett., 16(20)*, 895 (1966).
140. Aleksandrov, A. P., and Bredikhin, V. I., Measurement of the absolute value of the cross section for two-photon absorption in anthracene molecules, *Sov. Phys. Opt. Spect., 30(1)*, 37 (1971).
141. Kepler, R. G., Two-photon transitions to highly excited states in anthracene crystals, *Phys. Rev B, 9*, 4468 (1974).
142. Kaiser, W., and Maier, M., in *Laser Handbook* (F. T. Arecchi, and E. O. Schulz-Dubois, eds.) *2*, 1077, North-Holland, Amsterdam, 1972.
143. Blokhin, A. P., Povedalio, V. A., and Tolkachev, V. A., Polarisation of two-photon excited fluorescence of vapours of complex organic molecules, *Opt. Spect., 60(1)*, 37 (1986).
144. Bergman, A., and Jortner J., Consecutive two-photon absorption of azulene in solution utilizing dye lasers, *Chem. Phys. Lett., 20(1)*, 8–10 (1973).
145. McGraw, D. J., Michaekson, J., and Harris, J. M., Anharmonic forced Rayleigh scattering: A technique for study of saturated absorption in liquids, *J. Chem. Phys., 86(5)*, 2536 (1987).
146. Prior, Y., and Vogt, H., Measurements of UV two-photon absorption relative to known Raman cross sections, *Phys. Rev. B, 19*, 5388 (1979).
147. Fredrich, D. M., Van Alsten, J., and Walters, M. A., Polarization and concentration dependence of two-photon forbidden electronic origins: the $^1B_{2u}$ of benzene in solution, *Chem. Phys. Lett., 76(3)*, 504 (1980).
148. Twarowski, A. J., and Kliger, D. S., Multiphoton absorption spectra using thermal blooming, *Chem. Phys., 20*, 259 (1977).
149. Bree, A., Taliani, C., and Thirunamachandran, T., The polarised two-photon excitation spectrum of benzene monocrystals at 6 eV, *Chem. Phys., 56*, 123 (1981).
150. Razumova, T. K., and Starobogatov, I. O., Two-photon spectroscopy of liquid benzene and toluene, *Opt. Spect. (USSR), 49(6)*, 654 (1980).

151. Scott, T. W., Braun, C. L., and Albrecht, A. C., Multiphoton ionization in the organic condensed phase: a three photon study of liquid benzene, *J. Chem. Phys.*, *76(11)* 5195 (1982).
152. Hochstrasser, R. M., Meredith, G. R., and Trommsdorff, H. P., Resonant four wave mixing in molecular crystals, *J. Chem. Phys.*, *73(3)*, 1009 (1980).
153. Hochstrasser, R. M., Wessel, J. E., and Sung, H. N., Two-photon excitation spectrum of benzene in the gas phase and the crystal, *J. Chem. Phys.*, *60(1)*, 317 (1974).
154. Hochstrasser, R. M., Sung, H. N., and Wessel, J. E., Moderate resolution study of the two-photon spectrum of the $^1B_{2u}$ state of benzene-h_6 and benzene-d_6, *24(1)*, 7 (1974).
155. Levenson, M. D., and Bloembergen, N., Dispersion of the nonlinear optical susceptibilities of organic liquids and solutions, *J. Chem. Phys.*, *60*, 1323 (1974).
156. Zheng, B., Lin, M., Zhang, B., and Chen, W., Study of two-photon absorption cross section by multiphoton ionization spectroscopy, *Opt. Commun.*, *73(3)*, 208–212 (1989).
157. Boissel, P., Hanchercorne, G., Kerherve, F., and Meyer, G., *J. Phys. Lett.*, *39*, 319 (1978).
158. Monson, P. R., and McClain, W. M., *J. Chem. Phys.*, *53*, 29 (1970).
159. Wunsch, L., Neusser, H. J., and Schlag, E. W., *Chem. Phys. Lett.*, *31*, 633 (1975).
160. Hochstrasser, R. M., Meredith, G. R., and Trommsdorff, H. P., *J. Chem. Phys.*, *73*, 1009 (1980).
161. Gelwachs, G., and Wessel, J. F., *Opt. Eng.*, *16*, 197 (1977).
162. Krasinski, J., Majewski, W., and Glodz, M., Investigation of the two-photon absorption absorption spectrum of 3,4-benzopyrene by means of a tunable dye laser, *Opt. Commun.*, *14(2)*, 187 (1975).
163. Drucker, R. P., and McClain, W. M., Two-photon absorption in "double molecules", *Chem. Phys. Lett.*, *28(2)*, 255 (1974).
164. Drucker, R. P., and McClain W. M., Polarized two-photon studies of biphenyl and several derivatives, *J. Chem. Phys.*, *61(7)*, 2609 (1974).
165. Hochstrasser, R. M., Sung, H. N., and Wessel, J. E., High resolution two-photon excitation spectra, *J. Chem. Phys.*, *58(10)*, 4694 (1973).
166. Castellucci, E., and Salvi, P. R., and Foggi, P., Two-photon excitation spectra of the lowest electronic states of 2,2′-bipyridine, *Chem. Phys.*, *66*, 281 (1982).
167. Kennedy, S. M., Lytle, F. E., p-bis(o-methylstyryl)benzene as a power-squared sensor for two-photon absorption measurements between 537 and 694 nm, *Anal. Chem.*, *58(13)*, 2643 (1986).
168. Lotem, H., and Lynch, R. T., Destructive interference of imaginary resonant contributions to $\chi^{(3)}$, *Phys. Rev. Lett.*, *37(6)*, 334 (1976).
169. Maker, P. D., and Terhune, R. W., Study of optical effects due to an induced polarisation third order in the electric field strength, *Phys. Rev.*, *137(3A)*, A801 (1965).
170. Heumann, E., Kapp, I., Rentsch, S., and Triebel, W., Polarization dependent picosecond two-photon absorption spectra of diphenyl and chloronapthalene in solution, *Zh. Prikl. Spektros.*, *34(3)*, 481 (1981).
171. Eisenthal, K. B., Dowley, M. W., and Peticolas, W. L., Two-photon spectrum of a liquid, *Phys. Rev. Lett.*, *20(3)*, 93 (1968).
172. Liem, N. Q., Marconi, G., and Salvi, P. R., Two-photon vibronic transitions in crystalline chrysene, *Chem. Phys.*, *144*, 425–435 (1990).
173. Lotem, H., and de Araujo, C. B., Absolute determination of the two-photon absorption coefficient relative to the inverse Raman cross section, *Phys. Rev. B*, *16(4)*, 1711 (1977).
174. Singh, I. B., Rai, S. B., and Rai, D. K., Two-photon spectrum of O-dichlorobenzene, *J. Mol. Spectrosc.*, *163*, 364–370 (1994).

175. Bree, A., Leyderman, A., and Taliani, C., Competition between second harmonic generation and one- and two-photon absorption in the anthracene/ 9,10-dihydroanthracene mixed crystal, *J. Appl. Phys.*, *62(12)*, 4662–4667 (1987).
176. Bennett, J. A., and Birge, R. R., Two-photon spectroscopy of diphenylbutadiene. The nature of the lowest lying $^1A_g^{*-}$ $\pi\pi^*$ state, *J. Chem. Phys.*, *73(9)*, 4234 (1980).
177. Swofford, R. L., and McClain, W. M., New two-photon absorption spectrometer and its application to diphenylbutadiene, *Rev. Sci. Instrum.*, *46(3)*, 246 (1975).
178. Swofford, R. L., and McClain, W. M., *J. Chem. Phys.*, *59*, 5740 (1973).
179. Fang, H. L., Gustafson, T. L., and Swofford, R. L., *J. Chem. Phys.*, *78*, 1663 (1983).
180. Fang, H. L. -B., Thrash, R. J., and Leroi, G. E., Observation of the low energy ^1A_g state of diphenylhexatriene by two-photon excitation spectroscopy, *Chem. Phys. Lett.*, *57(1)*, 59 (1978).
181. Vasudev, R., and Brand, J. C. D., The two-photon excitation spectrum of fluorobenzene vapour, *J. Mol. Spect.*, *75*, 288 (1979).
182. Ravi Kumar, A. V., Padmaja, G., Nampoori, V. P. N., and Vallabhan, C. P. G., Two-photon absorption of formaldehyde using gas phase photoacoustic spectroscopy, in *Springer series in Optical Sciences* (J. C. Murphy, J. W. Maclachian-Spicer, L. Aamodt, and B. S. H. Royce, eds.), *62*, "Photoacoustic and photothermal phenomena II.", 360–362, Springer-Verlag, Berlin, 1990.
183. Twarowski, A. J., and Kliger, D. S., A search for a low lying excited ^1A state in 1,3,5-hexatriene, *Chem. Phys. Lett.*, *50(1)*, 36 (1977).
184. Nitti, S., Tan, H. M., Banfi, G., Degiorgio, V., Bailey, R. T., Cruickshank, F., Pugh, D., Shepherd, E. A, Sherwood J. N., and Simpson, G. S., Nonlinear susceptibility and damage thresholds of organic MBA-NP crystals, *J. Phys. D*: *Appl. Phys.*, *26*, B225–B229 (1993).
185. Penzkofer, A., and Falkenstein, W., Direct determination of the intensity of picosecond light pulses by two-photon absorption, *Opt. Commun.*, *17(1)*, 1–5 (1976).
186. Kleinschmidt, J., and Torpatschow, P., Experimental investigation of two-photon absorption in naphthalene, *Wiss. Z.*, *24*, 439 (1975).
187. Logothetis, E. M., Two-photon photoelectric spectroscopy in CsI, *Phys. Rev. Lett.*, *19*(26), 1470 (1967).
188. Mikami, N., and Ito, M., Two-photon excitation spectra of naphthalene and naphthalene-d_8, *Chem. Phys. Lett.*, *31(3)*, 472 (1975).
189. Galanin, M. D., and Chizikova, Z. A., Two-photon dichroism in nitrobenzene, *Sov. Phys. JETP Lett.*, *5*, 300 (1967).
190. Fuke, K., Nagakura, S., and Kobayashi, T., Two-photon absorption spectrum of (2,2)-paracyclophane, *Chem. Phys. Lett.*, *31*(2), 205 (1975).
191. Callis, P. R., Scott, T. W., and Albrecht, A. C., Polarized two-photon fluorescence excitation studies of pyrimidine, *J. Chem. Phys.*, *75(12)*, 5640 (1981).
192. Stachelek, T. M., Pazoha, T. A., and McClain, W. M., Detection and assignment of the "phantom" photochemical singlet of trans-stilbene by two-photon excitation, *J. Chem. Phys.*, *66(1)*, 4540 (1977).
193. Kleinschmidt, J., Rentsch, S., Tottleben, W., and Wilhelmi, B., Measurement of strong nonlinear absorption in stilbene-chloroform solutions, explained by the superposition of two-photon absorption and one-photon absorption from the excited state, *Chem. Phys. Lett.*, *24(1)*, 133 (1974).
194. Fuke, K., Sakamoto, S., Ueda, S., and Itoh, M., Two-photon absorption spectrum of trans stilbene. Trans-cis photoisomerisation via upper ^1A_g state, *Chem. Phys. Lett.*, *74(3)*, 546 (1980).
195. Gorelik, V. S., Sokolovskaya, A. I., Chernega, N. V., and Shcheglov, V. A., Stimulated two-photon excited luminescence in stilbene crystals, *Quantum Electron.*, *23(6)*, 505–507 (1993).

196. Katoh, R., and Kotani, M., Autoionization of higher excited states generated by two-photon absorption in p-terphenyl crystals, *Chem. Phys. Lett.*, *188(1,2)*, 80–84 (1992).
197. Samoc, M., Samoc, A., and Williams, D. F., Multiphoton processes of charge carrier generation in thianthrene crystals, *J. Chem. Phys.*, *76(7)*, 3768 (1982).
198. Chen, C. H., and McCann, M. P., Measurements of two-photon absorption cross-sections for liquid benzene and methyl benzenes, *J. Chem. Phys.*, *88*, 4671 (1988).
199. Worlock, J. M., *Laser Handbook*, (F. T. Arecchi, and E. O. Schulz-Dubois, eds.), North Holland, New York, 1972, Vol. 2, Chapter 3, p. 1364.
200. Fredericq, E., and Houssier, C., *Electric Dichroism and Electric Birefringence*, Clarendon Press, Oxford, 1973.

8
Z-Scan

Eric W. Van Stryland
CREOL (Center for Research and Education in Optics and Lasers)
University of Central Florida, Orlando, Florida

Mansoor Sheik-Bahae
University of New Mexico, Albuquerque, New Mexico

I. INTRODUCTION

There is considerable interest in finding materials having large yet fast nonlinearities. This interest, driven primarily by the search for materials for all-optical switching and sensor protection applications, concerns both nonlinear absorption (NLA) and nonlinear refraction (NLR). The database for nonlinear optical properties of materials, particularly organic, is in many cases inadequate for determining trends to guide synthesis efforts. Thus, there is a need to expand this database. Methods to determine nonlinear coefficients are discussed throughout this book. The Z-scan technique is a method which can rapidly measure both NLA and NLR in solids, liquids, and liquid solutions.[1,2] In this Chapter, we first present a brief review of this technique and its various derivatives. Simple methods for data analysis are then discussed for "thin" and "thick"[3,4,5,6] nonlinear media Z-scans, eclipsing Z-scan (EZ-scan),[7] two-color Z-scans,[8,9] time-resolved excite-probe Z-scans,[10,11] and top-hat-beam Z-scans.[12] Finally, an overview of the reported measurements of the nonlinear optical properties of organic materials as determined using these techniques will be presented.

The Z-scan method has gained rapid acceptance by the nonlinear optics community as a standard technique for separately determining the nonlinear changes in index and changes in absorption. This acceptance is primarily due to the simplicity of the technique as well as the simplicity of the interpretation. In most experiments the index change, Δn, and absorption change, $\Delta\alpha$, can be determined directly from the data without resorting to computer fitting. However, it must always be recognized that this method is sensitive to all nonlinear optical mechanisms that give rise to a change of the refractive index and/or absorption coefficient, so that determining the underlying physical processes present from a Z-scan is not in general possible. A series of Z-scans at varying pulsewidths, frequencies, focal geometries, etc., along with a variety of other experiments are often needed to unambiguously determine the relevant mechanisms. In this regard, we caution the reader that the conclusions as to the active nonlinear processes of any given reference using the Z-scan technique is often subject to debate.

II. TECHNIQUE AND SIMPLE RELATIONS

The standard "closed aperture" Z-scan apparatus (i.e., aperture in place in the *far field*) for determining nonlinear refraction is shown in Fig. 1. The transmittance of the sample through the aperture is monitored in the far field as a function of the position, Z, of the nonlinear sample in the vicinity of the linear optics focal position. The required scan range in an experiment depends on the beam parameters and the sample thickness L. A critical parameter is the diffraction length, Z_0, of the focused beam defined as $\pi w_0^2/\lambda$ for a Gaussian beam where w_0 is the focal spot size (half-width at the $1/e^2$ maximum in the irradiance). For "thin" samples (i.e., $L \leq n_0 Z_0$ where n_0 is the linear index), although all the information is theoretically contained within a scan range of $\pm Z_0$, it is preferable to scan the sample by at least $\pm 5Z_0$. This requirement, as we shall see, simplifies data interpretation when the sample's surface roughness or optical beam imperfections introduce background "noise" into the measurement system. In many practical cases where considerable laser power fluctuations may occur during the scan, a reference detector can be used to monitor and normalize the transmittance (see Fig. 1). To eliminate the possible noise due to spatial beam fluctuations, this reference arm can be further modified to include a lens and an aperture identical to those in the nonlinear arm.[9] The position of the

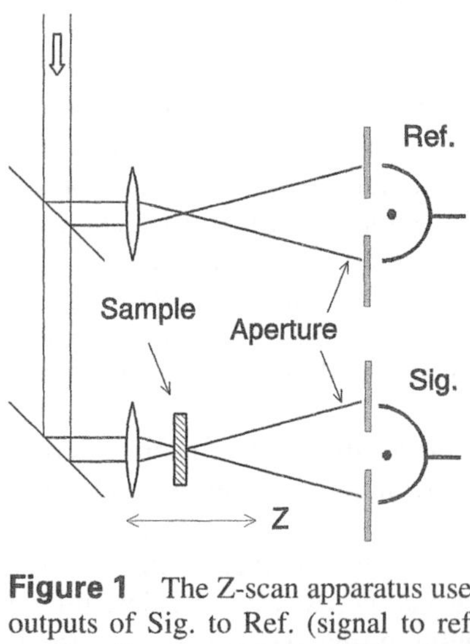

Figure 1 The Z-scan apparatus used to reduce the noise by monitoring the ratio of detector outputs of Sig. to Ref. (signal to reference). "Open aperture" Z-scans are obtained by removing the apertures (or disks for EZ-scan) shown in front of the signal and reference detectors and carefully collecting all of the transmitted light.

aperture is rather arbitrary as long as its distance from the focus, $d >> Z_0$. Typical values range from $20Z_0$ to $100Z_0$. The size of the aperture is signified by its transmittance, S, in the linear regime, i.e., when the sample has been placed far away from the focus at low energy. In most reported experiments, $0.1 < S < 0.5$ has been used for determining nonlinear refraction. Obviously, the $S = 1$ case corresponds to collecting all the transmitted light and therefore is insensitive to any nonlinear beam distortion due to nonlinear refraction. Such a scheme, referred to as an "open aperture" Z-scan, is suited for measuring nonlinear absorption ($\Delta\alpha$) in the sample. We will discuss this in more detail later.

Figure 2 shows a typical closed aperture Z-scan output for a thin sample exhibiting nonlinear refraction. For example, a self-focusing nonlinearity, $\Delta n > 0$ (solid line in Fig. 2), results in a valley followed by a peak in the normalized transmittance as the sample is moved away from the lens in Fig. 1 (increasing Z). The intensity is normalized so that the transmittance is unity for the sample far from focus where the nonlinearity is negligible (i.e., for $|Z| >> Z_0$). The positive lensing in the sample when it is placed before the focus moves the focal position closer to the sample resulting in a greater far field divergence and a reduced aperture transmittance. On the other hand, with the sample placed after focus, the

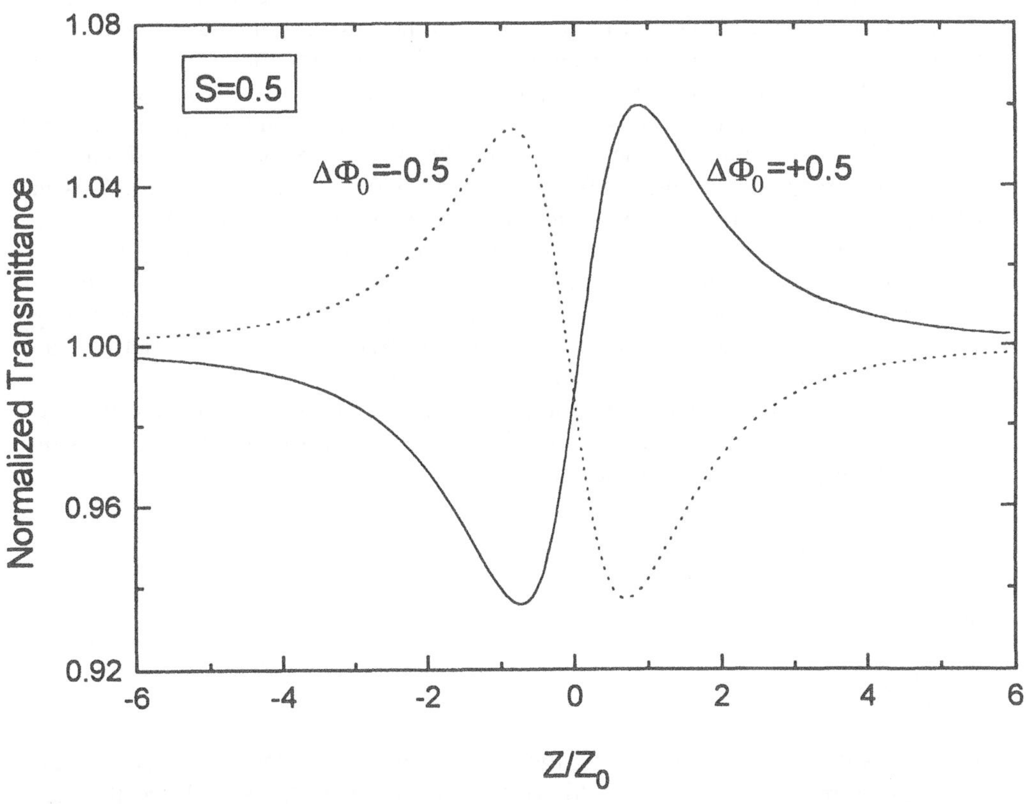

Figure 2 A typical Z-scan for positive (solid line) and negative (dashed line) third order nonlinear refraction.

same positive lensing reduces the far field divergence allowing for a larger aperture transmittance. The opposite occurs for a self-defocusing nonlinearity, $\Delta n < 0$ (Fig. 2, dashed line).

One of the attractive features of the Z-scan technique is the ease and simplicity by which the nonlinear optical coefficients can be determined with a high degree of accuracy. However, as is the case with most nonlinear optical measurement techniques, the measured quantities are the time-averaged nonlinearly induced $\langle \Delta n \rangle$ and/or $\langle \Delta \alpha \rangle$, where $\langle \quad \rangle$ denotes the time-average over a time corresponding to the temporal resolution of the detection system. Accurate determination of the nonlinear coefficients such as n_2 or β (defined in Eqs. 3 and 4) depend on how precisely the laser source is characterized in terms of its temporal and spatial profiles, power or energy content and stability. Furthermore, because different mechanisms respond on different time scales, time-averaged experiments often measure several competing mechanisms so that the results can depend strongly on the temporal profile of the laser pulse.

Once a specific type of nonlinearity is assumed (e.g., an ultrafast $\chi^{(3)}$ response), a Z-scan can be rigorously modeled for any beam shape and sample thickness by solving the appropriate Maxwell's equations. However, a number of valid assumptions and approximations will lead to simple analytical expressions, making data analysis easy yet precise. Aside from the usual SVEA (slowly varying envelope approximation), a major simplification results when we assume the nonlinear sample is "thin" so that neither diffraction nor nonlinear refraction cause any change of beam profile within the nonlinear sample. This implies that $L << n_0 Z_0$ and $L << Z_0/\Delta\Phi_0$, respectively, where $\Delta\Phi_0$ is the maximum nonlinearly induced phase change. The latter requirement assures "external self-action" and simply states that the effective focal length of the induced nonlinear lens in the sample should be much larger than the sample thickness.[13] In most experiments using the Z-scan technique, we find that this second criterion is automatically met since $\Delta\Phi_0$ is small. In Section IV we will analyze thick sample Z-scans ($L > Z_0$) and show that, for small enough phase distortions, simple expressions can still be derived for the Z-scan transmittance. Additionally, we will show that for the sample to be safely regarded as "thin", the first criterion for linear diffraction is more restrictive than it need be, and the condition $L < n_0 Z_0$ is sufficient.

The external self-action limit simplifies the problem considerably, and the amplitude $\sqrt{I}$ and phase $\Delta\phi$ of the electric field E are now governed in the SVEA by the following pair of simple equations:

$$\frac{d\Delta\phi}{dz'} = \frac{2\pi}{\lambda} \Delta n(I) \tag{1}$$

and

$$\frac{dI}{dz'} = -\alpha(I)I, \tag{2}$$

where z' is the propagation depth in the sample and $\alpha(I)$ in general includes linear and NLA terms. Note that z' should not be confused with the sample position Z.

For third order nonlinearities we take

$$\Delta n = \left\{ \frac{n_2}{2} |E|^2 \right\}_{\text{Gaussian}} = \{n_2 I\}_{\text{MKS}} \tag{3}$$

and

$$\Delta \alpha = \beta I, \tag{4}$$

where n_2 is the nonlinear index of refraction, E is the peak electric field (in Gaussian units), and I denotes the irradiance (in MKS units) of the laser beam within the sample. Here, β denotes the third order nonlinear absorption coefficient, which for ultrafast NLA is equal to the two-photon absorption (2PA) coefficient. n_2(Gaussian) and n_2(MKS) are related through the conversion formula, $n_2(\text{esu}) = (cn_0/40\pi)n_2(\text{MKS})$, where c (m/sec) is the speed of light in vacuum. We note, however, that while we are using n_2 here for *any* third order nonlinearity, it may not be the best description of cumulative nonlinearities. These occur, for example, in reverse saturable absorbing (RSA) dyes.[14] In such dyes, linear absorption is followed by excited state absorption (ESA) where the excited state cross-section is larger than the ground state cross-section. As the resulting change in absorption is best described by a cross-section and not by a two-photon absorption coefficient, the index change, here due to population redistribution, is better described by refractive cross-sections than by an n_2. Such an "n_2" (or β) would change with the laser pulsewidth.[15,16] This is discussed in more detail in Section V.

Once the amplitude and the phase of the beam exiting the sample are known, the field distribution at the far field aperture can be calculated using diffraction theory (Huygen's principle). We will briefly review this procedure in Section III for a Gaussian beam. Simple analytical or empirical relationships as obtained from those rigorous treatments will be presented in this Section. In most practical cases, these relationships present a convenient yet accurate method for estimating the nonlinear coefficients. In the remainder of this Chapter, n_2 always refers to n_2(MKS).

A. Nonlinear Refraction ($\Delta\alpha = 0$)

We define the change in transmittance between the peak and valley in a Z-scan as $\Delta T_{pv} = T_p - T_v$, where T_p and T_v are the normalized peak and valley transmittances as seen in Fig. 2. The empirically determined relationship between the induced phase distortion, $\Delta\Phi_0$, and ΔT_{pv} for a third order nonlinear refractive process in the absence of NLA is

$$\Delta T_{pv} \cong 0.406(1 - S)^{0.27}|\Delta\Phi_0|, \tag{5}$$

where

$$\Delta\Phi_0 = \frac{2\pi}{\lambda} n_2 I_0 L_{eff}, \tag{6}$$

with, $L_{eff} = (l - e^{-\alpha L})/\alpha$, and S is the transmittance of the aperture in the absence of a sample. $\Delta\Phi_0$ and I_0 are the on-axis ($r = 0$), peak ($t = 0$) nonlinear phase shift and the irradiance with the sample at focus ($Z = 0$), respectively. The sign of $\Delta\Phi_0$, and hence n_2, is determined from the relative positions of the peak and valley with

Z as shown in Fig. 2. Equation 5 is accurate to within $\pm 3\%$ for $\Delta T_{pv} < 1$. As an example, if the induced optical path length change due to the nonlinearity is $\lambda/250$, $\Delta T_{pv} \approx 1\%$ for an aperture transmittance of $S = 0.4$. Use of $S = 0.4$ is a good compromise between having a large signal which averages possible beam non-uniformities, thus reducing background signals, and reasonably high level of sensitivity.

A useful feature of the Z-scan trace is that the Z-distance between peak and valley, ΔZ_{pv}, is a direct measure of the diffraction length of the incident beam for a given order nonlinear response. In a standard Z-scan (i.e., using a Gaussian beam and a far field aperture), this relationship for a third order nonlinearity is given by:

$$|\Delta Z_{pv}| \approx 1.71 Z_0. \tag{7}$$

For small $\Delta\Phi_0$, peak and valley are equidistant ($\approx \pm 0.856 Z_0$) from the focus ($Z = 0$). As $\Delta\Phi_0$ increases, the peak and valley positions do not remain symmetric; the valley moves toward the focus and the peak away, but ΔZ_{pv} remains nearly constant as given above. Being independent of the irradiance, Eq. 7 is quite helpful in estimating Z_0, and hence the beam waist w_0, of the focused beam. We must re-emphasize that Eq. 7 is valid only for closed aperture Z-scans involving an n_2-type nonlinearity, a good quality Gaussian beam ($M^2 \approx 1$), and thin nonlinear samples. Any departure from these conditions will give rise to a different characteristic ΔZ_{pv}. Later, we briefly discuss cases involving thick samples, $\chi^{(5)}$-type nonlinearities, eclipsing Z-scans and top-hat-beam Z-scans.

The linear relationship between ΔT_{pv} and $\Delta\Phi_0$ makes it convenient to include a time-averaging factor for pulsed inputs which is not included in Eqs. 5 and 6. Inclusion of this temporal averaging reduces the measured ΔT_{pv} by a factor, A_τ, which generally depends on the pulse shape and the response time of the nonlinearity. For nonlinearities with response times much shorter than the pulsewidth (i.e., instantaneous nonlinearities), A_τ is given by[2]:

$$A_\tau = \frac{\int_{-\infty}^{+\infty} f^2(t)\, dt}{\int_{-\infty}^{+\infty} f(t)\, dt}, \tag{8}$$

where $f(t)$ denotes the dimensionless temporal profile of the incident laser pulse. For a Gaussian temporal shape, this gives $A_\tau = 1/\sqrt{2}$ while a $Sech^2$ pulse gives $A_\tau = 2/3$. In the other extreme, where the time response of the nonlinearity is much larger than the pulsewidth, $A_\tau = 1/2$ for a third order nonlinearity and is independent of the pulse shape.[2] Of course, in this case, the interpretation of n_2 changes. For example, in the case of reverse-saturable absorbers and under the approximations discussed in Ref. [16], $n_2 I_0$ is replaced by $(2\pi/\lambda)\alpha\sigma_r F(t)/\hbar W$, where σ_r is the excited state refractive cross-section and $F(t)$ is the time integral of the irradiance up to time, t, in the pulse (fluence). Cases involving higher order nonlinearities, and/or with response times that are comparable to the pulsewidth, require proper averaging of $\Delta\Phi_0(t)$ according to a generalization of Eq. 8, and will not be discussed here.

B. Higher Order Nonlinearities

Although many observed nonlinear optical effects give index changes proportional to the irradiance ($\Delta n \propto I$), we often encounter higher order effects where $\Delta n \propto I^{\eta}$, with $\eta > 1$. For example, a fifth order NLR, (a $\chi^{(5)}$-type nonlinearity where $\eta = 2$) becomes the dominant mechanism in semiconductors when Δn is induced by two-photon generated free-carriers.[17] For this type of nonlinearity, where $\Delta n = n_4 I^2$ is assumed, we can derive simple relations that accurately characterize the Z-scan data. For a Gaussian beam and far field aperture, these are given by:

$$\Delta T_{pv} \cong 0.21(1 - S)^{0.27}|\Delta\Phi_0|, \tag{9}$$

and

$$|\Delta Z_{pv}| \approx 1.2 Z_0, \tag{10}$$

where $\Delta\Phi_0 = kn_4 I_0^2 L'_{eff}$ with $L'_{eff} = [1 - \exp(-2\alpha L)]/2\alpha$. In certain cases where competing $\chi^{(3)}$ and $\chi^{(5)}$ processes are simultaneously involved, the data analysis becomes more complicated. A procedure is given in the literature for separating the two processes using a number of Z-scans at different irradiances.[17] This procedure makes use of simple relations of Eqs. 5 and 9 to estimate the nonlinear coefficients associated with both $\chi^{(3)}$ and $\chi^{(5)}$ processes.

C. Eclipsing Z-scan (EZ-scan)

As the Z-scan method relies on propagation of a phase distortion to produce a transmittance change, the minimum detectable signal is determined by how small a transmittance change can be measured. The surprising interferometric sensitivity comes about from the interference (diffraction) of different portions of the spatial profile in the far field. Recently, it was realized that this sensitivity could be greatly increased by looking at the outer edges of the beam in the far field rather than the central portion as in the Z-scan. This is accomplished by replacing the apertures in Fig. 1 with disks that block the central part of the beam. The light that leaks around the edges appears as an eclipse, thus the name EZ-scan for eclipsing Z-scan.[7] An analogous empirical expression to Eq. 9 for the EZ-scan technique is

$$\Delta T_{pv} \cong 0.68(1 - S)^{-0.44}|\Delta\Phi_0|, \tag{11}$$

which is accurate to within $\pm 3\%$ for $|\Delta\Phi_0| \leq 0.2$ and a disk linear transmittance rejection S in the range $0.98 > S > 0.995$, i.e., the fraction of light seen by the detector is $1 - S$. Figure 3 shows a comparison of an EZ-scan with a Z-scan for a phase distortion of 0.1 radian. The relative positions of peak and valley switch from the Z-scan since light that is transmitted by an aperture is now blocked by the disk and vise-versa. Evident from the above relation, as $S \to 1$ (large disks), the sensitivity increases significantly. Sensitivities to optical path length changes of $\cong\lambda/10^4$ have been demonstrated as compared to $\cong\lambda/10^3$ for Z-scan. For the range of S given above, the spacing between peak and valley, ΔZ_{pv}, is empirically found to be given by $\Delta Z_{pv} \approx 0.9$–$1.0 Z_0$, which grows to the Z-scan value of $\approx 1.7 Z_0$ as $S \to 0$. The enhancement of sensitivity in the EZ-scan, however, comes at the

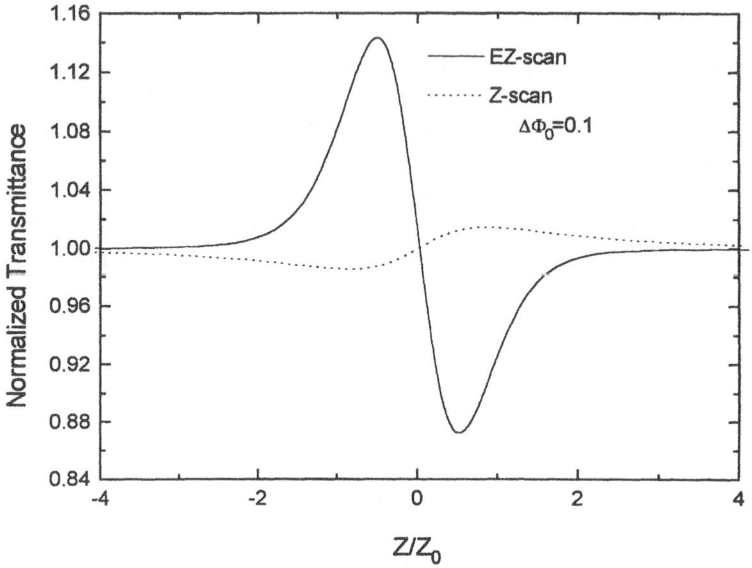

Figure 3 A comparison of EZ-scan data ($S = 0.98$) and Z-scan data ($S = 0.02$) for a self-focusing nonlinearity.

expense of signal photons as well as a reduction in accuracy and absolute calibration capability. This added uncertainty originates from the deviations of the actual laser beams from a Gaussian distribution, and the fact that we need to know S very accurately. We therefore recommend using this technique only when the added sensitivity is required and when a known reference sample is available to calibrate the system.

D. Nonlinear Absorption

While NLA can be determined using a two-parameter fit to a closed aperture Z-scan (i.e., fitting for both Δn and $\Delta\alpha$), it is more directly (and more accurately) determined in an open aperture Z-scan. For small third order nonlinear losses, i.e., $\Delta\alpha L = \beta I L_{\text{eff}} << 1$ with response times much less than the pulsewidth (e.g., two-photon absorption), and for a Gaussian temporal shape pulse, the normalized change in transmitted energy $\Delta T(=T(Z) - 1)$ becomes

$$\Delta T(z) \approx -\frac{q_0}{2\sqrt{2}}\frac{1}{[1 + Z^2/Z_0^2]}, \tag{12}$$

where $q_0 = \beta I_0 L_{eff}$ ($|q_0| << 1$). This mimics the Lorentzian distribution of the irradiance with Z for a focused Gaussian beam as seen in Fig. 4. If the response time of the material is much longer than the laser's temporal pulsewidth, the factor $2\sqrt{2}$ is replaced by 2. This is independent of the temporal pulse shape. Of course, in this case, the interpretation of β changes. For example, in the case of reverse-saturable absorbers and under the approximations discussed in Ref. [16], βI_0 is replaced by $\alpha\sigma F/2\hbar\omega$, where σ is the excited state absorption cross-section.

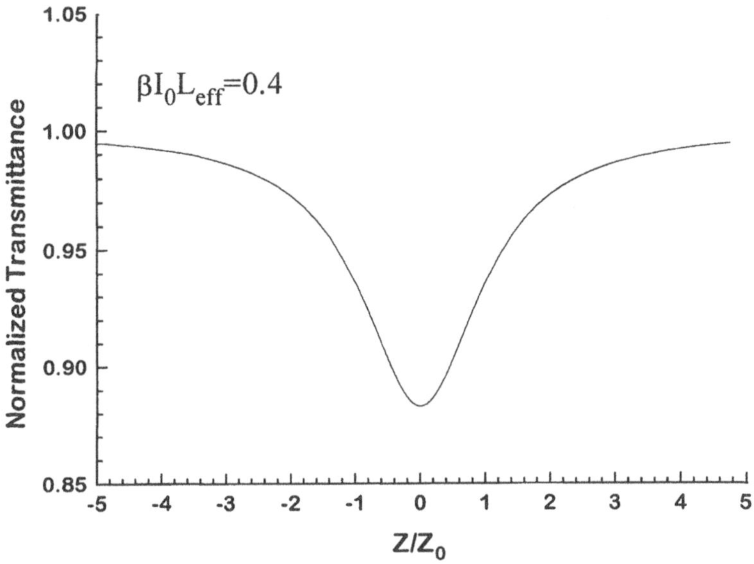

Figure 4 A typical open aperture Z-scan signal for third order nonlinear absorption.

E. Nonlinear Refraction in the Presence of Nonlinear Absorption ($\Delta\alpha \neq 0$)

We can also determine NLR in the presence of NLA. This can be done by fitting the Z-scan with a two-parameter fit or by separately measuring the NLA in a Z-scan performed with the aperture removed (i.e., open aperture Z-scan). This second method is more accurate since two single-parameter fits give a higher accuracy than one two-parameter fit. Within approximations elaborated in Ref. [2] (primarily that the Z-scan is not dominated by nonlinear absorption), a simple division of the curves obtained from the two Z-scans (closed/open) gives a curve that closely approximates what would be obtained with a closed aperture Z-scan on a material having the same Δn but with $\Delta\alpha = 0$. This greatly simplifies determining Δn. An example of this division process is shown in Fig. 5. In lieu of this division, with $\Delta\alpha$ known from the open aperture results, the Z-scan with aperture in place ($S < 1$) can be used to extract the remaining unknown, namely Δn.[17]

F. Excite-probe Z-scans

Excite-probe techniques in nonlinear optics have been commonly employed in the past to deduce information that is not accessible with a single beam geometry. The most significant application of such techniques concerns the ultrafast dynamics of the nonlinear optical phenomena. There has been a number of investigations that have used Z-scan in an excite-probe scheme.[8,9,10,11] The general geometry is shown in Fig. 6 where collinearly propagating excitation and probe beams are used. After propagation through the sample, the probe beam is then separated and analyzed through the far field aperture. Due to collinear propagation of the excitation and probe beams, we are able to separate them only if they differ in wavelength or

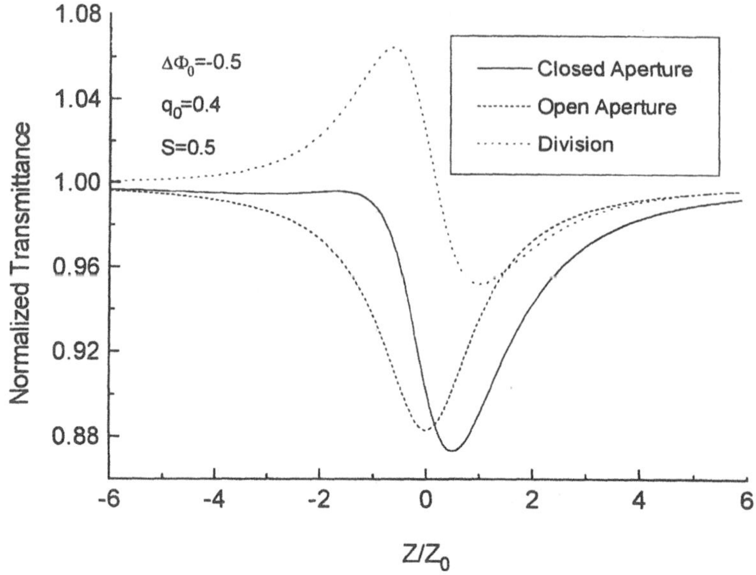

Figure 5 Calculations of closed and open aperture Z-scan data along with their ratio for self-defocusing accompanied by nonlinear absorption.

polarization. The former scheme, known as a 2-color Z-scan, has been used to measure the nondegenerate n_2 and β in semiconductors.[8,9] The time-resolved studies can be performed in two fashions. In one scheme, Z-scans are performed at various fixed delays between excitation and probe pulses. In the second scheme, the sample position is fixed (e.g., at the peak or the valley positions) while the transmittance of the probe is measured as the delay between the two pulses is varied. The analysis of the 2-color Z-scans is naturally more involved than that of a single beam Z-scan. The measured signal, in addition to being dependent on the parameters discussed for the single beam geometry, will also depend on parameters such as the excite-probe beam waist ratio, pulsewidth ratio and the possible focal separation due to chromatic aberration of the lens.[8,10]

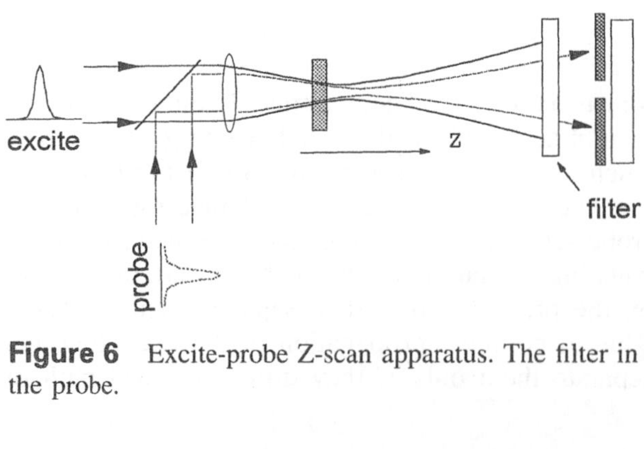

Figure 6 Excite-probe Z-scan apparatus. The filter in front of the detector transmits only the probe.

G. Z-scans with Non-Gaussian Beams

While Gaussian beams are extremely convenient since their propagation is particularly simple (e.g., a Gaussian beam remains Gaussian throughout a linear optical system in the absence of aberrations), the output of many lasers do not possess a Gaussian profile in space. Zhao and Palffy-Muhoray[12] derived the results of performing a Z-scan using a focused "top-hat" beam, where the profile at the initial focusing lens is approximately a step function (Heaviside function) in the radial coordinate r (i.e., $\Theta(r_0 - r)$ with r_0 a constant). In practice, one can produce this type of beam profile by sufficiently expanding any spatially coherent optical beam, and using a circular aperture to make the top-hat profile at the focusing lens. The lens focuses this beam to an Airy pattern in the absence of aberrations. The empirical expression relating ΔT_{pv} to $\Delta\Phi_0$ and aperture transmittance S is given by[12]:

$$\Delta T_{pv} \approx 2.8(1 - S)^{1.14} \tanh(0.37\ \Delta\Phi_0), \tag{13}$$

where $\Delta\Phi_0$ is the peak nonlinear phase shift at the center of the Airy disk at the focal plane. For $S \approx 0$ and small $\Delta\Phi_0$, the above expression gives $\Delta T_{pv} \approx 1.036\Delta\Phi_0$, indicating approximately a 2.5 times larger sensitivity than for a Gaussian beam Z-scan. This enhanced sensitivity is due to the steeper beam curvature gradients encountered by the nonlinear sample at Z positions near the focal plane.

It is also possible to use a sample of known nonlinearity as a reference to calibrate a system using a beam of arbitrary profile. Bridges and coworkers[18] show a way to use a reference sample to obtain the relative NLA and that is independent of the laser beam characteristics. This also allows violation of the thin sample approximation as long as the reference sample has the same thickness as the sample under measurement, and the irradiance is adjusted such that the ΔT_{pv}'s in both measurements are nearly equal. More generally, Z-scans using reference sample calibration are useful provided that the order of magnitude of each nonlinearity is the same (e.g., both are $\chi^{(3)}$ type) and conditions and parameters of both experiments are kept nearly the same.

H. Background Subtraction

In all the situations discussed above, it is often beneficial to perform experiments at high and low irradiance levels (low enough that the nonlinear response is negligible) and subtract the two sets of data.[2] This greatly reduces background signals due, for example, to sample inhomogeneities or sample wedge. A necessary condition for this background subtraction process to be effective is that the sample position be reproducible for both high and low irradiance scans (i.e., laterally, vertically, and along Z). It is also important that the data sets be normalized before subtraction such that $T(|Z| >> Z0)$ are made equal for high and low irradiance Z-scans. Experience shows that even when the signal is indistinguishable in a background that this subtraction can often uncover a usable signal.[2]

III. ANALYSIS OF Z-SCAN FOR A THIN NONLINEAR MEDIUM

While the above analysis gives the dependence of NLA on the sample position Z, the analysis for NLR was restricted to ΔT_{pv}. The Z dependence for NLR can be obtained by straightforward numerical techniques as outlined below.[2] We find it useful to fit the full Z dependence of the Z-scan signals since there is information regarding the order of the nonlinearity in this Z dependence as discussed in Section I.

The irradiance distribution and phase shift of the beam at the exit surface, I_L and $\Delta\phi$, of a sample exhibiting a third order nonlinear refractive index are obtained by simultaneously solving Eqs. 1 and 2:

$$I_e(Z,r,t) = \frac{I(Z,r,t)\exp^{-\alpha L}}{1 + q(Z,r,t)}, \tag{14}$$

and

$$\Delta\phi(Z,r,t) = \frac{kn_2}{\beta}\ln[1 + q(Z,r,t)], \tag{15}$$

where $q(Z,r,t) = \beta I(Z,r,t)L_{eff}$ and $I(Z,r,t)$ is the irradiance at the input face of the sample. Combining Eqs. 14 and 15, we obtain the complex field, E_e, at the exit surface of the sample to be[2,4]:

$$E_e = E(Z,r,t)e^{-\alpha L/2}(1 + q)^{(ikn_2/\beta - 1/2)}, \tag{16}$$

where $E(Z,r,t)$ is the incident electric field. The reflection losses can be safely assumed to be linear and hence will be ignored in this formalism. In evaluating the nonlinear coefficients, however, one should account for reflection loss of the first surface by taking the irradiance (i.e., I_0) to be that inside the sample.

In general for radially symmetric systems, a zeroth order Hankel transform of Eq. 16 will give the field distribution E_a at the aperture which is placed a distance d from the focal plane:

$$E_a(Z,r,t,d) = \frac{2\pi}{i\lambda d'}\exp\left(\frac{i\pi r^2}{\lambda d'}\right)\int_0^\infty r'dr'\, E_e(Z,r',t - d'/c) \exp\left(\frac{i\pi r'^2}{\lambda d'}\right) J_0\left(\frac{2\pi r r'}{\lambda d'}\right), \tag{17}$$

where $d' = d - Z$ is the distance from the sample to the aperture plane. The measured quantity is the pulse energy or average power transmitted through the far field aperture having a radius of r_a. The normalized transmittance is then obtained as:

$$T(Z) = \frac{\int_{-\infty}^{\infty} dt \int_0^{r_a} |E_a(Z,r,t,d)|^2\, r\, dr}{U}, \tag{18}$$

where U is the same as the numerator but in the linear regime (i.e., for $I \approx 0$). In the case of an EZ-scan, the limits of the spatial integral in Eq. 18 must be replaced by r_d to ∞ where r_d is the radius of the obscuration disk. It is generally more convenient to represent the aperture (or disk) size by the normalized transmittance (or rejection) S in the linear regime.

The formalism thus far presented is generally applicable to any radially symmetric beam. Here, however, we assume a $TEM_{0,0}$ Gaussian distribution for the incident beam as given by:

$$E(Z,r,t) = E_0(t)\frac{w_0}{w(Z)}\exp\left(-\frac{r^2}{w^2(Z)} + i\frac{\pi r^2}{\lambda R(Z)} + i\phi\right), \tag{19}$$

where $w(Z) = w_0(1 + Z^2/Z_0^2)^{1/2}$ and $R(z) = Z + Z_0^2/Z$. The radially invariant phase terms, contained in ϕ, are immaterial to our calculations, and hence will be ignored.

The integral in Eq. 17 can be analytically evaluated if we assume that $|q| < 1$ (in Eq. 16) and then perform a binomial series expansion of E_e in powers of q. Recalling that $q \propto I \propto \exp(-r^2/w^2)$, this expansion effectively decomposes E_e into a sum of Gaussian beams with varying beam parameters. This method of beam propagation known as Gaussian decomposition was first given by Wearie et al.[19] Following the expansion, we obtain:

$$E_e = E(Z,r,t)e^{-\alpha L/2}\sum_{m=0}^{\infty} F_m \exp\left(2mr^2/w^2(Z)\right), \tag{20}$$

where F_m, the factor containing the nonlinear optical coefficients, is given by:

$$F_m = \frac{(i\Delta\phi_0(Z,t))^m}{m!}\prod_{j=1}^{m}\left[1 + i\left(j - \frac{1}{2}\right)\frac{\lambda\beta}{2\pi n_2}\right], \tag{21}$$

with $F_0 = 1$ and $\Delta\phi_0(Z,t) = \Delta\phi(Z,r = 0,t)$ denoting the on-axis instantaneous nonlinear phase shift. The Hankel transform of E_e will then give the field at the aperture plane as a sum of Gaussian beams:

$$E_a(r,t) = E(Z,r = 0,t)e^{-\alpha L/2}\sum_{m=0}^{\infty} F_m \frac{w_{m0}}{w_m}\exp\left[-\frac{r^2}{w_m^2} + \frac{i\pi r^2}{\lambda R_m} + i\theta_m\right], \tag{22}$$

where the beam parameters of each Gaussian beam are as follows:

$$w_{m0}^2 = \frac{w^2(Z)}{2m + 1}, \qquad d_m = \frac{kw_{m0}^2}{2},$$

$$w_m^2 = w_{m0}^2\left[g^2 + \frac{d^2}{d_m^2}\right], \qquad R_m = d\left[1 - \frac{g}{g^2 + d^2/d_m^2}\right]^{-1},$$

and

$$\theta_m = \tan^{-1}\left[\frac{g}{d/d_m}\right],$$

where

$$g = 1 + d/R(Z). \tag{23}$$

Finally, the normalized transmittance can then be evaluated as given by Eq. 18. We should note here that with an incident Gaussian beam, the aperture transmittance can be given as $S = 1 - \exp(-2r_a^2/w_a^2)$ where $w_a = w_0(1 + d^2/Z_0^2)^{1/2}$ is the linear beam radius at the aperture.

It is worthwhile to analyze the implications of the above results under a number of further approximations. In the absence of nonlinear absorption (i.e., $\beta = 0$), $F_m = (i\Delta\phi_0(Z,t))^m/m!$ and the far field beam deformation will be the result of external self-action (self-focusing and self-defocusing). In that case we can write $\Delta\phi_0$ as:

$$\Delta\phi_0(Z,t) = \frac{2\pi}{\lambda} L_{eff} n_2 I(Z,t) = \Delta\Phi_0 \frac{f(t)}{1 + Z^2/Z_0^2}, \tag{24}$$

where $\Delta\Phi_0 = \Delta\phi_0(0,0)$ is the peak-on-axis nonlinear phase shift, and $f(t)$ represents the irradiance temporal profile of the incident pulse. We find that this Gaussian decomposition method is very useful for the small phase distortions detected with the Z-scan (or EZ-scan) method since only a few terms of the sum in Eq. 22 are needed. Figure 2 depicts calculated Z-scans for $\Delta\Phi_0 = \pm 0.5$ using the above formalism. The simple relations (Eqs. 9 and 11) given earlier were obtained by empirically fitting the calculated results using the equations derived in this Section. Such empirical relations are exact in the limit of small $\Delta\Phi_0$ where only one nonlinear term in the expansion (Eq. 22) is retained.[2]

With NLA present (i.e., $\beta \neq 0$), although no restriction is imposed on the magnitude of $\Delta\Phi_0$, the above formalism is valid only for $q_0 = |\beta I_0 L_{eff}| < 1$. Note that the coupling factor $\lambda\beta/2\pi n_2 = q_0/\Delta\Phi_0$ in Eq. 21 is twice the ratio of the imaginary to real parts of the third order nonlinear susceptibility, $\chi^{(3)}$ (i.e., $q_0/2\Delta\Phi_0 = \mathrm{Im}\{\chi^3\}/\mathrm{Re}\{\chi^3\}$). The 2PA figure-of-merit (FOM) for all-optical switching has been defined as 4π times this value.[20] Since the irradiance and effective length cancel in this ratio, this FOM can be deduced for third order nonlinearities without knowledge of the irradiance or sample length as long as the thin sample approximation is valid.

The Z-scan transmittance variations can be calculated following the same procedure as described previously. As is evident from Eqs. 21 to 22, the absorptive and refractive contributions to the far field beam profile, and hence to the Z-scan transmittance, are coupled. When the aperture is removed, however, the Z-scan transmittance is insensitive to beam distortion and is only a function of the nonlinear absorption. The total transmitted fluence (i.e., energy per unit area) in that case ($S = 1$) can be obtained by spatially integrating Eq. 14 without having to include the free-space propagation process. The resultant normalized transmittance for a pulse with a temporal profile $f(t)$ is[2]:

$$T(Z,S=1) = \frac{1 + Z^2/Z_0^2}{q_0} \frac{\int_{-\infty}^{+\infty} \ln\left[1 + q_0 \frac{f(\tau)}{1 + Z^2/Z_0^2}\right] d\tau}{\int_{-\infty}^{+\infty} f(\tau)d\tau}. \tag{25}$$

For $|q_0| < 1$, this transmittance can be expressed in terms of the peak irradiance in a summation form more suitable for numerical evaluation. Assuming a Gaussian temporal profile (i.e., $f(\tau) = \exp(-\tau^2)$), this can be written as:

$$T(Z,S=1) = \sum_{m=0}^{\infty} \frac{(-q_0)^m}{(1 + Z^2/Z_0^2)^m (m+1)^{3/2}}. \tag{26}$$

Thus, once an open aperture ($S = 1$) Z-scan is performed, β can be unambiguously deduced. With β known, the Z-scan with aperture in place ($S < 1$) can be used to extract the remaining unknown, namely, the coefficient n_2. Note that ΔT, as given by Eq. 12 in Section I, is simply the $m = 1$ term in Eq. 26.

IV. Z-SCAN FOR "THICK" SAMPLES

It is apparent from relations derived so far that a way to obtain larger Z-scan signals (ΔT_{pv}) is to increase $\Delta\Phi_0$ through either stronger focusing (shorter Z_0) or thicker samples (larger L). In either case, we recall that the validity of these relations becomes questionable once the *thin* sample criterion ($L << n_0Z_0$) is violated. In this Section we address this problem and analyze a general case in which no limitation is imposed on the sample length. The rigorous treatment of this problem involves numerical solutions to nonlinear wave equations and will not be discussed here.[6] In addition to numerical calculations, two types of approximate solutions, resulting in simple relationships, have been reported. One involves an "aberration-free" approximation of the nonlinear wave equation,[3,21] and the other treats the wave propagation exactly to first order in the nonlinear phase shift ($\Delta\phi$).[4,5] The latter approach requires that $\Delta\phi$ is small enough that no nonlinear beam distortion (self-action) occurs within the sample, although linear diffraction does occur. This condition is controllable and can be satisfied in an experiment.

Following Hermann and McDuff,[4] and Tian et al.,[5] the on-axis ($S \approx 0$) Z-scan transmittance of a *thick* nonlinear sample can be written as:

$$T(x) \approx 1 + \Delta\Phi_{z0}F(x,l), \tag{27}$$

where $\Delta\Phi_{z0} = (2\pi/\lambda)n_2I_0Z_0$ is the nonlinear phase shift occurring over the distance Z_0, $x = Z/Z_0$, and $l = L/(n_0Z_0)$ is the normalized length of the sample. $F(x,l)$ is given by[4]

$$F(x,l) = \frac{1}{4}\ln\left(\frac{[(x + l/2)^2 + 1][(x - l/2)^2 + 9]}{[(x - l/2)^2 + 1][(x + l/2)^2 + 9]}\right). \tag{28}$$

Plots of $F(x,l)$ are shown in Fig. 7 for $l = 1, 2, 5, 8$ and 10. The analysis here is strictly correct for $n_0 = 1$, but does not account for the longitudinal shift of the linear focus for $n_0 > 1$. Since no useful information (regarding the nonlinear optical measurement) exists in the absolute position of the transmittance peak and valley, this analysis is sufficient and simple. The position of peak and valley are obtained by evaluating $dF/dx = 0$, which gives:

$$X_{p,v} = \pm\sqrt{\frac{(l^2/2 - 10) + \sqrt{(l^2 + 10)^2 + 108}}{6}}. \tag{29}$$

The peak-valley separation, therefore, is given by $\Delta Z_{pv} = 2|X_{p,v}|Z_0$. As evident from Fig. 8 which shows $L_{eff}/(n_0Z_0)$ as a function of $L/(n_0Z_0)$, this separation approaches L/n_0 for $L >> n_0Z_0$.[3] All the above Equations reduce to the thin sample results of Section III when $l \rightarrow 0$. Moreover, as one would expect, it is seen in Fig. 8 that by increasing the sample thickness above $\approx 2n_0Z_0$, the signal (ΔT_{pv}) gradually levels off and ultimately becomes a constant. A useful quantity that illustrates this effect is the effective length of a thick nonlinear medium which is defined as the length that can be attributed to the sample if it were to be regarded as thin in the data analysis. Once we identify such an effective length, we can use the "thin" sam-

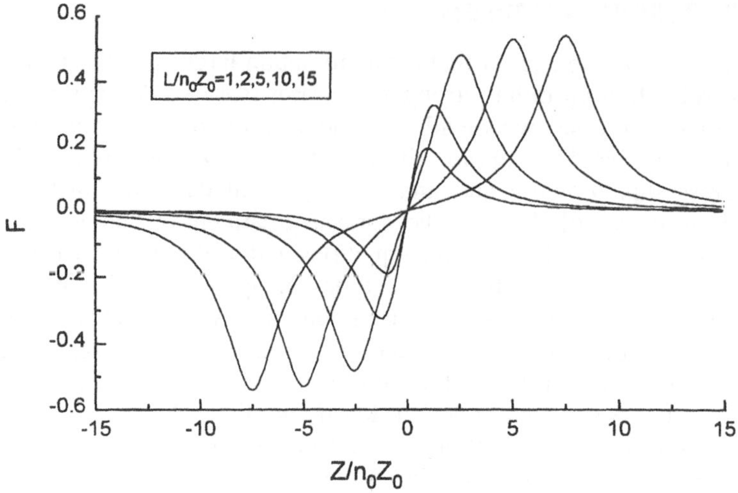

Figure 7 Normalized change in transmittance, F (see Eq. 28) for thick samples with L/n_0Z_0 =1, 2, 5, 8, 10 for $\Delta\Phi_{z0}$ = 0.01, as a function of Z/n_0Z_0.

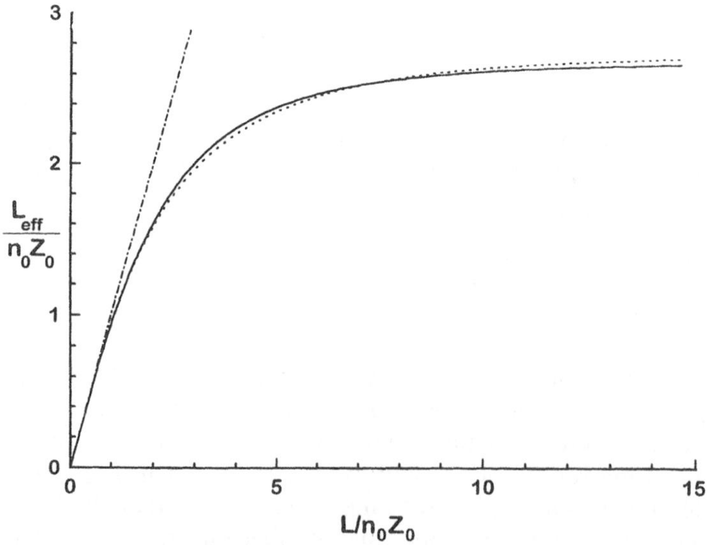

Figure 8 The effective sample length, l_{eff} (L_{eff}, in units of n_0Z_0) as a function of $l = L/n_0Z_0$ (solid line) and empirical fit to this line (dotted line). The dot-dashed line is a straight line of slope one.

ple formula given by Eq. 5 to quickly evaluate the n_2 coefficient for third order nonlinearities. We thus define this length as $l_{eff} = L_{eff}/(n_0 Z_0) = \Delta T_{pv}(\text{thick})/(0.406\Delta\Phi_{z0})$ which is evaluated from the above expressions as[3]:

$$l_{eff} = \frac{F(|X_{p,v}|,l)}{2 \times 0.406}. \tag{30}$$

This is plotted in Fig. 8 together with a simpler empirical fit given by:

$$l_{eff} = \frac{2.706[(l+1)^{1.44} - 1]}{l^{1.44} + 3.924}. \tag{31}$$

In an actual experiment where the refractive index is not unity, the above Equations are modified by replacing Z_0 by $n_0 Z_0$. This substitution, however, will not account for the longitudinal shift of the linear focus. Because the difference between the peak and valley do not change, this simple replacement yields the correct results.

V. INTERPRETATION

There are many physical processes which can lead to third order nonlinearities (i.e., effects proportional to the input irradiance, fluence, or energy). Ultrafast nonlinear absorption processes include multiphoton absorption,[22,23] stimulated Raman scattering,[24] and AC-Stark effects.[25,26] These lead via causality and Kramers-Kronig relations to the bound-electronic nonlinear refractive index, n_2.[25,26,27,28] Cumulative (i.e., slow) nonlinearities include population redistribution from linear absorption (this includes saturable and excited state or reverse saturable absorption and their refractive counterparts), reorientation of anisotropic molecules—which is dominant in liquids such as CS_2, thermal refraction, electrostriction, etc. The Z-scan measurement is sensitive to all of these nonlinearities including higher order effects and cannot simply be used by itself to distinguish these nonlinear processes or separate fast from slow nonlinearities.

A key to distinguishing these processes is to pay particular attention to the temporal response. Ultrafast nonlinearities are easily analyzed, as has been discussed. The use of pulsewidths much shorter than the decay times of excited states allows such cumulative nonlinearities to be more easily analyzed. As we shall show, in this regime, the excited state nonlinearities are fluence dependent, while the ultrafast effects remain irradiance dependent. The explicit temporal dependences of the nonlinearities can be obtained from, for example, degenerate four-wave mixing experiments[29] or time-resolved Z-scan experiments which can separate $\Delta n(t)$ and $\Delta\alpha(t)$.[10,11]

These time-resolved experiments can, in principle, separate slow and fast nonlinear responses. In addition, nondegenerate nonlinearities can be determined from pump-probe,[30] four-wave mixing,[29] or 2-color Z-scan,[8,9] etc., experiments. These nondegenerate nonlinear responses are also useful in distinguishing various contributing nonlinear mechanisms.

We illustrate the potential problems associated with interpreting nonlinear measurements with a single example of comparing excited state absorption (reverse saturable absorption) and two-photon absorption signals. The Equation describing 2PA in the presence of residual linear absorption is:

$$\frac{dI}{dz'} = -\alpha I - \beta I^2. \tag{32}$$

Excited states created by linear absorption in molecules are characterized by a $\Delta\alpha \approx \alpha\sigma/\eta\omega \int_{-\infty}^{t} I(t')dt'$.[15,16] By temporally integrating the resulting equation for dI/dz', we find the fluence F (energy per unit area) varies with z' as [16]:

$$\frac{dF}{dz'} = -\alpha F - \frac{\alpha\sigma}{2\eta\omega} F^2. \tag{33}$$

Notice that this Equation is exactly analogous to the Equation describing 2PA loss (Eq. 32) with the fluence replacing the irradiance and $\alpha\sigma/2\hbar\omega$ replacing β. Therefore, since in most experiments the pulse energy is detected, excited state absorption initiated by linear absorption and 2PA will give nearly identical results for loss as a function of input energy (microscopically ESA can be considered as the limit of 2PA with a resonant intermediate state). The difference between Eqs. 32 and 33 when determining the transmitted energy is in the temporal integral over the pulse for 2PA. For ESA, this integral has already been performed. In other words, in order to determine which of these nonlinearities is present, the temporal dependence must be measured in some way. An analogous problem exists with excited state refraction and the bound electronic n_2. Additionally, as seen in many semiconductors[17] and in some organic materials,[31,32] the excited states can be created by nonlinear absorption (e.g., 2PA) leading to fifth order nonlinear absorption and refraction, further confusing interpretation.

Equation 33 is only valid for low fluence where the changes in transmittance are small. For higher fluence, saturation of the groundstate absorption process (or even the excited state absorption[33]) can occur. In such cases, the best approach is to solve the system of rate equations to determine the level populations and then use these in the loss equation (Eq. 34) in terms of absorption cross-sections, σ_{ij}, or phase equation (Eq. 35) using refractive cross-sections $(\sigma_r)_{ij}$:

$$\frac{dI}{dz'} = -\sum_{i=1,j>i}^{N} \sigma_{ij}\,\Delta N_{ij}\, I, \tag{34}$$

and

$$\frac{d\phi}{dz'} = \sum_{i=1,j>i}^{N} (\sigma_r)_{ij}\,\Delta N_{ij}, \tag{35}$$

where ΔN_{ij} is the population difference between two levels $(N_i - N_j)$ coupled by an absorption cross-section σ_{ij}. The nonlinear refraction is due to the redistribution of level populations and the sign depends on the incident light's frequency relative to the resonance frequency as well as on whether the loss increases or saturates. For many materials (e.g., organic reverse saturable absorbers), ΔN_{ij} can be replaced by the population of the lower level N_i, since the upper level rapidly decays to an intermediate level (e.g., in the vibration/rotation band).[34] Temporal and spatial integrals of Eqs. 34 and 35 also need to be performed numerically. This procedure, of course, leads to a formalism where the loss or refraction are not described by the third order analysis given in this Chapter. This can also be said of the simple two-level saturation model which is only described by a third order response for small inputs.

It is important to accurately measure the laser mode and pulse parameters. For example, 2PA is irradiance dependent. Thus, given the pulse energy, we need to know both the beam area (i.e., spatial beam profile) and the temporal pulsewidth (i.e., temporal shape) in order to determine the irradiance. Any errors in the measurement of irradiance translate to errors in the determination of β as well as several other nonlinear coefficients.

There are several other papers that report methods or analysis for Z-scans that we have not yet mentioned. For example, Herman et al. discuss factors that affect optical limiting in thin samples with large nonlinearities related to Z-scans.[35] In Ref. 36 Hochbaum discusses the simultaneous determination of two or more nonlinear refractive constants in Z-scan measurements. Sutherland describes the effects of multiple internal reflections within a sample on the Z-scan signal.[37]

For some materials, the light permanently or temporarily changes the optical properties so that the sample properties change within the duration of a Z-scan experiment. Oliveira et al. discuss the analysis of such data.[38] Petrov et al. describe the use of a Z-scan in a reflection mode to determine changes of the complex dielectric function at surfaces.[39] Kershaw describes his analysis of EZ-scan measurements,[40] and Ma et al. propose a method to enhance the sensitivity of a 2-color Z-scan.[41]

VI. DATA

The following Table gives the results of Z-scan measurements on a variety of organic samples. There is no guarantee of the completeness of the given examples. Reported in Table 1 are values of nonlinear absorption coefficients, n_2's, third order nonlinear susceptibilities, $\chi^{(3)}$, and hyperpolarizabilities, as defined by the authors in the references cited. Other nonlinear coefficients such as thermal index changes and excited state cross-sections are also given. We also include our own measurements of CS_2 for reference.[42] We recommend consulting the original literature for numerical values and definitions. Indeed, this Table is primarily meant to be used as a guide to determine which references are of interest to the reader. A brief description of some of the findings from these measurements follows. As can be seen in this literature, the Z-scan technique has been used to measure many different nonlinear mechanisms in organic materials, some identified and some not. Table 2 lists a number of definitions of materials, solvents, techniques, etc., used both in Table 1 and in the descriptive text that follows.

A. Description of Measurements

In the following, we provide a description of many Z-scan measurements that were performed on a variety of materials and summarize the results.

(1) Winter et al.[43] and Oliver et al.[44] (same study) measured $\chi^{(3)}$ of several nickel dithiolene compounds (metal-sulfur ligand complexes) in a PMMA host with 100 ps pulses at 1064 nm. The SPIE proceedings data supersedes the Optics Communications publication.[45] Nonlinear refractive indices, n_2, calculated

Table 1 Nonlinear Properties of Organics as Measured by Z-Scan

Material	Solvent	γ (nm)	τ_p (nsec)	L	Beam radius	2PA Coefficient β (cm/GW)	n_2 (cm^2/W)	Ref.	α (cm^{-1})	Additional information
CS_2	Neat liquid	1064	0.04	1 mm		≈ 0	$3.1(\pm 0.2) \times 10^{-14}$	42	≈ 0	Reorientational Kerr effect
		532	0.028	1 mm	25 μm	≈ 0	$3.1(\pm 0.2) \times 10^{-14}$	2	≈ 0	
Ni:dithiolene polymer	PMMA	1064	0.10	5–188 μm		13 to 840 at high conc.	4.2×10^{-10} to 1.0×10^{-7} at high conc.	43,44	50	NLA increases faster than NLR with conc.
Ni:dithiolene polymer	DCM	1064	0.10	1 mm		0.55	-1×10^{-14}	43,44	50	
PTS	Single crystal	1600	0.065	210 μm	17 μm	<0.5	$2.2(+0.3) \times 10^{-12}$	48	≈ 1	See higher order NLR & NLA
PTS	Single crystal	1610–1120	0.065			Up to 200 at 1.05 eV	$(-2$ to $+5) \times 10^{-12}$	49	≈ 1	Spectra given
PTS	Single crystal	1064	0.035			100	$+5 \times 10^{-12}$	49	≈ 1	Spectra given
PTS	Single crystal	950–800	0.065			Up to 700 at 1.35 eV		49	≈ 1	Spectra given
PTS	Single crystal	1300	0.20	210 μm		20 ± 4	$3.2 \pm 0.3 \times 10^{-12}$	50	≈ 1	See higher order NLR & NLA
PTS	Single crystal	1064	0.035	210 μm	20 μm	100 ± 20	$5(\pm 1) \times 10^{-12}$	51	≈ 1	See higher order NLR & NLA
PTS	Single crystal	1064	0.07 & 0.01			65 ± 12	$-1.5(\pm 0.5) \times 10^{-11}$	52		Control rep. rate

DANS	50% by volume in polymer	780–1699	0.003	300 μm		Up to 5.5 at 920 nm		53		Peak 2PA at 920 mm
Phenylmethyl polysilane polymer	0.5 M/l THF	532	0.03	1 mm	≈18 μm	0.027	-2.1×10^{-14}	54	"≈0"	
Phenylmethyl polysilane polymer	0.5 M/l 1 THF	1064	≈0.04	1 mm	≈48 μm	"≈0"	$\approx +1.5 \times 10^{-15}$	54	"≈0"	
Thiophene oligomers nT $n = 2$–5	Dioxane 10^{-3} M/l	532	0.03	1 mm	26 μm	(1T) 0.026; (4T) 0.11	(1T) + 4.9; (3T) − 0.5 (4T) − 2.3; (5T) − 3.2 all $\times 10^{-15}$	55	"≈0"	Sat. of 2PA at 35 GW/cm^2
Thiophene oligomers nT $n = 6$	2,5-DCHEX	532	0.03	1 mm	26 μm	(6T) 0.23–0.34	(6T) − 3.5×10^{-15}	55	1.4	Sat. of 2PA at 35 GW/cm^2
Polythiophene	Thin film	532	0.03		50 μm	Saturation I_{sat} = 6.15 GW/cm^2	-9×10^{-11}	56	4×10^4	NLR from saturation of abs.
Pseudoazulene	Chloroform 1.2% weight	800	0.0001	1 mm		$\gamma_r = 1.7 \times 10^{-34}$ esu		57		
Disperse red 1	PMMA	1064	0.10 train	Few μm		≈0	-0.8×10^{-12}	58		2PA enhanced
HITCI	PMMA	1064	0.10 train			30	3.2×10^{-12}	58		2PA enhanced
Diphenylbutadiene	Neat liquid	532	0.035	1 mm	27 μm	2.65	6.3×10^{-12}	60		Molten at 154°C

Table 1 Continued

Material	Solvent	λ (nm)	τ_p (nsec)	L	Beam radius	2PA Coefficient β (cm/GW)	n_2 (cm^2/W)	Ref.	α (cm^{-1})	Additional information
BBTDOT	0.02 M/l THF	532	0.032	0.2 cm	19 μm	$\gamma_i = 4.8 \times 10^{-34}$ esu	$\gamma_r = 4.8 \times 10^{-34}$ esu	31	<0.1	3rd and 5th order
DDOS	5×10^{-3} M/l THF	532	0.032	0.2 cm	19 μm	$\gamma_i = 1.1 \times 10^{-33}$ esu	$\gamma_r = -4.5 \times 10^{-34}$ esu	31	<0.1	3rd and 5th order
Vitreous humor		532	0.060	1 mm	15 μm	$<4 \times 10^{-3}$	Rabbit = $(8.7 \pm 1.9) \times 10^{-16}$ Human = $(4.5 \pm 1.3) \times 10^{-16}$	61	≈0	
cis-$Mo(CO)_4(PPh_3)_2$	THF	532	20			≈0	$\chi^3 = 4.6 \times 10^{-9}$ esu	62	<0.1	Derived χ^3
trans-$Mo(CO)_4(PPh_3)_2$	THF	532	20			≈0	$\chi^3 = 1.4 \times 10^{-9}$ esu	62	<0.1	Derived χ^3
cis-$Mo(CO)_4(PPh_2Ome)_2$	THF	532	20			≈0	$\chi^3 = 2.6 \times 10^{-10}$ esu	62	<0.1	Derived χ^3
cis-$Mo(CO)_4(PPh_2Me)_2$	THF	532	20			≈0	$\chi^3 = 1.7 \times 10^{-10}$ esu	62	<0.1	Derived χ^3
$MO(CO)_5(PPh_3)$	THF	532	20			≈0	$\chi^3 = 8.2 \times 10^{-11}$ esu	62	<0.1	Derived χ^3
$MO(CO)_5(PPh_2NHMe)$	THF	532	20			≈0	$\chi^3 = 5.0 \times 10^{-11}$ esu	62	<0.1	Derived χ^3
HBT	n-hexane	1064	0.02				2.3×10^{-15} $\chi^3 = 4.7 \times 10^{-22} m^2/V^2$	63	≈0	#'s for solvent $n_2 > 0$ for solute

Azobenzene	PMMA	633	cw	100 μm		≈0	6 × 10^{-4} esu	64		Slow realign-ment
Bacteriorhodopsin	Dried film	633	cw	150 μm	50 μm		"−4.0 × 10^{-6}"	65	≈50	4–10 ms decay
Bacteriorhodopsin	Dried film	514	cw	150 μm	32 μm		"−3.6 × 10^{-6}"	65	≈50	4–10 ms decay
Bacteriorhodopsin	Dried film	488	cw	150 μm	31 μm		"−3.5 × 10^{-6}"	65	≈50	4–10 ms decay
Bacteriorhodopsin	Dried film	477	cw	150 μm	30 μm		"−3.5 × 10^{-6}"	65	≈50	4–10 ms decay
Bacteriorhodopsin	Dried film	633 & 476	cw			I_{sat} ≅ 3 mW/cm² at 633 ≅ 4 mW/cm² at 476	<0	66		Saturation of abs. + thermal defocusing
C_{60}	Thin film	633	cw	0.6 μm	48.4 μm	"(8 ± 3) × 10^{9}"	"0.16 × 10^{-9}"	71	1 × 10^{4}	2-step absorp-tion thermal focus
C_{60}	Toluene	527	30	1.0 cm	80 μm	ESA $\sigma_t \approx 3\sigma_g$	"<0"	72	0.17	NL scatter-ing
C_{60}	1-CloroNapth	532	6	50 μm		ESA	dn/dT	73		ESA + thermal lensing
C_{60}	Acetonitrile	532	7	0.1 cm	35 μm	ESA	≈0	70	≈3.6	
C_{70}	Toluene	532	0.035	1 mm		$\sigma_1 = 2.8 \times 10^{-17}$ cm²		74		ESA
CAP	10^{-3} M/toluene	532	0.029 & 0.061	0.2 cm	28 μm	$\sigma_1 = 2.3 \times 10^{-17}$ cm²	$\sigma_r = 1.8 \times 10^{-17}$ cm²	15.75	1.8	ESA and ESR
SINC	10^{-3} M/toluene	532	0.029 & 0.061	0.2 cm	28 μm	$\sigma_1 = 3.9 \times 10^{-17}$ cm²	$\sigma_r = 4.7 \times 10^{-18}$ cm²	15.75	1.8	ESA and ESR
TBP	THF	532	9	1 mm	21.4 μm	$\sigma_1 = 2.4 \times 10^{-16}$ cm²	$\sigma_r = (-4 \pm 1) \times 10^{-17}$ cm²	76.77	2.4	ESA and ESR

Table 1 Continued

Material	Solvent	λ (nm)	τ_p (nsec)	L	Beam radius	2PA Coefficient β (cm/GW)	n_2 (cm²/W)	Ref.	α (cm⁻¹)	Additional information
HITCI	33.5 μM methanol	532	15	1 mm	14.5 μm	$\sigma_1 = (38.4 \pm 0.4) \times 10^{-17}$ cm²		78,79,80	0.25	ESA
CuPC	30% in BaF_2 film	532	0.028	0.5 μm	29 μm	"5×10^3"	"4.3×10^{-11}"	81		ESA and ESR but quoted effective β and n_2
10^{-5} M Kiton red with polystyrene spheres	Colloid susp in H_2O	514	cw	100 μm	45 μm	"≈0"	"-1.3×10^{-7}"	82	3	
5CB	Nematic	532	7	25 μm		320	-2.4×10^{-12}	83		pol ∥ to alignment
Anthraquinone dye AQ1, AQ2	Nematic host	633	cw	35 μm	30 μm	"≈0"	n_2 not given, torque parameter η quoted	84	AQ2 = 142	$\eta = 238$ AQ2 homeotropic film
Anthraquinone dyes D4, D16, D27	E63	633	cw	35 μm	30 μm	"≈0"	n_2 not given, torque parameter η quoted	84	D4 = 275 D16 = 142 D27 = 105	D4: $\eta = -95$ D16: $\eta = -19$ D27: $\eta = -5$ homeotropic film
MNA	Chloroform	532	65	1 cm	33 μm	"≈0"	<0	85	0.036	Response time measurement
Chinese tea	Ethyl alcohol	633	cw	1 mm	Variable	"≈0"	$dn/dt = -3.4 \times 10^{-5}$	86	0.6	Thermal defocus
Chinese tea	Ethyl alcohol	633	cw	1 mm	Variable	"≈0"	$dn/dt = -3.4 \times 10^{-5}$	87	0.6	Thermal defocus

Black tea extract	H_2O	532	7	1 cm	44 μm	Saturation "−0.84"	-5.7×10^{-14}	88	1.4	Saturation of abs. + thermal defocusing
Toluene	Neat solvent	532	4.7	0.1 cm	22 μm	≈0	7×10^{-15}	7	<0.01	Demonstration of EZ-scan method
Toluene; ethanol and chloroform	Neat solvents	1064 or 532 + 633	6 at 532 nm 9 at 1064 nm	2 mm		1PA & 2PA induced thermal index change on cw probe	dn/dT	11		2-color time-resolved Z-scan
Diphenylbutadiene or R6G	Ethanol, H_2O or toluene	As above	As above	2 mm	60 μm + 40 μm	1PA & 2PA induced thermal index change on cw probe	dn/dT	11		2-color time-resolved Z-scan

Table 2 Definitions for Table 1

Spot sizes quoted as the half-width at the $1/e^2$ point of the maximum of the irradiance when known.
Pulsewidths quoted as the half-width at half maximum when known.
AQ1 = 1,8-dihydroxy, 4,5-diamino, 2,7-diisobutyl-anthraquinone
AQ2 = 1,8-dihydroxy, 4,5-diamino, 2,7-diisopentyl-anthraquinone
azobenzene = film of azobenzene-compound-doped poly(methyl methacrylate) with push group -NH-$C_{16}H_{23}$ and pull group -NO_2
BBTDOT = bisbenethiozole-substituted thiophene
CAP = Chloro-aluminum-phthalocyanine
CloroNapth = chloronaphthalene
CuPC = copper phthalocyanine
D4 = N,N′-(4-methylphenyl)-1,4-diamino-anthraquinone in E63
D16 = N-(4-nonyloxylphenyl)-1-amino-4-hydroxy-anthraquinone in E63
D27 = N-(4-dimethylaminophenyl)-1-amino-4-hydroxy-anthraquinone in E63
DANS = a di-alkyl-amino-nitro-stilbene side chain polymer
DCHEX = didecylhexathiophene
DCM = 4-(Dicyanomethylene)-2-methyl-6-(4-dimethylaminostyryl)-4H-pyran
DDOS = Didecyloxy substituted polyphenyl
derived χ^3 = is derived by extrapolating dilute data to the neat material
diphenylbutadiene = trans-trans 1,4-diphenyl-1,3-butadiene
ESA = excited state absorption
ESR = excited state refraction
E63 = a mixture composed mainly of biphenyls developed by British Drug House
few = a few or several
HBT = 2-(2′-hydroxyphenyl)benzothiazole
HITCI = 1,1′,3,3′,3′,-hexamethylindocarbocyanine iodide
nematic = nematic liquid crystal
Ni:dithiolene polymer = modified bis[1-ethyl-2-phenylethene-1,2-dithiolate(2-)-S,S′ nickel in PMMA
NLA = nonlinear absorption
NLR = nonlinear refraction
MNA = methyl nitroanaline
PC = phthalocyanine
PMMA = poly-methyl methacrylate
pol = polarization
Pseudoazulene = 1,2-5,6 dibenzoxalene
PTS = poly[2,4-hexdiyne-1,6-diol-bis-*p*-toluene-sulfonate]
Rep. Rate = repetition rate
sat = saturation
SINC = silicon naphthalocyanine
TBP = zinc meso-tetra(*p*-methoxyphenyl) tetrabenzporphyrin
THF = tetrahydrofuran
2PA = two-photon absorption
5CB = 4-cyano-4′-*n*-pentylbiphenyl
$\gamma = \chi^3\, 3^4/[(n^2 + 2)^4 N]$, where N is molecular density, n the linear index and γ is in units of $cm^5/statvolt^2$
σ = excited state absorption cross-section defined from $dI/dz = -\sigma NI$ where N is the density of excited states
σ_g = groundstate absorption cross-section defined from $dI/dz = -\sigma_g NI$ where N is the density in the groundstate
σ_r = excited state refractive cross-section defined from $d\phi/dz = \sigma_r N$ where N is the density of excited states and ϕ is the field phase
σ_t = triplet state absorption cross-section defined from $dI/dz = -\sigma_t NI$ where N is the density of triplet states

for neat solutions as high as 10^{-11} cm^2/W were found; however, as the concentration increased, the $\Delta\alpha$ increased more rapidly than Δn making them less desirable for all-optical switching applications. This was attributed to intermolecular interactions such as dimer and trimer complex formation.

(2) Underhill et al.[46] used DFWM and Z-scan to study NLA and NLR in PMMA films doped with Ni-dithiolene oligomers at 1064 nm to determine their suitability for all-optical switching applications. They report favorable figures-of-merit for some of the compounds, but no nonlinear coefficients are given.

(3) Gall et al.[47] incorporate a number of dyes into sol-gels and find that the nonlinear response does not change from the solution values using DFWM. They also report Z-scan measurements on CAP in a xerogel with 15 ns 532 nm pulses with $w_0 = 7$ μm, but give no values of the nonlinear coefficients.

(4) The study by Lawrence et al.[48] reports a large purely refractive nonlinear index, $n_2 = 2.2 \times 10^{-12}$ cm^2/W, of single crystal polydiacetylene paratoluene-sulfonate (PTS) at a wavelength of 1.6 μm. There is no measurable 2PA, and the linear absorption is ≈ 1 cm^{-1}, making this material useful for all-optical switching applications. A plot of the effective n_2 versus irradiance shows a straight line with a negative slope indicating that there is a fifth order nonlinearity reducing the total index change at higher irradiance levels.

(5) The same group[49] reports the measurement of the full 2PA spectrum of PTS from $0.8E_g$ to $1.6E_g$, where E_g is the bandgap energy of PTS. They also report dispersion of n_2 over a smaller wavelength range where the Z-scan signal is not overly dominated by nonlinear absorption. This is the only case known to the authors where the nonlinear absorption and nonlinear refraction of an organic material have been studied over such a large spectral range. Figure 9 shows the spectrum of the two-photon absorption coefficient (Fig. 9a) along with the dispersion of n_2 (Fig. 9b).

(6) In another study, Kim et al.[50] studied the usefulness of PTS as an all-optical switching material at 1.3 μm. These authors measure n_2 and β at 1.3 μm for 100 ps pulses in a sample of PTS 210 μm thick. They see higher order effects reducing the nonlinear refraction and raising the nonlinear absorption as the input irradiance is increased. However, they find the figures of merit are within the range necessary for all-optical switching.[20]

(7) A similar study of PTS at 1064 nm by Lawrence et al.[51] measures β and positive n_2, of PTS using 35 ps pulses. Again, a fifth order nonlinear absorption and refraction are observed. The fifth order nonlinear absorption reduces the loss as the irradiance increases, and the fifth order nonlinear refraction reduces the index, turning the refraction to self-defocusing above ≈ 7 GW/cm^2.

(8) Thakur et al.[52] measure PTS using Z-scan with 70 ps and 10 ps 1.06 μm pulse trains where the repetition rate and average power are controlled with a Pockels cell. They report a negative value of the intensity-dependent refractive index $n_2 = (-1.5 \pm 0.5) \times 10^{-5}$ cm^2/MW and $\beta = (65 \pm 12)$ cm/GW.

(9) In another series of experiments aimed at measuring the nonlinear spectrum of an organic material, Cha et al.[53] report β of a di-alkyl-amino-nitro-stilbene

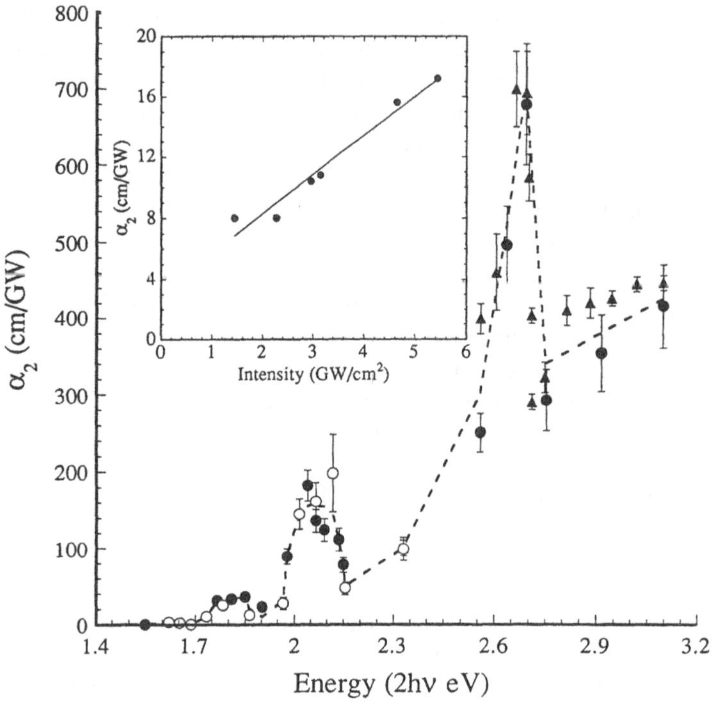

Figure 9a Two-photon absorption coefficient, $\beta = \alpha_2$ ($I = 0$), as a function of two-photon energy in PTS. The inset shows how the 2PA coefficient was determined from the low intensity intercept of the α_2 versus intensity curve. The slope is due to a fifth order nonlinear absorption process. The open and closed circles indicate data for two different samples, while the triangles show data taken with a second optical source. The dashed line represents a guide to the eye.

side chain polymer, DANS, from 780 to 1600 nm. They find a single 2PA peak at 920 nm with a maximum value of β = 5.5 cm/GW.

(10) Yang and coworkers[54] report measurements of self-focusing at 1064 nm and self-defocusing at 532 nm using 30 ps pulses in a THF solution of phenylmethyl polysilane. They attribute the change in sign at 532 nm to a two-photon absorption resonance transition.

(11) Hein and coworkers[55] measure both β and n_2 of thiophene oligomers for different numbers of repeat units (n = 2–6) for 10^{-3} molar solutions in dioxane at 532 nm using 30 ps pulses. They measure self-defocusing which increases in magnitude with chain length. They also find that β increases with chain length.

(12) Measurements of the self-defocusing and nonlinear absorption in polythiophene thin films are reported by Yang et al.[56] The nonlinearities at 532 nm using 30 ps pulses are attributed to saturation of the strong linear absorption ($\alpha \approx 4 \times 10^4$ cm^{-1}) at this wavelength.

(13) Samoc et al.[57] measured a derivative of 1,2-5,6 dibenzoxalene in chloroform (1.2 percent by weight) using 100 femtosecond pulses at 800 nm using Z-

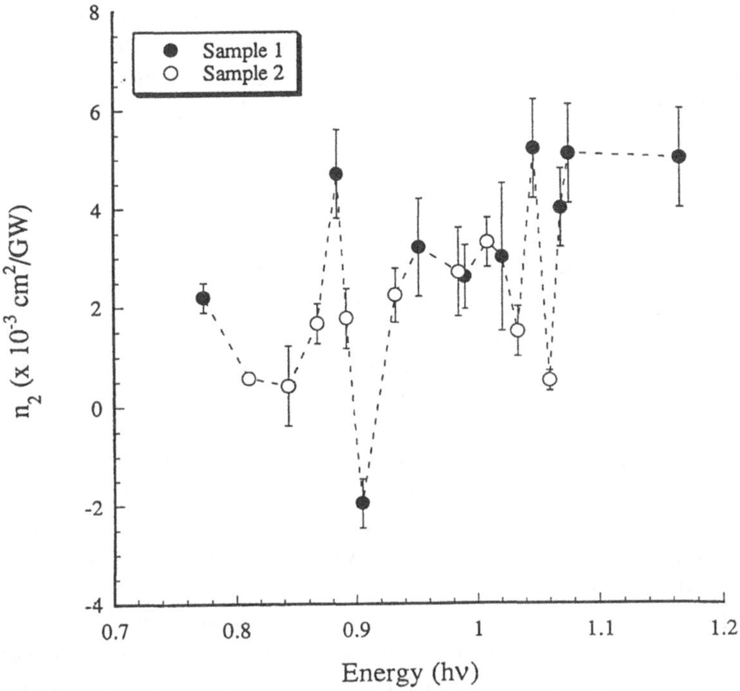

Figure 9b Nonlinear refractive index, n_2, as a function of photon energy for PTS (from Ref. [49] with permission). The open and closed circles show data for two different samples while the dashed line represents a guide to the eye.

scan to determine a hyperpolarizability of 1.7×10^{-34} esu which they claim is in reasonable agreement with theoretical predictions.

(14) Measurements of n_2 in Disperse Red 1 are described by Planas et al.[58] at a wavelength of 1064 nm which is below the 2PA resonance which occurs at 2×490 nm (980 nm) where they find $n_2 < 0$ ($n_2 = -0.8 \times 10^{-12}$ cm^2/W). Measurements at 1064 nm of n_2 in HITC (1,1′,3,3,3′,3′,-hexamethylindotricarbocyanine iodide) where the frequency is above the 2PA resonance at 2×755 nm (1510 nm) give $n_2 > 0$ ($n_2 = 3.2 \times 10^{-12}$ cm^2/W) while $\beta \cong 30$ cm/GW. Both materials were incorporated in thin (several μm thick) PMMA wafers and the laser used produced 100 ps Q-switched pulse trains. These results suggest that the large values of n_2 observed in these materials are associated with a 2PA resonance enhancement.

(15) Said et al. report[31] measurements of the irradiance dependence of the nonlinear response of THF solutions of a bisbenzthiozole-substituted thiophene compound, BBTDOT, and in a didecyloxy substituted polyphenyl compound, DDOS. In both compounds, third and fifth order responses were observed for the nonlinear refraction as well as for the nonlinear absorption. The third order responses are attributed to 2PA and bound electronic refraction, while the fifth order responses are attributed to absorption and refraction from the two-photon generated excited states. Hyperpolarizabilities, n_2, and excited

state cross-sections for both absorption and refraction are reported. Both compounds show increasing nonlinear absorption with irradiance, i.e., positive fifth order absorption, while the excited state refraction is self-defocusing in both materials. BBTDOT shows third order self-focusing, while DDOS shows third order self-defocusing. Van Stryland et al.[59] review some of the work of Said et al.[31] along with several other measurements.

(16) Fleitz et al.[60] report Z-scan measurements of molten diphenylbutadiene (*trans-trans* 1,4-diphenyl-1,3-butadiene) at 154 degrees Celsius in a 1 mm cell using 532 nm, 53 ps pulses. They interpret the increasing nonlinear loss with irradiance as 2PA followed by excited state absorption.

(17) Rockwell et al.[61] reports measurements of n_2 of water, salt water and vitreous humor (human and rabbit) at 532 nm using 60 ps pulses. The data show only minor differences between these materials within the measurement errors.

(18) Measurements reported by Zhai et al.[62] of $\chi^{(3)}$ at 532 nm with 20 ns pulses in various concentrations of a series of molybdenum-based metal-organic complexes in solution were made to determine the molecular hyperpolarizability, n_2. A close correlation is reported between the number of delocalized π electrons and the magnitude of n_2.

(19) Hofer et al.[63] present a study of the kinetics of intramolecular proton transfer using picosecond excitation followed by four-wave mixing studies. However, Z-scans of the solvent and solvent plus HBT were made in the nonresonant spectral range to determine the small but positive index change at 1064 and 532 nm.

(20) Fei et al.[64] use a cw HeNe laser at 633 nm to measure DFWM and Z-scan signals where they determine $n_2 = 6 \times 10^{-4}$ esu ($\chi^{(3)} \cong 10^{-4}$ esu) attributed to the slow realignment of the long chain molecules of azobenzene-compound-doped poly(methyl methacrylate) with a push group -NH-$C_{16}H_{23}$ and pull group -NO_2 attached. The sample was in the form of a 0.01 cm thick film.

(21) Song et al.[65] use continuous wave illumination at 633, 514, 488 and 477 nm of bacteriorhodopsin films to demonstrate self-defocusing index changes of $\approx 10^{-3}$ and anomalous nonlinear absorption.

(22) Song et al.[66] use the Z-scan at 633 nm and 476 nm (cw) to measure saturation and index changes due to population redistribution (two-level model) and thermal self-defocusing in chemically enhanced bacteriorhodopsin films. For example, they determine saturation intensities of $\cong$3 mW/cm^2 at 476 nm and $\cong$4 mW/cm^2 at 633 nm.

(23) Measurements of a series of inorganic metal cluster molecules are studied and described in the literature[67,68,69] by S. Shi et al. Using acetonitrile as the solvent, they measure NLA and NLR in the inorganic metal cluster molecules $C_{35}H_{72}N_5OS_6Cu_3Mo$, $WCu_2OS_3(PPh_3)_4$ and $MoCu_2OS_3(PPh_3)_3$.

(24) The same group[70] compares the nonlinearities of C_{60} to nonlinearities observed in other inorganic complexes and conclude that the nonlinear response is dominated by nonlinear absorption in both materials.

(25) Gu et al.[71] measure nonlinear absorption and self-focusing in C_{60} thin films using cw 633 nm light. They attribute the nonlinear absorption to excited triplet state absorption.

(26) In toluene solutions of C_{60}, Mishra et al.[72] report nonlinear absorption and self-defocusing with 30 ns, 527 nm pulses. They also attribute some loss to nonlinear scattering of a thermal origin.
(27) Justus et al.[73] attribute the nonlinear absorption they observe in C_{60} in 1-chloronaphthalene using 6 ns 532 nm pulses to excited state absorption. They attribute the self-lensing to photoinduced thermal refractive index changes.
(28) Yang et al.[74] measure the excited state absorption cross-section of C_{70} as $\sigma = 2.8 \times 10^{-17}$ cm^2 using 35 ps 532 nm pulses in a 1 mm cell.
(29) Measurements are reported of the third order excited state nonlinear absorption and refraction in chloro-aluminum phthalocyanine (CAP) solutions in ethanol at 532 nm using 30 ps pulses.[15,75] Solutions of SiNc, a silicon naphthalocyanine derivative, $Si(OSi(n\text{-hexyl})_3)_2Nc$, in toluene were also studied. By measuring the nonlinearities at two different pulsewidths (30 and 60 ps), they determined that the nonlinearities were both fluence dependent and, thus, due to populating the excited state. They report excited state absorption cross-sections, σ, and extinction coefficients. They also give excited state refractive coefficients defined by $d\phi/dz = \sigma_r N$, where ϕ is the induced phase distortion, σ_r the refractive cross-section and N the density of excited states. For CAP they report $\sigma = 2.3 \times 10^{-17}$ cm^2 or $\varepsilon = 6 \times 10^3$ liters M^{-1} cm^{-1}, and $\sigma_r = 1.8 \times 10^{-17}$ cm^2, while for SiNc, $\sigma = 3.9 \times 10^{-17}$ cm^2, $\varepsilon = 1.0 \times 10^4$ M^{-1} cm^{-1} and $\sigma_r = 4.7 \times 10^{-17}$ cm^2.
(30) Wood et al.[76,77] report Z-scan measurements of excited state absorption in zinc meso-tetra(p-methoxyphenyl) tetrabenzporphyrin called TBP dissolved in THF using 532 nm, 9 ns pulses. They use a quasi three-level model to determine $\sigma_{ex} = 2.4 \times 10^{-16}$ cm^2 and $\sigma_r = (-0.4 \pm 0.1) \times 10^{-16}$ cm^2, while the groundstate cross-section is $\sigma_{gr} = 8.0 \times 10^{-18}$ cm^2.
(31) Swatton et al.[78,79] and Welford et al.[80] report excited state absorption in HITCI in methanol using 15 ns, 532 nm pulses and describe the observed saturation of the Z-scan signal using a four-level model which also yields level lifetimes.
(32) The eclipsing Z-scan (EZ-scan) technique is used by Xia et al.[7] to measure nonlinear refraction in neat toluene as well as to compare the technique to Z-scan. The demonstrated sensitivity enhancement is $\approx$13 under the experimental conditions reported. The nonlinear refraction may have contributions from electrostriction because nanosecond pulses were used.
(33) Xia et al.[81] report the use of EZ-scan with picosecond pulses to measure nonlinear refraction and nonlinear absorption in thin films. They demonstrate the method by measuring purely nonlinear absorption in a 0.7 μm thick sample of CdSe clusters embedded in CuPC. They speculate that the nonlinearity is dominated by excited state absorption initiated by linear absorption (RSA). They also report measurements on a 0.5 μm thick BaF_2 film containing $\approx$30% CuPC showing a combination of NLR and NLA, again probably due to excited state processes.
(34) Herbert et al.[82] measure the thermal nonlinearity in a dye-doped colloidal crystal using Z-scan and use the distributed feedback structure for power limiting. The limiting is restricted to a narrow spectral bandwidth due to the diffractive nature of the limiting (i.e., Bragg scattering from the nonlinearly induced grating).

(35) Yuan et al.[83] measure β and n_2 in the pure nematic liquid crystal 5CB (4-cyano-4′-*n*-pentylbiphenyl) in the isotropic and nematic phase using cw 514 nm light (10 ms shuttered pulses) as a probe with 7 ns 532 nm excitation pulses in a 2-color Z-scan. They find large NLA and NLR for the optical electric field parallel to the director in the nematic phase using 25 μm thick aligned samples. For 7 ns, 532 nm pulses they find $\beta = 320$ cm/GW and $n_2 = -1.7 \times 10^{-9}$ esu (-2.4×10^{-16} m^2/W) for parallel polarizations of the beams (not measurable for perpendicular polarizations). Using 10 ms, 514 nm light they measured n_2 versus temperature for parallel and perpendicular polarizations and find large changes near the critical temperature for $n_2\|$ and $n_2\perp$. They observed no NLA for 10 ms pulses.

(36) Optical reorientation measurements in dye-doped nematic liquid crystal films are reported by Janossy et al.[84] The change of the optical path in a nematic film from the light-induced director reorientation and from thermally-induced refractive index variations are measured separately and the η parameter (ratio of dye-induced torque to the normal optical torque) determined. It was found that the dye-induced torque significantly exceeds the normal optical torque.

(37) The aperture transmission in a Z-scan measurement is temporally resolved by Toda et al.[85] to determine the speed of response of nonlinear refraction. This method is applied to observe the thermal nonlinearity of methyl nitroanaline.

(38) In two papers, Tian et al.[86,87] attribute the nonlinear refraction observed in Chinese tea in ethyl alcohol to linear absorption induced thermal self-defocusing.

(39) In a similar study, Gheung et al.[88] measure self-defocusing and absorption saturation of tea in water using 532 nm, 7 ns pulses. The index nonlinearity is again attributed to thermal lensing.

(40) The dual beam 2-color Z-scan is used by Kozich et al.[11] to time-resolve the thermal index changes due to linear or two-photon absorption in organic dyes. The 6 ns pulsed pump beam was at 1.06 μm or 532 nm and the nonlinear response was probed with a cw 633 nm laser. The temporal resolution allowed separation of the fast nonlinear response from the slower thermal index changes in toluene, ethanol and chloroform.

VII. CONCLUSION

There are a variety of methods and techniques for determining the nonlinear optical response, each with its own weaknesses and advantages. In general, it is advisable to use as many complementary techniques as possible over a broad spectral range in order to unambiguously determine the active nonlinearities. Z-scan is one of the simpler experimental methods to employ. Despite the wide range of available methods, it is rare that any single experiment will completely determine the physical processes behind the nonlinear response of a given material. A single measurement of the nonlinear response of a material, at a single wavelength and a single pulsewidth may give very little information on the material. In general, such limited data should not be used to judge the device performance of a material or to compare one material to another.

For someone who wants to measure a nonlinear refractive index, an important question to ask is what is the application for which the phase shift is to be used? For example, if optical limiting with nanosecond pulses is the purpose, so that materials having large nonlinear loss for 10 ns pulses is desirable, the pulsewidth to be used is 10 ns. However, in order to determine the physics behind the nonlinear loss it may be useful to look at this loss using shorter or longer pulses.[89]

Nonlinear absorption and refraction always coexist (although with different spectral properties) as they result from the same physical mechanisms. They are connected via dispersion relations similar to the usual Kramers-Kronig relations that connect linear absorption to the linear index (or, equivalently, relate the real and imaginary parts of the linear susceptibility).[25,26,27,28] The physical processes that give rise to NLA and the accompanying NLR include "ultrafast" bound electronic processes, "excited state" processes where the response times are dictated by the characteristic formation and decay times of the optically induced excited states, thermal refraction, etc. Ultrafast processes include multi-photon absorption,[22,23] stimulated Raman scattering,[24] and AC-Stark effects.[25,26] Excited state nonlinearities can be caused by a variety of physical processes including absorption saturation,[90] excited state absorption in atoms or molecules[15] or free-carrier absorption in solids,[17,91] photochemical changes,[92] as well as defect and color center formation.[93] The above processes can lead to increased transmittance with increasing irradiance (e.g., saturation, Stark effect) or decreased transmittance (e.g., multi-photon absorption, excited state absorption). A key to distinguishing these processes is to pay particular attention to the temporal response. One way of achieving this is the use of pulsewidths much shorter than the decay times of the excited states. In this regime, the excited state nonlinearities are fluence- (i.e., energy per unit area) dependent, while the ultrafast effects remain irradiance-dependent.

Z-scan has only recently been introduced as a useful technique for measuring nonlinearities and there are still relatively few measurements of organic materials using this technique. However, it is becoming more popular as both an absolutely calibrated method for determining standards and as a relative measurement method. The *Z*-scan signal as a function of irradiance and/or *Z* can give useful information on the order of the nonlinearity as well as its sign and magnitude.

VIII. ACKNOWLEDGMENT

We gratefully acknowledge the support of the National Science Foundation grant ECS#9510046, and the Naval Air Warfare Center Joint Service Agile Program Contract N66269-C-93-0256. The work presented in this Chapter represents many years of effort involving colleagues and many former and current students as well as post-doctoral fellows. We thank all those involved and acknowledge their contributions through the various referenced publications. We explicitly thank Edesly J. Canto-Said, J. Richard DeSalvo, Arthur Dogariu, David J. Hagan, David C. Hutchings, Edward Miesak, Ali A. Said, M. J. Soileau, Hermann Vanherzeele, Tai H. Wei, William E. Williams, Brian Wherrett, Milton A. Woodall, Yuen-Yen Wu and Tiejun Xia for their many contributions.

REFERENCES

1. Sheik-Bahae, M., Said, A. A., and Van Stryland, E. W., High Sensitivity, Single Beam n_2 Measurements, *Opt. Lett.*, *14*, 955–957 (1989).
2. Sheik-Bahae, M., Said, A. A., Wei, T. H., Hagan, D. J., and Van Stryland, E. W., Sensitive Measurement of Optical Nonlinearities Using a Single Beam, *JQE*, *QE26*, 760–769 (1990).
3. Sheik-Bahae, M., Said, A. A., Hagan, D. J., Soileau, M. J., and Van Stryland, E. W., Nonlinear Refraction and Optical Limiting in "Thick" Media, *Opt. Eng.*, *30*, 1228 1235 (1990).
4. Hermann, J. A., and McDuff, R. G., Analysis of spatial scanning with thick optically nonlinear media, *J. Opt. Soc. Am. B*, *10*, 2056–2064 (1993).
5. Tian, J.-G., Zang, W.-P., Zhang, C.-Z., and Zhang, G., Analysis of beam propagation in thick nonlinear media, *Appl. Opt.*, *34*, 4331–4336 (1995).
6. Chapple, P. B., Staromlynska, J., and McDuff, R. G., Z-scan studies in the thin- and the thick-sample limits, *J. Opt. Soc. Am. B*, *11*, 975–982 (1994).
7. Xia, T., Hagan, D. J., Sheik-Bahae, M., and Van Stryland, E. W., Eclipsing Z-Scan Measurement of $\lambda / 10^4$ Wavefront Distortion, *Opt. Lett.*, *19*, 317–319 (1994).
8. Sheik-Bahae, M., Wang, J., DeSalvo, J. R., Hagan, D. J., and Van Stryland, E. W., Measurement of Nondegenerate Nonlinearities using a 2-Color Z-Scan, *Opt. Lett.*, *17*, 258–260 (1992).
9. Ma, H., Gomez, A. S., and de Araujo, C. B., Measurement of nondegenerate optical nonlinearity using a two-color single beam method, *Appl. Phys. Lett.*, *59*, 2666 (1991).
10. Wang, J., Sheik-Bahae, M., Said, A. A., Hagan, D. J., and Van Stryland, E. W., Time-Resolved Z-Scan Measurements of Optical Nonlinearities, *JOSA B11*, 1009–1017 (1994).
11. Kozich, V. P., Marcano, A., Hernandez, F., and Castillo, J., Dual-beam time-resolved Z-scan in liquids to study heating due to linear and nonlinear light absorption, *Applied Spectroscopy*, *48*, 1506–1512 (1994). See also Castillo, J., Kozich, V., and Marcano, A., Thermal lensing resulting from one- and two-photon absorption studied with a two-color time-resolved Z-scan, *Opt. Lett.*, *19*, 171–173 (1994).
12. Zhao, W., and Palffy-Muhoray, P., Z-scan measurements of χ^3 using top-hat beams, *Appl. Phys. Lett.*, *65*, 673–675 (1994). See also Zhao, W., Kim, J. H., and Palffy-Muhoray, P., Z-scan measurements on liquid crystals using top-hat beams, *Appl. Phys. Lett.*, *65*, 673–675 (1994).
13. Kaplan, A. E., External Self-Focusing of Light by a Nonlinear Layer, *Radiophys. Quant Electron.*, *12*, 692–696 (1969).
14. Giuliano, C. R., and Hess, L. D., Nonlinear Absorption of Light: Optical Saturation of Electronic Transitions in Organic Molecules with High Intensity Laser Radiation, *IEEE J. Quant. Electron*, *QE-3*, 338–367 (1967).
15. Wei, T. H., Hagan, D. J., Sence, M. J., Van Stryland, E. W., Perry, J. W., and Coulter, D. R., Direct Measurements of Nonlinear Absorption and Refraction in Solutions of Phthalocyanines, *Applied Physics*, *B54*, 46–51 (1992).
16. Van Stryland, E., Sheik-Bahae, M., Said, A. A., and Hagan, D. J., Characterization of Nonlinear Optical Absorption and Refraction, *Prog. Crystal Growth and Charact.*, *27*, 279–311 (1993).
17. Said, A. A., Sheik-Bahae, M., Hagan, D. J., Wei, T. H., Wang, J., Young, J., and Van Stryland, E. W., Determination of Bound and Free-Carrier Nonlinearities in ZnSe, GaAs, CdTe, and ZnTe, *JOSA B*, *9*, 405–414 (1992).
18. Bridges, R., Fischer, G., and Boyd, R., Z-scan measurement technique for non-Gaussian beams and arbitrary sample thicknesses, *Opt. Lett.*, *20*, 1821–1823 (1995).

19. Weaire, D., Wherrett, B. S., Miller, D. A. B., and Smith, S. D., Effect of Low Power Nonlinear Refraction on Laser Beam Propagation in InSb, *Opt. Lett.*, *4*, 331–333 (1974).
20. Mizrahi, V., DeLong, K., Stegeman, G., Saifi, M., and Andrejco, M., Two-photon absorption as a limitation to all-optical switching, *Opt. Lett.*, *14*, 1140–1142 (1989). See also DeLong, K., Rochford, K., and Stegeman, G., Effect of two-photon absorption on all-optical waveguide devices, *Appl. Phys. Lett.*, *55*, 1823–1825 (1989).
21. Banerjee, P., Misra, R., and Maghraoui, M., Theoretical and Experimental Studies of Propagation of Beams Through a Finite Sample of a Cubically Nonlinear Material, *J. Opt. Soc. Am. B*, *8*, 1072 (1991).
22. Bechtel, J. H., and Smith, W. L., Two-Photon Absorption in Semiconductors with Picosecond Laser Pulses, *Phys. Rev. B*, *13*, 3515 (1976).
23. Van Stryland, E. W., Vanherzeele, H., Woodall, M. A., Soileau, M. J., Smirl, A. L., Guha, S., and Boggess, T. F., Two photon absorption, nonlinear refraction, and optical limiting in semiconductors, *Opt. Eng.*, *24*, 613 (1985).
24. See, for example, Shen, Y. R., *The Principles of Nonlinear Optics*, John Wiley and Sons, New York, 1984.
25. Sheik-Bahae, M., Hagan, D. J., and Van Stryland, E. W., Dispersion and Band-Gap Scaling of the Electronic Kerr Effect in Solids Associated with Two-Photon Absorption, *Phys. Rev. Lett.*, *65*, 96–99 (1989).
26. Sheik-Bahae, M., Hutchings, D. C., Hagan, D. J., and Van Stryland, E. W., Dispersion of Bound Electronic Nonlinear Refraction in Solids, *JQE*, *QE-27*, 1296–1309 (1991).
27. Hutchings, D. C., Sheik-Bahae, M., Hagan, D. J., and Van Stryland, E. W., Kramers-Kronig Relations in Nonlinear Optics, *Opt. Q.E.*, *24*, 1–30 (1992).
28. Bassani, F., and Scandolo, S., Dispersion Relations in Nonlinear Optics, *Phys. Rev.*, *B44*, 8446–8453 (1991).
29. Canto-Said, E. J., Hagan, D. J., Young, J., and Van Stryland, E. W., Degenerate Four-Wave Mixing Measurements of High Order Nonlinearities in Semiconductors, *JQE*, *QE27*, 2274–2280 (1991).
30. Von der Linde, D., Laubereau, A., and Kaiser, W., Molecular vibrations in liquids: direct measurement of the dephasing time; determination of the shape of picosecond light pulses, *Phys. Rev. Lett.*, *26*, 954 (1971).
31. Said, A. A., Wamsley, C., Hagan, D. J., Van Stryland, E. W., Reinhardt, B. A., Roderer, P., and Dillard, A. G., Third and Fifth Order Optical Nonlinearities in Organic Materials, *Chem. Phys. Lett.*, *228*, 646–650 (1994).
32. Lawrence, B. L., Chai, M., Torruellas, W. E., Stegeman, G. I., Etemad, S., and Baker, G., Z-scan Measurement of Third and Fifth Order Nonlinearities in Single Crystal PTS at 1064 nm, *Nonlinear Optics*, *10*, 193–205 (1995).
33. Wei, T.-H., Huang, T.-H., Lin, H.-D., and Lin, S.-H., Lifetime Determination for High-Lying Excited States Using Z-scan, *Appl. Phys. Lett.*, *67*, 2266 (1995).
34. Xia, T., Hagan, D., Dogariu, A., Said, A., and Van Stryland, E., Optimization of Optical Limiting Devices Based on Excited State Absorption, to be published in *Applied Optics* (1996).
35. Hermann, J., and Wilson, P., Factors affecting optical limiting and scanning with thin nonlinear samples, *Int. J. of Nonlinear Opt. Phys.*, *2*, 613–629 (1993).
36. Hochbaum, A., Simultaneous determination of two or more nonlinear refractive constants by Z-scan measurement, *Opt. Lett.*, *20*, 2261–2263 (1995).
37. Sutherland, R. L., Effects of multiple internal sample reflections on nonlinear refractive Z-scan measurements, *Appl. Opt.*, *33*, 5576–5584 (1994).
38. Oliveira, L. C., and Zilio, S. C., Single-beam time-resolved Z-scan measurements of slow absorbers, *Appl. Phys. Lett.*, *65*, 2121–2123 (1994).

39. Petrov, D. V., Gomes, A. S. L., and de Araujo, C. B., Reflection Z-scan technique for measurement of optical properties of surfaces, *Appl. Phys. Lett.*, *65*, 1067–1069 (1994).
40. Kershaw, S. V., Analysis of the EZ-scan measurement technique, *J. Mod. Opt.*, *42*, 1361–1366 (1995).
41. Ma, H., and de Araujo, C. B., Two-color Z-scan technique with enhanced sensitivity, *Appl. Phys. Lett.*, *66*, 1581–1583 (1995).
42. Present paper, see Table 1.
43. Winter, C., Manning, R., Oliver, S., and Hill, C., Measurement of the large optical nonlinearity of nickel dithiolene doped polymers, *Opt. Commun.*, *90*, 139–143 (1992).
44. Oliver, S. N., Winter, C. S., Manning, R. J., Rush, J. D., Hill, C., and Underhill, A. E., Nickel Dithiolene-PMMA guest-host polymers for all optical signal switching, *Proc. SPIE*, *1775*, Nonlinear Optical Properties of Organic Materials V, 110–120 (1992).
45. Manning, R., private communication.
46. Underhill, A., Hill, C., Charlton, A., Oliver, S., and Kershaw, S., Third-order NLO properties of PMMA films co-dispersed with metal dithiolene oligomers, *Synthetic Metals*, *71*, 1703–1704 (1995).
47. Gall, G., King, T., Oliver, S., Capozzi, C., Seldon, A., Hill, C., and Underhill, A., Third-order optical properties of metal dithiolene and phthalocyanine doped sol-gel materials, *SPIE*, *2288*, Sol-Gel Optics, 372–381 (1994).
48. Lawrence, B., Cha, M., Kang, J., Torruellas, W., Stegeman, G., Baker, G., Meth, J., and Etemad, S., Large purely refractive nonlinear index of single crystal P-toluene sulphonate (PTS) at 1600 nm, *Electron. Lett.*, *30*, 447–448 (1994).
49. Lawrence, B., Torruellas, W., Cha, M., Sundheimer, M., Stegeman, G., Meth, J., Etemad, S., and Baker, G., Identification of the role of two-photon excited states in a π-conjugated polymer, *Phys. Rev. Lett.*, *73*, 597–600 (1994).
50. Kim, D. Y., Lawrence, B., Torruellas, W., Stegeman, G., Baker, G., and Meth, J., Assessment of single crystal p-toluene sulfonate as an all-optical switching material at 1.3 μm, *Appl. Phys. Lett.*, *65*, 1742–1744 (1994).
51. Lawrence, B., Cha, M., Torruellas, W., and Stegeman, G., Measurement of the complex nonlinear refractive index of single crystal p-toluene sulfonate at 1064 nm, *Appl. Phys. Lett.*, *64*, 2773–2775 (1994).
52. Thakur M., and Quintero-Torres, R., The sign and magnitude of the off-resonant nonlinearities of polydiacetylene crystal measured by Z-scan, *SPIE*, *2025*, 446–449 (1993).
53. Cha, M., Torruellas, W., Stegeman, G., Horsthuis, W., Mohlmann, G., and Meth, J., Two-photon absorption of di-alkyl-amino-nitro-stilbene side chain polymer, *Appl. Phys. Lett.*, *65*, 2648–2650 (1994).
54. Yang, L., Dorsinville, R., Alfano, R., Zou, W., and Yang, N., Sign of $\chi^{(3)}$ in polysilane polymers, *Opt. Lett.*, *16*, 758–760 (1991).
55. Hein, J., Bergner, H., Lenzner, M., and Rentsch, S., Determination of real and imaginary part of $\chi^{(3)}$ of thiophene oligomers using the Z-scan technique, *Chem. Phys.*, *179*, 543–548 (1994).
56. Yang, L., Dorsinville, R., Wang, Q., Ye, P., and Alfano, R., Excited-state nonlinearity in polythiophene thin films investigated by the Z-scan technique, *Opt. Lett.*, *17*, 323–325 (1992).
57. Samoc, M., Swiatkiewicz, J., Samoc, A., and Luther-Davies, B., Third-order nonlinear optical properties of pseudoazulenes, *Acta Physica Polonica A*, *88*, 411–416 (1995).
58. Planas, S., Duarte, A., Mazzali, C., Palange, E., Fragnito, H., Cardosa, V., Sanches, M., Vallani, M., and Reggiani, A., Nonlinear refractive index of dye-doped organic polymers, *SPIE*, *2025*, 381–387 (1993).
59. Van Stryland, E. W., Sheik-Bahae, M., and Hagan, D. J., Characterization and Modeling of Nonlinear Optical Absorption and Refraction, in *Research Trends in Nonlinear and*

Quantum Optics, American Institute of Physics, Research Trends in Physics Series, Institute for Advanced Studies Press, 527–578 (1997).
60. Fleitz, P., Sutherland, R., and Bunning, T., Z-scan measurements on molten diphenylbutadiene in the isotropic liquid state, in *Mat. Res. Soc. Symposium Proceedings*, Vol. 374 (R. Crane, K. Lewis, E. Van Stryland and M. Koshnevisan, eds.), MRS, Pittsburgh, 1995, pp. 211–216.
61. Rockwell, B., Roach, W., Rogers, M., Mayo, M., Toth, C., Cain, C., and Noojin, G., Nonlinear refraction in vitreous humor, *Opt. Lett.*, *18*, 1792–1794 (1993).
62. Zhai, T., Lawson, C., Burgess, G., Lewis, M., Gale, D., and Gray, G., Nonlinear-optical studies of molybdenum metal organics, *Opt. Lett.*, *19*, 871–873 (1994).
63. Hofer, T., Kruck, P., Elsaesser, T., and Kaiser, W., Transient states of an intramolecular proton transfer cycle studied by degenerate four-wave mixing, *J. Phys. Chem.*, *99*, 4380–4385 (1995).
64. Fei, H., Wei, Z., Yang, Q., Che, Y., Shen, Y., Fu, X., and Qui, L., Low-power phase conjugation in push-pull azobenzene compounds, *Opt. Lett.*, *20*, 1518–1520 (1995).
65. Song, Q., Zhang, C., Gross, R., and Birge, R., Optical limiting by chemically enhanced bacteriorhodopsin films, *Opt. Lett.*, *18*, 775–777 (1993).
66. Song, Q., Zhang, C., Gross, R., and Birge, R., The intensity-dependent refractive index of chemically enhanced bacteriorhodopsin, *Opt. Commun.*, *112*, 297–301 (1994).
67. Shi, S., Ji, W., Kie, W., Chong, T. C., Zeng, H. C., Lang, J. P., and Kin, X. Q., The mixed metal cluster $(n\text{-}Bu_4N)_2[MoCu_3OS_3(NCS)_3]$: The first example of a nest-shaped compound with large third-order polarizability and optical limiting effect, *Mat. Chem. Phys.*, *39*, 298–303 (1995).
68. Shi, S., Ji, W., and Xin, X. Q., Synthesis and superior third-order nonlinear optical properties of the cluster $(n\text{-}Bu_4N)_4[Mo_8Cu_{12}O_8S_{24}]$, *J. Phys. Chem.*, *99*, 894–898 (1995).
69. Shi, S., Hou, H. W., and Xin, X. Q., Solid state synthesis and self-focusing and nonlinear absorptive properties of two butterfly-shaped clusters $WCu_2OS_3(PPh_3)_4$ and $MoCu_2OS_3(PPh_3)_3$, *J. Phys. Chem.*, *99*, 4050–4053 (1995).
70. Shi, S., Ji, W., Lang, J. P., and Xin, X. Q., New nonlinear optical chromophore: Synthesis, structures, and optical limiting effect of transition-metal clusters $(n\text{-}Bu_4N)_3[WM_3Br_4S_4]$ (W = Cu and Ag), *J. Phys. Chem.*, *98*, 3570–3572 (1994). See also Shi, S., Ji, W., and Tang, S. H., Synthesis and optical limiting capability of cubane-like mixed metal clusters $(n\text{-}Bu_4N)_3[MoAg_3BrX_3S_4]$ (X = Cl and I), *J. Am. Chem. Soc.*, *116*, 3615-3616 (1994).
71. Gu, G., Zhang, W., Zen, H., Du, Y., Han, Y., Zhang, W., Dong, F., and Xia, Y. Large nonlinear absorption in C_{60} thin films, *J. Phys. B: At. Mil. Opt. Phys.*, *26*, L451–L455 (1993).
72. Mishra, S. R., Rawat, H. S., Joshi, M. P., and Mehendale, S. C., The role of nonlinear scattering in optical limiting in C_{60} solutions, *J. Phys. B: At. Mil. Opt. Phys.*, *27*, L157–L163 (1994).
73. Justus, B., Kafafi, Z., and Huston, A., Excited-state absorption-enhanced thermal optical limiting in C_{60}, *Opt. Lett.*, *18*, 1603–1605 (1993).
74. Yang, L., Royer, E., Walser, A., and Dorsinville, R., Nonlinear optical responses of C_{70}-toluene solutions, *Chem. Phys. Lett.*, *239*, 390–394 (1995).
75. Perry, J. W., Khundkar, L. R., Coulter, D. R., Wei, T. H., Van Stryland, E. W., and Hagan, D. J., Excited State Absorption and Optical Limiting in Solutions of Metallo-phthalocyanines, proceedings of the NATO workshop on "Organic Materials for Nonlinear Optics and Photonics", La Rochelle, France, August 26–31, 1990.
76. Wood, G., Miller, M., and Mott, A., Investigation of tetrabenzporphrin by the Z-scan technique, *Opt. Lett.*, *20*, 973–975 (1995).

77. Wood, G., Mott, A., and Miller, M., Nonlinear optical properties of tetrabenzporphyrin, in *Mat. Res. Soc. Symposium Proceedings*, Vol. 374 (R. Crane, K. Lewis, E. Van Stryland and M. Koshnevisan, eds.), MRS, Pittsburgh, 1995, pp. 267–273.
78. Swatton, S. N. R., Welford, K., Till, S., and Sambles, J., Nonlinear absorption of a carbocyanine dye 1,1′,3,3,3′,3′-hexamethylindotricarbocyanine iodide using a Z-scan technique, *Appl. Phys. Lett.*, *66*, 1868–1870 (1995).
79. Swatton, S., Welford, K., Ray, S. C., and Till, S., Wavelength dependence of the non-linear transmission of HITCI using the z-scan technique, in *Mat. Res. Soc. Symposium Proceedings*, Vol. 374 (R. Crane, K. Lewis, E. Van Stryland and M. Koshnevisan, eds.), MRS, Pittsburgh, 1995, pp. 173–177
80. Welford, K., Swatton, S., Hughes, S., Till, S., Spruce, G., Hollins, R., and Wherrett, B. S., Nonlinear absorption in organic dyes, in *Mat. Res. Soc. Symposium Proceedings*, Vol. 374 (R. Crane, K. Lewis, E. Van Stryland and M. Koshnevisan, eds.), MRS, Pittsburgh, 1995, pp. 239–256.
81. Van Stryland, E. W., Sheik-Bahae, M., Xia, T., Wamsley, C., Wang, Z., Said, A. A., and Hagan, D. J., Z-scan and EZ-scan Measurements of Optical Nonlinearities, *International Journal of Nonlinear Optical Physics, Novel Nonlinear Optical Polymers and Organic Materials*, *IJNOP*, *3*, 489–500 (1994).
82. Herbert, C., Capinski, W., and Malcuit, M., Optical power limiting with nonlinear periodic structures, *Opt. Lett.*, *17*, 1037–1039 (1992).
83. Yuan, H. J., Li, L., and Palffy-Muhoray, P., Nonlinear birefringence of nematic liquid crystals, *SPIE*, *1307*, Electro-Optical Materials for Switches, Coatings, Sensor Optics and Detectors, 363–373 (1990).
84. Janossy, I., and Kosa, T., Influence of anthraquinone dyes on optical reorientation of nematic liquid crystals, *Opt. Lett.*, *17*, 1183–1185 (1992). See also Paparo, D., Maddalena, P., Abbate, G., Santamato, E., and Janossy, I., Wavelength Dependence of Optical Reorientation in Dye-Doped Nematics, *Mol. Cryst. Liq. Cryst.*, *251*, 73–84 (1994).
85. Toda, H., and Verber, C., Simple technique to reveal a slow nonlinear mechanism in a Z-scan like n_2 measurement, *Opt. Lett.*, *17*, 1379–1381 (1992).
86. Tian, J.-G., Zhang, C., Zhang, G., The origin of optical nonlinearities of Chinese tea, *Optik*, *90*, 1–4 (1992).
87. Tian, J.-G., Zhang, C., Zhang, G., and Li, J., Position dispersion and optical limiting resulting from thermally induced nonlinearities in Chinese tea liquid, *Appl. Opt.*, *32*, 6628–6632 (1993).
88. Cheung, Y., and Gayen, S., Optical nonlinearities of tea studied by Z-scan and four-wave mixing techniques, *J. Opt. Soc. Am.*, *B11*, 636–643 (1994).
89. Van Stryland, E. W., Application of Nonlinear Optics to Passive Optical Limiting, in *Nonlinear Optics of Organic Molecular and Polymeric Materials* (H. S. Nalwa and S. Miyata, eds.), 841–860, CRC Press, Boca Raton (1995).
90. See, for example, Shen, Y. R., *The Principles of Nonlinear Optics*, John Wiley and Sons, New York, 1984.
91. Boggess, T. F., Moss, S. C., Boyd, I. W., and Smirl, A. L., Nonlinear optical energy regulation by nonlinear refraction and absorption in silicon, *Opt. Lett.*, *9*, 291–293 (1984).
92. See, for example, Kosar, in *Light Sensitive Systems*, J. Wiley and Sons, New York, 1965. See also Bosomworth, D. R., and Gerritson, H. J., Thick holograms in photochromic materials, *Appl. Opt.*, *7*, 95 (1968).
93. Mansour, N., Mansour, K., Van Stryland, E., and Soileau, M. J., Diffusion of color centers generated by two-photon absorption at 532 nm in cubic zirconia, *J. Appl. Phys.*, *67*, 1475–1477 (1989).

9
Four Wave Mixing

R. A. Norwood
AlliedSignal Inc.
Morristown, New Jersey

I. INTRODUCTION

All third order nonlinear optical phenomena are fundamentally four wave mixing (FWM) events, meaning that the sum of the number of incoming and outgoing photons is four. This includes the generalization to quadratic electro optic phenomena in which two of the photons are at zero frequency. In other chapters of this volume, specific four wave mixing phenomena such as third harmonic generation, self-focusing, and the quadratic electro optic effect have been considered. Here we cover all other FWM techniques and measurements of organic materials and composites using them. A general FWM quantum electrodynamic diagram is given in Fig. 1, where we depict the virtual excitations that occur within a molecular system as it interacts with an electromagnetic field containing frequency components at ω_1, ω_2, ω_3, and ω_4. As is conventional, each vertex corresponds to a transition dipole matrix element with a corresponding resonant energy denominator.

Among the FWM phenomena we will discuss, two, the optical Kerr effect (OKE) and degenerate four wave mixing (DFWM), hold special significance since they relate most closely to possible commercial applications of third order nonlinear optics. The OKE was first observed in 1964 by Mayer and Gires[1] and is the optical analog of the well-known Kerr effect,[2] also known as the quadratic electro optic effect. The role of the strong dc field is replaced by an intense ac field, a pump laser beam. In materials with large electronic nonlinearities, the OKE can be used to make ultrafast light gates[3] or shutters[4] that could have use as optical processing elements.

DFWM in its most popular geometry (the phase conjugate configuration) was predicted and demonstrated in 1977 by Hellwarth.[5] It is essentially real-time holography in which a phase or amplitude grating is written in a nonlinear optical medium by two strong pumping beams, and detected by the diffraction of a third

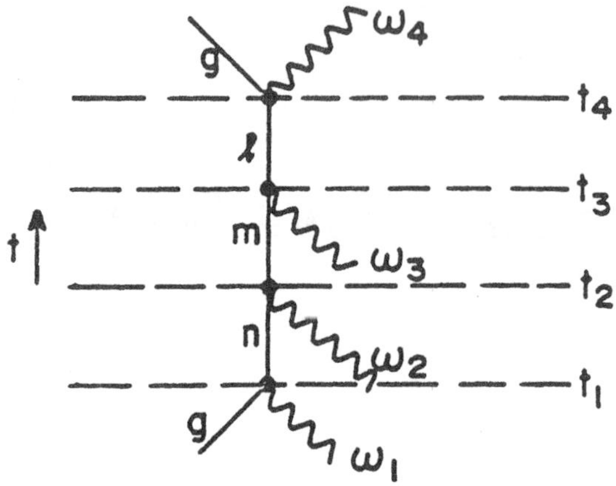

Figure 1 General quantum electrodynamic diagram of a four wave mixing interaction among fields at frequencies ω_1, ω_2, ω_3, and ω_4. Time increases up the page and virtual dipole interactions occur at the vertices between different virtual states.

weak probe beam. DFWM is especially favored for thin film studies owing to its high sensitivity and relatively large resonant enhancement (when performed at visible wavelengths). However, this sensitivity is purchased at the price of a multitude of mechanisms that can lead to DFWM signals, including electronic, orientational, and thermal effects. This contrasts with third harmonic generation (THG) for example (see Chapter 10), where the sensitivity of the experiment is relatively low, but only one mechanism can lead to the effect (electronic in this case).

In the following section we will discuss the conventions that prevail in the literature when discussing various four wave mixing type measurements, both for macroscopic and microscopic nonlinearities. We will then present the conventions that we will adopt in this work, striving to be consistent with the rest of this handbook. In Section III we discuss the various four wave mixing techniques, in each case giving the basic description in terms of Maxwell's equations, the microscopic origins of the effects, and some typical experimental configurations and methods. Note that it is not the intention of the author to give an exhaustive description of FWM techniques in terms of their ability to elucidate various physical phenomena, but rather to concentrate on the most prevalent techniques and how they are used to measure the third order nonlinearity. Section IV presents tables of n_{2B} (the intensity-dependent refractive index), $\chi^{(3)}$ (the molecular third order susceptibility) and γ (the bulk third order susceptibility), values for a variety of organic molecules and polymers. For each measurement we tabulate the material, the method used, the characteristics of the lasers used (wavelength, pulse length, repetition rate), as well as the relevant nonlinearity.

II. CONVENTIONS

A. Literature

1. *Macroscopic*

Units. Based on our review of the literature, the prevailing tendency is still to use Gaussian units for communicating the values of macroscopic optical nonlinearities obtained from four wave mixing measurements.

OKE. There are three prominent macroscopic conventions for quantifying the optical Kerr effect. The *optical Kerr constant*, B_λ, is given by:

$$B_\lambda = \frac{\delta n_\parallel - \delta n_\perp}{\lambda \langle E^2 \rangle}, \tag{1}$$

where $\delta n_\parallel$ and $\delta n_\perp$ are the changes in the refractive indices parallel and perpendicular to the intense pump field that are *induced* by that field at the probe wavelength λ, E is the intense pump electric field (assumed to be a plane wave), and the brackets denote a time average.

The second convention employs the familiar nonlinear refractive index n_{2B} through the relationship:

$$\delta n_\parallel - \delta n_\perp = \frac{1}{2} n_{2B} \langle E^2 \rangle. \tag{2}$$

Note that n_{2B} is a value explicitly obtained via an optically induced birefringence measurement, and may or may not relate directly to values of n_2 as obtained by other techniques. Clearly, $n_{2B} = 2\lambda B_\lambda$.

Finally, optical Kerr effect susceptibilities are often given directly in terms of the third order susceptibility; in this case, $\chi^{(3)}_{eff}$ is given by:

$$\chi^{(3)}_{eff}(-\omega_2;\omega_1,-\omega_1,\omega_2) = \frac{1}{2}\{\chi^{(3)}_{1212}(-\omega_2;\omega_1,-\omega_1,\omega_2) + \chi^{(3)}_{1221}(-\omega_2;\omega_1,-\omega_1,\omega_2)\}, \tag{3}$$

where we have included the explicit frequency-dependence of the susceptibilities at the probe frequency ω_2 and the pump frequency ω_1.

Note that each of the above definitions is primarily useful for isotropic media such as liquids and amorphous polymers. For crystalline materials or oriented polymer media, more complex expressions would obtain. In isotropic materials the following relationship exists between the unique nonvanishing components of $\chi^{(3)}$:

$$\chi^{(3)}_{1111} = \chi^{(3)}_{1212} + \chi^{(3)}_{1221} + \chi^{(3)}_{1122}, \tag{4}$$

and using Eq. 4 we can equivalently write Eq. 3 as:

$$\chi^{(3)}_{eff}(-\omega_2;\omega_1,-\omega_1,\omega_2) = \frac{1}{2}\{\chi^{(3)}_{1111}(-\omega_2;\omega_1,-\omega_1,\omega_2) - \chi^{(3)}_{1122}(-\omega_2;\omega_1,-\omega_1,\omega_2)\}. \tag{5}$$

DFWM. Literature values of DFWM nonlinearities are tabulated almost exclusively in terms of the various components of $\chi^{(3)}$. Measurements typically concentrate on the completely diagonal component, $\chi^{(3)}_{1111}$, as well as the off-diagonal components $\chi^{(3)}_{1221}$, $\chi^{(3)}_{1212}$, and $\chi^{(3)}_{1122}$. Occasionally, authors will relate their measured

$\chi^{(3)}$ tensor components to the nonlinear refractive index, n_2. In these cases, one must be careful to determine what definition of n_2 is being used, as pointed out in Chapter 1.

2. Microscopic

Units. In this case, there is an increasing tendency to use MKS units, but most past measurements used Gaussian units, resulting in the majority of results being in Gaussian.

General. The major difficulty in going from macroscopic third order susceptibility measurements using FWM to the second hyperpolarizability γ_{ijkl} is the proper isolation of the mechanisms involved in the macroscopic nonlinearity. We will discuss this in more detail below in the Section III (Techniques), but a few general comments can be made here. For measurements made on solid materials, the primary concerns in relating macroscopic measurements to microscopic quantities include:

1. local field effects
2. influence of orientational order
3. effects of aggregations/phase separation
4. excited state population effects
5. thermal effects.

Thermal effects are of particular concern in FWM measurements, especially those performed in the phase conjugate geometry with nanosecond pulses (see below). For liquid media, the role of optical field induced orientational order is large, and in fact in the majority of liquids composed of anisotropic molecules these effects dominate the nonlinear optical response.

B. This Work

Despite the prevailing tendencies in the literature, we will give all tabulated results in MKS units to be consistent with the optical engineering philosophy of this volume. In particular, we use the MKS(I) units of Shi and Garito, in which case $\chi^{(3)}$ has units of m^2V^{-2} and γ_{ijkl} has units of m^5V^{-2}.

III. TECHNIQUES

Four wave mixing encompasses a rich set of techniques that focus variously on measurement of optical nonlinearities, spectroscopy, and studies of atomic and molecular dynamics. As stated above, it is our intention to concentrate on the techniques that are best and most commonly used to measure third order optical nonlinearities. Nevertheless, we would be remiss not to direct the reader to a number of excellent references in the detailed polarization characteristics and dynamics of FWM.[6–15]

A. OKE

The OKE specifically refers to optical field induced birefringence but is often used more broadly to refer to any optically induced change in the refractive index of a material. In this section, we discuss some of the common experimental configurations for the OKE, give the macroscopic electric field expression for the OKE signal as a function of experimental parameters in the most common pulsed pump-cw probe configuration, and discuss the microscopic mechanisms leading to the OKE.

1. Configurations

A large number of the OKE studies that have been done have been performed using the experimental setup shown generically in Fig. 2, referred to as pump/CWprobe. An intense linearly polarized pulsed laser is used to induce the birefringence in the nonlinear medium, which is placed between crossed polarizers. This laser, referred to as the *pump*, is typically either Q-switched (ns pulses) or modelocked (ps) pulses and operating at repetition rates in the tens of Hz range. In general, shorter pulses are preferred, since a ps pulse of the same energy as a nanosecond pulse will have a thousand times the peak power, and for an instantaneously responding medium an OKE signal enhanced by six orders of magnitude. Low pulse energies are preferred to reduce the influence of thermally induced nonlinearities. Furthermore, the *beam shape* of the pump is important; it should be as close to Gaussian as possible to justify the standard plane-wave approximations. A weak linearly polarized CWprobe laser traverses the medium in a spatially coincident manner with the pump and it is the increased transmission of this probe beam through the second polarizer in the presence of the pump that constitutes the OKE signal. The probe laser should have excellent amplitude, pointing and frequency stability, and should also have a quasi-uniform beam shape. The small amount of probe light making it through the second polarizer is passed through filters to remove residual pump light and dumped onto a photomultiplier tube. Half wave plates are used to adjust the pump polarization to be at 45 degrees to the probe, which maximizes the OKE signal (see below). Often, the pump power is adjusted using filters or a half wave plate/polarizer attenuator to observe the power dependence of the OKE signal and assure that no other nonlinear effects such as stimulated Brillouin scattering, self-focusing, or stimulated Raman scattering are occurring.

Several minor changes can be made to the pump/CWprobe method that are advantageous for certain experimental objectives. The use of a *pulsed probe*[16] temporally coincident with the pump helps to reduce the problem of CWprobe light continually leaking through the second polarizer, thereby increasing the signal to noise ratio. However, this approach complicates the analysis of the nonlinear behavior since the overall effect is now a convolution of the pump temporal intensity profile, the probe temporal intensity profile, and the nonlinear response of the medium. A novel way to achieve the same objective of an increased signal to noise ratio without complicating the data analysis is to use an *optical gate*[17] whereby an additional portion of the pump beam is used to upconvert the OKE signal in a nonlinear mixing crystal positioned after the second polarizer (Fig. 3). Since the response of the nonlinear mixing crystal is electronic, the aforementioned convolution integral is considerably simplified. One can further reduce the susceptibility of pump/CWprobe to systematic errors such as pump beam temporal instabilities by providing a reference arm to the experiment, in which a standard material such as CS_2 is resident.

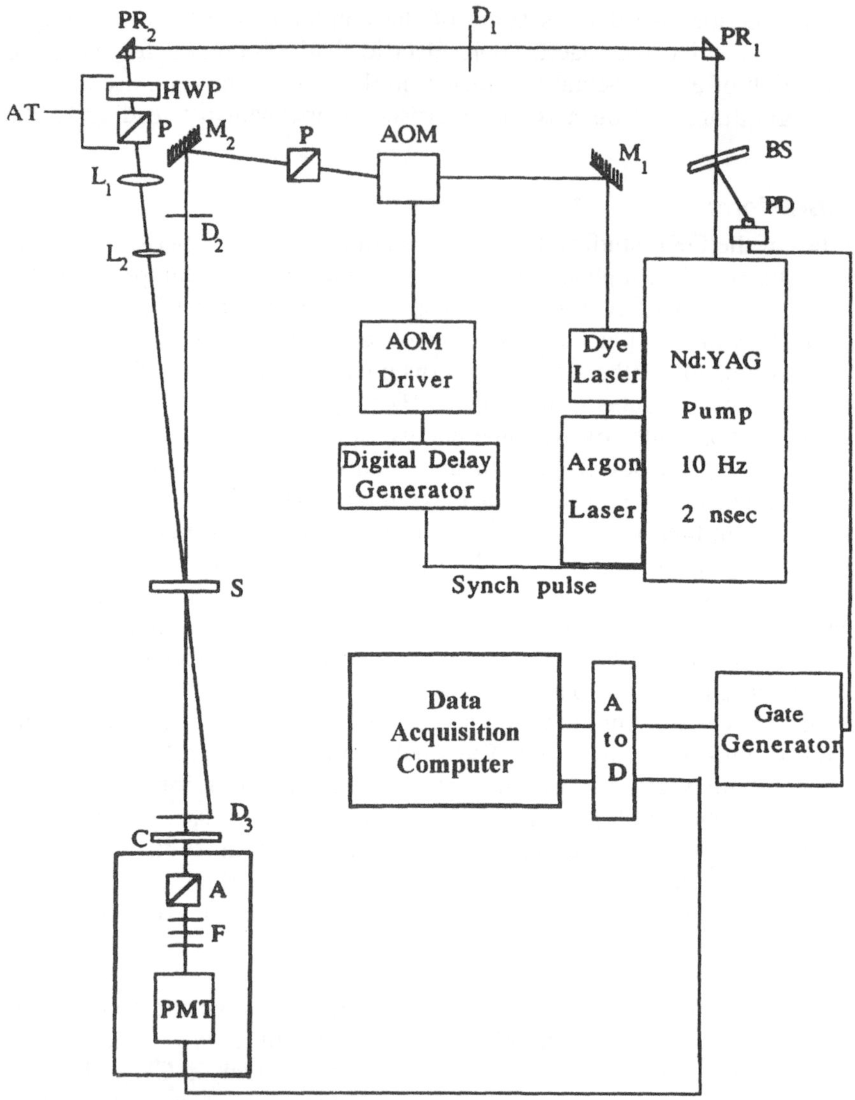

Figure 2 Standard optical Kerr effect experiment using a nanosecond pump laser at 1.06 μm and a tunable cw laser. Beamsplitter BS reflects a small part of the pump into photodiode PD, that provides a current pulse for data acquisition. Prism reflectors PR_1 and PR_2, along with diaphragm D_1, help to deliver the pump to the half-wave plate (HWP)/polarizer (P) combination used to attenuate (AT) the pump. Lenses L_1 and L_2 form a gently focusing beam at sample S and residual pump is dumped to diaphragm D_3. The probe is reflected from mirror M_1; the acousto-optic modulator (AOM) is used to reduce cw laser noise when the pump is not present. The probe is directed through diaphragms D_2 and D_3, through compensator C which nulls stray birefringence, analyzer A, and filters F before entering the detecting photomultiplier tube (PMT).

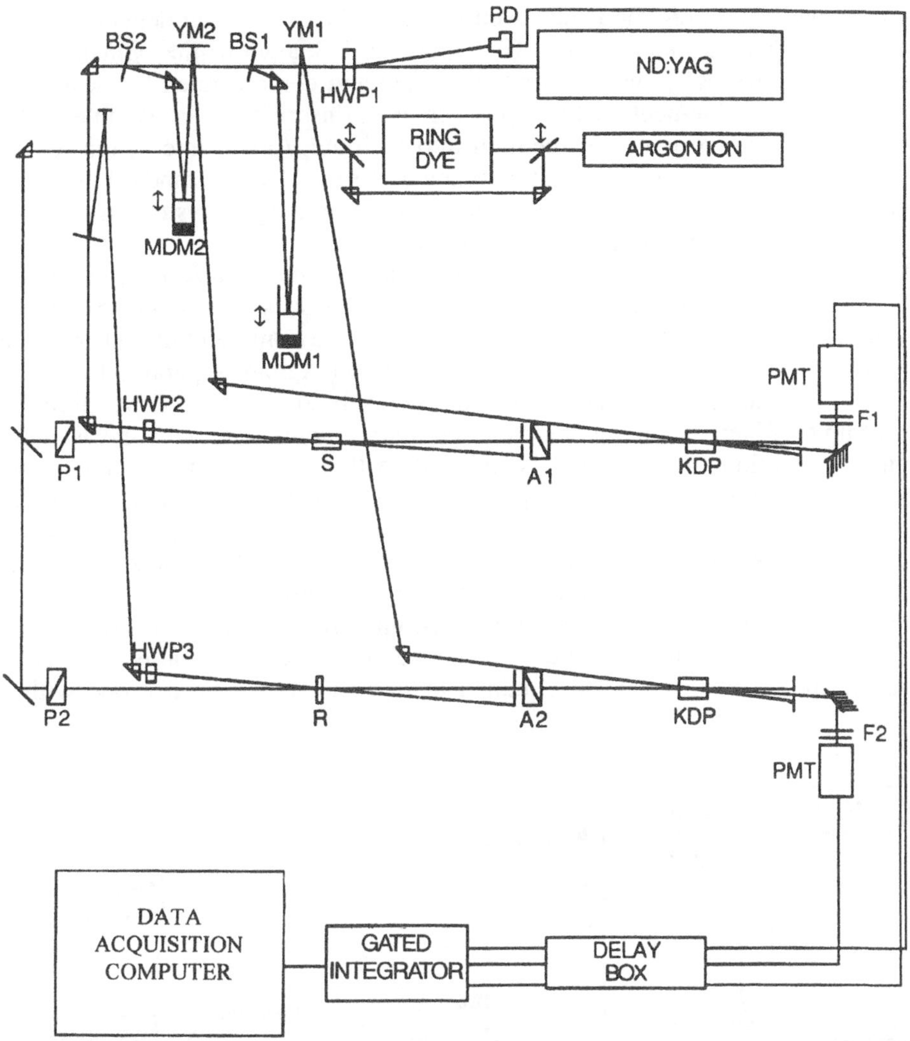

Figure 3 Fully referenced picosecond optical Kerr effect apparatus using an optical gate. The optical gates after the Kerr cells act to suppress background leakage of the cw probe beam through the crossed polarizer by dramatically reducing the averaging time of the low dark current. A1-A2: analyzers; BS1-BS2: beamsplitters; F1-F2: filters; HWP1-HWP3: half-wave plates; KDP: sum frequency generation crystal; MDM1-MDM2: motor driven mirrors; P1-P2: polarizers; PD: photodetector; PMT: photomultiplier tube; R: reference; S: sample; YM1-YM2: "YAG" mirrors.

As shown below, pump/CWprobe results in an OKE signal that is quadratically dependent on $\chi^{(3)}_{eff}$, which greatly compresses the dynamic range of the experiment as regards the measurement of materials with greatly differing nonlinearities. One approach to this problem is to slightly uncross the second polarizer in such a way that the measured signal now becomes the result of "interference" between the OKE signal and the probe light leaking through. By proper adjustment of the signal levels, an experimental regime is achieved in which the total signal depends nearly linearly on the material nonlinearity. This approach is known as the *optically heterodyned* OKE (OHOKE) and was first suggested and demonstrated by Scarpano, et al.[18] A further refinement to this approach is to insert both the sample to be measured and a known nonlinear medium into the pump/probe interaction region in pump/CWprobe.[19] The heterodyning is now performed by the standard material, and the interference of its signal with that of the sample again results in a signal that is linearly dependent on $\chi_{eff}^{(3)}$ of the sample.

Perhaps the most innovative configuration is the *Raman-induced* Kerr effect (RIKE).[20] The basic setup is shown in Fig. 4 and consists of two pulsed lasers, preferably individually tunable, one of which acts as the pump and the other as the probe. The frequencies of the two lasers, ω_{pump} and ω_{probe}, are separated by $\Delta\omega = |\omega_{pump} - \omega_{probe}|$, and this difference corresponds closely to a Raman active vibration of the sample. When the pump beam is *circularly polarized*, only the Raman-resonant contribution to the signal is induced, with the "ordinary" OKE being suppressed. By tuning the probe frequency, the Raman spectrum of the medium can be taken with great accuracy and zero background. By *elliptically* polarizing the pump beam contributions from the Raman resonance and the nonresonant OKE interfere. The lineshape that results (Fig. 5) contains information about both the Raman spectrum and $\chi_{eff}^{(3)}$, the latter being determined from the positions of minima and maxima in the spectrum. If the only objective of the measurement is to measure the Raman spectrum, this can be obtained in one shot by using a probe with a

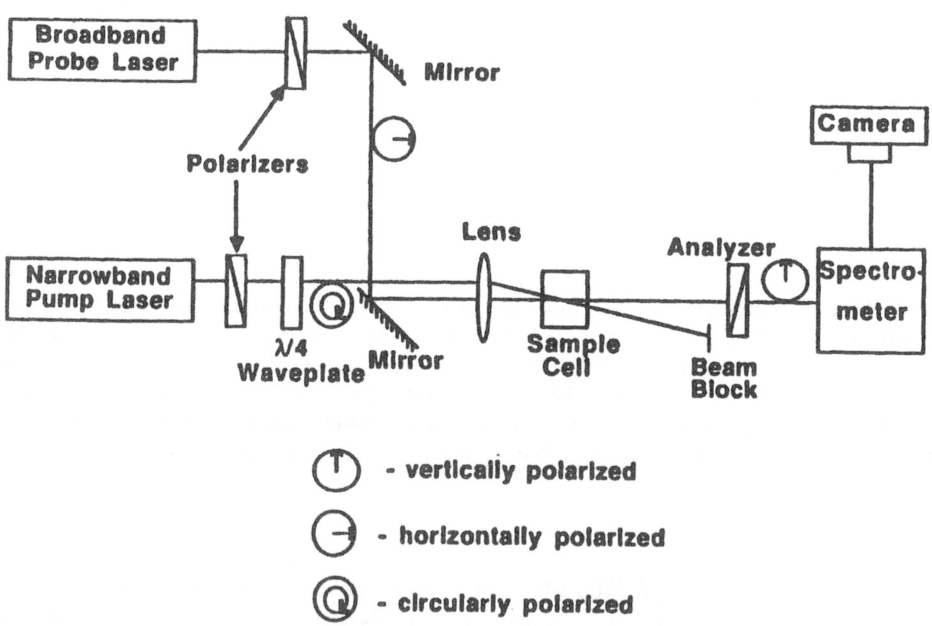

Figure 4 Schematic of a Raman-induced Kerr effect experimental apparatus. Note that the pump beam is circularly polarized, leading to no signal away from resonance. Nonresonant nonlinearities are measured by observing the interference between nonresonant and resonant nonlinearities.

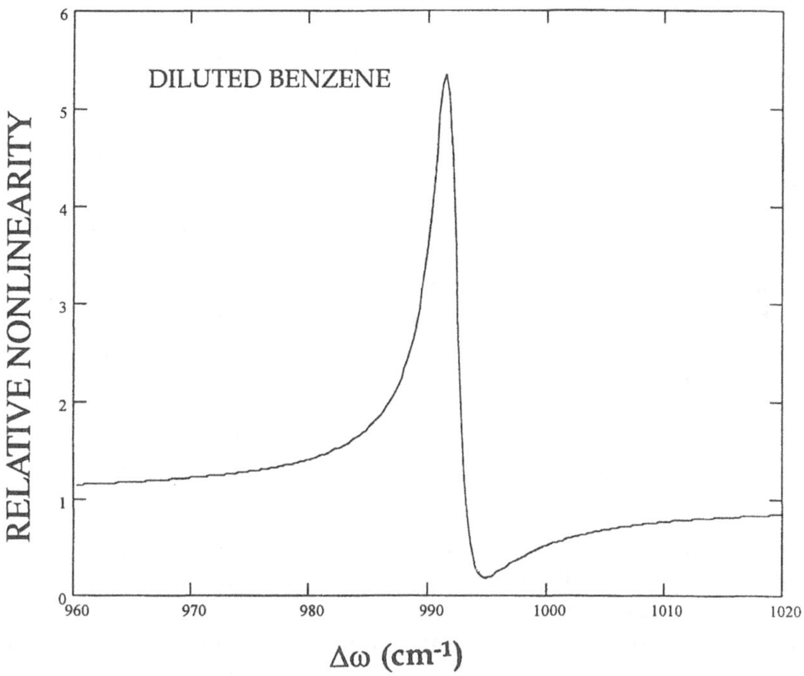

Figure 5 Lineshape obtained in a typical RIKES experiment. The positions of the minima and maxima are directly related to the nonresonant $\chi^{(3)}$ of the material.

bandwidth significantly larger than the Raman resonances to be measured. Surprisingly, this powerful technique has seen little use since its invention.

2. *Standard Pump-probe Analysis Result*

In the slowly varying amplitude approximation, assuming negligible absorption of both pump and probe lasers, and in the limit of small optically induced birefringence, the expression for the *instantaneous* probe power, P_{tran}, transmitted through the second polarizer in the pump/CWprobe configuration is given by[21]:

$$P_{tran} = \frac{KP_{probe}P_{pump}^2 L^2 \sin^2 2\theta}{\lambda_{probe}^2 n_0^3(\lambda_{probe}) n_0^2(\lambda_{pump}) \sigma_{pump}^2} (\chi_{eff}^{(3)})^2, \tag{6}$$

where:

K = numerical constants equal to $4 \cdot (96\pi^3)^2 c^{-2}$ in Gaussian units
P_{probe} = incident probe beam power
P_{pump} = pump beam power
L = sample length or length of pump/probe interaction region, which ever is shorter
θ = angle between the pump and probe polarization vectors
$\chi_{eff}^{(3)}$ = as defined in Eq. (3)
λ_{probe} = probe wavelength
λ_{pump} = pump wavelength

$n_0(\lambda_{probe})$ = the linear refractive index of the sample at λ_{probe}
$n_0(\lambda_{pump})$ = the linear refractive index of the sample at λ_{pump}
σ_{probe} = area of the probe beam
σ_{pump} = area of the pump beam.

Note that Eq. 6 is replaced with a time integral when the response is not instantaneous. In principle careful measurement of all of the quantities in Eq. 6 allows an absolute measurement of $\chi^{(3)}_{eff}$. In practice, investigators ordinarily reference measurements on unknown samples to a known material by consecutive measurements in the same apparatus with all conditions as similar as possible. An expression identical to Eq. 6 will obtain for the reference material, with the only sample specific quantities being the sample and reference lengths, refractive indices, and $\chi^{(3)}_{eff}$'s. It is then straightforward to show:

$$\frac{\chi^{(3)\,sample}_{eff}}{\chi^{(3)\,reference}_{eff}} = \frac{n_0^{sample}(\lambda_{pump})}{n_0^{reference}(\lambda_{pump})} \frac{L^{reference}}{L^{sample}} \left(\frac{n_0^{sample}(\lambda_{probe})}{n_0^{reference}(\lambda_{probe})}\right)^{3/2} \left(\frac{P^{sample}_{tran}}{P^{reference}_{tran}}\right)^{1/2}, \quad (7)$$

where the superscripts label the sample and reference. In this case, all that is required is knowledge of the refractive indices of the sample and reference at λ_{pump} and λ_{probe}, the reference and sample lengths, the reference $\chi^{(3)}_{eff}$, and the measured OKE signals. Changes to Eq. 7 in the presence of probe absorption are straightforward, in which case the reference and sample lengths are replaced by effective lengths obtained from Beer's law. Pump absorption will exhibit a more complex dependence; at any rate, pump absorption should be avoided since thermal effects will then become significant. The most common errors encountered in applying Eq. 7 result from differences between the extant experimental conditions and those used in the measurement being referred to, such as differences in pump beam pulse length, λ_{pump}, or λ_{probe}. The temporal response of the material must also be fast for both sample and reference if comparisons are to be meaningful. Every attempt should be made to refer to measurements employing experimental conditions as similar as possible to the extant conditions.

3. Mechanisms

In solid materials in which molecules are not free to rotate the predominant microscopic phenomena leading to the OKE is nonlinear distortion of the electron cloud, quantified by the second hyperpolarizability, γ_{ijkl}. If time-dependent quantum mechanical perturbation theory is used in the usual way, taking care to remove secular divergences, then for pump/CWprobe configuration we have[22]:

$$\gamma_{ijkl}(-E_{probe};E_{probe},-E_{pump},E_{pump})$$
$$= \frac{1}{4}P(E_{probe},-E_{pump},E_{pump})$$
$$\left[\sum_{rst}{}' \left\{\frac{\langle\mu_i\rangle_{gr}\langle\overline{\mu}_j\rangle_{rs}\langle\overline{\mu}_k\rangle_{st}\langle\mu_l\rangle_{tg}}{(E_{rg}-E_{probe})(E_{sg}-E_{probe}-E_{pump})(E_{tg}-E_{probe})}\right.\right.$$
$$+\frac{\langle\mu_j\rangle_{gr}\langle\overline{\mu}_i\rangle_{rs}\langle\overline{\mu}_k\rangle_{st}\langle\mu_l\rangle_{tg}}{(E^*_{rg}+(-E_{pump}))(E_{sg}-E_{probe}-E_{pump})(E_{tg}-E_{probe})}$$
$$+\frac{\langle\mu_l\rangle_{gr}\langle\overline{\mu}_k\rangle_{rs}\langle\overline{\mu}_i\rangle_{st}\langle\mu_j\rangle_{tg}}{(E^*_{rg}+E_{pump})(E^*_{sg}+E_{probe}+E_{pump})(E_{tg}-(-E_{pump}))}$$

$$+ \frac{\langle\mu_l\rangle_{gr}\langle\overline{\mu}_k\rangle_{rs}\langle\overline{\mu}_j\rangle_{st}\langle\mu_i\rangle_{tg}}{(E_{rg}^* + E_{pump})(E_{sg}^* + E_{probe} + E_{pump})(E_{tg}^* + E_{probe})}\Bigg\}$$

$$- \sum_{st}{}' \Bigg\{ \frac{\langle\mu_i\rangle_{gs}\langle\mu_j\rangle_{sg}\langle\mu_k\rangle_{gt}\langle\mu_l\rangle_{tg}}{(E_{sg} - E_{probe})(E_{sg} - (-E_{pump}))(E_{tg} - E_{probe})}$$

$$+ \frac{\langle\mu_i\rangle_{gs}\langle\overline{\mu}_j\rangle_{sg}\langle\overline{\mu}_k\rangle_{gt}\langle\mu_l\rangle_{tg}}{(E_{sg} - (-E_{pump}))(E_{tg}^* + E_{pump})(E_{tg} - E_{probe})}$$

$$+ \frac{\langle\mu_j\rangle_{gs}\langle\mu_i\rangle_{sg}\langle\mu_l\rangle_{gt}\langle\mu_k\rangle_{tg}}{(E_{sg}^* + E_{probe})(E_{sg}^* + (-E_{pump}))(E_{tg} + E_{probe})}$$

$$+ \frac{\langle\mu_j\rangle_{gs}\langle\mu_i\rangle_{sg}\langle\mu_l\rangle_{gt}\langle\mu_k\rangle_{tg}}{(E_{sg}^* + (-E_{pump}))(E_{tg} - E_{pump})(E_{tg}^* + E_{probe})}\Bigg\}\Bigg], \tag{8}$$

where:

$$E_{probe} = hc/\lambda_{probe}$$
$$E_{pump} = hc/\lambda_{pump}$$
$$\vec{\mu} = e\vec{r}$$
$$E_{ab} = E_a - E_b + i\Gamma_{ab}$$
$$\langle\overline{A}\rangle_{mn} = \langle A\rangle_{mn} - \langle A\rangle_{gg},$$

and where g is the groundstate and r, s, and t the excited states. The prime on the sums indicates that the summation excludes the groundstate, Γ_{ab} is a phenomenological width, and $P(E_{probe}, -E_{pump}, E_{pump})$ represents the average of all terms generated by permuting the three frequencies in the expression. Equation 8 can be used to predict γ_{ijkl} for molecules simple enough to be described by a Hermitian Hamiltonian. However, Eq. 8 is also useful in elucidating important symmetry rules concerning the importance of the contributions of different molecular states, as has been demonstrated by a number of workers.[23,24]

In liquids, another more significant mechanism can lead to the OKE, namely optical field induced orientation as a result of molecular anisotropy. The dipole that is induced by the interaction of the molecular polarizability, α_{ij}, with the field, can then interact with the field with an energy of orientation given by $\vec{p}^* \cdot \vec{E}$, where $\vec{p}^*$ is the dressed induced dipole (already proportional to $\vec{E}$). The birefringence induced is then proportional to $|E|^2$ as expected for a third order process. The general expression for the bulk anisotropy of the OKE susceptibility is in the low concentration limit calculated from the second term in Eq. 310 in Chapter 3 with the polarizabilities α_{ij}^* written in terms of a sum-over states model (described in Chapter 2):

$$B(0) = \frac{N}{15kT}\left\{\sum_n \left(\frac{|\langle\mu_{z'}\rangle_{ng}|^2 - |\langle\mu_{x'}\rangle_{ng}|^2}{E_{ng} - E_{probe}} + \frac{|\langle\mu_{z'}\rangle_{ng}|^2 - |\langle\mu_{x'}\rangle_{ng}|^2}{E_{ng}^* - E_{probe}}\right)\right\}$$
$$\times \left\{\sum_m \left(\frac{|\langle\mu_{z'}\rangle_{mg}|^2 - |\langle\mu_{x'}\rangle_{mg}|^2}{E_{ng} - E_{pump}} + \frac{|\langle\mu_{z'}\rangle_{mg}|^2 - |\langle\mu_{x'}\rangle_{mg}|^2}{E_{ng}^* - E_{pump}}\right)\right\} \tag{9}$$

where N is the number density of molecules, T is the temperature, and all quantities are as defined following Eq. 8. Also, we have assumed a uniaxial molecule with its highest symmetry axis along z'. Note that in the case of high molecule concentration, the transition dipole moments are replaced with dressed moments and the energy levels are shifted according to Eq. 58 in Chapter 3. Also note that there are no multiphoton terms in this expression, as would be expected from a contribution that can best be thought of as a direct product of one-photon Feynman diagrams. For anisotropic molecules, such as [S8-1] *carbon disulfide* and *nitrobenzene* [S8-2], (9) is the predominant source of the optical Kerr effect for pulse lengths larger than the reorientation time of the molecule (well-modeled by the Debye theory). We note here that a number of excellent studies have been done using time-domain measurements to elucidate the various molecular mechanisms contributing to the OKE and liquids.[25–27]

B. Nondegenerate Four Wave Mixing or CARS

In NDFWM, there are two incident laser beams of frequency ω_1 and ω_2, with the intensity at ω_1 being much higher than at ω_2; the generated radiation is then at frequency $\omega_3 = 2\omega_1 - \omega_2$. NDFWM has the requirement of phasematching to achieve measurable efficiencies, expressed in terms of the respective wave vectors as $2\vec{k}_1 - \vec{k}_2 = \vec{k}_{3s}$. The technique was first applied to the measurement of third order susceptibilities by Levenson et al.,[28] and was referred to as three wave mixing (TWM). These authors showed that the NDFWM signal comes from three terms, a nonresonant term directly related to γ_{ijkl}, a resonant Raman term, and a resonant two-photon term. Interference between these terms creates spectra that look like derivative spectra and whose features (amplitudes, widths, positions) can be used to directly determine the nonresonant portion if the resonant portion (i.e., the Raman resonances) has been otherwise characterized, being in this way similar to RIKE.

The requirements of multifrequency, synchronized pulsed lasers and phasematching has limited the use of NDFWM, except for transient species detection (combustion diagnostics).[29] NDFWM excels in these studies because of the extremely sharp resonances that can be observed often without significant background. Recent advances in laser technology have made nonresonant NDFWM measurements more feasible, enabling measurements at narrowly spaced frequencies in the near infrared (1 μm).[30] At these wavelengths material dispersion is often low, leading to long coherence lengths and, in some instances, eliminating the need for phasematching. The primary motivation for solving the experimental difficulties of NDFWM is that it and third harmonic generation are the only third order processes that are exclusively *electronic* in origin.

C. Degenerate Four Wave Mixing

DFWM occurs when all the frequencies in the four wave mixing interaction are the same. This substantially reduces the complexity of the experiment, since only one laser source is needed. Furthermore, DFWM is self-phasematched (at least in isotropic materials), which results in an efficient interaction. DFWM can be applied to samples in either film or solution form. The penalty paid for the ease of the experiment is the complexity of the data interpretation, since DFWM can result from a number of different mechanisms, such as nonresonant and resonant electronic motions, nuclear motion (orientation), and thermally induced densification and rarefaction. Separation of these different effects typically requires measurement of several different tensor elements of $\chi^{(3)}$, or time-resolved measurements which seek to separate the effects via their characteristic on-set and relaxation times.

1. Configurations

Phase Conjugate/Backward. Several early investigators recognized that DFWM in certain geometries had the ability to create the phase conjugate of one of the incident beams if certain limiting assumptions were obeyed, namely that the pump waves are homogeneous over the interaction region and that the cross-sectional area of the interaction region is small compared to the distance to the view region.[31] The phase conjugate or time-reversed beam has the property that it is a complete time-reversed replica of the incident beam, and therefore any distortions that are introduced on the incident beam are removed in the phase conjugate. First observed in *carbon disulfide*,[32] phase conjugate DFWM rapidly received attention as a fundamental nonlinear optical phenomenon,[33] and was soon applied to the measurement of nonlinear susceptibilities.[34]

A typical apparatus for measuring $\chi^{(3)}$ with DFWM is shown in Fig. 6; two of the incident beams (typically the counterpropagating ones) are considered pump beams and will be of approximately equal intensity, while the third beam (the probe) will be of less intensity. Note that factors such as the laser pulse width, energy per pulse, coherence length, and beam shape can significantly effect the results obtained in a DFWM measurement, owing to the plethora of mechanisms that can produce the effect. Other key experimental considerations include sample length and form, the sample scattering, pump beam polarizations, and the distances that each of the incident beams travels from the output of the laser. The latter can be a particular experimental difficulty, since for picosecond or femtosecond lasers the point at which the beams overlap must be correct to within a millimeter or less if the material response times are instantaneous.

Since many of the materials included in this handbook exhibit absorption at typical DFWM measurement wavelengths, we include an analysis of phase conjugate DFWM that explicitly takes account of absorption.[35] We let the incident fields be described by:

$$E_i(\vec{r},t) = \tfrac{1}{2}\{A_i(\vec{r}_i)\exp[i(\vec{k}_i \cdot \vec{r} - \omega t)] + c.c.\}, \tag{10}$$

where:

$$\vec{k}_i = \hat{k}_i(k_i' + i\alpha/2), \tag{11}$$

with $k_i' = n\omega/c$ and α the absorption coefficient. Fields 1 and 2 correspond to the counterpropagating pump fields, field 3 is the incident probe field, and field 4 is the phase conjugate. The nonlinear polarization is given by:

$$P(\vec{r})_i = \frac{3}{4}\sum_{jkl} \times \{\chi^{(3)}_{ijkl}(-\omega;\omega,\omega,-\omega)A_1(r_1)_j A_2(r_2)_k A_3^*(r_3)_l \exp\{i\vec{r}\cdot(\vec{k}_1 + \vec{k}_2 - \vec{k}_3^*)\}\}. \tag{12}$$

The phase conjugate reflectivity, the ratio of the phase conjugate intensity to the incident probe intensity, is given by:

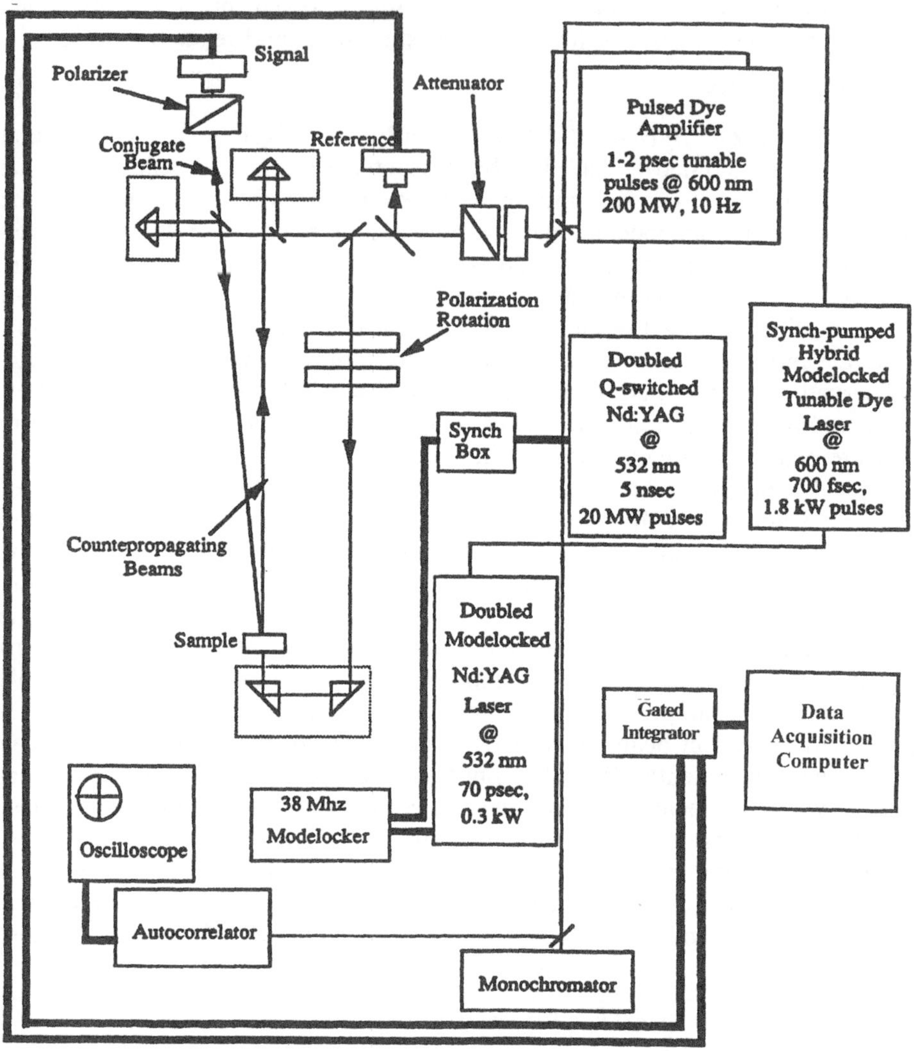

Figure 6 Standard picosecond degenerate four wave mixing experimental apparatus. Modelocked pulses, from a dye laser or Ti:sapphire laser, are amplified using a doubled Nd:YAG pumped dye amplifier to create high peak power (GW/cm^2) pulses for DFWM. The sample must be positioned so that all incident beams travel the same distance for maximum effect. Provisions are made for the systematic delay of various beams to investigate dynamics, as well as polarization control to allow the measurement of different tensor components.

$$R = \left| \frac{2Q \sin(\frac{1}{2} HL)}{H \cos(\frac{1}{2} HL) + \alpha \sin(\frac{1}{2} HL)} \right|^2, \tag{13}$$

with:

$$Q = \frac{-3\omega}{8cn} \chi^{3}_{eff}(-\omega;\omega,\omega,-\omega)A_1(0)A_2(L)\exp(-\alpha L/2), \tag{14}$$

and:

$$H = [4|Q|^2 - \alpha^2]^{1/2}. \tag{15}$$

In the limit of low reflectivity, this reduces to:

$$R \approx \frac{|Q|^2}{\alpha^2}[1 - \exp(-\alpha L)]^2, \tag{16}$$

where L is the effective geometrical sample length. We can see that the reflectivity is expected to be quadratically dependent on both the effective nonlinearity and the pump intensity, with a somewhat complex dependence on absorption and sample length. Equation 16 shows that in the limit of no absorption, R is also quadratic in the sample length.

Forward. While the phase conjugate geometry of DFWM is the most popular, a variety of forward scattering geometries have also been used to measure third order nonlinear optical responses. The simplest geometry is two wave mixing, in which two strong pump beams with wave vectors $\vec{k}_1$ and $\vec{k}_2$ mutually interact to generate a new beam at an angle that preserves momentum conservation. This is essentially the degenerate case of CARS-type four wave mixing discussed above, with the new beam emerging with wave vector $2\vec{k}_1 + \vec{k}_2$.

As with other DFWM geometries, real-time studies of grating dynamics can be performed with two wave mixing; because of the simplicity of the experiment, a straightforward expression for the generated signal as a function of delay, τ, between beams 1 and 2, obtains given by[36]:

$$I_4(\tau) = C\int_{-\infty}^{\infty} dt\, I_1(t)\left|\int_{-\infty}^{t} dt'\, K^{(3)}(t - t')E_1(t')E_2^*(t' - \tau)\right|^2, \tag{17}$$

where $K^{(3)}(t - t')$ is the impulse response function of the medium, $I_1(t)$ is the time dependent intensity of beam 1, and C is a constant. If the nonlinearity is primarily nonresonant electronic, the impulse response function will be nearly a delta function and Eq. 17 reduces to a cross-correlation function.

An alternative geometry for forward DFWM is the so-called folded BOXCARS geometry. In this case, there are three incident beams (as in phase conjugate DFWM) with a fourth beam generated with wave vector $\vec{k}_4 = \vec{k}_1 + \vec{k}_3 - \vec{k}_2$. This geometry provides great flexibility both in the choice of tensor component and for dynamical studies.

2. Determination of Nonlinearity

We can use Eq. 16 to arrive at an expression for the effective nonlinearity of an unknown sample, $\chi^{(3)}_{sam}$, determined with respect to that of a reference material, $\chi^{(3)}_{ref}$. If we assume that the reference material is completely transparent at the measurement wavelength, then we find the following expression for the ratio of the nonlinearities:

$$|\chi^{(3)}_{sam}| = |\chi^{(3)}_{ref}| \sqrt{\frac{S_{sam}}{S_{ref}} \left(\frac{n_{sam}}{n_{ref}}\right)^2 \left(\frac{l_{ref}}{l_{sam}}\right)} \frac{\alpha l_{sam}}{e^{-\alpha l_{sam}/2}(1 - e^{-\alpha l_{sam}})}, \tag{18}$$

where α is the absorption coefficient at the measurement wavelength of the sample, l_{sam} and l_{ref} are the sample and reference lengths, n_{sam} and n_{ref} are the sample and reference refractive indices, and S_{sam} and S_{ref} are the sample and reference four wave mixing signals. In the limit as α goes to zero, the last term in Eq. 18 approaches 1 and a very simple result obtains. Note the absolute value bars in Eq. 18, indicating that conventional four wave mixing measurements do not obtain information about the phase of $\chi^{(3)}$. Equation 18 is by far the most commonly used expression for the analysis of four wave mixing data and is applicable to both backward and forward geometries. Note that more careful experiments measure the signal as a function of incident intensity, and insert coefficients of a cubic polynomial fit into Eq. 18 rather than the signals at a given intensity. The merit of this approach is that if the signal is not strictly cubic in the incident intensity (prior to beamsplitting) it is an indication that phenomena other than nonresonant four wave mixing are contributing to the signal, such as excited state population gratings or higher order nonlinear phenomena.

3. Mechanisms

There are at least four microscopic mechanisms that can lead to macroscopic DFWM:

1. virtual electronic excitations
2. nuclear orientation
3. rarefaction and densification (thermal)
4. excited state population gratings.

For the electronic and nuclear mechanisms, Eqs. 8 and 9 can be essentially directly adopted (aside from normalization factors) in the limit as the probe and pump wavelengths become equal.

Thermal gratings are formed via the absorption of pump beam radiation and the subsequent propagation of the heat generated via acoustic waves. If the pump beams are given by $\vec{k}_1$ and $\vec{k}_2$, then the thermal grating has a wave vector given by:

$$\vec{q} = \vec{k}_2 - \vec{k}_1. \tag{19}$$

When the fractional rate of change in time of the amplitude of the grating is less than the sound frequency, which means that the acoustic waves equalize pressure everywhere in the medium instantaneously, then the reflectivity of the grating is given by:

$$R = \exp(-\alpha L) \left(\frac{\omega \left(\frac{\partial n}{\partial T}\right)_p}{\rho C_p} \right)^2 U_1 U_2 \eta, \tag{20}$$

where α is the absorption coefficient of the medium, L is the sample length, ω is the optical angular frequency of the pump beams, ρ is the density of the medium,

$(\partial n/\partial T)_p$ is the derivative of the refractive index with respect to temperature at constant pressure, U_1 and U_2 are the energy per unit area in the two pump beams, and η is a phase mismatch factor that influences overall efficiency. Thermal gratings with fairly high reflection efficiency (up to 10 percent) have been made in a number of media and frequently compete with gratings due to electronic excitations, even at nanosecond time scales. A convenient way to test whether a grating is thermal or not is to cross the polarizations of the pumping beams. For a thermal grating to form, the pumping beams must have the same polarization therefore a thermal grating will vanish when the pumping beam polarizations are crossed. Thermal gratings can also be avoided by using ultrashort pulse lasers (picosecond or less) or by assuring that the medium being measured is completely transparent at the measurement wavelength.

IV. TABLES

The results of various four wave mixing measurements are given in the accompanying tables. Data has been split up into three separate tables

1. optical Kerr measurements of n_2, primarily in organic liquids
2. four wave mixing measurements of $\chi^{(3)}$
3. four wave mixing measurements of γ.

This choice was motivated primarily by the trends in the literature and also by an interest in performing as little manipulation as possible with the existing data. For example, $\chi^{(3)}$ values can in some instances be converted to values of γ, but this usually involves a number of assumptions regarding mechanisms, local field corrections, and dispersion.

A. Methods

For Table 1, the optical Kerr effect was used to measure all of the entries. For Tables 2 and 3, the methods used are given in each table according to the following acronyms:

B-DFWM—folded BOXCARS degenerate four wave mixing
CARS—coherent anti-Stokes Raman scattering
CSRS—coherent Stokes Raman scattering
DFWM—phase conjugate degenerate four wave mixing
F-DFWM—forward degenerate four wave mixing
HOKE—heterodyne optical Kerr effect
OKE—optical Kerr effect
PCI—phase conjugate interferometry
RIKE—Raman-induced Kerr effect
TWM—three wave mixing.

B. Nonlinearities and Units

Table 1 tabulates the value of n_2, the nonlinear refractive index, for various organic liquids, as measured by the optical Kerr effect. Units are in m^2/V^2 and conversions

Table 1 Nonlinear Refractive Indices of Organic Liquids

Sample	λ (nm)	t_p (nsec)	$n_{2B}(10^{-19}\ m^2/V^2)$	Ref.
1,1,1-trichloroethane	$\lambda_{pump} = 1064$ $\lambda_{probe} = 532$	$\tau_{pump} = 0.010$ $\tau_{probe} = 0.009$	0.114	37
1,2-dichloroethane	$\lambda_{pump} = 1064$ $\lambda_{probe} = 532$	$\tau_{pump} = 0.010$ $\tau_{probe} = 0.009$	0.175	37
1-decanol	$\lambda_{pump} = 1064$ $\lambda_{probe} = 442$	10	0.118	38
1-decene	$\lambda_{pump} = 1064$ $\lambda_{probe} = 442$	15	0.184	39
1-decyne	$\lambda_{pump} = 1064$ $\lambda_{probe} = 442$	15	0.197	39
1-dodecanol	$\lambda_{pump} = 1064$ $\lambda_{probe} = 442$	10	0.570	38
1-heptanol	$\lambda_{pump} = 1064$ $\lambda_{probe} = 442$	10	0.0906	38
1-heptene	$\lambda_{pump} = 1064$ $\lambda_{probe} = 442$	15	0.190	39
1-heptyne	$\lambda_{pump} = 1064$ $\lambda_{probe} = 442$	15	0.213	39
1-hexanol	$\lambda_{pump} = 1064$ $\lambda_{probe} = 442$	10	0.0775	38
1-hexene	$\lambda_{pump} = 1064$ $\lambda_{probe} = 442$	15	0.164	39
1-hexyne	$\lambda_{pump} = 1064$ $\lambda_{probe} = 442$	15	0.203	39
1-nonanol	$\lambda_{pump} = 1064$ $\lambda_{probe} = 442$	10	0.112	38
1-nonene	$\lambda_{pump} = 1064$ $\lambda_{probe} = 442$	15	0.190	39
1-nonyne	$\lambda_{pump} = 1064$ $\lambda_{probe} = 442$	15	0.226	39
1-octanol	$\lambda_{pump} = 1064$ $\lambda_{probe} = 532$	$\tau_{pump} = 0.010$ $\tau_{probe} = 0.009$	0.0572	37
1-octanol	$\lambda_{pump} = 1064$ $\lambda_{probe} = 442$	10	0.101	38
1-octene	$\lambda_{pump} = 1064$ $\lambda_{probe} = 442$	15	0.164	39
1-octyne	$\lambda_{pump} = 1064$ $\lambda_{probe} = 442$	15	0.223	39

Table 1 Continued

Sample	λ (nm)	t_p (nsec)	$n_{2B}(10^{-19}\ m^2/V^2)$	Ref.
1-pentanol	$\lambda_{pump} = 1064$ $\lambda_{probe} = 442$	10	0.069	38
1-pentene	$\lambda_{pump} = 1064$ $\lambda_{probe} = 442$	15	0.069	39
1-pentyne	$\lambda_{pump} = 1064$ $\lambda_{probe} = 442$	15	0.246	39
1-propanol	$\lambda_{pump} = 1064$ $\lambda_{probe} = 532$	$\tau_{pump} = 0.010$ $\tau_{probe} = 0.009$	0.0377	37
1-propanol	$\lambda_{pump} = 1064$ $\lambda_{probe} = 442$	10	0.0576	38
1-undecanol	$\lambda_{pump} = 1064$ $\lambda_{probe} = 442$	10	0.124	38
1-undecene	$\lambda_{pump} = 1064$ $\lambda_{probe} = 442$	10–20	0.185	39
2,4-lutidine	$\lambda_{pump} = 1064$ $\lambda_{probe} = 632.8$	20	2.26	40
2,6-lutidine	$\lambda_{pump} = 1064$ $\lambda_{probe} = 632.8$	20	1.41	40
3,4-lutidine	$\lambda_{pump} = 1064$ $\lambda_{probe} = 632.8$	20	1.68	40
γ-picoline	$\lambda_{pump} = 1064$ $\lambda_{probe} = 632.8$	20	1.90	40
AC-SBAC	$\lambda_{pump} = 1064$ $\lambda_{probe} = 810$	6	0.277	41
acetic acid	$\lambda_{pump} = 1064$ $\lambda_{probe} = 441.6$	10	0.241	42
acetic acid	$\lambda_{pump} = 532$ $\lambda_{probe} = 441.6$	10	0.367	42
acetone	$\lambda_{pump} = 1064$ $\lambda_{probe} = 532$	$\tau_{pump} = 0.010$ $\tau_{probe} = 0.009$	0.108	37
acetonitrile	$\lambda_{pump} = 694.3$ $\lambda_{probe} = 488$	10	0.371	43
acetonitrile	$\lambda_{pump} = 1064$ $\lambda_{probe} = 532$	$\tau_{pump} = 0.010$ $\tau_{probe} = 0.009$	0.133	37
β-picoline	$\lambda_{pump} = 1064$ $\lambda_{probe} = 632.8$	20	1.62	40
BB-SBAC	$\lambda_{pump} = 1064$ $\lambda_{probe} = 810$	6	0.303	41
benzene	$\lambda_{pump} = 1064$ $\lambda_{probe} = 632.8$	20	0.915	40
benzene	$\lambda_{pump} = 694.3$ $\lambda_{probe} = 488$	10	0.862	43

Table 1 Continued

Sample	λ (nm)	t_p (nsec)	$n_{2B}(10^{-19}\ m^2/V^2)$	Ref.
benzene	$\lambda_{pump} = 1064$ $\lambda_{probe} = 532$	$\tau_{pump} = 0.010$ $\tau_{probe} = 0.009$	0.838	37
benzene	$\lambda_{pump} = 1064$ $\lambda_{probe} = 488$	70	0.122	4
benzene	$\lambda_{pump} = 694.3$ $\lambda_{probe} = 500$	20	0.977	1
benzoyl chloride	$\lambda_{pump} = 1064$ $\lambda_{probe} = 488$	70	0.367	4
benzoyl chloride	$\lambda_{pump} = 694.3$ $\lambda_{probe} = 500$	20	4.6	1
butyric acid	$\lambda_{pump} = 1064$ $\lambda_{probe} = 441.6$	10	0.204	42
butyric acid	$\lambda_{pump} = 532$ $\lambda_{probe} = 441.6$	10	0.254	42
carbon disulfide	$\lambda_{pump} = 694.3$ $\lambda_{probe} = 488$	10	5.16	43
carbon disulfide	$\lambda_{pump} = 1064$ $\lambda_{probe} = 532$	$\tau_{pump} = 0.010$ $\tau_{probe} = 0.009$	2.79	37
carbon disulfide	$\lambda_{pump} = 1060$ $\lambda_{probe} = 530$	0.005 0.005	2.79	3
carbon disulfide	$\lambda_{pump} = 694.3$ $\lambda_{probe} = 500$	20	5.45	1
carbon disulfide	$\lambda_{pump} = 694.3$ $\lambda_{probe} = 488$	20	5.72	45
carbon disulfide	$\lambda_{pump} = 694.3$ $\lambda_{probe} = 632.8$	30	7.58 ± 1.13	44
carbon disulfide	$\lambda_{pump} = 1064$ $\lambda_{probe} = 532$	$\tau_{pump} = 30$ $\tau_{probe} = 21.2$	6.66 ± 1.00	44
carbon disulfide	$\lambda_{pump} = 694.3$ $\lambda_{probe} = 441.6$	30	5.68 ± 0.79	44
carbon disulfide	$\lambda_{pump} = 1064$ $\lambda_{probe} = 488$	70	0.894	4
carbon tetrachloride	$\lambda_{pump} = 694.3$ $\lambda_{probe} = 488$	20	0.068	45
carbon tetrachloride	$\lambda_{pump} = 694.3$ $\lambda_{probe} = 488$	10	0.0354	43
carbon tetrachloride	$\lambda_{pump} = 1064$ $\lambda_{probe} = 532$	30	0.0628 ± 0.0094	44
chloroform	$\lambda_{pump} = 694.3$ $\lambda_{probe} = 488$	10	0.212	43

Table 1 Continued

Sample	λ (nm)	t_p (nsec)	$n_{2B}(10^{-19}\ m^2/V^2)$	Ref.
chloroform	$\lambda_{pump} = 694.3$ $\lambda_{probe} = 488$	20	0.218	45
chloroform	$\lambda_{pump} = 1064$ $\lambda_{probe} = 532$	$\tau_{pump} = 0.010$ $\tau_{probe} = 0.009$	1.48	37
cyclohexane	$\lambda_{pump} = 694.3$ $\lambda_{probe} = 488$	10	0.0797	43
cyclohexane	$\lambda_{pump} = 1064$ $\lambda_{probe} = 532$	$\tau_{pump} = 0.010$ $\tau_{probe} = 0.009$	0.0545	37
α-picoline	$\lambda_{pump} = 1064$ $\lambda_{probe} = 632.8$	20	1.43	40
decane	$\lambda_{pump} = 1064$ $\lambda_{probe} = 532$	$\tau_{pump} = 0.010$ $\tau_{probe} = 0.009$	0.214	37
ethanol	$\lambda_{pump} = 1064$ $\lambda_{probe} = 442$	10	0.0570	38
ethanol	$\lambda_{pump} = 1064$ $\lambda_{probe} = 532$	$\tau_{pump} = 0.010$ $\tau_{probe} = 0.009$	0.0349	37
ethyl iodide	$\lambda_{pump} = 1064$ $\lambda_{probe} = 532$	$\tau_{pump} = 0.010$ $\tau_{probe} = 0.009$	0.316	37
ethylene bromide	$\lambda_{pump} = 1064$ $\lambda_{probe} = 532$	$\tau_{pump} = 0.010$ $\tau_{probe} = 0.009$	0.479	37
formic acid	$\lambda_{pump} = 1064$ $\lambda_{probe} = 441.6$	10	0.275	42
formic acid	$\lambda_{pump} = 1064$ $\lambda_{probe} = 532$	$\tau_{pump} = 0.010$ $\tau_{probe} = 0.009$	0.140	37
freon 113	$\lambda_{pump} = 1064$ $\lambda_{probe} = 532$	30	0.241 ± 0.036	44
glycerin	$\lambda_{pump} = 1064$ $\lambda_{probe} = 532$	$\tau_{pump} = 0.010$ $\tau_{probe} = 0.009$	0.0349	37
heptanoic acid	$\lambda_{pump} = 1064$ $\lambda_{probe} = 441.6$	10	0.215	42
hexadecane	$\lambda_{pump} = 1064$ $\lambda_{probe} = 532$	$\tau_{pump} = 0.010$ $\tau_{probe} = 0.009$	0.216	37
hexane	$\lambda_{pump} = 694.3$ $\lambda_{probe} = 488$	20	0.0545	45
hexanoic acid	$\lambda_{pump} = 1064$ $\lambda_{probe} = 441.6$	10	0.212	42
m-bromotoluene	$\lambda_{pump} = 1064$ $\lambda_{probe} = 632.8$	20	2.13	40
m-dichlorobenzene	$\lambda_{pump} = 1064$ $\lambda_{probe} = 632.8$	20	2.22	40
m-nitrotoluene	$\lambda_{pump} = 1064$ $\lambda_{probe} = 532$	$\tau_{pump} = 0.010$ $\tau_{probe} = 0.009$	3.77	37

Table 1 Continued

Sample	λ (nm)	t_p (nsec)	$n_{2B}(10^{-19}\ m^2/V^2)$	Ref.
m-nitrotoluene	$\lambda_{pump} = 1064$ $\lambda_{probe} = 632.8$	20	2.65	40
m-xylene	$\lambda_{pump} = 1064$ $\lambda_{probe} = 632.8$	20	1.64	40
mesitylene	$\lambda_{pump} = 694.3$ $\lambda_{probe} = 632.8$	30	1.73 ± 0.26	44
mesitylene	$\lambda_{pump} = 694.3$ $\lambda_{probe} = 441.6$	30	1.28 ± 0.192	44
methanol	$\lambda_{pump} = 1064$ $\lambda_{probe} = 442$	10	0.0371	38
methanol	$\lambda_{pump} = 1064$ $\lambda_{probe} = 532$	$\tau_{pump} = 0.010$ $\tau_{probe} = 0.009$	0.031	37
methylcyclohexane	$\lambda_{pump} = 694.3$ $\lambda_{probe} = 488$	10	0.0974	43
n-decane	$\lambda_{pump} = 1064$ $\lambda_{probe} = 442$	15	0.117	39
n-decanol	$\lambda_{pump} = 1064$ $\lambda_{probe} = 532$	$\tau_{pump} = 0.010$ $\tau_{probe} = 0.009$	0.0545	37
n-dodecane	$\lambda_{pump} = 1064$ $\lambda_{probe} = 442$	15	0.133	39
n-heptane	$\lambda_{pump} = 1064$ $\lambda_{probe} = 442$	15	0.0854	39
n-hexane	$\lambda_{pump} = 1064$ $\lambda_{probe} = 442$	15	0.0777	39
n-nonane	$\lambda_{pump} = 1064$ $\lambda_{probe} = 442$	15	0.0965	39
n-octane	$\lambda_{pump} = 1064$ $\lambda_{probe} = 442$	15	0.0976	39
n-pentane	$\lambda_{pump} = 1064$ $\lambda_{probe} = 442$	15	0.0577	39
n-undecane	$\lambda_{pump} = 1064$ $\lambda_{probe} = 442$	15	0.120	39
nitrobenzene	$\lambda_{pump} = 1064$ $\lambda_{probe} = 532$	$\tau_{pump} = 0.010$ $\tau_{probe} = 0.009$	3.49	37
nitorbenzene	$\lambda_{pump} = 1060$ $\lambda_{probe} = 530$	0.005 0.005	1.04	3
nitrobenzene	$\lambda_{pump} = 694.3$ $\lambda_{probe} = 488$	20	3.95	45
nitrobenzene	$\lambda_{pump} = 694.3$ $\lambda_{probe} = 500$	20	4.61	1

Table 1 Continued

Sample	λ (nm)	t_p (nsec)	$n_{2B}(10^{-19}\ m^2/V^2)$	Ref.
nitorbenzene	$\lambda_{pump} = 694.3$ $\lambda_{probe} = 632.8$	30	6.10	44
nitrobenzene	$\lambda_{pump} = 1064$ $\lambda_{probe} = 488$	70	0.771 ± 0.108	4
nonanoic acid	$\lambda_{pump} = 1064$ $\lambda_{probe} = 441.6$	10	0.225	42
nonanoic acid	$\lambda_{pump} = 532$ $\lambda_{probe} = 441.6$	10	0.258	42
o-chlorotoluene	$\lambda_{pump} = 1064$ $\lambda_{probe} = 632.8$	20	2.25	40
o-dichlorobenzene	$\lambda_{pump} = 1064$ $\lambda_{probe} = 632.8$	20	2.34	40
o-nitrotoluene	$\lambda_{pump} = 1064$ $\lambda_{probe} = 632.8$	20	2.84	40
o-xylene	$\lambda_{pump} = 1064$ $\lambda_{probe} = 632.8$	20	1.69	40
octane	$\lambda_{pump} = 1064$ $\lambda_{probe} = 532$	$\tau_{pump} = 0.010$ $\tau_{probe} = 0.009$	0.0768	37
octanoic acid	$\lambda_{pump} = 1064$ $\lambda_{probe} = 441.6$	10	0.224	42
octanoic acid	$\lambda_{pump} = 532$ $\lambda_{probe} = 441.6$	10	0.276	42
p-chlorotoluene	$\lambda_{pump} = 1064$ $\lambda_{probe} = 632.8$	20	2.05	40
p-nitrotoluene	$\lambda_{pump} = 1064$ $\lambda_{probe} = 632.8$	20	2.42	40
p-xylene	$\lambda_{pump} = 1064$ $\lambda_{probe} = 632.8$	20	1.57	40
pentane	$\lambda_{pump} = 694.3$ $\lambda_{probe} = 488$	10	0.115	43
pentane	$\lambda_{pump} = 1064$ $\lambda_{probe} = 532$	$\tau_{pump} = 0.010$ $\tau_{probe} = 0.009$	0.0586	37
proprionic acid	$\lambda_{pump} = 532$ $\lambda_{probe} = 441.6$	10	0.337	42
proprionic acid	$\lambda_{pump} = 1060$ $\lambda_{probe} = 441.6$	10	0.229	42
pyridine	$\lambda_{pump} = 1064$ $\lambda_{probe} = 488$	70	0.184 ± 0.049	4
pyridine	$\lambda_{pump} = 1064$ $\lambda_{probe} = 632.8$	20	1.26	40
salol	$\lambda_{pump} = 1064$ $\lambda_{probe} = 532$	$\tau_{pump} = 0.010$ $\tau_{probe} = 532$	6.42	37

Table 1 Continued

Sample	λ (nm)	t_p (nsec)	$n_{2B}(10^{-19}\ m^2/V^2)$	Ref.
SBAC III	$\lambda_{pump} = 1064$ $\lambda_{probe} = 810$	6	0.433	41
succinonitrile	$\lambda_{pump} = 1064$ $\lambda_{probe} = 532$	$\tau_{pump} = 0.010$ $\tau_{probe} = 0.009$	0.349	37
sym-collidine	$\lambda_{pump} = 1064$ $\lambda_{probe} = 632.8$	20	2.46	40
tetrachloroethylene	$\lambda_{pump} = 694.3$ $\lambda_{probe} = 488$	10	1.87	43
toluene	$\lambda_{pump} = 694.3$ $\lambda_{probe} = 632.8$	30	1.61 ± 0.24	44
toluene	$\lambda_{pump} = 694.3$ $\lambda_{probe} = 441.6$	30	1.04 ± 0.16	44
toluene	$\lambda_{pump} = 1064$ $\lambda_{probe} = 488$	70	0.208 ± 0.041	4
toluene	$\lambda_{pump} = 694.3$ $\lambda_{probe} = 488$	20	1.27	45
toluene	$\lambda_{pump} = 694.3$ $\lambda_{probe} = 488$	10	1.17	43
toluene	$\lambda_{pump} = 1064$ $\lambda_{probe} = 632.8$	20	1.36	40
toluene	$\lambda_{pump} = 1064$ $\lambda_{probe} = 532$	$\tau_{pump} = 0.010$ $\tau_{probe} = 0.009$	0.838	37
trichloroethylene	$\lambda_{pump} = 1064$ $\lambda_{probe} = 532$	$\tau_{pump} = 0.010$ $\tau_{probe} = 0.009$	0.504	37
trichloroethylene	$\lambda_{pump} = 694.3$ $\lambda_{probe} = 488$	10	2.13	43
valeric acid	$\lambda_{pump} = 1064$ $\lambda_{probe} = 441.6$	10	0.216	42
valeric acid	$\lambda_{pump} = 532$ $\lambda_{probe} = 441.6$	10	0.272	42
water	$\lambda_{pump} = 1064$ $\lambda_{probe} = 532$	$\tau_{pump} = 0.010$ $\tau_{probe} = 0.009$	0.0182	37

AC-SBAC—symmetrically substituted benzylidene aniline dichloride with ethyl and carboxylic acid substitutions
BB-SBAC—2,5-dichloroterephthal-bis-(4-*N*,*N*-dibutyl aminoaniline)

Table 2 Third Order Nonlinear Optical Properties of Polymers and Bulk Organics as Measured by Four Wave Mixing Techniques

Sample	Method	λ (nm)	t_p (nsec)	$\chi^{(3)}_{1111}$ (10^{-19} m^2/V^2)	Ref.
α-[bis(ethylenedithio) tetrathiofulvalene] iodide (w/c* axis parallel to beam)	DFWM	675	0.5	7540	46
α-[bis(ethylenedithio) tetrathiofulvalene] iodide (w/c* axis perpendicular to beam)	DFWM	635	0.5	9770	46
1,1′-diethyl-2,2′-cyanine chloride pseudoisocyanine	DFWM	581	0.020	1400	47
2-methyl-10,12-docosadiyne-1,22-diyl ester 2-propenoic acid	DFWM	532	0.180	$\chi^{(3)}_{1221} = 0.98 \pm 0.28$	48
9,9′-(1,3,5-hexatriene-1,6-diyl) bis[10-(decylthio)-anthracene]	DFWM	1064	0.006	0.0265	49
BBB	DFWM	1064	0.035	0.77	50
BBL	DFWM	532	0.006	280	51
BBL	DFWM	1064	0.035	2.1	50
BBL (doped)	DFWM	1064	0.035	2.79	50
BDN in dichloromethane	DFWM	1064	0.100	$\chi^{(3)}_{1221} = 10.6$	52
benzene	CARS	$\lambda_1 \approx \lambda_2 \approx 550$	3	0.0024 ± 0.0002	53
benzene		$\lambda_1 \approx 545$	3	0.0026 ± 0.0001 $\chi^{(3)}_{1221} = 0.00075 \pm 0.00004$ $\chi^{(3)}_{1122} = 0.00091 \pm 0.00006$	54
benzene	RIKE	$\lambda_{pump} = 587$ $\lambda_{probe} = 555$	3	$\chi^{(3)}_{1122}{}^{NR} = 0.0047$	55
benzene	RIKE	$\lambda_{pump} = 587$ $\lambda_{probe} = 555$	3	$\chi^{(3)}_{eff} = 0.0059 \pm 0.0003$ $\chi^{(3)}_{elec} = 0.002 \pm 0.0001$	56
benzene	OKE	$\lambda_{pump} = 1064$ $\lambda_{probe} = 472$	0.030	$\chi^{(3)}_{eff} = 0.0080$	17

Table 2 Continued

Sample	Method	λ (nm)	t_p (nsec)	$\chi^{(3)}_{1111}$ (10^{-19} m^2/V^2)	Ref.
benzene	OKE	$\lambda_{pump} = 1064$ $\lambda_{probe} = 459$	0.030	$\chi^{(3)}_{eff} = 0.0080$	17
benzene	OKE	$\lambda_{pump} = 1064$ $\lambda_{probe} = 514.5$	0.030	$\chi^{(3)}_{eff} = 0.0082$	17
benzene	OKE	$\lambda_{pump} = 1064$ $\lambda_{probe} = 496$	0.030	$\chi^{(3)}_{eff} = 0.0095$	17
benzene	OKE	$\lambda_{pump} = 1064$ $\lambda_{probe} = 590$	0.030	$\chi^{(3)}_{eff} = 0.0098$	17
bis[1-ethyl, 2-phenyl, 1,2-ethenedithiolato-S,S′]nickel	DFWM	1064	0.100	$n_2 = 9.2 \times 10^{-9}$ cm^2/kw	57
bis[1-phenyl-1-dodecene-1,2-dithiolato(2-)-S,S′]-nickel	DFWM	1064	0.100	0.447 $\chi^{(3)}_{1221} = 0.036$	52
black tea in water	B-DFWM	532	7	0.279	58
BrMPPV film	F-DFWM	602	0.0004	125	59
BrMPPV/sol-gel	F-DFWM	602	0.0004	110	59
BuMPPV film	F-DFWM	602	0.0004	140	59
BuMPPV sol-gel	F-DFWM	602	0.0004	140	59
C_{60}	DFWM	675	0.0012	$\chi^{(3)\,eff}_{1111} = 11$	60
C_{60}-O_2	DFWM	675	0.0012	$\chi^{(3)\,eff}_{1111} = 13$	60
C_{60}-O_2 film	DFWM	590.5	0.0012	$\chi^{(3)\,eff}_{1111} = 73$	60
C_{60} film	DFWM	590.5	0.0012	$\chi^{(3)\,eff}_{1111} = 60$	60
C_{60} film	DFWM	1064	0.035	0.98	61
C_{60} film	DFWM	597	0.0012	53 ± 13	62
C_{60} film	DFWM	675	0.0012	11 ± 1	62
C_{70} film	DFWM	597	0.0012	293 ± 14	62
C_{70} film	DFWM	675	0.0012	89 ± 28	62
C_{60} film	DFWM	1064	0.035	$\chi^{(3)}_{1221} = 0.14$	63
C_{70} film	DFWM	1064	0.035	1.7 $\chi^{(3)}_{1221} = 0.14$	63

Cao1	DFWM	532	0.025	11.2 $\chi^{(3)}_{1221} = 4.13 \pm 0.78$ $\chi^{(3)}_{1212} = 5.16 \pm 0.65$	64
Cao2	DFWM	532	0.025	8.38 $\chi^{(3)}_{1221} = 3.09 \pm 0.59$ $\chi^{(3)}_{1212} = 3.36 \pm 0.18$	64
Cao3	DFWM	532	0.025	19.5	64
carbon disulfide	RIKE	$\lambda_{pump} = 587$ $\lambda_{probe} = 555$	3	$\chi^{(3)}_{eff} = 0.033 \pm 0.002$ $\chi^{(3)}_{elec} = 0.0083 \pm 0.0001$	56
charge transfer complex of C_{60} with TMPD*	DFWM	675	0.0012	$\chi^{(3)}_{1111}{}^{eff} = 0.0105$	60
chloronaphthalene	CSRS	1053	0.001	$\chi^{(3)}_{eff} = 0.040 \pm 0.009$	65
chloronaphthalene	CARS	1053	0.001	$\chi^{(3)}_{eff} = 0.044 \pm 0.010$	65
cobalt tetraphenyl porphyrin in toluene	DFWM	532	0.080	1.0 ± 0.5 (thermal)	66
copolymer of 3-methylthiophene and methylmethacrylate	DFWM	602	0.002	$\chi^{(3)\,yellow} = 0.98$ $\chi^{(3)\,red} = 12.2$	67
CS_2H_5Cl	RIKE	$\lambda_{pump} = 587$ $\lambda_{probe} = 555$	3	$\chi^{(3)}_{eff} = 0.011 \pm 0.001$ $\chi^{(3)}_{elec} = 0.0034 \pm 0.001$	56
cumylphenoxy platinum phthalocyanine in chloroform	DFWM	1064	0.010	0.41	68
DANS/MMA 60/40 (wt/wt)	DFWM	598	0.002	0.96 ± 0.14 $\chi^{(3)}_{1221} = 0.31 \pm 0.07$	69
DEANST in nitrobenzene (15 wt %)	OKE	$\lambda_{pump} = 590$ $\lambda_{probe} = 830$	0.004	$\chi^{(3)}_{eff} = 0.50$	70
Dennis1	DFWM	532	0.080	0.7 ± 0.3 $\chi^{(3)}_{1212} = 0.5 \pm 0.2$ $\chi^{(3)}_{1122} = 0.2 \pm 0.1$ $\chi^{(3)}_{1221} = 0$	71
diphenylbutadiene in DCE	PCI	532	5	$\chi^{(3)}_{1221} = 0.0037$	72
diphenylhexatriene in DCE	PCI	532	5	$\chi^{(3)}_{1221} = 0.00496$	72
Fe(III) tetraphenyl porphyrin	DFWM	532	15	11	73

Table 2 Continued

Sample	Method	λ (nm)	t_p (nsec)	$\chi^{(3)}_{1111}$ (10^{-19} m^2/V^2)	Ref.
Garvey1	OKE (Sagnac)	$\lambda_{pump} = 0164$ $\lambda_{probe} = 1064$	0.080	0.17 ± 0.10	74
Guha1	OKE	$\lambda_{pump} = 1064$ $\lambda_{probe} = 532$	0.030	$\chi^{(3)}_{eff} = 0.0043$	75
Guha2	OKE	$\lambda_{pump} = 1064$ $\lambda_{probe} = 532$	0.030	$\chi^{(3)}_{eff} = 0.013$	75
Guha3	OKE	$\lambda_{pump} = 1064$ $\lambda_{probe} = 532$	0.030	$\chi^{(3)}_{eff} = 0.0056$	75
Guha4	OKE	$\lambda_{pump} = 0164$ $\lambda_{probe} = 532$	0.030	$\chi^{(3)}_{eff} = 0.0098$	75
hexane	RIKE	$\lambda_{pump} = 588$ $\lambda_{probe} = 557$	3	$\chi^{(3)}_{eff} = 0.00078 \pm 0.00001$ $\chi^{(3)}_{elec} = 0.0012 \pm 0.0004$	56
magnesium octaphenyl tetraazaporphyrin 5% in PMMA	DFWM	598	0.003	$\chi^{(3)}_{1111} = 1.63$ $\chi^{(3)}_{1221} = 0.42$	76
Mates1	DFWM	598	0.005	0.43	77
Mates2	DFWM	598	0.005	0.37	77
Mates3	DFWM	598	0.005	0.30	77
MTPBT (extruded thin film)	F-DFWM	602	0.004	$\chi^{(3)\prime\prime}_{1111} = 64.5$ $\chi^{(3)}_{1111} = 59.3$	78
Nickel1	DFWM	1064	0.006	0.182	49
Nickel2	DFWM	532	0.006	0.0433	49
Nickel3	DFWM	1064	0.006	0.936	49
nitrobenzene	OKE	$\lambda_{pump} = 1064$ $\lambda_{probe} = 590$	0.030	$\chi^{(3)}_{eff} = 0.012$	17
nitrobenzene	OKE	$\lambda_{pump} = 1064$ $\lambda_{probe} = 459$	0.030	$\chi^{(3)}_{eff} = 0.018$	17
nitrobenzene	OKE	$\lambda_{pump} = 1064$ $\lambda_{probe} = 514.5$	0.030	$\chi^{(3)}_{eff} = 0.018$	17

nitrobenzene	OKE	$\lambda_{pump} = 1064$ $\lambda_{probe} = 496$	0.030	$\chi^{(3)}_{eff} = 0.020$	17
nitrobenzene	OKE	$\lambda_{pump} = 1064$ $\lambda_{probe} = 472$	0.030	$\chi^{(3)}_{eff} = 0.023$	17
nitrobenzene	CSRS	1053	0.001	$\chi^{(3)}_{eff} = 0.028 \pm 0.005$	65
nitrobenzene	CARS	1053	0.001	$\chi^{(3)}_{eff} = 0.029 \pm 0.007$	65
nitrobenzene	OKE (Sagnac)	1064	0.080	0.17 ± 0.03	74
ØNS/MMA 50/50 (wt/wt)	DFWM	598	0.002	0.28 ± 0.07	69
OONOOB	CARS	1053	0.001	$\chi^{(3)}_{eff} = 0.035\ (\pm 0.006)$	65
OONOOB	CSRS	1053	0.001	$\chi^{(3)}_{eff} = 0.037 \pm 0.005$	65
P(1APyre)	DFWM	1064	0.040	37.7 (extrapolated)	79
P(3APAz)	DFWM	1064	0.040	5.6 (extrapolated)	79
P(3APy)	DFWM	1064	0.040	3.1 (extrapolated)	79
P(3AQ)	DFWM	1064	0.040	5.2 (extrapolated)	79
P(4ABP)	DFWM	1064	0.040	20 (extrapolated)	79
P(4ABP)	DFWM	1064	0.040	18	80
P(5AQ)	DFWM	1064	0.040	2.7 (extrapolated)	79
P(6AQ)	DFWM	1064	0.040	3.6 (extrapolated)	79
P(8AQ)	DFWM	1064	0.040	3.2 (extrapolated)	79
P(Ani)	DFWM	1064	0.040	1.5 (extrapolated)	79
P(DPA)	DFWM	1064	0.040	1.8 (extrapolated)	79
P(DPA)	DFWM	1064	0.040	20	80
P(ITN)	DFWM	1064	0.040	7.0 (extrapolated)	79
P(NNPheBz)	DFWM	1064	0.040	1.3 (extrapolated)	79
P(NPhe2NaA)	DFWM	1064	0.040	0.56 (extrapolated)	79
p-ethoxy-benzylidene p-*N*-butylaniline	DFWM	532	0.026	0.21 ± 0.13	81
p-methoxy-benzylidene p-*N*-butylaniline	DFWM	532	0.026	0.056 ± 0.028	81
P3MT-electrochemically	Resonant	≈630	0.00008	$\chi^{(3)}_{1111} - \chi^{(3)}_{1122} = 5.4$	82

Table 2 Continued

Sample	Method	λ (nm)	t_p (nsec)	$\chi^{(3)}_{1111}$ (10^{-19} m^2/V^2)	Ref.
polymerized on glass	Kerr				
PBT-spin coated biaxial film	DFWM	585 or 604	0.008	0.75 $\chi^{(3)}_{2222} = 1$	83
PBTABQ	DFWM	532	0.025	6300	84
PBTBQ	DFWM	532	0.025	38000	84
PBTBQ solution in sulfuric acid	DFWM	532	0.025	38000	84
PDA-3BCMU (solution in formamide N,N-dimethyl)	DFWM	1060	0.033	$\chi^{(3)}_{1212} = 0.20$	85
PDA-3BCMU (blue film)	DFWM	1064	0.010	8.4 ± 0.4	36
PDA-3BCMU (blue gel)	DFWM	1064	0.010	3.8 ± 0.8	36
PDA-3BCMU (blue gel)	DFWM	1060	0.033	$\chi^{(3)}_{1212} = 1.82$	85
PDA-3BCMU film cast from chloroform	Resonant Kerr	630	0.00008	−360	82
PDA-4BCMU	B-DFWM	620	0.0001	28	86
PDA-4BCMU (cast film)	OKE	$\lambda_{pump} = 530$ $\lambda_{probe} = 1060$	0.008	$\chi^{(3)}_{1221} = 13$ $\chi^{(3)}_{eff} = 42$	87
PDA-4BCMU (red gel)	DFWM	1064	0.010	5.4 ± 0.8	36
PDA-4BCMU (red gel)	DFWM	1060	0.033	$\chi^{(3)}_{1212} = 1.26$	85
PDA-4BCMU (red solution)	DFWM	1064	0.010	5.9 ± 0.8	36
PDA-4BCMU (single xtal)	B-DFWM	681	0.001	140	88
PDA-4BCMU (single xtal)	B-DFWM	678	0.001	180	88
PDA-4BCMU (single xtal)	B-DFWM	758	0.001	28	88
PDA-4BCMU (single xtal)	B-DFWM	767	0.001	28	88
PDA-4BCMU (single xtal)	B-DFWM	749	0.001	32	88
PDA-4BCMU (single xtal)	B-DFWM	740	0.001	35	88
PDA-4BCMU (single xtal)	B-DFWM	675	0.001	290	88
PDA-4BCMU (single xtal)	B-DFWM	730	0.001	42	88
PDA-4BCMU (single xtal)	B-DFWM	722	0.001	56	88

PDA-4BCMU (single xtal)	B-DFWM	705	0.001	70	88
PDA-4BCMU (single xtal)	B-DFWM	696	0.001	84	88
PDA-4BCMU (single xtal)	B-DFWM	690	0.001	91	88
PDA-4BCMU (yellow solution)	DFWM	1064	0.010	0.59 ± 0.04	36
PDA-4BCMU film	B-DFWM	620	0.00009	28	89
PDA-4BCMU film	DFWM	1064	0.033	$\chi^{(3)}_{1212} = 2.8 \pm 0.1$, $\varphi = 80° \pm 20°$ $\chi^{(3)}_{1212} = 4.2 \pm 0.4$, $\varphi = 50 \pm 10°$ (High Intensity)	90
PDA-4BCMU (cast film) (Red)	DFWM	585	0.008	56	91
PDA-PTS (single xtal)	DFWM	691.5	0.006	63	34
PDA-PTS (single xtal)	DFWM	701.5	0.006	77	34
PDA-PTS (single xtal)	DFWM	675	0.006	290	34
PDA-PTS (single xtal)	DFWM	671	0.006	310	34
PDA-PTS (single xtal)	DFWM	667	0.006	320	34
PDA-PTS (single xtal)	DFWM	660	0.006	1000	34
PDA-PTS (single xtal)	DFWM	681.5	0.006	130	34
PDA-PTS (single xtal)	DFWM	651.5	0.006	1300	34
PDA-PTS (xtal)	DFWM	1064	0.010	56 ± 28	36
PDA-PTS single xtal	F-DFWM	700	0.006	70	92
PDA-PTS thermally polymerized	DFWM	689	0.0006	2800	93
PDA-PTS thermally polymerized	DFWM	645–667	0.0006	35000	93
PDTT in THF	B-DFWM	1064	≤0.015	4.2 ± 2.8	94
PDTT in THF	B-DFWM	630	≤0.015	180 ± 70	94
PDTT in THF	B-DFWM	532	≤0.015	1580 ± 280	94
PDTT in THF	B-DFWM	600	≤0.015	630 ± 140	94
PDTT in THF	B-DFWM	605	≤0.015	770 ± 140	94
PDTT in THF	B-DFWM	590	≤0.015	810 ± 140	94
PDTT in THF	B-DFWM	595	≤0.015	850 ± 140	94
PDTT in THF	B-DFWM	585	≤0.015	1080 ± 140	94

Table 2 Continued

Sample	Method	λ (nm)	t_p (nsec)	$\chi^{(3)}_{1111}$ (10^{-19} m^2/V^2)	Ref.
Ph_4 (vacuum sublimed)	DFWM	650	0.000140	≤0.022	95
Ph_5 (Quinque)	OKE	λ_{pump} = 1064 λ_{probe} = 4442	10–20 *cw probe*	<0.023	95
Ph_6 (Sexi)	OKE	λ_{pump} = 1064 λ_{probe} = 442	10–20 *cw probe*	0.11	95
Ph_7 (Septi)	OKE	λ_{pump} = 1064 λ_{probe} = 442	10–20 *cw probe*	0.18	95
Ph_8 (Octi)	OKE	λ_{pump} = 1064 λ_{probe} = 442	10–20 *cw probe*	0.24	95
PNO2STPV $FeCl_3$ doped → film	DFWM	602	0.010	24	96
poly (*p*-methoxyphenyl acetylene)	DFWM	532	10		97
Mo Catlyst				3.4	
Cr Cat.				3.1	
W Cat. in 1,2-dibromoethane				5.4	
poly(2-DDT) (electrochemical)	DFWM	602	0.00035	70	98
poly(2-DDT) (electrochemical)	DFWM	590	0.00035	100	98
poly(3-DDT)	DFWM	620	0.00006	7.7	99
poly(3-DDT) film (chemical)	DFWM	602	0.00035	46	98
poly(3-DDT) film (chemical)	DFWM	590	0.00035	63	98
poly(acetylene) (Ziegler-Natta)	B-DFWM	720	0.0001	5600	100
poly(hexylacetylene) in 1,2-dibromoethane	DFWM	532	10	6.1	97
poly(methoxyphenylacetylene) W catalyst in 1,2-dibromoethane	DFWM	532	10	1.8	97
poly(*N*-octylmethylsilane)	DFWM	532	10	0.4 ± 0.07	101
poly(*o*-methoxyphenyl-acetylene) MO catalyst in 1,2-dibromoethane	DFWM	532	10	2.3	97
PPMS film cast from THF	OKE	λ_{pump} = 694.3 λ_{probe} = 488	0.008	$\chi^{(3)}_{eff}$ = 0.28 ± 0.08	102

PPMS solution in toluene	DFWM	530	0.030	0.23 ± 0.03	102
poly(trimethylsilyl-phenylacetylene) W Catalyst in 1,2-dibromoethane	DFWM	532	10	17	97
poly-[2,5-dimethoxy paraphenylenevinylene] oriented film	DFWM	602	0.0004	560 ± 140	103
poly-C_{60}	DFWM	675	0.0012	$\chi^{(3)}_{1111}{}^{eff} = 12$	60
poly-C_{60}	DFWM	590.5	0.0012	$\chi^{(3)}_{1111}{}^{eff} = 120$	60
poly-hexyl-thiophene	B-DFWM	620	0.00090	100	89
polyacetylene (all-trans dispersion)	B-DFWM	720	0.0001	13800	100
polyacetylene film	B-DFWM	620	0.0001	7000	86
				$\chi^{(3)}_{1221} = 3400$	
polyacetylene film (Shirakawa)	B-DFWM	620	0.00090	7000	89
polyaniline film	B-DFWM	620	0.00090	110	89
polyaniline film	B-DFWM	620	0.0001	110	86
				$\chi^{(3)}_{1221} = 74$	
polydiethynylsilane	B-DFWM	620	0.0001	420	86
				$\chi^{(3)}_{1221} = 210$	
polydiethynylsilane film	B-DFWM	620	0.00090	420	89
polydiethynylsilane film	B-DFWM	620	0.00009	420	104
				$\chi^{(3)}_{1221} = 210$	
polyhexylthiophene	B-DFWM	620	0.0001	100	86
				$\chi^{(3)}_{1221} = 56$	
polyl(2-methoxy-1,4-phenylene vinylene) (thin film)	DFWM	602	0.0004	110	105
polylphenylene-ethynylene	DFWM	532	1	250 (extrapolated)	106
polystyrene in 1,2 dibromoethane	DFWM	532	10	0.1	97
polythiophene	B-DFWM	1064	0.015	4.2 ± 2.8	94
polythiophene	B-DFWM	630	0.015	100 ± 70	94
polythiophene	B-DFWM	590	0.015	420 ± 140	94
polythiophene	B-DFWM	605	0.015	420 ± 140	94
polythiophene	B-DFWM	600	0.015	450 ± 140	94
polythiophene	B-DFWM	595	0.015	530 ± 140	94

Table 2 Continued

Sample	Method	λ (nm)	t_p (nsec)	$\chi^{(3)}_{1111}$ (10^{-19} m^2/V^2)	Ref.
polythiophene	B-DFWM	585	0.015	700 ± 140	94
polythiophene	B-DFWM	532	0.015	920	94
polythiophene	B-DFWM	620	0.0001	84 $\chi^{(3)}_{1221} = 49$	86
polythiophene	B-DFWM	620	0.0009	84	89
PPA	DFWM	694.3	20	0.17	107
PPA (10% I-doped)	DFWM	694.3	20	0.249	107
PPA (20% I doped)	DFWM	694.3	20	0.321	107
PPA (40% I-doped)	DFWM	694.3	20	0.27	107
PPA (5% I-doped)	DFWM	694.3	20	0.191	107
PPA (60% I-doped)	DFWM	694.3	20	0.275	107
PPA (cat W) in 1,2-dibromoethane	DFWM	532	10	3.1	97
PPV	DFWM	602	0.0004	$\chi^{(3)\prime}_{1111} = -14$ $\chi^{(3)\prime\prime}_{1111} = 3.5$ $\chi^{(3)\parallel}_{2222} = 0.71$ $\chi^{(3)\perp}_{2222} = 0.14$ $\chi^{(3)}_{offdiag} = 0.70$ $\chi^{(3)\prime\prime}_{offdiag} = 8.4$	108
PPV	B-DFWM	620	0.0001	28 $\chi^{(3)}_{1221} = 14$	86
PPV film	F-DFWM	602	0.0004	56	59
PPV/sol-gel composite 1:1 film	OKE	$\lambda_{pump} = 602$ $\lambda_{probe} = 602$	0.00006	$\chi^{(3)}_{eff} = 3.5$	109
PPV/sol-gel composite 1:1 film	F-DFWM	602	0.00006	$\chi^{(3)}_{eff} = 6.3$	109
PPV/solgel silica composite	F-DFWM	602	0.0004	42	59
prepolymer A	DFWM	585	0.006	17	51
prepolymer A oxidized	DFWM	585	0.006	8.38	51
PSBAC	B-DFWM	595	0.0004	20.9	110
PSBAC	B-DFWM	586	0.0004	34.9	110

PSBAC	B-DFWM	620	0.0004	3.84	110
PSBAC	B-DFWM	580	0.0004	69.8	110
PSBAC	B-DFWM	610	0.0004	6.98	110
PTCDA vacuum deposited film	DFWM	602	0.0004	28	111
PTT	B-DFWM	1064	≤0.015	4.2 ± 2.8	94
PTT	B-DFWM	630	≤0.015	98 ± 38	94
PTT	B-DFWM	600	≤0.015	404 ± 280	94
PTT	B-DFWM	590	≤0.015	420 ± 140	94
PTT	B-DFWM	605	≤0.015	420 ± 140	94
PTT	B-DFWM	595	≤0.015	500 ± 140	94
PTT	B-DFWM	585	≤0.015	610 ± 140	94
PTT in THF	B-DFWM	532	≤0.015	830 ± 140	94
PVK:TNF 16:1 (cast film)	DFWM	602	0.002	0.47 ± 0.05	112
PVK:TNF 2:1 (cast film)	DFWM	602	0.002	2.8 ± 0.3	112
PVK:TNF 32:1 (cast film)	DFWM	602	0.002	0.28 ± 0.03	112
PVK:TNF 4:1 (cast film)	DFWM	602	0.002	1.7 ± 0.1	112
PVK:TNF 8:1 (cast film)	DFWM	602	0.002	1.0 ± 0.1	112
pyridine	RIKE	$\lambda_{pump} = 588$ $\lambda_{probe} = 557$	3	$\chi^{(3)}_{eff} = 0.0062 \pm 0.0004$ $\chi^{(3)}_{elec} = 0.0015 \pm 0.0001$	56
Saindon1	DFWM	598	0.005	1.7 ± 0.4	113
Saindon2	DFWM	598	0.005	3.6 ± 1.3	113
Saindon3	DFWM	598	0.005	0.21 ± 0.01	113
Saindon4	DFWM	598	0.005	0.28 ± 0.04	113
Sc bisphthalocyanine in chloroform	DFWM	1064	0.010	0.02	68
silica:poly (BrMPPV) composite	F-DFWM	602	0.0004	130	59
SiNc (30 wt% in PMMA)	DFWM	598	0.003	2.9 $\chi^{(3)}_{1221} = 0.9$	76
SiPc Langmuir-Blodgett film	DFWM	602	0.0004	280	114
SiPcMac/MMA 10/90 (wt/wt)	DFWM	598	0.003	13 $\chi^{(3)}_{1221} = 4.7$	76
SiPc/PMMA 10/90 (wt/wt) guest-host	DFWM	598	0.002	3.6 , $\chi^{(3)}_{1221} = 1.5$	69

Table 2 Continued

Sample	Method	λ (nm)	t_p (nsec)	$\chi^{(3)}_{1111}$ (10^{-19} m^2/V^2)	Ref.
SiPc bisacrylate/MMA 10/90 (wt/wt) copolymer	DFWM	598	0.002	13.1 $\chi^{(3)}_{1221} = 4.5$	69
SiPcMac/MMA 1/99 (wt/wt)	DFWM	598	0.002	1.4	69
SiPcMac/MMA 5/95 (wt/wt)	DFWM	598	0.002	8.4 ± 0.8	69
SiPcMac/MMA 10/90 (wt/wt)	DFWM	598	0.002	13.3 ± 1.4 $\chi^{(3)}_{1221} = 3.5 \pm 0.4$	69
SiPcMac/MMA 20/80 (wt/wt)	DFWM	598	0.002	22.3 ± 4.5	69
SiPcMac/MMA 40/60 (wt/wt)	DFWM	598	0.002	36.3 ± 10.9	69
SiPcMac/MMa 60/40 (wt/wt)	DFWM	598	0.002	27.9 ± 5.6	69
Spangler1	DFWM	532	0.030	$\chi^{(3)}_{neu} = 0.038$ $\chi^{(3)}_{dop} = 0.11$	115
Spangler1	DFWM	1064	0.030	$\chi^{(3)}_{neu} = 0.0075$ $\chi^{(3)}_{dop} = 0.020$	115
Spangler2	DFWM	532	0.030	$\chi^{(3)}_{neu} = 0.60$ $\chi^{(3)}_{dop} = 0.15$	115
Spangler2	DFWM	1064	0.030	$\chi^{(3)}_{neu} = 0.018$ $\chi^{(3)}_{dop} = 0.045$	115
Spangler3	DFWM	532	0.030	$\chi^{(3)}_{neu} = 3.60$ $\chi^{(3)}_{dop} = 0.195$	115
Spangler3	DFWM	1064	0.030	$\chi^{(3)}_{neu} = 0.092$ $\chi^{(3)}_{dop} = 0.038$	115
Spangler4	DFWM	532	0.030	$\chi^{(3)}_{neu} = 0.018$	115
Spangler5	DFWM	532	0.030	$\chi^{(3)}_{neu} = 0.133$	115
Spangler6	DFWM	532	0.030	$\chi^{(3)}_{neu} = 0.670$	115
Spangler7	DFWM	532	0.030	$\chi^{(3)}_{neu} = 2.37$	115
Spangler8	DFWM	532	0.030	$\chi^{(3)}_{neu} = 4.19$	115
Spangler9	DFWM	532	0.030	$\chi^{(3)}_{neu} = 13.9$	115
stilbene in dichloroethane	PCI	532	5	$\chi^{(3)}_{1221} = 0.0034$	72
succinonitrile	OKE	$\lambda_{pump} = 694.3$ $\lambda_{probe} = 397.2$	15	$\chi^{(3)}_{1111} - \chi^{(3)}_{1122} = 0.07$	116

tetramethylsilane	OKE	$\lambda_{pump} = 1064$ $\lambda_{probe} = D_2O$ continuum	0.033	$\chi^{(3)}_{1221} = 0.024$	117
tetraphenyl porphyrin in toluene	DFWM	532	0.080	4 ± 2 (Thermal)	66
Voglmann1	B-DFWM	720	0.001	14.0	88
Voglmann2	B-DFWM	720	0.001	195	88
Winter1	DFWM	1064	0.100	$n_2 = 2.1 \times 10^{-9}$ cm²/kw	57
Yt bisphthalocyanine in chloroform	DFWM	1064	0.010	0.012	68
Yu1	DFWM	585		136.5 ± 41.5 $\chi^{(3)}_{1221} = 25.1 \pm 6.98$ $\chi^{(3)}_{T} = 53 \pm 17$	51
Yu2	DFWM	585	0.006	124 ± 35 $\chi^{(3)}_{1221} = 26.5 \pm 7.0$ $\chi^{(3)}_{T} = 61.4 \pm 16.8$	51
Yu3	DFWM	585	0.006	103 ± 29 $\chi^{(3)}_{1221} = 22 \pm 7$ $\chi^{(3)}_{T} = 53 \pm 17$	51
Yu4		585	0.006	216 ± 71 $\chi^{(3)}_{1221} = 42 \pm 12$ $\chi^{(3)}_{T} = 108 \pm 28$	51
Yu5	DFWM	585	0.006	560	51
ZnTSPP-ZnTMPyP in H_2O/CH_3OH at 1.18×10^{-4}M	F-DFWM	1064	0.030	0.00022	118
ZnTSPP-ZnTMPyP in H_2O/CH_3OH at 1.18×10^{-4}M	F-DFWM (pumped)	1064 532 pump	0.030	0.000182	118
zinc tetraphenyl porphyrin in toluene	DFWM	532	0.080	2 ± 1 (thermal)	66

BrMPPV—poly (2-bromo-5-methoxy-*p*-phenylene vinylene)
BBB—poly(6,9-dihydro-6,9-dioxobisbenzimidazo [2,1b:1′2′]benzo[1mn]-[3,8]phenanthroline-3,2-diyl)
BBL—poly{(7-oxo-7,10H-benz[de]imidazo [4′,5′:5,6]benzimidazo[2,1-a]isoquinoline-3,4:10,11-tetra-10-carbonyl}
BDN—bis[1-[4-(dimethylamino)phenyl]-2-phenyl-1,2-ethenedithiolato(2-)-S,S′]-nickel
BuMPPV—poly(2-butyl-5-methoxy-*p*-phenylene vinylene)
DCE—1,2-dichloroethane
DANS—4-dialkylamino-4′-nitrostilbene
DEANST—4 (N,N′-diethylamino) nitrostyrene
DDT—dodecylthiophene
MMA—methylmethacrylate

MTPBT—poly[benzol 1,2-d:4,5-d′) bisthiazole-2,6-diyl]-2,5-thienylene
ONS—4-oxy-4′-nitrostilbene
OONOOB—4′-octyloxynaphthyl-1′-(4-decyoxybenzoate)
Ph—phenyl
P(1APyre)—poly(1-aminopyrene)
P(3AQ)—poly(3-aminoquinoline)
P(3APAz)—poly(3-aminopyraxole)
P(3APy)—poly(3-aminopyridine)
P(4ABP)—poly(4-amine biphenyl)
P(5AQ)—poly(5-aminoquinoline)
P(6AQ)—poly(6-aminoquinoline)
P(8AQ)—poly(8-aminoquinoline)
P(Ani)—poly(aniline)
PDA-PTS—poly(bis(*p*-toluene sulfonate) of 2,4 hexadiyne-1,6 diol)
PDA-3BCMU—poly[4,6-decadiyne-1,10-diol bis([(*n*-butoxycarbonyl)methyl]urethane)]
PDA-4BCMU—poly[5,7-dodecadiyne-1,12-diol bis([(*n*-butoxycarbonyl)methyl]urethane)]
P(DPA)—poly(diphenylamine)
PBT—poly(*p*-phenylenebenzobisthiazole)
PBTABQ—poly[γ-5,5^1-acetoxy benzylidene-block-γ-[5,5-bithiophene quindimethanediyl]
PBTBQ—poly[γ-[5,5^1-bithiophenediyl] benzylidene-block-γ-[5,5-bithiophene quindimethanediyl]
PDTT—poly(dithieno(3,2-b,2′,3′-d)thiophene)
P(ITN)—poly(isothianaphthene)
PMMA—poly(methyl methacrylate)
P(NNPheBz)—poly(N,N′-diphenylbenzidine)
P(NPhe2NaA)—poly(N-phenyl-2-naphthylamine)
PNO2STPV—poly [2-(3-{4-[2-)4-nitrophenyl) vinyl] phenoxy} propoxyl phenylenevinylene
PPA—polyphenylene acetylene
PPMS—poly(phenyl methylsilane)
PPV—poly(phenylenevinylene)
PSBAC—poly(symmetrically substituted benzylidene aniline dichloride)
PTCDA—perylene tetracarboxylic anhydride
PVK—poly-N-vinyl carbazole
PTT—poly(thieno(3,2-b)thiophene)
SiPc—silicon phthlalocyanine
SiPcMac—silicon phthalocyanine monoacrylate
SiNc—silicon naphthalocyanine
THF—tetrahydrofuran
TNF—2,4,7-trinitrofluorenone
TMPD—N,N,N′,N′-tetramethyl-1,4-phenylenediamine
ZnTSPP—zinc tetra (4-sulfonatophenyl) porphyrin
ZnTMPyP—zinc tetra (4-N-methyl-pyridyl) porphyrin

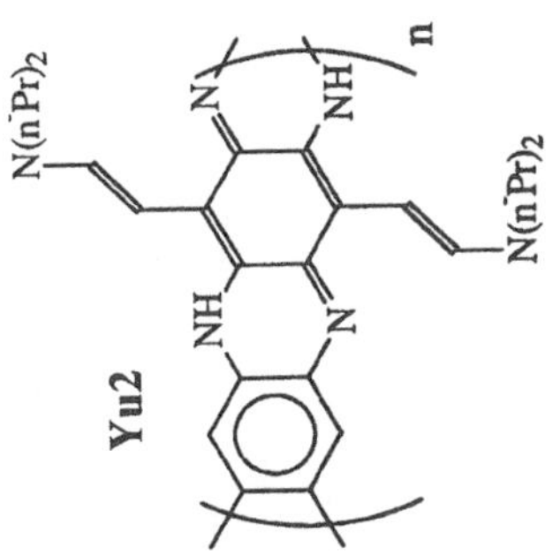

Yu2

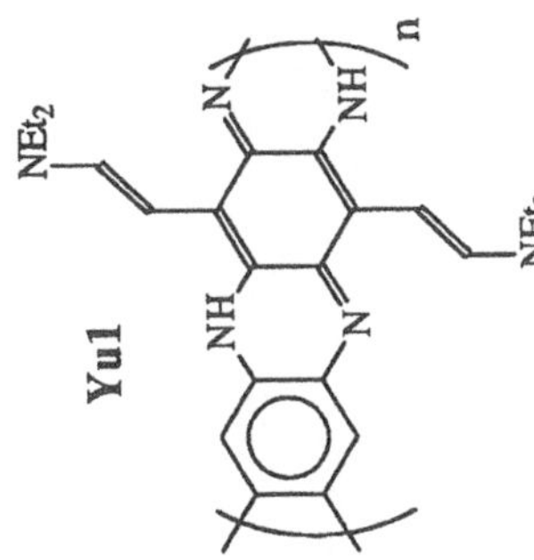

Yu1

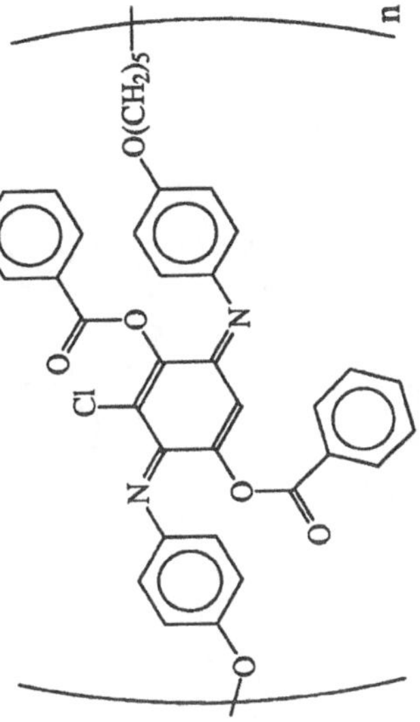

Yu5

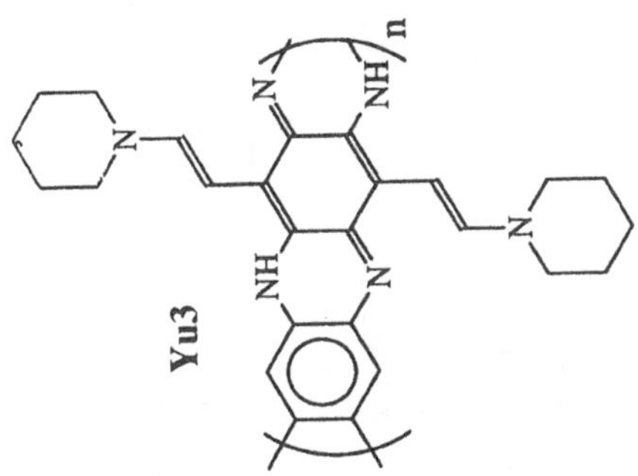

Yu3

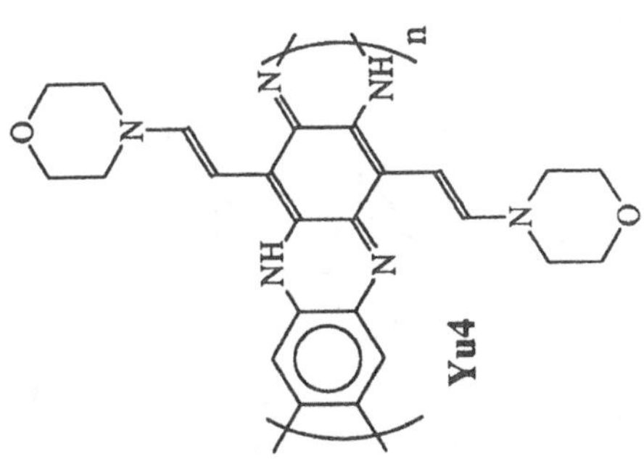

Yu4

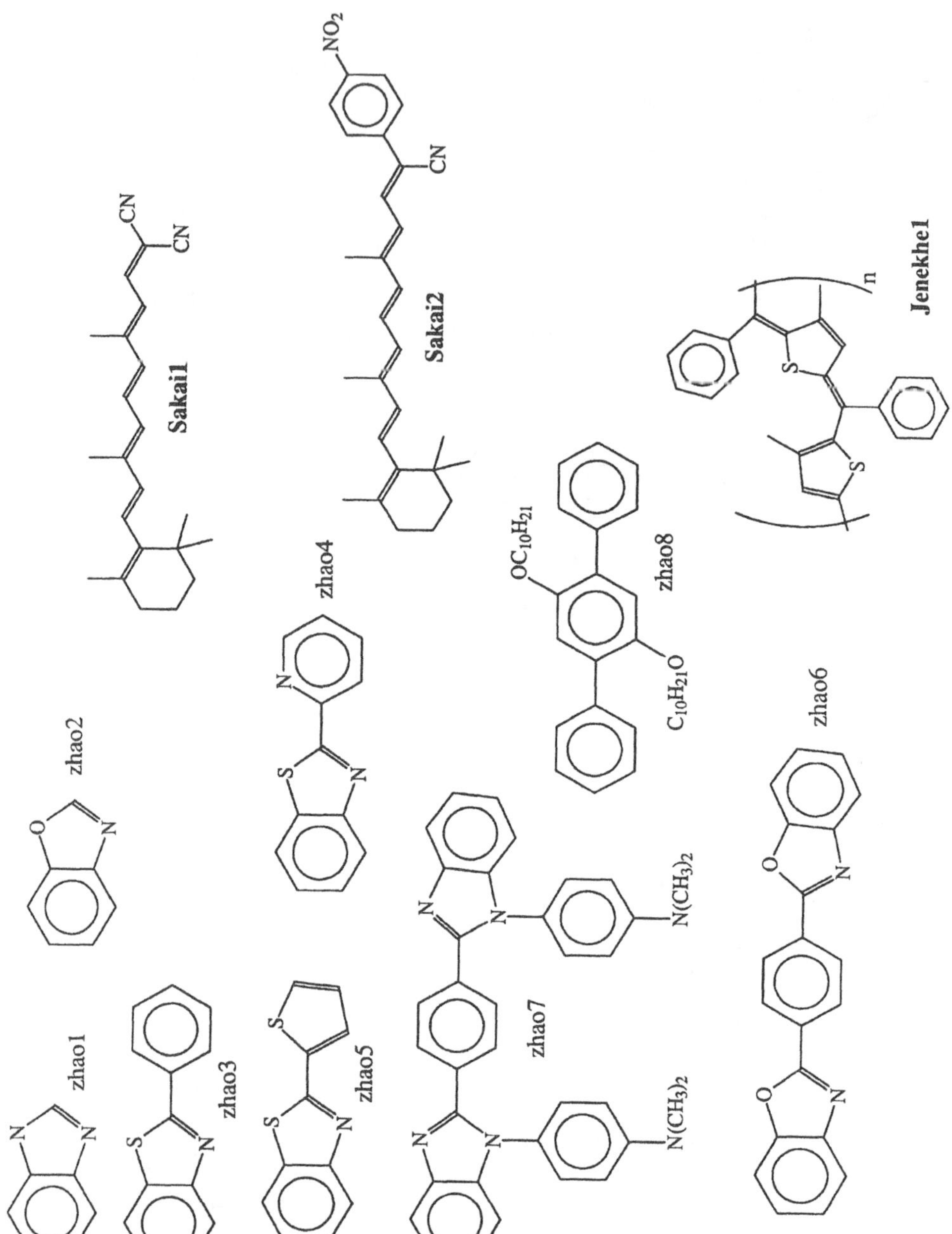

zhao1
zhao2
zhao3
zhao4
zhao5
zhao6
zhao7
zhao8
Sakai1
Sakai2
Jenekhe1

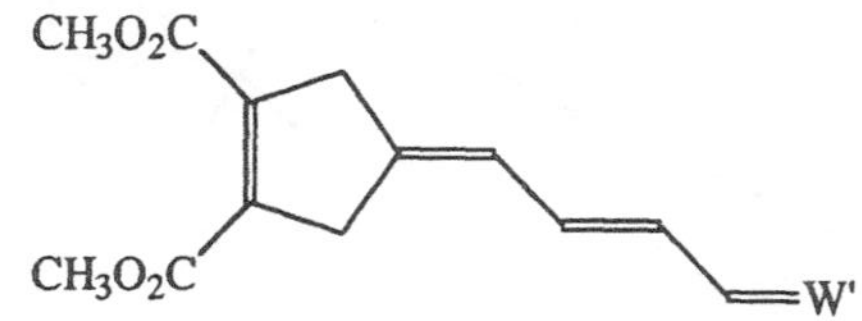

W'	name
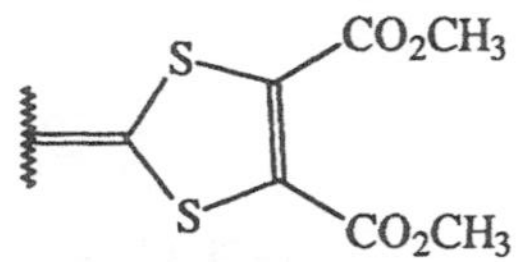	Saharovi1
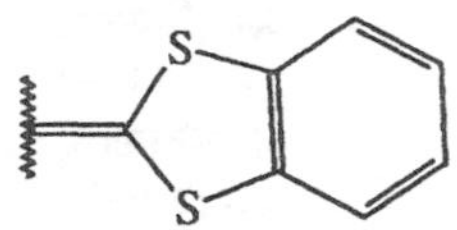	Saharovi2
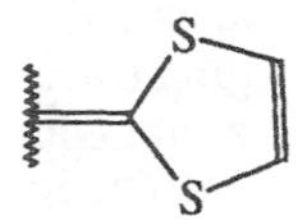	Sahaovi3
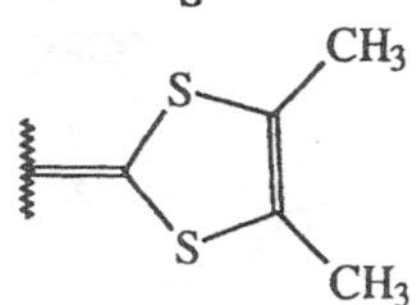	Saharovi4

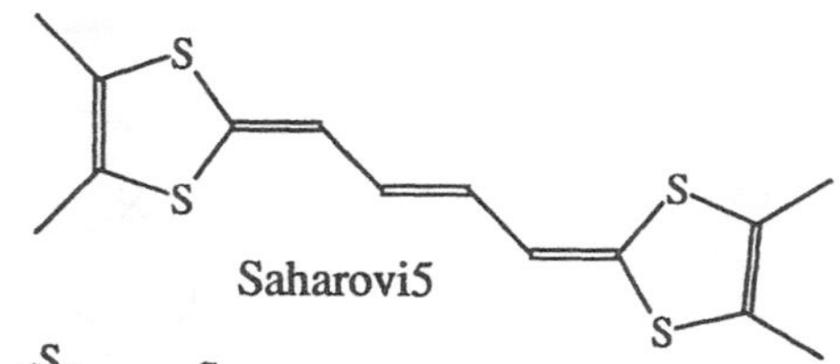

Saharovi5

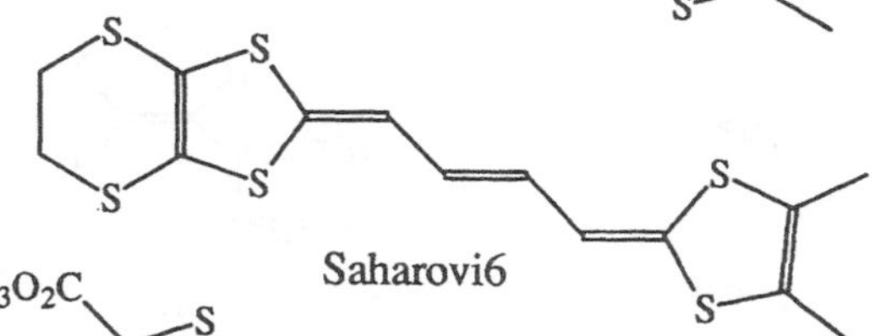

Saharovi6

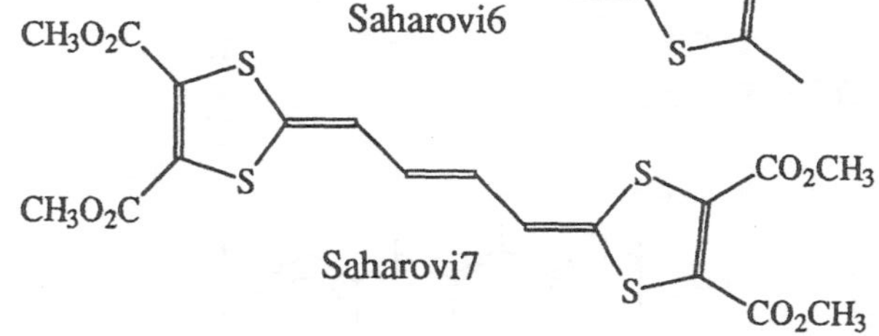

Saharovi7

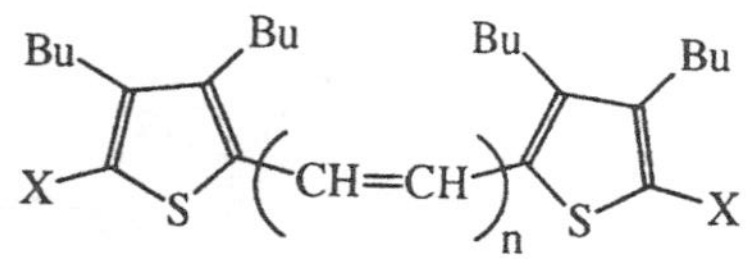

Name	X	n
Spangler1	H	5
Spangler2	H	7
Spangler3	H	8
Spangler4	H	3
Spangler5	H	4
Spangler6	H	5
Spangler7	H	6
Spangler8	H	7
Spangler9	H	8

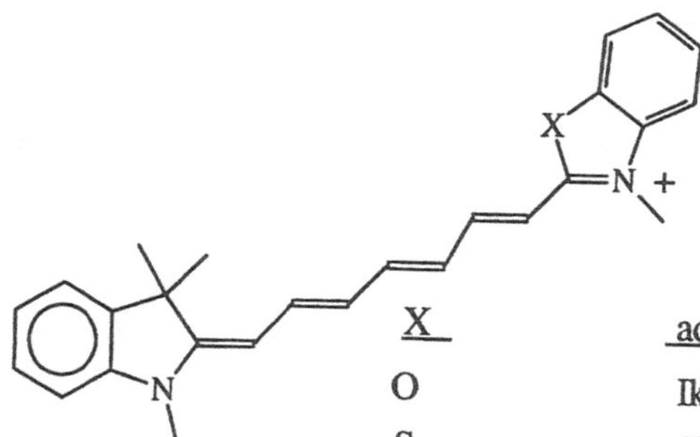

X	acronym
O	Ikeda1
S	Ikeda2

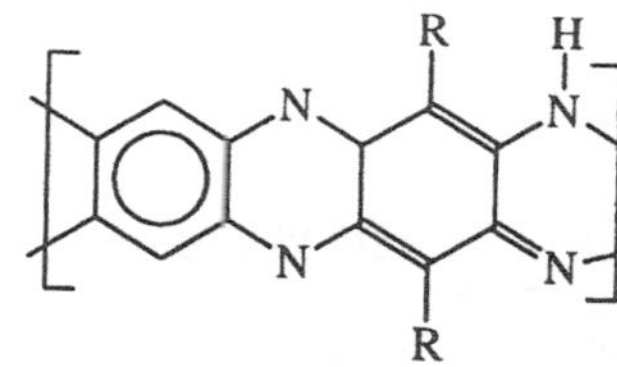

R	name/acronym
H	**Cao1**
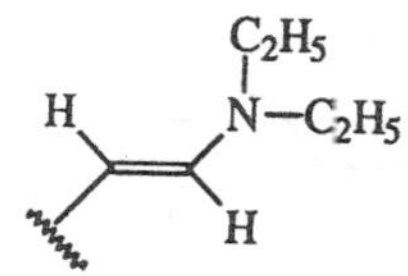	**Cao2**
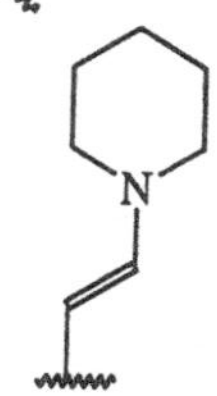	**Cao3**

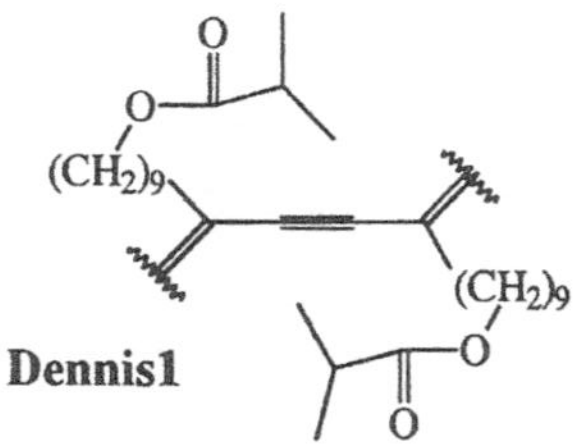

Dennis1

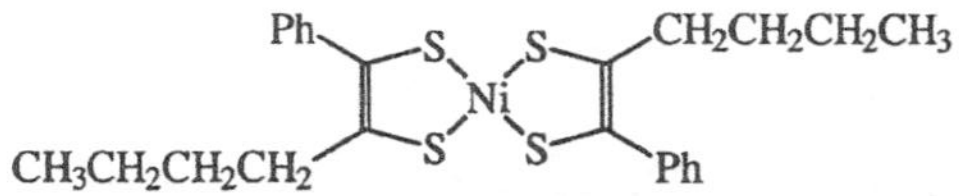

Winter1

Garvey1

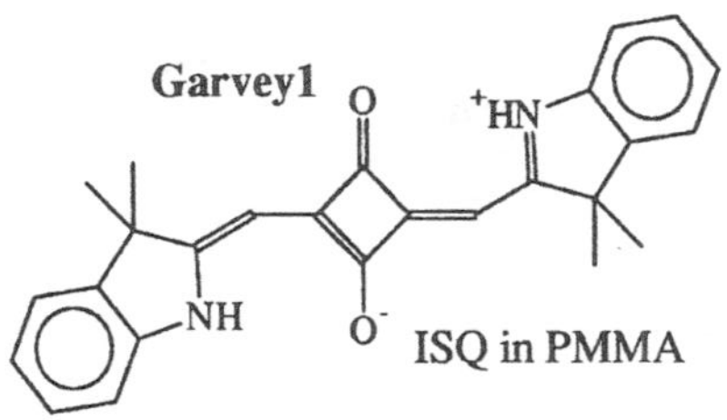

ISQ in PMMA

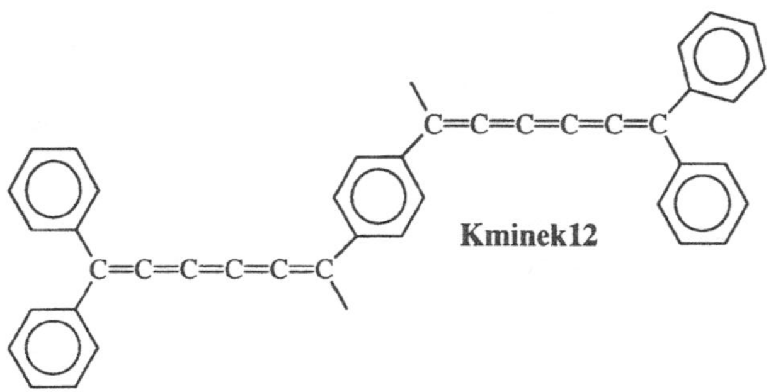

Kminek12

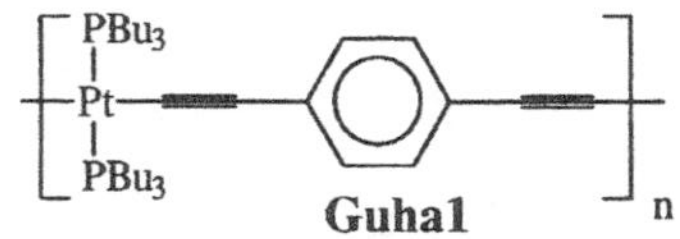

Guha1

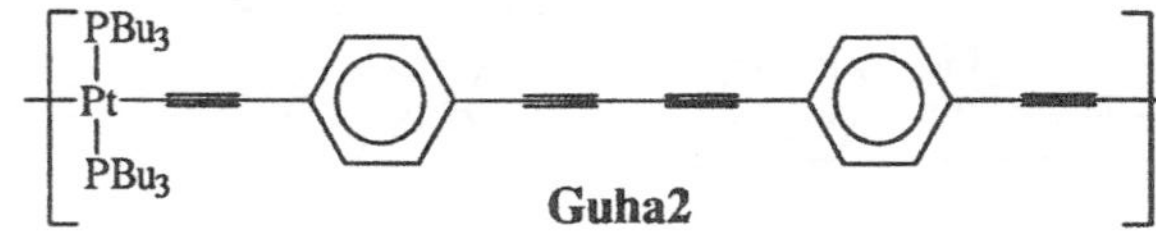

Guha2

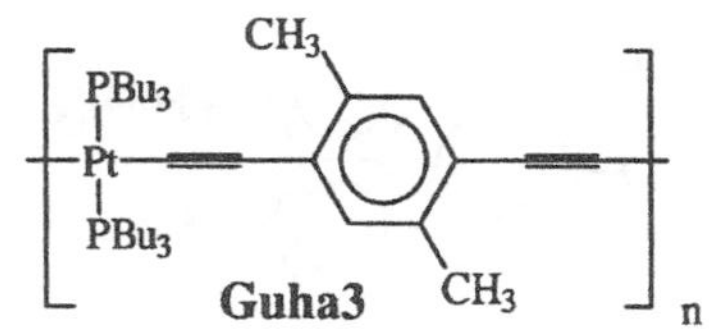

Guha3

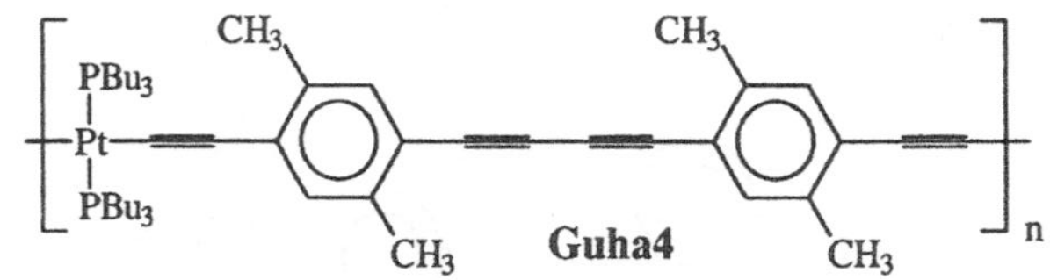

Guha4

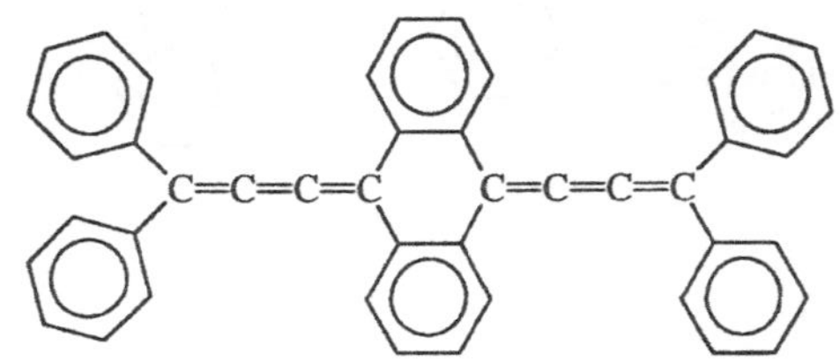

Kminek11

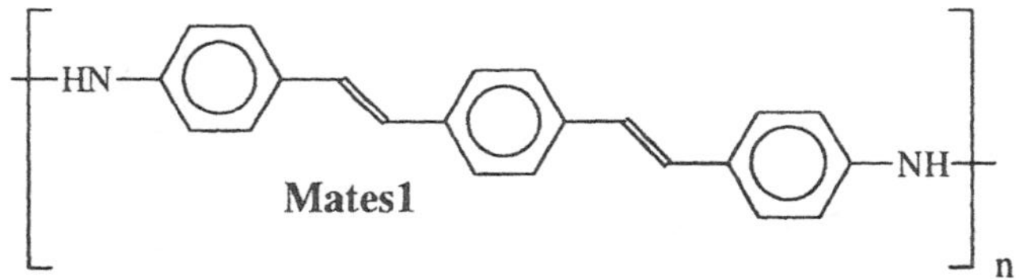

Mates1

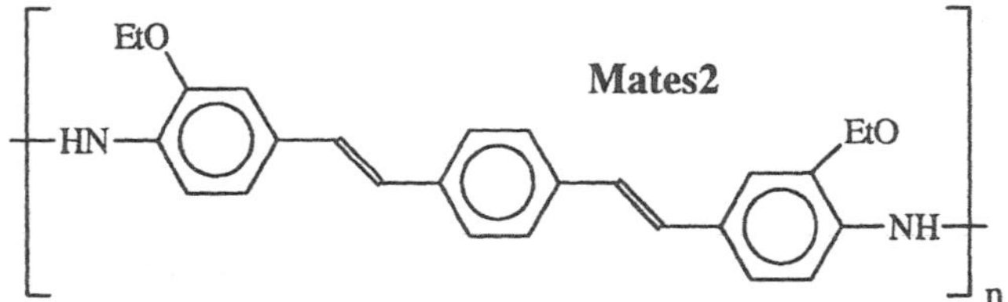

Mates2

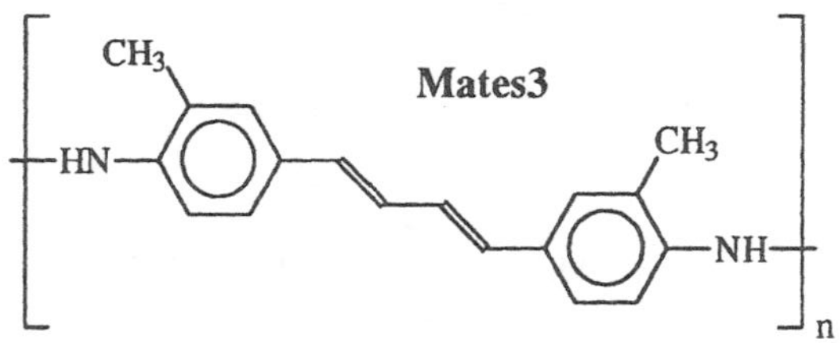

Mates3

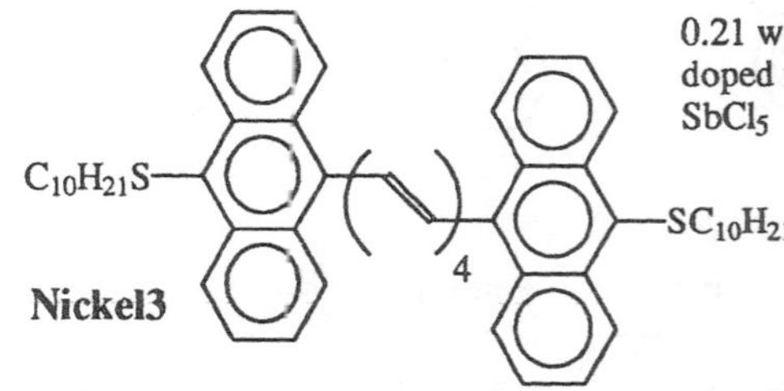

Nickel3

0.21 wgt,. % doped with $SbCl_5$

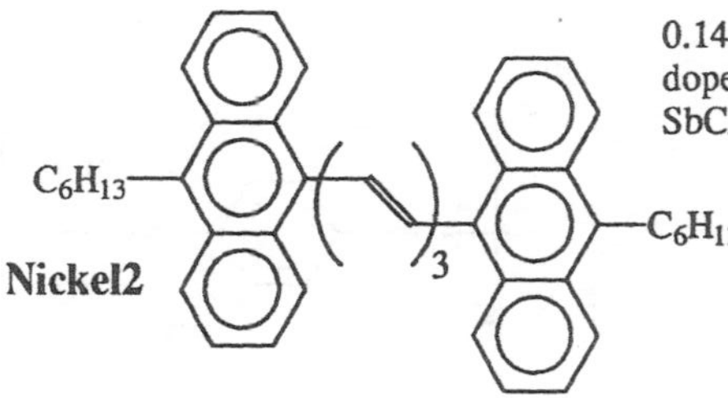

Nickel2

0.14 wgt. % doped with $SbCl_5$

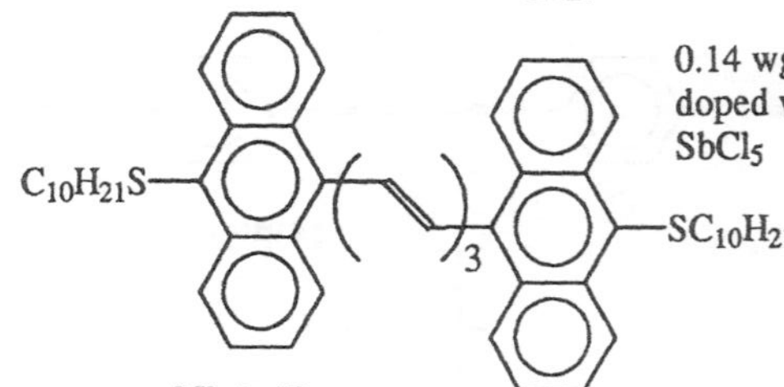

Nickel3

0.14 wgt. % doped with $SbCl_5$

NAME	R
Voglmann1	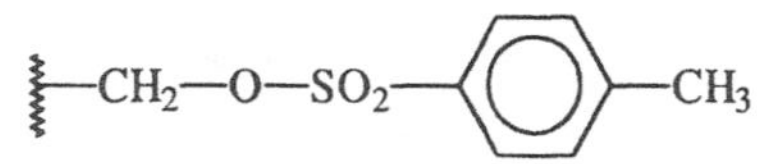
Voglmann2	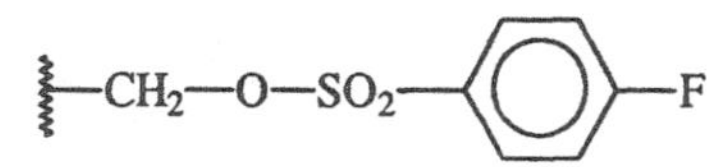

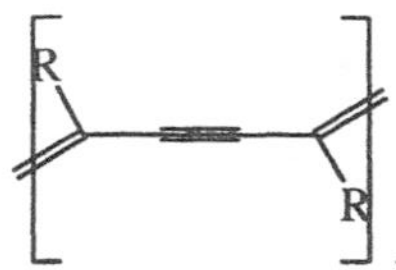

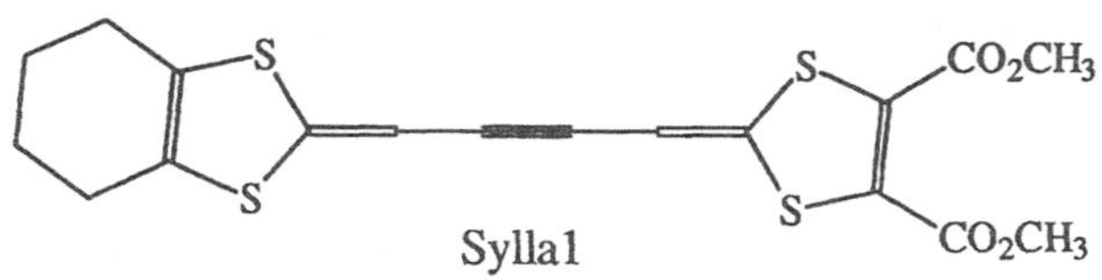

Sylla1

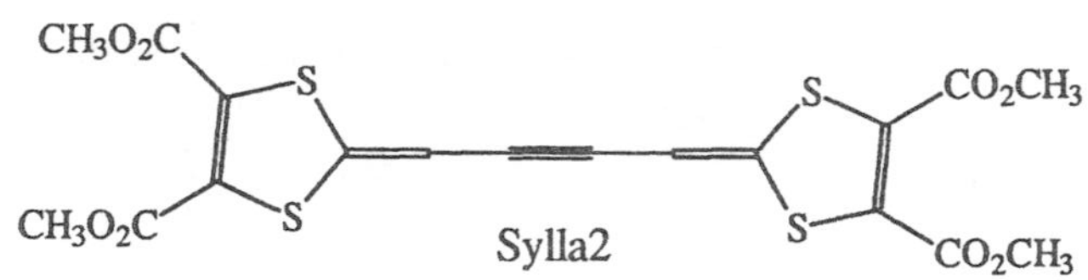

Sylla2

Sylla3

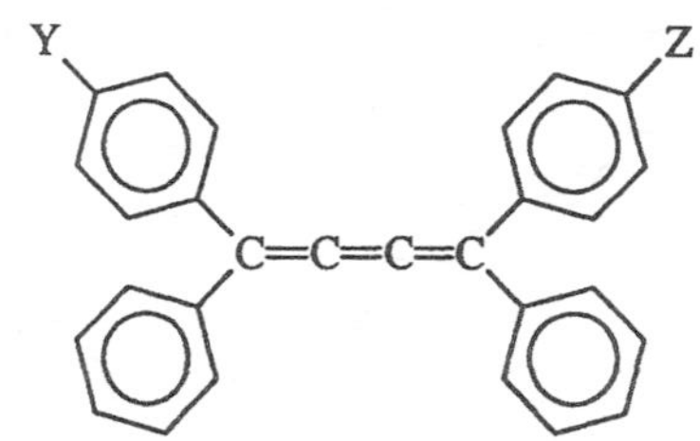

Y	Z	name
H	H	Kminek1
OCH_3	OCH_3	Kminek2
NO_2	$O(CH_2)_{11}CH_3$	Kminek3

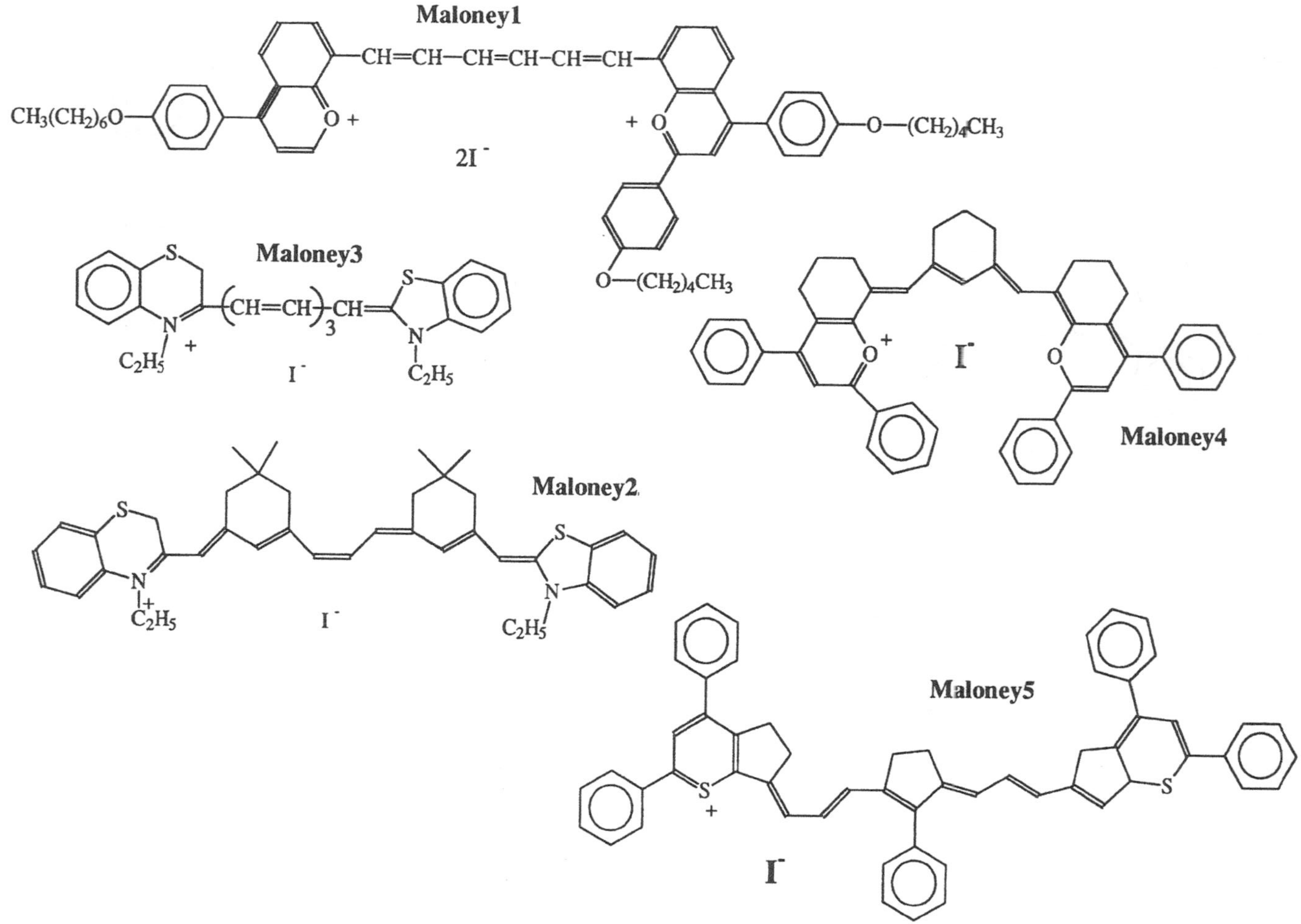

Maloney1

Maloney3

Maloney4

Maloney2

Maloney5

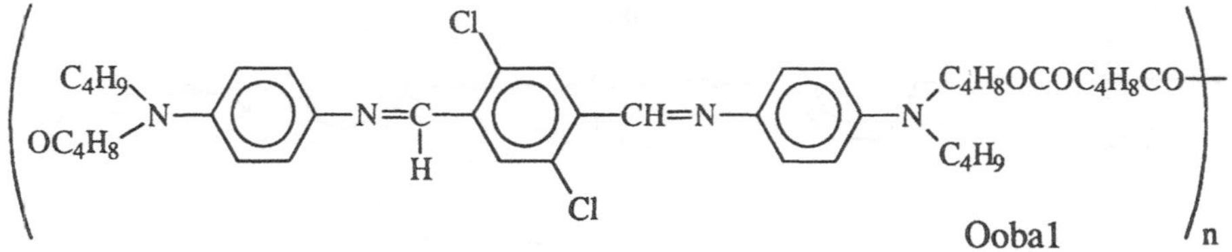

Ooba1

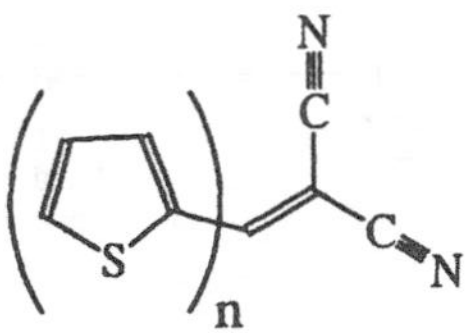

n	Name
1	Joshi1
2	Joshi2
3	Joshi3

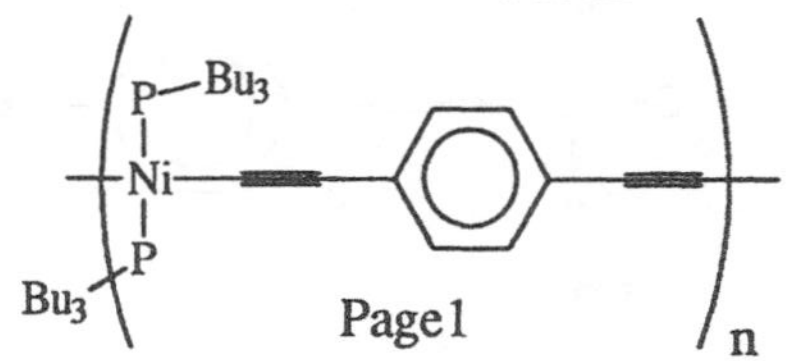

Page1

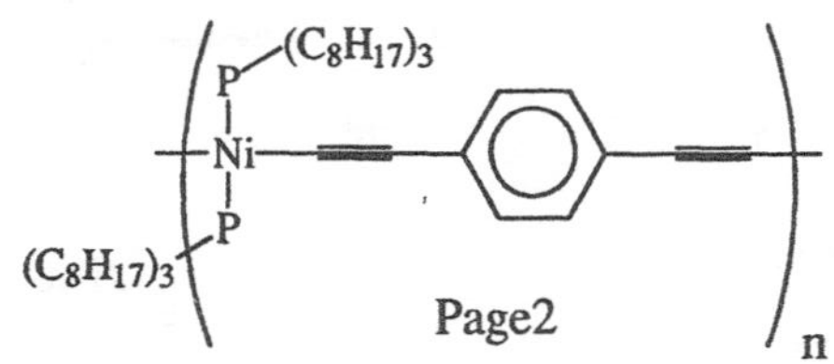

Page2

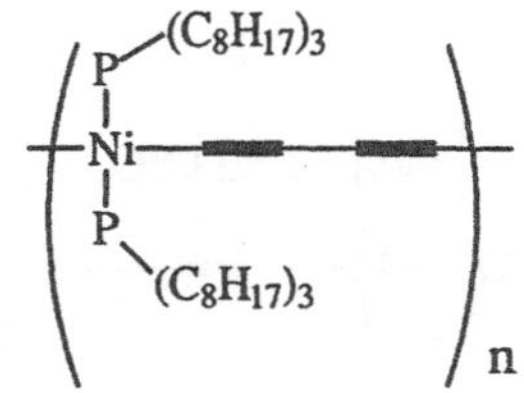

Page3

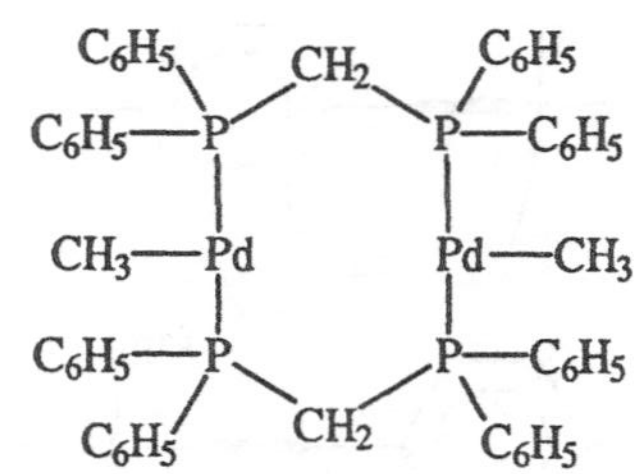

Page4

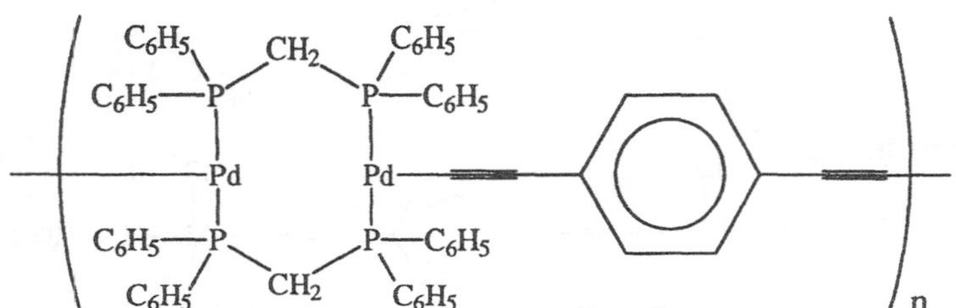

Page5

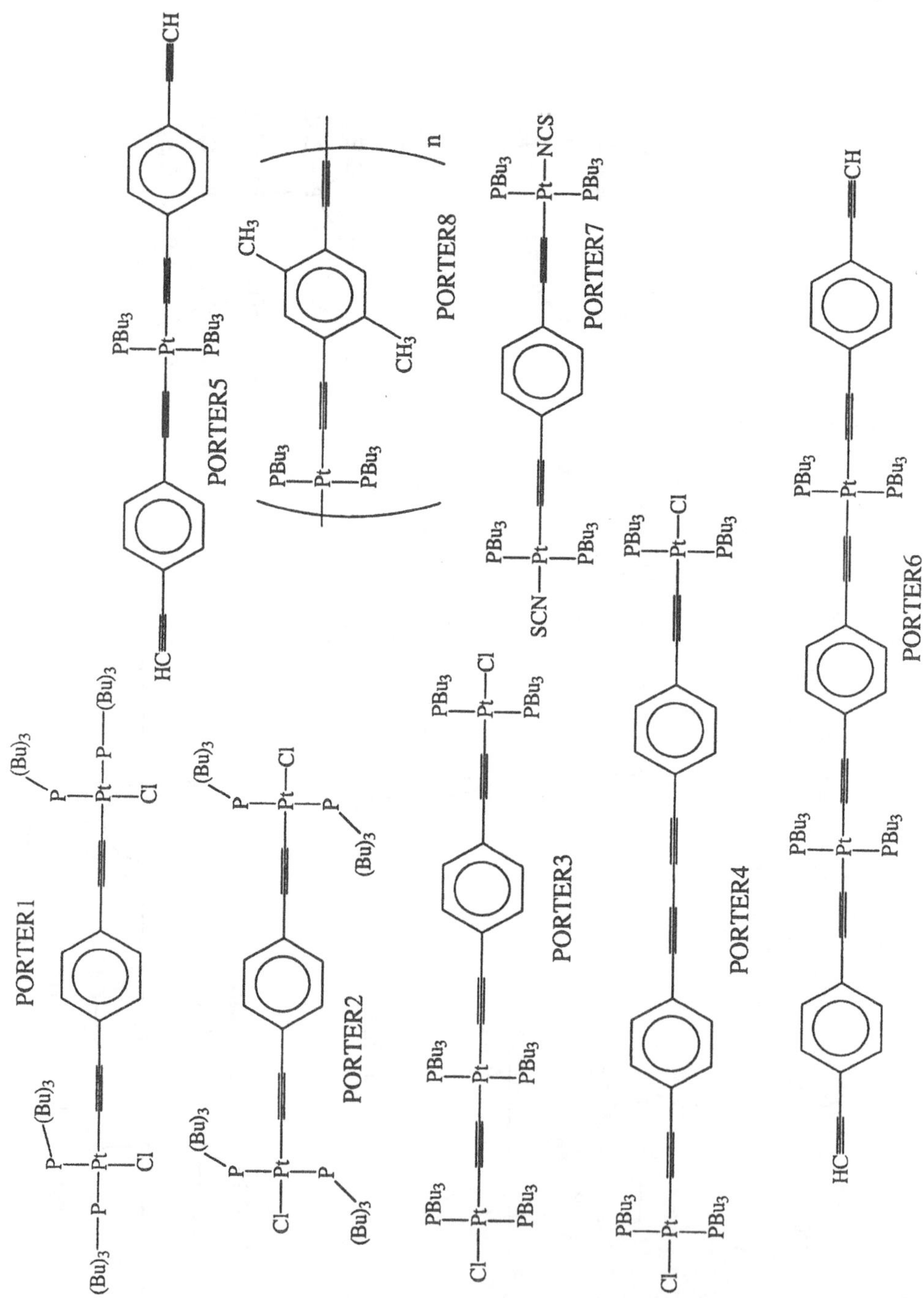
PORTER1
PORTER2
PORTER3
PORTER4
PORTER5
PORTER6
PORTER7
PORTER8

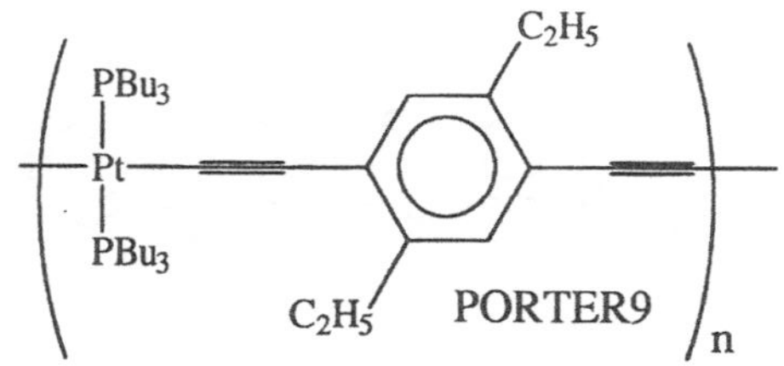

PORTER9

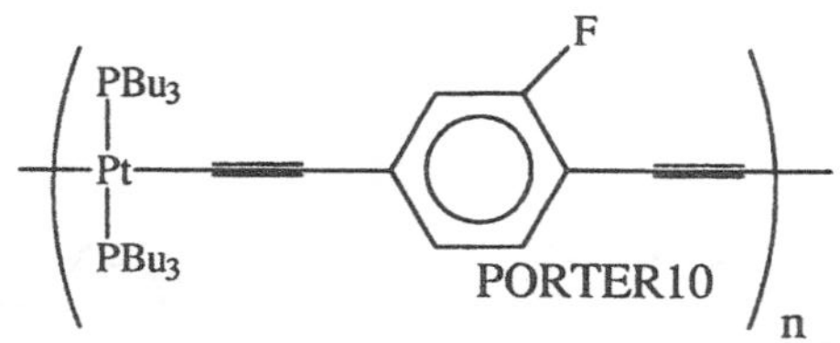

PORTER10

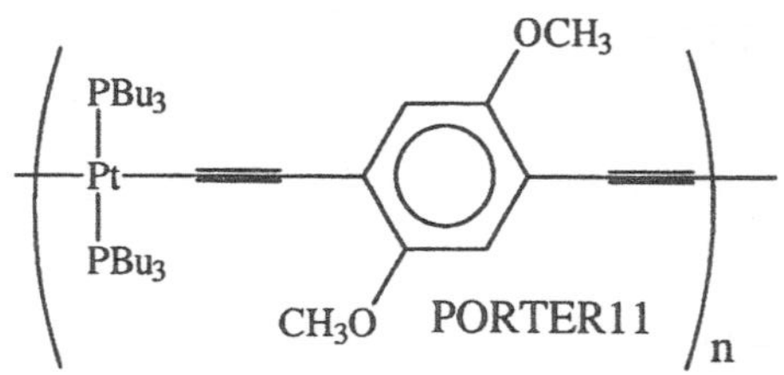

PORTER11

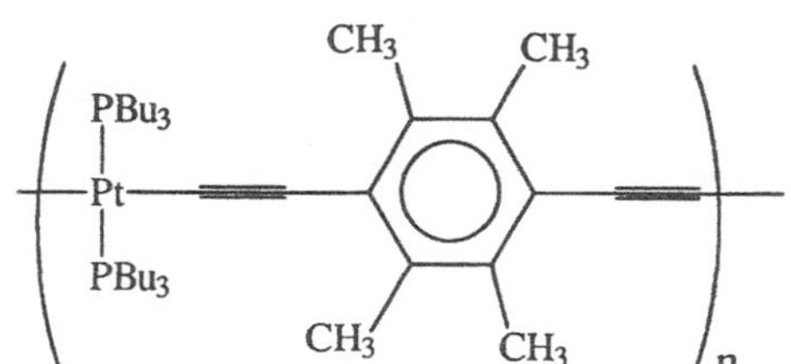

PORTER12

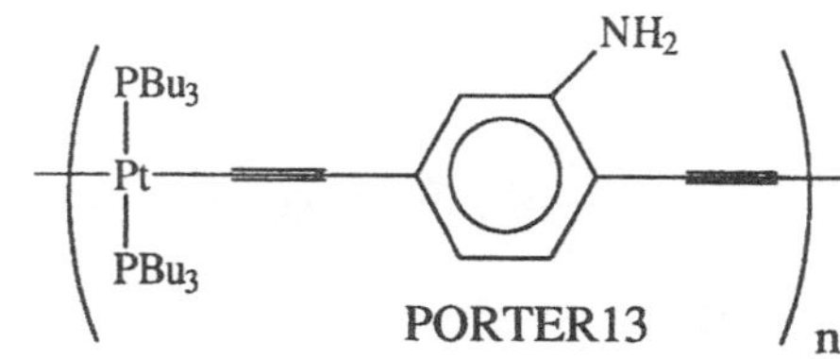

PORTER13

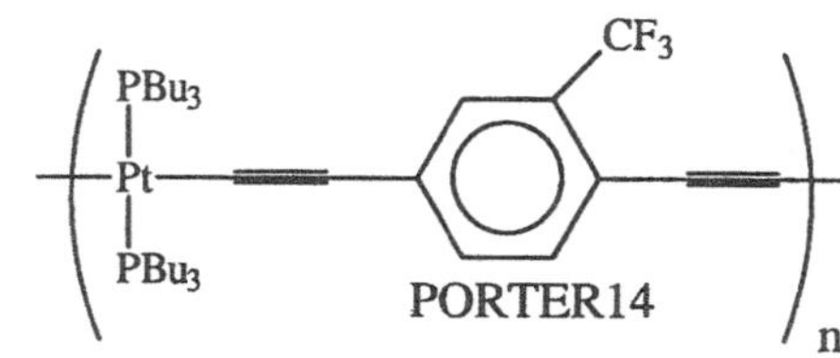

PORTER14

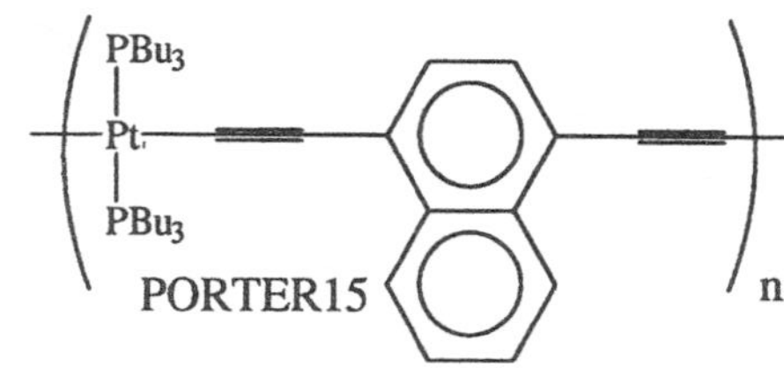

PORTER15

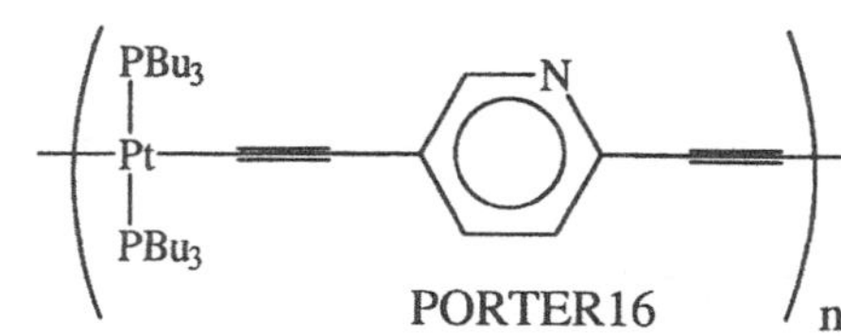

PORTER16

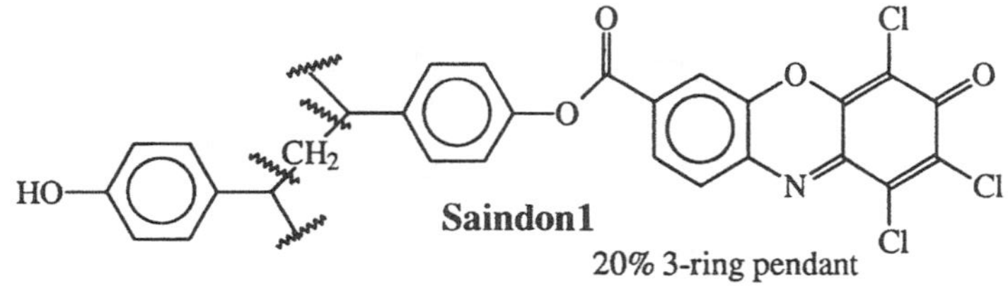

Saindon1

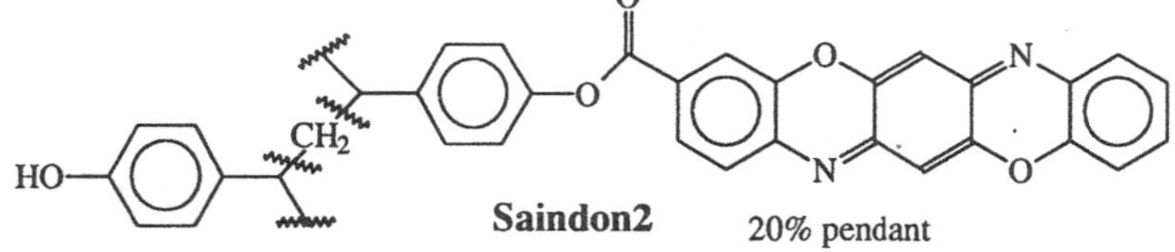

Saindon2

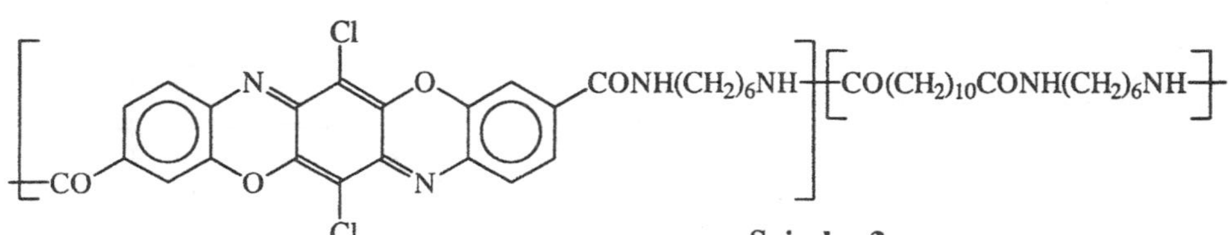

Saindon3

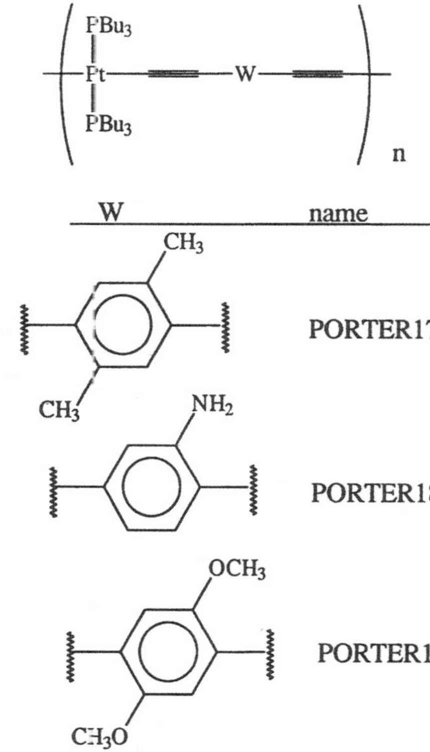

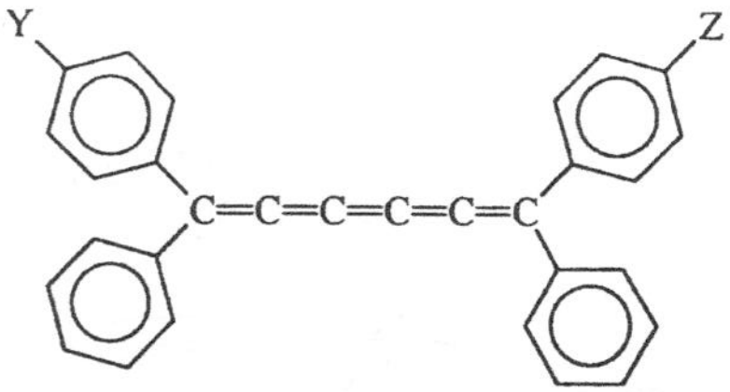

Y	Z	name
H	H	Kminek4
NO_2	NO_2	Kminek5
OCH_3	OCH_3	Kminek6
NO_2	OCH_3	Kminek7

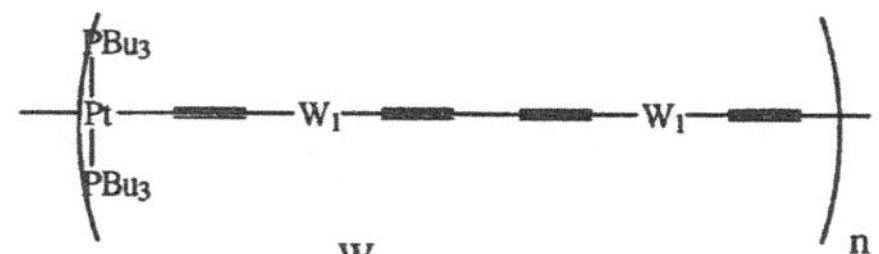

W_1

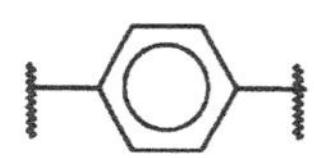

PORTER20

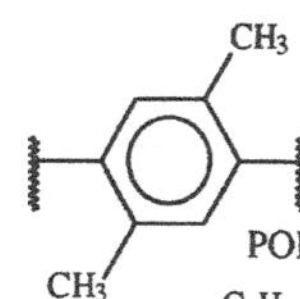

PORTER21

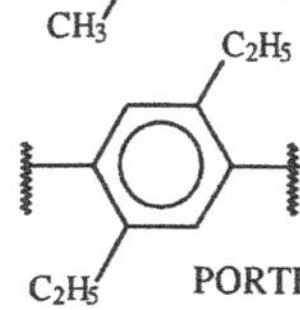

PORTER22

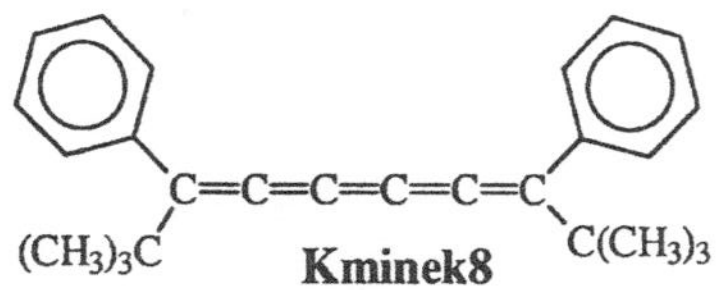

Kminek8

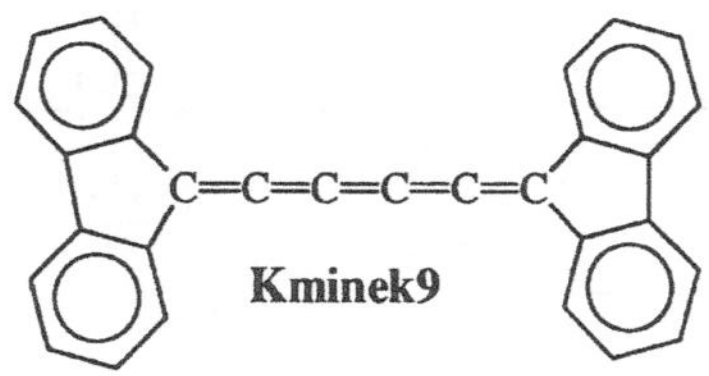

Kminek9

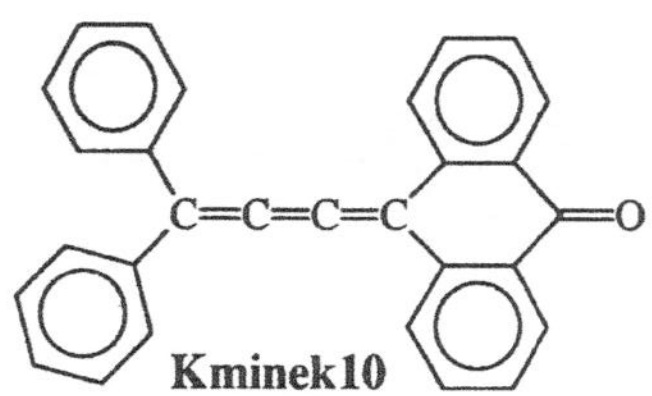

Kminek10

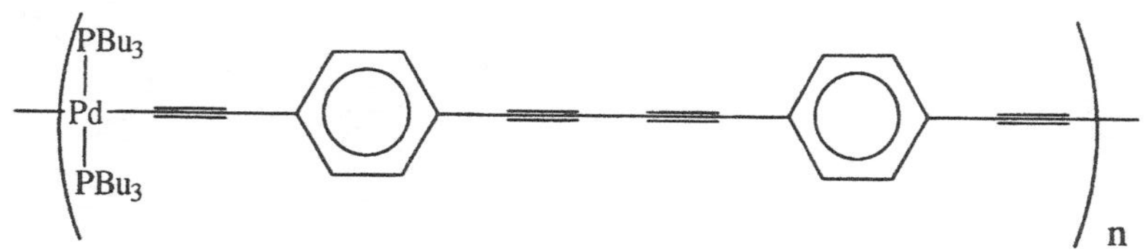

PORTER23

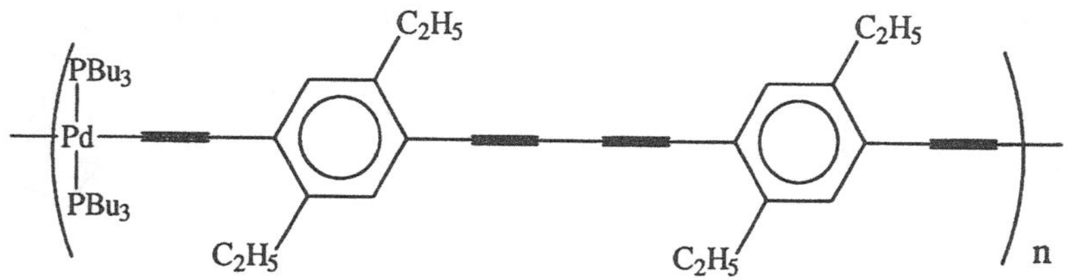

PORTER24

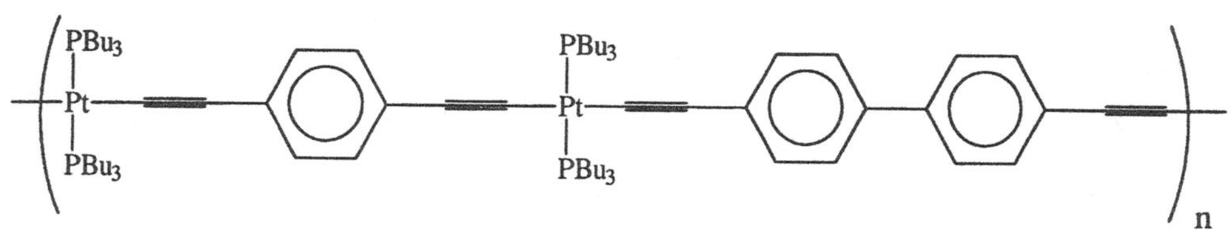

PORTER25

Table 3 Dressed Third Order Molecular Susceptibilities of Organic Molecules and Polymers Obtained by Four Wave Mixing Methods

Sample	Solvent	Method	λ (nm)	t_p (nsec)	$\langle\gamma^*\rangle$ (10^{-46} m^5/V^2)	Ref.
$[Ni(BDT)_2]$	chloroform	DFWM	1064	0.100	$\langle\gamma_{1111}\rangle = 14.4$	119
$[Ni(C_6H_6N_4)_2]$	chloroform	DFWM	1064	0.100	$\langle\gamma_{1111}\rangle = 21.4$	119
$[Ni(C_4H_2N_4)_2]$	chloroform	DFWM	1064	0.100	$\langle\gamma_{1111}\rangle = 5.86$	119
$[Ni(P(C_2H_5)_3)_2(C_4)_2]$	chloroform	F-DFWM	1064	—	-0.79, $\langle\gamma''\rangle = 1.7$	120
$[Ni(P(C_2H_5)_3)_2(C_4)_2]$	chloroform	F-DFWM	1064	—	-2.8, $\langle\gamma''\rangle = 1.5$	120
$[Ni(P(C_2H_5)_3)_2(C_4H_3S)_2]$	chloroform	F-DFWM	1064	—	-5.1, $\langle\gamma''\rangle = 0.46$	120
$[Ni(P(C_2H_5)_3)_2(C_8H_5S)_2]$	chloroform	F-DFWM	1064	—	-2, $\langle\gamma''\rangle = 1.0$	120
$[Ni(P(C_4H_9)_3)_2(C_4)]_n$	chloroform	F-DFWM	1064	—	-26.3, $\langle\gamma''\rangle = 24.1$	120
$[Ni(P(C_4H_9)_3)_2(C_4H_2S)]_n$	chloroform	F-DFWM	1064	—	-170, $\langle\gamma''\rangle = 200$	120
$[Ni(P(C_4H_9)_3)_2(C_8H_4S)]_n$	chloroform	F-DFWM	1064	—	-30, $\langle\gamma''\rangle = 10$	120
$[Ni(PBu_3)_2R]M$ R = thiophene	chloroform	F-DFWM	1064	0.070	-170, $\langle\gamma''\rangle = 200$	121
$[Ni(PBu_3)_2R]_m$ with R = isothionaphthene in	chloroform	F-DFWM	1064	0.070	-30, $\langle\gamma''\rangle = 10$	121
$[Pd(P(C_2H_5)_3)_2(C_4)_2]$	chloroform	F-DFWM	1064	—	-0.39, $\langle\gamma''\rangle = 0.092$	120
$[Pd(P(C_2H_5)_3)_2(C_4)_2]$	chloroform	F-DFWM	1064	—	-2.1, $\langle\gamma''\rangle = 0.34$	120
$[Pd(P(C_2H_5)_3)_2(C_4H_3S)_2]$	chloroform	F-DFWM	1064	—	-0.69, $\langle\gamma''\rangle = 1.3$	120
$[Pt(C_6H_2N_4)_2]$	chloroform	DFWM	1064	0.100	$\langle\gamma_{1111}\rangle = 6.70$	119
$[Pt(P(C_2H_5)_3)_2(C_4)_2]$	chloroform	F-DFWM	1064	—	-0.19, $\langle\gamma''\rangle = 0.077$	120
$[Pt(P(C_2H_5)_3)_2(C_4H_3S)_2]$	chloroform	F-DFWM	1064	—	-1.1, $\langle\gamma''\rangle = 0.22$	120
$[Pt(P(C_2H_5)_3)_2(C_4H_3S)_2]$	chloroform	F-DFWM	1064	—	-0.25, $\langle\gamma''\rangle = 0.60$	120
$[Pt(P(C_4H_9)_3)_2(C_4)]_n$	chloroform	F-DFWM	1064	—	-14.8, $\langle\gamma''\rangle = 17.4$	120
1:1:1 epoxy resin:epoxy hardner: rhodamine B	ethanol	DFWM	1064	4	60000	122
2,2-(1,3-phenylene)bis-thiophene	THF	DFWM	602	0.0004	0.014 ± 0.002	123
2,2′-(1,4-phenylene)bis-benzothiazole	THF	B-DFWM	602	0.0004	0.0265	124
2,2′-(2,5-pyridinediyl)bis[1-phenyl-1H-benzimidazole]	THF	B-DFWM	602	0.0004	0.1536 ± 0.0335	124

Table 3 Continued

Sample	Solvent	Method	λ (nm)	t_p (nsec)	$\langle\gamma^*\rangle$ (10^{-46} m^5/V^2)	Ref.
2,2′-(2,5-thiophenediyl)bis-benzothiazole	THF	B-DFWM	602	0.0004	0.0363 ± 0.0168	124
2,2′-[3,4-bis(decyloxy)-2,5-thiophenediyl]bis-benzothiazole	THF	B-DFWM	602	0.0004	0.0545 ± 0.0004	124
2-(2-phenylethenyl)-benzothiazole	THF	B-DFWM	602	0.0004	0.0108 ± 0.0025	124
2-[2-(2-thienyl)ethenyl]-benzothiazole	THF	B-DFWM	602	0.0004	0.0123 ± 0.0021	124
3,4-di-2-thienyl-1H-pyrrole	THF	DFWM	602	0.0004	0.0182 ± 0.0027	123
4-methoxy-4′-nitrostilbene	chloroform	PCI	1064	0.200	0.0124 ± 0.0005	72
5,5″-diiodo-2,2′:5′,2″-terthiophene	THF	DFWM	602	0.0004	0.070 ± 0.003	123
5,5″-dinitro-2,2′:5′,2″-terthiophene	THF	DFWM	602	0.0004	0.17 ± 0.045	123
5-BTP	methylene chloride	PCI	532	0.030	−350, $\langle\gamma''\rangle$ = −24	125
5-nitro-2,2′:5′2″-trethiophene	THF	DFWM	602	0.0004	1.00 ± 0.02	123
87% C_{60}, 13% C_{20}	toluene	DFWM	1064	10	$\langle\gamma_{1111}\rangle$ = 22.3 ± 7.0	126
β-carotene	CS_2	DFWM	1064	10	0.0011	127
β-carotene	toluene	DFWM	1064	0.160	<100	128
BDMAPEDTNi	toluene	DFWM	1064	0.160	8600	128
BDQP	THF	DFWM	602	0.0004	0.0096	129
BDSP	THF	DFWM	602	0.0004	0.0489	129
benzene	neat	OKE	λ_{pump} = 1064 λ_{probe} = 532	0.030	0.0005	75
benzoporphyrin I	THF	OKE	λ_{pump} = 1064 λ_{probe} = 1064	0.021	0.140 ± 0.070	130
benzoporphyrin II	THF	OKE	λ_{pump} = 1064 λ_{probe} = 1064	0.021	0.251 ± 0.126	130
benzothiazole	THF	B-DFWM	602	0.0004	0.0035 ± 0.0005	124

benzyl chloride	neat	TWM	$\lambda_1 \approx 561$ $\lambda_2 \approx 602$ $\lambda_3 \approx 525$	3	$\langle\gamma_{1111}\rangle = 0.00017 \pm 0.00006$	28
bis(TPOHBOBE) cobalt II	DCE	DFWM	1064	10	$\langle\gamma_{1111}\rangle = 182$	131
bis[3-hexene-3,4-dithiolato(2-)-S,S′]-nickel	THF	DFWM	602	0.0004	0.0265	124
bithiophene	THF	DFWM	602	0.0004	0.0031	132
C_{60}	toluene	DFWM	532	0.030	$\gamma_{1212} = 2300$	133
C_{60}	film	DFWM	1064	0.035	0.042	61
C_{60}	benzene	B-DFWM	602	0.0004	-0.700 ± 0.280, $\langle\gamma''\rangle = 1.26 \pm 0.14$	134
C_{70}	toluene	DFWM	532	0.030	$\gamma_{1212} = 8600$	133
canthaxanthin	THF	HOKE	620	0.00006	-15.36, $\langle\gamma''\rangle = 22.3$	135
carbon disulfide	neat	OKE	$\lambda_{pump} = 1064$ $\lambda_{probe} = 532$	0.030	0.00095	75
carbon disulfide	neat	TWM	$\lambda_1 \approx 561$ $\lambda_2 \approx 602$ $\lambda_3 \approx 525$	3	$\langle\gamma_{1111}\rangle = 0.00021 \pm 0.00003$	28
carbon tetrachloride	neat	OKE	$\lambda_{pump} = 1064$ $\lambda_{probe} = 532$	0.030	0.000085	75
carbon tetrachloride	neat	TWM	$\lambda_1 \approx 561$ $\lambda_2 \approx 602$ $\lambda_3 \approx 525$	3	$\langle\gamma_{1111}\rangle = 0.000067 \pm 0.000007$	28
cis-$Mo(CO)_4(PPh_2Me)_2$	THF	DFWM	532	20	0.296	136
cis-$Mo(CO)_4(PPh_2OMe)_2$	THF	DFWM	532	20	0.326	136
cis-$Mo(CO)_4(PPh_3)_2$	THF	DFWM	532	20	12.59	136
dithiophene	dioxan	HOKE	$\lambda_{pump} = 616$ $\lambda_{probe} = 616$	0.00008 0.00008	-3.9×10^{-34} esu	137
DMPEPM	DMSO	HOKE	$\lambda_{pump} = 796$ $\lambda_{probe} = 796$	0.00017	2.23, $\langle\gamma''\rangle = 2.93$	138
DMSO	neat	HOKE	$\lambda_{pump} = 796$ $\lambda_{probe} = 796$	0.00017	0.00021	138

Table 3 Continued

Sample	Solvent	Method	λ (nm)	t_p (nsec)	$\langle\gamma^*\rangle$ (10^{-46} m^5/V^2)	Ref.
DTCNBDA in DMSO	DMSO	DFWM	532	6.0	9.1	139
DTDCI	DMSO	HOKE	λ_{pump} = 796 λ_{probe} = 796	0.00017 0.00017	−82.4, $\langle\gamma''\rangle$ = 30.7	138
europium bis-phthalocyanine	chloroform	DFWM	1064	0.035	30.7	140
europium bis-phthalocyanine	chloroform	DFWM	1064	0.035	39.1	140
gadolinium bis-phthalocyanine	chloroform	DFWM	1064	0.035	30.7	140
gadolinium bis-phthalocyanine	chloroform	DFWM	1064	0.035	39.1	140
Garvey1	guest host in PMMA	OKE	λ_{pump} = 1064 λ_{probe} = 1064	0.080	106 ± 60	74
HDVPC	chloroform	OKE	λ_{pump} = 647 λ_{probe} = 647	0.000165	$\gamma_{1212} + \gamma_{1221}$ = 490	141
Ikeda1	DMSO	DFWM	532	1	41900	142
Ikeda2	DMSO	DFWM	532	1	75400	142
Jenehke1	DCE	DFWM	532	0.030	$\langle\gamma_{1111}\rangle$ = 0.168	143
Joshi1	chloroform	DFWM	1064	9	1.21	144
	pyridine	DFWM	1064	9	−0.88	
Joshi2	chloroform	DFWM	1064	9	0.31	144
Joshi3	chloroform	DFWM	1064	9	52	144
	pyridine	DFWM	1064	9	8.5	
Joshi4	chloroform	DFWM	1064	9	39	144
Kminek1	THF	F-DFWM	602	0.0004	−0.419	145
Kminek10	THF	F-DFWM	602	0.0004	−0.140	145
Kminek11	THF	F-DFWM	602	0.0004	−55.8	145
Kminek12	THF	F-DFWM	602	0.0004	−41.9	145
Kminek2	THF	F-DFWM	602	0.0004	−0.98	145
Kminek3	THF	F-DFWM	602	0.0004	−0.126	145
Kminek4	THF	F-DFWM	602	0.0004	−1.40	145
Kminek5	THF	F-DFWM	602	0.0004	−0.140	145

Kminek6	THF	F-DFWM	602	0.0004	−0.558	145
Kminek7	THF	F-DFWM	602	0.0004	−4.19	145
Kminek8	THF	F-DFWM	602	0.0004	−0.00014	145
Kminek9	THF	F-DFWM	602	0.0004	−2.79	145
lead phthalocyanine	chloroform	DFWM	1064	0.035	0.173	146
lutetium bis-phthalocyanine	chloroform	DFWM	1064	0.035	47.5	146
lutetium bisphthalocyanine	chloroform	DFWM	1064	0.035	30.7	140
magnesium octamethyl tetrabenzporphyrin	THF	DFWM	532	0.017	22.3	147
Maloney1	DCE	DFWM	1064	0.160	27000	128
Maloney2	methanol	DFWM	1064	0.160	3270	128
Maloney3	methanol	DFWM	1064	0.160	490	128
Maloney4	DCE	DFWM	1064	0.160	19000	128
Maloney5	DCE	DFWM	1064	0.160	12700	128
$Mo(CO)_5(PPh_2NHMe)$	THF	DFWM	532	20	0.044	136
$Mo(CO)_5(PPh_3)$	THF	DFWM	532	20	0.011	136
N,N-dimethyl-4-(2-phenyl-1H-benzimidazol-1-yl)-benzenamine	THF	B-DFWM	602	0.004	0.0155 ± 0.0047	124
naphthalene	benzene	TWM	$\lambda_1 \approx 561$ $\lambda_2 \approx 602$ $\lambda_3 \approx 525$	3	$\langle\gamma_{1111}\rangle = 0.00020 \pm 0.00010$	28
neodynium bis-phthalocyanine	chloroform	DFWM	1064	0.035	20.9	140
nigrosine	water	DFWM	1064	0.160	720	128
nitrobenzene	neat	TWM	$\lambda_1 \approx 561$ $\lambda_2 \approx 602$ $\lambda_3 \approx 525$	3	$\langle\gamma_{1111}\rangle = 0.00021 \pm 0.00011$ $\langle\gamma''_{1111}\rangle = 0.00017 \pm 0.00007$	28
nonether PPQ	chloroform	Pumped DFWM	$\lambda_{pump} = 532$ $\lambda_{probe} = 355$	0.035	$\gamma^{eff}_{1111} = 2300 \pm 420$	148
nonether PPQ	chloroform	DFWM	532	0.035	$\langle\gamma_{1111}\rangle = 1150 \pm 420$	148
octithiophene	dioxan	OKE	$\lambda_{pump} = 616$ $\lambda_{probe} = 616$	0.00008 0.00008	−0.700	137

Table 3 Continued

Sample	Solvent	Method	λ (nm)	t_p (nsec)	$\langle\gamma^*\rangle$ (10^{-46} m^5/V^2)	Ref.
Page1	chloroform	F-DFWM	1064	0.070	-10, $\langle\gamma''\rangle = 20$	149
Page2	chloroform	F-DFWM	1064	0.070	-40, $\langle\gamma''\rangle = 100$	149
Page3	chloroform	F-DFWM	1064	0.070	-40, $\langle\gamma''\rangle = 30$	149
Page4	chloroform	F-DFWM	1064	0.070	-1, $\langle\gamma''\rangle = 2$	149
Page5	chloroform	F-DFWM	1064	0.070	-20, $\langle\gamma''\rangle = 20$	149
paraphenylene oxide	benzene	TWM	$\lambda_1 \approx 561$ nm $\lambda_2 \approx 602$ nm $\lambda_3 \approx 525$ nm	3	$\langle\gamma_{1111}\rangle = 0.00223 \pm 0.00028$	28
PDA-4BCMU (yellow form)	chloroform	F-DFWM	1064	—	20	120
PDA-3BCMU (blue form)	chloroform	F-DFWM	1064	—	-40, $\langle\gamma''\rangle = 100$	120
PDA-4BCMU	chloroform	TWM	$\lambda_1 = 714$		$\gamma_T = -0.070$	150
PDA-4BCMU	chloroform	TWM	$\lambda_1 = 625$	3	$\gamma_T = -0.21$	150
PDA-4BCMU	chloroform	TWM	$\lambda_1 = 645$	3	$\gamma_T = -0.70$	150
PEBTBPPt	THF	OKE	$\lambda_{pump} = 1064$ $\lambda_{probe} = 1064$	—	0.0052, $\langle\gamma''\rangle = 0.2660$	151
phthalocyanine	chloroform	DFWM	1064	0.035	0.0173	146
platinum phthalocyanine	chloroform	DFWM	1064	0.035	1.21	146
poly (-3-tert-butyl) isothionaphthene (high M_w)	chloroform	F-DFWM	1064	0.050	-10, $\langle\gamma''\rangle = 30$, $\lvert\gamma\rvert = 40$	152
poly (2-1-butyl) isothionaphthene	chloroform	F-DFWM	1064	—	-90, $\langle\gamma''\rangle = 60$	120
poly(-3-tert-butyl) isothionaphthene (low M_w)	chloroform	F-DFWM	1064	0.050	90, $\langle\gamma''\rangle = 10$, $\lvert\gamma\rvert = 100$	152
poly(3-butyl) thiophene	chloroform	F-DFWM	1064	—	-5.7, $\langle\gamma''\rangle = 12$	120
poly(3-butyl) thiophene	chloroform	F-DFWM	1064	0.070	-6, $\langle\gamma''\rangle = 10$	121
polythiophene	chloroform	F-DFWM	1064	0.050	6, $\langle\gamma''\rangle = 10$, $\lvert\gamma\rvert = 10$	152
polyyne	THF	OKE	$\lambda_{pump} = 1064$ $\lambda_{probe} = 1064$	0.021	0.0168 ± 0.0042	130
Porter1	THF	OKE	$\lambda_{pump} = 1064$ $\lambda_{probe} = 1064$	—	0.0015, $\langle\gamma''\rangle = 0.0313$	151

Porter10	THF	OKE	$\lambda_{pump} = 1064$ $\lambda_{probe} = 1064$	—	0.0078, $\langle\gamma''\rangle = 0.176$	151
Porter11	THF	OKE	$\lambda_{pump} = 1064$ $\lambda_{probe} = 1064$	—	0.0091, $\langle\gamma''\rangle = 0.186$	151
Porter12	THF	OKE	$\lambda_{pump} = 1064$ $\lambda_{probe} = 1064$	—	0.0039, $\langle\gamma''\rangle = 0.185$	151
Porter13	THF	OKE	$\lambda_{pump} = 1064$ $\lambda_{probe} = 1064$	—	0.0025, $\langle\gamma''\rangle = 0.187$	151
Porter14	THF	OKE	$\lambda_{pump} = 1064$ $\lambda_{probe} = 1064$	—	0.0047, $\langle\gamma''\rangle = 0.300$	151
Porter15	THF	OKE	$\lambda_{pump} = 1064$ $\lambda_{probe} = 1064$	—	0.0027, $\langle\gamma''\rangle = 0.345$	151
Porter16	THF	OKE	$\lambda_{pump} = 1064$ $\lambda_{probe} = 1064$	—	0.0046, $\langle\gamma''\rangle = 0.316$	151
Porter17	THF	OKE	$\lambda_{pump} = 1064$ $\lambda_{probe} = 1064$	—	0.0027, $\langle\gamma''\rangle = 0.163$	151
Porter18	THF	OKE	$\lambda_{pump} = 1064$ $\lambda_{probe} = 1064$	—	0.0021 $\langle\gamma''\rangle = 0.245$	151
Porter19	THF	OKE	$\lambda_{pump} = 1064$ $\lambda_{probe} = 1064$	—	0.0031, $\langle\gamma''\rangle = 0.340$	151

Table 3 Continued

Sample	Solvent	Method	λ (nm)	t_p (nsec)	$\langle\gamma^*\rangle$ (10^{-46} m^5/V^2)	Ref.
Porter2	THF	OKE	$\lambda_{pump} = 1064$ $\lambda_{probe} = 1064$	—	0.0027, $\langle\gamma''\rangle = 0.1155$	151
Porter20	THF	OKE	$\lambda_{pump} = 1064$ $\lambda_{probe} = 1064$	—	0.0126, $\langle\gamma''\rangle = 0.636$	151
Porter21	THF	OKE	$\lambda_{pump} = 1064$ $\lambda_{probe} = 1064$	—	0.0162, $\langle\gamma''\rangle = 0.340$	151
Porter22	THF	OKE	$\lambda_{pump} = 1064$ $\lambda_{probe} = 1064$	—	0.0110, $\langle\gamma''\rangle = 0.690$	151
Porter23	THF	OKE	$\lambda_{pump} = 1064$ $\lambda_{probe} = 1064$	—	0.0092, $\langle\gamma''\rangle = 0.292$	151
Porter24	THF	OKE	$\lambda_{pump} = 1064$ $\lambda_{probe} = 1064$	—	0.0148, $\langle\gamma''\rangle = 0.487$	151
Porter25	THF	OKE	$\lambda_{pump} = 1064$ $\lambda_{probe} = 1064$	—	$\langle\gamma''\rangle = 0.624$	151
Porter3	THF	OKE	$\lambda_{pump} = 1064$ $\lambda_{probe} = 1064$	—	0.0063, $\langle\gamma''\rangle = 0.1670$	151
Porter4	THF	OKE	$\lambda_{pump} = 1064$ $\lambda_{probe} = 1064$	—	0.0123, $\langle\gamma''\rangle = 0.3026$	151
Porter5	THF	OKE	$\lambda_{pump} = 1064$ $\lambda_{probe} = 1064$	—	0.0074, $\langle\gamma''\rangle = 0.1060$	151

Porter6	THF	OKE	$\lambda_{pump} = 1064$ $\lambda_{probe} = 1064$	—	0.0092, $\langle\gamma''\rangle = 0.1854$	151
Porter7	THF	OKE	$\lambda_{pump} = 1064$ $\lambda_{probe} = 1064$	—	0.0042, $\langle\gamma\rangle = 0.1583$	151
Porter8	THF	OKE	$\lambda_{pump} = 1064$ $\lambda_{probe} = 1064$	—	0.0041, $\langle\gamma''\rangle = 0.168$	151
Porter9	THF	OKE	$\lambda_{pump} = 1064$ $\lambda_{probe} = 1064$	—	0.0060, $\langle\gamma''\rangle = 0.133$	151
quarterthiophene	dioxan	OKE	$\lambda_{pump} = 616$ $\lambda_{probe} = 616$	0.00008 0.00008	−0.127	137
quarterthiophene (γ)	THF	DFWM	602	0.0004	0.112	132
quinquethiophene	dioxan	OKE	$\lambda_{pump} = 616$ $\lambda_{probe} = 616$	0.0008 0.0008	−0.293	137
quinquethiophene (γ)	THF	DFWM	602	0.0004	0.363	132
retinal (molten)	—	DFWM	1064	10	0.0061	127
retinal (molten)	—	DFWM	1064	10	0.0014	127
retinal acetate	CS_2	DFWM	1064	10	0.0182	127
retinal acetate (molten)	—	DFWM	1064	10	0.0195	127
retinoic acid	CS_2	DFWM	1064	10	0.00038	127
Saharoui1	chloroform	DFWM	532	0.030	$\gamma^{el}_{1111} = 45.4$	153
Saharoui2	chloroform	DFWM	532	0.030	$\gamma^{el}_{1111} = 54.7$	153
Saharoui3	chloroform	DFWM	532	0.030	$\gamma^{el}_{1111} = 33.2$	153
Saharoui4	chloroform	DFWM	532	0.030	$\gamma^{el}_{1111} = 75.8$	153
Saharoui5	chloroform	DFWM	532	0.030	$\gamma^{el}_{1111} = -18.8$	153
Saharoui6	chloroform	DFWM	532	0.030	$\gamma^{el}_{1111} = -18.2$	153

Table 3 Continued

Sample	Solvent	Method	λ (nm)	t_p (nsec)	$\langle\gamma^*\rangle$ (10^{-46} m^5/V^2)	Ref.
Saharoui7	chloroform	DFWM	532	0.030	$\gamma^{el}_{1111} = 11.2$	153
Sakai1	DMSO	DFWM	532	6.0	251	139
Sakai2	DMSO	DFWM	532	6.0	2090	139
scandium bisphthalocyanine (2×10^{-4} M)	chloroform	DFWM	1064	0.035	67.0	140
scandium bisphthlocyanine	chloroform	DFWM	1064	0.035	28.0	140
sexithiophene	dioxan	OKE	$\lambda_{pump} = 616$ $\lambda_{probe} = 616$	0.00008 0.00008	−0.614	137
sexithiophene (γ)	THF	DFWM	602	0.0004	1.40	132
silicon phthlocyanine-TEA composite	—	DFWM	532	0.070	2.75 ± 0.31	154
Sylla1	chloroform	DFWM	532	0.025	$\langle\gamma_{1111}\rangle = 140$	155
Sylla2	chloroform	DFWM	532	0.025	$\langle\gamma_{1111}\rangle = 220$	155
Sylla3	chloroform	DFWM	532	0.025	$\langle\gamma_{1111}\rangle = 88$	155
TBP	THF	DFWM	532	0.017	69.8	147
terthiophene	dioxan	OKE	$\lambda_{pump} = 616$ $\lambda_{probe} = 616$	0.00008 0.00008	−0.073	137
terthiophene	chloroform	F-DFWM	1064	0.070	−0.34, $\langle\gamma''\rangle = 0.78$	121
terthiophene (γ)	THF	DFWM	602	0.0004	0.0223 ± 0.0028	123
terthiophene (γ)	THF	DFWM	602	0.0004	0.0223	132
THF	neat	OKE	$\lambda_{pump} = 1064$ $\lambda_{probe} = 532$	0.030	0.000078	75
thiophene	dioxane	OKE	$\lambda_{pump} = 616$ $\lambda_{probe} = 616$	0.00008 0.00008	0.00025	137
thiophene	THF	DFWM	602	0.0004	0.00057	132
thiophene	THF	DFWM	602	0.0004	0.00057	123
toluene	neat	OKE	$\lambda_{pump} = 1064$ $\lambda_{probe} = 532$	0.030	0.00071	75

TPCEPM	DMSO	DFWM	602	0.0004	11.3, $\langle\gamma''\rangle = 31.1$	138
TPOHBOBE	DCM	DFWM	1064	10	$\langle\gamma_{1111}\rangle = 6.4$	131
trans-[$(PBu_3)_2R_2Ni$] R = benzothiophene	chloroform	F-DFWM	1064	0.070	-2, $\langle\gamma''\rangle = 10$	121
trans-[$(PR_3)_2R_2M$]	chloroform					
R = thiophene, R′ = ethyl, and						
M = Ni		F-DFWM	1064	0.070	-5.1, $\langle\gamma''\rangle = 4.6$	121
M = Pd		F-DFWM	1064	0.070	-0.69, $\langle\gamma''\rangle = 1.3$	
M = Pt		F-DFWM	1064	0.070	-0.25, $\langle\gamma''\rangle = 0.6$	
trans-$Mo(CO)_4(PPh_3)_2$	THF	DFWM	532	20	2444	136
triphenyl	THF	DFWM	602	0.0004	0.019 ± 0.002	123
ytterbium bisphthalocyanine	chloroform	DFWM	1064	0.035	57.2	140
yttrium bis-phthalocyanine	chloroform	DFWM	1064	0.035	36.3	140
yttrium bisphthalocyanine	chloroform	DFWM	1064	0.035	30.7	140
Zhao1	THF	B-DFWM	602	0.0004	0.0028 ± 0.0004	124
Zhao2	THF	B-DFWM	602	0.0004	0.0022 ± 0.0004	124
Zhao3	THF	B-DFWM	602	0.0004	0.0060 ± 0.0008	124
Zhao4	THF	B-DFWM	602	0.0004	0.0066 ± 0.0007	124
Zhao5	THF	B-DFWM	602	0.0004	0.0075 ± 0.0002	124
Zhao6	THF	DFWM	602	0.0004	0.0126	124
Zhao7	THF	B-DFWM	602	0.0004	0.3072	124
Zhao8	THF	DFWM	602	0.0004	0.00312	129
zinc hexadecafluoro TBP	THF	DFWM	532	0.017	98	147
zinc meso-tetra-(*m*-fluorophenyl) TBP	THF	DFWM	532	0.017	531	147
zinc meso-tetra-(*p*-dimethyl-aminophenyl) TBP	THF	DFWM	532	0.017	1396	147

Table 3 Continued

Sample	Solvent	Method	λ (nm)	t_p (nsec)	⟨γ*⟩ (10^{-46} m^5/V^2)	Ref.
zinc meso-tetra-(*p*-methoxyphenyl) TBP	THF	DFWM	532	0.017	670	147
zinc meso-tetra-(*p*-methylphenyl) TBP	THF	DFWM	532	0.017	559	147
zinc meso-tetramethyl TBP	THF	DFWM	532	0.017	460	147
zinc meso-tetraphenyl TBP	DCM	DFWM	532	0.017	130	147

BDQP—2,5″-bis(decyloxy)-1,1′:4′,1″:4″,1‴:4‴,1′′′′′-quinquephenyl
BDSP—2‴,5‴-bis(decloxy)-1,1′:4′,1″:4″,1‴:4‴,1′′′′:4′′′′,1′′′′′,:4′′′′′,1′′′′′′-septiphenyl
BTP—bis-thienyl polyene
CS_2—carbon disulphide
DCE—1,2-dichloroethane
DCM—dichloromethane
DMPEPM—2,5-dimethoxy phenylene-1,5-bis14-ethenylene pyridimium methyl iodide
DMSO—dimethylsulfoxide
DTCNBDA—N′-[3,7-dimethyl-9-(2,6,6-trimethyl-1-cyclohexen-1-yl)-2,4,6,8-nonatetraenylidene]-N,N-dimethyl-1,4-benzenediamine
HDVPC—2,9,16,23-tetra-heptadecylamido-substituted vanadyl phthalocyanine
PDA—4BCMU—poly[5,7-dodecadiyne-1,12-diol bis([(*n*-butoxycarbonyl)methyl]urethane)]
PEBTBPPt—1-(1,4-phenylenedi-2,1-ethynediyl-C)bis(tributylphosphine)-platinum
pTT—polythieno(3,2-b)thiophene
TBP—tetrabenzporphyrin
THF—tetrahydrofuran
TPCEPM—triphenyl (phosphinium cyclopentodienylene-2,5-bis(4-ethenylene pyridinium methyl iodide)
TPOHBOBE—2,2,6,6-tetramethyl-pipiridine-4-ol-4N-oxy *p*-hydroxybenzaldehyde oxeme benzoic ester

have been made from esu units and from other measures of the nonlinear refractive index, such as B_0 or $\chi^{(3)}_{eff}$.

Table 2 tabulates the value of $\chi^{(3)}_{1111}$ as measured by four wave mixing techniques. These values are for organic media, primarily in thin film form, in units of m^2/V^2. Conversions have been made from Gaussian units. In cases where other tensor components have been measured, these are listed in the table explicitly. Note that in pump-probe measurements where the two colors used are the fundamental and second harmonic, the second harmonic pulse duration, τ_{SHG}, should be shorter than the fundamental pulse duration, τ_F, by an amount $\tau_{SHG} = \tau_F/\sqrt{2}$. If authors quote both durations as equal, then they are so tabulated.

Table 3 tabulates the value of $\langle\gamma^*\rangle$ the isotropic average of the dressed third-order molecular susceptibility γ^*_{ijkl} (assumed measured with all polarizations parallel in the four wave mixing), for various molecules in units of m^2/V^5. Note that in many measurements it is the absolute value of $\langle\gamma^*\rangle$ that is measured; no distinction is made here in those cases. In cases where both real and imaginary parts of $\langle\gamma^*\rangle$ have been measured, these are listed separately in the table.

C. Other Parameters

Along with names and nonlinearities, the tables provide other information such as the specific method used, the wavelengths used in the measurement, the pulse lengths of the pulsed lasers used, and the reference in which more detail about the measurement can be found, such as the linear optical properties of the materials—we have not included this information because all too often it is not present in the original source document. Note that if no pulse duration is listed for a wavelength, this indicates that the laser used was cw. Table entries are arranged in alphabetical order (as determined by word processing software).

D. Names

We have attempted to give a proper or commonly recognized name to each compound listed. In those cases where only alphanumeric chemical formulas were given by the original authors, these have been used. All acronyms used in a given table are listed in alphabetical order (by acronym) at the bottom of the table. Cases in which we have been unable to determine the proper chemical name of a compound that appeared in its original reference as a structure only are tabulated as structures at the end of the table. These compounds are named according to their author, and numerically within the reference. In one case, Ref. [113], the second author's name (Saindon) was used to avoid confusion with another reference.

ACKNOWLEDGEMENTS

I would like to acknowledge the help of Shari Stark in compiling the tables, Rob Johnson in naming the chemicals, and Chris Koeppen in hunting down many of the more difficult to find references. Most of all I would like to acknowledge my family for giving me the time to complete this task.

REFERENCES

1. Mayer, G., and Gires, F., *C. R. Acad, Sc. Paris*, *258*, 2039 (1964).
2. Buckingham, A. D., Birefringence resulting from the application of an intense beam of light to an isotropic medium, *Proc. Phys. Soc.*, *LXIX 3-B*, 344 (1955).
3. Duguay, M. A., and Hansen, J. W., An ultrafast light gate, *Appl. Phys. Lett.*, *15*, 192 (1969).
4. Coles, H. J., and Jennings, B. R., Laser induced birefringence in pure liquids, *Phil. Mag.*, *32*, 1051 (1975).
5. Hellwarth, R. W., Generation of time-reversed wave fronts by nonlinear refraction, *J. Opt. Soc. Am.*, *67*, 1 (1977).
6. Eichler, H. J., Günter, P., and Pohl, D. W., *Laser-Induced Dynamic Gratings*, Springer-Verlag, Berlin, 1986.
7. Meyers, A. B., and Hochstrasser, R. M., *IEEE J-QE*, *22*, 1482 (1986).
8. Etchepare, J., Grillon, G., Hamoniaux, G., Antonetti, A., and Orszag, A., *Revue. Phys. Appl*, *22*, 1749 (1987).
9. Etchepare, J., Grillon, G., Antonetti, A., and Orszag, A., *Physica Scripta*, *T23*, 191 (1988).
10. Etchepare, J., Grillon, G., Chambaret, J. P., Hamoniaux, G., and Orszag, A., *Opt. Commun.*, *63*, 329 (1989).
11. Deeg, F. W., and Fayer, M. D., *J. Chem. Phys.*, *91*, 2269 (1989).
12. Fishman, I. M., Marshall, C. D., Meth, J. S., and Fayer, M. D., Surface selectivity four-wave mixing: transient grating as a theoretical and experimental example, *J. Opt. Soc. Am.*, *B8*, 1880 (1991).
13. Fishman, I. M., Marshall, C. D., Tokmakoff, A., and Fayer, M. D., Transient grating diffraction from an interface between two materials: theory and experimental application, *J. Opt. Soc. Am.*, *B10*, 1006 (1993).
14. Fourkas, J. T., Trebino, R., Dugan, M. A., and Fayer, M. D., Extra resonances in time-domain four-wave mixing, *Opt. Lett.*, *18*, 781 (1993).
15. Dhar, L., Fourkas, J. T., and Nelson, K. A., Pulse-length-limited ultrafast pump-probe spectroscopy in a single laser shot, *Opt. Lett.*
16. Yang, L., Dorsinville, R., Ho, P. P., Zou, W. K., Yang, N. L., and Alfano, R. R., Intensity dependence of picosecond nonlinear response time of polydiacetylene, *Appl. Phys. Lett.*, *53*, 2009 (1988).
17. Kuzyk, M. G., Norwood, R. A., Wu, J. W., and Garito, A. F., Frequency dependence of the optical Kerr effect and third-order electronic nonlinear-optical processes of organic liquids, *J. Opt. Soc. Am.*, *B6*, 154 (1989).
18. Scarpano, M. A. F., Song, J. J., Lee, J. H., Cromer, C., and Levenson, M. D., Improved geometry for polarization-sensitive nonlinear spectroscopy, *Appl. Phys. Lett.*, *35*, 490 (1979).
19. Blaszczak, B., Concentration dependence of induced optical birefringence in carbon disulphide vapor, *Opt. Commun.*, *58*, 439 (1986).
20. Heiman, D., Hellwarth, R. W., Levenson, M. D., and Martin, G., Raman-induced Kerr effect, *Phys. Rev. Lett.*, *36*, 189 (1976).
21. Norwood, R. A., Optical Kerr effect and molecular symmetry in three Kerr liquids, Ph.D. Dissertation, University of Pennsylvania, 1988.
22. Kuzyk, M. G., Third order nonlinear optical processes in organic systems, Ph.D. Dissertation, University of Pennsylvania, 1985.
23. Wu, J. W., Heflin, J. R., Norwood, R. A., Wong, K. Y., Zamani-Khamiri, O., Garito, A. F., Kalyanaraman, P., and Sounik, J. R., Nonlinear optical processes in lower dimensional conjugated structures, *J. Opt. Soc. Am.*, *B6*, 707 (1989).
24. Dirk, C. W., Cheng, L.-T., and Kuzyk, M. G., *Int. J. of Quant. Chem.*, *43*, 27 (1992).

25. Feldstein, M. J., Vohringer, P., and Scherer, N. F., Rapid-scan pump-probe spectroscopy with high time and wave-number resolution: optical-Kerr-effect measurements of neat liquids, *J. Opt. Soc. Am.*, *B12*, 1500 (1995).
26. Etchepare, J., Grillon, G., Thomazeau, I., Migus, A., and Antonetti, A., *J. Opt. Soc. Am.*, *B2*, 649 (1985).
27. Lotshaw, W. T., McMorrow, D., Kalpouzos, C., and Kenney-Wallace, G., *Chem. Phys. Lett.*, *136*, 323 (1987).
28. Levenson, M. D., and Bloembergen, N., Dispersion of the nonlinear optical susceptibilities of organic liquids and solutions, *J. Chem. Phys.*, *60*, 1323 (1975).
29. Rahn, L. A., Zych, L. J., and Mattern, P. L., *Opt. Commun.*, *30*, 249 (1979).
30. Adair, R., Chase, L. L., and Payne, S. A., Nonlinear refractive index of optical crystals, *Phys. Rev. B*, *39*, 3337 (1989).
31. Martin, G., and Hellwarth, R. W., Infrared-to-optical image conversion by Bragg reflection from thermally induced index gratings, *Appl. Phys. Lett.*, *34*, 371 (1979).
32. Pepper, D. M., Fekete, D., and Yariv, A., Observation of amplified phase-conjugate reflection and optical parametric oscillation by degenerate four wave mixing in a transparent medium, *Appl. Phys. Lett.*, *33*, 41 (1978).
33. Gower, M. C., Diagrammatical analysis of phase conjugation using four and three-wave mixing processes in Kerr-like nonsaturated media, *IEEE J. Quant. Elec.*, *21*, 182 (1985).
34. Carter, G. M., Thakur, M. K., Chen, Y. J., and Hryniewicz, J. V., Time and wavelength resolved nonlinear optical spectroscopy of a polydiacetylene in the solid state using picosecond dye laser pulses, *Appl. Phys. Lett.*, *47*, 457 (1985).
35. Caro, R. G., and Gower, M. C., Phase conjugation by degenerate four wave mixing in absorbing media, *IEEE J. Quant. Elec.*, *18*, 1375 (1982).
36. Kajzar, F., Messier, J., Nunzi, J. M., and Raimond, P., Third-order materials: processes and characterization, in *Polymers for Lightwave and Integrated Optics* (L. A. Hornak, ed.), Marcel Dekker, New York, 1992, pp. 595–564.
37. Ho, P. P., and Alfano, R. R., Optical Kerr effect in liquids, *Phys. Rev.*, *A20*, 2170 (1979).
38. Harrison, N. J., and Jennings, B. R., Optical Kerr-effect measurement for a series of alcohols, *J. Appl. Phys.*, *73*, 8076 (1993).
39. Harrison, N. J., and Jennings, B. R., Optically induced Kerr constants for pure liquids: homologous series on n-alkanes, 1-alkenes and 1-alkynes, *J. Phys. Chem.*, *97*, 1551 (1993).
40. Blaszczak, Z., The influence of molecular interactions on optical orientation of pyridine methyl derivatives I. Optical birefringence and light scattering studies, *Acta Physica Polonica*, *A68*, 629 (1985).
41. Kanbara, H., Kobayashi, H., Kaino, T., Kurihava, T., Ooba, N., and Kubodera, K., Highly efficient ultrafast optical Kerr shutters with the use of organic nonlinear materials, *J. Opt. Soc. Am.*, *B11*, 2216 (1994).
42. Harrison, N. J., and Jennings, B. R., Laser-induced Kerr constant for alkanoic acids at 532 & 1064 nm inducing wavelengths, *Chem. Phys. Lett.*, *210*, 393 (1993).
43. Blaszczak, Z., The influence of molecular interaction on optical orientation of pyridine methyl derivatives. II. Effect of molecular structure, *Acta Physica Polonica*, *A71*, 601 (1987).
44. Martin, F. B., and LaLanne, J. R., Agreement between depolarized Rayleigh scattering and optical Kerr effect induced by Q-switched laser waves in some liquids, *Phys. Rev.*, *A4*, 1275 (1971).
45. Hellwarth, R. W., Effect of molecular redistribution on the nonlinear refractive index of liquids, *Phys. Rev.*, *152*, 156 (1966).

46. Huggard, P. G., Blau, W., and Schweitzer, D., Large third-order nonlinearity of the organic metal d-[bis(ethylenedithio)tetrathiofulvalene] iodide, *Appl. Phys. Lett.*, *51*, 2183 (1987).
47. Gadonas, R., Feller, K. H., and Pugzyls, A., Wavelength dependent nonlinear optical properties of pseudoisocyanine J-aggregates, *Opt. Commun.*, *112*, 157 (1994).
48. Dennis, W. M., Blau, W., and Bradley, D. J., Picosecond degenerate four-wave mixing in soluble polydiacetylene, *Appl. Phys. Lett.*, *47*, 200 (1985).
49. Nickel, E. G., Spangler, C. W., Tang, N., Hellwarth, R., and Dalton, L., Bipolaronic enhancement of $\chi^{(3)}$ in substituted bis-anthracenyl polyenes, *Nonlinear Optics*, *6*, 135 (1993).
50. Lindle, J. R., Bartoli, F. J., Hoffman, C. A., Kim, O. K., Lee, Y. S., Shirk, J. S., and Kafafi, Z. H., Nonlinear optical properties of benzimidazobenzophenanthraline type ladder polymers, *Appl. Phys. Lett.*, *56*, 712 (1990).
51. Yu, L., Polis, D. W., McLean, M. R., and Dalton, L. R., Polymeric materials for third-order nonlinear optical susceptibility, in *Electroresponsive Molecular and Polymeric Systems*, Vol. 2, (Terje A. Skotheim, ed.), Marcel Dekker, New York, 1991, pp. 113–186.
52. Winter, C. S., Hill, C. A. S., and Underhill, A. E., Near resonance optical nonlinearities in nickel dithiolene complexes, *Appl. Phys. Lett.*, *58*, 107 (1991).
53. Levenson, M. D., Feasibility of measuring the nonlinear index of refraction by third-order frequency mixing, *IEEE Jour. Quant. Elec.*, *QE-10*, 110 (1974).
54. Levenson, M. D., and Bloembergen, N., Dispersion of the nonlinear optical susceptibility tensor in centrosymmetric media, *Phys. Rev.*, *B10*, 4447 (1974).
55. Levenson, M. D., and Song, J. J., Raman-induced Kerr effect with elliptical polarization, *J. Opt. Soc. Am.*, *66*, 641 (1976).
56. Song, J. J., and Levenson, M. D., Electronic and orientational contributions to the optical Kerr constant determined by coherent Raman techniques, *J. Appl. Phys.*, *48*, 3496 (1977).
57. Winter, C. S., Oliver, S. N., Rush, J. O., Hill, C. A. S., and Underhill, A. E., Large third-order optical nonlinearities of nickel-dithiolene-doped polymethylmethacrylate, *J. App. Phys.*, *71*, 512 (1992).
58. Cheung, Y. M., and Gayon, S. K., Optical nonlinearities of tea studied by z-scan and four-wave mixing techniques, *J. Opt. Soc. Am.*, *B11*, 636 (1994).
59. Wung, C. J., Lee, K. S., Prasad, P. N., Kim, J. C., Jin, J.-I., and Shim, H. K., Study of third-order optical non-linearity and electrical conductivity of sol-gel processed silica: Poly(2-bromo-5-methoxy-p-phenylene vinylene) composite, *Polymer*, *33*, 4145 (1992).
60. Flom, S. R., Pong, R. G. S., Bartoli, F. J., and Kafafi, Z. H., Nonlinear optical properties of fullerenes and their derivatives, *Nonlinear Opt.*, *10*, 183 (1995).
61. Kafafi, Z. H., Lindle, J. R., Pong, R. G. S., Bartoli, F. J., Lingg, L. J., and Milliken, J., Off-resonant nonlinear optical properties of C_{60} studied by degenerate four-wave mixing, *Chem. Phys. Lett.*, *188*, 492 (1992).
62. Flom, S. R., Pong, R. G. S., Bartoli, F. J., and Kafafi, Z. H., Resonant nonlinear optical response of the fullernes C_{60} and C_{70}, *Phys. Rev. B*, *46*, 15598 (1992).
63. Lindle, J. R., Pong, R. G. S., Bartoli, F. J., and Kafafi, Z. H., Nonlinear optical properties of the fullerenes C_{60} and C_{70} at 1.064 μm, *Phys. Rev. B*, *48*, 9447 (1993).
64. Cao, X. F., Jiang, J. P., Bloch, D. P., Hellwarth, R. W., Yu, L. P., and Dalton, L., Picosecond nonlinear optical response of three rugged polyquinoxaline-based aromatic conjugated ladder-polymer thin films, *J. Appl. Phys.*, *65*, 5012 (1989).
65. Chuang, Y. H., Li, Z. W., Meyerhofer, D. D., and Schmind, A., Nonresonant $\chi^{(3)}$ obtained by nearly degenerate four-wave mixing using chirped-pulse technology, *Opt. Lett.*, *16*, 7 (1991).

66. Maloney, C., Byrne, H., Dennis, W. M., Blau, W., and Kelly, J. M., Picosecond optical phase conjugation using conjugated organic molecules, *Chemical Physics*, *121*, 21 (1988).
67. Nalwa, H. S., Third-order non-linear optical susceptibility measurements of thin films of 3-methylthiophene and methylmethacrylate copolymers, *Thin Solid Films*, *235*, 175 (1993).
68. Bartoli, F. J., Lindle, J. R., Shirk, J. S., Flom, S. R., Snow, A. W., and Boyle, M. E., Resonant third-order optical response in the metallo-phthalocyanines at 1064 nm, *Nonlinear Opt.*, *10*, 161 (1995).
69. Norwood, R. A., Sounik, J. R., Popolo, J., and Holcomb, D. R., Third order nonlinear optical characterization of side-chain copolymers, *Proc. SPIE*, *1560*, 54 (1991).
70. Kanbara, H., Kobayashi, H., Dubodera, K., Kurihara, T., and Kaino, T., Optical Kerr shutter using organic nonlinear optical materials in capillary waveguides, *IEEE Photonics Tech. Lett.*, *3*, 795 (1991).
71. Dennis, W. M., Blau, W., and Bradley, D. J., Optical phase conjugation in a soluble polymer, *Opt. Engin.*, *25*, 538 (1986).
72. Persoons, A., VanWonterghem, B., and Tackx, P., Measurements of second hyperpolarizabilities of diphenylpolynes by means of phase conjugate interferometry, *Proc. SPIE*, *1409*, 220 (1991).
73. Hao-Sheng, F., Li, H., Xi-cheng, A., Xui, Y., and Jia-cong, S., Reverse saturable absorption and third-order optical nonlinearity in porphyrin, *Chinese Science Bulletin*, *37*, 298 (1992).
74. Garvey, D. W., Li, Q., Kuzyk, M. G., Dirk, C. W., and Martinez, S., Sagnac interferometric intensity-dependent refractive-index measurements of polymer optical fiber, *Opt. Lett.*, *21*, 104 (1996).
75. Guha, S., Frazier, C. G., Porter, P. L., Kang, K., and Finberg, S. E., Measurement of the third-order hyperpolarizability of platinum poly-ynes, *Opt. Lett.*, *14*, 952 (1989).
76. Norwood, R. A., and Sounik, J. R., Third-order nonlinear optical response in polymer thin films incorporating porphyrin derivatives, *Appl. Phys. Lett.*, *60*, 295 (1992).
77. Mates, T. E., Ober, C. K., and Norwood, R. A., Conductivity and third order nonlinear optical measurements of polymers with distyrylbenzene and diphenyl butadiene segments, *Chem. Mat.*, *5*, 217 (1993).
78. Dotrong, M., Mehta, R., Balchin, G. A., Tomlinson, R. C., Sinsky, M., Lee, C. Y. C., and Evers, R. C., Synthesis, processing and third-order nonlinear optical properties of benzobisthiazole polymers containing thiophene, *J. Poly. Science.*, *A31*, 723 (1993).
79. Chandrasekhar, P., Thorne, J. R. G., and Hochstrasser, R. M., Structure-property relationship in off-resonant third-order nonlinearities of conducting polymers 1. Polyaromatic amines and ring-s-heteroatom containing polymers, *Synth. Metals*, *53*, 175 (1993).
80. Chandrasekhar, P., Thorne, J. R. G., Hochstrasser, R. M., Third-order nonlinear optical properties of polydiphenyl, aminyl and poly (4-amine biphenyl), novel processible conducting polymers, *Appl. Phys. Lett.*, *59*, 1661 (1991).
81. Jiang, J. P., and Rentzepis, P. M., Picosecond third-order nonlinear optical susceptibility in nematic liquid crystals, *J. Appl. Phys.*, *70*, 5125 (1991).
82. Kobayashi, T., Yoshizawa, M., Stamm, U., Taiji, M., and Hasegawa, M., Relaxation dynamics of photo-excitations in polydiacetylenes and polythiophene, *J. Opt. Soc. Am.*, *B7*, 1558 (1990).
83. Rao, D. N., Swiatkiewicz, Chopra, P., Ghoshai, S. K., and Prasad, P. N., Third order nonlinear optical interactions in thin films of poly-p-phenylenebenzobisthazole polymer investigated by picosecond and subpicosecond degenerate four wave mixing, *Appl. Phys. Lett.*, *48*, 1187 (1986).

84. Jenehke, S. A., Chen, W. C., Low, S., and Flom, S. R., Large third-order optical nonlinearities in organic polymer superlattices, *Appl. Phys. Lett.*, *57*, 126 (1990).
85. Nunzi, J. M., and Grec, D., Picosecond phase conjugation in polydiacetylene gels, *J. Appl. Phys.*, *62*, 2198 (1987).
86. Wong, K. W., and Vardeny, Z. V., Measurements of $\chi^{(3)}$ $(\omega; \omega, -\omega, \omega)$ in conducting polymers at λ = 620 nm, *Synth. Met.*, *49–50*, 13 (1992).
87. Ho, P. P., Yang, N. L., Jimbo, T., Wang, Q. Z., and Alfano, R. R., Ultrafast resonant optical Kerr effect in 4-butoxycarbonyl (methylurethane polydiacetylene, *J. Opt. Soc. Am.*, *B4*, 1025 (1987).
88. Vogtmann, Th., Schmidt, W., Fehn, Th., Baver, F., Baver, I., and Schwoerer, M., Nonlinear optics in polydiacetylene single crystals: anisotropy, spectral response and thermo-optic effect, *Synth. Metals*, *55–57*, 4018 (1993).
89. Wong, K. S., Hon, S. G., and Vardeny, Z. V., Studies of resonant and preresonant femtosecond degenerate four-wave mixing in unoriented conducting polymers, *J. Appl. Phys.*, *70*, 1896 (1991).
90. Nunzi, J. M., and Charra, F., Complex third-order phase conjugation nonlinearity of polymeric thin films, *Appl. Phys. Lett.*, *59*, 13 (1991).
91. Rao, D. N., Chopra, P., Ghoshai, S. K., Swiatkiewicz, J., and Prasad, P. N., Third-order nonlinear optical interaction and conformational transition poly-4-BCMU poly-diacetylene studied by picosecond and subpicosecond degenerate four wave mixing, *J. Chem. Phys.*, *48*, 7049 (1986).
92. Carter, G. M., Chen, Y. J., Rubner, M. F., Sandman, D. J., Thakur, M. K., and Tripathy, S. K., Degenerate third-order nonlinear optical susceptibility of polydiacetylenes, in *Nonlinear Optical Properties of Organic Molecules and Crystals*, Vol. 2 (D. S. Chemla and J. Zyss, eds.), Academic Press, Orlando, 1987, pp. 85–120.
93. Bolger, J., Hariey, T. G., Ji, W., Kar, A. K., Molyneux, S., Wherrett, B. S., Bloor, D., and Norman, P., Near resonant third-order optical nonlinearities in p-toluene sulfonate polydiacetylene, *J. Opt. Soc. Am.*, *B9*, 1552 (1992).
94. Dorsinville, R., Yang, L., Alfano, R. R., Zamboni, R., Daniels, R., Ruane, G., and Taliani, C., *Opt. Lett.*, *14*, 1321 (1989).
95. Marcy, H. O., Rasker, M. J., Warren, L. F., Reinhardt, B. A., Sinclair, M., and Segger, C. H., Time-resolved degenerate four-wave mixing studies of solid-state poly(p-phenylene) oligomers, *J. Chem. Phys.*, *100*, 3325 (1994).
96. Moon, K. J., Lee, K. S., and Ku Shim, H., Syntheses, electrical, and nonlinear optical properties of PPV derivatives containing alkoxynitrostilbene group, *Synth. Metals*, *71*, 1719 (1995).
97. Vijaya, R., Murti, Y. V. G. S., Vijayaraj, T. A., and Sundararajan, G., Optical nonlinearities in substituted conjugated polymers, *Opt. & Quant. Electronics*, *25*, 723 (1993).
98. Singh, B. P., Samoc, M., Nalwa, H. S., and Prasad, P. N., Resonant third-order nonlinear optical properties of poly(3-dodecylthiophene), *J. Chem. Phys.*, *92*, 2756 (1990).
99. Pang, Y., and Prasad, P. N., Photoinduced process and resonant third-order nonlinearity in poly(3-dodecylthiophene) studied by femtosecond time resolved degenerate four wave mixing, *J. Chem. Phys.*, *93*, 2201 (1990).
100. Schrof, W., Wunsch, J. R., Esser, A., Haas, K. H., and Naarman, H., Nonlinear optics of conjugated polyacetylene thin films produced by different preparation techniques, *Nonlinear Opt.*, *10*, 69 (1995).
101. McGraw, D. J., Siegman, A. E., Wallroff, G. M., and Miller, R. D., Resolution of the nuclear and electronic contributions to the optical nonlinearity in polysilanes, *Appl. Phys. Lett.*, *54*, 1713 (1989).

102. Yang, L., Wang, Q. Z., Ho, P. P., Dorsinville, R., Alfano, R. R., Zou, W. K., and Yang, N. L., Ultrafast time response of optical nonlinearity in polysilane polymers, *Appl. Phys. Lett.*, *53*, 1245 (1988).
103. Swiatkiewicz, J., Prasad, P. N., Karasz, F. E., Druy, M. A., and Glatkowski, P., Anisotropy of the linear and third-order nonlinear optical properties of a stretch-oriented polymer film of poly[2,5-dimethoxy paraphenylenevinylene], *Appl. Phys. Lett.*, *56*, 892 (1990).
104. Wong, K. S., Han, S. G., Vardeng, Z. V., Shinar, J., Pang, Y., Ijadi-Maghsoodi, S., Barton, T. J., Grigoras, S., and Parbhoo, B., Femtosecond dynamics of nonlinear optical response in polydiethenyl silane, *Appl. Phys. Lett.*, *58*, 1695 (1991).
105. Shim, H. K., Lee, K. S., Hwang, D. H., and Lee, J. I., Synthesis, electrical, and optical properties of o-symmetrically monoalkoxy-substituted PPV derivatives, *Synth. Metals*, *55–57*, 908 (1993).
106. Kondo, K., Okuda, M., and Fujitani, T., Synthesis and nonlinear optics of soluble polylphenyleneethynylene, *Macromolecules*, *26*, 7382 (1993).
107. Vijayo, R., Murti, Y. V. G. S., and Rao, T. A. P., Nonresonant third-order optical response of polyphenyl acetylene, *J. Appl. Phys.*, *69*, 3429 (1991).
108. Swiatkiewicz, J., Prasad, P. N., and Karasz, F. E., Anisotropy in the complex refractive index and the third-order nonlinear optical susceptibility of a stretch-oriented film of poly(p-phenylene vinylene), *J. App. Phys.*, *74*, 525 (1993).
109. Pang, Y., Samoc, M., and Prasad, P. N., Third-order nonlinearity and two-photon-induced molecular dynamics: Femtosecond time-resolved transient absorption, Kerr gate, and degenerate four-wave mixing studies in poly(*p*-phenylenevinylene)/sol-gel silica film, *J. Chem. Phys.*, *94*, 5282 (1991).
110. Ooba, N., Tomaru, S., Kurihara, T., Mori, Y., Shuto, Y., and Kaino, T., Resonant subpicosecond degenerate four wave mixing for a main-chain polymer with a definite short π-conjugation, *Chem. Phys. Lett.*, *207*, 396 (1993).
111. Samoc, M., and Prasad, P. N., Dynamics of resonant third order optical nonlinearity in perylene tetracarboxylic dianhydride studied by monitoring first- and second-order diffractions in subpsec DFWM, *J. Chem. Phys.*, *91*, 6643 (1989).
112. Ghoshal, S. K., Chopra, P., Singh, B. P., Swiatkiewicz, J., and Prasad, P. N., Picosecond degenerate four-wave mixing study of nonlinear optical properties of the poly-N-vinyl carbazole: 2,4,7-trinitrofluorenone composite polymer photo-conductor, *J. Chem. Phys.*, *90*, 5078 (1989).
113. Spangler, C. W., Saindon, M. L., Nickel, E. G., Sapochak, L. S., Polis, D. W., Dalton, L. R., and Norwood, R. A., Synthesis and incorporation of ladder polymer subunits in copolyamides, pendant polymers and composites for enhanced nonlinear optical response, *Proc. SPIE*, *1497*, 408 (1991).
114. Casstevens, M. K., Samoc, M., Pfleger, J., and Prasad, P. N., Dynamics of third-order nonlinear optical processes in Langmuir-Blodgett and evaporated films of phthalocyanines, *J. Chem. Phys.*, *92*, 2019 (1990).
115. Spangler, C. W., He, M., Liu, P., Nickel, E. G., Laquindanum, J., and Dalton, L. R., The design of new organic materials with enhanced nonlinear optical properties: incorporation of bipolaronic charge states, *Nonlinear Opt.*, *10*, 147 (1995).
116. Bischofberger, T., and Courtens, E., Measurement of the high frequency electro-optical Kerr effect in a plastic crystal: succinonitrile, *Phys. Rev.*, *A14*, 2278 (1976).
117. Pfeffer, N., Charra, F., and Nunzi, J. M., Phase and frequency resolution of picosecond optical Kerr nonlinearities, *Opt. Lett.*, *16*, 1987 (1991).
118. Si, J., Wang, Y., Yang, Q., Ye, P., Tian, H., Zhou, Q., and Xu, H., Dynamic studies of charge-separated-state enhancement of the optical nonlinearity of a porphyrin heterodimer, *Appl. Phys. Lett.*, *69*, 1832 (1996).

119. Dhindsa, A. S., Underhill, A. E., Oliver, S., and Kershaw, S., New $\chi^{(3)}$ materials for electro-optic and all-optical signal processing based on metal complexes, *Nonlinear Opt.*, *10*, 115 (1995).
120. Davey, A. P., Page, H., Blau, W., Byrne, H. J., and Cardin, D. J., Linear and third order nonlinear optical properties of one-dimensional organometallic systems, *Synth. Metals*, *55–57*, 3980 (1993).
121. Davey, A. P., Byrne, H. J., Page, H., Blau, W., and Cardin, D. J., Nonlinear optical studies of group 10 transition-metal thienyl systems, *Synth. Metals*, *58*, 161 (1993).
122. Rossi, B., Byrne, H. J., and Blau, W., Degenerate four-wave mixing in rhodamine doped epoxy waveguides, *Appl. Phys. Lett.*, *58*, 1712 (1991).
123. Zhao, M.-T., Samoc, M., Singh, B. P., and Prasad, P. N., Study of third-order microscopic optical nonlinearities in sequentially built and systematically derivatized structures, *J. Phys. Chem.*, *93*, 7916 (1989).
124. Zhao, M., Samoc, M., Prasad, P. N., Reinhardt, B. A., Unroe, M. R., Prazak, M., Evers, R. C., Kane, J. J., Jariwala, C., and Sinsky, M., Studies of third order optical nonlinearities of metal compounds containing benzothiazole, benzimidazole, and benzoxazole units, *Chem. Mater.*, *2*, 670 (1990).
125. Tang, N., and Partanen, J. P., Four-wave-mixing interferometer, *Opt. Lett.*, *21*, 1108 (1996).
126. Gang, Q., Sun, Y., Xia, Z., Zhou, Y. H., Gu, Z., Zhous, Z., and Qiang, D., Nonresonant third-order optical nonlinearity of all-carbon molecules C_{60}, *J. Appl. Phys.*, *71*, 3025 (1992).
127. Smith, P. W., Tomlinson, W. J., Eilenberger, D. J., and Maloney, P. J., Measurement of electronic optical Kerr coefficients, *Opt. Lett.*, *6*, 581 (1981).
128. Maloney, C., and Blau, W., Resonant third-order hyperpolarizabilities of large organic molecules, *J. Opt. Soc. Am.*, *B4*, 1035 (1987).
129. Zhao, M., Cui, Y., Samoc, M., Prasad, P. N., Unroe, M. R., and Reinhardt, B. A., Influence of two-photon absorption on third-order nonlinear optical processes as studied by degenerate four wave mixing: The study of soluble didecyloxy substituted polyphenyls, *J. Chem. Phys.*, *95*, 3991 (1991).
130. Guha, S., Kang, K., Porter, P., Roach, J. F., Remy, D. E., Aranda, F. J., and Rao, D. V. G. L. N., Third order-optical nonlinearities of metallo tetrabenzoporphyrins and platinum poly-yne, *Opt. Lett.*, *17*, 264 (1992).
131. Gong, Q., Xia, Z., Zou, Y. H., Li, Y., Wu, P., and Zhu, D., Third order nonlinear optical properties of some organic nitroxyls, *Appl. Phys.*, *B54*, 181 (1992).
132. Zhao, M. T., Singh, B. P., and Prasad, P. N., A systematic study of polarizability and microscopic third-order optical nonlinearity in thiophene oligomers, *J. Chem. Phys.*, *89*, 5535 (1988).
133. Zhang, Z., Wang, D., Ye, P., Li, Y., Wu, P., and Zhu, D., Studies of third-order optical nonlinearities in C_{60}-toluene solution, *Opt. Lett.*, *17*, 973 (1992).
134. Talopatra, G. B., Manickam, N., Samoc, M., Orczyk, M. E., Karna, S. P., and Prasad, P. N., Nonlinear optical properties of the C_{60} molecule: Theoretical and experimental studies, *J. Phys. Chem.*, *96*, 5206 (1992).
135. Orcyzk, M. E., Samoc, M., Swiatkiewicz, J., and Prasad, P. N., Dynamics of third-order nonlinearity of canthaxanthin caratenoid by the optically heterodyned phase-tuned femtosecond optical Kerr gate, *J. Chem. Phys.*, *98*, 2524 (1993).
136. Zhai, T., Lawson, C. M., Burgess, G. E., Lewis, M. L., Gale, D. C., and Gray, G. M., Nonlinear-optical studies for molybdenum metal organics, *Opt. Lett.*, *19*, 871 (1994).
137. Lap, D. V., Rentsch, S., and Naarmann, H., Femtosecond time resolved measurement of the nonlinear refractive index of thiophene oligomers in solution, *Nonlinear Opt.*, *10*, 79 (1995).

138. Prasad, P. N., Orczyk, M. E., Swiatkiewicz, J., Park, C. K., and Zhao, C. F., Third-order nonlinear optics of organic chromophores, *Nonlinear Opt.*, *10*, 37 (1995).
139. Sakai, T., Kawabe, Y., Ikeda, H., and Kawaski, K., Third-order nonlinear optical properties of retinal derivatives, *Appl. Phys. Lett.*, *56*, 411 (1990).
140. Shirk, J. S., Lindle, J. R., Bartoli, F. J., and Boyle, M. E., Third-order optical nonlinearities of bis(phthalocyanines), *J. Phys. Chem.*, *96*, 5847 (1992).
141. Yuan, P., Xia, Z., Zou, Y. H., Qiu, L., Shen, J., Shen, Y., and Xu, H., Ultrafast optical Kerr effect of phthalocyanine, *Chem. Phys. Lett.*, *224*, 101 (1994).
142. Ikeda, H., Sakai, T., and Kawasaki, K., Third-order optical nonlinearities of asymmetric carbocyanine dyes, *Chem. Lett.*, 1075 (1991).
143. Jenekhe, S. A., Lo, S. K., and Flom, S. R., Third-order nonlinear optical properties of a soluble conjugated polythiophene derivative, *Appl. Phys. Lett.*, *54*, 2524 (1989).
144. Joshi, M. V., Cava, M. P., Lakshmikantham, M. V., Metzger, R. M., Abdeldayem, H., Henry, M., and Venkateswarlv, P., Third-order nonlinear optical susceptibilities by laser self-trapping and degenerate four-wave mixing in oligothienyldicyano-vinylene, *Synth. Metals*, *55–57*, 3974 (1993).
145. Kminek, I., Klimovic, J., and Prasad, P. N., Third-order nonlinear optical response of some tetrasubstituted cumulenes, *Chem. Mater.*, *5*, 357 (1993).
146. Shirk, J. S., Lindle, J. R., Bartoli, F. J., Hoffman, C. A., Kafafi, Z. H., and Snow, A. W., Off-resonant third order optical nonlinearities of metal-substituted phthalocyanines, *APL*, *55*, 1287 (1989).
147. Rao, D. V. G. L. N., Aranda, F. J., Roach, J. F., and Remy, D. E., Third-order, nonlinear optical interaction of some benzporphyrins, *Appl. Phys. Lett.*, *58*, 1241 (1991).
148. Yan, J., Wu, J., Zhu, H., Zhang, X., Sun, D., Hu, Y., and Li, F., Excited-state enhancement of the third-order susceptibility of nonether polyphenylquinoxaline, *Opt. Lett.*, *20*, 255 (1995).
149. Page, H., Blau, W., Davey, A. P., Lou, X., and Cordin, D. J., Nonlinear optical properties of some mono- and bimetallic organometallic systems, *Synth. Metals*, *63*, 179 (1994).
150. Chance, R. R., Shand, M. L., Hogg, C., and Silbey, R., Three-wave mixing in conjugated polymer solutions: two photon absorption in polydiacetylenes, *Phys. Rev.*, *B22*, 3540 (1980).
151. Porter, P. L., Guha, S., Kang, K., and Frazier, C. C., Effects of structural variations of platinum and palladium poly-ynes on third-order nonlinearity, *Polymer*, *32*, 1756 (1991).
152. Borbridge, S. J., Page, H., Drury, A., Davey, A. P., Callaghan, J., and Blau, W., The third order nonlinear optical response of a soluble form of polyisonaphthalene, *Nonlinear Opt.*, *10*, 139 (1995).
153. Sahraoui, B., Sylla, M., Bourdin, J. P., Rivair, G., Zarenba, J., Nguyen, T. T., and Salle, M., Third-order nonlinear optical properties of ethylenic tetrathiafulvalene derivatives, *J. Mod. Opt.*, *42*, 2095 (1995).
154. Mandal, B. K., Bihari, B., Sinha, A. K., Karnath, M., and Chen, L., Third-order nonlinear optical response in a multilayered phthalocyanine composite, *Appl. Phys. Lett.*, *66*, 932 (1995).
155. Sylla, M., Zaremba, J., Chevalier, R., Rivoire, G., Khanous, A., and Gorgues, A., Third-order nonlinear optical susceptibility of acetylenic analogues of tetrathiafulvalene, *Synth. Metals*, *59*, 111 (1993).

10
Third Harmonic Generation

F. Kajzar
LETI, CEA, Technologies Advancées
DEIN/SPE, Saclay, France

I. INTRODUCTION

Third harmonic generation (THG) is a coherent four photon process in which three photons with angular frequency ω create one photon with tripled frequency 3ω through a material medium polarization. Although the principle of harmonic generation was described in the early sixties,[1] with the first experimental observation of second harmonic generation (SHG) reported in 1961,[2] the first experiments on THG have been published a few years later, first in a single crystal of calcite[3] and later in solution.[4–5] The latter publication has also reported a first demonstration of phase matching in a dye solution through anomalous refractive index dispersion. Subsequently, the technique has been utilized to determine the second molecular hyperpolarizabilities $\gamma(-3\omega;\omega,\omega,\omega)$ in solution and to determine different $\chi^{(3)}_{IJKL}(-3\omega;\omega,\omega,\omega)$ tensor components in single crystals. The technique had an important revival in the early seventies[6–7] and at the beginning of the eighties with the demonstration of the first THG experiments in thin films by Kajzar et al.[8] THG has been applied to measuring the electronic $\chi^{(3)}$ susceptibility as well as for nonlinear spectroscopy studies in solids[9] and in liquids[10-15] with the capability of determining both the magnitude and phase of $\chi^{(3)}$.[16-19] In contrast with second harmonic generation, the third harmonic generation process occurs in all materials, including the surrounding air. In the first THG experiments, the contribution due to air was not taken into account. As it has been shown by Kajzar and Messier[13] (see also discussion in Section II.A), air may contribute significantly to the third harmonic intensity of a sample resulting in a large error in determining the $\chi^{(3)}$ of the material.

Because the THG process is coherent, the harmonic photon must stay in step with the fundamental one if an appreciable amount of third harmonic is to be generated. Because third harmonic is generated only if the charges in the sample keep up with the optical field, it measures only the electronic mechanism of $\chi^{(3)}$. Although in practice one measures the macroscopic third order nonlinear optical susceptibility $\chi^{(3)}_{IJKL}(-3\omega;\omega,\omega,\omega)$, it is linked to the microscopic second hyperpolarizability $\gamma(-3\omega;\omega,\omega,\omega)$ through the following relation:

$$\chi^{(3)} = NF\gamma, \tag{1}$$

where N is the number of molecules per unit volume and F is the local field factor taking into account the fact that the molecule experiences a different electric field than the externally applied field, because of screening effect, called also the crystal field effect (see Chapter 3 for a general treatment). For THG, this factor is given by the following formula:

$$F = L_{3\omega}(L_{\omega})^3, \tag{2}$$

where for a spherically shaped molecule, $L_{\omega(3\omega)}$ is given by the Lorentz-Lorenz formula:

$$L_{\omega(3\omega)} = \frac{n^2_{\omega(3\omega)} + 2}{3}. \tag{3}$$

The molecular second hyperpolarizability tensor is a sum of vibrational (v), dipolar (d) and electronic (e) contributions:

$$\gamma = \gamma^v + \gamma^d + \gamma^e. \tag{4}$$

For an isotropic liquid or solid, the dipolar term is equal to zero.

As we already mentioned, the macroscopic third order nonlinear susceptibility $\chi^{(3)}_{IJKL}(-3\omega;\omega,\omega,\omega)$ and molecular susceptibility $\gamma_{ijkl}(-3\omega;\omega,\omega,\omega)$ are three-dimensional fourth rank tensors with $3^4 = 81$ components and the relation given by Eq. 1 must be generalized to tensor form. The obvious link between the macroscopic and microscopic hyperpolarizabilities are obtained by transforming the corresponding quantities from the microscopic ($ijkl$) to macroscopic ($IJKL$) reference frame by using the Wigners's rotation matrices a_{ij}. From Chapter 3, Eq. 58, the THG susceptibility is related to the microscopic susceptibility by:

$$\chi^{(3)}_{IJKL}(-3\omega;\omega,\omega,\omega) = L^{\omega}_{ii}L^{\omega}_{jJ}L^{\omega}_{kK}L^{3\omega}_{lL} \sum_n N^{(n)} \sum_{ijkl} a^{(n)}_{iI}a^{(n)}_{Jj}a^{(n)}_{Kk}a^{(n)}_{Ll}\gamma^{(n)}_{IJKL}(-3\omega;\omega,\omega,\omega), \tag{5}$$

where $N^{(n)}$ is the density of molecules in the n-th crystallographic sub-unit, $a^{(n)}_{iI}$ is a function of orientation of crystal subunit n, the Ls are local field tensors, and where summation convention applies. For a solution, due to the fact that all molecular axes are randomly distributed, Eq. 5 reduces to a much simpler form:

$$\chi^{(3)}_{IJKL} = NF\langle\gamma\rangle_{IJKL}, \tag{6}$$

where the diagonal component is:

$$\langle\gamma\rangle_{zzzz} = \tfrac{1}{5}(\gamma_{xxxx} + \gamma_{yyyy} + \gamma_{zzzz} + 2[\gamma_{xxyy} + \gamma_{xxzz} + \gamma_{yyzz}]). \tag{7}$$

the number of non-zero $\chi^{(3)}$ tensor components is, by symmetry of the isotropic solution, strongly reduced (to 9). In Eq. 6, it is assumed *a priori* that all molecules are identical.

Third harmonic generation can be viewed, in the dipolar approximation, as due to the nonlinear polarization created in the medium by the incident light at angular frequency ω. The link between the macroscopic polarization with the exciting field amplitude depends on the convention used, as discussed in chapter 1. Depending on the convention, different values for the nonlinear susceptibility $\chi^{(3)}$ and molecular hyperpolarizability γ will be obtained. We adopt the following convention for electric field and polarization amplitudes (in the case of a monochromatic beam):

$$\vec{E}(\vec{r},t) = \tfrac{1}{2}(\vec{E}(r)e^{i(\vec{k}\cdot\vec{r}-\omega t)} + c.c.), \tag{8}$$

where *c.c.* is the complex conjugate. The polarization term is analogously defined as:

$$\vec{p}(\vec{r},t) = \tfrac{1}{2}(\vec{p}(\vec{r})e^{i(\vec{k}\cdot\vec{r}-\omega t)} + c.c.).$$

In this convention, the third harmonic tensor introduced by Maker and Terhune[3] is given by:

$$C_{IJKL} = \chi^{(3)}_{IJKL}/4. \tag{9}$$

Consequently, the *i*-th component of the nonlinear polarization vector (in the frequency domain) describing the THG process is given by:

$$P^{I}_{NL}(3\omega) = \frac{\chi^{(3)}_{IJKL}}{4} E^{\omega}_{J}E^{\omega}_{K}E^{\omega}_{L}, \tag{10}$$

where, implicitly, Einstein's summation notation is used. The importance of which conventions and notations are used is appreciated when experimental data from different sources are compared, when experimental data are predicted by theoretical calculations, or when one nonlinear optical susceptibility is derived from another NLO process. These problems for second[20-22] and third order NLO processes[20,23-26] have been already addressed by several research groups (cf. also Chapter 1).

II. THIRD HARMONIC GENERATION IN HOMOGENOUS MEDIA

The general problem of harmonic generation of light in homogenous media was solved by Bloembergen and Pershan.[1] Let us consider here a simple case of a nonlinear medium immersed between two linear media and an incoming plane wave propagating in the *z*-direction (cf. Fig. 1). For the sake of simplicity, we consider a monochromatic plane wave. The case of focused Gaussian beams will be discussed later, when considering environmental effects; e.g., the surrounding air. In most practical cases, when the sample thickness is smaller than the beam waist and the sample is placed at the focus of a monochromatic laser beam, these assumptions are almost always fulfilled.

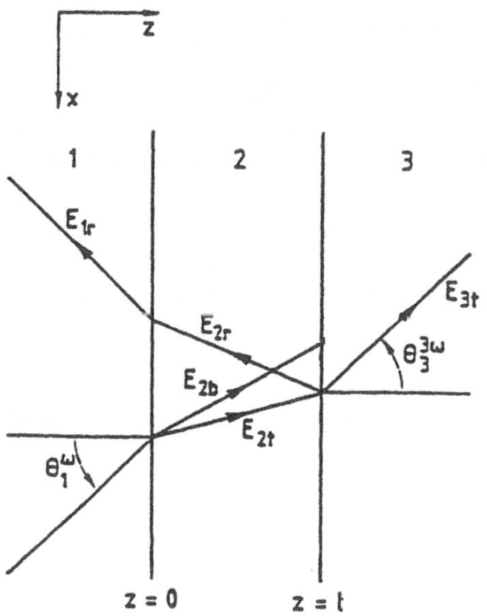

Figure 1

A. Third Harmonic Generation in a Non-absorbing Plane Parallel Slab

The expressions for harmonic intensities are obtained by considering Maxwell's equation in which the nonlinear polarization of the medium acts as a source term. Assuming the propagation equation for both the fundamental and harmonic frequencies have plane wave solutions, the wave equation in Gaussian units is:

$$\nabla^2 \vec{E}_{3\omega} + \frac{n^2(3\omega)}{c^2} \frac{\partial^2 E_{3\omega}}{\partial t^2} = -\frac{4\pi}{c^2} \frac{\partial^2 \vec{P}_{NL}(3\omega)}{\partial t^2}, \tag{11}$$

where the polarization in the driving term is given by Eq. 10.

The general solution of Eq. (11) is a free wave, $\vec{E}^f_{3\omega}(\vec{r},t)$:

$$\vec{E}^f_{3\omega}(\vec{r},t) = \tfrac{1}{2}[\vec{E}^f_{3\omega}(\vec{r})e^{i(\vec{k}_{3\omega}\cdot\vec{r}-3\omega t)} + c.c.], \tag{12}$$

and the particular solution of the homogenous equation is called the bound wave (because it follows the polarization) $\vec{E}^b_{3\omega}(\vec{r},t)$:

$$\vec{E}^b_{3\omega}(\vec{r},t) = \tfrac{1}{2}[\vec{E}^b_{3\omega}(\vec{r})e^{i(3\vec{k}^b_{\omega}\cdot\vec{r}-3\omega t)} + c.c.], \tag{13}$$

where

$$k^b_{\omega} = \frac{\omega\, n(\omega)}{c}, \tag{14}$$

and

$$k^f_{3\omega} = \frac{3\omega\, n(3\omega)}{c}. \tag{15}$$

By substituting the bound wave (Eq. 13) into Maxwell's equation (Eq. 11), one gets the bound wave amplitude:

$$\left[|k_\omega^b|^2 - \frac{n^2(3\omega)}{c^2}(3\omega)^2\right] E_{3\omega}^b = \frac{36\pi}{c^2} P_{NL}, \tag{16}$$

which yields

$$E_{3\omega}^b = \frac{4\pi P_{NL}}{\Delta\varepsilon}, \tag{17}$$

where $\Delta\varepsilon = n^2(\omega) - n^2(3\omega)$ is the dielectric constant dispersion. Consequently, the third harmonic wave propagating in the nonlinear medium is the sum of both free and bound waves,

$$\vec{E}_{3\omega}(\vec{r},t) = \tfrac{1}{2}[\vec{E}_{3\omega}^f(\vec{r})e^{i(\vec{k}_{3\omega}^f\cdot\vec{r}-3\omega t)} + \vec{E}_{3\omega}^b(\vec{r})e^{i(3\vec{k}_\omega^b\cdot\vec{r}-3\omega t)} + c.c.]. \tag{18}$$

Due to the refractive index dispersion, both waves propagate with different velocities and interfere giving rise to Maker fringes, which give rise to an intensity modulation in the material. The output third harmonic intensity is thus an oscillating function of the propagation distance.

The amplitude of the harmonic free wave is obtained by considering boundary conditions on interfaces between different linear and/or nonlinear media. We consider for the sake of simplicity the case of a nonlinear slab immersed between two linear media (cf. Fig. 1) and write boundary conditions for the harmonic fields. For all possible fundamental-harmonic beam configurations, it is sufficient to consider boundary conditions for the s- and p-polarized harmonic beams. The case of incident s-polarized light has been considered by Kajzar and Messier.[13] For the p-p (p-incident, p-transmitted) case, we will adapt the third order analogues of the formulas derived by Swalen and Kajzar[27] for the case of SHG.

For the s-polarized harmonic field, we have the following boundary conditions for electric and magnetic fields:

At interface $z = 0$ we have

$$E_r^{(1)} = E_t^{(2)} + E_r^{(2)} + E_b^{(2)} \tag{19}$$

for the electric field and

$$-N_\omega^{(1)}E_r^{(1)} = N_{3\omega}^{(2)}E_t^{(2)} - N_{3\omega}^{(2)}E_r^{(2)} + N_\omega^{(2)}E_b^{(2)} \tag{20}$$

for the magnetic field, respectively. Similarly, at $z = l$ interface the corresponding equations read:

$$E_t^{(2)}e^{i\varphi_2} + E_r^{(2)}e^{-i\varphi_2} + E_b^{(2)}e^{i\varphi_1} = E_t^{(3)} \tag{21}$$

for E field, and

$$N_{3\omega}^{(2)}E_t^{(2)}e^{i\varphi_2} - N_{3\omega}^{(2)}E_r^{(2)}e^{i\varphi_2} + N_\omega^{(2)}E_b^{(2)} = N_{3\omega}^{(3)}E_t^{(3)} \tag{22}$$

for the magnetic field. In Eqs. 19–22, the subscripts r, t, b refer to reflected (r), forward-propagating harmonic free (t) and harmonic bound (b) waves, respectively. $N_\omega^{(n)}$ is the effective index at ω and in layer n as defined later in Eq. 27 and the φ phase as defined later in Eq. 28. The superscripts refer to the medium under consideration. By solving simultaneously the set of four equations, one gets for the harmonic field in the linear medium (3) the following:

$$E_t^{(3)} = \frac{4\pi P_{NL}}{\Delta\varepsilon} e^{ik_{2r}^{3\omega}l}\{A_1^s[e^{i(\varphi_\omega-\varphi_{3\omega})} - 1] + A_2^s[e^{i(\varphi_\omega+\varphi_{3\omega})} - 1]\}/A_3^s, \tag{23}$$

where

$$A_1^s = \frac{N_{3\omega}^{(2)} + N_\omega^{(2)}}{N_{3\omega}^{(2)} + N_{3\omega}^{(2)}} \tag{24}$$

$$A_2^s = \frac{N_{3\omega}^{(2)} - N_\omega^{(2)}}{N_{3\omega}^{(3)} + N_{3\omega}^{(2)}} \frac{N_{3\omega}^{(1)} - N_{3\omega}^{(2)}}{N_\omega^{(1)} + N_{3\omega}^{(2)}} \tag{25}$$

$$A_3^s = 1 - \frac{N_{3\omega}^{(1)} - N_{3\omega}^{(2)}}{N_1^{3\omega} + N_2^{3\omega}} \frac{N_{3\omega}^{(3)} - N_{3\omega}^{(2)}}{N_3^{3\omega} + N_2^{3\omega}} e^{i(6\omega l N_2^{3\omega}/c)}, \tag{26}$$

with

$$N_{\omega,3\omega}^{(j)} = n_{\omega,3\omega}^{(j)} \cos\theta_{\omega,3\omega}^{(j)}, \tag{27}$$

and $\varphi_{m\omega} = 3\omega N_{m\omega}^{(2)} l/c$ ($m = 1,3$) are the phase mismatches for fundamental ($m = 1$) and harmonic ($m = 3$) waves in nonlinear medium (2), and

$$\Delta\varphi = \varphi_\omega - \varphi_{3\omega} = \frac{6\pi l}{\lambda}(n_\omega \cos\theta_\omega - n_{3\omega}\cos\theta_{3\omega}) \tag{28}$$

is a phase mismatch between fundamental and harmonic medium and depends on refractive index dispersion, medium thickness l and propagation angles $\theta_{\omega(3\omega)}^{(n)}$ with respect to slab normal in medium n. The subscripts $\omega(3\omega)$ refer to fundamental (harmonic) waves.

For a p-polarized harmonic wave, the corresponding boundary conditions are:

$$-E_r^{(1)}\cos\theta_{3\omega}^{(1)} = E_t^{(2)}\cos\theta_{3\omega}^{(2)} - E_r^{(2)}\cos\theta_{3\omega}^{(2)} + E_b^{(2)}\cos\theta_{3\omega}^{(1)}$$

and (29)

$$k_r^{(1)}E_r^{(1)} = k_t^{(2)}E_t^{(2)} + k_r^{(2)}E_r^{(2)} + k_b^{(2)}E_b^{(2)}$$

at $z = 0$, for the E and H fields, respectively. Similarly, at $z = l$ we have:

$$-E_t^{(2)}\cos\theta_{3\omega}^{(2)}e^{i\varphi_3} - E_r^{(2)}\cos\theta_{3\omega}^{(2)}e^{-i\varphi_3} + E_b^{(2)}\cos\theta_{3\omega}^{(2)}e^{i\varphi_1} = E_t^{(3)}\cos\theta_{3\omega}^{(3)}$$

and (30)

$$k_r^{(2)}E_r^{(2)}e^{-i\varphi_3} + k_t^{(2)}E_t^{(2)}e^{i\varphi_2} + k_b^{(2)}E_b^{(2)}e^{i\varphi_1} = k_t^{(3)}E_t^{(3)}.$$

Again, by solving simultaneously the set of Eqs. 29–30, one obtains for the harmonic p field the following:

$$E_t^{(3)} = \frac{4\pi P_{NL}}{\Delta\varepsilon} e^{i\varphi_{3\omega}}\{A_1^p[e^{i(\varphi_\omega-\varphi_{3\omega})} - 1] + A_2^p[e^{i(\varphi_\omega+\varphi_{3\omega})} - 1]\}/A_3^p, \tag{31}$$

where

$$A_1^p = \frac{N_{3\omega}^{(2)} + N_{\omega}^{(2)}}{N_{3\omega}^{(3)} + N_{\omega}^{(2)}} \tag{32}$$

$$A_2^p = \frac{(N_{3\omega}^{(2)} - N_{2\omega}^{(2)}}{(N_{3\omega}^{(2)} + N_{\omega}^{(2)})} \frac{(N_{3\omega}^{(1)} - N_{3\omega}^{(-2)}}{(N_{3\omega}^{(1)} + N_{3\omega}^{(-2)})} \tag{33}$$

$$A_3^p = 1 - \frac{(N_{3\omega}^{(1)} - N_{2\omega}^{(-2)})}{(N_{3\omega}^{(1)} + N_{\omega}^{(-2)})} \frac{(N_{3\omega}^{(-3)} - N_{3\omega}^{(2)})}{(N_{3\omega}^{(-3)} + N_{3\omega}^{(2)})} \tag{34}$$

$$N_{\omega,3\omega}^{(j)} = n_{\omega,3\omega}^{(j)} \cos \theta_{\omega,3\omega}^{(j)} \tag{35}$$

$$N_{3\omega}^{(j)} = n_{3\omega}^{(j)} \cos \theta_{3\omega}^{(j+1)} \tag{36}$$

$$N_{3\omega}^{(-j)} = n_{3\omega}^{(j)} \cos \theta_{3\omega}^{(j-1)}. \tag{37}$$

Equations 23 and 31 take account of multiple reflections of harmonic waves in the sample slab and account for simple transmission factors for the fundamental. Several approaches have been proposed in order to take into account the multiple reflections of fundamental beams.[28-31] They lead, however, to quite complicated formulae. The most appropriate approach is that given by Bethune for isotropic[28] and anisotropic media,[29] which consists of a simultaneous solution of propagation equations using matrix formalism. Van Beek et al.[43] proposed a simplified approach introducing into Eqs. 23 and 31 the fundamental field taking into account multiple reflections:

$$E_{\omega,t}^{(2)}(\vec{r}) = \frac{t_{\omega}^{12} E_{\omega}^{(1)}}{1 + r_{\omega}^{12} r_{\omega}^{23} e^{2i\varphi_{\omega}^{(2)}}} e^{i\vec{k}_{2t}^{\omega} \cdot \vec{r}} \tag{38}$$

$$E_{\omega,r}^{(2)}(\vec{r}) = \frac{t_{\omega}^{12} r_{\omega}^{23} E_{\omega}^{(1)} e^{2i\varphi_{\omega}^{(2)}}}{1 + r_{\omega}^{12} r_{\omega}^{23} e^{2i\varphi_{\omega}^{(2)}}} e^{i\vec{k}_{2r}^{\omega} \cdot \vec{r}} \tag{39}$$

where the subscript t denotes the transmitted field to medium (2) and r the reflected one (on interface 2-3), respectively. t^{ij} and r^{ij} are usual Fresnel transmission and reflection factors between media (i) and (j). $E_{\omega}^{(1)}$ in Eqs. 38–39 is the incident fundamental field (in medium 1). The resultant fundamental field in medium (2) is a sum of both Eq. 38 and 39 and the resultant harmonic field in medium (3) will depend now on different combinations of these two fields (for details see Ref. [43]).

As we mentioned already, the formulae taking full account of multiple fundamental and harmonic reflections are quite complicated and of limited practical utility, especially when fast screening of $\chi^{(3)}$ of many materials is required. Very often the studied materials are low refractive index materials like organics and the surface roughness decreases the multiple reflections so they can be neglected. This is also true when doing experiments as a function of incidence angle, as it is often the case. In that case, as discussed by Kajzar and Messier,[32] the multiple reflections give an oscillatory contribution to the overall THG fringes, as a function of incidence angle whose amplitude vanishes rapidly with increasing incidence angle, as it was observed also experimentally.[33] Multiply reflected beams in this type of experiment are spatially separated at large incidence angles. A good approximation for the harmonic field is then obtained by setting $A_2 = 0$ and $A_3 = 1$ (cf. Eqs. 23 and 31).

1. Graphical Representation of the Third Harmonic Field

For the case discussed above ($A_2 = 0$ and $A_3 = 1$), the harmonic field generated in a plane parallel slab has a simple form:

$$E_t^{(3)} = \frac{4\pi P_{NL}}{\Delta\varepsilon} A_1[e^{i\phi_\omega} - e^{i\phi_{3\omega}}]. \tag{40}$$

Because $e^{i\phi_\omega}$ and $e^{i\phi_{3\omega}}$ are complex functions of unit modulus, they can each be expressed as unit vectors in the complex plane. The third harmonic electric field, then, can be represented graphically in the complex plane (cf. Fig. 2) as the chord **AM** on a circle with radius given by $4\pi P_{NL}/\Delta\varepsilon\, A_1$ where the vectors OA and OM represent the unit vectors at angles $\phi_{3\omega}$ and ϕ_ω, respectively. The phase angle $\Delta\varphi = \varphi_\omega - \varphi_{3\omega}$, which is the phase mismatch between fundamental and harmonic waves in a nonlinear medium, varies with the incidence angle because of refractive index dispersion (cf. Eq. 28). Consequently, because AM is a cord of the circle, the difference between the two unit vectors spans from 0 to 2, so that $E_t^{(3)}$ varies between 0 ($\Delta\varphi = 2j\pi$; $j = 0, \pm 1, \pm 2, \ldots$) and $2 \times 4\pi P_{NL}/\Delta\varepsilon\, A_1(\Delta\varphi = (2j + 1)\pi$; $j = 0, \pm 1, \pm 2, \ldots)$.

B. Environmental Effects

Third harmonic generation takes place in any material medium and in particular in air. Although the THG susceptibility of air is significantly lower than that of any dense medium, the coherence length of air is much larger than the coherence length for most materials. Consequently, the third harmonic field generated in air may be comparable with that generated in the studied sample, giving rise to a complex interference between these two beams. For this reason, the THG experiments have

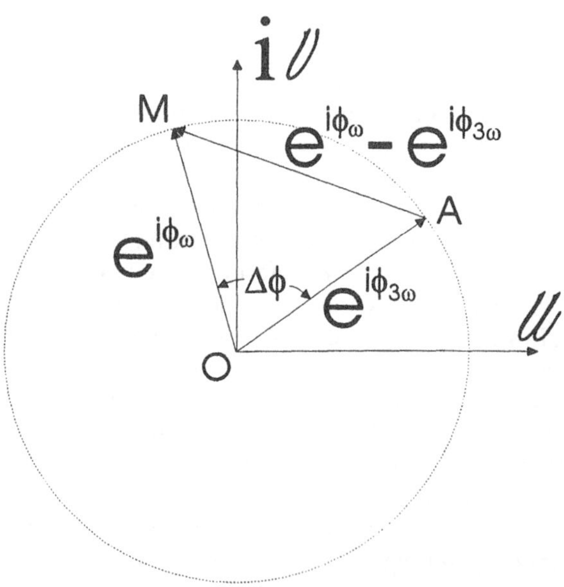

Figure 2

to be done with focused laser beam, otherwise the amplitude of harmonic field generated in air will be undetermined. All optical elements such as lenses, filters, polarizers, etc., will contribute to the third harmonic intensity, especially if all these components are exposed to the same intensity beam as is the case for an unfocused beam.

The problem of harmonic generation by focused laser beams was considered by Kleinman et al.[34] and applied in the case of gases by Ward and New[35] (see also Boyd and Kleinman[36]). The third harmonic field generated by a Gaussian beam, propagating in the z-direction at point (x,y,z) is given by:

$$E_{3\omega}(x,y,z) = \frac{4\pi P_{NL} I(z)}{\Delta\varepsilon} e^{ik_{3\omega}z} e^{-3(x^2+y^2)/W_0^2(1+i\tau)}, \tag{41}$$

where

$$I(z) = \frac{3\omega}{c} \Delta n \int_{-\infty}^{z} dz' \frac{e^{i\Phi}}{(1 + i\tau')^2}, \tag{42}$$

with

$$\Phi = \frac{3\omega}{c}(z - z'), \tag{43}$$

and

$$\tau' = \frac{2z'}{W_0^2 k_\omega}, \tag{44}$$

where W_0 is the beam waist (cf. Fig. 3), and Δn is the refractive index dispersion between the fundamental and third harmonic wavelengths. The function $I(z)$ (cf. Eq. 42) is an odd function ($I(z) = -I(-z)$) so:

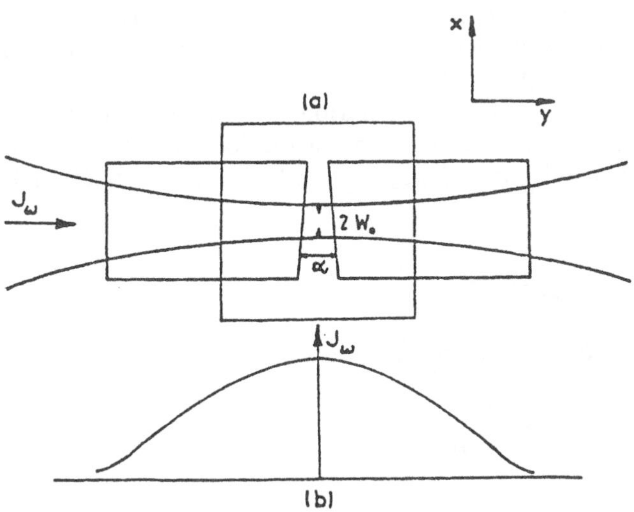

Figure 3

$$\int_{-\infty}^{z} dz' \frac{e^{i\Phi}}{(1 + i\tau')^2} = -\int_{z}^{\infty} dz' \frac{e^{i\Phi}}{(1 + i\tau')^2}, \tag{45}$$

or

$$\int_{-\infty}^{\infty} dz' \frac{e^{i\Phi}}{(1 + i\tau')^2} = 0. \tag{46}$$

Consequently, there is no resultant third harmonic field for a focused Gaussian beam at the output of an infinite medium. This is a very important result because it means that any THG experiment done with a focused laser beam in a thick medium will give no third harmonic beam at a distant output because of a cancellation of third harmonic fields generated before and after the focal point. However, as will be discussed later, introduction of a thin sample at the focus will give rise to a phase mismatch $\Delta\varphi$ between fundamental and harmonic fields, depending on sample thickness and refractive index dispersion (cf. Eq. 28) and the harmonic fields created before and after the focus will interfere. The result of interference will depend on the value of $\Delta\varphi$. As already mentioned, it is an important result which is the basis of the standard THG measurement technique in liquids as originally proposed by Kajzar and Messier.[14-15]

1. Graphical Representation

Similarly, as for THG process in a plane parallel slab and in the plane wave approximation, the harmonic field generated in an infinite medium can be represented graphically in the complex plane. However, in the present case, because of the change of fundamental field with propagation length, due to focusing the "elementary" harmonic fields (cf. Eq. 41) are described by a spiral (cf. Fig. 4) rather than a circle. From Fig. 4 it is seen that in an infinite medium, with no sample at the focal point, the harmonic fields generated before and after th focal points cancel. Introducing a plane parallel slab, with a thickness satisfying condition of the plane wave approximation for a fundamental beam inside the slab, the resultant harmonic field at the output of such system (e.g., air before sample, sample, air after sample) will be the sum of harmonic fields generated in these three media. This can be again represented graphically, as it is shown in Fig. 5. The resultant harmonic field is the sum of three vectors **OM** + **CA** + **AB**.

Figure 5 can also be used to show how air contributes to the third harmonic intensity from a plane parallel slab, placed in air and at the focal point of a converging lens. For the phase mismatch $\Delta\varphi = 2j\pi; j = 0, \pm 1, \pm 2, \ldots$, correspond-

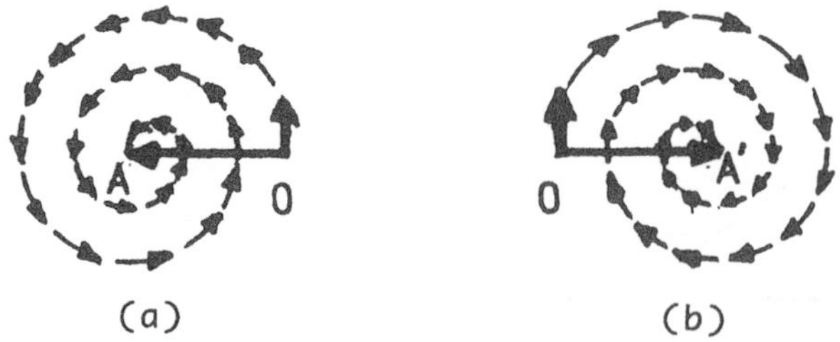

Figure 4

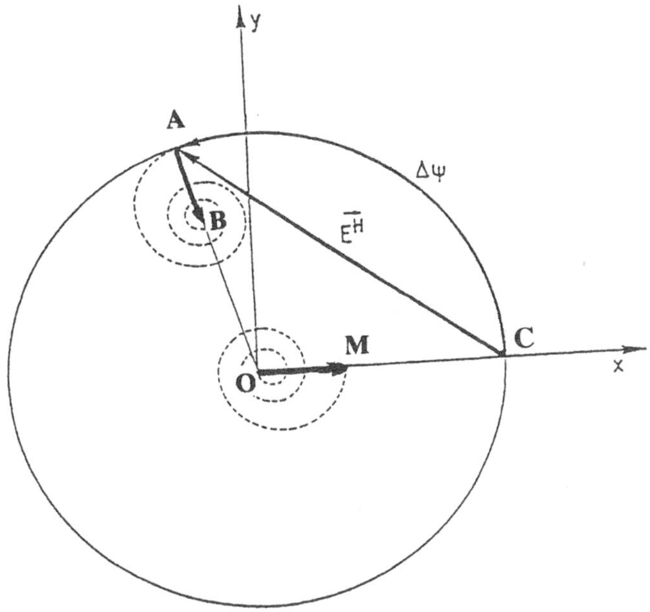

Figure 5

ing to the minima of Maker fringes (free and bound waves cancel) the air contribution from the air before the sample cancels the contribution from the air after the sample and no background air contribution is observed. The position of the Maker fringe minima are thus not affected by the air (cf. Figs. 6a and 6b). For a phase mismatch in the slab of $\Delta\varphi = (2j + 1)\pi$; $j = 0, \pm 1, \pm 2, \ldots$, corresponding to the maxima of a Maker fringe, the air contribution before the sample has the same sign as that generated after the sample, but opposite to the third harmonic generated in the sample. Consequently, the third harmonic electric field at a Maker fringe maximum will be reduced by two times the air's contribution to the third harmonic field. The resulting intensity is the square of this field. This explains the observed decrease of the amplitude of the Maker fringes due to the surrounding air. The effect can be quite important as it was observed by Kajzar and Mesier.[13] Figure 6 shows three different levels of relative air contribution: In Fig. 6a, the air contribution is smaller than the third harmonic field generated in the slab, in Fig. 6b, it is comparable, and in Fig. 6c, it is larger. In the first case, one observes a decrease in the amplitude of the third harmonic intensity relative to vacuum whose amplitude decreases at a faster rate than the envelope function in vacuum. In the second case (Fig. 6b), there is almost no third harmonic field, except at a large incidence angle. In the third case (Fig. 6c), the third harmonic intensity is lower but the envelope function increases with incidence angle. This behavior can be well understood in the light of the preceding remarks. The amplitude of the air contribution depends very weakly on the incidence angle, whereas the generated third harmonic fields in the sample depends much more strongly on an angle due to reflections. Thus, in the first case (Fig. 6a), the relative importance of the air contribution is increasing with increasing incidence angle (thus the difference is de-

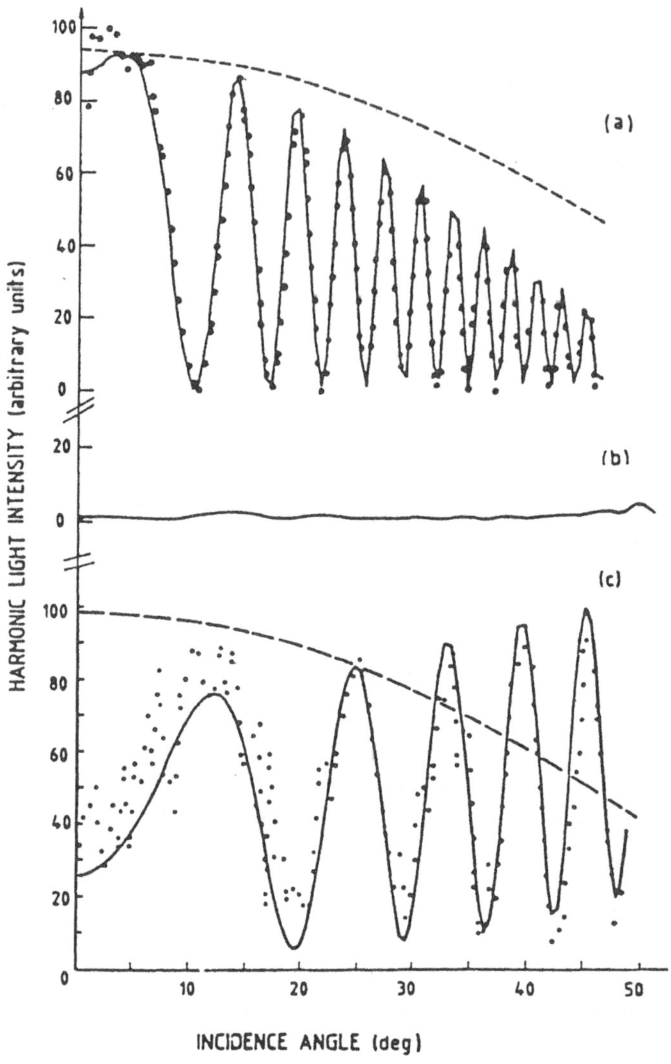

Figure 6

creasing). In the second case, we have almost complete cancellation at normal incidence. At larger incidence angles, this cancellation is weaker because the harmonic field generated in the slab is decreasing more rapidly with incidence angle than that generated in air. Consequently, in that case we substract a function that is decreasing with incidence angle (harmonic field generated in the slab) from almost a constant and larger field generated in air. The formalism developed by Kajzar and Messier[13] accounts analytically for the air contribution, as demonstrated by the agreement between the solid lines in Fig. 6 (depicting calculated intensities) and the data points.

Because the formalism developed by Kajzar and Messier[13] allows one to take into account the air contribution, it can also be used to measure the $\chi^{(3)}$ susceptibility of any gas that surrounds a sample. However, such measurements require that the data be fit to the theory with a large number of free parameters (the number of free parameters can be limited with independent THG measurements). Therefore, for practical reasons, it is more convenient to do THG measurements in vacuum. Some authors propose the use of tightly focused laser beams to avoid the air contribution.[37] In light of the above remarks, it is clear that this approach is not satisfactory because the plane wave approximation for THG in the sample will not be obeyed. Consequently, in that case, the formalism developed for third harmonic generation by a focused beam has to be used.

C. Third Harmonic Generation in Absorbing Media

The formalism for third harmonic generation described here is general and applies for absorbing media, provided that absorption is not too strong. This can be taken into account by introducing a complex refractive index ($n = n^r + i\kappa$, where n^r and κ represent the real and imaginary parts of n), a complex dielectric constant and a complex cubic susceptibility $\chi^{(3)}$. For an absorbing medium, the propagation angle θ is also complex. The necessity of a complex angle follows from the Descartes—Snell's law: $n \sin\theta = n_{\text{air}} \sin\theta^{air}$. When the surrounding medium is non-absorbing, $n \sin\theta$ must be real, so for complex n, θ must also be complex.

For small κ, the propagation direction of both the fundamental and third harmonic waves is close to what it would be in a non-absorbing medium. For a complex propagation angle θ_2 in the nonlinear medium (2), one can put formally:[38]

$$\cos\theta_2 = \cos\theta_2^r + i\,\frac{\kappa_2}{n_2^r \cos\theta_2^r} \tag{47}$$

Using Descartes—Snell's law, and assuming $n_1 = 1$, one obtains the following equation for the real part of the propagation angle:

$$\cos\theta_2^r = 1 - \frac{\sin^2\theta_2^r}{(n_2^r)^2} + \frac{\kappa_2^2}{2}\left[1 - \frac{\cos^2\theta_2^r}{(n_2^r)^2}\right] + O(\kappa_2^4), \tag{48}$$

where θ_2^r is the propagation angle (real) in medium (2). Equation 48 shows that for small extinction coefficients ($\kappa << 1$) the real part of the propagation angle in the absorbing medium is close to what it would be in a non-absorbing medium with the same real part of the refractive index. Consequently, the equation for a plane wave propagating in a nonlinear medium is given by:

$$E_\omega(z,t) = \frac{1}{2}\left[E_\omega(z)e^{-i(\omega t - k_\omega^r z)}e^{-\kappa l/\cos\theta^r} + c.c.\right], \tag{49}$$

where l is the thickness of the material.

In the graphical representation (cf. Ref. [39]), the absorption at fundamental and harmonic wavelengths results in a decrease of the OA and OM vectors (cf. Fig. 2), which are given by:

$$OA' = OAe^{-\kappa_{3\omega}(6\pi l/\lambda_\omega \cos\theta_{3\omega}^r)}. \tag{50}$$

Similarly,

$$OM' = OMe^{-\kappa_\omega(6\pi l/\lambda_\omega \cos\theta_\omega^r)}, \tag{51}$$

where the subscripts ω and 3ω refer to fundamental and third harmonic wavelength, respectively. Due to the large light intensities required for THG measurements, it is almost impossible to do these experiments in media that absorb an appreciable amount of light at the fundamental frequency. Consequently, the length of **OA** is constant in most experimental conditions. It is obvious that good contrast in Maker fringes requires that the free wave amplitude (or the length of **OM**′) cannot vanish over propagation lengths that are smaller than the material thickness. It imposes a true limit for medium absorption in THG experiments. Practically, it means that in the case of absorbing media the optical density along the longest propagation direction should be smaller than 2. Otherwise, Maker fringes are not observed.

III. EXPERIMENTAL TECHNIQUES

A. Oriented Systems

The simplest case of an oriented medium is a single crystal. In the general case, the relationship between macroscopic $\chi^{(3)}$ nonlinear optical susceptibility and microscopic hyperpolarizability γ is given by Eq. 5. The molecule symmetry as well as the symmetry of interacting fields limit significantly the number of non-zero γ_{ijkl} tensor components. The same is true for the macroscopic susceptibilities. Therefore it is necessary to know both molecule and single crystal symmetry. By doing experiments in various combinations of fundamental-harmonic field polarizations, it is possible to get different $\chi^{(3)}_{IJKL}$ tensor components. Because the three fundamental photons have the same electric field polarization, a limited number of tensor components can be measured in THG experiments, namely: $\chi^{(3)}_{IJJJ}$ $(I \neq J)$ and the diagonal components $\chi^{(3)}_{IIII}$. In some molecules, like conjugated quasi 1D polymers, the molecular second hyperpolarizability γ_{xxxx} is largest in the polymer chain direction x. When other γ_{ijkl} tensor component can therefore be neglected, the relationship between the macroscopic $\chi^{(3)}_{IJKL}$ and the microscopic γ_{xxxx} hyperpolarizability is much simpler:

$$\chi^{(3)}_{IJKL}(-3\omega;\omega,\omega,\omega) = F \sum_n N^{(n)} \cos^{(n)}(I,x) \cos^{(n)}(J,x) \cos^{(n)}(K,x) \cos^{(n)}(L,x)\gamma^{(n)}_{xxxx}(-3\omega;\omega,\omega,\omega), \tag{52}$$

where (I,z) is the angle between laboratory I and molecular z direction and F is the global field factor. For a mono-oriented system, composed of one molecular species, one gets a much simpler formula for the corresponding, non-zero tensor components of the macroscopic susceptibility $\chi^{(3)}$, assuming that the molecular x axis lies in the laboratory XY plane (cf. Refs. [40]–[41]):

$$\chi^{(3)}_{XXXX} = NF\gamma_{xxxx} \cos^4 \Theta \tag{53}$$

$$\chi^{(3)}_{YYYY} = NF\gamma_{xxxx} \sin^4 \Theta \tag{54}$$

$$\chi^{(3)}_{XXYY} = \chi^{(3)}_{YYXX} = \chi^{(3)}_{XYYXY} = 6NF\gamma_{xxxx} \sin^2 \Theta \cos^2 \Theta, \tag{55}$$

where Θ is the angle between the molecular x axis and the macroscopic X direction.

1. *Orientational Distribution*

In the case when the molecular axes are distributed in a regular way and this distribution can be described by a function $P(\Theta)$, where Θ is the angle between a privileged direction in the laboratory reference frame (e.g., X axis) and the molecular x axis, the macroscopic cubic susceptibility will be given by:

$$\chi^{(3)}_{IJKL}(-3\omega;\omega,\omega,\omega) = FN\langle \cos^{(4)}\Theta | P(\Theta)\rangle_{IJKL}\,\gamma_{xxxx}(-3\omega;\omega,\omega,\omega), \tag{56}$$

where the symbol <> denotes the orientational average weighted with orientation distribution function $P(\Theta)$.

Expanding the orientational distribution function $P(\Theta)$ into the series of complete set cos $n\Theta$ (cf. Ref. [42]) yields:

$$P(\Theta) = \sum_n O_n \cos n\,\Theta, \tag{57}$$

where O_n is an orientational order parameter of moment n.

Introducing this expression into Eq. 56, one obtains for the corresponding components of the cubic susceptibility the following:

$$\chi^{(3)}_{XXXX} = \frac{NF}{16}\gamma_{xxxx}(6 + 4O_2 + O_4) \tag{58}$$

$$\chi^{(3)}_{YYYY} = \frac{NF}{16}\gamma_{xxxx}(6 - 4O_2 + O_4) \tag{59}$$

$$\chi^{(3)}_{XXYY} = \chi^{(3)}_{YYXX} = \chi^{(3)}_{XYXY} = \frac{NF}{16}\gamma_{xxxx}(2 - O_4). \tag{60}$$

In general for a 1-dimensional molecule, the bulk response is related to the molecular third order response according to:

$$\langle\chi^{(3)}\rangle_{IJKL} = NF\gamma_{xxxx}\langle\cos^4\Theta\rangle_{IJKL}, \tag{61}$$

with

$$\langle\cos^4\Theta\rangle = \begin{cases} 1 & \text{for mono-orientation (all molecular axes parallel to a given direction)} \\ 3/8 & \text{for bidimensional disorder (all molecular axes parallel to a plane and randomly distributed within this plane)} \\ 1/5 & \text{for an isotropic distribution.} \end{cases}$$

In Section IV.B we show how one can make use of these formulae in nonlinear optical dichroism measurements applications.

B. Liquids

Different experimental techniques used for third harmonic generation in liquids have been reviewed extensively by Kajzar.[19] Any liquid has to be kept in a transparent container, which, in addition to the surrounding air, contributes to harmonic intensities. In treating the problem of harmonic generation in liquids, the approaches differ by the way the window contribution is treated. Thalhamer and Penzkofer[12] proposed to use a wedge-shaped liquid compartment with windows designed in such a way as to get their thickness exactly equal to an even number of coherence lengths. The cell is translated perpendicularly to the beam propagation direction. A simple Maker fringe pattern is obtained. This technique requires a window whose thickness must be determined to high precision, it works at one wavelength only, and the measurement must be performed in a vacuum.

Meredith et al.[10] proposed the use of a liquid cell with wedged windows and liquid compartments. The main disadvantage of this technique is the large number of free parameters that must be adjusted to fit, the model to the data as well as the necessity to do the THG experiments in a vacuum. Fully aware of this problem, in the next paper Meredith et al.[11] proposed a cell whose input window is wedge-shaped and the liquid compartment is a long cell (6 cm). The cell is translated perpendicularly to the beam propagation direction. Assuming that the THG experiments are done in a vacuum (otherwise the air contribution has to be taken into account), the harmonic field is a sum of harmonic fields generated at the window, which depends on translation, and in the liquid compartment. Because the principle of harmonic generation in an infinite medium can be applied to the liquid compartment, the harmonic field generated in it will not depend on translation. The main disadvantage of this technique is again the necessity of doing experiments in a vacuum and that it only works for highly transparent liquids. In a modification by Cheng and coworkers,[44] the input wedge-shaped window was replaced by a thick (2 cm) glass window and the solution occupied a long (3 cm) liquid compartment. This design avoids air contributions and doesn't require a vacuum, but again, due to the large liquid compartment thickness, the measurements are possible only in highly transparent solutions.

Kajzar and Messier[13] used a liquid cell with plane parallel glass windows. The optical pathlength variation is obtained by rotating the cell around an axis perpendicular to the beam propagation direction. For THG measurements done in a vacuum, the resultant harmonic field is a sum of harmonic fields generated in the three nonlinear media:

$$E_{3\omega} = T_1 E_{3\omega}^{G1} + T_2 E_{3\omega}^{L} + T_3 E_{3\omega}^{G2}, \tag{62}$$

where T_1, T_2, T_3 are overall transmission factors, and $E_{3\omega}^{G1}$, $E_{3\omega}^{G2}$, $E_{3\omega}^{L}$ are the harmonic fields generated in the input window, output window and in the liquid compartment, respectively, and are given by Eq. 23 or 31. Because these three fields vary independently and differently as a function of angle, and the two materials have different refractive index dispersion, thicknesses and nonlinear optical response, the resulting Maker fringe pattern, is complicated (see Fig. 7). This pattern is greatly simplified when the refractive index for liquid and window material are the same, and when the input and output window thicknesses are the same. In this case, Eq. 62 leads to the more simple expression for the third harmonic intensity:

$$I_{3\omega} = \frac{256\pi^4}{c^2} |T|^2 \left| \frac{\chi_L^{(3)}}{\Delta\varepsilon_L} \sin\frac{\Delta\varphi_L}{2} + 2\frac{\chi_G^{(3)}}{\Delta\varepsilon_G} \cos\frac{\Delta\varphi_L + \Delta\varphi_G}{2} \sin\frac{\Delta\varphi_G}{2} \right|^2 (I_\omega)^3, \tag{63}$$

where T is the overall transmission factor, I_ω is the fundamental beam intensity and the subscripts L and G refer to the liquid and glass, respectively, $\Delta\varepsilon$ is the corresponding dielectric constant dispersion and $\Delta\varphi$ is the corresponding phase mismatch (cf. Eq. 28).

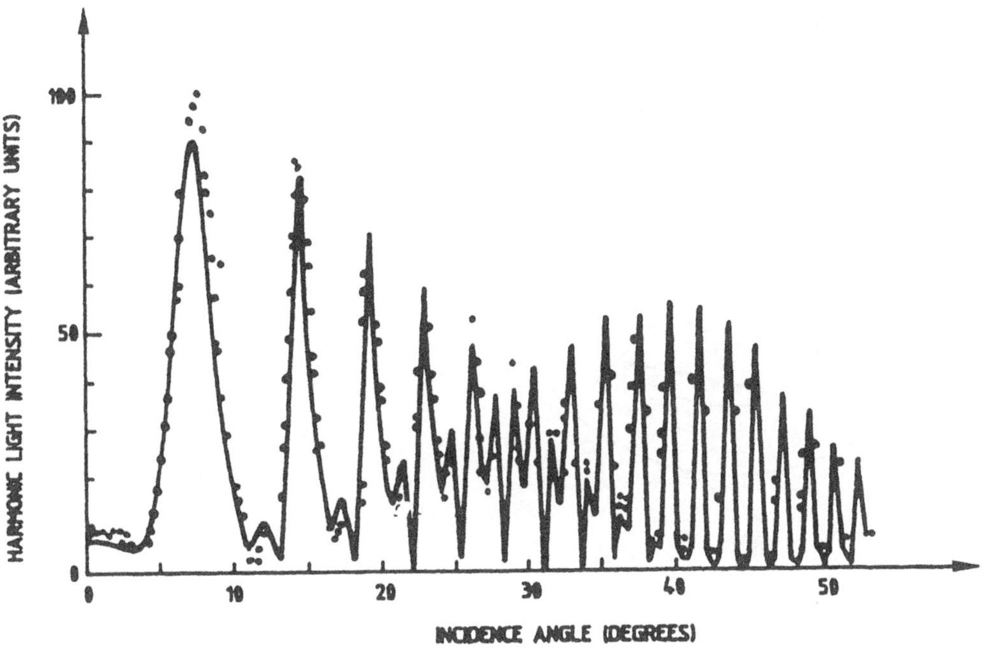

Figure 7

The quite complicated Maker fringe patterns from such a cell, shown in Fig. 7, can be fit to theory to determine all parameters of interest, such as the liquid $\chi^{(3)}$ susceptibility, its phase as well as refractive index dispersion.[13] Furthermore, the data can be directly calibrated against the window material. The technique suffers from a large number of free parameters, which number can be limited with additional THG experiments. The measurements have to be done in a vacuum, although the air contribution can be taken explicitly into account.[13] Figure 8 shows a graphical representation in the complex plane of harmonic fields generated in a plane parallel liquid cell, placed in vacuum (a) and in air (b). The resultant harmonic field is a sum of three vectors in the first case and of five vectors in the second one. Note that because the corresponding phase mismatches vary independently, this variation leads to complicated Maker fringe pattern, as shown in Fig. 7.

In order to decrease the number of free parameters and eliminate the air contribution, Kajzar and Messier[14-15] proposed another technique, based on the principle of harmonic generation by Gaussian laser beams in an infinite medium (cf. Section II.B) and applied it to the input and output windows, which are sufficiently thick (2–3 cm) to fulfill this condition (cf. Fig. 3). The liquid compartment is wedge-shaped and the cell is translated perpendicularly to the beam propagation direction. The harmonic field generated in the input window is given by:

$$E_{3\omega}^{G1} = \pi\{\chi^{(3)}/\Delta\varepsilon\}_G[E_\omega t_\omega]^3, \tag{64}$$

where

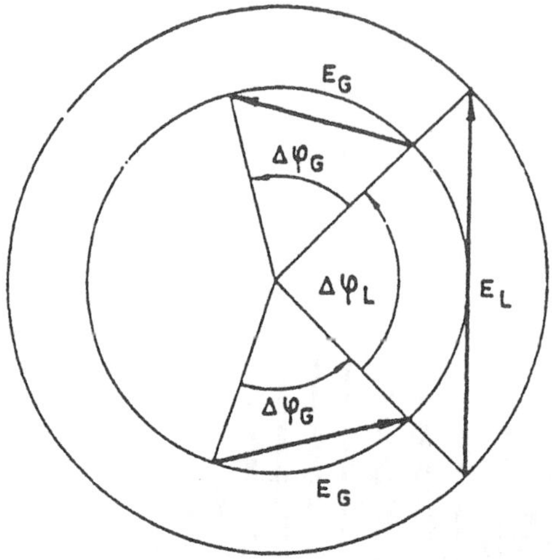

(a)

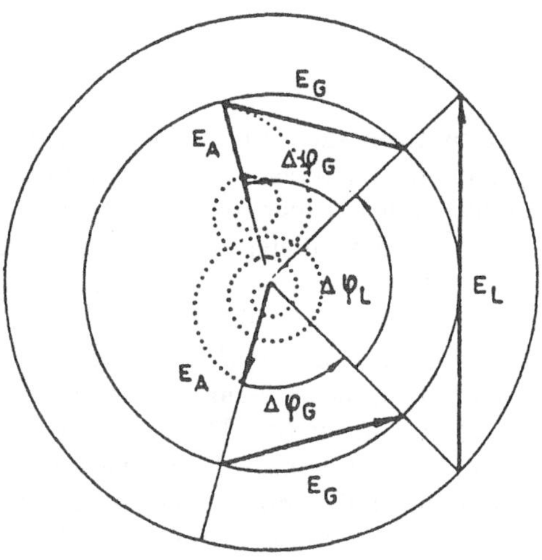

(b)

Figure 8

$$t_\omega = \frac{2}{1 + n_\omega^G} \tag{65}$$

is the transmission coefficient between air and the input window interface. The harmonic field generated in the output window differs from that generated in the input window by the transmission coefficients of the glass-liquid and liquid-glass interfaces as well as by the phase mismatch between fundamental and harmonic fields in the liquid compartment, and is given by:

$$E_{3\omega}^{G2} = -E_{3\omega}^{G1} e^{i\varphi_{\omega}^{L}} (t_{\omega}^{GL} t_{\omega}^{LG})^3.$$

Thus the resultant harmonic field outside the liquid compartment is the sum of the three harmonic fields generated in these three media and is given by (taking account of transmission losses as well as the corresponding phase mismatches):

$$E_{3\omega}^{R} = E_{3\omega}^{G1} e^{i\varphi_{3\omega}^{L}} t_{3\omega}^{GL} t_{3\omega}^{LG} t_{3\omega}^{GO} + E_L [t_{\omega} t_{\omega}^{GL}]^3 t_{3\omega}^{GO} - E_{3\omega}^{G2} e^{i\varphi_{\omega}^{L}} \{t_{\omega} t_{\omega}^{GL} t_{\omega}^{LG}\}^3 t_{3\omega}^{GO}, \tag{66}$$

where

$$t_{\omega(3\omega)}^{ij} = \frac{2n_{\omega(3\omega)}^{j}}{n_{\omega(3\omega)}^{i} + n_{\omega(3\omega)}^{j}}. \tag{67}$$

Consequently, the third harmonic intensity in air, behind the cell, is given by:

$$I_{3\omega} = \frac{64\pi^4}{c^2} \left| \left\{ \frac{\chi^{(3)}}{\Delta\varepsilon} \right\}_G e^{i\varphi_{3\omega}} A_G \right|^2 |t_{3\omega}^{GL} t_{3\omega}^{LG} - e^{i\Delta\varphi} \{t_{\omega}^{LG} t_{\omega}^{GL}\}^3 + \rho\{e^{i\Delta\varphi} - 1\} \{t_{\omega}^{GL} t_{\omega}^{LG}\}^3 t_{3\omega}^{GO}|^2 I_{\omega}^3, \tag{68}$$

where

$$\rho = \{\chi^{(3)} A_L / \Delta\varepsilon\}_L / \{\chi^{(3)} A_G / \Delta\varepsilon\}_G, \tag{69}$$

and A_G and A_L are factors arise from boundary conditions (cf. Eqs. 23 and 31).

The phase mismatch between fundamental and harmonic wave is given by Eq. 28, with the medium thickness depending on translation distance l and the wedge angle δ:

$$l(x) = l_0 + 2x \tan(\delta/2), \tag{70}$$

where l_0 is the initial thickness and x the distance of translation.

Equation 68 leads to regular Maker fringes, as a function of the translation x, as it is shown in Fig. 9. This can be easily visualized when the refractive index is the same for the window and for the liquid. In that case, the harmonic intensity is roughly given by:

$$I_{3\omega} \propto (1 - \rho)^2 I_{\omega}^3 \sin^2 \frac{\Delta\varphi}{2}, \tag{71}$$

and varies as the square of the sine function of half of the phase mismatch between the fundamental and the harmonic waves in the liquid compartment.

The present liquid cell can be also used for THG measurements of absorbing liquids, provided that the absorption is not too strong (see precedent discussion in Section III.C). In such cases, it is preferable to use a thinner liquid compartment design. Absorption at the harmonic (and/or fundamental) frequency manifests itself by non-zero minima (cf. Fig. 9). The analysis that leads to the resultant third harmonic field at the output of the liquid cell is analogous to the plane parallel

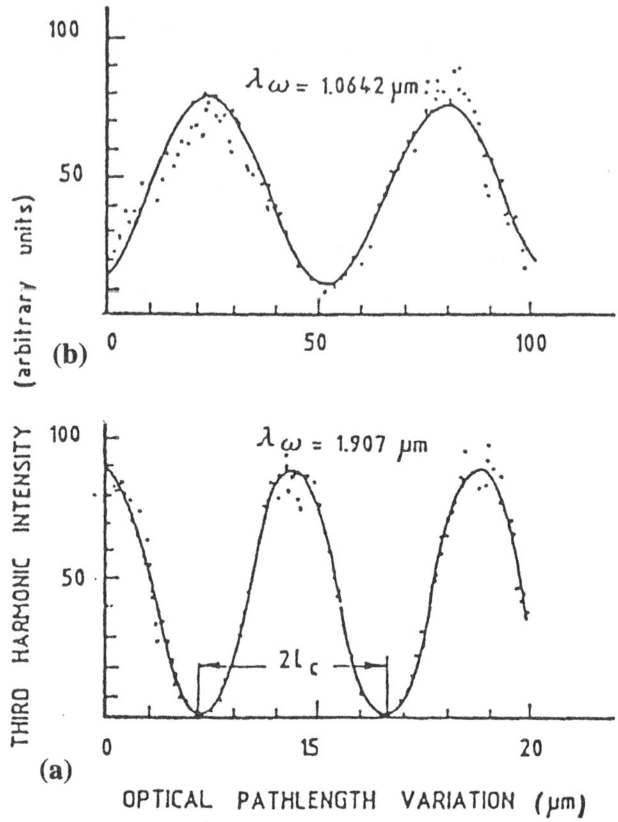

Figure 9

slab, sample in air problem (cf. Fig. 5). Some examples of $x^{(3)}$ susceptibilities by determination for different liquids are listed in Table 1.

For a solution, the THG susceptibility is given by:

$$\chi^{(3)} = N_p F_p \gamma_p + N_s F_s \gamma_s, \tag{72}$$

where N_p and N_s are the number density of solute (p) and solvent (s), respectively, γ's are the corresponding molecular second hyperpolarizabilities, and F's are global local field factors. The liquid $\chi^{(3)}$ susceptibility can be expressed in terms of average molecular density $d(C)$, solvent (M_s) and solute (M_p) molecular masses yielding:

$$\chi^{(3)} = \frac{F(C)d(C)N_A}{1 + C} \left\{ \frac{C\gamma_p}{M_p} + \frac{\gamma_s}{M_s} \right\}, \tag{73}$$

where $F(C)$ is the global local field factor at concentration C and N_A is Avagadro's number.

For practical reasons, it is convenient to calibrate the harmonic intensities of the solution against a pure solvent. The ratio of the amplitude of the harmonic

(*text resumes on p. 806*)

Table 1 Thin Film Real Part ($\chi'^{(3)}(-3\omega;\omega,\omega,\omega)$), Imaginary Part ($\chi''(-3\omega;\omega,\omega,\omega)$) and Modulus of Third Harmonic Susceptibility in Units of 10^{-14}esu

Material	Form	Fund wavelength (μm)	$\chi'^{(3)}$ 10^{-14}esu	$\chi''^{(3)}$ 10^{-14}esu	$\lvert\chi^{(3)}\rvert$ 10^{-14}esu	Remarks	References
			conjugated polymers				
PA	Isotropic film	1.907			$56(16) \times 10^4$	3ph, sp	5
PA	Oriented film	1.907			$1.7(7) \times 10^6$	3ph	6
PA	Oriented film	1.907			2.7×10^6	3ph	7
PDA-AFA	LB film (red)	1.22			$8.0(1) \times 10^2$	2ph, sp	1
PDA-AFA	LB film (blue)	1.35			$5.6(1) \times 10^3$	2ph, sp	2
PDA-AFA	LB film (blue)	1.907			$8.3(8) \times 10^3$	3ph, sp	2
PDA-AFA	Thin film (red)	1.22			$3.8(4) \times 10^3$	2ph, sp	3
PDA-AFA	Thin film (blue)	1.35			$1.2(1) \times 10^4$	2ph, sp	3
pDCH	Isotropic film	1.35			$7.0(7) \times 10^3$	2ph, sp	4
pDCH	Oriented film	1.907			$6.0(6) \times 10^4$	3ph, sp	8, 82
PDA-C_4UC_4	Oriented film	1.907			2.9×10^4 3×10^2	‖ chain, 3ph, sp ⊥ chain	9, 35
PDA-4BCMU	Spun film	1.6			$1.1(1) \times 10^4$	3ph, sp	11, 83
PDA-4BCMU	Oriented film	1.064			$4.9(5) \times 10^3$		12
PDA-4BCMU	Oriented film	1.907			$1.00(7) \times 10^4$		12
PDA-4BCMU	Spun film				$\times 10^4$	3ph, sp	13
PDA-4BCMU	Spun film	1.907			3.6×10^3	3ph, sp	14
PDA-DCHD	Thin single crystal	2.1			1.7×10^4	3ph, sp	31, 89
PDA-BTFP	Thin single crystal	2.1			1.8×10^4	3ph, sp	31
PDA-DFMP	Thin single crystal	2.1			0.45×10^4	5p	

Table 1 Continued

Material	Form	Fund wavelength (μm)	$\chi'^{(3)}$ 10^{-14}esu	$\chi''^{(3)}$ 10^{-14}esu	$\lvert\chi^{(3)}\rvert$ 10^{-14}esu	Remarks	References
			conjugated polymers continued				
PDA-PTS	Thin single crystal	2.1			0.4×10^4	3ph, sp	
PDA R_1—(C—C≡C—C)$_n$—R_2 $R=R_1=R_2$	Cast film	1.907			1.4×10^2	R═$C_6H_4NHCOC_{17}H_{35}$	32
″	Cast film	1.907			5.2×10^2	R═$(CH_2)_3$ $OCON(CH_2)_3$	32
″	Isotropic film	1.907			1.4×10^3	R═$CH_2OCONHC_4H_9$	33
″	Isotropic film	1.907			5.0×10^3	R_1═$C_{12}H_{25}$; R_2═$(CH_2)_8COOH$	32
″	Isotropic film	1.907			1.1×10^3	R═$(CH_2)_3OH$	32
″	Isotropic film	1.907			3.8×10^4	R═$(CH_2)_4OCONHC_3H_7$	34
	Cast film	2.16			$1.9 \times 10^{3(*)}$	R_1═$(CH_2)_4O$ CONH-Ph R_2═pBrPhS, sp	74
PDA-3BCMU	Unoriented film Bioriented film Oriented film	1.91			7×10^3 1.5×10^4 2.1×10^4	3ph	77

PPV	Isotropic film	1.45		1.4×10^4	3ph, sp	39
		1.336	$\phi = 227(15)°$	$1.60(17) \times 10^4$	3ph, sp	84
		1.064		2×10^3	3ph contr	40
PBT	Spun film	1.907		2.9×10^3	3ph, sp	14
PVK/TNF	Spun film	1.907		10^2	saturation value as function of TNF cont	93
PVK/DDQ	Spun film	1.907		2.25×10^2	saturation value as function of DDQ cont	93
PTT	Isotropic film	1.907		$2(2) \times 10^3$	3ph c, sp	10
PDTB	Isotropic film	1.907		$1.1(1) \times 10^3$	3ph c, sp	10
PDT	Spun film	1.907		8.0×10^2	3ph contr.	16
		1.0642	$\phi \sim 280°$	$\sim 1 \times 10^3$	3ph contr.	86
PMOT	″	2.4		4.4×10^4	3ph contr	43
PDDBT	Spun film	1.907		2.9×10^2	3ph	16
C_{60}	Isotropic film	1.064		$8.2(8) \times 10^3$	3ph, sp	23
		1.3		$6.1(6) \times 10^3$	2ph	
		1.907		$3.2(3) \times 10^3$	3ph contr	
		1.33		$2.7(4) \times 10^3$	2ph	96
		1.91		2×10^3	3ph contr	
		2.38		0.4×10^3		
		1.907		2.2×10^3	3ph contr	97
poly-C_{60}	Isotropic film	1.907		4.8×10^3	3ph contr	97
C_{60}:O_2	Isotropic film	1.907		4.6×10^3	3ph contr	97
C_{60}:TPN	Multilayer composite	1.907		6.7×10^2	3ph contr	97
				4.6×10^2	3ph contr	
C_{60}:TPP	Multilayer composite	1.907		7.7×10^2	3ph contr	97
				1.2×10^3	3ph contr	

Table 1 Continued

Material	Form	Fund wavelength (μm)	$\chi'^{(3)}$ 10^{-14}esu	$\chi''^{(3)}$ 10^{-14}esu	$\lvert\chi^{(3)}\rvert$ 10^{-14}esu	Remarks	References
			conjugated polymers continued				
C_{70}	Isotropic film	1.06			$3.0(3) \times 10^3$	2ph, sp	24
		1.42			$9(1) \times 10^3$	3ph	
		1.907			$2.0(2) \times 10^3$	3ph	
Rhodamine R6G		1.0642		$\phi = 241°$	3.4×10^2	3ph contr	85
PMPAZ	Isotropic film	1.5			8.0×10^2		25
PPV	Isotropic film	1.85			7.8×10^2		26
PPBBT	Isotropic film	2.4			$6.0(1.5) \times 10^2$		27
PPBBO	Isotropic film	2.4			8.1×10^2		28
PQ	Isotropic film	2.38			$2.2(4) \times 10^2$		29
PDPIM	Spun film	2.025			4.2×10^2	3ph contr, sp	99
PDPIHB	Spun film	2.025			2.0×10^3	3ph contr, sp	99
PPGN	Spun film	2.025			1.1×10^2	3ph contr, sp	100
PPV	Isotropic film	1.45			1.4×10^4	3ph, sp	39
		1.336		$\phi = 115(15)°$	$1.60(17) \times 10^4$	3ph, sp	84, 98
		1.064			2×10^3	3ph contr	40
PPA	Spun film	1.064		$\phi \sim 250°$	$\sim 1.2 \times 10^2$	R=H	86
PPA-0	Isotropic film	1.064		$\phi = 153(5)°$	$2.7(3) \times 10^2$	R=H	95
PPA-1				$\phi = 131(10)°$	$1.0(5) \times 10^2$	R=H	
PPA-2				$\phi = 170(10)°$	$2.5(5) \times 10^2$	R=H	
PPA-3				$\phi = 200(10)°$	$4.9(2) \times 10^2$	R=CH_5	
PPA-4				$\phi = 227(10)°$	$9.3(1.5) \times 10^2$	R=CH_5	
PPA-5				$\phi = 218(10)°$	$5.4(8) \times 10^2$	R=C_2H_5	
PPA-6				$\phi = 223(10)°$	$7.1(6) \times 10^2$	R=CH_5	

PPA-7				ϕ = 240(10)°	6.4(1) × 10^2	R=C_8H_{17}	
PPA-8				ϕ = 280(10)°	1.3(2) × 10^3	R=$Si(CH_3)_3$	
Phenylene vinylene oligomers OPV-n							
OPV-3	Isotropic film	1.064		ϕ = 102(15)°	1.61(20) × 10^3	sp	84, 98
OPV-4				ϕ = 101(15)°	1.70(20) × 10^3	sp	
OPV-5				ϕ = 123(15)°	1.72(20) × 10^3	sp	
Oligorylene Ory-n							
Ory-2		1.064		ϕ = 263°	4.8		90, 91
Ory-3				ϕ = 153°	8.1		
Ory-4				ϕ = 5°	15.7		
MOPP PPA-0V	"	1.6			1.5 × 10^4	3 ph, sp	39, 81
PAV	"	1.064			7.5 × 10^3		42
PTV	"	1.85			7.8 × 10^2		39
PDSHD (poly-4,4-diubstituted-1,6-heptadiyne	Spun film	1.907			6.5 × 10^3	3 ph, M_w = 9.7 × 10^4	94
					4.1 × 10^3	3 ph, M_w = 2.2 × 10^5	
					2.6 × 10^3	3 ph, M_w = 8.9 × 10^4	
β-carotene	Spun film	1.45			1.5 × 10^2	3ph, sp	61
β-Carotene (in polystyrene matrix)	Spun film	1.418			160(25)	3 ph, sp	54
		1.064		ϕ = 280(10)°	160	3ph contr, sp	88
Polysilane PMPS	Spun film	1.064	1.5 × 10^2	0(0.2)	1.5 × 10^2	3 ph contr	53
		1.064	6.5 × 10^2	3.1 × 10^2	7.2 × 10^2	3 ph contr	101
		1.907	4.2 × 10^2		4.2 × 10^2		
Polysilane PDHS	Spun film	1.064	10.3 × 10^2	4 × 10^2	11 × 10^2	3 ph c	101
		1.907	1.3 × 10^2		1.3 × 10^2		
Poly(di-n-hexyl-germane) PDHG	Spun film	1.064	3.1 × 10^2	1 × 10^2	3.3 × 10^2	3 ph c	101
		1.907	1.1 × 10^2		1.1 × 10^2		

Table 1 Continued

Material	Form	Fund wavelength (μm)	$\chi'^{(3)}$ 10^{-14}esu	$\chi''^{(3)}$ 10^{-14}esu	$\lvert\chi^{(3)}\rvert$ 10^{-14}esu	Remarks	References
			conjugated polymers continued				
Polysilane	LB film	1.2		1043(24)*		*related to substrate (silica)	52
Hemicyanine	LB film	1.064		ϕ = 48(20)°	140(20)	3ph, sp	47
		1.907			90(16)		48
			Functionalized polymers				
PDANB	Spun film	1.907			4.8×10^2	sp	15
PDANS	Spun film	1.907			7.3×10^2	sp	15
PDANDB	Spun film	1.907			4.4×10^2	sp	15
PDANDH	Spun film	1.907			4.4×10^2	sp	15
DRGMMA	Spun film	1.064			4.4×10^2	50 mol%	16
		1.907			1.4×10^3	2 ph, 50 mol%	
						25.9 mol%	
					1.26×10^2		80
DR#1	extrapolated	1.0642	1.38×10^2		1.38×10^2	2 ph contr.	66
	from solid sol	1.907	0.76×10^2		0.76×10^2	3 ph contr.	
DIAMBISA	Spun film	1.064			6.8×10^2	3 ph contr.	16
DIAMBISA	Spun film	1.907			3.8×10^2	2 ph	16
3DRCIN	Spun film	1.064			2.3×10^2	2 ph	16
3DRCIN	Spun film	1.907			6.1×10^2	3 ph contr.	16
PSS-HC	Cast film	1.5			1.7×10^3	3 ph, equimolar solution	36
		1.94			4.6×10^2		

2S-PAA	Spun film	1.9		0.54×10^2	sp	70
2A-PAA	Spun film	1.9		0.73×10^2	sp	70
3A-PAA	Spun film	1.9		2.7×10^2	sp	70
4AS-PAA	Spun film	1.9		1.2×10^3	sp	70
3A-PI	Spun film	1.9		2.8×10^2	sp	70
A-MA	Spun film	1.9		3.2×10^2	45.1 mol%, sp	70
S-MMA	Spun film	1.9		1.7×10^2	24 mol%, sp	70
A-MMA	Spun film	1.9		2.8×10^2	25.9 mol%, sp	70
PSTF	Spun film	1.6		5.9×10^3	63 wt%, 3 ph	78
		2.1		3.1×10^3		
TPN	Isotropic film	1.907		1.1×10^2	3 ph contr	97
TPP	Isotropic film	1.907		6.2×10^2	3 ph contr	97
			Phthalocyanines			
CuPc	Sublimed film	1.064		$3.8(4) \times 10^2$	3 ph contr	17
		1.907		$2.1(2) \times 10^2$	3 ph contr	
VONcR4		2.1		8.6×10^3	3 ph	18
InClNcR4		2.1		1.4×10^2	3 ph	18
Cl-GaPc		1.064		2.5×10^3	3 ph	19
F-AlPc		1.064		5×10^3	3 ph	19
Cl-InPc		1.907		1.3×10^4	3 ph	20
Cl-AlPc		1.907		1.5×10^3	3 ph	21
VOPc		2.1		4.0×10^3	3 ph	21
VOPc		1.907		1.85×10^4	3 ph	30
TiOPc		2.1		5.3×10^3	3 ph	21
CuPc		1.9071		1.5×10^2	3 ph	21
PtPc		1.5		7.6×10^1	3 ph	21
$CuPc(C_6H_{13}S)$		1.907		2.0×10^3	3 ph	21
$VOPc(C_6H_{13}S)$		2.1		3.1×10^3	3 ph	21
Opc (phthalocyana mito ruthenium complex)		1.064	$\phi = 155(10)°$	$3.7(1.5) \times 10^2$		87

Table 1 Continued

Material	Form	Fund wavelength (μm)	$\chi'^{(3)}$ 10^{-14}esu	$\chi''^{(3)}$ 10^{-14}esu	$\|\chi^{(3)}\|$ 10^{-14}esu	Remarks	References
			Phthalocyanines continued				
H_2Pc		1.907			3.0×10^2	3 ph	22
CoPc		1.907			7.5×10^2	3 ph	22
NiPc		1.907			2.3×10^2	3 ph	22
SnPc		1.907			4.0×10^3	3 ph	22
VOPc		1.907			9.3×10^3	3 ph	22
TBH_2Pc		1.907			1.9×10^2	3 ph	22
TBVOPc	Isotropic film	1.907			6.0×10^2	3 ph	22
		1.907			$5.0(8) \times 10^2$	3 ph, phase I	92
		1.907			$10(1) \times 10^2$	3 ph, phase II	
TBVOPc/PMMA	Spun film	1.907			$1.2(1) \times 10^2$	3 ph, phase I	92
39.2 w%					$6.0(7) \times 10^2$	3 ph, phase II	
$[t\text{-}Bu_4PcRu(dib)]_n$		1.064			3.7×10^2	3 ph	22
Multilayers: CuPc/ (NTCDA/CuPc)1.9/ KCl		1.9			$2.3 \times 10^{2(*)}$	3 ph	76
			Metallo-naphthalocyanines				
VNc	Spun film	2.1			8.6×10^3	3 ph	71, 72
CuNc	Spun film	2.1			2×10^2	3 ph	71
ZnNc	Spun film	2.1			1.88×10^2	3 ph	71
PdNc	Spun film	2.1			1.28×10^2	3 ph	71
NiNc	Spun film	2.1			1.59×10^2	3 ph	71

Organometals						
Metallocene derivatives R_1=CHNNH-φ-NO_2 R_2=CHNNH-φ-NO_2	Isotropic film	1.319		3.6×10^2	3 ph	73
R_1=CHNNH-φ-NO_2 R_3=CH_3	Isotropic film	1.319		3.5×10^2	3 ph	73
R=CHNNH-φ-NO_2 with Fe	Isotropic film	1.319		2.2×10^2	3 ph, sp	73
R=CHNNH-φ-NO_2 with Os	Isotropic film	1.319		2.4×10^2	3 ph	73
R=CHNNH-φ-NO_2 with Ru	Isotropic film	1.319		0.9×10^2	3 ph	73
R=CHNNH-φ-NO_2 with Mn	Isotropic film	1.319		1.4×10^2	3 ph	73
R=CHNNH-φ-NO_2	Isotropic film	1.319		0.9×10^2	3 ph	73
Metal-dionedioximes						
$Ni(dmg)_2$	Isotropic film	1.319		0.9×10^2	3 ph	73
$Pd(dmg)_2$	Isotropic film	1.319		3.1×10^2	3 ph	
$Pt(dmg)_2$	Isotropic film	1.319		6.4×10^2	3 ph	
$Ni(dpg)_2$	Isotropic film	1.319		1.0×10^2	3 ph	
$Pd(dpg)_2$	Isotropic film	1.319		1.6×10^2	3 ph	
$Pt(dpg)_2$	Isotropic film	1.319		2.6×10^2	3 ph	
$Ni(niox)_2$	Isotropic film	1.319		1.2×10^2	3 ph	
$Pd(niox)_2$	Isotropic film	1.319		3.2×10^2	3 ph	
$Pt(niox)_2$	Isotropic film	1.319		4.8×10^2	3 ph	
$Ni(bgd)_2$	Isotropic film	1.319		3.1×10^2	3 ph	

Table 1 Continued

Material	Form	Fund wavelength (μm)	$\chi'^{(3)}$ 10^{-14}esu	$\chi''^{(3)}$ 10^{-14}esu	$\lvert\chi^{(3)}\rvert$ 10^{-14}esu	Remarks	References
			Organometals continued				
$Pd(bgd)_2$	Isotropic film	1.319			5.2×10^2	3 ph	
Schiff base complexes:							
Pd-salen	Isotropic film	1.319			0.19×10^2	3 ph	73
Ni-salphen	Isotropic film	1.319			0.24×10^2	3 ph	
Pd-salphen	Isotropic film	1.319			0.22×10^2	3 ph	
Pd-dpsalen	Isotropic film	1.319			0.13×10^2	3 ph	
			Single crystals				
$CaCo_3$ (calcite)	R 3c	1.054			0.021	$\chi^{(3)}$eff	37
		1.054			0.37	$\chi^{(3)}$11	54
		1.054			$0.0(4) \times 10^{-4}$	$\chi^{(3)}$10	38
		1.054			−0.046	$\chi^{(3)}$35	38
PDA-PTS	$P2_1/c$	1.89			$85(50) \times 10^3$	∥ chain, 3 ph,	58
					<1000	⊥ chain	
		2.62			$16(10) \times 10^3$	∥chain	
					<200	⊥ chain	
PDA-TCDU	$P2_1/c$	1.89			$7(5) \times 10^3$	∥ chain, 3 ph	58
					<70	⊥ chain	
		2.62			$37(14) \times 10^2$	∥ chain	
					<40	⊥ chain	

β-BaB_2O_4	R3c	1.054		0.92	$\chi^{(3)}$eff	37
		1.054		0.49(20)	$\chi^{(3)}$eff	55
		1.054		0.76	C_{10}	
		1.054		0.32	C_{eff}	
α-quartz	R3c	1.91		3.81(15)	$\chi^{(3)}$xxxx	59
DEANS		1.9		1.0×10^4	$\chi^{(3)}\parallel$(mol.	50
				91	$\chi^{(3)}\perp$ axis)	
ABN	V_{2v}	1.318		19(6)	$\chi^{(3)}$eff	51
		1.0642		52(10)	$\chi^{(3)}$eff$^{(PM)}$	67
				88(4)	$\chi^{(3)}$yyyy	
				113	$\chi^{(3)}$zzzz	
			Liquid crystals			
MBBA		1.0642		70	$\chi^{(3)}$zzzz, at nematic phase	49
				11	$\chi^{(3)}$xxxx, at nematic phase	
			Amorphous materials			
silica		1.907	2.79	2.79		45
		1.9	2.8	2.8		57
		1.907	1.0	1.0		75
As_2S_3		1.9	220	220		57
$Cd_{0.9}Se_{0.1}$	doped glass	1.9	10	10		57
GeS_3		1.9	100	100		57
$90GeS_3$-$10Tl_2S$		1.9	70	70		57
$90As_2S_3$-10I		1.9	30	30		57
PMMA	1.0642	1.0642	7	7		66
		1.907	8.7	8.7		
BK7		1.907	4.67	4.67		46
		1.907	4.4	4.4		16
		1.0642	5.4	5.4		66

Table 1 Continued

Material	Form	Fund wavelength (μm)	$\chi'^{(3)}$ 10^{-14}esu	$\chi''^{(3)}$ 10^{-14}esu	$\lvert\chi^{(3)}\rvert$ 10^{-14}esu	Remarks	References
Liquids							
Alkanes:							
CH_3—$(CH_2)_{n-2}$—CH_3	n = 6	1.064	4.67		4.67		62
	n = 8	1.064	5.04		5.04		62
	n = 10	1.064	5.35		5.35		62
	n = 16	1.064	6.03		6.03		62
CH_3—$(CH_2)_{n-2}$—CH_2Cl	n = 6	1.064	5.57		5.57		62
	n = 10	1.064	6.10		6.10		62
	n = 12	1.064	6.47		6.47		62
	n = 14	1.064	6.66		6.66		62
CH_3—$(CH_2)_{n-2}$—CH_2Br	n = 6	1.064	7.47		7.47		62
	n = 10	1.064	6.87		6.87		62
	n = 14	1.064	6.93		6.93		62
CH_3—$(CH_2)_{n-2}$—CH_2I	n = 6	1.064	12.94		12.94		62
	n = 8	1.064	11.48		11.48		62
	n = 10	1.064	10.36		10.36		62
Benzene		1.0642	11.98		11.9		63
		1.907	10.64		10.64		63
		1.907	10.1		10.1		65
Carbone tetrachloride		1.0642	7.96		7.96		63
		1.907	7.37		7.37		63
		1.907	6.73		6.73		64
Chloroform		1.0642	7.00		7.00		63
		1.907	5.72		5.72		63
		1.907	5.78		5.78		64
Dichloromethane		1.0642	6.13		6.13		63
		1.907	4.77		4.77		63

Tetrachloroethylene	1.907	8.59	8.59	64
Cyclohexane	1.0642	6.47	6.47	63
	1.907	5.60	5.60	63
N-hexane	1.0642	4.42	4.42	63
	1.907	5.33	5.33	63
Acetone	1.0642	4.69	4.69	63
	1.907	4.10	4.10	63
Methanol	1.0642	2.98	2.98	63
	1.907	3.12	3.12	63
	1.907	2.86	2.86	65
Ethanol	1.0642	3.45	3.45	63
	1.907	4.13	4.13	63
	1.907	3.65	3.65	65
Propanol	1.907	4.11	4.11	65
2-Propanone	1.907	4.41	4.41	65
Dimethylformamide	1.0642	5.57	5.57	63
	1.907	5.05	5.05	63
Water	1.0642	2.80	2.80	63
Nitrobenzene	1.907	16.8	16.8	63
	1.907	13.7	13.7	63
Fluorobenzene	1.907	7.1	7.1	65
Chlorobenzene	1.907	10.6	10.6	65
Bromobenzene	1.907	14.0	14.0	65
Iodobenzene	1.907	23.5	23.5	65
Cyanobenzene	1.907	10.1	10.1	65
Aniline	1.907	16.8	16.8	65
Acetonitrile	1.907	2.54	2.54	65
Tetrahyrofuran	1.907	5.03	5.03	65
Methyl-cyclohexane	1.907	5.93	5.93	65
Pyridine	1.907	10.04	10.04	63
	1.907	9.91	9.91	63
CS_2	1.907	23	23	64
Toluene	1.907	9.81	9.81	65
p-Dioxane	1.064	6.03	6.03	68
		4.76	4.76	

Table 1 Continued

Material	Form	Fund wavelength (μm)	$\chi'^{(3)}$ 10^{-14}esu	$\chi''^{(3)}$ 10^{-14}esu	$\lvert\chi^{(3)}\rvert$ 10^{-14}esu	Remarks	References
			Gases				
He		0.6943	1.9×10^{-5} or 6.5×10^{-5}		1.9×10^{-5} or 6.5×10^{-5}		69
Ne		0.6943	4.8×10^{-5} or 11.6×10^{-5}		4.8×10^{-5} or 11.6×10^{-5}		69
Ar		0.6943	5.9×10^{-4} or 9.5×10^{-4}		5.9×10^{-4} or 9.5×10^{-4}		69
Kr		0.6943	1.65×10^{-3}		1.65×10^{-3}		69
Xe		0.6943	4.6×10^{-3}		4.6×10^{-3}		69
H_2		0.6943	3.8×10^{-4} or 9.4×10^{-4}		3.8×10^{-4} or 9.4×10^{-4}		69
CO_2		0.6943	7.3×10^{-4} or 2.4×10^{-3}		7.3×10^{-4} or 2.4×10^{-3}		69
N_2		0.6943	5×10^{-4} or 1.5×10^{-3}		5×10^{-4} or 1.5×10^{-3}		69
Air		0.6943	4.1×10^{-4} or 15.7×10^{-4}		4.1×10^{-4} or 15.7×10^{-4}		69
		1.055	4.9×10^{-5}		4.9×10^{-5}		79

He	0.6943	1.9×10^{-5} or 6.5×10^{-5}	1.9×10^{-5} or 6.5×10^{-5}	69

Numbers in brackets are standard deviations.
Data are usually calibrated with silica: $\chi^{(3)}(-3\omega;\omega,\omega,\omega) = 2.8 \times 10^{-14}$ esu (Ref. [45]).

(*)calibrated with silica: $\chi^{(3)}(-3\omega;\omega,\omega,\omega) = 1.0 \times 10^{-14}$ esu (Ref. [75])
Φ—phase of $\chi'^{(3)}(-3\omega;\omega,\omega,\omega)$
φ—phenyl ring
2 ph—two-photon resonance enhancement
3 ph—three-photon resonance enhancement
sp—$\chi^{(3)}$ spectrum can be find in original paper

For gases the results are uncertain within a factor of 3 (cf. Ref. [69]).
Abbreviations:

PDA—polydiacetylene
PA—polyacetylene
PPA—polyphenylacetylene
PT—polythiophene
PTT—polythieno(3,2)bithiophene
PDTB—poly[1,4-di-(2-thienyl)benzene)
PBT—poly(3-buthylthiophene)
PDT—poly(3-decylthiophene)
PDDBT—poly(4,4′-didecyl-2,2′)bithiophene
PMPAZ—propylmethylpolyazine
PPV—poly(p-phenylenevinilene)
PBBT—poly(p-phenylenebenzobisthiazole)
PBBT—poly(p-phenylenebenzobisoxazole)
PQ—polyquinoline
PAM—polyazomethines
BTFP—bis-(4-n-butyltetrafluorophenyl)butadiyne
DFMP—bis-(2,2′,5,5′-tetraxistrifluoromethylphenyl)butadiyne
PTS—poly-toluene sulphonate
PPV—poly (p-phenylene vinylene)
PAV—poly(arylene vinylene)
MOPPV—poly(2.5-dimethoxy-p-phenylene vinylene)
PTV—poly(2.5-thienylene vinylene)
PTS—poly-toluene sulphonate
TCDU—bis-(phenyl-urethane) of 5,7-dodecdiyne-1,2-diol

BCMU—butoxycarbonylurethane
TPN—5,6,11,12-tetraphenylnaphthacene (rubrene)
TPP—5,10,15,20-tetraphenyl-21H,23H-porphine
pDCH—poly bis[1,6-(9-carbazolyl)2-4 hexadiyne
PDPIM—poly-4,4′-diphenylimine methine
PDPIMHB—poly-4,4′-diphenylimine p-heptyloxybenzylidene
PPGN—polypernigraniline
PMMA—poly-methyl-methacrylate
TBVOPc—tert-butyl-substituted vanadyl phthalocyanine
PDANB, PDANS, PDANDB, PDANDB are functionalized polymers with 4-dialkylamino-nitrobenzene, 4-dialkylamino-4′-nitrostilbene, 4-dialkylamino-4′-diphenylbutadiene, 4-dialkylamino-4′-diphenylhexatriene as side groups
PSTF—fluorinated copolymer of polyurethane with a tris-azo dye
PVK—poly-n-vinyl carbazole
TNF—2,4,7-trinitrofluorenone
DDQ—2,3-dichloro-5,6-dicyano-1,4-benzoquinone
DRGMMA—copolymer of methyl-methacrylate (MMA) with Disperse Red (DR#1)
DIAM-BISA is a thermally crosslinked polymer of 2,4-diamino-4′-nitroazobenzene with an epoxy matrix: diglycidylether of Bisphenol A, 3
DR-CIN is a photo-crosslinked polymer of 4-N,N′-diethylamino-2-cinnamoyloxy-4″-nitrodiazobenzene with polyvinylcinnamate matrix
TB denotes tert butyl group
PSS-HC—hemicyanine in poly-styrenesulfonic acid
A-MA—copolymer of an azo dye with methacrylate
A-MMA—copolymer of an azo dye with methyl methacrylate
S-MMA—copolymer of stillbene dye with methyl methacrylate
A-PAA—copolymer of an azo dye with polyamic acid
S-PAA—copolymer of stillbene dye with polyamic acid
A-PI—copolymer of an azo dye with polyimide
DEANS—diethylamino-nitro-stilbene
ABN—4-amino-benzonitrile

(1) Kajzar, F., and Messier, J., *Polymer J.*, *19*, 275 (1987).
(2) Kajzar, F., and Messier, J., *Thin Sol. Films*, *132*, 11 (1985).
(3) Chollet, P. A., Kajzar, F., Messier, J., in *Nonlinear Optical and Electroactive Polymers* (P. N. Prasad and D. R. Ulrich, eds.), Plenum Press, New York, 1988, p. 121.
(4) Le Moigne, J., Thierry, A., Chollet, P. A., Kajzar, F., and Messier, J., *J. Chem. Phys.*, *88*, 6647 (1988).
(5) Fann, W. S., Benson, S., Madey, J. M. J., Etemad, S., Baker, G. L., and Kajzar, F., *Phys. Rev. Lett.*, *62*, 1492 (1989).
(6) Drury, M. R., *Solid State Commun.*, *68*, 417 (1988).
(7) Krausch, F., Winter, E., and Leising, G., *Phys. Rev. B*, *39*, 3701 (1989).

(8) Le Moigne, J., Thierry, A., and Kajzar, F., *Proc. SPIE*, *1125*, 9 (1990).
(9) Kanetake, T., Ishikawa, K., Hasegawa, T., Koda, T., Takeda, K., Hasegawa, M., Kubodera, K., and Kobayashi, M., *Appl. Phys. Lett.*, *54*, 2287 (1989).
(10) Kajzar, F., Ruani, G., Taliani, C., and Zamboni, R., *Synth. Metals*, *37*, 223 (1990); *Nonl. Optics*, *1*, 285 (1991).
(11) Torruellas, W. E., Rochford, K. B., Zanoni, R., Aramaki, S., and Stegeman, G. I., *Opt. Commun.*, *82*, 94 (1991).
(12) Hsu, C. C., Kawabe, Y., Ho, Z. Z., Peyghambarian, N., Polky, J. N., Krug, W., and Miao, E., *J. Appl. Phys.*, *67*, 7199 (1990).
(13) Townsend, P. D., Fann, W.-S., Etemad, S., Baker, G. L., Soos, Z. G., and McWilliams, P. C. M., *Chem. Phys. Lett.*, *62*, 1492 (1991).
(14) Torruellas, W. E., Neher, D., Zanoni, R., Stegeman, G. I., Kajzar, F., and Leclerc, M., *Chem. Phys. Lett.*, *175*, 11 (1990).
(15) Torruellas, W. E., Zanoni, R., Marques, M. B., Stegeman, G. I., Möhlmann, G. R., Erdhuisen, W. P., and Horsthuis, W. H. G., *Chem. Phys. Lett.*, *175*, 267 (1990).
(16) Kajzar, F., and Zagorska, M., *Nonl. Opt.*, *6*, 181 (1993).
(17) Chollet, P. A., Kajzar, F., and Le Moigne, J., *J. Mol. Eng.*, *1*, 35 (1991); see also *Nonlinear Optical Materials III* (P. Günter, ed.), Proceed. SPIE, *1273*, 87 (1990).
(18) Nalwa, H. S., Kobayashi, S., and Kakuta, A., *Nonl. Opt.*, *6*, 169 (1993).
(19) Ho, Z. Z., Yu, C. Y., and Hetherington, W. M., III, *J. Appl. Phys.*, *62*, 716 (1987).
(20) Shirk, J. S., Lindle, J. R., Bartoli, F. J., and Boyle, M. E., *J. Phys. Chem.*, *96*, 5847 (1992).
(21) Matsuda, H., Okada, S., Masaki, A., Nakanishi, H., Suda, Y., Shigehara, K., and Yamada, Y., *Proc. SPIE*, vol. *1337*, Bellingham, 1990, p. 105.
(22) Hosoda, M., Wada, T., Yamada, Y., Garito, A. F., and Sasabe, H., *Nonl. Opt.*, *3*, 183 (1992).
(23) Kajzar, F., Taliani, C., Danieli, R., Rossini, S., and Zamboni, R., *Chem. Phys. Lett.*, *217*, 418 (1994).
(24) Kajzar, F., Taliani, C., Danieli, R., Rossini, S., and Zamboni, R., *Phys. Rev. Lett.*, *73*, 1617 (1994).
(25) Nalwa, H. S., Hamada, T., Kakuta, A., and Mukoh, A., *Nonl. Opt.*, *6*, 155 (1993).
(26) Kaino, T., Kubodera, K., Tomaru, S., Kurihara, T., Saito, S., Tsutsui, T., and Tokito, S., *Electron. Lett.*, *23*, 1095 (1987).
(27) Vanherzeele, H., Meth, J. S., Jenekhe, S. A., and Roberts, M. F., *J. Opt. Soc. Am. B*, *9*, 524 (1992).
(28) Agrawal, A. K., Jenekhe, S. A., Vanherzeele, H., and Meth, J. S., *Chem. Mat.*, *3*, 765 (1991).
(29) Jenekhe, S. A., Osaheni, J. A., Vanherzeele, H., and Meth, J. S., *Chem. Mat.*, *4*, 683 (1992).
(30) Wada, T., Yamada, S., Matsuoka, Y., Grossman, C. H., Shigehara, K., Sasabe, H., Yamada, A., and Garito, A. F., Frontiers Materials for Organic Nonlinear Optics, in *Nonlinear Optics of Organics and Semiconductors* (T. Kobayashi, ed.), Springer Proceedings in Physics, vol. 36, Berlin, 1989, p. 292.
(31) Nakanishi, H., Matsuda, H., Okada, S., and Kato, M., Nonlinear Optical Properties of Polydiacetylenes with π-Conjugation Between the Main Chain and the Substituents, in *Nonlinear Optics of Organics and Semiconductors* (T. Kobayashi, ed.), Springer Proceedings in Physics, vol. 36, Berlin, 1989, p. 155. See also (89) Nakanishi, H., Matsuda, H., Okada, S., and Kato, M., *Polymers for Advanced Technology*, *1*, 75 (1990).
(32) Kubodera, K., and Kaino, T., $\chi^{(3)}$ Evaluation of Organic Nonlinear Optical Materials by THG Measurements, in *Nonlinear Optics of Organics and Semiconductors* (T. Kobayashi, ed.), Springer Proceedings in Physics, vol. 36, Berlin, 1989, p. 163.
(33) Tomaru, S., Kubodera, K., Zembutsu, S., Takeda, K., and Hasegawa, M., *Electron. Lett.*, *23*, 595 (1987).
(34) Tomaru, S., Kubodera, K., Kurihara, T., and Zembutsu, S., *Japan J. Appl. Phys.*, *26*, L1657 (1987).
(35) Kanetake, T., Ishikawa, K., Koda, T., Takeda, K., and Kubodera, K., Highly Oriented Polydiacetylene Films for Possible Nonlinear Optical (NLO) Application: Method of Fabrication, Morphology and Some NLO Properties, in *Nonlinear Optics of Organics and Semiconductors* (T. Kobayashi, ed.), Springer Proceedings in Physics, vol. 36, Berlin, 1989, p. 184.

(36) Matsuda, H., Okada, S., Nishiyama, T., Nakanishi, H., and Kato, M., Nonlinear Optical Properties of Ion-Dye Complexes, in *Nonlinear Optics of Organics and Semiconductors* (T. Kobayashi, ed.), Springer Proceedings in Physics, vol. 36, Berlin, 1989, p. 188.

(37) Penzkofer, A., Qiu, P., and Ossig, F., Picosecond Third Harmonic Generation in β-BaB_2O_4, in *Nonlinear Optics of Organics and Semiconductors* (T. Kobayashi, ed.), Springer Proceedings in Physics, vol. 36, Berlin, 1989, p. 312.

(38) Penzkofer, A., Qiu, P., and Ossig, F., *Appl. Phys. B*, *47*, 71 (1988).

(39) Kaino, T., and Kurihara, T., Third-Order Optical Properties of Polymeric Materials, in *Polymers for Lightwave and Integrated Optics: Technology and Applications* (L. A. Hornac, ed.), Marcel Dekker, Inc., New York, 1992, p. 50.

(40) McBranch, D., Sinclair, M., Heeger, A. J., Patil, A. O., Shi, S., Askari, S., and Wudl, F., *Synth. Mtals*, *29*, E85 (1989).

(41) Kaino, T., Kurihara, T., Mori, Y., Murata, H., Tsutsui, T., and Saito, S., *Chem. Phys. Lett.*, *183*, 534 (1991).

(42) Bradley, D. D. C., *Makromol. Chem. Makromol. Symp.*, *37*, 247 (1990).

(43) Houlding, V. H., Nahata, A., Yardley, T., and Elsebauer, R. L., *Chem. Mater.*, *2*, 169 (1990).

(44) Charra, F., Messier, J., Sentein, C., Pron, A., and Zagorska, M., Influence of Conformation on Two-Photon Spectra of Polyalkylthiophene, in *Organic Molecules for Nonlinear Optics and Photonics* (J. Messier, F. Kajzar and P. Prasad, eds.), NATO ASI Series, vol. E 194, Kluwer Acad. Publ., Dordrecht, 1991, p. 263.

(45) Meredith, G. R., Buchalter, B., and Hanzlik, C., *J. Chem. Phys.*, *78*, 1533 (1983).

(46) Buchalter, B., and Meredith, G. R., *Appl. Opt.*, *21*, 3221 (1982).

(47) Kajzar, F., Girling, I. R., and Peterson, I. R., *Electron. Lett.*, *22*, 1230 (1986).

(48) Kajzar, F., Girling, I. R., and Peterson, I. R., *Thin Sol. Films*, *160*, 209 (1987).

(49) Garito, A. F., and Wong, K. Y., *Polymer J.*, *19*, 51 (1987).

(50) Kobayashi, H., and Kubodera, K., *J. Appl. Phys.*, *69*, 3807 (1991).

(51) Pierre, M., Ehooman, J. B., Neerrgaar, G. R., Coutaz, J. L., and Ledoux, I., *Opt. Commun.*, *98*, 209 (1993).

(52) Mittler-Neher, S., Neher, D., Stegeman, G. I., Embs, F. W., and Wegner, G., *Chem. Phys.*, *161*, 289 (1992).

(53) Kajzar, F., Messier, J., and Rosilio, C., *J. Appl. Phys.*, *60*, 3040 (1986).

(54) Thalhammer, M., and Penzkofer, A., *Appl. Phys. B*, *32*, 137 (1983).

(55) Qiu, P., and Penzkofer, A., *Appl. Phys. B*, *45*, 225 (1988).

(56) Tomov, I. V., Van Wonterghem, B., and Rentzepis, P. M., *Appl. Optics*, *31*, 4172 (1992).

(57) Nasu, H., Ibara, Y., and Kubodera, K., *J. Non-Crystal. Solids*, *110*, 229 (1989).

(58) Sauteret, C., Hermann, J. P., Frey, R., Pradere, F., Ducuing, J., Baughman, R. H., and Chance, R. R., *Phys. Rev. Lett.*, *36*, 956 (1976).

(59) Meredith, G. R., *Phys. Rev. B*, *24*, 5522 (1981).

(60) Van Beek, J. B., Kajzar, F., and Albrecht, A. C., *J. Chem. Phys.*, *95*, 6400 (1991).

(61) Van Beek, J. B., Kajzar, F., and Albrecht, A. C., *J. Chem. Phys.*, *95*, 64000 (1991).

(62) Kajzar, F., and Messier, J., *J. Opt. Soc. Am. B*, *4*, 1040 (1987).

(63) Kajzar, F., and Messier, J., *Phys. Rev. A*, *32*, 2352 (1985).

(64) Meredith, G. R., Buchalter, B., and Hanzlik, C., *J. Chem. Phys.*, *78*, 1533 (1983).

(65) Meredith, G. R., Buchalter, B., and Hanzlik, C., *J. Chem. Phys.*, *78*, 1543 (1983).

(66) Morichere, D., Dumont, M., Levy, Y., Gadret, G., and Kajzar, F., Nonlinear Properties of Poled Polymer Films: SHG and Electrooptic Measurements. Proceed. SPIE, *1560*, 214 (1991).

(67) Moog, L., Thesis, Université Joseph Fourrier, December 1995.

(68) Burland, D. M., Walsh, C. A., Kajzar, F., and Sentein, C., *J. Opt. Soc. Am. B*, *8*, 2269 (1991).
(69) Ward, J. F., and New, G. H. C., *Phys. Rev.*, *185*, 57 (1969).
(70) Matsumoto, S., Kurihara, T., Kubodera, N., and Kaino, T., *Mol. Cryst. Liq. Crystals*, *182A*, 115 (1990).
(71) Nalwa, H. S., Kakuta, A., and Mukoh, A., *Chem. Phys. Lett.*, *203*, 109 (1993).
(72) Nalwa, H. S., Kakuta, A., and Mukoh, A., *J. Phys. Chem.*, *97*, 1097 (1993).
(73) Kamata, T., Fukaya, T., Kodzasa, T., Matsuda, H., Mizukami, F., Tachiya, M., Ishikawa, R., Uchida, T., and Yamazaki, Y., *Nonl. Optics*, *13*, 31 (1995).
(74) Shimada, S., Matsuda, H., Masaki, A., Okada, S., and Nakanishi, H., *Nonl. Optics*, *13*, 57 (1995).
(75) Mito, A., Hagimoto, K., and Takahashi, C., *Nonl. Optics*, *13*, 3 (1995).
(76) Imanishi, Y., Hattori, S., Hamada, T., Ishihara, S., Nalwa, H. S., and Kakuta, A., *Nonl. Optics*, *13*, 63 (1995).
(77) Hattori, Y., Mizoguchi, A., Uemiya, T., and Tanaka, G., *Nonl. Optics*, *13*, 73 (1995).
(78) Kaino, T., Tomaru, S., Kurihara, T., and Asobe, M., *Nonl. Optics*, *13*, 203 (1995).
(79) Thalhammer, M., and Penzkofer, A., *Appl. Phys. B*, *32*, 137 (1983).
(80) Matsumoto, S., Kubodera, K., Kurihara, T., and Kaino, T., *Appl. Phys. Lett.*, *51*, 1 (1987).
(81) Kaino, T., Kobayshi, H., Kubodera, K., Kurihara, T., Saito, S., Tsutsui, T., and Tokito, S., *Appl. Phys. Lett.*, *54*, 1619 (1989).
(82) Le Moigne, J., Kajzar, F., and Thierry, A., *Macromolecules*, *24*, 2622 (1991).
(83) Torruellas, W., Ph.D. Thesis, University of Arizona, 1991.
(84) Bubeck, C., *Nonl. Opt.*, *19*, 13 (1995).
(85) Grund, A., Mathy, A., Kaltbeitzel, A., Neher, D., Bubeck, C., and Wegner, G., in *Organic Materials for Nonlinear Optics II* (R. A. Hann and D. Bloor, eds.), Royal Soc. of Chem., Cambridge, 1991, p. 288.
(86) Neher, D., Wolf, A., Leclerc, M., Kaltbeitzel, A., Bubeck, C., and Wegner, G., *Synth. Met.*, *37*, 249 (1990).
(87) Grund, A., Kaltbeitzel, A., Mathy, A., Schwarz, R., Bubeck, C., Vermehren, P., and Hanack, M., *J. Phys. Chem.*, *96*, 7450 (1992).
(88) Aramaki, S., Torruellas, W., Zanoni, R., and Stegeman, G. I., *Opt. Commun.*, *85*, 527 (1991).
(89) Kistenmacher, A., Soczka, T., Baier, U., Ueberhofen, K., Bubeck, C., and Müllen, K., *Acta Polymer.*, *45*, 228 (1994).
(90) Schrader, S., Koch, K. H., Mathy, S., Bubeck, C., Müllen, K., and Wegner, G., *Synth. Met.*, *41–43*, 3223 (1994).
(91) Schrader, S., Koch, K. H., Mathy, S., Bubeck, C., Müllen, K., and Wegner, G., *Prog. Colloid. Polym. Sc.*, *85*, 143 (1994).
(92) Hosoda, M., Wada, T., Yamamoto, T., Kaneko, A., Garito, A. F., and Sasabe, H., *Japan J. Appl. Phys.*, *31*, 1071 (1992).
(93) Park, S. Y., Wada, T., and Sasabe, H., *Mol. Cryst. Liq. Cryst.*, *227*, 151 (1993).
(94) Park, S. Y., Cho, H. N., Kim, N., Park, J.-W., Jin, S.-H., Choi, S.-K., Wada, T., and Sasabe, H., *Appl. Phys. Lett.*, *65*, 289 (1994).
(95) Neher, D., Kaltbeitzel, A., Wolf, A., Bubeck, C., and Wegner, G., *J. Phys. D, Appl. Phys.*, *24*, 1193, 249 (1991).
(96) Meth, J. S., Vanherzeele, H., and Wang, Y., *Chem. Phys. Lett.*, *197*, 26 (1991).
(97) Okada-Shudo, Y., Kajzar, F., Meritt, C. D., and Kafafi, Z. H., Nonlinear Optical Properties of Multilayered Structures and Composites of C with Electron Donors, in *Fullerenes and Photonics II* (Z. Kafafi, ed.), Proc. SPIE, vol. 2854, in print.
(98) Mathy, A., Uebrhofen, K., Schenk, R., Gregorius, H., Garay, R., Müllen, K., and Bubeck, C., *Phys. Rev. B*, *53*, 4367 (1996).
(99) Chen, W.-C., Jenekhe, S. A., Meth, J. S., and Vanherzeele, H., *J. Pol. Sc. Part B, Polymer Physics*, *32*, 195 (1994).
(100) Osaheni, J. A., Jenekhe, S. A., Vanherzeele, H., Meth, J. S., Sun, Y., and MacDiarmid, A. G., *J. Phys. Chem.*, *96*, 2830 (1992).
(101) Baumert, J. C., et al. *Appl. Phys. Lett.*, *53*, 1147 (1988).

intensity from a solution with mass concentration C ($= m_p/m_s$) to the amplitude of the harmonic intensity generated in the pure solvent is:

$$I_{3\omega}^{N}(C) = I_{3\omega}(C)/I_{3\omega}(C = 0) = \frac{1}{\beta^2}\left(\left|\left\{A + B(C)\left[\frac{C\gamma_p'}{M_p} + \frac{\gamma_s}{M_s}\right]\right\}\right|^2 + C^2\left\{\frac{\gamma_p''}{M_p}B(C)\right\}^2\right) \Big/ \left\{A + B(0)\frac{\gamma_s}{M_s}\right\}^2, \tag{74}$$

where

$$A = t_{3\omega}^{GL}t_{3\omega}^{LG} - \alpha\beta e^{i\Delta\varphi}\{t_{\omega}^{GL}t_{\omega}^{LG}\}^3, \tag{75}$$

and

$$B(C) = \frac{F(C)d(C)N_A}{(1 + C)\chi_G^{(3)}}[\alpha\beta e^{i\Delta\varphi} - 1](t_{\omega}^{LG})^3 t_{3\omega}^{LG}\frac{(\Delta\varepsilon)_G}{(\Delta\varepsilon)_L}. \tag{76}$$

The coefficients α and β in Eqs. 75–76 take into account absorption at fundamental and harmonic frequencies, respectively, and are given by:

$$\alpha = e^{3\omega\kappa_{\omega}l/c} \tag{77}$$

and

$$\beta = e^{3\omega\kappa_{3\omega}l/c}. \tag{78}$$

In Eq. 74 we have introduced the complex molecular second hyperpolarizability of the solute $\gamma_p = \gamma_p' + i\gamma_p''$. The concentration dependence of the normalized third harmonic intensity can be used to obtain both the real (γ_p') and the imaginary part (γ_p'') of the solute second hyperpolarizability γ_p as it is seen in Fig. 10 (see also Ref. [18]). Usually the solute hyperpolarizability is significantly larger than that of the solvent. A negative γ_p' will result in a decrease of normalized harmonic intensity, with increased concentration. If $\gamma_p'' = 0$, the THG intensity equals zero when the solute and solvent contributions cancel. If γ_p' is sufficiently large so that the solute contribution to $\chi^{(3)}$ is larger than the solvent contribution before the solubility limit is reached, the third harmonic intensity then increases as the concentration is increased (cf. Fig. 10). This slope increase scales with γ_p''. Note that a decrease of normalized harmonic intensities with concentration may also be observed for absorbing liquids. The sign of γ_p'' in this case is generally undetermined because the solvent contribution is too small to act as an internal reference. In Table 2 we have listed γ values for selected molecules, as determined by THG measurements in solution. In some cases, both real and imaginary parts of γ were determined.

C. Thin Films

For the sake of simplicity, we consider the THG experiments in a vacuum. While the air's contribution can be accounted for using the procedure described by Kajzar and Messier,[13] for practical reasons it is more convenient to do experiments in vacuum.

Usually the thin film is supported by a substrate, which contributes also to the measured THG intensities. Although, again, the substrate $\chi^{(3)}$ susceptibility is often much lower than that for the thin film material, the substrate contribution to THG

(*text resumes on p. 820*)

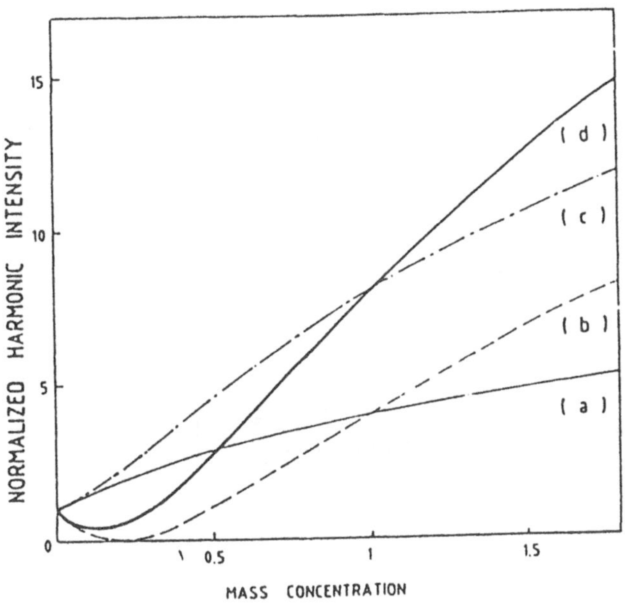
NORMALIZED HARMONIC INTENSITY
15
10
5
0
0.5
1
1.5
MASS CONCENTRATION
(d)
(c)
(b)
(a)

(a)

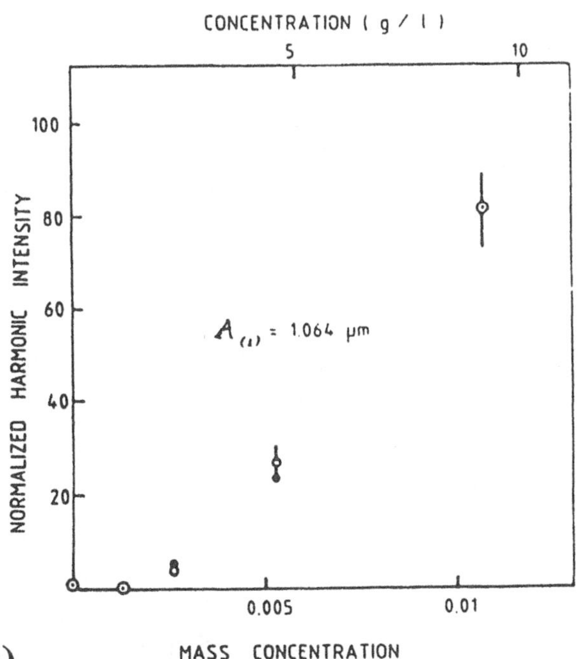
CONCENTRATION (g / l)
5
10
NORMALIZED HARMONIC INTENSITY
100
80
60
40
20
0.005
0.01
MASS CONCENTRATION

(b)

Figure 10

Table 2 Real Part ($\gamma'(-3\omega;\omega,\omega,\omega)$) Imaginary Part ($\gamma''(-3\omega;\omega,\omega,\omega)$) and Modulus of the Second Order Molecular Hyperpolarizability in Units of 10^{-36} esu, as Determined by THG

Molecule	Fundamental wavelength (μm)	γ' in 10^{-36} esu	γ'' in 10^{-36} esu	$\lvert\gamma\rvert$ in 10^{-36} esu	Remarks	Ref.
H_2Thp	1.34	−1000		1000	Fig. 1a	1
MnThp	1.34	−1500		1500	"	1
MnThp	1.907	−3000		3000	"	1
FeThp	1.34	−3100		3100	"	1
FeThp	1.907	−4600		4600	"	1
CoThp	1.34	−6100		6100	"	1
CoThp	1.907	−11000		11000	"	1
NiThp	1.34	−1500		1500	"	1
NiThp	1.907	−5000		5000	"	1
CuThp	1.34	−1100		1100	"	1
CuThp	1.907	−1000		1000	"	1
ZnThp	1.34	−900		900	"	1
$RgPcH_2$	1.34	−2100	±9000	9200	Fig. 1b	2
RgPcCu	1.34	−32000	±23000	39400	Fig. 1B	2
RgPcNi	1.34	−15400	±19300	24700	Fig. 1b	2
RgPcCo	1.34	−32200	±29900	43900	Fig. 1b	2
$[Pt(L)_2DEB]_n$	1.907	890	1300	1575	in THF	20
$[Pt(L)_2DEB]_n$	1.907	328	949	1004	in Benzene	20
$[Pt(L)_2DEX]_n$	1.907	471	464	661	in Benzene	20
$[Pt(L)_2(DEX)_2]_n$	1.907	1552	344	1590	in Benzene	20
$[Pt(L)_2C{\equiv}C{-}C{\equiv}C{-}]_n$	1.907	743	1296	1494	in THF	20
$[Pt(L)_2(C{\equiv}C{-}C{\equiv}C{-}H)_2]_n$	1.907	20	108	110	in THF	20
$trans[Pt(L)2(DEB)_2]_n$	1.907	245	221	3305	in Benzene	20
DEB	1.907	12	0	12	in THF	20

Thiophene oligomers	1.907			6	N = 1	3
	1.907			15	N = 2	3
	1.907			41	N = 3	3
	1.907			142	N = 4	3
	1.907			249	N = 5	3
	1.907			320	N = 6	3
	1.907			450	N = 7	3
	1.907			620	N = 8	3
Merocyanine	1.0642	−2300	±1600		in Ethanol	4
	1.907	800	0	800	″	4
	1.0642	−600	±2100		in Propanol	4
	1.907	400	0	400	″	4
PDA-TS-12	1.0642	−1700	±1100		in DMF	5
	1.907	1200	0	1200	″	5
DTDMP	1.064	−1.3(1)	1.5(2)	2.0(2)	in CH_2Cl_2	23
	1.907	0.40(4)		0.40(4)		
Polyene oligomers	1.064	$-0.31(5) \times 10^3$	0	$0.31(5) \times 10^3$	N = 1, γF	23
	1.907	$0.3(1) \times 10^3$	0	$0.3(1) \times 10^3$	product	
	1.064	$-13(1.5) \times 10^3$		$13(1.5) \times 10^3$	N = 4, ϕF	
	1.907	$4.1(5) \times 10^3$		$4.1(5) \times 10^3$	product	
	1.064	$24.9(2.5) \times 10^3$	$29.3(1.0) \times 1$	$38.5(2.7) \times 10^3$	N = 6, γF	
	1.907	$15(1.5) \times 10^3$	0^3	$15(1.5) \times 10^3$	product	
	1.064	$60.6(6.3) \times 10^3$		$60.6(6.3) \times 10^3$	N = 8, γF	
	1.907	$56.6(6.0) \times 10^3$		$56.6(6.0) \times 10^3$	product	

Table 2 Continued

Molecule	Fundamental wavelength (μm)	γ' in 10^{-36} esu	γ'' in 10^{-36} esu	$\|\gamma\|$ in 10^{-36} esu	Remarks	Ref.
Polyene oligomers	1.907	$8.1(1.5) \times 10^3$		$8.1(1.5) \times 10^3$	N = 28, THF	14
		$1.41(0.2) \times 10^4$		$1.41(0.2) \times 10^4$	N = 39	
		$2.47(0.2) \times 10^4$		$2.47(0.2) \times 10^4$	N = 50	
		$5.53(0.5) \times 10^4$		$5.53(0.5) \times 10^4$	N = 68	
		$1.025(0.15) \times 10^5$		$1.025(0.15) \times 10^5$	N = 88	
		$2.731(0.3) \times 10^5$		$2.731(0.3) \times 10^5$	N = 152	
		$3.794(0.5) \times 10^5$		$3.794(0.5) \times 10^5$	N = 240	
Asymmetric D-A	1.907	$2.0(5) \times 10^2$		$2.0(5) \times 10^2$	N = 2.5[a]	15
		$1.2(.12) \times 10^3$		$1.2(.12) \times 10^3$	N = 5.5[a]	
		$4.0(2) \times 10^3$		$4.0(2) \times 10^3$	N = 7.5[a]	
		$7.6(6) \times 10^3$		$7.6(6) \times 10^3$	N = 9.5[a]	
Symmetric A-A	1.34	$-7(1) \times 10^3$		$7(1) \times 10^3$	N = 10	
Symmetric D-D	1.34	$-4(.6) \times 10^4$		$4(.6) \times 10^4$	N = 10	
Hexatriene	1.907	1.04(16)	0	1.04	in Dioxane, sp	6
						19
						sp
β-Carotene	2.148	51.8(1.3)	0	51.8(1.3)	″	6
C_{60}	1.064	$-3.9(2.3) \times 10^3$	2.6(3.0)	$4.7(3.8) \times 10^3$	in Toluene	13
	1.5	$-1.3(0.8) \times 10^3$	$\times 10^3$	$1.3(0.8) \times 10^3$		
	2.0	$1.6(0.8) \times 10^3$	0	$1.6(0.8) \times 10^3$		
			0			
C_{70}	1.064	$-7.6(2.8) \times 10^3$	$57(4) \times 10^3$	$57.5(4.9) \times 10^3$	in Toluene	13
	1.5	$2.6(4.7) \times 10^3$	$22(15) \times 10^3$	$22(16.7) \times 10^3$		
	2.0	$-3.8(1.1) \times 10^3$	0	$3.8(1.1) \times 10^3$		
paranitroaniline (p-NA)	1.907	19		19	in Dioxane	10,11
MONS	1.0642			60	in Dioxane	12
OOCB	1.0642			79	in Dioxane	12

MNA	1.0642	−85(10)	0	−85(10)	in Chloroform	16
	1.907	33(3)		33(3)	in Chloroform	
	1.907	24(3)		24(3)	in Dioxane	
Monosubsituted benzenes						
φ—X, X = H	1.907	3.90		3.9	neat	18
X = CH_3		4.6		4.6	neat	
X = OH		5.0		5.0	in Dioxane	
X = SH		6.3		6.3	neat	
X = OMe		4.8		4.8	neat	
X = SMe		7.2		7.2	neat	
X = NH_2		5.4		5.4	neat	
X = NMe_2		8.1		8.1	neat	
X = F		3.5		3.5	neat	
X = Cl		4.5		4.5	neat	
X = Br		5.4		5.4	neat	
X = I		7.5		7.5	neat	
X = SO_2CH_3		3		3	in Dioxane	
X = SO_2F		4		4	neat	
X = CN		4.3		4.3	neat	
X = COH		5.3		5.3	neat	
X = $COCF_3$		5.3		5.3	in Dioxane	
X = NO		6.8		6.8	in Dioxane	
X = NO_2		5.7		5.7	in Dioxane	
X = $C_2H(CN)_2$		12		12	in Dioxane	
Monosubsituted stilbenes						18
φ—CH═CH—φ—X, X = H (trans)	1.907	26		26	in Chloroform	
X = H (cis)		12		12	in Chloroform	

Table 2 Continued

Molecule	Fundamental wavelength (μm)	γ' in 10^{-36} esu	γ'' in 10^{-36} esu	$\|\gamma\|$ in 10^{-36} esu	Remarks	Ref.
X = NH_2		31		31	in Dioxane	
X = NMe_2		64		64	in Dioxane	
X = NO_2		61		61	in Dioxane	
Disubsituted benzenes						
X-φ-Y, X=SO_2Me, Y=OH	1.907	3		3	in Dioxane	18
X=CN, Y=Me		5		5	neat	
X=CN, Y=Cl		5		5	in Dioxane	
X=CN, Y=Br		7		7	in Dioxane	
X=CN, Y=OPh		9		9	in Dioxane	
X=CN, Y=OMe		4		4	in Dioxane	
X=CN, Y=SMe		9		9	in Dioxane	
X=CN, Y=NH_2		6		6	in Dioxane	
X=CN, Y=NMe_2		10		10	in Dioxane	
X=CHO, Y=Me		7		7	neat	
X=CHO, Y=OPh		12		12	neat	
X=CHO, Y=OMe		8		8	neat	
X=CHO, Y=SMe		13		13	neat	
X=CHO, Y=NMe_2		18		18	in Dioxane	
X=i-$SO_2C_3F_7$, Y=OMe		5		5	in Chloroform	
X=$COCF_3$, Y=OMe		12		12	in Dioxane	
X=$COCF_3$, Y=OPh		7		7	in Dioxane	
X=$COCF_3$, Y=NMe_2		16		16	in Dioxane	
X=NO_2, Y=Me		8		8	in Dioxane	
X=NO_2, Y=OH		8		8	in Dioxane	
X=NO_2, Y=OPh		9		9	in Dioxane	
X=NO_2, Y=OMe		10		10	in Dioxane	
X=NO_2, Y=SMe		17		17	in Dioxane	

$X=NO_2$, $Y=N_2H_3$		9	9	in Dioxane	
$X=NO_2$, $Y=NH_2$		15	15	in Acetone	
Disubsituted stilbenes					18
Y-φ-CH=CH-φ-X	1.907	270	270	in NMP	
$X=NO_2$, $Y=N(Me)_2$					
$X=NO_2$, Y=COOMe		46	46	in CH_2Cl_2 for other substituents see Ref. 18	
Ortho- and Meta-Disubstituted Benzenes: X-φ-NO_2					18
Ortho: $X=NH_2$	1.907	8	8	in Dioxane	
Meta: $X=NH_2$		7	7	in Dioxane for other substituents see Ref. 18	
4-4′-Disubstituted Biphenyls and Fluorenes X-φ-φ-Y					24
$X=COCH_3$, Y=H	1.907	8	8	in Dioxane	
$X=NO_2$, $Y=NMe_2$		130	130	in Chloroform for other substituents see Ref. 24	
4- and β-Substituted Styrenes					24
X-φ-CH=CH-Y, X=OMe, Y=CN	1.907	11(5)	11(5)	in Chloroform	
$X=NMe_2$, Y=CHO		63(5)	63(5)	in Chloroform for other substituents see Ref. 24	

Table 2 Continued

Molecule	Fundamental wavelength (μm)	γ' in 10^{-36} esu	γ'' in 10^{-36} esu	$\lvert\gamma\rvert$ in 10^{-36} esu	Remarks	Ref.
4-4′-Nitropolyphenyl Oligomers						24
X-$(\varphi)_n$-NO_2, X=NH_2	1.907	21		21	n = 1, in NMP	
		96		96	n = 2	
		124		124	n = 3	
		133		133	n = 4	
X=OMe		10		10	n = 1	
		39		39	n = 2	
4-4′-Disubstituted Diphenylacetylenes: X-φ-C≡C-φ-Y						24
X=CO_2Me, Y=SMe	1.907	14		14	in Chloroform	
X=NO_2, X=NMe_2		151		151	in Chloroform for other substituents see Ref. 24	
4-4′-Disubstituted α,ω Diphenyl polyyne Oligomers: X-φ-$(C{\equiv}C)_n$-φ-Y, X=CN, Y=SMe	1.907	35		35	in Chloroform	24
X=NO_2, Y=NH_2		140		140	in NMP for other substituents see Ref. 24	
2′-Acceptor,2-Donor-Substituted α,ω-Diphenylpolyene Oligomers: X-φ-$(CH{=}CH)_n$-φ-Y,						24

X=2′-CN, Y=2-OMe	1.907	60	60	n = 2, in $CHCL_3$	
		147	147	n = 3, in $CHCL_3$	
X=2′-CN, Y=4-OMe		88	88	n = 2, in $CHCL_3$	
		165	165	n = 3, in $CHCL_3$	
X=2′-CN, Y=2-OMe		55	55	n = 2, in $CHCL_3$	
		149	149	n = 3, in $CHCL_3$ for other substituents see Ref. 24	24
Disubstituted α,ω Diphenyl polyene Oligomers:					
X-φ-$(CH=CH)_n$-φ-Y, X=CN, Y=OMe	1.907	54	54	n = 1, in $CHCL_3$	
		122	122	n = 2, in $CHCL_3$	
		234	234	n = 3, in $CHCL_3$	
X=NO_2, Y=Br		45	45	n = 1, in $CHCL_3$	
		146	146	n = 2, in $CHCL_3$	
		224	224	n = 3, in $CHCL_3$	
X=NO_2, Y=PMeO		93	93	n = 1, in $CHCL_3$	
		130	130	n = 2, in $CHCL_3$	
		230	230	n = 3, in $CHCL_3$ for other substituents see Ref. 24	24
4-Donor, β-Acceptor-Substituted α-Phenylpolyene Oligomers:					

Table 2 Continued

Molecule	Fundamental wavelength (μm)	γ' in 10^{-36} esu	γ'' in 10^{-36} esu	$\lvert\gamma\rvert$ in 10^{-36} esu	Remarks	Ref.
Y-φ-$(CH{=}CH)_n$-X, X=CHO, Y=MeO	1.907	28		28	n = 1, in $CHCL_3$	
		43		43	n = 2, in $CHCL_3$	
		120		120	n = 3, in $CHCL_3$	
X=CHO, Y=NMe_2		63			n = 1, in $CHCL_3$	
		140			n = 2, in $CHCL_3$	
		257			n = 3, in $CHCL_3$ for other substituents see Ref. 24	
			Gases			
He	1.055			9.14×10^{-4}	per atom	21
Ne	1.055			1.64×10^{-3}	per atom	21
Ar	1.055			2.14×10^{-2}	per atom	21
Kr	1.055			5.84×10^{-2}	per atom	21
Xe	1.055			0.172	per atom	21
He	1.055			1.93×10^{-2}	per atom	21
			Solvents			
Benzene	1.0642	4.4		4.4		7
	1.907	4.07		4.07		7
	1.907	3.85		3.85		9
Carbone tetrachloride	1.0642	3.63		3.63		7
	1.907	3.36		3.36		7
	1.907	3.09		3.09		8

solvents continued

Chloroform	1.0642	2.71	2.71	7
	1.907	2.27	2.27	7
	1.907	2.29	2.29	8
Dichloromethane	1.0642	1.94	1.94	7
	1.907	1.55	1.55	7
Cyclohexane	1.0642	3.32	3.32	7
	1.907	3.02	3.02	7
N-hexane	1.0642	3.43	3.43	7
	1.907	4.17	4.17	7
Acetone	1.0642	1.80	1.80	7
	1.907	1.88	1.88	7
Methanol	1.0642	0.83	0.83	7
	1.907	0.87	0.87	7
	1.907	0.80	0.80	9
Ethanol	1.0642	1.27	1.27	7
	1.907	1.49	1.49	7
	1.907	1.33	1.33	9
Propanol	1.0642	1.72	1.72	5
	1.907	1.82	1.82	5
	1.907	1.88	1.88	9
2-Propanone	1.907	2.04	2.04	9
Dimethylformamide	1.0642	2.2	2.2	7
	1.907	2.03	2.03	7
Water	1.0642	0.33	0.33	7
Nitrobenzene	1.907	6.57	6.57	7
	1.907	5.37	5.37	9
Fluorobenzene	1.907	3.19	3.19	9
Chlorobenzene	1.907	4.34	4.34	9
Bromobenzene	1.907	5.4	5.4	9
Iodobenzene	1.907	8.19	8.19	9
Cyanobenzene	1.907	4.11	4.11	9

Table 2 Continued

Molecule	Fundamental wavelength (μm)	γ' in 10^{-36} esu	γ'' in 10^{-36} esu	$\|\gamma\|$ in 10^{-36} esu	Remarks	Ref.
Aniline	1.907	5.71		5.71		9
Acetonitrile	1.907	0.87		0.87		9
Tetrahydrofuran	1.907	2.24		2.24		9
Methyl-cyclohexane	1.907	3.97		3.97		9
Pyridine	1.907	3.46		3.46		7
	1.907	3.36		3.36		9
CS_2	1.907	4.4		4.4		8
Xylene	1.907	5.2		5.2		18
Toluene	1.907	4.55		4.55		9
p-Dioxane	1.064	2.65		2.65		16
	1.907	2.15		2.15		16
	1.907	2.5		2.5		17

Numbers in brackets are standard deviations, N—number of double bonds; [a]number of equivalent double bonds; MONS—4,4′-methoxynitrostilbene, OCCB—4,4-octyloxycyanobiphenyl; φ-phenyl ring.

(1) Diaz-Garcia, M. A., Ledoux, I., Fernandez-Lazaro, F., Sastre, A., Torres, T., Agullo-Lopez, F., and Zyss, J., *J. Phys. Chem.*, *98*, 4495 (1994).
(2) Diaz-Garcia, M. A., Ledoux, I., Duro, J. A., Torres, T., Agullo-Lopez, F., and Zyss, J. *J. Phys. Chem.*, *98*, 8761 (1994).
(3) Cheng, L. T., Tour, J. M., Wu, R., and Bedworth, P. V., *Nonl. Opt.*, *6*, 87 (1993).
(4) Kajzar, F., Complex Hyperpolarizability and Solvatochromic Effects in Merocyanine Solutions, in *Nonlinear Optics of Organics and Semiconductors* (T. Kobayashi, ed.), Springer Proceedings in Physics, vol. 36, Berlin, 1989, p. 108.
(5) Kajzar, F., and Messier, J., in *Polydiacetylenes, Synthesis, Structure and Electronic Properties* (D. Bloor and R. R. Chance, eds.), NATO ASI Series, vol. E 102, Martinus Nijhoff Publ., Dordrecht, 1985, p. 325.
(6) Heflin, J. R., and Garito, A. F., Third-Order Optical Processes in Linear Chains: Electron Correlation Theory and Experimental Dispersion Measurements, in *Polymers for Lightwave and Integrated Optics: Technology and Applications* (L. A. Hornac, ed.), Marcel Dekker, Inc., New York, 1992, p. 50.
(7) Kajzar, F., and Messier, J., *Phys. Rev. A*, *32*, 2352 (1985).
(8) Meredith, G. R., Buchalter, B., and Hanzlik, C., *J. Chem. Phys.*, *78*, 1533 (1983).
(9) Meredith, G. R., Buchalter, B., and Hanzlik, C., *J. Chem. Phys.*, *78*, 1543 (1983).
(10) Sentein, C., Ph.D. Thesis, Universite Paris Sud, Orsay, France, 1993.
(11) Denovald, D. P., Ferguson, I., Hutchings, M. G., Kajzar, F., Messier, J., Malaney, K., and Sentein, C., *Nonl. Opt.*, *17*, 305 (1997).

(12) Gonin, D., Noël, C., and Kajzar, F., *Nonl. Opt.*, *8*, 37 (1994).
(13) Neher, D., Stegeman, G. I., Tinker, F. A., and Peyghambarian, N., *Opt. Lett.*, *17*, 1491 (1992).
(14) Samuel, I. D., Ledoux, I., Dhenaut, C., Zyss, J., Fox, H. H., Schrock, R. R., and Silbey, R. J., *Science*, *265*, 1070 (1994).
(15) Puccetti, G., Blanchard-Desce, M., Ledoux, I., Lehn, J.-M., and Zyss, J., *J. Phys. Chem.*, *97*, 9385 (1993).
(16) Burland, D. M., Walsh, C. A., Kajzar, F., and Sentein, C., *J. Opt. Soc. Am. B*, *8*, 2269 (1991).
(17) Cheng, L. T., Tam, W., Meredith, G., Rikken, G., and Meijer, E., Nonresonant EFISH and THG Studies of Nonlinear Optical Property and Molecular Structure Relations of Benzene, Stillbene and Other Arene Derivatives, in *Nonlinear Optical Properties of Organic Materials II* (G. Khanarian, ed.), Proceed. SPIE, vol. 1147, Bellingham, 1989, p. 61.
(18) Cheng, L.-T., Tam, W., Stevenson, S. H., Meredith, G. R., Rikken, G., and Marder, S. R., *J. Phys. Chem.*, *95*, 10631 (1991).
(19) Shi, R. F., Yu, Y. Z., Wu, M. H., Zhou, Q. L., and Garito, A. F., *Nonl. Optics*, *13*, 41 (1995).
(20) Marder, S. R., Metal-containing Materials for Nonlinear Optics, in *Inorganic Materials* (D. W. Bruce and D. O'Hare, eds.), Wiley & Sons, Inc., New York, 1992, p. 115.
(21) Lehmeier, H. J., Leupacher, W., and Penzkofer, A., *Opt. Commun.*, *56*, 67 (1985).
(22) Ward, J. F., and New, G. H. C., *Phys. Rev.*, *185*, 57 (1969).
(23) Messier, J., Kajzar, F., Sentein, C., Barzoukas, M., Zyss, J., Blanchard-Desce, M., and Lehn, J. M., *Nonl. Opt.*, 2, 53 (1992).
(24) Cheng, L.-T., Tam, W., Marder, S. R., Stiegman, A., Rikken, G., and Spangler, C. W., *J. Phys. Chem.*, *95*, 10643 (1991).

signal may be significantly larger if its coherence length is much larger than the thin film thickness. In many thin film experiments, the substrate is on the order of 1 mm thickness while the thin film is usually about 1 μm thick.

The harmonic intensity as a function of incidence angle θ from a thin film deposited on only one side of a substrate is given by[16]:

$$I_{3\omega}(\theta) = \frac{64\pi^4}{c^2} \left| \frac{\chi^{(3)}}{\Delta\varepsilon} \right|^2 |\exp [i(\varphi^s_\omega + \varphi^p_{3\omega})] \times \{T_1[1 - \exp(-i\Delta\varphi_s)] + \rho T_2 \exp (i\phi)[\exp (i\Delta\varphi_p) - 1]\}|^2 I^3_\omega, \quad (79)$$

where

$$\rho = \left(\frac{\chi^{(3)}}{\Delta\varepsilon}\right)_p \Big/ \left(\frac{\chi^{(3)}}{\Delta\varepsilon}\right)_s \quad (80)$$

is the ratio of the normalized thin film cubic susceptibility (p) to the substrate's normalized susceptibility where both susceptibilities are normalized to the dispersion in their respective dielectric constants. T's are transmission and boundary condition factors, whose exact form depends on experimental geometry (for details see Kajzar et al.[16]), Φ is the phase of the thin film cubic susceptibility assuming no optical loss (that of the substrate is assumed *a priori* to be real and positive), $\varphi^i_{\omega,3\omega}$ ($i = s,p$ i.e., s-polarization or p-polarization) is the phase of the fundamental (ω) and harmonic (3 ω) wave, respectively, and $\Delta\varphi_i$ ($i = s,p$) is the phase mismatch between fundamental and harmonic waves in the corresponding medium (cf. Eq. 28).

When a transparent thin film is deposited on both sides of a substrate (with identical thin film thicknesses on both sides) and when the refractive index of the thin film is close to that of the substrate, the resultant third harmonic intensity is given by Eq. 63 where the substrate plays the role of the liquid and the thin film plays the role of the window. For a thin film thickness less than the coherence length;

$$l < l_c = \frac{\lambda_\omega}{6(n_{3\omega} - n_\omega)}, \quad (81)$$

Eq. 63 reduces to the simple form

$$I_{3\omega} = \frac{4096\pi^5}{d^2} \left| \frac{\chi^{(3)}}{n_\omega + n_{3\omega}} \right|^2 |t^{01}_\omega|^6 |T^0 A_1(\theta)|^2 \left(\frac{l}{\lambda_\omega}\right)^2 I^3_\omega \cos^2 \frac{\Delta\varphi}{2}, \quad (82)$$

where for the sake of simplicity and for demonstration purposes we have neglected the substrate contribution to the resultant harmonic field. l in Eq. 81 and 82 is the thin film thickness, and $\Delta\varphi$ is the phase mismatch between the fundamental and the harmonic waves, created in the substrate (cf. Eq. 28) and depends on the substrate refractive indices and thickness only. The other quantities are related to thin film and have the usual meaning, as defined previously (cf. Eqs. 24, 32, and 67).

Equation 82 shows the interference between harmonic fields generated in the two thin films. This interference is governed by the phase mismatch $\Delta\varphi$ between fundamental and harmonic beams in the substrate itself. So, even if the substrate does not contribute significantly to harmonic intensities, because of a small cubic susceptibility value and harmonic field generated in each of the thin films vary smoothly and monotonically as a function of incidence angle, the resultant harmonic intensity will show interference fringes that look similar to Maker fringes (cf. Fig. 11). This effect is clearly illustrated in Fig. 11b. When the film on one side is removed, the interference pattern disappears.

The interference pattern, shown in Fig. 11a from a double-sided coated substrate, is sometimes erroneously called Maker fringes which leads to confusion because of its similarity to Maker fringes. The physics behind the fringes, however, is different. Maker fringes arise from interference between free and bound waves in the nonlinear medium whereas the interference fringes shown in Fig. 11a are due to interference between free waves created at the two different surfaces and separated by a dispersive medium.

When the thin film is deposited on only one side of the substrate, Maker fringes will be observed only if the third harmonic field created in the substrate is larger or comparable with that generated in the thin film. Figure 12 shows Maker fringes from a substrate and from a substrate coated with a thin film. In the latter, it is clearly seen that the thin film contributes a smoothly and monotonically varying harmonic field with incidence angle. From the difference in Maker fringe patterns between a substrate and a thin-film-coated substrate, it is possible to determine the phase ϕ of the thin film cubic susceptibility ($\chi^{(3)}(-3\omega;\omega,\omega,\omega) = |\chi^{(3)}(-3\omega;\omega,\omega,\omega)|e^{i\phi}$ as described in Ref. [16], provided that all necessary characteristics of the substrate are known, i.e., thickness, refractive indices at fundamental and harmonic frequencies, substrate thickness, and its THG susceptibility. These can be determined by doing independent THG measurements on the substrate alone by, e.g., removing *in situ* the measured film.

THG measurements can be also done in absorbing thin films, provided that the absorption is not too large. The effects of absorption can be minimized by decreas-

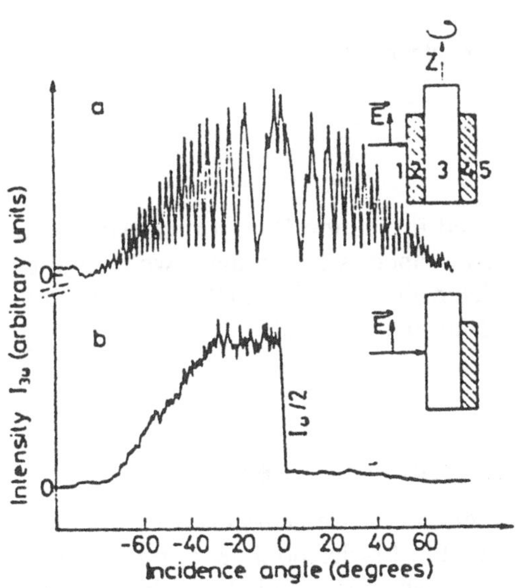

Figure 11

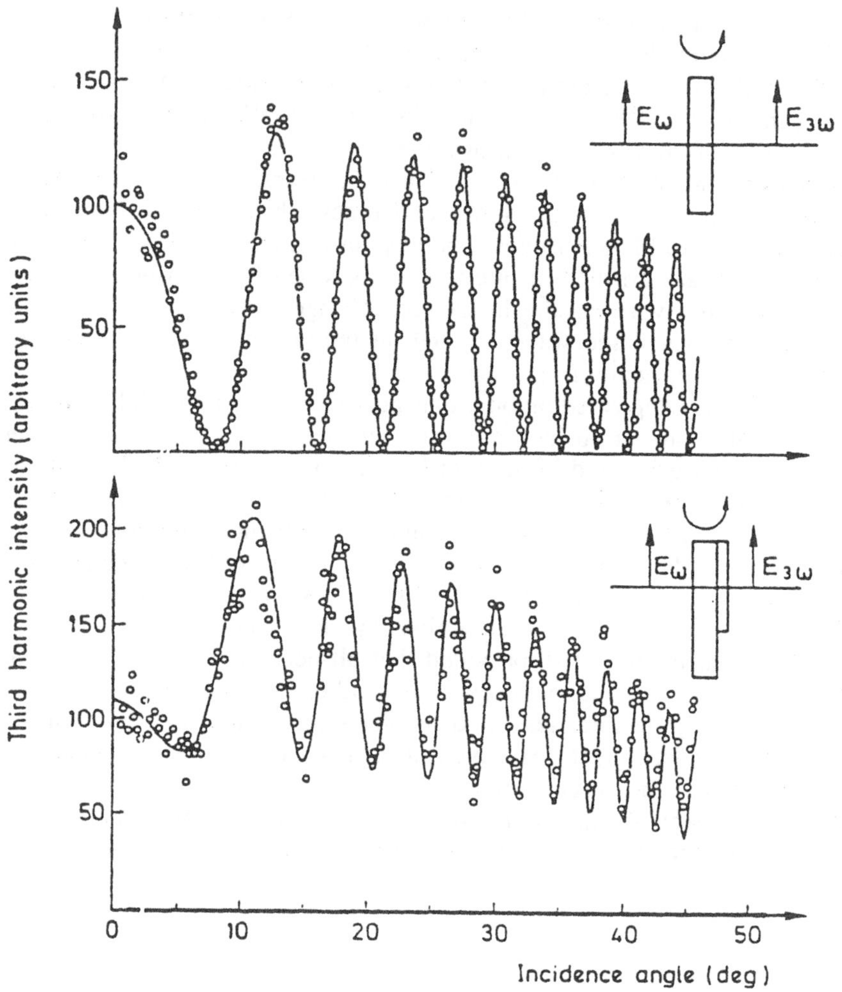

Figure 12

ing the thickness of the thin film. As discussed in Section II.C, it is possible to do such measurements on thin films whose absorbance is lower than two. Some examples of thin film cubic susceptibility determinations by THG for thin films, amorphous materials and for single crystals are listed in Table 1.

IV. APPLICATIONS

A. Characterization of Nonlinear Optical Properties

As previously discussed, third harmonic generation is a coherent process that measures the purely electronic contribution to the third order nonlinear optical susceptibility $\chi^{(3)}(-3\omega;\omega,\omega,\omega)$. Because the polarization must oscillate at the third harmonic frequency to generate light, the response time of the material must be at least as fast as the oscillation period, i.e., about 10^{-15} s. A typical experimental set up for wave-dispersed THG measurements is shown in Fig. 13. The light source is a pulsed neodymium-doped yttrium aluminium garnet laser, operating at 1.064 μm fundamental wavelength with a 13 ns pulse duration and 10 pps repetition rate. After frequency doubling in a phase-matched KDP single crystal, the 532 nm beam is used to pump a dye laser. The dye laser beam is subsequently shifted to the near infrared by stimulated Raman scattering in a high pressure (56 bars) hydrogen cell. This system allows the laser wavelength to be adjusted between 532 and 1907 nm with excellent beam quality (spectral width of 0.08 cm^{-1}). The beam is dispersed under 90 degrees in a Pellin-Broca prism and separated into two parts by using an ordinary glass slide. A lower intensity beam is sent to the reference arm where THG measurements are done simultaneously on a reference sample (a conjugated polymer) in order to correct for laser power fluctuations. The second beam, after parasitic light at non-fundamental wavelengths is removed by filtering, is focused

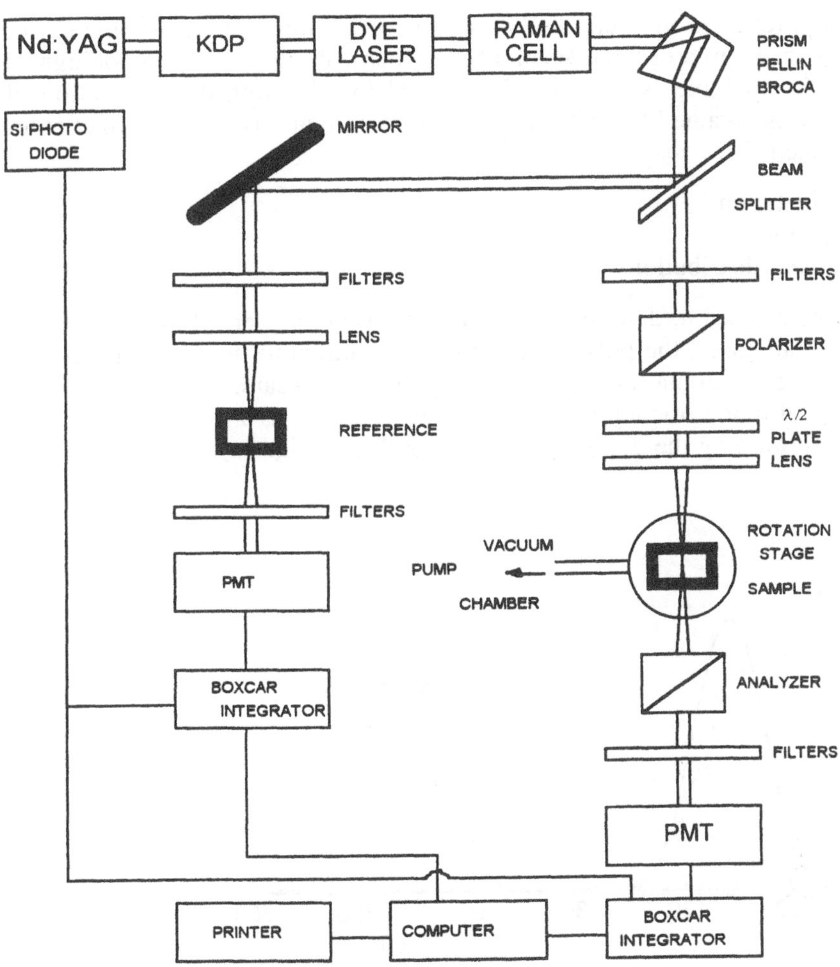

Figure 13

onto the sample that is mounted in a vacuum chamber to avoid the air contribution. The sample, mounted on a goniometer is rotated around an axis perpendicular and coincident to the beam propagation direction. The harmonic intensity is collected as a function of incidence angle. The experiment is fully computer controlled. This set-up can be used for THG measurements in solutions and in solids. In measurements of solids, a vacuum chamber is used. Figure 12 shows an example of the harmonic intensities as a function of incidence angle for a plane parallel silica plate and from a silica plate covered on one side by a polymeric thin film.

The solution and thin film THG measurements have been used for the nonlinear optical properties characterization of a large class of molecules. Data are listed in Table 1 for condensed phase third order nonlinear susceptibilities and in Table 2 for molecular hyperpolarizabilities, respectively. For example, Marder and coworkers used solution THG measurements extensively for testing the bond alternation model, by measuring γ as function of the solvent polarity.[46-49]

B. Nonlinear Optical Dichroism

The formalism presented in Section III.A and linking different macroscopic tensor components with the microscopic second hyperpolarizability γ_{xxxx} for a conjugated in one dimension polymer system has been used for checking the orientation of polymer thin films obtained by epitaxy.[41,50-52] More particularly, three distinct cases have been studied:

i. Mono-orientation
ii. Bi-orientation
iii. Orientational distributions

The measurements used to determine the orientations are presented in Figs. 14–16, respectively. In all cases, the polymer chains were parallel to the substrate and the thin film was rotated around an axis coincident with the beam.

In the case of mono-orientation, the harmonic intensity as a function of rotation angle Ψ is described by the eighth power of cos Ψ (cf. Fig. 14):

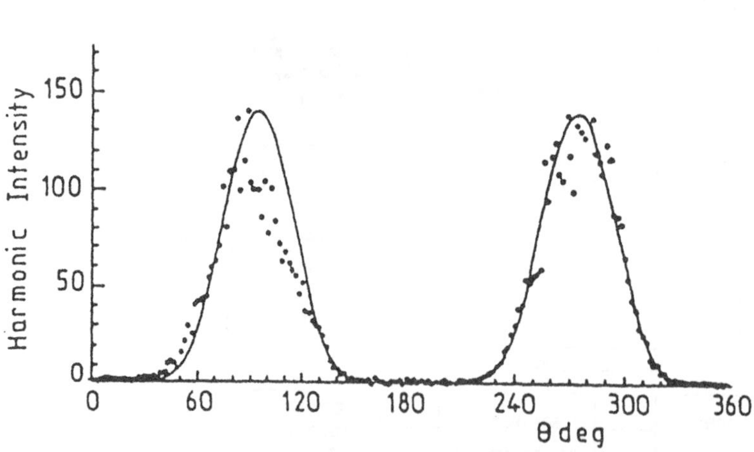

Figure 14

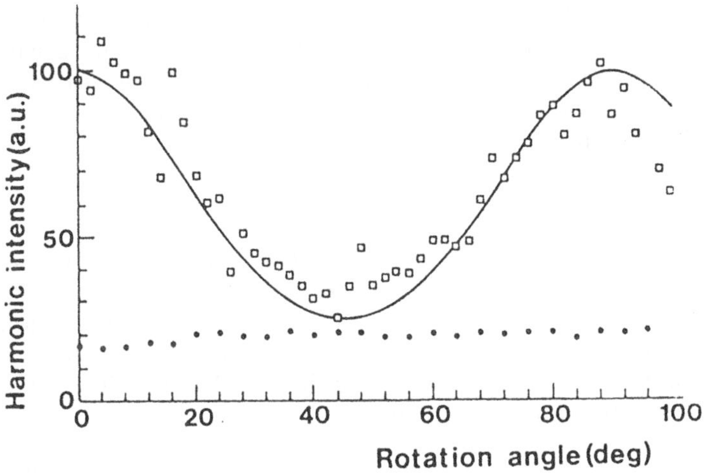

Figure 15

$$I_{3\omega} = AI_{\omega}^3 \cos^8 \Psi, \tag{83}$$

where A is a constant.

This dependence (Eq. 83) follows from Eq. 53 when Ψ is the angle between the X axis of the lab frame, perpendicular to the laser beam and the molecular axis.

For bi-orientation, the rotation angle dependence of the third harmonic intensity is described by the following expression (cf. Fig. 15):

$$I_{3\omega} = AI_{\omega}^3(\cos^4 \Psi + \sin^4 \Psi)^2. \tag{84}$$

Equation 84 can be again obtained from Eqs. 53 and 54 by considering two different, equally populated species, with equal molecular hyperpolarizabilities γ_{xxxx} and γ_{yyyy} (cf. Ref. [40]).

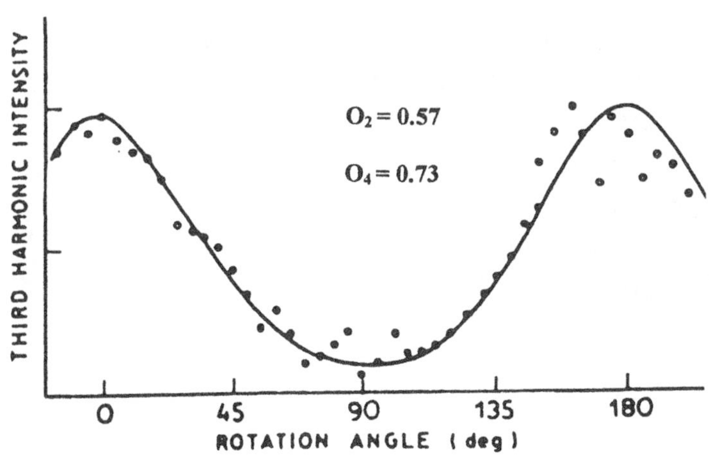

Figure 16

Figure 16 shows data for a polymer chain in which the orientation is described by a continuous orientational distribution function, $P(\Theta)$. By fitting the third harmonic intensities with the corresponding formulas for third harmonic intensities as given by Eq. 58–60, one can determine the order parameters O_2 and O_4 ($O_2 = 0.57$ and $O_4 = 0.73$ in this specific case, cf. Ref. [41]).

C. Scaling Laws

One important result in understanding the nonlinear optical response of organic molecules was the experimental verification of scaling laws which expresses the dependence of the nonlinear optical response on the electron's delocalization length. This interest was initiated by theoretical calculations by Rustagi and Ducuing[53] who showed that the second hyperpolarizability of electrons in a one-dimensional box varies strongly with the box length L:

$$\gamma \propto L^{a} \tag{85}$$

with $\alpha = 5$.

This result stimulated extensive theoretical[54-60] and experimental studies.[40,60-67] The calculated α_{th} and measured α_{exp} values for different conjugated quasi 1D systems are listed in Table 3. It is seen that the actual and the most realistic values of α are around 3 (where L represents the number of conjugated double bonds) and saturates for long chains.[40,66-67]

D. Nonlinear Spectroscopy

Third harmonic generation is a four photon process in which three incident photons with angular frequency ω create a photon with 3ω frequency. The off-resonant THG process can be represented by transition between virtual excited states, shown by the dashed lines in Fig. 17. If one of these virtual states coincides with an excited state of the unperturbed system (solid lines in Fig. 17) a resonance enhancement in the cubic susceptibility $\chi^{(3)}(-3\omega;\omega,\omega,\omega)$ will be observed. The frequency dependence of $\chi^{(3)}(-3\omega;\omega,\omega,\omega)$ is derived by time-dependent perturbation calculations[68] and is of the form:

$$\chi^{(3)}(-3\omega;\omega,\omega,\omega) = \frac{NF}{4\hbar^3}\sum_g \rho(g)\left\{\sum_{lmn}\left[\frac{\Omega_{gl}\Omega_{lm}\Omega_{mn}\Omega_{ng}}{(\omega_{lg}-3\omega)(\omega_{mg}-2\omega)(\omega_{ng}-\omega)}\right.\right.$$
$$+\frac{\Omega_{gl}\Omega_{lm}\Omega_{mn}\Omega_{ng}}{(\omega_{lg}^{*}+3\omega)(\omega_{mg}-2\omega)(\omega_{ng}-\omega)}$$
$$+\frac{\Omega_{gl}\Omega_{lm}\Omega_{mn}\Omega_{ng}}{(\omega_{lg}^{*}+\omega)(\omega_{mg}^{*}+2\omega)(\omega_{ng}-\omega)}$$
$$\left.+\frac{\Omega_{gl}\Omega_{lm}\Omega_{mn}\Omega_{ng}}{(\omega_{lg}^{*}+\omega)(\omega_{mg}^{*}+2\omega)(\omega_{ng}^{*}+3\omega)}\right]$$

Table 3 Theoretical Quantum Mechanical (α_{th}) and Measured (α_{exp}) Values of the Scaling Law Exponent. For Comparison, Experimental Values from Other than THG Measurements are Also Given

Material	α_{th}	Remark	Reference	α_{exp}	Remark	Ref.
Linear Polyene Oligomers	5	free electron model	1	2.82	CT polyenes ($\omega = 0$)	9
	3	INDO($\omega = 0$)	2	3.5	$\lambda_{\omega} = 1910$ nm	9
	3.27	SCF/STO-3G	3	2.3	Symmetric A-A THG ($\lambda_{\varphi} = 1910$ nm)	10
	3.14	SCF/3-21G	3			
	3.67	N = 2	4	3.3	Symmetric A-A	10
	3.98	N = 2 decrease for N > 6	4	2.6	N < 60, Saturation for N > 120	16
	3.8	DVB-PPP	5	3.6	EFISH in solution	18
	3.85	neutral solitons	6	4.65	Symmetric D-D	19
				2.3	Symmetric A-A	23
	6.32	charged polarons	6	3.6	Symmetric copolymers with discrete polyenes	20
	5.4	trans	7			
	4.7	cis	7, 14			
	6	N < 6, Saturation for N > 20	17			
	3.9	cis form, SDCI	22			
	4.0	trans form, SDCI				
Cyclic polyene oligomers				4.2	up N = 8	22
Thiophene Oligomers	4.5	CNDO N < 6 Saturation N > 6	8	4.05	DFM in solutions	11
				4.54	EFISH in solid solutions at 1060 nm Saturation N > 14	12
				3.4	THG	13
				2.8	THG in solution, N = 2, 4, . . . , 16, no saturation	14
Polyazine Oligomers				3.2	THG in thin films	21

N—number of double bonds

(1) Rustagi, K. C., and Ducuing, J., *Opt. Commun.*, *10*, 258 (1974).
(2) Pierce, B. M., *J. Chem. Phys.*, *91*, 791 (1989).
(3) Chopra, P., Carlacci, L., King, H. F., and Prasad, P. N., *J. Phys. Chem.*, *93*, 7120 (1989).
(4) Hurst, G., Dupuis, M., and Clementi, E., *J. Phys. Chem.*, *89*, 385 (1988).
(5) Soos, Z., and Ramesha, S., *J. Chem. Phys.*, *90*, 1067 (1989).
(6) De Melo, C. P., and Silbey, R., *J. Chem. Phys.*, *88*, 2567 (1988).
(7) Grossman, C., Heflin, J. R., Wong, K. Y., Zamani-Khamiri, O., and Garito, A. F., in *Organic Polymers for Nonlinear Optics* (J. Messier, F. Kajzar, P. N. Prasad and D. R. Ulrich, eds.), NATO ASI Series, Kluwer Academic Publ., Dordrecht, 1989, p. 61.

Table 3 Continued

(8) Charra, F., and Messier, J., in *Conjugated Materials: Opportunities in Electronics, Optoelectronics and Molecular Electronics* (J. L. Brédas and R. R. Chance, eds.), Kluwer Academic Publ., Dordrecht, 1990, p. 409.
(9) Messier, J., Kajzar, F., Sentein, C., Blanchard-Desce, M., and Lehn, J. M., *Nonl. Opt.*, *2*, 153 (1992).
(10) Blanchard-Desce, M., Lehn, J. M., Barzoukas, M., Ledoux, I., and Zyss, J., *Chem. Phys.*, *181*, 281 (1994).
(11) Zao, M. T., Singh, B. P., and Prasad, P. N., *J. Chem. Phys.*, *89*, 5535 (1988).
(12) Thienpont, H., Rikken, G., Meijer, E., Ten Hoeve, W., and Rynberg, H., *Phys. Rev. Lett.*, *17*, 2141 (1990).
(13) Kajzar, F., *Nonl. Opt.*, *5*, 329 (1993).
(14) Cheng, L. T., Tour, J. M., Wu, R., and Bedworth, P. V., *Nonl. Opt.*, *6*, 87 (1993).
(15) Garito, F., Heflin, J. R., Wong, K. Y., and Zamani-Khamiri, O., Nonlinear Optical Properties of Polyenes: Electron Correlation and Chain Conformation, in *Nonlinear Optical Properties of Polymers* (A. J. Heeger, J. Orenstein and D. R. Ulrich, eds.), MRS Proceed., vol. 109, Pittsburgh, 1988, p. 91.
(16) Samuel, I. D., Ledoux, I., Dhenaut, C., Zyss, J., Fox, H. H., Schrock, R. R., and Silbey, R. J., *Science*, *265*, 1070 (1994).
(17) Mukamel, S., and Wang, H. X., *Phys. Rev. Lett.*, *69*, 65 (1992).
(18) Craig, G. S. W., Cohen, R. E., Schrock, R. R., Silbey, R. J., Puccetti, G., Ledoux, I., and Zyss, J., *J. Am. Chem. Soc.*, *115*, 860 (1993).
(19) Puccetti, G., Blanchard-Desce, M., Ledoux, I., Lehn, J.-M., and Zyss, J., *J. Phys. Chem.*, *97*, 9385 (1993).
(20) Craig, G. S. W., Cohen, R. E., Schrock, R. R., Dhenaut, C., Ledoux, I., and Zyss, J., *Macromolecules*, *27*, 1875 (1994).
(21) Nalwa, H. S., Hamada, T., Kakuta, A., and Mukoh, A., *Nonl. Opt.*, *6*, 155 (1993).
(22) Shi, R. F., Yu, Y. Z., Wu, M. H., Zhou, Q. L., and Garito, A. F., *Nonl. Opt.*, *13*, 41 (1995).
(23) Blanchard-Desce, M., Lehn, J. M., Barzoukas, M., Runser, C., Fort, A., Pucetti, G., Ledoux, I., and Zyss, J., *Nonl. Opt.*, *10*, 23 (1995).

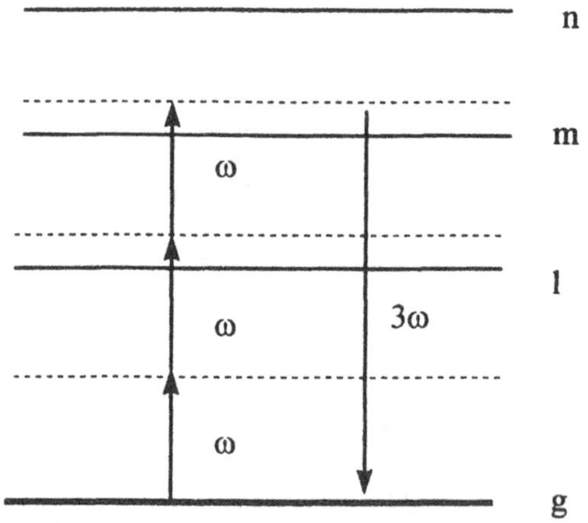

Figure 17

$$- \sum_{mn} \left[\frac{\Omega_{gm}\Omega_{mg}\Omega_{gn}\Omega_{ng}}{(\omega_{mg} - 3\omega)(\omega_{mg} - \omega)(\omega_{ng} - \omega)} \right.$$

$$+ \frac{\Omega_{gm}\Omega_{mg}\Omega_{gn}\Omega_{ng}}{(\omega_{mg} - \omega)(\omega_{ng}^* + \omega)(\omega_{ng} - \omega)}$$

$$+ \frac{\Omega_{gm}\Omega_{mg}\Omega_{gn}\Omega_{ng}}{(\omega_{mg}^* + 3\omega)(\omega_{mg}^* + \omega)(\omega_{ng}^* + \omega)}$$

$$\left. + \frac{\Omega_{gm}\Omega_{mg}\Omega_{gn}\Omega_{ng}}{(\omega_{mg}^* + \omega)(\omega_{ng} - \omega)(\omega_{ng}^* + \omega)} \right] \Bigg\}, \tag{86}$$

where $\Omega_{ij} = \langle i|e\mathbf{r}|j\rangle$ are dipolar transition matrix elements between i and j states, $\omega_{ij} = (E_i - E_j)/\hbar$ are the corresponding angular transition frequencies, and $\rho(g)$ is the density of fundamental states. N is the number density of molecules and F is the local field factor. Resonance line-widths are taken into account by introducing *a priori* a damping term Γ and consequently $\omega \Rightarrow \omega - i\Gamma$ and $\omega^* \Rightarrow \omega + i\Gamma$. $\chi^{(3)}$ is resonantly enhanced when the energy of the fundamental matches the energy of an excited state, one-half the energy of an excited state or one-third the energy of an excited state.

Equation 86 is the most general expression and includes the summation *a priori* over an infinite number of states. In practice, a good description of $\chi^{(3)}$ can be obtained by including only a limited number of states that are near the resonance condition. This approximation is sometimes called the "essential states model".

1. Two-level Model

When only the ground (g) and one excited state (n) is included in the sum in Eq. 86, the two-level model results:[26]

$$\chi^{(3)}(-3\omega;\omega,\omega,\omega) = \frac{NF}{4\hbar^3} |\mu_{gn}|^2$$

$$\left[|\Delta\mu|^2 \left\{ \frac{1}{(\omega_{ng} - 3\omega)(\omega_{ng} - 2\omega)(\omega_{ng} - \omega)} \right. \right.$$

$$+ \frac{1}{(\omega_{ng}^* + \omega)(\omega_{ng} - 2\omega)(\omega_{ng} - \omega)}$$

$$\left. + \frac{1}{(\omega_{ng}^* + \omega)(\omega_{ng}^* + 2\omega)(\omega_{ng} - \omega)} + \frac{1}{(\omega_{ng}^* + 2\omega)(\omega_{ng}^* + 3\omega)} \right\}$$

$$- |\mu_{gn}|^2 \left\{ \frac{1}{(\omega_{ng} - 3\omega)(\omega_{ng} - \omega)^2} + \frac{1}{(\omega_{ng}^* + \omega)(\omega_{ng} - \omega)^2} \right.$$

$$\left. \left. + \frac{1}{(\omega_{ng}^* + 3\omega)(\omega_{ng}^* + \omega)^2} + \frac{1}{(\omega_{ng}^* + \omega)^2(\omega_{ng} - \omega)} \right\} \right], \tag{87}$$

where $\Delta\mu = \mu_{nn} - \mu_{gg}$ is the difference of permanent dipolar moments between excited and fundamental states. Note that in centrosymmetric systems, $\Delta\mu = 0$. Note that in centrosymmetric systems $\Delta\mu = 0$. Consequently the two level model is inadequate for centrosymmetric molecules.

The two level model, first introduced for the frequency dispersion description of first hyperpolarizability $\beta(-2\omega,\omega,\omega)$ by Chemla and Oudar,[69] applies also for the THG susceptibility description, on the molecular or macroscopic level, for noncentrosymmetric molecules[26,70] and macroscopic structures.[71]

2. Three-level Model

The lowest approximation that includes all minimal symmetry and multiphoton state characteristics is the three-level model[63,95,96,105]:

$$\chi^{(3)}(-3\omega;\omega,\omega,\omega) = \frac{NF}{4\hbar^3}|\mu_{gn}|^2$$
$$\left[|\mu_{ng'}|^2\left\{\frac{1}{(\omega_{ng}-3\omega)(\omega_{ng}-2\omega)(\omega_{ng}-\omega)}\right.\right.$$
$$+\frac{1}{(\omega_{ng}^*+\omega)(\omega_{ng}-2\omega)(\omega_{ng}-\omega)}$$
$$+\frac{1}{(\omega_{ng}^*+\omega)(\omega_{ng}^*+2\omega)(\omega_{ng}-\omega)}$$
$$\left.+\frac{1}{(\omega_{ng}^*+\omega)(\omega_{ng}^*+2\omega)(\omega_{ng}^*+3\omega)}\right\}$$
$$-|\mu_{gn}|^2\left\{\frac{1}{(\omega_{ng}-3\omega)(\omega_{ng}-\omega)^2}+\frac{1}{(\omega_{ng}^*+\omega)(\omega_{ng}-\omega)^2}\right.$$
$$\left.\left.+\frac{1}{(\omega_{ng}^*+3\omega)(\omega_{ng}^*+\omega)^2}+\frac{1}{(\omega_{ng}^*+\omega)^2(\omega_{ng}-\omega)}\right\}\right]. \tag{88}$$

The three-level model has been successfully used for the description of the frequency dependence of the cubic susceptibility (or molecular second hyperpolarizability) of a number of conjugated molecules and polymers.[30,71-99] It allows a determination, in the framework of the validity of the model, of the excited state energy and width of forbidden two-photon levels. In centrosymmetric molecules, excited states can be classified into so-called *gerade* and *ungerade* states with opposite symmetry. Because a photon bears a spin of one, a transition between states of opposite symmetry are possible with one photon (or an odd numbers of photons). In contrast, transitions between states of the same symmetry are possible only with two (or an even number) photons. Thus transitions between states of the same symmetry will be not seen in the linear optical absorption spectrum (one-photon spectrum) but will be observable through a two-photon excitation. Figure 18 shows a THG spectrum of fullerene C_{60}. The points show experimental values of the cubic susceptibility $\chi^{(3)}(-3\omega;\omega,\omega,\omega)$ and the solid lines the calculated ones within a three-level model. Two resonance enhancements in $\chi^{(3)}(-3\omega;\omega,\omega,\omega)$ are seen: at about 1.06 μm fundamental wavelength (interpreted as a three-photon resonance) and at 1.2 μm fundamental wavelength (interpreted as a two-photon resonance). In Table 4 we have listed fundamental-excited and excited-excited states dipolar transition moments for selected molecules, as determined by THG measurements. Note that the THG technique is the only four-photon process which allows an observation of a three-photon resonance. Only nonlinear optical tech-

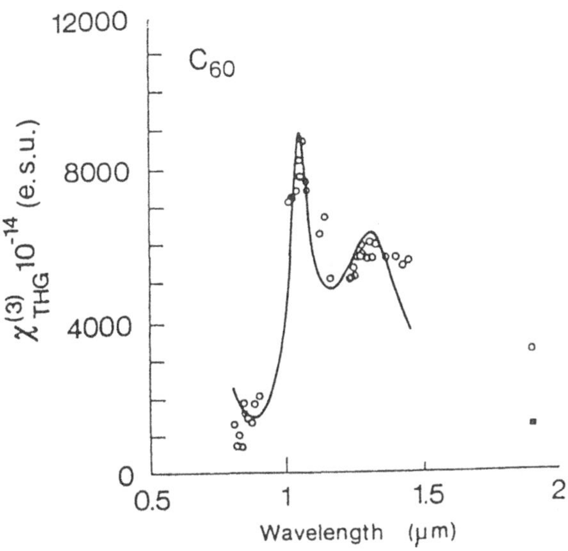

Figure 18

niques can give information about the transition dipolar moments between excited states.

V. CONCLUSIONS

The third harmonic generation technique is a useful tool for studies of nonlinear optical properties of materials in three different phases: gas, liquid and solid. Solution studies allow the determination of the molecular second hyperpolarizability γ. The techniques developed for THG measurements, in both solution and thin film, allow not only a precise determination of the modulus, but also their phase (real and imaginary parts).

Both thin film and solution measurements can be used for nonlinear spectroscopic studies. It gives information on the energy and the width of excited states as well as the transition dipolar moments between ground states and excited states as well as between excited states. Nonlinear spectroscopy is the only method available to get such information. On the other hand, these measurements are important not only from the fundamental point of view (knowledge of electronic structure and interaction) but also for practical applications, such as optical limiting.

The THG technique can be also used successfully, as it was presented and discussed here, in nonlinear optical dichroism studies. Although it applies to a limited number of molecules (with enhanced cubic hyperpolarizability tensor component in one direction, like conjugated polymers), it gives more information than the linear optical dichorism. As was discussed here with bi-oriented polydiacetylene thin films, it works even if the linear optical dichroism technique fails.

The THG technique can also be used directly for frequency tripling, as discussed by Kajzar and Messier.[39] This is of special interest in tripling very short femtosecond laser pulses, where the optical pathlength must be minimized because

Table 4 Transition Energies (ω_i) Between Fundamental (0) and Excited State (i), Dipolar Transition Moments μ_{ij} Between (i) and (j) States and Associated Dumping Terms Γ_{ij} as Determined by the Wave Dispersed THG Technique, Using a Quantum Three- or Four-Level Model

Compound	ω_1 cm^{-1}	Γ_{01} cm^{-1}	ω_2 cm^{-1}	Γ_{12} cm^{-1}	ω_3 cm^{-1}	Γ_{13} cm^{-1}	μ_{01} (D)	μ_{12} (D)	μ_{13} (D)	Model	Ref.
PDA-4BCMU	18350	4000	15350	1550	22850		13	25	35	4-level	1
Poly(3-butyl)	19610	3500	17610	3500[a]			17	21		3-level	1
thiophene	19610	3500	26610	3500[a]			17	33		3-level	
	19610	3000	18610	3000[a]	24610	3000[a]	17	15	23	4-level	
Poly(3-decyl)	19610	3000	18110	3000[a]			17	18		3-level	1
thiophene	19610	3000	26610	3000[a]			17	31		3-level	
	19610	3000	19110	3000[a]	24610	3000[a]	17	12	23	4-level	
β-Carotene	21050	2500	17050	2500	24550		11.5	12	35	4-level	1
	19890[b]	620	22200				13.5	31		3-level[b]	2
	19840	200	22200				12.8	31		3-level[c]	2
C_{60}	28570	814	14280	1178			25.3	26.3		3-level	3
C_{70}	21700	20100	1560	1650			29.0	30.4		3-level	4
Squaraines[a]:											
ISQ	15290(30)	363(24)	16532(80)	266(48)	26610(8060)	2900(970)	10.3(1.4)	4.8(1.0)	9.5(3.0)	4-level	5
C16-TSQ	15080(48)	339(24)	16290(80)	403(121)	29840(8060)	2016(967)	11.8(1.4)	4(1)	11(3)		

Note: In centrosymmetric structures, level "1" has opposite symmetry to the groundstate (0) and to levels (2) and (3).

[a]—assumed
[b]—time domain method
[c]—time domain method with inhomogeneous bradening
[d]—calibrated with BK-7 glass $\chi^{(3)}(-3\omega;\omega,\omega,\omega) = 0.58 \times 10^{-14}$ esu.

(1) Torruellas, W., Ph.D. Thesis, University of Arizona, 1991.
(2) Van Beek, J. B., Ph.D. Thesis, Cornell University, 1991; see also Van Beek, J. B., Kajzar, F., and Albrecht, A. C., *J. Chem. Phys.*, *95*, 6400 (1991).
(3) Kajzar, F., Taliani, C., Danieli, R., Rossini, S., and Zamboni, R., *Chem. Phys. Lett.*, *217*, 418 (1994).
(4) Kajzar, F., Taliani, C., Danieli, R., Rossini, S., and Zamboni, R., *Phys. Rev. Lett.*, *73*, 1617 (1994).
(5) Andrews, J. H., Khaydarov, J. D. V., Singer, K. D., Hull, D. L., and Chuang, K. C., *Nonl. Opt.*, *10*, 227 (1995).

of group velocity dispersion and temporal walkoff. The conversion efficiency for tripling by a thin film with thickness l is given by:

$$\eta = I_{3\omega}/I_{\omega} = \frac{2304\pi^6}{c^2}\left|\frac{\chi^3(-3\omega;\omega,\omega,\omega)}{1+n_{3\omega}}\right|^2\left(\frac{l}{\lambda_\omega}\right)^2 t_{01}^6 I_\omega^2, \tag{89}$$

where t_{01} is the transmission factor at fundamental frequency between air and the thin film and the other symbols have the same meaning as previously defined.

From Eq. 89 it is seen that the conversion efficiency increases rapidly with incident power and a high conversion efficiency can be obtained even for a thin conjugated polymer film (cf. Fig. 19).

A method to increase the THG susceptibility (and in general other cubic susceptibilities, involving at least three interacting photons) is through cascading of second order nonlinear optical susceptibilities $\chi^{(2)}(-2\omega;\omega,\omega)$. This effect has been studied by Meredith[99] in the case of a quartz single crystal and by Kajzar and Zagorska[100] with poled polymeric films.

Because tunable laser sources and sensitive electronic detection apparatus are common, THG experiments are easy to set up. A commercial apparatus fabricated by SOPRA[101] can be used for a rapid determination of $\chi^{(3)}(-3\omega;\omega,\omega,\omega)$ in solutions and in thin films. The technique allows a rapid determination of the electronic nonlinear optical susceptibility $\chi^{(3)}$ $(-3\omega;\omega,\omega,\omega)$ at one measurement frequency ω. Subsequently, the tensors for other third order processes can be derived from this single value via multi-level models, within given conventions and for a given frequency. Although such derivation is possible from a measurement at a single frequency, it requires knowledge of the excited state structure that is used in the multilevel model. Often it is necessary to measure the whole spectrum of $\chi^{(3)}$ $(-3\omega;\omega,\omega,\omega)$ to check the validity of the model parameters or to get a better estimation of the dipolar transition moments. For these reasons, the THG technique is widely used. However, there exist several sources of uncertainties when using this technique. These are:

1. *Conventions and calibration.* As discussed in the Introduction and in Chapter 1, two types of conventions for the Fourrier transform of the electric fields are

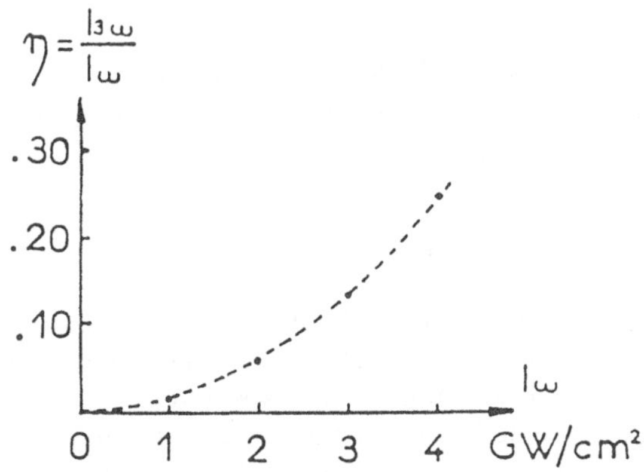

Figure 19

used (cf. Eq. 6) which lead to different definitions of the cubic susceptibility. Some authors also take explicit account of the degeneracy factors and others do not. Also, even in a given convention for $\chi^{(3)}$ $(-3\omega;\omega,\omega,\omega)$, the susceptibility of the silica standard used varies between researchers. We systematically use the value of $\chi^{(3)}(-3\omega;\omega,\omega,\omega) = 2.8 \times 10^{-14}$ esu, as determined by Meridith and Buchalter[102] at a 1.907 μm fundamental wavelength. A value of $\chi^{(3)}(-3\omega;\omega,\omega,\omega) = 1.1 \times 10^{-14}$ esu has been recently proposed by Mito et al.[103]

2. *Air contribution.* As shown by Kajzar and Messier,[13] air contributes significantly to the resulting THG intensities and alters the measured values of $\chi^{(3)}(-3\omega;\omega,\omega,\omega)$ if not taken correctly into account. The air contribution may be determined by doing experiments with, e.g., a silica slab (or any nonlinear optical slab) in air and in vacuum and following the procedure and the formulas as described in Ref. [13]. However, for practical reasons and for better accuracy, it is recommended that THG experiments be performed in a vacuum chamber. Some researchers are using strongly focused fundamental beams in order to eliminate the air contribution. This procedure diminishes the air contribution but it also violates the plane wave approximation condition. Consequently, the formulae derived within the plane wave approximation and commonly used for data screening (Eqs. 23–31) are no longer valid and those derived for harmonic generation in an infinite medium by a focused laser beam should be used (Eq. 41).
3. *Substrate contribution.* Equation 79 gives the THG intensity from a thin film and substrate assembly and neglects multiple reflections of the fundamental beam. Assuming that one harmonic field is small with respect to the second one (substrate or thin film) and assuming high transmission factors at the interfaces of the thin film and substrate as well as the factor arising from the boundary conditions, one can derive the harmonic intensity from a substrate alone or from a free-standing thin film when the thickness of the latter is smaller than its coherence length. Consequently, one gets the following expression for the harmonic field generated in an isotropic slab with thickness larger than the coherence length (as is often the case for the thin film substrate):

$$I^s_{3\omega} = \frac{256\pi^4}{c^2}\left[\frac{\chi^{(3)}(-3\omega;\omega,\omega,\omega)}{\Delta\varepsilon}\right]^2_s |T_s A_s|^2 \sin^2\frac{\Delta\varphi}{2}\,(I_\omega)^3, \qquad (90)$$

where $\Delta\varphi$ is the phase mismatch between fundamental and harmonic waves (cf. Eq. 28). The corresponding formula for a free-standing, non-absorbing thin film reads:

$$I^p_{3\omega} = \frac{1024^5}{c^2}\left[\frac{\chi^{(3)}(-3\omega;\omega,\omega,\omega)}{\Delta\varepsilon}\right]^2_p |T_p A_p|^2 \left(\frac{l}{\lambda_\omega}\right)^2 (I_\omega)^3. \qquad (91)$$

In Eqs. 90 and 91, A's are the factors arising from boundary conditions (cf. Eqs. 23 and 31) and T's are global transmission factors. The subscripts (superscripts) s and p refer to nonlinear slab and thin film, respectively.

By comparing both intensities (collected at the same conditions) and putting equal to 1 all transmission factors as well as the factors arising from boundary conditions and transmission, one gets the following expression for the thin film (p) cubic susceptibility, calibrated with the slab (s) susceptibility:

$$\chi_p^{(3)}(-3\omega;\omega,\omega,\omega) = \frac{2}{\pi}\left[\frac{I_{3\omega}^p}{I_{3\omega}^s}\right]^{1/2}\frac{l_p}{l_s^c}, \tag{92}$$

where l_p is the thin film thickness and l_s^c is the coherence length of the nonlinear slab. The harmonic intensities $I_{3\omega}^p$ and $I_{3\omega}^s$ are maximum intensities at the normal incidence (envelope value in the case of fringing patterns).

Equation 92 was originally derived by Tomaru et al.[104] and is frequently used by different authors in the screening of THG data. The derivation shown here illustrates the very limited applicability of Eq. 92. Indeed, it applies only if:

(i) refractive indices of thin film and of substrate are the same,
(ii) the harmonic field generated in thin film is much larger than that generated in the substrate so interference terms between both fields can be neglected,
(iii) thin film (as well as substrate) susceptibility is real, or
(iv) absence of absorption in the thin film at both fundamental and harmonic frequencies.

Some corrective formulae for absorption, valid in the low absorption limit, have been proposed together with a simplified formula to take account of the substrate contribution. The first applies only for small absorptions while the second one has been derived by neglecting an important term in the expression describing interference of harmonic fields generated in the thin film and in the substrate and its use is not justified (cf. discussion Kajzar and Zagorska[100]).

4. *Thin film absorption.* As mentioned above, corrections for absorption have been proposed, valid only at low absorption limit. The main problem is that usually when a thin film is absorbing, the $\chi^{(3)}$ $(-3\omega;\omega,\omega,\omega)$ susceptibility is complex. The formulae derived by Tomaru et al. assume $\chi^{(3)}$ to be real.

Note that Eq. 79 allows us to determine the thin film cubic susceptibility as well as its phase without ambiguity, provided that multiple reflection effects can be neglected. This is the case for substrates with low refractive index and close to that of the thin film, or substrates with rough surface. The problem of multiple reflections has been addressed by several authors (cf. Refs. [3,28–31]). Note that for practical reasons, the optical density at normal incidence should not exceed the value of about one in order to avoid a quasi-total extinction of the free wave with respect to the bound wave.

REFERENCES

1. Blombergen, N., and Pershan, P. S., *Phys. Rev.*, *128*, 606 (1962).
2. Franken, P. A., Hill, A. E., Peters, C. W., and Weinrech, G., *Phys. Rev. Lett.*, *7*, 118 (1961).
3. Maker, P. D., Terhune, R. W., and Savage, M., *Phys. Rev. Lett.*, *14*, 681 (1965); Maker, P. D., and Terhune, R. W., *Phys. Rev.*, *137*, A801 (1965).
4. Bey, P. P., Giuliani, J. F., and Rabin, H., *Phys. Rev. Lett.*, *19*, 819 (1967).
5. Chang, R. K., and Galbraith, R. K., *Phys. Rev.*, *171*, 993 (1968).
6. Hermann, J. P., Ricard, D., and Ducuing, J., *Appl. Phys. Lett.*, *23*, 178 (1973).

7. Hermann, J. P., and Ducuing, J., *J. Appl. Phys.*, *45*, 5100 (1974).
8. Kajzar, F., Messier, J., Zyss, J., and Ledoux, I., *Opt. Commun.*, *45*, 133 (1983).
9. Kajzar, F., and Messier, J., *Thin Sol. Films*, *132*, 11 (1985).
10. Meredith, G. R., Buchalter, B., and Hanzlik, C., *J. Chem. Phys.*, *78*, 1533 (1983).
11. Meredith, G. R., Buchalter, B., and Hanzlik, C., *J. Chem. Phys.*, *78*, 1543 (1983).
12. Thalhamer, M., and Penzkofer, A., *Appl. Phys. B*, *32*, 137 (1983).
13. Kajzar, F., and Messier, J., *Phys. Rev. A*, *32*, 2352 (1985).
14. Kajzar, F., and Messier, J., *Rev. Sci. Instrum.*, *58*, 2081 (1987).
15. Kajzar, F., and Messier, J., *J. Opt. Soc. Am. B*, *4*, 1040 (1987).
16. Kajzar, F., Messier, J., and Rosilio, C., *J. Appl. Phys.*, *60*, 3040 (1986).
17. Kajzar, F., and Messier, J., in *Polydiacetylenes, Synthesis, Structure and Electronic Properties* (D. Bloor and R. R. Chance, eds.), NATO ASI Series, vol. E 102, Martinus Nijhoff Publ., Dordrecht, 1985, p. 325.
18. Kajzar, F., Complex Hyperpolarizability and Solvatochromic Effects in Merocyanine Solutions, in *Nonlinear Optics of Organics and Semiconductors* (T. Kobayashi, ed.), Springer Proceedings in Physics, vol. 36, Berlin, 1989, p. 108.
19. Kajzar, F., Cubic Susceptibility of Organic Molecules in Solution, in *Nonlinear Optical Effects in Organic Polymers* (J. Messier, F. Kajzar, P. Prasad and D. Ulrich, eds.), NATO ASI Series, Series E: Applied Sciences, vol. 162, Kluwer Academic Publ., Dordrecht, 1989, p. 225.
20. Choy, M. M., and Byer, R. L., *Phys. Rev. B*, *14*, 1693 (1976).
21. Morley, J. O., and Pugh, D., *J. Chem. Soc. Faraday Trans.*, *87*, 3021 (1991).
22. Kajzar, F., Horn, K., Nahata, A. J., and Yardley, J., *Nonl. Opt.*, *8*, 205 (1994).
23. Burland, D. M., Welsh, C. A., Kajzar, F., and Sentein, C., *J. Opt. Soc. Am. B*, *8*, 2269 (1991).
24. Heflin, J. R., Cai, Y. M., and Garito, A. F., *J. Opt. Soc. Am. B*, *8*, 2132 (1991).
25. Willets, A., Rice, J. E., Burland, D. M., and Shelton, D. P., *J. Chem. Phys.*, *97*, 7590 (1992).
26. Gonin, D., Noël, C., and Kajzar, F., *Nonlinear Optics*, *8*, 37 (1994).
27. Swalen, J. D., and Kajzar, F., in *Organic Molecules for Waveguiding Nonlinear Optics: Science and Applications* (F. Kajzar and J. Swalen, eds.), Gordon and Breach Sc. Publ., Reading, 1996, p. 1.
28. Bethune, D. S., *J. Opt. Soc. Am. B*, *6*, 910 (1989).
29. Bethune, D. S., *J. Opt. Soc. Am. B*, *8*, 367 (1991).
30. Van Beek, J., Kajzar, F., and Albrecht, A. C., *J. Chem. Phys.*, *95*, 6400 (1991).
31. Neher, D., Wolf, A., Bubeck, C., and Wegner, G., *Chem. Phys. Lett.*, *163*, 116 (1989).
32. Kajzar, F., and Messier, J., in *Nonlinear Optical Properties of Organic Molecules and Crystals* (D. S. Chemla and J. Zyss, eds.), vol. 2, Academic Press, Orlando, 1987, p. 51.
33. Chollet, P. A., Kajzar, F., and Messier, J., in *Nonlinear Optical and Electroactive Polymers* (P. N. Prasad and D. R. Ulrich, eds.), Plenum Press, New York, 1988, p. 121.
34. Kleinman, D. A., Ashkin, A., and Boyd, G. D., *Phys. Rev.*, *145*, 338 (1966).
35. Ward, J. F., and New, G. H. C., *Phys. Rev.*, *185*, 57 (1969).
36. Boyd, G. D., and Kleinman, D. A., *J. Appl. Phys.*, *39*, 3597 (1968).
37. Krausch, F., Winter, E., and Leising, G., *Phys. Rev. B*, *39*, 3701 (1989).
38. Chemla, D. S., and Kupecek, P., *Rev. Phys. Appl.*, *6*, 31 (1971).
39. Kajzar, F., and Messier, J., *Third Order Nonlinear Optical Effects*, in *Conjugated Polymers: The Novel Science and Technology of Highly Conducting and Nonlinear Optically Active Materials* (J. L. Brédas and R. Silbey, eds.), Kluwer Academic Publ., Dordrecht, 1991, p. 509.
40. Kajzar, F., *Non. Opt.*, *5*, 329 (1993).

41. Kajzar, F., Lorin, A., Le Moigne, J., and Szpunar, J., *Acta Phys. Polonica*, *87*, 713 (1995).
42. Neher, D., Mittler-Neher, S., Cha, M., and Stegeman, G. I., in *Organic Materials V* (K. D. Singer, ed.), Proc. SPIE, vol. 1560, Bellingham, 1991, p. 335.
43. Van Beek, J. B., Kajzar, F., and Albrecht, A. C., *Chem. Phys.*, *161*, 299 (1992).
44. Cheng, L. T., Tam, W., Stevenson, S. H., Meredith, G. R., Rikken, G., and Marder, S. R., *J. Phys. Chem.*, *95*, 10631 (1991).
45. Cheng, L. T., Tam, W., Marder, S. R., Stiegman, A., Rikken, G., and Spangler, C. W., *J. Phys. Chem.*, *95*, 10631 (1991).
46. Bourhill, G., Cheng, L. T., Gorman, C. B., Lee, G., Marder, S. R., Perry, J. W., Perry, M. J., and Tiemann, B. G., Experimental Demonstration of the Relationship Between the Second- and Third-Order Polarizabilities of Conjugated Donor-Acceptor Molecules, Proceed. SPIE, vol. 2143, Bellingham, 1994.
47. Bourhill, G., Mansour, K., Tiemann, B. G., Gorman, C. B., Biddle, S., Marder, S. R., and Perry, J. W., Third-Order Polarizabilities of Symmetric and Non-symmetric Polyene and Cyanine-Like Organic Molecules, Proceed. SPIE, vol. 1853, Bellingham, 1993.
48. Gorman, C. B., and Marder, S., *Proc. Natl. Acad. Sci.*, *90*, 11297 (1993).
49. Marder, S. R., Gorman, C. B., Meyers, F., Perry, J. W., Bourhill, G., Brédas, J. L., and Pierce, B., *Science*, *265*, 632 (1994).
50. Le Moigne, J., Kajzar, F., and Thierry, A., *Macromol.*, *24*, 2622 (1991).
51. Le Moigne, J., Thierry, A., Chollet, P. A., Kajzar, F., and Messier, J., *J. Chem. Phys.*, *88*, 6647 (1988).
52. Kajzar, F., Le Moigne, J., and Thierry, A., Nonlinear Optics with Oriented Conjugated Polymer Thin Films, in *Electronic Properties of Polymers, Orientation and Dimensionality of Conjugated Systems* (H. Kuzmany, M. Mehring and S. Roth, eds.), Springer Verlag, Berlin, 1992, p. 202.
53. Rustagi, K. C., and Ducuing, J., *Opt. Commun.*, *10*, 258 (1974).
54. Agrawal, G. P., Cojan, C., and Flytzanis, C., *Phys. Rev. B*, *17*, 776 (1978).
55. Pierce, B. M., *J. Chem. Phys.*, *91*, 791 (1989).
56. Cha, M., Torruellas, W., Stegeman, G. I., Wang, H. X., Takahashi, A., and Mukamel, S., *Chem. Phys. Lett.*, *228*, 73 (1994).
57. Hurst, G., Dupuis, M., and Clementi, E., *J. Phys. Chem.*, *89*, 385 (1988).
58. Soos, Z., and Ramesha, S., *J. Chem. Phys.*, *90*, 1067 (1989).
59. De Melo, C. P., and Silbey, R., *J. Chem. Phys.*, *88*, 2567 (1988).
60. Grossman, C., Heflin, J. R., Wong, K. Y., Zamani-Khamiri, O., and Garito, A. F., in *Organic Polymers for Nonlinear Optics* (J. Messier, F. Kajzar, P. N. Prasad and D. R. Ulrich, eds.), NATO ASI Series, Kluwer Academic Publ., Dordrecht, 1989, p. 61.
61. Chopra, P., Carlacci, L., King, H. F., and Prasad, P. N., *J. Phys. Chem.*, *93*, 7120 (1989).
62. Fichou, D., Charra, F., Garnier, F., Kajzar, F., and Messier, J., Linear and Nonlinear Optical Properties of Thiophene Oligomers, in *Organic Materials for Nonlinear Optics* (R. A. Hann and D. Bloor, eds.), The Royal Soc. Chem. Special Publication, London, 1989, p. 176.
63. Messier, J., Kajzar, F., Sentein, C., Blanchard-Desce, M., and Lehn, J. M., *Nonl. Opt.*, *2*, 53 (1993).
64. Blanchard-Desce, M., Lehn, J. M., Barzoukas, M., Ledoux, I., and Zyss, J., *Chem. Phys.*, *181*, 281 (1994).
65. Zao, M. T., Singh, B. P., and Prasad, P. N., *J. Chem. Phys.*, *89*, 5535 (1988).
66. Thienpont, H., Rikken, G., Meijer, E., Ten Hoeve, W., and Rynberg, H., *Phys. Rev. Lett.*, *17*, 2141 (1990).

67. Samuel, I. D., Ledoux, I., Dhenaut, C., Zyss, J., Fox, H. H., Schrock, R. R., and Silbey, R. J., *Science*, *265*, 1070 (1994).
68. Orr, J., and Ward, J. F., *Mol. Phys.*, *20*, 513 (1971).
69. Oudar, J. L., and Chemla, D. S., *J. Chem. Phys.*, *66*, 2664 (1977).
70. Barzoukas, M., Fremaux, P., Josse, D., Kajzar, F., and Zyss, J., in *Nonlinear Optical Properties of Polymers* (A. J. Heeger, J. Orenstein and D. R. Ulrich, eds.), MRS Symp. Proceed., vol. 109, Pittsburgh, 1988, p. 171.
71. Charra, F., and Messier, J., in *Conjugated Materials: Opportunities in Electronics, Optoelectronics and Molecular Electronics* (J. L. Brédas and R. R. Chance, eds.), Kluwer Academic Publ., Dordrecht, 1990, p. 409.
72. Townsend, P. D., Fann, W.-S., Etemad, S., Baker, G. L., Soos, Z. G., and McWilliams, P. C. M., *Chem. Phys. Lett.*, *62*, 1492 (1991).
73. Baumert, J. C., Bjorklund, G. C., Jundt, D. H., Jurich, M. C., Looser, M. C., Miller, R. D., Rabolt, J., Sooryakumaran, R., Swalen, J. D., and Twieg, R. J., *Appl. Phys. Lett.*, *53*, 1147 (1988).
74. Kajzar, F., Messier, J., Nunzi, J. M., and Raimond, P., Third Order Materials: Processes and Characterization, in *Polymers for Lightwave and Integrated Optics: Technology and Applications* (L. A. Hornak, ed.), Marcel Dekker, Inc., New York, 1992, p. 595.
75. Kajzar, F., and Messier, J., *Polymer J.*, *19*, 275 (1987).
76. Kajzar, F., and Messier, J., *Thin Sol. Films*, *132*, 11 (1985).
77. Chollet, P. A., Kajzar, F., and Messier, J., in *Nonlinear Optical and Electroactive Polymers* (P. N. Prasad and D. R. Ulrich, eds.), Plenum Press, New York, 1988, p. 121.
78. Fann, W. S., Benson, S., Madey, J. M. J., Etemad, S., Baker, G. L., and Kajzar, F., *Phys. Rev. Lett.*, *62*, 1492 (1989).
79. Kanetake, T., Ishikawa, K., Hasegawa, T., Koda, T., Takeda, K., Hawegawa, M., Kubodera, K., and Kobayashi, M., *Appl. Phys. Lett.*, *54*, 2287 (1989).
80. Kajzar, F., Ruani, G., Taliani, C., and Zamboni, R., *Synth. Metals*, *37*, 223 (1990).
81. Torruellas, W. E., Rochford, K. B., Zanoni, R., Aramaki, S., and Stegeman, G. I., *Opt. Commun.*, *82*, 94 (1991).
82. Torrruellas, W. E., Neher, D., Zanoni, R., Stegeman, G. I., Kajzar, F., and Leclerc, M., *Chem. Phys. Lett.*, *175*, 11 (1990).
83. Kaino, T., and Kurihara, T., Third-Order Optical Properties of Polymeric Materials, in *Polymers for Lightwave and Integrated Optics: Technology and Applications* (L. A. Hornac, ed.), Marcel Dekker, Inc., New York, 1992, p. 50.
84. Vanherzeele, H., Meth, J. S., Jenekhe, S. A., and Roberts, M. F., *J. Opt. Soc. Am. B*, *9*, 524 (1992).
85. Agrawal, A. K., Jenekhe, S. A., Vanherzeele, H., and Meth, J. S. *Chem. Mat.*, *3*, 765 (1991).
86. Jenekhe, S. A., Osaheni, J. A., Vanherzeele, H., and Meth, J. S., *Chem. Mat.*, *4*, 683 (1992).
87. Kubodera, K., and Kaino, T., $\chi^{(3)}$ Evaluation of Organic Nonlinear Optical Materials by THG Measurements, in *Nonlinear Optics of Organics and Semiconductors* (T. Kobayashi, ed.), Springer Proceedings in Physics, vol. 36, Berlin, 1989, p. 163.
88. Heflin, J. R., and Garito, A. F., Third-Order Optical Processes in Linear Chains: Electron Correlation Theory and Experimental Dispersion Measurements, in *Polymers for Lightwave and Integrated Optics: Technology and Applications* (L. A. Hornac, ed.), Marcel Dekker, Inc., New York, 1992, p. 50.
89. Tomaru, S., Kubodera, K., Kurihara, T., and Zembutsu, S., *Japan J. Appl. Phys.*, *26*, L1657 (1987).

90. Tomaru, S., Kubodera, K., Zembutsu, S., Takeda, K., and Hasegawa, M., *Electron. Lett.*, *23*, 595 (1987).
91. Chollet, P. A., Kajzar, F., and Messier, J., Resonances of Cubic Nonlinear Optical Susceptibilities in Polydiacetylenes, in *Nonlinear Optics of Organics and Semiconductors* (T. Kobayashi, ed.), Springer Proceedings in Physics, vol. 36, Berlin, 1989, p. 171.
92. Torruellas, W., Ph.D. Thesis, University of Arizona, 1991.
93. Van Beek, J. B., Ph.D. Thesis, Cornell University, 1991.
94. Kajzar, F., Taliani, C., Danieli, R., Rossini, S., and Zamboni, R., *Chem. Phys. Lett.*, *217*, 418 (1994).
95. Kajzar, F., Taliani, C., Danieli, R., Rossini, S., and Zamboni, R., *Phys. Rev. Lett.*, *73*, 1617 (1994).
96. Andrews, J. H., Khaydarov, J. D. V., Singer, K. D., Hull, D. L., and Chuang, K. C., *Non. Opt.*, *10*, 227 (1995).
97. Mathy, A., Uebrhofen, K., Schenk, R., Gregorius, H., Garay, R., Müllen, K., and Bubeck, C., *Phys. Rev. B*, *53*, 4367 (1996).
98. Bubeck, C., Relations Between Structure and Third-Order Nonlinearities of Molecules and Polymers, in *Organic Thin Films for Waveguiding Nonlinear Optics* (F. Kajzar and J. Swalen, eds.), Gordon & Breach Sc. Publ., Reading, 1996, p. 137.
99. Meredith, G. R., *Phys. Rev. B*, *24*, 5522 (1981).
100. Kajzar, F., and Zagorska, M., *Nonl. Opt.*, *6*, 181 (1993).
101. Laser Focus World, December 1996.
102. Buchalter, B., and Meredith, G. R., *Appl. Opt.*, *21*, 3221 (1982).
103. Mito, A., Hagimoto, K., and Takahashi, C., *Nonl. Opt.*, *13*, 3 (1995).
104. Tomaru, S., Kubodera, K., Zembutsu, S, Takeda, K., and Hasegawa, M., *Electron. Lett.*, *23*, 595 (1987).
105. Kuzyk, M. G., and Dirk, C. W., *Phys. Rev. A*, *41*, 148 (1991).

Index